Concepts of
Physical Chemistry
Through Problems

for BSc (Hons) and all those preparing for examinations like IIT JAM, JEE Advanced, NET and GRE

Highlights

- More than 2100 selected questions with detailed explanations
- Quantitative and qualitative problems with solutions
- Auxiliary problems with answers for better understanding of the concepts
- University questions with solutions to enhance problem-solving skills
- Follows the basic pedagogy to help nurture constructive learning

Concepts of
Physical Chemistry
Through Problems

for **BSc (Hons)** and all those preparing for examinations like **IIT JAM, JEE Advanced, NET and GRE**

Dilip Kumar Khamrui PhD

Former Reader
Department of Chemistry
Seth Anandram Jaipuria College
Kolkata, India

CBS Publishers & Distributors Pvt Ltd

New Delhi • Bengaluru • Chennai • Kochi • Kolkata • Mumbai
Bhopal • Bhubaneswar • Hyderabad • Jharkhand • Nagpur • Patna • Pune
Uttarakhand • Dhaka (Bangladesh) • Kathmandu (Nepal)

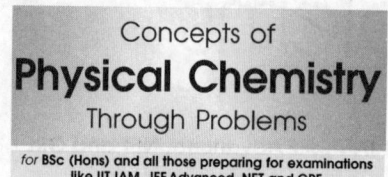

Concepts of
Physical Chemistry
Through Problems

for BSc (Hons) and all those preparing for examinations like IIT JAM, JEE Advanced, NET and GRE

ISBN: 978-93-89261-72-1

Copyright © Author and Publisher

First Edition: 2020

Published by Satish Kumar Jain and produced by Varun Jain for

CBS Publishers & Distributors Pvt Ltd
4819/XI Prahlad Street, 24 Ansari Road, Daryaganj, New Delhi 110 002, India.
Ph: 23289259, 23266861, 23266867   Fax: 011-23243014   Website: www.cbspd.com
e-mail: delhi@cbspd.com; cbspubs@airtelmail.in.

Corporate Office: 204 FIE, Industrial Area, Patparganj, Delhi 110 092
Ph: 4934 4934   Fax: 4934 4935   e-mail: publishing@cbspd.com; publicity@cbspd.com

Branches

- **Bengaluru:** Seema House 2975, 17th Cross, K.R. Road, Banasankari 2nd Stage, Bengaluru 560 070, Karnataka
  Ph: +91-80-26771678/79   Fax: +91-80-26771680   e-mail: bangalore@cbspd.com
- **Chennai:** 7, Subbaraya Street, Shenoy Nagar, Chennai 600 030, Tamil Nadu
  Ph: +91-44-26680620, 26681266   Fax: +91-44-42032115   e-mail: chennai@cbspd.com
- **Kochi:** 68/1534, 35, 36 Power House Road, Opp KSEB Power House, Ernakulam 682 018, Kochi, Kerala
  Ph: +91-484-4059061-65   Fax: +91-484-4059065   e-mail: kochi@cbspd.com
- **Kolkata:** 6/B, Ground Floor, Rameswar Shaw Road, Kolkata-700 014, West Bengal
  Ph: +91-33-22891126, 22891127, 22891128   e-mail: kolkata@cbspd.com
- **Mumbai:** 83-C, Dr E Moses Road, Worli, Mumbai-400018, Maharashtra
  Ph: +91-22-24902340/41   Fax: +91-22-24902342   e-mail: mumbai@cbspd.com

Representatives

- **Bhopal**   0-8319310552
- **Jharkhand**   0-9811541605
- **Pune**   0-9623451994
- **Kathmandu (Nepal)**   977-9818742655
- **Bhubaneswar** 0-9911037372
- **Nagpur**   0-9421945513
- **Uttarakhand** 0-9716462459
- **Hyderabad**   0-9885175004
- **Patna**   0-9334159340
- **Dhaka (Bangladesh)**   01912-003485

Printed at: India Binding House, Greater Noida, UP

to

My Guru

Professor Mihir Choudhury

Preface

$\mathbf{P}$roper understanding of chemistry requires clear concept of its principles which physical chemistry deals with. This can be achieved straight-forwardly through appropriately designed quiz problems. In the present book each chapter contains adequate number of such problems which are so arranged that they can be discussed in groups each concerned with some common physical quantity/quantities.

Qualitative problems which I have selected are mostly of fallacious nature. Regarding quantitative problems, those based on the same system under different conditions have been emphasised. For all such problems, often different working formula with different assumptions and approximations is required to be used in calculating the same physical quantity. One such problem is equivalent to several simple problems and is, therefore, highly effective in understanding the method of calculation through a limited number of problems. This purpose is pretty served also through selection of problems involving calculation of different quantities with reference to same system instead of different systems. In doing calculation, the data are put in the appropriate working formula using preferably the same system of units to get the result in the corresponding unit without any ambiguity. Although SI system of units is desirable, other units are used in some problems to make students familiar with such units and their advantages in certain cases. The necessity of numerical calculations is highlighted through interpretation and correlation of the results. They support the existing theories and at the same time, clear up the concept of different physical quantities by rooting out seeming contradictions.

Most problems end with a note regarding the importance of the concerned problem. This is useful in comparing the problems.

In producing this book, I am deeply indebted to my highly esteemed teacher Dr Dulal Chandra Mukherjee, President, Indian Chemistry Society, for his kind consent to be the editor of the book. I thank my bosom friend Professor (Dr) Bhabatarak Bhattacharya (Bhatnagar awardee), Bose Institute, for appreciating the nature of this book, my one-time colleague Professor (Dr) Debabrata Mondal, Calcultta University, for his valuable discussion about some confusing aspects, my former colleague Professor Bibekanjan Chatterjee for his assistance in proof-reading, and my ardent students Dr Dhiman Basu and Dr Joy Dutta for providing materials of academic importance. Lastly, I would like to thank all staff members of CBS Publishers & Distributors Pvt Ltd, whose active participation is deemed indispensable for publishing this book.

Any constructive criticism of this book will be gratefully acknowledged.

Dilip Kumar Khamrui

Preface

Proper understanding of chemistry requires clear concept of its principles which physical chemistry deals with. This can be achieved ...

... Dilip Kumar Khanna

Contents

11. Chemical Equilibrium 378–418

12. Ionic Equilibrium 419–467

13. Electrochemical Equilibrium 468–514

1

General Considerations

Preliminaries necessary for dealing with different aspects of physical chemistry are being considered here.

1.1 UNITS AND DIMENSIONS

Everything that can be numerically (quantitatively) expressed is termed a *quantity*. The quantities which are used to characterise matter or its observable change in nature are termed *physical* quantities.

Comparison of some quantity with a standard quantity is termed *measurement* and the standard quantity used is termed *unit of measurement* or simply *unit of the quantity*. The number of units considered in the measured quantity is termed the *numerical value of the quantity*. For example, when we say that a rod is 10 metres long, we mean that it is 10 times as long as a rod whose length has been defined to be 1 metre as unit.

Physical quantities are preferably defined in a way that reflects the operations that are to be carried out to measure the quantity. Such definitions are called *operational definitions*. For example, the speed of a body, defined as the distance travelled by it per unit time, can be obtained by measuring the distance travelled and the elapsed time, and dividing the first quantity by the second. It should be remembered that the physical quantities which cannot be operationally defined, e.g. entropy, wavefunction, etc. are also meaningful and have importance in the understanding of the operationally defined physical quantities.

The choice of a standard quantity rests on its high accessibility, permanency and reproducibility. It is for these factors that the unit originally defined for a quantity has been subsequently modified. Thus, the first international standard of length was provided by a bar of Pt-Ir alloy kept at the International Bureau of Weights and measured in Sèvres, near Paris. But due to lack of accessibility and inaccuracy in reproducibility of this bar, the standard of length has been changed (by international agreement in 1961) to an atomic constant, viz. the wavelength of the orange-red light emitted by the atoms of ^{86}Kr in the discharge tube. One metre has been redefined to be 1650763.73 times the wavelength of this light.

Sometimes modification of unit value for a physical quantity in a system of unit is done to make a simpler relation between the units of a physical quantity in different systems of measurement. Thus, common units of volume are cubic centimeter (cm^3), cubic meter (m^3) and liter (L). A liter was originally defined as the volume of 1000 g of water at 3.98 °C and 1 atm pressure, i.e. 1000.028 cm^3. However, in 1964, liter was redefined so that the following simple relations hold.

$$1 \text{ L} = 1000 \text{ cm}^3 = (10 \text{ cm})^3 = 1 \text{ dm}^3$$

Nowadays, the numerical value of a physical quantity is usually expressed in an international system of unit, called Système International (SI), preferably in the form $y10^x$, where x is a convenient integer (+ve or −ve) and y is a decimal number that rarely requires more than three digits (considering the limits of accuracy in a measurement) with only one numeral (nonzero) before the decimal point. The number of digits in y gives the *number of significant figures* in the approximate numerical value of the concerned physical quantity. The factor 10^x can be avoided using specific prefix (depending on the value of x) to the basic unit as given in Table 1.1.

Table 1.1: Standard SI prefixes

Factor (10^x)	10^{-18}	10^{-15}	10^{-12}	10^{-9}	10^{-6}	10^{-3}	10^{-2}	10^3	10^6	10^9	10^{12}	10^{15}	10^{18}
Prefix	atto	femto	pico	nano	micro	milli	centi	kilo	mega	giga	tera	peta	exa
Abbreviations	a	f	p	n	μ	m	c	k	M	G	T	P	E

On the basis of operational definition, a complex physical quantity can be expressed as product of some simple directly measurable quantities (or their inverse) which are listed along with their symbols and SI units, in Table 1.2. Such an expression (in terms of symbols) represents the *dimensions* of the concerned physical quantity which reflects its complex nature. Thus l/t represents the dimensions of speed which is defined as the length (l) traversed divided by the time (t) elapsed. The seven basic physical quantities given in Table 1.2 are dimensionally independent, i.e. they cannot be derived from one another by algebraic combinations.

Table 1.2: Basic physical quantities

Physical quantity	Symbol	SI unit	Abbreviation
Mass	m	kilogram	kg
Length	l	meter	m
Time	t	second	s
Electric current	I	ampere	A
Luminous intensity	I_v	candela	cd
Thermodynamic temperature	T	kelvin	K
Amount	n	mole	mol

Definitions of Unit Basic Quantities

Mass: The SI unit of mass is kilogram, which is mass of a cylinder of Pt–Ir alloy kept at the International Bureau of Weights and Measures in Sèvres.

Length: The SI unit of length is kilometre.

Time: The SI unit of time is second which was originally defined as 1/86400 part of the mean solar day. Subsequently (in 1967), the astronomical standard has been given up with the adoption of a more precise atomic standard. The second has been redefined as the duration of 9192631770 cycles of radiation caused by the transition between the two hyperfine energy levels in the ground state of ^{133}Cs atom (differing only in the spin of the outermost electron).

Electric current: The SI unit of electric current is ampere. It is the current that causes a force of 2×10^{-7} newton per metre between two infinitely long parallel wires (carrying this current) of negligible cross-section and placed one meter apart in vacuum.

Luminous intensity: The SI unit of luminous intensity is the candela. It is the quantity of light emitted per unit solid angle by a surface of 1/600000 square meters of a black-body at the melting temperature of Pt at a pressure of 101325 newtons per metre. (The term *luminous intensity* is somewhat misleading as the flux per unit area decreasing with increasing distance).

Thermodynamic temperature: The SI unit of thermodynamic temperature is kelvin. One kelvin is 1/273.16 of the thermodynamic temperature of the tripple point of water (at which liquid water, water vapour and ice are in equilibrium). Since the normal freezing point (0°C) of water is 0.01 degree below its triple point, any temperature which is t in celsius scale and T in kelvin scale will be related by

$$T/K = t/°C + 273.15$$
$$\approx t/°C + 273$$

Amount: The SI unit of amount is the mole which implies Avogadro number (i.e. 6.022×10^{23} which is the number of carbon atoms in 0.012 kg of ^{12}C) of elementary entities. Here specification must be made of elementary entity which may be an atom, a molecule, an ion, an electron, a photon, etc. or a specified group of such entities. The number of molecules (N_i) of the species i in a system is related to its amount (n_i) by $N_i = n_i N_A$, where N_A is the Avogadro constant having unit mol^{-1} (whereas Avogadro number is a pure number). It is correct to state that the amount of a species i in a system is n_i (and not n_i mol). Since n_i has a factor of 1 mol included in itself.

A mole is related to mass, provided the elementary entities possess mass. Thus:

  i. one mole of electron weighs 5.486×10^{-7} kg
  ii. one mole of H atom weighs 1.008×10^{-3} kg
  iii. one mole of H_2 molecule weighs 2.016×10^{-3} kg
  iv. one mole of $Fe_{0.95}O$ weighs 6.906×10^{-2} kg
  v. one mole of air containing 80 mole % N_2 and 20 mole % O_2 weighs 2.882×10^{-2} kg
  vi. one mole of a reaction mixture $H_2 + \frac{1}{2}O_2$ (which contains H_2 and O_2 in 2:1 mole ratio) weighs $\left(\frac{2}{3} \times 2.016 + \frac{1}{3} \times 32\right) \times 10^{-3}$ kg or 1.201×10^{-2} kg.

It is quite justifiable to define *one mole of a reaction*, e.g. $H_2 + \frac{1}{2}O_2 = H_2O$, as the reaction of one mole H_2 and half mole O_2. Chemists consider one mole of reaction as a unit reaction. This makes the unit $J \cdot mol^{-1}$ of ΔH for a chemical reaction obvious.

The amount of a substance has been introduced (by the general conference on Weights and Measures in 1971) into the set of basic physical quantities to deal with massless entities like photon (whose rest mass is zero) and also to serve the interest of the chemists. The quantity mass per amount of a substance (entity) is called its molar mass (M) whose SI unit is kg mol^{-1}. This renders the awkward term *molecular weight* obsolete.

Consistency of Dimension and Unit

Any equation relating to physical quantites must be dimensionally correct, i.e. same dimensions on both sides, as is found, for example, with the relation $\frac{1}{2}mc^2 = \frac{3}{2}PV$, where mc^2 has the dimensions $m(lt^{-1})^2$ and PV has the dimensions $(ml^{-1}t^{-2})$ (l^3) or ml^2t^{-2}. The dimensional consistency is useful in finding the dimensions of a complex quantity. This purpose will be readily served with the simplest relation involving the complex quantity concerned. Thus, the dimensions of molar gas constant R can be conveniently found from the ideal gas equation $PV = nRT$ or $R = PV/nT$ to be $(ml^{-1}t^{-2})$ $(l^3)/(n)$ (T) or ml^2t^{-2}/nT.

It should be remembered that from dimensional analysis, one cannot say whether a relation is physically meaningful or the constants involved are correct. Thus, molar entropy, molar heat capacity and molar gas constants all have same dimensions and SI unit ($JK^{-1}mol^{-1}$) but they are not of same kind, and hence addition and subtraction processes are not operative between them.

It is important to note that any physical quantity x in e^x, $\ln x$ and $\sin x$ must be dimensionless. This is in keeping with dimensional consistency. Thus, $\ln$ (x meter) carries no meaning, but $\ln$ (x/meter) or $\ln$ (x/cm) does have. To maintain the dimensional consistency, an expression is often written with reference to the unit choosen. For example, the variation of vapour pressure (p) in atmosphere of a liquid with temperature (T) in kelvin is more appropriately represented as

$$\ln (p/\text{atm}) = a - \frac{b}{T/K}$$

where a and b are dimensionless constants. Obviously, the values of a and b depends on the unit of p.

The dimensional consistency of an expression implies unit consistency. If all the quantities in a calculation are expressed in same system of units (say SI), the result will be in that system of units without involvement of any additional numerical factors. It is for this reason that the numerical calculations are preferably done using only one system of units.

A physical quantity may require to be converted from a given unit to some other unit. The conversion factor necessary for this can be found using the algebraic properties of units regarding multiplication, division and cancellation. For example, to convert time from min to sec, we have to use the relation 1 min = 60 sec or 1 = 60 sec min^{-1}. Then time in min can be converted into sec simply by multiplying with the conversion factor 60 sec min^{-1} (which amounts to multiplying by one). Hence, t min = t min (60 sec min^{-1}) = 60t sec, i.e. min, in effect, replaced by 60 sec. Common non-SI units and their SI equivalents are given in Table 1.3.

1.2 PRECISION AND APPROXIMATION

Numerical values of a physical quantity obtained from experimental observations are unlikely to be exact due to the deficiencies of the instrument and operator. The maximum instrumental error (called *precision*), unless mentioned on the instrument, is usually taken to be half of the smallest division on the instrument's scale, e.g. 0.5 mm in measuring length with a millimeter ruler. The personal error, which is random in nature, is less important but can be diminished using statistical means.

The precision of a numerical value can be expressed in relative form (often as percentage) or in absolute form. We can arrive at the significant figures of a numerical quantity from the specified error. Suppose in measuring length, one produces the result in form of absolute error as l = 14.31 cm $\pm$ 0.18 cm. Here there is no sense in writing out the hundredths, because the reported error is of the order of tenths of a centimeter. It is more correct to write l = 14.3 cm $\pm$ 0.2 cm. The last numeral in 14.3 is unreliable because it may be anything from 1 to 5. Since in this digit, the error is less than one half of a centimeter, we can disregard this and write l = 14 cm involving only two digits which are significant. Is there any difference between l = 14 cm and l = 14.0 cm? They differ in number of significant figures. 14.0 cm has three significant figures, where the zero after the decimal point implies that the tenths of a centimeter were measured but were not registered.

Table 1.3

Physical quantity	Name and symbol for SI unit	Definition of SI	Non-SI units and their SI equivalents
Volume	cubic meter (m^3)	m^3	1 liter = 10^{-3} m^3
Force	newton (N)	kg m s^{-2}	1 dyne = 10^{-5} N
Pressure	pascal (Pa)	Nm^{-2} = kg m^{-1}s^{-2}	1 atm (760 torr) = 1.01325×10^5 Pa
			1 bar = 10^5 Pa
Energy	joule (J)	Nm = kg m^{2}s^{-2}	1 erg = 10^{-7} J
			1 cal = 4.184 J
			1 electron volt (eV) = 1.602177×10^{-19} J
Power	watt (W)	Js^{-1} = kg m^{2}s^{-3}	1 horse power = 735.499 W
Electric charge	coulomb (C)	As	1 abcoulomb (emu) = 10 C
			1 statcoulomb (esu) = 3.336×10^{-10} C
Electric potential diff	volt (V)	JC^{-1} = kg m^{2}s^{-3}A^{-1}	1 abvolt = 10^{-8} V
			1 statvolt = 300 V
Electric resistance	ohm (Ω)	VA^{-1} = kg m^{2}s^{-3}A^{-2}	1 abohm = 10^{-9} Ω
Electric conductance	siemens (S)	Ω^{-1}	1 statohm = 9×10^{11} Ω
Dipole moment	Coulomb meter (C·m)	C·m	1 Debye = 3.335641×10^{-30} C·m
Magnetic flux	weber (Wb)	Vs = kg m^{2}s^{-2}A^{-1}	1 maxwell = 10^{-8} Wb
Magnetic flux density (magnetic induction)	tesla (T)	Wbm^{-2} = kg s^{-2}A^{-1}	1 gauss = 10^{-4}T
Luminous flux	lumen (lm)	cd·Sr*	
Illuminance	lux (lx)	lm·m^{-2}	
Luminance	nit (nt)	cd·m^{-2}	1 stibb (= 1 cd cm^{-2} = 10^{-4} nt)

* Sr is abbreviation of steradian which is SI unit of solid angle

Notes:

i. In writing units, words and symbols should not be mixed. Thus, it is improper to write N per sq metre; this should be written either as newton per sq metre or Nm^{-2}. Words in a unit should be in small letters. It is improper to write Newton per sq meter.

ii. It is safe to represent combination of units leaving a space in between. For instance, C·m stands for coulomb meter (unit of dipole moment) while cm for centimeter (unit of length).

iii. In writing units using slant line one should be cautious. Thus, the unit JK^{-1}mol^{-1} of R might be written as J/K mol but never J/K/mol which implies J mol/K.

iv. The word degree and the symbol (°) should not be used with kelvin (K).

v. The definition of mole refers to 12 g, instead of 12 kg (as per SI), of ^{12}C to make *molecular weight* numerically (though not dimensionally) equal to mass per amount of substance in unit of g mol^{-1}.

vi. Exponents have effect also on prefixes. For example

$$1 \text{ dm} = 10^{-1} \text{ m}$$
$$1 \text{ dm}^3 = (10^{-1} \text{ m})^3$$
$$= 10^{-3} \text{ m}^3$$
$$\neq 10^{-1} \text{ m}^3$$

Algebraic operations with uncertain numerical quantities will obviously give uncertain results which can, however, be obtained in good approximation using the following simple rule. Generally, a numerical result will have same number of significant figures as that of the component number having least number of significant figures. Thus, although 7.46 m^2/3.2 m = 2.33125 m, it is enough to take the approximate value 2.3 m. But if the first significant figure of the result is 1, then the number of significant figures should be one more. For instance, 4.58 m^2/3.2 m = 143125 m can be better approximated to 1.43 m (and not 1.4 m).

In this book, it has been assumed that the numerical values are precise to three or at most four significant figures, and thus the answers are stated to at most four significant figures after rounding off the result at the last stage of calculation (usually done with a scientific calculator).

1.3 GUIDELINES FOR SOLVING NUMERICAL PROBLEMS

Solution of numerical problems involves calculation of some quantity (ies) from the given data using appropriate relation between them under the condition specified in the problem. Selection of working formula and substitution of given data in it will be facilitated, if the given data are presented in a compact form diagramatically (as in problem 4.22 or 4.27).

In calculating more than one quantity in a given problem, it is better to calculate each quantity independently (to avoid transmission error). If it is not possible, the one(s) which is calculable directly from the given data is computed first and the result is used to compute others using the relation between them.

2

Gaseous State—Ideal

In gaseous state, matter behaves in a rather simple way compared to liquid and solid states.

2.1 GAS LAWS

Under ordinary condition, gases unlike liquids and solids, approximately obey some simple laws, such as

Boyle's law: $V \propto 1/P$ for constant n and T

Charles' law: $V \propto T$ for constant n and P

Avogadro's law: $V \propto n$ for constant P and T

where P stands for pressure, V for volume, T for temperature and n for amount (number of moles) of the gas. These laws can be put together in the following single expression.

$$\left. \begin{aligned} PV &= nRT \\ &= \frac{w}{M}RT \end{aligned} \right\} \tag{2.1a}$$

where w stands for mass and M for molar mass (formerly, molecular weight) of the gas (average molar mass in case of a mixture). This expression, which holds exactly only at extremely low pressure, represents an equation of state of an ideal gas [In general, for a system characterised by p independent properties (state variables), an equation of state is defined to be an algebraic expression that connects any $p + 1$ properties of the system. For example, a homogeneous system of fixed mass and composition (uninfluenced by any external field) is characterised by only two independent properties and hence a functional relation $V = f(P, T)$ or $f(P, V, T) = 0$ between three simple properties, P, V and T will constitute an equation state of such a system, called a PVT system]. In case of a mixture of ideal gases

$$P_i V = n_i RT \tag{2.1b}$$

where n_i is the number of moles of the ith component and P_i is its partial pressure such that

$$\sum P_i = P, \text{ the total pressure}$$

which is the Dalton's law of partial pressure.

In an attempt to explain the above equation of state a simple theory, called the *kinetic theory*, has been developed using the molecular concept of matter with necessary assumptions (mentioned in problem 2.10) regarding the behaviour of gas molecules, which leads to the following relation

$$PV = \tfrac{1}{3}mN\langle c^2 \rangle \tag{2.2a}$$

where N is the number of molecules each of mass m present in volume V, and $\langle c^2 \rangle$ is the mean square speed of the molecules.

The expression (2.2a) connects the macroscopic properties, P and V of a gas with a microscopic quantity $\langle c^2 \rangle$. It reveals the significance of P and T as statistical quantities, and also enables one to derive the above equation of state readily. It follows from this expression that total molecular kinetic energy of translation is

$$N \cdot \tfrac{1}{2}m\langle c^2 \rangle = \tfrac{3}{2}PV \tag{2.2b}$$

$$= \tfrac{3}{2}RT, \text{ for 1 mole}$$

i.e. T is determined by molecular kinetic energy due to random motion.

2.2 DISTRIBUTION OF MOLECULAR VELOCITIES

Molecules in a gas are continually undergoing change in their velocities due to collision among themselves and also with the walls of the container. Yet at equilibrium (often inappropriately called *steady state*), i.e. when the macroscopic properties (such as density) of the gas do not change with time, the number of molecules having velocity within certain specified range does not change with time and is given according to Maxwell by

or

$$\left.\begin{aligned}
\frac{dN_u}{N} &= \left(\frac{m}{2\pi kT}\right)^{1/2} e^{-mu^2/2kT}\, du \\[2mm]
\frac{1}{N}\cdot\frac{dN_u}{du} &= \left(\frac{m}{2\pi kT}\right)^{1/2} e^{-mu^2/2kT}
\end{aligned}\right\} \tag{2.3a}$$

where dN_u is the number of molecules having velocity component lying between u and $u + du$ in a particular direction and k is the Boltzmann constant. This is **Maxwell's law of distribution of molecular velocities** which is based on the rules of probability assuming that the molecules are independent. Its extension to the three-dimensional motion leads to the following expression for the distribution of molecular speed.

$$\frac{1}{N}\cdot\frac{dN_c}{dc} = 4\pi\left(\frac{m}{2\pi kT}\right)^{3/2} c^2 e^{-mc^2/2kT} \tag{2.3b}$$

where dN_c is the number of molecules having speed between c and $c + dc$.

The distribution curves corresponding to Eqs (2.3a) and (2.3b) are drawn in Fig. 2.1(a) and (b).

Average Kinetic Parameters

From the Maxwell distribution, the average value $\langle g \rangle$ of any function of speed $g(c)$ can be found using the relation

$$\langle g \rangle = \int_{c=0}^{c=\infty} g(c)\frac{dN_c}{N} \tag{2.4a}$$

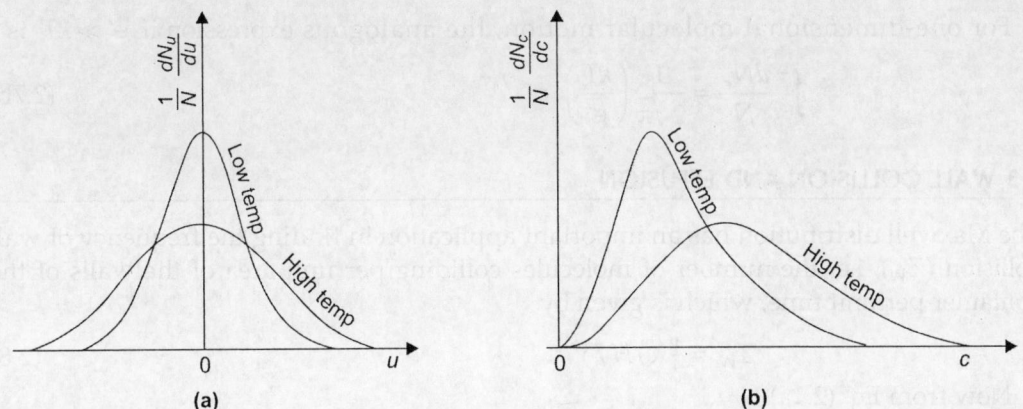

Fig. 2.1: (a) The distribution of the similar components of molecular velocities (b) The distribution of molecular speed

Thus we have the following two expressions

Average molecular speed
$$\langle c \rangle = \int_{c=0}^{c=\infty} c \frac{dN_c}{N} = \left(\frac{8kT}{\pi m} \right)^{1/2} \\ = \left(\frac{8RT}{\pi M} \right)^{1/2}$$

(2.4b)

Root mean square speed
$$\sqrt{\langle c^2 \rangle} = \left[\int_{c=0}^{c=\infty} c^2 \frac{dN_c}{N} \right]^{1/2} = \left(\frac{3kT}{m} \right)^{1/2} \\ = \left(\frac{3RT}{\pi M} \right)^{1/2}$$

(2.4c)

The most probable speed
$$C_{mp} = \left(\frac{2kT}{m} \right)^{1/2} = \left(\frac{2RT}{M} \right)^{1/2}$$

(2.4d)

The most probable speed is found by differentiating Eq. (2.3b) wrt c and putting the result to zero. The speed distribution Eq. (2.3b) can be transformed into the following energy distribution equation by putting $\frac{1}{2}mc^2 = \epsilon$, the molecular energy

$$\frac{1}{N} \cdot \frac{dN_\epsilon}{d\epsilon} = \frac{2}{\sqrt{\pi}} \left(\frac{1}{kT} \right)^{\frac{3}{2}} \epsilon^{\frac{1}{2}} e^{-\epsilon/kT}$$

(2.5)

This gives the following expression for the most probable molecular energy

$$\epsilon_{mp} = \frac{1}{2}kT$$

(2.6)

The fraction of molecules having energy above ϵ' is

$$\int_{\epsilon'}^{\infty} \frac{dN_\epsilon}{N} = \frac{2}{\sqrt{\pi}} \cdot \left(\frac{1}{kT} \right)^{\frac{3}{2}} \left[kT(\epsilon')^{1/2} e^{-\epsilon'/kT} + \frac{kT}{2} \int_{\epsilon'}^{\infty} \epsilon^{-\frac{1}{2}} e^{-\epsilon/kT} d\epsilon \right]$$

$$= \frac{2}{\sqrt{\pi}} \left(\frac{\epsilon'}{kT} \right)^{1/2} e^{-\epsilon'/kT}$$

(2.7a)

For one-dimensional molecular motion, the analogous expression if $\epsilon' \gg kT$ is

$$\int_{\epsilon'}^{\infty} \frac{dN_\epsilon}{N} = \frac{1}{2\sqrt{\pi}} \left(\frac{kT}{\epsilon'}\right)^{\frac{1}{2}} e^{-\epsilon'/kT} \tag{2.7b}$$

2.3 WALL COLLISION AND EFFUSION

The Maxwell distribution has an important application in finding the frequency of wall collision (Z_W), i.e. the number of molecules colliding per unit area of the walls of the container per unit time, which is given by

$$Z_W = \tfrac{1}{4}\langle c\rangle N/V \tag{2.8}$$

Now from Eq. (2.1a)

$$N/V = P/kT \tag{2.9}$$

where $k = R/N_A$ (N_A is Avogadro constant). Then it follows that effusion, i.e. the molecular flow through a small hole or porous vessel, will be governed by the following relation

$$\text{rate of effusion} \propto \frac{P}{\sqrt{T}}$$

which is Graham's law of effusion.

2.4 INTERMOLECULAR COLLISION AND MEAN FREE PATH

Although intermolecular collision is a more complex phenomenon than wall collision, the situation can be comfortably tackled treating the motion of a pair of colliding molecules as the motion of a single molecule, i.e. considering relative speed C_r instead of their actual speed C. This requires introduction of a quantity, called reduced mass μ, which is defined, for a pair of two types of molecules A and B, as

$$\frac{1}{\mu} = \frac{1}{m_A} + \frac{1}{m_B} = \frac{2}{m} \text{ when } A \equiv B$$

All these considerations lead to the following expression of collision frequency Z_1 for a single molecule in a pure gas (i.e. number of collisions made by a single molecule with other molecules per unit time)

$$Z_1 = \pi\sigma^2 \langle C_r\rangle \frac{N}{V} \tag{2.10a}$$

$$= \pi\sigma^2 \left(\frac{8kT}{\pi\mu}\right)^{\frac{1}{2}} \frac{N}{V}$$

$$= \sqrt{2}\pi\sigma^2 \left(\frac{8kT}{\pi m}\right)^{\frac{1}{2}} \frac{N}{V}$$

$$= \sqrt{2}\pi\sigma^2 \langle c\rangle \frac{N}{V} \tag{2.10b}$$

where σ is the collision diameter which is the closest distance of approach between two colliding molecules and $\pi\sigma^2$ is the collision cross-section.

The frequency of binary collision Z_{AA} between any two molecules in a single gas (or like molecules in a mixture) per unit volume is

$$Z_{AA} = \frac{1}{2} Z_1 \frac{N}{V}$$

$$= \frac{1}{2} \pi \sigma^2 \langle C_r \rangle \left(\frac{N}{V} \right)^2 \tag{2.11a}$$

$$= \frac{1}{\sqrt{2}} \pi \sigma^2 \langle C_r \rangle \left(\frac{N}{V} \right)^2 \tag{2.11b}$$

[The factor 1/2 is introduced otherwise every collision will be counted twice]

The frequency of binary collisions Z_{AB} between unlike molecules is

$$Z_{AB} = \pi \sigma_{AB}^2 \langle C_r \rangle \left(\frac{N}{V} \right)_A \left(\frac{N}{V} \right)_B \tag{2.12a}$$

$$\left. \begin{array}{l} = \pi \sigma_{AB}^2 \left(\dfrac{8kT}{\pi \mu} \right)^{1/2} \left(\dfrac{N}{V} \right)_A \left(\dfrac{N}{V} \right)_B \\[3mm] = \pi \sigma_{AB}^2 \sqrt{\langle C_A \rangle^2 + \langle C_B \rangle^2} \cdot \left(\dfrac{N}{V} \right)_A \left(\dfrac{N}{V} \right)_B \end{array} \right\} \tag{2.12b}$$

where $\sigma_{AB} = \frac{1}{2}(\sigma_A + \sigma_B)$ is the average collision diameter. The frequency of binary collision is an important factor determining the rate of a gaseous reaction.

The average distance a molecule moves between two successive collisions is called mean free path λ of the species concerned. It is given by

$$\left. \begin{array}{l} \lambda = \langle c \rangle / Z_1 \\[3mm] = \dfrac{1}{\sqrt{2} \pi \sigma^2 (N/V)} \end{array} \right\} \tag{2.13}$$

For a gas the mean free path is of vital importance in dealing with different transport phenomena, viz. diffusion (the migration of matter down a concentration gradient in absence of any bulk flow of the system), viscosity (the migration of linear momentum down a velocity gradient) and thermal conduction (the migration of energy down a temperature gradient). The corresponding transport coefficients (i.e. flux per unit value of the gradient) for an ideal gas can be expressed in the following forms

self-diffusion coefficient, $\quad D = \frac{1}{3} \langle c \rangle \lambda$ $\hfill (2.14)$

viscosity (coefficient), $\quad \eta = \frac{1}{3} \left(\dfrac{N}{V} \right) m \langle c \rangle \lambda$ $\hfill (2.15)$

thermal conductivity, $\quad K = \frac{1}{3} \left(\dfrac{N}{V} \right) C_V \langle c \rangle \lambda$ $\hfill (2.16)$

where $C_V = \dfrac{\overline{C_V}}{N_A}$

It is due to collision with the surrounding molecules that the intensity of a molecular beam falls from I_0 to I when it passes through a gas of thickness x

$$I = I_0 e^{-x/\lambda} \tag{2.17}$$

The factor $e^{-x/\lambda}$ is a measure of the chance for a molecule in the beam to travel a distance x in a gas without any collision with the gas molecules.

2.5 EQUIPARTITION PRINCIPLE AND HEAT CAPACITY OF GASES

According to the simple kinetic theory, a gas can store energy only through translational motion of its molecules which are assumed to be of point masses, i.e. structureless. But to account for heat capacity of gases, it has been subsequently assumed that the molecules have certain structure so that they can store energy also through their rotational and vibrational motion. The thermal energy of a molecule (i.e. one due to random molecular motion) is determined by its number of possible independent motions, called number of degrees of freedom, which is equal to the number or coordinates required to specify the positions of all the atoms in the molecule. For a molecule of atomicity x, total number of degrees of freedom is $3x$ of which 3 is due to translational motion, 2 or 3 is due to rotational motion according as the molecule is linear or not, and the rest is due to vibrational motion.

According to the law of equipartition of kinetic energy, in a system of molecules in thermal equilibrium at temperature T the mean kinetic energy is equally distributed between the various degrees of freedom and for each of them, it is equal to $\frac{1}{2}kT$ per molecule (where k is Boltzmann constant).

Any oscillating system has not only kinetic but also potential energy, and for a harmonic oscillator, the average value of each kind is same. Then assuming the molecules to behave as harmonic oscillators, the energy associated with a vibrational degree of freedom will be $2 \times \frac{1}{2}kT$ or kT per molecule.

Now molar heat capacity $\bar{C}_V = \left(\dfrac{\partial \bar{U}}{\partial T}\right)_V$, where $\bar{U}$ is the thermal energy per mole. Then according to equipartition principle

$$\bar{C}_V = 3N_A \cdot \tfrac{1}{2}k \quad \text{for monatomic species}$$
$$= \tfrac{3}{2}R \tag{2.18}$$
$$\bar{C}_V = 3N_A \cdot \tfrac{1}{2}k + 2N_A \cdot \tfrac{1}{2}k + (3x-5)N_A k, \text{ for linear molecules of atomicity } x$$
$$= \tfrac{3}{2}R + R + (3x-5)R \tag{2.19}$$
$$\bar{C}_V = 3N_A \cdot \tfrac{1}{2}k + 3N_A \cdot \tfrac{1}{2}k + (3x-6)N_A k, \text{ for non-linear molecules}$$
$$= \underset{\substack{\text{translational} \\ \text{contribution}}}{\tfrac{3}{2}R} + \underset{\substack{\text{rotational} \\ \text{contribution}}}{\tfrac{3}{2}R} + \underset{\substack{\text{vibrational} \\ \text{contribution}}}{(3x-6)R} \tag{2.20}$$

The equipartition principle cannot explain the facts: (i) $\bar{C}_V$ varies (increases) with temperature and (ii) the observed $\bar{C}_V$ equals the predicted one in case of monatomic but not polyatomic gases. In the latter case, the observed $\bar{C}_V$ is always less than the predicted one at ordinary temperature, but approaches the latter at very high temperature. The equipartition principle fails because it is based on the classical concept of energy as a continuous variable which nearly holds at ordinary temperature for monatomic species but only at much higher temperature for polyatomic molecules. This can be explained by the quantum theory according to which energy always appears in discrete packets called quanta which are quite small for translational energy but much higher with rotational energy and more so with vibrational energy. Consequently, the equipartition principle holds, though only approximately, at ordinary temperature for monatomic species which can undergo only translational motion, but at much higher temperature for polyatomic molecules due to their rotational and vibrational motions.

AUXILIARY PROBLEMS

2.1 In gaseous state, matter behaves in a rather simple way compared to liquid and solid states. Explain.

2.2 What is meant by an equation of state? Which of the following expression may be regarded as an equation of state of a gas?

    i. PV = constant

    ii. V/T = constant

    iii. PV/T = constant

    iv. $PV = nRT$

2.3 For a fixed amount (n) of a hypothetical gas, if PV is constant at constant T, will V/T be constant at constant P?

2.4 Unlike N_2 and O_2, NO_2 exhibits much variation of PV with P for fixed mass and temperature (i.e. much deviation from Boyle's law). Why?

2.5 The vapour density of an equilibrium mixture of N_2O_4 and NO_2 is 35. Find the degree of dissociation of N_2O_4.

2.6 One liter of a mixture of He and Ar weighs 0.5 g at STP. Find average molecular weight and composition of the gas mixture.

2.7 A 2 liter vessel contains H_2 at STP. Some He is introduced into this vessel till the mass density of the gas mixture becomes double. Find

    i. Amount of He introduced

    ii. Average molecular weight

    iii. Partial pressure of H_2

    iv. Total pressure of the gas mixture

2.8 A rubber balloon filled with D_2 at STP is placed in a box containing H_2 also at STP. What happens to the size of the balloon if the rubber is permeable (i) only to D_2 (ii) only to H_2 (iii) both to H_2 and D_2.

2.9 Suggest a simple experiment by which pure HD can be distinguished from an equimolecular mixture of H_2 and D_2.

2.10 Comment on the necessity and justification of the following assumptions of kinetic theory.

    i. A gas is made up of a large number of molecules moving at random

    ii. Molecules are like point masses

    iii. No forces act on the molecules except during their collision with one another or with walls of their container

    iv. Molecular collisions are elastic

2.11 The molecules of an ideal monatomic gas cannot remain at rest though a piece of diamond, which may be regarded as a macromolecule, can do. Explain. Can a piece of diamond be made to move translationally simply by heating?

2.12 The pressure of a gas results from the successive collisions of its molecules on the walls of the containing vessel. Then a steady gas pressure is unexpected. Discuss.

2.13 RMS speed is greater than the average speed of the gas molecules under usual conditions. Why? Under what condition do they become equal? Which one is of primary importance in dealing with pressure of a gas?

2.14 The relation $P = \frac{1}{3} m (N/V) \langle c^2 \rangle$ gives the instantaneous pressure of an ideal gas. Do you agree?

2.15 The relation $P = \frac{1}{3} m (N/V) \langle c^2 \rangle$ suggests a change in molecular speed (c) of an ideal gas with change in pressure (P). Is it?

2.16 Air gets heated on pumping into a tyre. Why? Will this happen with an ideal gas?

2.17 According to kinetic theory, temperature of a system is a measure of average translational kinetic energy of the molecular motion of the system. Then temperature should have no meaning with a solid whose constituent particles are devoid of any translational motion. Is it justified?

2.18 In a gas the molecules do not all have same speed. Is it justified to treat the molecules as hot or cold depending on their high or low speed?

2.19 If a vessel containing some gas moves, the molecular speed of the gas will increase and hence, according to kinetic theory, its temperature should increase. But this is contrary to our experience. Resolve this contradiction.

2.20 Comment on the validity of the equation of state $P\bar{V} = RT$ for an ideal gas under the following situations.
   I. One mol of a gas just after its introduction in a vessel
   II. One mol of a gas long after its introduction into a vessel
   III. The gas in the field of gravity
   IV. The gas in a free-falling container
   V. The gas consisting of only one molecule

2.21 An ideal gas expands when its volume becomes double and pressure becomes half. What is the change in kinetic energy of the gas molecules? Is any work available from this expansion?

2.22 Two vessels of equal volume are connected through a stop-cock. Initially both the vessels contain H_2 at $0°C$, but the gas pressure is 1 atm in one vessel and 2 atm in the other. If the stop-cock is opened what will be the final pressure? Also find the pressure if the temperature of the first vessel is lowered by $50°C$ and that of the second is raised by $50°C$.

2.23 A closed thermally insulated cylindrical vessel is divided into two parts by a freely movable partition which is initially at a fixed position. The two parts are filled with an ideal monatomic gas whose pressure (P), volume (V) and temperature (T) at the initial stage are indicated in the figure

$$\boxed{P_1, V_1, T_1 \mid P_2, V_2, T_2}$$

Find expressions for equilibrium pressure and temperature
(a) when the partition is removed
(b) when the partition is allowed to move assuming it to be a (i) conductor of heat (ii) non-conductor of heat.

2.24 Can a gas have all its molecules moving with same velocity or speed? If so, what crucial effect will it have in the atmosphere?

2.25 Maxwell's law for the distribution of molecular velocities does not refer to specified values of velocities but to ranges of velocities around them. Why?

2.26 The average value of the components of molecular velocities in a particular direction is zero. Why? Would the corresponding average value of energy be also zero?

2.27 (a) If the speed distribution function $P(c)$ is given by $P(c) \, dc = A c^m e^{-Bc^2} dc$, find dimensions of A and B. Also find a relation between A and B for $m = 2$.

(b) The exponential term in the distribution law has a –ve exponent. Does a +ve sign in it make any sense?

2.28 Express Maxwell's speed distribution law in a form independent of molecular mass and temperature. With the help of this, find the fraction of molecules having speed above (i) most probable speed (ii) mean speed (iii) rms speed.

Interpret the result. What fraction of molecules will have kinetic energy above the mean kinetic energy?

2.29 State whether Maxwell's distribution of speed or translational energy will hold in the following cases.

   i. An ideal gas at STP

   ii. A gas containing only a few molecules

   iii. A real gas at STP

   iv. A gas in gravitational field

   v. An effusing gas

   vi. A liquid

2.30 (a) 10 moles of an ideal gas each of molecular mass 6.64×10^{-24} g and speed 10^6 cm/s are initially confined in a thermally insulated vessel of volume 1 liter. Find

   i. $\langle c \rangle$ and $\sqrt{\langle c^2 \rangle}$, initially and also after a long time

   ii. equilibrium temperature

   iii. fraction of molecules having speed higher than 10^6 cm/s at equilibrium.

(b) Would the results be same if the number of molecules in the gas were just 10?

2.31 In an ideal gas 20% of the molecules have speed 100 m/s, 50% have speed 400 m/s and 30% have speed 200 m/s. Find mean speed and rms speed of the gas molecules. Would the calculated mean speed be same, less than or greater than that expected if the distribution of speed were Maxwellian?

2.32 A 10 L vessel contains 1 mole of H_2 molecules having average (Maxwellian) speed 5.5×10^5 cm/s. Find the gas pressure. Would the gas pressure be different, if the molecules all moved initially with the same speed equal to (i) $\sqrt{\langle c^2 \rangle}$ (ii) $\langle c \rangle$ at equilibrium? If so to what extent would it be so?

2.33 (a) Calculate the average speed of the H_2 molecules at atmospheric pressure, given the density of hydrogen equal to 0.00009 g/mL.

   (b) Also find the same for He molecules under the same condition of temperature and pressure.

   (c) Explain why earth's atmosphere contains insignificant amount of H_2 but significant amount of He (~10 times greater than that of H_2).

2.34 A 30 L vessel contains 100 g of O_2 under a pressure of 3×10^5 P_a. Find (i) C_{mp} (ii) $\langle c \rangle$ (iii) $\sqrt{\langle c^2 \rangle}$ (iv) mean square deviation $\langle (c - \langle c \rangle)^2 \rangle$ (v) mean kinetic energy (vi) most probable kinetic energy (vii) mean square deviation of kinetic energy of O_2 molecules.

2.35 In the context of the previous problem, find the fraction of O_2 molecules (i) whose speed lies within ± 1.0% of the most probable speed (ii) whose kinetic energy lies within ± 1.0% of the most probable kinetic energy (iii) whose kinetic energy lies above most probable kinetic energy.

2.36 A 2 L vessel contains He at 2 atm and a 5 L vessel contains Ne at 1 atm, both at same temperature. In which case will the gas have higher (i) molecular kinetic energy

(ii) most probable molecular speed (iii) most probable molecular momentum. Assume ideal behaviour of the gases.

2.37 Smoke particles in air, observed through microscope, are found to execute zig-zag motion, called Brownian motion. Find average translational energy and average speed of such particles at 27°C taking their masses of the order of 10^{-16} kg. Would these quantities depend on atmospheric pressure and size of the particle? Are these quantities same as those for oxygen?

2.38 In an experiment on Brownian motion, the particles each of mass 6.2×10^{-17} kg, suspended in a liquid at 27°C are found to move with a mean speed of 1.3×10^{-2} m/s. Calculate Avogadro constant. What would happen to Brownian motion if the Avogadro constant were infinite? Would the molar gas constant (R) and Boltzmann constant (k) be affected with change in Avogadro constant?

2.39 Calculate the probability that the velocity component of N_2 molecule lies between 999.5 and 1000.5 m/s at 300 K in a particular direction. Will this probability be same if all the molecules move in the chosen direction?

2.40 Consider some gas enclosed in a long, extremely narrow tube. On the basis of Maxwell's distribution law deduce expressions for (i) average molecular velocity (ii) average molecular speed (iii) mean square velocity (iv) mean square speed (v) most probable velocity and (vi) average molecular energy of translation. Depict the relevant velocity and speed distribution in the same plot. How will the expressions in question change if the gas is enclosed in a spherical vessel?

2.41 (a) Draw Maxwell energy distribution curve for three-dimensional motion of gas molecules and point out its differences from the corresponding speed distribution curve.

(b) How do the speed distribution curves for He and Ne differ? How do they look like at (i) $T = 0$ and (ii) $T = \infty$.

2.42 Write down Maxwell's law of distribution of molecular speed over a two-dimensional plane surface, and hence deduce expressions for (i) most probable molecular speed (ii) most probable molecular velocity (iii) average energy $\langle \epsilon \rangle$ of the molecules and (iv) fraction of molecules having energy $\geq \langle \epsilon \rangle$. Represent this speed distribution diagrammatically. How does it differ from that for three-dimensional motion?

2.43 A sample of He is heated to 1000 K in an oven. He molecules emerge through a small hole of the oven forming a beam. What will be the mean speed, mean velocity and most probable velocity of the He molecules (a) inside the oven (b) in the beam. Explain the difference in result.

2.44 For oxygen at STP, calculate (i) molecular density (n) (ii) average intermolecular distance (d) (iii) mean free path (λ) and (iv) collision frequency in the gas for a single molecule (Z_1). The bond distance in O_2 is 120 pm.

2.45 Recalculate the quantities in the previous problem under extreme vacuum condition of the order of 10^{-13} atm.

2.46 Consider an equimolecular gaseous mixture of ortho-H_2 and para-H_2. Here three types of binary collisions are possible. Find their relative frequency of occurrence.

2.47 A cylindrical vessel is partitioned into two equal compartments. Initially one compartment contains H_2 and the other contains D_2 at same pressure and temperature. If the partition is removed, what will happen to λ of H_2 and D_2? (Assume $\sigma_{D_2} = \sigma_{H_2}$).

2.48 For H_2 gas (molecular diameter 1 Å) at STP, find
    i. the average time between two successive collisions
    ii. the average time for a molecule to travel unit distance

2.49 What is the chance for a O_2 molecule to travel a distance (i) 10^{-6} mm (ii) 1 mm and (iii) λ (mean free path) without experiencing a collision, in oxygen gas at STP. Use the data in problem 2.44.

2.50 What distance a molecular beam of N_2 is expected to travel in an environment of nitrogen gas without scattering not more than 50% of the molecules initially forming the beam?

2.51 Consider an atomic beam of K passing through Ar at 7.5×10^{-5} torr and 27°C in a cell of length 2 cm. Find attenuation of the beam, taking K–Ar collision cross-section to be 6×10^{-14} cm^2.

2.52 How does mean free path of an ideal gas change when its pressure is increased
    i. by compression at constant temperature
    ii. by heating at constant volume

2.53 At very low pressure, the mean free path in a gas exceeds the dimensions of the container. Do any intermolecular collisions occur at this stage? Does the concept of mean free path work in this situation? Does Maxwell distribution law hold?

2.54 The kinetic derivation of the relation $P = \frac{1}{3}mn\langle c^2 \rangle$ is often done assuming that the gas molecules collide only with the walls of the container and not with each other. At what gas pressure at 27°C is this assumption justified for molecules of diameter 10^{-8} cm in a cubic vessel of volume 1 L?

2.55 Mean free path has no meaning in the upper atmosphere. Explain.

2.56 The atmosphere is cooler at the top of a mountain than at sea level, though hot air rises up. Explain.

2.57 What is the role of inert gas in the filament lamps?

2.58 Assuming D_2 and H_2 to have nearly the same collision diameter, which one should have higher thermal conductivity? Would you expect ^{4}He or ^{3}He to be better thermal insulator?

2.59 Molecules of a gas are moving at high speed, yet effusion of a gas is a slow process while propagation of sound through a gas is a fast process of the order of the molecular motion. Explain.

2.60 Can speed of sound in a gas exceed the average speed of gas molecules?

2.61 At ordinary temperature $\langle c \rangle$ of N_2 is 475 m/s. Find velocity of sound in N_2.

2.62 At every collision, gas molecules change their direction of motion and it may so happen that a gas molecule comes to its initial position. Hence diffusion of a gas would be most favourable in the absence of intermolecular collision. Do you agree?

2.63 Do you think that gases mix by diffusion more rapidly at higher pressure? Would the mixing rate be higher or lower at higher temperature?

2.64 The vapour pressure of water is 23.7 torr at 25°C. Assuming that H_2O molecules striking the surface of liquid water stick to it, find the rate of evaporation of water into dry air at 25°C.

2.65 A mixture of H_2 and He is prepared such that the frequency of wall collision of each gas is same. Will the frequency of collision between the like molecules be also same for the two gases? Which gas has the higher concentration?

2.66 Find (i) the most probable speed (ii) the average speed, of the gas molecules that collide with a small surface area.

2.67 (a) Show that the average relative kinetic energy with which bimolecular collisions occur in a gas is 2 kT. (b) Find the fraction of bimolecular collisions in which the relative kinetic energy of two colliding molecules exceeds ϵ'.

2.68 Show that rate of effusion of a gas $\propto P/\sqrt{mT}$, assuming ideal behaviour, and hence arrive at the Graham's law of effusion.

2.69 (a) A 10 L container with 0.1 mol of H_2 gas in it has a leakage hole of radius 0.001 mm. The root mean square speed of the H_2 molecules in the container is 400 m/s. How many H_2 molecules will initially escape through this hole per minute? How long would it take for (i) 0.20% of the H_2 (ii) 75% of the H_2 to escape? Mention the condition necessary for calculated values to be reliable.

(b) Find the pressure of the remaining gas after 10 minutes.

2.70 A mixture of Ne and Ar in 1:5 mol ratio effuses through a small orifice into an evacuated space. What is the composition of the effused gas mixture (i) initially (ii) after a long time?

2.71 In calculating degrees of freedom an atom is considered as a point mass though it contains nucleus and electrons separated by certain distance. How would you justify this?

2.72 Compare vibrational degrees of freedom for the molecules in each of the following pairs: (i) H_2O_2 and C_2H_2 (ii) BF_3 (planar) and NF_3 (pyramidal). Which will have higher heat capacity — H_2O_2 or C_2H_2?

2.73 Find, according to the principle of equipartition of energy, the high temperature limiting value of molar heat capacity at constant volume for an ideal gas consisting of polyatomic non-linear molecules. Why the heat capacity values rise rapidly with temperature for a polyatomic gas and reach a limiting maximum?

2.74 Using equipartition principle, set up an expression connecting atomicity of a gas with its $\gamma = C_P/C_V$.

2.75 For CO_2 and SO_2 gases at the same temperature and pressure, what will be their relative order of the following quantities.

  i. RMS speed of the molecules

  ii. Molar kinetic energy due to translational molecular motion

  iii. Molar kinetic energy due to all types of molecular motion

  iv. Molar thermal energy

  v. Molar heat capacity $\bar{C}_V$ introduced

  vi. $\bar{C}_P - \bar{C}_V$

2.76 What happens to $\bar{C}_V$ of gaseous chlorine when its temperature is raised?

ANSWERS

2.1 Because the intermolecular force (that determines potential energy) is negligible in gaseous state compared to that in liquid and solid states so that a gas may be treated as a collection of free particles, particularly at low pressure and high temperature.

Note: Gases exhibit simple behaviour (following the simple equation of state $P\overline{V} = RT$ at low molecular density) despite higher randomness of molecular motion in gaseous state. Because, properties of a molecule depend more on interaction with its surrounding molecules than on the molecular arrangement.

2.2 For a system characterised by p independent properties (state variables), an equation of state is defined to be an algebraic expression that connects any $(p + 1)$ properties of the system. For example, a homogeneous system of fixed mass and composition (uninfluenced by any external field) is characterised by only two independent properties and hence a functional relation $V = f(P, T)$ or $f(P, V, T) = 0$ between three simple properties—pressure (P), volume (V) and temperature (T) will constitute an equation of state of such a system (called a PVT system).

Then each of the given expression may be regarded as an equation of state for an ideal gas, of course under specified condition(s) which is constant amount (n) and constant T for (i), constant n and constant P for (ii), constant n for (iii), and not any for (iv). Generally (iv) is referred to an equation of state of an ideal gas.

Note: A heterogeneous system has no general equation of state in terms of P, V, T and composition, though each of its phase does have.

2.3 Not necessarily. Only if the involved constant (which is a function of T) is proportional to T (i.e. $PV \propto T$ as in an ideal gas), will V/T be constant for constant P and constant n.

Note: The abidance of Boyle's law is a necessary, but not sufficient condition for Charles law to hold.

2.4 For a gas to obey Boyle's law, its temperature (T) and number of moles (n) are required to be constant. NO_2, unlike N_2 and O_2, exhibits much deviation from this law due to variation of n with P (even for fixed mass) resulting from the following reaction

$$NO_2 \rightleftharpoons \tfrac{1}{2}N_2 + O_2$$

2.5 Starting with 1 mole of N_2O_4, we have, at equilibrium

$$N_2O_4 \rightleftharpoons 2NO_2 \quad \text{Total}$$

number of moles $1 - \alpha$ 2α $1 + \alpha$, α is the degree of dissociation

Now 2 vap density $= M_{av} = \dfrac{(1 - \alpha)\,M_{N_2O_4} + 2\alpha M_{NO_2}}{1 + \alpha}$

$$2 \times 35 = \frac{(1 - \alpha) \times 92.02 + 2\alpha \times 46.01}{1 + \alpha}$$

whence $\alpha = 0.3146$

2.6 $M_{av} = \dfrac{w}{PV}\,RT$ from Eq. (2.1)

$$= \frac{(0.5\,\text{g})\,(8.206 \times 10^{-2}\ \text{L atm K}^{-1}\,\text{mol}^{-1})\,(273.15\,\text{K})}{(1\,\text{atm})\,(1\,\text{L})} = 11.21\,\text{g mol}^{-1}$$

☐ Now $M_{av} = x\,M_{He} + (1 - x)\,M_{Ar}$, where x is the mole fraction of He

Then $11.21 = x \times 4 + (1 - x)\,40$

whence $x = 0.7997$

2.7 $$w = \frac{MPV}{RT}$$ from Eq. (2.1a)

i. Doubling of mass density at constant volume implies doubling of mass.

Then mass of He introduced = mass of H_2

$$= \frac{(2.016 \text{ g mol}^{-1})(1 \text{ atm})(2 \text{ L})}{(8.206 \times 10^{-2} \text{ L atm K}^{-1}\text{mol}^{-1})(273.15 \text{ K})}$$

$$= 0.1799 \text{ g}$$

ii. $$M_{av} = \frac{\text{mass of the gas mixture}}{\text{total number of moles}}$$

$$= \frac{2w}{\dfrac{w}{M_{H_2}} + \dfrac{w}{M_{He}}}$$

$$= \frac{2}{\dfrac{1}{2.016 \text{ g mol}^{-1}} + \dfrac{1}{4.003 \text{ g mol}^{-1}}} = 2.682 \text{ g mol}^{-1}$$

iii. Partial pressure of H_2 = pressure due only to H_2

$$= 1 \text{ atm}$$

iv. Total pressure $= \dfrac{w_{total}RT}{M_{av}V}$

$$= \frac{(2 \times 0.1799 \text{ g})(8.206 \times 10^{-2} \text{ L atm K}^{-1}\text{mol}^{-1})(273.15 \text{ K})}{(2.682 \text{ g mol}^{-1})(2 \text{ L})}$$

$$= 1.503 \text{ atm}$$

Alternatively

$$\frac{P_{total}}{P_{H_2}(=1 \text{ atm})} = \frac{n_{H_2} + n_{He}}{n_{H_2}} = 1 + \frac{n_{He}}{n_{H_2}} = 1 + \frac{(w/M)_{He}}{(w/M)_{H_2}}$$

$$= 1 + \frac{M_{H_2}}{M_{He}}, \text{ as } w_{He} = w_{H_2} \text{ (given)}$$

Then $$P_{total} = \left(1 + \frac{2.016}{4.003}\right) \times 1 \text{ atm} = 1.503$$

Note: Introduction of He causes x_{H_2} to decrease and P_{total} to increase such that the partial pressure $p_{He}\ (= x_{H_2}P_{total})$ remains unchanged with its value same as that of the pure gas pressure.

2.8 i. Here effusion, possible only with D_2, will reduce the number of gas molecules in the balloon leading to contraction of the balloon.

ii. Here the balloon will expand due to effusion only of H_2 that enhances the number of gas molecules in the balloon.

iii. Here H_2 will effuse into the balloon while D_2 will effuse out of it. But due to greater rate of effusion of H_2 (due to its lower molecular weight) there will be a net increase in number of gas molecules in the balloon leading to expansion of the latter.

2.9 The distinction cannot be made simply from determination of vapour density which is same for HD and an equimolecular $H_2 + D_2$ mixture. However, this can be done through effusion followed by measurement of vapour density of the effused gaseous sample. A change in vapour ensity will be observed with $H_2 + D_2$ mixture (but not with HD) due to change in composition resulting from the difference in rate with which H_2 and D_2 effuse.

2.10   i. The number of molecules should necessarily be large so that the frequency of intermolecular collision becomes high enough to make translational motion of the gas molecules quite random which is essential for uniform (equilibrium) distribution of the gas molecules.

   ii. Taking the assumption that the molecules are like point masses (i.e. have mass but no volume), the volume available for free movement of the gas molecules may simply be put equal to the volume of the container. Further, the total number of intermolecular collisions at any instant may be disregarded compared to the total number of molecules, and then the properties of a gas may be conveniently regarded as properties of free molecules.

   iii. Because the molecules (obeying Newton's law) will then move between any two successive collisions, in straight line (the simplest path) with constant speed.

   iv. The assumption that collisions between gas molecules (and also wall collisions) are elastic is necessary to avoid the possibility of a ridiculous gaseous state where the gas molecules have no translational kinetic energy (i.e. no translational motion) due to its conversion into other forms of energy such as rotational, vibrational, etc.

All these assumptions have been made in order that the behaviour of a gas can be represented in a mathematically simple way. But except (i), other assumptions have no rigid justification for real gases. (ii) and (iii) are fairly justified only at low molecular density. The assumption (iv) is not likely to be justified if the translational kinetic energy of the colliding molecules is comparable to the energy quantum for other types of molecular motion, viz. rotational, vibrational, etc. However, with large number of gas molecules at equilibrium, it is quite likely that on the average there is no net change of translational kinetic energy due to molecular collision.

2.11 This is justified by the kinetic theory which initially assumes that the molecules of gas are like point passes (i.e. structureless) without any interaction between them (except at the moment of their mutual collision). Then the inherent energy of such molecules will make their translational motion inevitable as they can neither rotate nor vibrate. But a piece of diamond need not move translationally for its energy content, as it can acquire energy through rotation and vibration possible due to its structure.

☐ No. Because this amounts to generation of directed macroscopic motion from random molecular motion (associated with heat) which cannot occur automatically. However, the converse happens naturally, e.g. when a falling body hits the ground. We can understand this if we remember that it is far more difficult to develop morality than to destroy it.

2.12 With enormous number of molecules assumed to be present in a gas, the frequency of wall collision at any instant is so high that its fluctuation with time occurs in imperceptible proportions making the gas pressure virtually steady.

If, however, the molecular density is not sufficiently high, the gas pressure will not become steady.

2.13   Average speed $\langle c \rangle = x_1 c_1 + x_2 c_2 + \ldots + x_n c_n$

   RMS speed $\sqrt{\langle c^2 \rangle} = \sqrt{x_1 c_1^2 + x_2 c_2^2 + \ldots + x_n c_n^2}$

where x_n is the fraction of molecules having speed c_n.

Since $x_n < 1, \sqrt{x_n c_n^2}$ (i.e. $\sqrt{x_n} \cdot c_n) > x_n c_n$ which makes $\sqrt{\langle c^2 \rangle} > \langle c \rangle$.

☐ $\sqrt{\langle c^2 \rangle} = \langle c \rangle$ becomes possible under the condition $c_1 = c_2 = \ldots = c_n$ which happens at absolute zero when speed of each molecule becomes zero.

☐☐ RMS speed because it is directly linked to the gas pressure.

2.14 No. Because $\langle c \rangle$, and hence P calculated from the given expression, is an invariable quantity for a gas at a fixed temperature (molecular collisions being assumed to be elastic). The calculated P would therefore correspond to the time averaged observed pressure and not to the variable instantaneous pressure unless molecular density is sufficiently high (vide problem 2.12).

2.15 From the given expression

$$\langle c^2 \rangle = \frac{3PV}{mN} = \frac{3kT}{m} \text{ since } PV = nkT \text{ for an ideal gas}$$
$$= \text{constant at constant } T$$

Then the given statement will not hold at constant T when $P \propto N/V$.

Note: Of the different factors involved in an expression, any two will mutually vary provided one of them does not bear any proportional relation with any of the rest factors.

2.16 Pumping of air into a tyre involves work done on air. This causes air molecules to have some directed motion which ultimately randomises (through intermolecular collisions) with consequent rise of temperature.

☐ Yes. Because pumping always involved work done on the gas, whether it is air or an ideal gas.

2.17 According to kinetic theory, temperature of a system is a measure of average kinetic energy of its random molecular motion which is not necessarily translational. The meaning of temperature remains also with a solid whose constituent particles execute random motion in form of vibration instead of translation. Therefore the given statement is not justified.

2.18 No. Because the terms *hot* and *cold* refer to high and low temperatures respectively which have meaning only with considerable amounts of the gas as a whole and not with its individual molecules, as temperature is a statistical quantity.

Note: Although the speed of a single molecule has meaning its temperature does not.

2.19 The temperature of a gas, according to kinetic theory, is a measure of the average kinetic energy of its random molecular motion. Here the additional molecular velocity due to motion of the container is same for all the molecules acting in the direction of motion of the container which not being random, has no effect on temperature. However, on stopping the container the temperature of the gas will raise due to randomisation of the directed molecular motion.

2.20 $P\bar{V} = RT$ will be valid for an ideal gas, provided the gas molecules are uniformly distributed throughout the system concerned. This requirement is fulfilled with (II), (III), ad (IV) but not with (I) and (V).

In case of (III), $P\bar{V} = RT$ is applicable to the system as a whole, if the latter is of ordinary size, otherwise it is applicable only to the individual layers of the system.

2.21 For an ideal gas, total KE of the gas molecules $= \frac{3}{2}PV$, by Eq. (2.26). In the given expansion PV does not change, and hence KE will not change.

□ Here the expression is isothermal, PV being constant. Work can be obtained by carrying out the expansion against an external pressure using a piston-cylinder device with the cylinder immersed in a heat reservoir at the prevailing temperature of the gas.

Note: Here work is obtained without any change in KE of the gas molecules. It results from transformation of heat supplied by the heat reservoir.

2.22 The first part of the problem may be depicted as

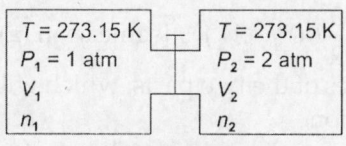

Final pressure $P = ?$

Here total number of moles $n = n_1 + n_2$ and total volume $V = V_1 + V_2$. Then using the ideal gas equation, we have, from $n = n_1 + n_2$

$$\frac{PV}{RT} = \frac{P_1 V_1}{RT} + \frac{P_2 V_2}{RT}$$

or
$$P = \frac{P_1 V_1 + P_2 V_2}{V}$$

$$= \frac{P_1 V_1 + P_2 V_2}{V_1 + V_2}$$

$$= \frac{1 \text{ atm} + 2 \text{ atm}}{1 + 1},$$

putting $V_1 = V_2$ and substituting the given values of pressure = 1.50 atm.

For the second part of the problem, the number of moles of the gas in the two vessels, n_1' and n_2' respectively, will be different to make the pressure P' equal in the two vessels with unequal temperatures T_1 and T_2. Since $n_1' + n_2' = n_1 + n_2$, we have

$$\frac{P' V_1}{RT_1} + \frac{P' V_2}{RT_2} = \frac{P_1 V_1}{RT} + \frac{P_2 V_2}{RT}$$

or
$$P' = \frac{P_1 + P_2}{T\left(\dfrac{1}{T_1} + \dfrac{1}{T_2}\right)} \quad \text{since } V_1 = V_2$$

$$= \frac{1 \text{ atm} + 2 \text{ atm}}{273.15\left(\dfrac{1}{-50 + 273.15} + \dfrac{1}{50 + 273.15}\right)} = 0.7249$$

Note: Here P' is different from P though temperature falls in one vessel and rises in the other by equal extent. This is due to change in translational energy of the gas molecules of the system as a whole, i.e. change of PV, V being same.

2.23 (a) If x is equilibrium pressure and T is equilibrium temperature, then

$$\frac{P(V_1 + V_2)}{RT} = \frac{P_1 V_1}{RT_1} + \frac{P_2 V_2}{RT_2} \quad \text{from conservation of number of moles}$$

Also $\dfrac{3}{2}P_1V_1 + \dfrac{3}{2}P_2V_2 = \dfrac{3}{2}P(V_1 + V_2)$

from conservation of molecular KE of translation

$$= \dfrac{3}{2}\left(\dfrac{P_1V_1}{T_1} + \dfrac{P_2V_2}{T_2} \right)T$$

which are the required expressions.

(b) i. At equilibrium the two parts will have same pressure and temperature, which are given by the same expression as in (a).

ii. The final pressure will be same for the two parts and is given by

$$\dfrac{3}{2}P_1V_1 + \dfrac{3}{2}P_2V_2 = \dfrac{3}{2}P(V_1 + V_2) \quad \text{[same as in (a)]}$$

But the final temperatures of the two parts, which will be different, cannot be found from the given information.

Note: The results depend on the nature of the partition but not on the nature of the gases. This will be true even if the two compartments contain different gases, provided they behave like ideal gases. Here atomicity is immaterial as it has no significant effect on molecular translational energy and hence on gas pressure.

2.24 No. Because the probability of a gas molecule to move exactly with a particular speed is exceedingly low since the number of different values of the speed is infinitely great while the number of molecules is finite. Obviously, it will be absolutely impossible for all the molecules to have exactly the same speed.

If all the molecules in the earth's atmosphere have same speed, they can rise only upto certain altitude (x, say) till their vertical components of velocity reduce to zero due to the action of the gravitational field. This would lead to the unrealistic situation that the atmosphere terminated abruptly at certain altitude (x).

2.25 Because the probability that a molecule moves with a particular velocity is virtually zero, but this is not so with a range of velocities around it.

Note: The range of velocities should necessarily be quite narrow, so that the probability density is practically same within a particular range and the distribution of velocities represented through various ranges becomes useful.

2.26 Because velocity is a vector quantity and distribution of molecular velocity is random.

☐ No. Because kinetic energy, which is related to the square of velocity, is a scalar quantity.

2.27 a. BC^2, being an exponent, is dimensionless. Then, the dimensions of B will be those of $1/C^2$, i.e. $l^{-2}t^2$.

Again, $P(c)dc$ (which implies probability) and e^{-BC^2} are both dimensionless. Then for dimensional consistency $AC^m dc$ will also be dimensionless, and hence the dimensions of A will be dimensions of $1/C^{m+1}$, i.e. $l^{-m-1}t^{m+1}$.

Integrating the given expression for $m = 2$

$$1 = \int_0^\infty P(c)dc = \int_0^\infty AC^2 e^{-BC^2} dc$$

$$= A \cdot \dfrac{1}{4}\left(\dfrac{\pi}{B^3} \right)^{1/2}$$

which is the required relation between A and B.

b. No. Because with +ve exponent of e, $\int_0^\infty P(c)dc$ (which represents total probability) will not be equal to 1 when $P(c)dc$ will lose its probabilistic significance.

Note: The speed distribution represented by $P(c)\Delta c = AC^m e^{-BC^2}\Delta c$ is conceptually incorrect, though it is dimensionally correct.

2.28 On replacing c in Maxwell's speed distribution law in Eq. (2.3b) by the speed (C_r) relative to the most probable speed, i.e. putting $C = C_r \cdot \sqrt{2kT/m}$, we get

$$\frac{1}{N}\frac{dNC_r}{dC_r} = \frac{4}{\sqrt{\pi}}C_r^2 e^{-C_r^2}$$

which is the desired expression in simplest form.

Note: Maxwell's distribution can be expressed without involving m and T by putting, in general, $C = A\sqrt{kT/m} \cdot C_r$ where A is a number. But the form thus obtained will be exponential unless $A = \sqrt{2}$, when C_r is the relative speed w.r.t. the most probable speed.

☐ i. The fraction of molecules having speed above the most probable speed (C_{mp}) will be

$$\int_{C_r=1}^{C_r=\infty} \frac{dNC_r}{N} = \frac{4}{\sqrt{\pi}}\int_1^\infty C_r^2 e^{-C_r^2}dC_r$$

$$= 0.57, \text{ i.e. more than half}$$

ii. The fraction of molecules with speed above the mean speed $(= 1.128\, C_{mp})$ will be

$$\frac{4}{\sqrt{\pi}}\int_{1.128}^\infty C_r^2 e^{-C_r^2}dC_r = 0.49, \text{ i.e. nearly half.}$$

iii. The fraction of molecules with speed above the rms speed $(= 1.225\, C_{mp})$ will be

$$\frac{4}{\sqrt{\pi}}\int_{1.225}^\infty C_r^2 e^{-C_r^2}dC_r = 0.41, \text{ i.e. appreciably less than half.}$$

☐☐ The fraction of molecules having speed above the mean speed $\langle c \rangle$, as calculated above, is found to be very nearly half. This is not unlikely with the mean value of a quantity. This would have been exactly half if the distribution curve were symmetrical relative to the average speed. Again, the fraction of molecules having speed above C_{mp} is appreciably higher than half while that having speed above C_{rms} is appreciably lower than half. Because $C_{mp} < C_{mean} < C_{rms}$.

☐☐☐ The fraction of molecules having kinetic energy above the mean kinetic energy will be same as that having speed above $\sqrt{\langle c^2 \rangle}$, i.e. 0.41. Because, a molecule having speed greater than $\sqrt{\langle c^2 \rangle}$ will also have its kinetic energy greater than the mean kinetic energy $\frac{1}{2}m\langle c^2 \rangle$.

2.29 Maxwell's distribution of molecular speed [Eq. (2.3b)] or translational energy [Eq. (2.5)], which is a particular form of Boltzmann distribution of energy, Eq. (21.13) will be applicable to a system, if

(1) the molecular speeds are independent in the sense that the total translational energy of the system is equal to the sum of those of its individual molecules.

(2) the system contains large number of molecules (as is required for any statistical law to hold)

These requirements are fulfilled in all the given situations excepting (ii).

Note: Gravitational field does not affect molecular translational energy and its distribution in a gas at equilibrium.

Effusion does not affect the speeds of the individual molecules nor does it affect the fraction of molecules having speed lying in a specified range, though the number of molecules in the effusing gas changes.

Maxwell's speed distribution law is relevant to any fluid, not just to an ideal gas. This can be quantum mechanically justified in terms of partition function.

2.30 a. i. Initially all the molecules have same speed and hence

$$\langle c \rangle = \sqrt{\langle c^2 \rangle} = \text{initial speed}$$

$$= 10^6 \text{ cm/s}$$

After a long time molecular speed will be randomised to a steady distribution (due to collision of the gas molecules with one another and also with the container's walls which are not smooth on molecular scale). Here $\langle c^2 \rangle$ suffers no change as translational energy of the molecules is conserved in an ideal gas, molecular collisions being assumed to be elastic. But $\langle c \rangle$ changes till a steady value is attained then, according to Maxwell's distribution

$$\langle c \rangle = \left(\frac{8}{3\pi} \langle c^2 \rangle \right)^{1/2}$$

$$= \left(\frac{8}{3 \times 3.143} \right)^{1/2} (10^6 \text{ cm/s}) = 9.211 \times 10^5 \text{ cm/s}$$

ii. Equilibrium temperature

$$= \frac{m\sqrt{\langle c^2 \rangle}}{3k} \quad \text{by Eq. (2.4c)}$$

$$= \frac{(6.64 \times 10^{-24} \text{ g})(10^6 \text{ cm/s})}{3(1.381 \times 10^{-16} \text{ erg K}^{-1})} = 1.603 \text{ K}$$

iii. Since 10^6 cm/s is the rms speed, the required fraction is 0.41 vide Prob 2.28 (iii).

b. Here the number of molecules (10) present in the system is so small that the statistical concept of temperature does not hold and the Maxwell's distribution is not applicable. Therefore the calculated values of the quantities in (a), will have no significance, except $\sqrt{\langle c^2 \rangle}$ and initial $\langle c \rangle$ which will be same.

2.31 Mean molecular speed,

$$\langle c \rangle = x_1 c_1 + x_2 c_2 + x_3 c_3$$

where x denotes the fraction of molecules having speed c indicated by the suffix

$$= \left(\frac{20}{100} \right)(100 \text{ m/s}) + \left(\frac{50}{100} \right)(400 \text{ m/s}) + \left(\frac{30}{100} \right)(200 \text{ m/s})$$

$$= 280 \text{ m/s}$$

RMS speed, $\sqrt{\langle c^2 \rangle} = \left[x_1 c_1^2 + x_2 c_2^2 + x_3 c_3^2 \right]^{\frac{1}{2}}$

$$= \left[\left(\frac{20}{100} \right)(100 \text{ m/s})^2 + \left(\frac{50}{100} \right)(400 \text{ m/s})^2 + \left(\frac{30}{100} \right)(200 \text{ m/s})^2 \right]^{-\frac{1}{2}}$$

$$= 306.6 \text{ m/s}$$

☐ Maxwellian distribution is characterised by a continuous variation of speed from 0 to ∞. Obviously, the given speed distribution is not Maxwellian. Now, for Maxwellian speed distribution,

$$\langle c \rangle = \left(\frac{8}{3\pi} \langle c^2 \rangle \right)^{1/2}$$

$$= \left(\frac{8}{3 \times 3.143} \right)^{1/2} (306.6 \text{ m/s}) = 282.5 \text{ m/s}$$

which is greater than that for the given distribution.

2.32
$$P = \frac{mN \langle c \rangle^2}{3V} \quad \text{by Eq. (2.2a)}$$

$$= \frac{\pi w \langle c \rangle^2}{8V}$$

where w is mass of the gas which is numerically equal to M for 1 mol.

$$= \frac{3.143 (2.016 \text{ g}) (5.5 \times 10^5 \text{ cm/s})^2}{8 (10 \times 10^3 \text{ cm}^3)} = 2.396 \times 10^7 \text{ dyn/cm}^2$$

☐ Yes.

☐☐ P would be less in case of (ii) due to lower molecular speed. It would be $8/3\pi$ times of that corresponding to (i), because $\langle c \rangle^2 = \frac{8}{3\pi} \langle c^2 \rangle$.

2.33   a.
$$\langle c \rangle = \left(\frac{8PV}{\pi w} \right)^{\frac{1}{2}} \quad \text{by Eqs (2.1a) and (2.4b)}$$

$$= \left(\frac{8P}{\pi \rho} \right)^{\frac{1}{2}} \quad \text{where } \rho \text{ is mass density}$$

$$= \left[\frac{8 (1.013 \times 10^6 \text{ dyn/cm}^2)}{3.143 (0.00009 \text{ g/cm}^3)} \right]^{\frac{1}{2}} = 1.693 \times 10^5 \text{ cm/s}$$

b.
$$\langle c \rangle_{He} = \left(\frac{M_{H_2}}{M_{He}} \right)^{\frac{1}{2}} \langle c \rangle_{H_2} \quad \text{by Eq. (2.4b) for same } T$$

$$= \left(\frac{2.016}{4.003} \right)^{\frac{1}{2}} (1.693 \times 10^5 \text{ cm/s}) = 1.201 \times 10^5 \text{ cm/s}$$

c. Although $\langle c \rangle$ calculated for H_2 is much smaller than the escape velocity for v_E (= 1.12×10^6 cm/s on the earth's surface), the fraction of H_2 molecules having speed higher than v_E is not quite insignificant (greater than that for He, since $m_{He} > m_{H_2}$) at the high temperature (~ 1000 K) of the upper atmosphere (where escape velocity is somewhat lower than 1.12×10^6 cm/s). Consequently, H_2 originally present in the earth's atmosphere long time ago, has now escaped almost entirely into outer space.

With He, the situation is different due to its continuous formation through radioactive decay from the earth's crust that makes He escaping at a steady rate. This is also favoured by lower mean speed of He molecules.

2.34 Here, pressure being rather low, oxygen is treated as an ideal gas

i.
$$C_{mp} = \left[\frac{2PV}{w}\right]^{\frac{1}{2}} \quad \text{by Eqs (2.4d) and (2.1a)}$$

$$= \left[\frac{2(3 \times 10^5\,Pa)\,(30 \times 10^{-3}\,m^3)}{(100 \times 10^{-3}\,kg)}\right]^{\frac{1}{2}} = 424.3 \text{ m/s}$$

ii.
$$\langle c \rangle = \left[\frac{8PV}{\pi w}\right]^{\frac{1}{2}}$$

$$= \frac{2}{\sqrt{\pi}}C_{mp} = \frac{2}{\sqrt{\pi}}(424.3 \text{ m/s}) = 478.7 \text{ m/s}$$

iii.
$$\sqrt{\langle c^2 \rangle} = \left[\frac{3PV}{w}\right]^{\frac{1}{2}}$$

$$= \sqrt{3/2}\,C_{mp} = \sqrt{3/2}\,(424.3 \text{ m/s}) = 519.7 \text{ m/s}$$

iv. $\langle (c - \langle c \rangle)^2 \rangle = \langle c^2 - 2c\langle c \rangle + \langle c \rangle^2 \rangle = \langle c^2 \rangle - 2\langle c \rangle \langle c \rangle + \langle c \rangle^2 = \langle c^2 \rangle - \langle c \rangle^2$

v. Mean molecular kinetic energy

$$\tfrac{3}{2}kT = \tfrac{1}{2}m\langle c^2 \rangle = \tfrac{1}{2}\cdot\frac{M_{O_2}}{N_A}\langle c^2 \rangle$$

$$= \frac{(32 \times 10^{-3}\,kg\ mol^{-1)}\,(519.7 \text{ m/s})^2}{2\,(6.022 \times 10^{23}\ mol^{-1})} = 7.176 \times 10^{-21}\,J$$

vi. Most probable molecular kinetic energy

$$= \tfrac{1}{2}kT \quad \text{by Eq. (2.6)}$$

$$= \tfrac{1}{3} \times \text{mean molecular kinetic energy}$$

$$= \tfrac{1}{3}(7.176 \times 10^{-21}\,J) = 2.392 \times 10^{-21}\,J$$

vii.
$$\langle (\epsilon - \langle \epsilon \rangle)^2 \rangle = \langle \epsilon^2 \rangle - \langle \epsilon \rangle^2$$

Now
$$\langle \epsilon^2 \rangle = \int_0^\infty \epsilon^2 \cdot \frac{2}{\pi^{1/2}}\left(\frac{1}{kT}\right)^{\frac{3}{2}} \epsilon^{\frac{1}{2}}\,e^{-\epsilon/kT}\,d\epsilon \quad \text{by Eq. (2.5)}$$

$$= \frac{2}{\pi^{1/2}}\left(\frac{1}{kT}\right)^{\frac{3}{2}}\int_0^\infty \epsilon^{\frac{5}{2}}\,e^{-\epsilon/kT}\,d\epsilon$$

$$= \frac{2}{\pi^{1/2}}\left(\frac{1}{kT}\right)^{\frac{3}{2}}\int_0^\infty x^6 e^{-x^2/kT}\,2dx \quad \text{putting } \epsilon = x^2$$

$$= \tfrac{15}{4}(kT)^2$$

and $\quad \langle \epsilon \rangle = \tfrac{3}{2}kT$

Therefore

$$\langle (\epsilon - \langle \epsilon \rangle)^2 \rangle = \tfrac{15}{4}(kT)^2 - \left(\tfrac{3}{2}kT\right)^2$$

$$= \tfrac{3}{2}(kT)^2$$

$$= \tfrac{3}{2}\left(\tfrac{2}{3} \times 7.176 \times 10^{-21}\,J\right)^2, \quad \text{from } (v)$$

Note:

1. $\epsilon_{mp} \neq \frac{1}{2} m C_{mp}^2$

2. $\langle \epsilon \rangle \neq \frac{1}{2} m \langle c \rangle^2$

3. In a gas, ϵ of a molecule continually vary, even at constant T, but $\langle \epsilon - \langle \epsilon \rangle)^2 \rangle$ does not.

2.35 i. The required fraction

$$= 4\pi \left(\frac{M}{2\pi RT} \right)^{\frac{3}{2}} C_{mp}^2 e^{-MC_{mp}^2 / 2RT} \, (0.02 C_{mp}) \; [\text{by Eq. (2.3b)}]$$

$$= 4\pi \left(\frac{M}{2\pi RT} \right)^{\frac{3}{2}} \left(\frac{2RT}{M} \right) e^{-1} \times 0.02 \left(\frac{2RT}{M} \right)^{\frac{1}{2}}$$

$$= \frac{0.08}{\sqrt{\pi} \, e} = 0.0166$$

ii. The required fraction

$$= \frac{2}{\sqrt{\pi}} \left(\frac{1}{kT} \right)^{\frac{3}{2}} \epsilon_{mp}^{1/2} \, e^{-\epsilon_{mp}/kT} \, (0.02 \, \epsilon_{mp}) \; [\text{by Eq. (2.5)}]$$

$$= \frac{2}{\sqrt{\pi}} \left(\frac{1}{kT} \right)^{\frac{3}{2}} \epsilon_{mp}^{3/2} \, e^{-\epsilon_{mp}/kT} \times 0.02$$

$$= \frac{2}{\sqrt{\pi}} \left(\frac{1}{kT} \right)^{\frac{3}{2}} \left(\tfrac{1}{2} kT \right)^{\frac{3}{2}} e^{-1/2} \times 0.02$$

$$= \frac{0.02}{\sqrt{2\pi}} e^{-\frac{1}{2}} = 0.004839$$

iii. The required fraction

$$= \frac{2}{\sqrt{\pi}} \left(\frac{\epsilon_{mp}}{kT} \right)^{\frac{1}{2}} e^{-\epsilon_{mp}/kT} \; [\text{by Eq. (2.7a)}]$$

$$= \sqrt{2/\pi} \, e^{-\frac{1}{2}} \qquad \text{putting } \epsilon_{mp} = \tfrac{1}{2} kT$$

$$= 0.4838$$

Note: Fraction of molecules with $\epsilon > \epsilon_{mp}$ differ appreciably from that with $C > C_{mp}$. But the fraction of molecules with $\epsilon > \langle \epsilon \rangle$ is same as that with $C > \langle c^2 \rangle$ (vide Problem 2.28) because $\epsilon_{mp} \neq \frac{1}{2} m C_{mp}^2$ but $\langle \epsilon \rangle = \frac{1}{2} m \langle c^2 \rangle$.

2.36 i. For an ideal gas total (translational) KE due to molecular motion $= \frac{3}{2} PV = \frac{3}{2} RT$ per mole. Then total molecular KE will be higher for Ne, PV being higher. However, KE per mole will be same for both the gases, T being same.

ii. Most probable molecular speed $C_{mp} = \left(\frac{2kT}{m} \right)^{\frac{1}{2}}$ will be higher for He due to its lower molecular mass (m).

iii. Most probable molecular momentum $m C_{mp} = m \left(\frac{2kT}{m} \right)^{\frac{1}{2}} = \sqrt{2mkT}$. Thus, it will be higher for Ne due to higher value of m.

Note: For an ideal gas:

(1) Total KE is determined by nT (i.e. by PV) but KE per mole is determined only by T (i.e. independent of P and V if T is fixed)

(2) $C_{mp} \propto 1/\sqrt{m}$, but most probable momentum $\propto \sqrt{m}$

2.37 Smoke particles exhibit zig-zag motion resulting from continual bombardment of randomly moving invisible air molecules on all sides of them. They may therefore, be supposed to have the same mean translational kinetic energy as the air molecules, just like the molecules of different ingradients in a gaseous mixture. Then

$$\text{average translational KE of smoke particles} = \tfrac{3}{2}kT$$

$$= \tfrac{3}{2}(1.381 \times 10^{-23}\,\text{JK}^{-1})(300.15\,\text{K})$$

$$= 6.124 \times 10^{-21}\,\text{J}$$

$$\text{average speed of smoke particles} = \left[\tfrac{3}{2}kT / \tfrac{1}{2}m\right]^{\frac{1}{2}} = \left(\frac{3kT}{m}\right)^{\frac{1}{2}}$$

$$= \left[\frac{6.214 \times 10^{-21}}{\tfrac{1}{2} \times 10^{-16}\,\text{kg}}\right]^{\frac{1}{2}}$$

$$= 1.115 \times 10^{-2}\,\text{m/s}$$

□ These quantities would not depend on atmospheric pressure, just as the energy and speed of gas molecules are independent of pressure at constant temperature. However, the speed of smoke particles, though not their energy, depends on their size, and hence mass, and varies as $1/\sqrt{m}$ at constant temperature, as is found with molecules of a gas.

□□ Obviously, average translational KE of the smoke particles, but not their average speed, will be same as that for oxygen.

Note: Brownian motion arises from imbalance of the impacts of molecular collisions with suspended particles. Hence it decreases with increasing size of the particles though their kinetic energy remains unchanged.

2.38 $$\langle c \rangle = \left(\frac{8RT}{\pi N_A m}\right)^{\frac{1}{2}} \quad \text{by Eq. (2.4b);} \quad \text{putting } k = \frac{R}{N_A}$$

Then, Avogadro constant

$$N_A = \frac{8RT}{\pi m \langle c \rangle^2}$$

$$= \frac{8(8.314\,\text{JK}^{-1}\text{mol}^{-1})(300.15\,\text{K})}{3.143(6.2 \times 10^{-17}\,\text{kg})(1.3 \times 10^{-2}\,\text{m/s})} = 6.06 \times 10^{23}\,\text{mol}^{-1}$$

□ If N_A is extremely high, approaching ∞, the molecular density in the medium, and hence the frequency of molecular collision with a suspended particle, will be so high that the impulse of molecular collisions on opposite sides of the particles becomes virtually equal. As a consequence, there will be no statistical imbalance, and hence no Brownian motion.

□ Since $R \propto N_A$ and $R = kN_A$, any change of N_A will affect R (proportionally) but not k, being the proportionality constant.

[That k is independent of number of particles can be understood on statistical ground, vide Sec. 21.2].

2.39 The probability for a velocity component u lie between u_1 and u_2 is

$$\int_{u_1}^{u_2} \left(\frac{M}{2\pi RT} \right)^{\frac{1}{2}} e^{-Mu^2/2RT} du \simeq \left(\frac{M}{2\pi RT} \right)^{\frac{1}{2}} \exp(-Mu^2/2RT) \Delta u \quad \text{by Eq. (2.3a)}$$

with $u = (u_1 + u_2)/2$, when $\Delta u = u_2 - u_1$ is quite small.

Then, to a good approximation, the required probability range will be

$$\left[\frac{28.02 \times 10^{-3} \text{ kg mol}^{-1}}{2 \times 3.143 (8.314 \text{JK}^{-1} \text{mol}^{-1})(300 \text{K})} \right]^{\frac{1}{2}} \exp\left[-\frac{(28.02 \times 10^{-3} \text{kg mol}^{-1})(1000 \text{ m/s})^2}{2(8.314 \text{JK}^{-1} \text{mol}^{-1})(300 \text{K})} \right] (1 \text{ m/s})$$

$$= 2.493 \times 10^{-9}$$

☐ Yes. Because a particle having velocity component lying between u and $u + du$ in a particular direction is equivalent to a particle moving with velocity lying between u and $u + du$ in that direction.

Note: In the above calculation, we could have used any value of u within the given narrow range getting practically the same probability density (vide problem 2.25). It is convenient to put $u = 1000$, being an integer.

2.40 Here the situation corresponds to one-dimensional molecular motion.

   i. Average molecular velocity

$$u_{av} = \int_{-\infty}^{+\infty} u \times \text{probability of velocity}$$

$$= \int_{-\infty}^{+\infty} u \frac{dN_u}{N}$$

$$= \int_{-\infty}^{+\infty} u \left(\frac{m}{2\pi kT} \right)^{\frac{1}{2}} e^{-mu^2/2kT} du \quad \text{by Eq. (2.3a)}$$

$$= 0, \text{ since the integral is an odd function of } u.$$

   ii. Average molecular speed

$$= \int_{0}^{+\infty} u \times \text{probability of speed}$$

$$= \int_{-\infty}^{+\infty} u \left(\frac{m}{2\pi kT} \right)^{\frac{1}{2}} e^{-mu^2/2kT} 2du$$

$$= 2 \left(\frac{m}{2\pi kT} \right)^{\frac{1}{2}} \int_{0}^{+\infty} u e^{-mu^2/2kT} du$$

$$= 2 \left(\frac{m}{2\pi kT} \right)^{\frac{1}{2}} \cdot \frac{1}{2m/2kT}$$

$$= \left(\frac{2kT}{\pi m} \right)^{\frac{1}{2}}$$

   iii. Mean square velocity

$$= \int_{-\infty}^{+\infty} u^2 \left(\frac{m}{2\pi kT} \right)^{\frac{1}{2}} e^{-mu^2/2kT} du$$

$$= 2 \int_{-\infty}^{+\infty} u^2 \left(\frac{m}{2\pi kT} \right)^{\frac{1}{2}} e^{-mu^2/2kT} du$$

$$= \frac{kT}{m}$$

iv. Mean square speed = Mean square velocity

v. Most probable velocity (v_{mp}), which corresponds to the maximum in the velocity distribution curve, is obtained on differentiating $\frac{1}{N}\frac{dN_u}{du}$, given by Eq. (2.3a), w.r.t. u and equating the result to zero, when we get

$$ue^{-mu^2/2kT} = 0$$

Then $u_{mp} = 0$; the other two solution $u = \pm\alpha$ corresponds to minimum in the distribution curve.

vi. Average molecular energy of translation

$$= \int_{-\infty}^{+\infty} \frac{1}{2}mu^2 \cdot \frac{dN_u}{N} \quad \text{by Eq. (2.4a)}$$

$$= 2\int_0^{+\infty} \frac{1}{2}mu^2 \cdot \left(\frac{m}{2\pi kT}\right)^{\frac{1}{2}} e^{-mu^2/2kT} du$$

$$= m\left(\frac{m}{2\pi kT}\right)^{\frac{1}{2}} \int_0^\infty u^2 e^{-mu^2/2kT} du$$

$$= m\left(\frac{m}{2\pi kT}\right)^{\frac{1}{2}} \cdot \frac{1}{4} \cdot \frac{\pi^{\frac{1}{2}}}{(m/2kT)^{\frac{3}{2}}} = \frac{1}{2}kT$$

This forms the basis of the law of equipartition of kinetic energy

The required velocity and speed distribution curves are shown in the adjoining diagram where the lower curve corresponds to velocity and the upper one to speed.

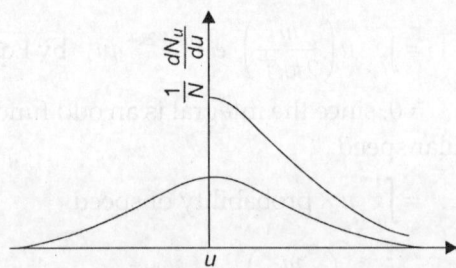

□ If the gas is enclosed in a spherical vessel, the molecular motion becomes three dimensional and the results will be changed accordingly, except the average molecular velocity (which is a vector quantity).

Here average molecular speed $= \left(\frac{8kT}{\pi m}\right)^{\frac{1}{2}}$

most probable velocity = most probable speed $= \left(\frac{2kT}{m}\right)^{\frac{1}{2}}$

average molecular energy of translation $= \frac{3}{2}kT$

Note:

1. $u_{mp} = u_{av}$ happens with one-dimensional, but not three-dimensional motion

2. Root mean square velocity in case of one-dimensional motion $(kT/m)^{\frac{1}{2}}$ is equal to the root mean square velocity in a particular direction for three-dimensional motion, $\sqrt{\frac{1}{3}\overline{C^2}}$.

3. Although the expression for molecular energy is different in two cases, the molecular energy is same. Because T is lower in case of three-dimensional

motion due to distribution of same amount of energy into larger number of degrees of freedom.

2.41 a. The Maxwell energy distribution curve corresponding to the energy distribution Eq. (2.5) is shown in the adjoining diagram. It has a vertical slope at the origin. Consequently it rises more quickly than the corresponding speed distribution curve which starts with a horizontal slope. After reaching the maximum, the energy distribution falls off more slowly than does the speed distribution.

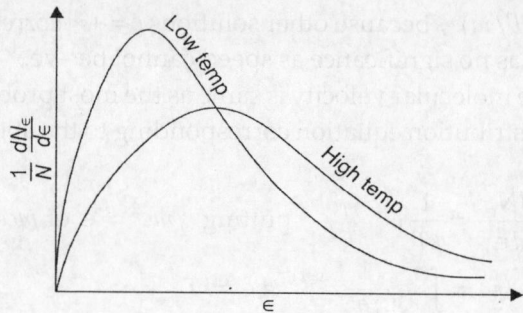

b. For Maxwellian speed distribution

$$C_{mp} = \left(\frac{2kT}{m} \right)^{\frac{1}{2}}$$

Then He, having lower molecular mass (m), will have higher C_{mp} and the speed distribution curve for He and Ne will be as shown in the adjoining diagram.

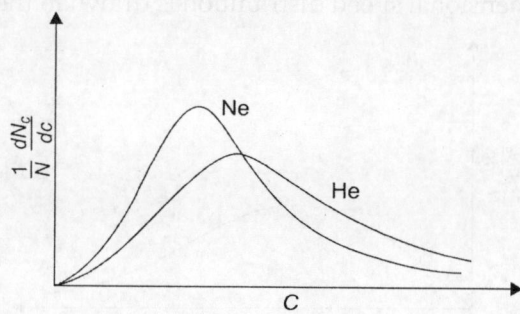

   i. At $T = 0$, $C_{mp} = 0$, then at this temperature, the distribution will be such that its maximum lies on the ordinate at infinite distance away from the origin.

   ii. At $T = \infty$, $C_{mp} = \infty$, then at this temperature, the distribution curve will have its maximum lying on the c-axis.

Note: It follows from distribution Eq. (2.3b) that m and T have opposite effect on probability distribution. A distribution curve for higher T will correspond to a distribution curve for lower m.

2.42 The probability that a molecule has two-dimensional speed between c and $c + dc$ is obtained on multiplying the probability density for the corresponding velocity distribution with an area of a strip of width dc and radius c, i.e.

$$\frac{dN_c}{N} = \left(\frac{m}{2\pi kT} \right)^{\frac{1}{2}} \left(\frac{m}{2\pi kT} \right)^{\frac{1}{2}} e^{-mc^2/2kT} 2\pi c dc$$

or $$\frac{1}{N} \frac{dN_c}{dc} = \frac{m}{kT} c e^{-mc^2/2kT}$$

where dN_c is the number of molecules having speed between c and $c + dc$. This represents the required Maxwell's law of distribution.

  i. Most probable molecular speed C_{mp}, which corresponds to the maximum in the distribution curve, is obtained on differentiating the above distribution equation w.r.t. c and equating the result to zero, when we get

$$\left(1 - \frac{mc^2}{kT}\right) e^{-mc^2/2kT} = 0$$

Then $C_{mp} = (kT/m)^{\frac{1}{2}}$, because other solutions $c = +\infty$ corresponds to a minimum while $c = -\infty$ has no significance as speed cannot be $-$ve.

 ii. Most probable molecular velocity is same as the most probable molecular speed.

iii. The energy distribution equation corresponding to the above speed distribution equation is

$$\frac{1}{N}\frac{dN_\epsilon}{d\epsilon} = \frac{1}{kT} e^{-\epsilon/kT}, \quad \text{putting } \tfrac{1}{2}mc^2 = \epsilon \text{ or } mcdc = d\epsilon$$

Then $$\langle\epsilon\rangle = \int_0^\infty \epsilon \cdot \frac{1}{kT} e^{-\epsilon/kT} d\epsilon = kT$$

Alternatively, $$\langle\epsilon\rangle = \tfrac{1}{2}m\int_0^\infty c^2 \cdot \frac{m}{kT} e^{-mc^2/kT} \cdot cdc$$

iv. The fraction of molecules having energy $\geq \langle\epsilon\rangle$ is

$$\frac{1}{kT}\int_{\langle\epsilon\rangle}^\infty e^{-\epsilon/kT} d\epsilon = e^{-\langle\epsilon\rangle/kT}$$

☐ The two-dimensional speed distribution is drawn in the adjoining diagram.

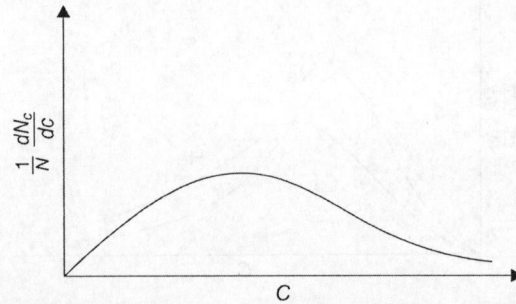

☐☐ This speed distribution curve is somewhat similar to the corresponding one for three-dimensional motion, with notable difference at $c = 0$, where the curve does not touch the c-axis.

Note: For two-dimensional energy distribution the fraction of molecules having energy above ϵ is equal to $e^{-\epsilon/kT}$. But this is not so far one-dimensional and three-dimensional energy distribution [(by equation (2.7b) and (2.7a)].

2.43 This problem is similar to problem 2.40, set in reverse way. Only difference is that the molecules in the beam are all moving along the same direction and in the same sense and hence the average molecular velocity is not zero but same as the average molecular speed. However, the latter (related to the thermal energy) is given by the same expression as in problem 2.40 where half of the molecules are moving in one direction while the other half in the opposite direction, with average molecular speed same for either half.

Note: The particles in the beam have some energy (in addition to their thermal energy) due to their directed motion imparted by the beam. This is unlike one-dimensional motion in problem 2.40. However, the thermal energy in the two cases can be treated similarly.

2.44 i. $$N/V = P/kT \quad \text{by Eq. (2.9)}$$

$$= \frac{(1.01 \times 10^5 \, P_a)}{(1.381 \times 10^{-23} \, \text{JK}^{-1})(273.15 \, \text{K})} = 2.685 \times 10^{25} \, \text{m}^{-3}$$

ii. Each molecule is associated on the average with a space V/N. Treating this space as a cube (but not sphere*), the length of such cube may be taken as average intermolecular distance (d), then

$$d = \left(\frac{1}{N/V}\right)^{\frac{1}{3}} = \left(\frac{1}{2.685 \times 10^{25} \, \text{m}^{-3}}\right)^{\frac{1}{3}} = 3.339 \times 10^{-9} \, \text{m}$$

[* Because a space can be divided entirely into elementary spaces of cubical symmetry but not of spherical symmetry]

iii. $$\lambda = \frac{1}{\sqrt{2} \times 3.143 \, (2 \times 120 \times 10^{-12} \, \text{m})^2 \, (2.685 \times 10^{25} \, \text{m}^{-3})}, \text{by Eq. (2.13)}$$

for O_2 molecule, being non-spherical, the collision diameter has been taken to be twice the O—O bond distance.

iv. $$Z_1 = \frac{\langle c \rangle}{\lambda} \quad \text{by Eq. (2.13)}$$

$$= \left(\frac{8RT}{\pi M}\right)^{\frac{1}{2}} = \left[\frac{8(8.314 \, \text{JK}^{-1}\text{mol}^{-1})(273.15 \, \text{K})}{3.143 \, (32 \times 10^{-3} \, \text{kg mol}^{-1})}\right]^{\frac{1}{2}} \Big/ 1.455 \times 10^{-7} \, \text{m}$$

$$= 2.921 \times 10^9 \, \text{s}^{-1}$$

2.45 Since $N/V \propto P$, at constant T

$$\frac{N}{V} = \frac{P}{P'}\left(\frac{N}{V}\right)'$$

the primed quantities refer to the data in the previous problem and the unprimed ones to the present problem

$$= \frac{10^{-13} \, \text{atm}}{1 \, \text{atm}} (2.685 \times 10^{25} \, \text{m}^{-3}) = 2.685 \times 10^{12} \, \text{m}^{-3}$$

Again $$d \propto \left(\frac{1}{N}\right)^{\frac{1}{3}} \propto \left(\frac{1}{P}\right)^{\frac{1}{3}}$$

Then $$d = \left(\frac{P'}{P}\right)^{\frac{1}{3}} d'$$

$$= \left(\frac{1 \, \text{atm}}{10^{-13} \, \text{atm}}\right)^{\frac{1}{3}} (3.339 \times 10^{-9} \, \text{m}) = 7.192 \times 10^{-5} \, \text{m}$$

$$\lambda = \frac{P'}{P} \cdot \lambda' \quad \text{since } \lambda \propto 1/P \text{ at constant } T$$

$$= \left(\frac{1 \, \text{atm}}{10^{-13} \, \text{atm}}\right) (1.455 \times 10^{-7} \, \text{m}) = 1.455 \times 10^6 \, \text{m}$$

$$Z = \frac{P}{P'} \cdot Z'$$

since $Z \propto P$ at constant T (because $\lambda \propto 1/P$ and $\langle c \rangle$ is independent of P)

$$= \left(\frac{10^{-13} \text{ atm}}{1 \text{ atm}} \right) (2.921 \times 10^9 \text{ s}^{-1}) = 2.921 \times 10^{-4} \text{ s}^{-1}$$

Note: At ordinary temperature and pressure, λ is small as compared with the usual dimensions of the container. Here a gas molecule collides with another at much higher frequency than with the container walls. Again, λ is not equal to the intermolecular distance but greater than this, because the colliding gas molecules may not be the nearest one.

At very low pressure, λ is much greater compared even with the dimensions of the container. Here the gas molecules collide far more often with the container walls than with one another.

At very high pressure, when the molecules almost touch each other, λ is zero. Here λ is less than the intermolecular distance.

2.46 $\qquad\qquad Z_{o-o} = Z_{p-p} = 1/2\, Z_{o-p}$

This follows from Eq. (2.11b) and (2.12b) on putting $\sigma_{o-H_2} = \sigma_{p-H_2}$, $\langle c \rangle_{o-H_2} = \langle c \rangle_{p-H_2}$ and $(N/V)_{o-H_2} = (N/V)_{p-H_2}$.

Note: Although $o-H_2$ and $p-H_2$ are different species, their kinetic behaviour is same.

2.47 For H_2, $\lambda (= \langle c \rangle / Z)$ changes by a factor

$$\frac{\lambda_f}{\lambda_i} = \frac{Z_i}{Z_f} = \frac{Z_{H_2-H_2}}{Z_{H_2-H_2} + Z_{H_2-D_2}}$$

$$= \frac{\dfrac{1}{\sqrt{2}} \left[\left(\dfrac{N}{V} \right)_{H_2} \right]_i^2 \Big/ \sqrt{M_{H_2}}}{\dfrac{1}{\sqrt{2}} \left[\left(\dfrac{N}{V} \right)_{H_2} \right]_f^2 \Big/ \sqrt{M_{H_2}} + \left[\left(\dfrac{N}{V} \right)_{H_2} \left(\dfrac{N}{V} \right)_{D_2} \right]_f \sqrt{\dfrac{1}{M_{H_2}} + \dfrac{1}{M_{D_2}}}}$$

by Eqs (2.11b) and (2.12b)

$$= \frac{4}{1 + \sqrt{3}} \qquad \because \left[\left(\frac{N}{V} \right)_{D_2} \right]_f = \left[\left(\frac{N}{V} \right)_{H_2} \right]_f = \frac{1}{2} \left[\left(\frac{N}{V} \right)_f \right]_i$$

$$= 1.464 \qquad\qquad \text{and } M_{D_2} = 2 M_{H_2}$$

Similarly, for D_2 $\dfrac{\lambda_f}{\lambda_i} = \dfrac{4}{1 + \sqrt{6}} = 1.16$

Note: λ for a gas is independent of its molar mass only when it is pure, otherwise it will depend on composition of the system and molar masses of its ingredients.

2.48   i. Average time between two successive collisions $(\lambda / \langle c \rangle)$

$$= \frac{1}{\text{average number of collisions suffered by a molecule per unit time}}$$

$$= \frac{1}{\sqrt{2} \pi \sigma^2 \langle c \rangle N/V} = \frac{1}{\sqrt{2} \pi \sigma^2 \left(\dfrac{8RT}{\pi M} \right)^{\frac{1}{2}} \dfrac{P}{kT}} = \frac{k\sqrt{MT}}{4\sqrt{\pi R} \cdot \sigma^2 P}$$

$$= \frac{(1.381 \times 10^{-23} \text{ JK}^{-1}) \sqrt{(4.003 \times 10^{-3} \text{ kg mol}^{-1})(273.15 \text{ K})}}{4\sqrt{3.143(8.314 \text{ JK}^{-1}\text{mol}^{-1})} \, (10^{-10} \text{ m})^2 \, (1.013 \times 10^5 \text{ Pa})} = 6.972 \times 10^{-10} \text{ s}$$

ii. For a molecule moving with speed c, the time required to travel unit distance is $1/c$ whose average is

$$\langle 1/c \rangle = \int_0^\infty \frac{1}{C} \frac{dN_c}{N} = \int_0^\infty \frac{1}{C} \cdot 4\pi \left(\frac{M}{2\pi RT} \right)^{\frac{3}{2}} c^2 e^{-MC^2/2RT} dc$$

$$= \left(\frac{2}{\pi} \right)^{\frac{1}{2}} \left(\frac{M}{RT} \right)^{\frac{3}{2}} \left(\frac{RT}{M} \right)$$

$$= \left(\frac{2M}{\pi RT} \right)^{\frac{1}{2}} \left[= \frac{4}{\pi} \frac{1}{\langle c \rangle} \right]$$

$$= \left[\frac{2(4.003 \times 10^{-3} \text{kg mol}^{-1})}{3.143(8.314 \text{ JK}^{-1}\text{mol}^{-1})(373.15 \text{ K})} \right]^{\frac{1}{2}} = 1.059 \times 10^{-3} \text{s}$$

Note: In (i) the distance travelled by the molecule between two successive collisions is not fixed. Here an average distance between two successive collisions have been considered supposing the molecules to move with a uniform speed $\langle c \rangle$.

In (ii) the distance travelled by the molecule is fixed. Here the time required for molecules of different speed is different and average of such times has been considered with the result $\langle \frac{1}{c} \rangle \neq \frac{1}{\langle c \rangle}$.

In (i), unlike (ii), the involved time varies with pressure. The quantities in the two cases become equal if the pressure is so chosen that $\lambda = 4/\pi$.

2.49 The chance for a molecule to travel a distance x without any collision is equal to $e^{-x/\lambda}$. Here $\lambda = 1.455 \times 10^{-7}$ m (from problem 2.44). Then

i. $e^{-x/\lambda} = \exp[-10^{-6} \times 10^{-3} \text{ m}/1.455 \times 10^{-7} \text{ m}] = 1.037 \times 10^{-3}$

ii. $e^{-x/\lambda} = \exp[-10^{-3} \text{ m}/1.455 \times 10^{-7} \text{ m}] \simeq 0$

iii. $e^{-x/\lambda} = e^{-1} = 0.368$

Note: The chance of moving without collision is higher for lower x. However, this chance for $x = \lambda$ happens to be only 0.368, i.e. 36.8% and not 100%. Because the mean free path does not imply a fixed distance λ, that a molecule moves between any two of its successive collisions.

2.50 The required distance x is given, according to the Eq. (2.17) by

$$\frac{x}{\lambda} = \ln \frac{I_0}{I}$$

or

$$x = \ln \frac{I_0}{I} \cdot \lambda$$

$$= \ln \frac{I_0}{\frac{1}{2} I_0} \cdot \lambda = 0.693 \lambda$$

2.51 Molecular density of Ar,

$$N/V = P/kT$$

$$= \frac{\left(\frac{7.5}{760} \times 10^{-5} \times 1.013 \times 10^6 \text{ dyn cm}^{-2} \right)}{(1.381 \times 10^{-16} \text{erg K}^{-1})(300.15 \text{ K})} = 2.412 \times 10^{12} \text{ cm}^{-3}$$

$$\lambda = \frac{\text{Average speed of K atom}}{\text{Frequency of collision of a K atom with Ar molecules}}$$

$$= \frac{\langle c \rangle_K}{\pi \sigma^2 \sqrt{\langle c \rangle_K^2 + \langle c \rangle_{Ar}^2} \cdot (N/V)_{Ar}}$$

$$= \frac{1}{\pi \sigma^2 \sqrt{1 + (M_K / M_{Ar}} \cdot (N/V)_{Ar}}$$

$$= \frac{1}{(6 \times 10^{-14} \text{ cm}^2 \sqrt{1 + \dfrac{39.1}{39.95}} (2.412 \times 10^{12} \text{ cm}^{-3})} = 4.912 \text{ cm}$$

Attenuation of the beam

$$\frac{I_0 - I}{I_0} = 1 - e^{-x/\lambda} = 1 - e^{-2/4.912} \quad \text{by Eq. (2.17)}$$

$$= 0.33450, \text{ i.e. } 33.45\%.$$

2.52
$$\lambda = \frac{1}{\sqrt{2} \pi \sigma^2 \dfrac{N}{V}}$$

i. Here λ decreases, because N/V increases due to increase of P at constant T, as $N/V = P/kT$.

ii. λ remains unchanged, because N/V does not change at constant volume.

Note: λ depends on P and T. But with change of P and T, λ will change provided N/V changes.

2.53 Since gas molecules have definite volume of their own, intermolecular collisions are inevitable, though at very low pressure, they occur at negligible frequency compared to wall collisions.

☐ Mean free path (λ) is defined to be the mean distance that a molecule travels between its consecutive collisions with other molecules. At very low pressure the frequency of collision is so low that λ thus defined loses its practical significance due to its fluctuating value.

☐☐ Here the Maxwell distribution law will not hold because of unattainability of steady distribution of molecular speed due to low molecular density.

2.54 The assumption mentioned in the question is justified only when the mean free path (λ) exceeds the length of the cubic vessel concerned, i.e. $(10^{-3} \text{ m}^3)^{\frac{1}{3}}$ or 0.1 m. The maximum gas pressure P_{max} (corresponding to $\lambda = 0.1$ m) at which this is expected can be found from the relation

$$P = \frac{kT}{\sqrt{2} \pi \sigma^2 \lambda} \quad \text{by Eq. (2.13) putting } N/V = P/kT$$

This gives
$$P_{max} = \frac{(1.381 \times 10^{-23} \text{ JK}^{-1})(300.15 \text{ K})}{\sqrt{2} \times 3.143 (10^{-10} \text{ m})^2 (0.1 \text{ m})} = 0.9327 \text{ Pa}$$

2.55 Because, at extreme upper atmosphere, the upwardly directed molecules either escape from the atmosphere if their velocity exceeds escape velocity for the earth or move circularly (like artificial satellites) without any intermolecular collisions.

2.56 Because, at the top of a mountain air has lower molecular density and hence lower rate of heat transfer to a body due to less number of collisions of air molecules with it.

Note: mperature has no necessary connection with sensation of heat. Thus, althou   the atmosphere at the mountain top is likely to have higher temperature, it appears cooler than at sea level.

2.57 Apart from providing a chemically inert atmosphere, the inert gas atoms minimise the condensation of the metal atoms evaporated from the filament on the wall of the lamp (forming a black coating) by reducing their free path through collision with them.

2.58 By Eq. (2.16), the thermal conductivity (K) of a gas is related to its molar mass (M) as

$$K \propto \langle c \rangle \text{ when } N/V, \bar{C}_V \text{ and } \lambda \text{ are constant}$$

$$\propto \frac{1}{\sqrt{M}}$$

For D_2 and H_2, $\bar{C}_v$ is same and λ is assumed to be same. Therefore, when N/V and T are same, H_2 will have higher K due to lower M.

☐ Under the same conditions of temperature and pressure, ^{4}He having higher M than ^{3}He, will have lower K and hence it will be a better thermal insulator.

2.59 Effusion involves increase in intermolecular distance which makes it a slow process despite high speed of gas molecules, due to low mean free path caused by random intermolecular collisions.

But propagation of sound through a gas does not involve any change of inter-molecular distance. Here the pulses of sound waves are transmitted from one region to another through intermolecular collisions, which is, therefore, expected to occur at a fast speed comparable to the average speed of the gas molecules.

2.60 No. Because sound waves are transmitted in a gas through intermolecular collisions.

Note: This mechanism has to be modified if the wavelength of sound is less than the mean free path. But our answer will not be affected thereby, because the velocity of sound is independent of its wavelength. [A more effective, but less obvious, mechanism of the propagation of sound waves (which are pressure waves) is that the gas molecules move to form necessary regions of different pressure]. Whatever might be the detailed mechanism of the propagation of sound waves, the gas molecules must move to form necessary regions of high and low pressure.

2.61. Velocity of sound $= \left(\dfrac{\gamma RT}{M} \right)^{\frac{1}{2}}$ where $\gamma = C_P/C_V$

$$= \left(\frac{\pi \gamma}{8} \right)^{\frac{1}{2}} \langle c \rangle \text{ by Eq. (2.4b)}$$

$$= \left[\frac{3.143 \times 1.40}{8} \right]^{\frac{1}{2}} (475 \text{ m/s}) = 352 \text{ m/s}$$

Note: The above expression of velocity of sound is based on the relation $PV^\gamma = $ constant which requires the gas to behave ideally and the process to be adiabatic and reversible. Such requirements are too stringent to be fulfilled in a real situation. Therefore, the velocity of sound provided by the above expression is liable to differ appreciably from the actual value which is more near to mean molecular speed than the rms speed.

2.62 No. Because the diffusion of a gas is a kind of molecular scattering and hence, it cannot occur without molecular collision*, though the latter makes it rather a slow process.

Note: * If there were no intermolecular collisions, as in a cluster of molecules each moving with same speed in the same direction, the relative positions of the molecules would remain unchanged and the cluster would move as such (somewhat similar to flying of a piece of cloud in the sky) without spreading.

2.63 Let us consider two gases A and B, separated by a partition, at same pressure P and temperature T. On removing the partition they will mix by diffusion. The rate of mixing r at the surface of separation is given by

$$r = D\left(= \tfrac{1}{3}\langle c \rangle \lambda\right) \times \text{concentration gradient across the surface of separation,}$$

by Eq. (2.14)

$$\propto \langle c \rangle \lambda \tfrac{P}{T} \text{ at the initial stage}$$

Since concentration gradient $\propto$ pressure gradient $\propto P$ for each gas

Then, with rise of P at constant T, r will not change because $\lambda \propto 1/P$ and $\langle c \rangle$ is independent of P. But with rise of T at constant P, r will rise because $\langle c \rangle$ rises, and $\lambda \tfrac{P}{T}$ does not change (as $\lambda \tfrac{T}{P}$).

2.64 At liquid-vapour equilibrium

rate of vaporisation of liquid

= rate of condensation of vapour assuming sticking probability to be one

$$= \frac{1}{k} \sqrt{\frac{R}{2\pi}} \frac{P}{\sqrt{MT}} \text{ by Eq. (2.8) putting } N/V = P/kT \text{ and } \langle c \rangle = \sqrt{\frac{8RT}{\pi M}}$$

$$= \frac{1}{(1.381 \times 10^{-23}\,\text{JK}^{-1})} \sqrt{\frac{8.314\,\text{JK}^{-1}\,\text{mol}^{-1}}{2 \times 3.143}} \cdot \frac{\frac{23.7}{760} \times 1.013 \times 10^5\,\text{Pa}}{\sqrt{18.016 \times 10^{-3}\,\text{kg mol}^{-1}}\,(298.15\,\text{K})}$$

$$= 3.936 \times 10^{25}\,\text{m}^{-2}\text{s}^{-1}$$

2.65 Since the frequency of wall collision, given by Eq. (2.8) is same

$$\langle c \rangle_{\text{He}} \left(\frac{N}{V}\right)_{\text{He}} = \langle c \rangle_{\text{H}_2} \left(\frac{N}{V}\right)_{\text{H}_2} \tag{A}$$

Then $$\frac{(N/V)_{\text{He}}}{(N/V)_{\text{H}_2}} = \frac{\langle c \rangle_{\text{H}_2}}{\langle c \rangle_{\text{He}}} = \left(\frac{M_{\text{He}}}{M_{\text{H}_2}}\right)^{\tfrac{1}{2}} > 1$$

Therefore, He has higher concentration

$$\frac{Z_{\text{He-He}}}{Z_{\text{H}_2\text{-H}_2}} = \frac{\sigma^2_{\text{He}}}{\sigma^2_{\text{H}_2}} \cdot \frac{\langle c \rangle_{\text{H}_2}}{\langle c \rangle_{\text{He}}} \text{ by Eq. (2.11b) using the relation (A)}$$

$$= \left(\frac{\sigma_{\text{He}}}{\sigma_{\text{H}_2}}\right)^2 \left(\frac{M_{\text{He}}}{M_{\text{H}_2}}\right)^{\tfrac{1}{2}} \simeq 1$$

i.e. $Z_{\text{H}_2\text{-H}_2} \simeq Z_{\text{He-He}}$

2.66 Total number of molecular collisions with a small surface area A

$$= \frac{1}{4} A \frac{N}{V} \langle c \rangle = \frac{1}{4} A \frac{N}{V} \int_0^\infty c\, P(c)\, dc$$

where $P(c)\, dc$ is the probability of molecular speed lying between c and $c + dc$.

Then the probability $P'(c)\, dc$ of collision with speed lying between c and $c + dc$ is

$$P'(c)\, dc = \frac{\frac{1}{4} A \frac{N}{V} c\, P(c)\, dc}{\frac{1}{4} A \frac{N}{V} \langle c \rangle} = \frac{c\, P(c)\, dc}{\langle c \rangle} = \frac{4\pi \left(\frac{m}{2\pi kT}\right)^{\tfrac{3}{2}} c^3 e^{-mc^2/2kT}}{\langle c \rangle}$$

i. Since $P' \propto c^3 e^{-mc^2/2kT}$, we have for maximum P' (where $dP'/dc = 0$)

$$3c^2 - c^4 \, m/kT = 0$$

or $$c = (3kT/m)^{\frac{1}{2}}$$

which is the required most probable speed. This is not same as the most probable speed $(2kT/m)^{\frac{1}{2}}$ in the gas phase.

ii. The average speed with which the gas molecules collide the surface

$$= \int_0^\infty cP'(c)\,dc$$

$$= \frac{4\pi \left(\dfrac{m}{2\pi kT}\right)^{\frac{3}{2}}}{\langle c \rangle} \int_0^\infty c^4 e^{-mc^2/2kT}\,dc$$

$$= 4\pi \left(\frac{m}{2\pi kT}\right)^{\frac{3}{2}} \cdot \frac{3}{8}\pi^{\frac{1}{2}} \left(\frac{2kT}{m}\right)^{\frac{5}{2}} \bigg/ \left(\frac{8kT}{\pi m}\right)^{\frac{1}{2}}$$

$$= \left(\frac{9\pi kT}{8m}\right)^{\frac{1}{2}}$$

This is not same as the average speed $\left(\frac{8kT}{\pi m}\right)^{\frac{1}{2}}$ in the gas phase.

Note: In both (i) and (ii), c is found to be greater than the corresponding value of c in the gas phase. Physically, this means that the molecules that strike a plane of given area are travelling at higher speeds than the molecules in a gas in general. Because the molecules having higher speed are more likely to strike the area in a given time.

2.67 The frequency of collisions between any two gas molecules

$$= \frac{1}{2}\pi\sigma^2 \langle C_r \rangle (N/V)^2 \quad [\text{(vide Eq. (2.11a)}]$$

$$= \frac{1}{2}\pi\sigma^2 \left(\frac{N}{V}\right)^2 \int_0^\infty C_r P(C_r)\,dC_r$$

Then following similar procedure as in the previous problem, the probability of collision with speed lying between C_r and $C_r + dC_r$ is

$$P'(Cr)dCr = \frac{C_r P(C_r)dC_r}{\langle C_r \rangle} = \frac{4\pi(\mu/2\pi kT)^{\frac{3}{2}} C_r^3 e^{-\mu C_r^2/2kT}\,dC_r}{(8kT/\pi\mu)^{1/2}}$$

Putting $$C_r = \left(\frac{2\epsilon_r}{\mu}\right)^{\frac{1}{2}} \text{ and } dC_r = \left(\frac{1}{2\mu\,\epsilon_r}\right)^{\frac{1}{2}} d\epsilon_r, \text{ we have}$$

$$P'(\epsilon_r)d\epsilon_r) = \left(\frac{1}{kT}\right)^2 \epsilon_r \, e^{-\epsilon_r/kT} d\epsilon_r$$

a. The required average relative kinetic energy

$$= \int_0^\infty \epsilon_r \, P'(\epsilon_r)d\epsilon_r$$

$$= \left(\frac{1}{kT}\right)^2 \int_0^\infty \epsilon_r^2 \, e^{-\epsilon_r/kT} d\epsilon_r$$

$$= \left(\frac{1}{kT}\right)^2 \cdot \frac{2!}{\left(\frac{1}{kT}\right)^{2+1}}$$

$$= 2kT \left[\because \int_0^\infty x^n e^{-ax} dx = \frac{n!}{a^{n+1}}\right]$$

b. The required fraction of bimolecular collision

$$= \int_{\epsilon'}^\infty P'(\epsilon_r) d\epsilon_r$$

$$= \left(\frac{1}{kT}\right)^2 \int_{\epsilon'}^\infty \epsilon_r\, e^{-\epsilon_r/kT} d\epsilon_r$$

$$= \left(\frac{1}{kT}\right)^2 \cdot (kT)^2 \left(1+\frac{\epsilon'}{kT}\right) e^{-\epsilon'/kT}$$

$$= \left(1+\frac{\epsilon'}{kT}\right) e^{-\epsilon'/kT}$$

$$= e^{-\epsilon'/kT} \text{ if } \epsilon' \ll kT$$

Note:

1. Average relative kinetic energy of gas molecules ($2kT$) is greater than average molecular kinetic energy $\left(\frac{3}{2}kT\right)$.

2. Fraction of molecular collisions with relative kinetic energy above ϵ' is somewhat greater than the fraction of molecules with kinetic energy above ϵ', which is $\dfrac{2}{\sqrt{\pi}} \left(\dfrac{\epsilon'}{kT}\right)^{\frac{1}{2}} e^{-\epsilon'/kT}$ [(Eq. (2.7a)]. However, both vary essentially as $e^{-\epsilon'/kT}$. In simple collision theory, the rate of a bimolecular gas-phase reaction is equated to the frequency of colliion per unit volume between the reactant molecules with relative kinetic energy above certain critical value (called energy of activation).

2.68 If a gas is confined to a container but open to a low pressure region through a small hole, the gas will flow through the hole until the pressure becomes equal on both sides. If the diameter of the hole is smaller than the mean free path, a molecule on reaching the hole is unlikely to collide with another but will pass right through the hole. Since in such free-molecule flow, called effusion flow, the random molecular motion is not affected, the Maxwell's law of speed distribution will hold for an effusion gas. Then the rate of effusion through a hole of area A will be equal to the rate at which the gas molecules happen to hit an area A of the hole and for the gas on either side of the hole it is given by

$$AZ_W = AP/(2\pi mkT)^{\frac{1}{2}} \text{ by Eq. (2.8)}$$

Putting $\qquad N/V = P/kT$ and $\langle c \rangle = \left(\dfrac{8kT}{\pi m}\right)^{\frac{1}{2}}$

i.e. $\qquad AZ_W \propto P/\sqrt{mT}$

Then, the rate of effusion $\propto 1/\sqrt{m}$, P and T remaining constant. This is Graham's law of effusion.

Note:

1. If the diameter of the hole exceeds the mean free path, the gas molecules will have fair chances to collide with other molecules during their passage through

the hole leading to a collective flow or directed mass motion towards the hole (as in case of viscous flow caused by pressure difference, vide problem 7.3). When this happens, Maxwell's distribution law and hence Graham's law will not hold.

2. Free flow of gas molecules through a capillary-like flow through a small hole, also occurs at a rate proportional to $1/\sqrt{m}$, but with proportionality constant depending on the length of the capillary apart from its diameter.

3. Effusion is closely related to diffusion in that both can be regarded as free expansion of a substance. Effusive flow occurs through a hole and has relevance only to gases where the molecules near the hole are mainly involved so that only pressure (and not the pressure gradient) is important in determining the rate of the process. In contrast, the diffusive flow occurs in a medium of constant pressure where the concentration gradient of a species determines its rate of diffusion (which is not restricted only to gases) following Fick's law, Graham's law, that governs effusion, has no relevance to diffusion.

2.69 a. Rate of escaping through the hole

$$= \text{area of the hole } (A) \times \text{frequency of wall collision}$$

$$= A \cdot \frac{1}{4} \langle c \rangle \frac{N}{V} \quad \text{by Eq. (2.8)}$$

where N is the number of molecules present in the container at the instant concerned

$$= \pi r^2 \cdot \frac{1}{4} \sqrt{\frac{8}{3\pi} \langle c^2 \rangle} \frac{N}{V}, \quad r \text{ is radius of the hole}$$

$$= \sqrt{\pi/6} \, r^2 \sqrt{\langle c^2 \rangle} \frac{N}{V}$$

$$= \sqrt{\frac{3.143}{6}} (0.001 \times 10^{-3} \, \text{m})^2 (400 \times 60 \, \text{m/min}) \left(\frac{0.1 \, \text{mol}}{10 \times 10^{-3} \, \text{m}^3} \right)$$

at the initial stage

$$= (1.737 \, \text{min}^{-1}) \, (0.1 \, \text{mol})$$

$$= 0.1737 \, \text{mol min}^{-1}$$

□ i. Required time $= \dfrac{\text{Number of molecules escaped}}{\text{Rate of escaping}}$

$$= \frac{0.1 \times \dfrac{0.20}{100}}{0.1737 \, \text{mol min}^{-1}} = 1.151 \times 10^{-3} \, \text{min}$$

Considering that the rate of effusion is very nearly same as that at the initial stage.

ii. Here molecular density of the gas remaining in the container, and hence the rate of escaping of the molecules, changes appreciably during effusion. Then, for correct calculation of the required time one has to apply integration on the following rate equation

$$- \frac{dN}{dt} = \sqrt{\frac{\pi}{6}} r^2 \sqrt{\langle c^2 \rangle} \frac{N}{V}$$

Then $- \displaystyle\int_{N_0}^{\frac{25}{100} N_0} \frac{dN}{N} = \sqrt{\frac{\pi}{6}} r^2 \sqrt{\langle c^2 \rangle} \cdot \frac{1}{V} \int_0^t dt$

or
$$t = \frac{-\int_{N_0}^{\frac{25}{100}N_0} \frac{dN}{N}}{\sqrt{\pi/6}\, r^2 \sqrt{\langle c^2 \rangle} \cdot \frac{1}{V}}$$

$$= \frac{\ln 4}{1.737 \text{ min}^{-1}} = 0.7979 \text{ min}$$

□□ The condition to be fulfilled here is that the diameter of the hole must not exceed the mean free path. Because only then, the gas can escape through a process of effusion when the working formula based on the Maxwell distribution can be effective.

b. Here
$$-\frac{dP}{P} = -\frac{dN}{N}$$

$$= \sqrt{\frac{\pi}{6}}\, r^2 \sqrt{\langle c^2 \rangle} \frac{1}{V} dt$$

Then
$$-\int_{P_0}^{P} \frac{dP}{P} = -\ln\left(\frac{P}{P_0}\right)$$

$$= \sqrt{\frac{\pi}{6}}\, r^2 \sqrt{\langle c^2 \rangle} \cdot \frac{1}{V} \int_0^t dt$$

or
$$P = \frac{n_0 M \langle c^2 \rangle}{3V} \text{Exp}\left[-\sqrt{\frac{\pi}{6}}\, r^2 \sqrt{\langle c^2 \rangle} \frac{1}{V} \int_0^{10} dt \right] \quad \because \langle c^2 \rangle = \frac{3P_0 V}{n_0 M}$$

$$= \frac{(0.1 \text{ mol})\,(2.016 \times 10^{-3} \text{kg mol}^{-1})\,(400 \text{ m/s})^2}{3(10 \times 10^{-3} \text{ m}^3)}$$

$$\exp[-1.737 \text{ min}^{-1} \times 10 \text{ min}] = 1.99 \times 10^{-15} \text{Pa}$$

2.70 i. In the effused gas mixture

$$\frac{\text{number of moles of Ne}}{\text{number of moles of Ar}} = \frac{\left[\frac{N}{V} \cdot \langle c \rangle\right]_{\text{Ne}}}{\left[\frac{N}{V} \cdot \langle c \rangle\right]_{\text{Ar}}}$$

$$= \frac{\left[\frac{N}{V}\right]_{\text{Ne}}}{\left[\frac{N}{V}\right]_{\text{Ar}}} \cdot \left(\frac{M_{\text{Ar}}}{M_{\text{Ne}}}\right)^{1/2}$$

$$= \frac{1}{5}\left(\frac{39.95}{20.18}\right)^{1/2}, \text{ initially} = 0.2814$$

ii. After a long time an equilibrium will be established between the effused part and the remaining part of the gas mixture when each part will have the given composition (just as in case of ordinary expansion of the gas mixture).

Note: The composition of the effused gas mixture is different in (i) and (ii) because it is kinetically controlled in (i) but thermodynamically controlled in (ii)

2.71 In kinetic theory, the concept of degrees of freedom was introduced to deal with heat capacity. Heat capacity of a molecular system is determined by the ability of its molecules to undergo various types of independent motions, called *degrees of freedom* in storing energy. Ordinarily, contribution of heat capacity is limited to translational, rotational and vibrational motions. It is quite unusual to find any contribution from electronic motion (transition) due to too high values of electronic energy quanta. This explains why atomic structure is disregarded in calculating degrees of freedom.

2.72 For a molecule of atomicity x

vibrational degrees of freedom,

f = total degree of freedom − translational degree of freedom
    − rotational degrees of freedom

= $3x - 3 - 2$ or 3 according as the molecule is linear or not

i. Here $x = 4$, then

for H_2O_2 molecule, which is non-linear

$$f = 3 \times 4 - 3 - 3 = 6$$

for C_2H_2 molecule, which is linear

$$f = 3 \times 4 - 3 - 2 = 7$$

ii. Here also $x = 4$ for both BF_3 and NF_3, and being non-linear both have $f = 6$ (like H_2O_2).

□ H_2O_2 and C_2H_2 have same atomicity and hence have total number of degrees of freedom same. But C_2H_2 having higher number of vibrational degrees of freedom will have higher heat capacity. Because a vibrational degree of freedom is associated with higher energy than both translational and rotational degrees of freedom.

Note: In applying classical equipartition principle the molecular structures other than linear and non-linear have no relevance. Hence in calculating $\bar{C}_V$, both BF_3 (planar) and NF_3 (pyramidal) are similarly treated as non-linear though they have structural difference having relevance to quantum theoretical calculations.

2.73 The high temperature limiting value of $\bar{C}_V$, predicted by the principle of equipartition of energy, is

$$\bar{C}_V = (3x - 3)R \quad \text{by Eq. (2.20)}$$

□ Explanation comes from quantum theory. At ordinary temperature, the translational contribution to $\bar{C}_V$ is fixed ($\frac{1}{2}R$ per degree of freedom) and the rotational contribution is nearly so. But the vibrational contribution, due to high values of vibrational quanta, is much lower than the classically predicted value (R per degree of freedom) and this contribution rises rapidly with temperature to a limiting value.

2.74 $\bar{C}_P$ (the molar heat capacity at constant pressure) is related to $\bar{C}_V$ by

$$\bar{C}_P - \bar{C}_V = \text{work done by the system against constant external pressure per mole per degree rise in temperature}$$

$$= P\left(\frac{\partial \bar{V}}{\partial T}\right)_P$$

$$= R, \text{ for an ideal gas for which } P\bar{V} = RT$$

Then
$$\gamma = \frac{\bar{C}_P}{\bar{C}_V} = \frac{\bar{C}_V + R}{\bar{C}_V}$$

$$= 1 + \frac{R}{\bar{C}_V}$$

Therefore the desired relation will be

$$\gamma = 1 + \frac{1}{3x - 5/2}, \text{ for linear molecules, by Eq. (2.19)}$$

$$\gamma = 1 + \frac{1}{3x - 3}, \text{ for non-linear molecules, by Eq. (2.20)}$$

2.75   i. RMS speed $\left(\frac{3RT}{M}\right)^{\frac{1}{2}}$: $CO_2 > SO_2$, since M is lower for CO_2.

ii. Molar KE due to translational molecular motion $\left(\frac{3}{2}RT\right)$: $CO_2 = SO_2$

iii. Molar KE due to translational, rotational and vibrational motion: $CO_2 = SO_2$. Since total number of degrees of freedom is same (3×3) for both the molecules and KE per degree of freedom for each type of motion is also same ($\frac{1}{2}RT$ per mole).

iv. Molar thermal energy $(\bar{U})$: $CO_2 [3 \times 3 - 5/2) RT] > SO_2 [(3 \times 3 - 3)RT]$
                                 linear   non-linear
                               molecule   molecule

v. Molar heat capacity $\left(\frac{\partial \bar{U}}{\partial T}\right)$: $CO_2 > SO_2$, following (iv)

vi. $\bar{C}_P - \bar{C}_V$ ($= R$, assuming ideal behaviour): $CO_2 = SO_2$

Note:

1. Although molecular KE (which is independent of molecular mass) is same, RMS speed (which depends on molecular mass) is different for the two gases.

2. Molar kinetic energy due to molecular motion is same but thermal energy (which includes potential energy of molecular vibration) is different for CO_2 and SO_2 due to difference in structure.

3. Although $\bar{C}_P$ (and also $\bar{C}_V$) is different for two gases, $\bar{C}_P - \bar{C}_V$, which is related only to the external work done by the gas (assuming ideal behaviour), is same for them.

2.76 At ordinary temperature, $\bar{C}_V$ of gaseous chlorine, which consists of Cl_2 molecules, will be less than its equipartition value due to partial vibrational contribution to $\bar{C}_V$. With rise in temperature this contribution will increase with consequent rise in $\bar{C}_V$. But at very high temperature when Cl_2 begins to atomise, $\bar{C}_V$ of the system will fall with further rise of temperature because Cl atoms can neither rotate nor vibrate.

UNIVERSITY QUESTIONS

2.1 What is meant by equation of state for a gas? Why is it called so?

(Calcutta BSC(H), 1996)

2.2 How can you depict Boyle's law by a straight line plot other than PV vs V (or P)?

(Burdwan BSC(H), 2001)

2.3 Show that for an ideal gas the cubic expansion coefficient and isothermal compressibility are respectively $1/T$ and $1/P$.   (Burdwan BSC(H), 2011)

2.4 Find the dimensions of RT/M.   (Burdwan BSC(H), 2011)

2.5 From the kinetic theory of gas arrive at a relation between pressure of a gas and its molecular velocity. From this equation justify Charle's law applied to the gas.

(Jadavpur BSC(H), 2012)

2.6 From the view point of kinetic theory give the concept of temperature and pressure.

(Calcutta BSC(H), 2008)

2.7 Two separate bulbs contain ideal gases A and B respectively at the same temperature. The density of gas A is thrice that of gas B and the molecular weight of A is half that of B. Calculate the ratio of the pressure of A to that of B.

(Calcutta BSC(H), 1994)

2.8 Two gases A and B have equal volume, equal number of moles and equal rms speed but unequal molar masses, $M_A > M_B$. Which gas has higher pressure.

(Burdwan BSC(H), 1994)

2.9 Calculate total pressure in a mixture of 4 g of oxygen and 3 g of hydrogen confined in a total volume of one liter at 0°C. (Madras BSc, 2008)

2.10 Explain the necessity of defining the rms speed. (Burdwan BSC(H), 1992)

2.11 RMS speed is appropriate in expressing average kinetic energy of gas molecules rather than average speed. Justify. (Burdwan BSC(H), 2008)

2.12 In calculating the average kinetic energy of a molecule one used the formula $E = \frac{1}{2}mc^2$ taking c as the average speed. Was he correct? It so, justify. If not, calculate the percentage error incurred in the calculation. (Calcutta BSC(H), 1992)

2.13 In a gaseous assembly 20% molecules possess a speed 490 cm s^{-1}, 30% possess 9 ms^{-1} and the rest 20 ms^{-1}. Find the average and rms speed.

(Calcutta BSC(H), 2008)

2.14 Oxygen has a density of 1.429 g/liter at STP. Calculate the rms and average speed of its molecules. (Delhi BSc, 2003)

2.15 Why should the rms speed be greater than the average speed?

(Burdwan BSC(H), 2001)

2.16 Why there is no air on the moon surface? (Calcutta BSc(H), 1995)

2.17 If an ideal gas at T K and P atm has an average speed C_a, by how many times C_a would change if (i) T is doubled at constant P and (ii) P is doubled at constant T?

(Calcutta BSC(H), 1993)

2.18 Calculate the value of mean square deviation $\langle (c - \langle c \rangle)^2 \rangle$ for O_2 at 27°C.

(Calcutta BSC(H), 1994)

2.19 What do you mean by the distribution of molecular speeds?

(Calcutta BSC(H), 1993)

2.20 State, with explanation, whether the distribution of molecular speeds is an example of static or dynamic equilibrium. (Calcutta BSC(H), 1995)

2.21 If out of N molecules of a gas at a given temperature, dN_u molecules have their x-component of velocity in the range u and $u + du$, then $\frac{1}{N} \cdot \frac{dN_u}{du}$ should be a function of u^2 and not u. Comment. (Burdwan BSC(H), 2009)

2.22 Maxwell distribution law of molecular velocity in one dimension is expressed as $\frac{1}{N} \cdot \frac{dN_u}{du} = Ae^{-mu^2/2kT}$. Find out the expression for A. (Calcutta BSC(H), 2011)

2.23 Draw schematically the one-dimensional velocity distribution plots at two different temperatures. Comment on the area under the curve and the average velocity of the gas molecules. (Calcutta BSc(H), 2009)

2.24 The average value of the x-component of molecular velocity of a gas is zero. Why?

(Burdwan BSc(H), 2001)

2.25 Justify or criticise the following: Since $\langle u \rangle = \langle v \rangle = \langle w \rangle = 0$, the average value of energy for each velocity component must be zero. (Burdwan BSc(H), 1996)

2.26 From the one-dimensional velocity distribution, find out the average kinetic energy of a molecule moving along one-dimension. (Calcutta BSc(H), 2011)

2.27 The expression for the distribution of molecular speeds (in three-dimensions) is $f(c) = 4\pi c^n \left(\frac{m}{2\pi kT}\right)^{\frac{3}{2}} e^{-mc^2/2kT}$. What does $f(c)\,dc$ signify? Using dimensional arguments show that $n = 2$ in the expression for $f(c)$. (Calcutta BSc(H), 2013)

2.28 a. Depict Maxwell's speed distribution curves at two different temperatures T_1 and T_2 $(T_2 > T_1)$.
   b. Explain their natures by analysing Maxwell's equation.
   c. How is it affected if the gas is changed from helium to argon?
   d. What is the significance of the area under the curve between any two different speeds? (Calcutta BSc(H), 2005)

2.29 How does the shape of the distribution curves change as $T \rightarrow 0$? (Calcutta BSc(H), 2007)

2.30 From the Maxwellian distribution of molecular speed, find the expression for calculating the most probable speed and the rms speed of a molecule. Which one is greater and why? Show the positions of the different types of speeds on the graphical representation. (Burdwan BSc(H), 1992)

2.31 For an ideal gas obeying the Maxwellian distribution of molecular speeds in three dimensions, find the maximum value of $\frac{1}{N} \cdot \frac{dN_c}{dc}$ for a gas of molar mass 4.0 g mol^{-1} kept at 127°C. (Calcutta BSc(H), 2012)

2.32 Find an expression for $\langle 1/c \rangle$ in case of a three-dimensional Maxwellian distribution of molecular speeds. (Calcutta BSc(H), 2012)

2.33 Show that the fraction of molecules of an ideal gas moving with speed between C_{mp} and 1.001 C_{mp} is constant for any gas at any temperature. (Calcutta BSc(H), 2013)

2.34 Starting from Maxwell's distribution formula for molecular speeds of a gas in three-dimensions, obtain the distribution formula for kinetic energy and hence derive an expression for the fraction of total molecules possessing kinetic energy in excess of ϵ. Point out the differences between the distribution curves for speed and kinetic energy. (Burdwan BSc(H), 2011)

2.35 In a hypothetical collection of N_2 gas at 27°C in an isolated vessel, each molecule has exactly the same kinetic energy to start with. Sufficient time is allowed to lapse such that total randomness is established. What should be
   i. the final temperature
   ii. the average and rms speed at the start and at the end
   iii. most probable velocity at the start and at the end. (Calcutta BSc(H), 1995)

2.36 Two-dimensional speed distribution function is $\frac{1}{N}\frac{dN_c}{dc} = \frac{m}{kT}c\,e^{-mc^2/2kT}$. From this expression, find out the fraction of molecules having kinetic energy $\geq \epsilon'$. (Calcutta BSc(H), 2011)

2.37 The molecules of an ideal gas are confined to move in a plane
   a. Derive the expression for rms speed
   b. Derive the expression for the most probable speed. (Calcutta BSc(H), 2014)

2.38 At 25°C and 1 atm, for oxygen molecules, calculate the number of wall collisions per sec per cm^2 and the rate of effusion through a hole of diameter 1 mm. (Calcutta BSc(H), 2008)

2.39 By how many times does the rate of effusion of a gas through a pin-hole into vacuum change when pressure is doubled and temperature is increased four times?

(Calcutta BSc(H), 2011)

2.40 What will be the ratio of final to initial wall collision frequency of an ideal gas if pressure is halved at constant gas density?   (Calcutta BSc(H), 2013)

2.41 Two separate bulbs are filled with two ideal gases A and B. The molecular weight of B is twice that of A. If B is at twice the absolute temperature and has half the density of A, what is the ratio of wall collision frequencies? (Calcutta BSc(H), 1993)

2.42 Calculate the number of binary collisions per cc of nitrogen gas per second at 2 atm and 30°C. The bond length of the gas molecule is 1.87Å.   (Calcutta BSc(H), 1999)

2.43 The binary collision frequency of the molecules of a gas is 10^{28} cm^{-3}s^{-1} at NTP. Find the same when the pressure is doubled temperature remaining analtered.

(Calcutta BSc(H), 1996)

2.44 The average speed and mean free path of oxygen molecules at 25°C and 1 atm are 444 ms^{-1} and 1.6×10^{-7} m respectively. Estimate the number of binary collisions between oxygen molecules per unit volume per second at 25°C and 1 atm pressure (assume ideal behaviour).   (Calcutta BSc(H), 2009)

2.45 Calculate the mean free path for oxygen molecules at 298 K and at a pressure of 500 torr (Given: molecular diameter 3.61×10^{-10} m).   (Calcutta BSc(H), 2014)

2.46 The mean free path of a gas is λ at pressure P and temperature T Kelvin. What will be the value of mean free path of the gas in terms of λ when
    i. the pressure is doubled keeping the temperature constant
    ii. the pressure is doubled and temperature is halved?   (Calcutta BSc(H), 2013)

2.47 Explain the effect of temperature rise on the mean free path of an ideal gas held at constant volume.   (Calcutta BSc(H), 2008)

2.48 What assumption of the kinetic model of gases has been modified in the equipartition principle?   (Burdwan BSc(H), 1992)

2.49 What is meant by 'degree of freedom' as referred to molecular motion?

(Calcutta BSc(H), 2000)

2.50 State the principle of equipartition of energy. Mention its advantages and limitations with explanations.   (Burdwan BSc(H), 1992)

2.51 Explain why the contribution of each translational and rotational mode to the average molecular energy is half of a vibrational mode.   (Calcutta BSc(H), 2008)

2.52   a. Find the total energy of all the molecules of He present in a vessel of volume V at pressure P, assuming ideal behaviour of the gas.
       b. Find the average value of energy for a H_2 molecule according to equipartition theorem.   (Burdwan BSc(H), 2006)

2.53 Using the principle of equipartition of energy, calculate $\bar{C}_v$ for a diatomic gas like chlorine in terms of R. At room temperature the experimental value of chlorine is $3R$. Explain the discrepancy.   (Calcutta BSc(H), 2013)

2.54 Explain why the equipartition principle predicts more correct value of heat capacity for He than HCl at room temperature.   (Calcutta BSc(H), 2011)

2.55 From the basic principle of equipartition of energy, show that γ of a polyatomic gas is less than that of a monatomic gas ($\gamma = C_P/C_V$).   (Jadavpur BSc(H), 2012)

2.56 What is the limiting value of γ as the number of atoms in the molecule becomes very large?   (Calcutta BSc(H), 2004)

2.57 The equipartition value of γ for a non-linear molecule A_nB is found to be 1.67, assuming ideal behaviour, find the value of n. (Calcutta BSc(H), 2009)

2.58 γ for gaseous H_2 is 1.42 at ordinary temperature but decreases as the temperature is increased. Explain. (Burdwan BSc(H), 1994)

2.59 Compare γ of CO_2 and SO_2 at high temperatures. (Burdwan BSc(H), 1998)

2.60 On the basis of kinetic theory of gases, show that for ideal gas $\bar{C}_p - \bar{C}_v = R$. (Punjab BSc, 2002)

KEY TO UNIVERSITY QUESTIONS

2.1 Vide Section 2.1.

□ An equation of state is called so because, from the knowledge of the state variables appearing in such an equation of a system, one can find out by means of it the other state variables of the system. For example, in case of an ideal gas, the coefficient of thermal expansion $\frac{1}{V}\left(\frac{\partial V}{\partial T}\right)_P = \frac{1}{T}$ (by $PV = nRT$) can be found from the knowledge of T.

2.2 Through a plot like P vs ρ (density) for a pure gas, or PV vs $1/M$ for a mixture of gases, at equilibrium (where M is average molar mass).

2.3 Cubic expansion coefficient

$$\frac{1}{V}\left(\frac{\partial V}{\partial T}\right)_P = \left(\frac{\partial \ln V}{\partial T}\right)_P$$

$$= 1/T, \text{ for an ideal gas}$$

where $PV = nRT$, or $\ln V = \ln T - \ln P + \ln(nR)$

Isothermal compressibility

$$-\frac{1}{V}\left(\frac{\partial V}{\partial P}\right)_T = -\left(\frac{\partial \ln V}{\partial P}\right)_T$$

$$= 1/P, \text{ for an ideal gas}$$

where $\ln V = \ln T - \ln P + \ln(nR)$

2.4 From ideal gas equation $PV = \frac{w}{M}RT$, RT/M will have the dimensions of PV/w, i.e. $(\text{ml}^{-1}\text{t}^{-2})(\text{l}^3)/(\text{m})$ or l^2t^{-2}.

2.5 To deduce the required relation let us consider N identical molecules of a gas confined in a cubic vessel of volume l^3. For a particular molecule, we can resolve its velocity c (a vector quantity) into three components u, v and w at right angles to one another and parallel to the edges of the cube. [This means each molecule moves as if with three velocities u, v and w simultaneously]. Each component has effect only on two opposite walls on which it acts normally. Here the molecule after collision with the wall will rebound with same velocity in the opposite direction, if collision is assumed to be elastic. If mass of the molecule is m, then for u component the change of its momentum per collision will be $-mu - mu = -2mu$, and hence the momentum, transferred to the wall will be $2mu$ (by conservation of linear momentum), and per unit time this will be $2mu$ multiplied by the frequency of collision (u/l, if molecules are like point masses and no forces act on them except at the moment of collision), i.e. $2\ mu^2/l$. The corresponding quantities for the components v and w will be $2mv^2/l$ and $2mw^2/l$ respectively. Therefore, the rate of transfer of momentum to the walls for the molecule due to its velocity c will be

$$\frac{2mu^2}{l} + \frac{2mv^2}{l} + \frac{2mw^2}{l} = \frac{2mc^2}{l}, \quad \text{since } c^2 = u^2 + v^2 + w^2$$

Then, for all the molecules, the rate of transfer of momentum, i.e. the force exerted on the wall will be

$$\frac{2m}{l}(N_1 C_1^2 + N_2 C_2^2 + ...) = \frac{2mN}{l} \sum \frac{N_i C_i^2}{N} = 2Nm\langle c^2 \rangle/l$$

where $\langle c^2 \rangle = \dfrac{1}{N} \sum N_i C_i^2$ is called the mean square velocity or speed of the gas.

Considering the random motion of the gas molecules, the pressure P of the gas on each wall of the container will be the same and the total force on all the six walls, each of area l^2, will be $6l^2 P$. Then

$$6l^2 P = 2Nm\langle c^2 \rangle/l$$

or $\qquad\qquad PV = \tfrac{1}{3}Nm\langle c^2 \rangle$, since $V = l^3$

Note: The expression thus deduced is valid for a vessel of any shape which can be regarded as consisting of large number of small cubes. Further, the intermolecular collisions, which are assumed to be absent, will not affect the final expression. Because any momentum due to be transferred to the walls by a particular molecule will be transferred by some other molecule, if the molecules are in large number and the collision is elastic.

☐ Kinetic theory relates temperature of a substance to its random molecular motion. According to this theory

average translational energy of a gas molecule $\propto T$

Then it follows from the kinetic expression of PV that

$V \propto T$ when N and P are constants

2.6 The expression $PV = \tfrac{1}{3}Nm\langle c^2 \rangle$, which follows from kinetic theory, implies that P of a gas is a statistical quantity since $\langle c^2 \rangle$ being an average, can have a meaningful steady value only when N is large.

T, like P, of a gas will also be a statistical quantity because of the following relation

$$PV \propto T$$

Then according to the kinetic theory, both P and T are statistical quantities. Originating from the molecular motion P relates to the molecular momentum while T relates to the molecular kinetic energy.

2.7 $\qquad\qquad P = \dfrac{w/V}{M}RT = \dfrac{\rho}{M}RT \qquad$ by Eq. (2.1a)

Then $\qquad\qquad \dfrac{P_A}{P_B} = \dfrac{\rho_A}{\rho_B} \cdot \dfrac{M_B}{M_A} \cdot \dfrac{T_A}{T_B} = 2 \times 2 \times 1$ from the given data

2.8 $\qquad \dfrac{P_A}{P_B} = \dfrac{n_A}{n_B} \cdot \dfrac{V_B}{V_A} \cdot \dfrac{T_A}{T_B} \qquad$ by Eq. (2.1a)

$\qquad\qquad = \dfrac{n_A}{n_B} \cdot \dfrac{V_B}{V_A} \cdot \dfrac{(\langle c^2 \rangle M)_A}{(\langle c^2 \rangle M)_B}$ by Eq. (2.4c)

$\qquad\qquad > 1; \quad$ from the given data

i.e $\qquad P_A > P_B$

2.9
$$P = (n_{O_2} + n_{H_2})RT/V$$

$$= \left[\left(\frac{w}{M}\right)_{O_2} + \left(\frac{w}{M}\right)_{H_2}\right]\frac{RT}{V}$$

$$= \left(\frac{4g}{32g\,mol^{-1}}\right) + \left(\frac{3g}{2.016g\,mol^{-1}}\right)\frac{(0.08206\,L\,atm\,K^{-1}mol^{-1})(273.15\,K)}{(1\,L)}$$

$$= 36.16\,atm$$

2.10 The necessity of defining the rms speed lies in the fact that even when the distribution of molecular speed in an isolated gas is not steady (e.g. with only small number of gas molecules), the rms speed (unlike $\langle c \rangle$ and C_{mp}) does not change with time (provided molecular collision is elastic) and is therefore, always a measure of molecular kinetic energy (vide problem 2.30).

2.11 Vide previous question.

2.12 No. Here it would be correct to take c as rms speed.

Now $\dfrac{\text{average speed}}{\text{rms speed}} = \left(\dfrac{8}{3\pi}\right)^{\frac{1}{2}}$ by Eqs (2.4b) and (2.4c)

Then, the % of error incurred

$$= \frac{\sqrt{8} - \sqrt{3\pi}}{\sqrt{3\pi}} \times 100 = -7.9$$

2.13 See problem 2.31.

2.14 See problem 2.33(a).

2.15 See problem 2.13.

2.16 See problem 2.33(c).

Although escape velocity is much lower for moon than for earth, it is still much higher than the average speed of the atmospheric molecules. But the fraction of molecules having speed higher than the escape velocity is not quite insignificant and this causes gradual escaping of the atmospheric molecules (once present in an appreciable concentration) ending in their insignificant concentration on the moon surface.

2.17 For an ideal gas

average speed $(C_a) = \sqrt{\dfrac{8RT}{\pi M}}$

Then i. C_a would change by a factor $\sqrt{2}$, if T is doubled.

ii. C_a (which depends only on T) would not change, if P is changed at constant T.

Note: $C_a = \sqrt{\dfrac{8P}{\pi\rho}}$ does not imply a change of C_a with P, because $P \propto \rho$ at constant T.

2.18 See problem 2.34(iv).

2.19 By the distribution of molecular speeds, we mean grouping of molecules belonging to different narrow specified range of speeds.

2.20 Dynamic equilibrium: Because at equilibrium, although the number of molecules having speeds within any specified range does not change, the speed of each molecule continually change through collision with other molecules.

2.21 Because, otherwise the probability density would not be same for both u and $-u$ for a random molecular motion.

2.22 Here
$$\int_{-\infty}^{+\infty} \frac{dN_u}{N} = \int_{-\infty}^{+\infty} A e^{-mu^2/2kT} \, du = A\left(\frac{\pi}{m/2kT}\right)^{\frac{1}{2}}$$

or
$$A = \left(\frac{m}{2\pi kT}\right)^{\frac{1}{2}}$$

2.23 The velocity distribution plots are drawn in Fig. 2.1a.

☐ The area under the curve is same for high and low temperatures, because it represents the total probability (equal to one) for the whole range of velocity $-\infty$ to $+\infty$.

The average velocity is zero for both high and low temperature, because at any temperature, the probability density is same around u and $-u$.

2.24 Because velocity is a vector quantity, and x-components of molecular velocity v_x and $-v_x$, have the same probability density for a random molecular motion.

2.25 No. Because the average values of energy for the respective velocity constants are proportional to $\langle u^2 \rangle, \langle v^2 \rangle$ and $\langle w^2 \rangle$ respectively, where $\langle u^2 \rangle = \langle v^2 \rangle = \langle w^2 \rangle \neq 0$.

2.26 See problem 2.40 (vi).

2.27 $f(c)\,dc$ signifies the probability of molecular speed to lie between c and $c + dc$.

Then
$$f(c)\,dc = 4\pi c^n \left(\frac{m}{2\pi kT}\right)^{\frac{3}{2}} e^{-mc^2/2kT} dc$$

will have no dimension for both sides. Now, $mc^2/2kT$, being an exponent, is dimensionless, and $m/2\pi kT$ has the dimensions of $1/c^2$. Therefore, RHS will be dimensionally equivalent to $C^n \left(\frac{1}{c^2}\right)^{\frac{3}{2}} dc$ having dimensions of C^{n-3+1} which will be dimensionless if $n - 3 + 1 = 0$ or $n = 2$.

2.28 a. As in Fig. 2.1(b).

   b. The peak of the distribution curves, which corresponds to C_{mp}, becomes sharper (i.e. the distribution is narrowed) and hence higher at lower temperature, because the area under the curve (which represents the total probability for the whole range of speed) is same (equal to unity) at all temperature. The peak also shifts to the left with lowering of temperature, due to lowering of speed.

   c. See problem 2.41(b).

   d. Mentioned in (b).

2.29 See problem 2.41(b).

2.30 Referred to Section 2.2 (average kinetic parameters).

The rms speed is greater that the most probable speed (C_{mp}), because, in the calculation of the former, all the molecules are considered where the molecules having speed above C_{mp} exceeds in number than the other.

2.31 The maximum value of $\frac{1}{N}\frac{dN_c}{dc}$ is obtained by putting $C = C_{mp} = \sqrt{\frac{2RT}{M}}$ in the distribution Eq. (2.3b) and this is equal to

$$4\pi\left(\frac{M}{2\pi RT}\right)^{\frac{3}{2}} \cdot \frac{2RT}{M} e^{-1} = \frac{4}{\sqrt{2\pi}} \sqrt{\frac{M}{RT}} \cdot e^{-1}$$

$$= \frac{4}{\sqrt{2\times3.143}} \sqrt{\frac{4.0\times10^{-3} \text{ kg mol}^{-1}}{(8.314 \text{ JK}^{-1} \text{ mol}^{-1})(400.15\text{K})}} \cdot \frac{1}{2.7183}$$

$$= 6.436 \times 10^{-12} \text{ m}^{-1}\text{s}$$

2.32 $$\left\langle\frac{1}{c}\right\rangle = \int_0^\infty \frac{1}{c}\frac{dNc}{N} = \int_0^\infty \frac{1}{c}\cdot 4\pi\left(\frac{M}{2\pi RT}\right)^{\frac{3}{2}} c^2 e^{-mc^2/2RT}\,dc$$

$$= \left(\frac{2}{\pi}\right)^{\frac{1}{2}}\left(\frac{M}{RT}\right)^{\frac{3}{2}} \int_0^\infty c e^{-mc^2/2RT}\,dc$$

$$= \left(\frac{2}{\pi}\right)^{\frac{1}{2}}\left(\frac{M}{RT}\right)^{\frac{3}{2}}\cdot\left(\frac{RT}{M}\right) = \frac{4}{\pi}\left(\frac{\pi M}{8RT}\right)^{\frac{1}{2}} = \frac{4}{\pi}\cdot\frac{1}{\langle c\rangle}$$

Note: $\left\langle\dfrac{1}{c}\right\rangle \neq \dfrac{1}{\langle c\rangle}$.

2.33 See problem 2.35(i).

2.34 Vide Section 2.2 (average kinetic parameter) and problem 2.41(a).

2.35 See problem 2.30.

2.36 See problem 2.42(iv).

2.37 (a) rms speed $= \displaystyle\int_0^\infty c^2 \frac{dN_c}{N} = \int_0^\infty c^2\frac{m}{kT}c e^{-mc^2/2kT}\,dc$

$$= \frac{m}{kT}\int_0^\infty c^3 e^{-mc^2/2kT}\,dc$$

$$= \frac{m}{kT}\frac{1}{2(m/2kT)^2} = 2kT/m$$

(b) See problem 2.42(i).

2.38 Frequency of wall collision

$$Z_W = \frac{1}{4}\langle c\rangle\frac{N}{V} \quad \text{by Eq. (2.8)}$$

$$= \frac{1}{4}\cdot\left(\frac{8RT}{\pi M}\right)^{\frac{1}{2}}\cdot\frac{P}{kT} \quad \text{by Eq. (2.9)}$$

$$= \frac{1}{k}\left(\frac{R}{2\pi MT}\right)^{\frac{1}{2}}_P$$

$$= \frac{1}{(1.381\times10^{-16}\,\text{erg K}^{-1})}\left[\frac{8.314\times10^7\,\text{erg K}^{-1}\text{mol}^{-1}}{2\times3.143(32\,\text{g mol}^{-1})(25+273.15)\text{K}}\right]^{\frac{1}{2}}(1.013\times10^6\,\text{dyn/cm}^2)$$

$$= 3.785\times10^{22}\,\text{cm}^{-2}\text{s}^{-1}$$

Rate of effusion through the hole of diameter d

$= $ Area of the hole $\times$ frequency of wall collision

$$= \pi\left(\frac{d}{2}\right)^2\cdot Z_W$$

$$= 3.143\left(\frac{0.1\,\text{cm}}{2}\right)^2(3.785\times10^{22}\,\text{cm}^{-2}\text{s}^{-1})$$

$$= 2.974\times10^{20}\,\text{s}^{-1}$$

2.39 Rate of effusion $\propto P/\sqrt{T}$ vide Sec. 2.3.

Then, if P is doubled, the rate of effusion will be doubled.

If T is increased four times, the rate of effusion will change by a factor $\frac{1}{\sqrt{4}}$, i.e. $\frac{1}{2}$.

2.40
$$Z_W = \frac{1}{4} \langle c \rangle \frac{N}{V}$$

$$= \frac{1}{4} \left(\frac{8P}{\pi \rho} \right)^{\frac{1}{2}} \cdot \frac{N_A}{M} \rho; \quad \text{since } \frac{N}{V} = \frac{w}{mV} = \frac{\rho}{M/N_A}$$

$$\propto P^{\frac{1}{2}}, \text{ at constant gas density } (\rho)$$

Then $\quad \dfrac{(Z_W)_{\text{final}}}{(Z_W)_{\text{initial}}} = \left(\dfrac{1}{2} \right)^{\frac{1}{2}}$, i.e. $\dfrac{1}{\sqrt{2}}$ if P is halved at constant density.

2.41
$$Z_W = \frac{1}{4} \langle c \rangle \frac{N}{V} = \frac{1}{4} \left(\frac{8RT}{\pi M} \right)^{\frac{1}{2}} \left(\frac{\rho}{M/N_A} \right)$$

$$\propto \frac{\rho T^{\frac{1}{2}}}{M^{\frac{3}{2}}}$$

$$\frac{(Z_W)_B}{(Z_W)_A} = \left(\frac{\rho_B}{\rho_A} \right) \left(\frac{T_B}{T_A} \right)^{\frac{1}{2}} \Big/ \left(\frac{M_B}{M_A} \right)^{\frac{3}{2}} = \left(\frac{1}{2} \right) (2)^{\frac{1}{2}} \Big/ 2^{\frac{3}{2}} = \frac{1}{4}$$

2.42 Binary collision frequency

$$= \frac{1}{\sqrt{2}} \pi \sigma^2 \langle c \rangle \left(\frac{N}{V} \right)^2 \quad \text{by Eq. (2.11b)}$$

$$= \frac{\pi \sigma^2}{\sqrt{2}} \left(\frac{8RT}{\pi M} \right)^{\frac{1}{2}} \left(\frac{P}{kT} \right)^2$$

$$= \frac{2\pi^{\frac{1}{2}}}{k^2} \left(\frac{R}{M} \right)^{\frac{1}{2}} \sigma^2 \frac{P^2}{T^{\frac{3}{2}}}$$

Here $\sigma = 2 \times$ O–O bond distance [vide problem 2.44(iii)].

2.43 Binary collision frequency $\propto \dfrac{P^2}{T^{\frac{3}{2}}}$

Then, on doubling P at constant T, binary collision frequency will be 2^2, i.e. four times of its initial value and is equal to $4 \times 10^{28} \text{cm}^{-3}\text{s}^{-1}$.

2.44 Binary collision frequency

$$= \frac{1}{2\lambda} \langle c \rangle \frac{N}{V} \quad \text{from Eqs (2.11b) and (2.13)}$$

$$= \frac{\langle c \rangle P}{2k\lambda T} \quad \text{putting } N/V = P/kT$$

2.45 See problem 2.44(iii).

2.46
$$\lambda = \frac{1}{\sqrt{2} \, \pi \sigma^2 (N/V)} \quad \text{by Eq. (2.13)}$$

$$= \frac{1}{\sqrt{2} \, \pi \sigma^2 (P/kT)}$$

$$\propto T/P$$

Then i. mean free path will be halved on doubling P at constant T
ii. mean free path will be one-fourth of its initial value.

2.47 See problem 2.52(ii).

2.48 According to the simple kinetic theory of matter, a gas can store energy only through translational motion of its molecules which are assumed to be of point masses and hence structureless. Subsequently, this assumption has been modified, particularly to account for the heat capacity of gases, by incorporating the idea that the molecules have the structure of their own. This enables the molecules to store thermal energy through their rotational and vibrational motion, apart from translational motion by following the equipartition principle.

2.49 By 'degrees of freedom' of a molecule, we mean various independent motions that the molecule can undergo, which, though involves some arbitrariness, are fixed in number determined by the molecular structure.

2.50 The principle of equipartition of energy states that, in a system of molecules in thermal equilibrium at temperature T, the mean kinetic energy is so distributed, that it is equal to $\frac{1}{2}kT$ per molecule per degree of freedom, irrespective of the nature of the molecular motion. Regarding the distribution of total energy ($KE + PE$) it has been further stated that the energy for vibrational degree of freedom (assuming the vibration to be harmonic) is $\frac{1}{2}kT$ ($\frac{1}{2}kT$ in the form of kinetic and $\frac{1}{2}kT$ in the form of potential).

The advantage of the equipartition principle lies in its simplicity in that it involves no molecular parameters, like bond length and bond angle.

The equipartition principle has the limitation that it cannot explain the following facts: (i) $\bar{C}_V$ varies (increases) with temperature and (ii) the observed $\bar{C}_V$ equals the predicted one only in case of monatomic but not in case of polyatomic gases. In the latter case the observed $\bar{C}_V$ is always less than the predicted one at ordinary temperature, but approaches the latter at very high temperature. The equipartition principle fails because it is based on the classical concept of energy as a continuous variable which is not tenable in case of molecules, particularly when vibrational energies are concerned.

2.51 A vibrational motion, unlike translational and rotational, entails potential energy (due to restoring force) along with kinetic energy, and that too in equal amounts in case of a harmonic oscillator. Again, according to the equipartition principle, the kinetic energy per degree of freedom is same for types of molecular motions. Then the contribution of each translational and rotational mode to the average molecular energy will be half of a vibrational mode (for harmonic vibration).

2.52   a. He molecules, being monatomic, possess only translational kinetic energy of which the total energy is given by

$$N \cdot \frac{1}{2}m \langle c^2 \rangle = \frac{3}{2}PV \qquad \text{by Eq. (2.2a)}$$

    b. H_2 molecules, being diatomic, possess translational energy for three translational degrees of freedom, rotational energy for two rotational degrees of freedom and vibrational energy for one vibrational degree of freedom. Then, by equipartition principle,

average energy of a H_2 molecule $= 3 \times \frac{1}{2}kT + 2 \times \frac{1}{2}kT + 1 \times kT = \frac{7}{2}kT$

2.53 For Cl_2 molecule, the atomicity is 2 and hence

$$\bar{C}_V = \frac{3}{2}R + R + (3 \times 2 - 5)R = \frac{7}{2}R \qquad \text{by Eq. (2.19)}$$

☐ The discrepancy arises from the partial contribution of vibrational mode to $\bar{C}_v$. This happens, according to quantum theory, due to rather high value of the vibrational energy quanta. Full contribution of vibrational mode occurs only at high temperature.

2.54 HCl molecules, unlike He, can have rotational and vibrational motion, apart from translational motion. The vibrational contribution to $\bar{C}_V$, expected from the equipartition principle, does not occur fully at ordinary temperature due to rather high value of vibrational energy quantum. Then $\bar{C}_V$ predicted by the equipartition principle for HCl is not so correct at ordinary temperature as for He.

2.55 For an ideal gas

$$\gamma = 1 + \frac{R}{\bar{C}_V} \quad \text{(vide problem 2.74)}$$

In case of monatomic gas

$$\bar{C}_V = \tfrac{3}{2}R \quad \text{by Eq. (2.18)}$$

$$\gamma = 1 + \tfrac{2}{3} = \tfrac{5}{3}$$

γ for a polyatomic linear molecules of atomicity x, which is greater than that for non-linear molecules of same atomicity, is given by

$$\gamma = 1 + \frac{1}{3x - 5/2} \quad \text{by Eq. (2.19)}$$

$$\leq \frac{9}{7}\left(<\frac{5}{3}\right) \quad \text{since } n \geq 2 \text{ for polyatomic molecules}$$

Then, γ for a polyatomic gas molecules will be less than that for a monatomic gas.

2.56 In case of polyatomic molecules of atmoicity x, we have

for linear structure $\quad \bar{C}_V = 3x - 5/2 \quad$ by Eq. (2.19)

for non-linear structure $\quad \bar{C}_V = 3x - 3 \quad$ by Eq. (2.20)

Then as $x \to \infty, \bar{C}_V \to \infty$ and $\bar{C}_P (= \bar{C}_V + R) \to \bar{C}_V$ whence $\gamma = \bar{C}_P / \bar{C}_V \to 1$.

2.57 From the equipartition principle, we have for non-linear polyatomic molecules of atomicity x

$$\gamma = 1 + \frac{1}{3x - 3} \quad \text{(vide problem 2.74)}$$

Then $\quad\quad\quad 1.67 = 1 + \dfrac{1}{3(n+1) - 3} \quad$ from the given data

or $\quad\quad\quad n \simeq 0.5$

2.58 The equipartition value of γ for linear molecules of atomicity x is

$$\gamma = 1 + \frac{1}{3x - 5/2} \quad \text{(vide problem 2.74)}$$

For H_2, $x = 2$ and then $\quad \gamma = 1 + \dfrac{1}{3 \times 2 - 5/2} = 1.29$

The experimental value (1.42) of γ $(= 1 + R/\bar{C}_V)$ at ordinary temperature is some-what higher than the equipartition value (1.29). This is due to partial vibrational contribution $\bar{C}_V$ with increase in temperature, the vibrational contribution will increase with consequent decrease in γ.

2.59 CO_2 and SO_2 have same atomicity and hence same (total) number of degrees of freedom. But the linear CO_2 molecule has one more vibrational degree of freedom and hence its $\bar{C}_V$ will be higher than SO_2.

2.60 On raising the temperature of a system by 1°C, its energies due to translational, rotational and vibrational molecular motion will all increase and these occur to the same extent whether it is achieved at constant volume or constant pressure. But in the latter case some extra amount of energy is absorbed by the system for doing work required for expansion against the surroundings. Then

$$\bar{C}_P - \bar{C}_V = P\left(\frac{\partial \bar{V}}{\partial T}\right)_P$$

$$= R, \text{ for an ideal gas [where } = P\bar{V} = RT]$$

3

Gaseous State—Real

The material behaviour of any real substance is too complicated, even in gaseous state to be effectively described by a simple equation except over a limited range of P, V and T.

3.1 THE VAN DER WAALS EQUATION

Real gases do not usually obey ideal gas equation $PV = nRT$ but approach the latter only at very low pressure and high temperature as it appears from the isothermal plot of compressibility factor Z (= PV/nRT) vs P (Fig. 3.1a) or P vs $\bar{V}$ curve (Fig. 3.1b).

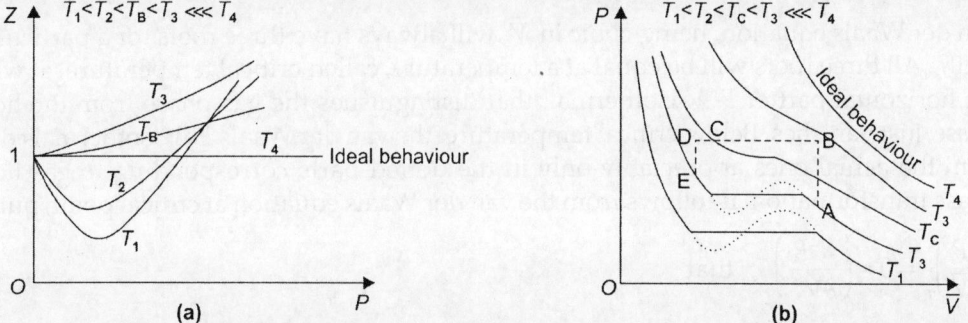

Fig. 3.1: Isothermals of a real gas showing deviation from ideal behaviour

A useful but simple equation of state for real gases was first theoretically deduced by van der Waals in the following form

$$\left(P + \frac{a}{\bar{V}^2}\right)(\bar{V} - b) = RT \tag{3.1a}$$

where $\bar{V} = V/N$ is the molar volume.

or $$\left(P + \frac{n^2 a}{V^2}\right)(V - nb) = nRT \tag{3.1b}$$

where a and b are van der Waals constants characteristic of a gas.

The van der Waals equation is based on the fact that the real gases donot obey most of the assumptions made for an ideal gas (listed in problem 2.10), particularly the assumption (ii) and (iii) due to their non-zero molecular volume and non-existence of intermolecular attraction. a is derived from intermolecular attraction while b from size of the gas molecules. b is given by

$$b = 4N_A \cdot \tfrac{4}{3}\pi \left(\frac{\sigma}{2}\right)^3 \tag{3.2}$$

where σ is the collision diameter (the shortest distance between two colliding molecules) and N_A is the Avogadro constant. σ would be same as the diameter of the molecules if they were like hard spheres when b would be four times the actual volume of the molecules per mole.

Boyle Temperature

The van der Waals equation leads to ideal gas equation when $\bar{V}$ is very high that happens at very low pressure and/or at very high temperature. The dip in the Z-P curve, due to $a/\bar{V}^2$, just vanishes at a temperature, called the Boyle temperature (T_B) where Boyle's law is obeyed for an appreciable range of pressure. The mathematical criterion for a gas to exhibit the Boyle temperature is

$$\lim\left(\frac{\partial Z}{\partial P}\right)_T = 0 \text{ only as } P \to 0 \tag{3.3}$$

Critical Constants

van der Waals equation, being cubic in $\bar{V}$ will always have three roots for a particular P and T. All three roots will be equal at a temperature, called critical temperature, at which the horizontal part of P-V isothermal, that distinguishes the gas phase from the liquid phase, just vanishes. Below critical temperature, the van der Waals P-V isothermals differ from the actual ones appreciably only in the dotted parts corresponding to gas-liquid phase transformation. It follows from the van der Waals equation at critical point, putting $\left(\dfrac{\partial P}{\partial \bar{V}}\right)_T = 0 = \left(\dfrac{\partial^2 P}{\partial \bar{V}^2}\right)_T$, that

$$\text{critical volume, } \bar{V}_c = 3b \tag{3.4}$$

$$\text{critical pressure, } P_c = \frac{a}{27b^2} \tag{3.5}$$

$$\text{critical temperature, } T_c = \frac{8a}{27Rb} \tag{3.6}$$

$$\text{critical compressibility factor, } \frac{P_c \bar{V}_c}{RT_c} = \frac{3}{8} \tag{3.7}$$

The van der Waals constant in terms of critical constants will then be

$$b = \frac{1}{3}\bar{V}_c = \frac{1}{8} \cdot \frac{RT_c}{P_c} \tag{3.8}$$

$$a = 3P_c \bar{V}_c^2 = \frac{27}{64} \cdot \frac{R^2 T_c^2}{P_c} \tag{3.9}$$

The van der Waals equation is quantitatively reliable only under the conditions where deviation from ideality is not large. Near critical point it fails utterly. Thus according to the van der Waals equation RT_c/P_cV_c is $\frac{8}{3}$ while its experimental value is found to vary from gas to gas in the range 2–3. The equation becomes invalid even qualitatively in the region of condensation where it predicts an absurd state for which volume increases with pressure at constant temperature.

Comparison between Real Gases

Of two real gases,

  i. one having higher value of b will have larger molecular size
  ii. one having higher a (more precisely a/b or T_c) is generally easier to liquefy
  iii. one having lower Z is easier to compress
  iv. one having Z nearer to 1 will behave more nearly ideally
  v. one having Z nearer to $\frac{3}{8}$ under critical condition will behave more nearly as van der Waals gas
  vi. one having higher intermolecular attraction will have lower Z and hence lower slope in the Z–P curve.

Now, for a van der Waals gas

$$Z \simeq 1 + \frac{P}{RT}\left(b - \frac{a}{RT}\right) \tag{3.10}$$

ignoring $ab/\bar{V}^2$ if P is not large when $P\bar{V} \simeq RT$

In low pressure region, b can be ignored when

$$Z \simeq 1 - \frac{aP}{(RT)^2}$$

provided $b \not> a/RT$, i.e. $T \not> a/Rb$ or $T \not> T_B$

In high pressure region, a can be ignored when

$$Z \simeq 1 + \frac{bP}{RT}$$

3.2 THE LAW OF CORRESPONDING STATES

Substitution of a, b and R in terms of critical constants in the van der Waals equation gives the following reduced equation of state

$$\left(P_r + \frac{3}{\bar{V}_r^2}\right)\left(\bar{V}_r - \frac{1}{3}\right) = \frac{8}{3}T_r \tag{3.11}$$

where $P_r = P/P_c$, $\bar{V}_r = \bar{V}/\bar{V}_c$ and $T_r = T/T_c$ are termed as reduced pressure, reduced volume and reduced temperature respectively.

Being guided by the universal nature of this equation, van der Waals proposed that all gases would have the same reduced equation of state. This is law of corresponding states which demands that

$$P_c\bar{V}_c/T_c = \text{constant} \quad \text{(same for all gases)} \tag{3.12}$$

3.3 OTHER EQUATIONS OF STATE

The Dieterici Equation

There are several other theoretical equations of state besides the van der Waals equation. The one proposed by Dieterici is

$$P = \frac{RT}{\bar{V} - b} e^{-a/RT\bar{V}} \tag{3.13}$$

This differs from the van der Waals equation only in the pressure correction which was arrived at on the basis of non-uniform distribution of gas molecules (due to intermolecular forces) using Boltzmann distribution of potential energy, instead of uniform distribution of molecules assumed by van der Waals.

General Equation of State

Since the equation of state of a real gas is too complicated to be known exactly, it will be safe for practical purposes to develop it empirically from the experimental data on P and $\bar{V}$ in the concerned ranges using the most general equation of state, called *virial* (Kammerling Onnes) *equation of state*

$$P\bar{V} = RT(1 + B/\bar{V} + C/\bar{V}^2 + ...) \tag{3.14}$$

Here B, C, ..., are the second, third, ..., virial coefficients which are characteristic constants for a gas depending only on temperature.

Any equation of state can be put in the virial form which is particularly useful for extrapolation purposes.

3.4 INTERMOLECULAR FORCES

In deriving equation of state, van der Waals emphasised on the existence of intermolecular forces in a real gas, which is evidenced by its liquefaction. Intermolecular forces exist also in case of an ideal gas but only as repulsion at the moment of collision between the molecules when they are at very short distance, called *collision diameter*.

In a real gas the forces of repulsion of course act at short distances between the molecules, but also long-range forces of attraction act between them. The latter forces are coulombic in origin and are called the van der Waals forces (because van der Waals first recognised them in formulating the equation of state for real gases) which are associated with molecular interactions of dipole-dipole, dipole-induced dipole and induced dipole-induced dipole type. But the short-range intermolecular repulsion can be explained only on the basis of quantum mechanics as mainly a consequence of the Pauli exclusion principle.

On theoretical and experimental grounds, it appears that both attractive and repulsive forces (F) are inversely proportional to certain power of the distance (r) between a pair of molecules

$$F \propto 1/r^p$$

For attraction F is –ve and $p \approx 7$, while for repulsion F is +ve and p takes on values from 9 to 19. Then with an increasing value of r the repulsive forces diminish much more rapidly than the attractive forces so that the former dominate, and hence perceived, only at very short distances.

Intermolecular interaction is often more conveniently put in the form of intermolecular potential energy (V) as

$$V = -\int_{\infty}^{r} F dx = \underset{\text{repulsion}}{\frac{B}{r^n}} - \underset{\text{attraction}}{\frac{A}{r^6}}$$

[where A and B are constants for a particular substance] which is called the Lennard-Jones ($n, 6$) potential; n is often conveniently chosen to be 12. The corresponding Lennard-Jones (12, 6) potential is usually put in the following more meaningful form

$$V = 4 \in \left[\left(\frac{\sigma}{r} \right)^{12} - \left(\frac{\sigma}{r} \right)^{6} \right] \tag{3.15}$$

where $\in$ is the depth of the potential well relative to the infinite separation and σ is the collision diameter, $\sigma \leq d$, the molecular diameter. This is represented diagramatically in Fig. 3.2.

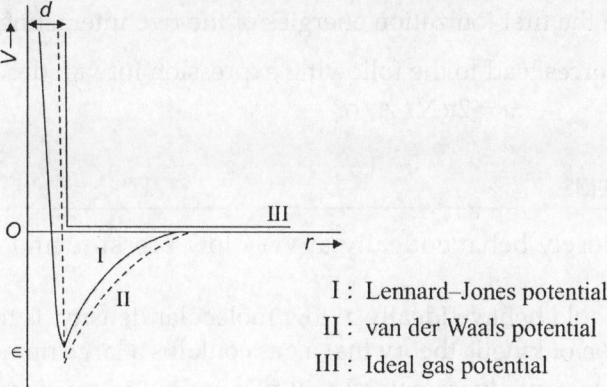

I : Lennard–Jones potential
II : van der Waals potential
III : Ideal gas potential

Fig. 3.2

Although the Lennard-Jones potential is fairly realistic, it is difficult to use. Ideal gas equation is based on the following crudely simplified potential

$$V = \begin{cases} \infty & \text{for } r < d \\ 0 & \text{for } r \geq d \end{cases} \tag{3.16}$$

The van der Waals equation is based on a hybrid of the ideal gas potential and the Lennard-Jones potential

$$V = \begin{cases} \infty & \text{for } r < d \\ 0 & \text{for } r = d \\ -\dfrac{A}{r^6} & \text{for } r > d \end{cases} \tag{3.17}$$

assuming the gas molecules to behave as hard spheres with forces of attraction always between them but repulsive forces appearing suddenly when $r < d$.

The average interaction energy due to dipole-dipole force, called the *Keesom force* is given by

$$V_{d-d} = -\frac{2\mu_1^2 \mu_2^2}{3kTr^6} \quad \text{(in CGS system)} \tag{3.18}$$

for SI, we have to divide this by $(4\pi\in_0)^2$

where μ_1 and μ_2 are dipole moments of a pair of polar molecules (e.g. H_2O, HCl, NH_3, etc.) rotating at a fixed separation r.

The average interaction energy due to dipole-induced dipole force, called *Debye force* is given by

$$V_{d-i} = -\frac{\alpha_2\mu_1^2 + \alpha_1\mu_2^2}{r^6} \tag{3.19}$$

where α_1 and α_2 are polarizability (i.e. distortability of electron cloud of a molecule) of two interacting molecules.

The interaction energy due to oscillating dipoles in the molecules (particularly non-polar molecules, e.g. He, H_2, CCl_4, etc.) is due to the force called *London force* or *dispersion force*

$$V_{disp} = -\frac{3I_1I_2}{2(I_1 + I_2)} \cdot \frac{\alpha_1\alpha_2}{r^6} \tag{3.20}$$

where I_1 and I_2 are the first ionization energies of the two interacting molecules.

van der Waals forces lead to the following expression for van der Waals constant a
$$a = 2\pi N_A A / \sigma^3$$

AUXILIARY PROBLEMS

3.1 Real gases closely behave ideally at very low pressure and high temperature. Explain.

3.2 Real gases should behave ideally at low molecular density. Is this contradictory to the assumption of kinetic theory that a gas contains a large number of molecules?

3.3 How many constants (parameters) will be required to construct a useful equation of state?

3.4 An equation of state should necessarily contain molar gas constant R. Comment.

3.5 Of the two equations of state one containing greater number of constants will be more accurate. Do you agree?

3.6 The equation of state of a real gas coincides with that of an ideal gas as $P \to 0$. Does this imply that all its properties will be same as those of an ideal gas in that limit? Discuss this with reference to molar volume.

3.7 Suppose that P and $\overline{V}$ are the pressure and molar volume of a real gas, and P_i and $\overline{V}_i$ are those for an ideal gas at the same temperature. Do you consider that $P_i \cdot \overline{V} = P\overline{V}_i = $ constant?

3.8 van der Waals equation is often used to represent approximately the behaviour of a real gas. This equation, being cubic in V, should always have three roots. But a gas is found to have only one value of V at a particular temperature and pressure. Resolve this paradox.

3.9 What is the maximum vapour pressure possible with a van der Waals gas having $\overline{V}_c = 0.065$ L mol^{-1} and $T_c = 33$ K?

3.10 The relation $P\overline{V} = RT$ holds for an ideal gas at any pressure and temperature, and also for a real gas when $P \to 0$. Does this relation hold for a real gas if P is not low? Under what condition $P\overline{V} > RT$ will hold for a real gas for any value of pressure?

3.11 The relation $P\bar{V} \propto T$ always holds for ideal gases. For what basis of measurement of P, $\bar{V}$ and T is similar relation possible for real gases? Will the proportionality constant be same in the two cases? Find its value for a van der Waals gas.

3.12 For different real gases $\bar{V}$ is remarkably different for fixed P and T. But V_r is more or less fixed for fixed P_r and T_r. How is this possible? For what molecular structure does this hold best?

3.13 Is collision diameter same as molecular diameter? Is it at all possible. Discuss.

3.14 Collision diameter should not vary with temperature. Do you agree? Give reason.

3.15 What makes van der Waals equation incorrect? Should a and b in the van der Waals equation vary with temperature?

3.16 (i) Assuming that the intermolecular potential energy for a pair of molecules (r apart) is A/r^6, derive a relation between a and σ, the collision diameter. (ii) How is b related to σ. (iii) On the basis of the derived relations explain the effect of T on a and b.

3.17 Do a and b follow additivity?

3.18 Is molecular density uniform throughout a gas as was considered by van der Waals? Is it more likely for molecular density to be higher or lower near the container wall than in the bulk?

3.19 Find the dimensions of the van der Waals constant a and b. For a gas $a = 6.55 \text{ L}^2$ atm mol^{-2} and $b = 0.032$ L mol^{-1}. Find their values in SI.

3.20 For approximate calculation using van der Waals equation, often a or b is put equal to zero. Under what condition will this be appropriate?

3.21 Comment on the possibility of the existence of a Boyle temperature for a van der Waals gas if (i) $a = 0$ (ii) $b = 0$ and (iii) $a = b = 0$.

3.22 Name a single parameter by which the non-ideality of real gases may be measured. What is its significance? How is it related to intermolecular attraction and repulsion?

3.23 At STP the molar volume of a gas is 20% less than the ideal volume. Find (i) compressibility factor and (ii) molar volume of the gas at STP.

3.24 The equation of state of a gas is $P(V - nb) = nRT$. Find compressibility factor if $\bar{V} = 9b$.

3.25 Assuming the intermolecular attraction to be negligible in case of He, and taking its molar volume at 0°C and 100 atm to be 1.107×10^{-2} times the volume at STP, calculate the van der Waals constant b of He.

3.26 For a gas $RT/\bar{V}$ is found to be 1.05 atm at 1 atm. If pressure is now halved at constant temperature will the new volume be doubled, less than or more than that?

3.27 Express van der Waals equation in virial form in terms of $1/\bar{V}$ and find the second and third virial coefficients in terms of van der Waals constant, hence find the Boyle temperature in terms of the latter.

3.28 Show that the compressibility factor of a van der Waals gas can be expressed in the following virial form

$$Z = 1 + \frac{1}{RT}\left(b - \frac{a}{RT}\right)P + \frac{a}{(RT)^3}\left(2b - \frac{a}{RT}\right)P^2 + \dots$$

At what temperature the slope of Z–P curve (at $P \to 0$) is numerically (i) lowest (ii) highest. What is the significance of temperature in case of (i)?

3.29 The compressibility factor of a real gas below its Boyle temperature is given by the relation $Z = \alpha + \beta P + \gamma P^2$. Show that (i) $\alpha = 1$ (ii) $\beta < 0$ and $\gamma > 0$ (iii) $4\gamma > \beta^2$

3.30 Do the constants, a and b, involved in the Dieterici equation have the same significance as in the van der Waals equation. State with reasons whether Z for a gas obeying Dieterici equation will be higher or lower than that expected for a van der Waals gas at same P and T.

3.31 Quantitatively Dieterici equation holds better than van der Waals equation for any real gas. But it is reliable in limited range of temperature and pressure. Why? Does it hold for an ideal gas?

3.32 Draw Z vs P plot on the same diagram for (i) ideal gas (ii) hydrogen and (iii) ammonia, at room temperature. Explain the difference between them.

3.33 From the law of corresponding states deduce the following relation for a real gas

$$P_c \bar{V}_c / T_c = \text{constant, which is same for all gases}$$

Does the law of corresponding states have any relevance to ideal gases?

3.34 Which of the following equations of state can be transformed into reduced form (i) van der Waals equation (ii) Dieterici equation (iii) Virial equation of state? Can same gas follow all three equations of state. Which of these conforms nearer to the law of corresponding states?

3.35 In the reduced equation of state, are the parameters characteristic of the gas completely removed or just hidden?

3.36 A real gas behaves ideally in an appreciable range of pressure at a characteristic temperature of the gas, called Boyle temperature. Is this ideality possible at some other temperature?

3.37 Boyle temperature of a gas is always greater than its critical temperature. Justify.

3.38 For an ideal gas, critical temperature is $0 \, \text{K}$. Comment.

3.39 Consider the following equations of state of real gases.

(I) $P\bar{V} = RT$

(II) $\left(P + \dfrac{a}{\bar{V}^2}\right)\bar{V} = RT$

(III) $P(\bar{V} - b) = RT$

(IV) $\left(P + \dfrac{a}{\bar{V}^2}\right)(\bar{V} - b) = RT.$

In which case(s) the gas can have (i) Boyle temperature (ii) critical temperature. In case the characteristic temperature exists, what will be its value? Also discuss the possibility of liquefaction for the given equations of state.

3.40 Discuss the possibility of the liquefaction of a gas, if its equation of state

  i. $Z = 1 + A/\bar{V} + B/\bar{V}^2$

  ii. $Z = 1 + A/\bar{V} + B/\bar{V}^2 + C/\bar{V}^3$

  iii. $Z = 1 + A/\bar{V} + B/\bar{V}^2 + C/\bar{V}^3 + D/\bar{V}^4$

  where A, B, C, D are characteristic parameters of the gas.

3.41 Under which of the following conditions a gas can be liquefied?

  (i) $T = T_c, P = P_c$ (ii) $T = T_c, P < P_c$ (iii) $T > T_c, P > P_c$ (iv) $T < T_c, P = P_c$ (v) $T < T_c, P < P_c$

3.42 Of the two gases, the one having higher T_c is easier to liquefy. Comment.

3.43 Which is easier to liquefy—H_2S ($T_c = 373.2 \, \text{K}$, $P_c = 88.2 \, \text{atm}$) or C_3H_8 ($T_c = 369.8 \, \text{K}$, $P_c = 41.9 \, \text{atm}$)?

3.44 a. From the P–V isothermals of a real gas arrive at the principle of continuity of states.

b. Is it possible to go from the gas side to the liquid side of an isothermal continuously? Justify your answer. Depict a path of such change.

3.45 a. A liquid may be regarded as a compressed gas, and a gas as a rarified liquid. Justify or criticise.

b. Does a solid–gas system have any temperature analogous to the critical temperature?

3.46 $P\bar{V}$ of a real gas above its Boyle temperature is same as that of an ideal gas. Which gas has higher temperature? Would your answer be different if the gas is kept below its Boyle temperature?

3.47 The van der Waals constants for two gases X and Y are as follows:

Gas	$a/L^2\ atm\ mol^{-2}$	$b/L\ mol^{-1}$
X	12	0.030
Y	6	0.028

Find the molecular volume of gas X and also its molar volume at STP. Which gas is (i) of higher T_c (ii) more nearly ideal in behaviour (iii) more compressible around STP? Will your answer in (ii) and (iii) be same for different regions of P and T?

3.48 A substance exhibits a maximum vapour pressure of 30 atm and this happens at 227°C. (a) Find its molecular radius. (b) At what maximum temperature is its transformation from gas phase to liquid phase possible? Also find its Boyle temperature. (c) In gaseous state will it be more compressible or less compressible around 1 atm and 400 atm, at (i) 1500°C (ii) 1000°C and (iii) 0°C?

3.49 One mole of gas A at 66 K and 3.2 atm and gas B at 608 K and 18.2 atm occupy a volume 1718 and 2782 cm³ respectively. Their critical constant values are shown in the following table. Find out the missing value in the table and state (i) which gas is more easily liquefiable and (ii) which gas is more close to the ideal behaviour around STP. Give the basis of your answer.

	T_c	P_c	$\bar{V}_c$
Gas A	33 K	12.8 atm	0.065 L mol^{-1}
Gas B	304 K	72.8 atm	–

Will gas A or B behave more closely as a van der Waals gas?

3.50 The second virial coefficient of gas is –0.145 L mol^{-1} at 0°C. Find $\bar{V}$ at STP. What is the nature of intermolecular force prevailing here?

3.51 In deriving equation of state van der Waals took into account only the intermolecular attraction, intermolecular repulsion was totally ignored. Comment.

3.52 Express the Lennard–Jones potential $V = \dfrac{B}{r^n} - \dfrac{A}{r^6}$ in terms of minimum potential ϵ and collision diameter σ (i.e. the finite value of r for which $V = 0$). Also relate A and B with ϵ, σ and n.

3.53 Using the common form of Lennard–Jones (12, 6) potential, find a relation between molecular diameter (d) and collision diameter (σ). Hence show that Lennard–Jones potential may also be expressed as

$$V = \epsilon\left[\left(\frac{d}{r}\right)^{12} - 2\left(\frac{d}{r}\right)^{6}\right]$$

3.54 Draw Lennard–Jones potential, van der Waals potential and ideal gas potential using same axes, and explain the notable difference between them.

3.55 Although hard-sphere potential is unrealistic, it works well at high temperatures. Why?

3.56 Explain the following order of boiling point
   a. $He < Ne < Ar < Kr$
   b. $H_2 < N_2 < O_2 < CH_4 < C_2H_6$
   c. $(CH_3)_3CH < (CH_3)_2C = CH_2 < (CH_3)_3N$
   d. $(CH_3)_2CH_2 < (CH_3)_2O < C_2H_4O$
   e. p-dichlorobenzene $\approx m$-dichlorobenzene $< o$-dichlorobenzene
   f. $CH_4 < NH_3 < HF < H_2O$

ANSWERS

3.1 An ideal gas is one where each molecule is not influenced by the others (except at the moment of collision), i.e. possesses no intermolecular potential energy. A real gas amply fulfils this requirement, i.e. closely behaves ideally, at very low pressure when the intermolecular distance is very high and also at very high temperature when the molecular kinetic energy is so high that the potential energy may be regarded as virtually absent. The situation is equivalent even at moderately low pressure, if the temperature is sufficiently high so that the molecular density is low.

3.2 No. Because, even with large number of molecules (required for statistical treatment), the molecular density may be low (required for low interaction) by making the volume of the gas large.

3.3 A useful equation of state, which always contains a limited number of terms, is customarily set up using the general virial equation of state [Eq. (3.14)]. The number of virial coefficient (parameters) to be considered depends on the concerned range of pressure and temperature and desired accuracy of calculation involved. For example, at very high temperature or at very low pressure when a real gas exhibits ideal behaviour, the equation of state $(P\bar{V} = RT)$ contains no characteristic parameter. On the other hand, even a two parameter equation, like van der Waals equation, is not useful at high pressure. Thus the given question has no unique answer.

Note: It is impossible to construct an equation of state containing limited number of terms that holds exactly at all ranges of temperature and pressure.

3.4 Yes. Because all equations of state of real gases virtually reduce to the ideal gas equation $(P\bar{V} = RT)$ at very high temperature or low pressure.

3.5 An equation of state designed to hold best for a specified range of temperature and pressure cannot be made significantly more accurate by including more characteristic parameter. However, the accuracy of the same equation for a wider range of temperature and pressure demands inclusion of more parameter(s) and then one containing greater number of parameters is likely to be more accurate.

3.6 No. This can be illustrated with reference to molar volume $\bar{V}$ which is given according to the virial equation of state, Eq. (3.14) by

$$\bar{V} = \frac{RT}{P} + \frac{B}{P\bar{V}/RT} \quad \text{as } P \to O$$

$$= \frac{RT}{P} + B \quad \text{Since } P\bar{V} = RT \text{ as } P \rightarrow O$$

Thus even at $P \rightarrow O$, $\bar{V}$ of a real gas is not equal to the ideal gas molar volume.

3.7 No. Because if the given relation were true, $P_i\bar{V} \cdot P\bar{V}_i$, i.e. $P_i\bar{V}_i \cdot P\bar{V}$ would have been constant. But this is not true, since at constant temperature $P_i\bar{V}_i$ is constant but $P\bar{V}$ is not.

3.8 Above T_C, two of the roots are imaginary. Below T_C, although three roots are all real only the highest one refers to the gaseous state (the lowest one to the liquid state and the intermediate one has no physical significance as it refers to a hypothetical state where P rises with V at constant T). At T_C, all three roots are real, but same.

Note: No solution of van der Waals equation is possible where two roots are real and one root is imaginary. Because the imaginary roots always occur in pair.

3.9 The maximum vapour pressure is the critical pressure P_C which, for a van der Waals gas, is given by

$$P_C = \frac{3}{8} \cdot \frac{RT_C}{\bar{V}_C} \quad \text{by Eq. (3.7)}$$

$$= \frac{3}{8} \cdot \frac{(0.08206 \text{ L atm K}^{-1}\text{mol}^{-1})(33\,K)}{(0.065 \text{ L mol}^{-1})} = 15.62 \text{ atm}$$

Note: For $P > P_C$ only one phase can exist, which is vapour or liquid according as $T > T_C$ or $T < T_C$.

3.10 Yes. For a particular temperature, the relation $P\bar{V} = RT$ holds for a real gas only for a particular pressure* (other than $P = 0$) corresponding to the ideal molar volume at the temperature and pressure concerned.

□ Above the Boyle temperature which is a/Rb for a van der Waals gas.

Note: At the Boyle temperature, $P\bar{V} = RT$ holds not just for a particular P, but for a range of P near zero.

3.11 Under critical condition similar relation, i.e. $P_C\bar{V}_C \propto T_C$, holds for real gases

$$\square \qquad \frac{P_C\bar{V}_C / T_C}{P\bar{V}/T} = \frac{1}{\lim\limits_{\substack{P \rightarrow 0 \\ T \rightarrow \infty}} f(P_r, T_r)} = \text{constant } (\neq 1),$$

for all real gases independent of their nature, by the law of corresponding states (vide problem 3.33)

Then the relevant proportionality constant will not be same for the ideal and real gases.

□□ For a van der Waals gas, $\dfrac{P_C\bar{V}_C}{T_c} = \dfrac{3}{8}R$. See Sec. 3.1 for critical constants.

3.12 For different real gases, the function $\bar{V} = f(P, T)$ is different. This is mainly because the measurement of P, $\bar{V}$ and T are based on some arbitrarily imposed fixed standards. But the arbitrariness disappears when P, $\bar{V}$ and T of a gas are measured in terms of their respective critical values for the same gas. Threfore, it is not unlikely that the function $V_r = f(P_r, T_r)$ is more or less same for all real gases.

□ The function $V_r = f(P_r, T_r)$ is expected to be same for different gases, if the interaction between two molecules is determined only by the distance between

---

* This pressure corresponds to the meeting point of the isotherms of the relevant real gas and that of an ideal gas at the temperature concerned.

them (as assumed by van der Waals in deriving the equation of state). This holds best for spherical molecules like He, Ne, etc.

Note: Interaction between two non-spherical molecules is determined not only by the distance between them but also by their relative orientation, particularly if they are polar.

3.13 No. It is not at all possible in case of real molecules because they behave not as hard particles, but as soft particles, being consisted of atomic nuclei surrounded by the electron cloud.

Note: The molecules of the inert gases (which are monatomic) act most nearly as rigid particles due to their tightly bound electronic structure.

3.14 No. Because, due to somewhat soft nature of the molecules, their collision diameter should decrease due to increase in vigour of the collision when the temperature is raised.

3.15 This is attributed primarily to the assumption of a faulty model in which the gas molecules are supposed to behave as hard spheres with forces of attraction always between them, but repulsive force appearing suddenly. In reality, however, molecules are somewhat soft with intermolecular repulsion, like attraction, growing gradually and not suddenly as the molecules approach each other. As a consequence, both 'a' and 'b' in the van der Waals equation, which depend on collision diameter, will vary with temperature making the equation incorrect.

Note: Instead of regarding 'a' and 'b' as characteristic constant of a gas (as was done by van der Waals), they should much better be regarded as adjustable parameters of the system.

3.16   i. Let us consider a large volume V of a gas containing N molcules. Each molecule may be supposed to be surrounded by spherical layers of the gas. The number of molecules dN in a layer of radius r and thickness dr is $dN = \frac{N}{V} \cdot 4\pi r^2 dr$. Considering all such layers, the average interaction energy per pair of molecules is

$$\langle E \rangle = \int E \cdot \frac{dN}{N}$$

$$= \int_{\sigma}^{\infty} -\frac{A}{r^6} \cdot 4\pi r^2 \frac{dT}{V}, \text{ where } \sigma \text{ is collision diameter}$$

$$= -\frac{4\pi A}{3\sigma^3 V}$$

Total interaction energy,

$$E = \text{Total number of pairs} \times \langle E \rangle$$
$$= \frac{1}{2} N(N-1) \langle E \rangle$$
$$= \frac{-2\pi N^2 A}{3\sigma^3 V} \quad \text{for } N \gg 1$$

[The factor $\frac{1}{2}$ appears to avoid counting each pair of molecules twice]

The energy per mole,

$$\bar{E} = N_A \cdot E/N$$
$$= -N_A \cdot 2\pi N \cdot A/3\sigma^3 V$$
$$= -2\pi N_A^2 A/3\sigma^3 \bar{V}, \quad \text{where } \bar{V} = N_A \cdot V/N$$

Then $\qquad \left(\dfrac{\partial \bar{E}}{\partial V}\right)_T = 2\pi N_A^2 \cdot A / 3\alpha^3 \bar{V}^2$

$\qquad\qquad\qquad = a / \bar{V}^2$, for a van der Waals gas, by Eq. (5.17)

Therefore, $\qquad a = 2\pi N_A^2 \cdot A / 3\sigma^3$

ii. Since the gas molecules have volume of their own, each molecule will exclude a volume $v = \frac{4}{3}\pi\sigma^3$ for free movement of the other molecules. If molecules are added one by one to a container of volume V, the volumes available to the 1st, 2nd, ... Nth molecule will be V, $V - v$, ..., $V(N-1)v$ respectively. The volume available to each molecule, on the average, will be

$$V - b = \tfrac{1}{N}[V + (V - v) + \ldots + V - (N-1)v]$$
$$= V - \tfrac{1}{2}(N-1)v$$
$$= V - \tfrac{1}{2}Nv \quad \text{for } N \gg 1$$

Then $\qquad b = \tfrac{1}{2}Nv$

$$= \tfrac{1}{2}N_A \cdot \tfrac{4}{3}\pi\sigma^3 \quad \text{for 1 mole}$$
$$= \tfrac{2\pi}{3}N_A\sigma^3$$

iii. With rise in temperature, b is always found to decrease. This follows from the above expression of b (which depends only on σ) due to decrease of σ.

In case of non-polar molecules, the rise of temperature affects a and b in reverse directions. Here, according to the above explosion of a, the effect of T on a is determined only by σ. (A being independent of T for non-polar molecules).

In case of polar molecules, generally a decreases with rise of temperature. Here, rising temperature causes both A and σ to decrease, but A is generally more affected.

Note: If molecules were like hard spheres, the volume v_0 of each molecule would be

$$v_0 = \tfrac{4}{3}\pi\left(\tfrac{\sigma}{2}\right)^3 = \tfrac{1}{8}v$$

and $\qquad b = \tfrac{1}{2}N \cdot 8v_0$

$\qquad\qquad = 4Nv_0$, i.e. four times the actual volume of the molecules

3.17 No. a and b have meaning only with pure gases. For a gaseous mixture, we cannot write $b = \sum n_i b_i / \sum n_i$, because of the involvement of more than one type of molecular collision and hence collision diameter. The same is true for a.

3.18 The molecular density near the wall is more likely to be somewhat lower than that in the bulk. This follows from the Boltzmann energy distribution law. Because the molecules near the wall possess excess potential energy over the interior molecules due to the existence of unbalanced intermolecular attraction.

Note: The non-uniformity of molecular distribution forms the basis of Dieterici equation.

3.19 It follows from the van der Waals equation that

a has the dimensions of $P(V/N)^2$, i.e. $(ml^{-1}t^{-2})(l^3)^2/(n)^2$ or $ml^5 l^{-2} n^{-2}$

b has the dimensions of V/n, i.e. $l^3 n^{-1}$

$$a = 6.55 \text{ L}^2 \text{ atm mol}^{-2}$$
$$= 6.55\,(10^{-3}\text{m}^3)^2\,(1.013 \times 10^5 \text{ Pa})\,\text{mol}^{-2}$$
$$= 6.635 \times 10^{-1}\text{m}^6\text{ Pa mol}^{-2}$$

$$b = 0.032 \text{ L mol}^{-1}$$
$$= 0.032 \, (10^{-3} \text{m}^3) \text{ mol}^{-1}$$
$$= 3.2 \times 10^{-5} \text{ m}^3 \text{ mol}^{-1}$$

3.20 Putting $a = 0$ has the justification at high pressure, because then $P \gg \frac{a}{\bar{V}^2}$.

Putting $b = 0$ has the justification at low pressure, because then $\bar{V} \gg b$.

Note: With increase of pressure, $a/\bar{V}^2$ also increases, but in the high pressure region $a/\bar{V}^2$ increases at lower rate (because $\bar{V}$ decreases at lower rate due to predominating effect of intermolecular repulsion over attraction).

3.21 For a real gas the Boyle temperature is a characteristic temperature at which

$$\lim \left[\frac{\partial(P\bar{V})}{\partial P} \right]_T = 0 \quad \text{only as } P \to 0$$

or $\qquad \lim \left[\frac{\partial(P\bar{V})}{\partial(1/\bar{V})} \right]_T = 0 \text{ only as } 1/\bar{V} \to 0$

Such requirement is not fulfilled by a van der Waals gas, i.e. the Boyle temperature is not possible for any of the given conditions

For $a = 0$, i.e. when $P(\bar{V} - b) = RT$

$$\left[\frac{\partial(P\bar{V})}{\partial P} \right]_T = b \neq 0 \text{ for any value of } P$$

For $b = 0$, i.e. when $\left(P + \frac{1}{\bar{V}^2} \right) \bar{V} = RT$

$$\left[\frac{\partial(P\bar{V})}{\partial(1/\bar{V})} \right]_T = -a \neq 0 \text{ for any value of } 1/\bar{V}$$

For $a = b = 0$, i.e. when $P\bar{V} = RT$

$$\left[\frac{\partial(P\bar{V})}{\partial P} \right]_T = 0 \text{ for any value of } P$$

Note: For a and/or b equal to zero, since the Boyle temperature does not exist, the relation $T_b = a/Rb$ will have no significance.

3.22 The appropriate parameter is the compressibility factor $Z = P\bar{V}/RT$.

☐ Z signifies the observed molar volume relative to the ideal molar volume (RT/P) at the prevailing temperature and pressure.

☐☐ At high pressures, all gases have $Z > 1$, signifying that they are more difficult to compress than an ideal gas. This corresponds to intermolecular repulsion dominating the attraction. At intermediate pressure, most of the gases have $Z < 1$, which corresponds to intermolecular attraction dominating the repulsion. At very low pressures, $Z \approx 1$ for all gases which then behave nearly ideally due to balancing of attractive and repulsive forces between the molecules.

Note: The variation Z with P (a macroscopic observation) discloses the simultaneous existence of attraction and repulsion between gas molecules.

3.23 i. $\qquad\qquad Z = \dfrac{\text{Observed molar volume}}{\text{Ideal molar volume}}$

$$= \frac{100 - 2}{100} = 0.98$$

ii. $$\bar{V} = ZRT/P$$

$$= \frac{(0.98)(0.08206 \text{ L atm K}^{-1} \text{ mol})(273.15 \text{ K})}{(1 \text{ atm})} = 21.97 \text{ L mol}^{-1}$$

3.24 $$Z = \frac{P\bar{V}}{RT}$$

$$= \frac{\bar{V}}{\dfrac{\bar{V} - nb}{n}}, \text{ from the given equation of state}$$

$$= \frac{\bar{V}}{\bar{V} - b}$$

$$= \frac{9b}{9b - b}, \text{ from the given data} = \frac{9}{8}$$

3.25 Neglecting intermolecular attraction, i.e. putting $a = 0$, the van der Waals equation becomes

$$P(\bar{V} - b) = RT$$

Then, at STP $1(\bar{V} - b) = 273.15R$

or $$\bar{V} = b + 273.15R$$

Again, at 0°C at 100 atm

$$100(1.107 \times 10^{-2}\bar{V} - b) = 273.15R$$

Then $1.107(b + 273.15R) - 100b = 273.15R$

or $$b = \frac{0.107 \times 273.15R}{98.893}$$

$$= \frac{0.107 \times 273.15 \times 0.08206}{98.893} \text{ L mol}^{-1} = 0.02425 \text{ L mol}^{-1}$$

3.26 $$Z = \frac{PV}{RT} = \frac{P}{RT/\bar{V}} = \frac{1}{1.05} < 1$$

Since $Z < 1$ at 1 atm which is rather a low pressure, the given data correspond to that part of Z vs P curve where Z rises with fall in pressure. Then on halving the pressure at constant temperature, the volume will be more than double.

Note: Even in the low pressure region the result will not be same if the experimental temperature lies above Boyle temperature when $Z > 1$.

3.27 The van der Waals equation is

$$P = \frac{RT}{\bar{V} - b} - \frac{a}{\bar{V}^2}$$

Then $$Z = \frac{P\bar{V}}{RT} = \frac{1}{1 - b/\bar{V}} - \frac{a}{RT\bar{V}}$$

$$= 1 + \left(b - \frac{a}{RT}\right)\frac{1}{\bar{V}} + b^2\left(\frac{1}{\bar{V}}\right)^2 + \dots$$

which is the van der Waals equation in the virial form.

Here, 2nd virial coefficient $= b - a/RT$

3rd virial coefficient $= b^2$

At the Boyle temperature

$$\lim \left(\frac{\partial Z}{\partial P}\right)_T = 0 \text{ as } P \to 0$$

$$\lim \left(\frac{\partial Z}{\partial(1/\overline{V})}\right)_T = 0 \text{ as } \frac{1}{\overline{V}} \to 0$$

Imposing this condition on the above virial equation, we have at the Boyle temperature, T_B

$$b - \frac{a}{RT_B} = 0$$

or $T_B = a/Rb$

3.28 For a van der Waals gas

$$Z = 1 + \left(b - \frac{a}{RT}\right)\frac{1}{\overline{V}} + b^2\left(\frac{1}{\overline{V}}\right)^2 + \dots \qquad \text{(vide problem 3.27)}$$

$$= 1 + \frac{B}{\overline{V}} + \frac{C}{\overline{V}^2} \dots \qquad \text{(I)}$$

or $$P = \frac{RT}{\overline{V}}Z = RT\left(\frac{1}{\overline{V}} + \frac{B}{\overline{V}^2} + \frac{C}{\overline{V}^3} + \dots\right)$$

Let $$Z = 1 + B'P + C'P^2 + \dots \qquad \text{(II)}$$

On substituting the expression for P, this becomes

$$Z = 1 + B'RT\left(\frac{1}{\overline{V}} + \frac{B}{\overline{V}^2} + \dots\right) + C'(RT)^2\left(\frac{1}{\overline{V}} + \frac{B}{\overline{V}^2} + \dots\right)^2 + \dots$$

Comparing this with (I)

$$B' = \frac{B}{RT} = \frac{1}{RT}\left(b - \frac{a}{RT}\right)$$

$$C' = \frac{C - B^2}{(RT)^2} = \frac{b^2 - \left(b - \frac{a}{RT}\right)^2}{(RT)^2} = \frac{a}{(RT)^3}\left(2b - \frac{a}{RT}\right)$$

Substituting B' and C' in II, we get

$$Z = 1 + \frac{1}{RT}\left(b - \frac{a}{RT}\right)P + \frac{a}{(RT)^3}\left(2b - \frac{a}{RT}\right)P^2 + \dots$$

Slope of Z vs P curve $= \dfrac{1}{RT}\left(b - \dfrac{a}{RT}\right)$ at $P \to 0$

☐ i. The lowest numerical value of this slope = 0 at $T = a/Rb$ and $T = \infty$.

ii. The temperature at which the slope is highest (here it is maximum) can be obtained by differentiating the slope with respect to T and equating the result to zero when we get $T = 2a/Rb$ ($T = \infty$ refers minimum slope).

☐☐ For $T = a/Rb$, $\left(\dfrac{\partial Z}{\partial P}\right)_T = 0$ and $Z = 1$ at $P \to 0$.

Then $T = a/Rb$ corresponds to the Boyle temperature where the ideal gas equation is obeyed at an appreciable range of pressure at $P \to 0$ only.

For $T = \infty$, $\left(\dfrac{\partial Z}{\partial P}\right)_T = 0$ and $Z = 1$, happen at all pressure, and the ideal gas equation is obeyed accordingly.

Note:

1. The slope of Z vs P curve (at $P \to 0$) varies with temperature exhibiting a minimum at $T = \infty$ but not at $T = a/Rb$ (though it is same at both the temperatures).

2. The slope is zero not only at the Boyle temperature but also at extremely high temperature.

3.29   i. Since $Z \to 1$ as $P \to 0$, we have from the given expression, $\alpha = 1$.

    ii. Since below Boyle temperature Z vs P curve passes through a minimum

$$\left(\frac{\partial Z}{\partial P}\right)_T = b + 2\gamma P = 0$$

and

$$\left(\frac{\partial^2 Z}{\partial^2 P}\right)_T = 2\gamma = +ve, \ \ e.g. \ \gamma > 0$$

Since P cannot be $-ve$, β and γ must have opposite sign. Then $\beta = -ve$, i.e. $\beta < 0$.

   iii. At the minimum of Z vs P curve the two roots of $Z = \alpha + \beta P + \gamma P^2$ are same, which requires $\sqrt{\beta^2 - 4\gamma(\alpha - Z)} = 0$.

or

$$\frac{\beta^2}{4\gamma} = \alpha - Z = 1 - Z$$

$$< 1, \ \ \text{since } Z < 1 \text{ corresponding to the minimum}$$

Then $4\gamma > \beta^2$.

3.30 At low value of P, when $\bar{V}$ is high, the Dieterici Eq. (3.13) becomes

$$P = \frac{RT}{\bar{V} - b}\left(1 - \frac{a}{RT\bar{V}}\right)$$

$$= \frac{RT}{\bar{V} - b} - \frac{a}{\bar{V}(\bar{V} - b)}$$

$$= \frac{RT}{\bar{V} - b} - \frac{a}{\bar{V}^2}, \ \ \text{Since } \bar{V}(\bar{V} - b) \simeq \bar{V}^2 \text{ considering } \bar{V} \gg b$$

i.e. the Dieterici equation transforms to the van der Waals equation at low pressure. This implies that a and b involved in this equation have the same significance as in the van der Waals equation.

□ In the Dieterici equation, the molecular density in the interior of the gas is considered to be higher than that near the surface with consequent higher attraction of the interior molecules over the surface molecules. For same P and T, $\bar{V}$ and hence $Z \ (= P\bar{V}/RT)$ calculated from this equation will, therefore, be lower than that given by the van der Waals equation.

3.31 The Dieterici equation is better than the van der Waals equation because the former, unlike the latter, takes care of non-uniformity in the distribution of molecules in a gas.

□ Because it is based on the same faulty hard sphere model as the van der Waals equation.

□□ Being applicable to a real gas (though in a limited range of P and T), the Dieterici equation will obviously hold for an ideal gas which is a mere simplified case of a real gas with $a = b = 0$.

3.32 The desired diagram is sketched in Fig. 3.3.

Here the initial slopes increase in the order NH_3 < ideal gas < H_2. This is due to the intermolecular attraction decreasing in the order NH_3 > ideal gas > H_2. In case of H_2 the intermolecular repulsion dominates over the attraction even at fairly low pressure due to relatively much smaller size of the molecules (and hence smaller dispersion force).

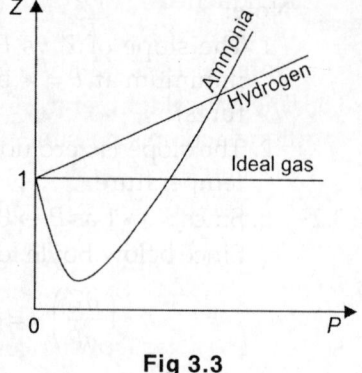

Fig 3.3

3.33 According to the law of corresponding states, $\bar{V}_r$ and hence $P_r\bar{V}_r/T_r$, of a real gas is a universal function of P_r and T_r

i.e. $$P_r\bar{V}_r/T_r = f(P_r, T_r)$$

Then $$P_c\bar{V}_c/T_c = \frac{P\bar{V}}{T} \cdot \frac{1}{f(P_r, T_r)}$$

Now, for a particular gas lhs is constant which is independent of P, $\bar{V}$ and T. Again, under the condition $P \rightarrow 0$, when $P_r \rightarrow 0$ and $T \rightarrow \infty$, when $T_r \rightarrow \infty$, all gases behave ideally, i.e. $P\bar{V}/T = R$ and the value of $f(P_r, T_r)$ is also same. Then it follows that $P_c\bar{V}_c/T_c$ = constant, which is same for all gases.

□ No. because P_r, $\bar{V}_r$ and T_r have no meaning with ideal gases as they cannot exhibit critical phenomenon.

Note: The law of corresponding states mere predicts $P_c\bar{V}_c/T_c$ to be a universal constant. Its value has to be found out by doing experiment on spherical molecules which conform to this law most. The value thus obtained is close to $0.3R$. As expected, the calculation based on a theoretical equation of state gives somewhat different value, depending on the equation of state used, e.g. $0.37R$ from the van der Waals equation and $0.27R$ from the Dieterici equation.

3.34 An equation of state can be transformed into the reduced form if it involves only two characteristic constants of the substance concerned. This requirement is fulfilled by (i) and (ii), but not by (iii) in general.

□ Yes. Because (i) is mere a particular case of (ii) in the low pressure region, and (ii) is a particular case of (iii) where the temperature dependent virial coefficients are restricted to be composed of only the van der Waals constant, a and b (apart from R).

□□ Since (iii) is not in general, relevant to the law corresponding states (ii), which is more general than (i), will conform nearer to this law.

Note: Although the general virial equation of state cannot be transformed into a reduced form, the van der Waals equation in virial form can be transformed.

3.35 The characteristic parameters of the gas remain just hidden. This is exposed by remarkable variation of $F_r V_r/T_r$ of a gas (even with spherical molecules) when its temperature is widely varied affecting its characteristic parameters appreciably.

Note: For a reduced equation to hold the critical constants of the gas, which are determined by its molecular parameter, must remain as such.

3.36 Yes, at $T \rightarrow \infty$ a real gas behaves ideally at any pressure (vide problem 3.28).

3.37 At the Boyle temperature (T_B) of a gas, its liquefaction is not possible because here $Z \nleq 1$, i.e. the intermolecular attraction does not exceed the repulsion. Hence T_B of a gas must always be greater than its T_C (the highest temperature at which a gas can be liquefied). For example, a van der Waals gas has $T_B = a/Rb$ more than three times of its $T_C = 8a/27\ Rb$).

Note: At T_B, $Z = 1$ at $P \to 0$ and $Z > 1$ at higher pressure.

3.38 The critical temperature of a gas (substance) is the highest temperature at which its liquefaction is possible. An ideal gas cannot be liquefied even at 0 K due to absence of intermolecular attraction. Therefore, the given statement is not justified.

3.39 i. The Boyle temperature is not possible in case of (I), (II) and (III) (see problem 3.21). But it is possible in case of (IV), where $\lim\left[\dfrac{\partial(P\bar{V})}{\partial(1/\bar{V})}\right]_T = 0$ only as $\dfrac{1}{\bar{V}} \to 0$ when $T = a/Rb$ (see problem 3.27).

   ii. For a gas (substance) to have critical temperature, its equation of state must be odd order in $\bar{V}$ and the order must not be less than three, otherwise it cannot have only one real root at high temperature (as in case of a van der Waals gas).

   This requirement is fulfilled only in case of (IV) which is of order three in $\bar{V}$. Here $T_c = 8a/27\ Rb$. This follows from the solution of the following three equations

$$P_C = \frac{RT_C}{\bar{V}_C - b} - \frac{1}{\bar{V}_C^2},$$

[which is the van der Waals equation at critical temperature]

$$\left(\frac{\partial P}{\partial \bar{V}}\right)_T = -\frac{RT_C}{(\bar{V}_C - b)^2} + \frac{2a}{\bar{V}_C^3} = 0$$

[Since the maximum and the minimum of P–V curve in the two phase region merge at T_C].

$$\left(\frac{\partial^2 P}{\partial \bar{V}^2}\right)_T = \frac{2RT_C}{(\bar{V}_C - b)^3} - \frac{6a}{\bar{V}_C^4} = 0$$

Only a gas having critical temperature, i.e. IV, can be liquefied. Liquefaction necessitates the simultaneous existence of both attraction and repulsion between the molecules as it happens with IV. For I the intermolecular force is absent, for II only intermolecular attraction (indicated by a) is present, and for III only intermolecular repulsion (indicated by b) is present. In case of II any attempt of liquefaction will lead to solidification (due to absence of intermolecular repulsion) in real gases.

Note: The existence of both the Boyle temperature and critical temperature requires the simultaneous existence of intermolecular attraction and repulsion in real gases.

3.40 The necessary condition for liquefaction of a gas, mentioned in the previous problem, leads to the following conclusions.

For (i), liquefaction is not possible because the equation of state is of even order (2) in $\bar{V}$.

For (ii), liquefaction is possible because the equation is of odd order three in $\bar{V}$.

For (iii), same as for (i), differing only in the value of the order which is four.

3.41 A gas can be liquefied always under the conditions (i) and (iv), and also under the condition (v), if P lies on the liquid side of the isotherm at the temperature concerned or above this value.

3.42 The liquefaction of a gas requires lowering of temperature below its T_C and raising of pressure to appropriate value. Here T_C is the prime factor, because above T_C liquefaction is not possible even if the applied pressure is higher than P_C. Hence, of the two gases the one having higher T_C is usually easier to liquefy, particularly when they differ much in T_C. But when the difference in T_C is low, the gas with lower P_C will be easier to liquefy. Hence the given statement is not always justified.

Note: In T_C ($\propto a/b$), a and b (i.e. intermolecular attraction and repulsion) are proportionately involved, while in P_C ($\propto a/b^2$), b is more involved than a.

3.43 Here the difference of T_C between the two gases is rather low, but the difference of P_C is quite high. Therefore C_3H_8 having lower value of P_C will be easier to liquefy, though it has lower T_C.

3.44 a. The difference between the gas and accompanying liquid phases, represented by the horizontal parts of the P–V isothermals, decreases with rise in temperature till it disappears at a temperature, called the critical temperature. The fact that, it is not always possible to distinguish between the gas and liquid phases of a substance leads to the principle of continuity of states which holds that there is no fundamental difference between these two phases.

b. Yes. This is justified by the principle of continuity of states.

The path of a phase transition of the indicated type is represented by the dashed line ABCDE in Fig. 3.1b. Here the gas phase passes continuously to the indistinguishable liquid phase only at the point C that lies just on the liquid side of the critical isothermal. [The same phase transition occurs discontinuously when brought about following the isothermal on which A and E lie. Because here the coexisting phases are distinguishable].

Note: A continuous phase transition is not possible isothermally (except through the critical isothermal). The path ABCDE is not an isothermal one. It involves three steps—isochoric heating, isobaric cooling and isochoric cooling, occurring successively.

3.45 a. The given statement is justifiable on the basis of the principle of continuity of states, if the gas and liquid phases do not coexist with a surface of separation between them when we are to treat the two phases differently, being part of the same system.

b. No. The critical temperature is one at which the liquid and gas phases of a substance are indistinguishable when they have same density. This arises from the same molecular arrangement in the liquid and gas phases under the stated condition, which does not happen with the solid and gas phases of a substance.

Note: A fluid is customarily called *liquid* if its temperature lies below T_C and density (ρ) lies above ρ_c, otherwise it is called *gas*.

3.46 For the same $P\overline{V}$, ZRT will be the same. Since $Z = 1$ for an ideal gas and $Z > 1$ for a real gas above its Boyle temperature, T will be higher for the ideal gas.

☐ Below the Boyle temperature, there are three possibilities (depending on pressure): (i) $Z = 1$ (ii) $Z < 1$ and (iii) $Z > 1$. Then the answer will be same or different according as (iii) holds or not.

3.47 Molecular volume $= \dfrac{b}{4N_A} = \dfrac{(0.030 \times 10^3 \text{ cm}^3\text{mol}^{-1})}{4(6.022 \times 10^{23} \text{ molecule mol}^{-1})}$

$$= 1.245 \times 10^{-23} \text{ cm}^3/\text{molecule}$$

According to the van der Waals equation $\overline{V} = \dfrac{RT}{P + a/\overline{V}^2} + b$

Being cubic in $\bar{V}$, it is difficult to solve for $\bar{V}$ exactly. However, a reasonably accurate value of $\bar{V}$ can be obtained through the following procedure. If the gas is assumed to behave ideally

$$\bar{V} = \frac{RT}{P} = \frac{(0.08206 \text{ L atm K}^{-1}\text{mol}^{-1})(273.15 \text{ K})}{(1 \text{ atm})}$$

$$= 22.41 \text{ L mol}^{-1}$$

Substituting this in the rhs of the above equation

$$\bar{V} = \frac{(0.08206 \text{ L atm K}^{-1}\text{mol}^{-1})(273.15 \text{ K})}{1 \text{ atm} + \dfrac{(12 \text{ L}^2 \text{ atm mol}^{-2})}{(22.41 \text{ L mol}^{-1})^2}}$$

$$= 21.92 \text{ L mol}^{-1}$$

Using this in the similar way and by further repeating the procedure, the result can be improved to the desired extent.

□ i. For a van der Waals gas, $T_C \propto a/b$. Since a/b is higher for X (12/0.030) than for Y (6/0.028), X will have higher T_C.

ii. For a van der Waals gas, $Z \approx 1 - aP/(RT)^2$ in the low pressure region. Then the gas Y having lower value of a will have $\approx$ nearer to 1, and hence it will be more nearly ideal in behaviour.

iii. The gas X having higher value of a will have lower value of Z and hence more (easily) compressible than Y.

□□ Answer (ii) will not be affected. Because in the low pressure region and also in the high pressure region (where $Z \approx 1 + Pb/RT$), the answer is determined only by the relative order of a and the relative order of b for the two gases, which are same in the given problem.

But the answer (iii) will be affected in high pressure region, though not in low pressure region.

Notes: 1. To find the volume of n moles of a van der Waals gas, it is better to calculate first $\bar{V}$ (using the above procedure) and then to use the relation $V = n\bar{V}$. With the van der Waals equation in the form $\left(P + \frac{n^2 a}{V^2}\right)(V - nb) = nRT$, the calculation will be somewhat lengthy.

2. The molecular volume can be conveniently expressed in vol/mol when it will be numerically equal to $b/4$.

3.48 Here $P_C = 30$ atm and $T_C = 227°C$

a. Molecular volume $= \frac{4}{3}\pi r^3 = \frac{b}{4N_A} = \frac{\frac{1}{8}RT_C/P_C}{4NA}$

Then molecular radius

$$r = \left[\frac{3RT_C}{128\pi N_A P_C}\right]^{\frac{1}{3}}$$

$$= \left[\frac{3(8.314 \text{ JK}^{-1}\text{mol}^{-1})(227 + 273.15)\text{K}}{128 \times 3.143 (6.022 \times 10^{23} \text{ mol}^{-1})(30 \times 1.013 \times 10^5 \text{Pa})}\right]^{\frac{1}{3}}$$

$$= 2.568 \times 10^{-10} \text{ m}$$

b. The maximum temperature in question is $T_C = 227°C$

$$T_B = \frac{a}{Rb} \qquad \text{(vide problem 3.27)}$$

$$= \frac{27}{8} T_C \qquad \text{by Eq. (3.6)}$$

$$= \frac{27}{8}(227 + 273.15)\,K = 1688\,K$$

c. (i) At 1500°C, which lies above T_B (1688 K), $Z > 1$ for any value of pressure. Then the gaseous state is less compressible in the given region of pressure.

ii. At 1000°C, which lies below T_C, $Z < 1$ at low pressure but $Z > 1$ at high pressure. Then the gaseous state is more compressible around 1 atm (low pressure region), but less compressible around 400 atm (high pressure region).

iii. At 0°C, same as (ii)

Note: 1. The maximum vapour pressure of a substance refers to its liquid-vapour (and not solid-vapour) equilibrium and this happens at T_C above which only gas phase can exist.

2. For compressibility, first think about temperature concerned and then pressure with the Amagat's isothermal in mind.

3. Low pressure region means $P < P_C$ and high pressure region $P > P_C$.

3.49 Since $(P_C\bar{V}_C/T_C)_A = (P_C\bar{V}_C/T_C)_B$, by the law of corresponding states (vide problem 3.33)

$$(\bar{V}_C)_B = \left(\frac{P_C\bar{V}_C}{T_C}\right)_A \cdot \left(\frac{T_C}{P_C}\right)_B$$

$$= \frac{(12.8\,\text{atm})(0.065\,\text{L mol}^{-1})(304\,K)}{(33\,K)\ (72.8\,\text{atm})} = 0.105\,\text{L mol}^{-1}$$

[Alternatively: Here $P_r = P/P_c$ is the same $\left(\frac{1}{4}\right)$ for A and B. Also $T_r = T/T_c = 2$ for both the gases. Then $\bar{V}_r$ will be the same for A and B.

i.e. $$(\bar{V}/\bar{V}_C)_A = (\bar{V}/\bar{V}_C)_B$$

or $$(\bar{V}_C)_B = \frac{(\bar{V}_C)_A}{\bar{V}_A} \cdot \bar{V}_B$$

$$= \frac{(0.065\,\text{L mol}^{-1})(2.782\,\text{L mol}^{-1})}{(1.718\,\text{L mol}^{-1})} = 0.105\,\text{L mol}^{-1}$$

i. Gas B having much higher T_C than A, is more easily liquefiable.

ii. Gas B, being more easily liquefiable, is more non-ideal than A.

Hence gas A is more close to the ideal behaviour.

☐ For A, $P_C\bar{V}_C/T_C$ = (12.8 atm) (0.065 L mol^{-1})/(33 K) = 0.0252 L atm K^{-1}mol^{-1}

For B, $P_C\bar{V}_C/T_C$ = (72.8 atm) (0.105 L mol^{-1})/(304 K) = 0.0251 L atm K^{-1}mol^{-1}

For a van der Waals gas $P_C\bar{V}_C/T_C$ = 0.308 L atm K^{-1}mol^{-1}

Then A and B are almost equally away from a van der Waals gas in their behaviour.

Note: The absolute values of P, V and T given in the problem are useful only for the alternative procedure which works here because P_r and also T_r are same for both the gases.

3.50 At low pressure $\quad \bar{V} = \dfrac{RT}{P} + B \quad$ vide problem 3.6

$$= \frac{(0.08206\,\text{L atm K}^{-1}\text{mol}^{-1})(273.15\,K)}{(1\,\text{atm})} - 0.145\,\text{L mol}^{-1}$$

$$= 22.3\,\text{L mol}^{-1}$$

☐ Here $\bar{V}$ is less than the ideal molar volume. Hence, the intermolecular force prevailing here is attractive in nature.

Note: The above procedure for the calculation of $\bar{V}$ is limited by the fact that the working formula used is a good approximation of the general Eq. (3.14) only at low pressure when $1/\bar{V}$ is quite low.

3.51 In deriving the equation of state of a real gas, van der Waals treated the gas molecules as hard spheres with the intermolecular attraction always between them but the repulsion suddenly appearing only at the moment of their collision. In the van der Waals equation, the constant a takes care of this attraction and b of repulsion that prevents the colliding molecules coming closer than a distance equal to their diameter. Then, the intermolecular repulsion is intimately connected to the molecular diameter which determines b as a measure of the excluded volume in the gas. Therefore, regarding the intermolecular repulsion, the given statement is not justified.

Note: Although the intermolecular repulsion is not ignored by van der Waals, it has been taken into account through an unrealistic hard sphere model (vide problem 3.15).

3.52 Here $V = 0$ for $r = \sigma$. Hence $A/\sigma^6 = B/\sigma^n = C\epsilon$, where C is a number.

Then
$$V = C\,\epsilon\left[\left(\frac{\sigma}{r}\right)^n - \left(\frac{\sigma}{r}\right)^6\right] \tag{I}$$

Now, for
$$V = \epsilon, \quad \frac{dV}{dr} = 0 \text{ whence } r^{n-6} = \frac{n}{6}\sigma^{n-6} \text{ or } r = \left(\frac{n}{6}\right)^{\frac{1}{n-6}}\sigma$$

Substituting this in (I), we have
$$C = \left[\left(\frac{6}{n}\right)^{\frac{n}{n-6}} - \left(\frac{6}{n}\right)^{\frac{6}{n-6}}\right]^{-1}$$

This, when used in (I), will give the desired expression for V.

Note: For $n = 12$ and $c = 4$, $V = 4\,\epsilon\left[\left(\frac{\sigma}{r}\right)^{12} - \left(\frac{\sigma}{r}\right)^6\right]$ which is the common form of Lennard–Jones potential.

3.53 Using $V = 4\,\epsilon\left[\left(\frac{\sigma}{r}\right)^{12} - \left(\frac{\sigma}{r}\right)^6\right]$, we have on putting $dV/dr = 0$ for $r = d$, $\sigma = d/2^{1/6}$.

Substitution of this in the expression of V gives
$$V = \epsilon\left[\left(\frac{d}{r}\right)^{12} - 2\left(\frac{d}{r}\right)^6\right]$$

3.54 Three potentials, defined below are depicted in Fig. 3.2.

$$V = \begin{cases} \infty \text{ for } r < d \\ 0 \text{ for } r \geq d \end{cases}$$

$$V = \begin{cases} \infty \text{ for } r < d \\ 0 \text{ for } r = d \\ -\dfrac{A}{r^6} \text{ for } r > d \end{cases}$$

$$\to \infty \text{ as } r \to 0$$
$$V = \begin{cases} 0 \text{ for } r = d/2^{\frac{1}{6}} \\ \dfrac{B}{r^{12}} - \dfrac{A}{r^6} \text{ for } r > d \end{cases}$$

Ideal gas potential   van der Waals potential   Lennard–Jones potential

Ideal gas potential has no minimum due to absence of intermolecular attraction, while van der Waals potential and Lennard–Jones potential exhibit minimum due to the presence of two opposing forces, attractive and repulsive, in real gas molecules. Although the minima for the two potential occur at virtually the same intermolecular distance equal to molecular radius, the minimum is less sharp with Lennard–Jones potential and lies slightly above that for the van der Waals potential due to flexibility of the molecules (arising from the diffused electron cloud around the molecules) considered in Lennard–Jones potential. For the same reason the V–r plot for $r < d$ in case of Lennard–Jones potential is slightly tilted towards the V-axis (and hence r for $V = 0$ is slightly lower) while the corresponding part of V–r plot is parallel to V-axis for van der Waals potential and also for ideal gas potential.

3.55 Because, at high temperature, the molecules travel with so much kinetic energy that the attractive part of Lennard–Jones potential is practically washed out.

3.56 (a) This is due to the intermolecular attraction (resulting solely from dispersion effect) increasing in the given order of the species in which their polarizability increases.

(b) Explanation same as in (a).

(c) This is ultimately due to the increase in polarity (dipole moment) of the species in the given order.

(d) Explanation same as in (c).

Here, unlike (c), the polarizability and hence dispersion effect decreases appreciably in the given order of the species. But we need not consider this effect which is likely to be less than the effect of polarity as the concerned molecules are quite small.

(e) Since the concerned species are isomeric, the dispersion effect is almost same for them. Hence the given order of their boiling point is same as that of their dipole moment. Here the molecules being quite large, the two polar C–Cl bonds interact appreciably only in ortho isomer leading to its higher dipole moment.

(f) This is attributed to the formation of hydrogen bonds (which are much stronger than van der Waals forces) that leads to molecular association of higher order in the given sequence of the species.

Although CH_4 has about 10% higher polarizability than NH_3, the former has lower boiling point. Because CH_4 is unable to form hydrogen bond.

Again, although HF has higher dipole moment and forms stronger hydrogen bond than H_2O, the former has lower boiling point. Because, on structural ground, a H_2O molecule is able to form greater number of hydrogen bonds than a HF molecule.

Notes:

1. The interaction energy of non-polar molecules is due only to the dispersion effect which depends largely on the polarizability of the molecules. As a general rule, the higher is the size of a molecule, the higher will be its polarizability and hence higher will be its dispersion energy and boiling point.

2. For polar molecules the dispersion effect, that arises inevitably, is important only when the molecules are large.

UNIVERSITY QUESTIONS

3.1 a. Draw clearly typical P vs V isothermals for real gases. Explain how these lead to the concept of critical state and mention the salient features of the critical state. (Calcutta BSc(H), 1994)

b. Discuss the principle of continuity of states from the plots.
(Burdwan BSc(H), 2007)

3.2 Why do gases fail to obey ideal gas equation at high pressure and low temperature? (Punjabi BSc, 2002)

3.3 Explain why beyond certain temperature gases cannot be liquefied whatever the pressure may be. (Guru Nanak Deb BSc, 2003)

3.4 a. Starting from the van der Waals equation for one mole of a gas arrive at the expression for n molar system. (Jadavpur BSc(H), 2012)

b. Justify the sign and magnitude of the correction terms used in the van der Waals equation. (Burdwan BSc(H), 1998)

3.5 Show that the excluded volume b is approximately four times the actual volume occupied by the molecules. (Burdwan, BSc(H), 2011)

3.6 In the van der Waals equation, what are the dimensions of the constants a and b?
(Burdwan BSc(H), 2004)

3.7 Write down the Dieterici equation and therefrom deduce the dimension of Dieterici a. Does the dimension differ from that of van der Waals a. (Calcutta BSc(H), 1995)

3.8 Demonstrate how can the Dielecterici equation for a real gas be reduced to the van der Waals equation and then to the ideal gas equation. What are the approximations employed in each stage? (Burdwan BSc(H), 1998)

3.9 Draw schematically a set of experimental as well as theoretical P–V isotherms based on the van der Waals equation. (Calcutta BSc(H), 1999)

3.10 Explain the basic differences between the van der Waals isotherms at $T \gg T_C$, $T = T_C$ and $T < T_C$. (Calcutta BSc(H), 2008)

3.11 What are the necessary conditions that are to be applied on the van der Waals equation to obtain the expression for critical temperature. (Calcutta BSc(H), 2012)

3.12 A gas obeys the equation of state: $PV = RT(1 + b/V)$
i. Would it have a critical temperature?
ii. Would it be possible to liquefy the gas? (Calcutta BSc(H), 2004)

3.13 How would you, in general, judge the quality of a real gas equation over the ideal gas equation? (Jadavpur BSc(H), 2012)

3.14 How does a general virial equation of state, in terms of $1/\bar{V}$ look like? What are virial coefficients? What do they depend on? Comment on their values for an ideal gas. (Burdwan BSc(H), 1993)

3.15 a. Define Boyle temperature (T_B). (Calcutta BSc(H), 1993)

b. What are the necessary conditions that are to be applied on the van der Waals equation to obtain an expression for T_B? (Calcutta BSc(H), 2012)

c. Write down the van der Waals equation of state in the form of virial equation of state. Identify the second virial coefficient and hence show that $T_B = a/Rb$.
(Burdwan BSc(H), 1995)

d. Remark whether T_B should be higher T_C. (Burdwan BSc(H), 2013)

3.16 Justify that a gas obeying the equation of state $\left(P + \frac{a}{\bar{V}^2}\right)\bar{V} = RT$ does not possess a Boyle temperature. Is is possible to liquefy the gas? (Calcutta BSc(H), 2007)

3.17 Explain how from Amagat's isothermal (Z vs P plot) below Boyle temperature it can be argued that gas molecules move under two opposing constants. (Calcutta BSc(H), 1995)

3.18 Draw PV vs P curves when the gas equation are (A) $P(\bar{V} - b) = RT$ and (B) $\left(P + \frac{a}{\bar{V}^2}\right)\bar{V} = RT$ with explanation. (Burdwan BSc(H), 2007)

3.19 Given below the plot of Z against P for three van der Waals gases CH_4, NH_3 and He at 0°C temperature. Identify, with justification, the gases corresponding to different curves in the figure. (Calcutta BSc(H), 2007)

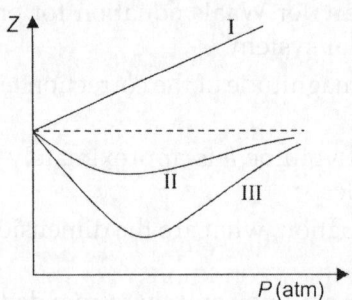

3.20 a. The compressibility factor Z for a van der Waals gas can expand as a power series in the pressure as follows
$$Z = 1 + A_1P + A_2P^2 + A_3P^3$$
Find the value of A_1 in the above equation. (Burdwan BSc(H), 2010)

   b. If one writes $Z = \alpha_0 + \alpha_1P$, what can you say about the magnitude of α_0 and signs of α_1 for different values of P? (Calcutta BSc(H), 2006)

3.21 Arrive at the reduced equation of state for a van der Waals gas. What is its significance? Explain whether such an equation is possible for an ideal gas?

3.22 The reduced equation of state for a van der Waals gas is $\left(\pi + \frac{3}{\phi^2}\right)(3\phi - 1) = 8\theta$ where the terms have their usual meaning. This equation is independent of a, b and R, so it is applicable to all gases. Justify or criticise. (Calcutta BSc(H), 2010)

3.23 Derive an expression for Z of a van der Waals gas in terms of its reduced quantities and draw the important conclusion therefrom. (Calcutta BSc(H), 2000)

3.24 Explain the significance of the van der Waals constants in terms of intermolecular forces. (Burdwan BSc(H), 1995)

3.25 The potential energy of attraction between neutral molecules is given by $U = A/r^n$. Remark on the sign of A and its dependence on molecular property. Comment on the value of n. (Calcutta BSc(H), 2006)

3.26 Draw neat diagrams showing plots of the van der Waals potential $U_v = A/r^6$ and the Lennard–Jones potential $U_L = -\frac{A}{r^6} + \frac{B}{r^n}$ for intermolecular interaction. (Burdwan BSc(H), 1996)

3.27 The Lennard–Jones potential $U_L = -\frac{A}{r^6} + \frac{B}{r^n}$ can be expressed in terms of U_m, the energy at the minimum, and r_0 the distance of separation at the minimum. Find A and B in terms of r_0 U_m and n. Write the potential in terms of the new parameters. If σ is the distance of separation when $U = 0$, find the relation between r_0 and σ. (Jadavpur BSc(H), 1996)

3.28 a. What is meant by van der Waals force? Why are they called so? Explain qualitatively the origin of this force.   (Burdwan BSc(H), 1997

b. Which of the three van der Waals attractive forces depend on

i. temperature of the gas

ii. dipole moment of the gas molecules.   (Calcutta BSc(H), 2013)

3.29 Identify the nature of attractive forces in each of the gases He, Ar, NH_3. How do such forces depend on temperature and size of species?   (Calcutta BSc(H), 2007)

3.30 Indicate the higher boiling substance and mention the major interaction responsible for it in the following pairs:

i. CH_3CH_2OH and CH_3—O—CH_3 (Jadavpur BSc(H), 2008)

ii. CH_3Cl and CCl_4 (Jadavpur BSc(H), 2009)

3.31 State with reasons, whether van der Waals constant a and b would depend on temperature or not.   (Calcutta BSc(H), 1993)

3.32 Will the pressure of the gas on the walls be greater or less than the ideal pressure if we take into account only the finite dimension of molecules.

(Calcutta BSc(H), 1997)

3.33 Two van der Waals gases have the same b but different a. Which one would occupy greater volume under identical conditions of temperature and pressure?

(Burdwan BSc(H), 2006)

3.34 A 10.0 L flask contains 64 g of oxygen at 27°C. Calculate its pressure using (i) van der Waals equation and (ii) ideal gas equation. Given: $a = 1.36$ L^2 atm mol^{-2} and $b = 0.037$ L mol^{-1}.   (Arunachal BSc(H), 2003)

3.35 For a van der Waals gas, find the reduced pressure and the pressure exerted by one mole of the gas taken in a 0.0804 L vessel at 37.5°C. Given: $P_c = 48.4$ atm, $T_c = 305.4$ K, $\bar{V}_c = 0.148$ L mol^{-1}.   (Calcutta BSc(H, 2013)

3.36 The critical temperature and pressure of CO_2 are respectively 31°C and 73 atm. Assuming that CO_2 obeys van der Waals equation, calculate the diameter of CO_2 molecule.   (Burdwan BSc(H), 2009)

3.37 Two van der Waals gases A and B are at corresponding states. The critical temperature and pressures of the gases are given below:

	P_C (atm)	T_C/K
A	48	105
B	33	125

Find out the volume of B at this corresponding state if the volume of A is 1.5 L.

(Calcutta BSc(H), 2007)

3.38 The behaviour of two gases A and B can be approximated by van der Waals equation. The critical constants of these gases are given below:

Gas	P_C (atm)	$\bar{V}_c$ (cm^{3}mol^{-1})	T_C/K
A	81.5	81.0	324.7
B	2.26	57.76	5.21

Explain (i) which gas has greater intermolecular force of attraction. (ii)Which gas obeys van der Waals equation more closely at the critical state.

(Calcutta BSc(H), 2010)

3.39 Gases NO and CCl_4 obeying van der Waals equations have T_C and P_C values as given below:

	T_C	P_C
NO	177 K	64 atm
CCl_4	550 K	45 atm

Which gas has (i) smaller value of a (ii) smaller value of b (iii) larger value of $\bar{V}_C$ and (iv) more nearly ideal behaviour at 300 K and 10 atm pressure? Give reasons for your answer. (Calcutta BSc(H), 1998)

3.40 If the temperature above which a van der Waals gas cannot be liquefied be 32.3°C and the minimum pressure to be applied at that temperature for liquefaction be 48.2 atm (i) find the diameter of gas molecule (ii) calculata a.

(Calcutta BSc(H), 1995)

3.41 Gases A, B, C and D obey the van der Waals equation, with a and b values as given (SI):

	A	B	C	D
a	0.6	0.6	0.2	0.005
$10^3 b$	0.025	0.15	0.10	0.02

Identify the gases which have the highest critical temperature, the largest molecules, the most nearly ideal behaviour at STP. (Calcutta BSc(H), 2003)

3.42 The compression factor is $Z = 1.00054$ at 0°C and 1 atm for a van der Waals gas. The Boyle temperature for that gas is 107 K. Estimate the values of a and b.

(Calcutta BSc(H), 2013)

3.43 The second virial coefficient of a gas is 13.7 L mol^{-1} at 273 K. Calculate the molar volume of the gas at STP. (Calcutta BSc(H), 2011)

KEY TO UNIVERSITY QUESTIONS

3.1 See Fig. 3.1b and problem 3.44.

Critical state refers to critical temperature (T_C) and critical pressure (P_C) which are characteristic constants of a substance such that when $T > T_C$ no liquid phase can exist (i.e. only gas phase can exist) and when $P > P_C$ no gas phase can exist in equilibrium with liquid phase (this does not necessarily mean that only liquid phase would exist).

Note: P_C of a substance equals the highest vapour pressure of its liquid phase.

3.2 Vide Sec. 3.1 and problem 3.1.

3.3 Because the liquefaction of a gas is not possible if its average molecular kinetic energy exceeds the highest value of the same for liquid state occurring at critical temperature.

3.4 a. The van der Waals equation for one mole of a gas is

$$\left(P + \frac{a}{V^2}\right)(V - b) = RT \quad \text{or} \quad \left(P + \frac{a}{\bar{V}^2}\right)(\bar{V} - b) = RT$$

Putting $\bar{V} = V/n$, we have

$$\left(P + \frac{n^2 a}{V^2}\right)(V - nb) = nR$$

which is the desired expression.

b. The non-zero volume of the molecules reduces the space available for their free movement. Then, for a real gas, V in the ideal gas equation should be replaced by $V - nb$, where b is called the excluded volume per mole, i.e. the volume correction for non-ideality is $-nb$ which is negative and numerically equal to four times the volume of the molecules taken together.

The intermolecular attraction makes the gas pressure lower than its ideal value. The pressure of a gas may be supposed to be proportional to the frequency and vigour of molecular collision with the container walls. Since both these factors are reduced by the intermolecular attraction by an amount proportional to the molar concentration n/V, the actual gas pressure will be less than the ideal gas pressure by an amount $a\,(n/V)^2$, where a is a positive constant characteristic of gas, i.e. the pressure correction for non-ideality is a $(n/V)^2$ which is positive.

3.5 See problem 3.16(ii).

3.6 See problem 3.19.

3.7 In the Dieterici equation $P = \frac{RT}{\bar V - b}\,e^{-a/RT\bar V}$. $a/RT\bar V$, being an exponent, has no dimension. Then a will have the dimension of $RT\bar V$, i.e. of $P\bar V^2$ (since RT is dimentionally equivalent to $P\bar V$) which is same as that of van der Waals a.

3.8 See problem 3.30.

3.9 Referred to Sec. 3.1.

3.10 See problem 3.8.

3.11 See problem 3.39(ii)

3.12 See problem 3.39(ii).

3.13 By calculating the critical compressibility factor $Z = P_C \bar V_C / RT_C$ from the concerned real gas equation and comparing it with the experimental value.

3.14 As given in Eq. (3.14). For an ideal gas, the virial coefficients are all zero, i.e. the equation of state is same for all ideal gases. But they do not exhibit universality in all of their properties. An example is offered by heat capacity.

3.15 For (a) and (b), refer to Boyle temperature in Sec. 3.1, (c) see problem 3.27 (d) see problem 3.37.

3.16 See problem 3.39.

3.17 See problem 3.22 for significance of Z.

Amagat's isothermals in Fig. 3.1a reflect that below the Boyle temperature Z is less than 1 (due to intermolecular attraction) at low pressure and Z is greater than 1 (due to intermolecular repulsion) at high pressure. Then, with rise in pressure, the molecules pass from attractive to repulsive interaction, i.e. they move under two opposing constraints.

3.18 PV vs P curves are sketched in the adjoining diagram.

The equation (A) implies that the actual volume ($\bar V$) is greater than the ideal gas volume by a fixed amount b. Hence $P\bar V$ becomes greater than its ideal value, i.e. RT (corresponding to ideal behaviour) and the difference $P\bar V - RT = bP$ increases linearly with P.

Regarding the Eq. (B), it implies that the actual pressure P is less than the ideal gas pressure by an

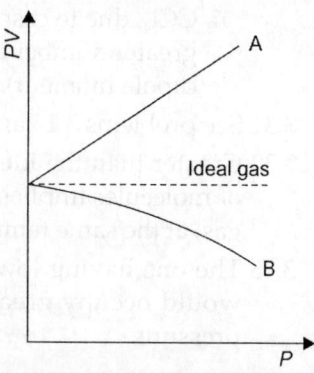

amount $a/\bar{V}^2$. Hence $P\bar{V}$ becomes less than its ideal value, i.e. RT and the difference $P\bar{V} - RT = -\frac{aP}{\bar{V}^2}$ varies with P, not linearly but at a greater rate due to decrease of V with increase of P.

3.19 Z and P plot will have lower initial slope for species having higher intermolecular attraction (vide problem 3.32). Then I, II and III will correspond to He, CH_4 and NH_3 respectively.

3.20 (a) At Boyle temperature $\lim\left(\frac{dZ}{dP}\right)_T = 0$, as $P \to 0$. Imposing this condition on the given expression, we have

$$A_1 = \text{Boyle temperature}$$
$$= a/Rb \text{ for a van der Waals equation}$$

(b) Experimentally, above Boyle temperature

$Z > 1$ and hence α_1 is +ve (since $\alpha_0 = 1$) for all pressures

below Boyle temperature

$Z < 1$ for low pressure region when α_1 is –ve

$Z > 1$ for high pressure region when α_1 is +ve

3.21 See Sec. 3.2. Since an ideal gas cannot exhibit critical phenomenon, a reduced equation is not possible for such a gas.

3.22 The given reduced equation follows from the van der Waals equation. Hence it is applicable only to the van der Waals gases.

3.23

$$Z = \frac{P\bar{V}}{RT} = \frac{P_C\bar{V}_C}{RT_C} \cdot \frac{P_r V_r}{T_r}$$

$$= \frac{3}{8} \cdot \frac{P_r V_r}{T_r} \qquad \text{for a van der Waals gas}$$

Then Z will be same for all van der Waals gases, since for the same P_r and T_r they have the same V_r.

3.24 See problem 3.51.

3.25 Referred to Sec. 3.4.

3.26 See problem 3.54.

3.27 See problems 3.52 and 3.53.

3.28 Referred to Sec. 3.4.

3.29 Referred to Sec. 3.4 and note on problem 3.56.

3.30   i. CH_3CH_2OH, due to hydrogen bonding which does not occur with CH_3-O-CH_3.

   ii. CCl_4, due to dispersion interaction which is much greater in CCl_4 (due to much greater number of electron) than in CH_3Cl (which, though polar, has only small dipole moment).

3.31 See problems 3.14 and 3.15.

3.32 Greater than the ideal pressure. Because the consideration of the finite dimension of molecules implies lower space for the movement of the molecules than the ideal gas, at the same temperature and pressure.

3.33 The one having lower value of a, i.e. involving lower intermolecular attraction, would occupy greater volume under identical conditions of temperature and pressure.

3.34   i. $P = \dfrac{RT}{V/n - b} - \dfrac{a}{(V/n)^2}$

$= \dfrac{(0.08206 \text{ L atm K}^{-1}\text{mol}^{-1})(300.15 \text{ K})}{\left(\dfrac{(10 \text{ L})}{\frac{64}{32} \text{ mol}}\right) - (0.032 \text{ L mol}^{-1})} - \dfrac{(1.36 \text{ L}^2 \text{ atm mol}^{-2})}{\left[\dfrac{10 \text{ L}}{\frac{64}{32} \text{ mol}}\right]^2} = 4.68 \text{ atm}$

ii. $P = \dfrac{RT}{V/n}$

$= \dfrac{(0.08206 \text{ L atm K}^{-1}\text{mol}^{-1})(300.15 \text{ K})}{\left(\dfrac{10 \text{ L}}{\frac{64}{32} \text{ mol}}\right)} = 4.92 \text{ atm}$

3.35 $\qquad P_r = \dfrac{\frac{8}{3} T_r}{\bar{V}_r - \frac{1}{3}} - \dfrac{3}{\bar{V}_r^2}$ by Eq. (3.11)

$= \dfrac{\frac{8}{3} T/T_c}{\bar{V}/\bar{V}_c - \frac{1}{3}} - \dfrac{3}{(\bar{V}/\bar{V}_c)^2}$

$= \dfrac{\frac{8}{3}(37.5 + 273.15) \text{ K}/(305.4 \text{ K})}{\dfrac{0.0804 \text{ L mol}^{-1}}{0.148 \text{ L mol}^{-1}} - \dfrac{1}{3}} - \dfrac{3}{\left(\dfrac{0.0804 \text{ L mol}^{-1}}{0.148 \text{ L mol}^{-1}}\right)^2} = 2.75$

$\qquad P = P_c P_r = (48.4 \text{ atm}) \times 2.75 = 133.1 \text{ atm}$

3.36 $\qquad 4 \cdot N_A \dfrac{4}{3}\pi \left(\dfrac{d}{2}\right)^3 = b = \dfrac{1}{8}\dfrac{RT_C}{P_C}$

or $\qquad d = \left[\dfrac{3R}{16\pi N_A} \cdot \dfrac{T_C}{P_C}\right]^{\frac{1}{3}}$

$= \left[\dfrac{3(8.314 \text{ JK}^{-1}\text{mol}^{-1})(31 + 273.15) \text{ K}}{16 \times 3.143(6.022 \times 10^{23} \text{ mol}^{-1})(73 \times 1.013 \times 10^5 P_a)}\right]^{\frac{1}{3}}$

$= 3.237 \times 10^{-10} \text{ m}$

3.37 Here $\qquad \dfrac{V_B}{V_A} = \dfrac{(\bar{V}_C)_B}{(\bar{V}_C)_A},$ Since A and B are in corresponding states

$= \dfrac{(T_C/P_C)_B}{(T_C/P_C)_A}$ Since $\dfrac{P_C \bar{V}_C}{T_C}$ = constant for a van der Waals gas

or $\qquad V_B = V_A \left(\dfrac{T_C}{P_C}\right)_B \cdot \left(\dfrac{P_C}{T_C}\right)_A$

$= (1.5 \text{ L}) \cdot \dfrac{125 \text{ K}}{33 \text{ atm}} \cdot \dfrac{48 \text{ atm}}{105 \text{ K}} = 2.60 \text{ L}$

Note: The two gases must have same equation of state which need not be of van der Waals type. Because $P_C \bar{V}_C/T_C$ is constant even if they follow some other equation of state provided it is transformable to the reduced form.

3.38   i. Of the two van der Waals gases, the one having higher value of a $(\alpha P_C \bar{V}_C^2)$ will have higher intermolecular force of attraction. Obviously it is A which has higher value of $P_C \bar{V}_C^2$ than B.

ii. See problem 3.49.

3.39   i. For a van der Waals gas, $a \propto T_C^2 / P_C$. Obviously, NO has smaller value of T_C^2 / P_C and hence smaller value of a.

ii. Since for a van der Waals gas $b \propto T_C/P_C$, NO having smaller value of T_C/p_C will have smaller value of b.

iii. Since $V_C \propto b$, CCl_4 having larger value of b will have larger value of $\bar{V}_C$.

iv. Here, 10 atm belongs to a low pressure region where $Z \simeq 1 - \frac{aP}{(RT)^2}$ for a van der Waals gas. Then NO having lower value of a will have Z nearer to 1 and hence it will have more nearly ideal behaviour at 300 K and 10 atm.

3.40   i. Similar to university question 3.36.

Here $T_C = 32.3°C$ and $P_C = 48.2$ atm.

ii. $a = \dfrac{27}{64} \cdot \dfrac{R^2 T_C^2}{P_C}$ for a van der Waals gas.

3.41 See problem 3.47.

3.42 For a van der Waals gas,
$$T_B = a/Rb$$

Also
$$Z = 1 + \frac{Pb}{RT}\left(1 - \frac{a}{RbT}\right) \quad \text{by Eq. (3.10)}$$

$$= 1 + \frac{Pb}{RT}\left(1 - \frac{T_B}{T}\right)$$

or
$$b = \frac{(Z-1)RT}{P(1 - T_B/T)}$$

$$= \frac{(1.00054 - 1)(0.08206 \text{ L atm K}^{-1}\text{mol}^{-1})(273.15 \text{ K})}{(1 \text{ atm})\left(1 - \dfrac{107 \text{ K}}{273.15 \text{ K}}\right)}$$

$$= 1.99 \times 10^{-2} \text{ L mol}^{-1}$$

$$a = RT_B b$$
$$= (0.08206 \text{ L atm K}^{-1}\text{mol}^{-1})(107 \text{ K})(1.99 \times 10^{-2}\text{L mol}^{-1})$$
$$= 0.175 \text{ L}^2\text{atm mol}^{-2}$$

3.43 Similar to problem 3.50.

4   First Law of Thermodynamics and Thermochemistry

Thermodynamics is the science of investigation based mainly on the consideration of energy. It has two main branches—classical and statistical. Classical thermodynamics is based on four phenomenological laws—zeroth, first, second and third, without any hypothesis (say molecular concept of matter) regarding the structure of a system.

4.1 RELEVANT TECHNICAL TERMS

The subject of any investigation is called *system* which is supposed to be separated from the rest of the universe, called *surroundings*, by a surface, called *boundary* (which may be real of imaginary).

Systems
- Open systems, which can exchange matter (and of course energy) with the surroundings
- Closed systems, which can exchange energy but not matter
- Isolated systems, which cannot exchange energy (and of course matter)

Properties of a system
- Microscopic properties, which refer to the constituent particles of the system (e.g. molecular speed)
- Macroscopic properties, which refer to the system as a whole. These are called extensive (e.g. mass, vol.) or intensive (e.g. density, pressure, temperature) according as they are affected or not on partitioning the system

$$\text{State of a system} = f(P, V, T, x) \tag{4.1a}$$

for a homogeneous system under ordinary conditions, where x is the mole fraction

$$= f(P, V, T), \text{ if } x \text{ is fixed} \tag{4.1b}$$
$$= f(P, V), f(P, T) \text{ or } f(T, V) \tag{4.1c}$$

for a hydrostatic system, i.e. one of fixed mass and composition.

The change in thermodynamic state of a system is called isobaric, isochoric, isothermal or physical according as P, V, T or x is constant.

A system, exposed to surroundings is said to be in 'thermodynamic equilibrium' if its macroscopic properties do not change with time (i) under exposed condition and also (ii) under isolated condition. If the condition (i) is fulfilled but (ii) is not, the system is said to be in 'steady state'.

If two systems A and B are each in thermal equilibrium with a third system C (i.e. no heat transfer in contact with C), then A will also be in thermal equilibrium with B. This is zeroth law of thermodynamics, which leads to the concept of temperature. Two bodies at thermal equilibrium have the same temperature.

The change in state of a system is effected by means of some operation, called process. A process is said to be reversible if (i) the system is at all times virtually in thermodynamic equilibrium, i.e. performed quasi-statically and (ii) it involves no dissipative effect (such as friction). It (i) or (ii) does not hold the process is called irreversible.

Heat and Work

Energy exchange between a system and its surroundings can occur in two forms—work and heat. Work is the energy exchange, due to the action of some organised macroscopic motion while heat is that due to the action of unorganised microscopic motion at the molecular level.

The most common type of work involved in thermodynamics is that due to volume change of the system. The work done (W) by a system due to its change in volume from V_i to V_f is given by

$$W = \int_{V_i}^{V_f} P_{ext} dV \tag{4.2a}$$

where P_{ext} is the external pressure

$$= \int_{V_i}^{V_f} P dV \tag{4.2b}$$

if the process is reversible at least mechanically

$$= \int_{V_i}^{V_f} \frac{nRT}{V} dV \tag{4.2c}$$

if the system comprises n mol of an ideal gas

$$= nRT \ln \frac{V_f}{V_i} \tag{4.2d}$$

if the process is isothermal.

For an adiabetic change of an ideal gas (where PV^γ = constant for a reversible process).

$$W = \frac{P_i V_i - P_f V_f}{\gamma - 1} \tag{4.3a}$$

$$= n\bar{C}_V (T_i - T_f) \tag{4.3b}$$

where $\bar{C}_V$ is the molar heat capacity of the gas at constant volume and $\gamma = C_P/C_V$.

According to international convention (recommended by IUPAC in 1970), *work* means work done on the system (i.e. work added to the system) which is negative of the work done by the system, just as *heat* means heat added to the system (so that heat and work can be treated as equivalent to each other).

4.2 FIRST LAW OF THERMODYNAMICS

The first law of thermodynamics is a particular form of the general law of the conservation of energy, because it primarily deals with the relation of heat (a particular form of energy exchange) with other forms of energy. Following two statements of the law are well-familiar

Clausius statement: Energy of the universe is constant.

Ostwald statement: A perpetual motion of the first kind (i.e. production of work without using any energy source) is impossible.

This law leads to the concept of a thermodynamic property, called internal energy of a system, U which is a single valued function of thermodynamic state and is extensive in nature.

If Q is the heat absorbed by a system and W is the work involved then ΔU of the system is given by

$$\Delta U = Q \pm W, \text{ if the system is closed} \tag{4.4a}$$

The positive sign refers to the work according to international convention (i.e. work done on the system) and negative sign refers to the work done by the system. For a specified change in state, Q and W depend on the path followed but ΔU does not. If $Q = 0$, the change is called adiabatic where $|W| = |\Delta U|$ when W, like ΔU, does not depend on the path of change. For an infinitesimal change the above expression becomes

$$dU = dQ \pm dW \tag{4.4b}$$

Here dU represents infinitesimal change but dQ and dW represent infinitesimal quantities.

Enthalpy, H (like U) is another thermodynamic property of a system, which is defined by

$$H = U + PV$$

It follows from this relation that

$$\Delta H = Q_P \tag{4.5}$$

in an isobaric process involving only PV-work.

[An isobaric process should necessarily be mechanically reversible (otherwise pressure of the system will have no meaning) during which the pressure of the system differs only infinitesimally from the externally applied pressure]

Further, heat capacity at constant pressure, $C_P = \frac{dQ_P}{dT} = \left(\frac{\partial H}{\partial T}\right)_P$ and heat capacity at constant volume, $C_V = \frac{dQ_V}{dT} = \left(\frac{\partial E}{\partial T}\right)_V$ if the system does only PV-work.

It can be established that

$$C_P - C_V = \left[\left(\frac{\partial U}{\partial V}\right)_T + P\right]\left(\frac{\partial V}{\partial T}\right)_P \tag{4.6}$$

$$= \left[-\left(\frac{\partial H}{\partial P}\right)_T + V\right]\left(\frac{\partial P}{\partial T}\right)_V \tag{4.7}$$

For an ideal gas $\left(\frac{\partial U}{\partial V}\right)_T = 0, \quad \text{i.e.} \, U = f(T)$

and $\left(\frac{\partial H}{\partial P}\right)_T = 0, \quad \text{i.e.} \, H = f(T)$

Then for such gas

$$C_P - C_V = nR \tag{4.8}$$

Note: Since classical thermodynamics disregards structure of matter, it cannot explain the origin of internal energy of a system. However, from the view point of statistical thermodynamics, which assumes molecular theory of matter, U of a system results from various motions and interactions that occurs within the system, and also due to change of mass (following Einstein equation). However, in ordinary chemical thermodynamics, the latter contribution is ignored so that we can suppose that mass is conserved and consider only the energy changes of various kinds occurring in equivalent amount. On doing so, we cannot apply the first law to the nuclear processes where mass in appreciable amount is converted into energy.

4.3 HEAT OF REACTION

This means heat absorbed by a system due to its chemical change (or phase change) which is equal to the accompanying ΔH or ΔU of the system according as the change occurs isobarically or isochorically (provided no work is involved other than that due to volume change).

Hess's Law

The law states that heat of a specified reaction is same whether it is brought about in one or more steps.

For example, the formation of 1 mol of CO_2 by combustion of carbon at a specified temperature and pressure would liberate the same amount of heat whether it occurs in a single step or in two steps involving CO as an intermediate, viz. $C(s) + \frac{1}{2}O_2(g) \rightarrow CO(g)$. This is justified on the ground that for a reaction like this; $Q_V = \Delta U$ and $Q_P = \Delta H$, and that ΔU and ΔH are independent of the path of change (vide problem 4.39).

ΔH and ΔU of a reaction are related as

$$\Delta H = \Delta U + \Delta(PV)$$
$$= \Delta U + \Delta v RT \tag{4.9}$$

where Δv is the sum of stoichiometric coefficients of the gaseous product minus that of the gaseous reactants, the gaseous components being assumed to obey ideal gas equation.

Kirchhoff's Equation

For a chemical reaction

$$v_A A + v_B B + \ldots = v_L L + v_M M + \ldots$$

$$\Delta H_{T_2} = \Delta H_{T_1} + \int_{T_1}^{T_2} \Delta C_P dT \tag{4.10a}$$

where $\Delta C_P = [v_L(\bar{C}_P)_L + v_M(\bar{C}_P)_M + \ldots] - [v_A(\bar{C}_P)_A + v_M(\bar{C}_P)_B + \ldots]$

$$\Delta U_{T_2} = \Delta U_{T_1} + \int_{T_1}^{T_2} \Delta C_V dT \tag{4.10b}$$

where ΔC_V is similar to ΔC_P.

These are two different forms of Kirchhoff's equation that represent variation of heat of reaction with temperature.

4.4 STANDARD ENTHALPY OF A REACTION

A thermochemical reaction is customarily represented quoting standard enthalpy of the reaction ΔH° (rather than ΔU) at a specified temperature (preferably 25°C). The zero superscript indicates the value of ΔH at the standard pressure which has been conventionally taken to be 1 bar (ordinarily 1 atm). In calculating ΔH° of a reaction the molar enthalpy $\bar{H}^\circ$ of every element has been arbitrarily assigned a zero value in its most stable state of aggregation at 25°C.

Thus gaseous hydrogen, liquid bromine, solid iodine, solid (graphite) carbon and solid rhombic sulfur all have $\bar{H}^\circ = 0$ at 25°C. From the value of $\Delta \bar{H}^\circ$ at 25°C, the value at some other temperature can be computed using Eq. (4.10a).

4.5 REMARKABLE FEATURES OF ΔH FOR COMMON THERMOCHEMICAL PROCESSES

The heat of reaction is suitably named according to the type of the process in question. For example, heat of formation, heat of combustion, heat of neutralisation, heat of solution, etc. All these are expressed per mole of the substance to which the process is referred.

The heat of formation of a substance is the enthalpy change when 1 mole of the substance is formed from its elements in their most stable states of aggregation at the temperature and pressure specified.

The heat of combustion is the enthalpy change per mole of a substance burnt in oxygen to stable products at the temperature and pressure specified. It is independent of the initial amounts of the reactants.

The heat of neutralisation is the enthalpy change per mole (or gm-equivalent) of an acid or base neutralised in solutions of specified dilution, e.g.

$$HCl(100H_2O) + NaOH(100H_2O) = NaCl(200H_2O) + H_2O(l), \ \Delta H^{\circ}_{298} = -27.15 \text{ kJ mol}^{-1}$$

The enthalpy change per mole of a solute due to dissolution of certain specified amount (n_2) of it in a solvent of specified amount (n_1) is called the integral heat of solution at the prevailing temperature and pressure. Here $\Delta H = \int (n_1, n_2)$. Then at constant T and P

$$d(\Delta H) = \left[\frac{\partial(\Delta H)}{\partial n_1}\right]_{n_2} dn_1 + \left[\frac{\partial(\Delta H)}{\partial n_2}\right]_{n_1} dn_2$$

$$= \Delta H_1 dn_1 + \Delta H_2 dn_2$$

where ΔH_2 is the differential heat of solution and ΔH_1 is the differential heat of dilution. For a fixed amount of solute, ΔH increases with increasing amount of solvent and approaches a limiting value at infinite dilution. This is unlike heat of combustion. With water as solvent, the dissolution process is often represented as

$$X + nH_2O \rightarrow X(nH_2O)$$

Note that $X(\propto H_2O)$ is represented by $X(aq)$ to mean an aqueous solution so diluted that further dilution produces no thermal effect.

4.6 BOND ENERGY AND BOND DISSOCIATION ENERGY

The bond-dissociation energy of a bond in a molecular species is the energy required to break that bond alone. It depends on the nature of the atoms linked by the bond and also on the nature of the species in which such bond is present. For example, atomisation of H_2O molecule that occurs in the following two steps:

1. $H-OH(g) = H(g) + OH(g)$
2. $O-H(g) = H(g) + O(g)$

Here the dissociation energy for O–H bond in H_2O molecules, corresponding to reaction 1, has somewhat higher value ($501.7 \text{ kJ mol}^{-1}$) than that for O–H bond in OH radical ($423.4 \text{ kJ mol}^{-1}$), corresponding to reaction 2.

The bond energy for a bond in a molecular species is the average energy per bond required to split all such bonds in the species. Thus in H_2O, O–H bond energy is the average of the energies involved in 1 and 2.

AUXILIARY PROBLEMS

4.1 Is there any system which can exchange matter but not energy with the surroundings? Can vacuum be treated as a thermodynamic system?

4.2 What is the function of boundary in thermodynamics? Is there any system having no real boundary? Consider certaim amount of a gas enclosed in a cylinder. Taking the gas as the thermodynamic system, what is the boundary? Find its dimensions.

4.3 Classify the following systems as open, closed, isolated or none (i) a barometer (ii) a thermometer (iii) a glowing electric bulb which is thermally insulated at its surface (iv) certain amount of a gas kept in a container such that the pressure of the gas remaining unchanged with change of temperature (v) a boiling solution under reflux (vi) a spontaneously freezing super-coded liquid enclosed in a Dewar flask (vii) human body (viii) an egg (ix) certain amount of uranium enclosed in a Dewar flask (x) sun (xi) moon (xii) Nux-30.

4.4 Classify the following properties as extensive, intensive or none (i) weight of a substance (ii) molecular weight (iii) molecular energy (iv) molar heat capacity (v) heat capacity (vi) molar gas constant (vii) acceleration due to gravity (viii) velocity of a system (ix) odour (x) solubility (xi) chemical reactivity (xii) equilibrium constant (xiii) emf of a galvanic cell (xiv) dipole moment.

4.5 Is is possible to express an intensive property only in terms of extensive properties? Is reverse possible? Discuss with an example (if any).

4.6 For an ideal gas $P = \frac{w}{MV} RT$. Here $P \propto w$ at constant V and T. Then P is an extensive property. Do you agree?

4.7 An extensive property is additive, at least approximately. Is reverse true? Discuss with examples.

4.8 Under what conditions we can write state of a system $= \int (P, V, T, x)$? Explain.

4.9 State with reasons whether the following cases correspond to equilibrium situation, steady state situation or none.
   a. One mole of a gas at STP. Here speed of each gas molecule does not remain constant.
   b. One molecule of a gas enclosed in a vessel.
   c. Liquid mercury and its vapour in a clinical thermometer (i) when it is entirely exposed to the atmosphere (ii) when it is partly within the body of a patient.
   d. A well-fed tiger in a cage, tiger's weight remaining constant.

4.10 Set up a relation between P and V for an adiabatic reversible process in case of (i) an ideal gas (ii) van der Waals gas.

4.11 Under what conditions the relation $PV^{\gamma} =$ constant will hold? Will $PV = nRT$ also hold simultaneously? On which of the following does this constant depend (i) mass (ii) molecular weight (iii) initial temperature (iv) molecular structure, of the gas? Is there any other factor that determines this constant?

4.12 Show that P–V curve for an ideal gas is steeper in adiabatic process than that in isothermal process. How do you explain this? Do you think that P–V curve for an adiabatic will always lie below that for an isothermal?

4.13 Discuss the possibility of the following:
   i. Expansion of an ideal gas is adiabatic as well as isothermal
   ii. Expansion of a real gas is adiabatic as well as isothermal
   iii. Expansion of any gas is isobaric as well as isothermal

4.14 State with reasons whether the following processes are reversible or not:
  i. Extremely slow expansion of a gas using ordinary piston-cylinder device
  ii. Free expansion of a gas
  iii. Transfer of heat frm a hot body to a cold body in extremely small instalments
  iv. Water freezes at STP
  v. Super-cooled water freezes at −10°C at 1 atm
  vi. Oscillation of a simple pendulum

4.15 Under what conditions the following relations will hold?
  i. $\Delta U = Q + W$

  ii. $W = -\int_{V_i}^{V_f} P dV$

  iii. $W = \dfrac{P_f V_f - P_i V_i}{\gamma - 1}$

  iv. $W = n\bar{C}_V (T_f - T_i)$
  v. $\Delta H = Q$

4.16 From the general expression, $\Delta U = Q + W$, for the first law arrive at the Clausius statement. The energy of the universe is a constant.

4.17 Consider a vessel containing certain amount of a gas connected to an empty vessel through a stop-cock. On opening the stop-cock the gas will flow into the empty vessel. Is the process reversible? Is it a free expansion?
  Assuming that the vessel and stop-cock are thermally insulated and the gas is ideal, answer the following:
  i. Whether any work will be involved
  ii. Whether U will change
  iii. Whether T will change
  Will the results be same if the gas were real?

4.18 a. One liter of an ideal gas initially at 2 atm is to be expanded to two liters at 1 atm without involving any work. How will you achieve this—(i) reversibly (ii) irreversibly? Find Q and ΔU in each case, and comment on their nature w.r.t. the path of change. Does your answer hold for any change?
  b. Is it possible to restore the gas to its initial state without involving any work isothermally or otherwise? Comment on the result.

4.19 Certain amount of an ideal gas undergoes reversible expansion till its volume becomes double in two different ways: (i) isothermally and (ii) adiabatically. In which case the work done by the gas will be lower? Will the final temperature in (ii) be lower or higher than that would be if the process were not reversible?

4.20 a. One liter of an ideal monatomic gas initially at 0°C and 10 atm expands isothermally at constant external pressure till its volume becomes double. Will the work involved be same, if (i) the process is adiabatic (ii) the gas is diatomic (iii) the gas is non-ideal (iv) P_{ext} is changed (v) initial temperature is changed.
  b. Will your conclusion in (a) change if one mole of the gas is considered instead of one liter?

4.21 For an ideal gas $\Delta H = C_p \Delta T$. Does this relation hold for a real system. If so under what condition? Give example of a physical and also of a chemical process (if any) where $\Delta H = 0$.

4.22 Deduce the following relation

$$C_P - C_V = \left[\left(\frac{\partial U}{\partial V}\right)_T + P\right]\left(\frac{\partial V}{\partial T}\right)_P$$

$$= \left[-\left(\frac{\partial H}{\partial P}\right)_T + V\right]\left(\frac{\partial P}{\partial T}\right)_V$$

Do you think that C_P will always be greater than C_V? Discuss with example (if any). Can heat capacity be (i) 0, (ii) ∞ (iii) –ve.

4.23 For an ideal gas, $C_P - C_V$ does not depend on atomicity though C_P/C_V does. Explain.

4.24 Justify or criticise the following:

 a. An isobaric process of expansion should necessarily be reversible.

 b. For a specified change of a system, the work obtained will be maximum if it is carried out reversibly.

 c. $\oint dW = 0$ in any cyclic process.

 d. Heat associated with a specified change of any system must be different for different path followed.

 e. Internal energy (U) is a state function. Therefore, any change of state of a system will be accompanied by a change in U of the system and vice versa.

 f. Internal energy of an isolated system is constant.

 g. H may be regarded as heat content for a system.

 h. For a constant pressure, $\Delta H = Q_P$. Then Q_P is a state function.

 i. C_P of an ideal gas is independent of pressure.

 j. If stirring work is done on an ideal gas in a closed system during a constant volume process, then $dQ \neq C_V dT$.

 k. C_P of a pure liquid in equilibrium with its vapour is infinite.

 l. C_V of a pure liquid in equilibrium with its vapour is infinite.

 m. If a system simultaneously attains mechanical equilibrium, thermal equilibrium and chemical equilibrium, then it will be in thermodynamic equilibrium.

4.25 An ideal monatomic gas is made to undergo a reversible cycle consisting of the following steps:

 State I (1 L, 1 atm) $\rightarrow$ State II (1 L, 3 atm)

 State II $\rightarrow$ State III (3 L, 1 atm)

 State III $\rightarrow$ State IV (125 mL, 32 atm)

 State IV $\rightarrow$ State I

 a. Indicate the isothermal and adiabatic steps (if any)

 b. Find W and Q for these steps

 c. Would W depend on mass of the gas

 d. Relate $\oint dW$ with $\oint dQ$

4.26 One mole of an ideal monatomic gas initially at 2 atm and 11.2 L undergoes reversible expansion in the following different ways till its volume becomes double: (i) isothermally (ii) adiabatically (iii) under the condition $P/V = $ constant (iv) under the condition $PT = $ constant. Calculate Q, W, ΔU and ΔH in each case.

4.27 One mole of an ideal gas $(\bar{C}_V = 3.0 \text{ cal K}^{-1}\text{mol}^{-1})$ undergoes the following reversible cycle:

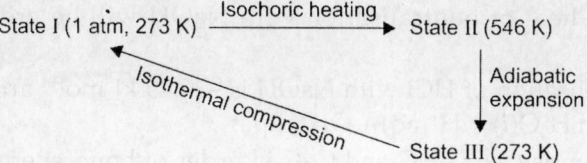

State I (1 atm, 273 K) —— Isochoric heating ——→ State II (546 K)

Isothermal compression

Adiabatic expansion

State III (273 K)

  a. Depict the cycle in a V vs T diagram
  b. Calculate Q, W and ΔU for each step and for the cycle
  c. Comment on the nature of heat, work and internal energy.

4.28 One liter of an ideal monatomic gas initially at 27°C and 10 atm undergoes adiabatic expansion against a constant external pressure of 1 atm till

  a. volume becomes double
  b. mechanical equilibrium is attained

Calculate W, ΔU and T_f in each case. Would the result be affected if the gas were non-ideal?

4.29 What fraction of heat absorbed by an ideal monatomic gas is transformed into work when heated (i) isothermally (ii) isobarically? Will the result be affected if the gas is diatomic?

4.30 One mole of an ideal gas is transformed from 27°C and 1 atm to 327°C and 20 atm. Find ΔH, ΔU and maximum value of Q possible for this change.

Given: $\bar{C}_P/(\text{JK}^{-1}\text{mol}^{-1}) = 20.5 + 0.045\,T/K$

4.31 a. One mole of an ideal monatomic gas initially at STP undergoes a reversible process in which volume is doubled, $Q = 400$ cal and $\Delta H = 500$ Cal. Calculate W and ΔU.

  b. Suppose that the gas were taken to the same final state by a process involving reversible isochoric and isothermal steps. Find Q, W, ΔU and ΔH for the simplest path of your choice.

4.32 Describe a reversible process by which one mole of water at 100°C can be converted into vapour at 0.5 atm isothermally. Calculate W, ΔU and ΔH in the process mentioning the assumption(s) involved. How would these quantities be affected if the operations were performed irreversibly? The latent heat of vaporisation of water = 540 cal g^{-1}.

4.33 Repeat the calculation of the quantities in the previous problem, if 1 mol of water vapour in equilibrium with liquid water is produced by introducing liquid water at 100°C into an initially evacuated vessel maintained at 100°C.

4.34 Iron (in very fine state of subdivision) is kept in an atmosphere of pure oxygen in a cylinder fitted with a frictionless piston that maintains the pressure of oxygen constant at 1 atm. Here following reaction occurs $2Fe(s) + \frac{3}{2}O_2(g) \rightarrow Fe_2O_3(s)$. For this reaction at constant temperature of 25°C, $\Delta H = -831.05$ kJ mol^{-1}. Find (a) Q, W and ΔU involved (b) the standard heat of combustion of iron (c) the standard heat of formation of Fe_2O_3 (d) the conventional enthalpy of Fe_2O_3, all at 25°C.

4.35 Account for the following:
  a. Heat of solution depends on the amount of solvent used but heat of combustion is independent of the amount of oxygen used.

b. Heat of neutralisation of any strong acid by a strong base is same. But this is not so for weak acid or weak base.

4.36 In which case heat of neutralisation with NaOH will be greater—CH_3CO_2H or HCO_2H?

4.37 Heat of neutralisation of HCl with NaOH is –26.45 kJ mol^{-1} around 25°C. Find ΔH for the reaction $H_2O(l) = H^+(aq) + OH^-(aq)$.

4.38 Arrange the bonds C–C, C=C and C≡C in order of bond energy. Would the order you suggest also represent the order of reactivity of the bonds?

4.39 According to Hess's law the heat of a reaction depends only upon the initial reactants and final products. Justify or modify this statement, citing appropriate example.

4.40 Zn is dissolved in (i) dil H_2SO_4 (ii) dil $CuSO_4$ solution at room temperature. Will more heat be evolved when the reaction is carried out in an open vessel or in a closed bomb?

4.41 When m mol of NaCl dissolves in 1 kg water at 25°C heat absorbed is given by $\Delta H / kJ = 3.86\ m + 1.99\ m^{3/2} - 3.04\ m^2 + 1.02\ m^{5/2}$. Find:
  i. heat of solution in the formation of one molal solution
  ii. heat of solution in the formation of an infinitely dilute solution
  iii. differential heat of solution for solution of unit molality
  iv. heat of dilution from molality 1 to 0.01

4.42 ΔH associated with mixing of 1 mol of H_2SO_4 with n mol of water at 25°C is given by $\Delta H / kJ = -\dfrac{75.5}{1 + 1.8/n}$. Find:
  i. Integral and differential heat of solution of H_2SO_4
  ii. Integral and differential heat of solution of H_2O for $n = 10$.

4.43 The standard heat of formation of HBr at 25°C is –8.7 kcal mol^{-1}. What do you mean by this? Heat of vaporisation of liquid bromine is 7.3 kcal mol^{-1} at 25°C. Molar heat capacities $(\bar{C}_P)$ of $H_2(g)$, $Br_2(g)$ and HBr(g) are of the form $\bar{C}_P = a + bT$ with following values of a and b.

Gas	a (cal K^{-1} mol^{-1})	$b \times 10^3$ (cal K^{-2} mol^{-1})
H_2	6.95	–0.20
Br_2	8.42	0.97
HBr	6.58	0.95

Calculate standard heat of formation of HBr at 100°C.

4.44 Heats of formation of CO(g) and $H_2O(g)$ are – 26.4 and – 58.08 kcal mol^{-1} respectively around 1000°C. Calculate ΔH for the water gas reaction
$$H_2O(g) + C(s) = H_2(g) + CO(g)$$
What should be the proportion by volume of air and steam which, if passed into a mass of coke at about 1000°C will maintain a constant temperature?

4.45 Find the enthalpy of the reaction $C_3H_8(g) + H_2(g) = C_2H_6(g) + CH_4(g)$ at 25°C using the given enthalpy of combustion values uner standard conditions:

Substance	$H_2(g)$	$CH_4(g)$	$C_2H_6(g)$	C (graphite)
H°/kJ mol^{-1}	–285.8	–890.0	–1560.0	–393.5

The standard enthalpy of formation of $C_3H_8(g)$ is –103.8 kJ mol^{-1}.

4.46 a. Heat of formation of $H_2O(g)$ at 25°C is -57.8 kcal mol^{-1}. Calculate the O–H bond energy from the following additional data.

$$H_2(g) = 2H(g), \quad \Delta H = 103 \text{ kcal mol}^{-1}$$
$$O_2(g) = 2O(g), \quad \Delta H = 34 \text{ kcal mol}^{-1}$$

b. A mixture of 2 mol H_2 and 11 mol O_2 is exploded adiabatically in a sealed bomb. Calculate

  i. ΔH and ΔU for the gaseous reaction. Also calculate these if the reaction were carried out isothermally at 25°C.

  ii. The maximum temperature that will be attained if the gas mixture is initially at 25°C.

  Molar heat capacities of $H_2(g)$ and $O_2(g)$ are both 6.5 cal K^{-1}mol^{-1} and that of $H_2O(g)$ is 7.5 cal K^{-1}mol^{-1}.

4.47 The polymerisation of ethylene to linear polyethylene is represented by the reaction

$$nCH_2 = CH_2 \rightarrow (-CH_2-CH_2-)_n$$

where n has a large integral value. Given that the average enthalpies of bond dissociation for C=C and C–C at 298 K are +590 and +331 kJ mol^{-1} respectively. Calculate ΔH of polymerisation per mol of ethylene at 298 K.

4.48 Compute the resonance energy of gaseous benzene from the following data:

$\in_{C-H} = 416.3$ kJ mol^{-1}, $\in_{C-C} = 331.4$ kJ mol^{-1}, $\in_{C=C} = 591.1$ kJ mol^{-1}

ΔH°_{sub} (C, graphite) = 718.4 kJ mol^{-1}

ΔH°_{diss} (H_2, g) = 435.9 kJ mol^{-1}

ΔH°_f (benzene, g) = 82.9 kJ mol^{-1}

ANSWERS

4.1 No. Because matter inevitably contains energy, and hence matter exchange must involve energy exchange (but reverse is not true).

☐ Yes. Because vacuum really consists of particles called photons.

4.2 Boundary mere helps in visualising a system.

☐ Yes, e.g. a mentally isolated portion of an object where boundary is imaginary.

☐☐ Here boundary is the interior surface of the cylinder.

☐☐☐ The dimension of boundary is that of area, i.e. l^2.

4.3 i. Open system, because here the surface of Hg is exposed open to the atmosphere and this makes evaporation of Hg into the atmosphere (surrounding) possible.

  ii. Closed system, because Hg enclosed in the thermometer can only exchange heat (energy) with the surroundings.

  iii. Closed system, because the electric bulb draws electrical energy from the surroundings though there is no heat exchange with the surroundings.

  iv. Open system, because of exchange of gas molecules with the surroundings to maintain the given condition.

  v. Open system, because the refluxing liquid loses some matter (though in small amount) to the atmosphere.

  vi. Isolated system, because here heat generated in freezing cannot go to the surroundings.

vii. Open system, because the exchange of matter with surroundings that occurs with a living human body (essential for its very existence) and also with a dead body (due to its natural decomposition).

viii. Open system, because of decomposition of egg into gaseous substance that effuse through the porous egg cell into the surrounding.

ix. None, because uranium undergoes radioactive disintegration that involves significant conversion of matter into energy and hence the given system cannot be regarded as thermodynamic one.

x. None, because of significant conversion of matter into energy due to nuclear fusion of ^1H into ^4He.

xi. Open system. This is unlike sun, there being no nuclear process to be considered here.

xii. None (considering Nux only), because of undetectably low concentration of Nux (10^{-30} M) in Nux-30.

4.4   i. Extensive.

ii. Intensive, being a macroscopic property independent of the size of the system.

iii. None, being a microscopic property considering the molecules individually (but total molecular energy is an extensive property).

iv. Intensive, being the ratio of two extensive quantities.

v. Extensive.

vi. None, being not a property of the system.

vii. None, being not a property of the system.

viii. None, being not a thermodynamic property.

ix. None, being not a thermodynamic property.

x. Intensive, being the ratio of two extensive quantities.

xi. None, being not a thermodynamic property.

xii. Intensive, being a macroscopic property independent of the size of the system.

xiii. Intensive, being a macroscopic property independent of the size of the cell.

xiv. None, being a microscopic property.

4.5 Yes, e.g. density = mass/volume.

☐ No, otherwise an extensive property will be independent of the size of the system, e.g. mass = volume × density, where volume is an extensive property.

4.6 No. Here, what is relevant is the expression of P and not the constancy of V and T. Since the expression for P involves only intensive variables, T and w/V, P is intensive.

Alternatively: Although w is generally extensive, w for a fixed volume is intensive and hence P is intensie, T being intensive.

4.7 An extensive property of a system is one that can be computed by summing up the values of this property for different volume elements having same intensive properties as the system. Again, an additive property of a system is one that can be computed by summing up independent contributions to this property due to different ingredients of the system. An extensive property will be additive provided there is no appreciable interaction between the ingredients of the system. For example, volume (an extensive property) of a mixture of ideal gas is equal to sum of the values of the component gases (when measured at the prevailing temperature and total pressure of the system).

But the reverse is not necessarily true. For example, pressure of a mixture of ideal gases is additive, being equal to sum of the partial pressures of its different ingredients. But pressure is not an extensive property.

4.8 The system is required to be homogeneous. For a heterogeneous system, the given function has meaning only with each phase of the system.

For example the volume of a system consisting of an ideal gas (where $x = 1$) can be ascertained from the knowledge of mass, temperature and pressure of the system. But this is not possible for a system consisting of ice in equilibrium with water. Because at equilibrium temperature and pressure, the volume of 10 g (say) of ice-water mixture may correspond to different amounts of ice and water having mass equal to 10 g.

4.9   a. Equilibrium situation. Because here the macroscopic properties do not all change with time even on isolating the system from the surroundings. The variation of molecular speed, which is a microscopic property, has no relevance to equilibrium.

   b. None. Because here the system is microscopic for which the macroscopic properties like pressure, temperature (which define equilibrium) have no meaning.

   c. (i) Equilibrium situation, the reason being same as in a. (ii) Steady state situation. Because the macroscopic property, temperature of the thermometer does not change with time so long as it is in contact with the patient's body but not when it is detatched from the latter.

   d. Tiger, as a living being, must exchange matter with the surroundings and hence it is an open system for which equilibrium is hardly possible (vide problem 4.24(m). Tiger may be related to a steady state situation, and that too only temporarily (so long as the tiger is alive).

Note: Tiger's weight fluctuates with time, remaining constant only on the average.

4.10   i. For any infinitesimal change in a closed system

$$dQ = dU + dW, \text{ where } dW \text{ is the work done by the system}$$

Then
$$0 = dU + dW, \text{ for an adiabatic process}$$

$$= dU + P_{ext}dV, \text{ for a system doing only } PV\text{-work}$$

$$= n\bar{C}_V dT + P_{ext}dV \text{ for an ideal gaseous system}$$

$$= n\bar{C}_V dT + PdV \text{ for a (mechanically) reversible process}$$

$$= n\bar{C}_V dT + \frac{nRT}{V}dV \text{ since } PV = nRT \text{ for an ideal gas}$$

$$= \frac{dT}{T} + \frac{R}{\bar{C}_V}\frac{dV}{V} \text{ dividing both sides by } n\bar{C}_V T$$

Integration gives

$$\ln T + (\gamma - 1) \ln V = \ln K, \text{ where } \gamma = C_P/C_V \text{ using } \bar{C}_P - \bar{C}_V = R$$

or $$TV^{\gamma-1} = K, \text{ constant}$$

or $$PV^\gamma = K', \text{ constant replacing } T \text{ by } PV/nR$$

Note: For the equation of state $P(V - nb) = nRT$, the relation $n\bar{C}_V dT + PdV = 0$ holds as in case of an ideal gas, since $(\partial U/\partial V)_T = 0$ for both, intermolecular attraction being absent. Then integration gives $P(V - nb)^\gamma = $ constant.

ii. Starting from the relation $dU + PdV = 0$, we have from the van der Waals equation

$$\left(P + \frac{n^2a}{V^2}\right)(V - nb) = nRT$$

$$\left(\frac{\partial U}{\partial T}\right)_V dT + \left(\frac{\partial U}{\partial V}\right)_T dV + \left(\frac{nRT}{V - nb} - \frac{n^2a}{V^2}\right)dV = 0$$

Now for a van der Waals gas, $\left(\frac{\partial U}{\partial V}\right)_T = \frac{n^2a}{V^2}$ [vide Eq. (5.17)]

Then $\quad n\bar{C}_V dT + \frac{n^2a}{V^2}dV + \frac{nRTdV}{V - nb} - \frac{n^2a}{V^2}dV = 0$

or $\qquad \frac{dT}{T} + \frac{R}{\bar{C}_V} \cdot \frac{dV}{V - nb} = 0$

Integration gives $\qquad \left(P + \frac{n^2a}{V^2}\right)(V - nb)^\gamma = K$, constant

4.11 As it appears from its derivation, the relation PV^γ = constant will hold, if the system consists of a fixed amount of an ideal gas and the process is adiabatic and reversible.
□ Yes. Because, the process being reversible, the system remains virtually in thermodynamic equilibrium at every stage where the equation of state $PV = nRT$ will hold.
□□ The constant in the relation PV^γ = constant depends on all the given factors.
□□□ Yes. The factor, apart from the given ones, is initial pressure or volume.
Note: For an ideal gas $PV = nRT$ holds at every stage of a reversible process, including isothermal and adiabatic. In case of an isothermal process, PV = constant (K) holds (because n and T remain constant) while in case of an adiabatic process $PV \neq$ constant (because T is not constant) but PV^γ = constant (K') holds. K and K' have different values. K, but not K', can be found only from the knowledge of n and T. To find K', we additionally require the knowledge P_i or V_i.

4.12 Slope of the reversible isothermal P–V curve

$$\frac{dP}{dV} = \frac{d(KV^{-1})}{dV}, \text{ where } PV = K$$

$$= -KV^{-2} = -\frac{P}{V}$$

Slope of the reversible adiabatic P–V curve

$$\frac{dP}{dV} = \frac{d(K'V^{-\gamma})}{dV}, \text{ where } PV^\gamma = K'$$

$$= -\gamma K'V^{-\gamma-1} = -\gamma\frac{P}{V}$$

Then $\qquad \left(\frac{dP}{dV}\right)_{ad} < \left(\frac{dP}{dV}\right)_{iso}$ since both the quantities are negative and $\gamma > 1$

or $\qquad \left|\left(\frac{dP}{dV}\right)_{ad}\right| > \left|\left(\frac{dP}{dV}\right)_{iso}\right|$ for same value of P and V

☐ Because in the isothermal process, the fall in pressure is determined only by increase in volume, but in the adiabatic process it is favoured also by the accompanying fall in temperature during expansion.

☐ ☐ No. The given statement is true for expansion but the reverse is true for contraction. This follows from the adjoining diagrams.

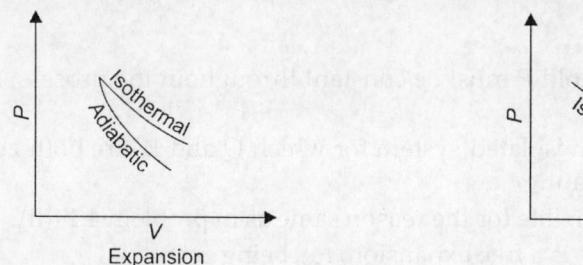

Expansion   Contraction

4.13   i. Possible simply through adiabatic free expansion of the gas, vide problem 4.17(iii).

ii. Possible through Joule-Thomson expansion of the gas at its inversion temperature.

iii. This happens when a gas (at pressure P and temperature T), enclosed in a thermally conducting cylinder by a frictionless piston, is subjected to an external pressure equal to P keeping the cylinder in a heat reservoir at $T + dT$.

[Here the gas absorbs heat till its temperature equals that of the reservoir with consequent rise of its pressure to $P_{ext} + dP$. This causes expansion of the gas till its pressure equals P_{ext} and temperature falls to T. These processes will continue till the desired expansion is achieved by arresting the motion of the piston.]

4.14   i. Irreversible, because of involvement of dissipation effect with ordinary piston-cylinder device that cannot operate without any friction.

ii. Irreversible, because the process is not quasi-static due to appreciable difference between the pressure of the system and the pressure imposed on the system.

iii. Irreversible, because the temperature difference between the two bodies that causes heat flow is appreciable.

iv. Reversible, because during freezing of water at STP (the freezing point of water) its liquid and solid phases remain virtually in thermodynamic equilibrium at all stages.

v. Irreversible, because during freezing of super-cooled water at –10°C (which is not the freezing temperature of water at 1 atm) its liquid and solid phases differ appreciably from being in thermodynamic equilibrium.

vi. Irreversible, because the oscillation of pendulum involves friction at the point of suspension.

4.15   i. The required condition is that the system must be closed, otherwise Q and W will have no meaning.

ii. The necessary conditions—the system must be closed capable of doing only PV-work, and the process must be (mechanically) reversible.

iii. This relation will hold if the system is closed, made up of ideal gases, and the process is adiabatic and (mechanically) reversible.

iv. Same as in iii except that the process need not be reversible.

v. The necessary conditions involved in deriving the relation are indicated below

$$dH = dU + PdV + VdP$$

$= dU + P_{ext}dV$, if the process is isobaric where $P = P_{ext}$ = constant

$= dU - dW$, if the system does only PV-work

$= dQ$

or $\qquad \Delta H = Q$

Note: For $\Delta H = Q_P$ to hold P must be constant throughout the process; mere $P_i = P_f$ is not enough.

4.16 Universe behaves as an isolated system for which Q and W are both zero, whence $\Delta U = 0$, i.e. U is consistant.

4.17 The process is not reversible for the reason same as in problem 4.14(ii).

□ Of course the process is a free expansion, P_{ext} being zero.

□□ i. $W = -\displaystyle\int_{V_i}^{V_f} P_{ext}dV = 0$, since $P_{ext} = 0$.

ii. Under the given condition, $Q = 0$, then $\Delta U = Q + W = 0 + 0$, i.e. U will not change.

iii. Since the gas is ideal $\Delta U = C_V\Delta T = 0$, and hence $\Delta T = 0$, i.e. T will not change.

□□□ The result will be same with i and ii but not with iii.

For a real gas $\Delta U \neq C_V\Delta T$, and hence although $\Delta U = 0$, $\Delta T \neq 0$, i.e. T will change.

4.18 a. Here work done by the gas, $W = \displaystyle\int_{V_i}^{V_f} P_{ext}dV$. Then W will be zero if P_{ext} or dV is zero for each stage of the process.

Now $\qquad T_f/T_i = P_fV_f/P_iV_i$, since the gas is ideal

$= 1$, from the given data.

or $\qquad T_f = T_i$, i.e. the change is isothermal.

ii. We can achieve the change irreversibly simply by connecting the vessel containing the gas, through a stop-cock, to another empty vessel of same volume keeping both the vessels in a thermostat at the temperature of the gas and then opening the stop-cock.

[**Note:** If $T_f \neq T_i$, then the thermostat would have to be maintained at T_f.]

i. We cannot achieve the change reversibly through a single step. However, this can be done through three reversible steps as indicated in the adjoining diagram. Here the first step is isochoric cooling of the gas till its pressure becomes zero. The second step is isobaric expansion at this pressure till volume becomes 2 liters. The third step is isochoric heating till pressure becomes 1 atm.

[Such a multi-step process if not practicable as pressure cannot be made zero exactly. Here the change is isothermal but the process is not.]

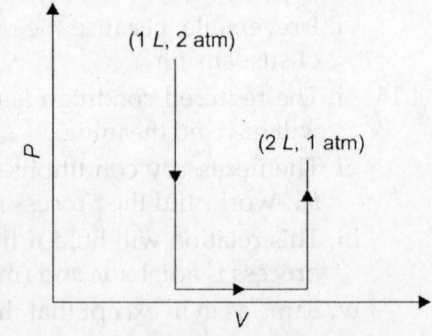

☐ In each case $\Delta U = C_V \Delta T$, since the system consists of ideal gas
$\qquad\qquad\qquad\qquad = 0$, since $\Delta T = 0$
Also, in each case $Q = \Delta U + W = 0 + 0$
Here the two paths have same ΔU and also same Q.

☐☐ For a specified change through different paths, ΔU will always be same, but Q will be same provided W is same as in the given problem.

  b. No. We can understand this considering the P–V diagram for (i) in reverse direction where least work will be involved. Here the second step, which is compression, must involve some work (though other two steps involve no work) because P_{ext} cannot be made exactly zero.

☐ The result leads us to conclude that the work obtained in any spontaneous change of a system is less than the work to be done in restoring the system to its original state.

4.19 Here work done by the gas, $W = \int_{V_i}^{V_f} P dV$.

Since volume change is same in the two cases, W will be lower in case of adiabatic due to lower value of P (for same value of V) resulting from the fall in temperature in the adiabatic expansion.

☐ Since $W = C_V(T_i - T_f)$ for adiabatic change of an ideal gas, T_f would be higher if the process were irreversible involving lower value of W.

4.20 Here work done by the gas, $W = P_{ext}(V_f - V_i)$.

  a. W will be same in all cases except (iv), because P_{ext}, V_i and V_f remain same.

  b. i. No, because P_{ext} and $(V_f - V_i)$ are same in isothermal and adiabetic though V_i and V_f change in each process due to change in amount of the gas.

    ii. No, because P_{ext}, V_i and V_f are not affected.

    iii. Yes, because $(V_f - V_i)$ changes (equation of states for ideal and real gas being different).

    iv. Yes. This is obvious from the expression of work.

    v. Yes. For the same reason as in (iii).

Note: 1. Here the process is not isobaric because pressure of the system does not remain constant, though P_{ext} is constant.

      2. For same volume change, W will be different for isothermal and adiabatic when the process is reversible but not when the process occurs against a constant external pressure.

4.21 The relation $\Delta H = C_P \Delta T$ holds also in case of a real system provided pressure of the system is constant. This follows just from the definition of C_P as $C_P = (\partial H/\partial T)_P$. For an ideal gas P need not be constant because here H is independent of P.

$\Delta H = 0$ for a physical process like Joule-Thomson expansion and chemical process like dextro-laevo transformation in case of optically active substance which involves only change in orientation of the bonds through breaking and making of the chemical bonds.

4.22 For systems doing only PV-work (when $dQ_P = dH$ and $dQ_V = dU$)

$$C_P = \left(\frac{\partial H}{\partial T}\right)_P$$

$$C_V = \left(\frac{\partial U}{\partial T}\right)_V$$

i. Now $\qquad H = U + PV$

or $\qquad dH = dU + PdV + VdP$

$$= \left(\frac{\partial U}{\partial T}\right)_V dT + \left(\frac{\partial U}{\partial V}\right)_T dV + PdV + VdP$$

if the system is hydrostatic, where $U = f(T, V)$.

Dividing both sides by dT at constant pressure, we have

$$\left(\frac{\partial H}{\partial T}\right)_P = \left(\frac{\partial U}{\partial T}\right)_V + \left[\left(\frac{\partial U}{\partial V}\right)_T + P\right]\left(\frac{\partial V}{\partial T}\right)_P$$

or $\qquad C_P - C_V = \left[\left(\frac{\partial U}{\partial V}\right)_T + P\right]\left(\frac{\partial U}{\partial T}\right)_P$

ii. Using the relation between dH and dU, we have

$$\left(\frac{\partial H}{\partial T}\right)_P dT + \left(\frac{\partial H}{\partial P}\right)_T dP = dU + PdV + VdP$$

if the system is hydrostatic where $H = f(T, P)$

Dividing both sides by dT at constant volume and then rearranging

$$C_P - C_V = \left[-\left(\frac{\partial H}{\partial P}\right)_T + V\right]\left(\frac{\partial P}{\partial T}\right)_V$$

☐ It follows from the expression $C_P - C_V = \left[\left(\frac{\partial U}{\partial V}\right)_T + P\right]\left(\frac{\partial V}{\partial T}\right)_P$ that C_P will not always be greater than C_V. C_P will be greater than, equal to or less than C_V according as $\left(\frac{\partial V}{\partial T}\right)_P >, =$ or < 0. Water offers examples of all three cases. Here $\left(\frac{\partial V}{\partial T}\right)_P >, =$ or < 0 according as temperature $>, =$ or $< 4°C$.

☐☐ Yes.

i. This happens with an adiabatically enclosed reaction system that undergoes exothermic reaction raising the temperature of the system.

ii. This happens at the point of any phase change where heat is absorbed without any change of temperature.

iii. This happens with a system that undergoes exothermic chemical change during rise in temperature of the system liberating more heat than that required for its temperature rise.

Note: C_P is greater than C_V for a gas, but not necessarily for a liquid or solid.

4.23 We cannot tackle such questions on the basis of classical thermodynamics which cannot explain the origin of internal energy of a system. The explanation comes from statistical thermodynamics (which, unlike classical thermodynamics, assumes molecular structure of matter) as due to various types of molecular motion (external and internal).

$$C_P - C_V = \left[\left(\frac{\partial U}{\partial V}\right)_T + P\right]\left(\frac{\partial V}{\partial T}\right)_P$$

$$= P\left(\frac{\partial V}{\partial T}\right)_P \text{ for an ideal gas where } \left(\frac{\partial U}{\partial V}\right)_T = 0$$

Now $P\left(\frac{\partial V}{\partial T}\right)_P$, which represents the increase in molecular kinetic energy for unit rise of temperature is determined only by translational degrees of freedom and hence it does not depend on atomicity and so does $C_P - C_V$.

C_V is determined by external as well as internal degrees of freedom and the latter is determined by atomicity. Then C_V and hence $\frac{C_P}{C_V} = 1 + \frac{R}{C_V}$ will depend on atomicity.

4.24   a. In an isobaric process of expansion, the pressure of the system remains constant and is at the same time greater than the external pressure. This is possible only if the external pressure is constant and differs only infinitesimally from the pressure of the system, i.e. if the process is mechanically reversible.

Note: The constancy of external pressure is a necessary, but not sufficient condition for a process to be isobaric.

b. The given statement has absolute justification only with respect to a particular way of change. To illustrate, let us consider the expansion of a gas from the initial state i to the final state f that can be achieved through the reversible path iaf or ibf as indicated in the adjoining diagram. The path ibf gives more work (measured by the area under the line) than the path iaf, and this happens even when the path ibf is considerably irreversible. But, for a particular sequence of isochoric and isobaric process, the work obtained is always greater if the operation is done reversibly when the expansion occurs against maximum pressure.

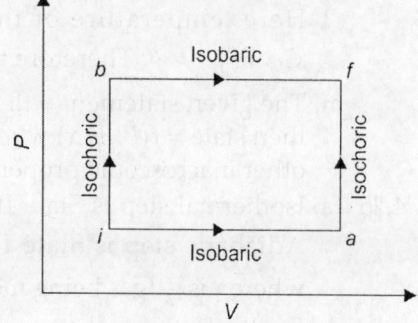

c. The given statement is not justified, because work (W) is not a state function. [But $\oint dW = 0$ cannot be ruled out for all cyclic processes, e.g. this happens with an isothermal reversible cycle and also with an adiabatic reversible cycle but not with their combination in a Carnot's cycle.]

d. Although heat Q depends on the path of change of a system, the given statement will be justified provided work W is different for different path (as it usually happens) because $\Delta U = Q + W$ is fixed for a specified change in state.

e. U is a single-valued function of state, but state is not a single-valued function of U. Hence a change in state may not be accompanied by change in U but the converse will always hold. For example, in the isothermal expansion of an ideal gas U does not change though state changes (due to change in volume). But when U changes state must change (since $\Delta U = C_V\Delta T$) due to change in T. Therefore, the first part of the given statement is not justified though the second part is.

f. In general, a system possesses macroscopic kinetic energy (due to motion of the system through space) and macroscopic potential energy (due to external field) apart from internal energy (determined by the internal state of the system). Then, for an isolated system sum total (E) of all these energies should actually be constant. However, in most applications of thermodynamics, we consider the system at rest ignoring the effect of external field, and only then the given statement will be justified.

g. It is not always justified to regard H as heat content. Because H is a measure of the absorbed heat only when the system is closed doing only PV-work and the process is isobaric.

h. The given statement is not justified because although H is a state function, ΔH is not since it refers to a change of state and not to a particular state.

i. The given statement is justified because $C_P = \left(\frac{\partial H}{\partial T}\right)_P$ and H is independent of P for an ideal gas.

j. The given statement is justified due to involvement of work (stirring work) other than PV-work when $dQ_V \neq dU$.

k. Here temperature of the system does not change on absorption of heat. Hence $C_P = \frac{dQ_P}{dT} = \infty$. Therefore, the given statement is justified.

l. Here temperature of the system changes on absorption of heat. Hence $C_V = \frac{dQ_V}{dT} \neq \infty$. Therefore the given statement is not justified.

m. The given statement will be justified only if the system is closed. Because only then state $= f(P, T, x)$ when constancy of P, T and x amounts to constancy of all other macroscopic properties of the system.

4.25   a. Isothermal step is State II → State III which obeys the relation $PV = $ constant. Adiabatic step is State IV → State I which obeys the relation $PV^\gamma = $ constant where γ is $\frac{5}{3}$ (gas being monatomic).

     b. For the isothermal step

$$W = -P_i V_i \ln \frac{V_f}{V_i}, \text{ according to IUPAC convention}$$

$$= -(3 \text{ atm})(1 \text{ L}) \ln \frac{3L}{1L} = -3.3 \text{ L atm}$$

$$Q = \Delta U - W$$
$$= C_V \Delta T - W, \text{ since the gas is ideal}$$
$$= -W, \quad \text{since } \Delta T = 0, \text{ the process being isothermal}$$

For the adiabatic step

$$W = \frac{P_f V_f - P_i V_i}{\gamma - 1}$$

$$= \frac{(1 \text{ atm})(1 \text{L}) - (32 \text{ atm})\left(\frac{125}{1000} \text{ L}\right)}{\frac{5}{3} - 1} = -6 \text{ L atm}$$

$$Q = 0$$

     c. No, because the quantities involved in the expression of W are all fixed.

     d. $\oint dQ + \oint dW = \oint dU$, by first law

$$= 0, \text{ since } U \text{ is a state function}$$

Note: Knowledge of temperature is not essential for calculation of W.

4.26   i. $W = -P_i V_i \ln \frac{V_f}{V_i}$

$$= -(2 \text{ atm})(11.2 \text{ L}) \ln 2$$
$$= -15.5 \text{ L atm}$$

$$\Delta U = n\bar{C}_V \Delta T$$

$$= 0, \text{ since } \Delta T = 0 \text{ the process being isothermal}$$

$$\Delta H = n\bar{C}_P \Delta T = 0$$

$$Q = \Delta U - W$$

$$= -W$$

ii. $\qquad\qquad Q = 0$

$$P_f = P_i \left(\frac{V_i}{V_f} \right)^\gamma$$

since the gas is ideal and the process is adiabatic and reversible

$$= (2 \text{ atm}) \left(\tfrac{1}{2} \right)^{\frac{5}{3}} = 0.63 \text{ atm}$$

$$W = \frac{P_f V_f - P_i V_i}{\gamma - 1}$$

$$= \frac{(0.63 \text{ atm}) (22.4 \text{ L}) - (2 \text{ atm}) (11.2 \text{ L})}{\frac{5}{3} - 1} = -12.4 \text{ L atm}$$

$$\Delta U = Q + W = 0 + W$$

$$\Delta H = \gamma \Delta U \quad \left(\text{since } \frac{\Delta H}{\Delta U} = \frac{C_P \Delta T}{C_V \Delta T} = \gamma \text{ for an ideal gas} \right)$$

iii. $\qquad W = -\int_{V_i}^{V_f} P dV$

$$= -\int_{V_i}^{V_f} KV dV, \text{ since } \tfrac{P}{V} = K, \text{ constant}$$

$$= -\tfrac{1}{2} |KV^2|_{V_i}^{V_f} = -\tfrac{1}{2} |PV|_{V_i}^{V_f} = \tfrac{1}{2}(P_i V_i - P_f V_f) \qquad\qquad \text{(I)}$$

$$= \tfrac{1}{2} \left(P_i V_i - \frac{P_i V_f^2}{V_i} \right)$$

$$= \tfrac{1}{2} [(2 \text{ atm}) (11.2 \text{ L}) - (2 \text{ atm}) (22.4 \text{ L})^2 / (11.2 \text{ L})]$$

$$\Delta U = n\bar{C}_V (T_f - T_i) = n \cdot \tfrac{3}{2} R(T_f - T_i) = \tfrac{3}{2}(P_f V_f - P_i V_i) = -3W \qquad \text{from (I)}$$

$$\Delta H = \gamma \Delta U = -\tfrac{5}{3}(3W) = -5W$$

$$Q = \Delta U - W = -3W - W = -4W$$

iv. Here, unlike (iii), P is given as a function of V; we have to derive the required function, thus

$$\left. \begin{array}{l} PT = \text{constant} \\ PV = nRT \end{array} \right\} \Rightarrow P = KV^{-\frac{1}{2}}, \text{ where } K \text{ is a constant}$$

Then proceeding as in (iii)

$$W = -\int_{V_i}^{V_f} P dV = 2(P_i V_i - P_f V_f)$$

$$\Delta U = n\bar{C}_V (T_f - T_i) = \tfrac{3}{2}(P_f V_f - P_i V_i) = -\tfrac{3}{4} W$$

Note: In case of (ii), W can also be calculated using the relation $W = n\bar{C}_V(T_f - T_i)$. But this is not convenient. Because we are to determine first T_i by relation $T_i = \frac{P_i V_i}{nR}$ and then T_f by $T_f = T_i \left(\frac{V_i}{V_f}\right)^{\gamma-1}$

4.27   a. The nature of the cycle in V vs T diagram is depicted below.

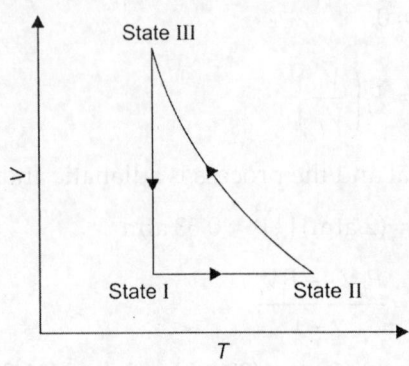

b. For the isochoric step, State I → State II
$$W = -P\Delta V = 0$$
$$\Delta U = n\bar{C}_V \Delta T$$
$$= (1 \text{ mol}) (3.0 \text{ cal K}^{-1}\text{mol}^{-1}) (546 \text{ K} - 273 \text{ K}) = 819 \text{ cal}$$
$$Q = \Delta U - W = (819 - 0) \text{ cal}$$
For the adiabatic step, State II → State III
$$\Delta U = n\bar{C}_V \Delta T$$
$$= (1 \text{ mol}) (3.0 \text{ cal K}^{-1}\text{mol}^{-1}) (273 \text{ K} - 546 \text{ K}) = -819 \text{ cal}$$
$$W = \Delta U - Q = (-819 - 0) \text{ cal}$$
$$\gamma = 1 + \frac{R}{\bar{C}_V} = 1 + \frac{2}{3} = \frac{5}{3}$$
$$P_{III} = P_{II}\left(\frac{T_{II}}{T_{III}}\right)^{\frac{\gamma}{1-\gamma}}$$
$$= (2 \text{ atm})\left(\tfrac{546}{273}\right)^{\frac{5/3}{1-5/3}} = 0.35 \text{ atm}$$

Using the relation $PT^{\gamma/1-\gamma} =$ constant. $P_{II} = 2$ atm, since $P \propto T$ at constant V.
For the isothermal step, State III → State I,
$$W = -nRT \ln\frac{P_i}{P_f} = -nRT_{III} \ln\frac{P_{III}}{P_I}$$
$$= -(1 \text{ mol}) (2 \text{ cal K}^{-1}\text{mol}^{-1}) (273 \text{ K}) \ln\tfrac{0.35}{1} = 573.3 \text{ cal}$$
$$\Delta U = n\bar{C}_V \Delta T = 0$$
$$Q = \Delta U - W = -W = 573.3 \text{ cal}$$
For the cycle $W =$ sum of W for the three steps
$$= [0 + (-819) + 573.3] \text{ cal} = -245.7 \text{ cal}$$
$Q =$ sum of Q for the three steps
$$= [819 + 0 + (-573.3)] \text{ cal} = 245.7 \text{ cal}$$
$\Delta U =$ sum of ΔU for three steps
$$= (819 - 819 + 0) \text{ cal} = 0$$

c. For the entire cycle $\Delta U = 0$ but $W \neq 0$ and also $Q \neq 0$.

This indicates that internal energy is a state function but work and heat are not.

4.28 Here the process is irreversible where V_f is known but P_f unknown in case of (a) while P_f is known but V_f are unknown in case of (b) which is treated first being more general in nature.

b.
$$W = -P_{ext}\Delta V, \text{ by IUPAC convention}$$

$$= P_f (V_i - V_f) \quad \text{(I)}$$

$$= P_f (V_i - V_f), \text{ since } P_{ext} = P_f = \text{constant}$$

$$\Delta U = n\bar{C}_V (T_f - T_i), \text{ since the gas is ideal}$$

$$= \tfrac{3}{2} nR (T_f - T_i) \quad \text{(II)}$$

Now $\Delta U - W = Q = 0$ (III)

Then $\tfrac{3}{2} nR (T_f - T_i) - P_f (V_i - V_f) = 0$

or $\dfrac{3}{2} nR (T_f - T_i) - P_f \left(\dfrac{nRT_i}{P_i} - \dfrac{nRT_f}{P_f} \right) = 0$

or
$$T_f = \frac{2}{5} \left(\frac{P_f}{P_i} + \frac{3}{2} \right) T_i$$

$$= \frac{2}{5} \left(\frac{1 \text{ atm}}{10 \text{ atm}} + \frac{3}{2} \right) (27 + 273) \text{ K} = 192 \text{ K}$$

$$\Delta U = \frac{3}{2} \left(\frac{P_i V_i}{T_i} \right) (T_f - T_i), \text{ from (II)}$$

$$= \frac{3}{2} \cdot \frac{(10 \text{ atm})(1 \text{ L})}{(27 + 273) \text{ K}} (192 - 300) \text{ K} = -5.40 \text{ L atm}$$

$$W = \Delta U \quad \text{from (III)}$$

$$= -5.40 \text{ L atm}$$

a.
$$W = P_{ext} (V_i - V_f), \quad \text{by (I)}$$

$$= -P_{ext} \cdot V_i, \quad \text{since } V_f = 2V_i$$

$$= -(1 \text{ atm})(1 \text{ L}) = -1 \text{ L atm}$$

$$\Delta U = W \quad \text{by (III)}$$

$$= -1 \text{ L atm}$$

$$T_f = T_i + \frac{2\Delta U}{3nR}, \quad \text{by (II)}$$

$$= T_i + \frac{2\Delta U T_i}{3P_i V_i}$$

$$= T_i \left(1 + \frac{2}{3} \cdot \frac{\Delta U}{P_i V_i} \right)$$

$$= (300 \text{ K}) \left[1 + \frac{2}{3} \cdot \frac{(-1 \text{ L atm})}{(10 \text{ atm})(1 \text{ L})} \right] = 280 \text{ K}$$

☐ In case of (a), W and ΔU will not be affected because the quantities involved in their expressions are all specified. But T_f will be different with non-ideal gas for which the relation (II) does not hold.

In case of (b), W, ΔU and T_f will all be different with non-ideal gas, V_f being different due to different equation of state.

Note: i. Here, the process being not reversible, T_f cannot be calculated using the relation $T_f = T_i(V_i/V_f)^{\gamma-1}$ in case of (a) and $T_f = T_i(P_i/P_f)^{\frac{\gamma-1}{\gamma}}$ in case of (b).

ii. In (a) calculations are done in the sequence W, ΔU and T_f while in (b) the sequence is reverse.

4.29 Available work $\quad W = P(V_f - V_i)$
$$= nR(T_f - T_i), \text{ since } PV = nRT, \text{ for an ideal gas}$$
$$\Delta U = n\bar{C}_V(T_f - T_i)$$

i. Here $\Delta U = 0$, since $T_f = T_i$

Then $\quad\dfrac{W}{Q} = \dfrac{W}{\Delta U + W} = \dfrac{W}{0 + W} = 1$

ii. Here $\quad\dfrac{W}{Q} = \dfrac{W}{\Delta U + W} = \dfrac{R}{\bar{C}_V + R}$

$$= \frac{1}{\frac{3}{2}+1} \quad \text{since } \bar{C}_V = \tfrac{3}{2}R, \text{ for an ideal monatomic gas}$$

$$= \tfrac{2}{5}$$

☐ With change in atomicity the result will be affected in case of (ii) due to change in $\bar{C}_V$, but not in case of (i)

Note: 1. Available work is maximum in case of isothermal heating.

2. In case of isobaric heating, the available work will decrease with increase in atomicity, because of increase in $\bar{C}_V$ due increase in degrees of freedom.

4.30 $$\Delta H = n\int_{T_i}^{T_f} \bar{C}_P dT$$

$$= (1 \text{ mol})\left[\int_{27+273}^{327+273} 20.5\, dT + 0.045\int_{27+273}^{327+273} T\, dT\right] \text{J mol}^{-1} \doteq 12225 \text{ J}$$

$$\Delta U = n\int_{T_i}^{T_f} \bar{C}_V dT = n\int_{T_i}^{T_f} \bar{C}_P - R)dT$$

$$= \Delta H - nR\int_{T_i}^{T_f} dT = \Delta H - nR(T_f - T_i)$$

$$= 12225 \text{ J} - (1 \text{ mol})(8.314 \text{ JK}^{-1}\text{mol}^{-1})(300 \text{ K}) = 9730.8 \text{ J}$$

Maximum possible $Q = \Delta U$, by Eq. (4.4a).

This corresponds to minimum value of work (W) done by the system [as in problem 18(a)].

4.31 Here the data are given in terms of thermodynamic quantities Q and ΔH.

a. $$\Delta U = \frac{\Delta H}{\gamma}, \quad \text{since the gas is ideal}$$

$$= \frac{500 \text{ cal}}{\frac{5}{3}} = 300 \text{ cal}$$

$$W = \Delta U - Q, \text{ by IUPAC convention}$$
$$= 300 \text{ cal} - 400 \text{ cal} = -100 \text{ cal}$$

b. The specified change can be achieved in innumerable ways under the specified condition. The simplest one is shown in the adjoining diagram by solid line which involves first isothermal expansion and then isochoric rise of temperature. Here work is involved only in the first step and is given by

$$W = -nRT \ln V_f/V_i$$
$$= -(1 \text{ mol}) (2 \text{ cal K}^{-1} \text{ mol}^{-1}) \ln 2$$
$$= -1.39 \text{ cal}$$

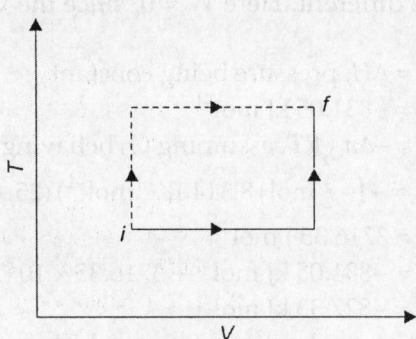

ΔU and ΔH, being independent of path, will be same as in (a).

But Q, being path dependent will be different and this will be

$$Q = \Delta U - W = 300 \text{ cal} - (-1.39 \text{ Cal})$$
$$= 301.39 \text{ cal}$$

Notes: i. Here the given change involves rise in temperature, ΔH being positive.

ii. If the sequence of the two steps in the new situation is reversed (shown by broken line), calculation of W, will be somewhat lengthy. Here, first we are to calculate T_f from $T_f - T_i = \frac{\Delta H}{C_p}$ and then W.

4.32 The specified change can be achieved through the following two reversible steps, using cylinder–piston device.

1. H_2O (l, 100°C, 1 atm) $\rightarrow$ H_2O (g, 100°C, 1 atm)
2. H_2O (g, 100°C, 1 atm) $\rightarrow$ H_2O (g, 100°C, 0.5 atm)

□ For step 1 $W_1 = -P(n\bar{V}_g - n\bar{V}_l)$, where $\bar{V}$ denotes molar volume

$$\approx -nP\bar{V}_g, \text{ since } \bar{V}_g \gg \bar{V}_l$$
$$= -nRT \quad \text{assuming water vapour to behave ideally}$$
$$= -(1 \text{ mol}) (2 \text{ cal K}^{-1}\text{mol}^{-1}) (100 + 273) \text{ K} = -746 \text{ cal}$$

$\Delta H_1 = Q_P = \text{Mass} \times \text{latent heat per unit mass}$
$$= (18 \text{ g}) (540 \text{ cal g}^{-1}) = 9720 \text{ cal}$$

$\Delta U_1 = Q + W = 9720 \text{ cal} - 746 \text{ cal} = 8974 \text{ cal}$

For step 2 $W_2 = -nRT \ln \dfrac{P_i}{P_f} = -(1 \text{ mol}) (2 \text{ cal K}^{-1}\text{mol}^{-1}) (373 \text{ K}) \ln \dfrac{1 \text{ atm}}{0.5 \text{ atm}} = -517 \text{ cal}$

$$\Delta U_2 = C_V\Delta T = 0$$
$$\Delta H_2 = C_P\Delta T = 0$$

Then for the overall process

$$W = W_1 + W_2 = -746 \text{ cal} + -517 \text{ cal}$$
$$\Delta U = \Delta U_1 + \Delta U_2 = 8974 \text{ cal} + 0$$
$$\Delta H = \Delta H_1 + \Delta H_2 = 9720 \text{ cal} + 0$$

□□ Since W and not ΔU and ΔH, depends on nature of the process, only W would be affected if the process were irreversible.

Note: $\Delta U = C_V \Delta T$ applies to the step (2) but not to (1)

$\Delta H = Q$ applies to step (1) but not to (2).

4.33 Here the vaporisation occurs irreversibly. However, since the vapour is produced at the same temperature (100°C), the accompanying ΔH and ΔU will be same as the corresponding ones in the previous problem, provided water vapour behaves ideally. But W will be different. Here $W = 0$, since the vaporisation occurs at zero external pressure.

4.34   a. Here
$$Q = \Delta H, \text{ pressure being constant}$$
$$= -831.05 \text{ kJ mol}^{-1}$$
$$W = -P\Delta V = -\Delta n_{O_2} RT, \text{ assuming } O_2 \text{ behaving ideally}$$
$$= -\left(-\tfrac{3}{2} \text{ mol}\right) 8.314 \text{ JK}^{-1}\text{mol}^{-1}) (25 + 273) \text{ K}$$
$$= 3716.35 \text{ J mol}^{-1}$$
$$\Delta U = Q + N = -831.05 \text{ kJ mol}^{-1} + 3716.35 \times 10^{-3} \text{ kJ mol}^{-1}$$
$$= -827.33 \text{ kJ mol}^{-1}$$

b. The standard heat of combustion of iron, which is ΔH per mole of Fe at 1 atm pressure and 25°C, is half of ΔH for the given reaction, i.e. $-831.05/2$ or $-415.53 \text{ kJ mol}^{-1}$.

c. The standard heat of formation of Fe_2O_3, which is ΔH for mole of Fe_2O_3 formed at 1 atm and 25°C, is same as ΔH for the given reaction, i.e. $-831.05 \text{ kJ mol}^{-1}$.

d. Since by convention $\overline{H}^0_{Fe} = 0$ and $\overline{H}^0_{O_2} = 0$ at 25°C, the conventional enthalpy of Fe_2O_3 for the given temperature and pressure will be the given ΔH, i.e. $-831.05 \text{ kJ mol}^{-1}$.

4.35   a. Because unlike chemical forces involved in combustion, the forces involved in dissolution (which is regarded as a physical process) are unsaturated.

b. Because in solution, strong acids and strong bases completely dissociate and exist in form of ions. Then neutralisation between them amounts to combination of H^+ (from acid) and OH^- (from base) to form water as represented below, for example in case of neutralisation between hydrochloric acid and sodium hydroxide

$H^+ + Cl^- + Na^+ + OH^- = Na^+ + Cl^- + H_2O$

or $H^+ + OH^- = H_2O$

But weak acids and weak bases are only partially dissociated. Here the neutralisation process involves two steps: (i) dissociation of the acid or base and (ii) combination of the resulting H^+ and OH^- ions. The heat liberated in (ii) is partly utilised in the process (i). Then the net heat evolved will be less than that expected for strong acid–strong base neutralisation by an amount determined by the characteristic heat of dissociation of the weak acid/base involved.

4.36 CH_3CO_2H. Because, being weaker than HCO_2H, it has higher heat of dissociation. Hence it liberates less heat on neutralisation making ΔH of the process less negative, i.e. greater.

4.37 Since HCl is a strong acid and NaOH is a strong base, heat of neutralisation between them will correspond to the reaction H^+ (ion) $+ OH^-$ (aq) $= H_2O$ (l), which is reverse of reaction for which ΔH has to be found out. Then, the required ΔH will be negative of the given heat of neutralisation, i.e. $26.45 \text{ kJ mol}^{-1}$.

4.38 The required order is C–C < C = C < C ≡ C, in which bond-order increases.

No. This is not unlikely. Because reactivity, which is not a thermodynamic 'property', has no necessary connection with bond energy.

4.39 Hess's law in form of the given statement is justified only for thermochemical reactions (where the involved work is only due to volume change). Because, only for such reaction $Q_V (= \Delta U)$ and $Q_P (= \Delta H)$ will be independent of the path followed. For example, the reaction $CuSO_4 (aq) + Zn \rightarrow ZnSO_4 (aq) + Cu$ can be brought about by direct contact between the reactants (when only PV-work is involved) or through Daniel cell (which involves electrical work along with small amount of PV-work). Here Q_P is different in the two cases. But when the statement is modified using ΔH or ΔU, instead of absorbed heat, its validity will not be limited only to the thermochemical reactions.

4.40 For a reaction involving a gas phase

$$Q_P = Q_V + \Delta v RT$$

where Δv is the increase in number of moles in the gas phase (assumed to behave ideally).

Since the reaction (i) involves a gas (H_2), $Q_P > Q_V$ (Δv being positive), and hence Q_V, which is heat evolved at constant volume, will be greater than Q_P which is heat evolved at constant pressure (i.e. in open vessel).

But the reaction (ii) involves no gas and hence $Q_P = Q_V$, i.e. heat evolved will be same whether the reaction is carried out in a closed bomb or in an open vessel.

4.41   i. Here mass of the solvent is 1 kg. Then $m = 1$. On putting this in the given expression, we have the following required value of ΔH.

$$\Delta H = 3.86 + 1.99 - 3.04 + 1.02 \text{ kJ mol}^{-1} = 3.83 \text{ kJ mol}^{-1}$$

ii. For infinitely dilute solution of 1 mol of NaCl

Integral heat of solution = differential heat of solution

$$= \left(\frac{\partial \Delta H}{\partial m} \right)_{composition} \quad \text{for } m \rightarrow 0$$

$$= 3.86 \text{ kJ mol}^{-1}$$

iii.   Differential heat of solution $= \left(\frac{\partial \Delta H}{\partial m} \right)_{composition} \quad \text{for } m = 1$

$$= \left(3.86 + \tfrac{3}{2} \times 1.99 - 2 \times 3.04 + \tfrac{5}{2} \times 1.02 \right) \text{kJ mol}^{-1}$$

$$= 3.315 \text{ kJ mol}^{-1}$$

iv. Required heat of dilution

= difference between the heats of solution for $m = 0.01$ and $m = 1$

$$= \tfrac{3}{2} \times 1.99 \left[(0.01)^{\frac{1}{2}} - 1^{\frac{1}{2}} \right] - 2 \times 3.04 \, [0.01 - 1] + \tfrac{5}{2} \times 1.02 \left[(0.01)^{\frac{3}{2}} - 1^{\frac{3}{2}} \right]$$

$$= 0.785 \text{ kJ mol}^{-1}$$

4.42   i. Integral heat of solution of H_2SO_4

$$= \frac{\Delta H \text{ for the given solution}}{\text{Number of moles of } H_2SO_4 \text{ mixed}}$$

$$= \frac{\left(-\frac{75.5}{1 + 1.8/10} \right) \text{kJ}}{1 \text{ mol}} \quad \text{putting } n = 10 \text{ in the given expression for } \Delta H$$

$$= -63.98 \text{ kJ mol}^{-1}$$

Differential heat of solution of H_2SO_4

$$= \left[\frac{\partial \Delta H}{\partial (1/n)} \right]_{composition}$$

$$= -\left[\frac{\partial}{\partial (1/n)} \left(\frac{1}{n} \cdot \frac{75.5}{1+1.8/n} \right) \right]_{composition}$$

(assuming that the solution of given composition contains $1/n$ mol H_2SO_4 and 1 mol H_2O)

$$= -\frac{75.5}{1+1.8/n} + \frac{75.5 \times 1.8}{n(1+1.8/n)^2}$$

$$= -\frac{75.5}{1+1.8/n} + \frac{75.5 \times 1.8}{n(1+1.8/10)^2}, \text{putting } n = 10$$

$$= -54.22 \text{ kJ mol}^{-1}$$

ii. Integral heat of solution of H_2O

$$= \frac{-\dfrac{75.5}{1+1.8/n}}{n}$$

$$= \frac{-\dfrac{75.5}{1+1.8/10}}{10 \text{ mol}} \text{ kJ} = -6.398 \text{ kJ mol}$$

Differential heat of solution of H_2O

$$= \left(\frac{\partial \Delta H}{\partial n} \right)_{composition}$$

$$= -\left[\frac{\partial}{\partial n} \left(\frac{75.5}{1+1.8/n} \right) \right]_{composition}$$

$$= -\frac{75.5 \times 1.8}{(1+1.8/n)^2 n^2}$$

$$= -\frac{75.5 \times 1.8}{(1+1.8/10)^2 10^2} = -0.976 \text{ kJ mol}^{-1}$$

Note: For a particular solution, the heat of solution will be different for different ingredients unless the ingredients are in same proportion by mole.

4.43 By the given statement, we mean that there will be a heat change (enthalpy change) of -8.7 kcal in the formation of 1 mole of HBr at 25°C and 1 atm (more precisely 1 bar) from its constituent elements H_2 (g) and Br_2 (l) at the temperature and pressure mentioned.

☐ From the given data at 25°C, we can write

i. $\dfrac{1}{2}H_2(g) + \dfrac{1}{2}Br_2(l) = HBr(g)$, $\Delta H^\circ_{(i)} = -8.7$ kcal/mol

ii. $\dfrac{1}{2}Br_2(l) = \dfrac{1}{2}Br_2(g)$, $\Delta H^\circ_{(ii)} = \dfrac{7.3}{2}$ kcal/mol

Since stable state of aggregation of bromine at 100°C is gaseous, the standard heat of formation ($\Delta H°$) of HBr at this temperature will correspond to the following equation

$$\tfrac{1}{2}H_2(g) + \tfrac{1}{2}Br_2(g) = 2HBr\,(g)$$

This equation can be obtained from the combination (i)–(ii). Hence

$$\Delta H° = \Delta H_{(i)} - \Delta H_{(ii)}$$
$$= (-8.7 - 7.3/2)\,kcal/mol$$
$$= -12.35\,kcal/mol\ at\ 25°C$$

Now, according to the Kirchhoff's equation, $\Delta H°$ for the concerned reaction at 100°C (373.15 K) is related to its value at 25°C (298.15 K) by

$$\Delta H°_{373.15} = \Delta H°_{298.15} + \int_{298.15}^{373.15}\left[(\bar{C}_P)_{HBr} - \tfrac{1}{2}(\bar{C}_P)_{H_2} - \tfrac{1}{2}(\bar{C}_P)_{Br_2}\right]dT$$

$$= \left[-12.35 \times 10^3 + \int_{298.15}^{373.15}\left(6.58 - \tfrac{6.95}{2} - \tfrac{8.42}{2}\right)dT +\right.$$

$$\left.\int_{298.15}^{373.15}\left(0.95 + \tfrac{0.20}{2} - \tfrac{0.97}{2}\right) \times 10^{-3}\,T dT\right]cal/mol$$

$$= -12425\,cal/mol$$

Note: In applying Kirchhoff equation, one must not forget that if any of substance(s) involved in a chemical reaction (like Br_2 in the above reaction) changes its state of aggregation in the temperature interval concerned, the corresponding change of enthalpy must be included.

4.44 From the given data, we can write

i. $C(s) + \tfrac{1}{2}O_2(g) = CO(g)$, $\Delta H_{(i)} \equiv 26.4\,kcal/mol$

ii. $H_2(g) + \tfrac{1}{2}O_2(g) = H_2O(g)$, $\Delta H_{(ii)} = -58.08\,kcal/mol$

Now (i) – (ii) gives, $H_2O(g) + C(s) = H_2(g) + CO(g)$

Then ΔH for this reaction will be given by

$$\Delta H = \Delta H_{(i)} - \Delta H_{(ii)}$$
$$= [-26.4 - (-58.08)]\,kcal/mol = 31.68\,kcal/mol$$

☐ From ΔH thus calculated, it follows that the given reaction will be accompanied by absorption of 31.68 kcal of heat per mole of H_2O reaction. To maintain a constant temperature, this amount of heat has to be produced by reaction (i), which is exothermic. Now production of 31.68 kcal of heat by (i) requires the reaction of $\tfrac{1}{2} \times 31.68/26.4$ mol of O_2 which is present in $5/2 \times 31.68/26.4$ or 3 mol of air (since air contains 20% by mol of O_2). Then the required proportion by volume of air and steam will be 3:1 (volume being proportional to mole number at constant T and P).

4.45 From the given data we can write

i. $\quad H_2(g) + \tfrac{1}{2}O_2(g) = H_2O(l)$ $\qquad\qquad\qquad \Delta H_{(i)} = -285.8\,kJ/mol$

ii. $\quad CH_4(g) + 2O_2(g) = CO_2(g) + 2H_2O(l)$ $\qquad \Delta H_{(ii)} = -890.0\,kJ/mol$

iii. $\quad C_2H_6(g) + \tfrac{7}{2}O_2(g) = 2CO_2(g) + 3H_2O(l)$ $\qquad \Delta H_{(iii)} = -1560\,kJ/mol$

iv. $\quad C\,(graphite) + O_2(g) = CO_2(g)$ $\qquad\qquad\qquad \Delta H_{(iv)} = -393.5\,kJ/mol$

v. $\quad 3C\,(graphite) + 4H_2(g) = C_3H_8(g)$ $\qquad\qquad \Delta H_{(v)} = -103.8\,kJ/mol$

We have to determine ΔH of the reaction

$$C_3H_8(g) + H_2(g) = C_2H_6(g) + CH_4(g)$$

This equation can be obtained from the combination–(v)–(iii)–(ii) + 3(iv) + 5(i). Then the required ΔH will be given by

$$\Delta H = -\Delta H_{(v)} - \Delta H_{(iii)} - \Delta H_{(ii)} + 3\Delta H_{(iv)} + 5\Delta H_{(i)}$$
$$= (103.8 + 1560.0 + 890.0 - 3 \times 393.5 - 5 \times 285.3) \, \text{kJ/mol}$$
$$= -55.7 \, \text{kJ/mol}$$

Note: Here the expression of ΔH cannot be so readily obtained as in previous problems. To solve such problems, first the principal chemical equation (whose ΔH is to be calculated) and the auxiliary chemical equation (whose thermochemical data are given) are set up. Then the reactants and products in the principal reaction are marked, which appear in only one of the auxiliary reactions. The corresponding auxiliary equations are then properly combined (considering the coefficients of the reactants and products). From the equation thus formed, the substances not involved in the princhpal chemical equation are removed using only the rest of the auxiliary chemical equations, and therefrom the required expression of ΔH is found. This general procedure is followed particularly when auxiliary equations are not small in number as in previous problems.

4.46   a. From the given data, we can write

   i. $H_2(g) + \frac{1}{2}O_2(g) = H_2O(g)$ $\qquad\qquad\qquad$ $\Delta H_{(i)} = -57.8 \, \text{kcal/mol}$

   ii. $H_2(g) = 2H(g)$ $\qquad\qquad\qquad\qquad$ $\Delta H_{(ii)} = 103 \, \text{kcal/mol}$

   iii. $O_2(g) = 2O(g)$ $\qquad\qquad\qquad\qquad$ $\Delta H_{(iii)} = 34 \, \text{kcal/mol}$

   O–H bond energy is half of ΔH for the reaction

   $$H_2O(g) = 2H(g) + O(g)$$

   This equation results from (ii) + $\frac{1}{2}$ (iii) – (i). Then

   $$\Delta H = \Delta H_{(ii)} + \frac{1}{2}\Delta H_{(iii)} - \Delta H_{(i)}$$
   $$= (103 + \tfrac{34}{2} + 57.8) \, \text{kcal/mol}$$
   $$= 177.8 \, \text{kcal/mol}$$

   Therefore, the O–H bond energy $= \Delta H/2 = 177.8/2 \, \text{kcal/mol}$

   b.   i. Here 2 mol of H_2 reacts (assuming the reaction to be complete) according to the equation

   $$2H_2(g) + O_2(g) = 2H_2O(g)$$

   Then required

   $$\Delta H = 2\Delta H_{(i)}, \text{ under isothermal condition}$$
   $$= 2 \times (-57.8) \, \text{kcal/mol} = -115.6 \, \text{kcal/mol}$$
   $$\Delta U = \Delta H - \Delta v \, RT$$
   $$= -115.6 \times 10^3 \, \text{cal/mol} - (-1)(2 \, \text{cal K}^{-1}\text{mol}^{-1})(25 + 273) \, \text{K}$$
   $$= -115004 \, \text{cal/mol}$$

   Under adiabatic condition

   $$\Delta U = Q_V = 0$$
   $$\Delta H = \Delta U + \Delta(PV)$$
   $$= \Delta U + R(n_f T_f - n_i T_i), \, T_f \text{ is obtainable from b(ii)}$$

ii. When the reaction occurs adiabatically, the heat evolved in the reaction raises the temperature of the gas mixture which consists of 2 mol H_2O (produced) and 10 mol O_2 (remaining unreacted)

Heat evolved $= -\Delta U$ under isothermal condition

$$= \left[2(\bar{C}_V)_{H_2O} + 10(\bar{C}_V)_{O_2}\right](T_f - T_i)$$

$$= \left[2(\bar{C}_P)_{H_2O} + 10(\bar{C}_P)_{O_2} - 12R\right](T_f - T_i)$$

Note: ΔU and ΔH of a reaction are different in isothermal and adiabatic conditions. At constant volume, $\Delta U = 0$ for all reactions under adiabatic condition. Normally, by ΔU of a reaction, we mean it under isothermal condition where it is a characteristic constant for the reaction.

4.47 In the given polymerisation process one C=C bond is replaced by two C–C bonds. Then the required ΔH is given by

$$\Delta H = -2\epsilon_{C-C} + \epsilon_{C=C}, \text{ where } \epsilon \text{ represents bond energy}$$
$$= (-2 \times 331 + 590) \text{ kJ/mol}$$

Note: From the given representation of the polymerisation reaction, it might seem that 1 C=C bond is replaced by 3 C–C bonds. But this is not really so, because the structural unit which is repeated here is CH_2–CH_2– and not –CH_2–CH_2–.

4.48 From the given data, we can write

i. $\quad C_6H_6 \text{ (g)} = 6C \text{ (g)} + 6H \text{ (g)} \quad \Delta H_{(i)} = 3\epsilon_{C-C} + 3\epsilon_{C=C} + 6\epsilon_{C-H}$

    Cyclohexatriene $\qquad\qquad\qquad = (3 \times 331.4 + 3 \times 591.1 + 6 \times 416.3) \text{ kJ mol}$

$\qquad\qquad\qquad\qquad\qquad\qquad\qquad = 5265.3 \text{ kJ/mol}$

ii. C (graphite)= C (g) $\qquad\qquad \Delta H_{(ii)} = 718.4 \text{ kJ/mol}$

iii. $\qquad H_2 \text{ (g)} = 2H \text{ (g)} \qquad\qquad \Delta H_{(iii)} = 435.9 \text{ kJ/mol}$

ΔH for formation of cyclohexatriene correspopnds to the reaction

$$6C \text{ (graphite)} + 3H_2 \text{ (g)} = C_6H_6 \text{ (g)}, \text{ benzene}$$

This reaction results from the combination 6(ii) + 3(iii) – (i)

Then $\qquad\qquad \Delta H = 6\Delta H_{(ii)} + 3\Delta H_{(iii)} - \Delta H_{(i)}$

$\qquad\qquad\qquad = (6 \times 718.4 + 3 \times 435.9 - 5265.3) \text{ kJ/mol}$

$\qquad\qquad\qquad = 352.8 \text{ kJ/mol}$

It is given that ΔH_f° (benzene, g) = 82.9 kJ/mol. This means benzene is more stable than cyclohexatriene by (352.8 – 82.9) or 269.9 kJ/mol which is the required resonance energy.

Note: Benzene is not same a cyclohexatriene. The former is more stable.

UNIVERSITY QUESTIONS

4.1 Classify the following systems as open, closed or idolated: Thermoflask, human body, a solution being boiled under reflux, clinical thermometer.

(Calcutta, BSC(H), 2008)

4.2 Justify or criticise: A system must be isolated if neither heat nor matter can enter or leave the system. (Calcutta BSC(H), 2002)

4.3 For which of the following systems the energy is conserved in every process?

(i) a closed system (ii) an open system (iii) an isolated system (iv) a system enclosed in adiabatic walls. (Calcutta BSc(H), 2004)

4.4 Classify each of the following as either intensive or extensive properties: Surface tension, molar volume, enthalpy, dielectric constant. (Jadavpur BSc(H), 2013)

4.5 Explain briefly with an example the meaning of a thermodynamically reversible process.

An ideal gas undergoes a free expansion into an evacuated vessel. State with reason whether this is a reversible or an irreversible process. (Calcutta BSc(H), 1993)

4.6 "A strictly thermodynamically reversible expansion of a gas in a cylinder fitted with an airtight piston is not possible." Justify. (Burdwan BSc(H), 2008)

4.7 Classify each of the following processes as reversible or irreversible:

(i) Freezing of water at 0°C and 1 atm (ii) Freezing of supercooled water at –10°C and 1 atm. (Calcutta BSc(H), 2007)

4.8 Explain with reason whether equilibrium exists in the following processes.

(i) Some water is boiling in a pressure cooker and the check-valve is closed (ii) Some water is boiling in an open beaker. (Burdwan BSc(H), 2008)

4.9 The work involved in an adiabatic expansion of a gas is independent of the path. Justify or criticise. (Calcutta BSc(H), 2014)

4.10 'Work done' may be a state function. Justify or criticise. (Calcutta BSc(H), 2008)

4.11 (a) What essential condition must be satisfied if the P-V work for a gas to be obtained by integration? (b) Is it possible to draw the path from State I to II if the change is brought about irreversibly. (Calcutta BSc(H), 1998)

4.12 What does P signify in PdV type of work? (Burdwan BSC(H), 2000)

4.13 Comment on 'irreversible work is always less than reversible work'. (Calcutta BSc(H), 2006)

4.14 State giving reasons whether the following statements are correct or incorrect:

(a) A certain gas adiabatically expands against zero pressure, and is found to cool down. Hence, the energy of the gas decreases. (b) When gaseous hydrogen and oxygen react to form water in a thermally insulated bomb, the energy of the system remains unchanged. (Calcutta BSc(H), 1986)

4.15 Define C_P of a thermodynamic system. "For any process occurring in the system $dH = C_P dT$." Justify or criticise. (Calcutta BSc(H), 2013)

4.16 Justify or criticise: $\Delta H = Q$ for a process in which P is not constant throughout, but for which the final and initial pressures are equal. (Calcutta BSc(H), 2002)

4.17 What is the dimension of enthalpy? (Burdwan BSc(H), 2001)

4.18 Write down the thermodynamic equation of state and identify the term called 'internal pressure'. Why is it so called? (Calcutta BSc(H), 2006)

4.19 For a closed system, show that $C_P - C_V = P\left(\dfrac{\partial V}{\partial T}\right)_P + \left(\dfrac{\partial U}{\partial V}\right)_T \left(\dfrac{\partial V}{\partial T}\right)_P$ where the terms have their usual meaning. Explain the physical significance of the two terms on the RHS, specially in case of a gaseous system. (Jadavpur BSc(H), 2012)

4.20 An ideal gas is in an initial state (P, V, T). It separately undergoes (i) isothermal and reversible expansion to (P_1, V_1, T) and (ii) adiabatic and reversible expansion to (P_2, V_2, T_2). Depict the changes on a P-V indicator diagram. Find the value of the ratio of the slopes of the P-V curves (adiabatic : isothermal) if the gas is diatomic. Explain the steps in your calculation. (Calcutta BSc(H), 2012)

4.21 Why P-V curves of adiabatic process is more steep than that of isothermal process?
(Burdwan BSc(H), 2012)

4.22 Which of the following parameters are state functions: Q, H, U and W. The terms have their usual meanings. Show that the work done in an isothermal (reversible) expansion of an ideal gas is greater than that of a van der Waals gas (for same volume change). (Kalyani BSc(H), 2003)

4.23 Calculate Q, W and ΔH for reversible isothermal expansion at 300 K of 5 moles of an ideal gas from 500 mL to 500 L. What would be the ΔU and W if the expansion occurs between the same initial and final states as before but is done by expanding the gas in vacuum? (Burdwan BSc(H), 1998)

4.24 A gas at 10 atm pressure occupies a volume of 10 liters at 300 K. It is allowed to expand at the constant temperature 300 K under a constant external pressure till volume equilibrates at 100 liters. Calculate the work done. (Punjab BSc(H), 2011)

4.25 If one mole of an ideal gas is expanded isothermally from T, P_1, V_1 to T, P_2, V_2 in two stages. First stage uses constant opposing pressure P_1 and second stage uses constant opposing pressure P_2. Show that the maximum value of the work produced

is given by $2RT\left[1 - \left(\frac{P_2}{P_1}\right)^{\frac{1}{2}}\right]$ and $P' = (P_1 P_2)^{\frac{1}{2}}$ (Calcutta BSc(H), 2003)

4.26 Show that the work produced in an adiabatic reversible expansion is

$$W = \frac{P_i V_i}{\gamma - 1}\left[1 - \left(\frac{P_f}{P_i}\right)^{\frac{\gamma - 1}{\gamma}}\right]$$ (Burdwan BSc(H), 2001)

4.27 Show that the expression for work (W) due to reversible isothermal (T_1) multistep expansion process of an ideal gas obeying the equation of state $PV^n = C$ (C is a

constant) is given by $W = \frac{RT_1}{1-n}\left[\left(\frac{P_2}{P_1}\right)^{\frac{n-1}{n}} - 1\right]$. (Burdwan BSc(H), 2012)

4.28 1 mole of an ideal monatomic gas at 298 K expands to double its volume at constant pressure. Calculate heat absorbed by the gas. (Calcutta BSc(H), 2006)

4.29 The pressure of a gas is represented by $20/V$ atm, where V is volume of the gas in liters. If the gas expands from 5 liters to 50 liters respectively and undergoes an increase in internal energy by 200 cal, how much heat will be absorbed in the process. (Calcutta BSc(H), 2014)

4.30 A gas expands against a constant pressure of 5 atm from 2 L to 6 L and in the process absorbs 500 J of heat. Find out the change in the internal energy of the gas.
(Jadavpur BSc(H), 2013)

4.31 2 mol of an ideal gas is confined under a constant pressure of 300 kPa. The temperature is changed from 127°C to 27°C at the same constant pressure. Calculate W, Q, ΔU and ΔH. Given $\bar{C}_V = 12.5$ JK^{-1} mol^{-1}. (Jadavpur BSc(H), 2012)

4.32 A 32 g sample of methane initially at 1 atm and 27°C is heated to 277°C. The empirical equation for molar heat capacity at constant pressure is $\bar{C}_P = 3 + 2 \times 10^{-2}T$ Cal mol^{-1}K^{-1}. Assuming ideal behaviour, calculate Q, ΔH and ΔU for an isobaric reversible process. (Vidyasagar BSc(H), 1997)

4.33 The coefficient of thermal expansion of water is 2×10^{-4} per °C. 200 gm of water at 25°C is heated to 50°C under 2 atm external pressure. Given the density of water at 25°C is 0.9970 gm/cc and $\bar{C}_P = 75.30$ J mol^{-1} K^{-1}. Calculate W, ΔH and Q.
(Calcutta BSc(H), 2001)

4.34 Calculate W and ΔU for the conversion of 1 mole of water into 1 mole of steam at a temperature of 373 K and 1 atm pressure. Latent heat of vaporisation of water is 540 cal g^{-1}. (Delhi BSc(H), 2001)

4.35 A cylinder is fitted with a frictionless piston and is kept in a thermostat. It contains 2 moles of an ideal gas at 27°C and 2 atm pressure. Following (i), (ii) and (iii) are three separate experiments carried out independently with the above.

   i. The piston is all on a sudden withdrawn to a position where pressure is reduced to 1 atm and equilibrium is restored.

   ii. Pressure is reduced at a single step from 2 to 1 atm.

   iii. Pressure is reduced slowly to 1 atm in such a way that the position of the piston remains unaltered if left to itself at any moment during the operation

   Calculate in each case ΔU, ΔH, Q and W. (Calcutta BSc(H), 1995)

4.36 2 moles of an ideal gas at 27°C are enclosed in a leak-proof cylinder fitted with a movable frictionless piston and thermally insulated from its surroundings. Justify whether this is an isolated system or not.

   The pressure on the piston is released very slowly to effect a quasistatic expansion to double its volume. Calculate

   i. the final temperature

   ii. the change in enthalpy of the gas

   iii. the work done by the gas

   Given $\bar{C}_V = 1.5R$ (Calcutta BSc(H), 2007)

4.37 One mole of a monatomic ideal gas at 300 K and 10 atm pressure is expanded adiabatically against a constant external pressure of 1 atm until it reaches equilibrium. Calculate the final temperature of the gas, W and ΔH.

(Calcutta BSc(H), 2006)

4.38 One mole of an ideal gas ($\bar{C}_V = 30$ cal K^{-1}mol^{-1}) undergoes the following reversible cycle:

   i. Depict the cycle in a V versus T diagram

   ii. Calculate Q, W and ΔU for each step and for the cycle.

State I ——— Isochoric heating ———→ State II
(1 atm, 273 K)   (546 K)

Isothermal compression ↑   State III (273 K)   Adiabatic expansion ←

(Calcutta BSc(H), 2009)

4.39 2 moles of an ideal gas are put through a cycle consisting of the following three reversible steps:

   i. Isothermal compression (at T_1) from (2 bar, 10 dm^3) to (20 bar, 1 dm^3)

   ii. Isobaric expansion at 20 bar to the original volume of 10 dm^3 with increase of temperature ($T_1 \rightarrow T_2$).

   iii. Cooling of the gas at constant volume to bring the gas to the original pressure and temperature.

   Depict the processes on a P-V diagram. Calculate ΔU, Q and W for each step of the cycle. Given $\bar{C}_V = 1.5R$. (Calcutta BSc(H), 2013)

4.40 The heats of solution of one mole of KCl in 200 moles of water under 1 atm pressure are

$$\Delta H = 4339 \text{ cal at } 21°C$$
$$\Delta H = 4260 \text{ cal at } 23°C$$

Determine ΔH at 25°C. State what assumption you have taken to work out the problem.

4.41 The standard heats of formation at 25°C are given as

$$Na_2SO_4 \text{ (aq), } \Delta H_f^{\circ} = -1387 \text{ kJ/mol}$$

$$Na_2SO_4, 10H_2O(s), \Delta H_f^{\circ} = -4324 \text{ kJ/mol}$$

$$H_2O \text{ (l), } \Delta H_f^{\circ} = -286 \text{ kJ/mol}$$

Calculate the heat absorbed in the process

$$Na_2SO_4 \cdot 10H_2O(s) \rightarrow Na_2SO_4 \text{ (aq)}$$ (Burdwan BSc(H), 2011)

4.42 Heats of combustion of hydrated copper sulphate and anhydrous copper sulphate are -2.80 and 15.89 kcal mol^{-1} respectively. Calculate the heat of hydration of copper sulphate. (Delhi BSc, 2004)

4.43 Define enthalpy of neutralisation. When 100 mL of 1 N HCl is neutralised by equivalent amount of NaOH, 5.273 kJ of heat is evolved. Calculate the heat of neutralisation of HCl. (Nagpur BSc, 2002)

4.44 How much heat is used up in dissociating HBO_2? Use the following data:

i. $HBO_2 + NaOH \rightarrow NaBO_2 + H_2O$ $\Delta H = -10.0$ kcal mol^{-1}

ii. $H^+ + OH^- \rightarrow H_2O$ $\Delta H = -13.4$ kcal mol^{-1}

State and explain the principle used in the calculation. (Burdwan BSc(H), 1996)

4.45 Calculate the heat of formation of acetic acid, its heat of combustion is -869.0 kJ mol^{-1}. The heats of formation of CO_2 (g) and H_2O (l) are -390 kJ mol^{-1} and -285.0 kJ mol^{-1} respectively. (Agra BSc, 2004)

4.46 The heats of combustion of C_2H_4 (g), C_2H_6 (g) and H_2 (g) and H_2 (g) are -1409 kJ mol^{-1}, -1558.3 kJ mol^{-1} and -285.65 kJ mol^{-1} respectively. Calculate the heat of hydrogenation of ethylene. (Assam BSc, 2005)

4.47 The standard heats of formation (ΔH_f°) of H_2O and H_2O_2 are x and y respectively. Evaluate the bond dissociation energy of the peroxide bond (O–O) in terms of x and y. (Calcutta BSc(H), 2013)

4.48 Define bond energy. Calculate ΔH for the reaction:

$$C_2H_4(g) + 3O_2(g) \rightarrow 2CO_2(g) + 2H_2O(g)$$

from the following values of bond energies:

Bond	C–H	O=O	C=O	O–H	C=C
Bond energy (kJ mol^{-1})	414	499	724	460	619

(Nagpur BSc, 2002)

4.49 2.0 g of C_6H_6 was burnt in excess of O_2 in a bomb calorimeter

$$C_6H_6(l) + 7\frac{1}{2}O_2(g) \rightarrow 6CO_2(g) + 3H_2O(l)$$

If the temperature rise is 40°C and the heat capacity of the system is 2.0 kJ K^{-1}, calculate the heat of combustion of C_6H_6 at constant volume.

(Guru Nanak Dev BSc, 2002)

KEY TO UNIVERSITY QUESTIONS

4.1 See problem 4.3.

4.2 A system must be isolated if any form of energy, and not just heat, cannot enter or leave the system. Therefore, the given statement is not justified.

4.3 (iii).

4.4 Here only enthalpy, which depends on the amount of a system, is extensive.

4.5 See Sec. 4.1 and problem 4.14(ii).

4.6 See problem 4.14(i).

4.7 See problem 4.14(iv) and (v).

4.8 Equilibrium does not exist in both the cases.

In (i), the system being close, the pressure and hence temperature of the system always change due to continuous formation of water vapour during boiling.

In (ii), the system being open, the vapour phase does not remain in equilibrium.

4.9 Here $W = \Delta U$ (Q being zero) and ΔU is independent of the path of change for specified initial and final state (U being a state function). Hence the given statement is justified.

4.10 Although work done (W) in some cases can be expressed only in terms of a state function (e.g. $W = U_f - U_i$ in case of an adiabatic change), it always refers to a change in state and not to a particular state. Hence the given statement is not justified.

4.11   a. The essential condition is that the pressure imposed on the system is either constant or expressable in terms of volume of the system as the only variable.

b. No, because an irreversible process, unlike a reversible one, cannot be described as a succession of well-defined equilibrium states.

4.12 P signifies the pressure imposed on the system when PdV gives the work done by the system against the imposed pressure.

4.13 See problem 4.24(b).

4.14   a. Here $Q = 0$ (since the process is adiabatic) and $W = 0$ (since expansion occurs against zero pressure). Here the energy change $\Delta U = Q + W = 0$, then the given statement is incorrect. Here the fall in temperature is due to conversion of kinetic energy to potential energy of the molecules resulting from the volume change against intermolecular attraction.

b. Here also $\Delta U = 0$ because $Q = 0$ (since the reaction vessel is thermally insulated and $W = 0$, since $\Delta V = 0$ (since the reaction vessel acts as a bomb). Then the given statement is correct.

Note: The interaction involved in the system is physical in (a) but chemical in (b). However, they are thermodynamically similar.

4.15 See problem 4.21.

4.16 See problem 4.15(v).

4.17 It follows from the definition of H as $H = U + PV$, that H will have the dimension of PV, i.e. $(ml^{-1}t^{-2})$ (l^3) or ml^2t^{-2}.

4.18 The commonly used thermodynamic equation of state is

$$\left(\frac{\partial U}{\partial V}\right)_T = T\left(\frac{\partial P}{\partial T}\right)_V - P \quad \text{(vide Eq. 5.16)}$$

Here the quantity $(\partial U/\partial T)_V$ which has the dimensions of pressure is called *internal pressure*. Because it originates from intermolecular interaction *–*.

4.19 Vide problem 4.22.

$P\left(\frac{\partial V}{\partial T}\right)_P$ signifies the energy absorbed by the system per unit rise of temperature for expansion against the external pressure, $\left(\frac{\partial U}{\partial V}\right)_T \left(\frac{\partial V}{\partial T}\right)_P$ signifies the corresponding quantity arising out of the internal pressure.

4.20 See problem 4.12.

☐ The equipartition value of the ratio of the slopes γ for a linear molecule of atomicity x is

$$\gamma = 1 + \frac{1}{3x - \frac{5}{2}} \quad \text{(vide problem 2.74)}$$

Then for diatomic molecule ($x = 2$) and
$$\gamma = 1 + \frac{1}{3 \times 2 - \frac{5}{2}} = 1.29$$

4.21 See problem 4.12.

4.22 State functions are U and H.

☐ The work done by a van der Waals gas

$$= \int_{V_i}^{V_f} P dV = \int_{V_i}^{V_f} \left(\frac{RT}{V-b} - \frac{a}{V^2} \right) dV$$

$$= RT \ln \frac{V_f - b}{V_i - b} + \left(\frac{1}{V_f} - \frac{1}{V_i} \right) \text{ for 1 mole}$$

This is less than the work done by an ideal gas $RT \ln \frac{V_f}{V_i}$ (for 1 mole).

Since $\frac{V_f - b}{V_i - b} < \frac{V_f}{V_i}$ and $\left(\frac{1}{V_f} - \frac{1}{V_i} \right)$ is –ve for expansion.

Note: But the same volume change against the same pressure (fixed), the work done by the gas will be the same for an ideal gas and a van der Waals gas.

4.23. Vide problem 4.26(i).

☐ ΔU being independent of path of change, would be unchanged. But W, being path dependent, would change to zero, since the external pressure is zero.

4.24
$$P_{\text{text}} = P_f = \frac{P_i V_i}{V_f} \qquad \text{(assuming the gas to be ideal)}$$

$$= (10 \text{ atm}) (10 \text{ L})/(100 \text{ L}) = 1 \text{ atm}$$
$$W = P_{\text{ext}} (V_i - V_f) \qquad \text{(by IUPAC convension)}$$
$$= (1 \text{ atm}) (10 \text{ L} - 100 \text{ L}) = -10 \text{ L atm}$$

Note: Here, unlike the previous question, the given temperature is of no use.

$$H = U + PV$$

Then
$$d_H = d_U + PdV + VdP$$

$$= \left(\frac{\partial U}{\partial T} \right)_V dT + \left(\frac{\partial U}{\partial V} \right)_T dV + PdV + VdP$$

for hydrostatic system where $U = f(T, V)$

Dividing both sides by dT at constant pressure

$$\left(\frac{\partial H}{\partial T} \right)_P = \left(\frac{\partial U}{\partial T} \right)_V + \left[\left(\frac{\partial U}{\partial V} \right)_T + P \right] \left(\frac{\partial V}{\partial T} \right)_P$$

or
$$C_P - C_V = \left[\left(\frac{\partial U}{\partial V} \right)_T + P \right] \left(\frac{\partial V}{\partial T} \right)_P$$

4.25 Let V' be the volume of the gas at pressure P'.

Then the work produced
$$W = P'(V' - V_1) + P_2(V_2 - V')$$
$$= P'V' - P'V_1 + P_2V_2 - P_2V'$$
$$= RT - P' \cdot \frac{RT}{P_1} + RT - P_2 \cdot \frac{RT}{P'}, \qquad \because P'V' = P_1V_1 = P_2V_2 = RT$$
$$= RT \left(2 - \frac{P'}{P_1} - \frac{P_2}{P'} \right)$$

$$\frac{dW}{dP'} = RT\left[-\frac{1}{P_1} + \frac{P_2'}{(P')^2}\right]$$

$$= 0 \quad \text{for maximum value of } W$$

whence $\qquad P = \sqrt{P_1 P_2}$

Substituting this in the above expression for W, we have

$$W_{max} = RT\left[2 - \frac{\sqrt{P_1 P_2}}{P_1} - \frac{P_2}{\sqrt{P_1 P_2}}\right]$$

$$= 2RT\left[1 - \left(\frac{P_2}{P_1}\right)^{\frac{1}{2}}\right]$$

4.26 $\qquad W = \dfrac{P_i V_i}{\gamma - 1}\left(1 - \dfrac{P_f V_f}{P_i V_i}\right) \qquad$ by Eq. (4.3a)

$$= \frac{P_i V_i}{\gamma - 1}\left[1 - \left(\frac{P_f}{P_i}\right)^{\frac{\gamma-1}{\gamma}}\right] \qquad \because \frac{V_f}{V_i} = \left(\frac{P_i}{P_f}\right)^{\frac{1}{\gamma}}$$

4.27 Here the P–V relation is similar to that in case of a reversible adiabatic change of an ideal gas (PV^γ = constant). Although the constant n in the given relation is not likely to be same as γ, W can be similarly calculated as in the previous question with the following result

$$W = \frac{P_i V_i}{n - 1}\left[1 - \left(\frac{P_f}{P_i}\right)^{\frac{n-1}{n}}\right]$$

$$= \frac{RT_1}{1 - n}\left[\left(\frac{P_f}{P_i}\right)^{\frac{n-1}{n}} - 1\right] \qquad \text{for 1 mol of the gas}$$

4.28 Here the process is reversible as it occurs at constant pressure of the system [vide problem 4.24(a)].

Now at constant pressure

$$T_f = T_i \cdot \frac{V_f}{V_i} \qquad \text{by ideal gas equation}$$

$$= 2T_i \qquad \because V_f = 2V_i$$

Then heat absorbed by the gas

$$= n\bar{C}_P(T_f - T_i) = n \cdot \tfrac{5}{2}RT_i$$

$$= (1 \text{ mol}) \cdot \tfrac{5}{2}(8.314 \text{ JK}^{-1}\text{mol}^{-1})(298 \text{ K}) = 6194 \text{ J}$$

4.29 Here $\qquad PV = K$, constant

$$W = -\int_{V_i}^{V_f} PdV = -K\int_{V_i}^{V_f}\frac{dV}{V} = K\ln\frac{V_i}{V_f}$$

$$= (20 \text{ L atm})\ln\tfrac{5}{50} = -46.06 \text{ L atm}$$

$$Q = \Delta U - W$$
$$= 200 \text{ cal} - (-46.06 \text{ L atm} \times 24.2 \text{ cal/L atm})$$
$$= 1314.6 \text{ cal}$$

Note: PV = constant does not necessarily imply that the gas is ideal and the process is isothermal. However, the calculation of W can be done similarly vide problem 26(i).

4.30
$$W = -P_{ext}(V_f - V_i)$$
$$= -(5 \times 1.013 \times 10^5 \text{Pa})(6-2) \times 10^{-3}\text{m}^3 = -2026 \text{ J}$$
$$\Delta U = Q + W = 500 \text{ J} - 2026 \text{ J} = -1526 \text{ J}$$

Note: Here, unlike question 4.28, the process if not reversible.

4.31
$$\Delta U = n\bar{C}_V \Delta T$$
$$Q = \Delta H = n\bar{C}_P \Delta T = n(\bar{C}_V + R)\,\Delta T$$
$$W = \Delta U - Q$$

4.32 $Q = \Delta H$, since the process is isobaric.

For calculation of ΔU and ΔH, proceed as in problem 4.30.

4.33
$$Q = \Delta H = n\bar{C}_P \Delta T$$
$$W = -P_{ext}\Delta V = -P_{ext}\,\alpha V \Delta T$$

where α is the coefficient of thermal expansion.

4.34 See problem 4.32.

4.35 In all the cases

$$\Delta U = n\bar{C}_V \Delta T = 0 \quad \text{since the gas is ideal and the process is isothermal}$$
$$\Delta H = n\bar{C}_P \Delta T = 0$$
$$Q = \Delta U - W$$

For (i), $P_{ext} = 0$ and hence $W = 0$

For (ii) $P_{ext} = 1$ atm

$$W = -P_{ext}\Delta V = -nRT \cdot P_{ext}\left(\frac{1}{P_f} - \frac{1}{P_i}\right)$$

For (iii) the process is reversible

$$W = -nR - \ln\frac{V_f}{V_i} = nRT \ln\frac{P_f}{P_i}$$

Note: W and Q, but not ΔU and ΔH are different in the three cases because of their path dependence.

4.36 The system is not isolated (vide university question 4.2) as it can exchange energy in form of work with the surroundings.

i.
$$T_f = T_i\left(\frac{V_i}{V_f}\right)^{\gamma-1}$$

by relation $TV^{\gamma-1}$ (found from combination of PV^γ = constant and $PV = nRT$)

$$\gamma = \frac{\bar{C}_P}{\bar{C}_V} = \frac{\bar{C}_V + R}{\bar{C}_V} = 1 + \frac{R}{\bar{C}_V}$$

ii.
$$\Delta H = n\bar{C}_P \Delta T$$

iii.
$$W = n\bar{C}_V(T_f - T_i)$$

4.37 For calculation of T_f see problem 4.28(b)

$$W = \Delta U \quad \text{since } Q = 0$$

$$= n\bar{C}_V \Delta T$$

$$= \tfrac{3}{2} nR\Delta T \quad \text{since the gas is monatomic}$$

$$\Delta H = n\bar{C}_P \Delta T$$

$$= n(\bar{C}_V + R)\, \Delta T$$

Note: Here $\Delta H \neq 0$, because pressure of the gas is not constant during the process though P_{ext} is constant. This is unlike question 4.31.

4.38 See problem 4.27.

4.39 Here $T_1 = P_1 V_1 / nR$ can be obtained from Step (i).

T_2 can be obtained from step (ii) where volume changes from 1 dm^3 to 10 dm^3. Then $T_2/T_1 = 10/1$ (since $T \propto V$ at constant T and n).

For calculation of ΔU, Q and W:

See problem 4.26(i) for step (i)

See problem 4.31 for step (ii)

See problem 4.27 for step (iii)

4.40 Assuming C_P to be independent of temperature in the region of temperature concerned, we have from Kirchhoff's Eq. (4.10a)

$$\Delta H_{t^\circ C} = \Delta H_{25^\circ C} + a(t - 25), \quad \text{where } a \text{ is a constant}$$

Then

$$\Delta H_{25^\circ C} = \Delta H_{23^\circ C} - \frac{\Delta H_{23^\circ C} - \Delta H_{21^\circ C}}{(23 - 21)^\circ C}(23 - 25)^\circ C$$

$$= 4260 \text{ cal} + (4260 - 4339) \text{ cal}$$

$$= 4181 \text{ cal}$$

4.41 Heat absorbed
$$\Delta H = \Delta H_f^\circ(Na_2SO_4 \text{ aq}) - \Delta H_f^\circ(Na_2SO_4 \cdot 10H_2O, s)$$
$$= (-1387 + 4324) \text{ kJ mol}^{-1}$$

Note: The given value of ΔH_f° (H_2O, l) is of no use.

4.42 From the given data we can write

i. $CuSO_4 \cdot 5H_2O(s) + \tfrac{1}{2}O_2(g) = CuO(s) + SO_3(g) + 5H_2O(g)$

$$\Delta H_{(i)} = -2.80 \text{ kcal mol}^{-1}$$

ii. $CuSO_4(s) + \tfrac{1}{2}O_2(g) = CuO(s) + SO_3(g) \qquad \Delta H_{(ii)} = 15.89 \text{ kcal mol}^{-1}$

We are to calculate DH for the reaction $CuSO_2(s) + 5H_2O(g) = CuSO_4 \cdot 5H_2O(s)$. This equation can be obtained by subtracting the Eq. (i) from Eq. (ii). Then the required DH will be given by

$$\Delta H = \Delta H_{(ii)} - \Delta H_{(i)}$$

$$= (15.89 + 2.80) \text{ kcal mol}^{-1}$$

$$= 18.69 \text{ kcal mol}^{-1}$$

4.43 Heat of neutralisation

$$= \frac{\text{Heat absorbed in the specified neutralisation}}{\text{Amount of neutralisation}}$$

$$= \frac{-5.273 \text{ kJ}}{\frac{10}{1000} \times 1 \text{ equivalent}} = -52.73 \text{ kJ/equivalent}$$

4.44 Heat used up in the dissociation

= Heat evolved in (ii) – heat evolves in (i)

= (13.4 – 10.0) kcal mol^{-1} [vide problem 4.35(b)]

4.45 The given data regarding the combustion of acetic acid may be represented as

$CH_3CO_2H(l) + 3O_2(g) = 2CO_2(g) + 2H_2O(l)$ $\Delta H = -869.0 \text{ kJ mol}^{-1}$

Then $\Delta H = 2\Delta H_f(CO_2, g) + 2\Delta H_F(H_2O, l) - \Delta H_f(CH_3CO_2H, l)$

                                              [vide problem 4.44]

or $\Delta H_f(CH_3CO_2H, l) = 2\Delta H_f(CO_2, g) + 2\Delta H_f(H_2O, l) - \Delta H$

                 = [2(–390.0) + 2(–285.0) + 869.0] kJ mol^{-1}

                 = –481 kJ mol^{-1}

4.46 The given data may be represented as

i. $C_2H_4(g) + 3O_2(g) = 2CO_2(g) + 2H_2O(l)$ $\Delta H_{(i)} = -1409 \text{ kJ mol}^{-1}$

ii. $C_2H_6(g) + \frac{7}{2}O_2(g) = 2CO_2(g) + 3H_2O(l)$ $\Delta H_{(ii)} = -1558.3 \text{ kJ mol}^{-1}$

iii. $H_2(g) + \frac{1}{2}O_2(g) = H_2O(l)$ $\Delta H_{(iii)} = -285.65 \text{ kJ mol}^{-1}$

We have to calculate ΔH for the reaction $C_2H_4(g) + H_2(g) = C_2H_6(g)$. This equation can be obtained from the combination (i) + (iii) – (ii). Then the required ΔH will be given by

$$\Delta H = \Delta H_{(i)} + \Delta H_{(iii)} - \Delta H_{(ii)}$$
$$= (-1409 - 285.65 + 1558.3) \text{ kJ mol}^{-1}$$
$$= -136.35 \text{ kJ mol}^{-1}$$

4.47 $\Delta H_f^{\circ}(H_2O) - \Delta H_f^{\circ}(H_2O_2) = 2\epsilon_{O-H} - (2\epsilon_{O-H} + \epsilon_{O-O}) = x - y$

whence $\epsilon_{O-O} = y - x$

4.48 $\Delta H = \underset{\text{for } H_2C=CH_2}{(\epsilon_{C=C} + 4\,\epsilon_{C-H})} + \underset{\text{for } O=O}{3\,\epsilon_{O-O}} - \underset{\text{for } O=C=O}{4\,\epsilon_{C=O}} - \underset{\text{for HOH}}{4\,\epsilon_{O-H}}$

                 = (619 + 4 × 414 + 3 × 499 – 4 × 724 – 4 × 460) kJ mol^{-1}
                 = –964 kJ mol^{-1}

4.49   Heat evolved = Heat capacity multiplied by rise of temperature

                            = 2.0 kJ K^{-1} × 40 K
                            = 80 kJ

Heat of combustion at constant volume

$$Q_V = \frac{\text{Heat absorbed}}{\text{Amount of } C_6H_6 \text{ burnt}}$$

$$= \frac{-80 \text{ kJ}}{(2/78 \cdot 108) \text{ mol}}$$

$$= -3124.32 \text{ kJ mol}^{-1}$$

5

Second Law of Thermodynamics

An important application of thermodynamics is to predict whether a system, under a given set of conditions, will undergo any spontaneous change or remain in equilibrium. This can be done, not through the first law, but through the second law of thermodynamics which is based on the general observation that all natural changes occur unidirectionally toward equilibrium. There are several statements of the second law. Thus, regarding heat flow, we can mention the Clausius statement: Heat cannot flow spontaneously in the diretion of higher temperature. However, it indirectly leads to other statements: Complete conversion of heat into work (through a cyclic process) is impossible (Kelvin–Plank statement), or a perpetual motion of the second kind (i.e. continuous conversion of heat into work using only one heat reservoir) is impossible (Ostwald statement).

5.1 THE CARNOT FORMULA

During a spontaneous (natural) process, it is possible to get some work through a device, called *engine*. The second law of thermodynamics originates from the study of the principles of the operation of a heat engine, a device that converts heat into work through a cyclic process using two heat reservoirs (bodies of infinite heat capacity).

If a heat engine absorbs $|Q_1|$ amount of heat from a reservoir (called *source*) at temperature T_1 and rejects $|Q_2|$ amount of heat to another reservoir (called *sink*) at lower temperature T_2, then the maximum work $|W|$ produced by it will be

$$|W| = |Q_1| - |Q_2| = |Q_1|\frac{T_1 - T_2}{T_1} \tag{5.1a}$$

whence

$$\frac{|Q_1|}{|Q_2|} = \frac{T_1}{T_2} \tag{5.1b}$$

and the efficiency of heat engine

$$\eta = \frac{|W|}{|Q_1|} = \frac{T_1 - T_2}{T_1} \tag{5.1c}$$

According to the IUPAC convention Q_1 is +ve, Q_2 is negative and W is –ve, so

$$-W = Q_1 + Q_2, \quad \text{since} \oint dU = W + Q_1 + Q_2 = 0$$

$$= \text{a +ve quantity}$$

When $T_1 < T_2$, i.e. transfer of heat in the direction of higher temperature, the device is called refrigerator or heat pump according as the purpose is to extract heat or to supply heat. Here, the heat flow being in the unnatural direction, there will be consumption (instead of production) of work, and we have the following expressions

The coefficient of performance of a refrigerator

$$\frac{|Q_1|}{-|W|} = \frac{T_1}{T_2 - T_1} \tag{5.1d}$$

The coefficient of performance of a heat pump

$$\frac{|Q_2|}{-|W|} = \frac{T_2}{T_2 - T_1} \tag{5.1e}$$

The expression (5.1a) was first deduced by Carnot through a hypothetical cycle, called the *Carnot cycle* which is a reversible cycle consisting of four parts, two isothermals and two adiabatics in alternate sequence.

5.2 ENTROPY

Carnot formula leads to the concept of entropy S as a single valued extensive state property of a system (just as U from the first law) that can be related to the heat change as

$$dS = dQ/T, \text{ for a reversible process or equilibrium situation}$$
$$> dQ/T \text{ for a spontaneous (irreversible) process}$$

or $$dS \geq dQ/T \tag{5.2}$$

which is the Clausius inequality. It demands that entropy of the universe (which behaves as an isolated system) tends to a maximum (due to the natural processes occuring in it). This is the Clausius statement of the second law of thermodynamics which gives the condition of spontanity often put in the following more familiar form.

$$dS_{universe} = +ve$$

Entropy being a state function, ΔS for a specified change in state of a system will be fixed (whether the process is reversible or not) and calculable from the knowledge of the parameters of the initial and final state of the system, but not from the accompanying heat change, unless the change occurs reversibly. Only in case of a reversible change, we can write

$$\Delta S = \int_i^f \frac{dQ}{T}$$

In case of irreversible change ΔS can be computed by devising a convenient reversible path for the specified change and applying this relation at every step. This can be done easily when both the initial and final states are equilibrium states. From such considerations, we can readily establish and understand the following simple relations.

For an ideal gas $$\Delta S = n\bar{C}_V \ln \frac{T_f}{T_i} + nR \ln \frac{V_f}{V_i} \tag{5.3a}$$

$$= n\bar{C}_P \ln \frac{T_f}{T_i} + nR \ln \frac{P_i}{P_f} \tag{5.3b}$$

For mixing of ideal gases

P_1, V_1, T n_1	P_2, V_2, T n_2	Isochoric mixing	P, V, T $n_1 + n_2$

$$\Delta S = \Delta S_1 + \Delta S_2$$

$$= n_1 R \ln \frac{P}{P_1} + n_2 R \ln \frac{P}{P_2}$$

$$= n_1 R \ln \frac{P}{P_1} + n_2 R \ln \frac{P}{P_2} \quad \text{when } P_1 = P_2 = P$$

$$= R \left(n_1 \ln \frac{1}{x_1} + n_2 \ln \frac{1}{x_2} \right)$$

$$= - nR(x_1 \ln x_1 + x_2 \ln x_2) \tag{5.4}$$

where $n = n_1 + n_2$ is the total number of moles.

It is important to note that in case of mixing the entropy change of each gas (ΔS_1 or ΔS_2) is not due to the presence of the other gas but due to its own expansion. Here the term *mixing* is somewhat confusing.

5.3 FREE ENERGY

Entropy provides the most fundamental criteria of equilibrium and spontaneity in the following general form

$$dS_{\text{universe}} \geq 0 \tag{5.5}$$

But the route is an inconvenient one due to involvement of the surroundings (being vast). To get simple route involving only the system, two additional functions A (called work function or Helmholtz free energy) and G (called Gibbs function or Gibbs free energy or Gibbs potential) have been introduced by defining them as follows:

$$A = U - TS \tag{5.6}$$

$$\left. \begin{array}{l} G = H - TS \\ = U + PV - TS \end{array} \right\} \tag{5.7}$$

Then for an infinitesimal change, we have (considering $dS \geq dQ/T$)

$$dA \leq - PdV - SdT \tag{5.8}$$

$$dG \leq VdP - SdT \tag{5.9}$$

if the system is closed and does only PV-work (i.e. expansion work)

or $\qquad dA_{T,V} \leq 0$ '=' gives the condition for equilibrium

$\qquad\qquad dG_{T,P} \leq 0$ and '<' gives the condition for spontaneity

G gives the most convenient criteria for spontaneity and equilibrium which refer only to the system (unlike S) at constant T and P at which most processes are carried out in a laboratory to collect precise data.

Dependence of Free Energy on Temperature

From Eqs (5.8) and (5.9)

$$\left. \begin{array}{l} \left(\dfrac{\partial A}{\partial T} \right)_V = - S \\[3mm] \left(\dfrac{\partial G}{\partial T} \right)_P = - S \end{array} \right\} \tag{5.10}$$

Then from Eq. (5.6) and (5.7)

$$
\left.
\begin{aligned}
A &= U + T\left(\frac{\partial A}{\partial T}\right)_V \\[2mm]
G &= H + T\left(\frac{\partial G}{\partial T}\right)_P
\end{aligned}
\right\}
\tag{5.11a}
$$

Rearranging

$$
\left.
\begin{aligned}
\left(\frac{\partial (A/T)}{\partial T}\right)_V &= -\frac{U}{T^2} \\[2mm]
\left(\frac{\partial (G/T)}{\partial T}\right)_P &= -\frac{H}{T^2}
\end{aligned}
\right\}
\tag{5.11b}
$$

or

$$
\left.
\begin{aligned}
\left(\frac{\partial (A/T)}{\partial (1/T)}\right)_V &= U \\[2mm]
\left(\frac{\partial (G/T)}{\partial (1/T)}\right)_P &= H
\end{aligned}
\right\}
\tag{5.11c}
$$

Equations (5.11a), (5.11b), (5.11c) and sometimes Eq. (5.10) are known as Gibbs-Helmholtz equations in different forms.

The chief application of the Gibbs-Helmholtz equation is in the calculation of ΔU and ΔH in an isothermal change using the following well familiar form of this equation

$$
\left.
\begin{aligned}
\Delta U &= \Delta A - T\left(\frac{\partial \Delta A}{\partial T}\right)_V \\[2mm]
\Delta H &= \Delta G - T\left(\frac{\partial \Delta G}{\partial T}\right)_P
\end{aligned}
\right\}
\tag{5.12}
$$

The knowledge of ΔA and ΔG and their temperature coefficients, available from emf measurements, thus enables one to avoid the cumbersome calorimetric method of determining ΔU (the heat of reaction at constant volume) and ΔH (the heat of reaction at constant pressure).

Dependence of Free Energy on Pressure

For any substance, we have from Eq. (5.9)

$$
\int_{G°}^{G} dG = \int_{P°}^{P} V dP \qquad \text{at constant temperature}
$$

or

$$
G = G°(T) + \int_{P°}^{P} V dP
\tag{5.13a}
$$

where $G°(T)$ is the standard free energy of the substance, i.e. the free energy under standard pressure $P°$ which is arbitrarily chosen to be 1 atm (more precisely 1 bar).

For an ideal gas, $\qquad V = nRT/P$

when we have $\qquad G = G°(T) + nRT \ln P$ $\qquad\qquad$ (5.13b)

or $\qquad\qquad \mu = \mu°(T) + RT \ln P$

where $\mu\ (= G/n)$ and $\mu°\ (= G°/n)$ are respectively the chemical potential and standard chemical potential of the gas.

For liquids and solids, V is almost independent of P when

$$G = G°(T) + V(P - 1) \tag{5.13c}$$
$$\approx G°(T) \quad \text{if } P \text{ is not very high}$$

Spontaneity of Different Types of Changes

For a change to occur spontaneously in a closed system, the accompanying $\Delta G = \Delta H - T\Delta S$ must be –ve at constant T and P.

Exothermic process		
Exothermic process	i. When ΔH = –ve, ΔS = +ve e.g. $2Na + H_2O = 2NaOH + H_2$	$\Delta G_{T,P}$ is always –ve and hence the change is expected to occur spontaneously at all temperatures. However, at very low temperature, the change may appear to be ceased due to extremely low rate
	ii. When ΔH = –ve, ΔS = –ve e.g. $2H_2 + O_2 = 2H_2O$ $CaO + CO_2 = CaCO_3$	$\Delta G_{T,P}$ = –ve at low temperature, and hence the change will occur spontaneously at sufficiently low temperature

Endothermic process	iii. When ΔH = +ve, ΔS = +ve i.e. reverse of (ii)	$\Delta G_{T,P}$ = –ve at high temperature, and hence the change will occur spontaneously only at sufficiently high temperature
	iv. When ΔH = +ve, ΔS = –ve e.g. electrolysis	$\Delta G_{T,P}$ is always +ve. Hence the change will never occur alone spontaneously. However, this can be brought about by coupling with another spontaneous change if $\Delta G_{T,P}$ for the coupled process is –ve. In case of electrolysis of a substance, ΔG of the substance electrolysed is +ve, but ΔG in the external source of emf is so –ve that total change of G is –ve

5.4 EXACT DIFFERENTIAL

A differential (very small change of a variable) is said to be exact (or perfect), if it always gives the same total quantity when integrated between two specified limits.

Suppose $\qquad dz = M(x, y)\, dx + N(x, y)\, dy$

Here dz will be an exact differential, if

$$\left(\frac{\partial M}{\partial y}\right)_x = \left(\frac{\partial N}{\partial x}\right)_y \tag{5.14}$$

This is Euler's criterion for exactness.

Again, when dz is an exact differential, the following relation will hold

$$\left(\frac{\partial x}{\partial y}\right)_z \left(\frac{\partial y}{\partial z}\right)_x \left(\frac{\partial z}{\partial x}\right)_y = -1 \tag{5.15}$$

This is cyclic rule.

5.5 MAXWELL RELATIONS

For any closed system that does only PV-work, we have, under equilibrium condition, the following four fundamental equations (the thermodynamic equation of state)

$$dU = TdS - PdV$$
$$dH = TdS + VdP$$
$$dA = -SdT - PdV$$
$$dG = -SdT + VdP$$

Now, dU, dH, dA and dG are all exact differentials as they are differentials of state functions. Then from these relations, we have, using Euler's criterion for exactness, the following four successive relations, called Maxwell relations

under adiabetic condition
$$\begin{cases} \left(\dfrac{\partial P}{\partial S}\right)_V = -\left(\dfrac{\partial T}{\partial V}\right)_S \\[4mm] \left(\dfrac{\partial V}{\partial S}\right)_P = \left(\dfrac{\partial T}{\partial P}\right)_S \end{cases}$$

under isothermal condition
$$\begin{cases} \left(\dfrac{\partial S}{\partial V}\right)_T = \left(\dfrac{\partial P}{\partial T}\right)_V \\[4mm] \left(\dfrac{\partial S}{\partial P}\right)_T = -\left(\dfrac{\partial V}{\partial T}\right)_P \end{cases}$$

Maxwell relations are useful in the calculation of complex thermodynamic quantities for a real system relating them to P, V and T.

Calculation of ΔU

For any hydrostatic system, we can write $U = f(T, V)$. Then

$$dU = \left(\frac{\partial U}{\partial T}\right)_V dT + \left(\frac{\partial U}{\partial V}\right)_T dV$$

$$= C_V dT + \left(\frac{\partial U}{\partial V}\right)_T dV$$

Again
$$dU = TdS - PdV$$

or
$$\left(\frac{\partial U}{\partial V}\right)_T = T\left(\frac{\partial S}{\partial V}\right)_T - P$$

$$= T\left(\frac{\partial P}{\partial T}\right)_V - P \qquad \text{by Maxwell relation} \tag{5.16}$$

Then
$$dU = C_V dT + \left[T\left(\frac{\partial P}{\partial T}\right)_V - P\right]dV$$

For a van der Waals gas, $P + \dfrac{n^2 a}{V^2} = \dfrac{nRT}{V - nb}$, when we have

$$\left(\frac{\partial U}{\partial V}\right)_T = \frac{nRT}{V - nb} - P = \frac{n^2 a}{V^2} \tag{5.17}$$

Then
$$dU = C_V dT + n^2 a \, dV/V^2$$

Integration gives $\quad \Delta U = \int_{T_i}^{T_f} C_V dT + n^2 a \int_{V_i}^{V_f} \dfrac{dV}{V^2}$

$$= C_V (T_f - T_i) + n^2 a \left(\dfrac{1}{V_i} - \dfrac{1}{V_f} \right) \qquad (5.18)$$

assuming C_V to be independent of T

Calculation of ΔS

$$dS = \left(\dfrac{\partial S}{\partial T} \right)_V dT + \left(\dfrac{\partial S}{\partial V} \right)_T dV$$

for a hydrostatic system where $S = f(T, V)$

$$= C_V \dfrac{dT}{T} + \left(\dfrac{\partial P}{\partial T} \right)_V dV$$

$$= C_V \dfrac{dT}{T} + nR \dfrac{dV}{V - nb}, \qquad \text{for a van der Waals gas}$$

Integration gives $\quad \Delta S = C_V \int_{T_i}^{T_f} \dfrac{dT}{T} + nR \int_{V_i}^{V_f} \dfrac{dV}{V - nb}$

$$= C_V \ln \dfrac{T_f}{T_i} + nR \ln \dfrac{V_f - nb}{V_i - nb} \qquad (5.19)$$

Note: i. a but not b appears in the expression ΔU while the reverse happens with ΔS.

ii. For an ideal gas, $a = 0$ and also $b = 0$ when $\Delta U = C_V(T_f - T_i)$ and

$$\Delta S = C_V \ln \dfrac{T_f}{T_i} + nR \ln \dfrac{V_f}{V_i} \qquad \text{that tallies with Eq. (5.3a)}$$

Calculation of $C_p - C_v$

$$C_P - C_V = \left[P + \left(\dfrac{\partial U}{\partial V} \right)_T \right] \left(\dfrac{\partial V}{\partial T} \right)_P$$

$$= T \left(\dfrac{\partial P}{\partial T} \right)_V \left(\dfrac{\partial V}{\partial T} \right)_P \qquad (5.20a)$$

Now $\qquad \left(\dfrac{\partial P}{\partial T} \right)_V = -\dfrac{(\partial V / \partial T)_P}{(\partial V / \partial P)_T}, \qquad \text{by cyclic rule}$

$$= \dfrac{\dfrac{1}{V} \left(\dfrac{\partial V}{\partial T} \right)_P}{-\dfrac{1}{V} \left(\dfrac{\partial V}{\partial P} \right)_T}$$

$$= \dfrac{\alpha, \text{ the coefficient of thermal expansion}}{K, \text{ the coefficient of compressibility}}$$

Then $\qquad C_P - C_V = T \cdot \dfrac{\alpha}{K} \cdot V \alpha$

$$= \dfrac{TV\alpha^2}{K} \qquad (5.20b)$$

In case of an ideal gas $PV = nRT$ when

$$\alpha = \frac{1}{V}\left(\frac{\partial V}{\partial T}\right)_P = \frac{1}{T}, K = -\frac{1}{V}\left(\frac{\partial V}{\partial P}\right)_T = \frac{nRT}{P^2V} = \frac{1}{P}$$

and then $C_P - C_V = nR$ by Eq. (5.20b).

5.6 JOULE-THOMSON PHENOMENON

In quest for existence of intermolecular force in a gas, Joule carried out an experiment in which a gas (air) was allowed to undergo expansion into a vacuum, called Joule expansion, using heat conducting vessels immersed in water bath (serving as calorimeter). He failed to detect any change of temperature in this process. It led him to the wrong conclusion that $\left(\frac{\partial U}{\partial V}\right)_T = 0$, i.e. $U = f(T)$ which is Joule's law that holds exactly for ideal gases and only approximately for real gases.

The existence of intermolecular force in a gas was clearly evidenced by the Joule–Thomson expansion in which a compressed gas is forced at constant pressure P_1 through a thermally insulated porous plug such that the exiting gas always remains at constant pressure P_2 using thermally insulated cylinder and double piston as shown in Fig. 5.1.

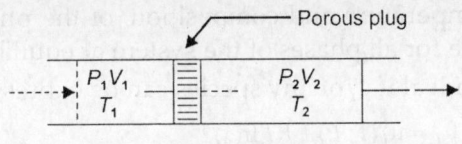

Fig. 5.1

If V_1 volume of gas at pressure P_1 becomes V_2 at pressure P_2 after passing through the plug, then

$$\Delta U + (P_2V_2 - P_1V_1) = Q = 0$$

or

$$\Delta U + \Delta(PV) = \Delta H = 0$$

Then Joule–Thomson expansion is isenthalpic but unlike Joule expansion, it is not a free expansion.

It has been found that on isenthalpic expansion a gas undergoes fall or rise in temperature (except at certain temperature called *inversion temperature* that depends on pressure). This phenomenon is known as Joule–Thomson effect. The relevant physical quantity $\left(\frac{\partial T}{\partial P}\right)_H$ is called Joule–Thomson coefficient (μ). Now considering the relation

$$dH = \left(\frac{\partial H}{\partial T}\right)_P dT + \left(\frac{\partial H}{\partial P}\right)_T dP = 0$$

we have

$$\mu = \left(\frac{\partial T}{\partial P}\right)_H = -\frac{1}{C_P}\left(\frac{\partial H}{\partial P}\right)_T \tag{5.21}$$

$(\partial H/\partial P)_T$ is called isothermal Joule–Thomson coefficient.

A comparable relation, that holds in case of a free adiabatic expansion of a gas, called Joule expansion (where U is supposed to remain constant) is

$$\left(\frac{\partial T}{\partial V}\right)_V = -\frac{1}{C_V}\left(\frac{\partial U}{\partial V}\right)_T \tag{5.22}$$

$\left(\frac{\partial U}{\partial V}\right)_T$ is called Joule coefficient.

5.7 FUNDAMENTAL EQUATIONS FOR OPEN SYSTEMS

To deal with open systems, particularly ones involving variable composition such as chemical systems, we need thermodynamic relations like

$$dU = TdS - PdV + \sum \mu_i dn_i \tag{5.23}$$

$$dH = TdS + VdP + \sum \mu_i dn_i \tag{5.24}$$

$$dA = -SdT - PdV + \sum \mu_i dn_i \tag{5.25}$$

$$dG = -SdT + VdP + \sum \mu_i dn_i \tag{5.26}$$

which are applicable to every phase of a system in a reversible process. Here n_i is the number of moles of the ith species in the phase concerned and μ_i is the chemical potential defined by

$$\mu_i = \left(\frac{\partial U}{\partial n_i}\right)_{S,V,n_{j(\neq i)}} = \left(\frac{\partial H}{\partial n_i}\right)_{S,P,n_{j(\neq i)}} = \left(\frac{\partial A}{\partial n_i}\right)_{T,V,n_{j(\neq i)}} = \left(\frac{\partial G}{\partial n_i}\right)_{T,P,n_{j(\neq i)}}$$

Obviously, μ is an intensive quantity (being a ratio of two extensive quantities) and hence, for a particular species in a particular system, it depends only on the intensive properties—pressure, temperature and composition, of the phase concerned. It can be established that μ_i is same for all phases of the system at equilibrium.

A system is said to be ideal if μ_i of any species can be expressed in the form

$$\mu_i = \mu_i^*(T, P) + RT \ln x_i \tag{5.27}$$

where $\mu_i^*(T, P)$ is the chemical potential of pure ith species ($x_i = 1$) in the same state of aggregation as the phase concerned at temperature T and pressure P. For an ideal gaseous system, this expression reduces to the following form

$$\mu_i = \mu_i^\circ(T) + RT \ln P_i \tag{5.28}$$

where P_i is the partial pressure of the ith species and μ_i° is the standard chemical potential corresponding to $P_i = 1$ atm (more precisely 1 bar) chosen arbitrarily as standard. [In fact Eq. (5.27) develops from Eq. (5.28)].

5.8 PARTIAL MOLAR QUANTITIES

The chemical potential is one of a family of properties called *partial molar quantities*. In general, any extensive property of a mixture can be expressed as a function of temperature (T), pressure (P) and mole number (n_i) of its constituents, and the partial molar property $\overline{X}_i$ of any species i of a mixture corresponding to an extensive property X of the mixture is defined by

$$\overline{X}_i = \left(\frac{\partial X}{\partial n_i}\right)_{T,P,nj(\neq i)}$$

Then
$$dX = \left(\frac{\partial X}{\partial T}\right)_{P,n_i} dT + \left(\frac{\partial X}{\partial P}\right)_{T,n_i} dP + \sum \overline{X}_i dn_i$$

On integrating this, keeping the intensive variables ($T, P, \overline{X}_i$) unchanged, we get

$$X = \sum n_i \overline{X}_i \tag{5.29}$$

Obviously, in case of a pure substance $X = n\bar{X}$ or $\bar{X} = X/n$, i.e. the partial molar quantities are identical with the corresponding molar quantities. For $X = G$, we have

$$G = \sum n_i \bar{G}_i$$

$$= \sum n_i \mu_i \tag{5.30a}$$

From Eq. (5.26) and (5.30a)

$$\sum x_i d\mu_i = 0 \quad \text{at constant } T \text{ and } P \tag{5.30b}$$

This is known as Gibbs-Duhem equation.

5.9 FUGACITY AND ACTIVITY

Real gases do not obey the Eq. (5.28). It is convenient to define a property fugacity (f) as a kind of fictitious pressure that measures the chemical potential of a real gas through an expression of the form analogous to Eq. (5.28)

$$\mu = \mu°(T) + RT \ln f \tag{5.31}$$
$$\lim f/P = 1$$
$$P \to 0$$

The quantity f/P is called *fugacity coefficient*. $\mu°$ is clearly the chemical potential of the gas at unit fugacity. Since f/P for most gases does not differ appreciably from unity except at pressures much greater than atmospheric the value of $\mu°$ is practically the same as the chemical potential at unit atmospheric pressure. f/P at pressure P is given by

$$\ln (f/P) = \int_0^P \frac{(Z-1)}{P} dP \tag{5.32}$$

where $Z(= P\bar{V}/RT)$ is the compressibility factor.

For a mixture of real (non-ideal) gases the fugacity f_i of the ith species is defined by

$$\mu_i = \mu_i°(T) + RT \ln f_i \tag{5.33}$$
$$\lim f_i/P_i = 1$$
$$P \to 0$$

Here fugacity f_i is analogous to partial pressure P_i of the gas. It is to be noted that f_i (unlike that of the pure gas) depends on the relative amount of the other gases present in the gas mixture. But $\mu_i°$ has precisely the same value as for the pure gas i at the same temperature.

The concepts of fugacity has importance only to gas phase. For all three phases, it is more convenient to use the concept of activity (a_i), as effective mole fraction (of a species i), that has been developed to treat real (non-ideal) systems through modification of the Eq. (5.27) to the following form

$$\mu_i = \mu_i^*(T, P) + RT \ln a_i \tag{5.34}$$
$$\lim a_i/x_i \to 1$$
$$x_i \to 1 \text{ or } 0, \text{ according as } x_i \text{ can be varied upto unity or not}$$

The quantity a_i/x_i, usually denoted by γ_i, is called activity coefficient of the ith species.

AUXILIARY PROBLEMS

5.1 When the sun shines on greenhouse, the temperature becomes higher inside than outside. Does this go against the Clausius statement of the second law of thermodynamics regarding heat flow?

5.2 In the isothermal expansion step of a Carnot cycle heat flows from a reservoir to the working substance (an ideal gas, say) of the engine though they are at same temperature. Here the Clausius statement regarding heat flow is violated. Comment.

5.3 Can heat be completely converted into work? Explain.

5.4 In a refrigerator, work is completely converted into heat which is equal to the heat extracted. Then the coefficient of performance of a refrigerator should always be unity. Comment.

5.5 What is the advantage of a heat pump over an ordinary electric heater? How do you explain that the coefficient of performance of a heat pump, unlike that of a refrigerator, is always greater than unity.

5.6 Draw Carnot cycle, with an ideal gas as the working substance, in P-V diagram and also in V-P diagram. What is the significance of the enclosed area in the two diagrams?

5.7 On which of the following factors does the efficiency of a heat engine depend?

(i) Material of the engine (ii) Working substance (iii) Temperature of source and sink (iv) Amount of heat withdrawn (v) Nature of the process, reversible or irreversible.

5.8 The efficiency of a reversible engine is independent of the working substance. Why then are we so much concerned about selection of a fuel for the engine? Can brick bats be used as fuel?

5.9 The efficiency of a Carnot engine is less than unity. Is this contradictory to the relation $\oint dQ = \oint dW$ (by the first law of thermodynamics)?

5.10 a. Is it possible to operate an engine following the Carnot cycle?

  b. In which case will the efficiency of a Carnot cycle be higher—when the temperature of the source is increased or when the temperature of the sink is decreased by the same amount $|\Delta T|$? Which one is the practical option for higher efficiency?

5.11 No practical heat engine is more efficient than the Carnot engine operating between the same two heat reservoirs. Establish this.

5.12 Carnot engine operates following a cycle. Does the efficiency of the engine depend on the starting point on this cycle? What happens to the efficiency of the engine if the adiabatic operations are done consecutively?

5.13 The efficiency of a heat engine would be 100% for its source at infinite temperature or sink at absolute zero. Should we expect the reverse, i.e. attainment of either temperature with such an engine?

5.14 Absolute zero is unattainable. Is it due to the fact that no engine is 100% efficient.

5.15 For cooling a room would you operate a refrigerator with its door open or use an ice box?

5.16 Carnot's formula contains both Kelvin statement and Clausius statement. Discuss.

5.17 Set up a relation between the efficiency of η of a Carnot engine and coefficient of performance ψ of the relevant refrigerator.

5.18 An ideal gas goes through a cycle consisting of alternate isothermal and adiabatic curves as shown in the diagram. The isothermal processes proceed at the temperatures T_1, T_2 and T_3. Find the efficiency of such a cycle, if in each isothermal expansion, the gas volume increases in the same proportion.

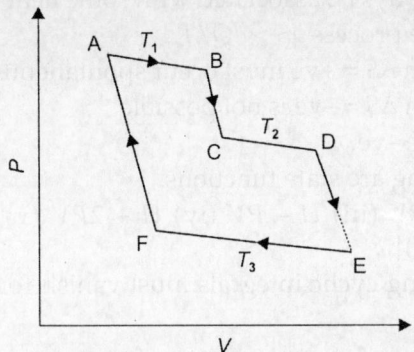

5.19 Three Carnot engines A, B and C are available. 'A' operates between the temperature, T_1 and T_2, 'B' operates between T_2 and T_3 and 'C' operates between T_1 and T_3. Show whether the efficiency of the combined action of A and B will be higher than, lower than or equal to that of C. Also find a relation between the efficiencies of A, B and C.

5.20 In a reversible heat engine, employing an ideal gas as the working substance, the adiabatic compression takes place from 20 L to 10 L. Calculate the efficiency of the engine. If the temperature of the source be 500°C, find that of the sink ($\gamma = 1.67$).

5.21 For a Carnot heat engine the operating temperature T_1 of source and T_2 of sink can vary within the range 300 K to 1000 K, but $T_1 - T_2$ is fixed at 100 K. Find the highest efficiency of the engine and T_1 corresponding to this.

5.22 A Carnot refrigerator receives 1000 J of heat from a heat reservoir at 100 K. How much heat does it deliver to the surroundings (sink) at 300 K? How much heat would be produced if the work consumed by the refrigerator in this operation is directly converted into heat? Is the principle of conservation of energy violated here?

5.23 A freezer is to be maintained at a temperature of 238 K on a summer day when the ambient temperature is 306 K. In order to maintain the freezer box at 238 K, it is necessary to remove heat from if at the rate of 1230 J/s. What is the maximum possible coefficient of performance of the freezer, and what is the minimum power that must be supplied to the freezer? Also find the heat discarded to the environment.

5.24 Calculate the minimum electrical energy required to produce one kg of ice from water by means of a refrigerator in a room at 27°C, if water is initially at (a) 0°C (b) 27°C. How much heat is given up to the surroundings in each case?

5.25 A Carnot engine operating between two reservoirs at 27°C and 0°C produces work which is used to run a Carnot refrigerator operating between two other reservoirs at 0°C and 27°C. How much heat will be absorbed by the heat engine for production of one kg of ice by the refrigerator? Will there be any production of ice if the two reservoirs are common to the heat engine and refrigerator?

5.26 Two identical bodies of finite heat capacity are available, one at high temperature T_1 and the other at low temperature T_2. They are connected through a Carnot engine which operates till both the bodies attain the same temperature T_3. Find T_3 and efficiency of the engine in terms of T_1 and T_2.

5.27 Justify or criticise the following:

   i. It follows from the defining equation of entropy, $ds = dQ/T$, that an entropy change should always be associated with some heat change.

   ii. In any irreversible process $dS > dQ/T$.

   iii. A change in which $\Delta S = +ve$ must occur spontaneously.

   iv. A change in which $\Delta S = -ve$ is not possible.

   v. $\Delta S_{universe}$ cannot be $-ve$.

5.28 Which of the following are state functions:

   (i) $Q + W$ (ii) $U + PV$ (iii) $U - PV$ (iv) $U + 2PV$ (v) G (vi) ΔG (vii) $(\partial G/\partial T)_P$ (viii) $[\partial^2(A/T)/\partial(1/T)]_V$.

5.29 Which of the following cyclic integrals must vanish for a closed system that does only PV-work?

   (a) $\oint VdV$ (b) $\oint PdV$ (c) $\oint(PdV + VdP)$ (d) $\oint(PdV - VdP)$ (e) $\oint dQ_{rev}$ (f) $\oint dQ_{rev}/T$

   (g) $\oint dQ_P$ (h) $\oint dW_{rev}/P$ (i) $\oint UdT$.

5.30 Mention the condition(s) for the following:

   i. $\Delta S = Q/T$.

   ii. An adiabatic change of a gas becomes isentropic.

   iii. An adiabatic change of a gas becomes isenthalpic.

5.31 Establish the following:

   i. $\Delta S_{uni} = 0$, for a reversible process

     $\Delta S_{uni} > 0$, for a natural process

   ii. $\Delta S_{surr} > 0$, for a natural cyclic process

   iii. An adiabatic irreversible cycle is not possible

   iv. Two reversible adiabates of a system will never meet

   v. A reversible adiabat and a reversible isothermal of a system can meet only at one point

   vi. If an isochore and an isobar of a gas for a reversible process are plotted in T-S diagram through a given point, the former will be steeper

   vii. For a system if enthalpy is independent of temperature, entropy will also be the same.

5.32 Draw Carnot cycle in T-S diagram corresponding to its P-V diagram. On the basis of such diagram, deduce an expression of efficiency of the Carnot engine and explain that the efficiency does not depend on the nature of the working substance. What is the significance of the enclosed area in this diagram. Will this area be same as that in P-V diagram?

5.33 An engine using ideal gas as the working substance is subjected to the following reversible cyclic process:

   (1) Adiabatic expansion when temperature falls from T_1 to T_2

   (2) Isothermal expansion at T_2

   (3) Adiabatic compression till temperature rises to T_1

   (4) Isothermal compression

   Represent the cycle in P-V and S-T diagrams. What does the enclosed area represent in each diagram. What type of engine it is?

5.34 The efficiency of a Carnot heat engine, using N_2 as working substance, is 80%. For the first step of isothermal expansion, it is found that $Q = 30$ kJ, $\Delta S = 100$ J/K and

$\Delta V = 20$ L per cycle. Find Q and ΔS for the third step. If N_2 is replaced by equal amount of He, will Q, ΔS and ΔV for the third step (source and sink remaining same) be same for the given data on the first step?

5.35 Using appropriate Maxwell relation, show that for an ideal gas S varies linearly with $\ln P$.

5.36 Show that if a gas behaves ideally (i.e. equation of state $PV = nRT$) then

$$\left(\frac{\partial U}{\partial V}\right)_T = \left(\frac{\partial H}{\partial P}\right)_T = 0. \text{ Is the converse true?}$$

5.37 State with basis whether $\left(\dfrac{\partial U}{\partial V}\right)_T$ and/or $\left(\dfrac{\partial H}{\partial P}\right)_T$ may be zero for any of the following equations of state:

(i) $\left(P + \dfrac{a}{V^2}\right)V = RT$ (ii) $P(V - b) = RT$ (iii) $\left(P + \dfrac{a}{V^2}\right)(V - b) = RT$.

5.38 $(\partial H/\partial P)_T = 0$ serves as a better criterian for ideality of a gas than $(\partial U/\partial V)_T = 0$. Justify or criticise.

5.39 Deduce the following relations
   a. $C_P = -T\,(\partial^2 G/\partial T^2)_P = 0$
   b. $C_P/C_V = (\partial V/\partial P)_T\,(\partial P/\partial V)_S$
   c. $\bar{C}_P - \bar{C}_V \simeq R\left(1 + \dfrac{2aP}{R^2 T^2}\right)$ for a van der Waals gas

5.40 Show that $(\partial C_V/\partial V)_T$ is same for an ideal gas and a van der Waals gas. How would you justify this.

5.41 Establish the relation $(\partial C_P/\partial P)_T = -T(\partial^2 V/\partial T^2)_P$. Hence show that for an ideal gas C_P is independent of pressure.

5.42 Under what condition the following relations will hold:
   (i) $G = H - TS$ (ii) $\Delta G = \Delta H - T\Delta S$

5.43 $G = f(H, S)$. Hence, change of G of a system cannot occur without any change of H or S. True or false.

5.44 Is there any meaning of G for the universe? What happens to S and G of the system (boat) and the universe when a boat drowns in a lake?

5.45 G for a gas is given by $\bar{G} = RT \ln P + A + BP + CP^2 + DP^3$ where A, B, C and D are temperature dependent constant. Derive an equation of state of the gas.

5.46 For a gas $\bar{A} = -\frac{a}{V} - RT \ln (\bar{V} - b) + c$, where a and b are characteristic constants for the gas, and c is a function of temperature. Derive an expression for pressure of the gas.

5.47 Give example of a process for which (a) $\Delta U = 0$, (b) $\Delta H = 0$ (c) $\Delta S = 0$ (d) $\Delta A = 0$ and (e) $\Delta G = 0$.

5.48 Which of the quantities W, ΔU, ΔH, ΔS, ΔA and ΔG, are zero for each of the following processes:
   a. Carnot cycle
   b. Joule-Thomson expansion of an ideal gas
   c. Joule-Thomson expansion of a real gas
   d. Vaporisation of water at 100°C and 1 atm
   e. Reaction between H_2 and O_2 with formation of H_2O at constant T and P
   f. Reaction between H_2 and O_2 in a thermally insulated bomb

5.49 An ideal gas expands isothermally at 27°C producing 4180 J of work. The entropy change is 10 cal/degree. Is the process reversible? What maximum work can be obtained from this change in state?

5.50 One mole of an ideal gas at 2 atm pressure expands irreversibly and adiabatically into vacuum to decrease its pressure to 1 atm. Calculate ΔS for the change.

5.51 Liquid water is isothermally compressed at 293 K:

(i) from 1 atm to 10 atm (ii) from 1 atm to 1000 atm (iii) from $V_{initial}$ to $V_{final} = 0.98$ $V_{initial}$. Calculate $\Delta \bar{S}$ for each case taking necessary assumptions and approximations. For water at 298 K and 1 atm cubic expensivity, $\alpha = 2.0 \times 10^{-4} K^{-1}$ and isothermal compressibility, $\beta = 4.53 \times 10^{-5}$ atm.

5.52 1 mole of a gas A at 300 K and 1 atm is adiabatically and isochorically mixed with 2 moles of another gas B at same temperature and pressure. Calculate ΔS assuming that A and B are ideal.

5.53 Equal amounts (n moles) of a liquid are adiabatically mixed at constant pressure. Find ΔS if initially the two quantities of the liquid have (i) same temperature (ii) different temperatures T_1 and T_2. Is your result consistent with the Clausius statement of the second law regarding entropy?

5.54 100 g of a chip of gold (sp heat 0.131 $Jg^{-1}K^{-1}$) at 127°C is put into 20 g of water at 27°C and the content is allowed to attain equilibrium under adiabatic condition. Find (i) ΔS_{Au} (ii) ΔS_{H_2O} (iii) ΔS_{system} (iv) $\Delta S_{universe}$.

5.55 a. Repeat the calculation of the quantities in the previous problem, if water is present in large amount and point out the difference in results.

b. How will the result change if the same amount of gold is dropped into water from a height of 10 m?

5.56 Boiling water at 100°C is connected to freezing water at 0°C through a metal rod and the entire system is thermally insulated. During a certain interval of time 100 g of ice melts. Find ΔS of (i) boiling water (ii) freezing water (iii) metal road (iv) universe.

5.57 a. An electric current of 2 amp is passed for 20 sec through a coil of resistance 10 ohm at a constant temperature of 25°C. Find entropy change of the system and also of the surroundings.

b. Also calculate the same quantities if the same current passes for the same duration through the same resistor under adiabatic condition [mass of the resistor is 1.5 g and sp heat is 0.13 $Jg^{-1}K^{-1}$].

5.58 10 mol of an ideal monatonic gas is compressed isothermally at 300 K from 10 kPa to 150 kPa. Calculate entropy change of the system of the surroundings and also of the universe, if (a) the process is mechanically reversible and the system is surrounded by a heat reservoir at 300 K (b) the process if mechanically reversible but the heat reservoir is at 200 K (c) the process is mechanically irreversible requiring 30% more work than in (b) and the heat reservoir is at 200 K.

Will the results be affected if the gas is diatomic?

5.59 1 L of an ideal gas initially at 10 atm is expanded isothermally against a constant pressure of 1 atm till mechanical equilibrium is attained. It is then isothermally but reversibly compressed to its original volume. Find W, Q and ΔS of the system and the surroundings for the entire process. Temperature of the surroundings is 27°C. Comment on the result.

5.60 1 L of an ideal monatomic gas initially at 10 atm and 300 K undergoes adiabatic expansion till its volume becomes double. Calculate ΔS for this change in case of (a) free expansion and (b) reversible expansion. Is it possible to achieve the change in (a) by a reversible adiabatic process? State how will you achieve this change reversibly? Will ΔS for the path you have choosen be same as for (a)?

5.61 2 mol of an ideal monatomic gas occupying 20 L at 27°C is (i) adiabatically expanded against a constant external pressure of 1 atm until volume is doubled and (ii) then reversibly and adiabatically compressed until the original temperature is reached. Calculate W, ΔS and ΔG of the gas as a result of the two-stage process.

5.62 a. 1 mole of an ideal gas A at STP is entirely transferred to a vessel containing 1 mole of another ideal gas B at STP. Find ΔS and ΔH due to mixing.

 b. How will the result differ if mixing is done under the condition of constant total pressure at constant T.

5.63 A one-liter bulb containing an ideal monatomic gas X at 1 atm and a two-liter bulb containing another ideal monatomic gas Y at 3 atm are connected through a stop-cock, the temperature of each gas being 300 K. The two gases are allowed to mix by opening the stop-cock isothermally. Calculate W, Q, ΔS and ΔG in the process if X and Y are (i) different (ii) same. Also calculate, in case of (i), the minimum work required to bring the gas mixture to the initial unmixed state.

5.64 1 mole of He initially at 298 K and 1 atm is isobarically heated till the volume becomes double. Assuming He to behave ideally, calculate ΔH, ΔS and ΔG for the process.

Here ΔG comes out negative. Does this imply that the process will occur spontaneously? [Absolute entropy of He at 298 K is 126.2 JK^{-1}mol^{-1}]

5.65 1 mole of water at 100°C is allowed to cool spontaneously to room temperature 25°C.

 a. Find the entropy change of the system and universe. Are the results qualitatively justified?

 b. Does the free energy of the system increase, decrease or remain unchanged? Does your conclusion contradict the second law.

5.66 A system undergoes isothermal isobaric change at STP with $Q = 20$ kJ and $|\Delta G| = 40$ kJ. Find ΔS. State whether $|\Delta S|$ is equal to, greater than or less than $|\Delta S_{surr}|$?

5.67 1 mol of water at 100°C and 1 atm is isothermally converted into vapour at 0.5 atm. Find ΔS and ΔG for this change in state. [Latent heat of vaporisation of water is 540 cal g^{-1}].

5.68 Find $\Delta \bar{S}$ and $\Delta \bar{G}$ for the following changes:

 a. H_2O (g, 1 atm, 100°C) $\rightarrow H_2O$ (l, 1 atm, 100°C)
 b. H_2O (g, 2 atm, 100°C) $\rightarrow H_2O$ (l, 2 atm, 100°C)

Take necessary assumption(s) and approximation(s).

5.69 1 mol of supercooled water freezes at –5°C at 1 atm.

 a. Find ΔS of (i) system and (ii) surroundings. Is the process spontaneous? [C_P of water and ice are 18 and 9 cal K^{-1}mol^{-1}]

 b. Find ΔG for the process. The vapour pressure of ice and supercooled water are 3.012 and 6.163 mm.

 c. ΔG and ΔA for the process from the densities of water (0.999 g cm^{-3}) and ice (0.917 g cm^{-3}), and C_P's as given in (a).

5.70 Find $\Delta G°$ of fusion of ice per mole at $-10°C$. Given $\overline{C}_{P_{H_2O(l)}} = 18$ Cal mol^{-1} and $\overline{C}_{P_{H_2O(s)}} = 9.0$ Cal mol^{-1}.

5.71 The variation of $\Delta H°$ of fusion of ice per mol between $0°C$ and $-25°C$ is given by the expression $\Delta H°/Cal = -1100 + 9.0\ T$. Calculate $\Delta G°$ per mole at $-10°C$.

5.72 $\Delta G°$ for a certain reaction at different temperatures is given by
$$\Delta G°/Cal\ mol^{-1} = -12640 - 5.4\ T \ln T + 104.7\ T$$
Find $\Delta H°$ for the reaction at 2000 K.

5.73 Give one example of chemical reaction in each of the following cases and indicate their possibility of occurrence at constant T and P.

  i. $\Delta H = -ve$, $\Delta S = +ve$

  ii. $\Delta H = -ve$, $\Delta S = -ve$

  iii. $\Delta H = +ve$, $\Delta S = +ve$

  iv. $\Delta H = +ve$, $\Delta S = -ve$

5.74 Some chemical reaction is possible only above a certain temperature T. Is the reaction exothermic or endothermic? Also find T, taking $|\Delta H| = 30$ kcal mol^{-1} and $|\Delta S| = 40$ e.u. for the reaction.

5.75 What are the thermodynamic implications of Joule expansion and Joule-Thomson expansion? Define Joule-Thomson inversion temperature. The analogous temperature is not found with Joule expansion. Why?

5.76 a. In ordinary adiabatic expansion, the temperature of a gas is found to fall, but in Joule-Thomson expansion, the temperature of a real gas may rise, fall or remain unchanged though $Q = 0$ for both. Explain.

  b. Why Joule-Thomson expansion but not ordinary adiabatic expansion is used for liquefaction.

5.77 Is it possible to bring about the same change in state as in Joule–Thomson expansion without using porous plug?

5.78 Justify/criticise Joule-Thomson expansion is a reversible isenthalpic process.

5.79 Define Joule coefficient (μ_J) and Joule-Thomson coefficient (μ_{JT}). Are they intensive or extensive quantities? Show that

$$\mu_J = \frac{1}{C_V}\left[P - T\left(\frac{\partial P}{\partial T}\right)_V\right]$$

$$\mu_{JT} = \frac{1}{C_P}\left[T\left(\frac{\partial V}{\partial T}\right)_P - V\right]$$

5.80 (a) Show that $\mu_{JT} = 0$ for an ideal gas.

  (b) Can μ_{JT} be zero for a real gas?

5.81 Show that for a particular pressure (below certain value) a van der Waals gas has two inversion temperatures. Find the highest and lowest possible inversion temperature for such a gas.

5.82 What are the conditions for Joule–Thomson cooling effect to be observed for a van der Waals gas?

5.83 Above certain pressure inversion temperature will have no real existence. Discuss this for a van der Waals gas. Will μ_{JT} be +ve, or −ve above this pressure? Give reasons for your answer.

5.84 Do you expect Joule–Thomson cooling of a van der Waals gas more likely at its critical temperature or at much lower or higher temperature? Can critical temperature be equal to inversion temperature?

5.85 Ordinarily H_2 and He, unlike other gases, exhibit heating effect in John–Thomson expansion. Explain.

5.86 BP and inversion temperatures at 1 atm for the gases A, B and C are given in the table below:

Gas	BP (K)	Lower inversion temp (K)	Higher inversion temp (K)
A	20.3	22.4	202
B	77.4	79	711
C	90	85	765

Discuss the possibility of liquefaction of these gases initially at 273 K using (i) only J-T expansion (ii) only ordinary adiabatic expansion.

5.87 Explain the following:

a. In Joule-Thomson expansion $\Delta H = Q$, though the overall process is not isobaric.

b. μ_{JT} of an ideal gas is zero.

c. As $P \to 0$ the equation of state of a real gas approaches that of an ideal gas, but μ_{JT} of the gas does not approach zero.

5.88 Chemical potential is defined as $\left(\dfrac{\partial G}{\partial n_i}\right)_{P,T,n_{j(\neq i)}}$ and not as $\left(\dfrac{\Delta G}{\Delta n_i}\right)_{P,T,n_{j(\neq i)}}$. Why?

5.89 Deduce the following expressions

$$dG = -SdT + VdP + \sum \mu_i dn_i$$

$$dU = -TdS - PdV + \sum \mu_i dn_i$$

Hence show that

$$\left(\frac{\partial G}{\partial n_i}\right)_{T,P,n_{j(\neq i)}} = \left(\frac{\partial U}{\partial n_i}\right)_{S,V,n_{j(\neq i)}}$$

What is the significance of $\sum \mu_i dn_i$?

5.90 Under what conditions the following relations will hold:

a. $dG = -SdT + VdP$

b. $dG = -SdT + VdP + \sum \mu_i dn_i$

c. $dG = \left(\dfrac{\partial G}{\partial T}\right)_{P,n_i} dT + \left(\dfrac{\partial G}{\partial P}\right)_{T,n_i} dP + \sum \left(\dfrac{\partial G}{\partial n_i}\right)_{T,P,n_j(\neq i)} dn_i$

5.91 Chemical potential can be defined in terms of U, H, A and G. But it is preferentially defined in terms of G. Why?

5.92 Establish the following:

a. It two phases of a system contains a common species i, then μ_i will be same in the two phases at equilibrium.

b. For a mixture of ideal gases, $\mu_i = \mu_i^\circ(T) + RT \ln P_i$.

5.93 An ideal gaseous mixture must be a mixture of ideal gases. Comment.

5.94 What does chemical potential of a species signify? Justify such name.

5.95 What is the advantage of μ over G in treating a thermodynamic system?

5.96 If V_1 vol of a substance is mixed with V_2 vol of another, the total vol V will be given by $V = V_1 + V_2$. Justify or criticise.

5.97 Establish the following:

a. $\left(\dfrac{\partial \mu_i}{\partial P} \right)_{T,n_i} = \bar{V}_i$

b. $\mu_i = \bar{H}_i + T \left(\dfrac{\partial \mu_i}{\partial T} \right)_{P,n_i}$

5.98 Fugacity has meaning only to gaseous system. Comment.

5.99 How is activity of a substance is related to its fugacity?

5.100 Most American authors use the following expression of chemical potential

$$\mu_i = \mu_i^\circ(T) + RT \ln a_i$$

[instead of $\mu_i = \mu_i^*(T, P) + RT \ln a_i$]

where μ° is the value of μ at the standard pressure of 1 atm. How far is it justified?

5.101 For an aqueous glucose solution what is the significance of $\mu_i^*(T, P)$ in $\mu_i = \mu_i^*(T, P) + RT \ln a_i$ for water and glucose at ordinary temperature and pressure?

ANSWERS

5.1 No. Because although the greenhouse is warmer than its immediate surroundings, its temperature is much too low compared to that of the source (sun) of radiant energy.

5.2 Initially there is no temperature difference between the working gas and the reservoir. Here the expansion of the gas, that occurs reversibly in a carnot cycle needs to be carried out by making the external pressure infinitesimally lower than the gas pressure. At the initial stage the gas expands infinitesimally just to offset the pressure difference. This involves work done by the gas at the cost of its own energy with consequent fall in its temperature. In an attempt to prevent this, the gas will absorb heat from the reservoir (which is now at higher temperature) without violating Clausius statement but with reappearance of the pressure difference. These two processes will, therefore, recur alternately till the gas attains the final equilibrium state.

Note: Here the change is isothermal but the process is only virtually so.

5.3 This is possible only with an ideal (hypothetical) gas through an isothermal process where $\Delta U = C_V \Delta T = 0$. But even with such a gas, the conversion of heat into work is only partially possible in a cyclic process due to non-availability of heat reservoir (acting as sink) at absolute zero (being unattainable).

We can justify our answer considering that heat, unlike work, originates from random molecular motion. Then conversion of heat into work amounts to transition from random motion to directed motion which, being an unnatural process, can be achieved only under certain conditions.

5.4. Coefficient of performance of a refrigerator, $\psi = \dfrac{\text{heat extracted}}{\text{work consumed}}$

Here heat extracted (Q) and heat equivalent of work consumed (Q^1) are different quantities. Q, unlike Q^1, depends on T_{source} and T_{sink}. Only when $Q = Q^1$ does ψ become unity.

5.5 The advantage is that, for same consumption of electrical energy (W), the heat produced (Q) is greater with a heat pump (where $Q > W$) than that with an ordinary electric heater (where $Q = W$).

□ Coefficient of performance of a heat pump

$$\psi^1 = \frac{\text{Output energy}}{\text{Input energy}}$$

$$= \frac{\text{heat extracted + electrical energy consumed}}{\text{electrical energy consumed}}$$

From this definition ψ appears to be always greater than unity. But the coefficient of performance of a refrigerator need not be greater than unity because here the output energy does not include the electrical energy consumed. ψ may be less than, equal to or greater than 1. This follows from Eq. (5.1d).

5.6 A P-V diagram ABCD of Carnot cycle turns into a V-P diagram through reflection on DB. Hence in the two diagrams (shown in the adjoining figure), the enclosed area will be same signifying the same net work procuced by the engine per cycle.

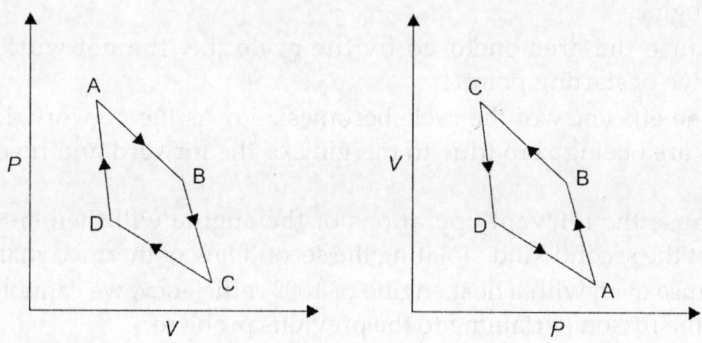

5.7 According to the Carnot theorem, the efficiency of a heat engine is determined only by the temperature of source and sink, and the extent to which it operates reversibly. Then the required factors are (iii) and (v).

5.8 For an engine, the working substance (which remains unchanged after a cycle of operation) should not be confused with a fuel (which serves as a source of energy input). Any material can in principle be used as working substance. But only substance of high calorific value are useful as fuel. Brick bats having no fuel property cannot be used as such.

5.9 Efficiency of a Carnot engine

$$= \frac{\text{Net work done by the engine per cycle } \oint dW \left(= \oint dQ\right)}{\text{Heat absorbed from the source, } Q} < 1$$

Obviously, no contradiction arises, as $\oint dQ < Q$.

5.10. a. No. Because a Carnot cycle refers to reversible operations which cannot be performed in practice.

b. Efficiency of a Carnot cycle $= \frac{T_1 - T_2}{T_1}$

Then for the specified values of T_1 (the temperature of source) and T_2 (the temperature of sink), the efficiency will be higher when the temperature of the sink is lowered, since $\frac{(T_1 + |\Delta T|) - T_2}{T_1 + |\Delta T|} < \frac{T_1 - (T_2 - |\Delta T|)}{T_1}$.

☐ However, the practical option for higher efficiency is to increase the temperature of the source, as the temperature of the available natural sinks (viz. rivers, oceans, atmosphere) is beyond our control due to their vastness.

5.11 A Carnot heat engine R operates reversibly while a practical engine S operates irreversibly. Since R is reversible, when it is operated in reverse direction, i.e. as heat pump, the heat changes and work involved will be numerically same with their sign reversed. Let us assume that S is more efficient than R when operated between the same two heat reservoirs. This means, for production of same amount of work S will absorb less amount of heat from hot reservoir and also release less amount of heat to the cold reservoir than R. If R is operated as heat pump by utilising the entire work produced by S as heat engine working between the same two heat reservoirs, then their combined action will be to transfer some heat from the cold reservoir to hot reservoir without consumption of any external energy. This goes against the Clausius statement of the second law. Hence our assumption is wrong, and we conclude that no practical heat engine is more efficiency than the Carnot engine.

5.12 No. Because the area enclosed by the cycle, i.e. the net work done is same irrespective of starting point.

☐ Here the efficiency of the cycle becomes zero, as the network done is zero, the enclosed area being zero (due to merging of the forward and reverse path of the cycle).

5.13 No. Because the relevant operations of the engine will then entail a perpetual motion of the second kind violating the second law of thermodynamics.

5.14 No. Because even with a heat engine of 100% efficiency, we cannot attain absolute zero for the reason pertaining to the previous problem.

5.15 Ice box. Operation of a refrigerator with its door oepn leads to warming instead of cooling. Because, here the room together with the refrigerator forms a closed system that gains certain amount of heat equivalent to the amount of electrical energy consumed by the refrigerator.

5.16 A heat engine absorbs heat Q at temperature T_1 and transfers a part of it at lower temperature T_2 producing work W according to the Carnot formula

$$W = Q \cdot \frac{T_1 - T_2}{T_1}$$

If, however, the engine absorbs heat at lower temperature T_2, W will be –ve, i.e. work will be consumed. This means transfer of heat from lower to higher temperature is possible only if work is supplied, i.e. spontaneous flow of heat to higher temperature is not possible, which is Clausius statement.

Again, according to the above equation, W will be +ve, so long as the temperature T_1 at which heat is absorbed by the engine is higher than the temperature T_2 at which heat is partially transferred. This means, no engine can produce work by cooling a body below the temperature of the coldest of the surrounding objects, which is Kelvin statement.

5.17 The efficiency η of a heat engine working with a source at temperature T_1 and sink at T_2 is

$$\eta = \frac{T_1 - T_2}{T_1}$$

The coefficient of performance ψ of the corresponding refrigerator (absorbing heat at T_2 and transforming it at T_1) is

$$\psi = \frac{T_2}{T_1 - T_2}$$

Then, the required relation is

$$\psi = \frac{1}{\eta} - 1$$

5.18 Considering the given cycle as combination of two Carnot cycles $A \to B \to X \to F \to A$ and $C \to D \to E \to X \to C$, we have

$$\frac{W}{Q} = \frac{T_1 - T_3}{T_1}$$

and

$$\frac{W'}{Q'} = \frac{T_2 - T_3}{T_2}$$

where

$$Q = RT_1 \ln\frac{V_B}{V_A} \text{ and } Q' = RT_2 \ln\frac{V_D}{V_C}$$

Then

$$\frac{Q}{Q'} = \frac{T_1}{T_2}, \text{ since } \frac{V_B}{V_A} = \frac{V_D}{V_C} \text{ from the given data.}$$

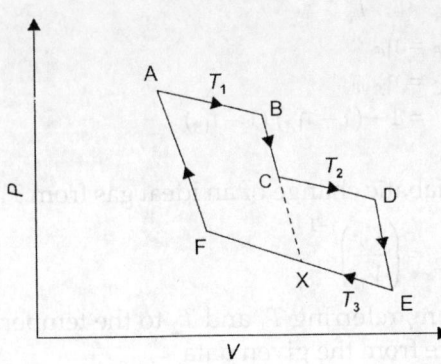

Efficiency of the cycle,

$$\eta = \frac{W + W'}{Q + Q'} = \frac{Q(T_1 - T_3)/T_1 + Q'(T_2 - T_3)/T_2}{Q + Q'}$$

$$= \frac{\dfrac{Q}{Q'}\dfrac{(T_1 - T_3)}{T_1} + \dfrac{(T_2 - T_3)}{T_2}}{\dfrac{Q}{Q'} + 1}$$

$$= \frac{\dfrac{T_1}{T_2}\dfrac{(T_1 - T_3)}{T_1} + \dfrac{(T_2 - T_3)}{T_2}}{\dfrac{T_1}{T_2} + 1}$$

$$= \frac{T_1 + T_2 - 2T_3}{T_1 + T_2}$$

Note: $\eta = \frac{T_2 - T_3}{T_2}$, when $T_1 = T_2$. This is quite expected because then the cycle reduces to a single Carnot cycle consisting of four usual steps.

5.19 Let the engine A absorbs Q_1 heat from a reservoir at T_1 transferring Q_2 heat to reservoir at T_2 and the engine B absorbs Q_2 heat from reservoir at T_2 transferring Q_3 heat to one at T_3. Then from the expression of efficiency (η) of a Carnot engine, we have

for A

$$\eta_A = \frac{Q_1 - Q_2}{Q_1} = \frac{T_1 - T_2}{T_1}, \text{ or } 1 - \eta_A = \frac{Q_2}{Q_1} = \frac{T_2}{T_1} \qquad (i)$$

for B

$$\eta_B = \frac{Q_2 - Q_3}{Q_2} = \frac{T_2 - T_3}{T_2}, \text{ or } 1 - \eta_B = \frac{Q_3}{Q_2} = \frac{T_3}{T_2} \qquad (ii)$$

(i) × (ii) gives

$$\frac{Q_3}{Q_1} = \frac{T_3}{T_1} = (1 - \eta_A)(1 - \eta_B) \qquad (iii)$$

Then the efficiency of the combined action of A and B

$$\eta_{A+B} = \frac{Q_1 - Q_3}{Q_1} = \frac{T_1 - T_3}{T_1}$$

Again, by Carnot formula, the efficiency of C,

$$\eta_C = \frac{T_1 - T_3}{T_1}$$

Hence $\eta_{A+B} = \eta_C$

Also from (iii) $\eta_C = \eta_{A+B}$

$$= 1 - (1 - \eta_A)(1 - \eta_B)$$

Note: $\eta_C \neq \eta_A + \eta_B$

5.20 In any reversible adiabatic change of an ideal gas from $T_1 V_1$ to T_2, V_2

$$T_1/T_2 = \left(\frac{V_2}{V_1}\right)^{\gamma - 1}$$

For the given problem, referring T_1 and T_2 to the temperatures of source and sink respectively, we have from the given data

$$T_1/T_2 = \left(\tfrac{20}{10}\right)^{1.67 - 1}$$

$$= 1.59 \qquad (I)$$

Then efficiency of the engine

$$= \frac{T_1 - T_2}{T_1} = 0.37$$

Again from (I) $T_2 = \dfrac{T_1}{T_1/T_2} = \dfrac{(500 + 273)\text{ K}}{1.59} = 486.2\text{ K}$

5.21 Efficiency of a Carnot engine, $\eta = \frac{T_1 - T_2}{T_1}$.

Here $(T_1 - T_2)$ being fixed (at 100 K), the highest value of η will correspond to the lowest value of $T_1 = (100 + 300)$ K $= 400$ K (the lowest value of T_2 being 300 K). Therefore the highest value of efficiency $= 100/400 = 0.25$.

Note: i. The upper limit of temperature (1000 K) is of no use.

ii. For highest value of $T_1 = 1000$ K, η is less than its highest value (0.25).

5.22 For a refrigerator, heat absorbed Q_1 at temperature T_1 and heat released Q_2 at temperature T_2 are related by

$$Q_2 = Q_1 \cdot \frac{T_2}{T_1} \quad \text{by Eq. (5.1b)}$$

$$= (1000 \text{ J}) \times \frac{300 \text{ K}}{100 \text{ K}} = 3000 \text{ J}$$

☐ Work consumed $= Q_2 - Q_1 = (3000 - 1000)$ or 2000 J.

This will produce equal amount of heat (i.e. 2000 J) on direct conversion.

☐☐ No. Because the delivered heat includes the extracted heat in addition to the heat produced by the consumed work.

5.23 Maximum coefficient of performance,

$$\psi = \frac{T_{source}}{T_{ambient} - T_{source}} \quad \text{by Eq. (5.1d)}$$

$$= \frac{238 \text{ K}}{306 \text{ K} - 238 \text{ K}} = 3.5$$

Minimum power required

$$= \frac{\text{rate of extraction of heat}}{\psi}$$

$$= \frac{1230 \text{ J/s}}{3.5} = 351.4 \text{ W}$$

∴ Heat discarded to the environment

$$= \text{heat extracted} + \text{power consumed}$$

$$= (1230 + 351.4) \text{ J/s} = 1581.4 \text{ J/s}$$

5.24 a. By Carnot's formula:

Minimum electrical energy required

$$= Q_1 \cdot \frac{T_2^* - T_1}{T_1} \quad \text{by Eq. (5.1d)}$$

$$= (80 \times 4.2 \text{ J} \times 10^3) \frac{(27 + 273) \text{ K} - (0 + 273) \text{ K}}{(0 + 273) \text{ K}} = 3.323 \times 10^4 \text{ J}$$

Heat given up to the surroundings

$$= \text{Energy consumed} + \text{heat extracted}$$

$$= 3.32 \times 10^4 \text{ J} + 80 \times 4.2 \times 10^3 \text{ J} = 3.69 \times 10^5 \text{ J}$$

[Alternatively, heat given up to the surroudings

$$= Q_1 \cdot \frac{T_2}{T_1}$$

$$= (80 \times 4.2 \times 10^3 \text{ J}) \cdot \frac{(27 + 273) \text{ K}}{(0 + 273) \text{ K}} = 3.69 \times 10^5 \text{ J}$$

b. For our purpose the formation of ice may be thought to occur through the following reversible steps

  i. Cooling of water from 27°C to 0°C

  ii. Freezing of water at 0°C

For step (i): Here the formula used in (i) is not applicable, because the temperature of water T is not constant till freezing starts. Here

Min. electrical energy required

$$= \int_{T_f}^{T_i} dQ \left(\frac{T_2 - T}{T} \right)$$

where dQ is the heat taken up by the refrigerator, i.e. given up by water at T which varies from T_i to T_f.

$$= \int_{T_f}^{T_i} C_V dT \left(\frac{T_2}{T} - 1 \right)$$

where C_V is the heat capacity of water

$$= C_V T_2 \int_{T_f}^{T_i} \frac{dT}{T} - C_V \int_{T_f}^{T_i} dT$$

$$= C_V T_2 \ln \frac{T_i}{T_f} - C_V (T_i - T_f)$$

$$= (4200 \text{ J/K}) (27 + 273) \text{ K} \ln \frac{27 + 273}{0 + 273} - (4200 \text{ J/K}) (27 \text{ K})$$

$$= 1.78 \times 10^4 \text{ J}$$

For step (ii): The energy consumed is same as in (a).

The required energy consumption

    = energy consumed in (i) + energy consumed in (ii)

Heat given up to the surroundings

    = energy consumed by refrigerator + heat given up by water in (i) + heat given up by water in (ii)

Note: On putting $T_1 = T_2 = 300$ K in the Carnot's formula, we arrive at the ridiculous result of zero energy consumption in ice formation. The fallacy lies in the variation of T before the start of freezing.

5.25 Here work produced by the heat engine is entirely consumed by the refrigerator. Then by Carnot's formula

$$Q_H \cdot \frac{T_H - T_H'}{T_H} = Q_R \cdot \frac{T_R' - T_R}{T_R}$$

where T_H and T_H' respectively denote the temperature of source and sink for heat engine, and T_R and T_R' denote the similar quantities for the refrigerator

or
$$Q_H = Q_R \cdot \frac{T_R' - T_R}{T_R} \cdot \frac{T_H}{T_H - T_H'}$$

$$= (80 \times 1000 \text{ cal}) \cdot \frac{(27 + 273) \text{ K}}{(0 + 273) \text{ K}}, \text{ since } T_R' - T_R = T_H - T_H'$$

$$= 87912 \text{ cal}$$

☐ No. Because in this case there will be no net loss or gain of heat by either reservoir due to compensating effect of heat engine and refrigerator regarding heat transfer.

This is in conformity with the Clausius statement (there being no external source of energy).

5.26 Here, unlike the problem 24(b), the temperature of both source and sink changes. During the operation of the engine if dQ is the heat change of the first body at temperature T and dQ' that of the second body at temperature T' at some stage, then

$$\frac{dQ}{T} + \frac{dQ'}{T'} = 0 \quad \text{by Eq. (5.1b)}$$

(or from $dS_{universe} = 0$ for a reversible process)

Integration gives

$$\int_{T_1}^{T_3} \frac{dQ}{T} + \int_{T_2}^{T_3} \frac{dQ'}{T'} = 0$$

or $\quad \int_{T_1}^{T_3} \frac{dT}{T} + \int_{T_2}^{T_3} \frac{dT'}{T'} = 0$

since $dQ = CdT$ and heat capacity C is same for both the bodies

or $\quad \ln \frac{T_3}{T_1} + \ln \frac{T_3}{T_2} = 0$

or $\quad \frac{T_3}{T_1} \cdot \frac{T_3}{T_2} = 1$

or $\quad T_3 = \sqrt{T_1 T_2}$

☐ Here the efficiency of the engine varies with time. The average efficiency (η_{av}) is given by

$$\eta_{av} = \frac{\text{Heat received from the first body} - \text{heat given up to the second body}}{\text{Heat received from the first body}}$$

$$= \frac{C(T_1 - T_3) - C(T_3 - T_2)}{C(T_1 - T_3)} = \frac{T_1 + T_2 - 2T_3}{T_1 - T_3}$$

$$= \frac{T_1 + T_2 - 2\sqrt{T_1 T_2}}{T_1 - \sqrt{T_1 T_2}}$$

$$= \frac{\sqrt{T_1} - \sqrt{T_2}}{\sqrt{T_1}}$$

Note: i. $\eta_{av} \neq \left[\eta_{initial} \left(= \frac{T_1 - T_2}{T_1} \right) + \eta_{final} (= 0) \right] \Big/ 2$

ii. $\eta_{av} = 1 - \sqrt{\dfrac{T_2}{T_1}} < 1 - \dfrac{T_2}{T_1} \left(\text{since } \sqrt{\dfrac{T_2}{T_1}} > \dfrac{T_2}{T_1} \right)$ i.e. $\eta_{av} < \eta_{initial}$

iii. Here $T_3 = \sqrt{T_1 T_2}$ (i.e. the geometric mean of T_1 and T_2).

If however, the two bodies were connected not through a heat engine but through a thermal conductor T_3 would have been $\frac{T_1 + T_2}{2}$ (i.e. the arithmatic mean of T_1 and T_2) which is greater than $\sqrt{T_1 T_2}$ because of transfer of heat from one body to the other completely (i.e. without any transformation into work).

5.27 i. The given statement is justified only in case of closed system for a reversible process. For an irreversible process $dS \neq \frac{dQ}{T}$, where $dS \neq 0$ does not necessarily imply $dQ \neq 0$, e.g. in case of irreversible adiabatic process $dS \neq 0$ but $dQ = 0$.

ii. The given statement is justified for a closed system only in case of natural irreversible process, but not in case of unnatural irreversible process where $dS < \frac{dQ}{T}$.

iii. The given statement is justified provided $\Delta G_{T,P} = \Delta H - T\Delta S = -$ve for the change concerned.

iv. The given statement will not be justified if ΔH is so $-$ve that $\Delta G_{T,P} = \Delta H - T\Delta S = -$ve for the change concerned.

v. According to the second law of thermodynamics, $S_{universe}$ increase with time. This must hold if we use *universe* to mean the system plus these parts of the world that interact with the system. If however, we use *universe* to mean everything that exists (i.e. the entire physical world that includes galaxies, intergalactic matter, electromagnetic radiation, etc.), there is no guarantee that this law will hold, i.e. $\Delta S_{universe} = -$ve cannot be ruled out.

According to the *big bang* model, the presently observed expanding universe results from the explosion of tiny ball of matter and radiation. If the universe contains enough matter (yet to be confirmed), the accompanying gravitational attraction will ultimately cause the universe to contact (after passing through a maximum expansion) making $\Delta S_{universe} = -$ve during the contraction phase.

5.28 State function of a system is one that refers to a single state of the system. Accordingly (ii), (iii), (iv), (v), (vii) and (viii) are state functions. Here (ii) $U + PV = H$, (v) G, (viii) $\left(\frac{\partial G}{\partial T}\right)_P = -S$ and (viii) $\left[\frac{\partial^2 \left(\frac{A}{T}\right)}{\partial \left(\frac{1}{T}\right)}\right]_V = U$ have real significance, but (iii) and (iv) do not.

(i) $Q + W = \Delta U$ and (vi) ΔG are not state functions as they refer to a change in state (and not to a single state). But they have real significance.

5.29 For the type of the system mentioned, a cyclic integral will always vanish only if the relevant differential is exact as found with a state function. This will then happen in the following cases:

a. $\oint V dV = \oint d\left(\frac{1}{2}V^2\right) = 0$, V^2 being a state function

c. $\oint (PdV + VdP) = \oint d(PV) = 0$, PV being a state function

f. $\oint \frac{dQ_{rev}}{T} = \oint dS = 0$, S being a state function

g. $\oint dQ_P = \oint dH = 0$, H being a state function

h. $\oint \frac{dW_{rev}}{P} = \oint dV = 0$, provided W is due only to volume change, V being a state function

i. $\oint U dT = \oint df(T) = 0$, if U is a function only of T, as in case of an ideal gas

Note:

1. PdV and also VdP are not exact differentials (so that $\oint PdV$ and $\oint VdP$ are both non-zero), but $PdV + VdP$ is an exact differential, though $PdV - VdP$ is not.

2. $\oint U dT \neq 0$ for a real system just as $PdV \neq 0$ for any system.

3. (i) is similar to (a).

4. dQ_{rev}/T is an exact differential though dQ_{rev} is not.

5. dW_{rev}/P is an exact differential though dW_{rev} is not.

5.30   i. The system must be closed and the process must be reversible and isothermal.

ii. The change must be brought about reversibly.

iii. The change must be brought about under throttling condition.

Note: For $dS = dQ/T$, unlike $\Delta S = Q/T$, the isothermal condition is not required.

5.31   i. For an infinitesimal change in a closed system

Then $\qquad dS \geq dQ/T$

'=' applies to reversible process and '>' to natural (spontaneous) process

Then $\qquad\qquad \geq 0$ for an isolated system, dQ being zero

This relation holds also for the universe since it behaves as an isolated system.

Then

$$\Delta S_{uni} = \int_i^f dS_{uni} = 0 \quad \text{for a reversible process}$$

$$> 0 \quad \text{for a natural process}$$

ii. $\qquad\qquad \Delta S_{surr} = \Delta S_{uni} - \Delta S \text{ (system)}$

$$> 0 \quad \text{for a cyclic process (where } \Delta S = 0)$$

occurring naturally (where $\Delta S_{uni} > 0$)

iii. For every step of a completely irreversible (natural) adiabatic process $dS > 0$ (in a closed system). Then $\int dS$ will not be zero for such an irreversible process occurring in a cycle. This contradicts the very concept of entropy as a state function. Hence such a cycle is not possible. This conclusion stands even if the cycle is not completely irreversible where $dS \geq 0$.

[**Note:** If a system undergoes irreversible adiabatic change, it cannot be restored to its initial state only through an adiabatic process.]

iv. In a reversible adiabatic process entropy of a system remains unchanged. Then if two reversible adiabats meet at any point, one will merge into the other denying their separate existence. This means two reversible adiabats of a system will never meet.

Note: Adiabats resemble electric or magnetic lines of force that never meet one another.

v. If a reversible adiabat and a reversible isothermal of a system meet twice then at the two meeting points entropy will be same if they are considered lying on the adiabat but different if they are considered lying on the isothermal leading to an absurdity. Then a reversible adiabat and a reversible isothermal cannot meet at more than one point.

vi. $\dfrac{\text{Slope of an isochoric T-S diagram}}{\text{Slope of an isobaric T-S diagram}} = \dfrac{(\partial T/\partial S)_V}{(\partial T/\partial S)_P} = \dfrac{T/C_V}{T/C_P}$

$$= \frac{C_P}{C_V} \quad \text{at a common point}$$

$$> 1$$

This establishes the given proposition.

vii. The given statement follows from the relation:

$$\left(\frac{\partial S}{\partial T}\right)_P = \frac{1}{T}\left(\frac{\partial H}{\partial T}\right)_P, \quad \text{for a hydrostatic system}$$

5.32 Carnot cycle in T–S diagram corresponding to a P–V diagram is shown in the adjoining figure.

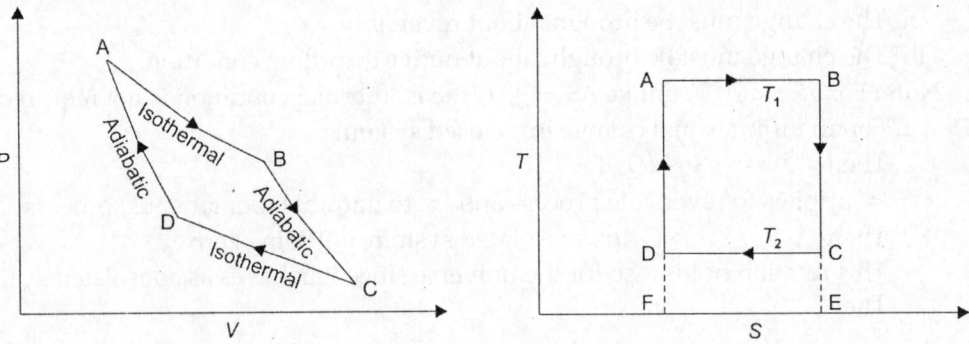

Efficiency of the Carnot's engine

$$\eta = \frac{\text{work produced (i.e. net heat absorbed) per cycle}}{\text{heat absorbed at } T_1}$$

$$= \frac{\text{Area ABCD}}{\text{Area ABEF}} = \frac{AB(T_1 - T_2)}{AB \times T_1}$$

$$= \frac{T_1 - T_2}{T_1}$$

Since Area ABCD/Area ABEF is independent of the working substance η will also be so.

The enclosed area $\oint T dS$ in T–S diagram represents the net heat absorbed (since $T dS = dQ$).

The area enclosed in the T–S diagram equals that in the P–V diagram (if expressed in the same unit) since $\oint T dS = \oint P dV$ in a Carnot's cycle.

5.33 The P–V diagram and S–T diagram are given in the adjoining figures.

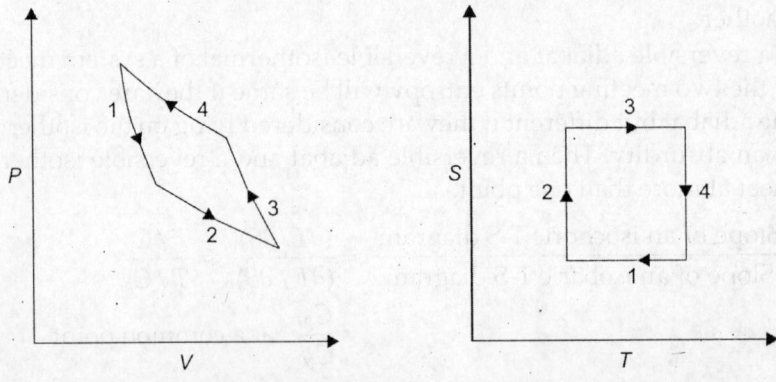

The engine in question acts as a refrigerator/heat pump following reversed Carnot cycle.

The enclosed area $\left(\oint P dV\right)$ in the P-V diagram represents the net work produced by the engine per cycle. And the enclosed area $\left(\oint T dS\right)$ over the S-axis in the S-T diagram represents the net heat absorbed per cycle. Then the two diagrams must involve equal enclosed area (by first law of thermodynamics) which is a measure of the electrical energy consumed.

Note: Here the engine does not start with absorption of heat. But it is immaterial. Because, the cycle being reversible, the enclosed area will not depend on the starting point of the cycle.

5.34 For the 1st step the temperature

$$T = \frac{Q}{\Delta S} = \frac{30 \text{ kJ}}{100 \text{ J/K}} = 300 \text{ K}$$

For the 3rd step the temperature T' is given by

$$\text{Efficiency} = \frac{T - T'}{T} = \frac{80}{100}$$

or $\qquad T' = T\left(1 - \tfrac{80}{100}\right) = (300 \text{ K})\left(1 - \tfrac{80}{100}\right) = 60 \text{ K}$

ΔS for the 3rd step $= -\Delta S$ for the 1st step,

T-S diagram of a Carnot cycle being rectangular (vide Problem 5.32)

$$= -100 \text{ J/K}$$

for the 3rd step $= T'\Delta S = (60 \text{ K}) (-100 \text{ J/K}) = -6000 \text{ J}$

☐ With He, Q and ΔS, but not ΔV, for the 3rd step will be same as found with N_2.

5.35 The appropriate Maxwell relation is

$$\left(\frac{\partial S}{\partial P}\right)_T = -\left(\frac{\partial V}{\partial T}\right)_P$$

$$= -nR/P \quad \text{for an ideal gas where } PV = nRT$$

Then $\qquad dS = -nR dP/P$ at constant temperature

Integration gives $\quad S = -nR \ln P + \text{constant}$

i.e. S varies linearly with $\ln P$ (for fixed amount of an ideal gas at constant T)

5.36 From the thermodynamic equation of state $dU = T dS - P dV$, we have

$$\left(\frac{\partial U}{\partial V}\right)_T = T\left(\frac{\partial S}{\partial V}\right)_T - P$$

$$= T\left(\frac{\partial P}{\partial T}\right)_V - P \quad \text{by Maxwell relation} \tag{I}$$

Similarly from $\quad dH = T dS + V dP$, we have

$$\left(\frac{\partial H}{\partial P}\right)_T = T\left(\frac{\partial S}{\partial P}\right)_T + V$$

$$= -T\left(\frac{\partial V}{\partial T}\right)_T + V \tag{II}$$

If follows from (I) and (II) that

$$\left(\frac{\partial U}{\partial V}\right)_T = \left(\frac{\partial H}{\partial P}\right)_T = 0$$

if $PV = nRT$ when the system is comprised of an ideal gas.

☐ The converse is not necessarily true. Because, for $\left(\frac{\partial U}{\partial V}\right)_T = \left(\frac{\partial H}{\partial P}\right)_T = 0$ to hold the necessary (and also the sufficient) condition is $PV/T = $ constant where the constant need not be equal to nR as in an ideal gas.

Note: The definition of an ideal gas provided by thermodynamics through certain energy relation $\left[\text{viz.} \left(\frac{\partial U}{\partial V}\right)_T = \left(\frac{\partial H}{\partial P}\right)_T = 0\right]$ is more general than that provided by kinetic theory through certain PVT relation (viz. $PV = nRT$).

5.37 For (i), $\qquad \left(\frac{\partial U}{\partial V}\right)_T \neq 0$

which follows from

$$\left(\frac{\partial U}{\partial V}\right)_T = T\left(\frac{\partial P}{\partial T}\right)_V - P, \text{ vide question 5.34.}$$

Also, $\qquad \left(\frac{\partial H}{\partial P}\right)_T \neq 0, \text{ from } \left(\frac{\partial H}{\partial P}\right)_T = -T\left(\frac{\partial V}{\partial T}\right)_P + V$

For (ii) $\qquad \left(\frac{\partial U}{\partial V}\right)_T = 0, \text{ but } \left(\frac{\partial H}{\partial P}\right)_T \neq 0$

For (iii) $\qquad \left(\frac{\partial U}{\partial V}\right)_T \neq 0$

and also $\qquad \left(\frac{\partial H}{\partial P}\right)_T \neq 0$ except at inversion temperature

5.38 When the deviation from ideality is only slight, a real gas generally follows an equation of state of the form

$$PV = RT + Pb \quad \text{for 1 mole}$$

[This follows from the virial form of the van der Waals equation at high temperature and low pressure]. For this equation of state $\left(\frac{\partial U}{\partial V}\right)_T = 0$ but $\left(\frac{\partial H}{\partial P}\right)_T \neq 0$ (as in the previous problem). This justifies the given statement.

Note: $\left(\frac{\partial H}{\partial P}\right)_T = 0$ but $\left(\frac{\partial U}{\partial V}\right)_T \neq 0$ is not unlikely for a gas as it appears from the equation of state $(P + a)V = RT$ (vide Question 5.34).

5.39 a. $\qquad C_P = \frac{dQ_P}{dT}$

$$= T\left(\frac{\partial S}{\partial T}\right)_P, \text{ for a reversible process in a closed system}$$

$$= T\left[\frac{\partial}{\partial T}\left(-\frac{\partial G}{\partial T}\right)_P\right]_P, \text{ since } dG = VdP - SdT$$

$$= -T\left(\frac{\partial^2 G}{\partial T^2}\right)_P$$

b. $\qquad \frac{C_P}{C_V} = \frac{dQ_P/dT}{dQ_V/dT} = \frac{T(\partial S/\partial T)_P}{T(\partial S/\partial T)_V}$

$$= \frac{(\partial S/\partial P)_T (\partial P/\partial T)_S}{(\partial S/\partial V)_T (\partial V/\partial T)_S} \text{ by cyclic rule}$$

$$= \left(\frac{\partial V}{\partial P}\right)_T \left(\frac{\partial P}{\partial V}\right)_S$$

c. $$\bar{C}_P - \bar{C}_V = T\left(\frac{\partial P}{\partial T}\right)_V \left(\frac{\partial \bar{V}}{\partial T}\right)_P$$

For a van der Waals gas

$$P\bar{V} - Pb + \frac{a}{\bar{V}} - \frac{ab}{\bar{V}^2} = RT$$

Then $$\left(\frac{\partial \bar{V}}{\partial T}\right)_P = \frac{R}{P - \dfrac{a}{\bar{V}^2} + \dfrac{2ab}{\bar{V}^3}}$$

$$\approx \frac{R}{P - \dfrac{a}{\bar{V}^2}}$$

$$\approx \frac{R}{\dfrac{RT}{\bar{V} - b} - \dfrac{2a}{\bar{V}^2}}$$

Also $$\left(\frac{\partial P}{\partial T}\right)_V = \frac{R}{\bar{V} - b}$$

Then $$\bar{C}_P - \bar{C}_V = \frac{RT}{\bar{V} - b} \cdot \frac{R}{\dfrac{RT}{\bar{V} - b} - \dfrac{2a}{\bar{V}^2}} = \frac{R}{1 - \dfrac{2a(\bar{V} - b)}{RT\bar{V}^2}}$$

$$\approx \frac{R}{1 - \dfrac{2a}{RT\bar{V}}} \text{ at ordinary pressure where } \bar{V} - b \approx \bar{V}$$

$$\approx R\left(1 + \frac{2a}{RT\bar{V}}\right) \text{ ignoring higher terms}$$

$$\approx R\left(1 + \frac{2aP}{R^2T^2}\right)$$

putting $\bar{V} = RT/P$ that holds reasonably at ordinary pressure

5.40 $$\left(\frac{\partial C_V}{\partial V}\right)_T = \left[\frac{\partial}{\partial V}\left(\frac{\partial U}{\partial T}\right)_V\right]_T$$

$$= \left[\frac{\partial}{\partial T}\left(\frac{\partial U}{\partial V}\right)_T\right]_V, \text{ since } dU \text{ is a perfect differential}$$

$$= 0 \text{ for an ideal gas where } \left(\frac{\partial U}{\partial V}\right)_T = 0$$

Similarly $\left(\frac{\partial C_V}{\partial V}\right)_T$ is also zero for a van der Waals gas where $\left(\frac{\partial U}{\partial V}\right)_T = \frac{n^2 A}{V^2}$ by Eq. (5.17). □ This is quite justified. Because C_V of a gas, which signifies the increase in molecular kinetic energy per unit rise of its temperature, does not depend on intermolecular attraction (existing in a van der Waals gas), volume remaining constant.

Note: C_P depends on intermolecular attraction because the latter determines the increase in volume, and hence increase in molecular kinetic energy, when heated at constant pressure.

5.41 $\left(\dfrac{\partial C_P}{\partial P}\right)_T = \left[\dfrac{\partial}{\partial P}\left(\dfrac{\partial H}{\partial T}\right)_P\right]_T = \left[\dfrac{\partial}{\partial P}\left\{T\left(\dfrac{\partial S}{\partial T}\right)_P\right\}\right]_T$

$$= T\left[\dfrac{\partial}{\partial T}\left(\dfrac{\partial S}{\partial P}\right)_T\right]_P$$

$$= T\left[\dfrac{\partial}{\partial T}\left(-\dfrac{\partial V}{\partial T}\right)_P\right]_P$$

$$= -T\left(\dfrac{\partial^2 V}{\partial T^2}\right)_P$$

□□ For an ideal gas $PV = nRT$, and hence $\left(\dfrac{\partial^2 V}{\partial T^2}\right)_P = 0$.

Then $\left(\dfrac{\partial C_P}{\partial P}\right)_T = 0$, i.e. C_P is independent of pressure.

Note: That C_P is independent of P for an ideal gas follows also from $\left(\dfrac{\partial H}{\partial P}\right)_T = 0$, following the procedure similar to the previous problem.

5.42 For (i), which refers to a state, there is no restriction regarding the system, i.e. holds for any system—closed or open.

(ii) which refers to a change, holds only under isothermal condition.

5.43 In general $G = f(T, H, S)$ and not $G = f(H, S)$. Hence the given statement is false.

5.44 In thermodynamics, the concept of free energy (G) was introduced to conveniently predict the reversibility and spontaneity of a process only from consideration of the system. This renders G of the universe useless though it is as meaningful as S of the university.

Considering that the state properties (P, V, T) of the boat and lake water do not change due to drowning of boat, H and S of the boat will not change. But its G will fall on hitting the bottom of the lake due to loss of its mechanical energy (arising from the gravitational field) which is part of its free energy.

H and S of lake water will, however, increase as it receives virtually the whole of heat produced due to friction of the drowning boat with water. Obviously, H and S of the universe (i.e. boat and lake together) will also increase. Here G of water will not change but G of the universe will fall.

Note: G unlike H and S, depends on the motion of the system and the external field acting on it. We must not forget this in calculating G from the usual thermodynamic relation $G = H - TS$.

5.45 $$\bar{V} = \left(\dfrac{\partial \bar{G}}{\partial P}\right)_T \quad \text{from Eq. (5.9)}$$

$$= \dfrac{RT}{P} + B + 2CP + 3DP^2 \quad \text{from the given expression}$$

5.46 $$P = -\left(\dfrac{\partial \bar{A}}{\partial V}\right)_T \quad \text{from Eq. (5.8)}$$

$$= -\dfrac{a}{\bar{V}^2} + \dfrac{RT}{\bar{V} - b} \quad \text{from the given expression}$$

which is the required expression for P.

5.47 a. Any isochoric adiabatic process (where $dU = dQ - P_{ext}dV$) in a system that can do only PV-work.

  b. Any isobaric adiabatic process in a system that can do only PV-work (where $dH = dQ - P_{ext}dV + PdV + VdP$).

  c. Any reversible adiabatic process (in a closed system) (where $dS = dQ/T$).

  d. Any reversible isochoric isothermal process in a closed system doing only PV-work (where $dA = -PdV - SdT$).

  e. Any (reversible) isobaric isothermal process in a closed system doing only PV-work (where $dG = VdP - SdT$).

Note: An adiabatic process occurring in a system always implies that a system the closed. Again an isobaric process implies that the process is mechanically reversible.

5.48 a. All except W. Because U, H, S, A and G are all state functions.

  b. $W, \Delta U$ and ΔH. Because the temperature of an ideal gas does not change in J-T expansion.

  c. Only ΔH. Because J-T expansion is a throttling process.

  d. Only ΔG. Because water is in equilibrium with its vapour at 100°C and 1 atm.

  e. None.

  f. W and ΔU. Because the system is closed doing only PV-work.

5.49 For a closed system the condition for reversibility of an isothermal process is
$$Q = T\Delta S$$

Now $\quad\quad\quad\quad\quad Q = \Delta U + W$, where W is the work produced
$$= C_V \Delta T + W, \text{ for an ideal gaseous system}$$
$$= W \quad \text{for an isothermal process}$$
$$= 4180 \text{ J}$$

Again $\quad\quad\quad T\Delta S = (27 + 273) \, K \, (10 \text{ cal}/K)$
$$= 3000 \text{ cal}$$
$$= 3000 \times 4.18 = 12540 \text{ J}$$

Since $\quad\quad\quad Q \neq T\Delta S$, the process is not reversible

Maximum work (W_{max}) will be obtained when the process is reversible.

Then $\quad W_{max} = Q_{rev} = T\Delta S$
$$= 12540 \text{ J}$$

5.50 $\quad\quad n\bar{C}_V \, (T_i - T_f) = \Delta U$
$$= Q - W \text{ (produced)}, \quad Q = 0 \text{ since the process is adiabatic}$$
$$= 0 - 0, \quad W = 0 \text{ since } P_{ext} = 0$$

Then $\quad\quad\quad\quad T_f = T_i$

$$\Delta S = nR \ln \frac{P_i}{P_f}$$
$$= (1 \text{ mol}) \, (8.314 \, JK^{-1}mol^{-1}) \ln \tfrac{2}{1}$$
$$= 5.762 \text{ JK}^{-1}$$

Note:

1. Here the process is adiabatic as well as isothermal. But $\Delta S \neq 0$, because the process is irreversible.

2. The expression for ΔS agrees with that found from Maxwell relation in problem 5.35.

5.51   i. $\left(\dfrac{\partial \bar{S}}{\partial P}\right)_T = -\left(\dfrac{\partial \bar{V}}{\partial T}\right)_P$,   Maxwell relation

Then $\Delta \bar{S} = -\alpha \bar{V}_i \Delta P$,

assuming $\bar{S}$ to vary linearly with P and ignoring the volume change, ΔP being not very high

$$= -(2.0 \times 10^{-4} K^{-1})\,(18 \times 10^{-6} m^3\,mol^{-1})\,(9 \times 1.013 \times 10^5 Pa)$$
$$= -3.28 \times 10^{-3}\,JK^{-1}mol^{-1}$$

   ii. Here the volume change is not ignorable, ΔP being quite high. However, we can use the above expression for $\Delta \bar{S}$ using the following approximation

$$\bar{V} = \frac{(\bar{V}_i + \bar{V}_f)}{2} = \bar{V}_i\left(1 + \frac{\beta \Delta P}{2}\right),$$

assuming α and β to be independent of V.

   iii. By Maxwell relation

$$d\bar{S} = \frac{\dfrac{1}{\bar{V}}\left(\dfrac{\partial \bar{V}}{\partial T}\right)_P}{-\dfrac{1}{\bar{V}}\left(\dfrac{\partial \bar{V}}{\partial T}\right)_T}\,d\bar{V} = \frac{\alpha}{\beta}\,d\bar{V}$$

or $\Delta \bar{S} = \dfrac{\alpha}{\beta}\,\Delta \bar{V}$, assuming α and β to be independent of V

$$= \frac{\alpha}{\beta}\,(-0.02\bar{V}_i)$$

$$= \frac{(2.0 \times 10^{-4}\,K^{-1})}{(4.53 \times 10^{-5} \times 1.013 \times 10^5\,P_a)} \cdot (-0.02 \times 18 \times 10^{-6}\,m^3 mol^{-1})$$

$$= -1.57 \times 10^{-9}\,JK^{-1}mol^{-1}$$

Note: If the liquid were completely incompressible $\Delta \bar{S}$ would be zero. Because, for a hydrostatic system $\bar{S} = f(T, V) = 0$ at constant T and V, and the Maxwell relation will then have no meaning. In (i), volume change actually occurs but it is ignored. It is due to too much lower compressibility of liquids that they exhibit very small variation of S with P compared with the gases (vide problem 5.50).

5.52 Here the temperature and pressure of the system will not change due to mixing

$$\Delta S_{mix} = -nR(x_A \ln x_A + x_B \ln x_B) \qquad \text{by Eq. (5.4)}$$

$$= -(1\,mol + 2\,mol)\,(8.314\,JK^{-1}mol^{-1})\left(\frac{1}{1+2}\ln\frac{1}{1+2} + \frac{2}{1+2}\ln\frac{2}{1+2}\right)$$

$$= 15.88\,JK^{-1}$$

5.53   i. $\Delta S = 0$, because none of the state properties of the system change on mixing.

   ii. Final temperature of the mixture will be $\dfrac{T_1 + T_2}{2}$

Then $\Delta S = nC_p \ln\dfrac{(T_1 + T_2)/2}{T_1} + nC_p \ln\dfrac{(T_1 + T_2)/2}{T_2}$ by Eq. (5.36)

$$= nC_p \ln \frac{(T_1 + T_2)^2}{4 T_1 T_2}$$

$$> 0 \quad \text{since } (T_1 + T_2)^2 = (T_1 - T_2)^2 + 4T_1T_2 > 4T_1T_2 \text{ for } T_1 \neq T_2$$

☐ Here $\Delta S_{universe} = \Delta S$, since $\Delta S_{surrounding} = 0$,
there being no heat exchange with surrounding

$$> 0$$

This is consistant with the Clausius statement.

5.54 Let T be the equilibrium temperature of the system (gold + water). Then equating heat lost by gold to heat gain by water

$(100 \text{ g}) (0.131 \text{ Jg}^{-1}\text{K}^{-1} (127 + 273 - T) \text{ K} = (20 \text{ g}) (4.18 \text{ Jg}^{-1}\text{K}^{-1}) (T - 273 - 27) \text{ K}$

whence $\quad\quad\quad\quad T = 313.5 \text{ K}$

i. $\quad\quad\quad \Delta S_{Au} = mc \ln \frac{T_f}{T_i}$ where m is mass of Au and c is its sp heat.

$$= (100 \text{ g}) (0.131 \text{ Jg}^{-1}\text{K}^{-1}) \ln \frac{343.5 \text{ K}}{(127 + 273)\text{K}} = -3.19 \text{ JK}^{-1}$$

ii. $\quad\quad\quad \Delta S_{H_2O} = mc \ln \frac{T_f}{T_i}$

$$= (20 \text{ g}) (4.18 \text{ Jg}^{-1}\text{K}^{-1}) \ln \frac{313.5 \text{ K}}{(27 + 273) \text{ K}} = 3.72 \text{ JK}^{-1}$$

iii. $\quad\quad\quad \Delta S_{system} = \Delta Au + \Delta S_{H_2O} = -3.19 \text{JK} + 3.72 \text{ JK} = 0.053 \text{ JK}^{-1}$

iv. $\quad\quad\quad \Delta S_{universe} = \Delta S_{system} + \Delta S_{surr}$

$$= 0.053 \text{ JK}^{-1} + 0$$

Note: Here Au is part of the system. $\Delta S_{Au} = -ve$ is not ruled out by the 2nd law which demands only ΔS_{univ} to be +ve as is found.

5.55 a. Here gold will finally attain initial temperature of water, i.e. 27°C. ΔS_{Au} can be calculated using the same expression as in the previous problem but it will have somewhat lower value due to lower T_f, i.e. $|\Delta S_{Au}|$ greater.

Entropy of water increases because it receives the heat lost by Au. But ΔS_{H_2O} cannot be calculated using the same expression as in the previous problem, since $T_i = T_f$. Here we are to use the relation

$$\Delta S_{H_2O} = \frac{\text{heat lost by Au}}{\text{temperature of } H_2O}$$

ΔS_{H_2O} thus calculated will be quite greater than that in the previous problem because of greater amount of heat received at lower temperature.

ΔS_{uni}, which is equal to ΔS of the system (Au + H_2O), will be more positive than in the previous problem.

b. ΔS_{Au}, which is determined by the initial and final thermodynamic state of Au, will be same as in (a).

But ΔS_{H_2O} will be somewhat greater, because H_2O receives some additional amount of heat (equivalent to mgh), equal to $(100 \times 10^{-3} \text{ kg}) (9.80 \text{ ms}^{-2} (10 \text{ m})$ or 9.80 J, resulting from the loss of mechanical energy of Au.

5.56 i. $\Delta S_{\text{boiling water}} = \dfrac{\text{Heat absorbed by boiling water for melting of 100 g of ice}}{\text{Boiling point of water}}$

$= -\dfrac{100 \times 80 \text{ cal}}{(100 + 273) \text{ K}} = -21.4 \text{ cal K}^{-1}$

ii. $\Delta S_{\text{freezing water}} = \dfrac{\text{Heat absorbed by freezing water}}{\text{Freezing point of water}}$

$= \dfrac{100 \times 80 \text{ cal}}{(0 + 273) \text{ K}}$

$= 29.3 \text{ cal K}^{-1}$

iii. $\Delta S_{\text{rod}} = 0$, since the metal rod absorbs no net heat

iv. $\Delta S_{\text{univ}} = \Delta S_{\text{boiling water}} + \Delta S_{\text{freezing water}} + \Delta S_{\text{rod}} + \Delta S_{\text{surr}}$

$= (-21.4 + 29.3 + 0 + 0) \text{ cal K}^{-1}$

$= +7.9 \text{ cal K}^{-1}$

Note: For boiling water the entropy change is due to the phase change $H_2O(g) \rightarrow H_2O(l)$ and for freezing water it is due to the phase change $H_2O(s) \rightarrow H_2O(l)$.

Entropy of the rod does not change due to flow of heat through it, just as the entropy of the resistor in case of isothermal conduction of electricity (vide problem 5.57).

5.57 a. Due to passage of electricity the thermodynamic state of the resistor does not change and hence its entropy, which is a state function, will not change.

To maintain a constant temperature, the heat produced by electricity goes entirely to the surroundings causing an entropy change of the latter.

$\Delta S_{\text{surr}} = \dfrac{\text{Heat produced, } i^2 rt}{T_{\text{surr}}}$

$= \dfrac{(2 \text{ amp})^2 (10 \text{ ohm}) (20 \text{ sec})}{(25 + 273) \text{ K}} = \dfrac{800 \text{ J}}{298 \text{ K}}$

$= 2.68 \text{ J K}^{-1}$

b. Here the heat produced by electricity causes temperature rise of the resistor leading to an increase in its entropy. The final temperature of the resistor is given by

$mc(T_f - T_i) = i^2 rt$

or $T_f = T_i + \dfrac{i^2 rt}{mc}$

$= 298 \text{ K} + \dfrac{800 \text{ J}}{(1.5 \text{ g}) (0.13 \text{ Jg}^{-1}\text{K}^{-1})}$

$= 4400 \text{ K}$

$\Delta S_{\text{resistor}} = mc \int_{T_i}^{T_f} \dfrac{dT}{T} = mc \ln \dfrac{T_f}{T_i}$

$= (1.5 \text{ g}) (0.13 \text{ Jg}^{-1}\text{K}^{-1}) \ln \dfrac{4400 \text{ K}}{298 \text{ K}} = 0.525 \text{ J K}^{-1}$

$\Delta S_{\text{surr}} = 0$, since surroundings is not affected

Note: In cases of both (a) and (b), $\Delta S_{\text{univ}} = +\text{ve}$, which is in agreement with the 2nd law, passing of electricity being a natural process.

5.58 a. Here, work done on the system

$$W = \int_{P_i}^{P_f} PdV = nRT \ln \frac{P_f}{P_i}$$

$$= (10 \text{ mol}) (8.314 \text{ JK}^{-1}\text{mol}^{-1}) (300 \text{ K}) \ln \frac{150}{100} = 10115 \text{ J}$$

$$\Delta U = C_V \Delta T = 0$$

$$Q = \Delta U - W = -W = -10115 \text{ J}$$

$$Q_{surr} = -Q = 10115 \text{ J}$$

Since the process is reversible and isothermal

$$\Delta S = \frac{Q}{T} = \frac{-10115 \text{ J}}{300 \text{ K}} = -33.72 \text{ JK}^{-1}$$

$$\Delta S_{surr} = \frac{Q_{surr}}{T_{surr}} = \frac{10115 \text{ J}}{300 \text{ K}} = 33.72 \text{ JK}^{-1}$$

$$\Delta S_{uni} = \Delta S + \Delta S_{surr} = 0$$

Note: $\Delta S_{uni} = 0$ is in accordance with the 2nd law, the process being reversible.

b. Here, due to appreciable difference in temperature between the system and the surroundings, the process is no longer completely reversible, the system being not in thermal equilibrium with the surroundings. However, the compression still occurs through a mechanically reversible process with the system always at the same temperature, and hence the system undergoes the same change in state by following the same path as in (a). Therefore, ΔS and also ΔU, W and Q will be same as in (a). However, ΔS_{surr}, and hence ΔS_{uni}, will be different due to different value of T_{surr}.

$$\Delta S_{surr} = \frac{Q_{surr}}{T_{surr}} = \frac{10115 \text{ J}}{200 \text{ K}} = 50.57 \text{ JK}^{-1}$$

$$\Delta S_{uni} = \Delta S + \Delta S_{surr} = -33.72 \text{ JK}^{-1} + 50.57 \text{ JK}^{-1}$$
$$= 16.85 \text{ JK}^{-1}$$

Note: The +ve value of ΔS_{uni} is in accordance with the 2nd law, the process being irreversible.

Here $\Delta S = Q/T$, though the process is not completely reversible. Because during the process, the system always remains in internal equilibrium which is enough for this relation to hold.

c. Here, the process is not mechanically reversible, although the change in state is same as in (a). Hence, ΔS and ΔU, which are independent of path, will be same as in (a), but W and Q will be different.

Here $$W = 10115 \text{ J} \times 1.3 = 13149 \text{ J}$$

$$Q = \Delta U - W = -W = -13149 \text{ J}$$

$$Q_{surr} = -Q = 13149 \text{ J}$$

$$\Delta S_{surr} = \frac{Q_{surr}}{T_{surr}} = \frac{13149 \text{ J}}{200 \text{ K}} = 65.74 \text{ JK}^{-1}$$

$$\Delta S_{uni} = \Delta S + \Delta S_{surr}$$
$$= -33.72 \text{ JK}^{-1} + 65.74 \text{ JK}^{-1}$$
$$= 32.02 \text{ JK}^{-1}$$

Note: Here $\Delta S \neq Q/T$, since the system does not remain in equilibrium even internally. But $\Delta S_{surr} = Q_{surr}/T_{surr}$ holds, because the surroundings always receive, or give up, heat reversibly (due to vastness of the surroundings).

☐ No. Because $\Delta U = C_V \Delta T$ is always zero for isothermal change of an ideal gas, though C_V depends on atomicity of the gas.

5.59 Since the gas is ideal and the expansion occurs isothermally,

$$P_i V_i = P_f V_f$$

or

$$V_f = P_i V_i / P_f,$$

$P_f = 1$ atm because the system finally attains mechanical equilibrium

$$= (10 \text{ atm}) (1 \text{ L})/(1 \text{ atm}) = 10 \text{ L}$$

For expansion $W_{exp} = -P_{ext} (V_f - V_i)$

$$= -(1 \text{ atm}) (10 \text{ L} - 1 \text{ L}) = -9 \text{ L atm}$$

For contraction $W_{cont} = -P_i V_i \ln \dfrac{V_f}{V_i}$

$$= -(1 \text{ atm}) (10 \text{ L}) \ln \frac{1 \text{ L}}{10 \text{ L}} = 23.03 \text{ L atm}$$

Since the expansion and contraction occur isothermally, the entire process becomes cyclic. Then for the entire process

$$\Delta S = 0$$

$$\Delta U = 0$$

and

$$-Q = W = W_{exp} + W_{cont} = -9 \text{ L atm} + 23.03 \text{ L atm} = 14.032 \text{ L atm}$$

$$\Delta S_{surr} = \frac{Q_{surr}}{T_{surr}} = \frac{-Q}{T_{surr}}$$

Here W and Q are not zero because of their path dependence.

5.60 a. Here $C_V(T_f - T_i) = \Delta U$, since the system consists of an ideal gas.

$$= W, \text{ since } Q = 0$$

$$= -P_{ext} \Delta V$$

$$= 0, \text{ since } P_{ext} = 0 \text{ in free expansion}$$

Then

$$T_f = T_i$$

Then

$$\Delta S = nR \ln \frac{V_f}{V_i}$$

$$= \frac{P_i V_i}{T_i} \ln \frac{V_f}{V_i}$$

$$= \frac{(10 \text{ atm}) (1 \text{ L})}{300 \text{ K}} \ln 2$$

b.

$$\Delta S = \int \frac{dQ_{rev}}{T} = 0$$

[**Note:** The relation $\Delta S = n\bar{C}_V \ln \frac{T_f}{T_i} + nR \ln \frac{V_f}{V_i}$ is not used (though it is applicable) to calculate ΔS. Because it involves the lengthy procedure of calculating first T_f by relation $T_f = T_i \left(\frac{V_i}{V_f}\right)^{\gamma-1}$. Moreover, ΔS thus calculated will not be exactly zero, though very nearly so.]

☐ No. Because if it were possible, ΔS for the two paths would be different (zero in reversible and non-zero in irreversible) which is contradictory to the concept of entropy as a state function.

☐☐ Referred to problem 4.18a.

☐☐☐ Yes. Because entropy being a state function, ΔS for the same change in state must be same irrespective of the path followed.

5.61 After the given two-stage process the system attains the initial temperature but not the original initial state. Because, here ΔS for the overall change of the system is not zero, the process being adiabatic and irreversible (where $\Delta S \neq 0$). Let T be the temperature at the end of step (i).

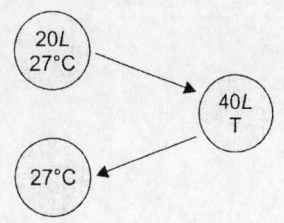

For step (i):

Work done on the system

$$W_1 = \Delta U, \ Q_1 \text{ being zero}$$
$$= n\bar{C}_V (T - 300)$$

$$\Delta S_1 = n\bar{C}_V \ln \frac{T_f}{T_i} + nR \ln \frac{V_f}{V_i}$$

T_f can be calculated as in problem 4.28a

$$\Delta G_1 = \Delta H_1 - (T_f S_f - T_i S_i), \quad \because \Delta G = \Delta H - \Delta(Ts)$$

Here ΔG_1 cannot be calculated because absolute entropy is not given.

For step (ii):

$$W_2 = n\bar{C}_V (300 - T)$$

$$\Delta S_2 = \int \frac{dQ_{rev}}{T} = 0, \ dQ_{rev} \text{ being zero}$$

ΔG_2 cannot be calculated for the same reason as ΔG_1.

For the overall process

$$W = W_1 + W_2 = 0$$
$$\Delta S = \Delta S_1 + \Delta S_2 = \Delta S_1 + 0$$
$$\Delta G = \Delta H - T\Delta S, \text{ since the overall change is isothermal}$$
$$= 0 - T\Delta S_1$$

Note: Here we can calculate ΔG for the overall isothermal change, though we cannot calculate if for each step. Because calculation of ΔG (and also ΔA), unlike other thermodynamic functions, requires the knowledge of absolute entropy unless the change is isothermal as in case of overall change in the given problem.

5.62 a. Here volume of the vessel is same as the initial volume of A. Since volume and temperature of either gas do not change in the given process, ΔS_A and ΔS_B will both be zero. Here

$$\Delta S_{mix} = \Delta S_A + \Delta S_B$$
$$= 0$$

Note: ΔS_{mix} is confusing in that ΔS of one goes is not due to the presence of the other gas.

b. Here each gas will have final volume double of its initial volume. Then

$$\Delta S_A = n_A R \ln \frac{V_f}{V_i} = R \ln 2$$

Likewise $\Delta S_B = R \ln 2$

Then $\Delta S_{mix} = \Delta S_A + \Delta S_B = 2R \ln 2$

Note: Compare this problem with problem 5.52.

5.63   i. Work done on the system $W = 0$, since $P_{ext} = 0$

$$\Delta U = \Delta U_X + \Delta U_Y$$
$$= (C_V)_X \Delta T + (C_V)_Y \Delta T$$
$$= 0, \ \Delta T \text{ being zero}$$
$$Q = \Delta U - W$$
$$= 0 - 0 = 0$$
$$\Delta S = \Delta S_X + \Delta S_Y$$
$$= \frac{(1\,\text{atm})(1\,\text{L})}{(300\,\text{K})} \ln \frac{1+2}{1} + \frac{(3\,\text{atm})(2\,\text{L})}{(300\,\text{K})} \ln \frac{1+2}{2}$$

$$= 0.1177\,\text{L atm/K} \quad \left[\text{Since for each gas } \Delta S = nR \ln \frac{V_f}{V_i} = \frac{P_i V_i}{T_i} \ln \frac{V_f}{V_i}\right.$$

$$\Delta G = \Delta H - T\Delta S$$
$$= 0 - T\Delta S,$$

since $\Delta H = \Delta H_X + \Delta H_Y$
$$= (C_P)_X \Delta T + (C_P)_Y \Delta T = 0$$

ii. Here $V_f \neq 3\,\text{L}$

Number of moles of gas in 1 liter vessel and 2 liter vessel are in the ratio $1 \times 1 : 2 \times 3$, i.e. 1:6 (since $n \propto PV$ at const T).

Here, for the gas in 1-liter vessel $V_f = \frac{1}{1+6}(1+2)\,\text{L}$ and for the gas in 2-liter vessel it is $\frac{6}{1+6}(1+2)\,\text{L}$.

Then $$\Delta S = \frac{(1\,\text{atm})(1\,\text{L})}{300\,\text{K}} \ln \frac{\frac{1}{1+6}(1+2)}{1} + \frac{(3\,\text{atm})(2\,\text{atm})}{300\,\text{K}} \ln \frac{\frac{6}{1+6}(1+2)}{2}$$

$$= 0.0022\,\text{L atm/K}$$

☐ Unmixing will require min work if it is carried out reversibly. Then,
$$\min W_{unmix} = \Delta U_{unmix} - Q_{unmix}$$
$$= 0 - T\Delta S_{unmix}$$
$$= T\Delta S_{mix}$$

Note: Here work obtained in natural mixing (which is zero) is less than the work to be done in reversing the change. This always holds for any natural process.

5.64 $$\Delta H = n\bar{C}_P(T_f - T_i)$$
$$= n\bar{C}_P T_i \quad \text{since } T_f = 2T_i \text{ for } V_f = 2V_i \text{ at const } P$$
$$= \tfrac{5}{2}nRT_i \quad \text{for a monatomic gas}$$

$$\Delta S = n\bar{C}_P \ln \frac{T_f}{T_i} = \tfrac{5}{2}nR \ln \frac{V_f}{V_i}$$

$$\Delta G = \Delta H - \Delta(TS) = \Delta H - (T_f S_f - T_i S_i) = \Delta H - [T_f(S_i + \Delta S) - T_i S_i]$$
$$= \Delta H - (T_f - T_i)S_i - T_f \Delta S$$

☐ No. For a closed system $\Delta G = -ve$, necessarily implies spontaneity of the involved change, provided T and P are constant, otherwise not as in the given situation where T is not constant.

5.65 a. Entropy change of water (system) is given by

$$\Delta S_{H_2O} = \int_i^f n(\bar{C}_P)_{H_2O} \frac{dT}{T} = n(\bar{C}_P)_{H_2O} \ln \frac{T_f}{T_i}$$

$$= (1 \text{ mol})(18 \text{ cal K}^{-1}\text{mol}^{-1}) \ln \frac{(25+273) \text{ K}}{(100+273) \text{ K}}$$

$$= -4.04 \text{ cal K}^{-1}$$

$$\Delta S_{surr} = \frac{Q_{surr}}{T_{surr}}$$

$$= \frac{\text{Heat given up by the system to the surroundings}}{T_{surr}}$$

$$= \frac{n(\bar{C}_p)_{H_2O} (T_i - T_f)}{T_f}$$

$$= \frac{18[(100+273)-(25+273)] \text{ cal}}{(25+273) \text{ K}}$$

$$= 4.53 \text{ cal K}^{-1}$$

$$\Delta S_{uni} = \Delta S_{H_2O} + \Delta S_{surr}$$

$$= -4.04 \text{ cal K}^{-1} + 4.53 \text{ cal K}^{-1}$$

$$= 0.49 \text{ cal K}^{-1}$$

□ Qualitative justification of the results lies in the Clausius inequality, $dS > dQ/T$. According to this, dS of a system, in a spontaneous process, is always +ve if $dQ = +$ve, but may be −ve if $dQ = -$ve as in the given process. Regarding the universe, $dS_{universe}$ is always predicted to be +ve, in a spontaneous process, as found above.

b. In any spontaneous process, for a closed system that does only PV-work (as the given system), we have from Eq. (5.9)

$$\left(\frac{\partial G}{\partial T}\right)_P < -S$$

Here $dT = -$ve and S is +ve (by 3rd law), and hence dG is +ve, i.e. G increases.
□ No. Because according to the 2nd law, G will necessarily decrease in a spontaneous process only if it is isothermal and isobaric, otherwise G may increase as in the given process which is not isothermal.

5.66 $$\Delta S = \frac{\Delta H - \Delta G}{T}$$

Here $\Delta H = Q$, since pressure is constant
          $= 20 \text{ kJ}$
          $\Delta G = -|\Delta G|$,

Since $\Delta G = -$ve, as $\Delta G \neq 0$ and the change occurs at constant T and P
          $= -40 \text{ kJ}$

Then $$\Delta S = \frac{20 \text{ kJ} - (-40 \text{ kJ})}{273 \text{ K}} = 0.22 \text{ kJ/K}$$

□ Since the given change occurs spontaneously, ΔG being −ve,
          $$\Delta S + \Delta S_{surr} > 0 \qquad\qquad \text{(I)}$$

Here $\Delta S = |\Delta S|$, as ΔS is positive

and $\Delta S_{surr} = -|\Delta S_{surr}|$, as ΔS_{surr} is $-ve$ due to loss of heat by the surroundings

Then, from (I) $|\Delta S| > |\Delta S_{surr}|$

5.67 ΔH and ΔG being path independent, for our purpose, we can regard the given change in state to occur through the following reversible steps:

1. H_2O (l, 100°C, 1 atm) → H_2O (g, 100°C, 1 atm)
2. H_2O (g, 100°C, 1 atm) → H_2O (g, 100°C, 0.5 atm)

For Step 1:

$$\Delta H_1 = (18 \text{ g}) (540 \text{ cal g}^{-1}) = 9720 \text{ cal}$$
$$\Delta G_1 = 0, \quad \text{by Eq. (5.9)}$$

Since liquid and vapour phases of H_2O remain in equilibrium at 100°C and 1 atm which remain constant.

For Step 2:

$$\Delta H_2 = n\bar{C}_P\Delta T, \quad \text{assuming water vapour tobehave ideally}$$
$$= 0, \text{ since } \Delta T = 0$$
$$\Delta G_2 = nRT \ln \frac{P_f}{P_i}$$
$$= (1 \text{ mol}) (1.98 \text{ cal K}^{-1}\text{mol}^{-1}) (378 \text{ K}) \ln \frac{0.5 \text{ atm}}{1 \text{ atm}}$$
$$= -511.96 \text{ cal}$$

Then for the given change
$$\Delta H = \Delta H_1 + \Delta H_2 = (9720 + 0) \text{ cal}$$
$$\Delta G = \Delta G_1 + \Delta G_2 = (0 - 511.96) \text{ cal}$$

Note: For Step 1, $\Delta G = 0$, while for Step 2, $\Delta H = 0$.

5.68 a. Here $H_2O(g)$ and $H_2O(l)$ are in equilibrium.

Then $\Delta \bar{S} = \frac{\Delta \bar{H}_{liq}}{T}$ at constant T and P

$$= -\frac{(18 \times 540)}{373 \text{ K}} \text{ cal mol}^{-1}$$
$$= -26.06 \text{ cal K}^{-1}\text{mol}^{-1}$$

$\Delta \bar{G} = 0$ as in the previous problem

b. For our purpose, the given irreversible change may be thought to occur through the following reversible steps:

1. H_2O (g, 100°C, 2 atm) → H_2O (g, 100°C, 1 atm)
2. H_2O (g, 100°C, 1 atm) → H_2O (l, 100°C, 1 atm)
3. H_2O (l, 100°C, 1 atm) → H_2O (l, 100°C, 2 atm)

For Step 1:

$$\Delta \bar{S}_1 = R \ln \frac{P_i}{P_f}$$
$$\Delta \bar{G}_1 = RT \ln \frac{P_f}{P_i}$$

For Step 2:

$\Delta \bar{S_2}$ and $\Delta \bar{G_2}$, same as in a.

For Step 3:

$$\Delta \bar{S_3} = \Delta \bar{G_3} = 0, \text{ assuming } H_2O(l) \text{ to be incompressible}$$

Because the given system is hydrostatic for which $S, G = f(T, V) = $ constant, at constant T and V (vide problem 5.51).

5.69 a. i. ΔS for a given change in state does not depend on the path followed. Then, for our purpose, we can regard the given change to occur through the following reversible steps:

1. Water (–5°C) $\xrightarrow{\text{1 atm}}$ Water (0°C)

2. Water (0°C) $\xrightarrow{\text{1 atm}}$ Ice (0°C)

3. Ice (0°C) $\xrightarrow{\text{1 atm}}$ Ice (–5°C)

$$\Delta S_1 = \int_i^f (C_P)_{water} \frac{dT}{T} = 18 \text{ cal K}^{-1} \int_{273-5}^{273+0} \frac{dT}{T} = 0.336 \text{ cal K}^{-1}$$

$$\Delta S_2 = \frac{\Delta H_{freezing}}{\text{freezing point}} = -\frac{80 \times 18 \text{ cal}}{(273+0) \text{ K}} = -5.275 \text{ cal K}^{-1}$$

$$\Delta S_3 = \int_i^f (C_P)_{ice} \frac{dT}{T} = 9 \text{ cal K}^{-1} \int_{273+0}^{273-5} \frac{dT}{T} = -0.168 \text{ cal K}^{-1}$$

Then

$$\Delta S = \Delta S_1 + \Delta S_2 + \Delta S_3$$
$$= -5.11 \text{ cal K}^{-1}$$

[**Note:** $\Delta S \neq -\frac{80 \times 18 \text{ cal}}{(273-5)\text{K}}$, since freezing of water does not occur reversibly at –5°C (but does at 0°C)].

ii.

$$\Delta S_{surr} = \frac{Q_{surr}}{T_{surr}}$$

$$= \frac{\text{Heat given up by the system to the surroundings}}{T_{surr}}$$

$$= \frac{[18(-5-0) + 18 \times 80 + 9(0+5)] \text{ cal}}{(273-5) \text{ K}} = 5.20 \text{ cal K}^{-1}$$

$$= 5.20 \text{ cal K}^{-1}$$

Here $\Delta S_{univ} = \Delta S_{system} + \Delta S_{surr} = (-5.11 + 5.20)$ or $+0.09$ cal K^{-1}

Since ΔS_{uni} is +ve, freezing of supercooled water is a spontaneous process.

b. Here

$$\Delta G_{T,P} = \underset{\substack{\text{at 1 atm}}}{G_{ice}} - \underset{\substack{\text{at 1 atm}}}{G_{water}}, \text{ at } -5°C$$

$$\simeq \underset{\substack{\text{at vap of ice}}}{G_{ice}} - \underset{\substack{\text{at vap of water}}}{G_{water}}$$

since G for solid and liquid state is almost independent of pressure

$$= \underset{\substack{\text{at vap of ice}}}{G_{H_2O(g)}} - \underset{\substack{\text{at vap of water}}}{G_{H_2O(g)}}$$

since water vap is in equilibrium with ice and water at their respective vap pr

$$= nRT \ln \frac{\text{vap pr of ice}}{\text{vap pr of water}}$$

by Eq. (5.13b) assuming $H_2O(g)$ to behave ideally

$$= (1 \text{ mol}) (1.98 \text{ cal K}^{-1}\text{mol}^{-1}) (273 - 5) \text{ K ln} \frac{3.012}{3.163}$$

$$= -25.9 \text{ cal K}^{-1}\text{mol}^{-1}$$

Since $\Delta G_{T,P}$ is –ve, freezing of supercooled water will occur spontaneously.

c. $\Delta G = \Delta H - T\Delta S$

$$\Delta H_{268} = \Delta H_{273} + [(C_P)_{ice} - (C_P)_{water}]\int_{273+0}^{273-5} dT, \quad \text{by Kirchhoff eqn.}$$

$$= [-18 \times 80 + (9 - 18)(-5 - 0)] \text{cal}$$

$$= -1395 \text{ cal}$$

Then

$$\Delta G_{268} = \Delta H_{268} - 268\Delta S_{268}$$

$$= -1395 \text{ cal} - (268 \text{ K})(-5.11 \text{ cal K}^{-1})$$

$$= -25.52 \text{ cal} \qquad \text{using } \Delta S \text{ calculated in (a)}$$

$$\Delta A \neq \Delta G - P\Delta V$$

$$P\Delta V = PM_{H_2O}\left(\frac{1}{\varrho_{ice}} - \frac{1}{\varrho_{water}}\right)$$

$$= (1 \text{ atm})(18 \text{ g})\left(\frac{1}{(0.917 \times 10^3 \text{g/L}} - \frac{1}{0.999 \times 10^3 \text{g/L}}\right) \times 24.2 \text{ cal/L atm}$$

$$= 3.46 \times 10^{-2} \text{ cal}$$

Then $\Delta A \approx \Delta G$

5.70 Here the change involved is $H_2O(s) \rightarrow H_2O(l)$

$$\Delta H_T^o = \Delta H_{273}^o + [(\bar{C}_P)_{H_2O(l)} - (\bar{C}_P)_{H_2O(g)}]\int_{273}^{T} dT, \quad \text{by Kirchhoff eqn}$$

$$= 18 \times 80 + (18 - 9)(T - 273)$$

$$= (-1017 + 9T/\text{K}) \text{ cal mol}^{-1}$$

Now $\left[\dfrac{\partial(\Delta G^o/T)}{\partial T}\right]_P = -\Delta H^o/T^2, \quad$ Gibbs–Helmholtz relation

On integrating

$$\int_{273+0}^{273-10} d\left(\frac{\Delta G^o}{T}\right) = -\int_{273+0}^{273-10} \frac{\Delta H^o}{T^2} dT$$

Then

$$\frac{\Delta G_{263}^o}{263} = \int_{263}^{273} \frac{(-1017 + 9T)}{T^2} dT$$

Since $\Delta G_{273} = 0$, $H_2O(s)$ being in equilibrium with $H_2O(l)$ at 273 K and 1 atm

or

$$\Delta G_{263}^o = 263\left|\frac{1017}{T} + 9\ln T\right|_{263}^{273} \text{ cal mol}^{-1}$$

$$= 51.1 \text{ cal mol}^{-1}$$

Note: The +ve value of ΔG_{263}^o implies that at 1 atm the concerned change will occur spontaneously only in the reverse direction, i.e. freezing instead of melting.

5.71 Similar to the previous problem. Here the expression of $\Delta H°$ is given.

5.72
$$\Delta H° = -T^2 \left[\frac{\partial(\Delta G°/T)}{\partial T} \right]_P, \quad \text{Gibbs–Helmholtz relation}$$

$$= -T^2 \left(\frac{12640}{T^2} - \frac{54 \text{ K}^{-1}}{T} \right) \text{cal mol}^{-1}$$

$$= (-12640 + 5.4 \times 2000) \text{ cal mol}^{-1}$$

$$= -1840 \text{ cal mol}^{-1}$$

Note: It is inconvenient to use the Gibbs–Helmholtz relation in the form

$$\Delta H° = \Delta G° - T \left(\frac{\partial \Delta G°}{\partial T} \right)_P, \quad \text{being lengthy.}$$

5.73 Vide Section 5.3 (spontaneity of different types of changes).

5.74 For any change to occur spontaneously, $\Delta G_{T,P}$ $(= \Delta H - T\Delta S)$ must be –ve. Then the given information implies that

$$\Delta H = +ve, \text{ i.e. the reaction is endothermic}$$

and $\Delta S = +ve$

Here $T = \dfrac{\Delta H}{\Delta S}$, which corresponds to $\Delta G_{T,P} = 0$

$$= \frac{30 \times 10^3 \text{ cal mol}^{-1}}{40 \text{ cal K}^{-1}\text{mol}^{-1}} = 750 \text{ K}$$

5.75 The thermodynamic implication of the Joule expansion is that U of a gas remains constant in such expansion and that of the Joule–Thomson expansion is that H of a gas remains constant in such expansion.

☐ The Joule–Thomson inversion temperature of a gas is one at which Joule–Thomson coefficient $(\partial T/\partial P)_H$ is zero.

☐☐ The Joule–Thomson inversion temperature corresponds to a situation where the thermal effect of the internal work done by a gas (due to intermolecular force) is off-set by that of the external work done by it. The analogous temperature does not arise with the Joule expansion because no external work is involved in such expansion (being a free expansion).

Note: $(\partial T/\partial P)_H = \mu$ of a real gas does not necessarily approach zero as the pressure is reduced even though the equation of state of the gas approaches that of an ideal gas. At very low pressure μ is zero only at certain temperature (depending on pressure) and not at all temperature (unlike an ideal gas) vide problem 5.87c.

5.76 a. In an ordinary adiabatic expansion, where $Q = 0$ but $\Delta H \neq 0$, the system does work (external and internal) at the expense of internal energy (U) and hence temperature falls. But in the Joule–Thomson expansion $Q = 0$ and also $\Delta H = 0$. It is due to this additional restriction on H (which is a function of P and T) that ΔU, and hence ΔT, may be –ve, +ve or zero in such expansion.

   b. Because with the Joule–Thomson expansion (i) there is no moving part in the device used and hence no problem of lubrication at low temperature and (ii) the lower the temperature, the larger the temperature drop for a given pressure drop.

5.77 Yes. The Joule–Thomson expansion using porous plug is but a convenient way of achieving an isenthalpic change.

5.78 Since pressure does not remain uniform throughout the system during Joule–Thomson expansion, the process is not reversible and cannot be regarded as isenthalpic, the system being in equation at the beginning and end of the process. Therefore, the given statement is not justified.

5.79 Joule coefficient and Joule–Thomson coefficient are defined by

$$\mu_J = \left(\frac{\partial T}{\partial V}\right)_U$$

$$\mu_{JT} = \left(\frac{\partial T}{\partial P}\right)_H$$

μ_J is an extensive quantity, being a ratio of an intensive quantity (∂T) and an extensive quantity (∂V). But μ_{JT} is an intensive quantity, being a ratio of two intensive quantities ∂T and ∂P.

Considering $\qquad H = f(T, P)$, for a hydrostatic system

We have

$$\left(\frac{\partial H}{\partial T}\right)_P dT + \left(\frac{\partial H}{\partial P}\right)_T dP = dH$$

$$= 0 \text{ for Joule–Thomson expansion}$$

or
$$\left(\frac{\partial T}{\partial P}\right)_H = -\frac{(\partial H/\partial P)_T}{(\partial H/\partial T)_P} = 0$$

$$= -\frac{1}{C_P}\left(\frac{\partial H}{\partial P}\right)_T$$

Again, the difference in enthalpy between two neighbouring equilibrium states of a closed system is given by

$$dH = TdS + VdP$$

or
$$\left(\frac{\partial H}{\partial P}\right)_T = R\left(\frac{\partial S}{\partial P}\right)_T + V$$

$$= -T\left(\frac{\partial V}{\partial T}\right)_P + V, \quad \text{by Maxwell relation}$$

Then
$$\mu_{J,T} = \frac{1}{C_P}\left[T\left(\frac{\partial V}{\partial T}\right)_P - V\right]$$

Likewise, considering $U = f(T, V)$ and $dU = 0$ for Joule expansion it can be shown that

$$\mu_J = \frac{1}{C_V}\left[P - T\left(\frac{\partial P}{\partial T}\right)_V\right]$$

5.80 a. This follows from either of the following relations

$$\mu_{JT} = -\frac{1}{C_P}\left(\frac{\partial H}{\partial P}\right)_T, \left(\frac{\partial H}{\partial P}\right)_T \quad \text{being zero for an ideal gas}$$

$$\mu_{JT} = \frac{1}{C_P}\left[T\left(\frac{\partial V}{\partial T}\right)_P - V\right], \quad \text{since } PV = nRT \text{ for an ideal gas}$$

  b. Yes, but only at certain temperatures, called inversion temperature, depending on the prevailing pressure.

5.81 For a van der Waals gas

$$P\bar{V} - Pb + \frac{a}{\bar{V}} - \frac{ab}{\bar{V}^2} = RT$$

Differentiation gives

$$\left(\frac{\partial \bar{V}}{\partial T}\right)_P = \frac{R}{P - \dfrac{a}{\bar{V}^2} + \dfrac{2ab}{\bar{V}^3}}$$

Then

$$\mu_{JT} = \frac{1}{C_P}\left[T\left(\frac{\partial \bar{V}}{\partial T}\right)_P - \bar{V}\right]$$

$$= \frac{1}{C_P}\left[\frac{RT}{P - \dfrac{a}{\bar{V}^2} + \dfrac{2ab}{\bar{V}^3}} - \bar{V}\right]$$

$$= \frac{1}{C_P}\left[\frac{-Pb + \dfrac{2a}{\bar{V}} - \dfrac{3ab}{\bar{V}^2}}{P - \dfrac{a}{\bar{V}^2} + \dfrac{2ab}{\bar{V}^3}}\right]$$

At inversion temperature, i.e. when $\mu_{JT} = 0$

$$-Pb + \frac{2a}{\bar{V}} - \frac{3ab}{\bar{V}^2} = 0$$

or

$$P = \frac{2a}{b\bar{V}} - \frac{3a}{\bar{V}^2}$$

Here P is maximum for $\dfrac{dP}{d\bar{V}} = \dfrac{6a}{\bar{V}^3} - \dfrac{2a}{b\bar{V}^2} = 0$ or $\bar{V} = 3b$ where $P_{max} = a/3b^2$. Again $\bar{V}$ has the highest value ∞ and lowest value $3b/2$, both for $P = 0$.

Then for a particular value of $P \le a/3b^2$, there are two values of V which correspond to two real inversion temperature (T_i) given by the van der Waals equation at T_i

$$\frac{2a}{b\bar{V}} - \frac{3a}{\bar{V}^2} + \frac{a}{\bar{V}^2}(\bar{V} - b) = RT_i$$

or

$$T_i = \frac{2a}{Rb}\left(1 - \frac{b}{\bar{V}}\right)^2$$

Then the highest inversion temperature, which corresponds to $\bar{V} = \infty$ will be $2a/Rb$ and the lowest inversion temperature, which corresponds to $\bar{V} = 3b/2$ will be $2a/9Rb$.

5.82 From the expression of μ_{JT} used in the solution of previous problem, we have for $\bar{V} \to \infty$

$$\mu_{JT} = \frac{1}{C_P}\left(\frac{2a}{P\bar{V}} - b\right)$$

$$= \frac{1}{C_P}\left(\frac{2a}{RT} - b\right) \text{ since } P\bar{V} = RT \text{ for } \bar{V} \to \infty$$

Then it follows that the highest inversion temperature $2a/9Rb$ is such that when $2a/Rb < T$ (the experimental temperature), $\mu_{JT} = $ –ve, i.e. Joule–Thomson heating effect and when $2a/Rb > T$, $\mu_{JT} = $ +ve, i.e. Joule–Thomson cooling effect.

Then regarding the lowest inversion temperature $2a/9Rb$, it follows (remembering that μ_{JT} changes its sign on passing through an inversion temperature) that when $T > 2a/9Rb$, $\mu_{JT} = $ +ve and when $T < 2a/9Rb$, $\mu_{JT} = $ –ve.

Again, as indicated in the solution of the previous problem, inversion temperature will be real only if $P \leq a/3b^2$.

Then for cooling effect (i.e. $\mu_{JT} = $ +ve) to be observed with a van der Waals gas, the necessary conditions are
$$P < a/3b^2$$
$$2a/9Rb < T < 2a/Rb$$

5.83 As indicated in the solution of problem 5.81 that an inversion temperature will be real only if $P \leq a/3b^2$, for a van der Waals gas, i.e. an inversion temperature will have no real existence above $a/3b^2$.

For $P > a/3b^2$, μ_{JT} will be –ve, i.e. only heating effect will be observed. Because under this condition of high pressure, intermolecular repulsion predominates over the attraction with consequent release of energy on expansion of the gas.

5.84 Joule–Thomson cooling of a van der Waals gas is quite possible at its critical temperature $T_C = 8a/27Rb$ and is observed when T_C lies between the two inversion temperatures associated with the prevailing gas pressure. This requirement is not fulfilled at the temperatures much lower than or higher than T_C as it appears from the lowest value $(2a/9Rb)$ and highest value $(2a/Rb)$ of the inversion temperature. The answer is now obvious.

☐ Yes. This is possible because the inversion temperature, unlike critical temperature of a gas is a pressure dependent variable quantity.

5.85 At ordinary pressure, the inversion temperature is near to $2a/Rb$ (the highest value of inversion temperature for a van der Waals gas). For H_2 and He, the value of a is much lower relative to other gases such that the value of $2a/Rb$ lies below the ordinary temperature and hence they ordinarily exhibit Joule–Thomson heating effect.

☐ Yes. H_2 and He too exhibit normal Joule–Thomson cooling effect when the experimental temperature is sufficiently low.

5.86 i. For A, liquefaction is not possible. Because the initial temperature (273 K) being greater than the highest inversion temperature (202 K), the gas will be heated on Joule–Thomson expansion.

For B, liquefaction is not possible, although cooling occurs. Because the gas will be cooled down only to the lower inversion temperature (79 K) when no further cooling to the boiling point (77.4 K) will occur.

For C, liquefaction is possible. Because here cooling will continue down to the boiling point (90 K) which is greater than the lower inversion temperature (84 K).

ii. Liquefaction of all three gases is possibly by ordinary adiabatic expansion which normally results cooling.

5.87. a. The relation $\Delta H = Q \neq 0$ is true for a gas only under isobaric condition. In Joule–Thomson expansion $\Delta H = Q$ holds without this restriction regarding pressure due to imposition of a different type of restriction, $\Delta H = 0$ (instead of $\Delta P = 0$).

b. Because the intermolecular force, which is responsible for Joule–Thomson effect, is absent in an ideal gas.

c. As $P \to 0$ although the equation of state of a real gas coincides with that of an ideal gas, the molar volume of a real gas $\bar{V}$ does not become same as the ideal gas molar vol RT/P but differ by a finite amount equal to the 2nd virial coefficient. Then

$$\mu_{JT} = \frac{1}{C_P}\left[T\left(\frac{\partial \bar{V}}{\partial T}\right)_P - \bar{V}\right]$$

$$= \frac{1}{C_P}\left[\frac{RT}{P} - \bar{V}\right] \text{ as } P \to 0$$

$$\neq 0 \quad \text{(vide problem 3.6)}$$

5.88 Chemical potential is a function of temperature, pressure and composition of the phase concerned. Due to this dependency on composition, its definition is restricted to infinitesimal addition of the species concerned. Because only then the addition will not virtually affect the composition.

Note: However, for a pure (single) substance, chemical potential (μ) can be defined by $\mu = \left(\frac{\Delta G}{\Delta n}\right)_{T,P}$, because here composition of the system is not affected on addition of any amount of the substance.

5.89 For a homogeneous system (or a phase) consisting of C number of different species, we can write

$$G = f(T, P, n_1, n_2, ..., n_c)$$

where n denotes the number of moles of the substance indicated by the suffix. Then

$$dG = \left(\frac{\partial G}{\partial T}\right)_{P,n_1,n_2,...} dT + \left(\frac{\partial G}{\partial P}\right)_{T,n_1,n_2,...} dP + \left(\frac{\partial G}{\partial n_1}\right)_{T,P,n_2,n_3,...} dn_1 + \left(\frac{\partial G}{\partial n_2}\right)_{T,P,n_1,n_3,...} dn_2 + ...$$

$$= \left(\frac{\partial G}{\partial T}\right)_{P,n_1,n_2,...} dT + \left(\frac{\partial G}{\partial P}\right)_{T,n_1,n_2,...} dP + \sum\left(\frac{\partial G}{\partial n_i}\right)_{T,P,n_j(\neq i)} dn_i$$

If the process is reversible and $dn_i = 0$, i.e. the system is closed, then the above expression must reduce to the following one

$$dG = -SdT + PdV$$

Then, for any reversible process, we can write

$$dG = -SdT + VdP + \sum \mu_i dn_i \tag{I}$$

where

$$\mu_i = \left(\frac{\partial G}{\partial n_i}\right)_{T,P,n_j(\neq i)}$$

Substitution of $G = U + PV - TS$ in (I) gives

$$dU = TdS - PdV + \sum \mu_i dn_i \tag{II}$$

☐ From (I) and (II), we have

$$\left(\frac{\partial G}{\partial n_i}\right)_{T,P,n_j(\neq i)} = \mu_i = \left(\frac{\partial U}{\partial n_i}\right)_{S,V,n_j(\neq i)}$$

□□ It appears from the expression of dU that $-\sum \mu_i dn_i$ represents the work (the chemical work) that a closed system can do due only to its change of composition in a reversible process.

5.90 a. Holds if the system is closed and the process is reversible.

b. Holds for any system if the process is reversible.

c. Holds for any system and any process (reversible or not).

Note: For $dG = -SdT + VdP$ to hold the system need not be a homogeneous one of fixed composition. Because this expression follows from $G = U + PV - TS$ with $Q = TdS$ as the only assumption that holds for a reversible process in any closed system (homogeneous or not) vide Sec. 5.3.

5.91 Because the definition of μ in terms of G refers to constancy of temperature and pressure that forms the most convenient condition in carrying out a reaction in laboratory.

5.92 a. Let only dn_i mol of the ith species is transferred from one phase (say α) to the other phase (say β) keeping the equilibrium temperature and pressure unchanged. Then for the overall system

$$0 = dG_{T,P} = dG^\alpha_{T,P} + dG^\beta_{T,P}$$

$$= \mu^\alpha_i (-dn_i) + \mu^\beta_i (dn_i), \quad \text{by Eq. (5.26)}$$

whence $\mu^\alpha_i = \mu^\beta_i$

b. For a single substance, we have from $dG = VdP - SdT$

$$\int_{G^\circ}^{G} dG = \int_{1\,atm}^{P\,atm} VdP, \quad \text{at constant } T$$

$$= \int_1^P \frac{nRTdP}{P}, \quad \text{for an ideal gas}$$

Then $\dfrac{G}{n} = \dfrac{G^\circ(T)}{n} + RT \ln P$

or $\mu = \mu^\circ(T) + RT \ln P$ (A)

If follows from this expression that the chemical potential of an ideal gas would not change on adding to it some other gas(s) at constant temperature and volume. Because in the resulting mixture, the pressure of the concerned gas (i), now called partial pressure (P_i), has the value same as that in the pure state. Then by replacing the pressure (P) of the pure gas with P_i in (A), we have the following (desired) expression of its chemical potential μ_i

$$\mu_i = \mu^\circ_i(T) + RT \ln P_i$$

5.93 The components of an ideal gaseous mixture (solution) should not necessarily be ideal gases obeying the ideal gas equation $P_iV = n_iRT$. By an ideal gaseous solution, we simply mean only the one having composition dependence of μ_i given by

$$\mu_i = \mu^*_i(T, P) + RT \ln x_i$$

where $\mu^*_i(T, P)$ is a constant for a component at a constant temperature and pressure but not necessarily equal to $= \mu^\circ_i(T) + RT \ln P$ as in case of ideal gas.

5.94 The chemical potential of a species signifies its tendency to escape from one region of the system to another to equalise this property.

□ Chemical potential owes its name to the flow of matter directed by it.

5.95 In thermodynamic treatment of a system, use of μ has the advantage over G in that the size of the system is of no concern, as μ (unlike G) is an intensive property.

5.96 The given statement is not absolutely justified unless the two substances concerned are same. Because $V = n_1 \bar{V}_1 + n_2 \bar{V}_2$ where $\bar{V}_1$ and $\bar{V}_2$, which represent the contributions to total volume per mol of the respective constituents, depend on the composition of the final mixture determined by n_1 and n_2, unless the substances concerned are same.

5.97 a.
$$\left(\frac{\partial \mu_i}{\partial P} \right)_{T, n_i} = \left[\frac{\partial}{\partial P} \left(\frac{\partial G}{\partial n_i} \right)_{T, P, n_{j(\neq i)}} \right]_{T, n_i}$$

$$= \left[\frac{\partial}{\partial n_i} \left(\frac{\partial G}{\partial P} \right)_{T, n_i} \right]_{T, P, n_{j(\neq i)}} \quad \text{Since } dG \text{ is a perfect differential}$$

$$= \left(\frac{\partial V}{\partial n_i} \right)_{T, P, n_{j(\neq i)}} , \quad \text{by Eq. (5.26)}$$

$$= \bar{V}_i$$

b. Similar procedure as in (a) leads to the relation
$$\left(\frac{\partial \mu_i}{\partial T} \right)_{P, n_i} = - \bar{S}_i$$

Now $\qquad G = H - TS$

Then $\qquad \mu_i = \left(\frac{\partial G}{\partial n_i} \right)_{T, P, n_{j(\neq i)}} = \left(\frac{\partial H}{\partial n_i} \right)_{T, P, n_{j(\neq i)}} - T \left(\frac{\partial S}{\partial n_i} \right)_{T, P, n_{j(\neq i)}}$

$$= \bar{H}_i - T \bar{S}_i$$

$$= \bar{H}_i + T \left(\frac{\partial \mu_i}{\partial T} \right)_{P, n_i}$$

5.98 Fugacity has meaning not only to gaseous systems but in general, it is a true measure of the escaping tendency of a component of any system—gaseous, liquid or solid. In a mixture, we can think of it as a kind of idealised partial pressure (for gaseous systems) or partial vapour pressure (for liquid and solid systems).

5.99 The activity of a substance is equal to the ratio of the fugacity of the substance for the specified state to its fugacity in a standard state.

5.100 It is not strictly justified. However, it leads only to minor difference in results since free energy does not vary appreciably unless pressure varies greatly in excess of atmospheric pressure, particularly for liquids and solids. Then, considering the simplicity of the given expression of chemical potential, it would always have been preferable had there been no need for precision of high degree.

5.101 $\mu_i^*(T, P)$ represents the chemical potential of a pure species i in the same state of aggregation as the solution concerned at the temperature T and pressure P. The given solution exists in liquid state of aggregation at ordinary temperature and pressure. Therefore, μ_{water}^* signifies the chemical potential of pure liquid water (that really exists at ordinary temperature), whereas $\mu_{glucose}^*$ signifies the chemical potential of pure, but hypothetical, liquid glucose at ordinary temperature and pressure (though pure glucose really exists as solid at ordinary temperature and pressure).

UNIVERSITY QUESTIONS

5.1 If V denotes the volume of n moles of an ideal gas, show that dV is a perfect differential. (Burdwan, BSC(H), 2007)

5.2 Which of the following are state functions?
(i) $Q + W$, (ii) Q_{rev}/T, (iii) $U + PV$, (iv) $U - PV$ (Burdwan, BSC(H), 2011)

5.3 Why is ΔS not a state function? (Burdwan, BSC(H), 2000)

5.4 Justify/critics: $[\partial(G/T)/\partial(1/T)]_P$ is a state function. (Burdwan, BSC(H), 2001)

5.5 Comment on: From first law of thermodynamics, we get $\oint dQ = \oint dW$. Hence, heat can be completely converted into work (by a cyclic process). (Burdwan, BSC(H), 2006)

5.6 Give an example of a process where heat is completely converted into work. (Burdwan, BSC(H), 2002)

5.7 Remark on the statement: 'heat cannot be completely converted into work' (as implied by Kelvin–Planck statement of second law). (Burdwan, BSC(H), 1999)

5.8 What do you mean by a perpetual motion of the second kind? Is this possible? Explain. (Jadavpur, BSC(H), 2000)

5.9 Show that $\oint dQ_{rev}/T = 0$ for a (reversible) Carnot engine (the working substance is an ideal gas). Hence find the efficiency of such an engine. (Calcutta, BSC(H), 2000)

5.10 Show that no engine is more efficient than a Carnot engine. (Burdwan, BSC(H), 2005)

5.11 One mole of an ideal gas is made to undergo the following reversible cycle involving the following steps:
 i. Isothermal expansion at temperature T, from state $P_1 V_1 \rightarrow P_2 \cdot V_2$.
 ii. Isochoric change at volume V_2 leading to change in state from $P_2, T_1 \rightarrow P_3, T_2$ where $P_3 < P_2$.
 iii. Adiabatic compression from $P_3, V_2, T_2 \rightarrow P_1, V_1, T_1$
 a. Represent the entire cycle on a properly labelled PV diagram.
 b. Obtain $|W|$ for each step.
 c. Show that $\eta = 1 - \dfrac{T_1 - T_2}{T_1 \ln \dfrac{T_1}{T_2}}$. (Calcutta, BSC(H), 2014)

5.12 A Carnot engine operates between two fixed temperatures and uses one mole of an ideal gas as the working substance. State how will the efficiency of the engine be affected when each of the following changes are done independently.
i. The amount of the gas is doubled, ii. The ideal gas is replaced by one mole of a van der Waals gas, iii. The engine is run in reverse cycle, iv. One step of the cycle is made irreversible. (Calcutta, BSC(H), 2000)

5.13 a. Prove that $\oint \dfrac{dQ}{T} \leq 0$ and from this expression, show that $dS \geq dQ/T$. (Calcutta, BSC(H), 2011)
 b. Comment on the values of ΔS of the universe for spontaneous changes occurring within it. (Jadavpur, BSC(H), 2012)
 c. Hot coffee cools spontaneously to room temperature. So for a spontaneous irreversible process entropy decreases. Is it true? (Calcutta, BSC(H), 2006)

    d. An endothermic process with ΔS = –ve is thermodynamically impossible. Justify or criticise. (Calcutta, BSC(H), 2012)

5.14 Differentiate a reversible cyclic process from an irreversible one in terms of net work done by the system and entropy change of the universe.

(Calcutta, BSC(H), 2007)

5.15 A non-ideal gas undergoes a Carnot cycle. Find the values of ΔU, ΔH, ΔS, $\Delta S_{universe}$, ΔA and ΔG. (Calcutta, BSC(H), 2000)

5.16 Draw Carnot cycle in T-S space, indicating clearly the steps involved. Derive efficiency of the engine. (Calcutta, BSC(H), 2008)

5.17 In the T-S diagram of a Carnot cycle, the end points of a diagonal have coordinates (S_2, T_1) and (S_1, T_2). Find the work done. (Burdwan, BSC(H), 2009)

5.18 Draw a Carnot cycle, with an ideal gas as the working substance in U-S plane.

(Calcutta, BSC(H), 2007)

5.19 Represent a Carnot cycle using an ideal gas in a H-T diagram with proper labels. Justify the diagram. (Calcutta, BSC(H), 2013)

5.20 The second law states that entropy of the universe (system + surroundings) increases in a spontaneous process: $\Delta S_{system} + \Delta S_{surr} > 0$. Give arguments that at constant T and P, ΔS_{surr} is related to the system enthalpy change by $\Delta S_{surr} = -\Delta H_{sys}/T$. Hence come to the conclusion that $-\Delta G > 0$ for the spontaneous process.

(Burdwan, BSC(H), 1997)

5.21 Justify or criticise: Since no engine is 100% efficient, absolute zero is unattainable.

(Burdwan, BSC(H), 1993)

5.22 State, giving reasons, whether the following statements are correct or incorrect?

    a. Entropy is a state function and hence does not depend on the amount of matter in the system. (Calcutta, BSC(H), 2001)

    b. In an irreversible cyclic process, the entropy of a closed system always increases.

    c. In any spontaneous process, the free energy of a closed system always decreases. (Calcutta, BSC(H), 1986)

    d. Free energy of a substance decreases with increase of temperature at constant pressure. (Calcutta, BSC(H), 1986)

5.23 Chemical potential of a gas is expressed as $\mu = \mu^0 + RT \ln f$. Comment on pressure and temperature dependence of μ and μ^0. Terms have their usual significance.

(Calcutta, BSC(H), 2014)

5.24 Point out whether for an ideal gas the following are an enthalpy or entropy effect.

    a. Increase of G with increase of P at constant T.

    b. Decrease of G with increase of T at constant P. (Burdwan, BSC(H), 1997)

5.25 Two isothermals of a system do not intersect. Why? (Burdwan, BSC(H), 2001)

5.26 What is the basic difference between Joule's and Joule–Thomson's experiments with gases? Mention one state function for each experiment which remains conserved in such experiments. Define their coefficients. Why throttled expansion of gas does not always produce cooling? (Burdwan, BSC(H), 1998)

5.27 Justify or criticise.

    a. Joule–Thomson expansion is an irreversible process. (Burdwan, BSC(H), 1997)

    b. Joule–Thomson expansion is an isenthalpic process as $H_f = H_i$.

(Calcutta, BSC(H), 2002)

5.28 Define Joule–Thomson coefficient μ_{JT}. Show that for an ideal gas $\mu_{JT} = 0$ and discuss the physical content of this equation. What is inversion temperature (T_i)? Obtain an expression for T_i in case of a van der Waals gas. (Burdwan, BSC(H), 2001)

5.29 Show that $\mu_{JT} = \frac{V}{C_P}(\alpha T - 1)$ where α is the temperature coefficient of volume expansion. (Calcutta, BSC(H), 2014)

5.30 What will happen if a gas obeying the equation of state $\left(P + \frac{n^2 a}{V^2}\right) = nRT/V$ undergoes Joule–Thomson expansion, and why? (Calcutta, BSC(H), 2006)

5.31 How Joule–Thomson (JT) cooling is different from adiabatic cooling? Find the expression of JT coefficient for a gas obeying the equation of state $P(\bar{V} - b) = RT$. What would be the JT effect for such a gas? (Calcutta, BSC(H), 2008)

5.32 Show that for a van der Waals gas $\left(\frac{\partial C_V}{\partial V}\right)_T = 0$. (Calcutta, BSC(H), 2012)

5.33 a. Evaluate $\left(\frac{\partial U}{\partial V}\right)_T$ and $\left(\frac{\partial H}{\partial P}\right)_T$ for an ideal gas. (Calcutta, BSC(H), 2003)

b. The internal energy and enthalpy of an ideal gas is a function of temperature only. Explain. (Burdwan, BSC(H), 1995)

5.34 Prove that if $\left(\frac{\partial U}{\partial V}\right)_T = 0$ for any gas, then it does not necessarily follow that $\left(\frac{\partial H}{\partial V}\right)_T = 0$ also. (Calcutta, BSC(H), 2011)

5.35 Show that for any system $C_P - C_V = T\left(\frac{\partial P}{\partial T}\right)_V\left(\frac{\partial V}{\partial T}\right)_P$, assuming thermodynamic equation of state. Hence explain the situation where (i) C_P tends to C_V (ii) $C_P < C_V$. (Calcutta, BSC(H), 2007)

5.36 With a suitable simplication for $\left(\frac{\partial V}{\partial T}\right)_P$ from the van der Waals equation, arrive at an expression for $(C_P - C_V)$ containing pressure and temperature. (Burdwan, BSC(H), 1997)

5.37 Derive $C_P - C_V = \left[P + \left(\frac{\partial U}{\partial V}\right)_T\right]\left(\frac{\partial V}{\partial T}\right)_P$. Show that for equation of state $P(V - b) = nRT$, C_P exceed C_V by the quantity nR. (Calcutta, BSC(H), 2010)

5.38 The difference in heat capacity $(C_P - C_V)$ of a thermodynamic system is given by the expression $C_P - C_V = TV\alpha^m/\beta^n$, where m, n are positive integers and other symbols have their own significance. Using dimensional arguments evaluate m and n. (Calcutta, BSC(H), 2013)

5.39 Show that

i. $C_P = -T\left(\frac{\partial^2 G}{\partial T^2}\right)_P$ (Calcutta, BSC(H), 1997)

ii. $C_V = -T\left(\frac{\partial^2 A}{\partial T^2}\right)_V$ (Calcutta, BSC(H), 2012)

iii. $C_P - C_V = \left(\frac{\alpha^2}{\beta}\right)TV$ (Calcutta, BSC(H), 2005)

5.40 Derive the following relations

i. $\left(\frac{\partial T}{\partial V}\right)_U = \frac{1}{C_V}\left(P - \frac{\alpha T}{\beta}\right)$ (Calcutta, BSC(H), 2008)

ii. $\left(\frac{\partial T}{\partial P}\right)_H = \frac{V}{C_V}(\alpha T - 1)$ (Calcutta, BSC(H), 2007)

iii. $\left(\dfrac{\partial T}{\partial P}\right)_S = \dfrac{V\alpha T}{C_P}$ (Calcutta, BSC(H), 2007)

iv. $\left(\dfrac{\partial H}{\partial P}\right)_T = V(1 - \alpha T)$ (Calcutta, BSC(H), 2005)

5.41 Assume that an isochore and an isobar have been plotted through a given point in a T-S diagram of a gaseous system. Show that the isochore will be steeper than the isobar. (Calcutta, BSC(H), 1979)

5.42 Derive the expression for $\left(\frac{\partial S}{\partial V}\right)_T$ and $\left(\frac{\partial S}{\partial T}\right)_V$, and show that they are always positive. What do the results signify? (Burdwan, BSC(H), 1999)

5.43 a. Combine the first and the second law of thermodynamics to derive the following expression for the entropy change for n mole of an ideal gas

$$dS = n\bar{C}_p d\ln V + n\bar{C}_V d\ln P$$

b. 1 mole of an ideal gas undergoes a change from $(P, V) \to \left(\frac{P}{m}, mV\right)$. Show that $\Delta S = R \ln (m)$. Is the change of state isothermal? (Calcutta, BSC(H), 2013)

5.44 Two blocks of same metal are of same size but are at different temperature T_1 and T_2. These are brought together and allowed to come to the same temperature. Show that the entropy change is given by $\Delta S = C_P \ln[(T_1 + T_2)^2 / 4T_1 T_2]$, if C_P is constant. How does this equation show that the change is spontaneous? (Burdwan, BSC(H), 2009)

5.45 Calculate the change in entropy (in SI units) when 50 gm of water at 50°C is mixed with 30 gms of water at 10°C. Will the result be same if the given two samples of water are mere thermally connected to each other without mixing? (Jadavpur, BSC(H), 2002)

5.46 An ideal gas undergoes compression under the condition $PV^\gamma = $ constant ($\gamma = C_P/C_V$). Comment on the change in entropy of the system, with justification. (Calcutta, BSC(H), 2012)

5.47 For a certain gas, assumed to be ideal, the entropy is expressed as

$$S = \frac{n}{2}\left(\alpha + 5R\ln\frac{U}{n} + 2R\ln\frac{V}{n}\right)$$

where n is the number of moles, α is a constant and the other symbols have usual significance. Calculate $\bar{C}_V$ and comment on the atomicity of the gas. (Calcutta, BSC(H), 2006)

5.48 If the Gibbs free energy varies with temperature according to $G/T = a + b/T + c/T^2$, how will the enthalpy and entropy vary with temperature. Check that these equations are consistent. (Burdwan, BSC(H), 2008)

5.49 The molar Gibbs frequency of a gas is given by $\bar{G} = RT\ln(P/\text{bar}) + a + bP + cP^2$, where a, b and c are constants. Obtain the equation of state for n mol of this gas. (Jadavpur, BSC(H), 2013)

5.50 For a gas, $A = -\left(\frac{a}{V}\right) - RT\ln(V - b) + f(T)$. Set up an expression for pressure. (Calcutta, BSC(H), 2013)

5.51 Starting from Clausius inequality arrive at $\Delta A_{T,V} < 0$ for a spontaneous change in a closed system. (Calcutta, BSC(H), 2007)

5.52 Show that the conditions for spontaneity of a process occurring in a closed system involving only expansion work are

i. $dG_{T,P} < 0$ (Calcutta, BSC(H), 2010)

    ii. $dS_{U,V} > 0$ (Calcutta, BSC(H), 2010)

    iii. $dU_{S,V} < 0$ (Calcutta, BSC(H), 2008)

5.53 To which external (maximal or minimal) condition must the state of a system reach at equilibrium?   (Burdwan, BSC(H), 2012)

5.54 Though entropy is a fundamental state function and free energy is a derived one, the latter can be used more conveniently. Justify the statement.

                                           (Calcutta, BSC(H), 1994)

5.55 Give examples of a process in each case in which for the system

    i. $\Delta G = 0; \Delta S > 0$

    ii. $\Delta G < 0; \Delta S < 0$ (Calcutta, BSC(H), 2005)

5.56 Show that

    i. $\left(\dfrac{\partial A}{\partial n_i}\right)_{T,V,n_{j(\neq i)}} = \left(\dfrac{\partial H}{\partial n_i}\right)_{S,P,n_{j(\neq i)}}$ (Calcutta, BSC(H), 2006)

    ii. $\left(\dfrac{\partial A}{\partial n_i}\right)_{T,V,n_{j(\neq i)}} = \left(\dfrac{\partial G}{\partial n_i}\right)_{T,P,n_{j(\neq i)}}$ (Calcutta, BSC(H), 2008)

    iii. $\sum n_i d\mu_i = 0$ at constant T and P (Calcutta, BSC(H), 2006)

    iv. $\left(\dfrac{\partial \mu_i}{\partial P}\right)_{T,n_i} = \bar{V}_i$ (Calcutta, BSC(H), 2008)

    v. $\left(\dfrac{\partial \mu_i}{\partial T}\right)_{P,n_i} = -\bar{S}_i$ (Calcutta, BSC(H), 2012)

5.57 Prove that the chemical potential of a component in every phase must be equal at equilibrium.   (Burdwan, BSC(H), 2010)

5.58 Elucidate the concept of chemical potential (μ) as escaping tendency and mention the significance with respect to the equilibrium state of the system.

                                           (Calcutta, BSC(H), 2013)

5.59 A person claims to have devised a cyclic engine which exchanges heat with reservoirs at 300 K and 540 K, and which can produce 450 J of work per 1000 J of heat extracted from the hot reservoir. Is the claim feasible?

                                           (Calcutta, BSC(H), 2011)

5.60 An engine operates between 100°C and 0°C and another engine operates between 100 K and 0 K. Find the efficiencies in two case. Comment on the result.

                                             (Calicut, BSC, 2009)

5.61 An engine takes up 5×10^4 J heat to complete the following cycle. Given $|\Delta P| = 10$ atm, $|\Delta V| = 20$ litres, find its efficiency.   (Calcutta, BSC(H), 2007)

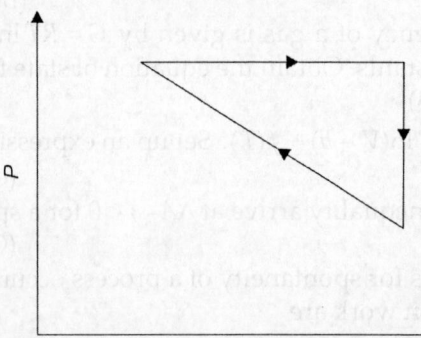

5.62 In a Carnot cycle, one mole of a gas ($\gamma = 1.4$) is taken as the thermodynamic working substance. The initial state of the gas is 600 K, 20 atm while the state of the gas at the end of second step becomes 300 K and 1 atm. Calculate the work involved in the first step of the cycle and the total work per cycle. (Jadavpur, BSC(H), 2012)

5.63 An ideal Carnot cycle operates on a temperature difference of 250°C. As a heat engine, its efficiency is 25%. Calculate the temperature of the source and the sink. Also find out the work done by the engine per cycle, given that 300 J of heat is released to the sink at the lower temperature per cycle. (Jadavpur, BSC(H), 2002)

5.64 A reversible Carnot refrigerator working between 0°C and 25°C is run for one minute using a 0.25 horse power electrical motor. Find

　i. the coefficient of performance of the refrigerator.

　ii. the amount of heat (in Joules) extracted from the water kept inside at 0°C.

　iii. the mass of ice produced. (Calcutta, BSC(H), 2010)

5.65 An electrical motor runs to operate a Carnot refrigerator. If outside temperature is 30°C and heat, leak into the refrigerator is 1200 Js^{-1}, find the coefficient of performance of the refrigerator and minimum power of the motor to maintain the temperature of –10°C inside the refrigerator. (Jadavpur, BSC(H), 2013)

5.66 How much work must be performed to freeze 1 kg of water at 0°C in a refrigerator kept in a room at 25°C? What will be the change in entropy of the room?
(Calcutta, BSC(H), 2006)

5.67 A Carnot refrigerator (using 1 mol of an ideal gas as the working substance) operates between 0°C and 27°C. It produces 400 J of work during isothermal expansion at the lower temperature. Find out the work expanded on the refrigerator per cycle of operation. How much heat will be released at the higher temperature per cycle of operation? (Jadavpur, BSC(H), 2000)

5.68 One mole helium (assumed ideal) initially at STP expands isothermally and irreversibly to 44.8 liter under condition such that $W = 100$ Cal. Calculate ΔS and ΔG and point out redundant data if any. (Calcutta, BSC(H), 1996)

5.69 0.5 mol of an ideal monatomic gas initially at 5 atm pressure and 0°C is allowed to expand against a constant external pressure of 0.5 atm. Conditions are such that the final volume is 10 times the initial volume, the final gas pressure equals the external pressure. Calculate ΔS and ΔG for the process. (Calcutta, BSC(H), 2005)

5.70 Consider the following cycle using 1 mol of an ideal gas, initially at 25°C and 1 atm pressure.

Step 1: Isothermal expansion against zero pressure to double the volume.

Step 2: Isothermal reversible compression from ½ atm to 1 atm.

　a.　i. Calculate the value of $\oint \frac{dQ}{T}$.

　　ii. Calculate ΔS for step 1.

　　iii. Show that ΔS for step 1 is not equal to the Q for step 1 divided by T.
(Calcutta, BSC(H), 2004)

　b. Calculate $\Delta S_{universe}$ for each of the steps. (Calcutta, BSC(H), 2014)

5.71 The fugacity coefficient of a certain gas at 200 K and 50 bar is 0.72. Calculate the difference of its chemical potential from that of a perfect gas in the same state.
(Calcutta, BSC(H), 2014)

5.72 5 mol of an ideal gas, initially at 50 atm and 300 K is expanded irreversibly where the pressure suddenly drops to 10 atm. The work involved is 4000 J. Show that the final temperature is greater than that of a reversible adiabatic expansion to the same pressure. If $C_V = 1.5R$, calculate the entropy change during the irreversible expansion. (Burdwan, BSC(H), 2002)

5.73 Calculate the final temperature and ΔS of 0.5 mol of an ideal gas ($\bar{C}_V = 1.5R$), initially at 3.0 atm and 300 K, when it undergoes an adiabatic expansion:
  i. reversibly to a final pressure of 1.0 atm.
  ii. against a constant external pressure of 1.0 atm till the final pressure of the gas becomes 1.0 atm. (Calcutta, BSC(H), 2009)

5.74 For a certain ideal gas, $\bar{C}_V = 7.05 + 17.8 \times 10^{-3}T - 308 \times 10^{-7}T^2$ Cal $K^{-1}mol^{-1}$. Calculate ΔS if 10 moles of the gas change state from 500 K, 50 atm pressure to 1000 K and 120 atm pressure. (Calcutta, BSC(H), 2014)

5.75 One mole of an ideal gas at 1 atm and 300 K undergoes free expansion adiabatically to double its initial volume. Find the final temperature, ΔS of the gas and ΔS of its surroundings. Comment on your result. Will the final temperature of the gas be same if the gas obeys van der Waals equation? (Calcutta, BSC(H), 2003)

5.76 Calculate the change in Gibbs free energy when 2 mol of an ideal diatomic gas is heated from 27°C to 177°C at a constant pressure of 2 atm. [Given: The absolute molar entropy of the gas, in $JK^{-1}mol^{-1}$ is given by the relation $S = 25.1 + 9.3T$]. (Calcutta, BSC(H), 2012)

5.77 At 373 K the entropy change for the transition of liquid water to steam is 109 $JK^{-1}mol^{-1}$. Calculate the enthalpy change ΔH_{vap} for the process. (Madurai, BSC, 2006)

5.78 Establish the condition for spontaneous vaporisation of water, given $\Delta H = 9590$ Cal mol^{-1} and $\Delta S = 26$ e.u. for the process. (Burdwan, BSC(H), 1982)

5.79 Calculate the entropy change when 1 kg of water at 27°C is converted into superheated steam at 200°C under constant atmospheric pressure. Given specific heat of liquid water = 4180 J/kg and specific heat of steam = $(1670 + 0.5T)$ J/kg at T K and latent heat of vaporisation of water = 23×10^5 J/kg. (Burdwan, BSC(H), 2010)

5.80 The heat of vaporisation, ΔH_{vap} of carbon tetrachloride at 25°C is 43 kJ/mol. If 1 mol of liquid carbon tetrachloride at 25°C has an entropy of 214 J/K, what is the entropy of 1 mole of the vapour in equilibrium with the liquid at this temperature. (Kalyani, BSC(H), 2005)

5.81 One mole of water vapour is compressed reversibly to liquid water at 373 K. Calculate ΔH, ΔS, ΔG and ΔA. (Calcutta, BSC(H), 2002)

5.82 Calculate the change in Gibbs potential when 36 g water initially at 100°C and 10 atm pressure is converted to vapour at 100°C and 0.01 atm pressure. (Calcutta, BSC(H), 1971)

5.83 Ice requires 6000 J mol^{-1} of heat to transform to water at 0°C. Find $\Delta \bar{S}$ for the system. If the surroundings is at +1.0°C, find $\Delta \bar{S}$ of the universe. Comment on the spontanity of the process. (Calcutta, BSC(H), 2007)

5.84 Calculate the latent heat of fusion of ice at –20°C if that at 0°C is 1440 Cal per mole, the sp heat of ice being 8.7 cal per mole. (Calcutta, BSC(H), 1998)

5.85 At –5°C the equilibrium vapour pressure of ice is 3.012 mm and that of supercooled liquid water is 3.163 mm. The latent heat of fusion of ice is 5.85 kJ mol^{-1} at –5°C.

Calculate ΔG and ΔS per mole for the transition H_2O (l, $-5°C$) $\rightarrow$ H_2O (s, $-5°C$). Comment on the spontaneity of the process. (Calcutta, BSC(H), 2004)

5.86 0.5 mol water at 1 atm pressure undergoes the process:

H_2O (l, $-10°C$) $\rightarrow$ H_2O (s, $-10°C$)

Compute ΔS and ΔG for the process from the following data:

Specific heat capacity of water and ice over the temperature range are 1.0 cal $deg^{-1}g^{-1}$ and 0.5 cal $deg^{-1}g^{-1}$ respectively, latent heat of fusion of ice is 80.0 cal g^{-1} at 0°C. (Calcutta, BSC(H), 2000)

5.87 Explain the following: Free energy (of formation) of an element at 1 atm and 298 K is assumed to be zero but entropy is not zero under the same conditions. (Gulbarga, BSC(H), 2004)

5.88 Calculate the entropy change in the surroundings for the formation of 1 mole of H_2O (l) under standard condition at 298 K.

$H_2(g) + \frac{1}{2}O_2(g) \rightarrow H_2O$ (l); $\Delta H^0 = -286$ kJ mol^{-1}

Also verify whether the reaction will be thermodynamically allowed or not under the given conditions:

At 298 K $\quad \bar{S}°(H_2, g) = 0.205$ kJ $K^{-1}mol^{-1}$

$\bar{S}°(O_2, g) = 0.130$ kJ $K^{-1}mol^{-1}$

$\bar{S}°(H_2O, l) = 0.070$ kJ $K^{-1}mol^{-1}$ (Calcutta, BSC(H), 2000)

5.89 Calculate the $\Delta G_f°$ for liquid water at 25°C. Given that $\Delta H_f°$ of H_2O (l) is -285.8 kJ mol^{-1} and standard molar entropies are $S°_{H_2O(l)} = 69.9$ $JK^{-1}mol^{-1}$, $S°_{H_2(g)} = 130.6$ $JK^{-1}mol^{-1}$ and $S°_{O_2(g)} = 205$ $JK^{-1}mol^{-1}$. (Vidyasagar, BSC(H), 2000)

5.90. From the following data, find the effect of temperature on ΔH and ΔS of the reaction

$A_2(g) + 2B_2(g) \rightarrow 2AB_2(g)$

$\bar{C}_P; A_2(g) = 32.0 + 1.04 \times 10^{-2}T$

$\bar{C}_P, B_2(g) = 45.2 + 2.06 \times 10^{-2}T$

$\bar{C}_P, AB_2(g) = 61.2 + 2.58 \times 10^{-2}T$ (all in $JK^{-1}mol^{-1}$) (Calcutta, BSC(H), 2009)

5.91 ΔG and ΔH valves for a reaction at 300 K are: -66.94 kJ mol^{-1} and -41.84 kJ mol^{-1} respectively. Calculate the free energy change at 330 K, assuming that ΔH and ΔS remain constant over this temperature range. (Nagpur, BSC(H), 2003)

5.92 For a reaction, ΔG (cal mol^{-1}) = $13580 + 16.1\ T \log_{10}(T/K) - 72.59\ T$. Find the ΔH and ΔS for the reaction at 300 K. (Jadavpur, BSC(H), 2013)

5.93 Calculate ΔH for isothermal compression of 1 mole of N_2 at 27°C from 1 atm to 500 atm pressure, assuming van der Waals behaviour of N_2 gas. Given JT coefficient,

$$\mu_{JT} = \left[\frac{2a}{RT} - b\right]\bigg/C_P$$

$a = 1.34$ L^2 atm mol^{-2}

$b = 0.039$ L mol^{-1} (Calcutta, BSC(H), 2008)

5.94 At 300°C and at pressures of 0 to 60 atm, the Joule–Thomson coefficient of nitrogen can be represented by the equation $\mu = 0.0142 - 2.60 \times 10^{-4}$ P. Assuming this equation to be temperature independent near 300°C, find the change in temperature on Joule–Thomson expansion of the gas from 60 to 20 atm. (Calcutta, BSC(H), 1996)

KEY TO UNIVERSITY QUESTIONS

5.1 For any PVT system (like an ideal gas), we can write:

$$dV = \left(\frac{\partial V}{\partial T}\right)_P dT + \left(\frac{\partial V}{\partial P}\right)_T dP, \qquad \text{for constant } n$$

Here the Euler's criterion for exactness of dV

Here $\left[\frac{\partial}{\partial P}\left(\frac{\partial V}{\partial T}\right)_P\right]_T = \left[\frac{\partial}{\partial T}\left(\frac{\partial V}{\partial P}\right)_T\right]_P$ [vide relation 5.14]

is fulfilled by the ideal gas equation $V = nRT/P$, each side being equal to $-nR/P^2$. Then dV is a perfect differential.

Note: Euler's criterian for exactness is always fulfilled by the differential of any state function. Thus dV for a van der Waals gas is also a perfect differential.

5.2 See problem 5.28.

5.3 Because ΔS refers to a change in state and not to a single state.

5.4 The given statement is justified. Because $\left[\frac{\partial (G/T)}{\partial (1/T)}\right]_P = H$, where H is a typical state function, called enthalpy.

5.5 Heat can be continuously converted into work only by means of a heat engine that operates through a cycle absorbing some heat (Q) from a reservoir and releasing a part of this heat (Q') to another reservoir. In a complete cycle $\oint dQ$ represents not Q but only the net heat ($Q - Q'$) converted into an equivalent amount of work $\oint dW$, by first law of thermodynamics. Hence, $\oint dQ = \oint dW$ does not at all imply that heat can be completely converted into work by a cyclic process.

5.6 See problem 5.3.

5.7 See problem 5.3.

5.8 By 'a perpetual motion of the second kind', we mean an operation (supposed to be cyclic for continuous running) that produces work through a machine using only one heat reservoir.

☐ No. Because under isothermal condition, no such device can produce work as implied by Carnot's formula for efficiency of a heat engine.

5.9 A Carnot engine follows a reversible cycle (called Carnot cycle) consisting of four steps—two isothermal and two adiabatic carried out alternately on a working substance (say n mol of an ideal gas) as represented in the adjoining diagram.

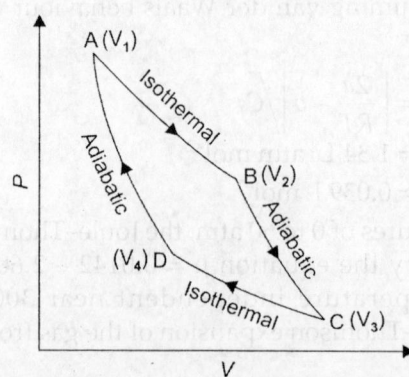

Here
$$\oint \frac{dQ}{T} = \int_A^B \frac{dQ\,(=dW)}{T} + \int_C^D \frac{dQ\,(=dW)}{T}$$

Since $dQ = dW$ for an ideal gas in an isothermal process where $dU = C_V dT = 0$

$$= nR \ln \frac{V_2}{V_1} + nR \ln \frac{V_4}{V_3}$$

$$= 0$$

Since
$$\frac{V_2}{V_1} = \frac{V_3}{V_4}$$

as $TV^{\gamma-1} = $ constant for an ideal gas in a reversible adiabatic process.

Efficiency of the engine

$$\frac{\oint dW}{\int_A^B dQ} = \frac{nRT \ln \frac{V_2}{V_1} + nRT' \ln \frac{V_4}{V_3}}{nRT \ln \frac{V_2}{V_1}}$$

$$= \frac{T - T'}{T}$$

5.10. See problem 5.11.

5.11. a. The given cycle is represented as P-V diagram in the adjoining figure.

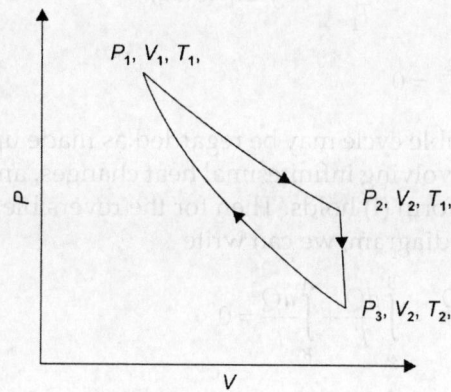

$P_1, V_1, T_1,$

$P_2, V_2, T_1,$

$P_3, V_2, T_2,$

b. For step (i) $\left| W_{(i)} \right| = RT_1 \ln \frac{V_2}{V_1}$

For step (ii) $\left| W_{(ii)} \right| = \int P dV = 0,$ dV being zero

For step (iii) $\left| W_{(iii)} \right| = \frac{P_1 V_1 - P_3 V_2}{\gamma - 1}$

c.
$$\eta = \frac{\oint dW}{Q_{(i)}}$$

$$= \frac{\left| W_{(i)} \right| + \left| W_{(ii)} \right| - \left| W_{(iii)} \right|}{\left| W_{(i)} \right|}$$

$$= 1 - \frac{(P_1V_1 - P_3V_2)/(\gamma - 1)}{RT_1 \ln \dfrac{V_2}{V_1}}$$

$$= 1 - \frac{(RT_1 - RT_2)/(\gamma - 1)}{\dfrac{RT_1}{\gamma - 1} \ln \dfrac{T_1}{T_2}}$$

since for an ideal gas $TV^{\gamma - 1} = $ constent in a reversible adiabatic process

$$= 1 - \frac{T_1 - T_2}{T_1 \ln \dfrac{T_1}{T_2}}$$

5.12 See problem 5.7.

In the reverse cycle (iii) the efficiency of the engine will remain unaffected. Because the work produced and the heat absorbed from the high temperature reservoir will be same as in the direct cycle only they have their sign reversed. Since Carnot cycle is a reversible one.

Note: On following reverse cycle the engine actually acts as a refrigerator with consumption, instead of production of heat.

5.13 a. In a Carnot cycle involving infinitesimal heat changes dQ_1 and dQ_2 at temperature T_1 and T_2 respectively

$$\frac{dQ_1 + dQ_2}{dQ_1} = \frac{T_1 - T_2}{T_1} \quad \text{by Eq. (5.1a)}$$

or $\quad \dfrac{dQ_1}{T_1} + \dfrac{dQ_2}{T_2} = 0$ \hfill (I)

Now any reversible cycle may be regarded as made up of an infinite number of Carnot cycles involving infinitesimal heat changes, and for each of the latter an equation of the form (I) holds. Then for the reversible cycle $A \to B \to A$, shown in the adjoining diagram, we can write

$$\oint \frac{dQ}{T} = \int_A^B \frac{dQ}{T} + \int_B^A \frac{dQ}{T} = 0 \hfill \text{(II)}$$

It appears from this relation that dQ/T represents the differential of some state property of the system, called entropy denoted by S

i.e. $\qquad\qquad dS = \dfrac{dQ_{\text{rev}}}{T}$ \hfill (III)

To obtain relation between the entropy change and heat change in an irreversible (natural) process, let us consider a Carnot-like cycle (which involves four steps similar to Carnot cycle except that the isothermal expansion is not reversible) involving irreversible heat change dQ_{irr} at temperature T_1 and reversible heatchange dQ_{rev} at temperature T_2. Here

$$\frac{dQ_{irr} + dQ_{rev}}{dQ_{irr}} < \frac{T_1 - T_2}{T_1}$$

since an irreversible cycle has lower efficiency than a reversible one

or $\quad \dfrac{dQ_{irr}}{T_1} + \dfrac{dQ_{rev}}{T_2} < 0 \qquad\qquad\qquad\qquad\qquad$ (IV)

Any irreversible cycle $A \xrightarrow{\;irr\;} B \xrightarrow{\;rev\;} A$ may be regarded as made up of an infinite number of appropriate small Carnot-like cycles for each of which the relation (IV) holds. Then for the entire cycle

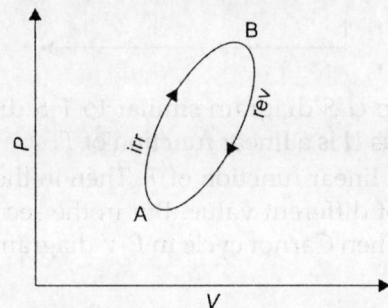

$$\oint \frac{dQ}{T} = \int_A^B \frac{dQ_{irr}}{T} + \int_B^A \frac{dQ_{rev}}{T} < 0 \qquad\qquad\qquad\qquad$ (V)

or $\qquad\qquad \displaystyle\int_A^B \frac{dQ_{irr}}{T} + \int_B^A dS < 0, \ $ by relation (III)

or $\qquad\qquad \displaystyle\int_A^B dS > \int_A^B \frac{dQ_{irr}}{T}$

or $\qquad\qquad dS > \dfrac{dQ_{irr}}{T} \qquad\qquad\qquad\qquad\qquad\qquad$ (VI)

Then $\oint \frac{dQ}{T} \le 0$, from (II) and (V), ('=' applies to reversible process and '<' applies to natural irreversible process)

and $\qquad\quad dQ/T \le dS$, from (III) and (VI)

b. See problem 5.27(v).

c. In the given spontaneous process, which is exothermic ($dQ = -ve$), S decreases (i.e. $dS = -ve$). But it may not be so for any spontaneous process as it appears from Clausius inequality $dS > dQ/T$.

d. We can justify this from Clausius inequality $dS > dQ/T$ which demands that dS must be +ve for an enothermic process, the associated dQ being +ve.

5.14 The two types of cyclic processes may be differentiated by their following characteristics

i. Work done by a system is greater in case of reversible cyclic process.

ii. ΔS_{uni} is zero for a reversible cyclic process but +ve for an irreversible one (vide problem 5.31).

5.15 See problem 5.48a.

In Carnot cycle ΔS_{uni} is zero because the cycle is reversible.

5.16 See problem 5.32

5.17 Work produced = Net heat absorbed

$$= \text{Enclosed area in } T\text{-}S \text{ diagram}$$

$$= (T_2 - T_1)(S_2 - S_1)$$

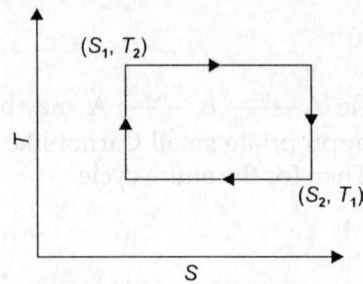

5.18 Carnot cycle will have U-S diagram similar to T-S diagram (as in problem 5.32) because for an ideal gas U is a linear function of T.

5.19 For an ideal gas H is a linear function of T. Then in the first and third steps H will remain constant but of different value. But in the second and fourth steps H will vary linearly with T. Then Carnot cycle in P-V diagram will reduce to a line in H-T diagram.

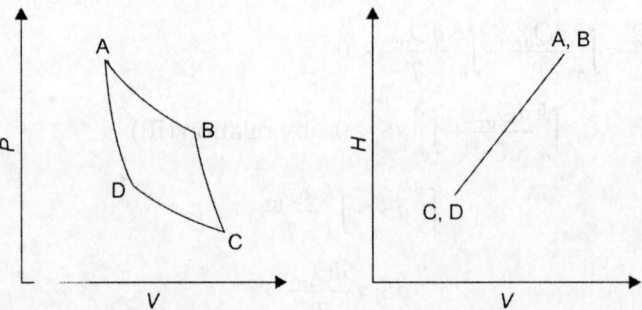

The area enclosed by H-T diagram of the Carnot cycle is zero. This is consistent with the fact that, for an ideal gas $\oint H dT = \oint df(T) = 0$.

5.20 Since surroundings always exchanges heat reversibly

$$\Delta S_{surr} = \frac{Q_{surr}}{T}$$

$$= \frac{-Q\,(\text{system})}{T}$$

$$= -\frac{\Delta H}{T}, \text{ at constant } T \text{ and } P, \text{ if the system does only } PV \text{ work}$$

Now by second law

$$\Delta S_{system} + \Delta S_{surr} > 0 \text{ for a spontaneous process}$$

Then $\qquad \Delta S_{system} - \dfrac{\Delta H}{T} > 0$

or $\qquad \dfrac{1}{T}(-\Delta H + T\Delta_{system}) > 0$

or $\qquad -\dfrac{\Delta G}{T} > 0$

Hence $\qquad -\Delta G > 0$, since $T > 0$

(This holds even if the system does some non-PV work).

5.21 See problem 5.14.

5.22 a. Incorrect. Because, although entropy is no doubt a state function, the state itself is determined by the amount of matter present in the system.

b. Incorrect. Because entropy is a state function and hence ΔS is always zero in a cyclic process for a closed system.

c. The given statement would be correct only at constant temperature and pressure, by relation (5.9).

d. Correct, by relation (5.9).

5.23. μ depends on pressure and temperature both. But $\mu°$, which refers to an arbitrarily fixed standard pressure, depends only on temperature (vide relation 5.31).

5.24 a. For an ideal gas (of fixed mass), $H = f(T)$. Then at constant T its change of $G\,(= H - TS)$ must be due only to change of S. Obviously the given change of G is an entropy effect.

b. With increase of T, both H and S increase. Here the decrease of $G\,(= H - TS)$ is certainly due to greater increase of TS. This points to an entropy effect.

5.25 Because if they intersect at a point, then at that point the two isothermals will have same temperature. This implies that at all other points of the two isothermals the temperature will be same and then the two isothermals cannot be treated as different. (Reasoning is similar to problem 5.31(iv)).

5.26 In the Joule experiment, the gas undergoes free expansion and hence no external work is involved. On the contrary, the Joule-Thomson experiment involves some external work.

□ U is conserved in the Joule experiment while H is consumed in the Joule–Thomson experiment.

□□ The relevant coefficients (μ) are defined as

$$\mu_J = \left(\frac{\partial T}{\partial V}\right)_U = -\frac{1}{C_V}\left(\frac{\partial U}{\partial V}\right)_T$$

$$\mu_{J\text{-}T} = \left(\frac{\partial T}{\partial P}\right)_H = -\frac{1}{C_V}\left(\frac{\partial H}{\partial P}\right)_T$$

□□□ Because in throttling of a gas $(\partial H/\partial P)_T$ may be +ve (i.e. heating effect) or –ve (i.e. cooling effect), except at inversion temperature, according as $\partial(PV)/\partial P$ is +ve or –ve (H being equal to $U + PV$). The possibility of such variation of $\partial(PV)/\partial P$ is corroborated by the observed $Z\,(= PV/nRT)$ vs P curves (Fig. 3.1a).

5.27 See problem 5.78.

5.28 See problem 5.75, problem 5.80 and problem 5.81.

5.29 See problem 5.79.

5.30 By similar procedure as in problem 5.81, we have

$$\mu_{JT} = \frac{2a/\bar{V}}{P - a/\bar{V}^2}$$
$$= + \text{ve}, \quad \text{since } P > a/\bar{V}^2$$

This means cooling effects will be observed in Joule–Thomson expansion. Because here only attractive force (indicated by a) exists between the molecules and the gas expands against this force.

5.31 First part: In Joule–Thomson cooling, the lower is the initial temperature of the gas, the larger is the temperature drop for a given pressure drop. This is unlike adiabatic/cooling.

Second part: This is similar to the previous question.

Here $\mu_{JT} = -b$
$$= -\text{ve}$$

This means the JT effect will be heating.

5.32 See problem 5.40.

5.33 a. See problem 5.36.

b. For a fixed amount of an ideal gas $PV \propto T$, and hence by relation (5.16) $\left(\frac{\partial U}{\partial V}\right)_T = 0$, i.e. U is a function of temperature only. This implies that $H = U + PV$ is also a function only of temperature for such gas for which PV depends only on temperature.

5.34 Vide problem (5.36).

It follows from the relation $\left(\frac{\partial U}{\partial V}\right)_T = T\left(\frac{\partial P}{\partial T}\right)_V - P$, that $\left(\frac{\partial U}{\partial V}\right)_T = 0$ will hold only if the equation of state is of the form

$$P/T = f(V) \quad \text{(for fixed amount)} \tag{I}$$

Again, it follows from the relation $\left(\frac{\partial H}{\partial P}\right)_T = -T\left(\frac{\partial V}{\partial T}\right)_P + V$, that $\left(\frac{\partial H}{\partial P}\right)_T$ or $\left(\frac{\partial H}{\partial V}\right)_T$ will be zero only if the equation of state is of the form

$$V/T = f(P) \quad \text{(for fixed amount)} \tag{II}$$

(I) does not necessarily imply (II). For example in case of $P(V - b) = nRT$, which is of the form (I), $\left(\frac{\partial U}{\partial V}\right)_T = 0$ but $\left(\frac{\partial H}{\partial V}\right)_T \neq 0$.

Note: For an ideal gas, $PV = nRT$, where (I) and (II) are both complied $\left(\frac{\partial U}{\partial V}\right)_T$ and $\left(\frac{\partial H}{\partial V}\right)_T$ are both equal to zero.

5.35 Vide university question 4.19 and apply Eq. (5.16) and (5.20a)

i. $C_P \rightarrow C_V$ when $\left(\frac{\partial V}{\partial T}\right)_P \rightarrow 0$, as in case of water around 4°C.

ii. $C_P < C_V$ when $\left(\frac{\partial V}{\partial T}\right)_T < 0$, as in case of water below 4°C.

5.36 See problem 5.39c.

5.37 Vide problem 5.39c.

5.38 C_P and C_V are dimensionally equivalent to energy/temperature i.e. $\frac{PV}{T}$, α is dimensionally equivalent to $1/T$.

β is dimensionally equivalent to $1/P$.

Then from the given expression, we have the following dimensional equivalency

$$\frac{PV}{T} \equiv TV\left(\frac{1}{T}\right)^m P^n$$

Equating the exponents of like quantities of both sides, we have

$$m = 2 \quad \text{and} \quad n = 1$$

5.39 i. See problem 5.39a.

ii. Similar to i. Here $dA = -PdV - SdT$.

iii. Vide relation (5.20a) and (5.20b).

5.40 For (i), (ii) and (iv), see problem 5.79.

(iii) $\quad\left(\dfrac{\partial T}{\partial P}\right)_S = \left(\dfrac{\partial V}{\partial S}\right)_P \quad$ by Maxwell relation

$$= \frac{V \cdot \dfrac{1}{V}\left(\dfrac{\partial V}{\partial T}\right)_P \cdot T}{dQ_P / dT} \quad \text{putting } dS = dQ/T$$

$$= \frac{V\alpha T}{C_P}$$

5.41 In a T-S diagram.

The slope of an isochore, $\dfrac{dT}{dS} = \dfrac{T}{C_V}$

The slope of an isobar, $\dfrac{dT}{dS} = \dfrac{T}{C_P}$

Since, for a gaseous system, $C_V < C_P$ the isochore will have greater slope, i.e. steeper, than the isobar at a common point.

5.42 $\qquad\left(\dfrac{\partial S}{\partial V}\right)_T = \left(\dfrac{\partial P}{\partial T}\right)_V, \quad$ vide Section 5.3 and 5.5

$$\left(\frac{\partial S}{\partial T}\right)_V = \frac{1}{T}\frac{dQ_V}{dT}, \quad \text{putting } dS = \frac{dQ}{T}$$

$$= \frac{C_V}{T}$$

$\left(\frac{\partial S}{\partial V}\right)_T$ is always +ve because $\left(\frac{\partial P}{\partial T}\right)_V$ is so.

Similarly $\left(\frac{\partial S}{\partial T}\right)_V$ is always +ve because C_V and T are so.

The results signify that entropy is a measure of randomness and increases with increase in spatial randomness and thermal randomness.

5.43 a. For a reversible change in state of a substance

$$dS = \frac{dQ}{T}$$

$$= \frac{dU + dW}{T}, \quad \text{from the first law, where } dW \text{ is the work done by the system}$$

$$= \frac{n\bar{C}_V dT + PdV}{T}, \quad \text{for } n \text{ mole of an ideal gas that does only } PV \text{ work}$$

$$= n\bar{C}_V \frac{dP}{P} + n\bar{C}_V \frac{dV}{V} + nR\frac{dV}{V} \quad \because PV = nRT \text{ for an ideal gas}$$

$$= n(\bar{C}_V + R)\frac{dV}{V} + n\bar{C}_V \frac{dP}{P}$$

$$= n\bar{C}_P d\ln V + n\bar{C}_V d\ln P$$

b. Here $P_iV_i = P_fV_f$ for the ideal gas. Hence the change of state is isothermal.

$$\Delta S = \int_i^f \frac{dQ}{T}$$

$$= \frac{RT \ln \dfrac{V_f}{V_i}}{T}, \quad \text{for isothermal change of 1 mol of an ideal gas}$$

$$= R \ln \frac{mV}{V} = R \ln m$$

5.44 See problem 5.53(ii).

5.45 First part is similar to the previous question.

Second part: Here the result will be same because the final state cannot be different in the to cases. [Final states would have been different if the two samples were made up of different kinds of substances].

5.46 The given conditions imply that the concerned process is an adiabatic reversible one. For such a process, the entropy of the system would not change.

5.47
$$\bar{C}_V = T \left(\frac{\partial \bar{S}}{\partial T} \right)_{\bar{V}}$$

$$= T \cdot \frac{5}{2} \frac{R}{\bar{U}} \left(\frac{\partial \bar{U}}{\partial T} \right)_{\bar{V}}, \quad \text{from the given expression}$$

$$= \frac{5}{2} RT \frac{\bar{C}_V}{\bar{U}}$$

Then
$$\bar{U} = \frac{5}{2} RT$$

Therefore
$$\bar{C}_V = \left(\frac{\partial U}{\partial T} \right)_{\bar{V}} = \frac{5}{2} R$$

$\bar{C}_V$ thus found is intermediate between the equipartition values of $\bar{C}_V$'s for a monatomic gas ($\bar{C}_V = 3/2\ R$) and a diatomic gas ($\bar{C}_V = 7/2\ R$). Probably, the given relation corresponds to a temperature region which is not sufficiently high for vibrational contribution to $\bar{C}_V$ to be significant (vide problem 2.73).

5.48
$$S = -\left(\frac{\partial G}{\partial T} \right)_P, \quad \text{since } dG = VdP - SdT \text{ under appropriate conditions}$$

$$= -a + \frac{C}{T^2} \quad \text{from the given expression}$$

$$H = \left[\frac{\partial(G/T)}{\partial(1/T)} \right]_P$$

$$= b + 2c/T$$

The given expression of G and the derived expressions of H and S are consistent with the relation $G = H - TS$.

5.49
$$V = n\bar{V} = n \left(\frac{\partial \bar{G}}{\partial P} \right)_T$$

$$= n \left(\frac{RT}{P} + b + 2CP \right)$$

5.50 See problem 5.46.

5.51 From the first law, we have in a reversible change

$$dU + PdV = dQ \text{ for a closed system doing only } PV \text{ work}$$
$$< TdS, \text{ from Clausius inequality (5.2) for a spontaneous change}$$

or $dU + PdV - TdS < 0$

or $d(U - TS)_{T,V} < 0$

i.e. $dA_{T,V} < 0$

5.52 From the relation, $dU + PdV - TdS < 0$, derived in previous question, we have:

   i. $d(U + PV - TS)_{T,P} < 0$

      i.e. $dG_{T,P} < 0$

   ii. $dS_{U,V} > 0$

   iii. $dU_{S,V} < 0$

Note:

1. The condition for spontaneity in terms of H is $dH_{S,P} < 0$ which follows from $dH - VdP = dQ < TdS$.

2. The condition for spontaneity $dG_{T,P} < 0$ holds even if the system does some non-pV work.

5.53 G provides the most useful condition for spontaneity in a closed system as

$$dG_{T,P} < 0$$

This implies that, at constant T and P, G of the system will decrease so long as the system undergoes spontaneous change, becoming minimum at equilibrium when $dG_{T,P} = 0$.

In terms of entropy, the equilibrium corresponds to a minimum when $dS_{U,V} = 0$.

5.54 Because the free energy, G provides the most convenient criteria for spontaneity and equilibrium as $dG_{T,P} \leq 0$ that does not involve surroundings (unlike S which offers the criteria $d_{uni} \geq 0$) and the constancy of T and P can be easily attained in the laboratory.

Note: Although G is conceptually more complex than S, it serve our purpose regarding spontaneity and equilibrium in a conveniently simple way.

5.55   i. This happens in any isothermal isobaric reversible process occurring in a closed system that does only PV work (when $\Delta G < 0$) with increase in randomness of the system (when $\Delta S > 0$), e.g. melting of ice at 0°C and 1 atm.

    ii. This happens in any isothermal isobaric natural process occurring in a closed system with decrease in randomness of the system, e.g. freezing of supercooled water at –10°C and 1 atm.

5.56 Similar to problem 5.89.

Starting from $\qquad dG = -SdT + VdP + \sum \mu_i dn_i$ $\qquad\qquad$ (I)

We have, on substituting $G = A + PV$ and $G = H - TS$

$$dA = -SdT - PdV + \sum \mu_i dn_i \qquad\qquad (II)$$

$$dH = TdS + Vdp + \sum \mu_i dn_i \qquad\qquad (III)$$

  i. From (II) and (III)

$$\left(\frac{\partial A}{\partial n_i} \right)_{T,V,n_{j(\neq i)}} = \mu_i = \left(\frac{\partial H}{\partial n_i} \right)_{S,P,n_{j(\neq i)}}$$

ii. From (I) and (II)

$$\left(\frac{\partial A}{\partial n_i}\right)_{T,V,n_{j(\neq i)}} = \mu_i = \left(\frac{\partial G}{\partial n_i}\right)_{T,P,n_{j(\neq i)}}$$

iii. Integration of (I), keeping the intensive variables (T, P, μ_i) unchanged, gives

$$G = \sum n_i \mu_i$$

Then

$$dG = \sum n_i d\mu_i + \sum \mu_i dn_i$$

Subtraction of this from (I) gives

$$\sum n_i d\mu_i = -SdT + VdP$$

$$= 0, \text{ at constant } T \text{ and } P$$

[This is Gibbs–Duhem equation that applies to every phase of a system when the phases are in equilibrium].

iv. See problem 5.97a.

v.

$$\left(\frac{\partial \mu_i}{\partial T}\right)_{P,n_i} = \left[\frac{\partial}{\partial T}\left(\frac{\partial G}{\partial n_i}\right)_{T,P,n_{j(\neq i)}}\right]_{P,n_i}$$

$$= \left[\frac{\partial}{\partial n_i}\left(\frac{\partial G}{\partial T}\right)_{P,n_i}\right]_{T,P,n_{i(\neq i)}} \text{, } dG \text{ being a perfect differential}$$

$$= -\left(\frac{\partial S}{\partial n_i}\right)_{T,P,n_{i(\neq i)}} \text{, since } dG = -SdT + VdP + \sum \mu_i dn_i$$

$$= -\bar{S}_i$$

Note: Chemical potential is often defined and interpreted in terms of G (rather than U, H, or A), because it refers to the constancy of T and P that forms the most convenient condition in carrying out a process in laboratory.

5.57 For our purpose, let us consider a multiphase multicomponent closed system at equilibrium. Let only dn_i mol of the ith species is transferred from one phase (say α) to another phase (say β) keeping the temperature and pressure unchanged. Then for the overall sstem

$$0 = dG_{T,P}$$

$$= dG^\alpha_{T,P} + dG^\beta_{T,P}$$

$$= \mu_i^\alpha(-dn_i) + \mu_i^\beta(dn_i),$$

by relation (I), in the previous problem, which holds for every phase of the system

or

$$\mu_i^\alpha = \mu_i^\beta$$

[Then, like pressure and temperature, the chemical potential of each species will have the same value in all the phases of an equilibriated system.]

5.58 Since it is the inherent tendency of every system to attain equilibrium, it follows that any species will tend to pass from a region of higher to a region of lower chemical potential to attain the same value in different phases (as shown in the previous problem), i.e. the chemical potential of a species is a measure of its escaping tendency in a non-equilibrium system.

5.59 The maximum work W_{max} that can be produced with the given two reservoirs is

$$W_{max} = 1000 \text{ J} \times \frac{(540 - 300)\text{K}}{540 \text{ K}}, \text{ by Eq. (5.1a)}$$

$$= 259.26 \text{ J}$$

Since the work claimed to be produced (450 J) exceeds W_{max}, the claim is not feasible.

5.60 By Eq. (5.1a), we have

for the first case, $\quad \eta = \dfrac{(273 + 100)K - (273 + 0) \text{ K}}{(273 + 100 \text{ K})} = 0.268$

for the second case, $\eta = \dfrac{100 \text{ K} - 0 \text{ K}}{100 \text{ K}} = 1$

□ Although $T_{source} - T_{sink}$ is same in the two cases, η's in the two cases are not in the ratio of the source temperatures (in Kelvin). Because when $T_{sink} = 0$ K, η is always same (equal to 1) independent of T_{source}.

5.61 Efficiency of the engine $= \dfrac{\text{Work produced (\equiv enclosed area in the diagram)}}{\text{Heat taken up}}$

$$= \frac{\frac{1}{2}(10 \text{ atm})(20 \text{ L}) \times 101.3 \text{ J/L atm}}{5 \times 10^4 \text{ J}}$$

$$= 0.203$$

5.62 For the second step, which is reversible adiabatic, the pressure at the beginning (P_2) is related to that at the end (P_3) by

$$P_2 = P_3 \cdot \left(\frac{T_3}{T_2}\right)^{\frac{\gamma}{1-\gamma}} \qquad \because PT^{\frac{\gamma}{1-\gamma}} = \text{const}$$

$$= (1 \text{ atm})\left(\frac{300}{600}\right)^{\frac{1.4}{1-1.4}}$$

$$= 11.31 \text{ atm}$$

In the first step, which is reversible isothermal, the pressure changes from $P_1 = 20$ atm to $P_2 = 11.31$ atm. Then in this step

work involved $= nRT_1 \ln \dfrac{P_1}{P_2}$, the working substance being an ideal gas

$$= (1 \text{ mol}) (8.314 \text{ JK}^{-1}\text{mol}^{-1}) (600 \text{ K}) \ln \frac{20}{11.31}$$

$$= 2843 \text{ J}$$

Since the working substance is an ideal gas, the heat absorbed will be equal to the work involved in the first step (being isothermal). Then the work produced per cycle $|W|$ by the engine operating between source temperature $T_1 = 600$ K and sink temperature $T_2 = 300$ K will be

$$|W| = (2843 \text{ J}) \frac{600 \text{ K} - 300 \text{ K}}{600 \text{ K}} \qquad \text{by Eq. (5.1a)}$$

$$= 1421 \text{ J}$$

5.63 Source temperature, $T_1 = \dfrac{T_1 - T_2}{\eta}$ by Eq. (5.1c)

$$= \frac{250 \text{ K}}{25/100} = 1000 \text{ K}$$

Sink temperature, $T_2 = T_1 - (T_1 - T_2)$

$$= 1000 \text{ K} - 250 \text{ K} = 750 \text{ K}$$

☐ Work done by the engine per cycle

$$|Q_1| - |Q_2| = |\dot{Q}_2| \cdot \frac{T_1 - T_2}{T_2}, \quad \text{by Eq. (5.1b)}$$

$$= (300 \text{ J}) \frac{(250 \text{ K})}{(750 \text{ K})}$$

$$= 100 \text{ J}$$

5.64 i. Coefficient of performance of the refrigerator,

$$\psi = \frac{T_1}{T_2 - T_1}, \quad \text{by Eq. (5.1d)}$$

$$= \frac{(0 + 273) \text{ K}}{(25 + 273) \text{ K} - (0 + 273) \text{ K}}$$

$$= 10.92$$

ii. Heat extracted $= \psi \times$ work consumed

$$= \psi \times \text{power} \times \text{duration}$$

$$= (10.92)(0.25 \times 735.5 \text{ Js}^{-1})(1 \times 60 \text{ s})$$

$$= 120475 \text{ J}$$

iii. Mass of ice produced $= \dfrac{\text{heat extracted}}{\text{latent heat of freezing of water}}$

$$= \frac{120475 \text{ J}}{80 \times 4.18 \text{ J/g}}$$

$$= 360.3 \text{ g}$$

5.65 Similar to the previous question.

Here, minimum power of the motor

$$= \frac{\text{Heat to be extracted per sec, i.e. heat leaked into the refrigerator}}{\psi}$$

5.66 Work to be performed = Heat to be extracted $\times \dfrac{T_{sink} - T_{source}}{T_{source}}$

$$= (10^3 \times 80 \times 4.18 \text{ J}) \times \frac{(25 + 273) \text{ K} - (0 + 273) \text{ K}}{(0 + 273) \text{ K}}$$

$$= 306 \times 10^4 \text{ J}$$

☐ Here $\Delta S_{univ} = \Delta S_{water} + \Delta S_{room} + \Delta S_{ref} = 0$, assuming rev operation of the refrigerator

Then $\Delta S_{room} = -\Delta S_{water}$, since $\Delta S_{ref} = 0$, as the refrigerator operates in cycles

$$= \frac{\text{heat lost by water}}{\text{temperature of water}}$$

$$= \frac{3.34 \times 10^5 \text{ J}}{(0 + 273) \text{ K}} = 1.22 \times 10^3 \text{ JK}^{-1}$$

Note: ΔS_{room} is independent of the room temperature, though heat received by the room is greater, greater is the room temperature. Here, we have avoided using the relation $\Delta S_{room} = \frac{\text{heat received by the room}}{\text{room temperature}}$, being a somewhat lengthy procedure.

5.67 Work consumed per cycle

$$-|W| = \text{Heat extracted} \times \frac{T_{sink} - T_{source}}{T_{source}}, \text{ by Eq. (5.1d)}$$

$$= (400 \text{ J}) \frac{(27 + 273)\text{K} - (0 + 273)\text{ K}}{(0 + 273)\text{ K}}$$

(since $\qquad Q = -W$ in isothermal expansion of an ideal gas)

$$= 39.56 \text{ J}$$

☐ Heat released to the sink = Heat absorbed $\times \dfrac{T_{sink}}{T_{source}}$, by Eq. (5.1b)

$$= (400 \text{ J}) \frac{(27 + 273)\text{ K}}{(0 + 273)\text{ K}} = 439.6 \text{ J}$$

5.68 $\Delta S = nR \ln \dfrac{V_f}{V_i}$, by Eq. (5.3a) under isothermal condition

$$= (1 \text{ mol}) (8.314 \text{ JK}^{-1}\text{mol}^{-1}) \ln \frac{44.8 \text{ L}}{22.4 \text{ L}}, \text{ since } \bar{V} = 22.4 \text{ L for an ideal gas at STP}$$

$$= 5.76 \text{ JK}^{-1}$$

$$\Delta G = nRT \ln \frac{P_f}{P_i}, \quad \text{by Eq. (15.13b)}$$

$$= nRT \ln \frac{V_i}{V_f} = T\Delta S$$

Here the given value of W is redundant. In fact, ΔS and ΔG, which are path independent, has no necessary connection with W which is path dependent.

5.69 Here $\quad P_i V_i (= 5 \times V_i) = P_f V_f (= 0.5 \times 10 V_i)$

Then the process is isothermal, the gas being ideal.

The problem is now similar to the previous one.

5.70 a. i. For step 1: $Q_1 = -W_1 = 0$, since $P_{ext} = 0$

For step 2: $Q_2 = -W_2 = nRT \ln \frac{P_i}{P_f}$

$$= (1 \text{ mol}) (8.314 \text{ JK}^{-1}\text{mol}^{-1}) (25 + 273)\text{K} \ln \frac{\frac{1}{2} \text{ atm}}{1 \text{ atm}}$$

$$= -1716.96 \text{ J}$$

$$\oint \frac{dQ}{T} = Q_1 + Q_2 = -1716.96 \text{ J}$$

ii. $\qquad \Delta S_1 = nR \ln \dfrac{P_i}{P_f}$

$$= (1 \text{ mol}) (8.314 \text{ JK}^{-1}\text{mol}^{-1}) \ln \frac{1 \text{ atm}}{\frac{1}{2} \text{ atm}}$$

$$= 1716.96 \text{ J}$$

iii. It appears from such calculations that $\Delta S_1 \neq Q_1/T$

Note: Here $\Delta S_1 > Q_1/T$, which is consistent with the Clausius inequality.

b. For step 1: $\Delta S_{surr} = 0$,

since the surroundings does not receive or absorb any heat (Q_1 being zero)

Then $\Delta S_{univ} = \Delta S_1$

For step 2: $\Delta S_{surr} = -\Delta S_2$, the process being reversible (for system)

Hence $\Delta S_{univ} = 0$

5.71 At particular T and P, the chemical potential of a real gas [given by Eq. (5.31)] differs from that of a perfect gas [given by Eq. (5.28)] by an amount equal to

$$RT \ln \frac{f}{p} = (8.314 \text{ JK}^{-1}\text{mol}^{-1})\,(200 \text{ K}) \ln 0.72$$

$$= -546.5 \text{ J mol}^{-1}$$

5.72 Here the irreversible expansion is assumed to be adiabatic (being quite fast). Then

$$W = \Delta U = n\bar{C}_V \, (T_f - T_i)$$

or $$T_f = T_i + \frac{W}{n\bar{C}_V}$$

$$= 300 \text{ K} + \frac{(-4000 \text{ J})}{(5 \text{ mol})\,(1.5 \times 8.314 \text{ JK}^{-1}\text{mol}^{-1})}$$

$$= 235.8 \text{ K}$$

In reversible adiabatic process

$$T_f = T_i \left(\frac{P_i}{P_f} \right)^{\frac{1-\gamma}{\gamma}} \qquad \because TP^{\frac{1-\gamma}{\gamma}} = \text{constant}$$

$$= T_i \left(\frac{P_i}{P_f} \right)^{-\frac{R}{\bar{C}_V + R}} \qquad \because \bar{C}_P = \bar{C}_V + R$$

$$= (300 \text{ K}) \left(\frac{50}{10} \right)^{-0.4}$$

$$= 157.6 \text{ K}$$

$$\Delta S = n\bar{C}_P \ln \frac{T_f}{T_i} + nR \ln \frac{P_i}{P_f}$$

$$= (5 \text{ mol}) \tfrac{5}{2} \times 8.314 \text{ JK}^{-1}\text{mol}^{-1}) \ln \tfrac{235.8}{300} + (5 \text{ mol})\,(8.314 \text{ JK}^{-1}\text{mol}^{-1}) \ln \tfrac{50}{100}$$

$$= 41.91 \text{ JK}^{-1} \qquad \left(\because C_P = C_V + R = \tfrac{5}{2}R \right)$$

5.73 i. Similar to previous question.

ii. Similar to problem (4.28b) regarding calculation of T_f.

5.74 $$\Delta S = \int_{T_i}^{T_f} \frac{n\bar{C}_P dT}{T} - \int_{P_i}^{P_f} \frac{nRdP}{P}$$

$$= (10 \text{ mol}) \int_{500}^{1000} \frac{(7.05 + 17.8 \times 10^{-3}T - 308 \times 10^{-7}T^2)dT}{T} \text{ JK}^{-1}\text{mol}^{-1}$$

$$- (10 \text{ mol})\,(8.314\text{JK}^{-1}\text{mol}^{-1}) \int_{50}^{120} \frac{dP}{P}$$

$$= -50.43 \text{ JK}^{-1}$$

5.75 Similar to question 5.70, the process being adiabatic as well as isothermal as
$$\Delta U = C_V \Delta T = 0$$

In case of van der Waals gas the final temperature will not be the same due to the work done by the gas against the intermolecular forces.

Here $\qquad \Delta U \neq C_V \Delta T$, but $\Delta U = 0$

5.76 $\qquad\qquad \Delta G = -\int_{T_i}^{T_f} n\bar{S}dT \quad$ by Eq. (5.9)

$$= -(2 \text{ mol})\int_{27+273}^{177+273} (25.1 + 9.3T) \, dT \text{ J mol}^{-1} = -1053780 \text{ J}$$

5.77 $\qquad\qquad \Delta S = \dfrac{\Delta H}{T}, \quad$ for a reversible change at constant T and P

Then $\qquad\qquad \Delta H_{vap} = T\Delta S_{vap}$

$$= (373 \text{ K})(109 \text{ JK}^{-1}\text{mol}^{-1}), \text{ for water at its normal bp (373 K)}$$
$$= 40657 \text{ J mol}^{-1}$$

5.78 $\qquad\qquad \Delta G_{vap} = \Delta H_{vap} - T\Delta S_{vap} \quad$ at constant T

$\qquad\qquad\qquad < 0$, for spontaneous vaporisation at constant T and P

or, $\qquad\qquad T > \dfrac{H_{vap}}{\Delta S_{vap}}$

or $\qquad\qquad T > \dfrac{9590 \text{ Cal mol}^{-1}}{26 \text{ Cal K}^{-1}\text{mol}^{-1}}$

or $\qquad\qquad T > 368.8 \text{ K}$

which is the required condition for spontaneous vaporisation of water.

5.79 The given change may be thought to occur through the following reversible steps.

$$H_2O(l, 27°C, 1 \text{ atm}) \xrightarrow{1} H_2O(l, 100°C, 1 \text{ atm}) \xrightarrow{2} H_2O(g, 100°C, 1 \text{ atm}) \xrightarrow{3} H_2O(g, 200°C, 1 \text{ atm})$$

$$\Delta S_1 = \int_{T_i}^{T_f} \frac{mC_p dT}{T} = (1 \text{ kg})(4180 \text{ JK}^{-1}\text{kg}^{-1})\int_{300}^{373} \frac{dT}{T} = 910.67 \text{ JK}^{-1}$$

$$\Delta S_2 = \frac{m\Delta H_{vap}}{T} = \frac{(1 \text{ kg})(23 \times 10^5 \text{J kg}^{-1})}{373 \text{ K}} = 6166.2 \text{ JK}^{-1}$$

$$\Delta S_3 = \int_{T_i}^{T_f} \frac{mC_p dT}{T} = (1 \text{ kg})\int_{373}^{473} \frac{(1670 + 0.5T) \, dT}{T} \text{ JK}^{-1}\text{kg}^{-1} = 446.91 \text{ JK}^{-1}$$

$$\Delta S = \Delta S_1 + \Delta S_2 + \Delta S_3 = 7523.8 \text{ JK}^{-1}$$

5.80 For reversible vaporisation of a liquid (at constant T and P)

$$\Delta \bar{S}_{vap} = \frac{\Delta \bar{H}_{vap}}{T}$$

$$= \frac{43 \text{ kJ mol}^{-1}}{(25 + 273) \text{ K}}, \quad \text{for CCl}_4$$

$$= 0.144 \text{ kJ K}^{-1}\text{mol}^{-1}$$

$$\bar{S}_{CCl_4(g)} = \Delta\bar{S}_{vap} + \bar{S}_{CCl_4(l)}$$

$$= (0.144 + 214 \times 10^{-3}) \text{ kJ K}^{-1}\text{mol}^{-1}$$

$$= 0.358 \text{ kJ K}^{-1}\text{mol}^{-1}$$

5.81 $\Delta H = -540 \text{ cal g}^{-1} \times 18 \text{ g} = -9720 \text{ cal}$

$$\Delta S = \frac{\Delta H}{T} = \frac{-9720 \text{ cal}}{373 \text{ K}} = -26.06 \text{ cal K}^{-1}$$

$\Delta G = 0$, since water is in equilibrium with its vap at its normal boiling point

$\Delta A = \Delta G - P\Delta V$

$\quad = \Delta G + nRT$

assuming water vapour to behave ideally and ignoring volume of liquid water

$\quad = 0 + (1 \text{ mol}) (1.987 \text{ cal K}^{-1}\text{mol}^{-1}) (373 \text{ K})$

$\quad = 741.2 \text{ cal}$

5.82 The given change may be thought to occur through the following reversible steps.

$$H_2O(l, 100°C, 10 \text{ atm}) \xrightarrow{1} H_2O(l, 100°C, 1 \text{ atm}) \xrightarrow{2} H_2O(g, 100°C, 1 \text{ atm}) \xrightarrow{3} H_2O(g, 100°C, 0.01 \text{ atm})$$

$\Delta G_1 = \int VdP = V_i(P_f - P_i)$ ignoring volume change

$\quad = (0.036 \text{ L}) (1 - 10) \text{ atm}$ taking density of water as 1 g/CC

$\quad = -0.324 \text{ L atm}$

$\Delta G_2 = 0$, since step 2 is isothermal isobaric and reversible

$\Delta G_3 = nRT \ln \dfrac{P_f}{P_i}$ assuming H_2O (g) to behave ideally

$\quad = \left(\dfrac{36}{18.016} \text{ mol} \right) (0.08206 \text{ L atm K}^{-1}\text{mol}^{-1}) \ln \dfrac{0.01 \text{ atm}}{1 \text{ atm}}$

$\quad = -0.7553 \text{ L atm}$

Then for the overall change,

$\Delta G = \Delta G_1 + \Delta G_2 + \Delta G_3$

$\quad = -1.079 \text{ L atm}$

Note: Here the effect of pressure on G of liquid is not negligible due to quite high pressure change.

5.83 Similar to problem 5.58b.

At 0°C and 1 atm, ice remains in equilibrium with water and ice-water system remains in internal equilibrium, though temperature of the surroundings is different. However, the melting will occur spontaneously as $\Delta S_{uni} = +ve$ (the system being not in equilibrium with the surroundings).

5.84 See problem 5.70.

5.85. For calculation of ΔG, see problem 5.69b.

Then use the relation

$$\Delta S = \frac{\Delta H - \Delta G}{T}$$

5.86 For calculation of ΔS, see problem 5.69a.

Then use the relation $\Delta G = \Delta H - T\Delta S$.

Here, unlike the previous question, ΔH at T is not given. It has to be calculated as in problem 5.70.

5.87 For a chemical reaction ΔH can be measured as Q_p of the reaction. But ΔS cannot be obtained from Q directly, it has to be obtained from the measurement of emf of an electrochemical cell where the reaction can be carried out reversibly. Unfortunately, this has been achieved only for few simple chemical reactions. Then the calculation of S of a compound from its entropy of formation ΔS_f has no practical importance like the calculation of enthalpy (H) from the measured ΔH_f taking H of each of the constituent elements of a compound to be zero at 1 atm and 25°C. This led to the introduction of a different type of convention for S based on the third law of thermodynamics where S of every substance (element or compound) has been assigned a zero value at absolute zero, which is most convenient.

5.88
$$\Delta S_{surr} = \frac{-\Delta H}{T},$$

Since heat released in the formation of H_2O is reversibly absorbed by the surroundings

$$= \frac{286 \text{ kJ mol}^{-1}}{298 \text{ K}}$$

$$= 0.9597 \text{ kJ K}^{-1}\text{mol}^{-1}$$

For the system $\quad \Delta S = \overline{S}^o_{(H_2O, l)} - \overline{S}^o_{(H_2, g)} - \frac{1}{2}\overline{S}^o_{(O_2, g)}$

$$= \left(0.070 - 0.205 - \frac{1}{2} \times 0.130\right) \text{ kJ K}^{-1}\text{mol}^{-1}$$

$$= -0.2 \text{ kJ K}^{-1}\text{mol}^{-1}$$

$$\Delta S_{uni} = \Delta S + \Delta S_{surr} = (0.9597 - 0.2) \text{ kJ K}^{-1}\text{mol}^{-1} = + \text{ve}$$

The +ve value of ΔS_{uni} implies that the reaction is thermodynamically allowed.

5.89 Here the relevant chemical reaction is

$$H_2(g) + \tfrac{1}{2}O_2(g) = H_2O \text{ (l)}$$

$$\Delta G^o_f(H_2O, l) = \Delta H^o_f \text{ (H}_2\text{O, l)} - T\Delta S^o_f \text{ (H}_2\text{O, l)}$$

$$= \Delta H^o_f \text{ (H}_2\text{O, l)} - T\left[(\overline{S}^o_{(H_2O, l)} - \overline{S}^o_{(H_2, g)} - \tfrac{1}{2}\overline{S}^o_{(O_2, g)}\right]$$

$$= -285.8 \text{ kJ mol}^{-1} - (298 \text{ K}) (69.9 - 130.6 - \tfrac{1}{2} \times 205) \text{ JK}^{-1}\text{mol}^{-1}$$

$$= 237.2 \text{ kJ mol}^{-1}$$

Note: However, for a particular species the relation
$$G° = H° - TS°$$
will hold only if the standard state chosen and the assigned conventional values are same for H, G and S. In general, however, the numerical values of H and G are based on the convention that at 25°C and 1 atm H and G of all elements are zero in their most stable state of aggregation at the stated temperature and pressure. But for S, a different convention, based on the third law, has been used where S of every substance (element or compound) has been assigned a zero value at 0 K.

5.90
$$\Delta H = 2\overline{H}_{AB_2} - \overline{H}_{A_2} - 2\overline{H}_{B_2}$$

Then $\quad \left(\dfrac{\partial \Delta H}{\partial T}\right)_P = 2\overline{C}_P(AB_2) - \overline{C}_P(A_2) - 2\overline{C}_P(B)_2, \quad$ since $\left(\dfrac{\partial H}{\partial T}\right)_P = C_P$

$$= 0, \text{ from the given data}$$

$$\Delta S = 2\overline{S}_{AB_2} - \overline{S}_{A_2} - 2\overline{S}_{B_2}$$

Then $\quad \left(\dfrac{\partial \Delta S}{\partial T}\right)_P = \dfrac{2\overline{C}_P(AB_2)}{T} - \dfrac{\overline{C}_P(A_2)}{T} - \dfrac{2\overline{C}_P(B_2)}{T}, \quad$ since $\left(\dfrac{\partial S}{\partial T}\right)_P = \dfrac{C_P}{T}$

$$= 0, \text{ from the given data}$$

Hence T has no effect on ΔH and ΔS.

Note: If enthalpy of a system is independent of temperature, entropy will also be the same since $\left(\frac{\partial S}{\partial T}\right)_P = \frac{1}{T}\left(\frac{\partial H}{\partial T}\right)_P$.

5.91

$$\Delta G_1 = \Delta H_1 - T_1 \Delta S_1$$

$$\Delta G_2 = \Delta H_2 - T_2 \Delta S_2$$

Then $\Delta G_2 = \Delta G_1 - (T_2 - T_1)\,\Delta S_1$, for $\Delta H_1 = \Delta H_2$ and $\Delta S_1 = \Delta S_2$

$$= \Delta G_1 - \frac{T_2 - T_1}{T_1} \cdot (\Delta H_1 - \Delta G_1)$$

$$= (-66.94 \text{ kJ mol}^{-1}) - \frac{(330 \text{ K} - 300 \text{ K})}{300 \text{ K}} \cdot (-41.84 + 66.94) \text{ kJ mol}^{-1}$$

$$= -69.45 \text{ kJ mol}^{-1}$$

5.92 Use the relations

$$\Delta H = -T^2 \left[\frac{\partial(\Delta G / T)}{\partial T} \right]_P$$

$$\Delta S = -\left(\frac{\partial \Delta G}{\partial T} \right)_P$$

5.93

$$\left(\frac{\partial H}{\partial P} \right)_T = -C_P \mu_{JT}, \quad \text{by Eq. (5.21)}$$

$$= b - \frac{2a}{RT}$$

Then $\Delta H = \left(b - \dfrac{2a}{RT} \right) \Delta P$

$$= \left[(0.039 \text{ L mol}^{-1}) - \frac{2(1.34 \text{ L}^2 \text{ atm mol}^{-2}}{(0.082 \text{ L atm K}^{-1}\text{mol}^{-1})\,(300 \text{ K})} \right] (500 - 1) \text{ atm}$$

$$\Rightarrow -34.9 \text{ L atm mol}^{-1}$$

5.94 Here $\left(\dfrac{\partial T}{\partial P} \right)_H = 0.0142 - 2.60 \times 10^{-4} P$

Then $\Delta T = \displaystyle\int_{60}^{20} (0.0142 - 2.60 \times 10^{-4} P)\, dP \cdot \text{K}$

$$= -0.152 \text{ K}$$

6

Liquid State—Stationary

Although the liquid state of a substance does not differ from its gaseous state at critical temperature, the two states differ markedly below this temperature.

6.1 STRUCTURE OF A LIQUID

A liquid (more appropriately liquid state of aggregation) physically behaves intermediate between a gas and a solid. Thus liquids (like gases, but unlike solids) are not characterised by any shape though (unlike gases, but like solids) exhibit ability to retain volume. These can be explained through the following structural model. A gas comprises mostly void space with the few molecules moving at random while reverse is the picture with a liquid. In a gas, the average intermolecular distance is very large compared with molecular diameter, and the intermolecular attractive forces, which decrease sharply with increase in intermolecular distance (r) (being proportional to r^{-7}), have no significant effect so that the molecules move disorderly. In a liquid, molecules are separated by a relatively short distance and hence intermolecular forces are strong enough to maintain a definite volume (though not shape) giving rise to some short-range order of molecular arrangement with a coordination number as high as 12. Here the molecular motion is less random since a liquid occupies a much smaller volume than a gas. In a liquid, there are vacancies of molecular size, though relatively a few on number and only a few molecules which are adjacent to such vacancies would move through (as in a gas) while others cannot (as in a solid). The low empty space left between the molecules in a liquid accounts for its low compressibility (a bulk property).

The properties of a liquid can be largely understood from the above molecular model. The most important properties which deserve special discussion are vapour pressure (connected to liquid vapour equilibrium), surface tension (a surface phenomenon) and viscosity (a transport phenomenon).

6.2 VAPOUR PRESSURE

If an evacuated vessel is partly filled with sufficient amount of a liquid, the molecules having energy higher than the cohesive energy of the liquid will evaporate till the vapour is in equilibrium with the remaining liquid (when formation of vapour and its condensation into liquid occur at the same rate). The pressure of the vapour thus developed is called vapour pressure of the liquid.

Any factor, such as temperature, that affects liquid-vapour equilibrium will affect the vapour pressure of a liquid. An equation showing the effect of temperature can be developed from the Boltzmann distribution law which when applied to molecular potential energy in vapour phase has the following form

$$n = Ae^{-\bar{U}/RT}$$

where n is the number density of the molecules having energy $\bar{U}$ per mole, A is a temperature dependent constant for the system and R is molar gas constant. For a liquid-vapour equilibrium $\bar{U}$ is high in the vapour phase (due to insignificant effect of intermolecular forces) compared with that in the liquid phase where we can put $\bar{U} = 0$. Then applying the above equation to the vapour phase and putting the vapour pressure $P = nkT$ (considering ideal behaviour), we have

$$P = A'e^{-\Delta\bar{U}_{vap}/RT}$$

where $A' = AkT$ and $\Delta\bar{U}_{vap} = \bar{U}_{vap} - \bar{U}_{liq}$ is the molar internal energy of vaporisation. Now $\Delta\bar{U}_{vap} = \Delta\bar{H}_{vap} - P\bar{V}_{vap} = \Delta\bar{H}_{vap} - RT$. Then the above equation becomes

$$P = Be^{-\Delta\bar{H}_{vap}/RT} \qquad \text{where } B = A'e$$

or $$\ln P = -\frac{\Delta\bar{H}_{vap}}{RT} + C \quad \text{where } C = \ln B \tag{6.1}$$

which relates the vapour pressure (P) of a liquid with its temperature (T).

This is of same form as Eq. (9.1c), which is based on classical thermodynamics, i.e. it contains Clausius–Clapeyron Eq. (9.1b) and Trouton's rule Eq. (9.1d).

Note: The Eq. (6.1) ignores the surface effect on vapour pressure, which is justified unless curvature of the liquid-vapour interface is high (vide Sec. 6.4B).

6.3 SURFACE TENSION

The surface molecules of a liquid differ energetically from those in the bulk. This gives rise to a number of phenomena observed with a liquid.

Let us consider a liquid in equilibrium with its vapour. Since the liquid phase has higher molecular density than the vapour phase, the liquid molecules at the surface (more precisely, interface), unlike those in the bulk, experience a pull toward the liquid phase because the intermolecular attractive forces acting on them are not balanced. This causes the surface layer to set up a pressure (called molecular pressure) on the bulk phase. As a consequence, the surface tends to contract and seems to behave as a stretched membrane. The tensional force that acts along the surface of a liquid perpendicular to unit length on the surface is called the surface tension of the liquid.

The term surface tension usually refers to interfacial tension when one of the phases forming the interface is gas. Strictly speaking, surface tension of a liquid refers to the interface formed by the liquid and its vapour in equilibrium. However, it is usually measured and recorded against air saturated with vapour of the liquid, because the result depends little on the presence of non-interacting gases at low or moderate pressure.

A. Surface Energy and Surface Free Energy

The minimum work, i.e. the reversible work, required to be done isothermally to increase unit surface area of a liquid against the force of surface tension (that binds parts of the surface together) is called surface free energy per unit area which is equal to the accompanying ΔA of the system at constant volume, or to accompanying ΔG of the system at constant pressure (for surface of constant curvature). The corresponding ΔU is a measure of surface energy. Then

$$\text{Surface free energy} = \gamma, \text{ the surface tension} \tag{6.2}$$
$$(\Delta A \text{ per unit area})$$

$$\text{Surface energy} = \gamma - T\left(\frac{\partial \gamma}{\partial T}\right)_V \quad \text{by Gibbs–Helmholtz equation} \tag{6.3}$$
$$(\Delta U \text{ per unit area})$$

Note: The relation $\Delta G = \gamma$ holds at constant T and P, provided the surface is flat because only then P can remain constant.

Better relation is $\Delta A = \gamma$ which holds at constant T even if the surface is not flat.

B. Angle of Contact and Shape of Liquid Surface

It is due to surface tension that a liquid drop assumes a spherical shape in absence of any external forces. This also happens if the surrounding field is uniform as in case of free falling liquid in gravity field or a liquid on an artificial satellite (because of no reaction force). But the situation is different when a liquid is held on a support which is stationary relative to the earth. Here, a force of reaction of the support acting normally on the liquid at the plane of contact flattens the liquid surface (by creating a hydrostatic pressure that results in spreading of the liquid in a horizontal direction). This is represented in Fig. 6.1.

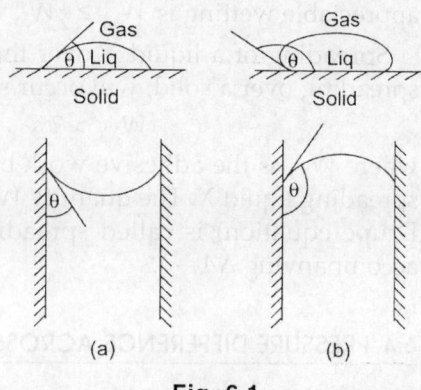

Fig. 6.1

The angle (θ) made by the tangent plane of the liquid-gas interface with that of the liquid-solid interface at a point of contact of the three phases within the liquid is called the angle of contact of the liquid on the solid involved. Balance of three interfacial tensions along a line of contact leads to the following expression, called Young equation

$$\gamma_{SG} = \gamma_{SL} + \gamma_{LG} \cos \theta \tag{6.4}$$

where γ_{SG}, γ_{SL} and γ_{LG} are three interfacial tensions of the solid-gas, solid-liquid and liquid-gas interfaces respectively. θ is same whether the solid surface is horizontal or vertical (in case of a tube). The situation with $\theta < 90°$ is represented by Fig. 6.1(a), where the surface of the liquid in the tube (i.e. meniscus) is concave (upward). The situation with $90° < \theta < 180°$ is represented by Fig. 6.1(b) where the meniscus is convex.

C. Cohesive Work, Adhesive Work and Spreading of a Liquid

Cohesive work (W_{co}) of a liquid (or any other phase) is the minimum work needed to tear off a column of unit cross-section of that liquid into two parts (without any lateral contraction). W_{co}, which is due to cohesive force (between molecules in same phase), is then given by

$$W_{co} = 2\gamma_{LG} \tag{6.5}$$

[The factor 2 arises due to formation of two liquid-gas interfaces.]

The adhesive work (W_{ad}) of a pair of solid and liquid (or of any two phases) is the minimum work needed to diminish the relevant solid-liquid interface by unit amount with simultaneous formation of solid-gas and liquid-gas interfaces each of unit area,

W_{ad}, which is due to adhesive force (between molecules in different phases) is then given by

$$W_{ad} = \gamma_{SG} + \gamma_{LG} - \gamma_{SL} \qquad (6.6a)$$

This is Drupe equation. In terms of contact angle, we can write

$$W_{ad} = \gamma_{LG} (1 + \cos \theta) \qquad (6.6b)$$

Then $\qquad \dfrac{W_{ad}}{\frac{1}{2} W_{co}} = 1 + \cos \theta$

So contact angle depends on the adhesive work and cohesive work, that is upon adhesive force and cohesive force. When $W_{ad} = W_{co}$, $\theta = 0°$, and the solid is then just completely covered by the liquid, i.e. completely wetted. Spreading or wetting will occur appreciably only upto $\theta = 90°$ which corresponds to $W_{ad} = \frac{1}{2} W_{co}$. Then the condition for appreciable wetting is $W_{ad} \geq \frac{1}{2} W_{co}$ or $2W_{ad} \geq W_{co}$.

Spreading of a liquid X over the entire surface of another immiscible liquid Y, like spreading over a solid, will occur spontaneously under the following condition

$$W_{XY} > 2\gamma_X \qquad (6.7)$$

where W_{XY} is the adhesive work between X and Y and $2\gamma_X$ is the cohesive work of the spreading liquid X. The quantity $W_{XY} - 2\gamma_X$, i.e. $\gamma_Y - \gamma_X - \gamma_{XY}$ (since $W_{XY} = \gamma_X + \gamma_Y - \gamma_{XY}$, by Drupe equation) is called spreading coefficient of X on Y (which is negative of the accompanying $\Delta A_{T,V}$).

6.4 PRESSURE DIFFERENCE ACROSS A CURVED SURFACE

For a curved surface, the concave side experiences higher pressure than the convex side. This pressure difference ΔP is given by the Laplace equation

$$\Delta P = \gamma \left(\frac{1}{r_1} + \frac{1}{r_2} \right) \qquad (6.8a)$$

where γ is the surface tension, and r_1 and r_2 are the radii of curvature of the interface at any point on it [r_1 and r_2 refer to the lines of intersection of the interface with two mutually perpendicular planes drawn through the point concerned].

For a spherical interface, the Laplace equation becomes

$$\Delta P = \frac{2\gamma}{r} \qquad (6.8b)$$

For a plane interface $\Delta P = 0$, r being ∞.

A. Capillary Action

If a capillary tube is partially immersed in a liquid, the liquid stands at different levels inside and outside the tube. This phenomenon is known as capillary action. This is a consequence of the fact that the liquid-gas interface is curved inside the tube though flat outside.

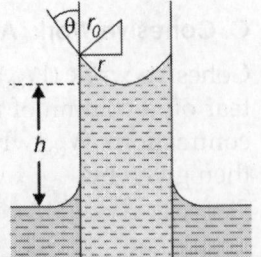

Fig. 6.2

If the liquid wets the capillary tube appreciably, the liquid-gas interface within the tube, called meniscus, will be concave and there will be capillary rise (Fig. 6.2). Because here the liquid within the tube is acted upon by the excess pressure (Laplace pressure) directed toward the centre of curvature of the meniscus, i.e. upward. This pressure will cause the meniscus to rise in the tube till it is balanced by the hydrostatic pressure due to the level difference (h) of the liquid inside and outside the tube. Then

$$2\gamma/r_0 = h(\rho - \rho')g,$$

ignoring the mass of the liquid above the minimum of the meniscus

or $$\gamma = r_0 h(\rho - \rho')g/2 \tag{6.9a}$$

where r_0 is radius of curvature of the meniscus.

In terms of radius (r) of the capillary tube, this becomes

$$\gamma = rh(\rho - \rho')g/2\cos\theta, \tag{6.9b}$$

since $r_0 = r/\cos\theta$, where θ is angle of contact

If $\theta \approx 0$, this equation virtually reduces to

$$\gamma = rh(\rho - \rho')g/2 \tag{6.9c}$$

i.e. $$h \propto \frac{1}{r} \tag{6.9d}$$

the other quantities involved being constant.

Note: The Eq. (6.9a) involving the radius of curvature is simpler than Eq. (6.9b) involving angle of contact.

If the liquid does not wet the capillary tube, capillary depression will occur. Because here the meniscus is convex of which the centre of curvature will be inside the liquid and hence the Laplace pressure will act on the liquid in the capillary in downward direction. However, the general expression for γ, Eq. (6.9b) still holds. Here h is negative as $\cos\theta$ is negative.

B. Vapour Pressure over a Curved Surface

As a consequence of Laplace pressure, the vapour pressure P of a liquid is higher or lower according as the liquid-vapour interface is convex (as with droplets) or concave (as with bubbles within a liquid), relative to the value P_0 for the plane interface. These are related by Kelvin equation

$$P = P_0 \exp[-2M\gamma/RT\rho r] \tag{6.10}$$

where r is radius of curvature of the liquid-vapour interface which is positive for concave and negative for convex, M is the molar mass of the liquid of density ρ. Although Kelvin equation is not exact, we can explain on this basis the existence of supersaturated vapour (as with clouds) as well as superheated liquid (i.e. one above its usual boiling point).

6.5 TEMPERATURE DEPENDENCE OF SURFACE TENSION

The surface tension of a liquid is mainly determined by the difference in properties of a substance, particularly the density in the liquid and gas phases. With rise in temperature this difference, and hence surface tension is expected to decrease till it becomes zero at critical temperature (T_c) when a liquid cannot be distinguished from its vapour of same density.

On somewhat theoretical ground, the temperature dependence of surface tension near critical temperature can be expressed by the following approximate relation (Eotvos equation)

$$-d[\gamma(M/\rho)^{\frac{2}{3}}]/dT = C, \text{ a universal constant}$$

or $$\gamma = C(T_c - T)(\rho/M)^{\frac{2}{3}} \tag{6.11a}$$

as $\gamma = 0$ at $T = T_c$.

If, however, T is fairly lower than T_c, the following relation is better

$$\gamma = C(T_c - T - 6)(\rho / M)^{\frac{2}{3}}$$ (6.11b)

The temperature coefficient of surface tension, i.e. $d\gamma/dT$, can then be expressed as

$$\frac{d\gamma}{dT} = -C(\rho / M)^{\frac{2}{3}}$$ (6.11c)

AUXILIARY PROBLEMS

6.1 For any substance, the liquid phase is supposed to be denser than the vapour phase at same temperature and pressure. Right or wrong?

6.2 Justify or criticise.
 i. A liquid of higher molecular density would have higher vapour pressure.
 ii. For any substance at its melting point the liquid phase will have same vapour pressure as the solid phase.

6.3 A supercooled liquid is unstable. Then the term *vapour pressure* will have no meaning with such state. Comment.

6.4 What happens to vapour pressure of a liquid if:
 i. the temperature is raised at constant volume.
 ii. the container is of large size.
 iii. the pressure of the system is increased by introducing argon into the container at constant volume and temperature.
 iv. the container is vigorously shaken all time at constant temperature.
 v. the container is packed with some shots of asbestos along with the liquid.

6.5 Find the vapour pressure of water at 99°C.

6.6 The vapour pressure of a liquid increases by 4% per degree rise of temperature around 25°C. Calculate heat of vaporisation and normal boiling point of the liquid. State the assumptions involved in such calculations. Are they justified?

6.7 Surface tension is a common property of solids, liquids and gases. Comment.

6.8 For most substances, the surface tension at the solid–gas interface is greater than that at the liquid–gas interface. Why?

6.9 Will the surface tension be same for different crystalline modification of a substance (e.g. rhombic and monoclinic sulfur)?

6.10 The force required to enlarge a liquid film, unlike a rubber sheet, does not depend on the amount of enlargement. Explain.

6.11 Under what condition (if any) the interfacial tension can be zero for *L-G*, *S-G* and *L-L* interfaces? Can it be negative?

6.12 The surface energy is more or less temperature independent, although the surface free energy is not. Establish.

6.13 Predict which substance in the following pair will have higher surface tension:
 i. hexane and benzene
 ii. dimethyl ether and ethanol
 iii. ethanol and water
 iv. glycerol and water
 v. mercury and water
 vi. molten calcium and mercury
 vii. Molten calcium and molten calcium chloride

6.14 CH_3OH and C_2H_5OH have nearly same surface tension around 20°C. Which one will have numerically higher temperature coefficient of surface tension? For such compounds $|d\gamma/dT|$ increases with increasing temperature. Explain.

6.15 Account for the following.

   a. Water wets glass but not polythene

   b. Rain drops do not wet inclined glass surfaces

   c. Hg does not wet most solids

6.16 Comment on the following.

   a. Capillary rise and capillary depression have no thermodynamic justification, because at particular T and P, they involve no change in volume and total interfacial area of the system and hence no change in thermodynamic property like entropy.

   b. The change in surface free energy (ΔA) per unit change in S-L interface will be ($\gamma_{SL} - \gamma_{SG}$) for capillary rise and ($\gamma_{SG} - \gamma_{SL}$) of opposite sign for capillary depression. Hence only in one of the cases ΔA can be negative where capillary action will be observed.

6.17 Account for the following.

   a. A freely falling drop of water is spherical in shape (having convex surface), but the meniscus of water in a glass capillary is concave in shape.

   b. When placed over a horizontal glass surface, a small quantity of mercury, unlike water, looks spherical but a lump of mercury looks ellipsoidal.

   c. Near melting point, a thin metallic wire becomes shorter.

   d. A piece of blotting paper soaks water but a polythene paper does not.

   e. Ploughing of soil before the onset of summer is necessary for crop production.

   f. Sprinkling of finely divided particles coated with thin film of a liquid on to the same liquid is associated with liberation of heat.

   g. In the liquid phase, CCl_4 and H_2O form two distinct layers but in the vapoour phase they get completely mixed up.

6.18 a. Determination of surface tension by drop weight method is based on the following relation between the surface tensions γ_1 and γ_2 of two liquids

$$\frac{\gamma_1}{\gamma_2} = \frac{\rho_1 n_2}{\rho_2 n_1}$$

where ρ denotes the density and n the number of drops per unit volume of the liquid.

Does this relation imply that higher is the density of a liquid, higher will be its surface tension?

   b. n is determined using a stalagmometer and not burette. Why? Would a stalagmometer made of polythene, serve the purpose?

6.19 In the determination of surface tension by capillary rise method, which of the following factors will affect the result?

   i. Material of the capillary

   ii. Radius of the capillary

   iii. Acceleration due to gravity

   iv. Atmospheric pressure

6.20 Will there be any change in the capillary rise if (a) temperature is raised (b) liquid level outside the capillary is gradually raised?

6.21 The vapour pressure of a liquid is higher when it has convex surface (as in case of droplets) and lower when it has concave surface (as in bubbles within a liquid), compared with that when the surface is flat. Explain.

6.22 Derive kelvin equation for vapour pressure of a liquid with curved surface.

6.23 a. Explain the existence of supersaturated vapours (whose pressure exceeds the vapour pressure of the liquid with flat surface).

   b. Rain in industrial areas is more likely on weekdays than at weekends. Justify.

6.24 a. A liquid generally boils above its true boiling point abruptly but boiling occurs smoothly in presence of porous bodies. Explain.

   b. In graduating a thermometer, it is more appropriate to put the thermometer-bulb into the steam over boiling water and not into the water. Justify.

6.25 Water vapour is suddenly cooled to 25°C to a supersaturated state containing microdroplets. Find the minimum pressure of the water vapour in absence of any foreign substance, required for condensation of such droplets if they are of radius 0.75 nm. The saturated vapour pressure of bulk phase water is 23.8 torr

$$\rho_{H_2O} = 10^3 \text{ kgm}^{-3}, \gamma_{H_2O} = 0.072 \text{ Nm}^{-1} \text{ at } 25°C$$

6.26 To what maximum extent will capillary rise/depression of water occur in the following cases?

   a. A glass capillary of internal diameter 0.2 mm.

   b. A polythene capillary of internal diameter 0.2 mm.

   c. Two vertical glass plates 0.2 mm apart.

   $[\gamma_{H_2O} = 72 \text{ dyn/cm}, \rho_{H_2O} = 1 \text{ g/cm}^3]$

6.27 In connection with problem 6.26(a):

   a. what minimum pressure is required to prevent capillary rise.

   b. if the capillary rise is restricted to half of its maximum value, what will be the angle of contact and radius of curvature

6.28 The limbs of a vertical U-tube have internal diameter 0.2 mm and 0.4 mm respectively. It is partly filled with water. What will be the difference of liquid levels in the two limbs $[\gamma_{H_2O} = 72 \text{ dyn/cm}, \rho_{H_2O} = 1 \text{ g/cm}^3]$.

   If the wider limb is just filled with water, will water overflow from the other limb?

6.29 In a vessel benzene and water form two layers. A capillary tube of diameter 0.2 mm is held vertically such that its one end is in benzene and the other end in water when water-benzene interface within the capillary is found to lie 3.95 cm above the water-benzene interface outside the capillary at equilibrium. State whether the water-benzene meniscus is concave or convex. Assuming it to be himispherical calculate water-benzene interfacial tension. $[\rho_{benzene} = 0.8 \text{ g/cm}^3, \rho_{H_2O} = 1 \text{ g/cm}^3]$.

6.30 What will be the excess pressure inside a bubble of diameter 0.2 mm (i) just below the surface of soap water and (ii) just above the surface of soap water ($\gamma_{soap \ water} = 150 \text{ dyn/cm}$).

   Would the excess pressure inside a soap bubble be same as that inside a drop of soap water of same size?

6.31 In the determination of surface tension by maximum bubble pressure method (as described in the adjoining figure), a capillary tube of diameter 0.1 cm is inserted into a dilute aqueous solution such that its open end is 10 cm below the surface.

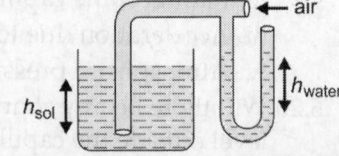

The air pressure just sufficient for the formation of bubble at the end of the tube, as measured by a water manometer, is found to be 12 cm. Calculate the surface tension of the solution. [$\rho_{soln} = 1.01$ g/cm^3]

6.32 A container has a circular hole of radius 0.03 mm at its bottom. To what maximum depth the container can be filled with water without dripping out through the hole? What assumption you have taken for solution? [$\gamma_{H_2O} = 72$ dyn/cm].

6.33 A thin metallic disc of mass 50 mg and radius 1.0 cm just floats in water. What minimum work will be needed to detatch it from the water surface. [$\gamma_{H_2O} = 72$ dyn/cm]?

6.34 A narrow metallic wire of mass 0.25 g and length 4 cm having a square shape is placed over water. Will it float or drown? [$\gamma_{H_2O} = 72$ dyn/cm].

What minimum force will be needed to detach it from the water surface (i) in horizontal position (ii) in vertical position? Does it depend on solid–liquid interfacial tension?

6.35 A drop of water breaks up into n microdrops of equal size. Show that the accompanying change in surface free energy increases linearly with $n^{\frac{1}{3}}$.

6.36 A spherical liquid drop of weight $W = 0.04$ g is dispersed into $n = 1500$ globules of radius $r = 0.02$ cm each. Find the resultant increase in surface free energy. ($\rho = 0.8$ g/cm^3, $\gamma = 27$ dyn/cm). Here the calculation using only the umber of globules (as in the problem 6.35) or only the size of the globules (in the problem 6.37) would lead to the same result. Then what is the justification of providing both the data?

6.37 At 293 K, the surface tension of water is 72.8 dyn/cm and its temperature coefficient is –0.15 dyn cm^{-1}K^{-1}. If 1.0 cm^3 of water is broken up into droplets, each of radius 10^{-5} cm, isothermally at 298 K, find the surface free energy and surface energy of the droplets relative to lump water. How much heat will be liberated in the reverse process?

6.38 Two globules of Hg, each of radius r coalesce to form a single globule. Obtain, an expression for the loss in free energy in the process in terms of r and γ_{Hg}.

6.39 At 20°C the interfacial tension between water and benzene is 3.50×10^{-2} N/m. If $\gamma_{benzene} = 2.88 \times 10^{-2}$ N/m and $\gamma_{H_2O} = 7.27 \times 10^{-2}$ N/m, calculate:
   i. the work of adhesion between water and benzene
   ii. the work of cohesion for benzene and for water
   iii. the spreading coefficient for benzene on water
Will benzene spread spontaneously on water? Will the reverse occur? Mention the assumption you have made.

ANSWERS

6.1 Right, except at critical point. When a liquid in equilibrium with its vapour is heated in a closed vessel, the density of the vapour phase increases (due to compression) while that of the liquid phase decreases (due to its expansion) till the density of the two phases becomes equal at the critical temperature.

6.2 i. Not justified. Because, although volatility of a liquid is favoured by higher molecular density, it is disfavoured by some other facor like cohesive force of the liquid, which depends on nature of the substance.

ii. For any substance, the solid and liquid phases can simultaneously remain in equilibrium with the vapour phase only at tripple point (which is a particular case of melting point). Therefore, only at tripple point, the term *vapour pressure* will have meaning for the two co-existing condensed phases and only then the given statement is justified.

6.3 No. Because although a supercooled liquid readily transforms to solid phase in presence of the latter, it can remain in equilibrium with the relevant vapour phase at certain pressure called its vapour pressure.

6.4   i. Vapour pressure will increase. Because at higher temperature, the average molecular energy is higher and hence liquid molecules find it easier to pass into the vapour phase overcoming stronger intermolecular attraction in the liquid phase.

ii. Vapour pressure will remain unchanged due to formation of requisite amount of vapour depending on the size of the container, provided the vaporisation is incomplete (otherwise the term vapour pressure will lose its meaning).

iii. Vapour pressure will increase. Because at higher pressure the liquid phase will be squeezed more with consequent higher tendency to vaporise.

iv. Vapour pressure will rise due to formation of small droplets having convex shape.

v. Vapour pressure will fall or rise according as the meniscus of the liquid in the capillaries of the asbestos is concave or convex (vide subsec. 6.4B).

6.5 The vapour pressure P_1 at T_1 is related to P_2 at T_2 by

$$\ln \frac{P_2}{P_1} = \frac{\Delta \bar{H}_{vap}}{R} \left[\frac{1}{T_1} - \frac{1}{T_2} \right],$$

from Eq. (6.1) ignoring temperature dependence of $\Delta \bar{H}_{vap}$ and C.

or

$$P_2 = P_1 \exp \left[\frac{\Delta \bar{H}_{vap}}{R} \left(\frac{1}{T_1} - \frac{1}{T_2} \right) \right]$$

$$= (760 \text{ torr}) \exp \left[\frac{(530 \times 18 \text{ cal mol}^{-1})}{(1.99 \text{ cal K}^{-1}\text{mol}^{-1})} \left(\frac{1}{373 \text{ K}} - \frac{1}{372 \text{ K}} \right) \right]$$

since $P_1 = 760$ torr for water at $T_1 = 373$ K, the normal b.p. of water

$$= 734.2 \text{ torr}$$

6.6 Heat of vaporisation,

$$\Delta \bar{H}_{vap} = RT^2 \frac{d \ln P}{dT}$$

by Eq. (6.1) ignoring temperature dependence of $\Delta \bar{H}_{vap}$ and C

$$= (1.99 \text{ cal} \cdot \text{K}^{-1}\text{mol}^{-1}) (298 \text{ K})^2 \left(\tfrac{4}{100} \text{ K}^{-1} \right)$$

$$= 7069 \text{ cal mol}^{-1}$$

Using Trouton rule [Eq. (9.1d)], which follows from Eq. (6.1), we have

$$\text{normal boiling point} = \frac{\Delta \bar{H}_{vap}}{\text{Trouton constant}}$$

$$= \frac{7069 \text{ cal mol}^{-1}}{21 \text{ cal K}^{-1}\text{mol}^{-1}} = 337 \text{ K}$$

The assumptions involved are (i) Boltzmann distribution law holds (ii) vapour obeys ideal gas equation (iii) $\Delta \bar{H}_{vap}$ is independent of temperature (iv) C in Eq. (6.1)is independent of temperature and nature of the liquid.

Except (i), the other assumptions are only hardly justified so that the quantities calculated above are quite approximate particularly for low-boiling liquids (like H_2) where (ii) is violated most and associated liquids (like water) where (iii) is not justified.

6.7 The term *surface tension* usually means an interfacial tension when one of the phases forming the interface is gas. Then it has meaning also with solids referring to solid–gas interface but not with gases, as two gases always mix forming a single phase.

6.8 Because the strength of the force field in solids is greater than that in liquids.

6.9 No. Because the different modifications of a substance have different surface density of their constituent particles.

6.10 A liquid film (even if very thin) can be considered as bulk liquid held between two closely spaced surface layers. The increase in area of the film due to an applied force results chiefly from sliding of one layer over the other. This is unlike a rubber sheet whose surface enlargement arises from elongation of the chemical bonds between the constituent atoms (following Hooke's law).

Note: In surface enlargement, the number of surface molecules increases in case of a liquid film but not in case of a membrane.

6.11 Zero value is possible with γ_{LG} at the critical temperature of the substance concerned (when the liquid and gas phases of same density cannot be distinguished) and also with γ_{LL} at the critical solution temperature (when two liquids have the same mutual solubility), but not with γ_{SG} as the difference between the solid and liquid phases of a substance never disappear.

☐ Interfacial tension cannot be negative, because the force responsible for this is considered to be directed towards the denser of the two phases involved.

6.12 This follows from Eotvos equation [Eq. (6.11a), which demands $\frac{\partial \gamma}{\partial T}$ = constant and $\frac{\partial^2 \gamma}{\partial T^2}$ = 0] together with Gibbs–Helmholtz equation [Eq. (6.3)].

Note: Eotvos equation is only an approximation violated most with substances forming hydrogen bonds.

6.13 A liquid having higher intermolecular attraction (determined mainly by molecular structure and nature of bonding) is likely to have higher surface tension. On this basis, we make the following predictions:

 i. benzene, for its higher polarisability due to unsaturation

 ii. ethanol, being able to form hydrogen bond

 iii. water, being able to form stronger hydrogen bond than ethanol

 iv. water, for the same reason as in iii.

 v. mercury, due to its metallic bond which is stronger than hydrogen bond

 vi. calcium, being able to form stronger metallic bond (indicated by higher melting point)

 vii. calcium chloride, due to interionic forces which are stronger than metallic bonds in calcium.

6.14 CH_3OH having lower molecular weight, by Eq. (6.11c).

☐ Because γ for such associated compounds depends much on their hydrogen bonding which breaks readily on raising temperature.

6.15 a. Because the adhesive work (W_{ad}) due to the interaction of dipoles (H_2O) with ions in glasses (which are ordinarily aluminosilicates) is greater than the cohesive work due to dipole–dipole interaction, which makes the contact angle $\theta < 90°$.

In case of polythene, W_{co} exceeds W_{ad} causing $\theta > 90°$.

b. Because the contact angle for rain drops on the sliiding side of the glass surface exceeds 90°.

c. Because mercury has so high value of W_{co} ($944 \times 10^{-3} J/m^2$) that its contact angle with most solid surfaces exceeds 90°.

6.16 a. No. Because, although total area of the interfaces, S–G, S–L and L–G, does not change, their proportions change in the two phenomena (which is equivalent to mixing) that can lead to requisite increase in randomness, i.e. entropy, for spontaneous occurrence of such phenomena.

b. No. ΔA is negative for capillary rise (where $\gamma_{SG} > \gamma_{SL}$) and also for capillary depression (where $\gamma_{SG} < \gamma_{SL}$, as $\theta > 90°C$).

Note: In both capillary rise and capillary depression L–G interfacial area does not change, but those of S–L and S–G interfaces both change though total interfacial area remain unchanged.

6.17 a. Because the shape of the liquid surface in case of a free falling drop of water is virtually determined only by the cohesive force, but that in case of water in a capillary is determined also by the solid–liquid and solid–gas adhesive forces such that the angle of contact for glass capillary is less than 90° and the meniscus appears concave.

b. Because the reaction force (due to the support) acting on the liquid is higher with lump of mercury which therefore, suffers greater distortion of spherical shape enough to assume ellipsoidal appearance.

c. This is due to surface tension. The effect is observable only near the melting point of the metal. Because at much lower temperature, the surface diffusion is not sufficiently fast.

d. Because blotting paper, but not polythene paper, is wetted by water. This enables blotting paper to soak water by capillary action of its pores.

e. Because this helps to retain subsoil water necessary for crop production by disrupting the capillaries in the soil, otherwise the underground water will rise through the capillaries to the surface and evaporate rapidly leaving the soil dry.

f. This is attributed to the disappearance of the liquid–air interface due to the film when the relevant interfacial energy appears as heat.

g. Mixing of two liquids, which involves breaking end making of bonds between like and unlike molecules respectively, will occur spontaneously only if the accompanying $\Delta G_{T,P}$ is negative. This requirement is not fulfilled in case of CCl_4 and H_2O in the liquid phase, because here ΔH is positive (due to high cohesive energy of H_2O) enough to make $\Delta G_{T,P}$ positive (despite ΔS being positive). But in the vapour phase, they get completely mixed up, because here $\Delta G_{T,P}$ is negative, being determined almost exclusively by ΔS which is positive.

6.18 a. The surface tension of a liquid is determined by its cohesive force and nature of the interface with its vapour phase. It has no necessary connection with ρ. In the given expression for γ, ρ appears as a consequence of the experimental procedure where the formation of drops is caused by their gravitational pull.

b. This is necessary so that the drops of a liquid can be of same size. A stalagmometer serves the purpose by holding up the liquid at its disc-shaped bottom part before drop formation.

□ No, if polythene is not wetted by the experimental liquid when the disc of the stalagmometer fails to serve its purpose.

6.19 The surface tension of a particular liquid, at a particular temperature, depends only on the nature of its interface with air (saturated with vapour of the liquid). Therefore, of the given factors, only (iv) will affect the obserrved value of surface tension (though only slightly), because the density of air depends on the atmospheric pressure.

Note: The quantities appearing on the right hand side of Eq. (6.9b) undergo mutual adjustment keeping γ virtually unaffected.

6.20 a. Capillary rise $h \simeq \frac{2\gamma \cos\theta}{rpg}$, by Eq. (6.9b). This implies that a rise in temperature will cause h to decrease as γ is more susceptible to temperature change than e.

b. h will remain same till the liquid meniscus reaches the upper end of the capillary. After this, h will gradually decrease ultimately to zero without any overflow due to simultaneous increase in θ is compliance with Eq. (6.9b).

Note:

1. h may be zero in (ii) but not in (i).
2. $h = 0$ corresponds to $\theta = 90°$ when the meniscus become flat.

6.21 This is a consequence of the pressure difference across a curved surface (having higher pressure on the concave side). We can understand this considering that the forces due to curvature of a liquid surface squeeze the liquid phase in case of convex surface leading to its higher tendency to vaporise, but has opposite effect in case of concave surface.

6.22 Let us imagine a closed vessel containing a liquid into which is a partly immersed capillary tube that is wetted by the liquid. Here the capillary forces will cause the liquid to rise in the tube to the height h given by

$$h \simeq \frac{2\gamma}{rpg}$$

where r is the radius of curvature of the concave meniscus in the capillary tube.

Since the vessel is closed, it will contain saturated vapour whose pressure P over the concave surface in the tube (at B) will be less than its pressure P_0 over the flat surface (at A), and they are related through barometric formula by

$$P = P_0 \exp(-Mgh/RT)$$

Then using the above expression for h, we have

$$P = P_0 \exp\left(-\frac{2M\gamma}{RT\rho r}\right)$$

which is Kelvin equation relating the vapour pressure P and P_0 of a liquid over its concave and flat surfact respectively.

The relevant relation in case of concave liquid meniscus can be obtained on replacing r by $-r$ in the above expression.

Note: The Kelvin equation, being free from g and h, will apply not only when the surface of a liquid is curved by a capillary, but also to such cases as a drop of a liquid (a convex surface) or a bubble in a liquid (a concave surface).

6.23. a. Supersaturated vapours are thermodynamically unstable. They have tendency to reduce their high free energy through condensation to a lump of liquid. But the process is very slow in absence of foreign particles. Because the microdroplets formed at the very beginning of the process have so high vapour pressure (due to their high curvature) that they mostly evaporate instead of growing, unless the initial pressure of the vapour is sufficiently higher than the bulk vapour pressure. However, the condensation is favoured in presence of foreign particles, such as dust particles and ions. Because the drops of liquid formed on them have appreciable radius from the very beginning of condensation which makes the vapour pressure over them only slightly greater than that over a flat liquid surface.

Note: The discussion made above constitutes the principle of artificial raining (by introducing dust particles). The cloud chamber used to determine the track of elementary particles works also on the same principle. Here the ions formed in their path act as nuclei for condensation and the trajectory of such particle is mapped out as a streak of condensed water.

b. The formation of rain drops from cloud (which is a clusture of microdroplets) requires appreciable concentration of atmospheric dust particles (which nucleate such process) which is more likely on week days due to running of factories producing dust. Then the given statement is justified.

6.24 a. Because, when the liquid is boiled alone most of the tiny bubbles formed at the initial stage of vaporisation condenses due to high Laplace pressure, apart from the external pressure, acting on them. Here the liquid gets superheated and boils abruptly when bubbles of vapour of sufficiently big size are formed by chance. Smooth boiling is due to big-sized bubbles formed by air liberated from the porous bodies on heating. Here air bubbles (due to their big size) facilitate evaporation, the prevailing Laplace pressure being quite low.

b. This is justified on the ground that the temperature of a boiling liquid is not an entirely definite quantity (it fluctuates within small limits) but the vapour above it is. Because no matter what happens inside a liquid, the vapour emerging from the bubbles bursting at its surface has a definite temperature (the true boiling point) at which the saturated vapour pressure equals the external pressure.

Note: Superheating of liquid (connected to vaporisation) and supersaturation of vapour (connected to condensation) are both based on Laplace pressure or Kelvin equation applied in reverse directions.

6.25 The minimum pressure of water vapour, which is equal to the vapour pressure of the droplets is

$$P = P_0 \exp\left[\frac{2M\gamma}{RT\rho r}\right] \qquad \text{by Eq. (6.10)}$$

$$= (23.8 \text{ torr}) \exp\left[\frac{2(18 \times 10^{-3}\,\text{kg mol}^{-1})(0.072\,\text{Nm}^{-1})}{(8.314\,\text{JK}^{-1}\text{mol}^{-1}((298\,\text{K})(10^3\,\text{kgm}^{-3})(0.75 \times 10^{-9}\,\text{m})}\right]$$

$$= 96.0 \text{ torr}$$

Note: Here only the quantities within [ ] is expressed in the same system of units, being convenient for this problem.

6.26 a. Here capillary rise will occur, because water wets glass

$$h_{max} = \frac{2\gamma}{r\,(\rho - \rho')g}, \text{ by Eq. (6.9b) taking maximum value of } \cos\theta = 1$$

$$\approx \frac{2\gamma}{r\rho g}$$

$$= \frac{2(72 \text{ dyn cm}^{-1})}{\left(\frac{0.2}{2} \times \frac{1}{10} \text{ cm}\right)(1 \text{ g}\cdot\text{cm}^{-3})(980 \text{ cm s}^{-2})}$$

$$= 14.7 \text{ cm}$$

   b. Here capillary depression will occur, because water does not wet polythene. The maximum depression will be same as h_{max} in (a)

   c. Here Laplace pressure $= \gamma\left(\dfrac{1}{r_1} + \dfrac{1}{\infty}\right)$, by Eq. (6.8a)

$$= \frac{\gamma}{r_1}$$

   which is half of the Laplace pressure in (a), r being same in the two cases. Then the capillary rise will be half of that in (a), i.e. 14.7/2 or 7.35 cm.

6.27 a. Minimum pressure required = Laplace pressure

$$= \frac{2\gamma}{r}$$

$$= \frac{2(72 \text{ dyn cm}^{-1})}{\dfrac{0.2}{2} \times \dfrac{1}{10} \text{ cm}} = 14400 \text{ dyn cm}^{-2}$$

   b. Radius of curvature

$$= \frac{2\gamma}{h\rho g}, \text{ by Eq. (6.9a) ignoring the density of the vapour phase}$$

$$= \frac{2\gamma}{\dfrac{h_{max}}{2}\rho g}$$

$$= 2r = \text{diameter of the capillary} \quad [\text{by Eq. (6.9c)}]$$

$$= 0.2 \text{ mm}$$

   Angle of contact θ is given by

$$\cos\theta = \frac{\text{radius of the capillary}}{\text{radius of curvature}} = \frac{1}{2}$$

   Then $\theta = 60°$.

Note:

1. Radius of curvature, like angle of contact, decreases with increase in capillary rise.

2. For h_{max}, that corresponds to $\theta = 0$, radius of curvature equals the radius of the capillary.

6.28 Level difference

$$= \frac{2\gamma}{\rho g}\left(\frac{1}{r_{narrow}} - \frac{1}{r_{wide}}\right), \quad \text{by Eq (6.9b) ignoring } \rho_{vap} \text{ and taking } \cos\theta \approx 1$$

$$= \frac{2(72 \text{ dyn/cm})}{(1 \text{ g/cm}^3)(980 \text{ cm/s}^2)}\left(\frac{1}{\frac{0.02 \text{ cm}}{2}} - \frac{1}{\frac{0.04 \text{ cm}}{2}}\right) = 7.35 \text{ cm}$$

No. Because θ changes (increases) properly so that overflowing is prevented. Vide problem 6.20b.

6.29 Concave (upward), as a consequence of capillary rise.

$$\text{Interfacial tension} = \frac{rh(\rho_{water} - \rho_{benzene})g}{2\cos\theta} \quad \text{by Eq. (6.9b)}$$

$$= \frac{\left(\frac{0.02}{2} \text{ cm}\right)(3.95 \text{ cm})(1 \text{ g cm}^{-3} - 0.8 \text{ g cm}^{-3})(980 \text{ cm/s}^2)}{2\cos 90°}$$

($\theta = 90°$ for hemispherical meniscus)

$$= 3.87 \text{ dyn/cm}$$

Note: The water–benzene interface is concave w.r.t. the lighter phase (benzene) forming the upper layer.

6.30 i. Here, there is only one liquid–gas interface and the excess pressure ΔP is given by

$$\Delta P = 2\gamma/r$$

$$= \frac{2(150 \text{ dyn/cm})}{\frac{0.02}{2} \text{ cm}} = 30000 \text{ dyn/cm}^2$$

ii. Here, there are two liquid–gas interfaces of virtually same radius and hence ΔP will be double of that in (i), i.e. 60000 dyn/cm².

□ The answer is 'yes' for the bubble in case of (i) but 'no' in case of (ii).

Note: The excess pressure, unlike hydrostatic pressure, is independent of density of the phase concerned.

6.31 For the nascent bubble

excess pressure $(2\gamma/r)$ = pressure within the bubble $(P + h_{water}\rho_{water}\, g)$
$\qquad\qquad\qquad\qquad$ – pressure outside the bubble $(P + h_{soln}\rho_{soln}\, g)$

where P is atmospheric pressure

Then $\quad \gamma = \frac{1}{2}r(h_{water}\rho_{water} - h_{soln}\rho_{soln})$

$$= \frac{1}{2}\left(\frac{0.1 \text{ cm}}{2}\right)[(12 \text{ cm})(1.0 \text{ g/cm}^3) - (10 \text{ cm})(1.01 \text{ g/cm}^3)](980 \text{ cm/s}^2)$$

assuming that the half-formed bubble is hemispherical with radius equal to that of the capillary (when the excess pressure is maximum, as the radius of curvature is then minimum)

$$= 46.55 \text{ dyn/cm}$$

6.32 The problem is similar to the previous one. Here the pressure difference $(2\gamma/r)$ across the liquid meniscus in the nascant (half-formed) drop is equal to the hydrostatic pressure at the bottom of the vessel ($h\rho g$). Then

$$h_{max} = \frac{2\gamma}{r\rho g}$$

$$= \frac{72 \text{ dyn/cm}}{\left(\frac{0.03}{10} \text{ cm}\right)\left(\frac{1.0 \text{ g}}{\text{cm}^3}\right)\left(\frac{980 \text{ cm}}{s^2}\right)} = 24.5 \text{ cm}$$

Here, it has been assumed that water does not wet the vessel so that the nascent drop is hemispherical with radius equal to that of the hole. It is under this condition, h is maximum as the radius of curvature is then minimum.

6.33   Work required = Area of the disc in contact with water × cohesive work

$$= \pi r^2 \times 2\gamma_{water}$$
$$= 3.14 \, (1 \text{ cm})^2 \times 2 \, (72 \text{ dyn/cm}) = 452.2 \text{ ergs}$$

Note:

i. Here cohesive work (but not adhesive work) is involved, because the metallic disc (which is wetted by water) is detatched together with a coating of water.

ii. Mass of the disc is of no use, because no work against gravitational force is involved.

6.34 Each side of the wire is of length 1 cm. If the wire is placed vertically on the liquid surface, the force that will act per unit length of the liquid surface is the weight of the wire (mg)

$$mg = (0.25 \text{ g}) \, (980 \text{ cm/s}^2) = 245 \text{ dyn}$$

which is higher than the surface tension of water (i.e. minimum force needed to cut through unit length of the liquid surface). Then the wire will not float in vertical position.

But in horizontal position, the force acting per unit length of the liquid surface is $\frac{0.25}{4} \times 980$ or 61.25 dyn, which is less than the surface tension of water. Then the wire will float in horizontal position.

☐ Since the metallic wire (wetted by water) is detatched along with a coating of water, the minimum force (F) needed for detatchment is

$$F = mg + 2\gamma l$$

where mg is the weight of the wire (due to gravitational force) and $2\gamma l$ is force due to surface tension of water. For vertical position $l = 1.0$ cm and for horizontal position, $l = 4.0$ cm.

[The number 2 appears in the expression of F because the liquid film associated with the detatchment of the wire has two surfaces].

☐☐ No. Because in the detatchment of wire, adhesive work is not involved.

Note:

1. Here the force, unlike work in the previous problem, depends on the weight of the wire.

2. If the wire were not wetted by the liquid, the results in (i) and (ii) would depend on solid–liquid interfacial tension due to involvement of adhesive work.

6.35 Let R be the radius of the drop that breaks into n microdrops each of radius r. Considering that

Volume of the initial drop $\left(\frac{4}{3}\pi R^3\right)$ = total volume of the microdrops $\left(n\frac{4}{3}\pi r^3\right)$

We have $r = \dfrac{R}{n^{\frac{1}{3}}}$.

Now, change in surface free energy = change in surface area × surface tension

$$= (n4\pi r^2 - 4\pi R^2)\gamma$$
$$= [4\pi n \, (R/n^{\frac{1}{3}})^2 - 4\pi R^2]\gamma$$
$$= 4\pi R^2 \, (n^{\frac{1}{3}} - 1)\gamma$$

which is a linear function of $n^{\frac{1}{3}}$.

6.36   Volume of the initial liquid drop $= \frac{4}{3}\pi R^3$, R is radius of the drop

$$= w/e$$

Then   surface area of this drop $= 4\pi R^2 = 4\pi \left(\frac{3w}{4\pi e}\right)^{\frac{2}{3}}$

The change in surface free energy

$$= (n \cdot 4\pi r^2 - 4\pi R^2)\gamma$$

$$= 4\pi \left[nr^2 - \left(\frac{3w}{4\pi e}\right)^{\frac{2}{3}}\right]\gamma$$

$$= 4 \times 3.14 \left[1500\,(0.02\text{ cm})^2 - \left\{\frac{3(0.04\text{ g}}{4 \times 3.14\,(0.8\text{ g/cm}^3}\right\}^{\frac{2}{3}}\right](27\text{ dyn/cm})$$

$$= 166.4\text{ ergs}$$

☐ This has no other justification then to suggest that the dispersion is not accompanied by considerable vaporisation as it follows from the given data that

volume of the initial liquid drop (w/ρ) = volume of the globules $\left(n\frac{4}{3}\pi r^3\right)$

$$= 0.05\text{ cm}^3$$

Note: The above calculation of surface free energy is valid even if the dispersion is accompanied by vaporisation. This is unlike the previous problem where the vaporisation is required to be absent.

6.37 Assuming that the surface tension varies linearly with temperature

$$\gamma_{T'} = \gamma_T + \frac{d\gamma}{dT}(T' - T$$

Then $\gamma_{298\text{ K}} = 72.8\text{ dyn/cm} - (0.15\text{ dyn cm}^{-1}\text{K}^{-1})\,(298\text{ K} - 293\text{ K})$

$$= 72.05\text{ dyn/cm}$$

Now, $\dfrac{\text{total volume of the droplets}}{\text{total surface area of the droplets}} = \dfrac{\frac{4}{3}\pi r^3}{4\pi r^2} = \dfrac{r}{3}$,

where r is radius of the droplets

Then total surface area of the droplets $= \dfrac{3(1\text{ cm}^3)}{(10^{-5}\text{ cm})} = 3 \times 10^5\text{ cm}^2$

(Considering that total volume of the droplet = Volume of initial lump water)

Therefore surface free energy of the droplets

$$= \text{Surface area of the droplets} \times \gamma\text{ water}$$

$$= (3 \times 10^5\text{ cm}^2)\,(72.05\text{ dyn/cm})$$

$$= 2.16 \times 10^7\text{ ergs at 298 K}$$

Surface energy of the droplets

$$= \text{Surface area of the droplets} \times \left[\gamma - T\left(\frac{\partial \gamma}{\partial T}\right)_V\right]$$

$$= (3 \times 10^5\text{cm}^2)\,[(72.05\text{ dyn/cm} - (298\text{ K})\,(-0.15\text{ dyn cm}^{-1}\text{K}^{-1})$$

$$= 8.20 \times 10^6\text{ ergs}$$

The surface free energy and surface energy relative to lump water will be practically same as their actual value calculated above as the surface area of the lump water is negligible compared to that of the droplets.

☐ In the condensation of the droplets the amount of heat liberated will be equal to their total surface energy (the surface energy of the resulting lump water being neglected), i.e. 8.2×10^6 ergs.

Note: Here we have considered (as in problem 6.35) that the total volume of the droplets is equal to the volume of the initial lump liquid, ignoring the vaporisation of the liquid. This is not required in case of previous problem where the size of the droplets and their number are both given.

6.38 Let R be the radius of the resulting globule. Considering that

Volume of resulting globule $\left(\frac{4}{3}\pi R^3\right)$

$$= \text{total volume of the smaller globules } \left(2 \times \frac{4}{3}\pi r^3\right)$$

We have $R = 2^{1/3}r$

The loss in free energy in the process = decrease in surface free energy

$$= \text{decrease in surface area} \times \gamma_{Hg}$$
$$= (2 \times 4\pi r^2 - 4\pi R^2)\,\gamma_{Hg}$$
$$= [2 \times 4\pi r^2 - 4\pi(2^{1/3}r)^2]\,\gamma_{Hg}$$
$$= 4\pi\,(2 - 2^{2/3})r^2\gamma_{Hg}$$

6.39 i. Work of adhesion between water and benzene $= \gamma_{water} + \gamma_{benzene} - \gamma_{water-benzene}$

$$= (7.27 + 2.88 - 3.50) \times 10^{-2}\,\text{N/m}$$
$$= 6.65 \times 10^{-2}\,\text{N/m}$$

ii. Work of cohesion for benzene $= 2\gamma_{benzene} = 2 \times 2.88 \times 10^{-2}\,\text{N/m}$

iii. Spreading coefficient for benzene on water $= \gamma_{water} - \gamma_{benzene} - \gamma_{benzene-water}$

$$= (7.27 - 2.88 - 3.50) \times 10^{-2}\,\text{N/m}$$
$$= 0.89 \times 10^{-2}\,\text{N/m}$$

Then, for spreading of benzene on water, $\Delta A_{T,V} = -\text{ve}$

Hence, benzene will spread spontaneously on water.

Again, for spreading of water on benzene

$$\Delta A_{T,V} = \gamma_{water-benzene} + \gamma_{water} - \gamma_{benzene}$$
$$= (3.5 + 7.27 - 2.88) \times 10^{-2}\,\text{N/m} = +\text{ve}$$

Then water will not spread on benzene spontaneously.

☐ Here it has been assumed that the benzene and water layers in contact with one another are pure. [More precisely, we should use γ of saturated solution of benzene in water instead of γ_{H_2O} and γ of saturated solution of water in benzene instead of $\gamma_{benzene}$].

UNIVERSITY QUESTIONS

6.1 Define surface tension. Why do liquids have surface tension?

(Guru Nanak Dev BSc, 2003)

6.2 At 25°C, the surface tension of ethanol is 21.8 dyn/cm whereas that of water is much higher, namely 72.8 dyn/cm. How would you account for this difference?

(Calcutta BSc(H), 1992)

6.3 Is there any difference between surface tension and surface energy?

(Jadhavpur BSc(H), 1992)

6.4 a. The surface tension of a liquid decreases as the temperature is raised. Explain the phenomenon qualitatively. (Calcutta BSc(H), 2010)

b. When should a liquid have a zero surface tension? (Calcutta BSc(H), 2002)

c. Comment on the temperature coefficient of surface tension in the case of associated liquids. (Calcutta BSc(H), 1997)

6.5 Why rain drops are spherical? (Arunachal BSc, 2003)

6.6 A liquid drop of radius R and surface tension γ breaks up into n tiny droplets of equal size. Show that the change in surface free energy is given by $4\pi R^2 \gamma (n^{1/3} - 1)$.

(Burdwan BSc(H), 2008)

6.7 A spherical drop of a liquid weighing 0.04 gm is dispersed into 1500 microglobules of radius 0.02 cm each by a suitable experimental device. Find the resultant increase in surface free energy. ($\rho_l = 0.8$ gm/cm^3, $\gamma = 27$ dyn/cm] (Calcutta BSc(H), 1998)

6.8 1 cc of water is broken into droplets having radius 10^{-5} cm. Surface tension of water is 72.8 dyn/cm at the same temperature. Calculate the surface free energy of fine droplets relative to that of water. (Calcutta BSc(H), 1994)

6.9 If a water drop of 1 mm radius is broken into a million droplets, calculate the increase in surface area and free energy. [$\gamma = 72.75$ erg cm^{-2}]

(Calcutta BSc(H), 2012)

6.10 In the determination of surface tension of a liquid A by the drop number method using a stalagmometer, equal volume of A and water formed 124 and 90 drops respectively at 25°C. Calculate the surface tension of A at 25°C.

[Given: Density of liquid A = 1.24 g/cm^3

Density of water = 0.996 g/cm^3

Surface tension of water = 72.8 dyn/cm, all at 25°C]

(Calcutta BSc(H), 2000)

6.11 Indicate with the help of sketch the angle of contact θ between liquid and solid. Obtain an expression for θ indicating the factors on which the value of θ depends. What does wetting of a solid surface by a liquid imply? (Calcutta BSc(H), 1994)

6.12 Show that the work of adhesion between two liquid phases α and β is given by $W_A^{\alpha\beta} = \frac{1}{2}(W_C^\alpha + W_C^\beta) - \gamma^{\alpha\beta}$ where W_C^α and W_C^β are the work of cohesion for α and β phases respectively. (Calcutta BSc(H), 2013)

6.13 Define spreading coefficient (ϕ) between two immiscible liquids. Show that for spontaneous spreading ϕ must be positive. (Calcutta BSc(H), 2006)

6.14 At 20°C the interfacial tension between water and benzene is 35 mN/m. If the surface tension $\gamma = 28.85$ mN/m for benzene and 72.75 mN/m for water, calculate:

i. the work of adhesion between water and benzene.

ii. the work of cohesion for benzene and water.

iii. the spreading coefficient for benzene and water. (Calcutta BSc(H), 2003)

6.15 Derive Laplace equation relating the pressure on the concave side of a cavity (P_{in}) with that on convex side of the cavity (P_{out}).

$P_{in} = P_{out} + 2\gamma/r$

where the terms have their usual significance. (Calcutta BSc(H), 2013)

6.16 Why it is difficult to blow a balloon initially, but becomes easier afterwards? (Consider the balloon as a bubble). (Calcutta BSc(H), 2014)

6.17 Why is the pressure on the concave side of a surface greater than that on the convex side? Use this to explain the phenomenon of capillarity. (Calcutta BSc(H), 1989)

6.18 Explain why one observes a capillary rise for water but for mercury there is a capillary depression when a glass capillary is partially dipped into the liquid.

(Burdwan BSc(H), 2009)

6.19 The surface tension of a liquid at 300 K is 27.1 dyn/cm and its density at this temperature is 0.988 g/cm³. What is the radius of the largest capillary that will allow the liquid to rise 2.0 cm? (Assume the angle of contact to be zero and $g = 981$ cm s⁻²). (Punjab BSc, 2005)

6.20 The internal diameter of a capillary tube is 0.30 mm, and that of another capillary is 0.50 mm. The two tubes are vertically placed side by side and partially dipped in a liquid of density 0.80 gm/cm³. The liquid level in one capillary is found to be 4.00 cm higher than the liquid level in the other capillary. Find the surface tension of the liquid. Assume zero contact angle. (Jadhavpur BSc(H), 1997)

6.21 The limbs of a vertical U-tube have internal diameters of 1 mm and 2 mm respectively. It is partially filled with a liquid of density 0.82 gm/cm³ and surface tension 50 dyn/cm. What is the difference in level of the liquid in the two limbs.

(Burdwan BSc(H), 2007)

6.22 Water rises to a height 10 cm when a certain capillary tube is dipped in water. If the same capillary tube is dipped in mercury, the level of mercury in the tube is depressed by 3.42 cm. If the density of water and mercury be 1.0 and 13.6 gm/cm³, the angles of contact for water and mercury are 0° and 135° respectively. Compare the surface tensions of water and mercury. (Burdwan BSc(H), 2009)

6.23 For water-air interface at 25°C and 1 atm, calculate the rise of water in a capillary tube of inside diameter 0.2 mm. The surface tension of water is 72 dyn/cm at 25°C. The densities of water and air at 25°C and 1 atm are respectively 0.997 g/cm³ and 0.001 g/cm³. Should the rise be lower at 40°C? Discuss very briefly.

(Burdwan BSc(H), 2001)

6.24 A spherical air bubble is created within a liquid of surface tension 72 dyn/cm. If the volume of the bubble is $\frac{\pi}{6}$ cm³, calculate the excess pressure inside the bubble.

(Calcutta BSc(H), 2003)

6.25 A spherical soap–bubble of volume $\frac{\pi}{6}$ cm³ stands suspended in air. What is the excess pressure inside the bubble? [Given: The interfacial tension for the soap solution–air interface is 27 dyn/cm]. (Calcutta BSc(H), 2000)

6.26 a. It is known that the vapour pressure of small drops of liquid is greater than that of the liquid present in bulk. How does surface tension measurement help in finding the relative magnitude of the two vapour pressures?

(Calcutta BSc(H), 1980)

   b. When a liquid is warmed, initially many small bubbles begin to form. Why?

(Guru Nanak Dev BSc, 2003)

6.27 Explain the phenomenon of boiling by bumping. (Burdwan BSc(H), 1995)

6.28 Calculate the vapour pressure of a water droplet at 25°C of radius 2.0 × 10⁻⁹ m. The vapour pressure of a flate surface of water is 3167 Pa at 25°C and the surface tension of water–air interface at 25°C is 0.072 Nm⁻¹. Density of water at 25°C is 10³ kg m⁻³.

(Calcutta BSc(H), 2008)

KEY TO UNIVERSITY QUESTIONS

6.1 Vide Section 6.3.

6.2 Vide problem 6.13(iii).

6.3 Surface tension is mathematically (i.e. numerically and dimensionally) equal to the surface free energy (which refers to the relevant ΔA) but not to surface energy (which refers to ΔU). This follows immediately if we remember that the surface free energy equals the force corresponding to the surface tension multiplied by unit distance through which the point of application of such force is moved. Vide Section 6.3.

Note: Now-a-days, surface tension is preferably expressed in the unit of energy, viz. J/m^2 (in SI).

6.4 a. Because a rise in temperature causes reduction of intermolecular forces (on which surface tension depends) due to increase in intermolecular distance.

b. At critical temperature, vide Section 6.5

c. Vide problem 6.14.

6.5 An isolated liquid drop assumes a spherical shape (which corresponds to minimimum surface area) to make its surface free energy minimum in compliance with thermodynamic requirement. The shape remains virtually same in the gravitational field (as with a free falling rain drop) because the latter imparts identical acceleration to all the particles of the drop.

Note: Slightly flattening of rain drops occurs due to the resistance of the air.

6.6 Vide problem 6.35.

6.7 Vide problem 6.36.

6.8 Vide problem 6.37.

6.9 Vide problem 6.35.

6.10
$$\frac{\gamma_A}{\gamma_{H_2O}} = \left(\frac{\rho}{n}\right)_A \bigg/ \left(\frac{\rho}{n}\right)_{H_2O} \qquad \text{(refer to problem 6.18a)}$$

or
$$\gamma_A = \gamma_{H_2O} \cdot \frac{\rho_A}{\rho_{H_2O}} \cdot \frac{n_{H_2O}}{n_A}$$

$$= (72.8 \text{ dyn/cm}) \frac{(1.24 \text{ g/cm}^3)}{(0.996 \text{ g/cm}^3)} \cdot \frac{(90)}{(124)}$$

$$= 65.8 \text{ dyn/cm}$$

6.11 Vide Section 6.3(B) and (C).

6.12 This follows from combination of Eqs (6.5) and (6.6a).

6.13 The spreading coefficient (ϕ) of a liquid X over another immiscible liquid Y is defined by

$$\phi = \text{adhesive work} - \text{cohesive work of } X$$
$$= W_{XY} - 2\gamma_X$$
$$= \gamma_Y - \gamma_X - \gamma_{XY} \quad \text{vide subsec. 6.3c}$$

Now, in the spreading of X over unit area of Y (which is accompanied by unit increase, in the liquid–vapour interface for X and unit decrease in the liquid–vapour interface for Y), $\Delta A_{T,V}$ (which is the minimum amount of work needed) is

$$\Delta A_{T,V} = \gamma_{XY} + \gamma_X - \gamma_Y$$
$$= -\phi$$

Then, for spontaneous spreading, ϕ must be positive so that $\Delta A_{T,V}$ is negative (a criterian for spontainity).

Note: $\phi = -\Delta A_{T,V}$

$\neq W_{XY}$

These are not unlikely if we remember that the expression of W_{XY} is based on the supposition that the total interface of both X and Y remain unchanged while the expression of ϕ is based on the supposition that the total interface of the spreading substance X increases but that of Y remains unchanged.

6.14 Vide problem 6.39.

6.15 Let us consider a spherical cavity of radius r in a liquid, we can regard the cavity as consisting of two equal hemispherical parts bound together through the force $2\pi r\gamma$ due to surface tension γ. This force prevents the hemispheres to be driven apart by the force $\pi r^2 \Delta P$ where ΔP is the excess pressure of the vapour inside the cavity. Then

$$\pi r^2 \Delta P = 2\pi r\gamma$$

or $$2\gamma/r = \Delta P$$

$$= P_{in} - P_{out}$$

whence $$P_{in} = P_{out} + 2\gamma/r$$

6.16 Because the excess pressure $\Delta P = 2\gamma/r$ inside the balloon (provided by blowing) gets reduced with growing r.

Note: Here the Laplace equation does not really hold, though we use it, being prompted by the apparent analogy between a balloon and a bubble vide problem 6.10.

6.17 Because the surface layer behaves as a stretched membrane due to intermolecular attraction.

□ Vide subsec. 6.4A.

6.18 Such phenomena are attributed to the curved nature of the meniscus (i.e. the liquid–vapour interface within a capillary) with the excess pressure (ΔP) directed toward the centre of curvature of the meniscus. Since glass is wetted by water but not by mercury, the meniscus is concave for water but convex for mercury causing ΔP to act in the upward and downward direction for the respective cases. The result is capillary rise for water but capillary depression for mercury. Vide Section 6.4A.

6.19 $$\gamma_{max} = \frac{2\gamma}{h\rho g} \text{ by Eq. (6.9c) ignoring the density of the vapour phase}$$

$$= \frac{2(27.1 \, \text{dyn/cm})}{(2.0 \, \text{cm})(0.9889 \, \text{g/cm}^3)(981 \, \text{cm/sec}^2)} = 0.28 \, \text{cm}$$

6.20 Level difference

$$\Delta h = -\frac{2\gamma}{\rho g}\left(\frac{1}{r_{narrow}} - \frac{1}{r_{wide}}\right), \text{ by Eq. (6.9c) ignoring density of the vapour phase}$$

or $$\gamma = \frac{\Delta h \rho g}{2\left(\dfrac{1}{r_{narrow}} - \dfrac{1}{r_{wide}}\right)}$$

$$= \frac{(4.0 \, \text{cm})(0.80 \, \text{g/cm}^3)(980 \, \text{cm/sec}^2)}{2\left(\dfrac{1}{\frac{1}{2} \times \frac{0.30}{10} \, \text{cm}} - 2\dfrac{1}{\frac{1}{2} \times \frac{0.50}{10} \, \text{cm}}\right)} = 58.8 \, \text{dyn/cm}$$

Note: This problem is essentially of same type as problem 6.28.

6.21 Vide problem 6.28.

6.22 $\quad \dfrac{\gamma_{Hg}}{\gamma_{H_2O}} = \dfrac{(h\rho/\cos\theta)_{Hg}}{(h\rho/\cos\theta)_{H_2O}}$, by Eq. (6.9b) ignoring the density of the vapour phase

$$= \dfrac{(-3.42\ \text{cm})\,(13.6\ \text{g/cm}^3)/\cos 135°}{(10\ \text{cm})\,(1.0\ \text{g/cm}^3)/\cos 0°} = 6.58$$

6.23 Capillary rise, $\quad h = \dfrac{2\gamma}{r(\rho_{water} - \rho_{air})g} \quad$ by Eq. (6.9b) taking $\cos\theta = 1$

$$= \dfrac{2(72\ \text{dyn/cm})}{\left(\frac{1}{2} \times \frac{0.2}{10}\ \text{cm}\right)(0.997\ \text{g/cm}^3 - 0.001\ \text{g/cm}^3)(980\ \text{cm/sec}^2)}$$

$$= 14.8\ \text{cm}$$

☐ Yes. Vide problem 6.20 (a).

6.24 The radius r of the bubble is given by

$$\tfrac{4}{3}\pi r^3 = \dfrac{\pi}{6}\ \text{cm}^3, \text{ the volume of the bubble}$$

whence $\qquad r = \tfrac{1}{2}\ \text{cm}$

Excess pressure $= \dfrac{2\gamma}{r}$

$$= \dfrac{2(72\ \text{dyn/cm})}{\frac{1}{2}\ \text{cm}} = 288\ \text{dyn/cm}^2$$

6.25 Excess pressure $= 2 \times 2\gamma/r$ vide problem 6.30

6.26 a. Vide problem 6.22.

   b. When a liquid is heated, its vaporisation is likely to begin with formation of bubbles of various size. But most of the tiny bubbles condenses due to high Laplace pressure acting on them. Vide problem 6.24a.

6.27 Vide problem 6.24a.

6.28 Vide problem 6.25.

7

Liquid State—Flowing Situation

An important characteristic of a liquid is its ability to flow.

7.1 FLUID FLOW AND VISCOSITY

Fluids (liquids and gases), unlike solids, cannot withstand a shear stress that always results in their flow, i.e. relative translational motion between their parts (layers); they are called so due to such characteristic. This flow is called laminar, if the velocity of flow at any point of the fluid does not vary with time and the velocity of fluid layers varies uniformly in the direction perpendicular to the direction of flow, otherwise the flow is called turbulent.

When a fluid flows the relative motion of its adjacent layers is opposed, the relevant property is called viscosity which may be regarded as the internal friction acting in a fluid.

A. Newton's Law

In case of laminar flow, the force (F) that opposes the relative motion of any two adjacent layers of a fluid is generally proportional to the area (A) of the interface between them and to the velocity gradient (dv/dx) between them in the direction x which is perpendicular to the direction of fluid flow, i.e.

$$F \propto A \frac{dv}{dx}$$

$$= -\eta A \frac{dv}{dx} \tag{7.1}$$

This is Newton's law of viscous flow. The proportionality constant η is called the coefficient of (Newtonian or dynamic) viscosity, or simply viscosity of the fluid.

In absence of external forces, the velocities of adjacent fluid layers level out (e.g. a storm abates with time) through transfer of impulse (momentum) from the rapidly moving to the slowly moving layer.

B. Poiseuille's Law

An important case of fluid flow is the flow through tubes. The volume (V) of a fluid flowing laminarly per unit time through a tube of (internal) radius R and length l under the pressure difference $P_1 - P_2$ is given by

$$V = \frac{\pi R^4 (P_1 - P_2)}{8 \eta l} \tag{7.2a}$$

which is Poiseuille's equation of viscous flow. Here V is measured at the mean pressure $(P_1 + P_2)/2$. If V is measured at some other pressure (P) then this expression has to be modified for gases (but not for liquids) as under, since the volume of a gas, unlike liquid, is very susceptible to pressure

$$V = \frac{\pi R^4 (P_1 - P_2)}{8 \eta l} \cdot \frac{P_1 + P_2}{2} \cdot \frac{1}{P}, \text{ by Boyle's law}$$

$$= \pi R^4 \frac{(P_1^2 - P_2^2)}{16 \eta l P} \tag{7.2b}$$

The flow through a tube involves flow of cylindrical layers of different radius ranging from R to O with velocity ranging from O to a maximum value (v_{max}) where

$$v_{max} = \frac{(P_1 - P_2) R^2}{4 \eta l} \tag{7.3a}$$

and average velocity of flow

$$\frac{V}{\pi R^2} = \frac{(P_1 - P_2) R^2}{8 \eta l} \tag{7.3b}$$

Whether the flow of a liquid is laminar or turbulant is determined by a combination of four factors, called Reynolds number, N_R defined as

$$N_R = \frac{\rho v d}{\eta} \tag{7.4}$$

where ρ and η are the density and viscosity of the fluid respectively flowing with an average velocity v. Experimentally, it has been found that the flow is laminar if $N_R < 2000$ and turbulent if $N_R > 3000$.

Although kinetics of flow is similar for liquids and gases, the kinetic molecular mechanism responsible for their viscosity is quite different.

7.2 VISCOSITY OF LIQUIDS

The liquid viscosity results mainly from intermolecular attractive forces (cohesive forces). In order to take part in the liquid flow a molecule must have to move from site to site (to maintain short-range order in molecular arrangement) against the attraction of its neighbours. Only those molecules will be involved which possess at least a minimum amount of energy E, called energy of activation, above the average. The number of such molecules, and hence the rate of flow may be supposed to be proportional to the Boltzmann factor $e^{-E/RT}$. Then the viscosity (η), i.e. resistance to flow, would be proportional to $e^{E/RT}$ (which is inverse of $e^{-E/RT}$) and we can write

$$\eta = \eta_0 e^{E/RT} \tag{7.5a}$$

where η_0 is a constant for a particular substance

or $\qquad \ln \eta = A + B/T \tag{7.5b}$

where $A = \ln \eta_0$ and $B = E/R$ are both constants for a particular liquid.

The above molecular picture of liquid flow is in agreement with the fact that viscosity of a liquid decreases with rise of temperature and increases with rise of pressure.

7.3 VISCOSITY OF GASES

For an ideal gas viscosity results from inter-layer jump of its molecules which is intimately connected to mean free path (λ). Kinetic theory leads to the following approximate relation

$$\eta = \tfrac{1}{3} mn\langle c\rangle \lambda \tag{7.6a}$$

where η is molecular density.

Then for an ideal gas η should be independent of pressure as $\langle c\rangle$ and $n\lambda \left(=\frac{1}{\sqrt{2}\pi\sigma^2}\right)$ are so. For real gases, the observed pressure dependence of η is due to the existence of intermolecular forces (as in case of liquids) that makes the above relation inapplicable.

Also it follows from the above relation that $\eta \propto \sqrt{T}$ (as $\langle c\rangle \propto \sqrt{T}$). For real gases, η rises with T, but somewhat different relation is followed

$$\eta = \frac{\eta_0 \sqrt{T}}{(1 + A/T)} \tag{7.6b}$$

where η_0 and A are constants for a particular gas.

7.4 EFFECT OF IMPURITY ON VISCOSITY

Dissolved ionic salts reduce cohesion and hence, in general, lower the viscosity of the medium. But the presence of macromolecules (or suspended matter) results in increase of viscosity; here the effect is large even at low concentrations, because the large molecules affect the surrounding fluid's flow over a long range. Then measurement of viscosity is expected to give information about the size and shape of macromolecules. The viscosity appropriate for this purpose is the intrinsic viscosity $[\eta]$ which is defined as

$$[\eta] = \lim_{c \to 0} \frac{1}{c}\left(\frac{\eta}{\eta_0} - 1\right) \tag{7.7a}$$

where η and η_0 are viscosities of the solution and the pure solvent respectively, and c is the mass concentration of the solute in solution. The quantity η/η_0 is referred to as relative viscosity and $\left(\frac{\eta}{\eta_0} - 1\right)$ as specific viscosity; these are types of reduced viscosity having no dimensions.

$[\eta]$ is related to the average molar mass M by an expression of the following form

$$[\eta] = KM^a \tag{7.7b}$$

where K and a are empirical constants for a solution which depends on the nature of the solute molecules and solvent involved. a is around 0.5 for nearly spherical molecules while it has higher value, as high as 2, for rod-like molecules.

7.5 STOKES LAW

When a body moves in a fluid, the layer of fluid in contact with the body moves along with it (no-slip condition), and a gradient of velocity develops in the surrounding fluid. This generates a viscous force (F) that resists the motion of the body. F increases proportionally with the velocity of the body. When F is balanced by the applied force, the body attains a constant velocity, called limiting or terminal velocity.

The viscous force F acting on a spherical body of radius r moving with velocity v through a Newtonian fluid of viscosity η is governed by Stokes law according to the following expression

$$F = 6\pi\eta rv \tag{7.8}$$

This expression holds only for spherical bodies provided (i) r is not less than the molecular radius of the fluid and it is much greater than the mean free path of the latter (arising for gaseous fluid) (ii) v is not very high.

Stokes law is used in the determination of viscosity of a fluid (by falling sphere method) and also the size of the particles moving through it. This is based on the fact that a sphere (of radius r and density ρ) falling in a viscous fluid (of density ρ' and viscosity η) attains a terminal velocity v when the weight of the sphere $\left(\frac{4}{3}\pi r^3\rho g\right)$ equals the viscous retarding force $(6\pi\eta rv)$ plus the force of buoyancy $\left(\frac{4}{3}\pi r^3\rho'g\right)$

i.e. $$\frac{4}{3}\pi r^3\rho g = 6\pi\eta rv + \frac{4}{3}\pi r^3\rho'g$$

or $$\eta = \frac{2}{9}\cdot\frac{r^2(\rho-\rho')}{v}g \tag{7.9}$$

The falling sphere method works well for fluids of quite high viscosity, such as glycerol. It is useless for fluids of low viscosity (where Ostwald viscometer based on Poiseuille's equation works well). Also it is unsuitable for fluids of very high viscosity, such as molten glass (where a rotating drum viscometer is most suitable).

AUXILIARY PROBLEMS

7.1 Viscosity may be regarded as an internal friction acting in a fluid. How does it differ from the external friction acting on a solid body sliding over another.

7.2 Viscosity of a stagnant fluid is zero, comment. Does the term viscosity have any relevance to raining?

7.3 In what important respects does viscous flow of a gas through a tube differ from effusive flow? Do you expect abating of storm to have cooling or heating effect?

7.4 Flowing of a gas through a tube violates Poiseuille's equation when diameter of the tube is high and also when it is very low. Why?

7.5 What is meant by (i) Newtonian fluid (ii) ideal fluid? Do they really exist?

7.6 Viscosity of a solution is found to decrease when its rate of flow is increased. What does this imply regarding this solution and solute?

7.7 Find the dimensions of ordinary viscosity and intrinsic viscosity. Which one is more complex quantity, and which one is more important to chemists?

7.8 Translate poise into SI unit.

7.9 Fluid flow through a tube offers simultaneously two different examples of transport processes. Discuss this mentioning the relevant laws.

7.10 Do you expect a substance to have higher viscosity—in liquid or gaseous state? Give reasons for your answer.

7.11 With rise in temperature, viscosity of a liquid falls but that of a gas rises. Explain. Can viscosity of a liquid be made to zero by heating?

7.12 Rise of gas viscosity with rise of temperature implies that energy of activation for gas flow is negative. Justify or criticise.

7.13 In which case a gaseous substance will have higher viscosity when it behaves as a real gas or an ideal gas? Also predict when the viscosity will be affected more with temperature?

7.14 State with reason which substance in each of the following pairs will have higher viscosity:
  i. *n*-pentane and neopentane
  ii. benzene and cyclohexane
  iii. mercury and kerosene
  iv. kerosene and glycerol

7.15 NaCl reduces the viscosity of water while protein does the reverse. Explain.

7.16 Deduce the Eq. (7.6a).

7.17 The expression $\eta = \frac{1}{3} mn\langle c \rangle \lambda$ for an ideal gas implies that its viscosity will be same at all pressure through molecular density in each layer is higher at higher pressure. What is the physical reason behind this?

7.18 The expression $\eta = \frac{1}{3} mn\langle c \rangle \lambda$ becomes invalid at very low pressure. Why?

7.19 The expression $\eta = \frac{1}{3} mn\langle c \rangle \lambda$ is based on the assumption that molecules are coming to a particular layer of the gas from the adjacent layers. How do you justify such assumption that seems to cause growing molecular density of the concerned layer? Again, although the molecules are jumping, and hence transferring momentum, in the direction perpendicular to the direction of flow, the associated force acts in the direction of flow. How?

7.20 Give reasons for the following.
  a. When a smooth-flowing stream of water comes out of a tap, it narrows as it falls, and if it falls far enough, it eventually breaks up into drops.
  b. Hot water running into a sink makes less noise and splashes less than cold water.
  c. Air escaping from the open end of a short pipe makes more noise than air escaping at the same volume rate from a long pipe.
  d. While falling, the path of a steel ball is straight in glycerine but not in water.

7.21 A solid body wetted by a liquid floats over the same liquid in open atmosphere. On which of the following factors, the force required to move it on the liquid surface at a uniform speed will depend?
  i. Shape, size and density of the body
  ii. Density of the liquid
  iii. Depth of the liquid
  iv. Viscosity of the liquid
  v. Surface tension of the liquid
  vi. Solid–liquid interfacial tension
  vii. Atmospheric pressure
  Will the dependent factor be same if the body is moved vertically?

7.22 Consider a solid body falling in a fluid. On which of the following characteristics of the body does the terminal velocity depend?
  (i) mass (ii) density (iii) size (iv) shape (v) roughness of the surface (vi) wettability.

7.23 Is there any terminal velocity of an air bubble in water?

7.24 Stokes law $F = 6\pi\eta rv$ fails under the following condions:
  i. *r* is less than molecular radius of the fluid or its mean free path (for gaseous fluid)

    ii. v is high

    iii. angle of contact greater than 90°

    Explain.

7.25 In connection with determination of viscosity of a liquid by the falling sphere method, explain the following:

    a. the sphere should be of quite small size

    b. the jar should have much higher radius than the sphere

    c. the depth of the fluid in the jar should be sufficiently high

7.26 The falling sphere method is useless for fluids of low viscosity and also for fluids of very high viscosity (such as molten glass). Why?

7.27 Ostwald viscometer is suitable only for liquids of low viscosity. Why?

7.28 Justify or criticise the following.

    a. Determination of viscosity by Ostwald viscometer is based on the relation $\eta_2/\eta_1 = \rho_2 t_2/\rho_1 t_1$ which implies higher viscosity (η) for higher density (ρ).

    b. Capillary portion of Ostwal viscometer is unnecessary as the viscosity found with the viscometer is independent of capillary radius.

7.29 In the determination of viscosity using Poiseuille's equation, find the error involved if that involved in the measurement of radius of the capillary is 2%.

7.30 The bottom of a cylindrical vessel of cross-sectional area 50 cm² ends in a horizontal capillary tube of length 15 cm and radius 0.08 cm. The vessel contains water ($\eta = 1.0$ cP) of depth 30 cm. If this water is allowed to pass through the capillary, how long would it take to drain out (i) half of water (ii) entire water.

    State the assumption implicity involved in your procedure, and establish the truth of the same.

7.31 Oil ($\eta = 0.3$ Pa s) is pumped through a pipeline 100 km long at a rate of 6 km³h⁻¹. The pipe has a diameter of 1.0 m. Find (i) the gauge pressure and (ii) minimum power required. $\rho_{oil} = 900$ kgm⁻³.

7.32 In Poiseuille's equation for liquid flow applicable also to gas flow?

    A gas ($\eta = 200$ µP) flows through a capillary tube 2.0 mm in diameter and 1.0 m in length. If 5.0 L of this gas passes through the tube every 10 sec, what must be the pressure head under which the gas is flowing when volume measured at (i) average pressure (ii) 1 atm pressure which is maintained at the exit end of the capillary. Also find the maximum velocity of flow.

7.33 What is the terminal velocity of a rain drop of radius 1.0 mm falling through air ($\rho = 1.30 \times 10^{-3}$ g cm⁻³, $\eta = 1.81 \times 10^{-4}$ P) at 20°C? How will it be affected with change in barometer reading? What is the Reynolds number in this case? Comment on the nature of the flow involved.

7.34 Find the velocity of an air bubble of radius 1.0 mm in water at room temperature, mentioning the assumption and approximation taken. Given $\eta_{water} = 1.0$ CP.

7.35 Viscosity coefficient of C_2H_5OH is 1.1 CP at 25°C. Find the value at 0°C. The activation energy for the viscous flow of C_2H_5OH is 3.23 kcal mol⁻¹.

7.36 For a certain liquid the effect of temperature (T) on viscosity (η) is such that the plot of $\log \eta$ vs $1/T$ yields a slope value of 600 K. Estimate the activation energy for flow.

7.37 Density of a gas is 1.5×10^{-3} g·cm⁻³ and its coefficient of viscosity is 1.03×10^{-4} poise. Calculate kinetic energy of gas molecules, if its mean free path is 10^{-5} cm. Molecular weight of the gas is 44.

7.38 Viscosity of H_2 is 8.53×10^{-5} P at STP. Find:
    a. molecular diameter and collision cross-section of H_2 molecules
    b. Viscosity of D_2 at STP
    State the nature of approximation involved.

ANSWERS

7.1 Viscosity depends largely on (i) the relative velocity of the adjacent fluid layers and (ii) the area of interface between them. In contrast, the friction acting externally between two solid bodies is almost independent of their relative velocity and contact area.

7.2 No. The term viscosity is not related to a stagnant fluid. It has meaning only with a flowing fluid as implied by its definition.

Viscosity has no relevance to rain which consists of water moving (but not flowing) in form of drops. However, the term is connected to the viscosity of air due to motion of the layer of air in contact with rain relative to the surrounding layers of air.

7.3 In viscous flow of a gas, there is transport of momentum from one layer to another through intermolecular collision, whereas in effusive flow gas molecules collide with the walls of the tube but not among themselves. A gas in a state of viscous flow, does not exactly obey Maxwell distribution law due to directed mass motion of the gas molecules (when molecular motion is not completely random). This is unlike effusive flow.

□ Storm involves viscous flow of air. When storm abates the energy associated with the directed mass motion of air molecules degrades into the energy of random molecular motion through transfer of momentum from fast moving to slow moving layers. We, therefore, expect a rise in temperature, i.e. heating effect.

7.4 Because, when diameter of the tube is high (compared with mean free path) the flow cannot be regarded as streamline, and then Poiseuille's equation, which is based on Newton law of viscous flow fails.

Again, for low diameter of the tube, the flow of gas cannot be regarded as viscous flow (being now an effusive flow).

7.5 i. A Newtonian fluid is one that obeys Newton's law of viscous flow under lamilarly flowing condition. This necessitates the particles of such a fluid to be spherical in shape so that their orientation has no affect on viscosity.

  ii. An ideal fluid is one whose viscosity is zero.

□ No. Because no real substance can completely fulfill the stringent conditions w.r.t. which a Newtonian fluid and an ideal fluid are defined.

Note: Even an ideal gas (that possesses viscosity) is not an ideal fluid.

7.6 The given data implies that the solution is a non-Newtonian fluid. Here the solute molecules are likely to have much-distorted spherical shape resembling rod or plate which enables them to be appropriately oriented to facilitate their flow.

7.7 Dimensions of ordinary viscosity, by Eq. (7.1) are those of (force) (time)/(area), i.e. $ml^{-1}t^{-1}$. But intrinsic viscosity has different dimensions which, by Eq. (7.7a) are those of (conc.)$^{-1}$, i.e. $l^3 n^{-1}$.

Then, dimensionally, ordinary viscosity is a more complex quantity. However, the intrinsic viscosity is more important to chemists because it gives a clue to the shape and size of the macromolecules.

7.8
$$1 \text{ poise} = 1 \text{ gm cm}^{-1}\text{s}^{-1}$$
$$\equiv 10^{-3} \text{ kg } (10^{-2}\text{m})^{-1}\text{s}^{-1} \text{ in SI unit}$$
$$= 10^{-1} \text{ kg m}^{-1}\text{s}^{-1}$$

Alternatively,
$$1 \text{ poise} = 1 \text{ dyne s cm}^{-2}$$
$$\equiv 10^{-5} \text{ N s } (10^{-2} \text{ m})^{-2} \text{ in SI unit}$$
$$= 10^{-1} \text{ N s m}^{-2}$$

7.9 A common feature of all transport processes is that they involve flow of some physical quantity (such as matter, momentum or electric charge) whose rate is proportional to the gradient of some other related physical quantity in the direction of flow. In fluid flow through a tube, bulk flow occurs along the tube whose rate is proportional to pressure gradient along the tube following Poiseuille's law. At the same time, momentum-flow occurs across the tube whose rate is proportional to the velocity gradient of fluid layers in that direction following Newton's law of viscous flow.

7.10 Liquid state. Because the fluid layers are denser in the liquid state that makes sliding one layer over the adjacent one more difficult in liquid state than in gaseous state.

7.11 Because, rise in temperature enhances interlayer exchange of gas molecules which causes higher gas viscosity. But in the liquid state, higher energy of molecules at higher temperature makes them more able to overcome interlayer attraction leading to lower viscosity.

☐ No. This follows from the temperature dependence of liquid viscosity according to the Eq. (7.5b).

Note: Even at very high temperature when a liquid is converted into gas, the viscosity will not be zero.

7.12 The given statement is not justified. Because the energy of activation has no meaning with the mechanism of gas viscosity which involves interlayer exchange of molecules with virtually no restriction regarding their energy.

7.13 Viscosity and also its temperature variation should both be higher when a gaseous substance behaves as a real gas. This is due to the existence of intermolecular forces that causes a real gas to lie between an ideal gas and a liquid when viscosity is concerned. Our conclusions are consistent with the facts that viscosity is higher and more susceptible to temperature change in liquid state compared to gaseous state of a substance (vide problem 7.10).

7.14 The viscosity of a fluid is determined not only by intermolecular forces but also by molecular size and shape.

   (i) *n*-pentane > neopentane, because the former has molecular shape more distorted from the spherical shape

   (ii) benzene > cyclohexane, because the former has higher cohesive force due to higher polarisability arising from unsaturation

   (iii) kerosene > mercury, for similar reason as in (i).

   (iv) kerosene > glycerol, the carbon chain being much longer with kerosene

Note: Molecular shape and size are more important in determining viscosity than surface tension. Thus kerosene has lower surface tension than mercury (which has higher cohesive energy due to metallic bonding) and glycerol (which has higher cohesive energy due to hydrogen bonding).

7.15 Because, hydration of ions of NaCl reduces cohesion of water molecules with consequent reduction of their internal friction. In contrast, macromolecules result in increase of viscosity by obstructing the surrounding fluid's flow over a long range due to their large size.

7.16 Let us imagine a gas flowing over a horizontal ground surface with a uniform velocity gradient q in the vertical direction. Let us consider a particular layer of gas at height x from the ground surface. If λ is the mean free path of the gas molecules, then on the average, all the molecules coming from a distance λ above and below this layer will make their first collisions, and hence their first transfer of momentum when they reach this layer. The average velocity of flow of the gas molecules in these top and bottom layers will be $(x + \lambda) q$ and $(x - \lambda) q$ respectively, assuming the gas layer adjacent to the ground surface to have zero velocity. It may be assumed, as an approximation, that one-third of the gas molecules are moving in each of the three directions at right angle, or one-sixth of the molecules are moving in the upward direction and one-sixth downward, at any isntant. If n is the number of molecules per unit volume and $\langle c \rangle$ is their average speed, $\frac{1}{6}n\langle c \rangle$ molecules will be moving upward, and equal number downward, through a, unit area in unit time. It follows, therefore, that the momentum (in the direction of flow) transferred to the layer under consideration (i.e. at distance x from the ground) per unit area per unit time will be $\frac{1}{6}mn\langle c \rangle(x + \lambda)q$ and $\frac{1}{6}mn\langle c \rangle(x - \lambda)q$ due to the molecules moving downward and upward respectively. The net downward transfer of momentum per second, i.e. the force acting per unit area is then the difference between these two quantities, i.e. $\frac{1}{3}mn\langle c \rangle\lambda q$. For unit velocity gradient (i.e. $q = 1$), this force becomes $\frac{1}{3}mn\langle c \rangle\lambda$ which is equal to the coefficient of viscosity (η), i.e.

$$\eta = \tfrac{1}{3}mn\langle c \rangle\lambda$$

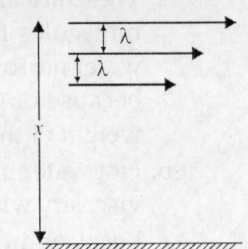

Note:

1. The factor $\frac{1}{3}$ is not quite correct. Because it arises from the assumption that molecular motion is completely random though this is not really so due to mass motion of the molecules in the direction of flow.

2. Here two types of motion have been considered—the directed motion of fluid layers which determines the momentum transferred by each of the jumping molecules and the random molecular motion which determines the inter-layer molecular jumps.

7.17 At higher pressure more molecules jump from layer to layer in the flowing gas, due to higher molecular density of the layers, but because of the shorter free paths each jump carries proportionally smaller momentum.

Note: The liquid viscosity, unlike gas viscosity, varies (increases) with pressure.

7.18 Because the concerned expression is based on the assumption that in gas flow the inter-layer jump of molecules is limited only to adjacent layers, which fails utterly at very low pressure (when the layers are so thinly populated that most of the molecules moving toward a particular layer will pass through it).

7.19 Justification lies on the ground that the gain of molecules by a particular layer is neutralised by its loss of molecules to the adjacent top and bottom layers.

☐ Here molecular momentum in the direction of flow is transferred from one layer to another and hence the associated force acts in the direction of momentum, i.e. the direction of flow (and not in the direction of transfer of momentum).

7.20 a. These are all due to the shearing effect of gravitational force that causes water molecules to move continuously relative to one another making the falling water narrowed down. Above certain length, it breaks into several parts (drops) because of poor tensile strength of liquid that makes it unable to support the weight of the entire length of falling water.

b. Hot water makes noise less due to lower density and splashes less due to lower viscosity which enhances the tendency to flow than to splash.

c. Because, in case of short pipe the gas exits at higher pressure, the pressure difference at the two ends of the pipe being smaller (by Poiseuille's law).

d. Because water has lower viscosity, so that the steel ball moves in it with higher speed causing much turbulence in the medium. The turbulence is much less in glycerol due to its higher viscosity.

7.21 Here the force that causes the motion of the body is just neutralised by the viscous force which is determined by the viscosity of the liquid, area of contact of the body with the liquid and the velocity gradient of the liquid layers in the vertical direction. Therefore, the dependent factors are (i) shape and size but not density (for just floating condition, (iii) depth of the liquid and of course (iv) the viscosity which depends slightly on the atmospheric pressure. The factor (vii) is absorbed in (iv).

☐ Yes, except for the depth of the liquid.

7.22 The terminal velocity depends on all the given factors, except mass, as expected from Stokes law. The roughness of surface, that makes surface area greater, has relatively minor effect.

Note: The Eq. (7.9), where mass (m) does not appear, implies that the terminal velocity is independent of m. Obviously, this will not happen with a non-spherical body to which the equation is not applicable.

7.23 Yes. There is no essential difference between the motion of a rising air bubble and that of a falling heavier body, both being associated with external gravitational force.

7.24 i. Stokes law is based on the assumption that momentum transfer happens only between the liquid layer adhered to the moving body and the adjacent liquid layer. This will hold only under the given condition.

ii. If v is high, the moving body will follow an undesirable zig-zag path due to enough of turbulence in the fluid behind it.

iii. For contact angle greater than 90°, the adhesion of liquid layer to the moving body is not sufficiently strong.

7.25 a. So that the sphere moves quite slowly causing least turbulance in the fluid.

b. So that wall of the jar has only insignificant effect on the motion of the sphere.

c. So that the speed of the sphere is only insignificantly affected by the bottom of the jar (due to its effect on fluid motion).

7.26 Because in the first case, the sphere moves with quite high speed following a zig-zag path (vide problem 7.24(ii).

In the second case, the experiment cannot be performed because it requires very high temperature which the glass apparatus cannot withstand.

7.27 Because liquids of low viscosity can flow through a capillary at a moderate rate suitable for measurement with a Ostwald viscometer.

7.28 a. The given working formula is based on Poiseuille's Eq. (7.2a) which does not involve density (ρ). This means viscosity has no necessary connection with density. Here ρ appears in the working formula because the force causing the fluid flow in the Ostwald viscometer is provided by means of level difference of the experimental fluid in the two limbs of the viscometer. Therefore, the given statement is not justified.

   b. Working of Ostwald viscometer is based on Poiseuille's equation which is valid only for laminar flow brought about by a capillary tube. Therefore, the given statement is not justified.

7.29 From Poiseuille's Eq. (7.2a)

$$\eta \propto R^4 \text{ (other factors involved in the equation remaining constant)}$$

whence $\qquad \dfrac{d\eta}{\eta} = 4\dfrac{dR}{R} = 4 \times \dfrac{2}{100}$

Then the % of error involved in $\eta = 4 \times \frac{2}{100} \times 100 = 8$.

7.30 Here rate of flow decreases with time due to decrease in pressure gradient caused by decrease in water depth. Let h be the initial water depth and x be its decrease in time t. Then from Poiseuille's Eq. (7.2a), we have

$$\frac{dV}{dt} = A\frac{dx}{dt} = \frac{\pi R^4 (h-x)\rho g}{8\eta l}$$

where A is the cross-sectional area of the vessel

i.
$$t = \int_0^t dt = \frac{8\eta l A}{\pi R^4 \rho g} \int_0^{\frac{h}{2}} \frac{dx}{h-x}$$

$$= \frac{8\eta l A}{\pi R^4 \rho g} \ln 2$$

$$= \frac{8(1.0 \times 10^{-2}\,P)(15\,\text{cm})(50\,\text{cm}^2)}{3.14\,(0.08\,\text{cm})^4\,(1\,\text{g/cm}^3)(980\,\text{cm/s}^2)} \cdot \ln 2$$

$$= 476\,\text{s·ln}\,2 = 330\,\text{s}$$

ii. If the above procedure is applied, t will be ∞ for $x = 0$ at $t = 0$ to $x = h$ at $t = t$. This happens because $\frac{dV}{dt} \to 0$ as $\frac{dx}{dt} \to 0$. However, practically entire water will be drained out in a finite period which can be attained by dividing the total volume of water with average volume rate of flow which is $\frac{dV}{dt} = \frac{\pi R^4 (P_{\text{initial}}/2)}{8\eta l}$.

Then the required time $= \dfrac{Ah}{\pi R^4 h\rho g/16\eta l}$

$$= \frac{2 \times 8A\eta l}{\pi R^4 \rho g} = 2\,(476\,\text{s}) = 952\,\text{s}$$

□ The above method of calculation implicity involves the assumption that water flows laminarly, otherwise the Poiseuille's equation used above will not hold. To verify the truth of this assumption, we are to calculate Reynolds number as done below.

Reynolds number

$$N_R = \frac{\rho v d}{\eta} = e\frac{(R^2 h\rho g)}{16\eta l}\, d/\eta$$

$$= \frac{\rho^2 R^3 hg}{8\eta^2 l}, \quad \text{putting } d = 2R$$

$$= \frac{(1\,\text{g/cm}^3)(0.08\,\text{cm})^3\,(30\,\text{cm})\,(980\,\text{cm/s}^2)}{8(0.01\text{P})^2\,(15\,\text{cm})} = 1254.4$$

Since the calculated value of N_R is less than 2000, the flow of water occurs laminarly, as assumed.

Note: Here v for average value of P has to be used.

7.31 i. Gauge pressure $P = \dfrac{8\eta l V}{\pi R^4}$ by Eq. (7.2a)

$$= \frac{8(0.3\text{Pas})(100 \times 10^3\,\text{cm})\,(6 \times 10^9\,\text{m}^3)/(60 \times 60\,\text{s})}{3.14\left(\frac{1}{2} \times 1.0\,\text{m}\right)^4}$$

$$= 2.04 \times 10^{12}\,\text{Pa}$$

ii. Minimmum power

$$= \tfrac{1}{2}\,(\text{rate of mass flow})\,(av\ \text{velocity of flow})^2$$

$$= \tfrac{1}{2}V\rho\left(\frac{V}{\pi R^2}\right)^2$$

$$= \frac{V^3\rho}{2\pi^2 R^4}$$

$$= \frac{\left(\frac{6 \times 10^9\,\text{m}^3}{60 \times 60\text{s}}\right)^3 (900\,\text{kg m}^{-3})}{2 \times (3.14)^2\left(\frac{1}{2} \times 1.0\,\text{m}\right)^4} = 3.38 \times 10^{21}\,\text{J/s}$$

Note: Gauge pressure depends on η and l, but minimum power does not.

7.32 Yes. Vide subsec. 7.1(B).

i. Pressure head

$$(P_1 - P_2) = \frac{8\eta l V}{\pi R^4}\quad \text{by Eq. (7.2a)}$$

$$= \frac{8(200 \times 10^{-6}\text{P})\,(1 \times 100\,\text{cm})\,(5 \times 10^3\,\text{cm}^3/10\,\text{s})}{3.14\left(\frac{1}{2} \times 2 \times 10^{-1}\,\text{cm}\right)^4}$$

$$= 2.55 \times 10^6\,\text{dyn/cm}^2$$

ii. Here $V = \pi R^4 \dfrac{(P_1^2 - P_2^2)}{16\eta l P_2}$ by Eq. (7.2b)

or $P_1 = \left[\dfrac{16\eta l P_2 V}{\pi R^4} + P_2^2\right]^{\frac{1}{2}}$

$$= \left[\frac{16(200 \times 10^{-6}\text{P})(1 \times 100\,\text{cm})(1.013 \times 10^6\,\text{dyn/cm}^2)(5 \times 10^3\,\text{cm}^3/10\text{s})}{3.14\left(\frac{1}{2} \times 2 \times 10^{-1}\text{cm}\right)^4} + (1.013 \times 10^6\,\text{dyn/cm}^2)^2\right]^{\frac{1}{2}}$$

$$= 1.24 \times 10^6\,\text{dyn/cm}^2$$

Therefore, pressure heat $(P_1 - P_2) = (1.242 - 1.013) \times 10^6$ or 0.229 dyn/cm^2.

☐ In either case, the maximum velocity of flow is given by Eq. (7.3a).

7.33 Terminal velocity, $v = \dfrac{2}{9} \cdot \dfrac{r^2 \rho g}{\eta}$ by Eq. (7.9),

ignoring density of air compared to that of water

$$= \tfrac{2}{9} \times \frac{\left(\tfrac{1}{10}\,\text{cm}\right)^2 \left(\frac{1\,\text{gm}}{\text{cm}^3}\right)\left(\frac{980\,\text{cm}}{\text{sec}^2}\right)}{(1 \times 10^{-4}\,\text{P})}$$

$$= 1.20 \times 10^4\,\text{cm/sec}$$

☐ Since viscosity and density of air do not significantly vary with change of atmospheric pressure, the latter does not practically affect v.

☐☐ Reynolds number $N_R = \dfrac{\rho_{\text{air}} v_{\text{drop}} d_{\text{drop}}}{\eta_{\text{air}}}$

$$= \frac{(1.3 \times 10^{-3}\,\text{gm/cm}^3)\,(1.2 \times 10^4\,\text{cm/sec})\left(\tfrac{2}{10}\,\text{cm}\right)}{(1.81 \times 10^{-4}\,\text{P})}$$

$$= 17237$$

☐☐☐ Since the calculated N_R exeeds 2000, the accompanying flow of fluid (air) will be turbulent.

Note: N_R refers to the motion of air layer in contact with the rain drop relative to the adjacent air layer.

7.34 Terminal velocity $= \tfrac{2}{9} \cdot \dfrac{r^2 \rho g}{\eta}$

ignoring the density of air compared to that of water assuming that the size and shape of air bubble do not alter during rising through water

$$= \tfrac{2}{9} \times \frac{\left(\tfrac{1}{10}\,\text{cm}\right)^2 \left(\frac{1\,\text{g}}{\text{cm}^3}\right)\left(\frac{980\,\text{cm}}{\text{sec}^2}\right)}{(1 \times 10^{-2}\,\text{P})}$$

$$= 217.8\,\text{cm/sec}$$

Note: Here, unlike the previous problem, the density of the moving body (air bubble) has been ignored. The working formula is, however, same in the two cases. Because, here weight of the air bubble and the frictional force acting on the bubble both act in the same direction (downward) opposite to that of buoyancy.

7.35 $\eta_{298\,\text{K}} = \eta_{273} \exp\left[\dfrac{E}{R}\left(\dfrac{1}{293\,\text{K}} - \dfrac{1}{273\,\text{K}}\right)\right]$, by Eq. (7.5a)

$$= (1.1\,\text{cP}) \exp\left[\frac{3.23 \times 10^3\,\text{cal mol}^{-1}}{1.99\,\text{cal K}^{-1}\,\text{mol}^{-1}}\left(\frac{1}{298\,\text{K}} - \frac{1}{273\,\text{K}}\right)\right] = 0.668\,\text{cP.}$$

7.36 $\dfrac{E}{2.303R} = \text{slope}$ by Eq. (7.5b)

or $\qquad E = 2.303\,R\,\text{slope}$

$$= 2.303\,(8.314\,\text{J K}^{-1}\,\text{mol}^{-1})\,(600\,\text{K})$$

$$= 11488\,\text{J mol}^{-1}$$

7.37 Kinetic energy of gas molecules per mol

$$= \tfrac{1}{2} M \overline{C}^2$$

$$= \tfrac{1}{2} M \cdot \frac{3\pi}{8} (\overline{C})^2$$

$$= \tfrac{1}{2} M \cdot \frac{3\pi}{8} \left(\frac{3\eta}{e\lambda}\right)^2 \quad \text{by Eq. (7.6a)}$$

$$= \frac{27\pi M\eta^2}{16 e^2 \lambda^2}$$

$$= \frac{27 (3.14) (44 \text{ g mol}^{-1}) (1.03 \times 10^{-4} P)^2}{16 (1.5 \times 10^{-3} \text{ g cm}^{-3})^2 (10^{-5} \text{ cm})^2}$$

$$= 2.04 \times 10^{10} \text{ ergs mol}^{-1}$$

Note: The speed involved in the expression of η is mean speed $\langle \overline{C} \rangle$ while that involved in the expression of KE is root mean square speed $\left(\sqrt{\overline{C}^2}\right)$.

7.38 a.
$$\eta = \tfrac{1}{3} \cdot \frac{M}{N} \frac{1}{\sqrt{2}\pi\sigma^2} \left(\frac{8RT}{\pi M}\right)^{\frac{1}{2}} \quad \text{by Eq. (7.6a)}$$

Since $\langle C \rangle = \left(\frac{8RT}{\pi M}\right)^{\frac{1}{2}}$ and $n\lambda = \frac{1}{\sqrt{2}\pi\sigma^2}$, where σ is molecular diameter assuming the molecules to behave as hard spheres.

whence $\quad \sigma = \left[\frac{2}{3\pi^{\frac{3}{2}}} \cdot \frac{\sqrt{RTM}}{N\eta}\right]^{\frac{1}{2}}$

$$= \left[\frac{2\sqrt{(8.31 \times 10^7 \text{ erg K}^{-1}\text{mol}^{-1}) (273 \text{ K}) (2.02 \text{ g mol}^{-1})}}{3(3.14)^{\frac{3}{2}} (6.02 \times 10^{23} \text{ mol}^{-1}) (8.53 \times 10^{-5} P)}\right]^{\frac{1}{2}}$$

$$= 2.23 \times 10^{-8} \text{ cm}$$

Collision cross-section

$$\pi\sigma^2 = 3.14 (2.23 \times 10^{-8} \text{ cm})^2$$

$$= 1.56 \times 10^{-15} \text{ cm}^2$$

b. From the above expression, we can write

$$\eta_{D_2} = \eta_{H_2} \sqrt{\frac{M_{D_2}}{M_{H_2}}}, \quad \text{taking } \sigma_{D_2} = \sigma_{H_2} \text{ as an approximation}$$

$$= (8.53 \times 10^{-5} P)\sqrt{\tfrac{4}{2}} = 12.06 \times 10^{-5} P$$

UNIVERSITY QUESTIONS

7.1 What is the difference in the nature of motion associated with (i) effusion, (ii) diffusion and (iii) viscosity of gases.   (Burdwan BSc(H), 1998)

7.2 State Newton's law of viscosity. Under what conditions, is it valid? What is fluidity and how is it connected with viscosity?   (Burdwan BSc(H), 2011)

7.3 Find the dimensions of viscosity coefficient and translate poise into SI units.
(Sambalpur BSc(H), 2004)

7.4 In the construction of Ostwald viscometer, the capillary part is unnecessary and it can be replaced by an ordinary glass tube. Justify or criticise.
(Calcutta BSc(H), 1974)

7.5 Why do you use the same viscometer for the liquid and water during the experimental determination of viscosity by Ostwald viscometer?
(Delhi BSc, 2000)

7.6 What is laminar flow of liquid? How one can characterise a liquid flow as turbulent or laminar? (Calcutta BSc(H), 2008)

7.7 Define the terms *relative viscosity, specific viscosity, reduced viscosity* and *intrinsic viscosity* as applied to the solutions of high polymers. (Burdwan BSc(H), 2003)

7.8 a. What is the difference between specific viscosity and intrinsic viscosity?
(Burdwan BSc(H), 2003)

   b. Explain how molecular weight of a polymer can be determined by viscometric method. (Burdwan BSc(H), 2011)

7.9 In the determination of viscosity coefficient (η) of a liquid by Poiseuille's method, what will be the percentage of error in η if the radius is measured with an error of -0.5%? (Calcutta BSc(H), 2012)

7.10 Should an ideal gas have viscosity? Why? (Burdwan BSc(H), 2000)

7.11 What is the phenomenon associated with transport of kinetic energy from one layer to another in a gas executing a laminar flow? (Burdwan BSc(H), 2006)

7.12 Viscosity and thermal conductivity of a gas must be related to one another. Why?
(Burdwan BSc(H), 2001)

7.13 In what respects is the viscosity of a gas different from that of a liquid?
(Calcutta BSc(H), 2007)

7.14 Why do liquids and gases exhibit opposite trends in the variation of viscosity with temperature? (Burdwan BSc(H), 2012)

7.15 The viscosity of (ideal) gases is independent of pressure at constant temperature but the independency fails at very low pressure. Explain. (Calcutta BSc(H), 2005)

7.16 The viscosity of H_2 gas at NTP is 8.4×10^{-5} poise and average speed of the molecules is 1.7×10^5 cm/sec. Calculate the mean free path. How is your mathematical formula valid at very low and very high pressure? (Burdwan BSc(H), 1997)

7.17 Viscosity coefficient of glycerol is much higher than that of *n*-propanol. Comment.
(Burdwan BSc(H), 1992)

7.18 What do you understand by activation energy of viscous flow? The viscosity coefficients of a liquid are 0.009 and 0.006 poise at 25°C and 50°C respectively. Calculate the energy of activation for the viscous flow of the liquid.
(Burdwan BSc(H), 2011)

7.19 The viscosity coefficient (η) of a liquid decrease by 2% per degree celsius rise in temperature. Show that if η is measured at 25°C and 75°C then $\eta_{75°C} : \eta_{25°C} = 1:e$
[*e* is the base of natural logarithm] (Calcutta BSc(H), 2010)

7.20 The coefficient of viscosity (η) of H_2 at 0°C and 1 atm is 8.53 µPas. Find η of D_2 at 0°C and 1 atm. Assume that $\sigma_{D_2} = \sigma_{H_2}$ (D = deuterium) (Calcutta BSc(H), 2012)

7.21 State Stokes law. A metallic sphere of diameter 1 cm takes 25 minutes to fall 1 cm through a liquid. If the densities of the metal and the liquid are 10 g/cc and 3 g/cc respectively, calculate the viscosity of the liquid. (Calcutta BSc(H), 2006)

7.22 Find out the terminal velocity of a rain drop of radius 0.01 cm falling through air of viscosity coefficient 1.85×10^{-4} poise. Neglect the density of air compared to that of water.   (Calcutta BSc(H), 2013)

7.23 A gas bubble of diameter 2 cm rises steadily through a solution of density 1.75 g/cm^3 at the rate of 0.35 cm/s. Calculate the coefficient of viscosity of the solution neglecting the density of the gas.   (Calcutta BSc(H), 2004)

7.24 One end of a capillary tube 10 cm in length is connected to water supply which has a pressure of 1.8 atm and the other end to a vessel open to air. At 25°C the tube delivers 0.2 lit min^{-1}. Calculate the diameter of the tube if η_{liq} at 25°C is 0.01 poise.
                                                              (Calcutta BSc(H), 2008)

7.25 Show how the time of flow of a liquid in a viscometer will change when the radius of the capillary is doubled.   (Burdwan BSc(H), 2010)

7.26 Water passes through a viscometer in 30 sec. The same volume of oil required 2263.7 sec. Calculate the viscosity of oil if its density is 1.1×10^3 kg m^{-3}. Density of water is 0.998×10^3 kg m^{-3}, viscosity of water is 0.00101 kg m^{-1}s^{-1}.
                                                                    (Kerala BSc, 2011)

KEY TO UNIVERSITY QUESTIONS

7.1 (i) Effusion refers to emergence of a gas from a container through a small hole (or a capillary). It involves no net force toward the hole. Here the molecules flow independently, i.e. without any collision between them.

   (ii) Diffusion refers to migration of matter down a concentration gradient through intermolecular collision. It involves no bulk flow (due to pressure difference).

   (iii) Viscosity of gases refers to migration of linear momentum (through inter-molecular collision) down a velocity gradient associated with a bulk flow.

   Note:
   1. Effusion is the simplest flow process.
   2. Maxwell distribution law holds for a gas undergoing effusive and diffusive flow but not visicoud flow (vide problem 7.3).

7.2 Vide Section 7.1(A).
   ☐ Vide problem 7.5
   ☐☐ Fluidity refers to the ease with which a fluid can flow. It is inversely connected with viscosity.

7.3 Vide problem 7.7 and 7.8.

7.4 Vide problem 7.28(b).

7.5 It follows from Poiseuille's equation [Eq. (7.2a)] that the time of flow for a fixed volume of liquid is proportional to η/ρ. The proportionality constant varies with radius and length of the capillary involved in the viscometer. Then for comparison of viscosities of two liquids using the working formula $\eta_2/\eta_1 = \rho_2 t_2/\rho_1 t_1$ experiment must be done with the same viscometer.

7.6 Vide Section 7.1.

7.7 Vide Section 7.4.

   Note:
   1. Intrinsic viscosity is the analogue of a virial coefficient considering the following expression

$$\eta = \eta_0 \left(1 + [n] c + ...\right)$$

2. The term viscosity is confusing for the quantities called relative viscosity and specific viscosity as they are dimensionless numbers.

7.8 a. Specific viscosity, unlike intrinsic viscosity, is a dimensionless number. It has meaning with a solution only for a particular concentration. In contrast, intrinsic viscosity has relevance to a solution irrespective of concentration.

b. This can be done on the basis of Eq. (7.7b). Experimentally, one must first determine K and a for the polymer and the solvent using polymer samples whose molecular weights have been found by some other methods (e.g. by osmotic pressure measurement). Once K and a are known, the molecular weight of a given sample of the polymer can be found by viscosity measurements.

Note: The experimentally found average molecular weight is not independent of the method of its determination. Light scattering measurements give weight average molecular weight $\left(\sum w_i M_i\right)$, osmometric measurements give number average molecular weight $\left(\sum x_i M_i\right)$ while viscometric measurements give a different type of molecular weight called viscosity average molecular weight $\left[\sum w_i M_i^a\right]^{\frac{1}{a}}$. Because the viscosity of polymer solution depends not only on the number and weight of the polymer molecules but also on their shape and compactness. Here, x_i stands for mole fraction and w_i for weight fraction of the polymeric species i.

7.9 Vide problem 7.29.

7.10 Yes. Because viscosity is connected to migration of linear momentum down a velocity gradient, which is not unlikely to occur with an ideal gas (even though it possesses no intermolecular forces) through inter-layer jump of its molecules.

7.11 Thermal conductivity.

7.12 Because in either phenomenon, the total physical quantity transported to a particular layer of the gas is equal to the number of molecules coming to this layer multiplied by the amount of the relevant physical quantity carried by each molecule which is linear momentum in case of viscosity and translational kinetic energy in case of thermal conductivity.

7.13 For a gas, viscosity increases with increase in temperature and is practically independent of pressure. In contrast, the viscosity of a liquid decreases with increase in temperature and increases with rise of pressure. Because the kinetic molecular mechanisms of transport of momentum are different in case of gas and liquid (vide problems 7.11 and 7.17).

7.14 Because the kinetic molecular mechanism responsible for viscosity is different in case of liquids and gases. Vide Sections 7.2 and 7.3 and problem 7.11.

7.15 Vide problems 7.17 and 7.18.

7.16 $$\lambda = \frac{3\eta}{\rho \langle c \rangle} \quad \text{by Eq. (7.6a)}$$

$$= \frac{3(8.4 \times 10^{-5}\,\text{P})}{(9 \times 10^{-5}\,\text{g cm}^{-3})(1.7 \times 10^{5}\,\text{cm s}^{-1})} = 1.6 \times 10^{-5}\ \text{cm}$$

□ The mathematical formula used is not sufficiently valid at very low pressure when the layers are thinly populated and also at very high pressure when the gas behaves quite non-ideally.

7.17 This is quite likely considering that glycerol (having though same carbon chain length as n-propanol) can form more hydrogen bonds with three OH groups per molecule. [Compare with problem 7.14(iv)].

7.18 Vide Section 7.2.

$$E = \frac{R \ln \dfrac{\eta_1}{\eta_2}}{\dfrac{1}{T_1} - \dfrac{1}{T_2}} \quad \text{by Eq. (7.5a)}$$

$$= \frac{(8.314 \times 10^7 \text{ erg K}^{-1}\text{mol}^{-1}) \ln \dfrac{0.009P}{0.006P}}{\dfrac{1}{298 \text{ K}} - \dfrac{1}{323 \text{ K}}} = 12976 \text{ J mol}^{-1}$$

7.19 Here

$$\frac{d \ln \eta}{dT} = -\frac{2}{100}$$

Integration gives

$$\ln \frac{\eta_{75\,^\circ C}}{\eta_{25\,^\circ C}} = -\frac{2}{100}(75 - 25)$$

whence $\eta_{75\,^\circ C} : \eta_{25\,^\circ C} = 1:e$

7.20 Vide problem 7.38(b).

7.21 Vide Section 7.5

$$\eta = \frac{2\left(\dfrac{1}{2}\text{ cm}\right)^2 \left(\dfrac{10g}{cc} - \dfrac{3g}{cc}\right)\left(\dfrac{980 \text{ cm}}{s^2}\right)}{9(1 \text{ cm}/25 \times 60\text{s})} \quad \text{by Eq. (7.9)}$$

$$= 571667 \text{ poise}$$

7.22 Vide problem 7.33.

7.23

$$\eta = \frac{2r^2 \rho g}{9v} \quad \text{by Eq. (7.9) ignoring the density of the gas}$$

$$= \frac{2\left(\dfrac{2\text{cm}}{2}\right)^2 \left(\dfrac{1.75 \text{ g}}{cm^3}\right)\left(\dfrac{980 \text{ cm}}{s^2}\right)}{9(0.35 \text{ cm/s})}$$

$$= 1089 \text{ poise}$$

7.24

$$R = \left[\frac{8V\eta l}{\pi(P_1 - P_2)}\right]^{\frac{1}{4}} \quad \text{by Eq. (7.2a)}$$

$$= \left[\frac{8\left(\dfrac{0.2 \times 10^3 \text{ cm}^3}{60 \text{ s}}\right)(0.01 \text{ poise})(10 \text{ cm})}{3.14(1.8 - 1) \times 1.013 \times 10^6 \text{ dyn/cm}^2}\right]^{\frac{1}{4}}$$

$$= 0.032 \text{ cm}$$

7.25 It follows from Poiseuille's equation that the time of flow (t) of a particular liquid will be proportional to $1/R^4$. Then according to the given data, t will change by a factor $1/2^4$.

7.26 Time (t) required for a fixed volume of two liquids to pass through the same viscometer will be related to their density (e) by

$$\frac{\eta_2}{\eta_1} = \frac{\rho_2 t_2}{\rho_1 t_1}$$

or

$$\eta_2 = \frac{\rho_2 t_2}{\rho_1 t_1} \cdot \eta_1$$

$$= \frac{(1.1 \times 10^3 \, \text{kg m}^{-3})(2263.7 \, \text{s})(0.00101 \, \text{kg m}^{-1}\text{s}^{-1})}{(0.998 \times 10^3 \, \text{kg m}^{-3})(30 \, \text{s})}$$

$$= 0.084 \, \text{kg m}^{-1}\text{s}^{-1}$$

8

Solid State—Crystallography

Solids are much less chaotic in their molecular motion, compared to liquids and gases, but their macroscopic characteristics are much more difficult to grasp.

8.1 CHARACTERISTICS OF SOLID STATE

The term solid is generally used to imply the form of matter having considerable rigidity and hence a definite shape and volume. The rigidity is caused by strong forces of attraction among the constituent particles of the solid, so much so that these particles can have mostly vibrational motion.

Solids are generally of two types—crystalline and amorphous. The crystalline substances or crystals, unlike the amorphous, have sharp melting point and regular shapes with plane surfaces, called faces, arranged symmetrically. Crystals split along well-defined planes while the amorphous solids have conchoidal fracture. Again, crystals are in general not isotropic, i.e. the directed properties, such as thermal or electrical conductivity, velocity of propagation of light, etc. within the crystal depend upon the direction of the crystal concerned.

The characteristics of crystalline state mentioned above results from long-range order of arrangement of the constituent particles of the crystal. In the amorphous state only short-range order exists.

8.2 CRYSTAL STRUCTURE AND SPACE LATTICE

A particular substance may be crystallised in different forms but they always have same angles between the corresponding crystal faces. This is first law of crystallography by Steno.

The regular external form of a crystal is due to the fact that a crystal is built up of identical smallest possible blocks, called *unit cells*, arranged in a regular order without any gap between them. This was first suggested by Hauy in connection with his law of rational indices (stated in Section 8.4). Since unit cells consists of atoms, molecules or ions, this idea of Hauy led to the concept that a crystal may be regarded as a space lattice, called *crystal lattice*, i.e. a regular three-dimensional arrangement of points each of which is connected to the site of an atom, or molecule, or ion, or a small group of such particles (called *basis*) constituting the crystal. Each point in the lattice has exactly the same environment as any other similar point (i.e. one representing the same kind of species). If the crystal is made up of a monatomic element (e.g. a metal) then there will be only one space lattice. But in case of crystals containing different kinds of species, the space lattice is conveniently regarded as the interpenetrating lattices each corresponding to one kind of species.

8.3 CLASSIFICATION OF CRYSTALS

A unit cell is formed in the shape of a parallelepiped by joining neighbouring lattice points. Based on the shape of the unit cell crystals are grouped into different types, called crystal systems, which are usually seven in number as defined in Table 8.1.

Table 8.1

Triclinic	$a \neq b \neq c$	$\alpha \neq \beta \neq \gamma \neq 90°$
Monoclinic	$a \neq b \neq c$	$\alpha = \beta = 90° \neq \gamma$
Orthorhombic	$a \neq b \neq c$	$\alpha = \beta = \gamma = 90°$
Hexagonal	$a = b \neq c\ (= 1.633a)$	$\alpha = \beta = 90°, \gamma = 120°$
Tetragonal	$a = b \neq c$	$\alpha = \beta = \gamma = 90°$
Trigonal (Rhombohedral)	$a = b = c$	$\alpha = \beta = \gamma$
Cubic	$a = b = c$	$\alpha = \beta = \gamma = 90°$

$$\text{Volume of a unit cell} = abc\,(1 - \cos^2\alpha - \cos^2\beta - \cos^2\gamma + 2\cos\alpha\cos\beta\cos\gamma)^{1/2} \qquad (8.1)$$

Here a, b and c are lengths of three vectors, called *primitive translation vectors*, represented by the three edges of the unit cell meeting at any corner such that the angle between b and c is α, that between c and a is β and that between a and b is γ. In all cases the edges of the cell are parallel to the edges or possible edges of the crystal. Then the system of a crystal may be determined by examining its appearance and measuring the interfacial angles.

Bravais, based purely on geometry, showed that at most fourteen different types of unit cells, called *Bravais lattices*, are necessary to generate all possible crystal lattices. The Bravais lattices include the seven crystal systems as given in Table 8.2.

Table 8.2

Crystal systems	Bravias lattices	Number of point groups	Number of space groups
Triclinic	Simple	2	2
Monoclinic	Simple and base-centred	3	13
Orthorhombic	Simple, body-centred, base-centred, face-centred	3	59
Hexagonal	Simple	7	27
Tetragonal	Simple and body-centred	7	68
Trigonal	Simple	5	25
Cubic	Simple, face-centred and body-centred	5	36
Total 7	14	32	230

Cells are called simple or primitive if the lattice points lie only at the corners of the cell. Other cells contain additional lattice point(s) lying at the position(s) indicated by the names.

A further classification of crystals can be made on the basis of symmetry. Crystals, unlike molecules, can have only some definite symmetry elements and possible combinations thereof. An element of symmetry is an operation which brings the crystal into coincidence with itself. The elements of symmetry found in crystals are (i) centre of symmetry (ii) plane of symmetry (iii) 2-, 3-, 4- and 6-fold axes of symmetry, and 4- and 6-fold axes of improper rotation or rotation–reflection. The prime rotational symmetry elements of seven crystal systems are given in Table 8.3.

The symmetry of a crystal corresponds to only 32 possible combinations, called point groups, of the above-mentioned symmetry elements. The name arises from the fact that the relevant operations leave at least one point in the crystal undisplaced. It is important to note that crystals, unlike molecules possess two translation symmetries—screw axis and glide plane. A screw axis corresponds to rotation around an axis through 180° followed by translation parallel to the axis, while a glide plane corresponds to reflection across a plane followed by translation parallel to the plane. A combination of these two symmetry elements with point group symmetry elements gives 230 different groups, called *space groups*. The distribution of point groups and space groups among the seven crystal systems are given in Table 8.2.

On the basis of the point group or space group, crystals have been conveniently classified into different types, called *crystal classes*. It is important to note that the crystal classes are based on the possible combination of the symmetry elements while the crystal systems are based on a particular type of symmetry element, viz. the prime rotational symmetry element of the crystals.

Table 8.3

System	Prime symmetries
Triclinic	None
Monoclinic	One C_2 axis
Orthorhombic	Three mutually perpendicular C_2 axes
Hexagonal	One C_6 axis
Tetragonal	One C_4 axis
Trigonal	One C_3 axis
Cubic	Four tetrahedrally oriented C_3 axes

8.4 CRYSTAL PLANES AND THEIR REPRESENTATION

A crystal contains innumerable net planes (ones containing lattice points) of varying reticular density (which is surface density of lattice points). The faces of a complete crystal are parallel to the net planes of highest reticular density.

It is always possible (due to regular arrangement of particles in a crystal) to find a set of axes of appropriate unit lengths (usually represented by a, b and c) on which the intercepts of a crystal plane will be given by rational numbers (called Weiss indices of the plane). This is Hauy's law of rational intercepts. A crystal plane is best represented by a set of three integers, called *Miller indices* (denoted by khl) which are the reciprocal intercepts (the numerical values) with all fractions cleared. To illustrate let us consider the crystal plane ABC that has intercepts $2a$, $3b$ and $4c$ on the x-, y- and z-axis respectively (Fig. 8.1). The reciprocals of these intercepts expressed in terms of respective unit

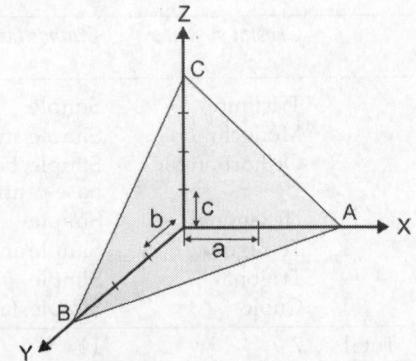

Fig. 8.1: Representation of crystal planes

lengths are $1/2$, $1/3$ and $1/4$. These, on multiplication with the least common multiple of their denominators, i.e. 12, give the Miller indices 6, 4, 3.

In connection with Miller indices, the following points are noteworthy:

i. The Miller indices of a plane are not equal to, but inversely proportional to, the intercepts of that plane on the chosen axes so that all equidistant parallel planes have the same Miller indices, i.e. Miller indices actually define the orientation of a set of equivalent equidistant parallel planes, one of which passes through the origin. Thus, by '643 planes' is meant a set of planes having intercepts $\frac{n}{6}a, \frac{n}{4}b, \frac{n}{3}c$ on the closen axes where n is an integer. Then 643 planes divide a into 6 equal parts, b into 4 equal parts and c into 3 equal parts. For $n = 1$, the corresponding plane is referred to as '(643) plane' having intercept $\frac{a}{6}, \frac{b}{4}, \frac{c}{3}$. The Weiss indices (now obsolete) of the 643 planes are $\frac{n}{6}, \frac{n}{4}$ and $\frac{n}{3}$. In cystalographic notation (hkl) refers to a crystal face (a particular plane) and hkl (without parenthesis) to a set of planes.

ii. The Miller indices for a plane may be negative. A negative index is written with a bar above the numeral, e.g. $(\bar{6}43)$ which implies that the plane has intercept on the negative side of x-axis.

iii. For a crystal system with $\alpha = \beta = \gamma = 90°$, called orthogonal system, the interplanar distance d_{hkl} for khl planes is given by

$$\frac{1}{d_{hkl}^2} = \frac{h^2}{a^2} + \frac{k^2}{b^2} + \frac{l^2}{c^2} \qquad (8.2a)$$

$$= \frac{1}{a^2}(h^2 + k^2 + l^2) \quad \text{for cubic system} \qquad (8.2b)$$

iv. Multiplication of the Miller indices hkl by an integer n reduces the interplanar spacing to d_{hkl}/n.

v. hkl type planes are to be distinguished from hkl actual planes or simply hkl planes whose interplanar distance is given by the actual Miller indices, e.g. by Eq. (8.2a) (for orthogonal crystal systems). The hkl type planes refer to hkl actual planes as well as the existing equivalent equidistant planes parallel to them. Thus in case of body-centred fubic crystals 100 type planes really mean 200 planes.

vi. For cubic crystals, the interplanar distances for 100, 110 and 111 type planes are in the following ratio

$$d_{100}{:}d_{110}{:}d_{111} = 1 : \frac{1}{\sqrt{2}} : \frac{1}{\sqrt{3}} \quad \text{for simple cubic } (sc) \text{ crystal}$$
$$1 : 0.707 : 0.577$$

$$d_{100}{:}d_{110}{:}d_{111} = 1 : \frac{2}{\sqrt{2}} : \frac{1}{\sqrt{3}} \quad \text{for body-centred cubic } (bcc) \text{ crystal}$$
$$1 : 1.414 : 0.577$$

$$d_{100}{:}d_{110}{:}d_{111} = 1 : \frac{1}{\sqrt{2}} : \frac{2}{\sqrt{3}} \quad \text{for face-centred cubic } (fcc) \text{ crystal}$$
$$1 : 0.707 : 1.154$$

Because in case of sc crystals, all three 100, 110 and 111 type planes are same as 100, 110 and 111 actual planes. But in case of bcc, only 110 (but not 100 and 111) type planes are same as 110 actual planes (because here 100 and 111 actual planes are interleaved by equivalent planes at just the half way). Similar is the case with fcc for 111 (but not 100 and 110) type planes. These planes are depicted in Fig. 8.2.

8.5 INTERNAL STRUCTURE OF CRYSTALS FROM X-RAY DIFFRACTION

The determination of actual positions of the constituent particles of a crystal cannot be done from measurement of interfacial angles which provides only the symmetry of crystal structure. For this, one has to make detailed exmination of the internal structure of the crystals using diffraction of x-ray (or electron or neutron).

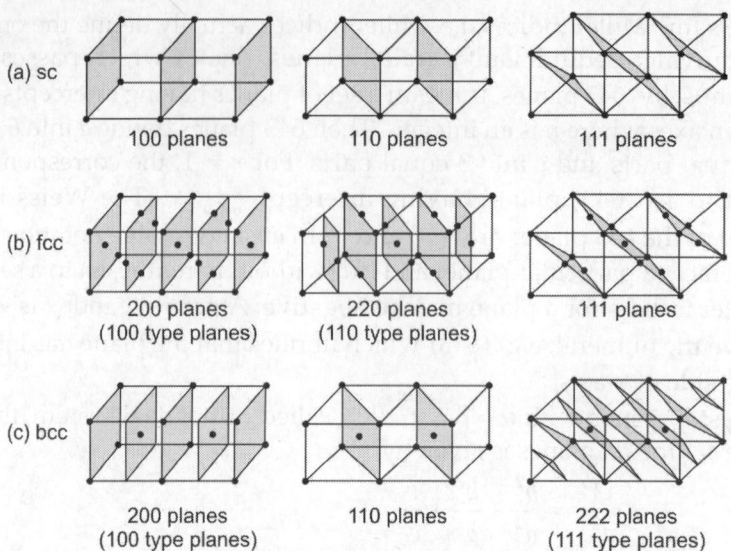

Fig. 8.2: Planes in cubic lattices. Shaded parts represent the specified planes for two adjacent cells

The separation d_{hkl} of the hkl planes (havig no in-between planes) of a crystal is given by the following Bragg equation

$$n\lambda = 2d_{hkl}\sin\theta_{hkl} \tag{8.3}$$

Here θ_{hkl} is the glancing angle (or Bragg diffraction angle) which is the angle between x-ray and khl planes, λ is wavelength of x-ray used and n is an integer called the order of reflection.

Higher order reflections may be looked as the first order reflection from planes of higher Miller indices in accordance with the Bragg equation. Thus, a second order reflection from 111 planes of a *fcc* crystal is equivalent to the first order reflection from 222 planes, for the same glancing angle.

8.6 INTENSITY OF DIFFRACTION AND STRUCTURE DETERMINATION

The problem of x-ray analysis lies in the indexing of the lines appearing in the diffraction pattern. The measurement of intensities of these lines throws much light in this regard.

Bragg equation deals with position of diffraction lines (determined by θ) and not with their intensity. If hkl planes have in-between planes, the diffraction lines produced by former may be stronger or weaker due to additional interference. In case the intensity falls to zero, the corresponding lines will be absent in the diffraction pattern. This happens when the in-between crystal planes lie mid-way between the consecutive planes concerned. An example of this situation is found with 100 planes of a cubic monatomic crystals of *bcc* or *fcc* type.

The absence of certain lines helps enormously in indexing the lines which do appear. From a complete study of line spacings and intensities in the diffraction pattern, it is possible to determine the shape and size of the unit cell and the arrangement of atoms within it.

To illustrate, let us consider the case of NaCl crystal which belongs to the cubic system. Here the interplanar distances for 100-, 110- and 111-type planes are in the following ratio

$$d_{100}{:}d_{110}{:}d_{111} \approx 1:0.707:1.154 \quad \left(\text{i.e.} \approx 1:\frac{1}{\sqrt{2}}:\frac{2}{\sqrt{3}} \right)$$

which corresponds to a *fcc* cubic lattice. This, therefore, forms the basis of NaCl crystal structure which really consists of a *fcc* lattice of Na^+ and a *fcc* of Cl^- interpenetrated in half-way. This is confirmed by intensity data. Thus, although intensity of reflection decreases as usual with order of reflection for 100- and 110-type planes (which contain both Na^+ and Cl^- with no in-between plane), this is not so with 111-type planes (which contain only one kind of ion, but with a in-between plane containing other kind of ion having different scattering power). For 111-type planes, the first order reflection is weaker than the second order reflection and third order weaker than the fourth order.

It is interesting to note that although NaCl and KCl have same crystal structure (isomorphous), the latter gives

$$d_{100}{:}d_{110}{:}d_{111} \approx 1:\frac{1}{\sqrt{2}}:\frac{1}{\sqrt{3}}$$
(for type planes)

which corresponds to simple cubic structure. This happens because K^+ and Cl^- are virtually equivalent to x-ray due to their almost same x-ray scattering power as both have 18-electron argon configuration. As a consequence, several diffraction lines (e.g. one produced by 111-type planes) of KCl, analogous to NaCl, are too weak to be observed. This makes KCl to have simpler diffraction pattern than NaCl in power protograph.

8.7 DIFFRACTION PATTERN OF CUBIC SYSTEM

From the Bragg equation

$$\sin^2\theta_{nkl} = \frac{\lambda^2}{4d_{hkl}^2}$$

$$= \frac{\lambda^2}{4a^2} \cdot (h^2 + k^2 + l^2) \quad \text{for cubic crystals} \qquad (8.4)$$

Then the possible values of $h^2 + k^2 + l^2$ can be calculated from the $\sin\theta$ values of different diffraction lines (appearing in the diffraction photograph of a powdered sample containing randomly oriented microcrystals) by dividing with the common factor $\lambda^2/4a^2$. As mentioned earlier, some of the diffraction lines observed with a *sc* structure will not appear in the diffraction pattern of *bcc* and *fcc* structures. These are now summarised in Table 8.4 for planes with simple Miller indices.

As reflected by this table, the diffractions are possible with all value of h, k and l for *sc*, only when $h + k + l = $ even for *bcc* and only when h, k, l are either all even or all odd (excluding zero) for *fcc*.

Table 8.4

Miller indices (khl) of the diffraction lines	Relevant cubic lattice type		
100	sc		
110	sc	bcc	
111	sc		fcc
200	sc	bcc	fcc
210	sc		
211	sc	bcc	
220	sc	bcc	fcc
221	sc		
300	sc		
310	sc	bcc	
311	sc		fcc
222	sc	bcc	fcc
320	sc		
321	sc	bcc	
400	sc	bcc	fcc

The possible values of $h^2 + k^2 + l^2$ are:

1, 2, 3, 4, 5, 6, 8, 9 (with 7 missing) for sc

2, 4, 6, 8, 10, 12, 14, 16 for bcc

3, 4, 8, 11, 16 (most irregular) for fcc

i.e. $h^2 + k^2 + l^2$, for first six lines, include 1 and 5 with sc but not with bcc and fcc.

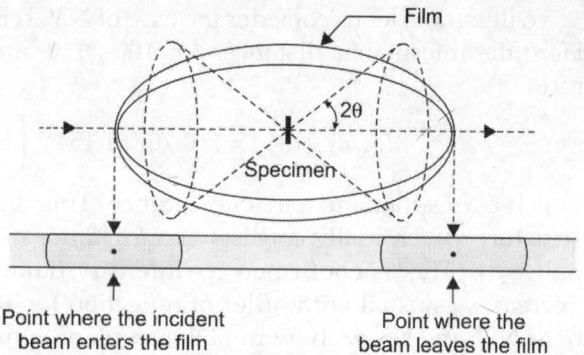

Fig. 8.3: Powder x-ray diffraction photography

A powder diffraction photograph is obtained using a camera with the powdered sample at the central position of the film wrapped round the circumference of the camera. For every value of the glancing angle θ a cone of diffracted x-ray beam will be formed on either side of the sample (shown in Fig. 8.3) producing two lines with same hkl on the film at equal distance D away from the central line due to incident x-ray beam. θ (rad) $= D/2R$ where R is the radius of the camera. This follows from the fact that the diffraction beam makes an angle 2θ with the incident beam. Formation of a cone for each value of θ arises from the different possible orientation of a particular khl plane forming the same angle θ with the incident beam.

8.8 DENSITY OF A CRYSTAL AND CLOSEST PACKING

The particle density of a crystal, which is determined solely by the arrangement of the particles, is often used for prediction or confirmation of crystal structure

$$\text{Particle density} = \frac{\sum n_i}{v} \qquad (8.5a)$$

$$\text{Bulk density} = \frac{\sum n_i M_i}{N_A v} \qquad (8.5b)$$

where n_i is the number of particles of the ith type per unit cell of volume v and M_i is molar mass.

The particle density is governed by the nature of binding in a crystal and size and shape of the constituent particles such that the potential energy of the crystal is minimum. The crystals are called metallic, ionic (e.g. NaCl), molecular (e.g. sugar), or covalent (e.g. SiC) according as the bending is largely metallic, electrostatic, van der Waals or covalent. A high coordination number is expected for the first three types of crystals where bonding has only slight directional character. The crystal structure of ionic compounds has to face the restriction that the distance between the ions of like charge must be greater than that between the ions of opposite charge, and this makes particle density of ionic compounds rather low. Highest particle density is expected with monatomic crystals like metals where the efficiency of the geometrical arrangement is the dominating factor. Here densest packing may be thought in terms of close-packed layers where each atom (supposed to be spherical) touches six others hexagonally. If the layers are stacked such that atoms of the top layers occupy only the hollows of the bottom layers then two simple types of compact structure will result where each atom has identical environment. These are called hexagonal close-packed or cubic close-packed according as the atoms lying on a vertical line are separated by one layer or two layers (shown in Fig. 8.4). The cubic close-packed structure is really face-centred cubic. However, the coordination number (12) and particle density are same for both types of structures.

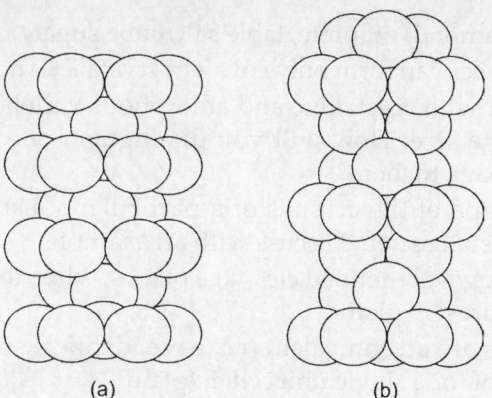

(a)   (b)

Fig. 8.4: Close-packed structures. (a) Hexagonal close packing (b) Cubic close packing

Even for a crystalline substance, the actual density is less than the ideal density. The latter is the density that would have been if there were no void interatomic space. Then

Ideal density = Actual density/fraction of space occupied

The compactness of three types of cubic systems are calculated below for monatomic crystals.

sc: Here $n = 1$, because each unit cell contains 8 corner atoms each of which is shared equally by 8 unit cells

$v = a^3$, where a is the length of the edge of the unit cell

In sc, atoms touch along edges of the unit cell, whence $a = 2r$, where r is radius of the atom.

$$\text{occupied space} = \frac{\frac{4}{3}\pi\left(\frac{a}{2}\right)^3}{a^3} \times 100 = 52.38\%$$

bcc: Here each unit cell contains 1 central atom and 8 corner atoms.

Then $n = 1 + \frac{1}{8} \times 8 = 2$.

In bcc atoms touch along body diagonals of the unit cells, whence $4r = \sqrt{3}a$, the length of the body diagonal. Then

$$\text{occupied space} = \frac{2 \times \frac{4}{3}\pi\left(\frac{\sqrt{3}a}{4}\right)^3}{a^3} \times 100 = 68.04\%$$

fcc: Here each unit cell contains 6 face atoms (each of which is shared equally by two unit cells) and 8 corner atoms. Then $n = \frac{1}{2} \times 6 + \frac{1}{8} \times 8 = 4$.

In fcc atoms touch along face diagonals of the unit cells, whence $4r = \sqrt{2}a$, the length of the face diagonal. Then

$$\text{occupied space} = \frac{4 \times \frac{4}{3}\pi\left(\frac{a}{2\sqrt{2}}\right)^3}{a^3} \times 100\% = 74.07\%$$

It appears from above calculations that fcc is the most compact form of the cubic system.

AUXILIARY PROBLEMS

8.1 Which of the following common substances are (i) crystalline (ii) amorphous (iii) none.

Copper, steel, diamond, graphite, table salt, cube sugar, ice, polyethylene, glass.

8.2 Covalent substances can form only covalent crystals. True or false.

8.3 You are supplied with crystalline and amorphous varieties of a substance in form of a cube of same size. How will you distinguish one from the other without causing any damage to them?

8.4 The size and shape of the crystals of a particular substance may vary with the conditions of crystallisation. Discuss with an example.

8.5 Crystals split along well-defined cleavage planes, while the amorphous solids have conchoidal fractures. Explain.

8.6 The lattice points are all equivalent (i.e. have identical surroundings) for a crystal as a whole but not for a single unit cell. Explain.

8.7 In a crystal, no atomic nucleus need lie at the relevant lattice points. Comment.

8.8 A crystal lattice cannot have a positive ion at one lattice point and a negative ion at another lattice point. Justify.

8.9 Unit cells of a crystal lattice cannot have spherical shape. Why?

8.10 Mention a symmetry element which is possible with molecules but not with crystals. Is there any symmetry element where the reverse happens?

8.11 In the crystallisation of molecules, can crystal symmetry exceed the molecular symmetry?

8.12 An axis of five-fold symmetry is not found with crystals, though it is not uncommon among molecules. Justify. Is axis of higher symmetry possible with a crystal.

8.13 There are three Bravais lattices belonging to the cubic system and only one belonging to the trigonal system. Which of these two crystal systems will be simpler?

8.14 Of the seven crystal systems which one is the simplest? Give reasons for your answer.

8.15 Mention the symmetry elements of a cubic system.

8.16 In representing crystal planes Miller indices are preferred to Weiss indices. Why?

8.17 Is Miller index system applicable to all seven crystal systems. For which system(s), the following expression for interplanar distance (d_{khl}) of khl planes is applicable?

$$\frac{1}{d_{hkl}^2} = \left(\frac{h}{a}\right)^2 + \left(\frac{k}{b}\right)^2 + \left(\frac{l}{c}\right)^2$$

How would you justify this expression considering that the Miller indices depend on the reference axes and origin chosen arbitrarily?

8.18 Find the Miller indices of the following planes having indicated intercepts on the respective crystallographic axes x, y and z.

(i) $6a$, $3b$ and $2c$ (ii) $2a$, b and $2/3a$ (iii) $2a$, b and ∞, (iv) $-6a$, $-3b$ and $-2c$ (v) $6a$, $-3b$ and $2c$.

Comment on the result.

8.19 Can any of the Miller indices of a plane be ∞? Is there any significance of $\infty\infty\infty$ planes.

8.20 How do the planes in the following pair differ?
    a. 110 and $\bar{1}\bar{1}0$ planes
    b. 111 and $\bar{1}\bar{1}1$ planes
    c. 111 and $\bar{1}\bar{1}\bar{1}$ planes

8.21 How many types of unit cells are possible for planar lattices? Draw them.

8.22 For a two-dimensional square lattice what will be the relation between planes in each of the following pairs:
    a. (11) and $(\bar{1}1)$ planes
    b. $(\bar{1}1)$ and $(1\bar{1})$ planes

8.23 Crystal structures are usually investigated using the diffraction of x-rays caused by the crystals. Would the purpose be served through ultraviolet ray, γ-ray, electron beam and neutron beam?

8.24 In elucidating crystal structure which is more effective tool to be used—x-ray diffraction or neutron diffraction? Give reasons for your answer.

8.25 Hydrogen atoms in a crystal structure cannot be located from x-ray diffraction study. Why? What is the way out?

8.26 First order Bragg reflection from 110 planes occurs with simple cubic crystals. Does this happen with body-centred and face-centred cubic crystals?

8.27 What type of cubic crystals will correspond to the following different statements?
    a. 100 planes and 110 planes contain all the lattice points
    b. 110 but not 100 planes contain all the lattice points
    c. 100 and 110 planes none contain all the lattice points

8.28 KCl (like NaCl) has face-centred cubic crystal structure. But to x-ray, KCl (unlike NaCl) behaves as a simple cubic crystal. Explain. Will KCl behave similarly towards neutron beams?

8.29 In power x-ray diffraction photograph, metallic crystals of *sc* type display larger number of lines than *fcc* crystals. But, although KCl and NaCl have same crystal structure (*fcc* type), KCl (which behaves towards x-ray as *sc*) displays fewer number of lines than NaCl and fewer more than metallic *sc* crystals. Thus 111 line is found with NaCl and also with *sc* but not with KCl. Explain.

8.30 The Eq. (8.4) is based on the Bragg equation for first order reflection. But this is used for indexing all lines arising also from higher order reflection in powder photograph. Is this justified?

8.31 A powder diffraction photograph contains even lines which correspond to the planes having no real existence, e.g. 200 planes of a *sc* system. Comment.

8.32 In the crystal analysis by powder method, the tiny crystals in the powdered sample should have dimensions of the order of microns. What's harm if the dimensions are too much smaller? Is power method effective for all crystal systems?

8.33 The interatomic distance in a body-centred cubic crystal of a metal is 5.2A. What is the maximum wavelength of radiation that can be used for analysis of this crystal? Using 0.8A x-rays, what will be the highest and lowest order of diffraction for (i) 110 planes and (ii) 111 planes?

8.34 A binary compound has cubic unit cell with A atoms only at the corners and B atoms only at the face-centres. Find the simplest formula of the compound.

8.35 Majority of metals crystallise with hexagonal close packing or cubic close packing. Why?

8.36 Which one will exhibit higher coordination number in a crystal—a metal or its salt?

8.37 Of the two crystals, one having closer packing will exhibit higher coordination number. True or false?

8.38 What is the highest coordination number that a crystalline substance can exhibit? Is this possible with crystals of the following substances?

Ar, Al, C, CO_2, CH_4 and NaCl

8.39 In close-packed structures of metallic crystals how many tetrahedral and octahedral holes are present per atom?

8.40 In a number of crystals containing two types of particles, one type of particles forms a close-packed structure and the other type of smaller size fits into the holes in this structure. Find the formula of the ionic crystals containing cation of element A and anion of element B forming the following well-known structures. Also find the highest radius ratio permitted in each case.

   a. *Rock salt structure*: Where anions form cubic close packing (CCP) and cations occupying octahedral holes.

   b. *Fluorite structure*: Where cations form CCP and anions, which are smaller than cations, occupying tetrahedral holes.

   c. *Antifluorite structure*: Where anions form CCP and cations occupying tetrahedral holes.

   d. *Zinc blende structure*: Where anions form CCP and half of the tetrahedral holes are occupied by cations.

   e. *Wurtzite structure*: Where anions form hexagonal close packing (HCP) and cations occupying half of the tetrahedral holes.

   f. *Corundum structure*: Where anions form *hcp* and cations occupying two third of octahedral holes.

8.41 a. Define cohesive energy of a crystal. How can it be determined?

   b. Cohesive energy of transition metals tends to be higher. Why?

   c. Covalent crystals have higher cohesive energy than metallic crystals. Comment.

8.42 Account for the following:

   a. Both Ni and NiO have face-centred cubic lattice. But nickel has higher coordination number in Ni crystal (12) than in NiO crystal (6).

   b. Graphite has higher melting point than diamond.

   c. Both CO_2 and SiO_2 are covalent compounds. But CO_2 is a gas at ordinary temperature while SiO_2 is a solid of very high melting point and boiling point.

   d. MgO has higher melting point than KF, though the difference in electronegativity between the constituent elements is greater in KF.

   e. Diamond has crystal lattice similar to zinc blende, yet diamond is very hard while zinc blende is rather soft.

   f. Hardness decreases in the order: diamond > steel > graphite.

   g. $CaCO_3$ crystal is harder than CaO crystal, though the latter is more compact.

8.43 For a hexagonal system, how are the lengths of the unit vectors (a, b and c) are related? Find the packing fraction of this system (i.e. the fraction of space occupied by the atoms).

8.44 a. An element has simple tetragonal crystal structure with $a = b = 4.5\text{Å}$ and $c = 7.5\text{Å}$. Find Weiss indices and Miller indices of its plane having intercepts a,

1.5*b* and 2*c* on the three choosen axes. Also find the interplanar distance of the group of planes represented by the Miller indices of this plane.

b. From the indicated plane if one proceeds towards the origin, how many equivalent planes are to cross per unit lattice distance along each axis?

8.45 a. The spacing between 222 planes of a *bcc* metallic crystal is 1.5Å. Find that for 110 planes.

b. For which planes (if at all) of this crystal, the spacing is (i) 3Å (ii) 0.75Å? Are these equivalent to 222 planes? How are they technically related?

c. Comment on 220 planes and 333 planes of this crystal.

d. What is the maximum wavelength of x-ray that can be used for diffraction of this crystal.

8.46 Is it possible to determine the coefficient of expansion of a solid using Bragg equaiton?

The first order reflection from a *khl* plane of a cubic crystal occurs at glancing angle θ at celcius temperature *t* and θ' at *t'*, the same x-ray source being used. Express the coefficient of cubic expression of the crystal in terms of θ, θ', *t* and *t'*.

8.47 When a metal of atomic radius *r* crystallises in cubic structure, its unit cell length *a*, the packing fraction *f* and coordination number *n* may be connected to the type of unit cell as listed below:

Crystal structure	a	f	n
sc	$2r$	$\pi/6$	6
bcc	$\dfrac{4}{\sqrt{3}}r$	$\dfrac{\pi}{8\sqrt{3}}$	8
fcc	$\dfrac{4}{\sqrt{2}}r$	$\dfrac{\pi}{3\sqrt{2}}$	12

Establish the given relations, mentioning the assumption(s) involved.

8.48 Polonium crystallises in a cubic system. It gives first order reflection from 100 planes. Is the unit cell premitive or not?

8.49 Cs (at wt 133) crystallises in a cubic structure. With x-rays of 80 pm, the $\sin\theta$ values for the first order reflection from 100-, 110- and 111-type planes are found to be 0.133, 0.094 and 0.230 respectively.

a. Which type of cubic structure is this?

b. Calculate (i) unit cell length (ii) radius Cs atom (iii) actual and ideal density of Cs metal (iv) reticular densities (number of particles per unit area) for the given planes.

8.50 An element (at wt 24) forms cubic crystal. Its unit cell contains atoms at positions (0, 0, 0), (1/2, 1/2, 1/2), (1, 0, 1) (1, 1, 0), (0, 1, 1). Considering various sets of planes containing all the atoms (i.e. having no interleaved planes) the highest interplanar distance is 1.5Å. Find the density of the element.

8.51 How much error will be involved in the determination of density of a cubic crystalline element by x-ray diffraction, if 1.0% error is involved in the determination of atomic radius?

8.52 A metal can crystallise in *sc*, *bcc* and *fcc* form.

a. Find the ratio of the densities of these three forms, mentioning the assumption, if any.

    b. Which form will be of closest packing in the three forms?

    c. The diffraction pattern has been studied for one of the modifications using x-rays of wavelength 50 pm. Strong first order reflections are found from 111 actual planes at $\sin\theta = 0.108$ and from 110-type planes at $\sin\theta = 0.175$. To which cubic form does the data correspond? Also find the unit cell length.

8.53 The smallest of observed Bragg diffraction angle for 111 planes of a potassium crystal is $6.613°$ when x-ray of wavelength 70.926 pm is used. Given that potassium exists as body-centred cubic lattice, determine the length of the unit cell (at wt of K = 39).

8.54 In a x-ray diffraction experiment using 155 pm x-ray a metal belonging to cubic system having molar volume 22.71 cc produces a first order reflection from 110 planes at glancing angle $15.04°$. Find out whether the metal has sc, bcc or fcc structure?

8.55 a. The metal titanium (at wt 48) forms hexagonal closest-packed crystals. Atomic radius of the metal is 146 pm. Find (i) coordination number (ii) number of atoms per unit cell (iii) unit cell dimensions (iv) density of crystal (v) packing fraction.

    b. Both hexagonal closest-packed and body-centred crystal have same number of atoms per unit cell. If Ti crystal were of body-centred type, what would happen to coordination number and density of the metal?

8.56 TiO_2 (rutile) has tegragonal lattice with $a = 4.59$ Å and $c = 2.96$ Å with all the lattice points occupied. Its density is 4.25 g cm^{-3}. Is the unit cell primitive or non-primitive?

8.57 When the NaCl crystal is investigated with x-rays of 58.5 pm, the first order Bragg diffraction occurs at $5.97°$ from the lattice planes parallel to the face of the face-centred cubic lattice. Find (a) the separation of these planes (b) the smallest distance between Na^+ and Cl^- ions (c) the smallest distance between Na^+ ions (d) the smallest distance between Cl^- ions (e) the Miller indices of the planes for which the first order reflections occurs at $8.4°$.

How do you justify that the minimum distance between Na^+ ions is same as that between Cl^- ions though Na^+ and Cl^- ions are of different radius.

8.58 MgO crystallizes in cubic structure with unit cell length 420 pm. The density of the crystal is 3.62 g cm^{-3}. Find the number of MgO unit per unit cell and hence predict the type of cubic structure does it have.

Which reflection will be weaker–first or second order, for (i) 100 planes (ii) 111 planes?

8.59 The molar volume of KCl is 1.3 times that of NaCl. If the glancing angle for first order Bragg reflection from 100 planes of NaCl is $5.9°$, calculate the same for KCl and also the angle of diffraction.

8.60 CsCl crystal is of body-centred type with Cs^+ ion at the centre and $8 Cl^-$ ions at the corners of the unit cell. Find

    a. density of CsCl relative to that of NaCl

    b. minimum value of $r+/r-$ for the existence of CsCl type of structure

    c. minimum value of $r+/r-$ for the existence of NaCl type of structure

[Given: $r_{Na^+} = 0.95$Å, $r_{Cs^+} = 1.69$Å, $r_{Cl^-} = 1.81$Å, Na = 23, Cs = 133, Cl = 35.5]

From which of the following three sets of planes should the first order reflection for CsCl would be most intense: 100, 110 or 111?

8.61 a. KCl crystallises in the NaCl-type structure. The ionic radius of Na^+ is 0.5 times that of Cl^-, and is 0.7 times of that of K^+. Calculate (i) the unit cell length of KCl (ii) the density of KCl, relative to that of NaCl.

b. What would happen to unit cell length and coordination number of constituent ions if NaCl crystals were of interpenetrating simple cubic instead of interpenetrating face-centred type. Is such crystal structure possible in reality?

8.62 Unit cell lengths for cubic crystals of NaCl, NaBr, KCl, KBr, CsCl and CsBr are 563, 596, 628, 659, 412 and 436 pm respectively. Do these data support the assertion that the ionic radii are independent of the counter ions?

8.63 a. In the transformation of a metal from one crystalline form to another, the molar vol should change in the inverse ratio of packing fraction. True or false.

b. Transformation of titanium form *hcp* to *bcc* structure entails 1.7% volume expansion. How do you account for this quantitatively?

8.64 ZnS (zinc blende) has cubic crystal structure of unit cell length 6 Å with Zn atoms in a *fcc* arrangement and the S atoms at the positions (1/4, 1/4, 1/4), (1/4, 3/4, 3/4), (3/4, 1/4, 3/4) and (3/4, 3/4, 1/4). Calculate

a. Zn-S bond distance

b. coordination number of Zn and also of S

c. density of ZnS crystal   [Zn = 65.4, S = 32]

8.65 Diamond has same space lattice as zinc blende except that all the atoms are identical. Its density is 3.52 g cm^{-3}. Calculate

a. unit cell length

b. C-C bond distance and bond angle

8.66 A salt A_xB_y has cubic unit cell with nucleus of A at each corner and also at the body centre, and nucleus of B only at two opposite face centres.

a. Find x and y.

b. What will be the coordination number of A and B, and the corresponding geometrical arrangement. Is such a structure possible in reality?

c. Predict whether the first order reflections will be weak or strong from 100, 110 and 200 planes.

8.67 A metal can crystallise in two cubic forms, A and B, whose x-ray diffraction patterns available from powder method are as follows:

Cubic form	Diffraction lines produced
A	111, 200, 220, 311, 222, 400, ...
B	110, 200, 211, 220, 310, 222, 321, 400, ...

To what type of unit cell do A and B belong?

8.68 The powder diffraction photograph of polonium (that forms cubic lattice) using Mo-x-rays gives consecutive lines at $\sin\theta$ = 0.149, 0.182, 0.211, 0.235, 0.259. Identify the kind of unit cell. Can cell dimension be calculated from the given data?

8.69 Ag crystallines in *fcc* form of unit cell length 408 pm. Find the glancing angles for the first three diffraction lines produced by Ag with x-ray of wavelength 154 pm.

8.70 MgO forms cubic crystal. Powder diffraction photograph of MgO using x-ray of wavelength 193 pm gives first three lines at $\sin\theta$ = 0.339, 0.461, 0.652. Find the type of unit cell and its dimension.

8.71 Ta (at wt 181) forms cubic crystal of density 16.69 g cm^{-3} at room temperature. Powder diffraction photograph of Ta with x-rays of wavelength 154.4 pm shows first three lines at $\theta = 19.31°$, $27.88°$ and $34.95°$. Determine the type of unit cell and its dimension.

8.72 A powder diffraction pattern of a cubic crystal contains lines which are at the following distances (in mm): 13.2, 18.4, 22.8, 26.2, 29.4, 32.42, 37.2, ..., from the central spot when x-rays of wavelength 70.8 pm were used in a camera of radius 5.76 cm:

  a. index the lines

  b. identify the kind of unit cell

  c. find unit cell length

8.73 The powder diffraction photograph of a cubic crystal gave first two lines at distances 12.05 mm and 27.16 mm from the central spot, when x-rays of wavelength 68.25 pm were used in a camera of radius 57.3 mm. The first line corresponds to the 100 planes

  a. index the other line

  b. identify the kind of unit cell

  c. determine the size of the unit cell

ANSWERS

8.1 (i) The crystalline substances are copper (a metallic crystal that involves metallic bonds), diamond (a covalent crystal that involves covalent bonds forming a three-dimensional network), graphite (a covalent crystal involving two-dimensional network), table salt (i.e. NaCl, which is an ionic crystal made up of Na$^+$ and Cl$^-$ ions), cube sugar (a molecular crystal composed of sucrose molecules held together by van der Waals forces and H-bonds) and ice (a hydrogen bonded molecular crystal).

(ii) The amorphous substances are glass (which has disordered and irregular structure usually involving Si-O bonds) and steel (which has no sharp melting point).

(iii) Polyethylene is partly crystalline and partly amorphous.

Note: A new kind of solid, called quasicrystal, has recently been discovered which has both short-range and long-range order but does not have symmetry property like a crystalline substance. Quasicrystals are formed when melts of certain alloys are rapidly cooled.

8.2 For most covalent substances, the atoms in the molecules are covalently bonded and the molecules interact through van der Waals forces or hydrogen bonding forming crystals called molecular crystals. Only in rare cases atoms in the crystal are all covalently bonded forming what is called a covalent crystal or a single giant molecule, e.g. diamond, Si etc. where a three-dimensional network of covalent bonds is involved. Then the given statement is not true.

8.3 From measurement of refractive index which will be same in all directions with amorphous variety and not with the crystalline variety.

8.4 This can result from the preferential growth of certain faces of a crystal depending on the conditions of crystallisation. For example, the crystal 'A' will take the shape of 'B' (shown in two dimensions) if the two faces marked by X, grow more rapidly than the other faces. As a specific example mention may be made of NaCl which usually crystallises as cubes from its aqueous solution but as octahedra from solutions containing a small amount of urea.

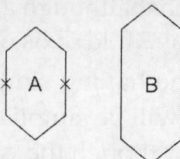

Note: Here each interfacial angle (i.e. the angle between the normals to the two intersecting faces) for the corresponding faces of the two crystalline forms remains unchanged.

8.5 In a crystal, the net planes having highest reticular density (the number of particles per unit area) are separated by highest distance (to make the bulk density uniform throughout the crystal) and hence they have minimum interaction between them. As a consequence, a crystal will naturally split along such well-defined planes.

In an amorphous form of a substance, due to random arrangement of the constituent particles, the reticular density is almost same for any surface imagined within the bulk. Here the natural fractures will necessarily be curved (conchoidal) to make the free energy (A) of the system minimum. This is similar to a liquid in its dispersion to droplets.

Note: An amorphous state may be regarded as an extremely viscous liquid state.

8.6 For a single unit cell, the lattice points occupying the corner positions will not be equivalent to those occupying the other positions due to difference in their surroundings. But such differentiation of lattice points becomes meaningless considering a crystal as a whole due to arbitrariness involved in the choice of a corner position.

8.7 This is not unlikely to happen considering the structural group (the basis) of a crystal.

8.8 The given statement is justified. Because the lattice points of an ionic crystal are determined jointly by both positive and negative ions and not by either ion alone. However, an ionic crystal like NaCl, is often conveniently represented by two interpenetrating lattices, the lattice points of one being occupied by positive ions and those of the other by negative ions (vide Sect. 8.6).

8.9 Because, with spherical cells a crystal cannot be built up without any gap between them.

8.10 An example is provided by five-fold axis of symmetry.
☐ Yes. This happens with the symmetry elements, screw axis and glide plane, which entail translational operation together with other operations appropriate to a finite figure. These are connected only to systems like crystals where there is a regular repetition of some pattern.

8.11 Yes. For example, if there are two unsymmetrical molecules in a unit cell, the crystal may have one or more two-fold axes. With more molecules in a unit cell, higher symmetries may also arise, even though the molecule itself is unsymmetrical.

8.12 Such peculiarity is not unlikely in view of the fact that that crystal symmetry, unlike molecular symmetry, is restricted by complete filling of space which is not possible with a figure of 5-fold axis of symmetry. This can be readily visualised in two dimensions. It is possible to tile a floor with parallelograms (C_2), equilateral triangles (C_3), squares (C_4) or regular hexagons (C_6) but not with regular pentagons (or heptagons) without leaving gaps in the tilling.

☐ It is quite interesting to note that although a 5-fold axis of symmetry is not found with crystals, 6-fold (though not 7-fold) axis of symmetry is found.

8.13 Of the two crystal systems, one having simplier relation between the primitive translation vectors, *a*, *b* and *c*, will be simpler. Then as it appears from Table 8.1, the cubic system will be simpler though the variety of Bravais lattices is more with this system.

Note:

i. Simplicity of a crystal system has no necessary connection with the number of Bravais lattices, number of point groups or number of space groups associated with it.

ii. Cubic system may be considered as a particular case of trigonal, if trigonal is defined as the one with $a = b = c$ and $\alpha = \beta = \gamma < 120°$.

8.14 Cubic system is the simplest because of simplest relation between its *a*, *b* and *c*.

8.15 A cubic system has all the symmetry elements, except C_6, possible for a unit cell (vide Section 8.3).

8.16 Because the Weiss indices refer to a single crystal plane whereas the Miller indices refer to a set of parallel equidistant equivalent planes with which crystallography is concerned most.

8.17 Yes.

☐ Only for orthogonal systems

☐☐ Although the origin and unit length for each axis are arbitrarily chosen, the interplanar distance found from the given expression is free from such arbitrariness. Because one of the planes represented by the Miller indices is assumed to pass through the origin chosen and the axes chosen are at right angle to each other, when d_{hkl} will be the distance of the origin from the nearest plane of Miller indices *hkl*.

Note: The unit lengths *a*, *b* and *c* are purely relative but their ratios (called axial ratios) are constant. *a*, *b* and *c* are either identical to or some simple multiple of the primitive translations. It is mere convenient to choose the axes required for Miller index system parallel to the three edges of the crystal, or unit cell thereof, that meet at a point.

8.18 (i) The inverse of the given intercepts are in the ratio $1/6 : 1/3 : 1/2$ or $1 : 2 : 3$ (in ratio of simple integers). Hence the required Miller indices will be 1, 2, 3.

(ii) Same as in (i).

(iii) The inverse of the intercepts are in the ratio $1/2 : 1 : 0$ or $1 : 2 : 0$. Hence the Miller indices will be 1, 2, 0.

(iv) Same as in (i), $\overline{1}\,\overline{2}\,\overline{3}$ planes being equivalent to 1 2 3 planes.

(v). $1\overline{2}3$

☐ The Miller indices are same for (i), (ii) and (iv), because the intercepts are in the same proportion in the three cases.

8.19 No. Because an infinite value of any Miller index implies that the relevant plane passes through the origin when the other two Miller indices will also be infinity. ∞∞∞∞ planes (in terms of Miller indices) have no significance because they do not represent a set of parallel planes having a definite orientation. However, a plane having intercepts 0, 0 and 0 has some significance in that it is supposed to be parallel to a set of parallel plans of specified Miller indices.

8.20 When all three Miller indices are of same sign, the same set of planes is represented by either sign and then negative signs are avoided. But in representing a single plane, the signs of the Miller indices are required to be mentioned to indicate whether the intercepts are on the positive or negative side of the chosen axes.

    a. In either case the planes are parallel to the z-axis with equal intercepts on the other two axes, and the two intercepts of each plane have same sign, +ve or −ve. Then 110 and $\bar{1}10$ planes belong to the same set.

    b. In either case, the intercepts on the three axes are equal. However, for 111 planes, the intercepts of each are all of same sign whereas for $\bar{1}11$ each plane has intercepts of same sign on x- and y-axes but of opposite sign on the z-axis.

    c. In either case the planes have equal intercepts on the three axes and the intercepts of each plane are of same sign. Then 111 and $\bar{1}\bar{1}\bar{1}$ planes form the same set.

Note: a and b are different, but a and c are similar, though reverse might appear at first glance.

8.21 For planar lattices, five types of unit cells are possible. These are sketched below.

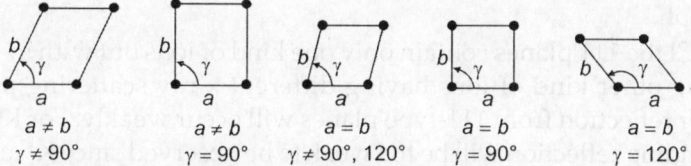

Fig. 8.5.

Note: Seven types of unit cells in three-dimensions follow from the above five types of planar unit cells considering an additional unit vector $\bar{c}$.

8.22 a. One is perpendicular to the other.

    (11) plane has +ve intercepts on both x- and y-axes, while $(\bar{1}1)$ plane has −ve intercept on x-axis and + intercept on y-axis.

    b. One is parallel to the other. $(\bar{1}1)$ plane has intercept −ve on x-axis and +ve on y-axes. The reverse happens with $(1\bar{1})$ plane.

8.23 Determination of crystal structure using x-rays is based on the ability of a crystal to behave as a diffraction grating, the interatomic spacings in crystals (usually of the order of 1 Å) being comparable to the wavelengths of the x-ray used. Neutrons of appropriate speed (of the order at 10^3 m/s) have de Broglie wavelengths comparable to x-rays and hence a neutron beam can also serve the same purpose as x-rays. But ultraviolet ray and γ-ray are unsuitable in this regard due to much lower wavelength with ultraviolet and more so with γ-ray. Again, although electrons having speed of the order of 10^7 m/s have de Brogli wavelength comparable to x-ray, they are unsuitable in studying the interior of a solid phase due to their strong coulombic interaction with electrons and nuclei of atoms. However, they are useful in studying molecules in the gas phase, on surfaces and in thin films.

8.24 Neutron diffraction. Because although neighbouring elements in the periodic table have almost identical x-ray scattering power and hence indistinguishable through x-ray diffraction, their neutron scattering strength may significantly differ.

8.25 Because of insufficient intensity with which x-rays are scattered by H-atoms due to their least electron content.

☐ The problem can be solved through neutron diffraction which, unlike x-ray diffraction, occurs through interaction with atomic nuclei. The neutron-scattering power of the H-nucleus is comparable to that of the other nuclei.

8.26 First order reflection from 110 planes will occur also from *bcc* but not from *fcc* crystals. Because 110 planes are not interleaved by any other planes for *bcc* while these are interleaved by equivalent planes at just the half way for *fcc*.

8.27   a. *sc* system, because here the given planes are not interleaved by any other planes.

b. *bcc* system, because here 100 but not 110 planes are interleaved by other planes.

c. *fcc* system, because here 100 and 110 planes are both interleaved by other planes.

8.28 Vide Section 8.6

☐ No. Because neutrons are scattered (following a mechanism different from x-ray scattering)with intensity determined mostly by atomic nuclei of the scattering species.

8.29 These are all linked to nearly same behaviour of K^+ and Cl^- ions towards x-ray. Although KCl behaves with x-rays as a simple cubic (*sc*) system, the unit cell is actually of face-centred cubic (*fcc*) type. The additional lattice points in *fcc* result in disappearance of several diffraction lines possible with *sc*, in powder diffraction photograph.

☐ For NaCl the 111 planes contain only one kind of ions but with in-between planes containing other kind of ions having different x-ray scattering power. Then the first-order reflection from 111-type planes will occur weakly. For KCl however, the corresponding reflection will be too weak to be observed since K^+ and Cl^- ions have almost same power of scattering x-rays. But the reflection from 111 planes of a *sc* crystal will be strong, there being no interleaved planes. The given facts are thus accountable.

8.30 Yes. Because a second order reflection from 110 planes, for example, occurs at the same angle as a first order reflection from 220 planes. Therefore, considering indices hkl of all values, the higher order reflections are automatically accommodated.

8.31 This can be justified on the ground that a second order reflection from the hypothetical 200 planes, e.g. is equivalent to, and hence can be treated as, a first order reflection from the real 100 planes.

8.32 This will cause the diffraction lines to be inconveniently broad. Because with decrease in size of the crystal, the number of faces increases with consequent decrease in orderly arrangement of lattice planes characteristic of a crystal.

☐ The powder method is effective only for the crystal systems that have not more than two lattice parameters to be decided (viz. cubic, trigonal, tetragonal and hexagonal systems).

8.33 For a *bcc* crystal, the highest interplanar distance is

$$d_{110}^* = \frac{a}{\sqrt{1^2 + 1^2 + 0^2}}$$

$$= \frac{2d}{\sqrt{3} \cdot \sqrt{2}} \quad \text{where } d \text{ is interatomic distance (Sect. 8.8)}$$

$$= \frac{2(5.2\text{Å})}{\sqrt{3}\sqrt{2}} = 4.25\text{Å}$$

[* d_{hkl} will be highest for planes that contain all the atoms and have least Miller indices]

$$\lambda_{max} = 2d_{110} \text{ by Eq. (8.3) putting } n = 1 \text{ and } \sin\theta = 1$$
$$= 2 \times 4.25\text{Å} = 850\text{Å}$$

i. The highest value of $n = \dfrac{2d_{110}}{\lambda}$ putting $\sin\theta = 1$ in Eq. (8.3)

$$= \frac{8.50\text{Å}}{0.8\text{Å}} = 10.6$$

$$= 11, \text{ taking the nearest higher integer}$$

The lowest value of $n = 1$, which is the lowest integer

ii. The highest value of $n = \dfrac{2d_{111}}{\lambda} = \dfrac{2d_{110}\sqrt{1^2 + 1^2 + 0^2} \Big/ \sqrt{1^2 + 1^2 + 1^2}}{\lambda}$

$$= 10.6\sqrt{2}/\sqrt{3} = 9 \quad \text{taking the nearest higher integer}$$

The lowest value of $n = 2$

Since the odd order reflections are absent as 111 planes are interleaves by equivalent planes at just the half-way.

Note:
1. The interplanar distance is lower than the interatomic distance.
2. λ_{max} is determined not by the interatomic distance but by the interplanar distance and is highest for planes with highest spacing.
3. Even for the same crystal the highest value of n is different for planes of different Miller indices, but the lowest value of n will be either 1 or 2 according as the specified planes are not interleaved or interleaved by equivalent planes at the half way.

8.34 Each unit cell contains, on the average, $1/8 \times 8$ or 1 'A' atoms and $1/2 \times 6$ or 3 'B' atoms which are in ratio 1:3. Hence the simplest formula of the compound is AB_3.

8.35 The forces of interaction which hold the atoms together in a metal crystal do not usually act in any preferred direction. For such crystals, the energetically most favourable structure is, therefore, expected to be the one in which each atom is surrounded by the greatest possible number of neighbouring atoms. This explains why most metals crystallise with hexagonal close packing or cubic close packing that gives highest coordination number. The departure from such packing exhibited by some metals (e.g. Fe, Mn, Mo, W, V, etc.) may be attributed to some covalent character (having directional property) of bonding that favours some other specific geometrical arrangement such as *bcc* structure.

Note: Crystal structure is likely to depend on temperature and pressure. But it seems unlikely that the highest packing is not always favoured at lower temperature. For example, at 1 atm crystalline iron has *fcc* structure in the temperature range 900–1400°C but *bcc* both above and below this range.

8.36 This question has no unique answer. For example, Cs and CsCl crystals exhibit same coordination number (8). However, in general a metal exhibits higher coordination number than its salts. Because a stable crystal structure (i.e. one having minimum energy) of a salt requires each ion to have only oppositively charged ions as its nearest neighbours, whereas the formation of a stable metal crystal primarily involves a spatial problem of accommodating highest number of particles surrounding each particle.

8.37 The given statement is true for crystals consisting of only one kind of particles but not necessarily so for other crystals. For example, CsCl crystl has less compact structure (*bcc*) but higher coordination number (8) compared to NaCl crystal of *fcc* structure with coordination number 6.

8.38 12 is the highest coordination number which is observed in a crystal for hexagonal close-packed (*hcp*) or cubic close-packed (*ccp*) structure.

☐ For highest coordination number to exhibit the crystal should be built up of only one kind of sphere-like particles held together by forces having no significant directional character. Such requirements are fulfilled by Ar (which forms molecular crystal where Ar atoms are held together by rather weak van der Waals forces), Al (which forms metallic crystal where Al atoms are held together by strong forces, called metallic bonds), CH_4 (which forms molecular crystal like Ar).

In crystalline carbon (a covalent crystal), the C atoms are held together by covalent bonds which have directional character. CO_2, like CH_4, forms molecular crystal but the molecules differ much from spherical shape. NaCl forms an ionic crystal which is composed of two different kinds of particles—Na^+ and Cl^- ions.

8.39 In building a close-packed structure, atoms (supposed to be spherical) of each layer rest only in half of the hollows of the bottom layer forming tetrahedral holes. The rest of the hollows form equal number of octahedral holes. Then it follows that per atom there will be two tetrahedral holes—one with bottom layer and another with top layer, and one octahedral hole (formed jointly by a hollow in the bottom layer and that in the top layer).

8.40 In a closed-packed structure (*ccp* and *hcp*) there are two tetrahedral holes and one octahedral hole for every constituent particle of the crystal. Then it follows that:
   a. A and B ions will be present in 1:1 mole ratio. Accordingly, the formula of the crystal will be AB.
   b. A and B ions will be present in 1:2 mole ratio. Hence the formula of the crystal will be AB_2.
   c. Likewise, the formula will be A_2B.
   d. The formula will be AB.
   e. The formula will be AB.
   f. A and B will be present in $\frac{2}{3}$:1 or 2:3 mol ratio.
   Then formula of the crystal will be A_2B_3.
   [**Note:**
   i. Fluorite and antifluorite structures are similar with cations and anions interchanged.
   ii. Zinc blende and wurtzite structures are equivalent.
   iii. It is not possible for both cations and anions to form close-packed structure.]
   ☐ A tetrahedral hole may be looked as a truncated cube of appropriate edge length (a) shown in Fig. 8.6. The anions of B forming close packing touch along the edges

of the tetrahedron of edge length $\sqrt{2}a$ (which is same as the length of a face diagonal of the cube). A cation of A occupies the mid-point of the body diagonal of length $\sqrt{3}a$, of the cube touching four ions of B each at the distance $\sqrt{3}a/2$ away. Then

$$\frac{\sqrt{2}a}{\sqrt{3}a/2} = \frac{2r_B}{r_A + r_B} \quad \text{whence} \quad \frac{r_A}{r_B} = 0.225$$

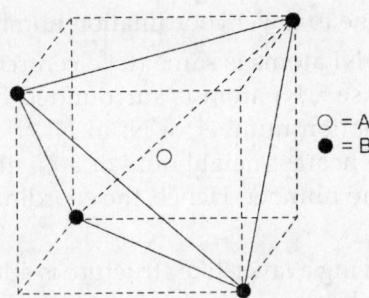

Fig. 8.6: Tetrahedron as a truncated cube

An octahedral hole may also be looked as a truncated cube shown in Fig. 8.7. Here the anions of B touch along the edges of the octahedron of edge length $a/\sqrt{2}$. A cation of A holding the central position of the cube touches six anions each at distance $a/2$ away. Then

$$\frac{a/\sqrt{2}}{a/2} = \frac{2r_B}{r_A + r_B} \quad \text{whence} \quad \frac{r_A}{r_B} = \sqrt{2} - 1 = 0.414$$

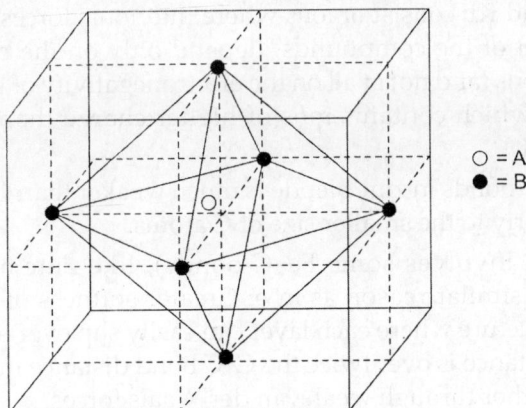

Fig. 8.7: Octahedron as a truncated cube

Then the required radius ratio, i.e. $r_{\text{smaller ion}}/r_{\text{larger ion}}$ will be 0.225 in (b), (c), (d), (e) and 0.414 in (a), (f).

8.41 a. The cohesive energy of a crystal is $\Delta H°$ for the isothermal conversion of the crystal into its ideal-gas-phase structural units at the temperature concerned. The structural units are isolated atoms for metallic and covalent crystals, molecules for molecular crystals, and ions for ionic crystals.

☐ Cohesive energy is not directly measurable. It has to be determined indirectly using the Born–Haber thermodynamic cycle.

b. This is due to covalent bonding involving d electrons.

c. The given statement is not always true. For example, W (a transition metal) has higher cohesive energy than even covalent solid like graphite. This happens due to its high coordination number in crystalline form together with covalent bonding involving d electrons.

However, for most covalent crystals, the given statement holds due to formation of strong covalent bonds in proper directions that compensates for lower interaction energy due to lower coordination number compared to metals.

8.42. a. The arrangement of Ni atoms is same (face-centred cubic) in both Ni and NiO crystals. In either case a Ni atom is surrounded by 12 other Ni atoms which provide the coordination number of Ni in Ni crystal but not in NiO crystal. Because in NiO, the nearest neighbours of a Ni atom are not Ni atoms but O atoms which are six in number. Hence, the coordination number of Ni in NiO is six.

b. Because graphite having a layer-like structure is additionally stabilised through delocalisation of π electrons which are involved in bonding C-atoms in the formation of two dimensional hexagonal ring structure in each layer.

c. Carbon forms small molecules of structure O=C=O while silicon, due to its larger atomic size prefers to form a giant molecule (a single crystal) through three dimensional network of Si–O single bonds. The vaporisation of such giant molecules, which involves breaking of strong covalent Si–O bonds, will then require quite a high amount of energy relative to vaporisation of carbon dioxide which involves breaking of much weaker van der Waals bonds between CO_2 molecules. This explains the given statement.

d. Both MgO and KF consist of ions where interionic forces, which determine the melting point of the compounds, depend only on the charge and size of the constituent ions (and not at all on the electronegativity of the relevant elements). Then MgO, which contains ions of higher charge than KF will have higher melting point.

e. Because Zn–S bonds in zinc blende is much weaker than C–C bonds in diamond due particularly to the smaller size of C atoms.

f. Steel (which involves some Fe–C bonds), like zinc blende, is softer than diamond for similar reason as in e. Greater softness of graphite is due to its layer-like structure where each layer can easily slip over other layer because the interlayer distance is over twice the C–C bond distance in a layer and the layers are held together through weak van der Waals forces.

g. This is due to involvement of directed covalent bonds that connect three O-atoms to a C-atom in a CO_3^{2-} ion. Such bonds reduce the ease of displacement of O (in CO_3^{2-}) with respect to Ca in $CaCO_3$ crystal compared to that in CaO crystal.

Note: Hardness is a mechanical property which is far from being defined quantitatively as it involves the concept of several properties (tensile strength, resistance to cleavage, etc.). It has no necessary connection with thermodynamic properties like cohesive energy. Thus diamond has higher hardness but lower cohesive energy (and hence lower melting point) than graphite.

8.43 A (primitive) hexagonal space lattice with a basis of two atoms constitutes the unit (shown in Fig. 8.8) through which a hexagonal close-packed structure can be generated. Then

$$a = b = 2r, \text{ assuming each atom to be spherical with radius } r$$

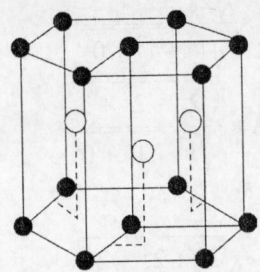

Fig. 8.8: A hexagonal unit cell (represented by broad lines) relating to *hcp* structure. Of the two atoms of the basis, one represented by ● occupies a lattice position but the other represented by O (lying within the unit cell) does not.

Again, in hexagonal close packing, each atom forms two tetrahedral holes—one with the top layer and another with the bottom layer. Then

$$c = 2 \times \text{height of a tetrahedron having all edges of same length } 2r$$

$$= \sqrt{(2r)^2 - (2r/\sqrt{3})^2}, \text{ from geometrical consideration}$$

$$= 4\sqrt{2/3} \cdot r$$

Then

$$c/a = 4\sqrt{2/3}\, r/2r$$

or

$$c = 1.633a$$

$$\text{Packing fraction} = \frac{\text{Volume of atoms per unit cell}}{\text{Volume of unit cell}} \times 100\%$$

$$= \frac{2 \times \dfrac{4}{3} \pi r^3}{a^2 c (1 - \cos^2 \gamma)^{1/2}} \times 100\% \qquad \text{by Eq. (8.1)}$$

$$= \frac{2 \times \dfrac{4}{3} \pi \times \left(\dfrac{a}{2}\right)^3}{a^2 \times 1.633a \sin 120°} \times 100\%$$

$$= 74.08\%, \text{ which is same as in case of } fcc$$

Note: The value of c/a found experimentally differ (though slightly) from 1.633 as calculated above assuming atoms to be spherical. This suggests non-spherical shape of an atom.

8.44 a. The Weiss indices of the given plane (which are multiples of a, b and c in the intercepts of the plane) are 1, 1.5, 2.

The inverse of the intercepts are in ratio 1/1 : 2/3 : 1/2 or 6 : 4 : 3 (in ratio of simplest integers). Hence the given plane has Miller indices (6 4 3).

□ Interplanar distance $= \left[\left(\dfrac{6}{4.5Å}\right)^2 + \left(\dfrac{4}{4.5Å}\right)^2 + \left(\dfrac{3}{7.5Å}\right)^2\right]^{-1/2}$ by Eq. (8.2a)

or

$$= 0.606Å$$

b. The consecutive planes represented by Miller indices 6 4 3 are separated by 1/6, 1/4 and 1/3 units along the respective chosen axes. Hence, the required number of planes to cross will be $1/(1/6)$, $1/(1/4)$ and $1/(1/3)$, i.e. 6, 4, 3 along the respective axes (same as the Miller indices).

8.45 a.

$$d_{110} = d_{222} \times \frac{\sqrt{2^2 + 2^2 + 2^2}}{\sqrt{1^2 + 1^2 + 0^2}} \qquad \text{by Eq. (8.2b)}$$

$$= 1.5\text{Å} \times \frac{\sqrt{2^2 + 2^2 + 2^2}}{\sqrt{1^2 + 1^2 + 0^2}}$$

$$= 3.67\text{Å}$$

b. $$\sqrt{h^2 + k^2 + l^2} = \frac{d_{222}}{d_{hkl}} \cdot \sqrt{2^2 + 2^2 + 2^2} \qquad \text{by Eq. (8.2b)}$$

i. $$\sqrt{h^2 + k^2 + l^2} = \frac{1.5\text{Å}}{3.0\text{Å}} \cdot \sqrt{2^2 + 2^2 + 2^2}$$

$$= \sqrt{1^2 + 1^2 + 1^2}$$

This corresponds to 111 planes

ii. $$\sqrt{h^2 + k^2 + l^2} = \frac{1.5}{0.75} \sqrt{2^2 + 2^2 + 2^2}$$

$$= \sqrt{4^2 + 4^2 + 4^2}$$

This corresponds to 444 planes which have no real existence for *bcc*, there being no plane between consecutive 222 planes.

111 planes (which exist) but not 444 planes (which do not exist) are equivalent to 222 planes (which exist).

222 planes, which include 111 planes and also in-between planes, are technically related to the latter as 111-type planes.

c. 220 planes, which have Miller indices double of 110 planes, do not exist in case of *bcc*, there being no in-between 110 planes. Again, 333 planes, which have Miller indices thrice of 111 planes, do not exist because there exists only one (but not two) plane between two consecutive 111 planes.

d. For *bcc* the maximum interplanar distance is d_{110}. Then the required maximum wavelength will be $2d_{110}$ (vide problem 8.33).

Note:

1. In a cubic crystal, planes with interplanar distance $= a/\sqrt{p^2 + q^2 + r^2}$ (where p, q and r are integers) may not exist (e.g. 333, 444-planes discussed above for *bcc*) but must not exist when this relation does not hold (as in case of interplanar distance $a/\sqrt{7}$).

2. Maximum wavelength to be used is determined by the interplanar distance of importance. Even for the same crystal, it varies with the crystal plane concerned. In the given problem any wavelength equal to $2d_{110}$ will be greater than $2d_{hkl}$ for any other (hkl) plane chosen for *bcc*.

3. Multiplication of Miller indices hkl of existing planes by n amounts to division of the space between consecutive hkl planes into n equal parts by introduction of additional planes. Only when these additional planes have real existence

would the Miller indices $nh\ nk\ nl$ be meaningful. Thus in case of *bcc* crystal, 111 planes exist but 333 and 444 planes do not, though 222 planes exist. It is not always an easy task to predict the real existence of planes defined by the Miller indices *hkl* unless h, k and l are of low value.

8.46 Yes, using the relation deduced below.

☐ Let d_{hkl} and d'_{hkl} be the interplanar distances at temperatures t and t' respectively. Then

$$\frac{d'_{hkl}}{d_{hkl}} = \frac{\sin\theta}{\sin\theta'} \qquad \text{by Bragg Eq. (8.3)}$$

Coefficient of linear expansion $\alpha = \dfrac{d'_{hkl} - d_{hkl}}{(t' - t)d_{hkl}}$

$$= \frac{\sin\theta - \sin\theta'}{(t' - t)\sin\theta'}$$

The coefficient of cubic expansion $= 3\alpha$

8.47 Coordination number (CN): For *sc*, each atom (shared by 8 unit cells) touches 6 other atoms along the 6 common sides. Hence CN is 6.

For *bcc*, each atom (conveniently, the body-centred one which belongs to only a single cell) touches 8 other atoms (at the corners) along four body diagonals. Hence CN is 8.

For *fcc*, each atom (conveniently, a face-centred one which is shared by only two cells) touches 12 other atoms (4 atoms at the corners of the common face and 8 atoms at the centres of 8 faces adjacent to the common face) at distance $a/\sqrt{2}$ (which is half the length of a face diagonal) away. Hence CN is 12.

Note: In deciding about CN, we can focus on any atom, cornered, body-centred or face-centred. Because atoms in a crystal are all equivalent (i.e. have identical surroundings), their geometrical position being determined arbitrarily by the unit cell imagined.

In each type of cubic crystal, the distance of an atom from the neighbouring atoms is same (atoms being assumed to be like hard spheres touching each other) although this is a with *sc*, $\sqrt{3}a/2$ with *bcc* and $a/\sqrt{2}$ with *fcc*. This is not unlikely in view of the fact that a is not same for *sc*, *bcc* and *fcc*, being determined by the number of atoms per unit cell (vide problem 8.52).

For a and f (vide Sect. 8.8).

8.48 The first order reflection from 100 planes implies that these planes are not interleaved by any other planes (vide Fig. 8.2). Then the unit cell is primitive.

8.49 a. The interplanar distances are in the ratio

$$d_{100}{:}d_{110}{:}d_{111} = \frac{1}{\sin\theta_{100}} : \frac{1}{\sin\theta_{110}} : \frac{1}{\sin\theta_{111}} \qquad \text{by Bragg equation}$$

$$= \frac{1}{0.133} : \frac{1}{0.094} : \frac{1}{0.230}$$

$$= 1 : 1.415 : 0.578$$

$$\simeq 1 : \frac{2}{\sqrt{2}} : \frac{1}{\sqrt{3}}$$

which corresponds to *bcc* type.

b. i. For a cubic crystal

$$d_{hkl} = \frac{a}{\sqrt{h^2 + k^2 + l^2}} \quad \text{which holds for } hkl \text{ actual planes}$$

$$= \frac{n\lambda}{2\sin\theta_{hkl}} \quad \text{by Bragg equation which holds for } hkl \text{ type planes}$$

Then $$a = \frac{n\lambda\sqrt{h^2 + k^2 + l^2}}{2\sin\theta_{hkl}}$$

which holds when hkl actual planes are same as type planes, i.e. when hkl planes contain all the atoms

$$= (1)(80 \text{ pm})\frac{\sqrt{1^2 + 1^2 + 0^2}}{2 \times (0.094)} \quad \text{using data for 110 planes which contain all the atoms}$$

$$= 601.8 \text{ pm}$$

ii. $$r = \frac{\sqrt{3}a}{4} \quad \text{[vide problem 8.47]}$$

iii. Actual density $$= \frac{\sum n_i M_i}{N_A a^3} \quad \text{by Eq. (8.5b)}$$

$$= \frac{(2)(133 \times 10^{-3} \text{ kg mol}^{-1})}{(6.02 \times 10^{23} \text{ mol}^{-1})(601.8 \times 10^{-12} \text{ m})^3}$$

$$= 2000 \text{ kg m}^{-3}$$

Ideal density = Actual density/packing fraction

$$= (2000 \text{ kg m}^{-3})/(0.6804) \quad \text{vide Section 8.8}$$

$$= 2939.5 \text{ kg m}^{-3}$$

iv. Reticular density = Particle density × Interplanar distance (d)

$$= \frac{\sum n_i}{a^3} \cdot d \quad \text{by Eq. (8.5a)}$$

For 110-type planes, $$d = \frac{a}{\sqrt{1^2 + 1^2 + 0^2}}$$

100-type planes, $$d = \frac{1}{2} \cdot \frac{a}{\sqrt{1^2 + 0^2 + 0^2}}$$

111-type planes, $$d = \frac{1}{2} \cdot \frac{a}{\sqrt{1^2 + 1^2 + 1^2}}, \text{etc.}$$

since 100- and 111-type planes are interleaved by equivalent planes (vide Sect. 8.8)

8.50 Here each unit cell contains 4 corner-atoms and 1 centre-atom. Hence number of atoms per unit cell (n) will be

$$n = 4 \times \frac{1}{8} + 1 = 1.5$$

Since the given unit cell is of body-centred type (with alternate corner points missing), the largest interplaner distance will correspond to 110 planes (with lowest Miller indices that contain all the atoms).

Then $\qquad a = d_{110}\sqrt{1^2 + 1^2 + 0^2} \qquad$ by Eq. (8.2b)

$$= (1.5\text{Å})\sqrt{1^2 + 1^2 + 0^2}$$

$$= 2.12\text{Å}$$

$$e = \frac{nM}{N_A a^3}$$

$$= \frac{(1.5)\,(24 \times 10^{-3}\text{ kg mol}^{-1})}{(6.02 \times 10^{23}\text{ mol}^{-1})\,(2.12 \times 10^{-10}\text{ m})^3}$$

$$= 6.28 \times 10^3 \text{ kg/m}^3$$

8.51 For a cubic crystal

$$e = \frac{nM}{N_A a^3} \quad n \text{ is number of particle per unit cell}$$

$$= \frac{B}{r^3} \quad \text{where } B \text{ is a constant for the crystal as } a \propto r, \text{ the atomic radius}$$

In logarithmic form

$$\ln e = \ln B - 3 \ln r$$

Differentiation gives

$$\frac{de}{e} = -3\frac{dr}{r}$$

Then error in $\qquad e = -3 \times$ error in r

$$= -3 \times 1\%$$

$$= -3\%$$

8.52 a. By relation $\qquad e = \dfrac{nM}{N_A a^3} \qquad$ from Eq. (8.5b)

$$e_{sc}:e_{bcc}:e_{fcc} = \left(\frac{n}{a^3}\right)_{sc} : \left(\frac{n}{a^3}\right)_{bcc} : \left(\frac{n}{a^3}\right)_{fcc}$$

$$= \frac{1}{(2r)^3} : \frac{2}{(4r/\sqrt{3})^3} : \frac{4}{(2\sqrt{2}r)^3} \qquad \text{(vide Sect. 8.8)}$$

$$= 1:1.299:1.414$$

This is based on the assumption that atoms are like hard spheres.

Note: For different crystalline forms of a particular metal, r is same but a is not. Because atoms touch each other along different directions for different forms.

b. fcc. This follows from a.

Note: For a particular substance, n and a both rise in the same direction

$$sc \rightarrow bcc \rightarrow fcc$$

But a rises in lower proportion than n, making e_{fcc} highest.

c. The strong first order reflection from 111 actual planes indicates that no intermediate planes are present between 111 planes (i.e. 111 actual planes are same as 111-type planes). Hence the concerned crystal will be either sc or fcc. Now, as in problem 8.49(b)

$$a = \frac{n\lambda\sqrt{h^2 + k^2 + l^2}}{2\sin\theta_{hkl}}$$

$$= \frac{(1)\,(50\ pm)\sqrt{1^2 + 1^2 + 1^2}}{2 \times (0.108)} \quad \text{using data for 111 plane}$$

$$= 400.9\ pm$$

Here calculation of a does not require the knowledge of the type of cubic structure.

If the crystal is assumed to be of sc type, 110 planes will be same as actual planes when

$$\sin\theta_{110} = \sin\theta_{111} \times \frac{d_{111}}{d_{110}}$$

$$= 0.108 \times \frac{\sqrt{1^2 + 1^2 + 1^2}}{\sqrt{1^2 + 1^2 + 0^2}}$$

$$= 0.132$$

Since $\sin\theta_{110}$ thus calculated differ much from the given value, the crystal is not of sc type as assumed. This indirectly implies that the crystal is of fcc type.

Note:

1. Distinguish between three types of cubic structure is possible only from $d_{110}{:}d_{111}$ (for 110- and 111-type planes) which is

$$d_{110}{:}d_{111} = 0.707{:}0.577 = 1{:}0.816 \text{ for } sc$$
$$= 1.414{:}0.577 = 1{:}0.408 \text{ for } bcc$$
$$= 0.707{:}1.154 = 1{:}1.632 \text{ for } fcc$$

For the given problem $d_{110}{:}d_{111} = \sin\theta_{111}{:}\sin\theta_{110} = 0.108{:}0.175 = 1{:}1.62$
This corresponds to fcc.
However, from $d_{100}{:}d_{110}$ distinction cannot be made between sc and fcc, while from $d_{100}{:}d_{111}$ distinction cannot be made between sc and fcc.

2. If $\sin\theta$ for 111-type planes (instead of 111 actual planes) were given, then first crystal type and then a had to be determined considering $\sin\theta$ for 111-type planes which are same as 111 actual planes in case of fcc.

8.53 Since the lattice is of bcc type, 111 (actual) planes will have equivalent in-between planes on the half way. Then odd order reflections from such planes will be absent, and the given data will correspond to the second order reflection. Then the Bragg equation $n\lambda = 2d_{hkl}\cdot\sin\theta_{hkl}$ (where $d_{hkl} = a/\sqrt{h^2 + k^2 + l^2}$), we have from the given data

$$(2)\,(70.926\ pm) = \frac{2a}{\sqrt{1^2 + 1^2 + 1^2}}\sin 6.613°$$

whence $\qquad\qquad a = 1067.3\ pm$

8.54 First order reflection from 110 planes indicates that the crystal structure will be either sc or bcc. Assuming it to be of sc type

$$\text{molar volume} = \frac{N_A a_{sc}^3}{n_{sc}} \quad \text{by Eq. (8.5a)}$$

$$= \frac{N_A}{n_{sc}} \left(\frac{n\lambda\sqrt{h^2 + k^2 + l^2}}{2\sin\theta_{hkl}} \right)^3$$

$$= \frac{(6.02 \times 10^{23} \text{ mol}^{-1})}{(1)} \left[\frac{(1)(1.55 \times 10^{-10} \text{ cm})\sqrt{1^2 + 1^2 + 0^2}}{2\sin 15.04°} \right]^3$$

$$\neq 22.71 \text{ cm}^3$$

Then the crystal is not of sc type. This implies that the metal has bcc structure.

Note: From $\bar{V} = N_A a^3/n$, n cannot be found from the given $\bar{V}$, because a (which depends on crystal type) is unknown. This led to the adoption of a trial procedure assuming a particular crystal type.

8.55 a. Vide problem 8.43

i. Here each atom has 6 nearest neighbours, 3 in the layer above it and 3 in the layer below it, i.e. coordination number is six.

ii. Two, vide Fig. 8.8.

iii. $a = b = c/1.633$

=2r, assuming each atom to be spherical with radius r

iv. Bulk density $= \dfrac{\sum n_i M_i}{N_A a^2 c (1 - \cos^2 \gamma)^{1/2}}$ by Eqs (8.1) and (8.5b)

$$= \frac{nM}{1.633 N_A a^3 \sin \gamma}$$

$$= \frac{nM}{1.633 N_A (2r)^3 \sin 120°}$$

$$= \frac{(2)(48 \times 10^{-3} \text{ kg mol}^{-1})}{1.633 (6.022 \times 10^{23} \text{ mol}^{-1})(2 \times 146 \times 10^{-12} \text{ m})^3 \cdot \sqrt{3}/2}$$

$$= 1.71 \times 10^3 \text{ kg/m}^3$$

v. Packing fraction = 74.08%

b. Coordination number would change to 8 for bcc (vide problem 8.47). Density (e) would change by a factor

$$\frac{e_{bcc}}{e_{hcp}} = \frac{v_{hcp}}{v_{bcc}} = \frac{1.633 a_{hcp}^3 \sin 120°}{a_{bcc}^3} = \frac{1.633(2r)^3 \sin 120°}{(4r/\sqrt{3})^3}$$

$$= 0.9185$$

This is just the ratio of the packing fractions for bcc and hcp.

Note: If a_{bcc} were equal to a_{hcp} then e_{bcc} would be greater than e_{hcp}.

8.56 Number of TiO_2 units per unit cell,

$$n = \frac{N_A a^2 c e}{M_{TiO_2}} \quad \text{by Eq. (8.5b)}$$

$$= \frac{(6.02 \times 10^{23} \text{ mol}^{-1})(4.59 \times 10^{-8} \text{ cm})^2 (2.96 \times 10^{-8} \text{ cm})(4.25 \text{ g cm}^{-3})}{(48 + 2 \times 16) \text{ g mol}^{-1}}$$

$$= 1.99 \simeq 2$$

Since $n > 1$, the unit cell is non-primitive.

8.57 a. Separation of the indicated (type) planes

$$d = \frac{(1)(58.6 \text{ pm})}{2\sin 5.97°} \quad \text{by Bragg Eq. (8.3)}$$

Such planes contain both Na^+ and Cl^- ions

b. In NaCl crystal Na^+ ions touch Cl^- ions along the edge of the unit cell. Then the required distance between Na^+ and Cl^- ions will be equal to d.

c. The smallest distance between Na^+ ions

$$= \frac{1}{2} \times \text{ face diagonal of the unit cell}$$

$$= \frac{1}{2} \times \sqrt{2}a \text{ because each type of ions are in } fcc \text{ arrangement}$$

$$= \sqrt{2}d, \text{ since } a = 2d$$

d. Same as c.

e. Reflection at 5.97° occurs from 100-type planes or 200-actual planes (which are also same as 200-type planes). If reflection at 8.4° occurs from hkl actual planes (which are same as type planes) then

$$\frac{d_{hkl}}{d_{200}} = \frac{\sin 5.97°}{\sin 8.4°} = \frac{\sqrt{2^2 + 0^2 + 0^2}}{\sqrt{h^2 + k^2 + l^2}}$$

or $\quad h^2 + k^2 + l^2 = (2^2 + 0^2 + 0^2) \cdot \dfrac{\sin^2 8.4°}{\sin^2 5.97°}$

$$= 8$$

Then the required Miller indices will be 220, or 202 or 022.

☐ The same value for c and d is not unjustified in view of the fact that in the crystal ions bearing like charge are not close enough to touch each other.

8.58 Number of MgO units per unit cell

$$n = \frac{N_A a^3 e}{M_{MgO}} \quad \text{(vide problem 8.5b)}$$

$$= \frac{(6.02 \times 10^{23} \text{ mol}^{-1})(4.20 \times 10^{-10} \text{ cm})^3 (3.62 \text{ g cm}^{-3})}{(24.3 + 16) \text{ g mol}^{-1}}$$

$$= 4.01 \simeq 4$$

This corresponds to fcc structure (with alkali halide crystals in mind).

☐ Considering the given planes to be the actual planes, the answer will be

  i. first order

 ii. first order

Considering the given planes to be the type planes, the answer will be

  i. second order

 ii. first order

The result is not same for different considerations. This is because, here 100 actual planes are different from 100 type planes (which are 200 actual planes) though 111 actual planes are same as 111 type planes.

Note: Type planes are to be equivalent to the actual planes (regarding composition). Although 111 actual planes of MgO have in-between planes, the latter are not equivalent to 111 actual planes. This is unlike 100 planes.

8.59 $\qquad \dfrac{\bar{V}_{KCl}}{\bar{V}_{NaCl}} = \left(\dfrac{a_{KCl}}{a_{NaCl}}\right)^3 , \text{ since } \bar{V} = \dfrac{N_A a^3}{n}$

[n being same for KCl and NaCl due to same crystal structure]

$$= \frac{(\sin \theta)_{\text{NaCl}}^3}{(\sin\theta)_{\text{KCl}}^3}$$

Since
$$a = \frac{\lambda \sqrt{h^2 + k^2 + l^2}}{\sin \theta_{khl}}$$

λ and $\sqrt{h^2 + k^2 + l^2}$ being same for KCl and NaCl

Then
$$(\sin \theta)_{\text{KCl}} = \left(\frac{\bar{V}_{\text{NaCl}}}{\bar{V}_{\text{KCl}}} \right)^{1/3} (\sin \theta)_{\text{NaCl}}$$

$$= \left(\frac{1}{1.3} \right)^{1/3} \sin 5.9°$$

$$= 0.09418$$

Therefore
$$\theta = 5.4°$$

Angle of diffraction $= 2 \times$ glancing angle (θ)
$$= 2 \times 5.4° = 10.8°$$

[glancing angle is sometimes called Bragg diffraction angle].

Note: Here '100 planes' imply '100 type planes' (i.e. 200 actual planes), because first order reflection does not occur from 100 actual planes for a *fcc* crystal like NaCl/ KCl. However, our result will not be affected since Miller indices and order of reflection do not appear in the final expression for calculation.

8.60 a.
$$\frac{e_{\text{CsCl}}}{e_{\text{NaCl}}} = \frac{(nM/a^3)_{\text{CsCl}}}{(nM/a^3)_{\text{NaCl}}}$$

$$= \frac{\left[nM \Big/ \left(\frac{2}{\sqrt{3}} \right)^3 (r_+ + r_-)^3 \right]_{\text{CsCl}}}{[nM / 2^3 (r_+ + r_-)^3]_{\text{NaCl}}}$$

[since Cs$^+$ and Cl$^-$ ions touch along body diagonals of unit cell, $\sqrt{3}a = 2(r_+ + r_-)$]

[since Na$^+$ and Cl$^-$ ions touch along the edges of the unit cell, $a = 2(r_+ + r_-)$]

$$= (\sqrt{3})^3 \frac{(nM)_{\text{CsCl}}}{(nM)_{\text{NaCl}}} \left(\frac{r_{\text{Na}^+} + r_{\text{Cl}^-}}{r_{\text{Cs}^+} + r_{\text{Cl}^-}} \right)^3$$

$$= (\sqrt{3})^3 \frac{1(133 + 35.3)}{4(23 + 35.5)} \left(\frac{0.95 + 1.81}{1.69 + 1.81} \right)^3 = 1.83$$

b. Here
$$\sqrt{3}a = 2(r_+ + r_-)$$

Again, the nearest distance between Cl$^-$ ions = edge length of the cell (a)

i.e.
$a \geq 2r_-$ '=' corresponds to touching of Cl$^-$ ions

Then
$$\sqrt{3} \cdot 2r_- \geq 2 (r_+ + r_-)$$

or
$$\frac{r_+ + r_-}{r_-} \geq \sqrt{3}$$

whence
$$\frac{r_+}{r_-} = \sqrt{3} - 1$$

Then the minimum value of $r_+/r_- = \sqrt{3} - 1 = 0.732$.

c. Here
$$a = 2(r_+ + r_-)$$

Again, the nearest distance between Cl^- ions

$$= \frac{1}{2} \times \text{length of the face diagonal} = \frac{a}{\sqrt{2}}$$

i.e.
$$\frac{a}{\sqrt{2}} \geq 2r_-$$

$$\frac{2(r_+ + r_-)}{\sqrt{2}} \geq 2r_-$$

or
$$\frac{r_+ + r_-}{r_-} \geq \sqrt{2}$$

whence
$$\frac{r_+}{r_-} \geq \sqrt{2} - 1$$

Then the minimum value of $r_+/r_- = \sqrt{2} - 1 = 0.414$.

☐ 110 (actual) planes. Because, unlike 110 planes, 100 and 111 planes are interleaved by other planes.

Note: The structure of ionic crystals changes with decrease in radius ratio $(\gamma) = r_{\text{smaller ion}}/r_{\text{larger ion}}$ (usually $r_{\text{cation}}/r_{\text{anion}}$) in the direction given below (radius ratio rule, due originally to Pauling):

cubic $\rightarrow$ Octahedral $\rightarrow$ Tetrahedral $\rightarrow$ Triangular $\rightarrow$ Linear
$\gamma \geq 0.732$ $\gamma \geq 0.414$ $\gamma \geq 0.225$ $\gamma \geq 0.155$ $\gamma < 0.155$

The values of γ mentioned are approximate. Because these are based on the over-simplified assumptions that the ions behave like rigid spheres and that oppositely charged ions are in physical contact with each other.

8.61 a. KCl has NaCl-type crystal structure such that the oppositely charged ions touch along the edges of the unit cell. Then

i.
$$\frac{a_{KCl}}{a_{NaCl}} = \frac{r_{K^+} + r_{Cl}}{r_{Na^+} + r_{Cl^-}} = \frac{0.5}{0.7}\frac{r_{Cl^-} + r_{Cl^-}}{0.5r_{Cl^-} + r_{Cl^-}} = 1.14$$

ii.
$$\frac{e_{KCl}}{e_{NaCl}} = \frac{M_{KCl}}{M_{NaCl}} \cdot \left(\frac{a_{NaCl}}{a_{KCl}}\right)^3, \quad \text{the number of ions per unit cell being same}$$

$$= \left(\frac{39.10 + 35.45}{22.99 + 35.45}\right)\left(\frac{1}{1.14}\right)^3 = 0.861$$

b. Here the crystal structure of lowest possible energy would correspond to one of *bcc* type where each ion touches 8 oppositely charged ions along body diagonals of the unit cell. This happens with lowering of cell length and highering of coordination number to 8.

☐ Such a structure is not permitted by the radius ratio being less than 0.732.

8.62 NaCl, NaBr, KCl and KBr all have crystal structure of *fcc* type where cell length a is related to the radii of the constituent cations and anions by
$$a = 2(r_+ + r_-)$$

If ionic radius were characteristic constant of an ion, then $(a_{NaBr} - a_{NaCl})$ would be equal to $(a_{KBr} - a_{NaBr})$, each being equal to $2(r_{Br^-} - r_{Cl^-})$. Now from the given data

$$a_{NaBr} - a_{NaCl} = 596 - 563 = 33 \text{ pm}$$

$$a_{KBr} - a_{KCl} = 659 - 628 = 31 \text{ pm}$$

For CsCl and CsBr, which form crystals of *bcc* type

$$\sqrt{3}a = 2(r_+ + r_-)$$

Here $\sqrt{3}(a_{CsBr} - a_{CsCl})$, instead of $(a_{CsBr} - a_{CsCl})$ would correspond to $2(r_{Br^-} - r_{Cl^-})$. Again, from the given data

$$\sqrt{3}(a_{CsBr} - a_{CsCl}) = \sqrt{3}(436 - 412) = 41.6 \text{ pm}$$

Such calculations do not support (rather oppose) the given assertion, particularly when the counter ion causes change of coordination number leading to greater polarisation of the concerned ions.

Note: Here $a_{CsCl} < a_{NaCl} < a_{KCl}$ is conclusive, though the given data are approximate.

8.63 a. The given statement holds true provided atomic size remains unchanged.

b. From the given data

$$\frac{\bar{V}_{bcc}}{\bar{V}_{hcp}} = 1.017$$

Again $\qquad \dfrac{f_{hcp}}{f_{bcc}} = \dfrac{0.7408}{0.6804} = 1.089 \qquad$ vide Section 8.8

Here $\bar{V}_{bcc} / \bar{V}_{hcp}$ significantly differ from f_{hcp}/f_{bcc}. This may be attributed to a change in atomic volume by a factor equal to

$$\frac{\bar{V}_{bcc}}{\bar{V}_{hcp}} \cdot \frac{f_{bcc}}{f_{hcp}} = \frac{1.017}{1.089} = 0.9339$$

Note:

1. Lower atomic size in *bcc*, compared to *hcp* is due to higher coordination number.
2. In crystallography, it is the usual procedure to assume that atoms are like hard sphere. The present problem points to the oversimplified nature of such assumption.

8.64 a. Zn–S bond distance (d) is the distance of a Zn atom, say at $(0, 0, 0)$ from the nearest S atom $(1/4, 1/4, 1/4)$. Then

$$d = \left[\left(\frac{1}{4}a - 0 \right)^2 + \left(\frac{1}{4}a - 0 \right)^2 + \left(\frac{1}{4}a - 0 \right)^2 \right]^{1/2}$$

$$= \frac{\sqrt{3}}{4}a = \frac{\sqrt{3}}{4} \times 6\text{Å} = 2.598\text{Å}$$

[Here d is 1/4th of a body diagonal of the unit cell].

b. Each S atom is surrounded (tetrahedrally) by four Zn atoms as nearest neighbour, and vice versa. Then the coordination number of both Zn and S will be 4.

c. Each unit cell contains four net S atoms and four net Zn atoms (8 corner atoms and 6 face-centred atoms). Then

$$e = \frac{n_{Zn} M_{Zn} + n_S M_S}{N_A^3 a^3} \qquad \text{by Eq. (8.5b)}$$

$$= \frac{(4)(65.4 \text{ g mol}^{-1}) + (4)(32 \text{ g mol}^{-1})}{(6.02 \times 10^{23} \text{ mol}^{-1})(6 \times 10^{-8} \text{ cm})^3} = 2.996 \text{ g cm}^3$$

Note: The formula ZnS of zinc blende found here tallies with that found from hole formalism in problem 8.40(d).

8.65 a.
$$a = \left(\frac{nM}{N_A e}\right)^{\frac{1}{3}} \qquad \text{by Eq. (8.5b)}$$

$$= \left[\frac{(8)(12.01 \text{ g mol}^{-1})}{(6.02 \times 10^{23} \text{ mol}^{-1})(3.52 \text{ g cm}^{-3})}\right]^{\frac{1}{3}}$$

$$= 3.567 \times 10^{-8} \text{ cm}$$

b. Here the unit cell may be looked as made up of 8 equivalent cells each of cell length $\frac{1}{2}a$. Considering a carbon atom A (1/4, 1/4, 1/4) bonded to a corner atom B (0, 0, 0) and a face-centred atom C (0, 1/2, 1/2), the bond angle θ may be represented by the following diagram.

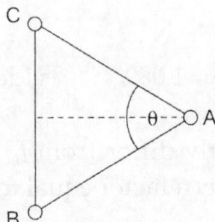

Here, bond distance

$$AB \text{ (or } AC) = \frac{1}{4} \times \text{length of a body diagonal of the unit cell}$$

$$= \frac{1}{4} \times \sqrt{3}a = \frac{\sqrt{3}}{4}(3.567 \times 10^{-8} \text{ cm}) = 1.545 \times 10^{-8} \text{ cm}$$

$$\sin\left(\frac{\theta}{2}\right) = \frac{OB}{AB} = \frac{\frac{1}{4} \times \text{length of a face diagonal of the unit cell}}{\frac{1}{4} \times \text{length of a body diagonal of the unit cell}}$$

$$= \frac{\frac{1}{4} \times \sqrt{2}a}{\frac{1}{4} \times \sqrt{3}a} = 0.8165$$

whence $\theta = 109°28'$

Note: Bond distance depends on cell length but bond angle does not.

8.66 Atom positions in the unit cell are shown in Fig. 8.9

a. Each unit cell contains net two A atoms (8 corner atoms and 1 central atom) and net one B atoms (2 face atoms). Then A and B are present in atomic ratio 2:1. Therefore, $x = 2$ and $y = 1$.

b. Here, there are two types of A atoms—those having two B atoms as nearest neighbours $a/2$ distance away and those having four nearest B atoms $a/\sqrt{2}$ distance away in square planes arrangement. Accordingly, A has coordination number (CN) 2 and 4.

● = A
○ = B

Each B atom has six neighbouring A atoms of which two are nearest at distance $a/2$ away while other four are

Fig. 8.9

at higher distance $a/\sqrt{2}$ away. Then CN of B is 2.

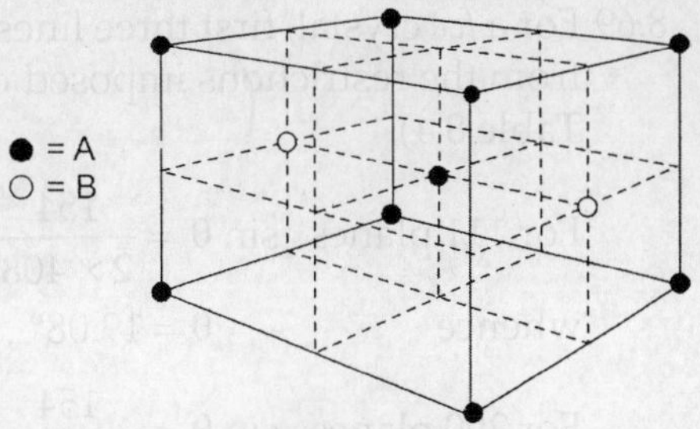

☐ Such a structure is not possible, for regular crystals where the constituent particles of same kind should necessarily have identical surroundings within the crystal.

[Note:

1. Here CN 2 of B happens with a figure of (equitorially) compressed octahedron. If this figure were an (equitorially) elongated octahedron, CN would be 4. Only when the octahedral arrangement would happen to be a regular one (with all six neighbouring A atoms equal distance away from a B atom) CN would be 6.

2. The formula of a crystalline compound relates to the number of constituent particles per unit cell but not necessarily to CN's of the particles. Thus in the present problem, the formula of the compound is same though CN of B is not same].

c. 100 planes containing A and B atoms are interleaved by planes containing only A atoms at the half-way, while 110 planes containing only A atoms are similarly interleaved by planes containing only B atoms. The partial destructive interference arising from the interleaved planes should then make the first order reflections from 100 and 110 planes weak, though not to the same extent (relative reflecting power of the interleaved planes being not same for two types of planes). For 200 planes the first order reflection should be strong, there being no interleaved planes.

8.67 For A, h, k and l are either all even or all odd (excluding zero). Hence A belongs to *fcc* type.

For B, $h + k + l$ is even for the given lines. Hence B belongs to *bcc* type.

8.68 From the given data

$\sin^2\theta$	0.0222	0.0331	0.0445	0.0552	0.0671
$h^2 + k^2 + l^2$	2	3	4	5	6

(dividing by the common factor 0.0111)

Since the values of $h^2 + k^2 + l^2$ are found to be consecutive integers, including 5, the given data corresponds to *sc* type of unit cell.

☐ Here $\dfrac{\lambda^2}{4a^2} = 0.0111$ by Eq. (8.4)

But a cannot be found as λ is not given.

8.69 For a *fcc* crystal, first three lines correspond to 111, 200 and 220 planes. This follows from the restrictions imposed on h, k and l for cubic systems (that form the basis of Table 8.4).

For 111 planes $\quad \sin \theta = \dfrac{154}{2 \times 408} \sqrt{1^2 + 1^2 + 1^2} \quad$ by Eq. (8.4)

whence $\qquad \theta = 19.08°$

For 200 planes $\quad \sin \theta = \dfrac{154}{2 \times 408} \sqrt{2^2 + 0^2 + 0^2}$

whence $\qquad \theta = 15.47°$

For 220 planes $\quad \sin \theta = \dfrac{154}{2 \times 408} \sqrt{2^2 + 2^2 + 0^2} \quad$ by Eq. (8.4)

whence $\qquad \theta = 32.25°$

8.70 Here (for a cubic crystal), $\Delta \sin^2 \theta$ between the first and second lines $(0.461^2 - 0.339^2 = 0.0976)$ is quite different from that between the second and third lines $(0.652^2 - 0.461^2 = 0.2126)$, i.e. consecutive lines do not differ in $h^2 + k^2 + l^2$ equally. This suggests that the unit cell is of *fcc* type.

Then $\sin \theta = 0.339$ for the first line will correspond to 111 planes having lowest $h^2 + k^2 + l^2$. Now a can be obtained from the following relation

$$a = \frac{\lambda}{2 \sin \theta} \sqrt{h^2 + k^2 + l^2} \quad \text{by Eq. (8.4)}$$

$$= \frac{193 \text{ pm}}{2(0.339)} \sqrt{1^2 + 1^2 + 1^2} = 493.1 \text{ pm}$$

8.71 Here $\quad \sin^2 27.88° - \sin^2 19.31° = 0.1099$

and $\quad \sin^2 34.95° - \sin^2 27.88° = 0.1101$

Since $\Delta \sin^2 \theta$ between consecutive lines is almost constant, the consecutive lines differ in $h^2 + k^2 + l^2$ equally. This suggests that the unit cell is either *sc* or *bcc*. Assuming it to be of *bcc* type the lowest θ, i.e. $19.31°$ would correspond to the lowest $h^2 + k^2 + l^2$, i.e. 110 planes, and a would be

$$a = \frac{\lambda}{2 \sin \theta} \sqrt{h^2 + k^2 + l^2}$$

$$= \frac{(154.4 \text{ pm}) \sqrt{1^2 + 1^2 + 0^2}}{2 \sin 19.31°} = 330.1 \text{ pm}$$

$$e = \frac{nM}{N_A a^3}$$

$$= \frac{(2)(181 \text{ g mol}^{-1})}{(6.022 \times 10^{23} \text{ mol}^{-1})(330.1 \times 10^{-10} \text{ cm})^3} = 16.73 \text{ g cm}^{-3}$$

The density thus calculated tallies with the given one. Hence the unit cell is of *bcc* type as assumed.

8.72 For any line glancing angle

$$\theta = \frac{\text{Distance } (D) \text{ of the line from the central spot}}{2 \times \text{radius of the camera}} \times \frac{360°}{2\pi}$$

we can then draw up the following table

D (mm)	13.2	18.4	22.8	26.2	29.4	32.42	37.2
θ (degree)	6.56	9.15	11.3	13.0	14.6	16.1	18.5
$\sin^2\theta$	0.0132	0.0255	0.0388	0.0506	0.0647	0.0782	0.101
$h^2 + k^2 + l^2$	1	2	3	4	5	6	8

(dividing by the common factor 0.0132)

a. The lines, corresponding to the values of $h^2 + k^2 + l^2$, are 100, 110, 111, 200, 210, 211 and 220.

b. Since the values of $h^2 + k^2 + l^2$ (up to 6) are consecutive integers, the unit cell is of *sc* type.

   [Alternatively, *fcc* is eliminated because indices are not of one type—odd or even. Again *bcc* is eliminated because values of $h + k + l$ are not all even].

c. $\left(\dfrac{\lambda}{2a}\right)^2 = 0.0132,$ the common factor in $\sin^2\theta$ values

   Then $a = \dfrac{\lambda}{2\sqrt{0.0132}}$

   $= \dfrac{70.8 \text{ pm}}{2\sqrt{0.0132}} = 308 \text{ pm}$

Note: This procedure for finding a is not safe if lines are few in number as in problems 8.70 and 8.71.

8.73 The expression for θ used in the previous problem gives θ = 6.022° for the first line and θ = 12.57° for the second line, the corresponding sin θ values are 0.1072 and 0.2352.

a. $\sqrt{h^2 + k^2 + l^2} = \dfrac{\sin\theta_{hkl}}{\sin\theta_{100}}\sqrt{1^2 + 0^2 + 0^2}$ by Eq. (8.4)

   $= \dfrac{0.2352}{0.1072} = 2,$ nearest integer

   This corresponds to 110 line.

   [**Note:** For a cubic crystal, if first line is 100, the second line must be 110 (vide Table 8.4).

b. The very appearance of the line 100 (where $h^2 + k^2 + l^2 = 1$) suggests that the unit cell is of *sc* type

c. $a = \dfrac{\lambda}{2\sin\theta_{hkl}}\sqrt{h^2 + k^2 + l^2}$

   $= \dfrac{68.25 \text{ pm}}{2(0.1072)}\sqrt{1^2 + 0^2 + 0^2} = 318.3 \text{ pm}$

Note: The procedure is applicable to any two lines, i.e. they need not be the first two or consecutive.

UNIVERSITY QUESTIONS

8.1 Differentiate between crystalline and amorphour solid. (Patna BSc, 2004)

8.2 Define crystal lattice and unit cell. (Kerala BSc, 2004)

8.3 a. What do you understand by crystal systems and crystal classes?

(Burdwan BSc(H), 1992)

b. How many crystal systems and Bravais lattices are there in crystalline solids?

(Burdwan BSc(H), 2011)

c. What are the dimensional characteristics of a triclinic bravais lattice?

(Burdwan BSc(H), 2012)

d. What are different Bravais lattice types of a cubic crystal?

(Vidyasagar BSc(H), 2003)

8.4 Enumerate various elements of symmetry of a cubic type of unit cell.

(Delhi BSc, 2002)

8.5 In crystals 5-fold and 7-fold rotational axes of symmetry do not exist. Justify.

(Jadavpur BSc(H), 2000)

8.6 a. Calculate the number of atoms in the unit cells of (i) a body-centred cube and (ii) a face-centred cube.

(Calcutta BSc(H), 1995)

b. Calculate the coordination number of an atom in the crystals (i) and (ii).

(Burdwan BSc(H), 2004)

c. Show that the body-centred cubic lattice is less economically packed compared to the face-centred lattice.

(Burdwan BSc(H), 1993)

8.7 a. What is law of rational indices?

(Delhi BSc, 2002)

b. What are Miller indices?

(Calcutta BSc(H), 2002)

c. Miller indices of a crystal face refer actually to a class of faces. Justify/criticise.

(Burdwan BSc(H), 1997)

8.8 Calculate:

a. Weiss and Miller indices for a plane perpendicular to one axis and parallel to the other two.

(Burdwan BSc(H), 2010)

b. Miller indices of a crystal plane which is cut through the crystal axes $2a, -3b, -c$.

(Guru Nanak Dev BSc, 2002)

8.9 Show that the distance of separation between the successive hk-planes in a two dimensional square lattice is $a/\sqrt{h^2 + k^2}$, where a is the unit distance along x- and y-axes.

(Calcutta BSc(H), 2014)

8.10 Show that in an orthorhombic unit cell, the separation of the hkl planes will be reduced by a factor of n if all three Miller indices are multiplied by that factor.

(Calcutta BSc(H), 2013)

8.11 a. The distance of two successive hkl planes in a cubic crystal is $a/\sqrt{3}$, where a is the length of the unit cell. Find the values of h, k, l.   (Calcutta BSc(H), 2007)

b. The distance between two successive crystal planes of a cubic lattice cannot be $a/\sqrt{7}$. Explain.

(Burdwan BSc(H), 2011)

8.12 What is the plane of closest packing in the (i) body-centredcubic and (ii) face-centred cubic structures?

(Calcutta BSc(H), 1991)

8.13 How is it that while x-rays can be used to obtain reflection patterns of crystal, neither UV nor γ-rays can be so used? For a given crystal, can you employ x-radiation of any wavelength? Explain.

(Burdwan BSc(H), 1991)

8.14 What is the minimum measurable value of spacing between the crystal planes when the wavelength 1.67Å is employed?

(Burdwan BSc(H), 2011)

8.15 The wavelengths of the characteristic K_α lines of Cr, Fe and Ni are 2.29, 1.94 and

1.66 Å respectively. Determine the suitability of these wavelengths for x-ray diffraction study of a crystal with an estimated interplanar spacing of 100 pm.

(Calcutta BSc(H), 2013)

8.16 nth order Bragg reflection from (hkl) planes of an orthorhombic cell (following Bragg equation $n\lambda = 2d_{hkl}\sin\theta$) can be thought of as first order reflection arising from its (nh nk nl) planes. Justify. (Jadavpur BSc(H), 1996)

8.17 Why is there a variation in intensity of the diffracted beam for different set of planes? (Burdwan BSc(H), 1993)

8.18 A crystal having simple cubic lattice has length a_0 pm of its unit cell. One of its planes show a first order Bragg reflection at an angle of 60°. Taking wavelength of x-ray as a_0 pm, find the Miller indices of the plane. (Burdwan BSc(H), 2005)

8.19 For identical experimental conditions, the first order Bragg reflection from a plane of a cubic crystal comes up at 5.9° and 5.85° respectively at 20°C and 50°C. Calculate the coefficient of cubic expansion of the solid. (Calcutta BSc(H), 2005)

8.20 A body-centred cubic element of density 10.3 g cm⁻³ has a cell edge of 314 pm. Calculate the atomic mass of element. (Madras BSc, 2007)

8.21 Sodium crystallises in body centred cubic structure with $a = 4.24$Å. Calculate (i) the theoretical density of Na, (ii) the radius of Na atom. (Burdwan BSc(H), 1994)

8.22 From the following data, determine the type of cubic lattice to which the system belongs

edge length	286 pm
density	7.86 g cm⁻³
molar mass	55.85 g mol⁻¹

(Calcutta BSc(H), 2014)

8.23 The density of lithium metal is 0.53 g cm⁻³ and the separation of 100 planes of metal is 350 pm. Determine whether the lattice is *fcc* or *bcc* (molar mass of lithium = 6.941 g mol⁻¹) (Baroda BSc, 2009)

8.24 At room temperature sodium has body-centred cubic lattice with a cell edge length 4.29Å. At 73 K, the density is only 4% larger but the cell edge is now 5.35Å. What type of cubic unit cell does sodium possess at 73 K? (Consider the nearest whole number). (Calcutta BSc(H), 2008)

8.25 The element polonium (at weight = 210) crystallises in the cubic system. Bragg first order reflections using x-rays of wavelength 0.154 nm occur at sin θ values of 0.225, 0.316 and 0.388 for reflections from (100), (110) and (111) type planes

i. Show whether the unit cell is simple, face-centred or body-centred.

ii. Calculate the value of a, the side of the unit cell.

iii. Calculate the density of polonium. (Calcutta BSc(H), 2007)

8.26 KCl has an *fcc* lattice. But from x-ray diffraction experiment, it appears to be simple cubic. Explain. (Calcutta BSc(H), 2013)

8.27 The molar volume of KCl is known to be 1.3 times that of NaCl. If the glancing angle for first order Bragg reflection from the 100 plane of NaCl is 5.9°, calculate the same for KCl. (Burdwan BSc(H), 2013)

8.28 Solid A has a face-centred cubic lattice with the length of the unit cube $a = 3.65$Å. Another solid B has a body-centred cubic lattice with $a = 2.90$Å. Calculate the ratio of the densities of the two solids. (Calcutta BSc(H), 1999)

8.29 Sodium chloride crystallises in face-centred cubic structure. Its density is 2.165 g cm⁻³. If the distance between Na⁺ and its nearest Cl⁻ is 281 pm, find out the Avogedro's number (Na = 23, Cl = 35.44). (Delhi BSc, 2010)

8.30 In a power diffraction photography, an element with simple cubic crystal structure gave a line corresponding to (111) planes at a distance of 21.0 mm from the central spot. The wavelength of the x-ray used is 71.0 pm and the camera radius is 57.3 mm. Calculate the molar volume of the element. (Jadavpur BSc(H), 1996)

8.31 The power diffraction pattern of a cubic crystal gave three lines at distances 12.05 mm, 27.16 mm and 34.55 mm respectively from the central spot when 68.25 pm w-x-rays were used in a camera of radius 57.3 mm. If the line at 12.05 mm corresponds to the (100) planes, index the planes for other two lines and determine the edge length of the unit cell. (Jadavpur BSc(H), 1999)

KEY TO UNIVERSITY QUESTIONS

8.1 Vide Sect. 8.1.

8.2 Vide Sect. 8.2.

8.3 Vide Sect. 8.3.

8.4 Vide Sect. 8.3.

8.5 Vide prob. 8.12.

8.6 Vide prob. 8.47.

8.7 Vide Sect. 8.4.

8.8 a. Let the plane has intercept p on the x-axis and parallel to the y and z axes. Then

Weiss indices will be p, ∞ and ∞

Miller indices will be 1 0 0

b. Vide problem 8.18.

8.9 A hk-plane AB nearest to the origin O will have intercept a/h on x-axis and $b/k = a/k$ on y-axis.

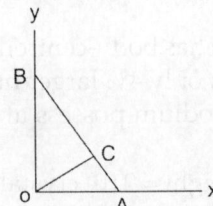

Now
$$\frac{1}{(a/h)^2} + \frac{1}{(a/k)^2} = \frac{1}{OA^2} + \frac{1}{OB^2}$$

$$= \frac{\sin^2 \angle OAC}{OC^2} + \frac{\sin^2 \angle OBC}{OC^2}$$

$$= \frac{1}{OC^2} (\sin^2 \angle OAC + \cos^2 \angle OAC)$$

$$= \frac{1}{OC^2}$$

Then the distance of separation of the successive planes

$$OC = \frac{1}{\sqrt{(h/a)^2 + (k/a)^2}}$$

$$= \frac{a}{\sqrt{h^2 + k^2}}$$

Note: One of the *hk*-planes passes through the origin so that the interplanar distance becomes equal to the length of the perpendicular drawn from the origin to the plane nearest to it.

8.10 For the orthorhombic system, interplanar distances of *hkl* planes and *nh nk nl* planes are given by

$$\frac{1}{d_{hkl}^2} = \frac{h^2}{a^2} + \frac{k^2}{b^2} + \frac{l^2}{c^2}$$

$$\frac{1}{d_{nhnknl}^2} = \frac{(nh)^2}{a^2} + \frac{(nk)^2}{b^2} + \frac{(nl)^2}{c^2}$$

Then on dividing one equation by the other

$$d_{nh\,nk\,nl} = \frac{1}{n} \cdot d_{hkl}$$

8.11 a. For a cubic system

$$d_{nkl} = \frac{a}{\sqrt{h^2 + k^2 + l^2}} \qquad (I)$$

$$= \frac{a}{\sqrt{3}}, \text{ given}$$

Then the relevant *h*, *k* and *l* are 1, 1 and 1 because the Miller indices can only be integer or zero.

b. It appears from Eq. (I) that the interplanar distance d_{hkl} cannot be $a/\sqrt{7}$, because 7 cannot be expressed as a sum of square of three numbers, integer or zero.

8.12 The planes of closest packing are those containing all the atoms (i.e. having no interleaved planes) with highest interplanar distance (to make the bulk density uniform throughout the crystal). Such planes will have lowest value of $h^2 + k^2 + l^2$ for cubic crystals. Then the relevant planes are 110 for (i) *bcc* and 111 for (ii) *fcc*.

Note: 100 planes having lowest value of $h^2 + k^2 + l^2$ are not closest packing for both *bcc* and *fcc*, because here 100 planes do not contain all the atoms. 100 planes are closest packing for *sc* crystals.

8.13 Vide problem 8.23.

☐ No. only those x-rays can be used whose wavelengths are comparable with the interatomic spacings of the crystal concerned. Because only then diffraction can occur effectively in compliance with the Bragg equation.

8.14 For a particular wavelength (λ) of radiation used, the minimum spacing (d_{min}) between the crystal planes is

$$d_{min} = \frac{\lambda}{2} \quad \text{by Bragg Eq. (8.3)}$$

$$= \frac{1.67\,\text{Å}}{2} = 0.835\,\text{Å}$$

8.15 For a given interplanar distance (*d*), the maximum wavelength (λ_{max}) that can be used is

$$\lambda_{max} = 2d$$
$$= 2\,(100\,\text{pm}) = 200\,\text{pm}$$
$$= 2\,\text{Å}$$

Then 2.29Å, which exceeds λ_{max}, is not suitable. The other two wavelengths, 1.94Å and 1.66Å, which are less than λ_{max}, may be used. However, 1.66Å, which is nearer to 100 pm (= 1Å) is more suitable.

8.16 Because, both correspond to the same diffraction angle and hence to the same line in the diffraction spectrum. This follows from the Bragg equation.

8.17 This is due to the difrrent reflecting power for different set of planes due to their difference in composition and reticular density.

8.18 For *hkl* planes containing all the atoms in a cubic crystal

$$\sqrt{h^2 + k^2 + l^2} = \frac{2a \sin \theta_{hkl}}{n\lambda} \quad \text{by Eqs. (8.2b) and (8.3)}$$

$$= \frac{2a_0 \sin 60°}{(1) a_0}$$

$$= \sqrt{3}$$

Then the relevant Miller indices are $h = 1$, $k = 1$ and $l = 1$.

8.19 Vide problem 8.46.

8.20 Atomic mass $\quad M = \dfrac{N_A a^3 \rho}{n} \quad$ by Eq. (8.5b)

$$= \frac{(6.02 \times 10^{23} \text{ mol}^{-1})(314 \times 10^{-10} \text{ cm})^3 (10.3 \text{ g cm}^{-3})}{(2)}$$

$$= 95.9 \text{ g mol}^{-1}$$

8.21 Vide problem 8.49(b).

8.22 Vide problem 8.58.

8.23 Assuming the lattice to be of *bcc* type, the density would be

$$e = \frac{nM}{N_A a^3}$$

$$= \frac{(2)(6.941 \text{ g mol}^{-1})}{(6.022 \times 10^{23} \text{ mol}^{-1})(350 \times 10^{-10} \text{ cm})^3}$$

since the separation of 100 planes is equal to a

$$= 0.537 \text{ g cm}^{-3}$$

This tallies with the actual density. Hence the lattice is of *bcc* type as assumed.

8.24 Although the lowering of temperature is quite high, the cell edge length rises notably by a factor 5.35/4.29, i.e. 1.25. This is possibly due to shifting to a *fcc* like form. Now, if there were no change of temperature

$$\frac{a_{fcc}}{a_{bcc}} = \frac{4r/\sqrt{2}}{4r/\sqrt{3}} \quad \text{[vide Sec. 8.8]}$$

$$= 1.22$$

This ratio is close to the actual one. Therefore, our conjecture is true. But rather low change of density suggests that the transformed structure is not exactly *fcc*.

8.25 Vide problem 8.49.

i. Here $d_{100}{:}d_{110}{:}d_{111} \approx 1 : \dfrac{1}{\sqrt{2}} : \dfrac{1}{\sqrt{3}}$

which corresponds to *sc* type

ii.
$$a = \frac{n\lambda\sqrt{h^2 + k^2 + l^2}}{2\sin\theta_{hkl}}$$

$$= \frac{(1)\,(0.154\text{ nm})\sqrt{1^2 + 0^2 + 0^2}}{2(0.225)} = 0.342\text{ nm}$$

iii.
$$e = \frac{(1)\,(210 \times 10^{-3}\text{ kg mol}^{-1})}{(6.02 \times 10^{23}\text{ mol}^{-1})\,(0.342 \times 10^{-9}\text{ m})^3}$$

$$= 7.455 \times 10^4\text{ kg m}^{-3}$$

8.26 Vide Sect. 8.6.

8.27 Vide problem 8.59.

8.28
$$\frac{\rho_A}{\rho_B} = \left(\frac{n_A}{n_B}\right)\left(\frac{a_B}{a_A}\right)^3\left(\frac{M_A}{M_B}\right) \quad \text{by Eq. (8.5b)}$$

$$= \left(\frac{4}{2}\right)\left(\frac{2.90}{3.65}\right)^3 \cdot \frac{M_A}{M_B}$$

8.29 In NaCl crystal Na^+ and Cl^- ions touch along the edge of the unit cell such that unit cell length, $a = 2 \times$ distance between a Na^+ ion and nearest Cl^- ion

$$\text{Avogedro number} = \frac{nM}{ea^3}$$

$$= \frac{(4)\,(58.44\text{ g mol}^{-1})}{(2.165\text{ g cm}^{-3})\,(2 \times 281 \times 10^{-10}\text{ cm})^3}$$

$$= 6.005 \times 10^{23}\text{ mol}^{-1}$$

8.30 For a spectral line, glancing angle

$$\theta = \frac{\text{Distance of the line from the central spot}}{2 \times \text{radius of the camera}} \times \frac{360°}{2\pi}$$

$$= \frac{21.0\text{ mm}}{57.3\text{ mm}} \times \frac{360°}{2\pi} = 21°$$

$$a = \frac{\lambda}{2\sin\theta_{hkl}}\sqrt{h^2 + k^2 + l^2} \quad \text{by Eq. (8.4)}$$

$$= \frac{(71.0\text{ pm})}{2\sin 21°}\sqrt{1^2 + 1^2 + 1^2} = 171.6\text{ pm}$$

$$\text{Molar volume} = \frac{N_A a^3}{n} \quad \text{by Eq. (8.5a)}$$

$$= \frac{(6.022 \times 10^{23}\text{ mol}^{-1})\,(171.6\text{ pm})^3}{(1)}$$

$$= 3.042 \times 10^{18}\text{ pm}^3\text{ mol}^{-1}$$

8.31 Vide problem 8.73.

9

Phase Equilibrium and Phase Rule

Physical changes are often accompanied by changes in state of aggregation or phase changes, though ordinarily they are not always visible. Such changes and relevant equilibria can be predicted by a simple mathematical equation called Gibbs' phase rule.

9.1 PHASE TRANSITION

A part or whole of a system is called a *phase*, if it is spatially uniform in its chemical composition and physical state. A system is said to be homogeneous or heterogeneous according as it comprises one or more phases.

Under a particular condition, a phase of a system is called *stable* if it does not undergo transition to another phase, otherwise it is called *unstable* or metastable according as the rate of transition is appreciable or not. During any phase transition of a system, when it occurs reversibly, T, P and G of the system do not change but the derivatives of G (wrt T or P) change. As originally proposed by Paul Ehrenfest, a phase change is said to be of nth order, if the nth derivative of G is discontinuous.

A phase transition for which the first-order derivative of G (wrt T or P) is discontinuous at the transition point is of first-order category. Examples are fusion, vaporisation and sublimation. Here $S = -\left(\frac{\partial G}{\partial T}\right)_P$ and $V = \left(\frac{\partial G}{\partial T}\right)_T$ (and hence U, H and A) undergo finite change at the transition point. But although $C_P = T\left(\frac{\partial S}{\partial T}\right)_P$ of either phase alone is finite, it becomes infinite when both the phases are present. In most cases, the high-temperature phase has lower C_P than the low-temperature phase.

There are phase transitions for which the first-order derivative of G is continuous but the second-order derivative is discontinuous at the transition point. Here S and V (hence U, H and A) do not change during the transition. But C_P undergoes a finite change during the transition. Such phase transitions are called second-order transitions. If, however, C_P does not change during the phase transition but attains an infinite value at the transition point, the relevant phase changes are called λ (*lambda*) *transitions*. Liquid helium offers example of both the occasions. The transition between liquid ^{3}He A (or ^{3}He B) and liquid ^{3}He N belongs to the second-order category while the transition between liquid ^{4}He I and liquid ^{4}He II belongs to λ-transition. Such transitions relate to some quantum effect (i.e. a phenomenon that can be explained only by quantum mechanics). The variation of G, S and C_P with temperature for three categories of phase transitions are indicated in Fig. 9.1.

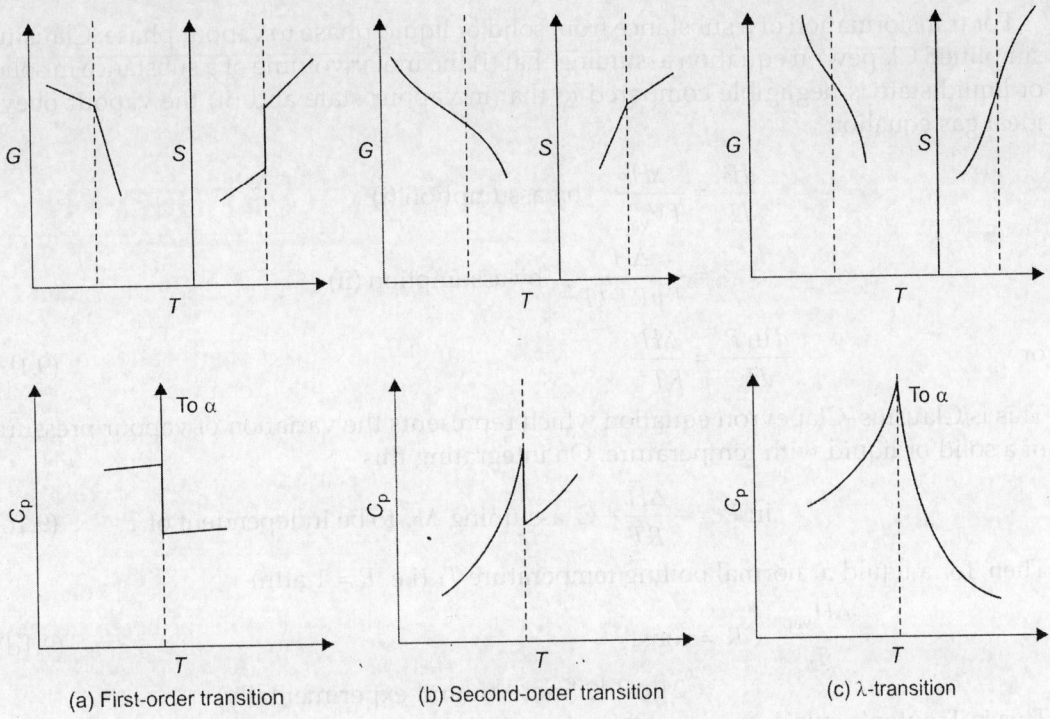

Fig. 9.1: The broken lines refer to transition points

In all three types of phase transitions G decreases continuously with rise of T at constant P, since $dG = VdP - SdT$. In case of first order transition $S [= -(\partial G/\partial T)_P]$ increases with T with a consequent gap between the values of S at the transition point while in case of second-order or λ-transition, S varies continuously with no gap in S at the transition point.

9.2 CLAPEYRON EQUATION: EFFECT OF PRESSURE ON TEMPERATURE OF A TWO-PHASE EQUILIBRIUM

If two phases of a substance exist in equilibrium at temperature T and pressure P, then the following relation will hold in case of first-order phase transition

$$\frac{dP}{dT} = \frac{\Delta H}{T\Delta V} \tag{9.1a}$$

where ΔH is the enthalpy change associated with the volume change ΔV resulting from the phase change at constant temperature and pressure. This is *Clapeyron equation*.

This follows from the Maxwell's relation $\left(\frac{\partial P}{\partial T}\right)_V = \left(\frac{\partial S}{\partial V}\right)_T$. Because in case of equilibrium between two phases of a substance $\left(\frac{\partial P}{\partial T}\right)_V = \frac{\partial P}{\partial T}$ (T being independent of V) and also $dS = \frac{d\theta_{rev}}{T} = \frac{dH}{T}$, where

$$\left(\frac{\partial S}{\partial V}\right)_T = \frac{1}{T}\left(\frac{\partial H}{\partial V}\right)_T$$

$$= \frac{1}{T}\frac{\Delta H}{\Delta V}, \quad \text{since } dH \propto dV$$

For transformation of a substance from solid or liquid phase to vapour phase, Clausius simplified Clapeyron equation assuming that (i) the molar volume of a substance in solid or liquid state is negligible compared to that in vapour state and (ii) the vapour obeys ideal gas equation

$$\frac{dP}{dT} = \frac{\Delta H}{TV_{vap}}, \quad \text{by assumption (i)}$$

$$= \frac{\Delta H}{TnRT/P}, \quad \text{by assumption (ii)}$$

or

$$\frac{d\ln P}{dT} = \frac{\Delta \bar{H}}{RT^2} \tag{9.1b}$$

This is Clausius–Clapeyron equation which represents the variation of vapour pressure of a solid or liquid with temperature. On integrating this

$$\ln P = -\frac{\Delta \bar{H}}{RT} + C \quad \text{assuming } \Delta \bar{H} \text{ to be independent of } T \tag{9.1c}$$

Then, for a liquid at normal boiling temperature T_b (i.e. $P = 1$ atm)

$$\frac{\Delta \bar{H}_{vap}}{T_b} \, CR = \text{const} \tag{9.1d}$$

$$\simeq 21 \text{ cal K}^{-1}\text{mol}^{-1} \text{ (found experimentally)}$$

This is Trouton's rule.

9.3 PHASE RULE

The equilibrium between different phases of a system is determined only by its intensive properties—temperature, pressure, composition (expressed in mole fraction) of each phase of the system. The minimum number of intensive variables that must be specified to describe the phase equilibrium of a system is called its number of degrees of freedom (F). This is the maximum number of intensive variables that can be independently varied without changing the number of phases of a system in equilibrium and is given according to phase rule by the following equation

$$F = C - P + 2 \tag{9.2}$$

where P is the number of (kinds of) phases in equilibrium and C is the number of components of the system.

A system can have only one gaseous phase (because the gases are mutually miscible in all proportions) but more than one liquid phase. Regarding solid phase, all different kinds of solids form different phases. Thus the system, $NH_4Cl(s) \rightleftharpoons NH_3(g) + HCl(g)$ consists of two phases (one gas and one solid), while the system $CaCO_3(s) \rightleftharpoons CaO(s) + CO_2(g)$ consists of three phases (one gas and two solids).

When a system is composed of more than one chemical species, the composition of different phases of he system is often different. The minimum number of chemical species required to describe the composition of each phase of a system in equilibrium is called the number of components (C) of the system. This is equal to the total number of constituents (N) minus the number of independent chemical equilibrium expressions (R) minus the number of independent additional mole fraction relations R' due to material balance (from stoichiometric equation) or charge balance, i.e.

$$C = N - R - R' \tag{9.3}$$

Then number of components ≤ number of constituents.

The equality sign holds in case of simple system involving no chemical interaction among the constituents.

Derivation of Phase Rule

Let us consider an equilibriated system containing C number of cemically inert constituents. Let there are P number of phases each of which is in contact with every other phases at same temperature and pressure. The state of each phase of the system can obviously be described by its temperature, pressure and mole fraction of each of its C constituents. But it is more convenient to use chemical potential (μ) instead of mole fraction (one being a function of the other), because then the description of each phase will amount to the description of the entire system and can be done with lower number of variables not exceeding $C + 2$, since each of these variables has the same value for all the phases. However, P number of these variables are fixed because in each phase there exists a relation $\sum x_i d\mu_i = 0$, at constant temperature and pressure [vide Eq. (5.30b)]. Then the number of degrees of freedom (F), which is equal to the total number of independent variables, of the system will be

$$F = C - P + 2$$

In case, there is chemical interaction among the constituents, let R be the independent chemical equilibrium relations. This will fix up R additional μ variables. Again μ variables are also fixed up by additional number of relations R' due to material balance or charge balance. Then from all these considerations

$$F = C - R - R' - P + 2$$
$$= C' - P + 2$$

where $C' = C - R - R'$ is defined to be the number of components so that the phase rule equation assumes, the same form whether the constituents undergo chemical reaction or not.

9.4 PHASE DIAGRAMS OF SIMPLE SYSTEMS

A phase diagram is one that gives the conditions of equilibrium between various phases of a system. Every point on such diagram represents state of the system in agreement with phase rule.

9.4.1 One-Component System

P vs T phase diagrams of some one-component systems (water, carbon dioxide and sulfur) are shown in Fig. 9.2. The lines on the phase diagram divide it into several regions (three in case of H_2O and CO_2 but four in case of S) each of which represents a single phase as indicated in the diagram. Each line (where slope is governed by Clapeyron equation) represents the equilibrium between a pair of phases. The point of intersection between any two lines represents the equilibrium between three phases and is called a *triple point*.

The phase diagrams of H_2O and CO_2 are similar except that solid-liquid line is inclined to the left in case of H_2O (ΔV_{fus} being −ve) while it is inclined to the right in case of CO_2 (ΔV_{fus} being +ve). Triple point pressure of H_2O (4.58 mm) is below 1 atm while it is much above 1 atm in case of CO_2.

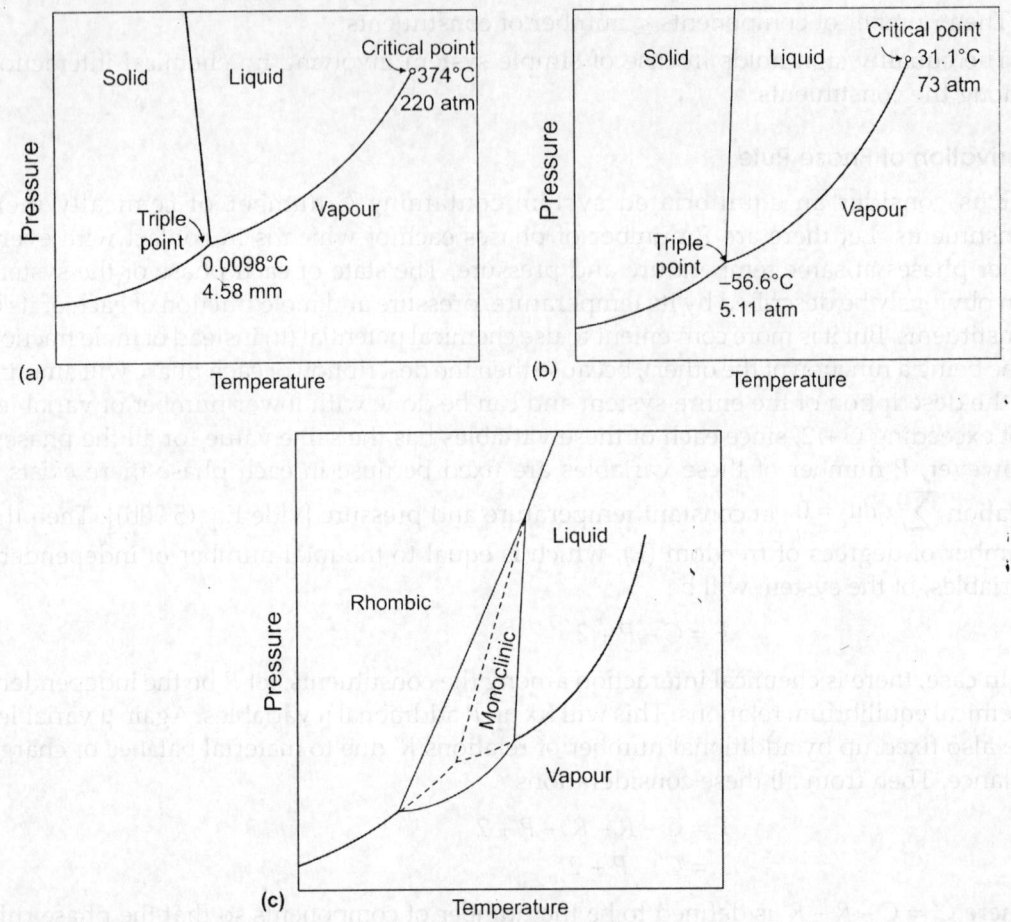

Fig. 9.2: Schematic phase diagrams of (a) H_2O (b) CO_2 and (c) S relating to common crystalline form(s)

Unlike H_2O and CO_2, sulfur can exist in two solid forms (rhombic and monoclinic) under ordinary condition. Accordingly, four triple points are possible with sulfur, though one in the monoclinic region is metastable.

9.4.2 Two-Component Liquid–Vapour System

A. Completely Miscible Liquid Pair

The vapour pressure (P) of a mixture of two liquids A and B obeying Raoult's law is given by

$$P = p_A^* + (p_B^* - p_A^*) x_B \qquad (9.4a)$$

$$= \frac{p_A^* p_B^*}{p_B^* + (p_A^* - p_B^*) y_B}, \quad \text{putting } x_B = y_B \cdot \frac{P}{p_B^*} \qquad (9.4b)$$

where x_B represents mol fraction of B in the liquid phase and y_B that in the vapour phase. The vapour pr vs composition diagram is shown in Fig. 9.3. Here the upper curve corresponds to the first equation and lower curve to the second equation. Above upper line only liquid phase exists, while below the lower line, only vapour phase exists. Any point a lying between the two lines represents the coexistence of liquid and vapour phase with their composition indicated by l and v respectively. The amount of liquid and vapour phases will be, according to the *lever rule*, in the ratio $av{:}al$, i.e. $(y_B - z_B){:}(z_B - x_B)$, where z_B is the overall composition of the system (i.e. if it were homogeneous) corresponding to the point a. Vide note on university question 9.36.

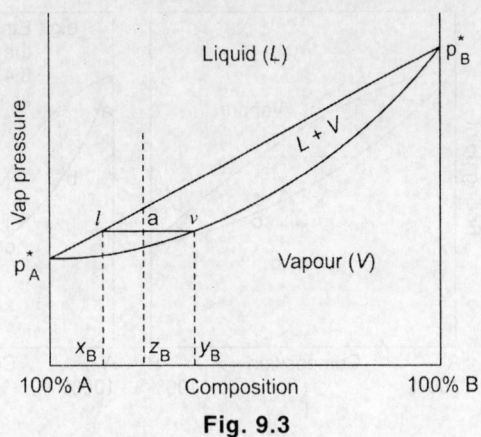

Fig. 9.3

However, for real liquid systems, due to their deviation from Raoult's law, the phase diagrams are usually of the following three types shown in Fig. 9.4.

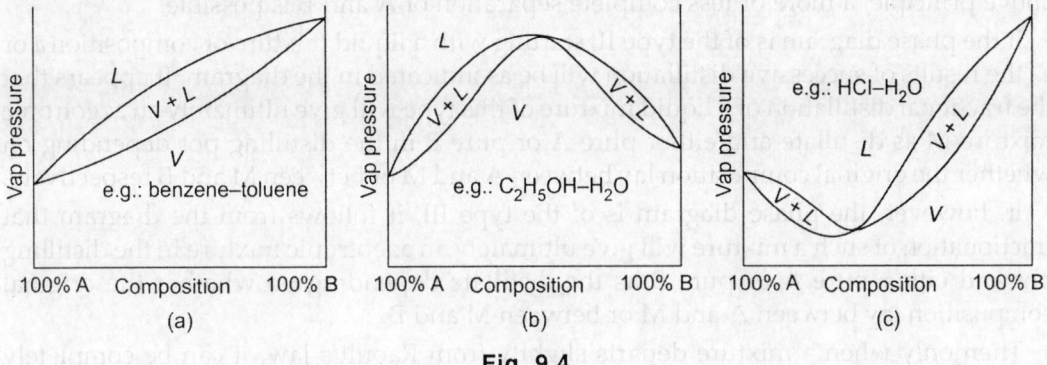

Fig. 9.4

Figure 9.4(a) corresponds to slight deviation from Raoult's law, 9.4(b) to profound +ve deviation and 9.4(c) to profound −ve deviation.

However, in all cases the composition of the liquid and vapour phases is usually different. The vapour phase is relatively richer in that component (not necessarily the more volatile one) whose addition to the system results in an increase in total vapour pr (or decrease in boiling point) of the system (Konowaloff's rule). Therefore, if we condense the vapour given off by a liquid mixture, the condensate will be richer in one of the components than the residual liquid mixture. The enrichment can be done to the desired extent by repeating the process over the successive condensates, unless the composition of the liquid and vapour phase in equilibrium is the same (as in case of mixture called azeotrope). This constitutes the principle of fractional distillation, which is usually done isobarically by heating instead of doing this isothermally by reduction of pressure.

The theory of fractional distillation can be readily understood considering the boiling point—composition diagram I, II and III in Fig. 9.5 corresponding to (a), (b) and (c) respectively (such diagrams resemble the vapour pressure composition diagrams turned upside down). Suppose, we distill a binary mixture having phase diagram I. Starting with a composition 'a', the first portion of the distillate would be of composition 'b'. Now if this is submitted to redistillation, the composition of the first portion of the new distillate will be 'c'. Thus by repeating the process a sufficient number of times, we can

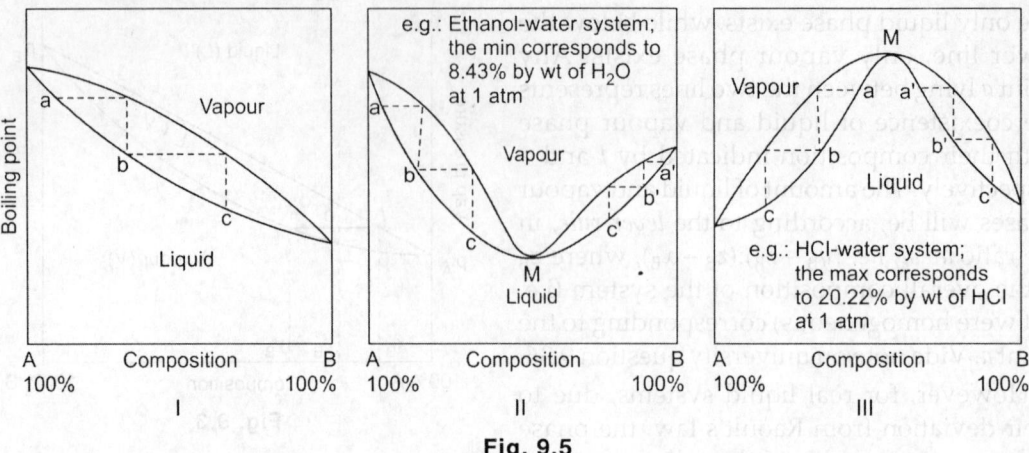

Fig. 9.5

come as close to pure B as we desire. Therefore, by fractional distillation, based on the above principle, a more or less complete separation of A and B is possible.

If the phase diagram is of the type II, starting with a liquid mixture of composition *a* or *a'*, the results of successive distillation will be as indicated in the diagram. It appears that the fractional distillation of a liquid mixture of this type will give ultimately an azeotropic mixture M as distillate and either pure A or pure B in the distilling pot depending on whether the original composition lay between A and M or between M and B respectively.

If, however, the phase diagram is of the type III, it follows from the diagram that fractionation of such a mixture will give ultimately an azeotropic mixture in the distilling pot and either pure A or pure B as the distillate depending on whether the original composition lay between A and M or between M and B.

Then only when a mixture departs slightly from Raoult's law, it can be completely separated (in principle) into their pure constituents by fractional distillation. Otherwise, it can be separated into only one pure constituent and an azotrophic mixture.

B. Immiscible Liquid Pair

Here, the total vapour pressure,

$$P = p_A^* + p_B^*$$

assuming A and B to obey ideal gas laws and ignoring the effect of pressure on the vapour pressure.

Since $P > p_A^*$ or p_B^*, the boiling point of the system will be lower than that of either of the pure components. The number of moles (n) and weights (w) of the two components in the vapour phase will be given by

$$w_A/w_B = n_A M_A/n_B M_B$$
$$= p_A^* M_A/p_B^* M_B \quad \text{by ideal gas equation} \tag{9.5}$$

Then the distillation of a pair of immiscible liquid will give a distillate in which the relative amounts of the components differ from that in the initial. This constitutes the principle of steam distillation whereby a liquid, virtually immiscible with water and having a relatively high boiling point, can be effectively steam distilled at a much lower temperature (usually by passing steam through it), if it has sufficient vapour pressure near the boiling point of water.

C. Partially Miscible Liquid Pair

Such systems are generally studied not at the vapour pr of the system (i.e. not under closed condition), but at constant pressure of atmosphere by exposing the system to the atmosphere. An example is phenol–water system. If a little phenol is added to water at ordinary temperature, then it will dissolve completely, but if the addition is continued, a point is reached when further dissolution will not occur and the two liquid layers called *conjugate solutions* are formed, of which one is saturated with phenol and the other with water. The phase diagram for this system

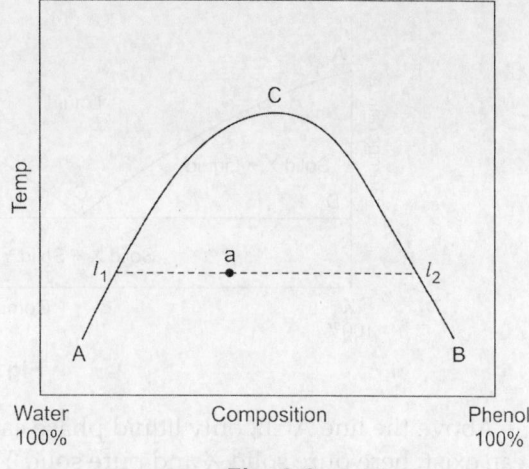

Fig. 9.6

exhibits a maximum at C (Fig. 9.6). Here AC represents the solubility curve of phenol in water, while BC that of water in phenol. With rise in temperature, the solubility of each component in the other increases and the two layers approach each other in composition till they become identical at C. The temperature at which the conjugate layers of a partially immiscible liquid pair have identical composition is called the *critical solution temperature* or the *consolute temperature*. Any point outside the curve ACB represents a homogeneous system while any point 'a' within the curve represents a heterogeneous system consisting of two liquid layers l_1 and l_2, their relative amount being given by the lever rule

$$\frac{\text{amount of } l_1}{\text{amount of } l_2} = \frac{al_2}{al_1}$$

The consolute temperature can occur not only at the maximum of the solubility curve but also at the minimum of such curve (e.g. in triethylamine–water system) and they are called *upper consolute temperature* and *lower consolute temperature* in the respective cases. Lower consolute temperature arises when the mutual solubility rises with lowering of temperature.

9.4.3 Two-Component Solid–Liquid System

Here we consider only the simplest case where the two components are completely miscible in the liquid state but completely immiscible in solid state. Such systems (e.g. Pb–Sb, KI–H_2O) are usually studied at fixed pressure by exposing them to the atmosphere.

If a liquid mixture of the two substances is cooled, a solid phase will appear at a definite temperature, called the *freezing point* or *saturation point* depending on whether the separated solid is solvent or solute. The phase diagram obtained on plotting the freezing point versus composition of a series of liquid mixture of two substancex X and Y is shown in Fig. 9.7. On cooling, the composition of the liquid mixture varies along AC when X freezes out. BC represents the situation when Y freezes out. The two curves meet at the point C, called the *eutectic point* where the liquid phase is in equilibrium with both solid solid X and Y. Obviously C represents the lowest freezing or melting point of the system.

[Slopes of the lines AC and BC are determined, in case of ideal mixture, by the equation

$$\ln x_1 = \frac{\Delta \bar{H}_{fus}}{R}\left(\frac{1}{T^\circ} - \frac{1}{T}\right)$$ related to depression of freezing point in Chapter 10]

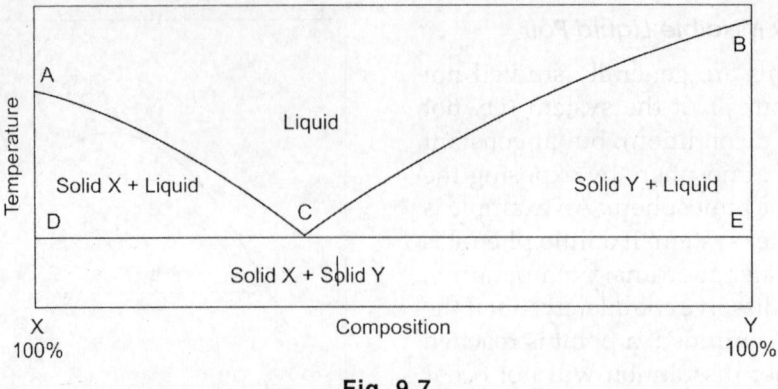

Fig. 9.7

Above the line ACB, only liquid phase can exist. Below the line DCE, no liquid phase can exist; here pure solid X and pure solid Y exist. Any point within ADC represents the coexistence of liquid mixture and solid X while any point within BEC represents the coexistence of liquid mixture and solid Y.

An eutectic mixture freezes at a fixed temperature (at a fixed pressure) without change of composition while other mixtures freeze over a range of temperature. Because of the sharp melting point, an eutectic mixture was formerly misrepresented as a chemical compound, called *cryohydrate* in case of aqueous system.

Note that at eutectic point, a liquid phase is in physical equilibrium with two solid phases. It differs from a *peritectic point* where a solid phase is in chemical equilibrium with a liquid phase and another solid phase (that happens, when a compound melts with decomposition). Both eutectic and peritectic points are tripple points, but for a two-component system (unlike one-component system), they are invariant only at constant pressure.

Freezing Mixtures

The production of freezing miture with common salt and ice can be readily understood from the phase rule. When the salt is added to ice at $0°C$, a part of it dissolves in the water accompanying the ice. The three phases of the system, solid salt, ice and solution, can be in equilibrium only at eutectic temperature ($-21°C$), which is much lower than the melting point of ice ($0°C$). To achieve this, the system will lower its temperature through melting of ice and dissolution of salt (both being endothermic process). If ice and salt are present in sufficient amount, eutectic temperature will be attained. The system will remain at this temperature until entire ice melts due to absorption of heat from the surroundings or entire salt dissolves. Disappearance of one of the phases makes the system univariant (at constant pressure) and the temperature will then rise. It should be mentioned that the eutectic point is formed here by solid $NaCl \cdot 2H_2O$ (not $NaCl$), ice and concentrated salt solution (unsaturated) coexisting in equilibrium at constant pressure.

9.4.4 Three-Component Liquid–Liquid System

Here, we consider only the most interesting situation where a substance (solute) is distributed in an immiscible liquid pair, e.g. I_2 between CCl_4 and H_2O.

Such systems are usually studied at constant temperature and pressure by exposing them to the atmosphere. Here, the degrees of freedom $F = C - P + 2 - 1 - 1 = 3 - 2 + 2 - 1 - 1 = 1$. This implies that if the conc of salute in one layer (C_1) is fixed that in the other layer (C_2) will also be fixed, i.e. $C_2 = f(C_1)$. Experiment shows that in the simplest situation

$$C_2/C_1 = K, \text{ a constant independent of } C_1 \text{ and } C_2$$

This is Nernst's distribution law which holds only if (i) the two solvents are completely immiscible (ii) solutions in the two layers behave ideally (iii) the conc of solute in each layer is law and (iv) the solute exists in the same molecular form in the two layers. All these follow from the thermodynamic derivation of the law based on the equality of chemical potential of the solute in the two solutions. K is called the *partition coefficient of the solute*.

An important application of the distribution law is in the recovery of a substance from a large volume of its solution by extraction with some better solvent which is immiscible with the first solvent.

Suppose V_1 volume of a solution is shaken with V_2 volume of the extracting solvent. If w_0 be the mass of the dissolved substance initially present in the first solvent and w_1 is that after first extraction, then according to distribution law

$$\frac{(w_0 - w_1)/V_2}{w_1/V_1} = K$$

the partition coefficient of the solute between the second and first solvent

or $\qquad\qquad\qquad w_1 = w_0 \left(\dfrac{V_1}{V_1 + KV_2} \right)$ $\qquad\qquad\qquad$ (9.6a)

If the extraction is done once more with V_2' volume of the extracting solvent, then the mass w_2 of the solute remaining in the first solvent will be given by

$$w_2 = w_1 \left(\frac{V_1}{V_1 + KV_2'} \right)$$

$$= w_0 \left(\frac{V_1}{V_1 + KV_2} \right)\left(\frac{V_1}{V_1 + KV_2'} \right) \qquad\qquad (9.6b)$$

$$= w_0 \left(\frac{V_1}{V_1 + KV_2} \right)^2, \text{ if } V_2 = V_2'$$

In general, if w_n be the mass of solute remaining in the first solvent after n successive extractions using equal volume V_2 of extracting solvent in each extraction, then

$$w_n = w_0 \left(\frac{V_1}{V_1 + KV_2} \right)^n \qquad\qquad (9.6c)$$

AUXILIARY PROBLEMS

9.1 Solid–solid transitions unlike ordinary liquid–liquid transitions, occur very slowly under normal condition. Discuss this with diamond→graphite transition as an example.

9.2 Metastable phases are unlikely to be gaseous or liquid. Comment.

9.3 For a first-order phase transition, C_P of the system becomes infinity though C_P of either phase is finite. Explain.

9.4 In both first-order phase transition and λ-transition, $C_P = \infty$ at the transition point. How do you account for this?

9.5 In a λ-transition, can two phases be distinguished at the transition point?

9.6 The Clapeyron equation holds for any phase change. True or false?

9.7 In which of the following cases Clapeyron equation is not applicable?
(i) open system (ii) closed system (iii) ideal gas (iv) real gas (v) supercooled liquid.

9.8 Trouton's rule fails for associated liquids and also for liquids having very low and very high boiling points. Explain.

9.9 Regarding entropy of melting at melting point, there is no general rule analogous to Trouton's rule. Why?

9.10 True or false.
a. For any substance, molar entropy of fusion cannot exceed that of boiling.
b. Entropy of fusion must always be positive.

9.11 Phase rule equation $F = C - P + 2$ will hold only if the phase regions are quite large. Justify.

9.12 What will be the number of phases in the following cases:
a. Water in equilibrium with two pieces of ice
b. Water together with ice and its vapour at 0°C and 1 atm
c. A non-volatile substance enclosed in a vessel under ordinary condition

9.13 Ordinary H_2O (a compound) can exist in three different phases while S (an element) can exist in one more phase. But in both cases, the maximum number of phases (P_{max}) that can coexist in equilibrium is same. Find P_{max} with justification. Also find the lowest and highest degrees of freedom in the two cases.

9.14 Water contains a large number of chemical species H_2O, $(H_2O)_2$, ... $(H_2O)_n$, and H_3O^+ and OH^-, yet it is a one-component system. On the contrary, a system containing H_2O, H_2 and O_2 is a three-component one. Justify.

9.15 Find the number of components in the following systems:
a. Aqueous solution of urea
b. Aqueous solution of $NaHSO_4$ (assuming no dissociation, incomplete dissociation, complete dissociation and saturated condition)
c. Aqueous solution of NH_4Cl
d. Aqueous $AlCl_3$

9.16 An aqueous solution contains NaCl (which dissociates into Na^+ and Cl^-) and KBr (which dissociates into K^+ and Br^-). Find the number of components of this system. How would it be affected on addition of (i) H_2O (ii) NaCl (iii) KCl (which dissociates into K^+ and Cl^-) to the system.

9.17 For the gaseous reaction system, $aA + bB = cC + dD$. What will be the number of components if the system initially consists of:
a. Arbitrary amounts of all four ingredients
b. Arbitrary amounts of A and B only
c. a mol of A and b mol of B only
d. a mol of A and c mol of C only
e. a mol of A and arbitrary amount of B only

9.18 Find the degrees of freedom in the following cases:
a. A pure substance at its normal boiling point
b. A pure substance at the critical point
c. A binary azeotropic mixture

9.19 What makes you sure that an azeotrobe is not a compound?

9.20 Find the degrees of freedom of the system consisting of water and air under the following conditions:

    a. when the system, enclosed in a cylinder by a frictionless piston, is exposed to the atmosphere

    b. air in (a) is disregarded

    c. the system is exposed open to the atmosphere disregarding air

    Comment on the results.

9.21 Comment on the applicability of phase rule to the following systems. Evaluate (if possible) the degrees of freedom in each case.

    a. Supersaturated water vapour in a closed container

    b. Supercooled water in a closed container

    c. Supercooled water exposed open to the atmosphere

    d. A solution in osmotic equilibrium

    e. A system in presence of a field

    f. A dispersed system

9.22 Discuss the possibility of distinguishing between a stable and a metastable system on the basis of phase rule alone, with S as an example.

9.23 Find the number of components and degrees of freedom of the following systems:

    a. (i) A mixture of $NH_4Cl(s)$, $NH_3(g)$ and $HCl(g)$ prepared by heating $NH_4Cl(s)$ (ii) some NH_3 is added to the mixture in (i)

    b. (i) A system consisting of $CaCO_3(s)$, $CaO(s)$ and $CO_2(g)$ prepared by heating $CaCO_3$ as starting material (ii) some $CO_2(g)$ is added to (i) and (iii) some $CaO(s)$ is added to (i)

    c. (i) A system consisting of $Pb(NO_3)_2(s)$, $PbO(s)$, $NO_2(g)$ and $O_2(g)$ prepared by heating $Pb(NO_3)_2$ as starting material (ii) Some $PbO(s)$ is added to (i) and (iii) Some O_2 is added to (i)

    Which are the components in case of (b)?

9.24 Find the number of degrees of freedom of the following systems:

    a. Diamond and graphite at ordinary temperature and pressure

    b. $O_3 + O_2$

    c. $o–H_2 + p–H_2$

    d. Phenol + water

    e. Aqueous solution containing dextro- and laevo-rotatory lactic acid

    f. Aqueous solution containing racemic- and meso-tartaric acid

9.25 Consider the equilibrium $A \rightleftharpoons B + C$. Find the number of components and degrees of freedom when:

    a. only vapour phase exists

    b. vapour plus a liquid phase consisting of pure A exists

    c. vapour plus a liquid phase each comprising all the species exists

9.26 Account for the following.

    a. Snowfall does not occur in coastal areas

    b. On heating at 1 atm ice melts but solid CO_2 sublimes

    c. Melting point of rhombic sulfur depends on rate of heating. If heated rapidly, it melts at lower temperature

d. Crystals of rhombic sulfur are found in nature though liquid sulfur ordinarily solidifies to monoclinic form

e. With rise of pressure melting point of a substance may increase or decrease but boiling point always increases

9.27 Justify or criticise.

a. Melting and boiling points of a substance must always differ

b. Sublimation is possible with all substances

c. Triple point of any system is an invariant

d. Vapour phase in equilibrium with a binary solution will be richer in more volatile component

e. On boiling, a dilute aqueous solution of HCl becomes concentrated and a concentrated solution becomes dilute

f. A binary azeotropic mixture boils at fixed temperature and hence it will freeze also at fixed temperature, if pressure if fixed.

9.28 During boiling at constant pressure, the temperature of ethanol–water system changes till it becomes constant. But similar boiling of nitrobenzene–water system occurs all time at constant temperature. Explain. What is expected with phenol–water system? In which case the boiling point of the system will be lower, if phenol–water system is (i) completely immiscible or (ii) completely miscible?

9.29 Which of the following substances can be steam distilled?

(a) Acetone (b) Aniline (c) Mercury.

9.30 The partition coefficient of a substance between two immiscible solvents may be regarded asthe ratio of its solubilities in the two solvents. True or false.

9.31 Give example (if any) of a binary system that exhibits both upper and lower consolute temperature. What is the degrees of freedom of a binary system at its consolute temperature?

9.32 What is the chemical implication of the lower consolute temperature? Is there any binary system that exhibits no consolute temperature, upper or lower? Give examples (if any) with reasons.

9.33 Do you expect consolute temperature to increase or decrease when a salt is added to the system? Cite examples.

9.34 What is the essential difference between eutectic and peritectic points in a binary system? What will be the degrees of freedom in the two cases?

9.35 Is eutectism limited to binary systems? What about azeotropism?

9.36 Does the lowest temperature of an ordinary freezing mixture (NaCl + ice) depend on the relative amount of NaCl and ice present in the system? Here NaCl and ice are ordinarily taken in 2:7 weight ratio. Why?

9.37 Find (a) freezing point and (b) boiling point of water at 3 atm.

9.38 The vapour pressure of solid SO_2 (sp gr 2.93) is given by

$$\log P \,(\text{mmHg}) = -\frac{1871}{T/K} + 10.6$$

and that of liquid SO_2 (sp gr 1.43) is given by

$$\log P \,(\text{mmHg}) = -\frac{1426}{T/K} + 8.3$$

Find (a) molar heat of sublimation (b) molar heat of fusion (c) normal boiling point (d) tripple point (e) normal freezing point (f) Trouton constant.

9.39 At what pressure, diamond (sp gr 3.52) will be in equilibrium with graphite (sp gr 2.25). $\mu_{diamond}^{0}$ = 685 cal mol^{-1} at 25°C.

9.40 Normal boiling point of iodine is 457.5 K. The vapour pressure of iodine is 100 torr at 390 K and that of solid iodine is 1 torr at 312 K. Find triple point of iodine, if its enthalpy of fusion is 15.5 kJ mol^{-1}.

9.41 An ideal mixture of two liquids A and B having composition x_A is distilled to half of its content when the boiling point becomes T. At this temperature vapour pressure (P) of the condensate is 800 torr, p_A^* = 950 torr and p_B^* = 450 torr. Find (a) composition of the residual liquid mixture (x_A') (b) composition of the distillate (x_A'') (c) percent of A distilled off.

9.42 Steam distillation of an oil takes place at 99°C under 1 atm. The collected distillate is 9.5% by weight in organic compound. Calculate the molecular weight of the organic compound. The vapour pressure of water is 730 torr at 99°C.

9.43 A gaseous mixture containing 0.2 mol of H_2O and 0.8 mol of benzene has a total vapour pressure of 1492 torr at 100°C. The mixture is isochorically cooled to 80°C, the boiling point of pure benzene. Assuming benzene and water to be completely immiscible, predict whether any condensation will occur and if so, find the nature of the liquid formed and its amount. Also find the total vapour pressure of the system at 80°C. The vapour pressure of water is 355 torr at 80°C.

9.44 A gaseous mixture containing benzene and water in 1:1 mol ratio is slowly compressed isothermally at 80°C (the bp of benzene)

a. which component will start condensing first and at what pressure does this occur?

b. What should be the composition of the initial gaseous mixture so that both the components will start condensing simultaneously?

c. If the vapour of composition you determine is compressed at 100°C, will the condensation of each component begin simultaneously? The vapour pressure of water is 355 torr at 80°C.

9.45 A and B are substances which can form ideal liquid solution. A gaseous mixture of A and B in 1:4 mol ratio is slowly compressed isothermally at temperature T.

a. Find the pressure at which liquid phase will first appear and composition of this new phase.

b. What is the composition of the solution whose normal boiling point is T

Given p_A^* = 0.6 atm and p_B^* = 1.5 atm, at T.

9.46 Water and phenol are partially miscible at 323 K. When these two liquids are mixed at 323 K and 1 atm, then at equilibrium, one phase is 89% water by weight and the other is 37.5% water by weight. If 6.00 g of phenol and 4.00 g of water are mixed at 323 K and 1 atm, find the mass of water and the mass of phenol in each phase at equilibrium.

9.47 Apply phase rule to show whether a system comprising $CuSO_4 \cdot 5H_2O(s)$, $CuSO_4 \cdot 3H_2O(s)$, $CuSO_4 \cdot H_2O(s)$ and $H_2O(g)$ can be in equilibrium or not? If 0.0006 mol of $CuSO_4 \cdot 5H_2O$ is put into a closed vessel of volume 10 L and is heated to 50°C, what will be the composition of the solid at equilibrium? The vapour pressure of the penta-, tri- and mono-hydrates of copper sulfate are 47, 30 and 4.5 torr respectively at 50°C.

9.48 Succinic acid has to be extracted with ether from 200 cm³ of its aqueous solution containing 50 g of the acid. Calculate:

a. the extraction expected in a single operation with 100 cm³ of ether.

b. the maximum extraction expected in two-stage operation with a total volume of 100 cm³ of ether.

c. minimum number of extraction operations each with 100 cm³ of ether required for at least 80% extraction of the acid.

The partition coefficient of succinic acid between ether and water is 3.0. Comment on the results.

ANSWERS

9.1 Such contrast is quite likely in view of the fact that the mobility of molecules in liquids is freezed in solids due to much stronger intermolecular forces prevailing in solids. Thus thermodynamically favourable transition of diamond to graphite of lower free energy occurs too slowly to be observed.

9.2 Metastable phases are kinetically stable though they are thermodynamically unstable. Such phases are mostly solids. Because, ordinarily liquid molecules are quite mobile and gas molecules are more so. However, apart from molecular mobility, some other factors like surface tension, might have significant role in determining metastability of a phase. The relevant examples of metastable phases are supercooled liquids (ones below their freezing points), superheated liquids (ones above their boiling points) and supersaturated vapours. Thus, water can be rapidly cooled to as low as –40°C and rapidly heated to as high as 280°C without any phase change in absence of dust particles (which have catalytic effect).

Note: Generally, a solid cannot be superheated. On heating, a solid usually begins to melt below its melting point to give a thin liquid-like surface film whose thickness increases as the melting point is approached.

9.3 C_P of a two-phase system is obviously finite at any temperature other than transition temperature, C_P of each phase being finite. But at transition point for a first order phase transition, the absorption or evolution of heat by the system occurs without any change of temperature; this makes C_P of the system infinite.

9.4 For a first-order transition, the reason is given in the previous problem. For a λ-transition, the reason is different, dQ being zero. Here, on approaching the transition point, C_P of each phase (individually), and hence of the overall two-phase system approaches infinity.

9.5 No. Because here the macroscopic properties (including C_P) are all same for the two phases at the transition point (just as in case of critical point).

Note: This is in contrast with a first-order phase transition.

9.6 False. Because, for phase-transition of higher than first order, the Clapeyron equation $dP/dT = (1/T) (\Delta H/\Delta V)$ becomes meaningless, ΔH and ΔV being both zero.

9.7 Clapeyron equation, which deals only with first-order phase change, will not be applicable to

(i) if the concerned phases are not in equilibrium.

(iii) because an ideal gas cannot undergo any phase change.

(v) if liquid–solid phase change is concerned that does not occur reversibly for this system.

Note:

1. The Clapeyron equation applies to open systems only for specific phase changes, e.g. melting.

2. Even for supercooled liquids (metastable systems), the Clapeyron equation applies to liquid–vapour and solid–vapour equilibria.

9.8 For high boiling liquids, particularly the associated (i.e. hydrogen-bonded) ones, the intermolecular forces are high that causes the molecules to have less random arrangement. As a consequence, the entropy change (i.e. increase in randomness) due to vaporisation becomes significantly higher than that normally provided by the Trouton's rule. But this rule fails also for liquids having very low boiling point (though they have low intermolecular forces) due to much non-ideal behaviour of their vapour arising from higher molecular concentration (c) at very low temperature (that appears from $P = cRT$).

Note: The anomaly with low boiling liquids has been considerably managed by comparing $\Delta \bar{S}_{vap}$ of the liquids for their vapour at same concentration and not at same pressure of one atmosphere (corresponding to normal bp). This modification of Trouton's rule was originally due to Hildebrand who suggested a vapour concentration of $1/22.4$ mol/L for this purpose.

9.9 Because melting involves two condensed phases whose entropies depend considerably on the molecular shape. $\Delta \bar{S}_{fus}$ varies usually from 2 cal K^{-1} mol^{-1} for nearly spherical molecules to as high as 30 cal K^{-1} mol^{-1} for long molecules due to their change of regular orientation in solid to a great many different orientations in melt through coiling.

9.10 a. Melting involves much lower change in volume than boiling. But $\Delta \bar{S}_{fus}$ maynot be less than $\Delta \bar{S}_{vap}$. Because, in melting the entropy change due to small volume change may be accompanied by much greater contribution due to structural change through coiling as mentioned in the previous problem. Then, the given statement is not true.

   b. The general truth of the given statement is impaired by fusion of solid helium at 0 K, for which $\Delta \bar{S}_{fus} = 0$ (vide problem 21.70).

   Note: At $T = 0$ K, P has no effect on S which is determined only by T.

9.11 Phase rule equation is based on the supposition that the chemical potential (μ) of each chemical species in a phase is determined only by pressure, temperature and composition of the phase. Here, it has been implicitly assumed that the contribution of surface energy to μ is small enough to be ignored and this necessitates phase regions to be quite large (so that total interfacial area for a system of fixed mass becomes sufficiently low).

9.12 a. In phase rule concerning phase equilibria, the phrase 'number of phases' is used to mean 'number of kinds of phases'. Since the equilibrium of a system is not affected by subdivision of phases, the number of phases will be same whether water is in equilibrium with one or more pieces of ice, and this is 2, viz. 1 liquid and 1 solid.

   b. Only the phases in equilibrium have relevance to the phase rule. At 0°C and 1 atm, only water and ice can be in equilibrium. Hence the number of phases is 2.

Note:

1. On keeping, the vapour phase will disappear without affecting solid–liquid equilibrium.

2. Phase rule may be applicable to a part of the system, even when it does not apply to the whole system.

c. Even a non-volatile solid (e.g. Fe) or liquid (e.g. Hg) is always accompanied by its vapour (though in trace amount) which remains in equilibrium under the given condition. Hence the required number of phases is 2.

Note: It is equilibrium, and not the concentration, of a phase that has relevance to the phase rule.

9.13 Both H_2O and S belong to one-component system for which P_{max}, which corresponds to $F = 0$, i.e.

$$P_{max} = C - P + 2 = 1 - 0 + 2 = 3$$

The same value of P_{max} for both the system is justified. Because, the complexity of the two systems (determined by c) being same, the complexity they can exhibit in their appearance (determined by P_{max}) is also same.

☐ In both cases, the lowest value of F, which corresponds to P_{max} is zero. The highest value of F, which correspond to lowest value of P (i.e. $P = 1$) is

$$C - P + 2 = 1 - 1 + 2 = 2.$$

9.14 Here $(n + 2)$ specified species are connected by the following n rapid equilibria

$$H_2O + H_2O \rightleftharpoons H_3O^+ + OH^-.$$

$$H_2O + H_2O \rightleftharpoons (H_2O)_2$$

$$\overline{(H_2O)_2 + H_2O \rightleftharpoons (H_2O)_3}$$

$$\overline{(H_2O)_{n-1} + H_2O \rightleftharpoons (H_2O)_n}$$

Also there is one restricting equation $x_{H_3O^-} = x_{OH^-}$ which is always established in water in absence of any added acid or alkali. Then, number of components

$$= N - R - R' \quad \text{by Eq. (9.3)}$$
$$= (n + 2) - n - 1 = 1$$

☐ The system containing H_2O, H_2 and O_2 is a three-component one. Because, here the relevant chemical equilibrium $H_2O \rightleftharpoons H_2 + \dfrac{1}{2}O_2$ is established extremely slowly making $R = 0$. Also $R' = 0$, as x_{H_2} and x_{O_2} are not connected by any fixed relation in general.

Note: The so-called pure water does not consist of only one species.

9.15. a. Here dissociation of water is the only chemical reaction that occurs significantly, and that too can be disregarded as it has no effect on number of components (shown in the previous problem). Then $R = 0$ and also $R' = 0$, and

$$C = N - R - R' = 2 - 0 - 0 = 2$$

b. Assuming no dissociation: $C = 2$, as in a.

Assuming incomplete dissociation: Here there are six constituent species (i.e. $N = 6$), viz. H_2O, $NaHSO_4$, HSO_4^-, Na^+, H^+ and SO_4^{2-}, two independent chemical equilibria (i.e. $R = 2$)

$$NaHSO_4 \rightleftharpoons Na^+ + HSO_4^-$$

$$HSO_4^- \rightleftharpoons H + SO_4^{2-}$$

and two independent mole fraction relating equations

$$x_{Na^+} = x_{HSO_4^-} + x_{SO_4^{2-}}$$
$$x_{H^+} = x_{SO_4^{2-}}$$

[**Note:** The equation $x_{H^+} + x_{Na^+} = x_{HSO_4^{2-}} + 2x_{SO_4^{2-}}$, arising from charge balance is not considered as it is not an independent one which can be obtained from addition of the above two mole fraction relations.]

Then $\qquad C = 6 - 2 - 2 = 2$

Assuming complete dissociation: Here there are four constituent species (i.e. $N = 4$), viz. H_2O, Na^+, H^+ and SO_4^{2-}, but no chemical equilibria ($R = 0$), though there are two independent mole fraction relating equations, ($R' = 2$)

$$x_{Na^+} = x_{HSO_4^-} + x_{SO_4^{2-}}$$
$$x_{H^+} = x_{SO_4^{2-}}$$

Then $\qquad C = N - R - R' = 4 - 0 - 2 = 2$

Saturated condition: $C = 6 - 2 - 2 = 2$, as in case of incomplete dissociation with which it differs (without any effect) only in the chemical equilibrium (I) which is now heterogeneous, i.e. $NaHSO_4(s) \rightleftharpoons Na^+ + HSO_4^-$.

c. Here, disregarding dissociation, there are four constituent species (i.e. $N = 4$) viz. H_2O, NH_4Cl, NH_4OH and HCl, one chemical equilibrium (i.e. $R = 1$)

$$NH_4Cl + H_2O = NH_4OH + HCl$$

and one mole fraction relating equation (i.e. $R' = 1$)

$$x_{NH_4OH} = x_{HCl}$$

Then $\qquad C = N - R - R' = 4 - 1 - 1 = 2$

d. Here, disregarding dissociation, there are four constituent species (i.e. $N = 4$), viz. H_2O, $AlCl_3$, $Al(OH)_3$ and HCl, one chemical equilibrium (i.e. $R = 1$)

$$AlCl_3 + 3H_2O = Al(OH)_3(s) + 3HCl$$

but no mole fraction relating restricting equations (i.e. $R' = 0$) as some of the resulting $Al(OH)_3$ precipitates.

Then $\qquad C = N - R - R' = 4 - 1 - 0 = 3$

Note: So long as no new phase appears due to reaction between constituents of a system, no matter whatever we assume to take place in the system, the number of constituents will remain same.

9.16 Here there are five constituent species (i.e. $N = 5$), viz. H_2O, Na^+, K^+, Cl^- and Br^-. $R = 0$ and $R' = 2$ due to the equations

$$x_{Na^+} = x_{Cl^-}$$
$$x_{K^+} = x_{Br^-}$$

[**Note:** The equation $x_{Na^+} + x_{K^+} = x_{Cl^-} + x_{Br^-}$ is not an independent equation as it can be obtained from addition of the above equations]

Then $\qquad C = N - R - R' = 5 - 0 - 2 = 3$

☐ i. No effect

ii. No effect

iii. Here $R = 0$ but $R' = 1$ due to the equation $x_{K^+} = x_{Cl^-}$ (but $x_{Na^+} \neq x_{Cl^-}$ and $x_{K^+} \neq x_{Br^-}$). Then $C = 5 - 0 - 1 = 4$.

Note: The results would be the same if NaCl, KBr and KCl were undissociated.

9.17 a. $\qquad$ $N = 4, R = 1, R' = 0$

Then $\qquad$ $C = N - R - R' = 4 - 1 - 0 = 3$

b. $\qquad$ $N = 4, R = 1$, but $R' = 1$ due to the relation $x_C/c = x_D/d$

Then $\qquad$ $C = N - R - R' = 4 - 1 - 1 = 2$

c. $\qquad$ $N = 4, R = 1$ but $R' = 2$ due to the following relations

$$\frac{x_A}{a} = \frac{x_B}{b}$$

$$\frac{x_C}{c} = \frac{x_D}{d}$$

Then $\qquad$ $C = N - R - R' = 4 - 1 - 2 = 1$

d. $\qquad$ $N = 4, R = 1$, but $R' = 0$

Then $\qquad$ $C = N - R - R' = 4 - 1 + 0 = 3$

e. Same as in d

Note: R' refers to a (temperature independent) relations between mole fractions which are not affected during the reaction. No such relation exists in case of (a), (d) and (e) and hence C is same in all three cases though they are experimentally different.

9.18 a. At normal boiling point, two phases $(P = 2)$, liquid and vapour, of a pure substance $(C = 1)$ remain in equilibrium at a fixed pressure of 1 atm. Then

$$F = C - P + 2 - 1, \quad -1 \text{ is due to fixed pressure}$$
$$= 1 - 2 + 2 - 1 = 0$$

b. At critical point, two phases $(P = 2)$, liquid and vapour, of a pure substance $(C = 1)$ are indistinguishable. Then

$$F = C - P + 2 - 1, -1 \text{ is due to indistinguishability of phases}$$
$$= 1 - 2 + 2 - 1 = 0$$

Note:

1. Here we cannot take $P = 1$ though the two phases are indistinguishable (vide problem 9.31).

2. F is zero for a substance both at its critical point and triple point.

c. An azeotropic mixture is one that passes from liquid to vapour phase without any change of composition. Here

$$F = C - P + 2 - 1, -1 \text{ is due to same composition in the two phases}$$
$$= 2 - 2 + 2 - 1 = 1 \text{ for binary azeotrope } (C = 2)$$
$$= 1$$

9.19 At fixed pressure we have $F = 0$ for an azeotrope (and also for a pure liquid), i.e. intensive variables of the system are all fixed. As a consequence, the composition of an azeotrope depends on the pressure at which it boils. This makes us sure that an azeotrope is not a compound.

Note: At fixed pressure $F = 0$ happens with any azeotrope, even if it is not binary.

9.20 a. Here $\qquad$ $F = C - P + 2 - 1, \quad -1$ is due to fixed pressure of atmosphere
$$= 2 - 2 + 2 - 1 = 1$$

b. $\qquad$ $F = C - P + 2 + 1 - 1$
$$= 1 - 2 + 2 = 1$$

Here -1 is due to fixed pressure and $+1$ is due to difference in pressure of water vapour and that acting on the condensed phase (water). [On disregarding air, the vapour phase (containing only water vapour) and the liquid phase will be at different pressure].

c.
$$F = C - P + 2 - 1$$
$$= 1 - 1 + 2 - 1 = 1$$

Here the gaseous phase has been disregarded as it is not in equilibrium.

☐ F is same in all three cases. The systems involving condenses phase(s) can, therefore, be conveniently studied by exposing them open to the atmosphere. F calculated under this condition will be same as that for the closed system. Vide 9.21(b) and (c).

9.21 a. The phase rule is applicable even to a metastable system like the given one. Here
$$F = C - P + 2$$
$$= 1 - 1 + 2 = 2$$

b. Here also the system is metastable which consists of two phases, water (liq) and its vapour
$$F = C - P + 2$$
$$= 1 - 2 + 2 = 1$$

c. The phase rule is applicable to this system in the following form
$$F = C - P + 2 - 1$$
$$= 1 - 1 + 2 - 1 = 1 \quad \text{vide problem 9.20(c)}$$

d. The given system consists of two liquid phases which are in equilibrium but at different pressure. Here, the phase rule is applicable not in the usual form $(F = C - P + 2)$ but in the modified form $F = C - P + 2 + 1$. Accordingly
$$F = 2 - 2 + 2 + 1 = 3$$

e. The phase rule, in form of simple equation $F = C - P + 2$, is not applicable to a system where chemical potential is significantly affected by the applied field. Because this equation is based on the assumption that the chemical potential is determined only by temperature, pressure and composition of the system concerned.

f. Here the phase rule is not effective in its usual form $(F = C - P + 2)$. Because a dispersed system involves non-planar interfaces of large (total) area whose effect on chemical potential cannot be ignored.

9.22 There is no such possibility. Because phase rule mere concerns with the equilibrium of a system and not the stability of the equilibrium.

Thus by phase rule alone we can predict the existence of $^4C_3$ or four triple points of sulfur (which can exist in four different phases) but not their relative stability. Only by phase diagram, we can ascertain that one of these triple points is metastable and this corresponds to the coexistence of rhombic, liquid and vapour sulfur [indicated in Fig. 9.2(c) by the meeting point of three broken lines].

9.23 a. i. Here $N = 3$, $R = 1$ due to the following chemical equilibrium

$$NH_4Cl(s) \rightleftharpoons NH_3(g) + HCl(g)$$

and $\qquad R' = 1$ due to relation between x_{NH_3} and x_{HCl}

Then $\qquad C = N - R - R' = 3 - 1 - 1 = 1$
$$F = C - P + 2 = 1 - 2 + 2 = 1$$

Note: The vapour pressure of NH_4Cl system (having $F = 1$) depends on T and not on V of the system.

ii. Here, unlike (i), $R' = 0$ as no fixed relation now exists between x_{NH_3} and x_{HCl}. Then

$$C = N - R - R' = 3 - 1 - 0 = 2$$
$$F = C - P + 2 = 2 - 2 + 2 = 2$$

b. i. Here $N = 3$, and $R = 1$ due to the following chemical equilibrium

$$CaCO_3(s) \rightleftharpoons CaO(s) + CO_2(g)$$

But. unlike a(i), $R' = 0$ as there is no fixed (temperature independent) relation between the mole fractions of the constituents (viz. $CaCO_3$, CaO and CO_2) in the vapour phase. Then

$$C = N - R - R' = 3 - 1 - 0 = 2$$
$$F = C - P + 2 = 2 - 3 + 2 = 1$$

ii. Same as in (i).

iii. Same as in (i).

Note: In (b) the vapour phase is not composed of pure CO_2 due to the presence of traces of $CaCO_3(g)$ and CaO (g) varying in amount with temperature of the system.

c. i. Here $N = 4$, $R = 1$ due to the following chemical reaction

$$2Pb(NO_3)_2(s) \rightleftharpoons 2PbO(s) + 4NO_2(g) + O_2(g)$$

and $R' = 1$ due to the relation $x_{NO_2} = 4x_{O_2}$

Then $C = N - R - R' = 4 - 1 - 1 = 2$
$$F = C - P + 2 = 2 - 3 + 2 = 1$$

ii. Same as in (i).

iii. Now $R' = 0$, there being no fixed relation between x_{NO_2} and x_{O_2}. Then

$$C = N - R - R' = 4 - 1 - 0 = 3$$
$$F = C - P + 2 = 3 - 3 + 2 = 2$$

☐ In dealing with phase equilibria, the term *constituent* is used to mean a specified chemical species existing as such in a system. The term *component* is used to mean a chemically independent species that may not exist as such in the system concerned, but it must contain one or more constituent atoms of the system. Then the question "which are the components of a system" is quite awkward. In case of (b), any two of the constituents $CaCO_3$, CaO and CO_2 may be taken as the components. Even we can designate any two of the three kinds of constituent atoms of $CaCO_3$ as the components. Because if we fix the percentages of these two kinds of atoms in any phase, the percentage of the other kind of atom in the same phase becomes fixed.

Note: It is the number, and not the nature, of components which has relevance to the phase rule.

9.24 a. Here $C = 1$ and $P = 2$ (solid and vapour) because graphite (the more stable solid phase) is in equilibrium with the vapour phase (extremely dilute) though the two solid phases are not in equilibrium under the given condition. Then

$$F = C - P + 2 = 1 - 2 + 2 = 1$$

Note: Considering the system exposed open to the atmosphere of fixed pressure, the result will be same [vide problem 9.20(c)]. Here $C = 1$, $P = 1$ (considering only the more stable solid phase) and hence

$$F = C - P + 2 - 1 = 1 - 1 + 2 - 1 = 1$$

b. Here $C = 2$ (the two constituents O_2 and O_3 being not in equilibrium) and $P = 1$. Then $F = C - P + 2 = 2 - 1 + 2 = 3$

Note: Here O_2 and O_3 cannot be treated as the same constituent as they belong to the same phase. But in (a) diamond, graphite and carbon vapour are treated as the same constituent as they belong to different phases.

c. Here $C = 1$ (o-H_2 and p-H_2 being in equilibrium) and $P = 1$

Then $\qquad F = C - P + 2 = 1 - 1 + 2 = 2$

d. Here $C = 2$ and $P = 2$ for one vapour phase and only one liquid phase, but $P = 3$ for one vapour phase and two liquid phases

When $P = 2$, $\qquad F = C - P + 2 = 2 - 2 + 2 = 2$

and when $P = 3$, $\quad F = C - P + 2 = 2 - 3 + 2 = 1$

e. $C = 2$, viz water and lactic acid (as d-lactic acid and l-lactic acid together corresponds to only one component, the two optical isomers being of some thermodynamic properties), and $P = 2$ (liquid and vapour).

Then $\qquad F = C - P + 2 = 2 - 2 + 2 = 2$

f. $C = 3$ (racemic- and meso-modifications being of different thermodynamic properties, like chemical potential), and $P = 2$.

Then $\qquad F = C - P + 2 = 3 - 2 + 2 = 3$

9.25 a. Here $N = 3$, $R = 1$ and in general $R' = 0$

Then, in the general case

$$C = N - R + -R' = 3 - 1 - 0 = 2$$
$$F = C - P + 2 = 2 - 1 + 2 = 3$$

In case the system is prepared from A as starting material, $R' = 1$ due to the relation $x_B = x_C$ established in the system.

Then, in this special case

$$C = N - R + -R' = 3 - 1 - 1 = 1$$
$$F = C - P + 2 = 1 - 1 + 2 = 2$$

b. C, but not F, is same as in (a)

Here $\qquad F = C - P + 2 = 2 - 2 + 2 = 2$ in general case

and $\qquad F = 1 - 2 + 2 = 1$ in the special case

c. Here $R' = 0$ in the general case. But this happens also in special case due to natural difference in the (temperature dependent) solubility of B and C in A that spoils any fixed (temperature independent) relation between x_B and x_C in the system. Therefore, in both cases $C = 2$ and $F = 2$.

9.26 a. For snowfall to occur the partial pressure of water vapour in the atmosphere must be less than 4.58 mm, the triple point pressure of H_2O. This requirement is not fulfilled in the coastal areas.

b. Because the triple point pressure lies below 1 atm for H_2O but above 1 atm for CO_2.

c. On heating rhombic sulphur rapidly, the system having no sufficient time to form a new solid phase (monoclinic sulfur) of different structure, moves on to a metastable triple point where rhombic sulphur begins to melt. But on heating sufficiently slowly, rhombic sulfur changes to monoclinic form and the system passes on to a higher stable triple point where monoclinic sulfur begins to melt at higher temperature. The path leading to metastable triple point is represented by a broken line in the phase diagram of sulfur (Fig. 9.2).

Note: A change to a metastable state should necessarily be rapid.

d. The availability of rhombic sulfur in nature is attributed to its crystallisation from the molten sulfur at very high pressure (above 1290 atm) provided by the earth during its solidification. Though at ordinary pressure, solidification to monoclinic form results. These are justified by the phase diagram of sulfur.

· **Note:** Although under ordinary condition, rhombic sulfur is less stable than monoclinic, the transition to the latter form occurs at an extremely slow rate.

e. For both boiling and melting of a substance, the sign of dT/dP depends only on the accompanying ΔV (by Clapeyron Eq. (9.1a), because ΔH_{vap}, ΔH_{melt} and T are all positive. Now ΔV is always positive for boiling but may be positive or negative for melting. Then with rise of pressure, boiling point of a substance will always increase but its melting point may increase or decrease.

9.27 a. The given statement is not justified at the triple point which is a particular case of melting as well as boiling point.

b. Sublimation is a process in which a substance changes directly from solid to gaseous state below certain pressure. This does not happen with helium. Because solid helium cannot be in equilibrium with its vapour in any condition. Therefore, the given statement is not justified.

Note: We can go from solid He to gaseous He only through the liquid He (He I or He II).

c. For triple point ($P = 3$) to be invariant,
$$F = C - P + 2 = C - 3 + 2 = 0$$
or $\qquad C = 1$

Then the given statement is justified only for one-component systems.

d. The given statement may not hold. Because although the concentration of an ingredient in the vapour phase of a binary mixture is favoured by volatily of this (pure) ingredient, it is disfavoured by concentration of the other ingredient in the liquid phase (by Raoult's law).

e. The given statement is justified. Because on boiling a dilute aq HCl, the residual solution approaches the composition of the relevant azeotropic mixture where the concentration of HCl is much higher (~ 6N). Again, on boiling a solution having concentration of HCl higher than 6N the residual solution approaches the composition of the same azeotropic mixture whereby it gets diluted.

[**Note:** The formation of azeotropic HCl–H_2O mixture is used in inorganic analysis where reaction with concentrated HCl is brought about without boiling while boiling is often preferred for reaction with dilute HCl].

f. For a binary system, the composition of the liquid mixture for azeotropism (that relates to fixed b.p.) is unlikely to be same as that for eutectism (that relates to fixed f.p.). Hence the given statement is not justified.

9.28 At boiling point $P = 2$ (one vapour phase and one liquid phase) for ethanol–water system, but $P = 3$ (one vapour phase and two liquid phases) for nitrobenzene–water system. Then at constant pressure $F = (C - P + 2 - 1)$ is one for the first system but zero for the second. Here lies the explanation of the given observation.

□ With phenol–water system (partially miscible), P may be 2 (when there is one liquid phase) or 3 (when there are two liquid phases) if the system is closed. Accordingly, during boiling the temperature of this system will change in the first case but not in the second case.

□□ The partial vapour pressure of each component is equal to its vapour pressure (p^*) in pure state in case of (i) but less than p^* in case of (ii), by Raoult law. Then in case of (i) the total vapour pressure will be higher and hence boiling point will be lower.

Note: During boiling each liquid phase of nitrobenzene–water system remain pure (i.e. suffers no change in composition), but the relative amounts of the two components change if two liquid phases are considered together.

9.29 Here only aniline can be effectively steam distilled because it is almost immiscible with water and has sufficient vapour pressure near the boiling point of water.

9.30 False. Because the idea of partition coefficient of a substance as a constant ratio of its concentrations in two immiscible solvents is based on certain restrictions (vide Section 9.4.4) which are not fulfilled in case of concentrated solutions associated with most solubility equilibria. Increasing concentration of solutions causes not only their deviation from ideality but also affects immiscibility of the solvents concerned.

9.31 Nicotine–water system.

☐ At consolute temperature, the composition of the two liquid phases becomes same. Due to this additional restriction, $C = N - 1 = 2 - 1 = 1$.

Then $\qquad\qquad F = C - P + 1$

since the given system is equivalent to a condensed system (having only two liquid phases) exposed open to the atmosphere [vide problem 9.20(c)].

$$= 1 - 2 + 1 = 0$$

Note: At consolute temperature the two liquid phases have same composition but we cannot take them as same phase (because of phase boundary yet to disappear). Here we are to take $P = 2$ and $C = 1$ (to consider equality of composition of the two liquid phases). We cannot take $P = 1$ and $C = 2$ giving $F = 2$.

9.32 The lower consolute temperature suggests the formation of some weak complex between the two components of the system, presumably through hydrogen bonding, which is favoured at lower temperature.

☐ Yes. For example, in case of ether–water system. Although mutual solubility increases with decreasing temperature, water freezes before lower consolute temperature is reached, and hence unattainable. Again, for chloroform–water system, although the mutual solubility increases with rise of temperature, critical temperature of chloroform is reached before the theoretical upper consolute temperature is attained.

Note: Although the expected consolute temperature is not always attainable with any liquid pair, the analogous temperature is sometimes observed with a pair of gases, e.g. NH_3–CH_4 and He–Ne in the regions of high liquid-like densities. This warns us aginst making the general statement that gases are miscible in all proportions.

9.33 If the added salt is insoluble in both the liquid phases, the consolute temperature will not be affected.

If the salt dissolves only in one of the liquid phases, it is likely to be solvated only by the major constituent of that phase with salting-out of the oher constituent. As a result, mutual solubility of the two constituents will be equalised at higher temperature, i.e. consolute temperature will rise. For example, consolute temperature of phenol–water system is raised by KCl which dissolves only in water-rich layer.

If, however, the salt dissolves in both the liquid phases, it will be solvated by both the constituents. Here salting-in effect, instead of salting-out, will occur. This will cause an increase in mutual solubilities with consequent lowering of consolute temperature. For example, in presence of sufficient LiI, the consolute temperature of aniline–water system is so lowered that aniline and water become completely miscible in all proportions at ordinary temperature. Here LiI dissolves in both aniline and water.

9.34 Vide Subsect. 9.43.

□ As triple point of a binary system, F for both eutectic and peritectic point will be same which is

$$F = C - P + 2 = 2 - 3 + 2 = 1$$

9.35 Practically so, because the chance for more than two substances melting at a particular temperature is virtually nil.

However, an azeotrope is not limited to a binary system, as the transition of a homogeneous mixture from liquid to vapour phase or reverse involves no severe restrictions.

9.36 No, because here the lowest temperature corresponds to an eutectic system (where solid salt, its solution and ice coexist in equilibrium) which is invariant at constant pressure.

□ The temperature of freezing mixture will remain unchanged until entire ice melts due to absorption of heat from the surroundings or entire salt dissolves. In preparing freezing mixture NaCl and ice are usually taken in 1:3.5 ratio by weight so that they disappear at nearly the same time.

9.37 a. For a pure liquid, the variation of freezing point ΔT from its normal freezing point due to variation of pressure ΔP from normal pressure is given by

$$\frac{\Delta T}{\Delta P} \simeq \frac{dT}{dP} \quad \text{when } \Delta P \text{ is not high}$$

$$= \frac{T\Delta V}{\Delta H} \quad \text{by Clapeyron Eq. (9.1a)}$$

Then $$\Delta T = \frac{T\Delta P\Delta V_{fus}}{\Delta H_{fus}}$$

$$= \frac{(273.15 \text{ K}) (2 \text{ atm})\left(-\dfrac{1}{11} \times 10^{-3} \text{L g}^{-1}\right)}{80 \text{ cal g}^{-1} \times \dfrac{1}{24.2} \text{ L-atm/cal}}$$

for water that freezes at 273.15 K at 1 atm

$$= -0.015 \text{ K}$$

Therefore the required freezing point is (273.15 – 0.015) K, i.e. 273.135 K or –0.015°C

b. The boiling points T_1 at P_1 and T_2 at P_2 are related by

$$\ln\frac{P_2}{P_1} = \frac{\Delta\bar{H}_{vap}}{R}\left(\frac{1}{T_1} - \frac{1}{T_2}\right) \quad \text{by Eq. (9.1c)}$$

For water, $T_1 = 373.15$ K at $P_1 = 1$ atm, and then T_2 at $P_2 = 3$ atm will be given by

$$\ln\frac{3}{1} = \frac{540 \times 18 \text{ cal mol}^{-1}}{1.987 \text{ cal K}^{-1}\text{mol}^{-1}}\left(\frac{1}{373.15} - \frac{1}{T_2}\right)$$

whence $T_2 = 407.33$

9.38 a. By Eq. (9.1c)

$$\log P \text{ (mm)} = -\frac{\Delta \bar{H}_{sub}}{2.303 \, RT} + \text{constant} \qquad \text{(I)}$$

Comparison of this with the given vapour pressure equation for solid SO_2 gives

$$\frac{\Delta \bar{H}_{sub}}{2.303 \, R} = 1871 \text{ K}$$

or

$$\Delta \bar{H}_{sub} = 2.303 \, (1871 \text{ K}) \, R$$
$$= 2.303 \, (1871 \text{ K}) \, (1.987 \text{ cal K}^{-1} \text{mol}^{-1})$$
$$= 8562 \text{ cal mol}^{-1}$$

b. The vapour pressure equation for liquids will be of the following form, similar to (I)

$$\log P \text{ (mm)} = -\frac{\Delta \bar{H}_{vap}}{2.303 \, R} + \text{constant}$$

The given vapour pressure equation for liquid SO_2 being of this form

$$\Delta \bar{H}_{vap} = 2.303 \, (1426 \text{ K}) \, R$$
$$= 2.303 \, (1426 \text{ K}) \, (1.987 \text{ cal K}^{-1} \text{mol}^{-1})$$
$$= 6525 \text{ cal mol}^{-1}$$
$$\Delta \bar{H}_{fus} = \Delta \bar{H}_{sub} - \Delta \bar{H}_{vap}$$
$$= (8562 - 6525) \text{ cal mol}^{-1}$$
$$= 2037 \text{ cal mol}^{-1}$$

c. Normal boiling point T_b corresponds to $P = 760$ mm. Then from the given expression of vapour pressure for liquid SO_2

$$\log 760 = -\frac{1426 \text{ K}}{T_b} + 8.3$$

whence $T_b = 263.1$ K

d. At triple point the solid and liquid places of a substance have same vapour pressure. Then from the given two vapour pressure equations, the triple point temperature T is given by

$$-\frac{1871 \text{ K}}{T} + 10.6 = -\frac{1426 \text{ K}}{T} + 8.3$$

whence $T = 193.5$ K

Triple point pressure can be obtained by substituting this in either of the given equations. Then

$$\log P \text{ (mm)} = -\frac{1871 \text{ K}}{193.5} + 10.6 \qquad \text{from the first given equation}$$

whence $P = 8.53$ mm

e. Triple point may be regarded as f.p $T_1 = 193.5$ K at $P_1 = 8.53$ mm. Then normal f.p. T_2 at $P_2 = 760$ mm will be given, as in previous problem, by

$$T_2 - T_1 = \frac{T_1 (P_2 - P_1) \bar{V}_{fus}}{\Delta \bar{H}_{fus}}$$

$$= \frac{T_1(P_2 - P_1)\left(\dfrac{1}{\rho_{liq}} - \dfrac{1}{\rho_{solid}}\right)M}{\Delta\bar{H}_{fus}} \quad \text{where } \rho \text{ is density and } M \text{ is molar mass}$$

$$= \frac{(193.5 \text{ K})\left(\dfrac{760 - 8.53}{760}\right)\text{atm}\left(\dfrac{1}{1.43} - \dfrac{1}{2.93}\right)\times 10^{-3}\text{Lg}^{-1}\,(64.07 \text{ g mol}^{-1})}{2037 \text{ cal mol}^{-1} \times \dfrac{1}{24.2}\text{ L atm cal}^{-1}}$$

$$= 52.1 \text{ K}$$

Then $T_2 = T_1 + 52.1 = (193.5 + 52.1)\text{ K} = 245.6 \text{ K}$

f. Trouton constant $= \dfrac{\Delta\bar{H}_{vap}}{T_b} = \dfrac{6525 \text{ cal mol}^{-1}}{263.1 \text{ K}}$

$$= 24.8 \text{ cal K}^{-1}\text{ mol}^{-1}$$

9.39 For solids $\qquad \mu = \mu° + (P-1)\bar{V} \quad$ by Eq. (5.13c)

Diamond will be in equilibrium with graphite when they have same chemical potential (μ) and then

$$(P-1)(\bar{V}_{graphite} - \bar{V}_{diamond}) = \mu°_{diamond} - \mu°_{graphite}$$

or $\qquad\qquad P - 1 = \dfrac{\mu°_{diamond} - \mu°_{graphite}}{M\left(\dfrac{1}{\rho_{graphite}} - \dfrac{1}{\rho_{diamond}}\right)}$

where ρ is density and M is molar mass of carbon

$$= \frac{(685 - 0)\text{ cal mol}^{-1} \times \dfrac{1}{24.2}\text{ L-atm cal}^{-1}}{(12 \text{ g mol}^{-1})\left(\dfrac{1}{2.25 \text{ g cm}^{-3}} - \dfrac{1}{3.52 \text{ g cm}^{-3}}\right)\times 10^{-3}\text{ L cm}^{-3}} \quad [\text{at 25°C when } \mu°_{graphite} = 0]$$

$$= 14715 \text{ atm}$$

Then $P = 14716$ atm

Note: Graphite is more stable than diamond at ordinary temperature and pressure but not at very high pressure when the order of stability becomes reversed. This is the basis of preparation of artificial diamond first successfully done by Moisan.

9.40 Liquid iodine has vapour pressure $P_1 = 100$ torr at $T_1 = 390$ K and $P_2 = 760$ torr at $T_2 = 457.5$ K (the normal boiling point).

Then $\qquad\qquad \Delta\bar{H}_{vap} = \dfrac{R\ln(P_2/P_1)}{(1/T_1) - (1/T_2)}$

$$= \frac{8.31 \text{ J K}^{-1}\text{mol}^{-1}\ln(760/100)}{(1/390 \text{ K}) - (1/457.5 \text{ K})} = 44586 \text{ J mol}^{-1}$$

$$\Delta\bar{H}_{sub} = \Delta\bar{H}_{fus} + \Delta\bar{H}_{vap}$$
$$= 15.5 \times 10^3 \text{ J mol}^{-1} + 44586 \text{ J mol}^{-1}$$
$$= 60086 \text{ J mol}^{-1}$$

Since triple point (T, P) is common to liquid–vapour and solid–vapour curves

$$\Delta\bar{H}_{vap} = \frac{R\ln(P/P_1)}{(1/T_1) - (1/T)} \quad \text{considering liq–vap curve}$$

$$\Delta \bar{H}_{sub} = \frac{R \ln(P / P_1')}{(1 / T_1') - (1 / T)} \quad \text{considering solid–vap curve}$$

where the vap pr $P_1' = 1$ torr at temp $T_1' = 312$ for solid iodine

From these two equations T and P can be found out.

9.41 By Eq. (9.4a) $760 = p_B^* + (p_A^* - p_B^*) x_A' = 450 + (950 - 450) x_A'$

(Here by *boiling point*, we mean *normal boiling point* when vap pr is 760 torr)

and $800 = 450 + (950 - 450) x_A''$

Therefore $x_A' = 0.62$

$x_A'' = 0.70$

Here we need to know x_A which is provided by lever rule

$$\frac{\text{moles of condensate}}{\text{moles of residual liquid}} = \frac{x_A - x_A'}{x_A'' - x_A}$$

Then $1 = \dfrac{x_A - 0.62}{0.72 - x_A}$

or $x_A = 0.66$

% of A distilled off $= \dfrac{(n/2) x_A''}{n x_A} \times 100$,   where n is the moles of initial liquid mixture

$$= \frac{(0.70 / 2)}{0.66} \times 100 = 53.03$$

Note: The % of A (more volatile component) distilled off is found to be more than 50.

9.42 By Eq. (9.5)

$$M_{org} = M_{water} \cdot \frac{w_{org}}{w_{water}} \cdot \frac{P_{water}^*}{P - P_{water}^*}$$

where P is total vap pressure, i.e. external pressure

$$= 18 \times \frac{9.5}{100 - 9.5} \times \frac{730}{760 - 730} = 46$$

9.43 At 100°C, $P_{H_2O} = \left(\dfrac{0.2}{0.2 + 0.8}\right)(1492 \text{ torr}) = 298.4 \text{ torr}$, by Dalton's law $p_i = x_i P_{total}$

$$P_{benzene} = \left(\frac{0.8}{0.2 + 0.8}\right)(1492 \text{ torr}) = 1193.6 \text{ torr}$$

At 80°C, assuming no condensation

$$P_{H_2O} = 298.4 \text{ torr} \times \frac{80 + 273.15}{100 + 273.15} = 282.4 \text{ torr}, \quad \text{assuming } P \propto T$$

$$P_{benzene} = 1193.6 \text{ torr} \times \frac{80 + 273.15}{100 + 273.15} = 1129.6 \text{ torr}.$$

At 80°C the calculated value of $P_{benzene}$ exceeds 760 torr, the saturated vap pressure of benzene at 80°C (b.p. of benzene being 80°C). But this does not happen with H_2O, whose saturated vap pressure is 355 torr at 80°C. Hence benzene, and not H_2O, will condense on cooling to 80°C.

The amount of benzene condensed

$$= 0.8 \text{ mol} \times \frac{1129.6 - 760}{1129.6} \qquad \text{assuming } n \propto p$$

$$= 0.2618 \text{ mol}$$

☐ Total vapour pressure at 80°C

$$= p_{H_2O} + p_{benzene} \text{ (saturated value)}$$

$$= 282.4 \text{ torr} + 760 \text{ torr} = 1042.4 \text{ torr}$$

Note: On cooling, condensation happens with the more volatile component, benzene (being in much higher mole fraction).

9.44 a. The components being in same relative amount, the less volatile one, i.e. H_2O will start condensing first. This will happen when the partial pressure of H_2O attains its saturation value, i.e. 355 torr at 80°C. This corresponds to the total pressure, $P_{total} = P_i/x_i = 355/(1/2)$ or 710 torr.

b. For both the components to start condensing simultaneously, their amounts must be in the ratio of their saturated vapour pressure, i.e.

$$n_{benzene} : n_{water} = 760 : 355$$

c. No, because the ratio of saturated vapour pressure is not same at 80° and 100°C.

9.45 a. By Eq. (9.4b) $P = \dfrac{p_A^* p_B^*}{p_A^* + (p_B^* - p_A^*) y_A}$

$$= \frac{(0.6 \text{ atm})(1.5 \text{ atm})}{0.6 \text{ atm} + (1.5 \text{ atm} - 0.6 \text{ atm}) \dfrac{1}{1+4}} = 1.15 \text{ atm}$$

By Eq. (9.4a) $x_A = \dfrac{P - p_B^*}{p_A^* - p_B^*}$ $\qquad\qquad$ (I)

$$= \frac{1.15 - 1.5}{0.6 - 1.5} = 0.39$$

b. At normal boiling point, $P = 1$ atm. Then by Eq. (I)

$$x_A = \frac{1 - 1.5}{0.6 - 1.5} = 0.56$$

Note: The previous problem involves two immiscible liquids where only one of the components condenses on cooling a gas mixture, unless the latter has a particular composition. The present problem differs in that it involves two completely miscible liquids where the condensed phase does not consist of a pure component due to mixing with the other component.

9.46 % of H_2O in the overall system

$$= \frac{4}{6+4} \times 100 = 40$$

By lever rule

$$\frac{\text{mass of water-rich layer } (X)}{\text{mass of phenol rich layer } (Y)} = \frac{40 - \% \text{ of } H_2O \text{ in } Y}{\% \text{ of } H_2O \text{ in } X - 40}$$

Then $\qquad\qquad$ mass of X = mass of $(X + Y) \dfrac{40 - \% \text{ of } H_2O \text{ in } Y}{\% \text{ of } H_2O \text{ in } X - \% \text{ of } H_2O \text{ in } Y}$

$$= (6+4) \text{ g} \times \frac{40 - 37.5}{89 - 37.5} = 0.485 \text{ g}$$

In X $\qquad$ mass of $H_2O = 0.485 \text{ g} \times \dfrac{89}{100} = 0.432 \text{ g}$

$\qquad$ mass of phenol $= 0.485 \text{ g} - 0.432 \text{ g} = 0.053 \text{ g}$

In Y $\qquad$ mass of $H_2O = 4.00 \text{ g} - 0.432 \text{ g} = 3.568 \text{ g}$

$\qquad$ mass of phenol $= 6.00 \text{ g} - 0.053 \text{ g} = 5.947 \text{ g}$

9.47 Here the relevant chemical equilibria are

$$CuSO_4{\cdot}5H_2O(s) \rightleftharpoons CuSO_4 \cdot 3H_2O(s) + 2H_2O(g) \qquad \text{(I)}$$
$$CuSO_4{\cdot}3H_2O(s) \rightleftharpoons CuSO_4 \cdot H_2O(s) + 2H_2O(g) \qquad \text{(II)}$$
$$CuSO_4{\cdot}H_2O(s) \rightleftharpoons CuSO_4(s) + H_2O(g) \qquad \text{(III)}$$

Assuming that all three equilibria exist

$$C = N - R = 5 - 3 = 2$$

Since $\qquad F = C - P + 2 - 1$ at constant temperature

the maximum number of phases, which corresponds to $F = 0$, will be

$$P = C - F + 2 - 1 = 2 - 0 + 2 - 1 = 3$$

Therefore, the four specified phases cannot be in equilibrium

☐ Here the equilibrium can exist between $H_2O(g)$ and only two solid phases.

Now, the number of moles of H_2O (g), $n = PV/RT$, required to produce the given vapour pressure in a 10 L vessel at 50°C will be as under.

For $P = 47$ torr, $\qquad n = \dfrac{\left(\dfrac{47}{760} \text{ atm}\right)(10 \text{ L})}{(0.082 \text{ L-atm K}^{-1}\text{mol}^{-1})(50 + 273) \text{ K}} = 0.0233 \text{ mol}$

For $P = 30$ torr, $\qquad n = 0.0233 \times \dfrac{30}{47}$ or 0.0149 mol

For $P = 4.5$ torr, $\qquad n = 0.0233 \times \dfrac{4.5}{47}$ or 0.0022 mol

The attainment of equilibrium (I) is not possible starting with 0.006 mol of $CuSO_4{\cdot}5H_2O(s)$ because its conversion to $CuSO_4{\cdot}3H_2O$ (s) cannot provide minimum amount (0.0233 mol) of H_2O required to be present in the system.

The equilibrium (III) is also not possible because the amount of H_2O produced in the conversion of 0.006 mol of $CuSO_4{\cdot}5H_2O$ to $CuSO_4{\cdot}H_2O$ much exceeds the maximum amount (0.0022 mol) of H_2O required to be present in the system.

Then the equilibrium (II) is likely to be established. Let x mol of $CuSO_4{\cdot}3H_2O(s)$ and y mol of $CuSO_4{\cdot}H_2O(s)$ remain in the system at equilibrium. Then, by material balance

$$2x + 4y = 0.0149$$
$$x + y = 0.006$$

whence $\qquad x = 0.00185 \text{ mol}$

$\qquad y = 0.00415 \text{ mol}$

Note: The composition of the solid residue at equilibrium depends on the volume of the vessel at constant temperature. It will vary from pure $CuSO_4{\cdot}5H_2O(s)$ for very low volume to pure $CuSO_4(s)$ for very high volume of the vessel.

9.48 a. By Eq. (9.6a), the fraction of solute expected to remain unextracted

$$\frac{w_1}{w_0} = \frac{V_1}{V_1 + KV_2}$$

$$= \frac{200}{200 + 3 \times 100} = \frac{2}{5}$$

Then the fraction expected to be extracted $= 1 - (2/5) = (3/5)$

b. Let $V_2 = x$ ml of ether is used in the first extraction and hence $V_2' = (100 - x)$ ml used in the second extraction. Then, by Eq. (9.6b)

$$\frac{w_2}{w_0} = \left(\frac{V_1}{V_1 + KV_2} \right)\left(\frac{V_1}{V_1 + KV_2'} \right)$$

$$= \left(\frac{200}{200 + 3x} \right)\left[\frac{200}{200 + 3(100 - x)} \right]$$

$$= \frac{40000}{122500 - 9(50 - x)^2}$$

w_2/w_0 will be minimum for maximum value of the denominator that happens with $x = 50$. Then the minimum value of w_2/w_0 will be

$$\left(\frac{200}{200 + 3 \times 50} \right)^2 = 0.33$$

Then the maximum fraction expected to be extracted

$$= 1 - 0.33 = 0.67$$

c. The number of operations (n) required for 80% extraction, i.e. for 20% remaining unextracted, is given, according to Eq. (9.6c), by

$$\frac{20}{100} = \left(\frac{200}{200 + 3 \times 100} \right)^n$$

$$= \left(\frac{2}{5} \right)^n$$

Then $\qquad n = \dfrac{\log(1/5)}{\log(2/5)} = 1.7$

Therefore, the required number of operations

$$= 2 \text{ (the higher integer nearest to 1.7)}$$

☐ It is evident from the results found in (a) and (b) that the efficiency of extraction with a fixed amount of an extracting solvent is greater when it is used in parts with separate operation for each part.

UNIVERSITY QUESTIONS

9.1 What do you mean by 'degrees of freedom'? Can it be negative? Explain.

(Burdwan BSc(H), 1992)

9.2 Calculate in each of the following examples, the number of phases, components and degrees of freedom:

a. $NH_4Cl(s) \rightleftharpoons NH_3(g) + HCl(g)$

b. $CaCO_3(s) \rightleftharpoons CaO(s) + CO_2(g)$ (Burdwan BSc(H), 2008)

9.3 Find out the number of degrees of freedom in the following systems:

a. $Na_2SO_4 \cdot 10H_2O(s) \rightleftharpoons Na_2SO_4(s) + 10H_2O(g)$ (Allahabad BSc, 2002)

b. Saturated solution of NaCl (Allahabad BSc, 2002)

     c. Unsaturated solution of sucrose kept in an open beaker.

(Burdwan BSc(H), 2005)

9.4 Calculate the number of components and degrees of freedom for a system of sodium chloride solution in water containing undissolved salt in equilibrium with water vapour.   (Baroda BSc, 2005)

9.5 For a substance (e.g. H_2O), three phases (solid, liquid and vapour) coexist at the freezing point, and also at the triple point. But the triple point is invariant whereas the freezing point is variable. How would you account for this?

(Calcutta BSc(H), 1976)

9.6 Discuss the salient features of phase diagram of sulphur system. Why can four phases of this system not exist at equilibrium?   (Madras BSc, 2010)

9.7 a. Find the maximum number of degrees of freedom of a one-component system.

(Burdwan BSc(H), 2002)

    b. How many three and two phase sulphur systems are possible? Compute their degrees of freedom.   (Burdwan BSc(H), 1994)

9.8 Justify the following:

    a. Rhombic sulfur is more abundant in nature than monoclinic ones.

(Jadavpur BSc, 2013)

    b. On heating, sulphur melts but iodine sublimes, at ordinary pressure.

(Calcutta BSc(H), 1986)

9.9 Define first order phase transition.   (Calcutta BSc(H), 2006)

9.10 Write down the differences between normal phase transition and lambda transition.   (Jadavpur BSc(H), 2003)

9.11 Starting from Clapeyron equation, deduce Clausius–Clapeyron equation, mentioning the necessary assumptions.   (Calcutta BSc(H), 2014)

9.12 a. The Clausius–Clapeyron equation is a special case of the van't Hoff equation for liquid vapour equilibrium. Justify or criticise.   (Burdwan BSc(H), 2008)

    b. Explain, on its basis, whether the boiling point of a liquid can ever decrease with rise of pressure.   (Burdwan BSc(H), 2005)

9.13 Give reasons why:

    a. The slope of P vs T line is greater for solid–gas curve than for liquid gas curve at a triple point.   (Burdwan BSc(H), 1998)

    b. A solid which sublimes under open condition can be melted in closed container.

(Jadavpur BSc(H), 2003)

9.14 In the phase diagram for water:

    a. what is the upper limit of liquid–vapour equilibrium line?

    b. why is solid–liquid equilibrium line for water almost vertical and slightly tilted to the left?   (Punjab BSc, 2002)

9.15 What is Trouton's rule? Indicate the cases where it fails.   (Calcutta BSc(H), 2006)

9.16 Water and carbon tetrachloride have widely different values of Trouton's constant. Why?   (Burdwan BSc, 2001)

9.17 Starting from the appropriate form of the Duhem–Margules equation obtain Konowaloff's rule and use this to construct the boiling point-composition curve to explain the distillation of binary liquid pairs with minimum boiling point?

(Calcutta BSc(H), 1997)

9.18 a. What condition must be satisfied for a binary liquid mixture to show azeotropism? (Burdwan BSc(H), 2000)

   b. A pair of completely miscible liquids A and B cannot show azeotropism if they obey Raoult's law through the entire range of mole fractions. Justify or criticise. (Burdwan BSc(H), 2012)

9.19 a. Calculate degrees of freedom of the following systems:
   i. A liquid at critical temperature, ii. A binary azeotrope. (Mysore BSc, 2000)

   b. An azeotrope is not a pure compound. Explain. (Burdwan BSc(H), 2009)

9.20 Liquid A boils at 25°C, while liquid B at 80°C and a mixture containing one mole of A and two moles of B exhibits a maximum boiling point of 100°C, all under 1 atm pressure.

   a. Draw the boiling point-composition diagram for A and B, and label the diagram

   b. Describe the results of continuous heating of a mixture containing 2 moles of A and 1 mole of B. If this mixture is distilled, what would be the compositions of the initial distillate and the final residue? (Calcutta BSc(H), 1985)

9.21 a. It is not possible to get pure ethanol from 50% aqueous ethanol by distillation only. Why (Burdwan BSc(H), 2001)

   b. How can water–ethanol azeotropism be avoided in purifying ethanol. (Burdwan BSc(H), 2005)

9.22 Liquid–liquid phase diagram of a partially miscible liquid pair can intersect with the liquid–vapour one. Explain. (Burdwan BSc(H), 2001)

9.23 a. What is consolute temperature? Can we have systems with (i) upper, (ii) lower and (iii) upper and lower consolute temperatures? If so, give one example for each. (Calcutta BSc(H), 1993)

   b. The composition of a mixture of partially miscible liquids at consolate temperature is always fixed. Justify or criticise. (Burdwan BSc(H), 1998)

9.24 a. Each of two phenol–water mixtures with 40% and 60% by mass of phenol respectively becomes just homogeneous at the same temperature. Calculate the amounts of the two layers when a phenol–water mixture containing 5 g of each of the components is equilibriated at this temperature. (Jadavpur BSc(H), 2012)

   b. Is there any affect of addition of a little NaCl tothe system? (Burdwan BSc(H), 2007)

9.25 a. A high boiling liquid can be distilled off at much lower temperature even under normal atmospheric pressure. Explain. (Jadavpur BSc(H), 2003)

   b. Aniline–water mixture boils at 371.4 K and 1 atm. At this temperature, vapour pressure of aniline is 42 mmHg. If the mass ratio of aniline to water in the distillate is 0.13, calculate the molar mass of aniline. (Burdwan BSc(H), 2009)

9.26 How many degrees of freedom are there in the following system?
   I_2 dispersed between liquid water and liquid carbon tetrachloride at 1 atm with no solid I_2 present. (Calcutta BSc(H), 1991)

9.27 Establish the Nernst distribution law. What is partition coefficient? Does it vary with temperature? Explain why and how the distribution law would change in case a solute partially dimerises in one of the solvents. How does the result simplify if complete dimerisation is assumed? (Burdwan BSc(H), 1993)

9.28 Show that, in a solvent extraction process, it is more economical to use a small volume of solvent in stages than the total amount of solvent at a time. (Burdwan BSc(H), 2005)

9.29 The distribution coefficient of a substance between a solvent A and water is double of that between another solvent B and water at a particular temperature. Find out which one of the following is more efficient for extraction of this substance from a certain volume of its aqueous solution:

a. A single stage process using V cc of A as extractant.

b. A two stage process using V cc of B as extractant at each stage.

(Jadavpur BSc(H), 2002)

9.30 An organic compound is extracted from an aqueous solution with successive quantities of 25 ml of chloroform. The volume of the aqueous solution is 200 ml. The distribution coefficient of the compound between chloroform and water is 20. Calculate the number of extraction needed for 99% recovery.

(Calcutta BSc(H), 1987)

9.31 Two metals, which are completely soluble in the liquid state but show no solid–solid solubility, form an eutectic. Apply phase rule to find whether the eutectic point is invariable or not. How can it be verified that the eutectic mixture is not a compound?

(Calcutta BSc(H), 1977)

9.32 Solid X has the melting point 630°C and Y has the melting point 346°C. X and Y exhibit a simple eutectic at 246°C with eutectic composition being 30 weight % of X. Draw and explain the cooling curves of (i) liquid having the composition where the weight % of X = 50 and (ii) liquid having eutectic composition.

(Calcutta BSc(H), 2007)

9.33 What modifications in the phase diagram of a simple eutectic system take place when one of the components exists in two allotropic forms? Show that the transition point is an invariant point.

(Jadavpur BSc(H), 2012)

9.34 Draw the phase diagram (T vs mole % B) of a system consisting of solids A and B forming a stable compound A_2B with congruent melting point. Show the different phases present in the different regions of the diagram [Given: Melting point of A_2B < melting point of A < melting point of B].

(Calcutta BSc(H), 2006)

9.35 Distinguish between eutectic and peritectic points.

(Delhi BSc, 2001)

9.36 In Na-K system, a compound Na_2K is formed and the peritectic point is found to be at 56 weight % of K and 7°C. If 10 g of the peritectic solution is mixed with 5 g of pure solid Na, find out the composition of the phases in equilibrium at just below 7°C.

(Jadavpur BSc(H), 2001)

9.37 Explain:

a. An aqueous solution can be formed from a solid by evaporation of water.

(Jadavpur BSc(H), 2003)

b. Ice-common salt mixture can be used as freezing point.

c. Tin–lead mixture can be used for soldering.

(Calcutta BSc(H), 1986)

9.38 At 0°C the heat of fusion of ice is 333.5 J/g and the densities of water and ice are 0.9998 g·cm^{-3} and 0.9168 g·cm^{-3} respectively. Show that an increase in pressure by 1 bar lowers the freezing point of water by 0.0075 K.

(Burdwan BSc(H), 1976)

9.39 The vapour pressure of 2,2,4-trimethylpentane at 20.7°C and 29.1°C are 40 and 60 torr respectively. Calculate the enthalpy of vaporisation of this compound.

(Delhi BSc, 2006)

9.40 The vapour pressure of water at 25°C is 23.76 mmHg. Calculate the average value of molar enthalpy of vaporisation (ΔH_{vap}) of water over the temperature range 25°C to 100°C.

(Jadavpur BSc(H), 2001)

KEY TO UNIVERSITY QUESTIONS

9.1 Vide Section 9.3.

☐ No. Because the definition of 'degrees of freedom' does not make any sense for negative values.

9.2 a. Here $P = 2$.

In the general case of arbitrary composition of the gas phase $C = 2$ and then $F = C - P + 2 = 2 - 2 + 2 = 2$.

In the special case, when the system is prepared starting from NH_4Cl, $C = 1$ and $F = C - P + 2 = 1 - 2 + 2 = 1$.

Vide problem 9.23(a).

b. Here, in both general and special cases, $P = 3$, $C = 2$ and $F = 1$. Vide problem 9.23(b).

9.3 a. $P = 3$, $C = 2$ and hence $F = 1$, similar to the previous question 9.2(b).

b. Here $C = 2$. Vide problem 9.15(b).

$P = 3$ (solid, liquid and gas). Then $F = C - P + 2 = 2 - 3 + 2 = 1$.

c. $C = 2$, $P = 1$ disregarding the vapour phase which is not in equilibrium.

$F = C - P + 2 - 1$, -1 is due to fixed pressure of atmosphere

$\quad = 2 - 1 + 2 - 1 = 2$

9.4 Same as the previous question (b).

9.5 For any substance ($C = 1$), the vapour phase is in equilibrium at triple point (and hence $P = 3$) but not at freezing point (where $P = 2$). Then $F = C - P + 2$ will be zero for triple point but not for freezing point.

9.6 Vide Section 9.4.

☐ Because $P = 4$ corresponds to $F = C - P + 2 = 1 - 4 + 2 = -1$, which, being negative, is absurd.

9.7 a. Here $F_{max} = 2$ vide problem 9.13.

b. S can exist in 4 single phases. The number of ways in which it can exist as three-phase system is $^4C_3$, i.e. 4 and as two-phase system is $^4C_2$, i.e. 6.

☐ For $P = 3$, $F = C - P + 2 = 1 - 3 + 2 = 0$

For $P = 2$, $F = 1 - 2 + 2 = 1$

Note: The degrees of freedom is not determined by the number of possible ways of combining phases.

9.8 a. Vide problem 9.26(d).

b. Because the triple point pressure of iodine, unlike sulfur, lies above 1 atm.

9.9 Vide Section 9.1.

9.10 Vide Section 9.1.

9.11 Vide Section 9.2.

9.12 a. The van't Hoff equation, which refers usually to chemical equilibrium, is

$$\frac{d \ln K_p}{dT} = \frac{\Delta \bar{H}}{RT^2} \quad \text{where } K_p \text{ is equilibrium constant}$$

When applied to liquid–vapour equilibrium (a physical equilibrium which is simplier than chemical equilibrium), this equation becomes

$$\frac{d \ln P}{dT} = \frac{\Delta \bar{H}_{vap}}{RT^2}$$

[where P is the vapour pressure of the liquid, which is a measure of activity of the vapour (assumed to behave ideally), considering liquid activity as constant], which is Clausius–Clapeyron equation. Therefore, the given statement is justified.

b. Since $\Delta \bar{H}_{vap}$ and T cannot be negative, it follows from the Clausius–Clapeyron equation that $d\ln P/dT$ is always positive, i.e. the boiling point (T) of a liquid cannot decrease with rise of applied pressure.

9.13 a. It follows from the Clausius–Clapeyron equation (9.1b) that at triple point, which is common to both solid–gas and liquid–gas curves

$$\left(\frac{dP}{dT}\right)_{sub} \Big/ \left(\frac{dP}{dT}\right)_{vap} = \frac{\Delta \bar{H}_{sub}}{\Delta \bar{H}_{vap}} > 1$$

The given statement follows from this relation.

b. For a solid, which sublimes on heating under open condition, the triple point pressure lies above 1 atm which is not possible for the resulting vapour to attain. But it is possible for the accumulated vapour in a closed container with consequent rise of pressure.

9.14 a. Critical point. Because above critical temperature only vapour phase can exist.

b. Here the low variation of transition temperature with pressure, given by the Clapeyron Eq. (9.1a), arises from low volume change in the transition between the condensed phases. Again, due to negative ΔV for $H_2O(s) \rightarrow H_2O(l)$ transition, the line representing this transition in P–T phase diagram is tilted to the left.

9.15 Vide Section 9.2.
□ Vide problem 9.8.

9.16 This arises from molecular association (through hydrogen bonding) occurring in water but not in carbon tetrachloride.

9.17 For a binary solution of A and B the Duhem–Margules Eq. (10.7) may be written as

$$\frac{x_A}{P_A} \cdot \frac{dP_A}{dx_A} = \frac{x_B}{P_B} \cdot \frac{dP_B}{dx_B}$$

or $\quad \dfrac{x_A}{P_A} \cdot \dfrac{dP_A}{dx_A} + \dfrac{x_B}{P_B} \cdot \dfrac{dP_B}{dx_A} = 0 \quad$ since $dx_A = -dx_B$

Then $\quad \dfrac{dP}{dx_A} = \dfrac{d(P_A + P_B)}{dx_A} = \dfrac{dP_B}{dx_A}\left(1 - \dfrac{x_B P_A}{x_A P_B}\right) \quad$ (I)

Since $\quad \dfrac{dP_B}{dx_A} = -\dfrac{dP_B}{dx_B}$ is negative, $\dfrac{dP}{dx_A}$ will be positive

if $\quad x_B P_A > x_A P_B \quad$ or $\quad \dfrac{P_A}{P_B} > \dfrac{x_A}{x_B}$

Then the vapour phase is relatively richer in that component (A) whose addition to the system results in an increase in total vapour pressure or decrease in boiling point. This is *Konowaloff's rule*.

On the basis of this rule, the boiling point-composition curve that is obtained for a binary liquid pair with minimum boiling point will be of the type shown in Fig. 9.5(II). Such a phase diagram implies that distillation of the solution will give ultimately an azeotropic mixture (M) as distillate and either pure A or pure B in the distilling pot depending on whether the original composition lay between A and M or between M and B respectively.

9.18 a. Azeotropism refers to the phenomenon of boiling of a solution without change of composition for a fixed applied pressure. The necessary condition is that the solution has to exhibit enough deviation from Raoult's law so that the vapour pressure curves with respect to liquid and vapour phase composition touch each other at some minimum or maximum point.

  b. The given statement is justified. Because touching of vapour pressure curves with respect to liquid and vapour phase composition, which is necessary for azeotropism, is not possible if Raoult's law is obeyed.

9.19 a. Vide problem 9.18(b) and (c).

  b. Vide problem 9.19.

9.20 a. Vide Fig. 9.5 III.

  b. Since the solution is initially richer in A than the azeotrope, the initial distillate will be richer in A than the solution of given composition.

  The final residue would be an azeotropic mixture containing A and B in 1:2 mol ratio.

Note: Although phase rule demands that the composition of the initial distillate is fixed under given conditions, it cannot be found out quantitatively. Because here, no definite relations between vapour pressure and composition exist due to non-ideal nature of the solution (compare with problem 9.41). However, it is possible to find exactly the composition of the final residue, because it is same as that of azeotrope of given composition.

9.21 a. Because ethanol–water system forms a minimum boiling azeotrope containing 8.43% by weight of H_2O (at 1 atm). The solution of given composition being richer in water than the azeotrope, the fractional distillation will ultimately give the azeotrope as the distillate and pure water as the final residue.

  b. This can be done conveniently (i) by distillation with a third substance like benzene (ii) by chemical reaction with some suitable substance like CaO that removes water.

  [In (i) water is removed as ternary azeotrope and excess benzene as binary azeotrope formed with ethanol.]

9.22 This is not unlikely because in case of coexistence of three phases, one vapour and two liquid phases, of such two-component system (under closed condition), the degrees of freedom (+1) does not become negative.

Note: The given statement does not have any meaning for the open system where the vapour phase, being not in equilibrium, has to be disregarded.

9.23 a. Vide Section 9.4.2(c) and problem 9.31.

  b. The given statement is justified, because here $F = 0$. Vide problem 9.31.

9.24 a. Vide problem 9.46.

  b. Vide problem 9.33.

9.25 a. This is possible by distillation of the high boiling liquid (A) with another liquid (B), immiscible with it, as in steam distillation. Here total pressure $P = p_A^* + p_B^*$ (assuming A and B to obey ideal gas law). Since $P > p_A^*$ or p_B^*, the boiling point of the system will be lower than that of either of the pure components.

  b. Vide problem 9.42.

9.26 $F = C - P + 2 - 1$, for fixed pressure

$= 3 - 2 + 2 - 1 = 2$, vide Section 9.4.4.

9.27 When a solute, more appropriately 'chemical species', is distributed between two immiscible solvents 1 and 2, its chemical potential (μ) in the two layers becomes same at equilibrium

i.e. $\mu_1^* + RT \ln a_1 = \mu_2^* + RT \ln a_2$

where a_1 and a_2 are activities of the solute in the respective solvents, and μ_1^* and μ_2^* are constants at constant temperature and pressure. Then

$$\frac{a_1}{a_2} = \text{constant, at constant temperature and pressure}$$

Again $\qquad \dfrac{a_1}{a_2} = \dfrac{x_1}{x_2}$, if the solutions are ideal

$$= \frac{c_1}{c_2}, \text{ if the solutions are dilute}$$

Then $\qquad \dfrac{c_1}{c_2} = \text{constant } (K),$

at constant temperature and pressure it the solute exists in the form of same species in the two solvents

This is Nernst distribution law in its original form that holds under the specified conditions.

The constant K is called the partition coefficient of the dissolved species between solvent 1 and 2. As implied by its nature, K should vary with temperature.

If the solute species (A) dimerises in solvent 2 (say)

$$2A \xrightarrow{Ka} A_2$$

the distribution law will be applicable only to the species A which is common to both the solvents. Then the above distribution law will now change to the following form

$$\frac{C_1}{(C_2 \alpha / 2K_a)^{1/2}} = K$$

C_2 now represents the concentration of the solute if it were entirely in the form A; α is the degree of association.

If the dimerisation is assumed to be complete (i.e. $\alpha = 1$), then

$$\frac{C_1}{\sqrt{C_2}} = \text{constant}$$

Note:

1. In general, in case of association (assumed to be complete

$(C_1 / C_2)^{1/n} = \text{constant}$ and in case of dissociation into n species

$(C_1 / C_2)^n = \text{constant}$. The distribution law may, therefore, be used to get an idea of the form in which a dissolved substance exists.

2. Partition coefficient (which refers to a particular chemical species) will be same as the distribution coefficient (which refers to the solute as a whole) provided the solute exists in form of only one species in the solutions concerned.

3. In the distribution of a solute between two immiscible solvents, the two liquid layers may individually behave as ideal solution, though the system as a whole is non-ideal (as the involved solvents are immiscible).

9.28 In the recovery of a substance from V_1 volume of its solution, let n successive extraction operations are done with an extracting solvent using V_2 volume in each operation. The fraction remaining unextracted f_n and that after a single extraction operation with all nV_2 volume f_1 will be

$$f_1 = \frac{V_1}{V_1 + KnV_2} \quad \text{by Eq. (9.6a)}$$

$$f_n = \left(\frac{V_1}{V_1 + KV_2}\right)^n \quad \text{by Eq. (9.6c)}$$

Then
$$\frac{f_1}{f_n} = \frac{[1 + (KV_2/V_1)]^n}{1 + Kn(V_2/V_1)}$$

$$= \frac{1 + Kn(V_2/V_1) + \text{positive terms}}{1 + Kn(V_2/V_1)} > 1$$

The given statement follows from this relation.

Note: Washing out of impurities resembles in principle with solvent extraction.

9.29 Here (a) is equivalent to a single stage extraction with $2Vcc$ of B. Now a solvent extraction gives better results when it is done in stages using the allotted solvent in parts. Then (b) will be more efficient than (a).

9.30 The number of operations (n) needed for 99% recovery, i.e. for 1% remaining unextracted is given, according to Eq. (9.6c), by

$$\frac{1}{100} = \left(\frac{200}{200 + 20 \times 25}\right)^n = \left(\frac{2}{7}\right)^n$$

Then
$$n = \frac{\log(1/100)}{\log(2/7)} = 3.7$$

Therefore, the number of extraction needed = 4 (the higher integer nearest to 3.7).

9.31 A system of this type, involving condensed phases, is usually studied by exposing it to the atmosphere (vide problem 9.20). At eutectic point $P = 3$ (two solid phases of pure metal and one liquid phase containing both the metals). Then

$$F = C - P + 2 - 1, \quad -1 \text{ is due to the fixed pr of atmosphere}$$
$$= 2 - 3 + 2 - 1 = 0$$

Therefore, the eutectic point is invariable.

□ At fixed pressure, the eutectic point, like melting point of a pure substance, is invariable. However, an eutectic is not a compound on the ground that (i) eutectic composition is a function of pressure (ii) an eutectic appears heterogeneous under the microscope (iii) the physical properties of an eutectic, e.g. density, heat of solvation, are mean of the values for their components.

9.32 One way of detecting phase change in a solid–liquid system is by thermal analysis that takes advantage of the effect of enthalpy change during a first-order transition. The technique consists in withdrawing heat from a melt at a steady rate and recording the temperature of the system as a function of time. The slope of the cooling curve thus found is approximately proportional to the system's heat capacity C_p. The first appearance (or disappearance) of a phase gives rise to a *break* and appearance of one more phase to a *halt*. (i.e. no change of temperature) in the cooling curve. The relative lengths of different parts of the cooling curve depends on initial composition of the melt. The horizontal part of the cooling curve is of highest length for a melt of eutectic composition.

The cooling curves for (i) and (ii) are shown in Fig. 9.8.

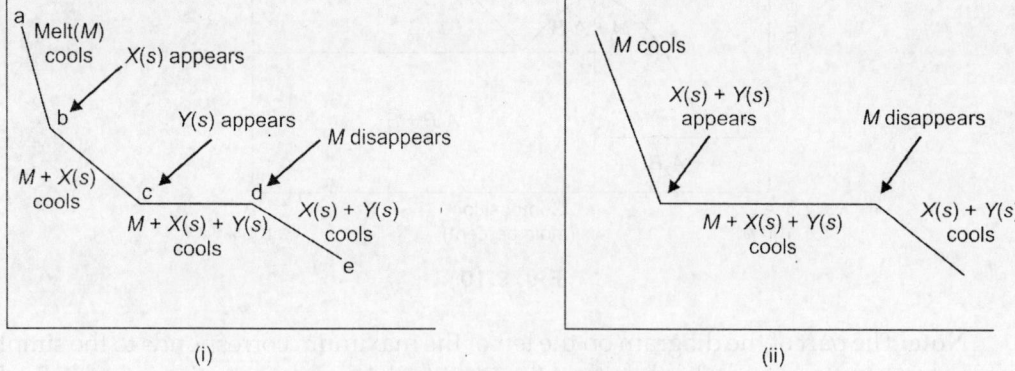

Fig. 9.8

In (i) the new phase that appears first is $X(s)$ because the starting melt is richer in X than the eutectic melt. The slope of cooling curve decreases in the order $ab > bc > de$ because the heat capacity of the system increases in the order

$$C_p(M) < C_p \text{ (M + solid X)} < C_p \text{ (solid X + solid Y)}$$

The horizontal part of the cooling curve is a consequence of the invariability (i.e. $F = 0$) of the system at the eutectic point. The horizontal part is longer in (ii) because starting with same amount of melt of different composition, the amount freezing at eutectic temperature is greater when the melt is of eutectic composition.

9.33 To fit to the given situation, the phase diagram of a simple eutectic has to be modified through introduction of an additional line that takes care of different solubility of the two allotropes α and β of A in molten B, shown in Fig. 9.9.

☐ At the transition point X, $P = 4$, and then
$$F = C - P + 2 - 1, \text{ when exposed to the atmosphere}$$
$$= 3 - 4 + 2 - 1 = 0$$

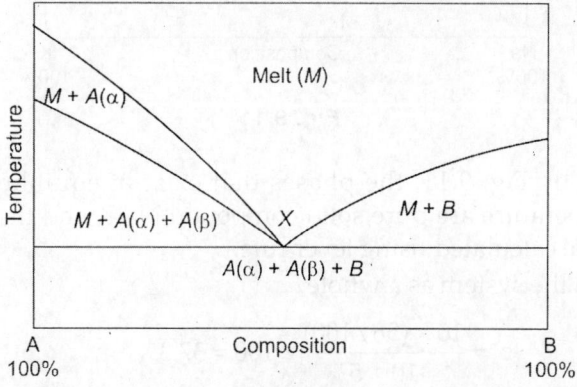

Fig. 9.9

9.34 The required diagram is shown in Fig. 9.10.

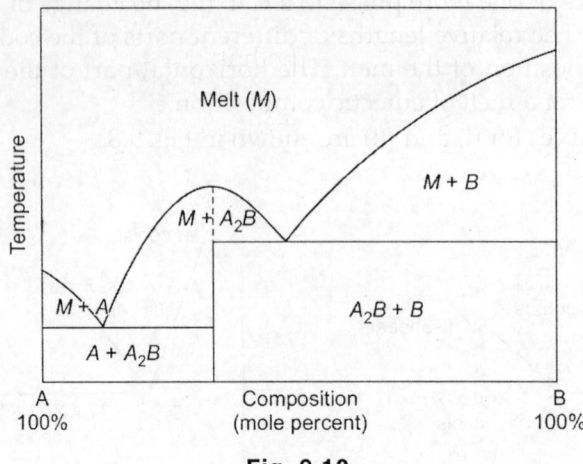

Fig. 9.10

Note: The part of the diagram on the left of the maximum corresponds to the simple eutectic system $A – A_2B$ while the other part to the simple eutectic system $A_2B – B$. The overall system is however, a two-component one.

9.35 Vide Section 9.4.3.

9.36 Here, the relevant phase diagram is of the type shown in Fig. 9.11.

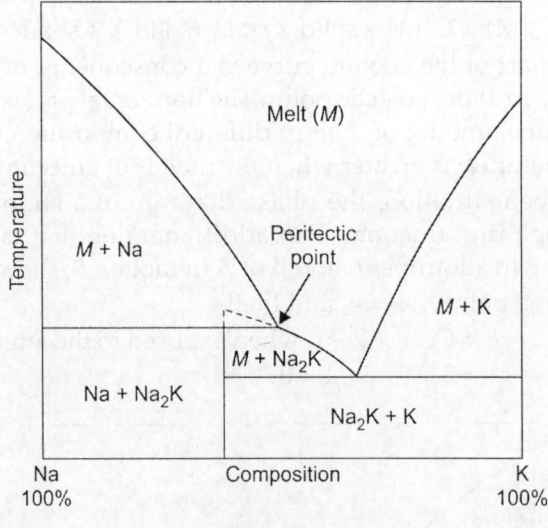

Fig. 9.11

As suggested by Fig. 9.11, the phases that exist in equilibrium just below the peritectic temperature are pure solid compound Na_2K and the melt (M) having % of K that can be calculated using lever rule.

Here % of K in the system as a whole,

$$z = \frac{10 \times (56/100)}{10 + 5} \times 100 = 37.3$$

Now $\quad \dfrac{\text{mass of } M \text{ (final)}}{\text{mass of } Na_2K \text{ formed}} = \dfrac{z - \text{weight \% of K in } Na_2K}{\text{weight \% of K in } M \text{ (final)} - z}$, by lever rule

If x g of K comes out of M, then

$$\frac{10 - x}{(85/39)x} = \frac{37.3 - 45.9}{100\left(\dfrac{5.6 - x}{10 - x}\right) - 37.3} \quad \text{since 39 g of K} \equiv 85 \text{ g of } Na_2K$$

whence $\qquad x = 4.35$

The required weight % of K in the melt $= 100\left(\dfrac{5.6 - x}{10 - x}\right)$

$$= 100\left(\frac{5.6 - 4.25}{10 - 4.25}\right) = 23.5$$

Note:

1. Here the involved compound melts with decomposition (at the peritectic temperature) while in question 9.34, the compound melts without decomposition.

2. The lever rule is not limited to mole % scale. Because we can arrive at this rule simply from material balance as follows. If n is the total number of moles of a two-phase system of overall composition z_i (i.e. if the system were homogeneous), n_1 is that of the phase of composition x_i and n_2 is that of the other phase of composition y_i, then

$$\left.\begin{array}{r} nz_i = n_1 x_i + n_2 y_i \\ nz_i = n_1 z_i + n_2 z_i \end{array}\right\} \Rightarrow \frac{n_1}{n_2} = \frac{z_i - y_i}{x_i - z_i}$$

It does not concern with state of aggregation of the phases, nor does it concern with number of components of the system, as it concerns only with distribution of a particular component i between the phases.

9.37   a. This is possible with a hydrated substance that melts without decomposition (e.g. $FeCl_3 \cdot 12H_2O$), if the prevailing temperature does not lie below the eutectic temperature of the system formed by this hydrate with other lower hydrate(s). Here, below certain % of H_2O determined by the maximum in the phase diagram, lowering of H_2O content by isothermal evaporation will lead to the formation of an aqueous solution.

     b. Because ice and salt forms an eutectic system with eutectic temperature much below the melting point of ice.

     c. Because tin and lead forms an easily fusable eutectic. Toughness of tin, together with its non-toxicity, has the added advantage.

9.38 At freezing point T, at pressure P

$$\Delta T = \frac{T \Delta P \, \Delta V_{\text{fus}}}{\Delta H_{\text{fus}}} \quad \text{vide problem 9.37(a)}$$

$$= \frac{(273.15 \text{ K})\,(10^5 \text{Pa})\left[\left(\dfrac{1}{0.9998} - \dfrac{1}{0.9168}\right) \times 10^{-6} \text{ m}^3\text{g}^{-1}\right]}{(333.5 \text{ J g}^{-1})}$$

$$= 0.0075 \text{ K}$$

9.39 $$\Delta \bar{H}_{vap} = \frac{R \ln(P_2/P_1)}{(1/T_1) - (1/T_2)} \quad \text{by Eq. (9.1c)}$$

assuming ΔH to be independent of T

$$= \frac{(8.314 \text{ JK}^{-1} \text{mol}^{-1}) \ln \dfrac{60}{40}}{\dfrac{1}{(20.7 + 273.15) \text{ K}} - \dfrac{1}{(29.1 + 273.15) \text{ K}}}$$

$$= 3.371 \times 10^4 \text{ J mol}^{-1}$$

9.40 Vide previous question.

Here $P_2 = 760$ mmHg at $T_2 = 373$ K, the normal boiling point of water.

10

Ideal Solutions—
Colligative Properties

A solution is a homogeneous multicomponent system. In this chapter, we concentrate on liquid solutions in evaluating some of their simply correlated thermodynamic properties.

10.1 IDEAL SOLUTIONS

An ideal solution is one of which the chemical potential (μ_i) of any constituent species i can be expressed as

$$\mu_i = \mu_i^*(T, P) + RT \ln x_i \qquad \text{(vide Section 5.7)} \qquad (10.1)$$

The macroscopic properties commonly attributed to an ideal solution, can all be deduced from this relation. Thus on mixing the ingredients of an ideal solution, the accompanying change in thermodynamic properties will be

$$\Delta G_{mix} = G_{final} - G_{initial}$$

$$= \sum n_i \mu_i - \sum n_i \mu_i^* = RT \sum n_i \ln x_i \qquad (10.2)$$

$$\Delta S_{mix} = -\left[\frac{\partial \Delta G_{mix}}{\partial T}\right]_{P, n_i} = -R \sum n_i \ln x_i \qquad (10.3)$$

$$\Delta H_{mix} = \Delta G_{mix} + T \Delta S_{mix} = 0$$

$$\Delta V_{mix} = \left[\frac{\partial \Delta G_{mix}}{\partial P}\right]_{T, n_i} = 0$$

$\Delta H_{mix} = 0$ implies that in an ideal solution, the energies of interaction between like molecules and unlike molecules are same. Again $\Delta V_{mix} = 0$ implies that molecules of the system are of same size and shape.

The above expression of chemical potential serves as a simple mathematical model for solutions just as $PV = nRT$ for gases.

10.2 COLLIGATIVE PROPERTIES

These are properties associated with solutions which depend primarily on the mole fraction of the solute particles but not on their nature. Well familiar examples are lowering of vapour pressure, elevation of boiling point, depression of freezing point and osmotic pressure. These properties are all bound together through their common origin—the lowering of chemical potential of solvent by a solute, and hence they are called colligative (Latin *colligatus* means tied). Purely colligative properties are expected only with ideal solutions. The basis of their treatment is the equilibrium between two appropriate phases, i.e. equality of chemical.potential of the common constituent (i.e. solvent) between the relevant two phases.

10.3 LOWERING OF VAPOUR PRESSURE OF A SOLVENT BY A SOLUTE

Raoult's Law

Due to mixing, the vapour pressure of an ingredien i is lowered from p_i^* in pure state to p_i in solution of mole fraction x_i such that

$$p_i = p_i^* x_i \text{ approximately} \tag{10.4a}$$

This is Raoult's law. If this relation is obeyed exactly, the solution is called ideal (on experimental ground).

For a binary solution, using the suffix 1 for solvent and 2 for solute

and
$$\left. \begin{array}{l} p_1 = p_1^* x_1 \\ p_2 = p_2^* x_2 \end{array} \right\} \tag{10.4b}$$

Then relative lowering of vapour pressure

$$\frac{p_1^* - p_1}{p_1^*} = 1 - x_1 = x_2 \tag{10.4c}$$

where x_2 is mole fraction of the solute.

Total vapour pressure

$$\begin{aligned} P &= p_1 + p_2 = p_1^* x_1 + p_2^* x_2 \\ &= p_1^* + (p_2^* - p_1^*) x_2 \text{ since } x_1 + x_2 = 1 \end{aligned} \tag{10.4d}$$

Then Raoult's law demands that a plot of P vs x_2 should be linear. Linearity is also expected for $1/y$ vs $1/x$ plot, where x is the mole fraction of either ingredient in the liquid phase and y is that of the same ingredient in the vapour phase.

As expected from its derivation, the Raoult's law is an ideal law which holds at all concentrations only for ideal solutions. For real solutions, Raoult's law serves as a limiting law that may be written as

$$p_i \rightarrow p_i^* x_i \text{ as } x_i \rightarrow 1$$

At ordinary concentration, real solutions exhibit either negative or positive deviations. However, if one of the constituents in a binary solution exhibits negative deviation, the other will also exhibit the same. This is also observed in case of positive deviation.

Henry's Law

In real solutions of low concentrations, the solvent obeys Raoult's law but the solute does not. It has been found that the vapour pressure of a volatile (non-electrolyte) solute in dilute solution is proportional to its mole fraction, i.e.

$$p_i = K_i x_i \quad K_i \neq p_i^*, \text{ the vapour pressure of pure solute} \tag{10.5}$$

This is Henry's law.

We can arrive at this law thermodynamically on the basis of the following relations

$$\mu_{i\,(liq)} = \mu_{i\,(vap)}, \text{ at equilibrium}$$

$$\mu_{i\,(\text{liq})} = \mu^*_{i(\text{liq})}(T, P) + RT \ln x_i, \text{ if the solution is ideal}$$

$$\mu_{i\,(\text{liq})} = \mu^*_{i(\text{vap})}(T) + RT \ln p_i, \text{ if vapour behaves as an ideal gas}$$

From all these, we have

$$\frac{p_i}{x_i} = \exp\left[\frac{\mu^*_{i(\text{liq})} - \mu^*_{i(\text{vap})}}{RT}\right] \tag{10.6}$$

$$= K_i, \text{ the Henry's law constant}$$

It appears from this derivation that Henry's law is applicable only when the solution behaves ideally. Further K_i is independent of composition of the solution and practically independent of pressure due to negligible dependence of $\mu^*_{i(\text{liq})}$ on pressure. Then in case of ideal solution, $K_i = p_i^*$ (since $p_i = p_i^*$ for $x_i = 1$) and Henry's law becomes Raoult's law, i.e. both the laws become applicable to the same constituent, solvent or solute, at all concentrations.

For real solutions, however, Henry's law serves as a limiting law that can be expressed as

$$p_i \rightarrow K_i x_i \text{ as } x_i \rightarrow 0$$

Further, this law is now applicable only to one of the constituents of a solution in a small concentration range (0 to x_2 say), and Raoult's law only to the other constituent in the corresponding concentration range (i.e. 1 to $1 - x_2$), for a binary solution.

A solution is said to be ideal-dilute if the solute(s) obeys Henry's law (but not Raoult's law) and the solvent obeys Raoult's law (Fig. 10.1).

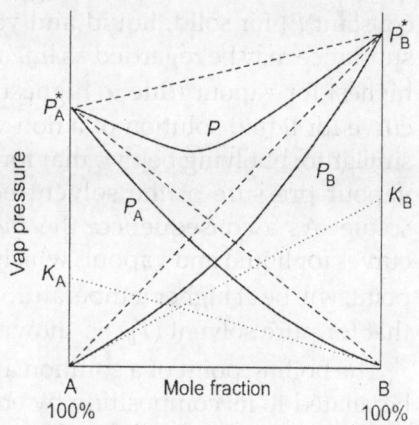

Fig. 10.1: Vapour pressure—composition diagram for a binary solution of A and B. Broken lines correspond to Raoult's law and dotted lines correspond to Henry's law.

Duhem–Margules Equation

The chief difficulty in the study of solutions is that thermodynamics does not provide detailed information concerning the dependence of chemical potential on the composition. The only guidance comes from the Gibbs–Duhem Eq. (5.30b) which for a binary solution reads

$$x_1 d\mu_1 + x_2 d\mu_2 = 0$$

Then

$$x_1 \frac{d\mu_1}{dx_1} = -x_2 \frac{d\mu_2}{dx_1}$$

$$= x_2 \frac{d\mu_2}{dx_2}, \quad \text{since } x_1 + x_2 = 1 \text{ and hence } dx_1 = -dx_2$$

Now, for any constituent i of a liquid mixture

$$\mu_i = \mu_i^\circ(T) + RT \ln p_i$$

where p_i is the partial vapour pressure of the ith constituent assuming it to behave as an ideal gas

Differentiation with respect to x_i at constant temperature and total pressure gives

$$\frac{d\mu_1}{dx_1} = RT \frac{d\ln p_1}{dx_1} \quad \text{(for binary solutions)}$$

$$\frac{d\mu_2}{dx_2} = RT\frac{d\ln p_2}{dx_2}$$

Then in combination with the above relation

$$x_1\frac{d\ln p_1}{dx_1} = x_2\frac{d\ln p_2}{dx_2}$$

or
$$\frac{d\ln p_1}{d\ln x_1} = \frac{d\ln p_2}{d\ln x_2} \qquad (10.7)$$

which is Duhem–Margules equation.

10.4 ELEVATION OF BOILING POINT

If the variation of T is not large, the μ vs T curve (at constant P) for solid, liquid and vapour state of a pure substance may be regarded as linear with negative slope highest for vapour (due to highest $\bar{S}$ for vapour). Such curve for liquid solution of a non-volatile solute will be similar to, but lying below, that for the pure solvent, the vapour pressure of the solvent being lowered by the solute. As a consequence, the meeting point of such curves for liquid and vapour, which represent the boiling point, will be at higher temperature (T_b') for solution than that for pure solvent (T_b) as shown in Fig. 10.2.

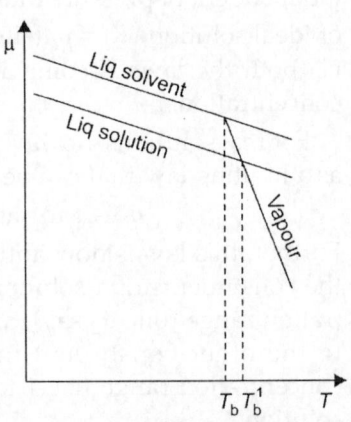

Fig. 10.2

The boiling point of a solution at fixed pressure P can be related to its composition by equating the chemical potential of solvent in the liquid phase to that in the vapour phase. In case of non-volatile solute, we have, using suffix 1 for solvent

$$\mu_1^*(g,T,P) = \mu_1(l,T,P,x_1)$$
$$= \mu_1^*(l,T,P) + RT\ln x_1, \text{ if the solution is ideal}$$

or
$$\ln x_1 = \frac{\mu_1^*(g,T,P) - \mu_1^*(l,T,P)}{RT}$$

$$= \frac{\Delta\bar{G}_{vap}}{RT} \qquad (10.8)$$

where $\Delta\bar{G}_{vap}$ is the molar free energy of vaporisation of pure solvent at T and P

Then
$$\left(\frac{\partial\ln x_1}{\partial T}\right)_P = \frac{1}{R}\left[\frac{\partial(\Delta\bar{G}_{vap}/T)}{\partial T}\right]_P$$

$$= -\frac{\Delta\bar{H}_{vap}}{RT^2} \qquad (10.9)$$

by Gibbs–Helmholtz relation Eq. (5.11b) where $\Delta\bar{H}_{vap}$ is the molar enthalpy of vaporisation of pure solvent.

On integrating, $\displaystyle\int_1^{x_1} d\ln x_1 = -\frac{\Delta\bar{H}_{vap}}{R}\int_{T_b}^{T_b'}\frac{dT}{T^2}$, assuming $\Delta\bar{H}_{vap}$ to be independent of T

or
$$\ln x_1 = \frac{\Delta \bar{H}_{vap}}{R}\left(\frac{1}{T_b} - \frac{1}{T_b^*}\right) \tag{10.10a}$$

$$= \frac{\Delta \bar{H}_{vap}}{R}\left(\frac{1}{T_b^* + \Delta T_b} - \frac{1}{T_b^*}\right) \tag{10.10b}$$

where T_b^* is boiling temperature of solvent and T_b is that of solution of composition x_1 at same pressure P, and $\Delta T_b = T_b - T_b^*$ is the elevation of boiling point.

In case of binary solution of high dilution, when mole fraction of the solute $x_2 \ll 1$, $\ln x_1 = \ln(1 - x_2) \simeq -x_2$ and $T_b \simeq T_b^*$, the equation (10.10a) becomes

$$x_2 = \frac{\Delta \bar{H}_{vap}}{R} \cdot \frac{\Delta T}{T_b^{*2}}$$

if the solute is neither associated nor dissociated

or
$$\Delta T_b = \frac{RT_b^{*2}}{\Delta \bar{H}_{vap}} \cdot x_2 \tag{10.11a}$$

$$= \frac{RT_b^{*2}}{\Delta \bar{H}_{vap}} \cdot \frac{m}{1000/M_1} \tag{10.11b}$$

where m is molality of the solution and M_1 is the molecular weight of the solvent

$$= K_b m \tag{10.11c}$$

where
$$K_b = \frac{RT_b^{*2}}{1000 \Delta \bar{H}_{vap}/M_1} \tag{10.12a}$$

K_b is the molal boiling-point-elevation constant, or ebullioscopic constant, of the solvent. If M_1 is expressed in kg mol^{-1}, then

$$K_b = \frac{RT_b^{*2}}{\Delta \bar{H}_{vap}/M_1} \tag{10.12b}$$

In case of volatile solute, depression, instead of elevation, of boiling point may occur. Here

$$\mu_1(g, T, P, y_1) = \mu_1(l, T, P, x_1)$$

where x_1 and y_1 are mole fractions of solvent in the liquid and vapour phases respectively. Then

$$\mu_1^*(g, T, P) + RT\ln y_1 = \mu_1^*(l, T, P) + RT\ln x_1$$

or
$$\ln \frac{x_1}{y_1} = \frac{\Delta \bar{G}_{vap}}{RT} = \frac{\Delta \bar{H} - T\Delta \bar{S}}{RT}$$

which leads to the relation

$$\ln \frac{x_1}{y_1} = \frac{\Delta \bar{H}_{vap}}{R} = \left(\frac{1}{T_b} - \frac{1}{T_b^*}\right) \tag{10.13}$$

ignoring temperature variation of $\Delta \bar{H}$ and $\Delta \bar{S}$ and putting $\Delta \bar{S} = \Delta \bar{H}/T_b^*$

It follows from this expression that if

$$x_1 < y_1 \text{ then } T_b > T_b^*, \text{ i.e. elevation of boiling point}$$

and if
$$x_1 > y_1 \text{ then } T_b < T_b^*, \text{ i.e. depression of boiling point}$$

This conclusion holds though T_b^* now corresponds actually to an azeotrope where $x_1 = y_1$.

10.5 DEPRESSION OF FREEZING POINT

On cooling a solution, the solid phase that appears may contain (i) only solvent (ii) only solute or (iii) both solvent and solute, depending on composition of the solution. Freezing refers to the situation (i).

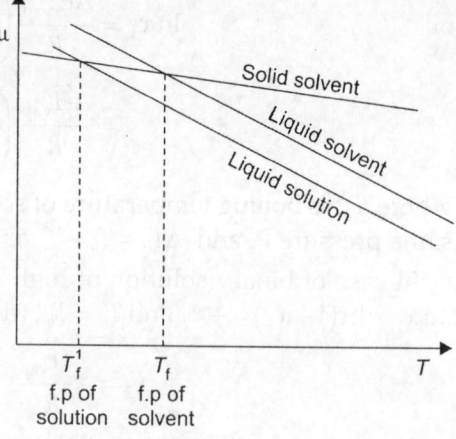

Fig. 10.3

As in case of elevation of boiling point, the usual depression of freezing point of a solvent by a solute can be understood from $\mu - T$ curves (Fig. 10.3), remembering that the meeting point of such curves for solid and liquid phases represents the freezing point.

The freezing point (T) of a solution can be related to its composition (x_1) by equating the chemical potential of solvent in the solid phase to that in the liquid phase, i.e.

$$\mu_1^* (S, T, P) = \mu_1 (l, T, P, x_1)$$

when pure solid solvent is in equilibrium with liquid solution [S denotes solid state]

$$= \mu_1^* (l, T, P) + RT \ln x_1 \text{ if the solution is ideal}$$

or
$$\ln x_1 = \frac{\mu_1^* (l, T, P) - \mu_1^* (S, T, P)}{RT}$$

$$= -\frac{\Delta \overline{G}_{fus}}{RT} \tag{10.14}$$

where $\Delta \overline{G}_{fus}$ is the molar free energy of fusion of pure solvent

[The negative sign implies that while the boiling point is normally elevated, the freezing point is depressed]. Then, by following similar mathematical procedure as for the boiling-point elevation, we can arrive at the following relations:

$$\ln x_1 = \frac{\Delta \overline{H}_{fus}}{R} \left(\frac{1}{T_f^*} - \frac{1}{T_f} \right) \tag{10.15a}$$

$$= \frac{\Delta \overline{H}_{fus}}{R} \left(\frac{1}{T_f^*} - \frac{1}{T_f^* - \Delta T_f} \right) \tag{10.15b}$$

where $\Delta T_f = T_f^* - T_f$ is the depression of freezing point

and
$$\Delta T_f = \frac{RT_f^{*2}}{\Delta \overline{H}_{fus}} \cdot x_2 \tag{10.16a}$$

$$= K_f m, \text{ for dilute solution of molality } m \tag{10.16b}$$

where K_f is the molal freezing-point-depression constant, or cryoscopic constant of the solvent given by

$$K_f = \frac{RT_f^{*2}}{1000 \Delta H_{fus} / M_1} \tag{10.17a}$$

$$\equiv \frac{RT_f^{*2}}{\Delta \overline{H}_{fus} / M_1} \text{ if } M_1 \text{ is expressed in kg mol}^{-1} \tag{10.17b}$$

If the solid phase, in equilibrium with the liquid solution, is a solid solution, instead of pure solvent, then there might be so-called freezing-point elevation. Here

$$\mu_1\,(S,\,T,\,P,\,y_1) = \mu_1\,(l,\,T,\,P,\,x_1)$$

This leads to the relation

$$\ln\frac{x_1}{y_1} = \frac{\Delta\bar{H}_{fus}}{R}\left(\frac{1}{T_f^*} - \frac{1}{T_f}\right)\tag{10.18}$$

It follows from this relatin that if $x_1 < y_1$, then $T_f < T_f^*$ which corresponds to freezing-point depression and if $x_1 > y_1$, then $T_f > T_f^*$ which corresponds to freezing-point elevation.

The Eq. (10.18) is similar to Eq. (10.13). Here T_f^* corresponds actually to a solid solution where $x_1 = y_1$.

Solubility

Suppose a solid substance remains in equilibrium with its solution at the saturation point. The situation may be looked as freezing of a solution with the terms *solute* and *solvent* interchanged. Then in case of a saturated solution, the Eq. (10.15a) transforms into the following one

$$\ln x_2 = \frac{\Delta\bar{H}_{fus}}{R}\left(\frac{1}{T^*} - \frac{1}{T}\right)\tag{10.19}$$

where x_2 is the mole fraction of the dissolved substance in its saturated solution, i.e. solubility at temperature T, $\Delta\bar{H}_{fus}$ is its molar enthalpy of fusion and T^* is its melting point. This equation implies that a particular substance will have same solubility in all solvents with which it forms ideal solutions. Further, the solubility will be favoured by low T^* and high $\Delta\bar{H}_{fus}$.

10.6 OSMOSIS AND OSMOTIC PRESSURE

If two solutions, containing same solvent and solute but at different concentration, are kept at same temperature and pressure separated by a membrane permeable only to the solvent, the latter will flow through the membrane toward the more concentrated solution (where concentration of the solvent is lower). This phenomenon is called osmosis.

Osmotic pressure of a solution is the pressure difference required to just prevent the flow of solvent molecules from pure solvent to the solution separated by a semipermeable membrane.

Let a solution at pressure P is separated from the pure solvent at pressure P' by a membrane permeable only to the solvent. At equilibrium, the chemical potential of solvent will be same on both sides of the membrane, i.e.

$$\mu_1^*\,(l,\,T,\,P') = \mu_1\,(l,\,T,\,P,\,x_1)$$

$$= \mu_1^*\,(l\,T,\,P) + RT\ln x_1,\ \text{if the solution is ideal}$$

or $\qquad RT\ln x_1 = \mu^*(l,\,T,\,P') - \mu^*(l,\,T,\,P)$

$$= \int_P^{P'} \bar{V}_1^*\,dp$$

where $\overline{V}_1^*$ is the molar volume of pure liquid solvent at temperature T and pressure P

$$= \overline{V}_1^* (P' - P)$$

(assuming $\overline{V}_1^*$ to be independent of pressure)

$$= -\overline{V}_1^* \pi$$

where $\pi = P - P'$ is the osmotic pressure of the solution

or $\qquad \overline{V}_1^* \pi + RT \ln x_1 = 0$

In case of binary ideal solution of high dilution, $\ln x_1 = \ln(1 - x_2) \simeq -x_2 \simeq -n_2/n_1$, where n_1 is the number of moles of solvent and n_2 that of solute present in the solution

$$\pi = \frac{n_2 RT}{n_1 \overline{V}_1^*} \simeq \frac{n_2 RT}{V}, \text{ where } V \text{ is volume of the solution}$$

$$= CRT \tag{10.21}$$

where C is molar concentration of the solute in solution.

Equation (10.21) represents the van't Hoff's law.

10.7 VAN'T HOFF FACTOR

For ideal solutions of low concentration, colligative property $\propto$ mole fraction of solute (x_2). For different colligative property (X), the proportionality constant is different but the following ratio is almost same

$$\frac{X_{\text{observed}}}{X_{\text{calculated, disregarding any association or dissociation of the solute}}} = i$$

i, called van't Hoff factor, will have value depending on the molecular form of the solute.

$$i = 1 + (n - 1)\alpha, \text{ in case of dissociation} \tag{10.22a}$$

$$= 1 + \left(\frac{1}{n} - 1\right)\alpha, \text{ in case of association} \tag{10.22b}$$

where α is the degree of dissociation/association and n is the order of dissociation/association. It should be remembered that the Van't Hoff factor will have significance, only if the solution is dilate; otherwise the different colligative properties are to be connected through x_1 as follows

$$\ln \frac{p_1}{p_1^*} = \ln x_1 = \frac{\Delta \overline{H}_{\text{vap}}}{R} \left(\frac{1}{T_b^* + \Delta T_b} - \frac{1}{T_b^*}\right) = \frac{\Delta \overline{H}_{\text{fus}}}{R} \left(\frac{1}{T_f^*} - \frac{1}{T_f^* - \Delta T_f}\right) = -\frac{\overline{V}_1^* \pi}{RT} \tag{10.23}$$

Another term relevant with colligative properties is osmotic coefficient (ϕ) of the solvent which is defined by the relation

$$\mu_1 = \mu_1^*(T, P) + \phi RT \ln x_1$$

where $\phi \to 1$ as $x_1 \to 1$.

10.8 DETERMINATION OF MOLECULAR WEIGHT

Colligative properties can be conveniently used in the determination of molecular weight using the expressions for dilute solutions. Cryoscopic method is more reliable than the ebullioscopic method which again is better than that based on the vapour pressure measurement. However, these methods are not considered suitable if molecular weight is very high (exceeding 5000, as in case of carbohydrates and proteins) when the measurement of osmotic pressure is preferred.

Determination of molecular weight by osmometry, i.e. from the measurement of osmotic pressure is based on the following virial-like expression (due to deviation from ideality)

$$\frac{\pi}{\rho} = \frac{RT}{M}\left(1 + \frac{B}{M}\rho + ...\right) \tag{10.24}$$

where π is the osmotic pressure of the solution, ρ is its density and M is molecular weight of the solute, and B is osmotic virial coefficient. From π/ρ vs ρ plot (if linear), the intercept RT/M (at $\rho = 0$) and hence M can be found out.

Now-a-days molecular weight is determined accurately using mass spectroscopy.

10.9 NONIDEAL SOLUTIONS

Here the expressions for different properties of an ideal solution have to be modified for non-ideality of the liquid phase by replacing mole fraction with activity (and pressure with fugacity, if the vapour phase deviates appreciably from the ideal gas laws) using the concept developed in Section 5.9. It should be kept in mind that in the expression $\mu_i = \mu_i^* (T, P) + RT \ln a_i$, μ_i^* represents the chemical potential of the pure constituent species i in a state depending on the standard state chosen for $\gamma_i = (a_i/x_i) = 1$. The choice rests on either of the following two conventions

Convention I: $\gamma_i \to 1$ as $x_i \to 1$, based on Raoult's law
Convention II: $\gamma_i \to 1$ as $x_i \to 0$, based on Henry's law

For either convention, the activity of a species is unity in the standard state. This is obvious for convention I, where both γ_i and x_i are unity but not for convention II where $\gamma_i = 1$ but $x_i \neq 1$.

According to the convention I, the standard state of a constituent species of a solution is the pure species at the same temperature and pressure as the solution concerned. But in convention II, the standard state of the species is its hypothetical pure state with chemical environment of its infinitely dilute solution. The convention I is customarily used for that constituent of the solution whose mole fraction can be varied up to unity without change of phase. Otherwise the convention II is used.

Then, in terms of activity and activity coefficient, we have

Raoult's law: $p_i = a_i p_i^* = \gamma_i x_i p_i^*$, by convention I
Henry's law: $p_i = a_i' K_i = \gamma_i' x_i K_i$, by convention II
Obviously $\gamma_i' \neq \gamma_i$ as $K_i \neq p_i^*$.

10.10 REGULAR SOLUTIONS

In an ideal solution, the activity coefficient of each species is same (equal to unity) for all composition. But, for a real solution, the matter is not so simple. However, so-called regular solutions exhibit following regularity in the variation of activity coefficient (γ_i) with composition x_i

$$\log \gamma_i = - A x_i^2$$

where 'A' is a parameter that has the same value for each species. A regular solution may be thought of as one in which the constituent species are distributed randomly (as in an ideal solution) but have different energies of interaction with each other.

AUXILIARY PROBLEMS

10.1 Show that the formation of an ideal mixture from its pure ingredients always occur spontaneously. Find ΔH_{mix} and ΔV_{mix} for such process. What do the results suggest regarding the molecules of such systems?

10.2 Which of the following solutions will be more nearly ideal?

a. Aq NaCl

b. Aq H_2SO_4

c. Solution of toluene in benzene

d. Solution of o-xylene in p-xylene

10.3 A mixture of $CHCl_3$ and $CDCl_3$ should behave as an ideal solution exactly. True or false.

10.4 Would you consider a homogeneous mixture of dextro- and laevo-rotatory CHDClBr as an ideal solution?

10.5 Show that for an ideal solution ΔG_{mix} will be minimum if the constituents are present in equimolecular amount.

10.6 Which of the following properties are colligative?

a. Partial pressure of a gas

b. Vapour pressure of a liquid

c. Lowering of vapour pressure of a solvent by a solute

d. Osmotic pressure

e. Coefficient of diffusion in a solution

f. Solubility

10.7 The vapour pressure of a liquid depends on temperature. Hence the relative lowering of vapour pressure will also do the same? Is it affected by an inert gas present in the system?

10.8 What is the cause of vapour pressure lowering?

10.9 Lowering of vapour pressure of a liquid by a non-volatile solute is due to arresting of solvent molecules through solvation. Justify or criticise.

10.10 Show that in an ideal mixture of two volatile liquids, the plot of $1/y$ vs $1/x$ is linear where x is the mole fraction of either constituent in the liquid phase and y is that of the same constituent in the vapour phase. Mention the assumptions involved.

10.11 Justify Raoult's law from the molecular view point.

10.12 An ideal solution is better defined in terms of chemical potential by

$$\mu_i = \mu_i^* (T, P) + RT \ln x_i$$

than by Raoult's law. Justify or criticise.

10.13 In a mixture of two kinds of molecular species A and B, if A exhibits negative deviation, B will also do the same. Same is the rule with positive deviation. Justify.

10.14 It is possible that a constituent in a binary solution shows negative deviation from Raoult's law in dilute solution but positive deviation is concentrated solution?

10.15 What physicochemical effects are culminated in the deviation from Raoult's law?

10.16 Henry's law constant is dependent of pressure. True or false?

10.17 In a solution of two liquids A and B, the Henry's law constant of A exceeds its vapour pressure in pure state. What is expected with B. Justify your answer.

10.18 Henry's law is fairly obeyed for aqueous solution of oxygen but not for aqueous solution of HCl even when the solution is dilute. Explain.

10.19 Point out the differences between the following terms:
    a. ideal solution
    b. ideally dilute solution
    c. infinitely dilute solution

10.20 Explain the difference in behaviour of the solute and solvent in a solution of low concentration.

10.21 In a binary solution, if Henry's law holds for one of the constituents in a limited range, Raoult's law will hold for the other constituents in the corresponding range. Establish this.

10.22 What are the significances of the slopes of the vapour pressure composition curve of a binary solution at its two ends.

10.23 Ebullioscopic constant is the rise in boiling point of a solvent associated with a solution of unit molality. Comment.

10.24 Would the ebullioscopic constant be negative for depression of boiling point?

10.25 The boiling point of a solution should always be higher than that of the pure solvent, since the solute lowers the vapour pressure of the solvent. Justify or criticise.

10.26 For a specified amount of a solvent, the elevation of boiling point depends on the amount of solute dissolved in it. Hence, the boiling point elevation is an extensive property. True or false.

10.27 Usually a solute causes elevation of boiling point but depression of freezing point. Why? Which one is greater for a particular solution. Do they depend on the applied pressure? How?

10.28 Show that for a solution with a non-polar solvent at a particular concentration the ratio of the boiling point elevation to the boiling point of the solution is independent of the nature of the solvent. State the assumptions involved.

10.29 On cooling an aqueous solution of urea, what will be the composition of the solid phase that first appears in the following two cases? (a) Dilute solution (b) Concentrated solution. Mention the assumption involved. Would you consider water as solvent in the two cases?

10.30 The solubility x of a substance at temperature T is given by

$$\ln x = \frac{\Delta \bar{H}_{fus}}{R} \left(\frac{1}{T^*} - \frac{1}{T} \right)$$

where T^* is the melting point of the substance and $\Delta \bar{H}_{fus}$ is its heat of fusion. Then, the solubility of a particular substance should be same in all solvents. Comment. Which will have higher solubility in benzene—naphthalene or anthracene?

10.31 Is it always possible to produce a saturated solution? Give reason for your answer.

10.32 Explain depression of freezing point from the molecular view point. Show that the depression of freezing point of an ideal solution is proportional to the decrease in chemical potential of the solvent by solute.

10.33 Explain whether and when an ordinary thermometer (accuracy 0.1°C) can be used for molecular weight determination (to ordinary precision) by Cryoscopic method?

10.34 Explain whether the same Beckmann thermometer can be used for both cryoscopic and ebullioscopic experiments.

10.35 Molecular weight can be determined more accurately by the cryoscopic method than by ebullioscopic method. Why? Does the cryoscopic method always work?

10.36 For the determination of molecular weight exceeding 5000 osmometry is preferred to cryoscopic method. Why? What type of average molecular weight is provided by this method?

10.37 In deducing the relation between the elevation of boiling point of a solvent and the composition of solution it is often assumed that the solute is (a) non-volatile and (b) non-electrolyte. Why? How would the relation be modified if these assumptions are not fulfilled? Will the modification be of same nature in case of depression of freezing point?

10.38 Impurity always lowers the melting point of a pure substance. True or false?

10.39 Concentrations of two solutions can be equalised through osmosis or diffusion. How do the two phenomena originate and differ? Would ΔG be same in the two cases? Diffusion does not involve transportation of solvent. Why?

10.40 a. Does the osmotic pressure of a solution depend on the nature of the semi-permeable membrane used? Does the mechanism of osmosis depend?
   b. Will the osmotic pressure of a 40% aqueous C_2H_5OH be same if it is measured using a membrane permeable only to (i) H_2O and (ii) C_2H_5OH? Explain.

10.41 An ideal solution refers to Raoult's law but not to van't Hoff equation. Why?

10.42 The van't Hoff equation for osmotic pressure has got the resemblance with ideal gas equation. Then osmotic pressure has got the same origin as the gas pressure and may therefore be ascribed to the bombardment of solute molecules on the semipermeable membrane concerned. Do you agree?

10.43 Are osmotic pressure and van't Hoff equation limited only to liquid solutions?

10.44 A solution should be made isotonic (i.e. of same osmotic pressure) with the contents of the blood cell before injecting into the blood stream. Why?

10.45 A vessel can withstand a maximum pressure of 5 atm. Will it crack when a solution of osmotic pressure 6 atm is poured on it.

10.46 Arrange $(N/10)$ NaCl, $(N/10)$ $MgSO_4$ and $(N/10)$ $MgCl_2$ in the increasing order of colligative properties and van't Hoff factor. Comment on the results.

10.47 The vapour pressures of $CHCl_3$ and CCl_4 are 1040 and 620 mm at 70°C. Find the composition of the mixture having normal boiling point 70°C and the composition of the vapour in equilibrium with the liquid solution at 70°C.

10.48 Liquids A and B form an ideal solution. The vapour pressure of A is double of that of B at a certain temperature. If a vapour–liquid equilibrium is established at this temperature in a system prepared by mixing 2 moles of A with 3 moles of B, where the number of moles of A is half of that of B in the liquid phase, calculate the total number of moles in the liquid.

10.49 Henry's law is obeyed by dilute solutions of C_2H_5Br in C_6H_6. The normal boiling point of such a solution containing 1.0 mole percent solution of C_2H_5Br is 79.4°C and that of C_6H_6 is 80.0°C. Find the vapour pressure of a 1.5 mole percent solution of C_2H_5Br in C_6H_6 at 79.4°C. Does the expression $\Delta T_6 = K_b m$ work for this solution. [$\Delta \overline{H}_{vap} = 2.37$ K cal mol^{-1} for C_6H_6].

10.50 10 g of cane sugar (mol weight 342) is dissolved in 1000 g of water. 2 L of dry air is drawn slowly through a series of bulbs containing the solution at 35°C. Calculate the weight of water vapour removed by air. Vapour pressure of water is 42 mm at 35°C.

10.51 Vapour of ether (b.p. 34.6°C) is passed into nitrobenzene (b.p. 210°C) up to saturation point at 35.6°C. Find the concentration of the resulting solution. [$\Delta H_{vap} = 90$ cal g^{-1} for ether].

10.52 A 1.25 weight % solution of a substance in benzene has a vapour pressure of 752 mm at 80°C and normal boiling point of 80.5°C. The normal boiling point of benzene is 80°C. Assuming the solute to be non-volatile, calculate:

a. molecular weight of the substance

b. $\Delta \bar{H}_{vap}$ of benzene

c. $\Delta \bar{G}_{vap}$ of benzene at 1 atm at (i) 80°C and (ii) 80.5°C.

10.53 Acetic acid molecules (mol wt 60) associate in benzene to form dimeric molecules. 1.65 g of the acid when dissolved in 100 g of benzene raised the boiling point of the solvent by 0.36°C. Calculate:

a. the Van't Hoff factor

b. the degree of association. $[K_b = 2.6 \text{ K kg mol}^{-1} \text{ for benzene}]$

10.54 A solution containing 0.684 g cane sugar (mol wt 342) in 100 g water freezes at −0.037°C while a 0.01 molal solution of $K_3Fe(CN)_6$ freezes at −0.062°C. Calculate

a. molal freezing point depression constant for water

b. van't Hoff factor

c. the percentage dissociation of $K_3Fe(CN)_6$.

10.55 A solution containing 10 g of a dibasic acid in 1000 g of water freezes at −0.15°C. 12 ml of (N/10) NaOH is required for neutralisation of 10 ml of this solution. Calculate Van't Hoff factor for the acid and its molecular weight. $[K_f = 1.86 \text{ K kg mol}^{-1} \text{ for water}]$.

10.56 An equimolecular mixture of two liquids, A and B having same freezing point, is slowly cooled. Which species will freeze out first? $\Delta \bar{H}_{fus}$ is lower with A.

10.57 An ideal mixture of A and B freezes at 5°C and boils at 140°C. Pure A melts at 15°C and boils at 130°C and has $\Delta \bar{H}_{vap} = 9 \text{ kcal mol}^{-1}$. Calculate composition of the solution, $\Delta \bar{H}_{fus}$ and $\Delta \bar{H}_{sub}$ of A.

10.58 A and B are two non-volatile solids. A is dissolved in a solvent to give a dilute solution in which it undergoes dissociation of order two (with degree of dissociation α). In the same solvent, when dissolved in the same molecular proportion, B is dimerised (with degree of dimerisation β). The elevation of boiling point in the solution of A is twice that of the solution of B. If α be 80%, what should be β?

10.59 Osmotic pressures of two aqueous solutions of urea are 2 atm and 0.2 atm at 27°C. Find ΔG per mole of H_2O due to its transfer from the concentrated to the dilute solution at the same temperature.

10.60 An aqueous solution of KCl which is 95% dissociated exhibits an osmotic pressure of 20 atm at 27°C. Find (a) the concentration of KCl and (b) the freezing point of the solution. Mention the assumption involved. Given, density of water is 0.99 g cm^{-3} at 27°C and K_f for water is 1.86 K kg mol^{-1}.

10.61 An aqueous solution of a carbohydrate containing 34 g per 100 cm^3 of the solution is found to be isotonic with another aqueous solution 100 cm^3 of which contains 2.92 g of NaCl (mol wt 58.4). Find

a. molecular weight of the carbohydrate

b. normal freezing point of either solution

c. vapour pressure of the solution at 100°C

Will the freezing point of two isotonic solutions be always same?

ANSWERS

10.1 In the mixing process, $\Delta G_{mix} = RT \sum n_i \ln x_i$. by Eq. (10.1)

Here ΔG_{mix} is negative (since $x_i < 1$) which indicates that the mixing will occur spontaneously (at constant T and P).

Vide Section 10.1 for the other parts of the question.

10.2 (d) Because here the necessary condition for a solution to be ideal, $\Delta V_{mix} = 0$, is fulfilled most due to nearly same molecular size of the two isomers.

10.3 False. Because the difference in isotopic mass leads to a difference in molecular size and molecular interaction, though very slightly.

10.4 Such a mixture cannot be considered as an ideal solution which is generally defined on thermodynamic ground. Even it cannot be regarded as a solution on the same ground, because the dextro- and laevo-rotatory species having same chemical potential are identical thermodynamically.

Note: A mixture of dextro- and laevo-rotatory species has to be treated as a solution, if the properties (e.g. optical property) other than thermodynamic properties are concerned. Then the traditional definition of solutions and their classification as ideal and non-ideal are not always effective.

10.5 For a binary solution

$$\Delta G_{min} = \frac{RT}{n_1 + n_2} [x_1 \ln x_1 + x_2 \ln x_2]$$

$$= \frac{RT}{n_1 + n_2} [x_1 \ln x_1 + (1 - x_1) \ln(1 - x_1)]$$

Differentiating with respect to x_1 and equating the result to zero, we have

$$1 + \ln x_1 - 1 - \ln(1 - x_1) = 0$$

or $\qquad x_1 = \frac{1}{2}$, i.e. $n_1 = n_2$ for ΔG_{mix} to be minimum

In case of multicomponent solutions ΔG_{mix} will be minimum if this requirement is fulfilled by each pair of constituents, i.e. if the constituents are all present in equimolecular amount.

10.6 As per definition of colligative property, only (c) and (d) belong to this category.

a. Partial pressure of a gas is not a colligative property. Because, although it refers to a solution the latter is in gas phase where chemical potential of any species cannot be changed by addition of other species at constant volume (assuming the gas mixture to obey ideal gas laws).

b. Vapour pressure of a pure liquid is not a colligative property because the system consists of only one substance.

e. Here the property depends on the nature of the two species involved.

f. Solubility of a substance is not a colligative property. Because it depends on the nature of the solvent involved. [in contrast with freezing-point depression (a colligative property) which is independent of the nature of the solute particles].

10.7 No, if the vapour phase is kinetically ideal and liquid phase is thermodynamically ideal. With variation of temperature, P^* and P both change by a temperature dependent common factor. This follows from the Clausius–Clapeyron euation, ΔH_{vap} being same for pure solvent and ideal solution. (Since $\Delta H_{mix} = 0$ for an ideal solution).

☐ Yes. In presence of an inert gas P^* and P change in same amount so that $P^* - P$ remains unchanged and $(P^* - P)/P^*$ changes.

Note: Inert Inert gas affects vaporisibility of the liquid solvent but not the condensibility of the solvent vapour by affecting $\mu*(l)$ and not $\mu^*(g)$, while temperature affects both.

10.8 It is due to lowering of the chemical potential of the solvent by the solute. For an ideal solution, it is an entropy effect, ΔH_{mix} for an ideal solution being zero.

From molecular view point, the vaporisation of a liquid is a manifestation of its tendency to achieve greater randomness. This tendency is reduced in the presence of a solute that makes the liquid phase more random, and hence vapour pressure is lowered.

From kinetic stand point, the lowering of vapour pressure by a solute results from lowering of surface density of solvent molecules in the liquid phase with consequent lowering of rate of vaporisation and hence lowering of vapour pressure.

Note: All colligative properties arise due to the entropy effect rather than the enthalpy effect.

10.9 Not justified (as it appears from solution of the previous problem) on the ground that lowering of vapour pressure of a liquid by a non-volatile solute is a common affair with all solutions including ideal solutions where the interaction between unlike molecules is not different from that between like molecules. However, arresting of solvent molecules through solvation (if it occurs) will favour lowering of vapour pressure.

10.10
$$p_2 = p_2^* x_2 \quad \text{by Eq. (10.4b)}$$
$$= y_2 P \quad \text{by Dalton's law}$$

Then
$$\frac{p_2^*}{y_2} = \frac{P}{x_2}$$
$$= \frac{p_1^*}{x_2} + (p_2^* - p_1^*) \quad \text{by Eq. (10.4d)}$$

This transforms to
$$\frac{p_1^*}{y_1} = \frac{p_2^*}{x_1} + (p_1^* - p_2^*) \quad \text{through interchange of suffixes}$$

Then plot of $1/y$ vs $1/x$ of either constituent will be linear.

The assumptions involved are:

i. Dalton's law holds good

ii. p_1^* and p_2^* are independent of total pressure of the system which depends on the composition of either phase.

10.11 The rate of vaporisation of solvent from solution ∝ Surface density of solvent molecules, and hence mole fraction of solvent (x_1), in the liquid phase

The rate of condensation of solvent molecules ∝ Their concentration in the vapour phase

∝ Partial pressure (p_1) of solvent in the vapour phase, if the ideal gas laws are obeyed

At equilibrium, the two rates are equal, and hence

$$p_1 \propto x_1$$

Then $\qquad\qquad p_1 = p_1^* x_1$

[Since $p_1 = p_1^*$ (the vapour pressure solvent) for $x_1 = 1$]

which is Raoult's law.

10.12 The given statement is justified. Because the expression $\mu_i = \mu_i^*(T, P) + RT \ln x_i$ does not involve any assumption regarding the vapour. But Raoult's law requires the vapour to obey ideal gas laws, as it appears from its thermodynamic and kinetic derivation.

10.13 The negative deviation suggests that A–B interaction involves more attractive force than A–A and B–B interactions. Since attraction is mutual, if A exhibits negative deviation. B will also do the same.

Regarding positive deviation, the explanation can be done similarly in terms of repulsive force.

10.14 Such possibility cannot be ruled out in view of the fact that the nature of interaction between molecules varies from attractive to repulsive as they come very close to each other.

10.15 Extreme negative deviation results in compound formation while extreme positive deviation results in immiscibility, i.e. separation between the species concerned.

10.16 Not absolutely true. Vide Henry's law, Section 10.3.

10.17 The same is expected with B. Because the given information regarding A implies its positive deviation from Raoult's law which should necessarily happen with B.

10.18 For any constituent of a solution to obey Henry's law, it must exist in the same molecular form in both liquid and vapour phases (which follows from its derivation). The requirement is fulfilled by oxygen (O_2) and not by hydrogen chloride which exists in form of HCl molecule in the vapour phase but largely as hydrated H^+ and Cl^- ions in the liquid phase. This explains the given statement.

Note: The validity of Henry's law depends on both solute and solvent. Although this law is not valid for HCl in water, it is valid for HCl in benzene. Further, the Henry's law constant for a particular substance is different in different solvents.

10.19 They differ as follows:

In (a), solute, as well as solvent, obey Raoult's law.

In (b), only solvent obeys Raoult's law. Here solute obeys Henry's law.

In (c), the qualities of the solution (like vapour pr, mole fraction etc.) are not affected on addition of solvent.

10.20 In a solution of low concentration, each molecule of the solvent, and also of solute, is surrounded only by the solvent molecules as nearest neighbours. Then the solvent would behave much like a pure liquid, but the solute would behave entirely differently from its pure state, unless the solvent and solute molecules happen to be very similar (as in an ideal solution), in which case the solute would also obey Raoult's law.

10.21 Let Henry's law is obeyed by the solute in the concentration range 0 to x_2, i.e.

$$p_2 = K_2 x_2$$

where the suffix 2 refers to the solute in a binary solution

or $\qquad\qquad \dfrac{d \ln p_2}{d \ln x_2} = 1$

Then, by Duhem–Margules Eq. (10.7), we have for solvent

$$\frac{d\ln p_1}{d\ln x_1} = 1$$

Integration gives $p_1 = K_1 x_1$

$$= p_1^* x_1 \quad \text{since } P_1 = p_1^* \text{ for } x_1 = 1 \text{ (when } x_2 = 0 \text{)}$$

i.e. Raoult's law is obeyed by the solvent in the concentration range $1 - x_2$ to 1.

Note: $K_2 \neq p_2^*$, since $x_2 \neq 1$. This means the solute does not obey Raoult's law. In ideal solutions however, Raoult's law (and also Henry's law) is obeyed by either constituent of the solution in all ranges of concentration.

10.22 For either constituent, the slope of the curve at one end where its mole fraction is zero signifies its Henry's law constant while that at the other end signifies its vapour pressure in pure state (vide Fig. 10.1).

10.23 Such a definition of K_b is faulty, because it is based on the relation $\Delta T_b = K_b m$ which is exact only for a dilute solution having molality (m) much less than unity.

Note: K_b should be defined as

$$K_b = \lim_{\Delta m \to 0} \frac{\Delta T_b}{\Delta m}$$

10.24 Boiling point depression, unlike elevation, does not refer to a pure liquid (solvent). Hence it cannot be expressed, by analogy with elevation, as $\Delta T = Km$ with a negative value of K. However, the meaning of ebullioscopic constant for a pure liquid, as defined by Eq. (10.12), stands and it is always positive.

10.25 The given statement is justified in case of non-volatile solute but not so if the solute is volatile. In the latter case, although the solute lowers the vapour pressure of the solvent, the total vapour pressure (which is sum of the vapour pressures of solvent and solute), on which the boiling point of a solution depends, may not be lower (i.e. the boiling point of the solution may not be higher) than that of the pure solvent.

10.26 False. Here elevation of boiling point is essentially due to increase in mole fraction of the solute and not due to increase in moles of the system. This is supported by the fact that the boiling point of a solution of a particular composition does not depend on the amount involved.

10.27 This is due to (i) lowering of chemical potential of solvent (ii) higher negative slope of μ vs T plot for the vapour phase than that for the liquid and solid phases of a substance (solvent).

☐ Depression of freezing point. This is due to (ii).

☐☐ At ordinary pressure μ is almost independent of P for liquid and solid phases but not for the vapour phase (vide Eq. 5.13b). With increase in P, the slope of μ vs T curve becomes significantly more negative for the vapour phase but remains almost unchanged for the liquid and solid phases. As a consequence, the boiling point elevation is significantly reduced while freezing point depression remains almost same.

Note: With change in pressure, boiling point is more affected than freezing point. Then the elevation of boiling point will change more than the depression of freezing point for same change of pressure.

10.28 From Eq. (10.10a) $\quad \dfrac{\Delta T_b}{T_b} = -\dfrac{RT_b^*}{\Delta \bar{H}_{vap}} \ln x_1$

$\qquad\qquad\qquad\qquad$ = constant, for a solution of particular composition x_1

assuing that $\Delta \bar{H}_{vap}/T_b^*$ is constant for a non-polar liquid (by Trouton's rule).
The assumptions involved:

   i. the solution is ideal $\qquad\qquad\qquad$ ii. the solute is non-volatile

   iii. ΔH_{vap} is independent of T $\qquad\qquad$ iv. Trouton's rule is valid

10.29 Pure ice for (a) and pure urea for (b).

☐ Here it has been assumed that urea does not form any hydrate.

☐☐ Generally, the species which is of primary importance in a binary solution is designated as solute and the other species as solvent that serves as a medium (which necessitates the pure solvent to exist in the same state of aggregation as the solution at the prevailing temperature and pressure). Then, if urea is to emphasise in its aqueous solution, it has to be considered as solute and hence water as solvent is (a) and also in (b).

However, in dealing with colligative properties, it is convenient to consider the species common to the relevant two phases as solvent. From this viewpoint, solvent is water in (a) but urea in (b).

10.30 The given expression is valid provided the solution is ideal, and only then the given statement will be justified.

☐ For real solutions, the given expression which is not valid quantitatively, can only be used for qualitative purposes. Considering that naphthalene has much lower melting point (T^*) than anthracene, it is expected to have higher solubility (x_2) in benzene.

10.31 No. Because in case of ideal solution, for example of a liquid substance in a liquid solvent

$$\ln x_2 = \frac{\mu_2^*(l, T, P) - \mu_2^*(l, T, P)}{RT} = 0 \quad \text{vide Eq. (10.8)}$$

or $x_2 = 1$, i.e. the dissolved substance can have concentration in all ranges and hence never forms a saturated solution. This will happen whenever two substances in the same state of aggregation mix forming an ideal solution, but not necessarily when the solution is non-ideal, even with a gaseous solution (vide note on problem 9.32). But two substances with different states of aggregation can always form a saturated solution under ordinary condition.

Note: The formation of saturated solution is not limited to a pair of substances with different states of aggregation. For example, saturated solution is formed by NaCl(s) with $H_2O(l)$ and also by phenol (l) with H_2O (l) [though not by C_2H_5OH (l) with $H_2O(l)$ under ordinary condition].

10.32 The dissolution of a solute in a solvent enhances the molecular randomness of the system, and this opposes the separation of a pure solid from the solution, i.e. freezing. At lower temperature, this randomness is reduced and then equilibrium between the solid and liquid phases becomes possible, i.e. freezing point is lowered.

☐ This follows from Eq. (10.15a) which transforms to

$$T_t^* - T_f = -\frac{T_f^*}{\Delta \bar{H}_{fus}} \cdot RT_f \ln x_1$$

$\qquad\qquad \propto RT_f \ln x_1$ [the decrease in chemical potential of the solvent at the freezing point of solution.]

Note: The changes of properties of a system are ultimately controlled by the change in free energy (ΔG). In case of colligative properties (i.e. properties of solution relative to solvent) which are associated with mixing of solute with solvent, ΔG is is determined by ΔS alone, if the solution is ideal for which $\Delta H_{mix} = 0$. Then colligative properties of solutions in general are due to entropy effect rather than the enthalpy effect.

10.33 The cryoscopic method is based on the Eq. (10.16b) for the depression of freezing point (ΔT_f) of a dilute solution. Only when K_f is sufficiently high so that ΔT_f is much higher than 0.1°C that an ordinary thermometer can be used.

10.34 Remembering that a Beckmann thermometer is a short-range differential thermometer and the freezing point depression is quite greater than the boiling point elevation for same solution, the same Beckmann thermometer should not be used both for cryoscopic and ebullioscopic experiments for accurate measurement.

10.35 Because

(i) the depression of freezing point of a solution, being greater than the elevation of its boiling point for the same solute concentration, can be more precisely measured.

(ii) the freezing point of a liquid and its depression are much less susciptible to change in atmospheric pressure than the boiling point nd its elevation.

(iii) the experimental substance has less chance of decomposition at the freezing point of the solution which is much less than the boiling point.

□ No, because the cryoscopic method is subject to the condition that the solution must be such that only pure solvent can freeze out.

10.36 Because

(i) for a substance of high molecular weight, the freezing point depression in a dilute solution is too low to be measured accurately.

(ii) membranes which are semipermeable to big solute molecules of such weight are easily available.

□ Number average molecular weight. Because osmotic pressure is a colligative property that depends primarily on the mole fraction of the solute molecules in solution and not on their nature.

10.37 Such assumptions are made to treat the matter in a simplified way.

In case of elevation of boiling point, if the assumption (a) is not fulfilled, Eq. (10.10a) has to be modified to Eq. (10.13), and for failure of the assumption (b) the Eq. (10.11a) for dilute solution has to be modified as

$$\Delta T_b = \frac{RT_b^{*2}}{\Delta \bar{H}_{vap}} \cdot i x_2 \quad \text{due to association or dissociation of the electrolyte}$$

where x_2 is the mole fraction of the solute that would have been, if it were neither associated nor dissociated, and i is the Van't Hoff factor.

In dealing with depression of freezing point, we are concerned only with solid–liquid equilibrium. Hence the assumption (b), but not (a) is relevant. Here due to failure of this assumption, the Eq. (10.16a) would require modification similar to elevation of boiling point.

10.38 False. Because the temperature at which an impure substance melts or freezes, may be higher than that of the pure substance depending on the composition of the solid and liquid phases involved, as it appears from Eq. (10.18).

10.39 These two phenomena originate from the same basic cause—the inequality of chemical potential (due to difference in concentration) of a common constituent in the two solutions.

They differ only in the constituent species that passes from one solution to the other. It is solvent in case of osmosis but solute in case of diffusion.

☐ Yes, because the initial and final states, on which ΔG depends, are same in the two cases.

☐☐ Because the transportation of solvent is inconvenient in diffusion as it involves displacement of greater amount of matter than the transportation of solute.

10.40 a. No, because osmotic pressure is a property only of the solution (system) that is manifested under certain condition involving a semipermeable membrane.

☐ Yes. The mechanism of osmosis, even of same system, is not same with different semipermeable membranes. This is not contradictory to the thermodynamic principle that the change of a state function does not depend on the path of change.

b. No. Here the different osmotic pressure is due to different final states, corresponding to equality of chemical potential of H_2O in (i) and of C_2H_5OH in (ii), rather than due to different membranes which mere create the relevant situations.

Note:

1. The nature of membrane matters when the membranes are permeable to different species, but not when they are permeable to the same species.
2. The value of osmotic pressure of a solution will not be meaningful unless the solvent (or solute) is mentioned or implied.

10.41 Because Raoult's law is obeyed by all ideal solutions while van't Hoff equation is obeyed only by dilute ideal solutions.

Note: This problem has some relevance to problem 10.12.

10.42 No. Because if the given statement were true then, like ideal gas equation $P = cRT$, van't Hoff's equation would have been valid at all ranges of concentration (c) instead of being limited only to dilute solutions.

10.43 Osmotic pressure has meaning also with gaseous solutions (and even with solid solutions, though useless). However, the osmotic pressure of a gas mixture is simply the partial pressure* of the molecular species considered as solute (i.e. one to which the relevant semipermeable membrane is not permeable). Hence the measurement of osmotic pressure of a gas does not require a semipermeable membrane and the term osmotic pressure is not used with gases.

Further, van't Hoff's equation for osmotic pressure is also applicable to gases though at all concentrations, if the ideal gas laws are obeyed.

[* Because pressure of the pure gas P_1 is equal to its partial pressure in the gas mixture at total pressure $P_1' = P_1 / x_1$. Then osmotic pressure

$$P_1' - P_1 = (1 - x_1)P_1' = x_2 P_1']$$

10.44 If this is not done, osmosis might damage the blood cells by causing them to swell abnormally in solutions of lower osmotic pressure. With solutions of higher osmotic pressure shrinking of the cells results.

10.45 The vessel will crack only if the pressure exerted by the solution exceeds 5 atm. But this cannot be ascertained from the given value of osmotic pressure which mere signifies a pressure difference necessary for attainment of equilibrium between the solution and the pure solvent separated by a suitable semipermeable membrane. Therefore, no definite answer can be provided.

10.46 Assuming complete dissociation of the given electrolytes, i.e. $\alpha = 1$, the van't Hoff factor (i), given by Eq. (10.22a), will have the following order

$$\left(\frac{N}{10}\right)\underset{i=2}{NaCl} = \left(\frac{N}{10}\right)\underset{i=2}{MgSO_4} < \left(\frac{N}{10}\right)\underset{i=3}{MgCl_2}$$

The colligative properties will be in the order of ic where c is formal molarity or molality of the electrolyte. Then the required order of colligative property will be as under

$$\left(\frac{N}{10}\right)MgSO_4 < \left(\frac{N}{10}\right)MgCl_2 < \left(\frac{N}{10}\right)NaCl$$

$$ic = 2 \times \frac{1}{2 \times 10} \qquad ic = 3 \times \frac{1}{2 \times 10} \qquad ic = 2 \times \frac{1}{10}$$

□ Although $(N/10)\,MgSO_4$ and $(N/10)\,NaCl$ have same i and same concentration (in normality), they do not have the same value of a colligative property. Because, the number density of the particles (ions) for the two solutions is not the same.

Note: In dealing with colligative properties, it will be convenient if the concentrations of the dissolved particles are provided in units (like mole fraction, molalrity or molality) which are directly linked to number of particles while in dealing with chemical property normality is preferred.

10.47 $$P = \overset{*}{p}_{CHCl_3} + (\overset{*}{p}_{CHCl4} - \overset{*}{p}_{CHCl_3})\, x_{CCl_4} \quad \text{by Eq. (10.4d)}$$

AT normal boiling point (70°C), the total vapour pressure of the mixture, $P = 760$ mm. Then from the given data

$$760 = 1040 + (620 - 1040)\, x_{CCl_4}$$

whence $\qquad x_{CCl_4} = 2/3$

Mole fraction in the vapour phase

$$y_{CCl_4} = \frac{y_{CCl_4} \cdot \overset{*}{p}_{CCl_4}}{P} \quad \text{vide problem (10.10)}$$

$$= \frac{(2/3) \times 620}{760} = 0.544$$

10.48 In the liquid phase, the mole fraction of A, $x_A = 1/3$, from the given data. Now, the mole fraction of A in the vapour phase y_A is related to x_A as

$$\frac{\overset{*}{p}_A}{y_A} = \frac{\overset{*}{p}_B}{x_A} + (\overset{*}{p}_A - \overset{*}{p}_B) \quad \text{vide problem (10.10)}$$

Then $\qquad \dfrac{2\overset{*}{p}_B}{y_A} = \dfrac{\overset{*}{p}_B}{1/3} + (2\overset{*}{p}_B - \overset{*}{p}_A) \quad \text{since } \overset{*}{p}_A = 2\overset{*}{p}_B$

whence $\qquad y_A = \dfrac{1}{2}$

Let z_A be the overall mole fraction of A in the system, if it were homogeneous. Then

$$\frac{\text{number of moles in the liquid phase}}{\text{number of moles in the vapour phase}} = \frac{z_A - x_A}{y_A - z_A}$$

by lever rule (vide note on University question 9.36)

or number of moles in the liquid phase

$$= \text{Total number of moles of the system} \times \frac{z_A - x_A}{y_A - x_A}$$

$$= (2 + 3) \times \frac{2/(2+3) - 1/3}{(1/2) - (1/3)} = 2$$

10.49 Since C_2H_5Br obeys Henry's law, C_6H_6 must obey Raoult's law. Then total vapour pressure (P) of their mixture will be given by

$$P = K_{C_2H_5Br} \, x_{C_2H_5Br} + p^*_{C_6H_6} \, x_{C_6H_6} \tag{I}$$

To find P by this equation, we need to know the Henry's law constant $K_{C_2H_5Br}$ and $p^*_{C_6H_6}$ at the temperature concerned.

$p^*_{C_6H_6}$ at 79.4°C can be calculated using Clausius–Clapeyron equation

$$\ln \frac{P_2}{P_1} = \frac{\Delta \bar{H}_{vap}}{R} \left(\frac{1}{T_1} - \frac{1}{T_2} \right)$$

where $P_1 = 1$ atm at $T_1 = (273 + 80.0)$ K and P_2 is the vapour pressure at $T_2 = (273 + 79.4)$ K, and $\Delta \bar{H}_{vap} = 2.37$ k cal mol^{-1} (given).

$K_{C_2H_5Br}$ at 79.4°C (the normal b.p. of C_2H_5Br) can be calculated from (I) putting $P = 1$ atm, $x_{C_2H_5Br} = \frac{1}{100}$, and $x_{C_6H_6} = 1 - \frac{1}{100}$

P can now be calculated using (I) for $x_{C_2H_5Br} = \frac{1.5}{100}$

☐ No, because of volatility of C_2H_5Br.

10.50 Vapour pressure of cane-sugar solution

$$p_{H_2O} = p^*_{H_2O} \, x_{H_2O}$$

$$= 42 \text{ mm} \times \frac{1000/18.02}{(1000/18.02) + (10/342)} = 41.98 \text{ mm}$$

Air will remove the H_2O vapour till it becomes saturated when partial pressure of H_2O vapour is 41.98 mm.

Then weight of H_2O vapour removed

$$\equiv 2 \text{ L of } H_2O \text{ vapour at 41.98 mm pressure and at 35°C}$$

$$= \frac{\left(\frac{41.98}{760} \text{ atm} \right) (2 \text{ L}) (18.02 \text{ g mol}^{-1})}{(0.08206 \text{ L atm K}^{-1}\text{mol}^{-1}) (273 + 35) \text{ K}} = 0.0787 \text{ g}$$

10.51 Since nitrobenzene has much higher boiling point than ether, the resulting solution (b.p. 35.6°C) may be regarded as one of non-volatile solute. Further, the boiling point elevation of ether $\Delta T = (35.6 - 34.6)$ or 1 K being quite low, the solution may be treated as dilute having concentration given by

$$m = \frac{1000 \, \Delta \bar{H}_{vap}/M_1}{RT^{*2}} \cdot \Delta T, \text{ from Eq. (10.11b)}$$

$$= \frac{1000 \, (90 \text{ cal g}^{-1})}{(1.99 \text{ cal K}^{-1}\text{mol}^{-1}) [(273 + 34.6) \text{ K}]^2} (1 \text{ K}) = 0.478 \text{ molal}$$

Note: Here neither $\Delta \bar{H}_{vap}$ nor M_1, but only their ratio is supplied. This prevents us from using more exact relation given by Eq. (10.10b).

10.52 a. Mole fraction of solvent (benzene)

$$x_1 = \frac{p_1}{p_1^*} \quad \text{assuming the solution to be ideal}$$

Then from the given data at 80°C

$$\frac{(100-1.25)/78}{(100-1.25)/78+(1.25/M)} = \frac{752}{760} = 0.989$$

whence M, the molecular weight of the substance $= 88.8$

b.
$$\Delta \bar{H}_{vap} = \frac{R \ln x_1}{(1/T)-(1/T^*)} \quad \text{by Eq. (10.10a)}$$

$$= \frac{(8.314 \text{ J K}^{-1}\text{mol}^{-1}) \ln(752/760)}{\dfrac{1}{(273+80.5) \text{ K}} - \dfrac{1}{(273+80) \text{ K}}} = 21824 \text{ J mol}^{-1}$$

c. i. Since liquid benzene is in equilibrium with its vapour at 1 atm and 80°C (b.p.), $\Delta \bar{G}_{vap} = 0$.

 ii. At $T = 80.5$°C pure benzene does not remain in equilibrium with its vapour at 1 atm.

Here $\Delta \bar{G}_{vap} = RT \ln x_1$, by Eq. (10.8)

$$= (8.31 \text{ JK}^{-1}\text{mol}^{-1})(273+80.5) \text{ K} \ln 0.989 = 14.1 \text{ J mol}^{-1}$$

Note: The temperature variation of $\Delta \bar{H}_{vap}$, unlike ΔG_{vap}, is small enough to be ignored.

10.53 a. Elevation of boiling point being small, the expression for the dilate solute can be used

$$i = \frac{\Delta T_b}{K_b m}$$

where m is formal molality of the solution [vide problem (10.37) and Eq. (10.11c)]

$$= \frac{(0.36 \text{ K})}{(2.6 \text{ K kg mol}^{-1})\left(\dfrac{1.65/60}{100/1000} \text{ mol kg}^{-1}\right)} = 0.503$$

b.
$$\alpha = \frac{i-1}{(1/n)-1} \quad \text{by Eq. (10.22b)}$$

$$= \frac{0.503}{(1/2)-1} = 0.994$$

10.54 a.
$$K_f = \frac{\Delta T_f}{im}$$

$$= \frac{(0.037 \text{ K})}{(1)\left(\dfrac{0.684/342}{100/1000}\right) \text{ mol kg}^{-1}} = 1.85 \text{ K kg mol}^{-1}$$

Substituting the given data for cane sugar solution for which $i = 1$

[Here change in f.p. is same as f.p. itself of the solution, f.p. of water being 0°C].

b.
$$i = \frac{\Delta T_f}{K_f m}$$

$$= \frac{(0.062 \text{ K})}{(1.85 \text{ K kg mol}^{-1})(0.01 \text{ mol kg}^{-1})} = 3.35$$

Substituting the given data for $K_3Fe(CN)_6$

c.
$$\alpha = \frac{i - 1}{n - 1} \quad \text{by Eq. (10.22a)}$$

$$= \frac{3.35 - 1}{4 - 1} = 0.783$$

[$n = 4$ for $K_3Fe(CN)_6$, since 1 mole of it gives 4 moles on complete dissociation.] ·

10.55 Concentration of the acid solution
$$= \frac{12 \times 0.1 \text{ N}}{10} = 0.12 \text{ N}$$

Then formal molality (m) of the dilute dibasic acid in solution (being same as molarity in dilute aqueous solution)
$$= \frac{0.12}{2} = 0.06$$

$$i = \frac{\Delta T_f}{K_f m}$$

$$= \frac{(0.15 \text{ K})}{(1.86 \text{ K kg mol}^{-1})(0.06 \text{ mol kg}^{-1})} = 1.34$$

$$\text{Molecular weight of the acid} = \frac{\text{Amount of acid in g/L}}{\text{Molarity}}$$

$$= \frac{10}{0.06} = 166.7$$

10.56 From Eq. (10.15a), using given data, if follows that T_f (the freezing point of the solution) will be lower with A as solvent. This implies that T_f will be higher with B as solvent.

Hence B will freeze out first.

Note: Although the two species have same freezing point (when pure), they do not start freezing simultaneously even from solution where they have same mole fraction.

10.57 From the given data

$$\ln x_A = \frac{9 \times 10^3 \text{ cal mol}^{-1}}{1.99 \text{ cal K}^{-1} \text{mol}^{-1}} \left(\frac{1}{(273 + 140) \text{ K}} - \frac{1}{(273 + 130) \text{ K}} \right) \quad \text{by Eq. (10.10a)} \quad \text{(I)}$$

$$= -0.27111 \text{ or } x_A = 0.763$$

$$\text{Also } \ln x_A = \frac{\Delta H_{fus}}{1.99 \text{ cal K mol}^{-1}} \left(\frac{1}{(273 + 15) \text{ K}} - \frac{1}{(273 + 5) \text{ EK}} \right) \quad \text{by Eq. (10.15a)} \quad \text{(II)}$$

(II) ÷ (I) gives $\Delta \bar{H}_{fus} = (9 \times 10^3 \text{ cal mol}^{-1}) \times \dfrac{(1/413) - (1/403)}{(1/288) - (1/278)}$

$$= 4.3 \times 10^3 \text{ cal mol}^{-1}$$

$$\Delta \bar{H}_{sub} = \Delta \bar{H}_{fus} + \Delta \bar{H}_{vap}$$
$$= 4.3 \times 10^3 + 9.0 \times 10^3 \text{ or } 13.3 \times 10^3 \text{ cal mol}^{-1}$$

Note: Calculation of composition using Eq. (10.11a) for dilute solution is inaccurate, the given solution being not dilute as indicated by its high variation of b.p. and f.p.

10.58 In dilute solution, the elevation of boiling point (ΔT) of a particular solvent for a particular formal concentration of a solute is given by

$$\Delta T \propto i \quad \text{vide problem 10.37}$$

Then by Eqs (10.22a) and (10.22b), we have from the given data

$$2 = \frac{1 + (2-1) \times (80/100)}{1 + [(1/2)-1]\beta}$$

whence $\beta = 0.2$, i.e. 20%

10.59 Due to transfer per mole of water from solution of composition (x_{H_2O}) to one of composition (x'_{H_2O})

$$\Delta G = (\mu^*_{H_2O} + RT \ln x'_{H_2O}) - (\mu^*_{H_2O} + RT \ln x_{H_2O})$$
$$= RT \ln x'_{H_2O} - RT \ln x_{H_2O}$$
$$= \bar{V}^*_{H_2O} (\pi - \pi') \quad \text{by Eq. (10.20)}$$
$$= (18 \text{ cm}^3 \text{ mol}^{-1}) (2 \text{ atm} - 0.2 \text{ atm})$$
$$= 32.4 \text{ cm}^3 \text{ atm}$$

Note: Concentration of a solution usually refers to solute. In the given transfer process x_{H_2O} increases with consequent increase in G.

10.60 a. $\qquad \ln x_1 = -\dfrac{\pi \bar{V}^*}{RT} \quad$ by Eq. (10.20)

$$= -\frac{(20 \text{ atm}) \left(\dfrac{18.02}{0.99} \times 10^{-3} \text{ L mol}^{-1} \right)}{(0.082 \text{ L atm K}^{-1} \text{ mol}^{-1})(273 + 27) \text{ K}} = -0.0148$$

whence $\qquad x_1 = 0.985$

Again $\qquad x_1 = \dfrac{1000/18.02}{im + (1000/18.02)},$ where m is formal molality of solute

$$= \frac{1000/18.02}{1.95 \text{ m} + 1000/18.02} = 0.985$$

whence $\qquad m = 0.422$

b. $\qquad \ln x_1 = \dfrac{M_1 T_f^{*2}}{K_f} \left(\dfrac{1}{T_f^*} - \dfrac{1}{T_f} \right) \quad$ by Eqs (10.15a) and (10.17b)

$$= \frac{(18.02 \times 10^{-3} \text{ kg mol}^{-1})(273 \text{ K})^2}{(1.86 \text{ K kg mol}^{-1})} \left(\frac{1}{273 \text{ K}} - \frac{1}{T_f} \right) = -0.0148$$

whence $\qquad T_f = 271.5$

☐ Here it has been assumed (apart from ideality of the solution) that the degree of dissociation and hence x_1 does not change on cooling.

Note: Here the % of dissociation is of no use in calculating T_f.

10.61 a. By Eq. (10.20), two isotonic solutions with same solvent will hae same x_1, provided the solutions are ideal. Then from the given data

$$x_1 = \frac{100/18.02}{34/M_2 + 100/18.02} \quad \text{for carbohydrate solution}$$

$$= \frac{100/18.02}{2 \times \dfrac{2.92}{58.4} + \dfrac{100}{18.02}} = 0.982$$

for NaCl solution assuming complete dissociation of NaCl

whence $\qquad M_2 = 340$

b. By Eq. (10.15a)

$$\ln 0.982 = \frac{80 \times 18.02 \text{ cal mol}^{-1}}{1.99 \text{ cal K}^{-1} \text{mol}^{-1}} \left(\frac{1}{273 \text{ K}} - \frac{1}{T_f} \right)$$

whence $\qquad T_f = 271.3$ K

c. Vapour pressure of the solution

$$= P_1^* x_1 \quad \text{assuming Raoult's law to hold}$$
$$= (1 \text{ atm}) (0.982)$$
$$= 0.982$$

☐ It follows from Eq. (10.23) that the freezing points of two isotonic ideal solutions will be same only when they are made up of same solvent (when $\bar{V}_1^*$, $\Delta \bar{H}_{fus}$ and T_f^* will be same)

Note: Any two solutions (even of different solvents) with same number density of solute particles will be isotonic, provided they are dilute [vide Eq. (10.21)].

UNIVERSITY QUESTIONS

10.1 Define an ideal solution. (Kerala BSc, 2000)

10.2 Find expressions for the ΔG_{mix} and ΔS_{mix} for an ideal binary solution and hence show that ΔV_{mix} and ΔH_{mix} are zero for such solutions. (Calcutta BSc (H), 2013)

10.3 Show that ΔS_{mix} will be maximum for equilimolecular binary mixture.
(Calcutta BSc (H), 2014)

10.4 Why does a real solution deviate from ideality? (Calcutta BSc (H), 2000)

10.5 Which of the two solutions, aqueous solution of NaCl or an alcoholic solution of NaCl of same concentration, will show more non-ideality and why?
(Burdwan BSc (H), 2002)

10.6 An ideal solution need not be a dilute solution. Comment.
(Burdwan BSc (H), 1991)

10.7 Both solute and solvent tend to behave ideally with increasing dilution. Discuss.
(Jadavpur BSc (H), 2001)

10.8 Is it possible for one component of a solution to behave ideally but not the other component? (Jadavpur BSc (H), 2012)

10.9 What are colligative properties? Why they are so called? (Calcutta BSc (H), 1998)

10.10 Colligative properties of a solution arise due to the entropy effect rather than the enthalpy effect. Elucidate/justify. (Jadavpur BSc (H), 2012)

10.11 Explain qualitatively the reason as to why the vapour pressure of the solution of a non-volatile solute is lower than that of a pure solvent.   (Calcutta BSc (H), 1979)

10.12 a. Lowering of vapour pressure of a liquid by a non-volatile solute is due to arresting of solvent molecules through solvation. Justify or criticise.

(Calcutta BSc (H), 1995)

b. Lowering of vapour pressure of a liquid does not occur when a volatile solute is dissolved in it. Justify or criticise.   (Calcutta BSc (H), 2013)

10.13 Deduce Raoult's law of relative lowering of vapour pressure from thermodynamic stand point. Under what conditions is the law independent of temperature?

(Calcutta BSc (H), 1987)

10.14 a. Two liquids A and B obeying Raoult's law form an ideal mixture in all proportions. Following Raoult's law, write expressions for the partial vapour pressures P_A, P_B and the total vapour pressure P. Plot these quantities against the mole fraction x_B of one compound in the liquid mixture. Do you expect that the mole fraction y_B of component B in the vapour phase mixture will be the same (x_B) as in the liquid phase mixture? If it is not so, indicate graphically the variation of total vapour pressure with y_B.   (Burdwan BSc (H), 1997)

b. Apply the Duhem Margules equation to a binary liquid mixture to show that (i) if component A obeys Raout's law at all compositions, the other component B will also obey it and (ii) if there is positive deviation in one component, the other component must also show a positive deviation.

(Burdwan BSc (H), 1997)

10.15 Why do positive deviations occur?   (Guru Nanak Dev BSc, 2003)

10.16 Show that Raoult's law is a special case of Henry's law.   (Delhi BSc, 2000)

10.17 State Henry's law. Show that, if ideal behaviour is assumed, Henry's law is equivalent to Nernst distribution law.   (Burdwan BSc (H), 1991)

10.18 Show thermodynamically that when the solvent follows Raoult's law, the solute must obey Henry's law in sufficiently dilute solution.   (Jadavpur BSc (H), 2002)

10.19 Can a liquid solute obey Raoult's law in one solvent and Henry's law in another? Explain.   (Jadavpur BSc (H), 2003)

10.20 Why is the boiling point of a solution (of non-volatile solute) greater than that of a pure solvent?   (Calcutta BSc (H), 2008)

10.21 Define ebullioscopic constant. Mention whether it is a property of solute or solvent.

(Burdwan BSc (H), 2010)

10.22 How does the boiling point elevation ΔT_b depend on the mole fraction of solute? Plot ΔT_b vs ΔP (ΔP = vapour pressure lowering) for dilute solution. Explain the observation. Remark on the slope.   (Burdwan BSc (H), 1991)

10.23 Show that for a solution of a solute in a non-polar solvent at a particular concentration, $\Delta T_b / T_b$ is independent of the nature of the solvent. State the assumption involved.   (Burdwan BSc (H), 1994)

10.24 Explain how the boiling point of a solution may be lower than that of the pure solvent. Derive mathematical expression which shows such behaviour quantitatively.   (Jadavpur BSc (H), 2002)

10.25 It may not be possible to boil a saturated solution of a certain substance under normal atmospheric pressure. Explain.   (Jadavpur BSc (H), 2003)

10.26 Normal boiling points and enthalpies of vaporisation of two solvents benzene and carbon tetrachloride are (80.1°C, 7.64 k cal mol^{-1}) and (76.8°C, 7.45 k cal mol^{-1})

respectively. Find out which one will be more suitable for estimating the molecular weight of a substance by boiling point elevation technique.

(Jadavpur BSc (H), 2001)

10.27 What is the reason behind depression of freezing point of a solvent?

(Burdwan BSc (H), 1998)

10.28 In a single graph, draw the μ–T plots for solid and liquid phases of (i) pure solvent and (ii) solvent in solution. Mention the thermodynamic basis for the depression of freezing point.   (Calcutta BSc (H), 2008)

10.29 a. Deduce thermodynamically expression relating freezing point depression of a solvent of a non-volatile non-electrolyte with its molal concentration.

b. How should it be modified if the solute be an electrolyte? Is there any advantage in using molality instead of molarity as concentration unit?

(Calcutta BSc (H), 1994)

10.30 Explain whether and when an ordinary thermometer (instead of Beckmann thermometer) can be used for molecular weight determination by the cryoscopic method.   (Calcutta BSc (H), 1985)

10.31 Is the freezing point depression method of determination of molecular weight applicable in all cases? Give reasons for your answer.   (Calcutta BSc (H), 1983)

10.32 What conditions favour the largest elevation in the boiling point and biggest depression in the freezing point of a solvent containing a solute at a particular concentration?   (Burdwan BSc (H), 1995)

10.33 The freezing point of a solvent may be elevated after addition of a solute. Justify/ criticise.   (Jadavpur BSc (H), 2013)

10.34 Compare the processes of osmosis and diffusion. Both are consequences of the second law of thermodynamics". Explain.   (Calcutta BSc (H), 1996)

10.35 Explain what is meant by the osmotic pressure of a solution and put forward possible mechanism for osmosis.   (Calcutta BSc (H), 1992)

10.36 How is relative lowering of vapour pressure of solutions of non-volatile solute and volatile solvent related to osmotic pressure? Deduce the relationship.

(Madras BSc, 2009)

10.37 Osmometry is better than cryoscopic method for the determination of molecular weight of polymers. Comment.   (Calcutta BSc (H), 2014)

10.38 Write Van't Hoff equation for osmotic pressure of a solution. How can it be modified to determine molecular weight of polymers?   (Punjab BSc, 2010)

10.39 Gas pressure and osmotic pressure obey same type of equation (Ideality is assumed in both cases). Comment.   (Jadavpur BSc (H), 2013)

10.40 What is Van't Hoff factor? What information can be obtained from it?

(Jadavpur BSc (H), 2000)

10.41 Draw a relationship between the Van't Hoff factor (i) and the degree of dissociation of a solute, M_xM_y dissociating as: $M_xA_y \rightleftharpoons xM^{z+} + yA^{z-}$.   (Burdwan BSc (H), 2007)

10.42 Van't Hoff factor for a weak electrolyte solution depends on concentration, while it is practically independent of concentration for a strong electrolyte. Discuss.

(Jadavpur BSc (H), 2001)

10.43 The following aqueous solutions at 300 K are given: (a) 1 M NaCl, (b) 1 M sucrose, (c) 1 M $BaCl_2$ and (d) phenol. Assuming that electrolytes are completely ionised while phenol is completely dimerised, arrange the solutions in the order of increasing colligative properties, giving justification for the same.

(Calcutta BSc (H), 1987)

10.44 Which of the following two solutions will have lower vapour pressure?

(i) 1.0 molal aqueous solution of cane sugar

(ii) A cane sugar solution in which the mole fraction of cane sugar is 0.1

[Given the density of water to be 1 g/cc] (Calcutta BSc (H), 1977)

10.45 The vapour pressure of a solution containing 6.69 g of $Mg(NO_3)_2$ dissolved in 100 g of water is 747 torr at 100°C. Calculate the degree of dissociation of the salt.

(Burdwan BSc (H), 2012)

10.46 The total vapour pressure at 25°C of a mixture of benzene and toluene in which the two mole fractions are equal, is 62 mmHg. The vapour pressure of pure benzene at 25°C is 95 mmHg. Calculate the mole fraction of benzene in the vapour in equilibrium with the liquid mixture (assume ideal behaviour of the mixture).

(Calcutta BSc (H), 2001)

10.47 The vapour pressure of A is 939.4 mm of Hg and that of B is 495.8 mm of Hg at 140°C. Assuming that they form an ideal solution, what will be the composition of a mixture which boils at 140°C under 1 atm. What will be the composition of the vapour at this temperature? (Calcutta BSc (H), 2004)

10.48 Equal volumes of two aueous solutions, one containing 36 g of glucose ($C_6H_{12}O_6$) in 1000 g of water and the other 36 g of urea in 1000 g of water, are mixed. What will be the freezing point of the resulting solution? The heat of fusion of ice at 0°C is 80 Cal g^{-1}. (Calcutta BSc (H), 1976)

10.49 At 100°C the vapour pressure of a solution of 6.5 g of a solute in 100 g of water is 732 mm. What is the boiling point of the solution? (Given $K_b = 0.52$ K kg mol^{-1}).

(Burdwan BSc (H), 1996)

10.50 Normal boiling point of heavy water is 102°C and its ebullioscopic constant is 10% higher than that of water. Calculate the latent heat of evaporation of heavy water. Latent heat of evaporation of water is 540 Cal g^{-1} at 100°C at 1 atm.

(Calcutta BSc (H), 2013)

10.51 Calculate molal boiling point elevation constant (K_b) of water at 2 atm ($\Delta H_{vap} = 40.5$ kJ mol^{-1}). Assume ideal behaviour of water vapour. (Calcutta BSc (H), 2000)

10.52 A solution containing 5.0 g of KCl per liter of water boils at 100.065°C at 760 mm pressure. Determine the degree of dissociation of KCl (K_b for water is 0.54 K kg mol^{-1}). (Delhi BSc, 2008)

10.53 What will be the boiling point of solution containing 0.06 g of urea (molecular mass 60) in 2.5 g of water? (Punjab BSc , 2011)

10.54 Acetic acid associates in benzene to form double molecule. 1.65×10^{-3} kg of acetic acid when dissolved in 100×10^{-3} kg of benzene raised the boiling point by 0.36 K. Calculate the Van't Hoff factor and degree of association of acetic acid in benzene (K_b for benzene = 2.57 K kg mol^{-1}) (Delhi BSc, 2003)

10.55 Pure benzene freezes at 5.4°C and a solution of 0.223 g of phenylacetic acid ($C_6H_5CH_2COOH$) in 4.4 g of benzene freezes at 4.47°C. The molar enthalpy of fusion of benzene is 9.89 kJ mol^{-1}. Calculate the relative molar mass of phenylacetic acid and comment on the result. (Jadavpur BSc (H), 2012)

10.56 A certain solution of benzoic acid in benzene has a freezing point of 3.1°C and a boiling point of 82.6°C at 1 atm. Explain these observations and suggest structure

of the solute particles at two temperatures. [Normal freezing point of benzene = 5.5°C, normal boiling point of benzene = 80.1°C, K_f = 5.12 K kg mol^{-1} and K_b = 2.53 K kg mol^{-1}]
(Calcutta BSc (H), 2007)

10.57 A solution containing 0.684 g cane sugar in 100 g water freezes at –0.037°C, while a solution containing 0.585 g NaCl in 100 g water freezes at –0.342°C. Calculate (a) the normal freezing point depression constant of water, (b) the Van't Hoff factor, and (c) the percentage dissociation of NaCl in solution. (Calcutta BSc (H), 1982)

10.58 An aqueous solution of sucrose freezes at –0.2°C. Calculate the molality of an aqueous solution of NaCl (assuming completely ionized) having the same boiling point of the sucrose solution. [Given: K_f = 1.86 K kg mol^{-1} and K_b = 0.52 K kg mol^{-1} for water]. (Jadavpur BSc (H), 2013)

10.59 0.1 g of a solute with molecular mass 80 is dissolved in 50 ml water. If the osmotic pressure of the solution is measured to be 0.5 atm at 27°C, find out with reasons whether the solute has significantly dissociated or associated in the solution.
(Jadavpur BSc (H), 2003)

10.60 At 233 K, osmotic pressure of a solution of urea is 500 mm of Hg. The solution is diluted and the temperature is raised to 298 K when the osmotic pressure is found to be 105.3 mmHg. Determine the extent of dilution. (Burdwan BSc (H), 2008)

10.61 What would be the concentration in g litre^{-1} of a solution containing a non-electrolyte (MW = 180) isotomic with a decinormal solution of NaCl? The apparent dissociation of solute is 85%. (Burdwan BSc (H), 2010)

10.62 Calculate osmotic pressure of human blood at body temperature (36.9°) which shows a freezing depression of 0.558°C (assumed to contain no associating or dissociating substances). Molal depression constant of water is equal to 1.86 K kg mol^{-1}. (Andhra BSc, 2002)

KEY TO UNIVERSITY QUESTIONS

10.1 An ideal solution is best defined to be the one of which the chemical potential of each constituent species can be expressed in the form of Eq. (10.1).

Note: This definition is not limited only to liquid solution.

10.2 Vide Section 10.1.

10.3 For an ideal binary mixture

$$\Delta S_{mix} = -\frac{R}{n_1 + n_2}[x_1 \ln x_1 + x_2 \ln x_2], \quad \text{by Eq. (10.3)}$$

$$= -\frac{R}{n_1 + n_2}[x_1 \ln x_1 + (1 - x_1)\ln(1 - x_1)]$$

$$= +ve$$

For maximum $\Delta S_{mix} = \dfrac{\partial \Delta S_{mix}}{\partial x_1} = 0$

when we have $1 + \ln x_1 - 1 - \ln(1 - x_1) = 0$

whence $\qquad x_1 = \dfrac{1}{2}$

which corresponds to an equimolecular mixture.

10.4 The thermodynamic definition of an ideal solution, according to Eq. (10.1), imposes the restriction that the solute and solvent particles are to be very similar regarding their size (so that $\Delta V_{mix} = 0$) and mutual interaction (so that $\Delta H_{mix} = 0$), vide Section 10.1. Due to such stringent restriction, which can hardly be fulfilled, a real solution deviates from ideality.

10.5 An aqueous NaCl will show more non-ideality than an alcoholic NaCl due mainly to higher solute–solvent (ion–dipole) interaction compared to solvent–solvent (dipole–dipole) interaction.

10.6 An ideal solution, being hypothetical, need not be dilute. However, for a real solution to behave ideally, it should necessarily be dilute so that $\Delta V_{mix} = 0$ and $\Delta H_{mix} = 0$.

10.7 Because, with increasing dilution, the activity coefficient of both solvent and solute tends to one, though concentration of solvent rises and that of the solute falls. This is a consequence of the different standard states chosen for solvent (based on Raoult's law) and solute (based on Henry's law).

10.8 This possibility is ruled out by Duhem–Margules equation that represents the inevitable connection between the partial vapour pressures and the relative amounts of the two components in a binary system at equilibrium. Vide question 10.14(b).

10.9 Vide Section 10.2.

10.10 Vide note on problem 10.32.

10.11 Vide problem 10.8.

10.12 a. Vide problem 10.9.

b. The given statement is not justified. The vapour pressure of a solvent is always lowered by addition of even a volatile solute, though here, the total vapour pressure rises (unless no azeotrope is formed).

10.13 Vide Henry's law, Section 10.3

☐ Vide problem 10.7. The required conditions are: (i) the vapour phase need to be kinetically ideal (ii) the liquid phase need be thermodynamically ideal.

10.14 a. Vide Section 10.3 and Section 9.4.2(A)

b. i. If $\dfrac{d\ln p_1}{d\ln x_1} = 1$ at all composition

then $\dfrac{d\ln p_2}{d\ln x_2} = 1$ also at all composition

ii. If $\dfrac{d\ln p_1}{d\ln x_1} > 1$ holds

then $\dfrac{d\ln p_2}{d\ln x_2} > 1$ will also hold.

10.15 Vide problem 10.13.

10.16 Vide Henry's law, Section 10.3.

10.17 Vide Henry's law, Section 10.3.

Since partial pressure of any species (i) in an ideal gas mixture is proportional to molar concentration $(C_i)_{vap}$ and mole fraction of any species in its dilute solution is also proportional to its molar concentration $(C_i)_{liq}$, the Eq. (10.6) transforms.

Equation (10.6) transforms to the following form

$$\frac{(C_i)_{vap}}{(C_i)_{liq}} = constant$$

which represents the distribution of any species i in a system between vapour and liquid phases. This is equivalent to the Nernst distribution law regarding distribution of any species between two liquid phases.

10.18 Vide problem 10.21.

10.19 Yes. The solute (i) will obey the two laws under the following conditions

Raoult law, if $x_i \rightarrow 1$

Henry's law, if $x_i \rightarrow 0$

Note: The term *solute* is rather misfit when $x_i \rightarrow 1$, which is possible if solvent and solute are miscible in all proportions.

10.20 This is due to lowering of vapour pressure of solvent by solute, and hence of total vapour pressure of the solution, the solute being non-volatile.

10.21 Vide Section 10.4.

K_b is a property of solvent. Because the defining Eq. (10.12) for K_b contains constant which are all characteristic of solvent.

Note: K_b is a constant of a liquid (solvent), but it has meaning only with respect to a solute.

10.22 For dilute solution, the dependence of ΔT_b on x_2 is given by Eq. (10.11a).

☐ ΔT_b vs ΔP plot is linear.

☐☐ This follows from the relation

$$\Delta T_b = \left(\frac{RT_b^{*2}}{\Delta H_{vap} p_1^*} \right) \Delta P, \quad \text{by Eqs (10.4c) and (10.11a)}$$

where the quantity within ( ) is a characteristic constant for the solvent.

☐☐☐ The slope of ΔT_b vs ΔP plot depends on the solvent involved.

10.23 Vide problem 10.28.

10.24 This can happen if the solute is volatile so that lowering of vapour pressure of the solvent is more than compensated by the contribution of the solute to the total vapour pressure of the solution with consequent depression of boiling point.

☐ The relevant quantitative relation is Eq. (10.13)

10.25 For a liquid to boil its vapour pressure must be equal to the external pressure. On raising temperature, the solubility of the solute, and hence its concentration, may fall with consequent rise in vapour pressure of the solution exceeding the normal atmospheric pressure making boiling impossible at this pressure.

10.26 For pure liquids $K_b \propto \dfrac{MT_b^{*2}}{\Delta \overline{H}_{vap}}$ by Eq. (10.12)

For the given two liquids T_b^* and $\Delta \overline{H}_{vap}$ differ by small amounts. Then CCl_4 having much higher molecular weight (154) than benzene (78) will have higher K_b. Hence for same solute concentration, CCl_4 having higher K_b than C_6H_6 will entail higher elevation of boiling point that can be measured more accurately. Therefore, CCl_4 will be more suitable for the given purpose.

10.27 The thermodynamic reason behind depression of freezing point is the lowering of chemical potential of solvent by a solute.

The molecular version of this reason is the tendency of a system to maximise its randomness. Vide problem 10.32.

10.28 Vide Section 10.5.

10.29 a. Vide Section 10.5.

   b. If the solute is an electrolyte, the Eq. (10.16b) should be modified thus
$$\Delta T_f = i\,K_f m$$
where i is the Van't Hoff factor.

   □ The chief advantage of molality, over molarity is that it is temperature independent.

10.30 Vide problem 10.33.

10.31 Vide problem 10.35.

10.32 The required condition are that the solvent used should have K_b and K_f as large as possible in the respective cases concerned.

10.33 Vide Section 10.5.

10.34 Vide problem 10.39.

10.35 The mechanism of osmosis is not same for all semipermeable membranes. Often a membrane acts as mere sieve allowing smaller solvent molecules to pass through its pores while blocking the larger solute particles.

Some membranes dissolve pure solvent and release it to the solution. This mechanism is particularly suitable when solvent molecules are larger than solute molecules.

Note: The actual mechanism of osmosis has to be ascertained from kinetic study. Thermodynamics is of no use for this purpose, because it deals only with the equilibrium which is same for all membrane.

10.36 The required relation is

$$\ln\left(1 - \frac{p_1^* - p_1}{p_1^*}\right) = \ln\frac{P_1}{P_1^*} = \frac{\bar{V}_1^* \pi}{RT}, \quad \text{by Eqs. (10.4b) and (10.20)}$$

   □ Vide Section 10.6.

10.37 Vide Problem 10.36.

10.38 The required modified Van't Hoff equation is Eq. (10.24).

10.39 Vide problem 10.42.

10.40 Vide Section 10.7.

   □ i provides informations regarding association or dissociation of solute in a solution.

10.41 Let us consider one mole of solute of which α mol dissociates forming $x\alpha$ mol of M^{z+} and $y\alpha$ mol of A^{z-} ions. Then

$$i = \text{Total number of moles of particles (molecules + ions)}$$
$$= (1 - \alpha) + x\alpha + y\alpha$$
$$= 1 + (x + y - 1)\,\alpha$$

which is the desired relation.

10.42 In solution, the degree of dissociation (α) of a weak electrolyte (that dissociates incompletely) depends on its concentration while α of a strong electrolyte (that dissociates almost completely) is practically independent of its concentration. Then i, given by Eq. (10.22a), will have concentration dependence as in the given statement.

10.43 Colligative properties increase with concentration of particles due to solute, *ic*. Here *c*, the formal molarity is same for all the given solutions. Then colligative properties will increase in the increasing order of *i* [given by Eq. (10.22)] as under

$$\text{Phenol} < \text{Sucrose} < \text{NaCl} < \text{BaCl}_2$$
$$i = (1/2) \qquad i = 1 \qquad i = 2 \qquad i = 3$$

10.44 In solution (i), the mole fraction of $H_2O = \dfrac{1000/18}{(1000/18)+1} = 0.982$

In solution (ii), the mole fraction of $H_2O = 1 - 0.1 = 0.900$

Then, by Raoult's law, the solution (ii) having lower mole fraction of H_2O will have lower vapour pressure.

10.45
$$x_{H_2O} = \frac{p_{H_2O}}{p_{H_2O}^*}$$

Then from the given data

$$\frac{100/18}{i \times (6.69/148.3) + (100/18)} = \frac{747}{760} \quad \text{Molecular weight of } Mg(NO_3)_2 = 148.3$$

whence $\qquad i = 2.14$

Degree of dissociation $= \dfrac{i-1}{n-1}$ by Eq. (10.22a)

$$= \frac{2.14-1}{3-1} = 0.57$$

Note: α is found to be much less than 1 due to formation of ion pairs, like $MgNO_3^+$, where oppositely charged ions are held together by electrostatic forces.

10.46 Mole fraction of benzene in the vapour $= \dfrac{x_{C_6H_6} P_{C_6H_6}^*}{P_{total}}$ vide problem (10.10)

$$= \frac{\left(\frac{1}{2}\right)(95 \text{ mm})}{(62 \text{ mm})} = 0.77$$

10.47 Vide problem 10.47.

10.48 Assuming that the two solutions are of same density equal to that of water, 2000 g of the resulting solution will contain 36 g of glucose and 36 g of urea. Then by Eq. (10.15a)

$$\ln \frac{\dfrac{2000}{18}}{\dfrac{36}{180} + \dfrac{36}{60} + \dfrac{2000}{18}} = \frac{80 \times 18 \text{ cal mol}^{-1}}{1.99 \text{ cal K}^{-1}\text{mol}^{-1}} \left(\frac{1}{273 \text{ K}} - \frac{1}{T_f} \right)$$

whence $\qquad T_f = 272.3 \text{ K}$

10.49
$$\ln \frac{p_1}{p_1^*} = \frac{\Delta \overline{H}_{vap}}{R} \left(\frac{1}{T_b} - \frac{1}{T_b^*} \right) \quad \text{by Eq. (10.23)}$$

$$= \frac{M_1 T_b^{*2}}{K_b} \left(\frac{1}{T_b} - \frac{1}{T_b^*} \right) \quad \text{by Eq. (10.12b)}$$

Then $\qquad \ln\dfrac{732}{760} = \dfrac{(18\times10^{-3}\text{ kg mol}^{-1})(373\text{ K})^2}{(0.52\text{ K kg mol}^{-1})}\left(\dfrac{1}{T_b}-\dfrac{1}{373\text{ K}}\right)$

whence $\qquad T_b = 374.1$ K

Note: The given composition of the solution is of no use.

10.50 $\qquad (\Delta\bar{H}_{\text{vap}})_{\text{D}_2\text{O}} = \dfrac{(K_b)_{\text{H}_2\text{O}}}{(K_b)_{\text{D}_2\text{O}}}\cdot\dfrac{M_{\text{D}_2\text{O}}}{M_{\text{H}_2\text{O}}}\cdot\dfrac{(T_b^*)^2_{\text{D}_2\text{O}}}{(T_b^*)^2_{\text{H}_2\text{O}}}\cdot(\Delta\bar{H}_{\text{vap}})_{\text{H}_2\text{O}}$ by Eq. (10.12a)

$\qquad\qquad = \dfrac{100}{110}\cdot\dfrac{20}{18}\cdot\dfrac{(375\text{ K})^2}{(373\text{ K})^2}\cdot(540\text{ cal g}^{-1}) = 551.3\text{ cal g}^{-1}$

10.51 $\qquad\qquad K_b = \dfrac{M_1 R T_b^{*2}}{\Delta\bar{H}_{\text{vap}}}$ by Eq. (10.12b)

where T_b^* is the boiling point of water at 2 atm, which can be calculated from the normal boiling point 100°C of water using the Clausius–Clapeyron equation.

10.52 As in case of problem 10.53

$$i = \dfrac{\Delta T_b}{K_b m}$$

[where m is the formal molality which is approximately equal to molarity for aqueous solution]

$$= 1 + (n-1)\alpha, \text{ by Eq. (10.22a)}$$

Then $\qquad\qquad \alpha = \dfrac{1}{n-1}\left[\dfrac{\Delta T_b}{K_b m}-1\right]$

$\qquad\qquad = \dfrac{1}{2-1}\left[\dfrac{(0.065\text{ K})}{(0.54\text{ K kg mol}^{-1})\left(\dfrac{5.0}{74.6}\text{ mol kg}^{-1}\right)}-1\right] = 0.796$

10.53 Here, unlike the previous problem, the solution cannot be treated as dilute. Using Eq. (10.10a), we have

$$\ln\dfrac{\dfrac{2.5}{18}}{\dfrac{0.06}{60}+\dfrac{2.5}{18}} = \dfrac{540\times18\text{ cal mol}^{-1}}{1.99\text{ cal K}^{-1}\text{ mol}^{-1}}\left(\dfrac{1}{T_b}-\dfrac{1}{373\text{ K}}\right)$$

whence $\qquad T_b = 373.2$ K

10.54 Vide porblem 10.53.

10.55 $\qquad\qquad \Delta T_f = \dfrac{M_1 R T_f^{*2}}{1000\Delta\bar{H}_{\text{fus}}}\cdot m$ by Eqs (10.16b) and (10.17a)

Then from the given data

$$(5.4-4.47)\text{ K} = \dfrac{(78\text{ g mol}^{-1})(8.31\text{ JK}^{-1}\text{mol}^{-1})(278.4\text{ K})^2}{1000(9.89\times10^3\text{ J mol}^{-1})}\left(\dfrac{0.223}{M_2}\times\dfrac{1000}{4.4}\right)$$

whence M_2, the molar mass $= 276.8$ g mol^{-1}

☐ Compared to the M_2 corresponding to the molecular formula $C_6H_5CH_2COOH$ (136.1), the value calculated above is

i. almost double due to dimerisation of the acid caused by hydrogen bonding

ii. Slightly more than double due to non-ideality.

Note: The present problem points to the limitation of the cryoscopic method

10.56 For the given purpose, we can treat the solution as dilute

Then
$$m = \frac{\Delta T_b}{K_b} \quad \text{by Eq. (10.11c)}$$

$$= \frac{2.5 \text{ K}}{2.53 \text{ K kg mol}^{-1}} \approx 1 \text{ mol kg}^{-1}$$

Again,
$$m = \frac{\Delta T_f}{K_f} \quad \text{by Eq. (10.16b)}$$

$$= \frac{2.4 \text{ K}}{5.12 \text{ K kg mol}^{-1}}$$

$$\approx \frac{1}{2} \text{ mol kg}^{-1}$$

It appears that molality (m) of the solution at the freezing point is almost half of that at the boiling point. This suggests that in solution at freezing point benzoic acid has the structure which is dimeric of that at boiling point.

10.57 Vide problem 10.54.

10.58
$$\Delta T_b = \frac{K_b}{K_f} \cdot \Delta T_f \quad \text{by Eqs (10.11c) and (10.16b)}$$

$$= K_b \cdot 2m'$$

where m' is the formal molality of NaCl assuming completely ionised

or
$$m' = \frac{\Delta T_f}{2K_f}$$

$$= \frac{(0.2 \text{ K})}{2(1.86 \text{ K kg mol}^{-1})} = 0.0538 \text{ mol kg}^{-1}$$

Note: Here the given K_b is of no use.

10.59
$$\pi = CRT \quad \text{Eq. (10.21)}$$

$$= \left(\frac{1}{80} \text{ mol} \middle/ \frac{50}{1000} \text{ lit}\right)(0.082 \text{ lit atm K}^{-1}\text{mol}^{-1})(300 \text{ K})$$

$$= 6.15 \text{ atm}$$

The observed osmotic pressure is much lower than the calculated one. This suggests significant association of the solute in the solution.

10.60
$$\frac{C}{C'} = \frac{\pi}{\pi'} \cdot \frac{T'}{T} \quad \text{by Eq. (10.21)}$$

$$= \frac{500 \text{ mm}}{105.3 \text{ mm}} \times \frac{298 \text{ K}}{233 \text{ K}} = 6.07$$

Therefore, the extent of dilution is 6.07.

10.61

$$\pi = CRT \quad \text{for the non-electrolyte}$$
$$= iC'RT \quad \text{for NaCl}$$

Then $\qquad c = ic'$

$$= (1.85)\,(0.1\ \text{molar}) = 0.185\ \text{molar}$$
$$\equiv 0.185 \times 180 \ \text{or} \ 33.3\ \text{g litre}^{-1}$$

10.62 Considering molarity equal to molality (the solution being aqueous)

$$c = \frac{\pi}{RT} \quad \text{by Eq. (10.21)}$$

$$= \frac{\Delta T_f}{K_f} \quad \text{by Eq. (10.16b)}$$

Then $\qquad \pi = \dfrac{RT\Delta T_f}{K_f}$

$$= \frac{(0.0821\ \text{L atm K}^{-1}\text{mol}^{-1})\,(309.9\ \text{K})\,(0.558\ \text{K})}{(1.86\ \text{K kg mol}^{-1}} = 7.63\ \text{atm}$$

11

Chemical Equilibrium

In a closed system, a chemical reaction, unlike phase change, cannot go to completion due to the development of what is called a *chemical equilibrium*. The possibility of a chemical reaction and the extent to which it can lead to, can be predicted thermodynamically. However, due to formation of new substances in a chemical reaction, it is much more difficult to deal with a chemical equilibrium than a physical equilibrium.

11.1 REVERSIBLE NATURE OF CHEMICAL REACTIONS

Any chemical reaction occurring in a system (called chemical system) is always accompanied by a reverse reaction. These two opposing processes would occur ultimately at the same rate in a closed system due to its tendency to attain a chemical equilibrium when the composition becomes constant and uniform in·each phase of the system.

Thermodynamically, an equilibrium state corresponds to a minimum free energy (G) of the system at fixed temperature and pressure when $\Delta G_{T,P} = 0$. This results from a compromise between two tendencies of a system—the tendency to attain lowest enthalpy (roughly the energy) and the tendency to attain highest entropy (i.e. randomness).

If the reaction system is not in equilibrium, a net reaction will occur in the direction in which $\Delta G_{T,P} = -ve$ (vide Section 5.3).

The net amount of a reaction occurring in a system is expressed in terms of *advancement of reaction* denoted usually by ξ. For a chemical reaction represented stoichiometrically by $aA + bB + ... = lL + mM + ...$, ξ at any instant is defined as

$$\xi = (n_A^o - n_A)/a = (n_B^o - n_B)/b = ... \quad \text{in terms of reactants}$$
$$= (n_L - n_L^o)/l = (n_M - n_M^o)/m = ... \quad \text{in terms of products}$$

where n's represent the mole numbers of the species, indicated by the suffix, present in the system at the instant concerned, and n^o's denote the initial mole members. $\xi = 1$, corresponding to reaction of the species in stoichiomeric amounts, forms a convenient unit, viz. 'one mol of a reaction'.

11.2 FREE ENERGY CHANGE IN CHEMICAL REACTIONS

Let us consider a closed system where the following chemical reaction occurs
$$aA + bB + ... = lL + mM + ...$$

At any stage of the reaction, the Gibbs free energy change due to infinitesimal advancement $d\xi$ of the reaction at constant temperature and pressure is given by

$$dG = \sum \mu_i dn_i$$
$$= (\mu_L l d\xi + \mu_M m d\xi + ...) - (\mu_A a d\xi + \mu_B b d\xi + ...)$$

where the μ's denote the chemical potentials of the species indicated by the suffix. Then, the reaction free energy ΔG (at constant T and P), which is really $(\partial G/\partial \xi)_{T,P}$, is given by

$$\Delta G = (l\mu_L + m\mu_M + ...) - (a\mu_A + b\mu_B + ...) \tag{11.1}$$

Now
$$\mu_i = \mu_i^*(T, P) + RT \ln a_i$$

where μ_i^* is the chemical potential of the pure ith species which depends only on temperature and pressure, and a_i is the activity of the ith species in the system. Then

$$\Delta G = (l\mu_L^* + m\mu_M^* + ...) - (a\mu_A^* + b\mu_B^* + ...) + RT \ln \frac{a_L^l a_M^m ...}{a_A^a a_B^b ...}$$

$$= \Delta G^*(T, P) + RT \ln \frac{a_L^l a_M^m ...}{a_A^a a_B^b ...} \tag{11.2}$$

where $\Delta G^*(T, P)$ represents reaction free energy per unit advancement of the reaction under unmixed condition of reactants and products at the prevailing temperature and pressure. ΔG is similar to ΔG^* except that it refers to reaction in a mixture of specified composition. For standard pressure of 1 atm (more precisely 1 bar), ΔG^* is replaced by ΔG° called standard reaction free energy.

At equilibrium $\qquad \Delta G = 0$, at constant T and P

and hence $\qquad \Delta G^*(T, P) = -RT \ln \dfrac{a_L^l a_M^m}{a_A^a a_B^b}$

$$= -RT \ln K_a \tag{11.3}$$

where $\qquad K_a = \dfrac{a_L^l a_M^m}{a_A^a a_B^b} \tag{11.4}$

K_a is a constant, called *equilibrium constant of the reaction system* at constant temperature and pressure, as ΔG^* is a similar constant.

Then $\qquad \Delta G = -RT \ln K_a + RT \ln \dfrac{a_L^l a_M^m}{a_A^a a_B^b} \tag{11.5}$

which is Van't Hoff reaction isotherm in general form. For heterogeneous chemical system a's corresponding to pure solids and liquids are deleted (taking them to be constant) from the expressions of ΔG and K at ordinary pressure.

For ideal reaction systems

$$\mu_i = \mu_i^*(T, P) + RT \ln x_i$$

Again, for ideal gaseous systems

$$\mu_i = \mu_{i(P)}^\circ(T) + RT \ln (P_i / P^\circ)$$
$$= \mu_{i(c)}^\circ(T) + RT \ln (C_i / C^\circ)$$

where the prefix indicates the value for standard state. Such expressions of chemical potential lead to the following three specific forms of equilibrium constant for ideal systems

$$K_x = \frac{x_L^l x_M^m ...}{x_A^a x_B^b ...} \tag{11.6}$$

$$K_p = \frac{(P_L/P^o)^l \, (P_M/P^o)^m \, ...}{(P_A/P^o)^a \, (P_B/P^o)^B \, ...} \tag{11.7}$$

$$K_c = \frac{(C_L/C^o)^l \, (C_M/C^o)^m \, ...}{(C_A/C^o)^a \, (C_B/C^o)^B \, ...} \tag{11.8}$$

The corresponding expressions of Van't Hoff isotherm are

$$\left.\begin{aligned} \Delta G &= \Delta G^*(T,P) + RT \ln \frac{x_L^l x_M^m \, ...}{x_A^a x_B^b \, ...} \\[2mm] &= - RT \ln K_x + RT \ln \frac{x_L^l x_M^m \, ...}{x_A^a x_B^b \, ...} \end{aligned}\right\} \tag{11.9}$$

$$\left.\begin{aligned} \Delta G &= \Delta G_{(P)}^o(T) + RT \ln \frac{(P_L/P^o)^l \, (P_M/P^o)^m \, ...}{(P_A/P^o)^a \, (P_B/P^o)^b \, ...} \\[2mm] &= - RT \ln K_P + RT \ln \frac{(P_L/P^o)^l \, (P_M/P^o)^m \, ...}{(P_A/P^o)^a \, (P_B/P^o)^b \, ...} \end{aligned}\right\} \tag{11.10}$$

$$\left.\begin{aligned} \Delta G &= \Delta G_{(C)}^o(T) + RT \ln \frac{(C_L/C^o)^l \, (C_M/C^o)^m \, ...}{(C_A/C^o)^a \, (C_B/C^o)^b \, ...} \\[2mm] &= - RT \ln K_C + RT \ln \frac{(C_L/C^o)^l \, (C_M/C^o)^m \, ...}{(C_A/C^o)^a \, (C_B/C^o)^b \, ...} \end{aligned}\right\} \tag{11.11}$$

$\Delta G_{(P)}^o$ is the standard reaction free energy per unit advancement of the reaction with unmixed reactants and unmixed products each at the specified temperature and standard pressure. $\Delta G_{(c)}^o$ represents analogous quantity with different standard state which is 1 mol L^{-1}.

$$\left.\begin{aligned} \Delta G_{(p)}^o - \Delta G_{(c)}^o &= - RT (\ln K_p - \ln K_c) \\[1mm] &= - RT \ln (K_p / K_c) \\[1mm] &= \Delta v RT \ln (0.082 \, T / K \end{aligned}\right\} \tag{11.12}$$

where $\Delta v = (l + m + ...) - (a + b + ...)$

The reaction isotherm provides answer to the question of foremost importance to the chemists regarding the possibility and direction of chemical reaction in a system. A reaction in a particular direction is possible if the accompanying $\Delta G_{T,P}$ is –ve and its extent is related to the magnitude of ΔG^* or ΔG^o involved.

For a reaction system if G (at constant T and P) is plotted against ξ, we will get a curved line instead of a straight line as shown in Fig. 11.1. The minimum of this curve corresponds to equilibrium at which $(\partial G/\partial \xi)_{T,P} = 0$. The diagram reflects the reversible nature of chemical reaction. Thus although the final products have lower free energy than the initial reactants, the former spontaneously react to some extent, ΔG for the process being –ve. This is due to the free energy of mixing of the resulting substances with

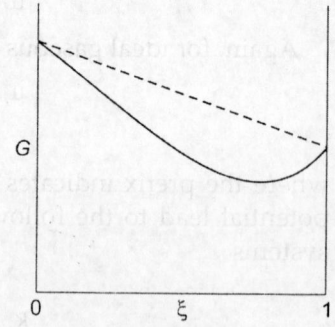

Fig. 11.1

the unreacted substances, which is always −ve while the free energy due to the chemical reaction alone may not be so.

11.3 RELATION BETWEEN DIFFERENT EQUILIBRIUM CONSTANTS

K_p, K_c and K_x have meaning also with (ideal) systems which are heterogeneous for which these are conventionally defined only with respect to the gaseous species appearing in the stoichiometric equation (vide question 11.13). In general, they are inter-related as

$$K_p = K_c \, (RTC^\circ/P^\circ)^{\Delta v} \tag{11.13}$$
$$= K_x \, (P/P^\circ)^{\Delta v} \text{ (since } P_i = C_i RT = x_i P) \tag{11.14}$$

For heterogeneous reaction systems Δv has relevance only to gaseous reactants and gaseous products.

If the stoichiometric equation representing a chemical reaction can be obtained by algebraic summation of those for simple reactions then the corresponding equilibrium constants will be related in form of product or quotient. This is due to additivity of reaction free energy. To illustrate let us consider the following reactions

(1) $CO_2 + H_2O = CO + H_2O$
(2) $2H_2 + O_2 = 2H_2O$
(3) $2CO + O_2 = 2CO_2$

Since (2)–(3) gives 2(1)

$$\Delta G_2^\circ - \Delta G_3^\circ = 2\Delta G_1^\circ$$

Then $-RT \ln K_2 + RT \ln K_3 = -2RT \ln K_1$

or $$K_1^2 = \frac{K_2}{K_3}$$

11.4 VARIATION OF EQUILIBRIUM CONSTANT WITH TEMPERATURE

For ideal gaseous reaction system

$$\ln K_p = -\Delta G^\circ / RT$$

Then $$\frac{d \ln K_p}{dT} = -\frac{1}{R} \frac{d(\Delta G^\circ / T)}{dT}$$
$$= \Delta H^\circ / RT^2, \text{ by Gibbs–Helmholtz relation}$$
$$= \Delta H / RT^2 \tag{11.15}$$

since H is independent of P for an ideal gas

Equation (11.15) is one form of what is known as van't Hoff equation (also sometimes called reaction isobar, though it was termed as *isochore* by van't Hoff). In terms of ΔU

$$\frac{d \ln K_c}{dT} = \frac{\Delta U}{RT^2} \tag{11.16}$$

this is sometimes called *reaction isochore*.

A more general form of van't Hoff equation is

$$\left[\frac{\partial \ln K_x}{\partial T} \right]_P = \frac{\Delta H}{RT^2}$$

which holds for both ideal gaseous and liquid reaction systems.

However, the most general form of van't Hoff equation is

$$\left[\frac{\partial \ln K_a}{\partial T}\right]_P = \frac{\Delta H}{RT^2} \qquad (11.17a)$$

which holds for any system. This is consistent with van't Hoff reaction isotherm.

The suffix P is unnecessary with H° as the latter is independent of pressure.

Assuming ΔH to be independent of temperature, integration of Van't Hoff equation gives

$$\ln \frac{K'}{K} = \frac{\Delta H}{R}\left(\frac{1}{T} - \frac{1}{T'}\right) \qquad (11.17b)$$

11.5 VARIATION OF EQUILIBRIUM CONSTANT WITH PRESSURE

For ideal gaseous reaction system, K_p and K_c are both independent of pressure, but K_x depends on pressure. Since $K_x = K_p P^{-\Delta v}$ unless unless $\Delta v \neq 0$.

But, for real systems the equilibrium constant (K_a) always depends on pressure. Since $K_a = e^{-\Delta G^*/RT}$, where ΔG^* is a function not only of T but also of P. However, for condensed systems, this dependency on pressure is negligible because the pressure variation of chemical potential is negligible for a solid or liquid. This is also true for systems involving real gases, unless the pressure is large.

It is important to note that in case of real systems only K_f (the equilibrium constant in terms of fugacity) and K_a are true (thermodynamic) equilibrium constants. Because these two (unlike other equilibrium constants) are not composition dependent. We can understand this if we remember that $f_i = P_i \times$ fugacity coefficient and $a_i = x_i \times$ activity coefficient, both the coefficients being dependent on pressure and composition.

11.6 LE CHATELIER PRINCIPLE

If the equilibrium of a system is disturbed by changing any of its equilibrium controlling factors (viz. temperature, pressure, concentrations of the constituents) the system will tend to attain a different equilibrium state to oppose the imposed change. This is Le Chatelier principle whereby the shift in equilibrium can be qualitatively predicted, thus:

i. Heating of a system at constant volume will shift the equilibrium state in the endothermic direction whereby the change in properties of the system caused by addition of heat can be minimised.

ii. Compression of a system at constant temperature will shift the equilibrium in the direction in which the system contracts.

iii. Addition of any reacting substance to a system at constant volume and temperature will shift the equilibrium in the direction in which the added substance reacts.

11.7 EVALUATION OF SHIFT OF EQUILIBRIUM

Le Chatelier principle gives only qualitative information regarding the shift of equilibrium. To get quantitative imformation, we are to use appropriate expression involving equilibrium constant which, for an ideal gaseous reaction $aA + bB + ... = lL + mM + ...$, is

$$K_P = \left[\frac{n_L^l n_M^m \cdots}{n_A^a n_B^b \cdots} \right] \left(\frac{P/P°}{\sum n_i} \right)^{\Delta v} \tag{11.18}$$

Let us discuss the effect of addition of inert gas to the system on the basis of this equation. Here $\sum n_i$ includes the moles of the added inert gas.

If the inert gas is added at constant volume and temperature, $P/\sum n_i$ (i.e. RT/V) will not be affected. Then K_P remaining unchanged (at constant temperature), the quantity within [ ] will remain unaffected, i.e. equilibrium will not be disturbed whatever might be the value of Δv.

If, however, the inert gas is added at constant (total) pressure and temperature, $P/\sum n_i$ will decrease and the following three cases might arise

i. when $\Delta v = +$ ve, $(P/\sum n_i)^{\Delta v}$ decreases on addition of inert gas. Then K_P remaining unaffected, the quantity within [ ] will increase, i.e. equilibrium will shift in the forward direction.

ii. when $\Delta v = -$ve, reverse will happen.

iii. when $\Delta v = 0$, nothing will happen, because here $\left(\dfrac{P/P°}{\sum n_i} \right)$ is always equal to 1.

The effect of addition of any reactant or product can also be predicted from the above equation. If addition is made at constant volume, $P/\sum n_i$ will not change (as in case of addition of inert gas) but quantity within [ ] will momentarily increase or decrease, depending on whether a product or reactant is added and accordingly, the equilibrium will be shifted in the backward or forward direction, K_P remaining unaffected.

For reaction having $\Delta v = 0$, the affect of addition of any reactant or product at constant pressure will be same as that at constant volume. But this is not so when $\Delta v \neq 0$. Let us consider the case with $\Delta v = +$ve. Here due to addition of any reactant or product to the system at constant pressure $(P/\sum n_i)^{\Delta v}$ will decrease (as in case of addition of inert gas). Again, if a reactant (A, say) is added, the quantity within [ ] will momentarily decrease. Then, K_P remaining unchanged, the quantity within [ ] must increase subsequently, i.e. equilibrium will be shifted in forward direction. But if a product is added, the quantity within [ ] will momentarily increase which may be in the same, greater or lower proportion than the decrease of $(P/\sum n_i)^{\Delta v}$. Accordingly the equilibrium will remain unchanged, shift in backward direction or shift in forward direction, K_P being unchanged.

Curiously, when $\Delta v = +$ve, the addition of any product to the reaction system at constant pressure and temperature may result in the shift of equilibrium toward the direction of the added substance. Apparently, this seems to be a violation of Le Chatelier principle.

Relation Between Equilibrium Constant and Advancement of Reaction

Let us consider the chemical reaction $aA + bB + ... = lL + mM + ...$ If the system initially contain p mol A, q mol B, ... but no product, then

$$K_n = \frac{n_L^l n_M^m ...}{n_A^a n_B^b ...} \tag{11.19}$$

$$= \frac{(l\xi)^l (m\xi)^m ...}{(p - a\xi)^a (q - b\xi)^b ...} \tag{11.20}$$

where ξ is the advancement of reaction

For ideal reaction system

$$K_P = K_n \left(\frac{P/P^\circ}{\sum n_i} \right)^{\Delta v} \tag{11.21}$$

If $\Delta v = 0$, $K_P = K_n$, when ξ (which is a meanure of equilibrium position) will be determined only by equilibrium constant. But this is not so when $\Delta v \neq 0$. If $\Delta v = +ve$, ξ will be lower the higher is the applied pressure. Reverse will happen when $\Delta v = -ve$.

AUXILIARY PROBLEMS

11.1 A system will be in chemical equilibrium if its composition does not change with time. True or false?

11.2 In a bunsen flame, composition does not change with time. Is it an example of chemical equilibrium?

11.3 In which of the following cases, chemical equilibrium is possible?
   i. Strongly heated $CaCO_3$ in an open vessel.
   ii. Strongly heated $BaCO_3$ in a closed vessel.
   iii. $RaCO_3$ enclosed in a vessel.

11.4 Is chemical equilibrium possible in a system which is not in mechanical equilibrium?

11.5 Whereas in the laboratory Na_2CO_3 and $CaCl_2$ react to form $NaCl$ and $CaCO_3$, the reverse reaction is found to occur on the shore of Egyption lakes where deposits of Na_2CO_3 are found. How would you explain this?

11.6 a. The idea of chemical reversibility fails for chemical reactions, like $NH_4NO_2 \rightarrow N_2 + 2H_2O$, which are known to occur only in one direction. Therefore, such reactions, often called irreversible, should not be characterised by an equilibrium constant. Comment.

   b. In a closed system, a phase change can go to completion but a chemical change cannot. Justify or criticise.

11.7 The reaction $Ca_3(PO_4)_2 + SiO_2 \rightarrow CaSiO_3 + P_2O_5$ occurs at high temperature though it involves liberation of an acid (P_2O_5) by a weaker acid (SiO_2). Explain.

11.8 A chemical system reacts so as to minimise its energy, hence no endothermic reaction is expected to occur. True or false?

11.9 If ΔG° of a reaction is +ve, it will not occur spontaneously. True or false.

11.10 Is it possible to bring about a reaction if its ΔG is +ve?

11.11 The decomposition of ozone ($2O_3 \rightarrow 3O_2 +$ heat) is favoured by rise of temperature. Here Le Chatelier principle is violated. Comment.

11.12 The dissolution of NaOH in water is exothermic. Yet contrary to Chatelier principle, the solubility of sodium hydroxide increases with rise of temperature. Resolve this paradox.

11.13 For the reaction $N_2 + O_2 = 2NO$, with NO initially absent, $K_P = \left(\dfrac{2\xi}{1-\xi}\right)^2$. Then the reaction system should entail two equilibrium positions corresponding to two different values of ξ (the advancement of reaction). Comment.

11.14 Justify or modify the statement: K_P for the reaction $N_2 + O_2 = 2NO$ should depend only on temperature (and not on pressure).

11.15 For a real system, which are the true equilibrium constants—K_p, K_c, K_f, K_x or K_a?

11.16 For an ideal liquid phase reaction, which one is more fundamental—K_x or K_c? Is there any meaning of K_p for such reaction systems?

11.17 An equilibrium constant should be a dimensionless quantity. Justify.

11.18 On which of the following does equilibrium constant depend?
   i. Stoichoimetric representation of the reaction concerned.
   ii. Standard state chosen.
   iii. Initial composition of the reaction system.

11.19 Is the relation $K_p = K_c$ valid for the reaction $N_2 + O_2 = 2NO$?

11.20 The relation $K_p = K_c$ is not valid for the reaction $N_2 + 3H_2 = 2NH_3$ where $\Delta\nu \neq 0$. Comment.

11.21 ΔG^* for a chemical reaction is an extensive quantity as it depends on stoichiometric representation of the reaction. Then, bythe relation $\Delta G^* = -RT\ln K$, K should be an extensive quantity. Comment.

11.22 Equilibrium in a molecular system is dynamic in nature microscopically but static in effect macroscopically. Justify this adducing experimental evidence.

11.23 How will you distinguish between a true and false equilibrium? What happens to the composition of a reaction system in true equilibrium if it is cooled
   i. very rapidly
   ii. very slowly.

11.24 The reaction $CaCO_3(s) = CaO(s) + CO_2(g)$ occurs appreciably in the forward direction at high temperature, but in the reverse direction at low temperature. Explain.

11.25 Does equilibrium pressure of CO_2 for the system is the previous problem depend on volume of the reaction vessel and amount of $CaCO_3$?

11.26 The law of mass action was originally formulated in terms of active mass which was supposed to be a variable in case of gaseous substances but constant for pure solids and liquids in a heterogeneous reaction system. Is this justified? What is the thermodynamic significance of active mass?

11.27 To which thermodynamic quantity is the term *chemical affinity* relevant–U, H, S or G?

11.28 Is the law of mass action justified on thermodynamic ground?

11.29 Of the two reactions having $\Delta G_{T,P}$ equal to –1 unit and –2 units respectively, which one will be faster?

11.30 A and B are two isomeric compounds of which A is more stable. If pure A is kept isolated, will it remain in pure state?

11.31 In which direction the shift of equilibrium (if any) will occur in the following systems when compressed at constant temperature?

    i. $CO(g) + H_2O(g) = CO_2(g) + H_2(g)$
    ii. $C(s) + H_2O(g) = CO(g) + H_2(g)$
    iii. $C(diamond) = C(graphite)$

11.32 Is Le Chatelier principle applicable to any system?

    a. Apply this principle to predict the direction in which the position of equilibrium of the reaction $N_2 + 3H_2 = 2NH_3$ + heat will shift when total pressure of the system is increased:

        i. by reducing the volume of the system at constant temperature
        ii. by increasing the temperature at constant volume
        iii. by introducing some He at constant volume and temperature
        iv. by introducing N_2 at constant volume and temperature

    b. Also predict the direction in which the equilibrium will be shifted if some (i) N_2 (ii) NH_3 is introduced to the system at constant total pressure and constant temperature.

11.33 Arrange the oxide of nitrogen in order of stability from the following data regarding standard enthalpy of formation (ΔH_f°) and standard free energy of formation (ΔG_f°) at 25°C.

	N_2O	NO	NO_2	N_2O_4	N_2O_5
ΔH_f° (kJ mol^{-1})	82.05	90.25	33.18	9.16	11.3
ΔG_f° (kJ mol^{-1})	104.2	86.55	51.31	97.89	113.9

Will the order of instability be same as the order of reactivity?

11.34 a. 1.588 g of N_2O_4 gives a total pressure of 1 bar when partially dissociated into NO_2 in a 500 cm^3 vessel at 25°C. Find K_p and K_c.

    b. If the vessel initially contains one mole each of N_2O_4 and NO_2 at the same temperature in which direction the net reaction will occur? Find the minimum amount of N_2O_4 required to be introduced into the vessel to produce NO_2 at 1 atm at 25°C.

11.35 0.1 mol of H_2 and 0.2 mol of CO_2 are introduced in a vacuum flask at 450°C. The reaction $H_2 + CO_2 \rightarrow H_2O + CO$ occurs and at equilibrium, the pressure is 0.5 atm. The equilibrium mixture contains 10 mol percent of steam. Find equilibrium constant at 450°C.

11.36 A mixture of N_2 and H_2 in 1:3 volume ratio is allowed to attain equilibrium $N_2 + 3H_2 \rightleftharpoons 2NH_3$ at 10 atm pressure at certain temperature. Find K_p for the reaction if the equilibrium mixture contains (i) 20% by volume of NH_3 (ii) 20% by weight of NH_3 (iii) 20% by volume of N_2 (iv) 20% of the initial amount of N_2. How much H_2 will remain unreacted in each case.

11.37 Show that the yield of NH_3 will be maximum if N_2 and H_2 are initially taken in 1:3 volume ratio.

11.38 In a study of the equilibrium, $SO_3 \rightleftharpoons SO_2 + \frac{1}{2}O_2$, it has been found that when 8 g of SO_3 is introduced into a vacuum flask at 600°C, the equilibrium pressure is 1.8 atm and density 1.6×10^{-3} g cm^{-3}. Find the degree of dissociation and K_c of the system.

11.39 a. Find the equilibrium constant for the dissociation of phosgene gas according to the equation $COCl_2 = CO + Cl_2$, if the reaction system initially contains only CO and Cl_2 at pressure 5001 torr of each and the equilibrium pressure of the system is 5002 torr at 27°C.

b. Also find the fraction of phosgene dissociated at the same temperature when 2 mol of it is placed in a 100 L vessel which initially contains (i) nothing (ii) 1 mol of N_2 (iii) 1 mol of Cl_2 (iv) Cl_2 at 1 atm pressure.

11.40. a. A vessel contains an equilibrium mixture comprising 2 mol of $PCl_5(g)$, 2 mol of $PCl_3(g)$ and 2 mol of $Cl_2(g)$. Will the equilibrium be disturbed if 1 mol each of the three species is removed from the vessel (temperature and volume of the vessel remaining unchanged)? If so, in which direction the system will tend to react?

b. How many moles of (i) He and (ii) Cl_2 are to be introduced into the vessel at constant temperature to make the total pressure of the system double of the initial?

11.41 In a study of the equilibrium, $H_2(g) + I_2(g) \rightleftharpoons 2HI(g)$, let x mol of HI is formed when 1 mol of H_2 and 3 mol of I_2 are introduced into a flask at a constant temperature. On introducing 2 additional mol of H_2 at same temperature, the amount of HI formed is $2x$. Find K_p.

11.42 a. A gaseous equilibrium mixture of three isomers A, B and C is obtained at certain temperature T starting only with 1 mol of A. If the equilibrium constant for the reaction $A \rightleftharpoons B$ is $\frac{1}{2}$ and that of $B \rightleftharpoons C$ is $\frac{2}{3}$, find that for the reaction $A \rightleftharpoons C$ and the number of moles of each isomer.

b. Will the result be affected if B is replaced by some other isomer?

11.43 4.4 g of CO_2 is introduced into a 1 L vessel containing excess of solid carbon at 1000°C, when the equilibrium, $CO_2(g) + C(s) \rightleftharpoons 2CO(g)$ is reached the average molecular weight in the gas phase is 36.

a. Calculate equilibrium pressure, K_c and K_p.

b. How many mol of CO_2 will be required to produce 0.2 mol of CO if the reaction is carried out in the 1 L vessel at 1000°C?

c. What will be the effect of addition of (i) CO (ii) C, at constant volume and constant pressure to the system at equilibrium.

11.44 a. Find K_p for the reaction $PCl(g) = PCl_3(g) + Cl_2(g)$ at 25°C. Given that the standard free energy of formation at 25°C is –267.8 kJ mol^{-1} for $PCl_3(g)$ and –305.0 kJ mol^{-1} for $PCl_5(g)$.

b. Also calculate $\Delta H°$ and $\Delta S°$ for the reaction at 25°C, if the degree of dissociation of PCl_5 increases by 6% at constant pressure around this temperature.

c. If the eror in the determination of $\Delta S°$ is 1%, find that involved in the determination of K_p.

11.45 A price of Ni is introduced into $H_2O(g)$ containing 1% H_2 at 1000 K in a closed vessel. Will there be any formation of NiO? Given the following $\Delta G°$'s at this temperature

$$2Ni(s) + O_2(g) = 2NiO(s) \qquad \Delta G° = 296.2 \text{ kJ mol}^{-1}$$
$$2H_2(g) + O_2(g) = 2H_2O(g) \qquad \Delta G° = -381.6 \text{ kJ mol}^{-1}$$

11.46 a. The dissociation vapour pressure of NH_4Cl is 608 kPa at 700 K and 1131 kPa at 733 K. Calculate (i) the equilibrium constant, (ii) $\Delta G°$ (iii) $\Delta H°$ and (iv) $\Delta S°$ all at 700 K.

b. If 0.5 mol of $NH_4Cl(s)$ is introduced into a one-litre vessel at 700 K, find the fraction of it that will decompose. Also find the number of mol of NH_3 that would have to be added to the vessel to reduce the decomposition of NH_4Cl to 1%.

    c. If NH_4Cl dissociates in the same vessel at 700 K which initially contains NH_3 at 100 kPa, what will be the total pressure of the system equilibrium?

11.47 The equilibrium constant (K) of a reaction varies with temperature as

$$\ln K = -1.15 - \frac{986\ K}{T} + \frac{2.50 \times 10^4\ K^2}{T^2}$$

in 400 K–600 K range. Calculate $\Delta H°$ and $\Delta S°$ at 500 K.

ANSWERS

11.1 True, provided the composition of the system remains unaffected on isolating the system.

11.2 No. The flame is an example of chemical steady state, because it will disappear if the supply of fuel is cut off.

11.3 Chemical equilibrium is possible only if the relevant process is chemically reversible, as in (ii).

In (i) chemical reaction occurs unidirectionally as $CaCO_3(s) \rightarrow CaO(s) + CO_2(g)$ due to escaping of CO_2 from the system being not closed. Again (iii) involves a nuclear reaction occurring unidirectionally.

11.4 Yes, because the two equilibria have no necessary connection between them, e.g. a free falling vessel containing the reaction system $H_2 + Cl_2 = 2HCl$ of fixed composition.

11.5 For both the occasions, the relevant chemical equilibrium $Na_2CO_3 + CaCl_2 \rightleftharpoons CaCO_3 + 2\ NaCl$ is same. But the position of equilibrium is much more on the side of Na_2CO_3 in case of Egyptian lakes. This is due to much higher concentration of NaCl in lake water compared with the concentration of the usual laboratory reagents ($\sim N/10$). This explains the contrast between the two observations cited.

11.6 a. It is a thermodynamic necessity that all reaction systems under closed condition will attain chemical equilibrium. However, for the given system the equilibrium position lies on the far side of the products so much so that the amount of the reactant remaining at equilibrium is undetectably small, and it seems that the reverse reaction does not occur.

Obviously, such reactions, like all other reactions, should have characteristic equilibrium constants, but of very high value.

  b. Like chemical change, the phase change also occurs reversibly, i.e. incompletely in a closed system, but it is considered as such only up to certain minimum amount of either phase in compliance with the definition of a phase as a macroscopic region (to make the phase rule equation simple). Therefore, the given statement is not completely justified.

11.7 Because P_2O_5, which is volatile at high temperature, escapes from the system, if the system is open. As a consequence, the reaction occurs only in the forward direction due to the unsuccessful attempt of the system to attain equilibrium.

Note: If the temperature is not sufficiently high P_2O_5 will not volatilise off, when the reverse reaction is the likely one with $\Delta G_{T,P} = \Delta H - T\Delta S < 0$. Here G of the reaction system is entropy dominated ($\Delta S > 0$) at high temperature but enthalpy dominated ($\Delta H < 0$) at low temperature.

11.8 False. A chemical system reacts so as to minimise its free energy G (at constant T and P) and not necessarily total energy (roughly H). Hence an endothermic reaction ($\Delta H > 0$) is not unlikely to occur if it is accompanied by $\Delta G_{T,P} = (\Delta H - T\Delta S) < 0$.

Note: Naturally occurring endothermic reactions are not very few in number. A notable example is $N_2 + O_2 = 2NO$ that occurs during lightning.

11.9 False. Even if $\Delta G° > 0$ for a reaction, it will occur spontaneously if ΔG_{mix} due to mixing of reactants and products is so negative that $\Delta G = (\Delta G° + \Delta G_{mix}) < 0$.

Note: The possibility of a reaction is determined by the sign of ΔG (at constant T and P), while its extent is determined by the magnitude of $\Delta G°$ or ΔG^* (which contains equilibrium constant).

11.10 Yes. This is possible by coupling the given reaction system with another system so that for the coupled process $\Delta G_{T,P}$ is negative. For example, in case of electrolysis of a substance, ΔG of the substance electrolysed is +ve but ΔG in the external source of emf is so negative that the total change of G is –ve.

11.11 The Chatelier principle is not applicable to the given system because of very slow attainment of $2O_3 \rightleftharpoons 3O_2$ equilibrium. Here the phrase 'violation of the principle' is irrelevant.

11.12 The exothermic nature of the dissolution of anhydrous NaOH in water is due to the formation of hydrates. However, such hydrates dissolve endothermically at ordinary temperature and hence their solubility increases with rise of temperature according to the Chatelier principle. This clears up the seeming paradox.

11.13 Of the two roots of the given equation, only one (which is positive and less than 1) will have the real significance representing the equilibrium position corresponding to the minimum free energy (G) of the reaction system. Then given statement is not true.

Note: This is consistent with the concept of G as a single valued function of state of the system.

11.14 The given statement would have been justified if the reaction system were ideal. For non-ideality of the system, the given statement should be modified by replacing K_p with K_f (the equilibrium constant in terms of fugacity).

Note: For ordinary pressure, the given statement may be regarded as true.

11.15 Only K_f and K_a. Vide Section 11.5.

Note: In case of ideal gaseous system K_x, K_p and K_c are also justified as true equilibrium constants, being independent of composition for such systems.

11.16 K_x. Because here K_x is always independent of composition of the reaction system but K_c is not. Only when the reactants and products are present in very diluted form, K_c will bear the meaning of equilibrium constant, being the independent of initial composition of the reaction system, because only under this diluted condition $C_i \propto x_i$ and hence $\mu_i = \mu^*_{i(c)} (T, P) + RT \ln C_i$.

☐ Yes. Because even for a condensed system, the same chemical equilibrium exists also in the (inevitably) accompanying vapour phase for which K_p is meaningful but difficult to determine.

Note: On freezing, a gaseous reaction system, like $N_2 + 3H_2O \rightleftharpoons 2NH_3$, will not be devoid of K_p.

11.17 Equilibrium constant, defined thermodynamically in terms of reaction free energy (ΔG) by Van't Hoff isotherm or Eq. (11.3), should not have any dimension, being involved within logarithm.

Note: From thermodynamic view point equilibrium constants should have no unit, though sometimes they are expressed in suitable units following the law of mass action.

11.18 For any reaction system (ideal or real), the equilibrium constant K_a is given thermodynamically by

$$\ln K_a = -\frac{\Delta G^*}{RT} \quad \text{by Eq. (11.3)}$$

It follows from this relation that, like ΔG^*, K_a should depend on (i) and (ii) but not on (iii).

11.19 For the given reaction, although $\Delta v = 0$, $K_p = K_c$ is not strictly valid due to non-ideality of the reaction system.

11.20 It follows from Eq. (11.13) that the relation $K_p = K_c$ is valid for any gas phase reaction at $T = 1/R$ provided the reaction system behaves ideally. It is due to non-ideality of any real system that K_p will not be exactly equal to K_c for the reaction cited.

Note: For ideal reaction systems, $K_p = K_c$ happens for a particular type of reaction $(\Delta v = 0)$ at any temperature and for any type of reaction at a particular temperature $(T = 1/R)$.

11.21 Although G of a system is an extensive quantity, ΔG^* for a chemical reaction, which signifies the free energy change of the reaction system per mol at reaction under particular condition, is intensive. Hence by the given relation K is intensive as T is so. Therefore the given statement is wrong.

11.22 Because even at equilibrium, the reaction in both forward and backward direction proceeds but with equal rate that causes the net reaction nil, i.e. no macroscopic effect.

The dynamic nature of chemical equilibrium can be established by the following simple experiment. If some $^{14}CO_2$ is added to a closed vessel containing an equilibrium mixture of $Ca^{12}CO_3$, CaO and $^{12}CO_2$ keeping the pressure of carbon dioxide unchanged, by an arrangement shown in Fig. 11.2, the equilibrium of the system remains unaffected but calcium carbonate becomes radioactive. The obvious conclusion from this result is that the back reaction occurs even at equilibrium.

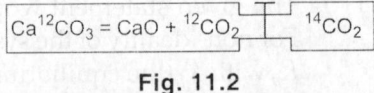

$$Ca^{12}CO_3 = CaO + {}^{12}CO_2 \qquad {}^{14}CO_2$$

Fig. 11.2

11.23 In case of true equilibrium, the composition of the system does not change with time due to equality of the forward and backward reaction rates while in case of false equilibrium, the composition seems to be unchanged due to slowness of the reaction rates in both directions. The two can be distinguished by application of a catalyst or a shock (i.e. sudden supply of energy) that enhances the reaction rates; if the system is not in true equilibrium, the composition will change.

☐ i. Composition of the system remains almost unchanged due to lack of sufficient time for change.

ii. Composition of the system changes due to shift of equilibrium in the exothermic direction, according to Le Chatelier principle.

11.24 For the forward reaction (which is decomposition) both ΔH and ΔS are positive and hence $\Delta G = \Delta H - T\Delta S$ is negative at high temperature but positive at low temperature. Therefore, the reaction will occur in this direction at high temperature, but in the reverse direction at low temperature.

Note:

1. Here the reaction accompanying decrease in G is entropy dominated at high temperature but enthalpy dominated at low temperature.

2. In the forward direction, the entropy, i.e. randomness of the system increases tending to be maximum while in the backward direction, the enthalpy (roughly

the energy) decreases tending to be minimum. The two tendencies are compromised at equilibrium when $\Delta G_{T,P} = \Delta H - T\Delta S = 0$. The reversible nature of a chemical reaction is thus linked to the natural tendency of the reaction system to attain equilibrium.

11.25 No, provided the system contains $CaCO_3(s)$ and $CaO(s)$ in equilibrium and the pressure dependence of their chemical potential and the non-ideality of CO_2 have no consequences. This follows from the relation $\mu_{CaCO_3} = \mu_{CaO} + \mu_{CO_2}$ at equilibrium which leads to $RT \ln p_{CO_2} = \mu^{\circ}_{CaCO_3} - \mu^{\circ}_{CaO} - \mu^{\circ}_{CO_2}$ under the stated conditions.

Note: The phase rule leads to the same conclusion. Since the system $CaCO_3(s) = CaO(s) + CO_2(g)$ is univariant, the pressure and composition of the system, and hence p_{CO_2}, will be fixed when its temperature is fixed.

11.26 Originally, by active mass of a reactant or product in a reaction system was meant its molar concentration (C), i.e. mol per unit volume. For gaseous substances, unlike liquid and solid substances in a heterogeneous reaction system, the volume is same as that of the whole system concerned and hence C will change during a chemical reaction at constant volume (decreases in case of a reactant and increases in case of a product). But with pure liquid and solid substances C does not change because their amount changes proportionally with their volume during the reaction. The idea of active mass is thus justified.

☐ The thermodynamic significance of active mass of any substance is its activity [defined by Eq. (5.34)] which is equal to its molar concentration in a homogeneous reaction system, provided the system is ideal and dilute. In heterogeneous reaction system, the activity of a pure liquid or solid substance is constant depending on the prevailing temperature and pressure.

11.27 Chemical affinity is relevant to G which gives the most convenient criterian for spontaneous occurrence of a chemical reaction. Mutual affinity of reacting substances for a particular chemical reaction is measured by $-\Delta G_T^{\circ}$ associated with that reaction under standard condition.

11.28 The law of mass action deal with the effect of reactant concentration of the reaction rate. According to this law, the rate (r) of a reaction $aA + bB = cC + dD$ in the forward and backward direction will be

$$r_f = k_c C_A^a C_B^b \text{ and } r_b = k_b C_C^c C_D^d$$

where k_f and k_b are corresponding rate constants and at equilibrium when $r_f = r_b$

$$\frac{C_C^c C_D^d}{C_A^a C_B^b} = \frac{k_f}{k_b} = K$$

which is a constant for the system.

Also, according to thermodynamics

$$\frac{C_C^c C_D^d}{C_A^a C_B^b} = K$$

But on thermodynamic ground, $C_A^a C_B^b$ cannot be related to r_f nor $C_C^c C_D^d$ to r_b. Because thermodynamics deals with the possibility of a change and equilibrium in a system but not with any time-related quantity like reaction rate, as time is no variable in thermodynamics. Therefore, the law of mass action cannot be justified thermodynamically.

11.29 The −ve sign of $\Delta G_{T,P}$ of a reaction mere implies that it will occur spontaneously in the direction concerned. But more −ve value does not necessarily imply higher reaction rate due to unpredictability of the reaction rate on thermodynamic ground. Then the question asked has no unique answer.

Note: For any reaction the magnitude of ΔG is not so important as its sign. The converse may be said about $\Delta G°$ (vide note on problem 11.9).

11.30 No, because of tendency of the process A → B which is accompanied by negative ΔG due to negative ΔG_{mix}. Vide Fig. 11.1.

Note: Here $\Delta G^* = +ve$ but $\Delta G = -ve$.

11.31 To compress a system means to reduce the fraction of space remaining unoccupied by the particles of the system. The system will tend to oppose this through self contraction. Then, shift in equilibrium:

– will not occur in (i), volume change being zero in either direction making the system unable to reduce the strain through shift of equilibrium.

– will occur in (ii) in the backward direction in which system's volume decreases.

– will occur in (iii) in the backward direction in which system's volume decreases.

These are in accordance with the Chatelier principle.

Note: All three reactions involve no change in number of moles, but regarding shift of equilibrium they differ. Because the shift of equilibrium is determined by change of volume of the system rather than by change of number of moles. The heterogeneous reactions (ii) and (iii), unlike the homogeneous gas reaction (i), undergo change in volume.

11.32 a. No. This principle is applicable only to closed systems which can attain equilibrium rapidly.

According to the Chatelier principle, the shift of equilibrium in the different cases are stated below:

i. In the forward direction in which the system undergoes self contraction to reduce the strain caused by compression.

ii. In the (endothermic) backward direction so that a part of the added heat (responsible for rise of pressure) is consumed.

iii. No shift of equilibrium, there being no change in equilibrium controlling factors, like temperature, concentration of the reactants or products.

iv. In the forward direction so that a part of the added N_2 (which creates strain) is consumed.

b. i. In the given process, volume of the system increases, i.e. the system is decompressed. This will tend to shift the equilibrium in the backward direction in which the system can undergo self-expansion. Again, the added N_2 will tend to shift the equilibrium in the forward direction in which a part of it is consumed. If the relative amount of the added N_2 is so specific that the two opposing tendencies nullify each other, there will be no shift of equilibrium, otherwise the equilibrium will be shifted in the forward or backward direction according as the relative amount of the added N_2 is higher or lower than the specific amount.

ii. Here added NH_3 and enhanced volume, both tend to shift the equilibrium in the same direction—the backward direction. This results in shift of equilibrium in the said direction.

Note:

1. In all cases of (a) the system is same and its pressure is raised. But the shift in equilibrium is not same, because the rise in pressure is not due to the same cause. This reminds us that in applying the Chatelier principle, we must not forget the cause behind the imposed change.
2. For (b) (i), unlike (b) (ii), the answer is not definite.
3. The effect of addition of N_2 regarding the shift of equilibrium in (a) (iv) (at constant volume) is different from that in (b) (i) (at constant pressure).

11.33 The stability of a substance refers to its ability to remain unchanged when kept isolated. The stabilities of the oxides will be in the order of their negative ΔG_f° (not necessarily ΔH_f°) and this is

$$N_2O_5 < N_2O < N_2O_4 < NO < NO_2$$

☐ Not necessarily. The reactivity of a substance toward another substance refers to the readiness of the relevant reaction. Being a time-related quantity, it cannot be predicted thermodynamically.

Note: The thermodynamic stability should not be confused with thermal stability. The latter refers to the ability of a substance to resist its decomposition into constituent elements when heated alone and it increases with increase in negative ΔH_f° in the following order for the given oxides

$$NO < N_2O < NO_2 < N_2O_5 < N_2O_4$$

which is quite different from the above order of negative ΔG_f°. It is interesting to note that compared to the lowest oxide (N_2O), the highest oxide (N_2O_5) has less thermodynamic stability but more thermal stability though N_2O_5 is more susceptible to decomposition by heat. The seeming paradox is resolved when we remember that the thermal stability refers to the decomposition to constituent elements and not to any other species. On heating N_2O_5 readily decomposes to lower oxides but not to its elements.

11.34 a. 1.588 g of $N_2O_4 \equiv \dfrac{1.588}{92.02}$ or 0.01726 mol.

Let x mol of N_2O_4 react at equilibrium, when the number of moles of different species will be as under

$$\begin{array}{ccc} N_2O_4 = 2NO_2 & & \text{Total} \\ 0.01726 - x = 2x & & 0.01726 + x \end{array}$$

Then assuming ideal gas equation

$$n = \frac{PV}{RT} \text{ to be applicable}$$

$$(0.01726 + x)\,\text{mol} = \frac{\left(\dfrac{1}{1.013}\,\text{atm}\right)\left(\dfrac{500}{1000}\,\text{L}\right)}{(0.0821\ \text{L atm K}^{-1}\text{mol}^{-1})\,(25 + 273)\,\text{K}}$$

whence $\quad x = 0.002914$

$$K_p = \left[\frac{n_{NO_2}^2}{n_{N_2O_4}}\right]\left(\frac{P/P^\circ}{\sum n_i}\right)^{\Delta v} \qquad \text{by Eq. (11.18)}$$

$$= \left[\frac{(2x)^2}{0.01726 - x}\right]\left(\frac{P/P^\circ}{0.01726 + x}\right)^{2-1}$$

$$= \left(\frac{(2 \times 0.002914)^2}{0.01726 - 0.002914} \right) \left(\frac{1/1}{0.01726 + 0.002914} \right) \text{ taking } P^\circ = 1 \text{ bar}$$

$$= (0.002362)(49.58) = 0.1171$$

$$K_c = \left[\frac{n_{NO_2}^2}{n_{N_2O_4}} \right] \left(\frac{1}{VC^\circ} \right)^{\Delta v}$$

$$= (0.002362 \text{ mol}) \left[\frac{1}{\dfrac{500}{1000} \text{ L } (1 \text{ mol L}^{-1})} \right]^{2-1}$$

$$= 0.004724$$

Alternatively, K_p can be calculated from K_c by Eq. (11.13)

b.

$$\frac{(C_{NO_2}/C^\circ)^2}{C_{N_2O_4}/C^\circ} = \frac{n_{NO_2}^2}{n_{N_2O_4}} \cdot \frac{1}{VC^\circ}$$

$$= 2, \text{ for the given mixture}$$

$$> K_c$$

Then the net reaction will occur in the backward direction.
□ In the equilibrium mixture

$$n_{NO_2} = \frac{PV}{RT}$$

$$= \frac{(1 \text{ atm}) \left(\dfrac{500}{1000} \text{ L} \right)}{(0.0821 \text{ L atm K}^{-1}\text{mol}^{-1})(25 + 273) \text{ K}} = 0.02044 \text{ mol}$$

$$n_{N_2O_4} = \frac{n_{NO_2}^2}{K_c VC^\circ}$$

$$= \frac{(0.02044 \text{ mol})^2}{(0.004724) \left(\dfrac{500}{1000} \text{ L} \right) (1 \text{ mol L}^{-1})} = 0.1769 \text{ mol}$$

The required amount of N_2O_4
$$= \text{amount of } N_2O_4 \text{ reacted} + \text{amount of } N_2O_4 \text{ unreacted}$$

$$= \frac{n_{NO_2}}{2} + n_{N_2O_4}$$

$$= \frac{0.02044}{2} \text{ mol} + 0.1769 \text{ mol}$$

$$= 0.1871 \text{ mol}$$

Note: The amount of N_2O_4 required is greater than the amount of N_2O_4 reacted. Because the equilibrium requires some unreacted N_2O_4 whose amount depends on the volume and temperature of the vessel.

11.35 In the given reaction the number of mols of the system does not change. Hence total number of moles = initial number of moles = (0.1 + 0.2) or 0.3 mol.

Then, the number of moles of the different species at equilibrium will be as under

$$\underset{0.1-0.03=0.07}{H_2} + \underset{0.2-0.03=0.17}{CO_2} = \underset{0.3\times\frac{10}{100}=0.03}{H_2O} + \underset{0.03}{CO}$$

$$K_p = K_c = K_n = \frac{n_{H_2O}\ n_{CO}}{n_{H_2}\ n_{CO_2}}$$

$$= \frac{(0.03)\ (0.03)}{(0.07)\ (0.17)} = 0.076$$

Note: The given pressure is of no use.

11.36 Here N_2 and H_2 are initially present in volume ratio (same as mol ratio) 1:3 in which they react. Hence this ratio will remain same at all stages of the reaction including equilibrium.

i. Let the system contains 100 moles of mixture at equilibrium where

$$n_{NH_3} = 20\ \text{mol}$$

$$n_{N_2} = (100 - 20) \times \frac{1}{1+3} \text{ or } 20\ \text{mol}$$

$$n_{H_2} = 3n_{N_2} = 3 \times 20 \text{ or } 60\ \text{mol}$$

$$K_p = \left[\frac{n_{NH_3}^2}{n_{N_2} n_{H_2}^3}\right]\left(\frac{P/P_o}{\sum n_i}\right)^{\Delta v} \qquad \text{by Eq. (11.18)}$$

$$= \left(\frac{20^2}{20^1 \cdot 60^3}\right)\left(\frac{10\ \text{atm}/1\ \text{atm}}{100}\right)^{2-1-3}$$

$$= 9.26 \times 10^{-3}$$

☐ 20 mol of $NH_3 \equiv 30$ mol of H_2.

Then $(30 + 60)$ or 90 mol of H_2 was initially present and hence the amount of H_2 remaining unreacted is $\frac{60}{90} \times 100$ or 66.7%.

ii. Let the system contains 100 g of mixture at equilibrium, where

$$n_{NH_3} = \frac{20}{17}\ \text{mol} = 1.18\ \text{mol}$$

$$n_{N_2} = (100 - 20)\ \frac{28}{28+6} \times \frac{1}{28} \text{ or } 2.35\ \text{mol}$$

Since N_2 and H_2 react in $28:(28 + 3 \times 2)$ weight ratio

$$n_{H_2} = 3n_{N_2} = 3 \times 2.35 \text{ or } 7.05\ \text{mol}$$

$$\sum n_i = (1.18 + 2.35 + 7.05) \text{ or } 10.58\ \text{mol}$$

$$K_p = \left(\frac{1.18^2}{2.35 \times 7.05^3}\right)\left(\frac{10\ \text{atm}/1\ \text{atm}}{10.58}\right)^{-2} \qquad \text{by Eq. (11.18)}$$

$$= 1.89 \times 10^{-3}$$

☐ Since mass of the system does not change due to chemical reaction

% of H_2 reacted = % by weight of NH_3 in the reaction system

$$= 20$$

Hence the amount of H_2 remaining unreacted = $(100 - 20)$ or 80%.

Of course the common procedure used in (i) is applicable here.

iii. Let the system contains 100 mol of mixture at equilibrium where

$$n_{N_2} = 20 \text{ mol}$$

$$n_{H_2} = 3n_{N_2} = 60 \text{ mol}$$

$$n_{NH_3} = 100 - n_{N_2} - n_{H_2} = 20 \text{ mol etc.}$$

☐ Similar to (i).

iv. Let the system initially contain 100 mol of N_2. Then at equilibrium

$$n_{N_2} = 20 \text{ mol}$$

$$n_{H_2} = 3n_{N_2} = 60 \text{ mol}$$

$$n_{NH_3} = 2 \times \text{mol of } N_2 \text{ reacted}$$

$$= 2 \times (100 - 20) \text{ or } 160 \text{ mol}$$

$$\sum n_i = n_{N_2} + n_{H_2} + n_{NH_3}, \text{etc.}$$

☐ % of H_2 remaining unreacted

$$= \% \text{ of } N_2 \text{ remaining unreacted}$$

$$= 20$$

Note: In all cases some convenient amount of the system is considered depending on the given condition. This will not affect the results as K_p is an intensive quantity.

11.37 Let the system contains N_2 and H_2 in $1{:}r$ mol ratio at equilibrium. Considering the Stoichiometric equation $\frac{1}{2}N_2 + \frac{3}{2}H_2 = 2NH_3$

$$K_p = \frac{P_{NH_3}}{P_{N_2}^{\frac{1}{2}} P_{H_2}^{\frac{3}{2}}} = \frac{P_{NH_3}}{r^{\frac{3}{2}} P_{N_2}^2} = \frac{P_{NH_3}}{r^{\frac{3}{2}} \left(\dfrac{P - P_{NH_3}}{1+r} \right)^2}$$

or

$$\frac{r^{\frac{3}{2}}}{(1+r)^2} = \frac{1}{K_p} \cdot \frac{P_{NH_3}}{(P - P_{NH_3})^2} = f(P_{NH_3})$$

Differentiation with respect to r gives

$$-\frac{2r^{\frac{3}{2}}}{(1+r)^3} + \frac{\frac{3}{2}r^{\frac{1}{2}}}{(1+r)^2} = f'(P_{NH_3}) \cdot \frac{dP_{NH_3}}{dr}$$

$$= 0, \text{ for max yield of } NH_3 \text{ when } \frac{dP_{NH_3}}{dr} = 0$$

or $r = 3$, i.e. the equilibrium mixture contains N_2 and H_2 in $1{:}3$ mol ratio. Since this is the mol ratio in which N_2 and H_2 react, they will also be present initially in the same ratio. Then for maximum yield N_2 and H_2 must be taken in $1{:}3$ mol ratio or volume ratio.

11.38 8 g of $SO_3 \equiv \dfrac{8}{80}$ or 0.1 mol of SO_3.

Let α be the degree of dissociation of SO_3, then the number of mol of different species will be as under

$$\begin{array}{cccc} SO_3 & = SO_2 & + \frac{1}{3}O_2 & \text{Total} \\ 0.1(1-\alpha) & 0.1\alpha & \dfrac{0.1\alpha}{2} & 0.1\left(1 + \dfrac{\alpha}{2}\right) \end{array}$$

Assuming ideal gas equation

$$P = \frac{\rho}{M_{av}} RT \text{ to be applicable}$$

$$= \frac{\rho RT}{\dfrac{w}{0.1\,(1 + \alpha/2)}}$$

or $\quad 0.1\,(1 + a/2)\ \text{mol} = \dfrac{Pw}{\rho RT}$

$$= \frac{(1.8 \text{ atm})\,(8 \text{ g})}{(1.6 \times 10^{-3} \times 10^3 \text{g L}^{-1})\,(0.082 \text{ L atm K}^-\text{mol}^{-1})\,(600 + 273)\,\text{K}}$$

$$= 0.126 \text{ mol}$$

whence $\alpha = 0.52$

$$K_c = \left[\frac{n_{SO_2}\, n_{O_2}^{\frac{1}{2}}}{n_{SO_3}}\right]\left(\frac{1}{VC^\circ}\right)^{\Delta v}$$

$$= \left[\frac{(0.1\alpha)\,(0.1\alpha/2)^{\frac{1}{2}}}{0.1(1 - \alpha}\right]\left[\frac{P}{0.1(1 + \alpha/2)\,RTC^\circ}\right]^{1 + \frac{1}{2} - 1}$$

$$= \left[\frac{(0.1 \times 0.52)\,(0.1 \times 0.52/2)^{\frac{1}{2}}}{0.1(1 - 0.52)}\right]\left[\frac{1.8 \text{ atm}}{(0.126)\,(0.082 \text{ L atm K}^{-1}\text{mol}^{-1})\,(873 \text{ K})}\right]^{\frac{1}{2}} (1 \text{ mol L}^{-1})$$

$$= 4.05 \times 10^{-3}$$

11.39 a. The partial pressures of different species at equilibrium will be as under

$$\underset{p}{COCl_2} = \underset{5001 - p}{CO} + \underset{5001 - p}{Cl_2}$$

Total pressure $= p + (5001 - p) + (5001 - p) = 5002$

or $\qquad\qquad p = 5000$ torr

$$K_p = \left(\frac{P_{CO} P_{Cl_2}}{P_{COCl_2}}\right)\frac{1}{P^\circ}$$

$$= \frac{(1 \text{ torr})\,(1 \text{ torr})}{(5000 \text{ torr})} \cdot \frac{1}{760 \text{ torr}}$$

$$= 2.63 \times 10^{-7}$$

b. i. Let α be the fraction of $COCl_2$ dissociated at equilibrium

$$\underset{2 - 2\alpha}{COCl_2} = \underset{2\alpha}{CO} + \underset{2\alpha}{Cl_2}$$

Then $\qquad K_p = \left[\frac{(2\alpha)\,(2\alpha)}{(2 - 2\alpha)}\right]\left(\frac{RT}{VP^\circ}\right)^{2 - 1} \qquad$ by Eq. (11.18)

$$\simeq 2\alpha^2 \frac{RT}{VP^\circ} \qquad \text{taking } \alpha \ll 1$$

$$\text{or } \alpha = \left[\frac{K_p P^\circ V}{2RT}\right]^{\frac{1}{2}}$$

$$= \left[\frac{(2.63 \times 10^{-7)} \,(1 \,\text{atm})\,(100 \,\text{L})}{2\,(0.082 \,\text{L atm K}^{-1}\text{mol}^{-1})\,(27+273)\,\text{K}}\right]^{\frac{1}{2}}$$

where 2 actually implies 2 mol

$$= (1.07 \times 10^{-6}/2)^{\frac{1}{2}}$$

$$= 7.31 \times 10^{-4}$$

Note: The value of α thus calculated justifies the approximation taken. This approximation is taken to avoid solution of quadrantic equation.

ii. Same as in (i). Because N_2, which acts as an inert gas, does not affect the equilibrium at constant volume whether present in the vessel initially or introduced subsequently.

iii. Initially present Cl_2 will lower α. Here $n_{Cl_2} = 2\alpha + 1 \approx 1$ and then

$$K_p \approx \alpha \cdot \frac{RT}{VP^\circ}$$

$$\text{or} \qquad \alpha = \frac{K_p P^\circ V}{RT}$$

$$= 1.07 \times 10^{-6}, \text{ as calculated in (i)}$$

iv. Amount of Cl_2 initially present

$$= \frac{PV}{RT}$$

$$= \frac{(1 \,\text{atm})\,(100 \,\text{L})}{(0.082 \,\text{L atm K}^{-1}\text{mol}^{-1})\,(27+233 \,\text{K})} = 4.06 \,\text{mol}$$

Here $\qquad n_{Cl_2} = 2\alpha + 4.06 \approx 4.06$

Then $\qquad K_p \approx 4.06\alpha \cdot \dfrac{RT}{VP^\circ}$

or $\qquad \alpha \approx \dfrac{1}{4.06}(1.07 \times 10^{-6})$ from (iii)

$$= 2.63 \times 10^{-7}$$

11.40 a. For the equilibrium mixture

$$\frac{C_{PCl_3}C_{Cl_2}}{C_{PCl_5}} = \frac{(2/V)\,(2/V)}{(2/V)} = \frac{2}{V} = K_c, \text{numerically}$$

On removing one mol of each species

$$\frac{C_{PCl_3}C_{Cl_2}}{C_{PCl_5}} = \frac{(1/V)\,(1/V)}{(1/V)} = \frac{1}{V}(\neq K_c) < K_c$$

This means the equilibrium will be disturbed.

☐ Here there will be a net reaction in the forward direction.

b.   i. He, being an inert gas, will not affect the equilibrium when added to the system under the given condition. Then the number of moles of He to be introduced will be equal to the initial total number of moles of the system, i.e. $2 + 2 + 2$ or 6.

ii. Due to introduction of Cl_2, the equilibrium $PCl_5 \rightleftharpoons PCl_3 + Cl_2$ will be shifted in the backward direction. Let x mol of Cl_2 is introduced of which y mol reacts at equilibrium. Then the number of mol of different species at equilibrium will be as under

$$PCl_5 = PCl_3 + Cl_2$$

Mol number before introduction of Cl_2 $\quad$ 2 $\quad$ 2 $\quad$ 2

Mol number after introduction of Cl_2 $\quad$ $2+y$ $\quad$ $2-y$ $\quad$ $2+x-y$

Then, numerically

$$K_c = \frac{(2/V)(2/V)}{(2/V)}, \text{ before introduction of } Cl_2$$

$$= \frac{\dfrac{(2-y)}{V}\dfrac{(2+x-y)}{V}}{\left(\dfrac{2+y}{V}\right)}, \text{ after introduction of } Cl_2$$

whence $\qquad (2-y)(2+x-y) = 2(x+y)$ $\hfill$ (I)

Again total number of moles $= (2+y) + (2-y) + (2+x-y) = 12$

or $\qquad\qquad\qquad\qquad x - y = 6$ $\hfill$ (II)

From (I) and (II), $x = 7.2$ and $y = 1.2$.

Note: Here the position of equilibrium shifts by 1.2 units of (net) reaction in the backward direction.

11.41 The number of moles of different species present at equilibrium will be as under

$$H_2(g) \quad + \quad I_{2(g)} \quad = \quad 2HI(g)$$

In the first stage $\quad$ $1 - x/2$ $\quad$ $3 - x/2$ $\quad$ x

In the second stage $\quad$ $3 - x$ $\quad$ $3 - x$ $\quad$ $2x$

$$K_p = \frac{x^2}{(1 - x/2)(3 - x/2)} \quad \text{for the first stage}$$

$$= \frac{(2x)^2}{(3 - x)(3 - x)} \quad \text{for the second stage}$$

whence $\qquad\qquad x = 1.5$

$$K_p = 1.44$$

11.42 a. At equilibrium, the number of moles of different species will be related as under

$$\frac{n_B}{n_A} = \frac{1}{2} \text{ or } n_A = 2n_B \quad \text{considering } A \rightleftharpoons B$$

and $\qquad\qquad \dfrac{n_C}{n_B} = \dfrac{2}{3} \text{ or } n_C = \dfrac{2}{3}n_B \quad \text{considering } B \rightleftharpoons C$

Again $\qquad n_A + n_B + n_C = 1$

Then $\qquad 2n_B + n_B + \frac{2}{3}n_B = 1$

whence $\qquad\qquad n_B = \dfrac{3}{11}$

$$n_A = 2 \times \frac{3}{11} = \frac{6}{11}$$

$$n_C = \frac{2}{3} \times \frac{3}{11} = \frac{2}{11}$$

For $A \rightleftharpoons C$ $\qquad K = \frac{n_C}{n_A} = \frac{n_C}{n_B} \cdot \frac{n_B}{n_A} = \frac{2}{3} \cdot \frac{1}{2} = \frac{1}{3}$

Note: To find K, we need not calculate n_A and n_C.

b. Here the equilibrium constant for $A \rightleftharpoons C$ will not be affected. Because $\Delta G°$, and hence equilibrium constant is independent of the intermediate involved. But the number of moles of each isomer will be affected, because the other two equilibrium constants are affected.

Note: The equilibrium constant for $A \rightleftharpoons B$ is not affected due to the equilibrium $B \rightleftharpoons C$.

11.43 a. 4.4 g of $CO_2 \equiv \frac{4.4}{44}$ or 0.1 mol of CO_2.

Let x mol of CO_2 reacts when equilibrium is attained. Then

$$CO_2(g) + C(s) = 2CO(g) \qquad \text{Total} \left(\sum n_i \right) \text{ in the gas phase}$$

Number of mol at equilibrium $\quad 0.1 - x \qquad\qquad 2x \qquad\qquad 0.1 + x$

Then $\qquad M_{av} = \dfrac{(0.1 - x) M_{CO_2} + 2x M_{CO}}{0.1 + x}$

$$= \frac{(0.1 - x) 44 + 2x \times 28}{0.1 + x} = 36$$

or $\qquad\qquad x = 0.033$

Equilibrium pressure $= \dfrac{\sum n_i RT}{V}$

$$= \frac{(0.1 + 0.033) \text{ mol } (0.082 \text{ L atm K}^{-1}\text{mol}^{-1})(1000 + 273) \text{ K}}{1 \text{ L}}$$

$$= 13.88 \text{ atm}$$

$$K_c = \left[\frac{n_{CO}^2}{n_{CO_2}} \right] \left(\frac{1}{VC°} \right)^{\Delta v}$$

$$= \left[\frac{(2x)^2}{(0.1 - x)} \right] \left(\frac{1}{VC°} \right)^{2-1}$$

$$= \left[\frac{(2 \times 0.033 \text{ mol})^2}{(0.1 - 0.033) \text{ mol}} \right] \left[\frac{1}{(1 \text{ L})(1 \text{ mol L}^{-1})} \right] = 0.065$$

$$K_p = K_c \left(\frac{RTC°}{P°} \right)^{\Delta v}$$

$$= (0.065) \left[\frac{(0.082 \text{ L atm K}^{-1}\text{mol}^{-1})(100 + 273) \text{ K } (1 \text{ mol L}^{-1})}{(1 \text{ atm})} \right]^{2-1} = 6.78$$

b. Let y mol of CO_2 is required. Then

$$CO_2(g) + C(s) = 2CO(g)$$

Number of moles at equilibrium $\quad y - 0.1 \qquad\qquad\qquad 0.2$

$$K_c = \left[\frac{(0.2 \text{ mol})^2}{(y - 0.1) \text{ mol}}\right]\left[\frac{1}{(1 \text{ L})(1 \text{ mol L}^{-1})}\right] = 0.065$$

or $\qquad\qquad y = 0.715$

Note: y is lower with vessel of higher volume, the forward reaction being then favourable.

c.   i. On addition of CO at constant volume, the equilibrium will be shifted in the backward direction so that a part of the added CO (which creates strain in the system) is consumed.

The addition at constant pressure will involve increase in volume of the system. This will oppose the backward shift of equilibrium. It may so happen that for certain amount of added CO, there will be no shift of equilibrium. Above this, the equilibrium will shift in the backward direction while below this, it will shift in the opposite direction. Vide problem 11.32(b).

ii. On addition of C at constant volume, its concentration is not affected but pressure of the system is increased, i.e. the system is compressed. The system will oppose this by self contraction through the shift of equilibrium in the backward direction. Vide problem 31(ii).

On the contrary, the addition at constant pressure will cause increase in volume, i.e. decompression of the system, with consequent shift of equilibrium in the forward direction.

However, the addition of C will affect the system only insignificantly, because the effect arises indirectly from the change in pressure or volume of the system and that too is small enough to be ignored under ordinary condition.

11.44 a.

$$\Delta G^\circ = (\bar{G}_f^\circ)_{PCl_3} + (\bar{G}_f^\circ)_{PCl_2} - (\bar{G}_f^\circ)_{PCl_5}$$

$$= [-267.8 + 0.0 - (-305.0)] \text{ kJ mol}^{-1}$$

$$= 37.2 \text{ kJ mol}^{-1}$$

$$K_p = \exp\left[-\frac{\Delta G^\circ}{RT}\right]$$

$$= \exp\left[-\frac{(37.2 \times 10^3 \text{ J mol}^{-1}}{(8.31 \text{ JK}^{-1}\text{mol}^{-1})(25 + 273) \text{ K}}\right] = 0.985$$

b. If α be the degree of dissociation of PCl_5, then

$$K_p = \left(\frac{\alpha^2}{1 - \alpha}\right)\left(\frac{P/P^\circ}{1 + \alpha}\right)^{1+1-1}$$

$$= \left(\frac{\alpha^2}{1 - \alpha^2}\right)\left(\frac{P}{P^\circ}\right)$$

Then $\qquad \ln K_p = 2\ln\alpha - \ln(1 - \alpha^2) + \ln(P/P^\circ)$

Differentiation at constant pressure gives

$$\frac{d \ln K_p}{dT} = \frac{2d \ln \alpha}{dT} + \left(\frac{2\alpha}{1-\alpha^2}\right)\frac{d\alpha}{dT}$$

$$= \frac{2d \ln \alpha}{dT} + \left(\frac{1}{1-\alpha^2}\right)$$

$$\approx \frac{2d \ln \alpha}{dT} \quad \text{ignoring } \alpha^2 \text{ compared to 1}$$

[We can arrive at this expression taking this approximation in the expression for K_p]

Then using van't Hoff Eq. (11.15)

$$\frac{\Delta H^\circ}{RT^2} = \frac{2d \ln \alpha}{dT}$$

or

$$\Delta H^\circ = 2RT^2 \frac{d \ln \alpha}{dT}$$

$$= 2(8.31 \text{ JK}^{-1}\text{mol}^{-1})(298 \text{ K})^2 \left(\frac{6}{100} \text{ K}^{-1}\right) = 88.6 \text{ kJ mol}^{-1}$$

$$\Delta S^\circ = \frac{\Delta H^\circ - \Delta G^\circ}{T}$$

$$= \frac{88.6 - 37.2}{298} = 0.172 \text{ kJ K}^{-1}\text{mol}^{-1}$$

c. From the relation $-RT \ln K_p = \Delta G^\circ = \Delta H^\circ - T\Delta S^\circ$, if error lies in the determination of ΔS° and not in ΔH°, then

$$-RT d \ln K_p = -Td\Delta S^\circ$$

or

$$\frac{dK_p}{K_p} = \frac{d\Delta S^\circ}{R} = \frac{\Delta S^\circ \times \frac{1}{100}}{R}$$

or

$$\frac{dK_p}{K_p} \times 100 = \frac{\Delta S^\circ}{R} = \frac{172 \text{ JK}^{-1}\text{mol}^{-1}}{8.31 \text{ JK}^{-1}\text{mol}^{-1}} = 20.7$$

Then the error in the determination of $K_p = 20.7\%$

11.45 Given that

(1) $2Ni(s) + O_2(g) = 2NiO(s)$ $\Delta G_1^\circ = 296.2 \text{ kJ mol}^{-1}$

(2) $2H_2(g) + O_2(g) = 2H_2O(g)$ $\Delta G_2^\circ = -381.6 \text{ kJ mol}^{-1}$

(1)–(2) gives

$$2Ni(s) + 2H_2O = 2NiO(s) + 2H_2(g) \quad \Delta G^\circ = \Delta G_1^\circ = \Delta G_2^\circ = 677.8 \text{ kJ mol}^{-1}$$

For this reaction $\Delta G = \Delta G^\circ + RT \ln \dfrac{p_{H_2}^2}{p_{H_2O}^2}$

$$= \Delta G^\circ + 2RT \ln \frac{p_{H_2}}{p_{H_2O}}$$

$$= 677.8 \text{ kJ mol}^{-1} + 2(8.31 \times 10^{-3} \text{ kJ K}^{-1}\text{mol}^{-1})(1000 \text{ K}) \ln \frac{1}{99} > 0$$

The positive value of ΔG implies that the gas mixture will not react with Ni to form NiO.

Note: The oxidation of Ni to NiO is possible only if $H_2O(g)$ is almost free from H_2 when ΔG for the reaction becomes negative. This requirement is fulfilled when $H_2O(g)$ is passed over Ni.

11.46 a. The relevant chemical equation is

$$NH_4Cl(s) = NH_3(g) + HCl(g)$$

i.
$$K_p = \frac{P_{NH_3}}{P°} \cdot \frac{P_{HCl}}{P°}$$

$$= \left(\frac{\frac{1}{2}P}{P°}\right)^2 \quad \because P_{NH_3} = P_{HCl} = \tfrac{1}{2}P \quad \text{where } P \text{ is the vap pr of } NH_4Cl$$

$$= \left(\frac{1}{2} \times \frac{608\,kPa}{100\,kPa}\right)^2 \quad \text{taking } P° = 1\,bar \equiv 100\,kPa$$

$$= 9.24$$

ii.
$$\Delta G° = -RT \ln K_p$$

$$= -(8.31 \times 10^{-3}\,kJK^{-1}mol^{-1})(700\,K)\ln 9.24$$

$$= -1.29\,kJ\,mol^{-1}$$

iii.
$$\Delta H° = \frac{R\ln(K_p'/K_p)}{\dfrac{1}{T} - \dfrac{1}{T'}} \quad \text{by Eq. (11.17b)}$$

$$= \frac{2R\ln(P'/P)}{\dfrac{1}{T} - \dfrac{1}{T'}} \quad \because K_p \propto P^2$$

$$= \frac{2(8.31 \times 10^{-3}\,kJK^{-1}mol^{-1})\ln(1131/608)}{\dfrac{1}{700\,K} - \dfrac{1}{733\,K}}$$

$$= 160.4\,kJ\,mol^{-1}$$

iv.
$$\Delta S° = \frac{\Delta H° - \Delta G°}{T}$$

$$= \frac{160.4 + 1.29}{700} = 0.231\,kJK^{-1}mol^{-1}$$

☐ The required expression is

$$\frac{d(\Delta G°/T)}{dT} = -\frac{\Delta H°}{RT^2} \quad \text{by Gibbs–Helmoltz Eq. (5.11b)}$$

$$= -\frac{160.4}{8.31 \times 10^{-3}} \cdot \frac{1}{T^2}$$

$$= -\frac{19.3 \times 10^3}{T^2}$$

b. Let α be the fraction of solid NH_4Cl decomposed, then the gas phase contains NH_3 and HCl each of $n\alpha$ mol, where n is the number of mol of NH_4Cl initially present. Then

$$\text{total number of mol} = 2n\alpha = \frac{PV}{RT}$$

or

$$\alpha = \frac{PV}{2nRT} = \frac{(608 \times 10^3 \, Pa)\,(10^{-3}\, m^3)}{2\,(0.5\, \text{mol})\,(8.31\, JK^{-1}mol^{-1})\,(700\, K)}$$

$$= 0.104$$

□ Let the fraction of NH_4Cl decomposed be α' in presence of x mol (say) of added NH_3. Then

$$C^{\circ 2}K_c = \left(\frac{n\alpha}{V}\right)\left(\frac{n\alpha}{V}\right) = \left(\frac{n\alpha' + x}{V}\right)\left(\frac{n\alpha'}{V}\right)$$

or

$$x = n\left(\frac{\alpha^2}{\alpha'} - \alpha'\right)$$

$$= (0.5\, \text{mol})\left[\frac{(0.104)^2}{0.01} - 0.01\right]$$

$$= 0.536\, \text{mol}$$

c. Let the partial pressure of HCl be p_{HCl} in absence of added NH_3 and p'_{HCl} when NH_3 is initially present at 100 kPa. Then

$$p^{\circ 2}K_p = p_{HCl} \cdot p_{NH_3} = p'_{HCl}\,(p'_{HCl} + 100\,\text{kPa})$$

Using the given data

$$(304\, \text{kPa})\,(304\, \text{kPa}) = p'_{HCl}\,(p'_{HCl} + 100\,\text{kPa})$$

whence

$$p'_{HCl} = 258\, \text{kPa}$$

$$\text{Total pressure} = p'_{HCl} + p'_{NH_3} = p'_{HCl} + p'_{HCl} + 100\,\text{kPa}$$

$$= (258 + 258 + 100)\, \text{or}\, 616\, \text{kPa}$$

11.47

$$\Delta H^\circ = \frac{d(\Delta G^\circ / T)}{d(1/T)} \quad \text{by Gibbs–Helmholtz Eq. (5.11c)}$$

$$= -R\frac{d\ln K}{d(1/T)}$$

$$= -R[-986\, K + 2 \times 2.50 \times 10^4\, K^2/T]$$

Then at 500 K $\quad \Delta H^\circ = -(8.31\, JK^{-1}mol^{-1})\left[-986\, K + \frac{2 \times 2.50 \times 10^4\, K^2}{(500\, K)}\right]$

$$= 7.36\, \text{kJ mol}^{-1}$$

$$\Delta S^\circ = -\frac{d\Delta G^\circ}{dT} \quad \text{by Eq. (5.10)}$$

$$= \frac{d}{dT}(RT\ln K)$$

$$= R\left[\frac{d}{dT}(-1.15T - 986\,K + 2.50 \times 10^4\,K^2\,/T)\right]$$

$$= R[-1.15 - 2.50 \times 10^4\,K^2\,/T^2]$$

Then at 500 K, $\Delta S° = (8.31\,JK^{-1}mol^{-1})[-1.15 - 2.5 \times 10^4\,K^2\,/(500\,K)^2$

$$= -10.4\,JK^{-1}mol^{-1}$$

Note:

1. It is more convenient to calculate $\Delta S°$ directly than to do it using the relation

$\Delta S° = \dfrac{\Delta H° - \Delta G°}{T}$ as in the previous problem.

2. $\Delta H°$ and $\Delta S°$, like $\Delta G°$, do not depend on pressure.

3. $\Delta G° = -RT\ln K_a$ holds for any system but $\Delta G° = -RT\ln K_p$ holds only for ideal systems.

UNIVERSITY QUESTIONS

11.1 Draw free energy (G) verses extent of reaction (ξ) plot for the following gas phase reaction

$$A(g) \rightleftharpoons B(g)$$

What will be the nature of the plot if we assume that no mixing of product with reactant takes place during the reaction? Explain your answer.

(Calcutta BSc(H), 2013)

11.2 A reaction of the type $A(g) + A(g) \rightarrow A_2(g)$ is expected to be exothermic. Explain.

(Burdwan BSc(H), 2000)

11.3 Give qualitative arguments to explain the fact that for the reaction $N_2(g) \rightleftharpoons 2N(g)$, $\Delta G°$ at very low T is positive while its value is negative at very high T.

(Burdwan BSc(H), 1992)

11.4 Consider the reaction: $H_2(g) + I_2(g) = 2HI(g)$.

If there are 1 mole of H_2, 1 mole of I_2 and O moles of HI present before the reaction starts, express the free energy of the reaction in terms of degree of advancement ξ. If standard free energy of a reaction is positive, can the reaction proceed in forward direction. Comment.

(Calcutta BSc(H), 2008)

11.5 For the reaction $2SO_2(g) + O_2(g) \rightleftharpoons 2SO_3(g)$, $K_p = 10$ at 960 K. For a reaction mixture with partial pressures $P_{SO_2} = 1.0 \times 10^{-3}$ bar, $P_{O_2} = 0.02$ bar and $P_{SO_3} = 1.0 \times 10^{-4}$ bar, determine in which direction the reaction would be spontaneous and why [Take $P° = 1$ bar]

(Calcutta BSc(H), 2007)

11.6 For a general chemical reaction at equilibrium prove that

i.

$$\left(\frac{\partial \xi}{\partial T}\right)_P = \frac{\Delta H}{T}\bigg/\left(\frac{\partial^2 G}{\partial \xi^2}\right)_{equilibrium}$$

ii.

$$\left(\frac{\partial \xi}{\partial P}\right)_T = -\Delta V\bigg/\left(\frac{\partial^2 G}{\partial \xi^2}\right)_{equilibrium}$$

All terms have their usual significance.

(Calcutta BSc(H), 2014)

11.7 Prove that for an ideal gas reaction

$$\left(\frac{\partial \ln K_n}{\partial P}\right)_T = -\frac{\Delta v}{P}$$

(Calcutta BSc(H), 2013)

11.8 State whether the equilibrium $2SO_2 + O_2 \underset{650°C}{\overset{\text{Pt catalyst}}{\rightleftharpoons}} 2SO_3$ will be disturbed and if so, in which direction when

    i. the volume of the reaction vessel is increased at constant temperature.

    ii. the catalyst is somehow removed from the vessel, without removing any gas, at constant volume and temperature.

    iii. the reaction vessel is suddenly cooled to 0°C.

    Briefly indicate the reasons for your answer.   (Calcutta BSc(H), 1973)

11.9 a. For the ideal gas reaction $PCl_5\,(g) \rightleftharpoons PCl_3\,(g) + Cl_2\,(g)$, state with reasons, how the equilibrium is affected when each of the following changes is made in the above equilibrium mixture at 25°C.

    i. He (g) is added at constant T and V.

    ii. He (g) is added at constant T and P.   (Calcutta BSc(H), 2014)

  b. Show the quantitative effect of introducing x moles of inert gas at constant pressure into the equilibriated system

$$PCl_3\,(g) \rightleftharpoons PCl_3\,(g) + Cl_2\,(g)$$ (Calcutta BSC(H), 1976)

11.10 Consider the equilibrium $A + 3B \rightleftharpoons 2C$ (ΔH negative). Do you think that the increase of concentration of A, decrease of temperature and presence of a catalyst will change the equilibrium constant of the reaction? Consider the effects separately and justify your answer.   (Calcutta BSc(H), 1993)

11.11 Does the equilibrium constant of a chemical reaction depend on the following?

    i. Standard states chosen for the reactants and products.

    ii. The Stoichiometric representation of the reaction.

    Justify your answer.   (Calcutta BSc(H), 2005)

11.12 The standard free energy $\Delta G°$ at 500 K for an ideal gas reaction $A + 2B \rightleftharpoons 2C$ is -478.0 Cal mol^{-1}. Find the value of the equilibrium constant K_p for the reaction $\frac{1}{2}A + B \rightleftharpoons C$ at this temperature. Will the value of K_p change with change of pressure.   (Calcutta BSC(H), 1993)

11.13 $CaCO_3$ is heated in a closed container. Obtain an expression for K_p. Does the expression imply that K_p depends on pressure?   (Burdwan BSC(H), 1992)

11.14 Consider the following gaseous reaction

$$AB\,(g) \rightleftharpoons A\,(g) + B\,(g); \quad \Delta H = -ve$$

    i. Derive a suitable equilibrium constant expression to explain the effect of pressure on the reaction equilibrium.

    ii. Starting from van't Hoff's reaction, isotherm obtain the integrated form of van't Hoff's equation at constant pressure and explain the effect of temperature on the position of equilibrium of the above reaction (assume heat of reaction to be independent of temperature).

    iii. With a graphical representation, explain how ΔH of the reaction can be experimentally determined.

    iv. The relation $K_p = K_c RT$ for the reaction implies that K_p/K_c has the dimension of energy. Comment on this statement.   (Calcutta BSc(H), 2000)

11.15 Find the temperature for which K_p and K_c values of the following reaction are same

$$A\,(g) = 2B\,(g)$$

(Assume all gases in equilibrium behave ideally)   (Jadavpur BSc(H), 2013)

11.16 K_c for the following reaction at 1173 K is 0.28

$$CS_2(g) + 4H_2(b) \rightleftharpoons CH_4(g) + 2H_2S(g)$$

Calculate the value of K_p at this temperature. (Nagpur BSC, 2005)

11.17 Under what condition would the equilibrium constant not change with temperature? (Burdwan BSc(H), 1999)

11.18 K_p of a reaction varies with temperature (T) as

$$\ln K_p = -1.04 - \frac{1088}{T} + \frac{1.51 \times 10^5}{T^2} \text{ [in 300 K–600 K range]}$$

Calculate $\Delta S°$ at 400 K. (Calculate BSc(H), 2007)

11.19 K_p for the reaction $N_2(g) + 3H_2(g) \rightleftharpoons 2NH_3(g)$ at 400°C is 1.6×10^{-4}.

a. What will be the equilibrium constant at 500°C if the heat of the reaction in this temperature range is 10.18 kJ? (Delhi BSc, 2003)

b. Calculate the values of the standard free energy change $(\Delta G°)$ with respect to K_p and K_c, and interpret the values of $\Delta G°$ in the two cases if these are found to be different. (Calcutta BSc(H), 1991)

11.20 A certain amount of NOCl(g) is introduced into an evacuated flask and at equilibrium the total pressure is 1 atm and the partial pressure of NOCl(g) is 0.6 atm at 200°C.

Calculate K_p for the reaction $2NOCl(g) = 2NO(g) + Cl_2(g)$ at this temperature. K_p increases by 2% per degree celsius rise in temperature around 200°C. Calculate $\Delta H°$ and $\Delta S°$ for the reaction at this temperature.

Also calculate the average molecular weight of the mixture at 200°C and 1 atm total pressure. (Calcutta BSc(H), 1999)

11.21 For the equilibrium $COCl_2(g) \rightleftharpoons CO(g) + Cl_2(g)$, K_p is 8×10^{-9} at 127°C.

a. Calculate the degree of dissociation of phosgene and $\Delta H°$ for the for the reaction at that temperature.

Given that total pressure is 2 atm and $\Delta S°_{400 K} = 300$ Cal deg^{-1}mol^{-1}. (Calcutta BSC(H), 2012)

b. Find the fraction of phosgene dissociated at this temperature when 1 mole of phosgene is placed in a 100 litre vessel containing nitrogen at a partial pressure of 1 atm. (Calcutta BSc, 1978)

11.22 At 2155°C and 1 atm pressure $H_2O(g)$ is 1.18% decomposed into $H_2(g)$ and $O_2(g)$ in accordance with the equation $2H_2O(g) = 2H_2(g) + O_2(g)$. Calculate K_p for the process. (Calcutta BSc(H), 1997)

11.23 In a gas phase reaction $2A(g) + B(g) \rightleftharpoons 3C(g) + D(g)$, it was found that when 1.2 and 1 mol of A, B and D respectively were mixed and allowed to come to equilibrium at 300 K, the resulting mixture contained 0.90 mol of C at a total pressure of a bar. Calculate K_p. (Jadavpur BSc(H), 2012)

11.24 1 mol of H_2 and 0.2 mol of CO_2 introduced in a vacuum flask at 450°C, reacts as

$$H_2(g) + CO_2(g) \rightarrow H_2O(g) + CO(g)$$

At equilibrium the pressure is 0.5 atm and the amount of steam is 10 mole percent. Find the equilibrium constant at 450°C. (Calcutta BSc(H), 2010)

11.25 N_2 and O_2 combine at a given temperature to produce NO. At equilibrium the yield of NO is x% by vlume. If $x = \sqrt{Kab} - \dfrac{K(a+b)}{4}$, where K is the equilibrium constant of the reaction at the given temperature and a and b are the volume percentages of N_2 and O_2 respectively in the initial pure mixture, what would be the initial composition of the reaction mixture in order that the maximum yield of NO is ensured. (Calcutta BSc(H), 2013)

11.26 a. When N_2 and H_2 are mixed in the proportion 1:3 (by volume) at 650°C under 50 atm pressure equilibrium is established when NH_3 is produced to the extent 25% by weight. Calculate the equilibrium constant. (Burdwan BSc(H), 2010)

b. Prove that yield of NH_3 is maximum in the reaction $N_2 + 3H_2 = 2NH_3$, when N_2 and H_2 are mixed in the ratio 1:3 (by volume). (Burdwan BSc(H), 2008)

11.27 A sample of $CaCO_3(s)$ is introduced into a sealed container of volume 0.5 litre and heated to 800 K until equilibrium is reached. The equilibrium constant (K_p) for the reaction $CaCO_3(g) \rightleftharpoons CaO(g) + CO_2(g)$ is 3.9×10^{-2} at this temperature. Calculate the amount of CaO present at equilibrium. (Delhi BSc, 2005)

11.28 The vapour pressure of solid NH_4HS at 25°C is 50 cm of Hg.

a. Assuming complete dissociation of the vapour into NH_3 and H_2S, calculate total pressure when solid NH_4HS is allowed to dissociate at 25°C in a vessel containing NH_3 at a pressure of 32 cm of Hg. (Calcutta BSc(H), 1976)

b. If 0.06 mol of solid NH_4HS is introduced into a 2.4 litre flask at the said temperature

i. calculate the percentage of solid that will get decomposed into NH_3 and H_2S at equilibrium.

ii. Calculate the number of moles of NH_3 that would have to be added to the flask to reduce decomposition of solid to 1% (assume all the gases to be ideal). (Calcutta BSc(H), 2003)

c. Once equilibrium is reached, does the addition of solid NH_4HS affect the equilibrium? (Calcutta BSc(H), 2008)

11.29 For the reaction $PCl_5(g) \rightleftharpoons PCl_3(g) + Cl_2(g)$, find $\Delta G°$ and $\Delta H°$ at 27°C. Given standard free energy of formation

$\Delta G_f^°$ for PCl_5 at 27°C = –77.47 kCal/mol

for PCl_3 at 27°C = –68.39 kCal/mol

Standard heat of formation

$\Delta H_f^°$ for PCl_5 = –95.35 kCal/mol

for PCl_3 = –73.22 kCal/mol

$\Delta H_f^°$ values assumed independent of temperature. (Calcutta BSc(H), 1994)

11.30 Calculate $\Delta G_f^°$ for liquid water at 25°C. Given that $\Delta H_f^°$ of $H_2O(l)$ is –285.5 kJ mol^{-1} and standard molar entropies are $S_{H_2O(l)}^° = 69.9$ JK^{-1}mol^{-1}, $S_{H_2(g)}^° = 130.6$ JK^{-1}mol^{-1} and $S_{O_2(g)}^° = 205$ JK^{-1}mol^{-1}. (Vidyasagar BSc(H), 2007)

11.31 A two-step reaction has stepwise equilibrium constants K_1 and K_2. From free energy consideration show that the overall equilibrium constant is K_1K_2. (Burdwan BSc(H), 2012)

11.32 Calculate the equilibrium constant (K_p) for the reaction: $C(s) + CO_2(g) \rightleftharpoons 2CO(g)$ at 1300 K from the following data

$C(s) + 2H_2O(g) \rightleftharpoons CO_2(g) + 2H_2(g)$, $K_p = 3.9$

$H_2(g) + CO_2(g) \rightleftharpoons CO(g) + H_2O(g)$, $K_p = 0.7$ (Burdwan BSc(H), 2007)

11.33 $\Delta G°$ values for the hydrolysis of glucose-1-phosphate and glucose-6-phosphate are -21 kJ/mol and -14 kJ/mol, respectively. Calculate the equilibrium constant for the following equilibrium at 25°C.

Glucose-1-phosphate $\rightleftharpoons$ Glucose-6-phosphate   (Calcutta BSc(H), 2012)

KEY TO UNIVERSITY QUESTIONS

11.1 Vide Fig. 11.1, Section 11.2.

11.2 A reaction will occur spontaneously only if $\Delta G_{T,P} = \Delta H - T\Delta S$ is negative. For a combination reaction of the given type, randomness decreases, i.e. ΔS is negative. Then to make $\Delta G_{T,P}$ negative ΔH of the reaction should necessarily be negative, i.e. the reaction is expected to be exothermic, provided the decrease in G due to mixing does not dominate.

Note: Not all combination reactions occur with a negative ΔH. For example, $N + O = NO$ is exothermic but $\frac{1}{2}N_2 + \frac{1}{2}O_2 = NO$ is endothermic, because the latter, unlike the former, involves also bond breaking.

11.3 For the given dissociation reaction which involves only bond breaking $\Delta H°$ and $\Delta S°$ are both positive. Then for this reaction $\Delta G° = \Delta H° - T\Delta S°$ will decrease with rise of temperature, and it is quite likely that $\Delta G°$ is positive at very low T but negative at very high T.

11.4 Let ξ be the advancement of the reaction represented below. Then from the given condition the number of moles of the different species will be as under

$$H_2(g) + I_2(g) = 2HI(g)$$
$$1-\xi \quad 1-\xi \quad 2\xi$$

Here $\Delta G_{T,P} = 2\mu_{HI} - \mu_{H_2} - \mu_{I_2}$

If the system is assumed to be ideal, $\mu_i = \mu_i° + RT \ln \dfrac{C_i}{C°}$, and then

$$\Delta G_{T,P} = (2\mu_{\mu I}° - \mu_{H_2}° - \mu_{I_2}°) + RT \ln \frac{(C_{HI}/C°)^2}{(C_{H_2}/C°)(C_{I_2}/C°)}$$

$$= \Delta G° + RT \ln \frac{(n_{HI}/V)^2}{(n_{H_2}/V)(n_{I_a}/V)}$$

$$= \Delta G° + RT \ln \frac{2\xi}{(1-\xi)(1-\xi)}$$

☐ Yes. This happens if the second term in the expression for ΔG is so negative that ΔG becomes negative.

Note: Even for the same composition of a chemical system, ξ and hence ΔG, depends on the stoichiometric representation of the reaction involved. For the given reaction represented as $\frac{1}{2}H_2(g) + \frac{1}{2}I_2(g) = HI(g)$

$$\Delta G = \Delta G° + RT \ln \left[\frac{\xi}{(1-\xi/2)(1-\xi/2)} \right]^{\frac{1}{2}}$$

11.5 For the given reaction mixture

$$\frac{P_{SO_3}/P°)^2}{(P_{SO_2}/P°)^2 (P_{O_2}/P°)} = \frac{(1.0 \times 10^{-4})^2}{(1.0 \times 10^{-3})^2 (0.02)} = 0.5 < K_p (= 10)$$

Then a net reaction would occur spontaneously in the forward direction, because by this the LHS becomes equal to K_p at equilibrium.

Note: Equilibrium constant of a chemical reaction is a measure of the ability of a system of specified composition to undergo that reaction.

11.6 For a chemical reaction, the reaction free energy $\Delta G = \left(\frac{\partial G}{\partial \xi}\right)_{T,P}$ is a function of T, P and ξ, and hence

$$d\Delta G = \left(\frac{\partial \Delta G}{\partial T}\right)_{P,\xi} dT + \left(\frac{\partial \Delta G}{\partial P}\right)_{T,\xi} dP + \left(\frac{\partial \Delta G}{\partial \xi}\right)_{T,P} d\xi$$

$$= -\Delta S dT + \Delta V dP + G'' d\xi \quad \text{(remembering that } dG = VdP - SdT)$$

If the variation of T, P and ξ occur infinitesimally so that the system always remain virtually in equilibrium when $\Delta G = 0$, and hence $d\Delta G = 0$ and also $\Delta S = \Delta H/T$ ($\because \Delta H - T\Delta S = \Delta G$), then by above equation

$$-\frac{\Delta H}{T} dT + \Delta V dP + G'' d\xi = 0$$

This leads to the following expressions

i.
$$\left(\frac{\partial \xi}{\partial T}\right)_P = \frac{\Delta H}{TG''} = \frac{\Delta H}{T} \bigg/ \left(\frac{\partial^2 G}{\partial \xi^2}\right)_{\text{equilibrium}}$$

ii.
$$\left(\frac{\partial \xi}{\partial P}\right)_T = -\frac{\Delta V}{G''} = -\Delta V \bigg/ \left(\frac{\partial^2 G}{\partial \xi^2}\right)_{\text{equilibrium}}$$

Note: i and ii are thermodynamic formulation that implicitly justify Chatelier principle with regard to the effect of temperature and pressure on equilibrium. Since $(\partial^2 G/\partial \xi^2)_{T,P} = +ve$ at equilibrium (i) implies that an increase in T at constant P will shift the equilibrium in the direction of increasing H, and (ii) implies that an increase in pressure at constant T will shift the equilibrium in the direction ot decreasing V.

It is risky to apply Chatelier principle unless the system changes slowly.

11.7 For an ideal gas reaction $aA + bB + ... = lL + mM + ...$

$$K_p = \frac{(p_L/P°)^l (P_M/P°)^m ...}{(P_A/P°)^a (P_B/P°)^b ...}$$

$$= \left[\frac{n_L^l n_M^m ...}{n_A^a n_B^b ...}\right] \left(\frac{P}{P° \sum n_i}\right)^{\Delta v} \quad \text{Since } P_i = \frac{n_i}{\sum n_i} P$$

$$= K_n \left(\frac{P}{P° \sum n_i}\right)^{\Delta v} \quad \text{where } \Delta v = (l + m + ...) - (a + b + ...)$$

Then $\quad \ln K_n = \ln K_p - \Delta v \ln P + \Delta v \ln\left(P° \sum n_i\right)$

Differentiating with respect to P at constant T

$$\left(\frac{\partial \ln K_n}{\partial P}\right)_T = \left(\frac{\partial \ln K_p}{\partial P}\right)_T - \frac{\Delta v}{P} + 0$$

Since $P°$ is an arbitrary constant and $\sum n_i$ may be taken as constant as K_p does not depend on $\sum n_i$

$$= 0 - \frac{\Delta v}{P}$$

Since K_p does not depend on P, for an ideal gas reaction.

Note: Even for an ideal gas reaction, K_n is not independent of P or V (unless $\Delta v = 0$ or $RT = 1$), though (like other equilibrium constants) it is independent of initial composition of such reaction systems.

11.8 i. The equilibrium will be shifted in the backward direction in which the system undergoes self-expansion to reduce the strain caused by decompression [vide problem (11.32a(i)].

ii. The equilibrium will not be disturbed. Because a catalyst does not determine the equilibrium position. It mere hastens up the attainment of equilibrium by enhancing both forward and backward reaction rates and has, therefore no role to play after the equilibrium is attained by the reaction system.

iii. The system will not remain in equilibrium, though the composition of the system will remain almost unchanged due to lack of sufficient time for change (vide problem 11.23).

Note: The Chatelier principle is not applicable to (iii), the system being not in equilibrium.

11.9 a. According to the Chatelier principle the qualitative effect on equilibrium will be as follows:

  i. No shift of equilibrium. Vide problem 11.32a(iii).

  ii. Shift of equilibrium in the forward direction. Vide problem 11.32(b).

Alternatively, we can arrive at these conclusions using Eq. (11.18), vide Section 11.7.

b. Due to forward shift in equilibrium, the change in composition of the system may be represented as follows:

$$PCl_5(g) = PCl_3(g) + Cl_2(g)$$

Mol no. at the first equilibrium	a	b	c
Mol no. at the second equilibrium	$a-y$	$b+y$	$c+y$

Then, by Eq. (11.18)

$$K_p = \left[\frac{(b)(c)}{(a)}\right]\frac{P/P°}{(a+b+c)}, \text{ for the first equilibrium}$$

$$= \left[\frac{(b+y)(c+y)}{(a-y)}\right]\frac{P/P°}{(a+b+c+y+x)}, \text{ for the second equilibrium}$$

where y can be calculated from the knowledge of a, b, c and x.

11.10 Due to concentration change, the equilibrium constant (K) will not change, because ΔG^* (or $\Delta G°$), that determines K by Eq. (11.3) is independent of composition of the system.

With change in temperature K will be affected. Because ΔG^*, that determines K (by $\Delta G^* = -RT \ln K$) is a function of T.

In pressure of a catalyst K will not virtually change, because at the end of the specified reaction, a catalyst remains unchanged in mass and chemical composition though not necessarily in physical form. Hence a catalyst cannot significantly affect $\Delta G^* = RT \ln K$, the effect being only due to its minor physical change.

Note: The factors which affect K will also affect equilibrium position, but the converse may not be true (as with reactant concentration).

11.11 Vide problem 11.18.

11.12 Unit advancement of the second reaction amounts to half of that of the first reaction.

Hence ΔG° of the second reaction $= \frac{1}{2}$ of ΔG° of the first

$$= -\frac{478 \text{ cal mol}^{-1}}{2} = -239 \text{ cal mol}^{-1}$$

The required $\quad K_p = \text{Exp}\left(-\frac{\Delta G^\circ}{RT}\right)$

$$= \text{Exp}\left(-\frac{-239 \text{ cal mol}^{-1}}{(1.99 \text{ cal K}^{-1}\text{mol}^{-1})(500 \text{ K})}\right) = 1.27$$

☐ No, because $\Delta G^\circ = -RT \ln K_p$ is independent of P for an ideal gaseous reaction system.

11.13 The vapour phase of the reaction system $CaCO_3 = CaO + CO_2$ contains CO_2 along with traces of $CaCO_3$ and CaO. At equilibrium

$$\mu_{CaCO_3} = \mu_{CaO} + \mu_{CO_2} \qquad\qquad (I)$$
$$\mu_{CaCO_3 (g)} = \mu_{CaCO_3 (s)}$$
$$\mu_{CaO(g)} = \mu_{CaO(s)}$$

Obviously it is immaterial whether μ_{CaCO_3} and μ_{CaO} in (I) are referred to the species in the solid or vapour phase. Considering the second alternative and assuming the vapour phase to behave ideally $\left[\text{when } \mu_i = \mu_i^\circ(T) + RT \ln\left(\frac{P_i}{P^\circ}\right)\right]$, we have

$$-RT \ln K_p = \mu_{CaO(g)}^\circ + \mu_{CO_2(g)}^\circ - \mu_{CaCO_3(g)}^\circ \qquad\qquad (II)$$

where $\quad K_p = \dfrac{P_{CaO}/P^\circ \cdot P_{CO_2}/P^\circ}{P_{CaCO_3}/P^\circ}$ is the equilibrium constant

Since P_{CaO} and P_{CaCO_3} represent saturated vapour pressures at a particular temperature, they may be taken as fixed quantities (ignoring their variation with total pressure of the system). Hence it is more convenient to write

$$K_p' = \frac{P_{CO_2}}{P^\circ} = K_p \frac{P_{CaCO_3}}{P_{CaO}}$$

It follows from (II) that K_p depends on temperature and not on pressure of the system. The same is virtually true also for the conventional equilibrium constant K_p'. The expression $K_p' = \dfrac{P_{CO_2}}{P^\circ}$ does not really represent the pressure dependence of K_p'; it simply represents temperature dependence through P_{CO_2} (which is a function of T).

11.14 i. Vide question 11.7.

ii. For the given reaction, the van't Hoff reaction isotherm is

$$\Delta G = -RT \ln K + RT \ln K \frac{a_A a_B}{a_{AB}} \quad \text{by Eq. (11.5)}$$

or

$$\ln K = -\frac{\Delta G}{RT} + \ln \frac{a_A a_B}{a_{AB}}$$

Then

$$\left(\frac{\partial \ln K}{\partial T}\right)_P = -\frac{1}{R}\left[\frac{\partial(\Delta G/T)}{\partial T}\right]_P + 0, \text{ since } \frac{a_A a_B}{a_{AB}} \text{ is an arbitrary quantity}$$

$$= \frac{\Delta H}{RT^2} \quad \text{by Gibbs–Helmholtz relation}$$

Integration gives

$$\ln K = -\frac{\Delta H}{RT} + \text{constant, assuming } \Delta H \text{ to be independent of } T.$$

iii. From the slope $(= -\Delta H/R)$ of the linear plot of $\ln K$ against $1/T$ (for short range of T), ΔH can be calculated as shown in Fig. 11.3.

iv. In consistent with dimensions, we should write

$$K_p = K_n \left(\frac{P/P^\circ}{\sum n_i}\right)^{\Delta v}$$

(assuming that the reactants and products are all ideal gases)

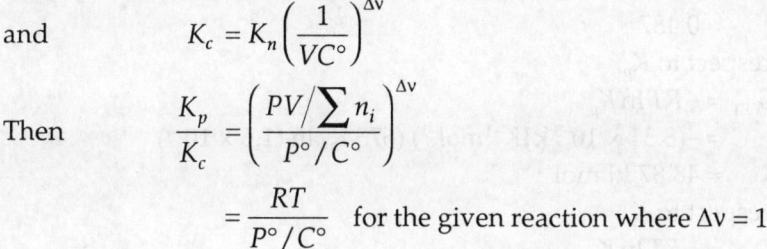

Fig. 11.3

and

$$K_c = K_n \left(\frac{1}{VC^\circ}\right)^{\Delta v}$$

Then

$$\frac{K_p}{K_c} = \left(\frac{PV/\sum n_i}{P^\circ/C^\circ}\right)^{\Delta v}$$

$$= \frac{RT}{P^\circ/C^\circ} \quad \text{for the given reaction where } \Delta v = 1.$$

Since $\dfrac{RT}{P^\circ/C^\circ}$ has no dimensions, the given statement is not justified.

Note: The relation $\dfrac{K_p}{K_c} = (RT)^{\Delta v}$ is correct numerically, though not dimensionally.

11.15 For the given reaction

$$K_p = K_c \cdot \frac{RTC^\circ}{P^\circ}$$

Then K_p and K_c will have same value for $RTC^\circ/P^\circ = 1$

or

$$T = \frac{P^\circ}{RC^\circ} = \frac{1}{0.08206} \quad \text{or } 12.19 \text{ K}$$

11.16 For the given reaction

$$K_p = K_c \left(\frac{RTC^\circ}{P^\circ}\right)^{(1+2)-(1+4)}$$

assuming reactants and products behaving ideally

$$= 0.28 \left[\frac{(0.082 \text{ L atm K}^{-1}\text{mol}^{-1}) (1173 \text{ K}) (1 \text{ mol L}^{-1})}{(1 \text{ atm})} \right]^{-2} = 3.02 \times 10^{-5}$$

11.17 The required condition is that ΔH of the reaction must be zero. This follows from the Van't Hoff Eq. (11.17a).

11.18 From dimensional consideration, the given expression should be more appropriately written as

$$\ln K_p = -1.04 - \frac{1088}{T/K} + \frac{1.51 \times 10^5}{(T/K)^2}$$

Vide problem 11.47.

11.19 a. $K_P' = K_p \exp\left[\frac{\Delta H}{R} \left(\frac{1}{T} - \frac{1}{T'} \right) \right]$ by Eq. (11.17b)

$$= (1.6 \times 10^{-4}) \exp\left[\frac{(10.18 \text{ kJ mol}^{-1})}{(8.31 \times 10^{-3} \text{kJ K}^{-1}\text{mol}^{-1})} \left(\frac{1}{673 \text{ K}} - \frac{1}{773 \text{ K}} \right) \right]$$

$$= 2.025 \times 10^{-4}$$

b. Here $K_c = K_p \left(\frac{P^\circ}{RTC^\circ} \right)^{-2}$

$$= (1.6 \times 10^{-4}) \left[\frac{(1 \text{ atm})}{(0.082 \text{ L atm K}^{-1}\text{mol}^{-1}) (673 \text{ K}) (1 \text{ mol L}^{-1})} \right]^{-2} \quad \text{at } 400°C$$

$$= 0.487$$

With respect to K_p,

$\Delta G_{(P)}^\circ = -RT \ln K_p$

$= -(8.31 \times 10^{-3} \text{kJK}^{-1}\text{mol}^{-1}) (673 \text{ K}) \ln (1.6 \times 10^{-4})$

$= 48.87 \text{ kJ mol}^{-1}$

With respect to K_c,

$\Delta G_{(C)}^\circ = -RT \ln K_c$

$= -(8.31 \times 10^{-3} \text{kJK}^{-1}\text{mol}^{-1}) (673 \text{ K}) \ln 0.487$

$= 4.023 \text{ kJ mol}^{-1}$

Both $\Delta G_{(P)}^\circ$ and $\Delta G_{(C)}^\circ$ are standard reaction free energy, but they have different values because they refer to different standard states, 1 atm for $\Delta G_{(P)}^\circ$ but 1 mol L^{-1} for $\Delta G_{(C)}^\circ$.

11.20 At equilibrium

$$x_{NOCl} = \frac{p_{NOCl}}{P} = \frac{0.6 \text{ atm}}{1 \text{ atm}} = 0.6 \quad \text{assuming ideal behaviour of gases}$$

$$x_{Cl_2} = (1 - x_{NOCl}) \times \frac{1}{3}$$

$[\because x_{NOCl} + x_{NO} + x_{Cl_2} = x_{NOCl} + 2x_{Cl_2} + x_{Cl_2} = 1]$

$$= (1 - 0.6) \times \frac{1}{3} = \frac{0.4}{3}$$

$$x_{NO} = 2x_{Cl_2} = \frac{0.8}{3}$$

$$K_p = K_x \left(\frac{P}{P^\circ} \right)^{\Delta v}$$

$$= \frac{x_{NO}^2 \, x_{Cl_2}}{x_{NOCl}^2} \cdot \left(\frac{P}{P^\circ} \right)^{2+1-2}$$

$$= \frac{(0.8/3)^2 \, (0.4/3)}{(1-0.6)} \cdot \left(\frac{1\,atm}{1\,atm} \right)^1 = 0.237$$

$$\Delta H^\circ = RT^2 \frac{d \ln K_p}{dT}$$

$$= (8.31 \times 10^{-3} \, kJ K^{-1} mol^{-1})(473\,K)^2 \left(\frac{2}{100} \, K^{-1} \right) = 3.72 \, kJ\,mol^{-1}$$

$$\Delta S^\circ = \frac{\Delta H^\circ - \Delta G^\circ}{T} = \frac{\Delta H^\circ + RT \ln K_p}{T}$$

$$= \left[\frac{3.72 \, kJ\,mol^{-1} + (8.31 \times 10^{-3} \, kJ\,mol^{-1})(473\,K) \ln (0.237)}{473\,K} \right]$$

$$M_{av} = (xM)_{NOCl} + (xM)_{NO} + (xM)_{Cl_2}$$

$$= 0.6 \times 65.5 + \frac{0.8}{3} \times 30.0 + \frac{0.4}{3} \times 70.9 = 56.8$$

11.21 a. Let α be the degree of dissociation of phosgene, then the relative number of moles of the different species at equilibrium will be as under

$$COCl_2 = CO + Cl_2$$
$$1 - \alpha \quad \alpha \quad \alpha$$

$$K_p = \left(\frac{n_{CO} n_{Cl_2}}{n_{COCl_2}} \right) \left(\frac{P/P^\circ}{\sum n_i} \right)^{\Delta v}$$

$$= \left(\frac{\alpha \cdot \alpha}{1 - \alpha} \right) \left(\frac{P/P^\circ}{1 + \alpha} \right)^{1+1-1}$$

$$\simeq \alpha^2 \frac{P}{P^\circ} \qquad \qquad \text{taking } \alpha \ll 1$$

or $$\alpha = \frac{P^\circ K_p}{P} = \frac{(1\,atm)(8 \times 10^{-9})}{(2\,atm)} = 4 \times 10^{-9}$$

$$\Delta H^\circ = T\Delta S^\circ + \Delta G^\circ = T\Delta S^\circ - RT \ln K_p$$

b. Vide problem 11.39.

11.22 In terms of degree of dissociation (α), the relative mole number of different species at equilibrium will be

$$2H_2O(g) = 2H_2(g) + O_2(g) \qquad Total$$
$$1 - \alpha \qquad \alpha \qquad \alpha/2 \qquad 1 + \alpha/2$$

Then
$$K_p = \left[\frac{\alpha^2 \cdot \alpha/2}{(1-\alpha)^2}\right]\left(\frac{P/P^\circ}{1+\alpha/2}\right)^{2+1-2}$$

$$= \frac{P/P^\circ}{\left(\frac{1}{\alpha}-1\right)^2 \left(\frac{2}{\alpha}+1\right)}$$

$$= \frac{1\,\text{atm}/1\,\text{atm}}{\left(\frac{100}{1.18}-1\right)^2 \left(2\times\frac{100}{1.18}+1\right)} \quad \text{taking } P^\circ = 1 \text{ atm}$$

Note: In this problem, unlike the previous one, α is not ignorable compared to 1.

11.23 According to the Stoichiometric representation of the reaction, the number of moles of the different species at equilibrium will be as under

$$
\begin{array}{ccccccc}
2A\,(g) & + & B\,(g) & = & 3C\,(g) & + & D\,(g) & \text{Total} \left(\sum n_i\right) \\
1-\dfrac{2\times0.9}{3}=0.4 & & 2-\dfrac{0.9}{3}=1.7 & & 0.9 \text{ given} & & 1+\dfrac{0.9}{3}=1.3 & 4.3
\end{array}
$$

$$K_p = \left[\frac{n_C^3\, n_B}{n_A^2\, n_B}\right]\left(\frac{P/P^\circ}{\sum n_i}\right)^{\Delta v}$$

$$= \left[\frac{(0.9)^3\,(1.3)}{(0.4)^2\,(1.7)}\right]\left(\frac{1\,\text{bar}/1\,\text{bar}}{4.3}\right)^{(3+1)-(2+1)} \quad \text{Taking } P^\circ = 1 \text{ bar}$$

$$= 3.48$$

11.24 Here the total number of moles, which does not change during the reaction, is $(1 + 0.2)$ or 1.2. Then the number of moles of the different species at equilibrium will be as under:

$$
\begin{array}{ccccccc}
H_2(g) & + & CO_2(g) & = & H_2O(g) & + & CO(g) \\
1-0.12 & & 0.2-0.12 & & 1.2\times(10/100) & & 0.12 \\
=0.88 & & =0.08 & & =0.12 & &
\end{array}
$$

$$K_p = \frac{n_{H_2O}\cdot n_{CO}}{n_{H_2}\cdot n_{CO_2}}, \quad \Delta v \text{ being zero}$$

$$= \frac{(0.12)\,(0.12)}{(0.88)\,(0.08)} = 0.205$$

Note: Here the given pressure and temperatre are of no use.

11.25 Here $a + b = 100$. Therefore, the given expression may be rewritten as

$$x = \sqrt{K\,a(100-a)} - \frac{100\,K}{4}$$

Then
$$\frac{dx}{da} = \frac{\sqrt{K}}{2}\cdot\frac{(100-2a)}{\sqrt{a(100-a)}} = 0, \quad \text{for yield to be maximum}$$

Then for maximum yield $a = 50$, and hence $b = 50$, i.e. initially N_2 and O_2 must be in 1:1 mol ratio. This is confirmed by negative d^2x/da^2 for $a = 50$.

11.26 a. Vide problem 11.36.

b. Vide problem 11.37.

Note: The procedure used in (b) is most convenient and of course, it can be applied to the previous problem.

11.27 Here K_p is numerically equal to P_{CO_2}.

Then, the amount of

$$CaO \equiv 0.500 \text{ L } CO_2 \text{ at } 3.9 \times 10^{-2} \text{ atm and } 800 \text{ K}$$

$$= \frac{(3.9 \times 10^{-2} \text{ atm}) (0.500 \text{ L})}{(0.082 \text{ L atm K}^{-1}\text{mol}^{-1}) (800 \text{ K})} = 2.97 \times 10^{-4} \text{ mol}$$

11.28 (a), (b), vide problem 11.46.

(c) vide problem 11.43(c)

11.29 Vide problem 11.44(a)

$$\Delta G° = \Delta G_f° (PCl_3) - \Delta G_f° (PCl_5)$$

$$= -68.39 \text{ kcal mol}^{-1} + 77.47 \text{ kcal mol}^{-1}$$

$$= 9.08 \text{ kcal mol}^{-1}$$

$$\Delta H° = \Delta H_f° (PCl_3) - \Delta H_f° (PCl_5)$$

$$= -73.22 \text{ kcal mol}^{-1} + 95.35 \text{ kcal mol}^{-1}$$

$$. = 22.13 \text{ kcal mol}^{-1}$$

Note:

1. In calculating $\Delta G°$ of a reaction in terms of free energies of formation, only the compounds involved are considered. However, this is completely justified only at 25°C.

2. $\Delta G_f° (PCl_5) \neq G°(PCl_5)$ because the temperature is not 25°C.

These two quantities become equal only at 25°C when $G°$'s of the elements are conventionally taken to be zero.

But $\Delta H_f° (PCl_5) = H°(PCl_5)$, since the given data regarding enthalpy applies to 25°C.

11.30 $\Delta G_f° (H_2O(l)) = \Delta H_f° (H_2O(l)) - T\Delta S_f° (H_2O(l))$

$$= \Delta H_f° (H_2O(l)) - T[S°(H_2O(l)) - S°(H_2(g)) - \tfrac{1}{2}S°(O_2(g))]$$

$$= -285.5 \text{ kJ mol}^{-1} - (298 \text{ K})\left[69.9 - 130.6 - \tfrac{1}{2} \times 205\right] \times 10^{-3} \text{kJ mol}^{-1}$$

$$= -216.3 \text{ kJ mol}^{-1}$$

Note: Values of H, S and G for a substance, but not their changes, depend on the standard state chosen. Hence, although $G = H - TS$ is meaningless, unless same standard state is used for H, S and G, $\Delta G = \Delta H - T\Delta S$ is meaningful.

11.31 Vide Section 11.3.

For the first step $\Delta G_1° = -RT \ln K_1$

For the second step $\Delta G_2° = -RT \ln K_2$

For the overall reaction,

$$\Delta G° = \Delta G_1° + \Delta G_2°$$

Then $-RT \ln K = -RT \ln K_1 - RT \ln K_2$

or $K = K_1 K_2$

11.32 From the given data

(1) $C(s) + 2H_2O(g) = CO_2(g) + 2H_2(g),$ $\Delta G_1 = -RT \ln 3.9$

(2) $H_2(g) + CO_2(g) = CO(g) + H_2O(g),$ $\Delta G_2 = -RT \ln 0.7$

   (1) + (2) gives

$$C(s) + CO_2(g) = 2CO(g) \qquad\qquad \Delta G = -RT \ln K_p$$

Then $\Delta G = \Delta G_1 + 2\Delta G_2$

So $-RT \ln K_p = -RT \ln 3.9 - 2RT \ln 0.7$

Hence $K_p = 3.9\,(0.7)^2$

           $= 1.91$

11.33 The reaction concerned is equivalent to the difference between the given two hydrolysis reactions which may be denoted by 1 and 2 respectively. Then $\Delta G°$ for this reaction will be

$$\Delta G° = \Delta G_1° - \Delta G_2° = -21 \text{ kJ mol}^{-1} - (-14 \text{ kJ mol}^{-1})$$

$$= -RT \ln K = -(8.314 \times 10^{-3} \text{ k JK}^{-1}\text{mol}^{-1})\,(298 \text{ K}) \ln K$$

whence $K = 16.86$

Ionic Equilibrium

Ionic equilibrium is a particular case of chemical equilibrium where charged molecular species, called ions, are involved. This can be effectively treated quantitatively only through the concept of activity with some additional assumption(s).

12.1 ACTIVITY IN CASE OF ELECTROLYTES

The problem of defining activities is somewhat more complicated in electrolytic solutions than in non-electrolytic ones because of non-separability of positive and negative ions formed by the electrolytes.

Let us consider a strong electrolyte MX that undergoes virtually complete dissociation in solution as

$$MX = M^+ + X^-$$

Then at equilibrium (as in any chemical equilibrium), we have the following relation between the chemical potential μ of MX, μ_+ of M^+ ion and μ_- of X^- ion

$$\mu = \mu_+ + \mu_-$$

In terms of activity

$$\mu^* + RT \ln a = \mu_+^* + RT \ln a_+ + \mu_-^* + RT \ln a_-$$

Here a, that refers to undissociated electrolyte, is conventionally regarded as the activity of the electrolyte as a whole.

If activities are so defined that

$$\mu^* = \mu_+^* + \mu_-^* \quad \text{(which means } \Delta G^* = 0\text{)}$$

then $\qquad a = a_+ \cdot a_- = a_\pm^2$ where $a_\pm$ is the mean ionic activity

In general, for an electrolyte $M_{v_+} X_{v_-}$ dissociating as

$$M_{v_+} X_{v_-} = v_+ M^{Z+} + v_- X^{Z-}$$

$$a = a_+^{v_+} \cdot a_-^{v_-} = a_\pm^{v} \quad \text{(where } v = v_+ + v_-\text{)} \tag{12.1}$$

For an electrolytic solution, the activities of the dissolved species are conveniently expressed as product of molal concentration (m) and activity coefficient (γ)

$$a = m\gamma$$
$$a_+ = m_+ \gamma_+$$
$$a_- = m_- \gamma_-$$

where m is the molality of the undissociated electrolyte and not the total molality of the electrolyte.

Then $\qquad\qquad a = (m_+ \gamma_+)^{v_+} (m_- \gamma_-)^{v_-} = (m_\pm)^v (\gamma_\pm)^v \tag{12.2}$

where mean ionic molality

$$m_{\pm} = (m_+^{v_+} \cdot m_-^{v_-})^{\frac{1}{v}}$$ (12.3)

mean ionic activity coefficient

$$\gamma_{\pm} = (\gamma_+^{v_+} \cdot \gamma_-^{v_-})^{\frac{1}{v}}$$ (12.4)

[Strictly speaking for any species i, $a_i = \dfrac{m_i \gamma_i}{m^{\circ}}$

where m° is the standard value of molality (1 mol kg^{-1} of solvent)]

It is important to note that individual ion activity coefficients, and hence the relevant activities, cannot be determined experimentally, though they can be calculated theoretically using Dabye–Huckel theory (because it is not possible to get a solution of only positive or only negative ions). However, $\gamma_{\pm}$ can be found experimentally. Here lies the usefulness of the Eq. (12.2).

12.2 DEBYE–HUCKEL THEORY

Arrhenius theory (i.e. the idea of ions in solution in equilibrium with parent molecular species) is more or less satisfactory in explaining the equilibrium properties and conducting behaviour only of weak electrolytes (that dissociate only to a small extent at moderate concentrations). In order to account for the behaviour of strong electrolytes Debye and Huckel put forward a theory by attributing the deviation of ionic solutions from ideality to the interionic coulombic forces (which Arrhenius failed to consider though it is appreciable in case of strong electrolytes). The key feature of their theory is that in an electrolytic solution, each ion (supposed to be spherical and of negligible size) is surrounded by a spherically symmetrical atmosphere of equal but opposite charge (due to excess of oppositely charged mobile ions that tend to preferentially surround each ion). It is this ionic atmosphere that controls the equilibrium properties and conducting behaviour of an electrolytic solution. On computing the electrical energy of the ion due to its atmosphere (supposing distribution of ions in the ionic atmosphere to be of Boltzmann type) and equating the result to $RT \ln \gamma_i$ (which measures the deviation from ideality), they arrived at the following expression for activity coefficient (γ_i) of an ionic species i.

$$-\log \gamma_i = \frac{1}{2.303} \cdot \frac{F^2}{8\pi N_A \, \epsilon_o \epsilon_r \, RT r_D} \cdot Z_i^2$$ (12.5a)

$$= \frac{1}{2.303} \cdot \frac{F^3}{4\pi N_A} \left(\frac{e}{2 \, \epsilon_o^3 \epsilon_r^3 \, R^3 T^3} \right)^{\frac{1}{2}} Z_i^2 \sqrt{I}$$ (12.5b)

$$\left. \begin{array}{l} = A Z_i^2 \sqrt{I} \\[2mm] = 0.509 \, Z_i^2 \, \sqrt{I/m^{\circ}} \end{array} \right\} \quad \text{for aqueous solution at 25°C}$$ (12.5c)

Here Z_i is the valency of the ion; A is a constant depending only on temperature T, dielectric constant (relative permittivity) ϵ_r and density e of the solvent; r_D is Debye length, which is an approximate measure of the thickness of the ionic atmosphere (i.e. the distance over which the electrostatic field of an ion is appreciable); I is ionic strength of the solution.

The expression (12.5c) of γ_i (which is non-measurable) leads to the following expression of $\gamma_\pm$ (which is measurable)

$$-\log\gamma_\pm = AZ_+Z_- \sqrt{I} \qquad (12.5d)$$

where Z_+ is the valency of the positive ion and Z_- that of the negative ion of the electrolyte concerned. The Eq. (12.5d) represents what is called the Debye–Huckel limiting law. This is called so because it holds only in the limit $I \to 0$.

For I exceeding 0.01 mol kg^{-1}, $\gamma_\pm$ is more approximately given by the Debye–Huckel extended law

$$-\log\gamma_\pm = \frac{AZ_+Z_- \sqrt{I}}{1 + B\sqrt{I}} \qquad (12.5e)$$

where B (*like* A) is another constant that arises out of consideration of the definite size of the ions (which has been disregarded in the limiting law).

I, appearing in the Debye–Huckel equations, is defined as

$$I = \tfrac{1}{2}\sum m_i Z_i^2 \qquad (12.6)$$

Then I is such a quantity that $\sqrt{I}$ is roughly a measure of the concentration of the ionic charge in the solution concerned.

12.3 SOLUBILITY PRODUCT

In saturated solution of an electrolyte $M_{v_+} X_{v_-}$, strong or weak, both physical and chemical equilibria are involved as represented below

$$M_{v_+} X_{v_-} \text{ (pure)} \rightleftharpoons M_{v_+} X_{v_-} \text{ (soln)} \rightleftharpoons v_+ M^{z+} + v_- X^{z-}$$

Then considering the equilibrium constant together with constant activity of a pure substance

$$a_{M^{z+}}^{v_+} \cdot a_{M^{z-}}^{v_-} = K_{th} \qquad (12.7)$$

where K_{th} is the thermodynamic (or activity) solubility product which is a constant for a particular electrolyte at a particular temperature and pressure. K_{th} is basically an equilibrium constant, though it has meaning only with an electrolyte in its saturated solution (and should therefore be better called solubility constant). The corresponding quantity in unsaturated solution is called the *ionic product of the electrolyte*. A new phase of pure electrolyte will appear (i.e. precipitation will occur) only when its ionic product exceeds the solubility product.

Variation of Solubility of an Electrolyte Due to Presence of Other Electrolytes

From Eq. (12.7) $\qquad K_{th} = [M^{z+}]^{v_+} [X^{z-}]^{v_-} \gamma_{M^{z+}} \gamma_{X^{z-}}$

$$= K_{sp} \gamma_\pm^{v} \qquad (12.8)$$

where [ ] denotes the concentration of the enclosed species expressed in molality (often in molarity for aqueous solutions); $\gamma_{M^{z+}}$ is the activity coefficient of the positive ion in the corresponding concentration scale used and $\gamma_{X^{z-}}$ is that of the negative ion. The quantity K_{sp} is often called the concentration solubility product

$$K_{sp} \to K_{th} \text{ as } \gamma_\pm \to 1, \text{ i.e. } I \to 0$$

In presence of a foreign electrolyte, K_{th} is not affected, but K_{sp} is affected due to change in $\gamma_\pm$ resulting from change in I. The effect depends on the type of foreign electrolyte.

(1) If the foreign electrolyte has no ion in common with the concerned electrolyte, it will enhance K_{sp}, and hence solubility of the latter due to decrease in $\gamma_\pm$.

(2) If the foreign electrolyte has an ion in common with the concerned electrolyte, the effect of I will be same as in 1 but the effect of common ion will be to reduce solubility (provided there is no complex formation) due to shift of the solubility equilibrium to the undissociated side with K_{sp} unchanged. Usually in such cases, the solubility first decreases, reaches a minimum and then decreases as the concentration of the foreign electrolyte changes from low to high value.

Knowledge of K_{sp} of an electrolyte $M_{v_+} X_{v_-}$ enables one to calculate its solubility S_0 in water and S in x molal/molar solution of a foreign electrolyte AX (which dissociates completely) using the following relation

$$K_{sp} = (v_+ S_0)^{v_+} (v_- S_0)^{v_-} \tag{12.9a}$$
$$= (v_+ S)^{v_+} (v_- S + x)^{v_-} \tag{12.9b}$$
$$\simeq (v_+ S)^{v_+} (x)^{v_-}, \text{ if } x \gg v_- S \tag{12.9c}$$

12.4 pH SCALE

Hydrogen ion plays a central role in many chemical processes. To express its concentration in aqueous solution in a convenient way Sorensen introduced a scale of pH by defining the latter as

$$pH = -\log C_{H^+} \tag{12.10a}$$

where C_{H^+} is the total molarity of the hydrogen ion that exists practically entirely in the hydrated form H_3O^+.

Since thermodynamic properties of a solution depends not on the analytical concentration but on the activity of the dissolved species, pH was subsequently defined in terms of hydrogen ion activity as

$$pH = -\log a_{H^+} \tag{12.10b}$$

But the problem lies with the undefined single-ion activity.

At present pH of a solution is defined operationally with much adherence to the measurement of emf of a concentration cell using the concerned solution and some other reference solution in its two electrodes (vide note on problem 13.12).

In ordinary calculation of pH, a_{H^+} is taken to be equal to C_{H^+} though it is legitimate only for solutions of low ionic strength when molality and molarity are almost same for aqueous solution.

It is important to note that pH is not a good index of acidity (i.e. tendency of a solution to transfer proton to a base) of an aqueous solution of high concentration. A more satisfactory acidity function is H_0 which has been defined as

$$H_0 = -\log \left(\frac{a_{H^+} \gamma_B}{\gamma_{BH^+}} \right) \tag{12.11}$$

where a_{H^+} is the total activity of the hydrogen ion in the solution, and γ_B / γ_{BH^+} is the ratio of the activity coefficients of a base B and its conjugate acid BH^+. Since γ_B / γ_{BH^+} has the same value for all bases in a given solvent, H_0 is a property of the solvent and approaches pH in dilute aqueous solution where γ_B / γ_{BH^+} approaches unity.

12.5 ACID AND BASE DISSOCIATION CONSTANTS

In aqueous medium, the dissociation constants K_a of an acid HA and K_b of a base B usually mean the equilibrium constant for the following reaction

$$HA + H_2O \rightleftharpoons H_3O^+ + A^-, \quad K_a = a_{H^+} \cdot a_{A^-} / a_{HA} \tag{12.12}$$

(or simply $HA \rightleftharpoons H^+ + A^-$) where $a_{H^+} \equiv a_{H_3O^+}$

$$B + H_2O \rightleftharpoons BH^+ + OH^-) \quad K_b = a_{BH^+} \cdot a_{OH^-} / a_B \tag{12.13}$$

In each case it has been assumed that the activity of H_2O is constant (assuming the aqueous solution to be dilute) and absorbed into K_a and K_b.

Similar consideration of autodissociation of water

$$H_2O + H_2O \rightleftharpoons H_3O^+ + OH^-$$

leads to a constant K_w, called ionic product of water defined by

$$K_w = a_{H^+} \cdot a_{OH^-} \tag{12.14}$$
$$= 1.008 \times 10^{-14} \text{ at } 25°C$$

The dissociation constants K_a of an acid HA and K_B of its conjugate base A^- are related by

$$K_a K_b = K_w \text{ (in aqueous solution)} \tag{12.15}$$

This follows from the following equilibrium relations

$$HA + H_2O \rightleftharpoons H_3O^+ + A^-, \quad K_a = a_{H^+} \cdot a_{A^-} / a_{HA}$$
$$A^- + H_2O \rightleftharpoons HA + OH^-, \quad K_b = a_{HA} \cdot a_{OH^-} / a_{A^-}$$

For low ionic strength, the activities may be replaced by concentrations.

Calculation of pH

This is simplest in case of solution of strong acids (because of their complete dissociation) and the acid is monobasic HA

$$pH = -\log C_{H^+} \quad \text{(ignoring activity coefficient effect)}$$
$$= -\log F_{HA} \tag{12.16a}$$

where F is formality (i.e. total molarity of HA).
For strong monoacidic base

$$pH = -\log\left(\frac{K_w}{C_{OH^-}}\right) = -\log\left(\frac{K_w}{F_B}\right)$$
$$= PK_w + \log F_B \tag{12.16b}$$

Difficulty arises when the acid or base involved is weak. This happens also with hydrolysable salts. Due to hydrolysis, the solution becomes acidic or basic according as the resulting acid has higher or lower strength than the resulting base.

Let us consider the general case of a solution of weak acid HA and its soluble salt MA having formalities F_{HA} and F_{MA} respectively. Here

$$F_{HA} + F_{MA} = C_{HA} + C_{A^-} \quad \text{material balance equation}$$
$$C_{H^+} + C_{M^+} = C_{A^-} + C_{OH^-} \quad \text{charge balance equation}$$

or
$$C_{A^-} = F_{MA} + C_{H^+} - C_{OH^-}, \quad \because C_{M^+} = F_{MA}$$
and
$$C_{HA} = F_{HA} + F_{MA} - C_{A^-}$$

Then
$$C_{H^+} = K_a \frac{C_{HA}}{C_{A^-}} \quad \text{from Eq. (12.12)}$$

$$= K_a \cdot \frac{F_{HA} - C_{H^+} + C_{OH^-}}{F_{MA} + C_{H^+} - C_{OH^-}} \tag{12.17a}$$

The corresponding expression for a mixture of weak base B and its salt is

$$C_{OH^-} = K_b \frac{F_B - C_{OH^-} + C_{H^+}}{F_{BHA} + C_{OH^-} - C_{H^+}} \tag{12.17b}$$

The Eq. (12.17a) will be reduced in the following different case:

i. For acidic solution C_{OH^-} may be ignored. Again if the solution is not very acidic (pH ≮ 4) C_{H^+} may also be ignored, and then

$$C_{H^+} = K_a \frac{F_{HA}}{F_{MA}}$$

or
$$pH = pK_a + \log \frac{F_{MA}}{F_{HA}} \tag{12.18a}$$

Similarly, for a mixture of weak base and its salt, if the mixture is not very alkaline (pH ≯ 10)

$$pOH = pK_b + \log \frac{F_{BHA}}{F_B} \tag{12.18b}$$

The Eqs (12.18a) and (12.18b) are known as Henderson equation for weak acid and weak base respectively which are good approximation in the pH range 4 to 10.

ii. For pH considerably lower than 4, C_{H^+} should be calculated using the relation

$$C_{H^+} = K_a \frac{F_{HA} - C_{H^+}}{F_{MA} + C_{H^+}} \tag{12.19a}$$

For pH considerably greater than 10, one should use the relation

$$C_{OH^-} = K_b \frac{F_B - C_{OH^-}}{F_{BHA} + C_{OH^-}} \tag{12.19b}$$

iii. For solution containing only acid (HA), $F_{MA} = 0$ and C_{OH^-} is negligible, then

$$K_a = \frac{C_{H^+}^2}{F_{HA} - C_{H^+}}$$

$$= \frac{C_{H^+}^2}{F_{HA}} \quad \text{if } F_{HA} \gg C_{H^+}$$

or
$$pH = \tfrac{1}{2}(pK_a - \log F_{HA}) \tag{12.20a}$$

The corresponding relation for a base is

$$\left. \begin{array}{l} pOH = \tfrac{1}{2}(pK_b - \log F_B) \\ pH = pK_w - \tfrac{1}{2}(pK_b - \log F_A) \end{array} \right\} \tag{12.20b}$$

iv. For a solution containing only salt (MA), $F_{HA} = 0$, and if the salt is of strong base and weak acid (when $C_{OH^-} > C_{H^+}$)

$$C_{H^+} = K_a \frac{C_{OH^-}}{F_{MA} - C_{OH^-}}$$

$$\simeq K_a \frac{C_{OH^-}}{F_{MA}}$$

$$\simeq K_a \cdot \frac{K_w / C_{H^+}}{F_{MA}}$$

or $\qquad pH = \frac{1}{2}(pK_w + pK_a + \log F_{MA})$ (12.21a)

v. For a solution containing only salt (BHA) of weak base and strong acid (when $C_{OH^-} < C_{H^+}$), Eq. (12.17b) reduces to

$$C_{OH^-} = K_b \frac{C_{H^+}}{F_{BHA} - C_{H^+}}$$

$$\simeq K_b \frac{C_{H^+}}{F_{BHA}}$$

whence $\qquad C_{H^+}^2 = \frac{K_w F_{BHA}}{K_b}$

or $\qquad pH = \frac{1}{2}(pK_w - pK_b - \log F_{BHA})$ (12.21b)

For the solution of a salt of weak base (B) and weak acid (HA) the expression of pH cannot be derived from Eq. (12.17a) or (12.17b). However, this can be obtained as follows

$$C_{H^+} = \frac{K_a C_{HA}}{C_{A^-}}$$

$$= K_a \left(\frac{K_w}{K_a K_b} \right)^{\frac{1}{2}}$$

Since $\qquad \dfrac{C_{HA}}{C_A'} = \dfrac{C_B}{C_{BH^+}} = \left(\dfrac{K_w}{K_a K_B} \right)^{\frac{1}{2}}$ (vide problem 12.35)

Then $\qquad \left. \begin{aligned} pH &= \frac{1}{2}(pK_w + pK_a - pK_b) \\ &= \frac{1}{2}(pK_a + pK_a') \end{aligned} \right\}$ (12.21c)

where pK_a' corresponds to the conjugate acid BH^+ of the base B.

The last equation is applicable to half neutralisation point of an equimolecular mixture of two weak monobasic acids having dissociation constant K_a and K_a', or of a dibasic acid with first dissociation constant K_a and sound dissociation constant K_a'.

For a solution containing two weak acids HA_1 and HA_2 and their salts produced by b equivalent of a base added per litre of the solution, C_{H^+} is given by

$$C_{H^+} = \frac{F_1 K_1}{C_{H^+} + K_1} + \frac{F_2 K_2}{C_{H^+} + K_2} + C_{OH^-} - b$$ (12.22)

This follows from the consideration of the material balance equation $(F = C_{HA} + C_{A^-})$ and dissociation equilibrium $(HA \rightleftharpoons H^+ + A^-)$ for each acid and the charge balance equation for the solution.

12.6 ACID–BASE INDICATOR

An idea of pH of a solution can be obtained from the colour imparted to it by small amount of some foreign substance called acid–base indicator. Colour of such indicators changes with pH due to their structural change. They may be supposed to exist predominantly in form of an acid (In_A) at low pH but in form of a base (In_B) at high pH. In solution, the equilibrium between the two forms may be represented as

$$In_A \rightleftharpoons H^+ + In_B$$

This entails an equilibrium constant, called indicator constant (K_{In}).

The chief characteristic of acid–base indicators is that the change from a predominantly *acid* colour to a predominantly *basic* colour takes place gradually though it can be perceived by our eye only within a limited range of pH (usually not more than two pH units around pK_{In}), termed the colour-change interval of the indicator. In acid–base titrations, the pH at which the colour change of an indicator is most prominent is called pT, or titration exponent, of an indicator.

In order that an indicator may be used to detect the end point of a titration, the change of pH near the equivalence point should not be very gradual. As a general rule, if there is a change of at least two units of pH around the equivalence point due to addition of a small amount, say $x\%$ of the amount of the titrant needed at equivalent point then an indicator can be used in that titration with an error within $x\%$. If pH change is satisfactory, only then an indicator can be selected that exhibits a distinct colour change near the equivalent point. Greater is the concentration of the acid or alkali titrated and greater is the $\%$ of error allowed, larger will be the number of effective indicators for a particular titration (Table 12.1).

Table 12.1: pT and colour-change intervals of some indicators

Indicator	Colour-change	Range	pT
Methyl orange	Red in acid $\leftrightarrow$ yellow in alkali	3.1–4.4	4
Methyl red	Red in acid $\leftrightarrow$ yellow in alkali	4.2–6.3	5.5
Phenolphthalein	Colourless in acid $\leftrightarrow$ red in alkali	8.3–10	9
Litmus	Red in acid $\leftrightarrow$ blue in alkali	5–8	

12.7 BUFFER SOLUTION

The solutions having appreciable capacity to resist the change in its pH due to addition of acid or base are called buffer solutions. Usually they consist of a mixture of a weak acid or base and its salt.

Buffer capacity (β) of a solution is defined by

$$\beta = \frac{db}{d(pH)}$$

where $d(pH)$ is the change in pH of the solution due to addition of db equivalent of a strong base per litre of the solution.

Buffer capacity of an acid (or base) changes during its neutralisation. Let b equivalent of base is added per liter of a strong acid HA of formality F. Then

$$pH = -\log(F - b) = -\frac{1}{2.303}\ln(F - b)$$

So $$\beta = \frac{db}{d(pH)} = \frac{1}{d(pH)/db} = 2.303\,(F - b) \qquad (12.23a)$$

Here β decrease linearly with b and becomes zero when $b = F$, i.e. when completely converted into salt.

But if HA is a weak acid

$$pH = pK_a + \log\frac{b}{F - b}, \text{ by Henderson equation}$$

$$= pK_a + \frac{1}{2.303}\ln\frac{b}{F - b}$$

So $$\beta = \frac{1}{d(pH)/db} = 2.303b(F - b)/F \qquad (12.33b)$$

Here, during neutralisation, β first rises, reaches a maximum and then falls to zero when $b = F$. The maximum value of β corresponds to

$$\frac{d\beta}{db} = \frac{2.303}{F}(F - 2b) = 0$$

or $$b = F/2$$

i.e. at half neutralisation stage when pH = pKa.

Then the best way to prepare an acid buffer is to half neutralise a weak acid having pK_a equal to the desired pH. If, however, such an acid is not available, then an acid having pK_a close to the desired pH is selected and then neutralised to the required extract as per Henderson equation. But in this case β will be less than its maximum value $(2.303\,F/4)$ possible with an acid of formality F. It can be shown that if $|pH - px_a| > 1$, then β will be less than one-third of its maximum value and the solution of desired pH will have no value as buffer solution.

AUXILIARY PROBLEMS

12.1 ΔG° for the dissociation of any strong electrolyte is conventionally considered to be zero. Is it justified?

12.2 For a strong electrolyte MX, dissociating as $MX \rightleftharpoons M^+ + X^-$, which of the following expressions will be meaningful?

    i. $a = m\gamma$, where m is the molality and γ the activity coefficient of the undissociated species.

    ii. $a = a_\pm^2$

    iii. $a = (m_\pm)^2\gamma_\pm^2$

    iv. $\gamma = \gamma_\pm^2$

    $\gamma_\pm$ has practical significance but γ_+ or γ_- does not. Comment.

12.3 Ionic strength of a solution is defined as $I = \frac{1}{2}\sum m_i z_i^2$. why?

12.4 State and explain the nature of variation of the thickness of ion atmosphere with ionic strength and temperature.

12.5 Discuss qualitatively the effect of temperature and dielectric constant of a medium on the activity coefficient of an ion.

12.6 For which of the following substances is the term solubility product meaningful? $AgCl$, $NaCl$, $NaOH$, $NaHSO_4$, CH_3Cl, $C_6H_5CO_2H$, H_2S and NH_3.

12.7 By concentrating an aqueous solution of $NaCl$ and KBr, which substance(s) can be crystallised out—$NaCl$, KCl or $NaBr$? Which one will be crystallised first.

12.8 When a solution containing two isomorphous salts is evaporated, a mixed crystal is formed though the solution is saturated with respect to only one of the salts. Explain.

12.9 An aqueous solution of $MgCl_2$ gives a precipitate of $MgCO_3$ with $NaHCO_3$ but a precipitate of basic magnesium carbonate with Na_2CO_3, while $FeCl_3$ gives a precipitate only of $Fe(OH)_3$ with both. Explain.

12.10 $PbSO_4$ is soluble in NH_4AC solution but not in NH_4Cl or HA_C. Why?

12.11 Solubility product of $CaCO_3$ is not equal to the square of its solubility in pure water. Explain.

12.12 Of the two salts, one having higher solubility product will have higher solubility. Comment.

12.13 Discuss the effect of $FeSO_4$, HCN, HCl and $NaCl$ on the solubility and solubility product of $Fe(CN)_2$.

12.14 Will the ionic product of water change due to addition of C_2H_5OH and $NaCl$?

12.15 What is the advantage of Sorensen's pH scale over the molarity scale in expressing the concentration of hydrogen ion?

12.16 Can pH of a solution be negative or exceed 14? Is there any limitation regarding the lowest and highest value of pH?

12.17 What happens to pH of water when it is completely vaporised or freezed?

12.18 Is the idea of pH limited only to aqueous medium?

12.19 True or false.

   i. pH of a solution in a sealed bottle remains same through out the year.

   ii. On heating, pH of water decreases, and hence water becomes more acidic at higher temperature.

   iii. If pH of water does not change on addition of a salt, then the latter is not hydrolyzed.

   iv. An aqueous solution of a salt can never act as a buffer.

   v. For the same formal concentration, a strong acid has much higher buffer capacity than a weak acid. Therefore the former can be used as a buffer.

12.20 Which solution will have higher concentration of H^+ ion—$(N/10)$ HCl or $(N/10)$ CH_3CO_2H? Will the amount of alkali required to neutralise liquid volume of the two solutions be same or different?

Can concentration of H^+ ion in an acid solution be determined by direct titration with an alkali?

12.21 Explain why

   i. most of the common indicators can be used in the strong acid–strong base titration.

   ii. practically no single indicator is suitable for weak acid–weak base titration.

iii. in an acid–base titration pH varies at a higher rate near the equivalence point.

iv. a mixture of acid and its salt can act as a buffer only if the acid is weak.

v. $CH_3CO_2NH_4$ can act as a buffer but $NaHCO_3$ does not (though pH does not depend on concentration of the solution for both the salts).

12.22 Find ionic strength of the aqueous solution which is 0.01 m in each of the solutes— NaCl, KNO_3, $K_3Fe(CN)_6$ and urea. Also find the mean ionic activity of NaCl and $K_3Fe(CN)_6$ in the solution. Will the activity coefficient of each of the ions in this solution be same.

12.23 Find the value of A in the Debye–Huckel equation for aqueous solution at 25°C. Dielectric constant of water is 78.54 and its density 997 kg m^{-3} at 25°C.

12.24 The solubility of $PbCl_2$ in water is 5.35×10^{-2} mol L^{-1} at 18°C and 6.00×10^{-2} mol L^{-1} at 25°C. Find ΔH of solution for $PbCl_2$.

12.25 The ionic product of water is 0.68×10^{-14} at 20°C and 1.47×10^{-14} at 30°C. Find (i) pH of water at 20°C (ii) ΔH of neutralisation of HCl by NaOH. How will the values of ΔH differ if NH_3 is used instead of NaOH?

12.26 Find the pH of (i) 10^{-5} N HCl solution (ii) 10^{-7} N HCl solution and (iii) 10^{-9} N HCl solution.

12.27 What is the pH of a 1.0 F acetic acid solution? To what extent must this solution be diluted (i) to double the pH (ii) to double OH$^-$ ion concentration. $K_a(CH_3COOH) = 1.8 \times 10^{-5}$.

12.28 An aqueous solution of acetic acid which is 0.05 F in sodium acetate has pH = 5.0 at 25°C. Find the exact concentration of H$^+$ in the solution.

12.29 Calculate the concentrations of all the species present in a 0.20 F H_2SO_4 solution

Given $HSO_4^- \rightarrow H^+ + HSO_4^-$; strong

$HSO_4 \rightleftharpoons H^+ + SO_4^{2-}$; $K_2 = 1.3 \times 10^{-2}$

12.30 Find the concentration of H$^+$ in an aqueous solution of 0.1 F succinic acid ($K_1 = 6.5 \times 10^{-5}$ and $K_2 = 3.3 \times 10^{-6}$).

12.31 A solution which is 0.08 F in HCl, 0.09 F in HA and 0.1 F in CH_3CO_2H has a pH = 1. If K_a of CH_3CO_2H is 10^{-5}, find that of HA.

12.32 A buffer solution contains 0.002 mol of lactic acid ($pK_a = 3.87$) and 0.01 mol of sodium lactate per liter. Calculate (a) the pH of this buffer (b) the change in pH due to addition of 2 ml of 0.5 N HCl per liter of this buffer (c) the expected pH change if this quantity of acid is added to 1 liter of a solution of a strong acid of the same initial pH.

12.33 Aqueous solution of a weak acid HA is titrated with 0.1 F NaOH and the equivalence point is reached when 30 mL of the base is added. Then 20 ml of 0.1 F HCl is added to the solution and the pH is then found to be 5. Find pK_a of HA.

12.34 A 40 ml solution of a weak base is titrated with 0.1 N HCl solution. The pH of the solution is found to be 10 and 9 after the addition of 5 ml and 20 ml of the acid respectively. Find pK_b of the base.

12.35 What is the pH of an aqueous solution of $CH_3CO_2NH_4$ at 25°C. What fraction of acetic acid will remain unneutralised when it is mixed with an equal volume of aqueous NH_3 of same strength [$K_a(CH_3CO_2H) = K_b(NH_3) = 1.8 \times 10^{-5}$].

12.36 Find the pH of the resulting solution obtained on mixing a 0.1 F NaOH with 0.25 F CH_3CO_2H in the volume ratio (i) 1:5 (ii) 2.5:1 (iii) 5:1. [$pK_a(CH_3CO_2H) = 4.7$]

12.37 What volume of 0.1 F HCl and 0.1 F aqueous NH_3 are to be mixed to prepare 100 ml of a solution of pH equal to (i) 3 (ii) 7 (iii) 9 $[pK_b(NH_3) = 4.7]$.

12.38 You are to prepare a buffer solution of pH = 5. Two monobasic acids having pK_a equal to 4 and 5.5 are available. Which one is preferable for this purpose and why? What will be the buffer capacity of the prepared buffer if the acid used is of concentration 0.1 F?

12.39 An acid HA of formal concentration IF is titrated with a base BOH of same concentration. Find expressions of pH (i) before $x\%$ of the equivalence point (ii) at the equivalence point and (iii) after $y\%$ of the equivalence point, when (a) HA and BOH are both strong (b) HA is strong but BOH is weak (c) HA and BOH are both weak.

12.40 pT's of methyl red and phenolphthalein are 5.5 and 9 respectively. Which one is suitable for titration of 0.1 F CH_3CO_2H with 0.1 F NaOH. Find the percent of error involved in using the other indicator. Will the error be same if the titration is done in the reverse way? $[pK_a(CH_3CO_2H) = 4.75]$

12.41 Equal volume of 0.04 F HA_C and 0.02 F NaCN are mixed up at 25°C. Find the concentrations of all the species present in the resulting solution. $[K_a(HA_C) = 1.8 \times 10^{-5}$ and $K_a(HCN) = 8 \times 10^{-10}]$

12.42 How much acid is to be added to a 0.01 F Na_2CO_3 solution to make pH of the resulting solution 6, if the acid is (a) HCl (b) CH_3CO_2H. K of CH_3CO_2H is 1.8×10^{-5}, and K_1 and K_2 of H_2CO_3 are 4.2×10^{-7} and 4.8×10^{-11} respectively. Mention the approximation(s) involved.

12.43 Some solid NaOH is added to 1 liter of 0.02 F aqueous H_2S. Find (a) the initial pH (b) pH when the added NaOH is (i) 0.01 mol (ii) 0.02 mol and (iii) 0.04 mol. K_1 and K_2 of H_2S are 10^{-7} and 10^{-15} respectively.

12.44 A solution which is 0.1 F in HCl and 0.1 F in CH_3CO_2H is titrated with 0.1 F NaOH solution. Calculate the pH (a) at the start (b) at the first equivalence point (c) at the second equivalence point. $[pK_a(CH_3CO_2H) = 4.75]$

12.45 Repeat the calculations for different stages of titration (a), (b) and (c) in the previous problem with HCO_2H instead of HCl $[pK_a (HCO_2H) = 3.75]$

12.46 0.1 F NaOH is added to an equal volume of a solution which is 0.1 F in NH_4Cl and 0.1 F in CH_3CO_2H. Find the pH of the resulting solution. $[pK_a (CH_3CO_2H) = pK_b (NH_3) = 4.7]$

12.47 The thermodynamic solubility product of BaOx (Ox = oxalate) is 1.6×10^{-7} and K_1 and K_2 for H_2Ox are 6.0×10^{-2} and 6.4×10^{-5} respectively. Calculate (a) solubility of BaOx in water (b) pH of the saturated solution of BaOx. Mention the assumption/ approximation taken.

12.48 K_{th} of PbF_2 is 4×10^{-9}. Calculate the solubility of PbF_2 in
   i. Water ii. 0.1 F NaCl iii. 0.1 F NaF iv. 0.1 F HCl (neglecting activity coefficient effect and taking $K_a(HF) = 2 \times 10^{-2}$)

12.49   a. State whether $PbCl_2$ ($K_{sp} = 10^{-3}$) will be precipitated in the following cases:
   i. 1 ml of 1 F $CaCl_2$ is added to 100 ml of 1 F $Pb(NO_3)_2$
   ii. 50 ml of 1 F $CaCl_2$ is added to 50 ml of 1 F $Pb(NO_3)_2$
   Also find the amount of precipitate (if any).
   b. How much $PbCl_2$ can be dissolved in 500 ml of 0.1 F $CaCl_2$ solution?

12.50 One litre each of 0.06 F of $Pb(NO_3)_2$, 0.03 F K_2SO_4 and 0.04 F KIO_3 are mixed together. Calculate the amount of the substance(s) precipitated and the concentrations of all the species present in the solution in equilibrium. Take the solubility products of $PbSO_4$ and $Pb(IO_3)_2$ to be 10^{-8} and 10^{-13} respectively.

12.51 A 0.1 F solution of NaCl is titrated with 0.1 F $AgNO_3$ solution. Find the concentration of Cl^- ion in the solution (a) before 1% of the equivalence point (b) at the equivalence point (c) after 1% of the equivalence point. Solubility product of AgCl is 1.5×10^{-10} (neglect activity coefficient effect).

12.52 a. To a mixture of $AgNO_3$ and $Pb(NO_3)_2$, KIO_3 is gradually added. If the mixture is initially (i) 0.1 F (ii) 0.001 F in each of the nitrates, which iodate $AgIO_3$ or $Pb(IO_3)_2$, will be precipitated first? In case of (i) what percent of metal ion remains to be precipitated when the other ion just starts precipitating?

b. For what concentration of a solution containing equimolar amount of $AgNO_3$ and $Pb(NO_3)_2$ will $AgIO_3$ and $Pb(IO_3)_2$ start precipitating simultaneously? If 0.001 mol of KiO_3 is added to one litre of this solution, what will be the composition of the precipitate and concentration of IO_3^- ion in the solution? [Solubility products of $AgIO_3$ and $Pb(IO_3)_2$ are 5.3×10^{-8} and 3.2×10^{-13} respectively].

12.53 To a suspension of 0.2 mol of (i) $PbCl_2$ (ii) PbF_2 in 1 liter of water solid $(HH_4)_2SO_4$ is gradually added until $PbCl_2/PbF_2$ is just completely converted into $PbSO_4$. Calculate the number of moles of $(NH_4)SO_4$ added in each case. Explain the difference in result.

Take $K_{Sp}(PbCl_2) = 10^{-3}$ and $K_{Sp}(PbF_2) = K_{Sp}(PbSO_4) = 10^{-8}$.

12.54 Calculate the amount of HCl justsufficient to convert (i) 0.03 mol of Na_2OX dissolved in 1 liter of water into chloride (ii) 0.03 mol of CaOX suspended in 1 liter of water into chloride. Solubility product of CaOX is 2.6×10^{-9} and K_1 and K_2 of oxalic acid are 6.0×10^{-2} and 6.4×10^{-5} respectively.

12.55 Calculate how much HCl is to be added per liter of a 0.001 F lead salt to just prevent precipitation when saturated with H_2S. Solubility product of PbS is 10^{-28}, K_1 and K_2 for H_2S are 10^{-7} and 10^{-15} and solubility of H_2S is 0.1 F in acid solution.

12.56 Find the lowest pH at which 0.01 mol of $Zn(OH)_2$ will go into solution (1 liter) as $Zn(OH)_4^{2-}$ and the highest pH at which it will dissolve as Zn^{2+}? Also find the lowest solubility of $Zn(OH)_2$.

Given $Zn(OH)_2(s) \rightleftharpoons Zn^{2+}(aq) + 2OH^-(aq)$, $K_1 = 1.2 \times 10^{-17}$

$Zn(OH)_2(s) + 2OH^-(aq) \rightleftharpoons Zn(OH)_4^{2-}(aq)$, $K_2 = 0.13$

12.57 $HgCl_2$ dissolves in water and remains largely undissociated. However, small amount of $HgCl^+$, Hg^{2+} and Cl^- are formed as under:

1. $HgCl_2 \rightleftharpoons HgCl^+ + Cl^-$ $K_1 = 3.3 \times 10^{-7}$

2. $HgCl^+ \rightleftharpoons Hg^{2+} + Cl^-$ $K_2 = 1.8 \times 10^{-7}$

Find the concentration of Hg^{2+} in 0.1 F $HgCl_2$ solution.

12.58 The solubility product of $CaCO_3$ is not equal to the square of the solubility but is somewhat less due to the hydrolysis of the carbonate ion. How would you find out the true solubility product of $CaCO_3$?

ANSWERS

12.1 ΔG for a chemical process consists of two parts: $\Delta G°$ due to formation of new species and ΔG_{mix} due to their mixing. The stated convention is justified for strong electrolytes because they exist not as neutral molecules but as combination of ions, and hence their dissociation does not give rise to formation of any new species.

Note: For weak electrolytes $\Delta G° \neq 0$.

12.2 All except (iv), because $m\gamma = (m_\pm)^2\,(\gamma_\pm)^2$ and $m \neq (m_\pm)^2$. Vide Section 12.1.

Note: For a strong electrolyte a corresponds to a hypothetical molecular species.

12.3 Because in an ionic solution the intensity of the electric field, on which the activity coefficient of an ion depends, increases with ionic concentration and is higher with electrolytes containing ions of higher charge.

12.4 The thickness of ionic atmosphere, called Debye length r_D, varies inversely with square root of ionic strength (I). Because $I^{\frac{1}{2}}$ may be regarded as proportional* to the concentration of charge in an ionic solution and hence as it increases the counterions crowd more closely around the central ion with consequent reduction of r_D.

On the other hand, r_D varies proportionally with $T^{\frac{1}{2}}$ because with rise in T the counter-ions, whose speed may be regarded as proportional** to $T^{\frac{1}{2}}$, can move away from the ionic atmosphere more freely.

Note: * This is analogous to the relation between molecular speeds in ideal gases where rms speed is proportional to the mean speed.

** Analogous to that assumed with ideal gases.

12.5 Non-ideality due to interionic forces becomes less intense with increase in dielectric constant (due to lowering of such forces) and also with increase in temperature (due to rise of kinetic energy of the ions), and consequently activity coefficient rises.

12.6 Thermodynamic solubility product has meaning with all the given substances except for CH_3Cl (which is a non-electrolyte) and NH_3 (which behaves as a non-electrolyte in aqueous medium).

Note:

1. The term *solubility* has meaning with all substances but *solubility product* has meaning only with electrolytes.

2. For $NaHSO_4$ solubility products, viz. $a_{Na^+} \cdot a_{HSO_4^-}$ and $a_{Na^+} \cdot a_{H^+} \cdot a_{SO_4^{2-}}$, are possible just as more than one dissociation constant of a polybasic acid.

3. The concept of solubility product is not limited only to salts.

4. For a substance to have solubility product, it need not be a solid at ordinary temperature, e.g. H_2S, for which the solubility product $a_{H^+}^2 \cdot a_{S^{2-}}$ corresponds to the equilibrium $H_2S(g) \rightleftharpoons 2H^+(aq) + S^{-2}(aq)$.

12.7 All three substances.

□ One for which the ionic product first reaches the solubility product will be crystallised first. This will not necessarily happen with the one having least solubility product, because the composition of the solution has also a role to decide it.

Note: From such solution NaBr may be crystallised though it is not used in preparing the solution. Here the separation of NaBr, unlike NaCl or KBr, would have to be regarded as a chemical producess in view of the starting materials used to prepare the solution.

12.8 Due to dilution, the solubility product of either electrolyte in the mixed crystal is less than that referred to its pure crystal. Hence, for the formation of mixed crystal, the activities of the constituent ions of either electrolyte in the relevant solution are not required to be as high as in case of formation of pure crystal of the electrolyte from the solution containing only this electrolyse. This explains the given statement.

12.9 Aqueous solution of both $NaHCO_3$ and Na_2CO_3 contain OH^- ions (due to hydrolysis) and CO_3^{2-} ions (due to dissociation). In both the solutions, the concentration of OH^- is enough for precipitation of $Fe(OH)_3$ having very low solubility product ($\approx 10^{-38}$). In $NaHCO_3$ solution, the concentration of OH^- is not enough to precipitate $Mg(OH)_2$ ($K_{th} \approx 10^{-12}$) though the concentration of CO_3^{2-} is enough to precipitate $MgCO_3$ ($K_{th} \approx 10^{-5}$) as $C_{CO_3^{2-}} \gg C_{OH^-}$. However in Na_2CO_3 solution, the concentrations of both OH^- and CO_3^{2-} are enough to precipitate both $Mg(OH)_2$ and $MgCO_3$ as mixed crystal, called basic carbonate.

Note: Alternative explanation of such precipitation reactions as a consequence of neutralisation process is not quite satisfactory.

12.10 Because in presence of NH_4Ac, the solubility equilibrium $PbSO_4(s) \rightleftharpoons Pb^{2+}(aq) + SO_4^{2-}(aq)$ shifts in the forward direction due to removal of Pb^{2+} ions in form of undissociated $Pb(Ac)_2$. This does not happen with NH_4Cl (as $PbCl_2$ is more ionizable than $PbSO_4$) or HAc (as HAc cannot liberate much stronger acid H_2SO_4).

Note: NH_4Ac affects solubility equilibrium but not solubility product constant.

12.11 This is due to hydrolysis of CO_3^{2-} ions that causes $C_{CO_3^{2-}} \neq$ solubility of $CaCO_3$ measured by $C_{Ca^{2+}}$.

Note: Due to hydrolysis, solubility of $CaCO_3$ slightly increases resulting from partial removal of CO_3^{2-} ions as HCO_3^- or H_2CO_3, but the solubility product does not change (decrease in $C_{CO_3^{2-}}$ being associated with increase in $C_{Ca^{2+}}$).

12.12 This is not necessarily true, because the solubility of a salt is determined not only by its solubility product but also by its formula. For example, AgCl has higher solubility product (1.5×10^{-10}) than Ag_2CrO_4 ($K_{th} = 2.4 \times 10^{-12}$), but AgCl has less solubility (1.2×10^{-5} mol L^{-1}) than Ag_2CrO_4 (8.4×10^{-5} mol L^{-1}).

12.13 In the saturated solution of $Fe(CN)_2$ following equilibria are involved

$$Fe(CN)_2(s) \rightleftharpoons Fe(CN)_2\,(aq) \rightleftharpoons Fe^{2+}(aq) + 2CN^-(aq)$$

Solubility of $Fe(CN)_2$

– decreases in presence of $FeSO_4$ due to the effect of the common ion Fe^{2+} that shifts the above equilibria to the left.

– increases in presence of HCN due to formation of the complex $H_3Fe(CN)_6$ that shifts the above equilibria to the right (though concentration of Fe^{2+} decreases*).

– increases in presence of HCl due to removal of CN^- ion as HCN that shifts the equilibrium to the right.

– increases in presence of NaCl due to increase in ionic strength of the solution that lowers $\gamma_\pm$ with consequent rise of K_{SP} [by Eq. (12.8)].

In all cases the activity solubility product is not affected though the concentration solubility product increases due to decrease in $\gamma_{\pm}$ of $Fe(CN)_2$ resulting from the increase in ionic strength.

Note: * The solubility of an electrolyte can be measured by the concentration of either of its constituent ions only if the dissolved electrolyte exists almost entirely in the form of such ions.

12.14 Yes. K_w $(= a_{H^+} \cdot a_{OH^-})$ changes due to change in mole fraction (or molality) and activity coefficient of the involved ions on addition of both C_2H_5OH and NaCl. However, the effect will not be appreciable if the concentration of the added substance is very low. In case of C_2H_5OH, the change in activity coefficient arises from the change in dielectric constant of the solution, but with NaCl this is mainly due to change in ionic strength of the solution.

12.15 The concentration of (hydrated) H^+ ions in aqueous solutions range from $\simeq 1$ mol L^{-1} (e.g. in 1 N HCl) to $\simeq 10^{-14}$ mol L^{-1} (e.g. in 1 N NaOH). Sorensen's pH scale has the advantage in that it reduces this wide span of H^+ ion concentration into a much shorter range 0 to 14 approximately.

12.16 Yes. For example, pH of 2N HCl (aq) is approximately –0.3. Negative pH is not unlikely because it merely indicates $a_{H^+} > 1$. Similar is the situation with negative pOH which corresponds to pH > 14, as pH + pOH = 14 at 25°C

☐ There is some practical (though not theoretical) limitation regarding the lowest and highest values of pH. Thus a_{H^+} corresponding to pH = –1 is very high and is rarely encountered, and a_{H^+} corresponding to pH = –2 is simply unrealisable. Similar limitation prevails regarding pH = 15 which corresponds to pOH = –1.

12.17 pH will have no meaning with vaporised water or frozen water, because they do not contain any (mobile) H^+ ion.

12.18 No. However, in non-aqueous medium, the value of pH provided by its operational definition deviates much from its thermodynamic significance.

12.19   i. False. Because pH of an aqueous solution is ultimately related to the equilibrium $H_2O + H_2O \rightleftharpoons H_3O^+ + OH^-$ which is affected with natural temperature variation during the year.

 ii. False, because in pure water C_{H^+} and C_{OH^-} are always equal through they increase on heating.

 iii. True, except when the relevant acid and base of the salt are equally weak, as in case of NH_4Ac [NH_3 and HAc being equally weak, i.e. $K_b(NH_3) = K_a(CH_3COOH)$].

 iv. False. A salt can act as a buffer if it produces ions of appreciable acidic character together with those of basic character. For example, NH_4Ac (where NH_4^+ acts as an acid and Ac^- as a base) and potassium hydrogen tartrate, KHTa (where HTa^- acts as an acid and Ta^{2-} as a base).

 v. False, because a solution containing only an acid (strong or weak) cannot function as a buffer and therefore, the value of buffer capacity does not matter here.

Note: For same formal concentration, a strong acid has much higher buffer capacity (due to high C_{H^+}) than a weak acid, but for same pH the reverse happens.

12.20 N/10 HCl.

☐ Same, because H+ ions that react directly with an alkali are produced in equal amount from equal volume of N/10 HCl and N/10 CH$_3$CO$_2$H (when completely dissociated). In case of the latter H+ ions removed by the alkali are replenished by further dissociation of CH$_3$COOH molecules till they are completely exhausted.

☐☐ No, unless H+ ions contained in the system are produced by virtually complete dissociation of the acid.

12.21 i. Because in a strong acid–strong base titration, the range of pH near the equivalence point is remarkably broad (roughly 4 to 10) to which pT's of most of the common indicators belong.

ii. Because unlike (i), the range of pH is so narrow that no common (single) indicator can effectively serve the purpose.

iii. This is a consequence of the logarithmic relation $pH = -\log C_{H^+}$ for low concentration of H+ ions near equivalence point.

iv. Because the anion of only a weak acid can effectively act as a base.

v. Because in the interaction of NaHCO$_3$ with an acid most of the liberated H$_2$CO$_3$ escapes as CO$_2$ gas.

12.22 Ionic strength

$$= \tfrac{1}{2}\Big[m_{K^+} \cdot Z^2_{K^+} + m_{Na^+} \cdot Z^2_{Na^+} + m_{Cl^-} \cdot Z^2_{Cl^-} + m_{NO_3^-} \cdot Z^2_{NO_3^-} + m_{Fe(CN)_6^{3-}} \cdot Z^2_{Fe(CN)_6^{3-}} \Big]$$

$$= [(\underset{\underset{KNO_3}{\uparrow}}{0.01} + \underset{\underset{K_3Fe(CN)_6}{\uparrow}}{3 \times 0.01}) \times 1^2 + 0.01 \times 1^2 + 0.01 \times 1^2 + 0.01 \times 1^2 + 0.01 \times 3^2] \, mol\, kg^{-1}]$$

$$= 0.08 \; mol\, kg^{-1}$$

Urea, being non-electrolyte, does not contribute to the ionic strength.

☐ For NaCl $\qquad \log \gamma_{\pm} = -0.509 Z_{Na^+} \cdot Z_{Cl^-} \sqrt{\dfrac{I}{m^\circ}}$ at 25°C

$$= -0.509(1)(1)\sqrt{0.08}$$

$$= -0.144$$

whence $\qquad\qquad \gamma_{\pm} = 0.718$

Mean ionic activity

$$a_{\pm} = m_{\pm} \cdot \gamma_{\pm} = (m_{Na^+} \cdot m_{Cl^-})^{\frac{1}{2}} \gamma_{\pm}$$

$$= (0.01 \times 0.01)^{\frac{1}{2}} \times 0.718 \quad \text{using only numerical values of } m$$

$$= 0.00718$$

For K$_3$Fe(CN)$_6$,

$$\log \gamma_{\pm} = -0.509 Z_{K^+} \cdot Z_{Fe(CN)_6^{3-}} \sqrt{\dfrac{I}{m^\circ}}$$

$$= -0.509(1)(3)\sqrt{0.08} = -0.432$$

whence $\qquad\qquad \gamma_{\pm} = 0.370$

$$a_{\pm} = \left(m^3_{K^+} \cdot m_{Fe(CN)_6^{3-}} \right)^{\frac{1}{1-3}} \gamma_{\pm}$$

$$= [\underset{\underset{KCl}{\uparrow}}{0.01} + \underset{\underset{K_3Fe(CN)_6}{\uparrow}}{3 \times 0.01})^3 \times 0.01]^{\frac{1}{4}} \times 0.370 = 0.0103$$

☐ On the basis of the Debye–Huckel limiting law the activity coefficients of all the monovalent ions (cation or anion) will be same and greater than that of the trivalent ion $Fe(CN)_6^{3-}$. However, due to difference in sizes of the monovalent ions (which is not considered in the limiting law) their activity coefficients will not really be the same.

Note: In calculating $m_\pm$ of $K_3Fe(CN)_6$, we have to consider total molality of K^+ ion in the solution and not just the contribution from $K_3Fe(CN)_6$.

12.23 $\quad A = \dfrac{F^3}{9.212\pi N_A}\left(\dfrac{e}{2\,\epsilon_o^3\,\epsilon_r^3\,R^3T^3}\right)^{\frac{1}{2}}$ by Eq. (12.5b)

$$= \dfrac{(9.6485\times10^4\,C\,mol^{-1})^3}{9.212\times3.143(6.022\times10^{23}\,mol^{-1})}$$

$$\times\left[\dfrac{(997\;kg\;m^{-3})}{2(8.854\times10^{-12}\,C^2N^{-1}m^{-2})^3\,(78.54)^3\,(8.314\;JK^{-1}mol^{-1})^3\,(298.15\;K)^3}\right]^{\frac{1}{2}}$$

$$= 0.509\;kg^{\frac{1}{2}}mol^{-\frac{1}{2}}$$

Note: A is not unitless. In Debye–Huckel Eq. (12.5c), the units of A and $\sqrt{I}$ are cancelled out. Therefore, in calculating activity coefficient, A is often treated as unitless with the Debye–Huckel equation written as

$$\log\gamma_\pm = AZ_+Z_-\sqrt{\dfrac{I}{m^\circ}}$$

12.24 For $PbCl_2$ the solubility equilibrium may be represented as

$PbCl_2(s)\,[\rightleftharpoons PbCl_2(aq)]\rightleftharpoons Pb^{2+}(aq) + 2Cl^-(aq)$

Assuming ΔH of solution to be temperature independent, it is given, according to Van't Hoff equation, by

$$\ln\dfrac{K'_{th}}{K_{th}} = \dfrac{\Delta H}{R}\left(\dfrac{1}{T}-\dfrac{1}{T'}\right)$$

or $\qquad \Delta H = \dfrac{R\ln\dfrac{K'_{th}}{K_{th}}}{\dfrac{1}{T}-\dfrac{1}{T'}}$

$$= \dfrac{R\ln\dfrac{K'_{sp}}{K_{sp}}}{\dfrac{1}{T}-\dfrac{1}{T'}}\qquad \text{taking } K_{sp}\approx K_{th}$$

$$= \dfrac{3R\ln\dfrac{S'}{S}}{\dfrac{1}{T}-\dfrac{1}{T'}}\quad \text{by Eq. (12.9a)}$$

where S is solubility at T and S' and T'

$$= \dfrac{3(8.31\;JK^{-1}mol^{-1})\ln\dfrac{1.4\times10^{-14}}{0.68\times10^{-14}}}{\dfrac{1}{293\;K}-\dfrac{1}{303\;K}} = 159.5\;kJ\;mol^{-1}$$

Note: In case of non-electrolytes, the dissolved molecules remaining as such,

$$\Delta H = \frac{R \ln \dfrac{S'}{S}}{\dfrac{1}{T} - \dfrac{1}{T'}}$$

12.25 **i.** $\qquad$ $pH = pOH = \frac{1}{2} PK_w$

$$= -\tfrac{1}{2} \log (0.68 \times 10^{-14}) = 7.08$$

ii. ΔH of neutralisation of HCl (a strong acid) by NaOH (a strong base) corresponds to the net reaction

$$H^+(aq) + OH^-(aq) = H_2O(l)$$

Then $\qquad\qquad$ $\Delta H = \dfrac{\ln \dfrac{K'}{K}}{\dfrac{1}{T} - \dfrac{1}{T'}}$

$$= \dfrac{\ln \dfrac{K_w}{K'_w}}{\dfrac{1}{T} - \dfrac{1}{T'}}, \quad \text{since } K = \dfrac{1}{K_w} \text{ etc.}$$

☐ Here ΔH will be numerically lower than that calculated above. Because, NH_3 being a weak base, a part of the heat liberated by the above reaction is utilised in the dissociation process

$$NH_3 + H_2O = NH_4^+(aq) + OH{-}(aq)$$

12.26 pH of an aqueous solution is defined approximately as $pH = -\log C_{H^+}$. In an acid solution C_{H^+} includes contributions from the acid and water.

i. Since HCl dissociates completely, its contribution to C_{H^+} of the solution is 10^{-5} mol L^{-1}. Let the contribution from water (that dissociates only slightly) be x. Then

$$(x + 10^{-5}) x = 10^{-14}, \text{ since } C_{H^+} \cdot C_{OH^-} = K_w$$

or $\qquad\qquad$ $x = \left[10^{-14} + \left(\dfrac{10^{-5}}{2} \right)^2 \right]^{\frac{1}{2}} - \dfrac{10^{-5}}{2} \approx 0$

Then $\qquad$ $C_{H^+} = x + 10^{-5} \approx 10^{-5} \text{ mol } L^{-1}$

Therefore $\qquad$ pH = 5.

ii. The same procedure as in (i) gives

$$x = 0.618 \times 10^{-7}$$

Then $\qquad$ $C_{H^+} = x + 10^{-7} = 1.618 \times 10^{-7} \text{ mol } L^{-1}$

Therefore $\qquad$ pH = 6.79

iii. Here $\qquad$ $x \approx 10^{-7} - \dfrac{10^{-9}}{2}$

Then $\qquad$ $C_{H^+} = x + 10^{-9} \approx 10^{-7} \text{ mol } L^{-1}$

Therefore $\qquad$ pH = 7 (and not 9)

Note: The contribution of H_2O to C_{H^+} is insignificant in (i) but dominant in (iii). However, in (ii) both HCl and H_2O have comparable (but not equal) contributions to C_{H^+}. These are all due to the effect of common ion (H^+) that decreases in the order (i) > (ii) > (iii) whereby dissociation of H_2O is suppressed in the same order. The present problem reminds us that it is risky to calculate pH of a solution of an acid (or base) with only the acid (or base) in mind, particularly when the solution is dilute.

12.27 For acetic acid (HAc)

$$K_a = \frac{C_{H^+} \cdot C_{AC^-}}{F_{HAc} - C_{H^+}} = \frac{C_{H^+}^2}{F_{HAc} - C_{H^+}} \tag{I}$$

$$\simeq \frac{C_{H^+}^2}{F_{HAc}} \quad \text{ignoring } C_{H^+} \text{ compared to } F_{HAc} \tag{II}$$

or

$$C_{H^+} = (K_a F_{HAc})^{\frac{1}{2}}$$
$$= [(1.8 \times 10^{-5})(1)]^{\frac{1}{2}} = 4.24 \times 10^{-3} \text{ M}$$

Therefore $\quad$ pH $= -\log(4.24 \times 10^{-3}) = 2.37$

[The calculated value of C_{H^+} justifies the use of the approximate expression II]

i. For the diluted solution

$$\text{pH} = 2 \times 2.37 = 4.74$$

Hence $\quad C_{H^+} = 10^{-4.74}$
$$= 1.8 \times 10^{-5}$$

Then by Eq. (I)

$$1.8 \times 10^{-5} = \frac{(1.8 \times 10^{-5})^2}{F_{HAc} - 1.8 \times 10^{-5}}$$

Whence $\quad F_{HAc} = 3.6 \times 10^{-5} \text{ M}$
Therefore, the dilution factor

$$\frac{1M}{9.6 \times 10^{-5}} = 2.8 \times 10^4$$

ii. On doubling C_{OH^-}, C_{H^+} becomes half (K_w remaining same)

i.e. $\quad C_{H^+} = \frac{1}{2} \times 4.24 \times 10^{-3} \text{ M} = 2.12 \times 10^{-3} \text{ M}$

Then by Eq. (II)

$$F_{HAc} = \frac{C_{H^+}^2}{K_a}$$

$$= \frac{(2.12 \times 10^{-3})^2}{(1.8 \times 10^{-5})} = 0.25 \text{ M}$$

Note: The approximate expression (II) is not good enough for (i) to be used.

12.28 $\qquad I = m_{Na^+} + m_{H^+} \quad \text{by Eq. (12.6)}$
$$\simeq 0.05 \text{ mol kg}^{-1}$$

Now $\qquad \log m_{H^+} = \log a_{H^+} - \log \gamma_{H^+} \quad$ since $a_{H^+} = m_{H^+} \cdot \delta_{H^+}$

$$= -pH + A Z_{H^+}^2 \sqrt{I}$$

$$= -5 + 0.509\sqrt{0.05}$$

Whence $\qquad m_{H^+} = 1.30 \times 10^{-5} \text{ mol kg}^{-1}$

Here recalculation with m_{H^+} thus found will lead to an improvement of the result only insignificantly.

12.29 Let the concentration of SO_4^{2-} ion at equilibrium be x M. Then the concentration of different ionic species at equilibrium will be as under

$$\underset{(0.20-x)\,\text{M}}{HSO_4^-} \rightleftharpoons \underset{(0.20+x)\,\text{M}}{H^+} + \underset{x\text{M}}{SO_4^{2-}}$$

Now $\qquad K_2 = \dfrac{C_{H^+} \cdot C_{SO_4^{2-}}}{C_{HSO_4^-}}$

Then $\qquad 1.3 \times 10^{-2} = \dfrac{(0.20+x)\,(x)}{(0.20-x)}$

Whence $\qquad x = 0.012$

Therefore $\qquad C_{H^+} = (0.20 + 0.012) \text{ or } 0.212 \text{ M}$

$\qquad\qquad C_{HSO_4^-} = (0.20 - 0.012) \text{ or } 0.188 \text{ M}$

$\qquad\qquad C_{SO_4^{2-}} = 0.012 \text{ M}$

$\qquad\qquad C_{H_2SO_4} = 0$

12.30 For aqueous succinic acid (H_2A) following relations are useful

$$C_{HA^-} = K_1 \frac{F_{H_2A}}{C_{H^+}} \quad \text{due to dissociation equilibrium } H_2A \rightleftharpoons H^+ + HA$$

$$C_{A^{2-}} = K_1 K_2 \frac{F_{H_2A}}{C_{H^+}^2} \quad \text{due to dissociation equilibrium } H_2A \rightleftharpoons 2H^+ + A^{2-}$$

$$C_{H^+} = C_{HA^-} + 2C_{A^{2-}} + C_{OH^-} \quad \text{due to electroneutrality of the solution}$$

Combination of these relations gives

$$C_{H^+} = \frac{F_{H_2A}}{C_{H^+}} \cdot K_1 \left(1 + \frac{2K_2}{C_{H^+}} \right) + C_{OH^-}$$

Here the solution is likely to be sufficiently acidic for C_{OH^-} to be ignored. Then assuming $\dfrac{2K_2}{C_{H^+}} \ll 1$

$$C_{H^+} = \sqrt{K_1 F_{H_2A}} \qquad \text{same as Eq. (12.20a)}$$

$$= \sqrt{(6.5 \times 10^{-5})\,(0.1)} = 2.55 \times 10^{-3} \text{ M}$$

The value of C_{H^+} thus found justifies the assumption taken.

Note: For a weak dibasic acid, the second dissociation (corresponding to K_2) can be ignored if $\dfrac{K_2}{C_{H^+}}$ or $K_2 \sqrt{FK_1} \ll 1$. But disregarding of second dissociation will not be justified if the acid is strong, as in the previous problem.

12.31 For HAc $K_a = \dfrac{C_{H^+} C_{Ac^-}}{C_{HAc}}$

or $\dfrac{C_{Ac^-}}{C_{HAc}} = \dfrac{K_a}{C_{H^+}}$

$$= \dfrac{1.8 \times 10^{-5}}{10^{-1}} = 1.8 \times 10^{-4}$$

The low value of this ratio implies that the contribution of HAc to C_{H^+} in the solution may be ignored.

HCl, being a strong acid, will contribute 0.08 M to C_{H^+}. Then contribution by HA, and hence C_{A^-} will be (0.1 – 0.08) or 0.02 M.

Therefore

$$K_a(\text{HA}) = \dfrac{C_{H^+} \cdot C_{A^-}}{C_{HA}} = \dfrac{C_{H^+} C_{A^-}}{F_{HA} - C_{A^-}}$$

$$= \dfrac{(0.1)(0.02)}{0.09 - 0.02} = 2.86 \times 10^{-2}$$

12.32 a. By Henderson Eq. (12.18a)

$$\text{pH} = 3.87 \ \log \dfrac{0.01}{0.002} = 4.57$$

The pH thus obtained is reasonably accurate as it lies within the range of pH (4 to 10) where the Henderson equation is valid.

b. 2 ml of 0.5 N HCl ≡ 0.001 mol of HCl

HCl, being strong acid, will react with sodium lactate producing virtually 0.001 mol of lactic acid. So in the resulting solution

$F_{\text{lactic acid}} = (0.002 + 0.001)$ or 0.003 M, ignoring volume change

$F_{\text{Na-lactate}} = (0.01 - 0.001)$ or 0.009 M

Then by Henderson equation

$$\text{pH} = 3.87 + \log \dfrac{0.009}{0.003} = 4.35$$

Therefore, the required change in pH = 4.35 – 4.57 = 0.22.

c. In the acid solution of pH = 4.57, $C_{H^+} \doteq 10^{-4.57}$ or 2.7×10^{-5} M when 2 ml of 0.5 N HCl or 0.001 mol of HCl is added to 1 L of a solution of strong acid, the resulting solution will have

$$C_{H^+} = 0.001 + 2.7 \times 10^{-5} \approx 0.001 \text{ M}$$

Then   pH = –log 0.001 = 3

Therefore, the change in pH = 3 – 4.57 = –1.57.

Note: Due to addition of small amount of acid, a buffer solution suffers only a small change in pH compared to other solutions of same pH.

12.33 Amount of NaA produced ≡ 30 mL of 0.1 F NaOH ≡ $\dfrac{30}{1000} \times 0.1$ or 0.003 mol.

Amount of HA liberated by HCl ≡ 20 mL of 0.1 F HCl ≡ $\dfrac{20}{1000} \times 0.1$ or 0.002 mol.

Then the solution will finally contain 0.002 mol of HA and (0.003 – 0.002) or 0.001 mol of NaA and by Henderson Eq. (12.18a)

$$5 = pK_a + \log \frac{0.001}{0.002}$$

whence $\qquad pK_a = 5.3$

12.34 Since the aqueous solution of a salt weak base and strong acid is acidic, i.e. pH < 7, the pH 10 and 9 will correspond to the solution containing the free base and its salt.
For pH = 10, i.e. pOH = 4, the amount of salt present

$$\equiv 5 \text{ ml of } 0.1 \text{ N HCl}$$

$$= \left(\frac{5}{1000} \times 0.1 \right) \text{ or } 5 \times 10^{-4} \text{ mol}$$

For pH = 9, i.e. pOH = 5, the amount of salt present

$$\equiv 20 \text{ ml of } 0.1 \text{ N HCl}$$

$$= \left(\frac{20}{1000} \times 0.1 \right) \text{ or } 20 \times 10^{-4} \text{ mol}$$

If the amount of base initially present in the solution is x mol, then by Henderson Eq. (12.18b)

$$4 = pK_b + \log \frac{5 \times 10^{-4}}{x - 5 \times 10^{-4}} \qquad\qquad \text{(I)}$$

$$5 = pK_b + \log \frac{20 \times 10^{-4}}{x - 20 \times 10^{-4}} \qquad\qquad \text{(II)}$$

(II) – (I) gives $\qquad 1 = \log \left(\frac{x - 5 \times 10^{-4}}{x - 20 \times 10^{-4}} \times 4 \right)$

whence $\qquad x = 30 \times 10^{-4} \text{ mol}$

Substituting this in II, we have

$$pK_b = 5 - \log \frac{20 \times 10^{-4}}{30 \times 10^{-4} - 20 \times 10^{-4}} = 4.7$$

12.35 By Equation (12.21c)

$$pH = \tfrac{1}{2}[14 + \log(1.8 \times 10^{-5)} - \log(1.8 \times 10^{-5})] = 7$$

☐ In aqueous solution of NH_4Ac, the hydrolysis equilibrium may be represented as

$$\underset{C(1-\alpha)}{NH_4^+} + \underset{C(1-\alpha)}{Ac^-} \overset{H_2O}{\rightleftharpoons} \underset{C\alpha)}{NH_3} + \underset{C\alpha)}{HAc}$$

where c is the total concentration (formality) of NH_4Ac and α is the degree of hydrolysis.

Then $\qquad \dfrac{C_{NH_3} \cdot C_{HAc}}{C_{NH_4^+} \cdot C_{Ac^-}} = \dfrac{C_{HAc}}{C_{H^+} C_{A^-}} \cdot \dfrac{C_{NH_3}}{C_{NH_4^+} C_{OH^-}} \cdot C_{H^+} \cdot C_{OH^-}$

$$= \frac{K_{av}}{K_a K_b} = \frac{\alpha^2}{(1-\alpha)^2}$$

or $\qquad \dfrac{\alpha}{1-\alpha} = \left(\dfrac{K_w}{K_a K_b} \right)^{\frac{1}{2}} = \left[\dfrac{10^{-14}}{(1.8 \times 10^{-5})(1.8 \times 10^{-5})} \right]^{\frac{1}{2}}$

whence $\qquad \alpha = 0.0055$

The fraction of HAc that remains unneutralised, which is just α, is then 0.0055.

12.36 i. Here CH_3CO_2H (HAc) will not be completely neutralised. In the resulting mixture

$$F_{HAC} = \frac{1}{5+1}(5 \times 0.25 - 1 \times 0.1) \text{ or } 0.1916 \text{ M}$$

$$F_{NaAc} = \frac{1}{5+1}(0.1) \text{ or } 0.0167 \text{ M}$$

Then by Henderson Eq. (12.18a)

$$pH = 4.7 + \log\frac{0.0167}{0.1916} = 3.7$$

The pH thus calculated is somewhat less than 4, and hence it cannot be regarded as accurate. For more precise value of pH, we are to calculate first C_{H^+} from the following equation and then pH

$$C_{H^+} = 10^{-4.7} \times \frac{0.1916 - C_{H^+}}{0.0167 + C_{H^+}} \quad \text{by Eq. (12.19a)}$$

ii. Here the acid and base are in equivalent amount and the mixture corresponds to a solution of NaAc whose pH is

$$pH = \frac{1}{2}\left[14 + 4.7 + \log\frac{0.1 \times 2.5}{2.5+1}\right], \text{ by Eq. (12.21a)}$$

$$= 8.8$$

iii. Here NaOH is in excess and this dissociates completely. Then

$$C_{OH^-} = \frac{1}{5+1}(5 \times 0.1 - 1 \times 0.25) \text{ or } 0.0416 \text{ M}$$

$$pH = 14 - pOH$$
$$= 14 + \log 0.0416$$
$$= 12.6$$

12.37 If NH_3 and HCl are mixed in 1:1 volume ratio, the mixture will correspond to a solution of NH_4Cl with $F_{NH_4Cl} = \frac{0.1}{2}$ or 0.05 M whose pH is given, according to Eq. (12.21b) by

$$pH = \frac{1}{2}(14 - 4.7 - \log 0.05) = 5.3$$

Let x volume of NH_3 and $(100 - x)$ volume of HCl are mixed.

i. Since the required pH is less than 5.3 the mixture must contain an excess of HCl that dissociates completely. Then in the mixture

$$C_{H^+} = F_{HCl} = [(100 - x) - x] \times \frac{0.1}{100} M = 10^{-3} M$$

whence $\qquad x = 49.5$

ii. Since the required pH is greater than 5.3, the mixture must contain an excess of NH_3 where

$$F_{NH_4Cl} \propto (100 - x), \text{ the vol. of } NH_3 \text{ neutralised} \equiv \text{vol. of HCl consumed}$$
$$F_{NH_3} \propto [x - (100 - x)] = 2x - 100$$

Then $\quad pOH = PK_w - pH = 14 - 7$

$$= 4.7 + \log\frac{100 - x}{2x - 100} \quad \text{by Eq. (12.18b)}$$

whence $\qquad x = 50.1$

iii. Here $\quad 14-9 = 4.7 + \log\dfrac{100-x}{2x-100}$

whence $\quad x = 60.0$

Note:

1. In (ii), pH = 7 (as in pure water) though the solution contains free base. Here excess NH_3 just nullifies the acidity due to NH_4Cl.
2. Although pH lies in the neutral region in (ii) and in the basic region in (iii), the same expression for pH holds in both cases due to similar composition of the solution.

12.38 The acid with $pK_a = 5.5$ which is nearer to the desired pH of the buffer is preferable, because this acid can provide a buffer of higher buffer capacity than the other acid. To produce the buffer of pH = 5, the amount of base, say b equivalent, required to be added per liter of the acid is given, according to the Henderson Eq. (12.18a) by

$$5 = 5.5 + \log\dfrac{b}{0.1-b}$$

whence $b = 0.024$. Substituting this in the Eq. (12.23b) of buffer capacity β, we have

$$\beta = 2.303\,(0.024)\,(0.1-0.024)/0.1$$
$$= 0.042$$

12.39 a. i. $\quad C_{H^+} = \dfrac{100}{100+(100-x)} \cdot \underbrace{\dfrac{x}{100}}_{\text{Dilution factor}} M = \dfrac{x}{200-x}\,M$

ii. $\quad \text{pH} = \tfrac{1}{2}PK_w$

iii. $\quad C_{OH^-} = \dfrac{100}{100+(100+y)} \cdot \dfrac{y}{100}\,M = \dfrac{y}{200+y}\,M$

b. i. Same as in a(i).

ii. $\quad \text{pH} = \dfrac{1}{2}\left(pK_w - pK_b - \log\dfrac{F}{2}\right)$ by Eq. (12.21b) $F_{BHA} = \dfrac{1}{2}F_{HA}$ due to dilution)

iii. $\quad \text{pOH} = pK_b + \log\dfrac{100}{y}\quad$ by Eq. (12.18b)

c. i. $\quad \text{pH} = pK_a + \log\dfrac{100-x}{x}\qquad$ by Eq. (12.18a)

ii. $\quad \text{pH} = \tfrac{1}{2}(pK_w + pK_a - pK_b)\quad$ by Eq. (12.21c)

iii. Same as in b(iii)

Note: Knowledge of pH close to the equivalence point before and after this point is useful in selecting suitable indicator in an acid–base titration.

12.40 pH at the equivalence point which corresponds to a solution of NaAc of formality 0.1/2 will be

$$\text{pH} = \tfrac{1}{2}\left(14 + 4.75 + \log\tfrac{0.1}{2}\right)\quad \text{by Eq. (12.21a)}$$
$$= 8.72$$

This is nearer to the PT of phenolphthalein and hence this indicator is the suitable one for the titration concerned.

□ With methyl red as indicator, the end point is indicated quite earlier before the arrival of equivalence point. This causes a considerable (−ve) error, say x%. At this end point pH of the solution is

$$pH = 4.75 + \log \frac{100 - x}{x}$$

$$= 5.5 \text{ (the pT of methyl red)}$$

whence $\qquad x = 15.1$

□□ In reverse titration the error will be numerically same but with opposite sign.

12.41 Since HCN is a weak acid the CN⁻ ion will act as fairly strong base and as such CN⁻ ions coming from NaCN will react almost completely with HAc (acting as fairly strong acid)

$$HAc + CN^- = Ac^- + HCN$$

Then in the resulting solution

Concentration of HAc $= \frac{1}{2}(0.04 - 0.02) = 0.01 \text{ F}$

Concentration of HCN $= \frac{1}{2} \times 0.02 = 0.01 \text{ F}$

Concentration of NaAc $= 0.01 \text{ F}$

Considering the equilibrium

$$HAc \rightleftharpoons H^+ + Ac^-$$

$$C_{H^+} = K_a \frac{C_{HAc}}{C_{Ac^-}} \simeq K_a \frac{F_{HAc}}{F_{NaAc}}$$

$$= (1.8 \times 10^{-5}) \times \frac{0.01}{0.01} = 1.8 \times 10^{-5} \text{ M}$$

Again, considering the equilibrium $HCN \rightleftharpoons H^+ + CN^-$

$$C_{CN^-} = K_a \frac{C_{HCN}}{C_{H^+}} \simeq K_a \frac{F_{HCN}}{C_{H^+}}$$

$$= (8 \times 10^{-10}) \times \frac{0.01}{1.8 \times 10^{-5}} = 4.4 \times 10^{-7} \text{ M}$$

Obviously $\qquad C_{Na^-} = F_{NaAc} = 0.1 \text{ M}$

Lastly $\qquad C_{OH^-} = \dfrac{K_w}{C_{H^+}} = \dfrac{10^{-14}}{1.8 \times 10^{-5}} = 5.5 \times 10^{-10} \text{ M}$

12.42 a. Suppose the amount of HCl added per litre is x mol. Now, from the ionisation equilibrium $H_2CO_3 \overset{K_1}{\rightleftharpoons} H^+ + HCO_3^-$

$$\frac{C_{HCO_3^-}}{C_{H_2CO_3}} = \frac{K_1}{C_{H^+}}$$

$$= \frac{4.2 \times 10^{-7}}{10^{-6}} = 0.42$$

or $\qquad C_{HCO_3^-} = \dfrac{0.42}{1.42}(C_{H_2CO_3} + C_{HCO_3^-})$ $\qquad\qquad$ (I)

Again, from the ionisation equilibrium $HCO_3^- \xrightleftharpoons{K_2} H^+ + CO_3^{2-}$

$$\frac{C_{CO_3^{2-}}}{C_{HCO_3^-}} = \frac{K_2}{C_{H^+}}$$

$$= \frac{4.8 \times 10^{-11}}{10^{-6}} = 4.8 \times 10^{-5}$$

Obviously $C_{CO_3^{2-}}$ is insignificant.

Further $C_{H_2CO_3} + C_{HCO_3^-} + C_{CO_3^{2-}} = F_{NaCO_3}$, from material balance.

Then, ignoring $C_{CO_3^{2-}}$, we have from the given data

$$C_{H_2CO_3} + C_{HCO_3^-} = 0.01$$

Hence $\quad C_{HCO_3^-} = \dfrac{0.42}{1.42} \times 0.01 \quad$ from (I)

$$= 2.96 \times 10^{-3} \text{ M}$$

Substituting this value of $C_{HCO_3^-}$, $C_{Na^+} = 0.02$ and $C_{Cl^-} = x$ in the following charge balance equation, and neglecting C_{H^+}, C_{OH^-} and $C_{CO_3^{2-}}$

$$C_{H^+} + C_{Na^+} = C_{Cl^-} + C_{HCO_3^-} + 2C_{CO_3^{2-}} + C_{OH^-}$$

We have $\quad 0.02 = x + 2.96 \times 10^{-3}$

or $\quad\quad\quad x = 0.017$

b. HAc, unlike HCl, dissociates incompletely as $HAc \xrightleftharpoons{K} H^+ + Ac^-$. Here similar procedure as in (a) will give $C_{Ac^-} = 0.017$ M (corresponding to C_{Cl^-})

Now $\quad \dfrac{C_{Ac^-}}{C_{HAc}} = \dfrac{K}{C_{H^+}} = \dfrac{1.8 \times 10^{-5}}{10^{-6}} = 18$

Then, moles of HAC to be added per litre

$$= C_{HAC} + C_{Ac^-} \quad \text{(by material balance)}$$

$$= \frac{18 + 1}{18} \times C_{Ac^-}$$

$$= \frac{19}{18} \times 0.017 = 0.0179$$

Approximation involved:

1. The volume change due to addition of acid is ignored
2. Activity coefficient effects have been ignored.

Note: HAC is required to be added in greater amount than HCl to maintain additional equilibrium.

12.43 a. Since K_2 is much too smaller than K_1 only the first dissociation step need be considered as the source of H^+ (vide note on problem 12.30).

Then $\quad\quad pH = \frac{1}{2}(pK_1 - \log F_{H_2S}) \quad$ by Eq. (12.20a)

$$= \frac{1}{2}(7 - \log 0.02) = 4.65$$

b. i. This corresponds to a stage before the first equivalence point

Then $\quad pH = pK_1 + \log \dfrac{C_{HS^-}}{C_{H_2S}} \quad$ by Eq. (12.18a)

$$= 7 + 0 \quad \text{since } C_{HS^-} \approx C_{H_2S}$$

ii. This corresponds to the first equivalence point

Then $pH = \frac{1}{2}(pK_1 + pK_2)$ by Eq. (12.21c)

$= \frac{1}{2}(7 + 15) = 11$

iii. This corresponds to the second equivalent point

$pH = \frac{1}{2}(pK_w + pK_2 + \log F_{Na_2S})$ by Eq. (12.21a)

$= \frac{1}{2}(14 + 15 + \log 0.02) = 13.6$

12.44 a. HCl dissociates completely but CH_3CO_2H remains almost entirely in the undissociated form as an effect of the common ion H^+, then

$pH = -\log F_{HCl}$ by Eq. (12.16a)

$= -\log 0.1 = 1$

b. At the first equivalence point, only HCl, which is a strong acid, is neutralised. The resulting solution corresponds to one of weak acid CH_3CO_2H of formality $0.1/2$ or 0.05 (due to doubling of volume). Then

$pH = \frac{1}{2}(4.75 - \log 0.05)$ by Eq. (12.20a)

$= 3.7$

c. At the second equivalence point both the acids will be neutralised and the resulting solution will be $0.1/3$ or 0.033 F with respect to NaCl and CH_3CO_2Na. Since CH_3CO_2Na (but not NaCl) hydrolyses, we have

$pH = \frac{1}{2}(14 + 4.75 + \log 0.033)$ by Eq. (12.21a)

$= 10.6$

12.45 a. From Eq. (12.22), ignoring C_{OH^-} and also the dissociation constant compared to C_{H^+}, we have

$$C_{H^+} = \left[(FK_a)_{CH_3CO_2H} + (FK_a)_{HCO_2H}\right]^{\frac{1}{2}}$$

$$= [(0.1)(10^{-4.75}) + (0.1)(10^{-3.75})]^{\frac{1}{2}} \, M$$

$$= 4.45 \times 10^{-3} \, M$$

Therefore $pH = -\log(4.45 \times 10^{-3})$

$= 2.35$

b. Here also C_{OH^-} may be ignored. Then putting $b = F_1 = F_2$ and ignoring all the third order terms, we have from Eq. (12.22)

$$C_{H^+} = \left[(K_a)_{CH_3COOH} \cdot (K_a)_{HCO_2H}\right]^{\frac{1}{2}}$$ [which follows also from Eq. (12.21c)]

or $pH = \frac{1}{2}[PK_{a(CH_3CO_2H)} + pK_{a(HCO_2H)}]$

$= \frac{1}{2}(4.75 + 3.75) = 4.25$

c. At the second equivalence point $b = F_1 + F_2$ and C_{H^+} is negligible compound to the dissociation constants when we have from Eq. (12.22)

$$C_{H^+} = C_{OH^-}$$

or $C_{H^+} = K_w^{\frac{1}{2}}$

Therefore $pH = \frac{1}{2}pK_w = 7$

Note:

1. In (a) C_{H^+} is less than $\left(\sqrt{(F \cdot K_a)_{CH_3CO_2H}} + \sqrt{(FK_a)_{HCO_2H}}\right)$ where each term represents the concentration of H^+ in the solution only of the indicated acid. This is due to mutual suppression of dissociation by the effect of common ion (H^+).

2. In (c) pH = 7, although pH of the solution containing only CH_3CO_2Na or HCO_2Na is above 7. This is due to mutual suppression of hydrolysis by the OH^- ion which is the common product.

12.46 The resulting solution corresponds to half neutralisation of a mixture of two weak acids HAc and NH_4^+ of equal formality.

Then

$$pH = \tfrac{1}{2}[pK_a (HAc) + pK_a (NH_4^+)] \quad \text{by Eq. (12.21c)}$$

$$= \tfrac{1}{2}[pK_a (HAc) + pK_w - pK_b (NH_3)]$$

$$= \tfrac{1}{2}(4.7 + 14 - 4.7) = 7$$

Note: Here the resulting mixture is equivalent to one of NH_3 and CH_3CO_2H in equal amount (vide problem 12.35).

12.47 a. Ignoring hydrolysis and activity coefficient effects,

Solubility of BaOx in water $= \sqrt{K_{SP}} \quad$ by Eq. (12.9a)

$$= \sqrt{1.6 \times 10^{-7}} = 4 \times 10^{-4} \text{ M}$$

b. Slight hydrolysis [ignored in (a)] will make actual solubility (S) slightly greater than that calculated in (a). Further, the solution will be alkaline. Here

$$pH = \tfrac{1}{2}(pK_w + pK_2 + \log S) \quad \text{by Eq. (12.21a)}$$

$$= \tfrac{1}{2}(14 - \log (6.4 \times 10^{-5}) + \log(4 \times 10^{-4})] = 7.4$$

12.48 i. Let the solubility of PbF_2 be S in molarity. Then neglecting the activity coefficient effect

$$S(2s)^2 = K_{sp} \quad \text{by Eq. (12.9a)}$$

$$= 4 \times 10^{-9}$$

whence $\qquad S = 10^{-3} \text{ M}$

For precise value of S, we are to calculate I (using the approximate value of S), $\gamma_\pm$ and K_{sp}, as in (ii).

ii. Considering that the ionic strength of the solution is practically due to NaCl

$$I = 0.1 \quad \text{by Eq. (12.6)}$$

The mean ionic activity coefficient ($\gamma_\pm$) of PbF_2 is given, according to Eq. (12.5d), by

$$\log \gamma_\pm = -0.509 \,(2)\,(1)\,\sqrt{0.1}$$

whence $\qquad \gamma_\pm = 0.476$

Now $S(2s)^2 = K_{sp} = \dfrac{K_{th}}{\gamma_\pm^3}$

$$= \frac{4 \times 10^{-9}}{(0.476)^3} = 3.71 \times 10^{-8} \quad \text{by Eq. (12.8)}$$

whence $\qquad S = 2.1 \times 10^{-3} \text{ M}$

iii. Here I and hence $\gamma_{\pm}$ and K_{sp} will be essentially same as in (ii)

Then $3.71 \times 10^{-8} = (S)(2S + 0.1)^2$ by Eq. (12.9b)

$$\simeq (S)(0.1)^2 \because S \ll 0.1$$

or $\qquad S = 3.71 \times 10^{-6}$ M

iv. Of the total fluorine content of the solution (due to PbF_2), let the fraction remaining as F^- ion be α. Then from the dissociation equilibrium

$$\underset{2S(1-\alpha)M}{HF} \rightleftharpoons \underset{\simeq 0.1 M}{H^+} + \underset{2S\alpha M}{F^-}$$

$$\frac{(0.1)(2S\alpha)}{2S(1-\alpha)} = K_n = 2 \times 10^{-2}$$

whence $\qquad \alpha = \frac{1}{6}$ and $C_{F^-} = 2S\alpha = S/3$

Then $\qquad S(S/3)^2 = 4 \times 10^{-9}$

whence $\qquad S = 3.3 \times 10^{-3}$ M

Note: Both NaCl and HCl donot contain any ion in common with PbF_2. But HCl enhances the solubility more than NaCl by removing F^- ions largely as HF molecules.

12.49 a. i. If there were no precipitation, the concentration of Pb^{2+} and Cl^- ions would be

$$C_{Pb^{2+}} = \frac{100}{100+1} \times 1 \text{ M} = 0.99 \text{ M}$$

$$C_{Cl^-} = \frac{100}{100+1} \times 2 \text{ M} = 0.02 \text{ M}$$

Then ionic product of $PbCl_2 = (0.99)(0.02)^2$ or 3.96×10^{-4} which is less than the solubility product (10^{-3}). Therefore, $PbCl_2$ will not be precipitated.

ii. Here, by following the same procedure as in (i), we have ionic product

$$= \left(\frac{50}{100}\right)\left(\frac{2 \times 50}{100}\right)^2 > K_{sp} (10^{-3})$$

Therefore, $PbCl_2$ will be precipitated.

☐ Let x moles of $PbCl_2$ precipitate at equilibrium. Then the concentration of Pb^{2+} and Cl^- ions remaining in the solution will be

$$C_{Pb^{2+}} = \left(\frac{\frac{50 \times 1}{1000} - x}{\frac{50+50}{1000}}\right) \text{M} = (0.5 - 10x) \text{ M}$$

$$C_{Cl^-} = \left(\frac{\frac{50 \times 2}{1000} - 2x}{\frac{50+50}{1000}}\right) \text{M} = 2(0.5 - 10x) \text{ M}$$

Now, for saturated solution at equilibrium
ionic product = solubility product
Then $(0.5 - 10x)[2(0.5 - 10x)]^2 = 10^{-3}$
whence $\qquad x = 0.044$

b. Let the solubility of $PbCl_2$ in 0.1 F $CaCl_2$ be S in molarity. Then

$$\underset{\text{ionic product}}{S(2S + 2 \times 0.1)^2} = \underset{\text{solubility product}}{10^{-3}} \qquad (I)$$

If S within ( ) is ignored, then we have

$$S = \frac{10^{-3}}{(2 \times 0.1)^2} \text{ or } 0.025 \text{ M}$$

It appears from the calculated value of S that $2s$ is not small enough to be ignored compared to 2×0.1. We can get better result putting $S = 0.025$ within ( ) in I when

$$S = \frac{10^{-3}}{(2 \times 0.025 + 2 \times 0.1)^2} \text{ or } 0.016 \text{ M}$$

The amount of $PbCl_2$ dissolved

$$= \text{solubility} \times \text{volume of medium}$$

$$= (0.016 \text{ M}) \left(\frac{500}{1000} \text{ L} \right) = 0.008 \text{ mol}$$

12.50 From the preliminary analysis of the given data, it appears that the concentration of Pb^{2+} is enough to precipitate SO_4^{2-} and IO_3^- ions almost completely producing essentially 0.03 mol of $PbSO_4(s)$ and 0.04/2 or 0.02 mol of $Pb(IO_3)_2(s)$. In the resulting 3 L solution

$$C_{Pb^{2+}} = \frac{\text{amount of remaining } Pb^{2+}}{\text{Volume of the resulting solution}}$$

$$= \frac{(0.06 - 0.03 - 0.02) \text{ mol}}{3 \text{ L}} = 0.00333 \text{ M}$$

$$C_{NO_3^-} = \frac{2 \times 0.06}{3} \text{ or } 0.04 \text{ M}$$

$$C_{K^+} = \frac{(2 \times 0.03 + 1 \times 0.04)}{3} \text{ or } 0.0333 \text{ M}$$

$$C_{SO_4^{2-}} = \frac{K_{sp} (PbSO_4)}{C_{Pb^{2+}}} = \frac{10^{-8}}{0.00333} \text{ or } 3.0 \times 10^{-6} \text{ M}$$

$$C_{IO_3^-} = \left[\frac{K_{sp} (Pb(IO_3)_2)}{C_{Pb^{2+}}} \right]^{\frac{1}{2}} = \left[\frac{10^{-13}}{0.00333} \right]^{\frac{1}{2}} \text{ or } 5.48 \times 10^{-6} \text{ M}$$

12.51 a. Neglecting the contribution of AgCl to the concentration of Cl^- ion in the solution before 1% of the equivalence point

$$C_{Cl^-} = 0.1 \times \frac{1}{100} \times \frac{100}{100 + 99} \text{ or } 5.02 \times 10^{-4} \text{ M}$$

b. At the equivalence point

$$K_{sp (AgCl)} = C_{Ag^+} . C_{Cl^-} = C_{Cl^-}^2$$

$$= 1.5 \times 10^{-10}$$

whence $C_{Cl^-} = 1.22 \times 10^{-5} \text{ M}$

c. Neglecting the contribution of AgCl to the concentration of Ag^+ ion in the solution after 1% of the equivalence point

$$C_{Ag^+} = 0.1 \times \frac{1}{100} \times \frac{100}{100+101} \quad \text{or } 4.98 \times 10^{-4} \text{ M}$$

Then

$$C_{Cl^-} = \frac{K_{sp}(AgCl)}{C_{Ag^+}} = \frac{1.5 \times 10^{-10}}{4.98 \times 10^{-4}} = 3.01 \times 10^{-7} \text{ M}$$

Note: Cl^- ions are not completely precipitated even after equivalence point. However, the amount of Cl^- ions remaining unreacted is only $\left(\frac{200 \times 1.22 \times 10^{-5}}{100 \times 0.1} \times 100\right)$ or 0.024%. This justifies the titrimetric estimation.

12.52 a. That iodate will precipitate first for which the necessary concentration of IO_3^- ion is smallar.

    i. $AgIO_3$.

    ii. $Pb(IO_3)_2$

□ When precipitation of $Pb(IO_3)_2$ starts, the concentration of IO_3^- ions in the solution is

$$C_{IO_3^-} = \left(\frac{3.2 \times 10^{-13}}{0.1}\right)^{\frac{1}{2}} \quad \text{or } 1.79 \times 10^{-6} \text{ M}$$

For this concentration of IO_3^- the concentration of Ag^+ ions remaining in the solution is

$$C_{Ag^+} = \frac{5.3 \times 10^{-8}}{1.79 \times 10^{-6}} \quad \text{or } 2.96 \times 10^{-2} \text{ M}$$

Then % of Ag^+ ions that remains to be precipitated

$$= \frac{2.96 \times 10^{-2}}{0.1} \times 100 = 29.6$$

b. Let x be the concentration of $AgNO_3$ and $Pb(NO_3)_2$ when $AgIO_3$ and $Pb(IO_3)_2$ precipitate simultaneously. Here the minimum concentration of IO_3^- required for precipitation is

$$C_{IO_3^-} = \frac{5.3 \times 10^{-8}}{x} \quad \text{for precipitation of } AgIO_3$$

$$= \left(\frac{3.2 \times 10^{-13}}{x}\right)^{\frac{1}{2}} \quad \text{for precipitation of } Pb(IO_3)_2$$

whence $x = 8.78 \times 10^{-3}$ M

□ Suppose a mol of $AgIO_8$ and b mol of $Pb(IO_3)_2$ are precipitated. Then in the solution

$$C_{Ag^+} = 8.78 \times 10^{-3} - a$$

$$C_{Pb^{2+}} = 8.78 \times 10^{-3} - b$$

$$C_{IO_3^-} = \frac{5.3 \times 10^{-8}}{8.78 \times 10^{-3} - a} = \left[\frac{3.2 \times 10^{-13}}{8.78 \times 10^{-3} - b}\right]^{\frac{1}{2}} \tag{I}$$

Also $a + 2b + C_{IO_3^-} = 0.001$, from material balance.

Now from preliminary analysis, it appears that most of the IO_3^- ions will be precipitated and so $C_{IO_3^-}$ may be ignored.

Then $\qquad a + 2b = 0.001$ (II)

From (I) and (II), we have

$$a = 2 \times 10^{-4}$$
$$b = 4 \times 10^{-4}$$

Then from (I)

$$C_{IO_3^-} = \frac{5.3 \times 10^{-8}}{8.78 \times 10^{-3} - 2 \times 10^{-4}} \text{ or } 6.18 \times 10^{-6} \text{ M}$$

This justifies ignoring of $C_{IO_3^-}$.

Note: In problem 12.50 two different types of ions are precipitated simultaneously and almost completely through their reaction with a common ion which is present in excess. But in the present problem, the precipitation of the two ions occurs successively and competitively with the common precipitating ion added in less amount than the total equivalents of other ions concerning the precipitation. This provides the basis of the principle of fractional precipitation whereby a desired ion can be precipitated from a mixture of several ions.

12.53 i. When $PbCl_2$ is just entirely converted into $PbSO_4$, the solution will have

$$C_{Cl^-} = 2 \times 0.2 \text{ or } 0.4 \text{ M}$$

$$C_{Pb^{2+}} = \frac{K_{sp}(PbCl_2)}{C_{Cl^-}^2} = \frac{10^{-3}}{(0.4)^2} \text{ or } 0.00625 \text{ M}$$

$$C_{SO_4^{2-}} = \frac{K_{sp}(PbSO_4)}{C_{Pb^{2+}}} = \frac{10^{-8}}{0.00625} \text{ or } 0.0000016 \text{ M}$$

Required amount of $(NH_4)_2SO_4$

$\equiv$ moles of $PbSO_4(s)$ formed + moles of SO_4^{2-} (aq) (unreacted)

$= (0.2 - 0.00625) + 0.0000016$

≈ 0.194 mol

ii. When PbF_2 is just entirely converted into $PbSO_4$, the solution will have

$$C_{F^-} = 2 \times 0.2 \text{ or } 0.4 \text{ M}$$

$$C_{Pb^{2+}} = \frac{K_{sp}(PbF_2)}{C_{F^-}^2} = \frac{10^{-8}}{(0.4)^2} \text{ or } 6.25 \times 10^{-8} \text{ M}$$

$$C_{SO_4^{2-}} = \frac{K_{sp}(PbSO_4)}{C_{Pb^{2+}}} = \frac{10^{-8}}{6.25 \times 10^{-8}} \text{ or } 0.16 \text{ M}$$

Required amount of $(NH_4)_2SO_4$

$\equiv$ moles of $PbSO_4(s)$ formed + moles of SO_4^{2-} (aq)

$= (0.2 - 6.25 \times 10^{-8}) + 0.16$

$\approx 0.2 + 0.16$ or 0.36 mol

❑ The amount of $(NH_4)_2SO_4$ is lower than the initial amount of $PbCl_2$ but higher than the initial amount of PbF_2. Because $PbCl_2$ has much higher solubility than PbF_2

so that in aqueous suspension, the amount of the undissolved salt is lower with $PbCl_2$.

Note: If the system were homogeneous, the amount of $(NH_4)_2SO_4$ required would not be lower than the initial amount of $PbCl_2$ (vide next problem).

12.54 i. The relevant reaction is

$$Na_2OX(aq) + 2HCl(aq) = 2NaCl(aq) + H_2OX(aq)$$

It appears from the Stoichiometry of this reaction that 0.03 mol of Na_2OX will require 2×0.03 or 0.06 mol of HCl for conversion to NaCl (the back reaction being insignificant).

ii. The relevant reaction is

$$CaOX(s) + 2HCl(aq) = CaCl_2(aq) + H_2OX(aq)$$

Here apart from HCl consumed in this reaction, some additional HCl is required to keep the oxalate ion concentration just sufficient to maintain calcium ion in solution. When CaOX just dissolves the solution will have

$$C_{Ca^{2+}} = 0.03 \text{ M ignoring the volume change due to addition of HCl}$$

$$C_{OX^{2-}} = \frac{2.6 \times 10^{-9}}{0.03} \quad \text{or} \quad 8.67 \times 10^{-8} \text{ M}$$

$$C_{H_2OX} = 0.03 \text{ M}$$

assuming that the resulting oxalic acid is mostly present as H_2OX molecule

Then

$$C_{H^+} = \left[\frac{K_1 K_2 CH_2OX}{C_{OX^{2-}}} \right]^{\frac{1}{2}} \quad \text{from } H_2OX \rightleftharpoons 2H^+ + OX^{2-}$$

$$= \left[\frac{(6 \times 10^{-2})(6.4 \times 10^{-5})(0.03)}{8.67 \times 10^{-8}} \right]^{\frac{1}{2}} \quad \text{or } 1.33 \text{ M}$$

Therefore, the total amount of HCl required

$$= 0.06 \text{ mol (for the above reaction)} + 1.33 \text{ mol (additional)}$$
$$= 1.39 \text{ mol, since HCl is completely dissociated}$$

The assumption that the resulting oxalic acid is mostly present as H_2OX molecule can be justified by the low value of $\dfrac{C_{HOX^-}}{C_{H_2OX}} = \dfrac{K_1}{C_{H^+}}$ for quite high value of C_{H^+} found.

12.55 Precipitation of PbS will be just prevented when the solution has

$$C_{S^{2-}} = \frac{K_{sp}(PbS)}{C_{Pb^{2+}}} = \frac{10^{-28}}{0.001} \quad \text{or } 10^{-25} \text{ M}$$

$$C_{H^+} = \left[\frac{K_1 K_2 C_{H_2S}}{C_{S^{2-}}} \right]^{\frac{1}{2}}, \quad \text{from } H_2S \rightleftharpoons 2H^+ + S^{2-}$$

$$= \left[\frac{(10^{-7})(10^{-15})(0.1)}{10^{-25}} \right]^{\frac{1}{2}} \quad \text{or } 10 \text{ M}$$

Therefore, the required amount of HCl = 10 mol.

Note: The prevention of precipitation by adjustment of acidity is an important technique used particularly in analytical chemistry.

12.56 a. For the required lowest pH, that relates to K_2,

$$C_{OH^-} = \left[\frac{C_{Zn(OH)_4^{2-}}}{K_2} \right]^{\frac{1}{2}}$$

$$= \left(\frac{0.01}{0.13} \right)^{\frac{1}{2}} \text{ or } 0.277 \text{ M}$$

Therefore pH = 14 – pOH

$$= 14 + \log 0.277$$

$$= 13.4$$

For the required highest pH, that relates to K_1,

$$C_{OH^-} = \left[\frac{K_1}{C_{Zn^{2+}}} \right]^{\frac{1}{2}}$$

$$= \left(\frac{1.2 \times 10^{-17}}{0.01} \right)^{\frac{1}{2}} \text{ or } 3.46 \times 10^{-8}$$

Therefore pH = 14 – pOH

$$= 14 + \log (3.46 \times 10^{-8})$$

$$= 6.54$$

b. The lowest solubility of $Zn(OH)_2$ relates to K_1 for $C_{OH^-} = 3.46 \times 10^{-8}$ and this is

$$\frac{1.2 \times 10^{-17}}{(3.46 \times 10^{-8})^2} \text{ or } 0.01 \text{ M}.$$

12.57 (1) + (2) gives

$$HgCl_2 = Hg^{2+} + 2Cl^-, \quad K = K_1 K_2$$

Then $C_{Hg^{2+}} \cdot C_{Cl^-}^2 = K C_{HgCl_2} \approx K F_{HgCl_2}$

or $$C_{Hg^{2+}} = \left[\frac{K F_{HgCl_2}}{4} \right]^{\frac{1}{3}} \text{ since } C_{Cl^-} = 2 C_{Hg^{2+}}$$

$$= \left[\frac{(33 \times 10^{-7})(1.8 \times 10^{-7})(0.1)}{4} \right]^{\frac{1}{3}} = 2.46 \times 10^{-5} \text{ M}$$

12.58 This is possible by studying the solubility equilibrium (1) and hydrolytic equilibrium (2) using a soluble carbonate

1. $CaCO_3 (s) + H_2O \rightleftharpoons Ca^{2+}(aq) + HCO_3^-(aq) + OH^-(aq) \quad K_1$

2. $CO_3^{2-}(aq) + H_2O \rightleftharpoons HCO_3^-(aq) + OH^-(aq) \quad K_2$

(1)–(2) gives

$$CaCO_3(s) = Ca^{2+}(aq) + CO_3^{2-}(aq) \quad K = \frac{K_1}{K_2}$$

Then $$K_{sp} = C_{Ca^{2+}}(aq) C_{CO_3^{2-}}(aq) = \frac{K_1}{K_2}$$

UNIVERSITY QUESTIONS

12.1 Define mean ionic activity coefficient ($\gamma_\pm$). How is log $\gamma_\pm$ related to non-ideality of an electrolyte solution. (Burdwan BSC(H), 2006)

12.2 Suppose 4 water molecules are attached to Na^+ ions and 2 to Cl^- ions in a 1 mole/kg solution of the salt. Calculate the number of moles of free water per kg of solvent. Referring to activity coefficients, what does this result imply? What would have happened in dilute solution? (Burdwan BSC(H), 1996)

12.3 Express μ of $ZnCl_2$ in terms of μ_+ and μ_-. Express a of $ZnCl_2$ in terms of m_i and $\gamma_\pm$. μ is chemical potential of the electrolyte as a whole and a is its activity. Other terms have their usual significance. (Calcutta BSc(H), 2011)

12.4 Outline the essential features of Debye–Huckel theory of interionic attraction. Discuss qualitatively the effect of temperature and dielectric constant on the activity coefficient of an electrolyte. (Calcutta BSc(H), 1989)

12.5 According to Debye–Huckel model, the electrical potential at a distance r from the central positive ion of charge Ze is given by

$$\frac{Ze}{\epsilon_r r} - \frac{Ze}{\epsilon_r r_D}$$

Interpret the terms. Find the work done for transferring the charged ion from the solution in infinite dilution to that in a given concentration. Find dimension of r_D. (Burdwan BSc(H), 2008)

12.6 If Debye–Huckel potential at a distance r from a central positive ion is given by

$$\psi = A\frac{e^{-r/r_D}}{r}, \text{ calculate } A. \text{ Explain the significance of } r_D. \quad \text{(Burdwan BSc(H), 1994)}$$

12.7 The A factor in Debye–Huckel limiting law is unitless. Comment. (Burdwan BSc(H), 1998)

12.8 Starting from the relation $\log \gamma_i = -AZ_i^2 \sqrt{I}$, arrive at a formula the $\log \gamma_\pm$. Draw $\log \gamma_\pm$ versus $\sqrt{I}$ for a $1-1$ electrolyte stating clearly the value of slope. How does the slope vary with temperature. (Burdwan BSc(H), 2005)

12.9 An extended form of Debye–Juckel limiting law equation for a 1:1 electrolyte is given by $\log \gamma_+ = A\sqrt{I}/(1+B\sqrt{I})$ where A and B are constants. Suggest suitable linear plot fo find A and B. (Burdwan BSc(H), 2006)

12.10 Explain the idea of ion atmosphere.

Calculate the thickness of the ion atmosphere for a 0.01 M $MgCl_2$ solution at 298 K. Given that the thickness of the ion atmosphere for 0.1 M NaCl solution is 0.96 nm at 298 K. (Burdwan BSc(H), 2000)

12.11 Calculate mean ionic activity coefficient $\gamma_\pm$ of (i) 0.01 M KCl and (ii) 0.01 M $BaCl_2$ solution at 298 K using Debye–Huckel limiting law ($A = 0.51$ at 298 K). What will be the value $\gamma_\pm$ of KCl solution at 308 K? Assuming that the dielectric constant does not change appreciably. (Burdwan BSc(H), 1995)

12.12 Calculate $a_\pm$ for 0.1 molar solution of H_2SO_4 at 25°C where $\gamma_\pm = 0.265$. (Burdwan BSc(H), 1996)

12.13 Calculate the ionic strength of a solution obtained by mixing 50 ml of 0.2 M $Al_2(SO_4)_3$, 25 ml of 0.2 M $K_4Fe(CN)_6$ and 25 ml 0.1 M urea solution at a given temperature. (Jadavpur BSc(H), 2003)

12.14 If the solubility product of PbI_2 is determined at a given temperature from the product of ionic concentration in a saturated solution of PbI_2 in KI solution of varying concentration, is the solubility product actually constant? Justify your answer.

Discuss how the thermodynamic solubility product of PbI_2 can be obtained?

(Calcutta BSc(H), 1994)

12.15 Do you expect the solubility of CaF_2 in aqueous KNO_3 solution to be the same as lower than or higher than its solubility in pure water at the same temperature. Give reasons for your answer. (Jadavpur BSc(H), 2001)

12.16 A double salt may not separate out on evaporation of its unsaturated aqueous solution. Explain. (Jadavpur BSc(H), 2003)

12.17 Calculate the solubility product of silver chromate if its solubility in pure water be 2.5×10^{-2} gm/litre. Find out its solubility in 0.001 molar potassium chromate solution. The activity coefficient may be taken as unity. (Burdwan BSc(H), 2011)

12.18 Solubility of CaF_2 at 25°C in water is 2.04×10^{-4} mol L^{-1}. Calculate the thermodynamic solubility product and solubility in 0.1 M NaF solution. Assume Debye–Huckel limiting law to be valid with $A = 0.509$ $L^{\frac{1}{2}}$ $mol^{-\frac{1}{2}}$ at 25°C.

(Jadavpur BSc(H), 2012)

12.19 The solubility product of lead bromide is 8.0×10^{-5}. If the salt is 80% dissociated in the saturated solution, find the solubility of the salt. (Madurai BSc, 2010)

12.20 Equal volume of 0.08 N $CaCl_2$ and 0.02 N Na_2SO_4 are mixed at room temperature. Will there be any precipitation? The solubility product is 2.4×10^{-5}.

(Calcutta BSc(H), 1996)

12.21 The solubility product of PbI_2 is 7.47×10^{-9} at 15°C and 1.39×10^{-8} at 25°C. Calculate molar heat of solution of PbI_2. (Calcutta BSc(H), 2005)

12.22 Assuming heat of solution to be independent of temperature, the molar solubility (s) of $BaSO_4$ in water at a given temperature is given by

$$\log_{10}S = \frac{2.60 \times 10^2}{T/K} - 5.87$$

Find (i) the heat of solution of $BaSO_4$ in water and (ii) the thermodynamic solubility product of $BaSO_4$ at 25°C. (Given, Debye–Huckel constant of water at 25°C = 0.51).

(Calcutta BSc(H), 2000)

12.23 Which quantity do we use in place of pH to measure the acidity of a concentrated aqueous solution. Show that it reduces to pH in dilute solution.

(Burdwan BSc(H), 2000)

12.24 Justify/criticise pH + pOH = 14 is always true. (Burdwan BSc(H), 1996)

12.25 Do you expect the pH of pure water at 100°C to be less than 7, or more than 7. Explain your answer. (Nagpur BSc, 2000)

12.26 Justify/criticise

(a) pH of a 10^{-8} M HCl solution (aq) is 8. (Burdwan BSc(H), 2005)

(b) For an infinitely dilute acid solution pH = α. (Burdwan BSc(H), 1992)

12.27 What will be the pH of a sample of pure water at 25°C and 50°C? Given $K_w = 1 \times 10^{-14}$ at 25°C and heat of neutralisation of a strong acid by a strong base, $\Delta H = -13.7$ kCal per g equivalent. From the calculated value of pH at 50°C, do you think that water is acidic? (Calcutta BSc(H), 1989)

12.28 All acid–base neutralisations do not occur at exactly pH = 7. Justify with reasons and give two examples, one for each case (pH > 7 and pH < 7). Mention the indicators suitable for such titration. (Burdwan BSc(H), 2010)

12.29 Salt hydrolysis is a manifestation of the Le Chatelier principle on the dissociation equilibrium of water. Explain.

12.30 The hydrolysis constant of KCN is 1.39×10^{-5}. Calculate the degree of hydrolysis of KCN in (i) 10^{-1} M and (ii) 10^{-2} M, solution of KCN. (Burdwan BSc(H), 2010)

12.31 Calculate the degree of hydrolysis of ammonium acetate. The dissociation constant of NH_4OH is 1.8×10^{-5} and that of acetic acid is 1.8×10^{-5} and $K_w = 1.0 \times 10^{-14}$.
(Rajasthan BSc, 2008)

12.32 Aqueous solutions of two salts BA' and B'A at equal formality have equal pH. The acid HA' and the base B'OH are strong while the acid HA and base BOH are weak. Establish a relation between pK values of the weak acid and weak base in terms of salt concentration. (Burdwan BSc(H), 2010)

12.33 A salt of a strong acid HA and weak base B shows pH = 6.2 when dissolved in water to make a 0.1 N solution. Find the dissociation constant of B and explain why that of HA cannot be determined from this observation. (Burdwan BSc(H), 2005)

12.34 Calculate the pH of a 0.01 M solution of a weak acid ($K_a = 10^{-5}$) when the solution is 1 M with respect to NaCl also. (Burdwan BSc(H), 2006)

12.35 What should be the essential condition for a compound to be able to serve as an acid–base indicator? Find out the range of such an indicator.
(Calcutta BSC(H), 1998)

12.36 Why weak acid–weak base titration is difficult to carry out. (Madras BSc, 2010)

12.37 a. Calculate the pH of the mixture when 1, 20, 50 and 60 ml of 0.1 N NaOH are added to a solution of 50 ml of 0.1 N HCl.

   b. Repeat the same calculations for the mixtures when same alkali is added to 50 ml 0.1 N acetic acid ($K_a = 1.82 \times 10^{-5}$).

From the result of above two sets, select the proper indicator to be used to find the equivalence point in each case. (Burdwan BSc(H), 2006)

12.38 Suppose you are titrating 50.0 ml of a mixture of 0.1 M hydrochloric acid and 0.1 M acetic acid with 0.2 M sodium hydroxide. Calculate the pH of the resulting solutions that correspond to the following volumes in ml of titrant added (i) 0.0, (ii) 10,0, (iii) 25,0, (iv) 35.0, (v) 50.0 and (vi) 60.0. Draw the rough sketch of the titration curve. (K_a of acetic acid = 1.75×10^{-5}). (Burdwan BSc(H), 1997)

12.39 A solution was known to contain only Na_2CO_3 and $NaHCO_3$ with initial pH 10.63. When a 25 ml solution was titrated with 0.1 M HCl, pH of the solution after adding 21 ml acid was 6.35. Find the amount of each component in original solution.
[Given $pK_1 = 6.35$, $pK_2 = 10.33$ for H_2CO_3] (Burdwan BSc(H), 2007)

12.40 10 cc of 0.1 NH_4OH is titrated against 0.1 N HCl. Calculate the pH of the solution (i) at the start, (ii) at the half-neutralisation point and (iii) at the equivalence point.
[Given $pK_b(NH_4OH) = 4.74$] (Calcutta BSc, 1987)

12.41 a. What is a buffer solution? (Madurai BSc, 2000)

   b. Why a mixture of NaCl and NaOH does not act as a buffer?
(Burdwan BSc(H), 2001)

12.42 How will you explain the buffer action of aqueous solution of ammonium acetate?
(Punjabi BSc, 2000)

12.43 How buffer capacity of a solution is described? Show that buffer capacity is maximum when $pH = pK_a$. (Burdwan BSc(H), 2012)

12.44 You are supplied with two weak acids A and B of pK values 4.8 and 5.8 respectively. State with reasons, which one you prefer to prepare a buffer solution of pH 5.1. (Calcutta BSc(H), 1993)

12.45 You are supplied with 0.1 N NH_4OH^- and 0.1 N HCl solutions. Show how you will use these solutions to prepare 100 ml of a buffer solution of pH =9.0. (Calcutta BSc(H), 1998)

12.46 Calculate the volume of 0.2 N NaOH solution that has to be added to 10 ml of 0.1 M NaH_2PO_4 solution to prepare a phosphate (Na_2HPO_4–NaH_2PO_4) buffer of maximum buffer capacity at 25°C. What will be the pH of the resulting buffer?

Which of the following indicators will be suitable for the determination of the pH of this buffer by colorimetric method?

Bromophenol blue ($pK_{In} = 4.0$), Bromocresol green ($pK_{In} = 4.7$), Bromothymol blue ($pK_{In} = 7.0$).

[Given, at 25°C for H_3PO_4, $K_1 = 7.5 \times 10^{-3}$, $K_2 = 6.2 \times 10^{-8}$ and $K_3 = 4.8 \times 10^{-13}$]. (Calcutta BSC(H), 2000)

12.47 What will be the pH of a buffer solution containing 0.06 mole of acetic acid and 0.04 mole of sodium acetate per liter? Calculate the changes in pH resulting from the addition of (a) 0.05 mole of a strong acid, and (b) 0.03 mole of strong base per liter of the buffer solution? [Given pK for acetic acid = 4.76] (Calcutta BSc(H), 1989)

12.48 What will be the pH of a buffer solution containing 0.06 mole of acetic acid and 0.04 mole of sodium acetate per liter? Calculate the changes in pH resulting from the addition of (a) 0.05 mole of a strong acid, and (b) 0.03 mole of strong base per liter of the buffer solution? [Given pK for acetic acid = 4.76] (Calcutta BSc(H), 1989)

12.49 A reaction is giving out H_3O^+ and a product which is stable only at pH = pK_a of acetic acid. Explain why a mixture of ammonium acetate and acetic acid is best suited as the medium to carry out the reaction. (Burdwan BSc(H), 2008)

12.50 The dissociation constant of NH_4OH is 1.8×10^{-5}. The solubility product of $Mg(OH)_2$ is 1.22×10^{-11}. How many gms of solid NH_4Cl must be added to a mixture of 50 cc of (N) NH_4OH solution and 50 cc of $MgCl_2$ solution so that the precipitate of $Mg(OH)_2$ just disappear? It is assumed that the volume of the solution is not changed by dissolving solid NH_4Cl and that the dissociation of the neutral salt is complete. (Burdwan BSc(H), 2001)

KEY TO UNIVERSITY QUESTIONS

12.1 Vide Section 12.1 and 12.2.

12.2 Number of moles of free H_2O molecules per kg of pure water = 1000/18 or 55.5. Number of moles of bound H_2O molecules per kg of water in the given solution is = (4 + 2) or 6.

Therefore, the required number of moles of free water = (55.5 – 6) or 49.5.

□ Activity coefficient is unity for pure water where each H_2O molecule is surrounded by identical (free) molecules. But in the given solution surrounding H_2O molecules are not all free, a considerable portion being bound to Na^+ and Cl^-

ions. This implies that the activity coefficient of H_2O in the given solution is less tan unity.

□□ With increasing dilution activity coefficient will approach unity because of increase in fraction of free H_2O molecules in the solution.

12.3
$$\mu = \mu_{Zn^{2+}} + 2\mu_{Cl^-}$$

$$a = (m_{Zn^{2+}} \cdot m_{Cl^-}^2) \gamma_{\pm}^3$$

12.4 Vide Section 12.2.

□ Vide problem 12.5.

12.5 The first term, which decreases with r, is due to the central ion alone and the second term, which is independent of r, is due to its oppositely charged surrounding atmosphere of radius r_D.

□ The required work is $-\dfrac{(Ze)^2}{2\epsilon_r r_D} + kT \ln x_i$.

The first term corresponds to the energy of ion (i) due to its charge Ze in the field of its ionic atmosphere and the second term is due to the increase in ionic concentration (mole fraction, x_i) of the solution treated as ideal.

□□ The first and second terms in the question must have the same dimension. Hence r_D will have the dimension of r, i.e. l.

12.6 From the given expression

$$\psi = A\left(\frac{1 - r/r_D}{r}\right); \text{ ignoring the higher terms}$$

Now, in very dilute solution, the potential at short distance r from an ion of charge Ze is due to the ion alone, i.e. $Ze/\epsilon_r r$. Then, $A = Ze/\epsilon_r$.

□ In moderately dilute solution $\psi = \dfrac{Ze}{\epsilon_r r} - \dfrac{Ze}{\epsilon_r r_D}$.

Here the second term corresponds to the contribution to ψ due to the ionic atmosphere. r_D signifies the mean radius of the ionic atmosphere such that if its total charge $(-Ze)$ where placed at this distance from the central ion, the potential produced at it would be equal to $-Ze/\epsilon_r r_D$. r_D equals roughly to the average distance of the innermost counterions from their central ion.

Note: An ionic atmosphere shields the central ion from interactions with other ions lying outside it.

12.7 Vide note on problem 12.23.

12.8
$$\log \gamma_{\pm} = \frac{1}{v_+ + v_-}(v_+ \log \gamma_+ + v_- \log \gamma_-) \quad \text{from Eq. (12.4)}$$

$$= -A\left(\frac{v_+ Z_+^2 + v_- Z_-^2}{v_+ + v_-}\right)\sqrt{I} \quad \text{using the given relation}$$

$$= -A Z_+ Z_- \sqrt{I}$$

introducing $v_+ Z_+ = v_- Z_-$ (due to electroneutrality of the solution)

□ $\log \gamma_{\pm}$ versus $\sqrt{I}$ plot will be linear with slope $-AZ_+Z_-$ which is $-A$ (0.509 for aqueous solution at 25°C) for 1 – 1 electrolyte.

□□ The slope will vary inversely with $T^{\frac{3}{2}}$, by Eq. 12.5(b).

12.9 From the given relation $\dfrac{1}{\log \gamma_\pm} = \dfrac{1}{A\sqrt{I}} + \dfrac{B}{A}$. Then, A and B can be found from the

plot of $\dfrac{1}{\log \gamma_\pm}$ verses $\dfrac{1}{\sqrt{I}}$ (where $I =$ molality for $1 - 1$ electrolyte) which is linear

with slope $= 1/A$ and intercept $= B/A$.

12.10 Vide Section 12.2.

☐ Thickness of the ion atmosphere, $r_D \propto \dfrac{1}{\sqrt{I}}$, from Eq. (12.5a) and (12.5b).

Then $\qquad (r_D)_{MgCl_2} = (r_D)_{NaCl}\sqrt{\dfrac{I_{NaCl}}{I_{MgCl_2}}}$

$$= (0.96 \text{ nm})\sqrt{\dfrac{\frac{1}{2}(0.1 \times 1^2 + 0.2 \times 1^2)}{\frac{1}{2}(0.01 \times 2^2 + 0.02 \times 1^2)}} \quad \text{by Eq. (12.6)}$$

$$= 1.75 \text{ nm}$$

Note: In an electrolytic solution r_D, like interionic distance, refers to the solution as a whole and not to the type of the ions present in the solution.

12.11 Vide problem 12.22

☐ A and hence $\log \gamma_\pm$, varies inversely with $T^{\frac{3}{2}}$. Then for KCl

$$\log (\gamma_\pm)_{308} = \log (\gamma_\pm)_{298} \times \left(\dfrac{298}{308}\right)^{\frac{3}{2}}$$

12.12 $\quad a_\pm = (m_{H^+}^2 \cdot m_{SO_4^{2-}})^{\frac{1}{3}} \gamma_\pm \quad$ for the dissociation $H_2SO_4 = 2H^+ + SO_4^{2-}$ by Eq. (12.3)

$$= [(0.2)^2 \cdot (0.1)]^{\frac{1}{3}} (0.265) = 0.519$$

12.13 In the resulting mixture of $(50 + 25 + 25)$ or 100 ml

$$\dfrac{C_{Al^{3+}}}{2} = \dfrac{C_{SO_4^{2-}}}{3} = 0.2 \text{ M} \times \dfrac{50}{100} = 0.1 \text{ M}$$

$$\dfrac{C_{K^+}}{4} = C_{Fe(CN)_6^{4-}} = 0.2 \text{ M} \times \dfrac{25}{100} = 0.05 \text{ M}$$

$$I = \tfrac{1}{2}\left(C_{Al^{3+}} \cdot Z_{Al^{3+}}^2 + C_{SO_4^{2-}} \cdot Z_{SO_4^{2-}}^2 + C_{K^+} \cdot Z_{K^+}^2 + C_{Fe^*(CN)_6^{4-}} \cdot Z_{Fe(CN)_6^{4-}}^2\right)$$

$$= \tfrac{1}{2}[(2 \times 0.1 \text{ M}) \times 3^2 + (3 \times 0.1 \text{ M}) \times 2^2 + (4 \times 0.05 \text{ M}) \times 1^2 + (0.05 \text{ M}) \times 4^2]$$

$$= 2.0 \text{ M} \approx 2.0 \text{ mol kg}^{-1}$$

since for dilute aqueous solution molarity $\approx$ molality

Urea, being a non-electrolyte, does not directly contribute to the ionic strength. However, the urea solution indirectly causes change in I through volume change of the resulting mixture. Note the difference from problem 12.22.

12.14 For $PbI_2 \qquad \log K_{sp} = \log K_{th} - \log \gamma_\pm^3 \quad$ by Eq. (12.8)

$$= \log K_{th} + 3A Z_{Pb^{2+}} \cdot Z_{I^-} \sqrt{I}$$

Since K_{th} is a real constant for PbI_2 at a given temperature, K_{sp} will vary with I in varying concentration of KI.

☐ From the intercept ($\log K_{th}$) of the linear plot of $\log K_{sp}$ versus $\sqrt{I}$ we can find out K_{th}.

12.15 In aqueous KNO_3, which has higher ionic strength than pure water, $\gamma_\pm$ of CaF_2 will be lowered. Therefore, CaF_2 will have higher K_{sp}, and hence higher solubility in aqueous KNO_3 solution than in pure water.

Note: With increasing concentration of the foreign electrolyte above certain high value, $\gamma_\pm$ of the concerned electrolyte is likely to pass through a minimum leading ultimately to a decrease in the solubility of the latter. Such salting-out effect can be explained only through proper modification of Eq. (12.5e), such as

$$\log \gamma_\pm = -0.51 Z_+ Z_- \left[\frac{\sqrt{I}}{1 + \sqrt{I}} - 0.30 I \right] \text{ in } H_2O \text{ at } 25°C$$

which is proposed by Davies.

12.16 This may be explained as due to the appreciable salting-out effect of one of the component salts over the other.

12.17 $\dfrac{2.5 \times 10^{-2} \text{ g}}{\text{litre}}$ of $Ag_2CrO_4 = \dfrac{2.5 \times 10^{-2}}{2 \times 108 + 52 + 4 \times 16}$ or 7.53×10^{-5} mol/litre

$$K_{sp}(Ag_2CrO_4) = (2 \times 7.53 \times 10^{-5})^2 (7.53 \times 10^{-5}) \quad \text{by Eq. (12.9a)}$$
$$= 1.71 \times 10^{-12}$$

☐ The solubility S in 0.001 M K_2CrO_4 is given by

$$1.71 \times 10^{-12} = (2S)^2 (S + 0.001) \quad \text{by Eq. (12.9b)}$$
$$\simeq (2S)^2 (0.001) \quad \text{ignoring } s \text{ compared to } 0.001$$

whence $\qquad S = 2.07 \times 10^{-5}$ mol/liter

12.18 For CaF_2 $\qquad \log \gamma_\pm = A Z_+ Z_- \sqrt{I}$
$$= -A Z_+ Z_- \sqrt{3S}$$

$[\because I = \frac{1}{2}(S \times 2^2 + 2S \times 1^2) = 3S$, where S is the solubility of CAF_2]

$$= -(0.509)(2)(1) \sqrt{3(2.04 \times 10^{-4})}$$

whence $\qquad \gamma_\pm = 0.94$

$$K_{th} = K_{SP} \gamma_\pm^3$$
$$= (S)(2S)^2 (0.94)^3$$
$$= 4S^3 (0.94)^3$$
$$= 4(2.04 \times 10^{-4})^3 (0.94)^3$$
$$= 2.83 \times 10^{-11}$$

☐ Vide problem 12.48 (iii).

12.19 The relevant solubility equilibrium is

$$PbBr_2(s) \rightleftharpoons PbBr_2 \text{ (aq)} \rightleftharpoons Pb^{2+} \text{ (aq)} + 2Br^- \text{(aq)}$$
$$\qquad\qquad\qquad S(1-0.8) \qquad\quad 0.8S \qquad\quad 2 \times 0.8S$$

where S is solubility of $PbBr_2$

Then $\qquad\qquad K_{sp} = C_{Pb^{2+}} \cdot C_{Br^-}^2$
$$= (0.8S)(2 \times 0.8S)^2$$
$$= 8.0 \times 10^{-5}$$

whence $\qquad S = 0.0339$ mol L^{-1}

Note: Here the concentration of electrolyte in the undissociated form is not small enough to be ignored (as is usually done in dealing with solubility problem) compared to its solubility. As a consequence $K_{sp} \neq (S)(2S)^2$, the concentration of neither ion being a measure of solubility. $K_{sp}(PbBr_2)$ would have been greater if there were no undissociated $PbBr_2$.

12.20 If there were no precipitation, the resulting mixture would have

$$C_{Ca^{2+}} = \frac{0.08}{2} N = \frac{0.08}{2 \times 2} \text{ or } 0.02 \text{ M}$$

$$C_{SO_4^{2-}} = \frac{0.02}{2} N = \frac{0.02}{2 \times 2} \text{ or } 0.005 \text{ M}$$

Then the ionic product of $CaSO_4$ is $(0.02)(0.005)$ or 1.0×10^{-4} which is greater than the solubility product 2.4×10^{-5}. Therefore precipitation of $CaSO_4$ will occur.

Note: Here concentrations of the ions are expressed in molarity rather than in normality. This is done for understanding an equilibrium situation readily in terms of number density of the particles in the system.

12.21 Vide problem 12.24.

12.22 i. $\Delta H_{soln} = RT^2 d \ln K_{sp}/dT$ by van't Hoff equation

$= 2 \times 303 \, RT^2 \, d \log S/dT$, since $K_{sp} = S^2$

$= 2 \times 2.303 \, (-2.60 \times 10^2 K) \, R$, from the given expression

$= 2 \times 2.303 \, (-2.60 \times 10^2 K) \, (8.31 \text{ JK}^{-1}\text{mol}^{-1})$

$= -9.95 \times 10^3 \text{ J mol}^{-1}$

ii. $\log S = \dfrac{2.60 \times 10^2}{298} - 5.87$ from the given expression at 25°C

whence $S = 1.78 \times 10^{-5}$

$\log \gamma_{\pm} = AZ_+ Z_- \sqrt{I}$

$= AZ_+ Z_- \sqrt{4S} \qquad \because I = \tfrac{1}{2}(S \times 2^2 + S \times 2^2)$

$= (0.51)(2)(2\sqrt{4 \times (1.78 \times 10^{-5}}$

whence $\gamma_{\pm} = 0.96$

$K_{th} = K_{sp}\gamma_{\pm}^2$

$= S^2 \gamma_{\pm}^2$

$= (1.78 \times 10^{-5})^2 (0.96)^2$

$= 2.92 \times 10^{-10}$

12.23 Vide Section 12.4.

12.24 From Eq. (12.14)

$$pH + pOH = pK_w = \text{constant at a constant temperature}$$
$$= 14, \text{ at 25°C only}$$

Then the given statement is not justified.

12.25 At higher temperature pH of water becomes lower due to higher concentration of H_3O^+ produced due to autodissociation, $H_2O + H_2O \rightleftharpoons H_3O^+ + OH^-$. Since pH of pure water is 7 at 25°C, it will be less than 7 at 100°C.

12.26 a. For an aqueous solution $pH \approx -\log C_{H_3O^+}$ where $C_{H_3O^+}$ includes contributions from solute and water. In 10^{-8} M HCl, the contribution from HCl is 10^{-8} M but that from H_2O is near 10^{-7} M, i.e. $C_{H_3O^+} > 10^{-8}$ M. Hence the given statement is not justified.

b. For an infinitely dilute acid solution although the contribution of the acid to $C_{H_3O^+}$ of the solution is virtually zero, the contribution of water is near 10^{-7} M, i.e. $C_{H_3O^+} \approx 10^{-7}$ M. Hence the given statement is not justified.

12.27 Vide problem 12.25.

☐ Vide problem 12.19(ii).

12.28 A neutralisation reaction involving weak acid/weak base will not result in a solution of pH = 7 unless pK_a of the acid is equal to pK_b of the base. This is due to the hydrolysis of the conjugate base and conjugate acid in the respective cases, occurring in appreciable extent depending on their relative strength. A weak acid, (e.g. HAc) neutralised by a strong base (e.g. NaOH) results in pH > 7 due to hydrolysis of the conjugate base of the acid (with formation of OH⁻ ions in excess of H_2O^+), as it acts as a base of appreciable strength (by Bronsted concept). On the contrary, a strong acid (e.g. HCl) neutralised by a weak base (e.g. NH_3) results in pH < 7 due to hydrolysis of the conjugate acid of the base, as it acts as an acid of appreciable strength.

For HAc–NaOH titration an indicator like phenolphthalein having pK_{In} (or pT) lying in the basic region is suitable while for HCl–NH_3 titration and indicator like methyl orange having pK_{In} lying in the acid region is suitable.

Note: pH = 7 of an aqueous salt does not necessarily imply that there is no hydrolysis of the salt. For example, aqueous solution of NH_4Ac has pH = 7. Here NH_4^+ and Ac^- both hydrolyse, but to the same extent producing H_3O^+ and OH⁻ ions in equal amount.

12.29 On the basis of the Chatelier principle salt hydrolysis may be thought to occur through the partial removal of H_3O^+, or OH⁻ ions by the ions of the salt with consequent shift of the equilibrium $2H_2O \rightleftharpoons H_3O^+ + OH^-$ to the right and growing in concentration of the Co ions. For example, in the hydrolysis of sodium acetate (NaAc), H_3O^+ ions are partly removed by the Ac^- ions with the formation of undissociated HAc molecules. As a result OH⁻ ions exceed H_3O^+ ions in concentration and the solution becomes alkaline. This is consistent with the Bromsted acid–base concept that suggests Ac^- ion to act as a base of appreciable strength (i.e. appreciable tendency to be protonated) being the conjugate base of the weak acid HAc.

12.30 In aqueous solution of KCN the hydrolytic equilibrium may be represented as

$$CN^- + H_2O \rightleftharpoons HCN + OH^-$$
$$C(1-\alpha) \qquad\qquad C\alpha \qquad C\alpha$$

where C is the total concentration of KCN and α is the degree of hydrolysis. The hydrolysis constant K_h is given by

$$K_h = \frac{C_{HCN} \cdot C_{OH^-}}{C_{CN^-}}$$

$$= \frac{(C\alpha)(C\alpha)}{C(1-\alpha)} = \frac{C\alpha^2}{1-\alpha} \qquad\qquad\qquad (I)$$

$$\approx C\alpha^2 \quad \text{ignoring } \alpha \text{ compared to 1}$$

or
$$\alpha = \sqrt{\frac{K_h}{C}}$$

i.
$$\alpha = \sqrt{\frac{1.39 \times 10^{-5}}{10^{-1}}} = 0.0118$$

Here α is found in good approximation and it is quite small compared to 1. However, the result can be improved on putting $\alpha = 0.0118$ in the denominator of (I) and recalculating α.

ii. Proceeding as with (i)

$$\alpha = \sqrt{\frac{1.39 \times 10^{-5}}{10^{-2}}} = 0.0373$$

Here the result is considerably inaccurate, because α thus found is not small enough to be ignored compared to 1. For precise value of α we are to solve the quadratic equation (I).

12.31 Vide problem 12.35.

12.32 BA' is a salt of weak base and strong acid for which pH is given by Eq. (12.21b).

$B'A$ is a salt of strong base and weak acid for which pH is given by Eq. (12,.21a).

Then under the given conditions

$$pK_a + pK_b = -\log F_{BA'} - \log F_{B'A}$$

which is the required relation.

12.33 The dissociation constant (K_a) of a weak acid HA of formality F is given by

$$K_a = \frac{C_{H^+} \cdot C_{A^-}}{C_{HA}} = \frac{C_{H^+} \cdot C_{A^-}}{F_{HA} - C_{H^+}}$$

$$\simeq \frac{C_{H^+}^2}{F_{HA}} \quad \text{ignoring } C_{H^+} \text{ compared to } F_{HA}$$

Now, BH^+ ion of the salt acts as an acid of appreciable strength (being the conjugate acid of the weak base B) with dissociation constant equal to K_w/K_b [by Eq. (12.15)] where K_b is the dissociation constant of the base B. Then

$$\frac{K_w}{K_b} = \frac{C_{H^+}^2}{F_{BHA}}$$

or
$$K_b = \frac{K_w F_{BHA}}{C_{H^+}^2}$$

$$= \frac{(10^{-14})(0.1)}{(10^{-6.2})^2} = 2.5 \times 10^{-3}$$

Analogous procedure, treating A^- ion of the salt as a conjugate base of the acid HA, will not work if HA is a strong acid for which A^- ion does not virtually act as a base.

Note: The hydrolytic equilibrium of a salt of weak base (B) and strong acid (HA) is really the dissociation equilibrium of the acid BH^+, and that of a salt of strong base and weak acid is same as the dissociation equilibrium of the base A^-.

12.34 For a weak acid HA

$$K_a = \frac{C_{H^+}^2}{F_{HA}} \quad \text{vide previous problem}$$

Then
$$pH = -\tfrac{1}{2}[\log K_a + \log F_{HA}]$$

$$= -\tfrac{1}{2}[\log(10^{-5}) + \log(0..01] = 3.5$$

NaCl will definitely raise the pH by lowering γ_{H^+} caused by increase in ionic strength. But it is not easy to calculate. In our simple calculation of pH, the effect of ionic strength has been ignored. For improved value, we are to add to such calculated pH a quantity $-\log \gamma_{H^+}$ given by the Davies equation (vide note on question 12.15) with I due only to NaCl (being present in relatively much higher concentration).

12.35 Vide Section 12.6.

12.36 Vide problem 12.21(ii).

12.37 a. When 1 ml of NaOH is added, the resulting solution (51 ml) contains excess of HCl (that dissociates completely) for which

$$C_{H^+} = \frac{1}{51}(50 \times 0.1 - 1 \times 0.1) \text{ or } 0.096 \text{ M}$$

Then
$$pH = -\log(0.096) = 1.02$$

For 20 ml of added NaOH, pH can be similarly calculated.

For 50 ml of added NaOH, HCl and NaOH are both completely neutralised and the resulting solution contains NaCl which is a salt of strong acid and strong base. Then

$$pH = \tfrac{1}{2}pK_w$$

For 60 ml of added NaOH, the resulting solution contains excess of NaOH, vide problem 12.36(iii).

  b. Vide problem 12.36.

☐ For (a) pH near equivalence point lies around 7 in quite a broad range, hence most of the common indicators, having pT lying below and above 7, e.g. methyl red (pT = 5.5) and phenolphthalein (pT = 9), will be suitable.

For (b) pH near equivalence point lies above 7 and hence only indicators having pT above 7, e.g. phenolphthalein, will be suitable.

12.38   i. Corresponds to starting point of the titration.

iii. Corresponds to neutralisation of only HCl (first equivalence point).

  v. Corresponds to neutralisation of both HCl and CH_3CO_2H (second equivalence point.

These are all similar to problem 12.44.

Again

ii. Corresponds to partial neutralisation only of HCl (i.e. before first equivalence point).

In the resulting solution (60 ml),

amount of unneutralised HCl $\equiv (50 - 2 \times 10)$ or 30 ml of 0.1 M HCl

$$(\because 10 \text{ ml of } 0.2 \text{ M NaOH} \equiv 2 \times 10 \text{ ml } 0.1 \text{ M NaOH})$$

Then $\quad C_{H^+} = F_{HCl^-} = \dfrac{30 \times 0.1}{60}$ or 0.05 M

$$pH = -\log C_{H^+} = -\log 0.05 = 1.3$$

iv. Corresponds to complete neutralisation of HCl (consuming 25 ml of NaOH) and partial neutralisation of CH_3CO_2H (consuming 10 ml of NaOH), i.e. between the first and second equivalence point. In the resulting solution

amount of NaAc $\equiv$ 10 ml of 0.2 M NaOH

amount of HAc $\equiv (50 - 2 \times 10)$ or 30 ml of 0.1 M HAc

$$pH = pK_a + \log\frac{F_{NAAc}}{F_{HAc}}$$

$$= -\log(1.75 \times 10^{-5}) + \log\frac{10 \times 0.2}{90 \times 0.1} = 4.58$$

vi. Corresponds to the stage of titration after second equivalence point. In the resulting solution (110 ml)

amount of excess NaOH $\equiv (60 - 50)$ or 10 ml of 0.2 M NaOH

Then $$C_{OH^-} = F_{NaOH} = \frac{10 \times 0.2}{110} \text{ or } 0.0182 \text{ M}$$

$$pH = pK_w - pOH = 14 + \log 0.0182 = 12.3$$

☐ The titration curve is shown in Fig. 12.1.

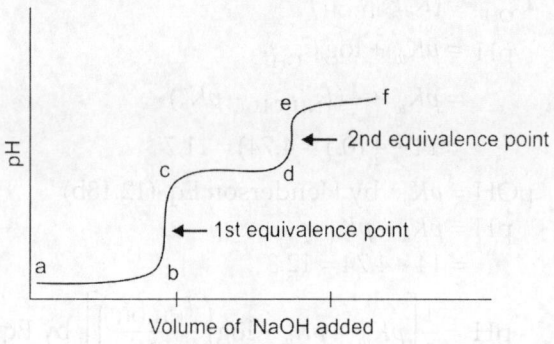

Fig. 12.1

abc corresponds to neutralisation of HCl (strong acid) while *cde* corresponds to neutralisation of CH_3CO_2H (weak acid). For neutralisation of strong acid pH continuously rises with slightly increasing rate (*ab*) up to equivalence point, while for weak acid it rises (along cd) first with decreasing rate up to half-equivalence point and then with increasing rate. Near equivalence point, pH rises rapidly though it involves much wider range with strong acid (*bc*) than with weak acid (*de*). All these arise from logarithmic nature of pH ($pH \approx -\log C_{H^+}$).

12.39   pH of $NaHCO_3$ $= \frac{1}{2}(pK_1 + pK_2)$

     solution

$$= \frac{1}{2}(6.35 + 10.33) = 8.34$$

Then the resulting solution, having lower pH (= 6.35), will contain excess of HCl.
Let the given solution contain x mol of Na_2CO_3 and y mol of $NaHCO_3$

Then $\underbrace{\dfrac{21}{1000} \times 0.1}_{\text{amount of HCl added}} = \underbrace{(2x+y)}_{\text{amount of HCl reacted}} + \underbrace{\dfrac{21+25}{1000} \times 10^{-6.35}}_{\text{amount of HCl unreacted}}$

$\simeq 2x + y$

Again initial pH = 10.63

$= pK_1 + \dfrac{x}{y}$ by Henderson equation

$= 6.35 + \dfrac{x}{y}$

Solving these two equations

$x = 0.00094$ mol

$y = 0.00022$ mol

12.40 i. For the dissociation equilibrium, $NH_4OH \rightleftharpoons NH_4^+ + OH^-$

$$K_b = \dfrac{C_{NH_4^+} \cdot C_{OH^-}}{C_{NH_4OH}}$$

$$= \dfrac{C_{NH_4^+} \cdot C_{OH^-}}{F_{NH_4OH} - C_{OH^-}}$$

$$\simeq \dfrac{C_{OH^-}^2}{F_{NH_4OH}}, \text{ ignoring } C_{OH^-} \text{ compared to } F_{NH_4OH}$$

or $\quad C_{OH^-} = (K_b F_{NH_4OH})^{\frac{1}{2}}$

Then $\quad pH = pK_w + \log C_{OH^-}$

$= pK_w + \tfrac{1}{2}(F_{NH_4OH} - pK_b)$

$= 14 + \tfrac{1}{2}(0.1 - 4.74) = 11.7$

ii. $\quad\quad pOH = pK_b$ by Henderson Eq. (12.18b)

Then $\quad pH = pK_w - pK_b$

$= 14 - 4.74 = 12.3$

iii. $\quad\quad pH = \dfrac{1}{2}\left[pK_w - pK_b - \log\left(\dfrac{F_{NH_4OH}}{2}\right)\right]$ by Eq. (12.21b)

$= \dfrac{1}{2}\left[14 - 4.74 - \log\left(\dfrac{0.1}{2}\right)\right] = 5.28$

12.41 a. Vide Section 12.7.
   b. A solution acts as a buffer through its reaction with both an acid and a base. The given solution cannot act as such because NaCl (and of course NaOH) cannot act as an acid.

12.42 Because NH_4Ac can act as an acid through NH_4^+ ion (being the conjugate acid of the weak base NH_3) as well as a base through Ac^- ion (being the conjugate base of the weak acid HAc).
   Note: Although NH_4Ac is a normal salt, it exhibits appreciable buffer action.

12.43 Vide Section 12.7.

12.44 Vide problem 12.38.

12.45 Vide problem 12.37(iii).

12.46 Maximum buffer capacity corresponds to pH at the half neutralisation of the acid NaH_2PO_4 concerned here. The volume x of NaOH solution required is then given by

$$x \times 0.2 = \frac{10}{2} \times 0.1$$

or

$$x = 2.5 \text{ ml}$$

☐
$$pH = pK_a \text{ of } NaH_2PO_4 \text{ (i.e. } pK_2)$$
$$= -\log (6.2 \times 10^{-8})$$
$$= 7.2$$

☐☐ For pH of a solution to be determined by colorimetric method, it must lie within the visible colour change interval, i.e. $pKin \pm 1$, of the indicator chosen. Then bromothymol blue ($pK_{in} = 7$) will be suitable indicators.

12.47 Vide problem 12.32.

12.48 Because a buffer of particular pH = pK_a of the acetic acid (HAc) will have maximum buffer capacity as it corresponds to half neutralisation of HAc.

12.49 In the given mixture

$$F_{NH_4OH} = \frac{50}{50 + 50} \times 1 \text{ or } 0.5 \text{ M}$$

$$C_{Mg^{2+}} = \frac{50}{50 + 50} \times \frac{1}{2} \text{ or } 0.25 \text{ M}$$

The precipitate of $Mg(OH)_2$ will just disappear when the solution has

$$C_{OH^-} = \left[\frac{K_{sp} Mg(OH)_2}{C_{Mg^{2+}}} \right]^{\frac{1}{2}}$$

$$= \frac{K_b C_{NH_4OH}}{C_{NH_4^+}} \quad \text{considering the equilibrium } NH_4OH \rightleftharpoons NH_4^+ + OH^-$$

$$= \frac{K_b F_{NH_4OH}}{F_{NH_4Cl}}$$

Then $F_{NH_4Cl} = K_b F_{NH_4OH} \cdot \left(\frac{C_{Mg^{2+}}}{K_{sp}} \right)^{\frac{1}{2}}$

$$= (1.8 \times 10^{-5}) (0.5) \sqrt{\frac{0.25}{1.22 \times 10^{-11}}} = 1.29$$

Therefore, the amount of NH_4Cl to be added

$$= 1.29 \times \frac{100}{1000} \text{ or } 0.129 \text{ mol}$$

13

Electrochemical Equilibrium

Treatment of ionic equilibria becomes complicated in presence of electric field, as in an electrochemical cell, because it needs modification of the concept of chemical potential. Study of such equilibria is particularly important in view of the fact that the most effective way to bring about a chemical reaction reversibly is to carry out it through an electrochemical cell.

13.1 ELECTROCHEMICAL POTENTIAL

A difference in electric potential always arises at the junction of two different electrically conducting phases. Consequently, an electrical work, in addition to chemical work, is involved when charged species are transferred from one phase to another. Here the escaping tendency of a charged species i in an electric field will be determined not by its chemical potential μ_i but by its electrochemical potential $\bar{\mu}_i$ defined below

$$\bar{\mu}_i = \mu_i + Z_i F \phi \qquad (13.1)$$

where Z_i is the charge number (including sign) of the species, ϕ is the electric potential in the region of its location and F is the Faraday constant. Here μ_i (that depends only on chemical environment) represents the chemical contribution to the escaping tendency, while $Z_i F \phi$ (that depends on electrical environment) is the electrical contribution to the escaping tendency. In all thermodynamic equations for chemical system μ_i is required to be replaced by $\bar{\mu}_i$ to obtain the corresponding equations for the electrochemical systems.

It should be kept in mind that although $\bar{\mu}_i$ is conveniently thought to consist of two parts, they cannot be separated. Because there is no way of separating electricity from matter. What we can measure is only $\bar{\mu}_i$.

13.2 ELECTROCHEMICAL CELL

An electrochemical cell is a device to convert free energy of a physical or chemical change into electrical (free) energy or reverse, and accordingly it is called galvanic or electrolytic cell.

An electrolytic cell functions through simultaneous oxidation (i.e. loss of electron) and reduction (i.e. gain of electron) occurring in its two different parts, called electrodes or half-cells. An electrode is called cathode or anode according as it serves as a source or sink for electron.

The simplest galvanic cell comprises two metal electrodes immersed in an electrolyte solution. In some cases, e.g. in Daniel cell, two electrolyte solutions are involved which are in contact with one another. Such liquid-junction contributes to the emf of the cell.

This can be minimised (but not completely eliminated) by connecting the two electrolyte solutions through a salt bridge which is a bent glass tube filled with a concentrated solution of KCl or NH_4NO_3 in agar gel (that allows migration of ions under applied electric field, but prevents mixing of the electrolytes).

A cell is diagramatically represented by writing in order each successive phase that makes up the electric circuit of the cell, using a single vertical line for a phase boundary, dashed vertical line for a liquid-junction and a double vertical lines for a salt bridge. Different species present in the same phase are represented by their usual formula with a comma between consecutive species. For example, a Daniel cell, which consists of Zn-rod immersed in $ZnSO_4$ solution and Cu-rod in $CuSO_4$ solution, is represented as

$Zn|ZnSO_4(aq)\vdots CuSO_4(aq)|Cu,$ cell with liquid junction potential

$Zn|ZnSO_4(aq)\|CuSO_4(aq)|Cu,$ cell without liquid junction potential

The terminals other than the electrodes are usually omitted from the cell diagram.

A cell reaction is conveniently expressed in terms of two hypothetical half-cell processes or hall-reactions which, according to the IUPAC conventions, involve oxidation at the left electrode (treated as anode) and reduction at the right electrode (treated as cathode) in the cell diagram used.

reaction at the left electrode   : $Zn \rightarrow Zn^{2+}(aq) + 2e$

reaction at the right electrode : $Cu^{2+}(aq) + 2e \rightarrow Cu$

Net cell reaction   : $Zn + Cu^{2+}(aq) \rightarrow Zn^{2+}(aq) + Cu$

The emf E of the cell is defined as

$$E = \phi_R - \phi_L \tag{13.2}$$

where ϕ_R and ϕ_L are the respective open-circuit electric potentials of the terminals on the right side and left side of the cell diagram. A +ve emf for a cell diagram implies that the corresponding cell reaction occurs spontaneously in the forward direction when the cell produces electric current.

A cell (or an electrode) is said to be reversible thermodynamically, if it operates in such a way that the process occurring in the cell (or on electrode surface) can be reversed by reversing the direction of current in the cell through only an infinitesimal change in the externally applied potential difference. A chemically irreversible cell will of course be thermodynamically irreversible. But even when a cell is chemically reversible (e.g. Daniel cell), it can hardly become thermodynamically reversible. One source of irreversibility is the liquid junction. Strictly speaking, no cell can as a whole remain in thermodynamic equilibrium, even in an open-circuit condition.

13.3 DERIVATION OF CELL EMF FROM CELL REACTION

Let n faraday of electricity passes through a galvanic cell of emf E during its operation. If the cell reaction is

$$v_A A + v_B B = v_L L + v_M M$$

the accompanying ΔG is given by

$$\Delta G = \Delta G^\circ + RT \ln \frac{a_L^{v_L} a_M^{v_M}}{a_A^{v_A} a_B^{v_B}}, \text{ by reaction isotherm}$$

$$= - \text{ work other than } PV\text{-work obtained from the cell}$$

$$= -nFE, \text{ if the cell operates thermodynamically reversibly} \quad (13.3)$$

Then
$$E = E^\circ - \frac{RT}{nF} \ln \frac{a_L^{v_L} a_M^{v_M}}{a_A^{v_A} a_B^{v_B}} \Bigg\}$$
$$= E^\circ - \frac{0.059}{n} \ln \frac{a_L^{v_L} a_M^{v_M}}{a_A^{v_A} a_B^{v_B}} \Bigg\}, \text{ at } 25^\circ C \quad (13.4)$$

where $E^\circ = \Delta G^\circ / nF$ is the standard emf, which is emf of the cell when the chemical species involved in the cell reaction are all at standard state of unit activity at a pressure of 1 bar. This is Nernst equation of cell emf.

Emf of a reversible cell can be conveniently expressed as the difference of two analogous quantities, called electrode potentials, one corresponding to each electrode. Electrode potential is the electric potential difference that arises at the electrode surface due to charge separation. Being unable to measure it, we express it relative to that of a standard hydrogen electrode (the reference electrode) taken as zero at $25^\circ C$, and by IUPAC convention, equate it to the emf of the cell that has standard hydrogen electrode on the left of the cell diagram and the concerned electrode on the right. Electrode potential can therefore, be expressed similar to the emf of a cell as given by Nernst equation.

Instead of deducing the Nernst equation using reaction isotherm, we can deduce it in a more fundamental way using the following equation for an electrochemical system

$$\Delta G = \sum v_i \bar{\mu}_i = 0 \text{ at equilibrium at constant } T \text{ and } P$$

where v_i's are the stoichiometric coefficients in the cell reaction.

In case of half cell process, M^{n+} (soln) + ne (metal) = M (metal), we have at equilibrium

$$\bar{\mu}_M - \bar{\mu}_{M^{n+}} - n\bar{\mu}_e = 0$$

or $\quad \mu_M - \mu_{Mn^{n-}} - nF\phi_{soln} - (n\mu_e - nF\phi_{metal}) = 0$

where ϕ_{soln} is the electric (galvanic) potential in solution and ϕ_{metal} is that in metal

or $\quad -\mu_{Mn^{n-}} = -nF(\phi_{metal} - \phi_{soln})$

[by convention $\mu_{element} = 0$ at 1 bar at $25^\circ C$, $\mu_e = 0$ in every metal, because electronic potential is measured with wires of same composition]

$$= -nFE, \text{ where } E = \phi_{metal} - \phi_{soln} \text{ is the electrode potential}$$

or $\quad E = E^\circ + \frac{RT}{nF} \ln a_{M^{n+}}$

where E° is the standard electrode pot (i.e. when $a_{M^{n+}} = 1$)

Note:

1. For any ion in aq soln
$$\mu^\circ = \pm nFE^\circ \quad (13.5)$$
[where E° is the standard electrode potential for the equilibrium between the ion and the corresponding element, + sign applies to cations and – sign to anions]
But $\mu_e^\circ = 0$ in every metal

2. For any ion in pure element
$$\mu^\circ = 0,$$
E° being zero in the same phase

13.4 TYPES OF ELECTRODES

A. Metal–Cation Electrode

It consists of a bar or strip of a metal M dipping in a solution containing its ion M^{n+}.

Symbol $M^{n+}|M$, e.g. $Zn^{2+}(aq)|Zn(s)$, as cathode

Electrode process $Zn^{2+}(aq) + 2e = Zn(s)$

Electrode potential $E = E° + \dfrac{RT}{2F} \ln a_{Zn^{2+}}$

Active metal like Ca can be used as electrode not in pure form but in the form of amalgam, and such electrode is represented as $Ca^{2+}(aq)/Ca(Hg)$.

Hydrogen electrode belongs to this category because it forms cation in aq medium. Here

Symbol $H^+(aq)|H_2(g)|Pt$ as cathode

Electrode process $2H^+(aq) + 2e = H_2(g)$

Electrode potential $E = E° - \dfrac{RT}{2F} \ln \dfrac{f_{H_2}/P°}{a_{H^+}^2}$ (13.6)

In elementary work, the fugacity f is replaced by the pressure P.

B. Nonmetal–Anion Electrode

It consists of an inert metal dipping in a solution saturated with a non-metal and containing ion of it.

Symbol $Cl^-(aq)|Cl_2(g)|Pt$, as an example

Electrode process $Cl_2(g) + 2e = 2Cl^-(aq)$

Electrode potential $E = E° - \dfrac{RT}{2E} \ln \dfrac{a_{Cl^-}^2}{f_{Cl_2}/P°}$

Note: Electrode symbol indicates mere order of arrangement of the involved substances in a cell and not necessarily their order in the electrode process. The electrode process is oxidation or reduction according as the inert metal is on the left or right of the electrode symbol.

C. Metal-Insoluble Salt-Anion Electrode

It consists of a metal in contact with (or layered with) its suitable sparingly soluble salt in an electrolytic solution containing a common anion.

Symbol $Cl^-(aq)|AgCl(s)|Ag(s)$, as an example

Electrode process $AgCl(s) + e = Ag(s) + Cl^-(aq)$

Electrode potential $E = E° - \dfrac{RT}{F} \ln a_{Cl^-}$ $E° = 0.222$ V at 25°C

Another example of this type of electrode is calomel electrode that involves Hg_2Cl_2 (known as calomel)

Symbol $Cl^-(aq)|Hg_2Cl_2(s)|Hg(l)$

Electrode process $Hg_2Cl_2(s) + 2e = 2Hg(l) + 2Cl^-(aq)$

Electrode potential $E = E° - \dfrac{RT}{F} \ln a_{Cl^-}$ $E° = 0.268$ V at 25°C

Being easy to prepare and handle, calomel and silver-silver chloride electrodes are often used as auxiliary reference electrodes, instead of standard hydrogen electrode which is the basic reference electrode (whose potential has been conventionally taken to be zero).

Metal-insoluble hydroxide (or oxide) electrodes belong to this category

Symbol $\qquad$ $OH^-(aq)|Ag_2O(s)|Ag(s)$, for example

Electrode process $\qquad$ $Ag_2O(s) + H_2O(l) + 2e = 2Ag(s) + 2OH^-(aq)$

Electrode potential $\quad E = E° - \dfrac{RT}{F} \ln a_{OH^-}$

D. Oxidation-Reduction Electrode

It consists of an inert metal dipping in a solution containing ions of an element in different valence state.

Symbol $\qquad$ $Fe^{3+}(aq), Fe^{2+}(aq)/Pt$, as an example

Electrode process $\qquad$ $Fe^{3+}(aq) + e = Fe^{2+}(aq)$

Electrode potential $\quad E = E° - \dfrac{RT}{F} \ln \dfrac{a_{Fe^{2+}}}{a_{Fe^{3+}}}$

The name of this type of electrode (historically given) cannot be justified when we think that oxidation or reduction is common to any other type of electrodes, though it is retained in absence of any better name.

E. Glass Electrode

It is a kind of membrane electrode where a membrane allows passing essentially only some selective positive ion like Na^+. Usually, it consists of a thin-walled bulb of special quality of glass containing a solution of constant pH (usually 0.1 N HCl solution) with an immersed reference electrode (usually $Ag/AgCl/Cl^-$ electrode) and surrounded by the experimental solution during experiment. A typical glass electrode may then be represented as

$$Ag\,|AgCl(s)|\,HCl\,(aq)\,|Glass|\,\text{Experimental solution}$$

The potential difference across the glass membrane, called glass electrode potential E_G, results from the exchange of alkali metal ions (mainly Na^+) of glass membrane with H^+ ions of the solution

$$Na^+ (glass) + H^+ (soln) \rightleftharpoons Na^+ (soln) + H^+ (glass)$$

Assuming that only the surface ions of the membrane are involved in this equilibrium, it can be deduced that

$$E_G = E_G° + \frac{RT}{F} \ln a_{H^+ (soln)}, \text{ provided } a_{H^+ (soln)} \gg a_{Na^+ (soln)}$$

$$= E_G° - 0.059\,\text{pH} \quad \text{at 25°C (similar to H-electrode)} \quad (13.7)$$

where $E_G°$ is a constant for the electrode at constant temperature.

From the measurement of emf (E) of the cell obtained on coupling the glass electrode with a reference electrode, like calomel electrode, the pH of different solutions can be compared using the relation

$$E = E_{Cal} - E_G$$

However, once this cell is calibrated by means of solutions of known pH, it can be conveniently used for measurement of pH of any unknown solution. Such as arrangement is known as pH meter. Due to high resistance of the glass membrane ($10^7 - 10^9 \Omega$), emf of such a cell cannot be measured using ordinary potentiometer circuit, an electronic voltmeter (that draws only a negligible current from the cell) can, however be used instead of potentiometer.

13.5 CONCENTRATION CELL

In an electrochemical cell oxidation occurs at one electrode and reduction at the other. But the net process occurring in the cell may be chemical (when it is called chemical cell) or physical (when it is called concentration cell). In a concentration cell electrical energy is produced due to a net process that amounts to the transfer of matter from higher to lower concentration either in the electrodes or in the electrolytic solutions.

Let us consider the electrolyte concentration cell

$$Pt\,|H_2(P)|\,HCl\,(m_1): HCl\,(m_2)\,|H_2(P)|\,Pt$$

Such a cell with liquid junction is called *cell with transfer* due to possible transfer of ions across the junction. When one faraday of electricity passes through the cell from left to right (i.e. when the cell produces electrical energy), the following processes will occur in the cell

Anode process $\qquad \frac{1}{2}H_2(P) = H^+(m_1) + e$

Cathode process $\qquad H^+(m_2) + e = \frac{1}{2}H_2(P)$

Transfer process $\qquad t_{H^+}H^+(m_1) = t_{H^+}H^+(m_2)$

$\qquad\qquad\qquad\quad t_{Cl^-}Cl^-(m_2) = t_{Cl^-}Cl^-(m_1)$

Net process $\qquad H^+(m_2) + t_{H^+}H^+(m_1) + t_{Cl^-}Cl^-(m_2)$

$\qquad\qquad = H^+(m_1) + t_{H^+}H^+(m_2) + t_{Cl^-}Cl^-(m_1)$

or $t_{Cl^-}H^+(m_2) + t_{Cl^-}Cl^-(m_2) = t_{Cl^-}H^+(m_1) + t_{Cl^-}Cl^-(m_1)$, putting $t_{H^+} = 1 - t_{Cl^-}$

[Less appropriately, $t_{Cl^-}HCl(m_2) = t_{Cl^-}HCl(m_1)$]

For this net process

$$\Delta G = t_{Cl^-}\,RT\ln\frac{(a_{H^+}\cdot a_{Cl^-})_1}{(a_{H^+}\cdot a_{Cl^-})_2}$$

$$= -FE \quad \text{where } E \text{ is emf of the cell}$$

Then $\qquad\qquad E = t_{Cl^-}\dfrac{RT}{F}\ln\dfrac{(a_{H^+}\cdot a_{Cl^-})_2}{(a_{H^+}\cdot a_{Cl^-})_1}$ $\qquad\qquad$ (13.8a)

$$= 2t_{Cl^-}\frac{RT}{F}\ln\frac{(a_\pm)_2}{(a_\pm)_1} \qquad\qquad (13.8b)$$

$$= 2t_{Cl^-}\frac{RT}{F}\ln\frac{(m_\pm\gamma_\pm)_2}{(m_\pm\gamma_\pm)_1} \qquad\qquad (13.8c)$$

E will be +ve if $(a_\pm)_2 > (a_\pm)_1$, or less appropriately if $m_2 > m_1$, and only then the cell reaction as written above will occur spontaneously.

For the cell without transfer, the emf results only from the process occurring at the electrodes, i.e.

$$H^+(m_2) = H^+(m_1)$$

The emf E' associated with this process is given by

$$E' = \frac{RT}{F} \ln \frac{(a_{H^+})_2}{(a_{H^+})_1}$$

$$= \frac{RT}{F} \ln \frac{(a_\pm)_2}{(a_\pm)_1} \tag{13.9}$$

Then emf of the cell with transfer,

$$E = 2t_{Cl^-} E'$$

Also

$$E = E' + E_j \tag{13.10}$$

Then from the measurement of emf of a cell with transfer and without transfer, it is possible to find transport number t of the ions of the electrolyte and liquid junction potential E_j.

13.6 APPLICATION OF EMF MEASUREMENT

Electrochemical techniques provide methods for accurate determination of a wide variety of quantities.

Thermodynamic quantities $\Delta G°$, $\Delta S°$ and $\Delta H°$ of a chemical reaction occurring in a cell can be calculated from the measured standard emf $E°$ of the cell and its temperature variation using the following relations

$$\Delta G° = - nFE° \tag{13.11}$$

$$\Delta S° = -\left(\frac{\partial \Delta G°}{\partial T}\right)_P$$

$$= nF\left(\frac{\partial E°}{\partial T}\right)_P \tag{13.12}$$

$$\Delta H° = \Delta G° + T\Delta S°$$

$$= - nFE° + nFT\left(\frac{\partial E°}{\partial T}\right)_P \tag{13.13}$$

where n is the number of faraday of electricity passing through the cell due to the reaction represented by the stoichiometric equation used. Further, from $\Delta G°$ we can find the equilibrium constant (K), from the relation $\Delta G° = -RT \ln K$, and other related quantities.

Other quantities which can be best determined from emf measurement are ionic activity and activity coefficient, pH and transport number. All these are kept for illustration through solution of relevant problems.

A sophisticated application of emf measurement in analytical chemistry is the potentiometric titration of one electrolyte by another, i.e. by measuring at every step, the emf of a galvanic cell of which the solution being titrated is a part. The equivalence point corresponds to the maximum slope of the curve obtained by plotting the measured emf versus volume of the titrant added.

AUXILIARY PROBLEMS

13.1 Justify or criticise the following:
   i. Single electrode potential represents the electric potential at a particular part of the electrode.

ii. Chemical potential of a charged species must be same at all the phases of a system at equilibrium.

iii. Electrode potentials are additive properties.

iv. Emf of a cell is the algebraic sum of its electrode potentials.

v. The expression $E = -\Delta G/nF$ for the emf of a cell indicates E to be an extensive quantity, considering ΔG to be extensive.

vi. In an electrochemical cell, oxidation occurs at the anode and reduction at the cathode. Hence only a red-ox reaction can be brought about in a cell.

vii. μ_e is taken to be zero in every metal.

viii. For a galvanic cell, the electrical energy produced should be equal to the chemical energy consumed.

ix. $\Delta H \ne Q_P$ in an electrochemical cell.

x. Since $E = E° - \dfrac{RT}{nF} \ln \dfrac{[Red]}{[Ox]}$, E should vary linearly with T.

13.2 Give reasons for the following:

i. A difference in electric potential always arises at the junction of two different phases.

ii. It is not possible to measure a purely electric potential difference between a metal and an aqueous solution of its salt.

iii. Phases involved in a galvanic cell must all be electrical conductors of which at least one must be an electrolytic conductor.

iv. Electricity flows from high to low potential outside a galvanic cell but from low to high potential inside the cell.

v. In the reaction $Zn + CuSO_4 = Cu + ZnSO_4$, energy is liberated mainly in form of heat when occurs directly but in electrical form when brought about in Daniel cell.

vi. All reversible cells are not suitable as strong cell.

vii. KCl and not NaCl is used in salt bridge.

viii. Nernst equation is not applicable to any cell.

ix. A potentiometer and not an ordinary voltmeter should be used for measuring emf of a galvanic cell.

x. No electrochemical cell can as a whole remain in thermodynamic equilibrium even in an open-circuit condition.

xi. Half-cell potential for reduction of periodates is greater than that of permanganates. But permanganates oxidise aldehydes rapidly though periodates do not.

xii. The standard potential for the half-reaction $Fe(CN)_6^{3-} + e = Fe(CN)_6^{4-}$ is lower than that for $\tfrac{1}{2}I(s) + e = I^-$, yet liberation of I_2 from iodides is possible with $Fe(CN)_6^{3-}$ in acid medium and also in presence of Zn^{2+} ions.

13.3 Give example (if any) of a cell for which (i) $E° = 0$ and $E = 0$ (ii) $E° = 0$ but $E \ne 0$ (iii) $E° \ne 0$ but $E = 0$.

13.4 Write net reaction for the following cell
$$Zn \mid H_2SO_4(aq) \mid Cu$$
Emf of this cell does not conform to the Nernst equation. Explain. Suggest a cell which has net reaction same as this cell but conforming to the Nernst equation.

13.5 According to IUPAC convention the cell reaction for the Daniel cell represented as $Zn|ZnSO_4(aq)||CuSO_4(aq)|Cu$ is $Zn + Cu^{2+}(aq) \rightarrow Zn^{2+}(aq) + Cu$ but when represented as $Cu|CuSO_4(aq)||ZnSO_4(aq)|Zn$ the cell reaction is reversed. Which is the actual cell reaction?

13.6 Write the cell reaction and expression of emf for the Weston cell represented below.

$$Pt|Cd(Hg)|\text{satd. } CdSO_4 \cdot \tfrac{8}{3}H_2O|Hg_2SO_4(s)|Hg|Pt$$

This is widely used as standard cell (i.e. one whose emf is stable and accurately known). Why?

13.7 Will the emf be same for the following pairs of cells? If not, which one will have higher emf?

a. $Zn|ZnCl_2(m_1)||ZnCl_2(m_2)|Zn$ and $Zn|ZnSO_4(m_1)|ZnSO_4(m_2)|Zn$

b. $Zn|ZnCl_2(m_1):ZnCl_2(m_2)|Zn$ and $Zn|ZnSO_4(m_1):ZnSO_4(m_2)|Zn$

13.8 Find the expressions for cell emf and liquid junction potential for the cell in the previous problem 13.7b.

13.9 Set up galvanic cells in which the following processes occur.

i. $K_2Cr_2O_7 + 6KI + 14HCl = 2CrCl_3 + 3I_2 + 8KCl + 7H_2O$

ii. $\qquad Ag(s) + \tfrac{1}{2}Cl_2(g) = AgCl(s)$

iii. $\qquad H_2(g) + \tfrac{1}{2}O_2(g) = H_2O(l)$

iv. $\quad HCl(aq) + NaOH(aq) = NaCl(aq) + H_2O(l)$

v. $\qquad H^+(aq) + OH^-(aq) = H_2O(l)$

vi. $\qquad 2Na(s) + 2H_2O(l) = 2NaOH(aq) + H_2(g)$

vii. $\qquad HgO(s) + H_2(g) = Hg(l) + H_2O(l)$

viii. $\qquad HCl(0.1\ m) \rightarrow HCl(0.05\ m)$

13.10 Calculate emf of a cell at 25°C which comprises a saturated calomel electrode coupled through a salt bridge, with a platinum black electrode dipping into solution A and H_2 gas (1 atm pressure) is bubbled over the platinum surface. The solution A is prepared by mixing 10 mL (N/10) NaOH with 90 mL (N/100) HCl solution ($E_{\text{saturated calomel}} = 0.242$ V at 25°C).

13.11 Two weak acid solutions HA_1 and HA_2 each with the same concentration and having pK values 3 and 5 respectively, are placed in contact with two hydrogen electrodes (1 atm pressure at 25°C) and are interconnected through a salt bridge. Find the emf of the resulting cell. What assumption you have to make in order to solve this problem?

13.12 The cell $Ag(s)|AgCl(s)/HCl(0.1\ N)|Glass|buffer|$ saturated calomel electrode gives an emf 0.1120 V at 25°C when the pH of the buffer is 4. The emf is 0.3865 V with a buffer of unknown pH. Calculate the pH of the unknown buffer.

13.13 What are the advantages and disadvantages of the operational definition of pH on the basis of emf measurement?

13.14 Regarding glass electrode, answer the following:

a. What is the function of the inner reference electrode.

b. What is the origin of the electrode potential? What makes you sure that it is not due to the difference in H^+ ion concentration on the two sides of the glass membrane as in a concentration cell?

c. Glass electrode gives accurate pH only in the range 1–9.

d. An ordinary potentiometer circuit is not suitable for the measurement of emf of a cell involving glass electrode.

13.15 Is hydrogen electrode reversible with respect to OH^- ion? Find the standard potential of the electrode $Pt|H_2(g)|OH^-(aq)$. Do you expect H_2 to act as a better reducing agent in acid or alkaline medium?

13.16 Calculate standard potentials for the following half-reactions at 25°C

i. $\frac{1}{2}O_2(g) + H_2O(l) + 2e = 2OH^-(aq)$

ii. $\frac{1}{2}O_2(g) + 2H^+(aq) + 2e = H_2O(l)$

Given $H_2(g) + \frac{1}{2}O_2(g) + H_2O(l) = 2H^+(aq) + 2OH^-(aq)$, $\Delta G^\circ_{25°C} = -77.7\ kJ/mol$. Which species is expected to have more corrosive action on metals, $-H^+$ or O_2? Will such corrosions be favoured by an acid or alkali?

13.17 Calculate the stability constant for the complex ion formed in the reaction

$$Zn^{2+}_{(aq)} + 4NH_3 = Zn(NH_3)_4^{2+}(aq)$$

Given $Zn + 4NH_3 = Zn(NH_3)_4^{2+}(aq) + 2e$, $E^\circ = 1.03\ V$

$Zn = Zn^{2+}(aq) + 2e$, $E^\circ = 0.763\ V$

13.18 The emf of the following cell is 0.0062 V at 25°C.

$$H_2(P)|HCl(a_\pm = 0.01) : HCl(a_\pm = 0.02)|H_2(P)$$

Calculate the transport number of H^+ ion, and the value of the liquid junction potential. Which side of the liquid junction is positive?

13.19 For the following cell:

$$Cl_2(P_1 = 0.01\ atm)|NaCl(m_1 = 0.001) : NaCl(m_2 = 0.01|Cl_2(P_2 = 1\ atm)$$

Calculate emf and liquid junction potential

i. if $t_{Na^+} = 0.4$.

ii. if the two solutions are separated by a membrane which is permeable only to Cl^- ions.

iii. if the membrane is permeable only to Na^+ ions.

13.20 Mention the ion(s) wrt which the electrodes $I^-(aq)|AgI(s)|Ag$ and $Ag^+(aq)|Ag$ are reversible. Set up a relation between E°'s of the two electrodes.

13.21 Find (a) net reaction (b) emf, for each of the following cells

i. $Pb|PbCl_2(s)|PbCl_2(aq)|Pb$

ii. $Pb|PbCl_2(s)|KCl(0.1\ m)|Pb$

iii. $Pb|PbCl_2(s)|Pb(NO_3)_2(0.1\ m)|Pb$

iv. $Pb|PbCl_2(satd\ soln)||Pb(NO_3)_2(0.1\ m)|Pb$

Also find E° of the cell (iv) $[K_{sp}(PbCl_2) = 10^{-4}]$.

13.22 Derive an expression for the emf of the following cell:

$$Pt|H_2(P)|HCl(m_1)|AgCl(s)|Ag|AgCl(s)|HCl(m_2)|H_2(P)|pt$$

Explain why this is called a cell without transfer. What other information would you need to obtain the transport number of H^+ and Cl^- ions from emf measurements?

13.23 For the cell $Pt|H_2(P)|HCl(m_1) : HCl(m_2)|H_2(P)|Pt$, $m_2 > m_1$ which side of the liquid junction is positive? How will you determine exact emf of the corresponding cell without transfer.

13.24 For certain cell, the measured values of E at different temperatures are as follows:

T/K	293.15	298.15	303.15
E/V	1.0477	1.0486	1.0494

Find ΔH for the cell reaction per faraday of electricity when the cell produces electrical energy at 298.15. If the measured E is accurate to $\pm$ 0.001 V and T to ± 0.01 K, how precise will be the calculated ΔH?

13.25 Consider the following Daniel cell $Zn|ZnSO_4$ (1 m)$||CuSO_4$ (1 m)$|$ Cu where each solution contains 1 kg of water. Calculate the quantity of electricity delivered by this cell until the potential drops to 1.0 V. How long the original cell could deliver electricity at the rate of 0.3A? Given $E^{\circ}_{Zn^{2+}/Zn} = -0.76$ V and $E^{\circ}_{Cu^{2+}/Cu} = 0.34$ V.

13.26 Assuming the temperature coefficient of emf to be zero for the Daniel cell in the previous problem calculate ΔG°, ΔS°, ΔH° and Q° for the reaction that occurs in the cell under the following conditions:

   i. when the cell operates reversibly producing 1 F of electricity.

   ii. when equivalent amount of chemical reaction occurs by direct reaction between Zn and $CuSO_4$.

   iii. when the cell is charged reversibly with 1 F of electricity.

   iv. when the charged cell is short-circuited.

13.27 Consider the following simplest but most useful fuel cell (where cell reaction amounts to combustion of a fuel)

$$H_2 \text{ (g)}|\text{electrolyte}| O_2 \text{ (g)}$$

where the electrolyte may be an acid or alkali. For acidic electrolyte, write half-cell reactions and net cell reaction, and also find E° at 25°C. Would E° be same for both acidic and alkaline electrolyse? Given $\mu^{\circ}_{H_2O(l)} = -237.13$ kJ mol^{-1}.

Fuel cells have high thermodynamic efficiency (theoretically 100%) in the conversion of Gibbs free energy into mechanical energy. But the efficiency of an internal combustion engine is much lower. Explain.

13.28 Find E° of the fuel cell that uses gaseous ethane as fuel at 25°C indicating the half-cell reaction involved. Given

$$\mu^{\circ}_{C_2H_6(g)} = -32.82 \text{ kJ mol}^{-1}, \mu^{\circ}_{CO_2(g)} = -394.36 \text{ kJ mol}^{-1}, \mu^{\circ}_{H_2O(l)} = -237.13 \text{ kJ mol}^{-1}.$$

What is the main problem with such fuel cells (using usual kind of fuel)? Can coal be used in fuel cells?

13.29 Devise a cell whose net reaction is

$$AgCl(s) \rightarrow Ag^+(aq) + Cl^-(aq)$$

Find i. E° of the cell

   ii. K for the cell reaction

   iii. Solubility of AgCl

   iv. Standard Gibbs free energy of formation of AgCl(aq)

Given $\mu^{\circ}_{AgCl(s)} = -109.79$ kJ mol^{-1}

   $\mu^{\circ}_{Ag^+(aq)} = +77.11$ kJ mol^{-1}

   $\mu^{\circ}_{Cl^-(aq)} = -131.23$ kJ mol^{-1}

13.30 Write net reaction and an expression of emf for the following cell

$$H_2 \text{ (1 atm)}|HCl \text{ (m)}| AgCl(s)| Ag$$

Emf measurements for the cell at 25°C were made using two HCl solutions of different molalities (m) with the following results

Molality (m)	0.0032	0.1238
Emf	0.5205	0.3420

Assuming infinite dilution for the first solution, find $E°$ of the cell and the mean ionic activity coefficient of HCl in the second solution. Also calculate $\Delta G°$ for the cell reaction.

13.31 For the following cell $E° = -0.627$ V at 25°C

$$Ag\,|Ag_2SO_4(s)|H_2SO_4\,(0.1\,m)\,|H_2(1\,atm)|Pt$$

a. write cell reaction.

b. calculate E at 25°C (i) neglecting activity coefficient (effect) and (ii) taking $\gamma_\pm = 0.7$ for H_2SO_4. If the cell reaction you write spontaneous?

c. solubility product of Ag_2SO_4

Given $E_{Ag^+/Ag}^° = 0.799$ V.

13.32 Design a cell suitable for the determination of ionic product of water from the measurement of its emf that does not depend on the concentration of the electrolyte involved. Write down the relevant cell reaction and the expression for the cell emf.

13.33 Find the ionic product of water from the $E°$ values of the following cells at 25°C.

$$Pt|H_2\,(g)|\,NaOH\,(aq),\,NaCl\,(aq)|AgCl(s)|Ag \qquad E° = 1.0486\,V$$
$$Pt|H_2\,(g)|\,HCl\,(aq),\,|AgCl(s)|Ag \qquad\qquad E° = 0.2223\,V$$

13.34 At 25°C the solubility of I_2 is 1.33×10^{-3} mol dm^{-3} in water. Find the same in 0.5 M KI. Given $E_{I_2/I^-}^° = 0.535$ V and $E_{I_3^-,I^-}^° = 0.536$ V

13.35 Given $E°$ values for the following half-reaction

$$Cu^{2+}\,(aq) + 2e = Cu, \quad E° = 0.34\,V$$
$$Cu^{2+}\,(aq) + e = Cu^+\,(aq), \quad E° = 0.16\,V$$
$$\tfrac{1}{2}I_2(s) + e = I^-(aq), \quad E° = 0.53\,V$$

Find

a. i. $E°$ for the half-reaction $Cu^+\,(aq) + e = Cu$

ii. $\mu°$ of Cu^+ and Cu^{2+} in aqueous solution and in pure Cu metal

iii. $E°$ for the half-reaction $Cu^{2+}\,(aq) + I^-\,(aq) + e = CuI(s)$, $K_{sp}(CuI) = 4 \times 10^{-12}$

iv. Equilibrium constant for the reaction $Cu + Cu^{2+}\,(aq) = 2Cu^+\,(aq)$

b. i. Would Cu^+ ion disproportionate in aqueous medium?

ii. Can you set up a cell where this reaction can occur?

iii. Can Cu^+ ion liberate I_2 from KI solution?

13.36 Given the following $E°$ values

$MnO_4^-\,(aq) + 8H^+\,(aq) + 5e = Mn^{2+}\,(aq) + 8H_2O\,(l)$	$E° = 1.51$ V
$MnO_2(s) + 4H^+(aq) + 2e = Mn^{2+}\,(aq) + 2H_2O\,(l)$	$E° = 1.23$ V
$Tl^{3+}\,(aq) + 2e = Tl^+(aq)$	$E° = 1.25$ V

Find $E_{MnO_4^-/MnO_2}^°$ in acid medium. Can Tl^+ ion reduce MnO_4^- to Mn^{2+}?

13.37 a. In the potentiometric titration of $FeSO_4$ with $Ce(SO_4)_2$ do the measured potentials at different stages of the titration correspond to the iron couple or cerium couple? What is the value of the potential at the equivalence point?

b. If $FeSO_4$ and $Ce(SO_4)_2$ are of same concentration, find the percent of Fe^{2+} remaining unreacted at the following steps of the titration.

i. before 1% of the equivalence point

      ii. at the equivalence point

      iii. after 1% of the equivalence point

   c. Draw the titration curve.

     Given $E^{\circ}_{Fe^{3+},Fe^{2+}} = 0.77$ V, $E^{\circ}_{Ce^{4+},Ce^{3+}} = 1.61$ V.

13.38 For the storage cell $Pb|PbSO_4(s)|H_2SO_4(a_{\pm})|PbSO_4(s)|PbO_2(s)|Pb$, $E^{\circ} = 2.04$ V and $E = 2.016$ V at 25°C. Write the cell reaction and find $a_{\pm}$ of H_2SO_4.

13.39 At 25°C, $E^{\circ} = 1.27$ V for the following alkali storage cell

$$Fe|FeO(s)|KOH(aq)|Ni_2O_3(s)|NiO(s)|Ni$$

   a. Write cell reaction.

   b. Here all the species involved in the cell reaction are solids. Then why not $K = 1$ and $E^{\circ} = 0$?

   c. Find the emf of the cell.

   d. If a small amount of H_2SO_4 is added to the cell, what effect it will have on the emf of the cell?

   e. What are the advantages and disadvantages of such alkali storage cell over the acid storage cell in the previous problem?

ANSWERS

13.1   (i) The given statement is not justified. The so-called single electrode potential actually represents a potential difference resulting from the formation of an electrical double layer between the phases involved in the electrode process.

     (ii) For a charged species, it is the electrochemical potential and not the chemical potential, which is same at all the phases of a system at equilibrium. This is a consequence of the existence of the interphase potential.

     [However, chemical potential of a charged species is same everywhere within the same phase since electric potential is insignificant everywhere within a phase (due to insignificant deviation from electroneutrality].

     (iii) The given statement follows from the very definition of electrode poptential.

     (iv) The given statement will be justified provided there is no other potential difference in the cell except at the electrodes.

     (v) The given statement is not justified, because here ΔG, which signifies free energy change per mole of specified cell reaction, is not extensive but intensive. Vide problem 11.21.

     (vi) In an electrochemical cell although oxidation occurs at the anode and reduction at the cathode, the net cell reaction may not be of oxidation–reduction type. Even the net cell process may be physical instead of chemical as in case of a concentration cell. Therefore the given statement is not justified.

     (vii) The stated convention is justified in dealing with emf of an electrochemical cell where μ_e is cancelled out, the cell terminals being made up of a material of same composition (usually Cu).

     (viii) For a galvanic cell functioning reversibly, the electrical energy produced and chemical energy consumed are respectively equal to $-\Delta G$ and $-\Delta H$ of the cell reaction at constant temperature and pressure. Since $\Delta G \neq \Delta H$ unless $\Delta S = 0$ (by $\Delta G = \Delta H - T\Delta S$), the given statement is not justified.

(ix) For $\Delta H = Q_P$ to hold the system must not do any work other than PV-work in any process. This requirement is not fulfilled when an electrochemical cell operates due to involvement of electrical work. Therefore, the given statement is justified.

(x) The given statement is not justified because $E°$ varies with temperature.

13.2 (i) This is due to one or more of the following reasons: (a) change transfer between phases, (b) orientation of polar molecules, (c) polarisation of molecules, and (d) unequal adsorption of ions, in the interphase region.

Note: Interphase potential does not constitute an emf (i.e. cannot result in an electric current in a closed circuit) under isothermal condition.

(ii) Electric potential is defined with reference to the movement of a test charge in an electric field. Bringing it from one region to another of different field intensity and different chemical composition entails not only electrical work (by which electric potential is measured) but also some chemical work (due to inevitable matter content of the test charge). Since there is no way of experimentally separating these two work terms (because there is no way of separating electricity from matter), it is impossible to measure a purely electric potential difference between two points in media of different chemical composition, for instance between a metal and an aqueous solution of its salt.

Note: Measurement of electric potential difference between two points is possible only when these are located in the regions of same chemical composition. The terminals of an electrochemical cell are made of the same metal (such as Cu) so that we can measure the cell emf which is the open-circuit electric potential difference between the terminals of same composition.

(iii) Phases involved in a galvanic cell must all be electrical conductors to make the internal resistance of the cell minimum and continuous flow of current possible. But all the phases must not be electronic conductors, at least one must be an electrolytic conductor (through which electrons cannot pass) so that the cell terminals can have different potential under open-circuit condition.

(iv) Because in the outer part of the cell, the flow of electricity is due to electrons while inside the cell, it is due to ions. Electrons generated at the –ve electrode (at low potential) move toward the +ve electrode (at high potential) outwardly (being unable to pass through an electrolyte), which, by convention, amounts to electricity passing from high to low potential. This restricts the flow of electricity to occur through the electrolyte in the opposite direction to avoid accumulation of electricity in any part of the circuit concerned.

(v) Because in the given reaction, the electron-transfer from Zn to Cu^{2+} occurs haphazardly in direct reaction but through organised motion of electron when brought about in Daniel cell.

(vi) Because of rather high dissipation of energy in charging and discharging most cells due to their appreciably high internal resistance.

(vii) The success of a salt bridge is attributed to (1) the high concentration of ions in it that enables them to conduct most of the current near its ends (2) the nearly equal mobility of the cation and anion in the salt bridge that can result in only small junctional potentials of opposite sign at the two ends of it, leading almost to their cancellation.

The reason for the use of KCl in salt bridge is that KCl, unlike NaCl, contains cation and anion of nearly same mobility.

(viii) The Nernst equation for emf (E) of a cell is based on the following expression of ΔG of the cell reaction for n faraday (F) of electricity

$$\Delta G = -nFE$$

Since this expression holds only for the reactions which can be carried out thermodynamically reversibly the Nernst equation is limited to such reactions.

(ix) The emf of a cell is measured by equating it to the potential difference between the terminals of the cell under the condition that no (appreciable) current passes through the cell. This condition is fulfilled with a potentiometer but not with an ordinary voltmeter.

Note: The emf of a cell can be measured with an electronic voltmeter which, unlike ordinary voltmeter, draws only a negligible current from the cell.

(x) Because the direct reaction between the species involved in the cell reaction cannot be completely arrested.

Note: This is true even with a standard cell like Weston cell where chemical reaction between the dissolved Hg_2SO_4 and Cd (both in trace amount) cannot be denied, though it is too small to be considered.

(xi) From the potential data, we can predict only the possibility of a red-ox reaction and not its rate. The higher reduction potential of $IO_4^- - IO_3^-$ system mere implies that any substance oxidised by permagnate will of course be oxidised by periodate. But it may so happen, as in case of oxidation of aldehyde, that the rate of oxidation by periodate is much slower than that with permanganate.

(xii) H^+ combines with $Fe(CN)_6^{4-}$ to a greater extent than $Fe(CN)_6^{3-}$. As a consequence, in solution of low pH, the potential for the half reaction $Fe(CN)_6^{3-} + e \rightarrow Fe(CN)_6^{4-}$ becomes higher than that for $\frac{1}{2}I_2(s) + e \rightarrow I^-$. Hence $Fe(CN)_6^{3-}$ undergoes reduction to oxidise I^- to I_2.

Zn^{2+} acts in a way similar to H^+.

13.3　(i) Any concentration cell where no electrode reaction can occur, e.g.
$$H_2(P)|HCl(aq)|H_2(P)$$

(ii) Any concentration cell where electrode reaction can occur

e.g.　　　　$H_2(P_1)|HCl(aq)|H_2(P_2), \quad P_1 \neq P_2$

(iii) Any chemical cell that reaches equilibrium, i.e. cannot undergo any reaction spontaneously.

13.4 When the cell produces current, the chemical reaction that occurs is
$$Zn + 2H^+(aq) \rightarrow Zn^{2+}(aq) + H_2(g) \qquad \text{(I)}$$

□ The given cell is not a reversible one, because on reversing the direction of current, the cell reaction is not reversed,, instead the reaction $Cu + 2H^+(aq) \rightarrow Cu^{2+}(aq) + H_2$ occurs. Hence the Nernst equation is not applicable to this cell.

□□ The following reversible cell having cell reaction (I) is the appropriate one
$$Zn|ZnSO_4(aq)||H_2SO_4(aq)|H_2(g)|Cu$$

13.5 The actual cell reaction is one for which emf, according to the Nernst equation, is positive. This is determined by the concentrations of $ZnSO_4$ and $CuSO_4$ in the cell.

However, for comparable concentrations, the cell reaction would be $Zn + Cu^{2+} \rightarrow$ $Zn^2 + Cu$ which corresponds to higher potential for the half reaction $Cu^{2+} + 2e \rightarrow$ Cu, than that for $Zn^{2+} + 2e \rightarrow Zn$.

13.6 Anode reaction $\qquad Cd\,(Hg) = Cd^{2+}(aq) + 2e$

Cathode reaction $\qquad Hg_2SO_4(s) + 2e = 2Hg(l) + SO_4^{2-}\,(aq)$

Net cell reaction $\quad Cd(Hg) + Hg_2SO_4(s) = Cd^{2+}\,(aq) + SO_4^{2-}\,(aq) + 2Hg(l)$

Gross cell reaction:

$$Cd(Hg) + Hg_2SO_4(s) + \tfrac{8}{3}H_2O(l) = CdSO_4 \cdot \tfrac{8}{3}H_2O(s) + 2Hg(l)$$

$$E = E^\circ - \frac{RT}{2F}\ln\frac{a(CdSO_4 \cdot \tfrac{8}{3}H_2O(s))\,a^2\,(Hg(l))}{a(Cd)\,a(Hg_2SO_4(s))\,a_{(H_2O\,(l))}^{\tfrac{8}{3}}}$$

$$\simeq \text{constant at constant temperature}$$

☐ The Weston cell is the most widely used standard cell because of its following advantages:

   i. No liquid–liquid junction is involved.

   ii. Due to insolubility of Hg_2SO_4 direct reaction between the dissolved Hg_2SO_4 and Cd is insignificant and the problem of maintaining the cell does not arise.

   iii. Sufficiently low temperature coefficient of the cell emf.

13.7 a. Emf of both the cells (being without transfer) is due only to the processes occurring at the electrodes leading to the same net process, i.e.

$$Zn^{2+}\,(m_2) \rightarrow Zn^{2+}\,(m_1) \quad \text{assuming } m_2 > m_1$$

Then emf will be same for both.

   b. Here emf (E) of each cell is given by

$$E = E' + E_j$$

where E' is the emf of the corresponding cell without transfer and E_j is the liquid junction potential. For the two cells E' is same yet E is different as E_j is different due to different mobility (u) of Cl^- and SO_4^{2-} ions.

Since $(u_{Cl^-} - u_{Zn^{2+}}) < (u_{SO_4^{2-}} - u_{Zn^{2+}})$, E_j and hence E of the second cell will be greater.

13.8 **For the first cell:** When 2F of electricity passes through the cell from left to right (i.e. when the cell produces electrical energy), the following net process will occur $t_{Cl^-}ZnCl_2\,(m_2) = t_{Cl^-}ZnCl_2\,(m_1)$, assuming $m_2 > m_1$ vide Section 13.5 or more appropriately

$$t_{Cl^-}\,Zn^2\,(m_2) + 2t_{Cl^-}\,Cl^-\,(2m_2) = t_{Cl^-}\,Zn^2\,(m_1) + 2t_{Cl^-}\,Cl^-\,(2m_1)$$

Then $\qquad\qquad\qquad E = t_{Cl^-}\dfrac{RT}{2F}\ln\dfrac{(a_{Zn^{2+}} \cdot a_{Cl^-}^2)_2}{(a_{Zn^{2+}} \cdot a_{Cl^-}^2)_1}$

$$= t_{Cl^-}\frac{3RT}{2F}\ln\frac{(a_\pm)_2}{(a_\pm)_1}$$

Emf without transfer, $\qquad E' = \dfrac{RT}{2F}\ln\dfrac{(a_\pm)_2}{(a_\pm)_1}$

$$E_j = E - E'$$

$$= (3t_{Cl^-} - 1)\frac{RT}{2F}\ln\frac{(a_\pm)_2}{(a_\pm)_1}$$

For the second cell: By following similar procedure, we have

$$E = t_{SO_4^{2-}} \frac{RT}{2F} \ln \frac{(a_{Zn^{2+}} \cdot a_{SO_4^{2-}})_2}{(a_{Zn^{2+}} \cdot a_{SO_4^{2-}})_1}$$

$$= t_{SO_4^{2-}} \frac{RT}{F} \ln \frac{(a_{\pm})_2}{(a_{\pm})_1}$$

$$E' = \frac{RT}{2F} \ln \frac{(a_{\pm})_2}{(a_{\pm})_1}$$

$$E_j = E - E'$$

$$= (2t_{SO_4^{2-}} - 1) \frac{RT}{2F} \ln \frac{(a_{\pm})_2}{(a_{\pm})_1}$$

13.9 Here we are to search for appropriate cathode and anode reactions keeping in mind the net cell reaction which is obtained on writing the concerned reaction in ionic form excluding the species common to both sides.

(i) The relevant cell is

$$Pt|I_2(s) \, KI(aq) \| K_2Cr_2C_7, \, CrCl_3, \, HCl(aq)| \, Pt$$

Anode reaction $\qquad\qquad 6I^-(aq) = 3I_2(s) + 6e$

Cathode reaction

$$Cr_2O_7^{2-}(aq) + 14H^+(aq) + 6e = 2Cr^{3+}(aq) + 7H_2O(l)$$

Net cell reaction

$$Cr_2O_7^{2-}(aq) + 6I^-(aq) + 14H^+(aq) = 2Cr^{3+}(aq) + 3I_2(s) + 7H_2O(l)$$

Overall reaction

$$K_2Cr_2O_7(aq) + 6KI(aq) + 14HCl(aq) = 2CrCl_3(aq) + 8KCl(aq) + 3I_2(s) + 7H_2O(l)$$
(adding K^+ and Cl^- on both sides)

(ii) $\qquad\qquad\qquad Ag|AgCl(s)|Cl^-(aq)|Cl_2(g)| \, Pt$

Anode reaction $\qquad\qquad Ag + Cl^-(aq) = AgCl(s) + e$

Cathode reaction $\qquad\qquad \frac{1}{2}Cl_2(g) + e = Cl^-(aq)$

Net cell reaction $\qquad\qquad Ag + \frac{1}{2}Cl_2(g) = AgCl(s)$

(iii) $\qquad\qquad Pt|H_2(g)|H^+(aq)|O_2(g)|Pt \,$ [or, $Pt|H_2(g)|OH^-(aq)|O_2(g)|Pt$]

Anode reaction $\qquad\qquad\qquad H_2(g) = 2H^+(aq) + 2e$

Cathode reaction $\quad \frac{1}{2}O_2(g) + 2H^+(aq) + 2e = H_2O(l)$

Net cell reaction: $\qquad\qquad H_2(g) + \frac{1}{2}O_2(g) = H_2O(l)$

[or Anode reaction $\qquad H_2(g) + 2OH^-(aq) = 2H_2O(l) + 2e$

Cathode reaction $\quad \frac{1}{2}O_2(g) + H_2O(l) + 2e = 2OH^-(aq)$]

(iv) $\qquad\qquad Pt|H_2(g)| \, NaOH(aq)\|HCl(aq)\|H_2(g)|Pt$

Anode reaction $\qquad \frac{1}{2}H_2(g) + OH^-(aq) = H_2O(l) + e$

Cathode reaction $\qquad\qquad H^+(aq) + e = \frac{1}{2}H_2(g)$

Net cell reaction $\qquad H^+(aq) + OH^-(aq) = H_2O(l)$

Overall reaction $\qquad HCl(aq) + NaOH(aq) = NaCl(aq) + H_2O(l)$
$\qquad\qquad\qquad\qquad$ (adding Na^+ and Cl^- on both sides)

(v) $Pt|H_2(g)|OH^-(aq)||H^+(aq)|H_2(g)|Pt$

(vi) $Pt|Na(Hg)|NaOH(aq)|H_2(g)|Pt$

Anode reaction	$2Na(Hg) = 2Na^+(aq) + 2e$
Cathode reaction	$2H_2O(l) + 2e = 2OH^-(aq) + H_2(g)$

Net cell reaction	$2Na(Hg) + 2H_2O(l) = 2Na^+(aq) = 2OH^-(aq) + H_2(g)$
or	$2Na(s) + 2H_2O(l) = 2NaOH(aq) + H_2(g)$

(vii) $Pt|H_2(g)|OH^-(aq)|HgO(s)||Hg(l)|Pt$

Anode reaction	$H_2(g) + 2OH^-(aq) = 2H_2O(l) + 2e$
Cathode reaction	$HgO(s) + H_2O(l) + 2e = Hg(l) + 2OH^-(aq)$

Net cell reaction	$HgO(s) + H_2(g) = Hg(l) + H_2O(l)$

(viii) $Pt|H_2(P)|HCl(0.05\,m):HCl(0.1\,m)|H_2(P)|Pt$ or $Ag|AgCl(s)|HCl(0.1\,m):HCl(0.05\,m)|AgCl(s)|Ag$

Anode process:
$Ag(s) + Cl^-(0.1\,m) \rightarrow AgCl(s) + e$
Cathode process:
$AgCl(s) + e \rightarrow Ag(s) + Cl^-(0.05\,m)$
Transfer process:
$t_{H^+}H^+(0.1\,m) \rightarrow t_{H^+}H^+(0.05\,m)$
$t_{Cl^-}Cl^-(0.05\,m) \rightarrow t_{Cl^-}Cl^-(0.1\,m)$

Net process:
$t_{Cl^-}HCl(0.1\,m) \rightarrow t_{Cl^-}HCl(0.05\,m)$
vide Section 13.5

Net process:
$t_{H^+}HCl(0.1\,m) \rightarrow t_{H^+}HCl(0.05\,m)$
putting $t_{Cl^-} = 1 - t_{H^+}$

Note: In (iv), (v) and (viii), the overall process is not of redox type. Here electrode processes involve some species not appearing in the overall process.

(iv) is a particular case of (v) with NaOH as alkali and not HCl as acid.

For (viii) the relevant two cells do not involve same amount of HCl transferred due to the same amount of electricity passing from one solution to the other (in opposite direction).

13.10 For the solution A

$$C_{H^+} = \frac{\left(10 \times \frac{1}{10} - 90 \times \frac{1}{100}\right)}{10 + 90} \text{ or } 10^{-3}\,M$$

Emf of the cell $= E_{calomel} - E_{hydrogen}$

$\qquad\qquad\qquad = E_{calomel} - 0.059 \log C_{H^+}$ by Eq. (13.6) taking $a_{H^+} = C_{H^+}$

$\qquad\qquad\qquad = 0.242 - 0.059 \log(10^{-3})$

$\qquad\qquad\qquad = 0.419\,V$

13.11 For a weak acid HA of formality F

$\qquad$ dissociation constant $K = \dfrac{C_{H^+} \cdot C_{A^-}}{F_{HA} - C_{A^-}}$ $\qquad$ ignoring activity coefficient affect

$\qquad\qquad\qquad\qquad = \dfrac{C_{H^+}^2}{F_{HA}}$ $\qquad$ ignoring H^+ from H_2O and C_{A^-} compared to F_{HA}

or $\qquad\qquad\qquad C_{H^+} = \sqrt{KF_{HA}}$

$\qquad$ Emf of the cell = Difference between the potentials for two hydrogen electrodes one containing HA_1 and the other containing HA_2

$$= 0.059 \log \frac{(C_{H^+})_{HA_1}}{(C_{H^+})_{HA_2}} \quad \text{by Eq. (13.6)}$$

$$= 0.059 \log \frac{\sqrt{K_{HA_1} F_{HA_1}}}{\sqrt{K_{HA_2} F_{HA_2}}}$$

$$= 0.059 \times \tfrac{1}{2}(pK_{HA_2} - pK_{HA_1}), \quad \text{since } F_{HA_1} = F_{HA_2}$$

$$= 0.059 \times \tfrac{1}{2}(5 - 3) = 0.059 \text{ V}$$

Note: The given cell may be treated as an electrolyte concentration cell with different concentration of H^+ in the two electrolyte HA_1 and HA_2.

13.12 The difference in pH between two solutions is related to the difference between the relevant glass electrode potentials (E_G) by

$$0.059 \, (pH_2 - pH_1) = (E_G)_1 - (E_G)_2$$
$$= E_2 - E_1 \quad \text{where } E = E_{cal} - E_G$$

or
$$pH_2 = pH_1 + \frac{E_2 - E_1}{0.059}$$

$$= 4 + \frac{0.3865 - 0.1120}{0.059} = 8.65$$

Note: At present pH of a solution is defined operationally in terms of emf of the cell consisting of a hydrogen electrode, or more conveniently a glass electrode, and a saturated calomel reference electrode. If emf of this cell is E_S with a standard solution S of preassigned pH(s) and E_X with a solution X of unknown pH(X), then

$$pH(X) = pH(s) + \frac{E_X - E_S}{2.303 \, RT/F}$$

The most convenient primary standard is saturated aqueous solution of potassium hydrogentartrate, which has pH = 3.557 at 25°C.

13.13 The operational definition of pH has the advantage in that it is easy to implement with high reproducibility. It suffers from the disadvantage in that the value of the defined quantity found experimentally is not precisely equal to $-\log a_{H^+}$ (found from theoretical calculation of γ_{H^+}). Because the liquid junction potential associated with the relevant cell is not adequately eliminated by the salt bridge and the assigned values of pH lack precision.

Note: It appears awkward that the current definition of pH lays more emphasis on the simplicity of the method of its measurement and reproducibility of the result than to the thermodynamic significance of the result.

13.14 a. This merely provides electrical connection with the inner side of the glass membrane. Liquid Hg with Pt terminal can serve the same purpose.

   b. The origin lies in the following equilibrium pertaining to the exchange of alkali metal ions (mainly Na^+) of glass membrane with H^+ ions of the surrounding solution

$$Na^+ \text{ (glass)} + H^+ \text{ (soln)} \rightleftharpoons Na^+ \text{ (soln)} + H^+ \text{ (glass)}$$

That the origin is not due to the difference in H^+ ion concentration on the two sides of the membrane is ensured by the fact that the membrane used in the glass electrode is permeable to Na^+ ions but not to H^+ ions.

c. Because beyond this range of pH, the usual expression of glass electrode potential, Eq. (13.7) does not work. This is due to appreciable violation of the underlying assumptions (i) that the relevant ion exchange process involves only the surface ions of the membrane (if pH $\ll 1$) and (ii) that $a_{H^+ \text{(soln)}} \gg a_{Na^+ \text{(soln)}}$ (if pH $\gg 10$) due to appreciable interaction of glass with OH^- ions.

d. Because the resistance of a glass electode is so high (10^7 to 10^9 ohms) that the galvanometer used in the ordinary potentiometer circuit cannot readily detect the extremely small currents involved. Electronic voltmeter can, however be used.

Note: The membrane of a glass electrode can be replaced by a crystal of a water-insoluble salt having significant ionic conductivity at room temperature, e.g. LaF_3.

13.15 By hydrogen electrode, we usually mean one with the electrode process

$$\tfrac{1}{2}H_2 \, (g) = H^+ \, (aq) + e \quad \text{as anode}$$

$$H^+ \, (aq) + e = \tfrac{1}{2}H_2 \, (g) \quad \text{as cathode}$$

Hydrogen electrode is reversible wrt H^+ ion as this ion is involved in the electrode process. Also it is reversible wrt OH^- ion with which H^+ ion can establish rapid equilibrium. Since electrode potential is only indirectly affected by OH^-, hydrogen electrode is often conveniently thought to be reversible wrt H^+.

The standard potential $E°$ of the given electrode corresponds to the following half-reaction

$$\tfrac{1}{2}H_2 \, (g) + OH^- \, (aq) = H_2O \, (l) + e \qquad\qquad \Delta G° = - FE°$$

This can be obtained by adding the following two half-reactions

1. $\qquad\qquad \tfrac{1}{2}H_2 \, (g) = H^+ \, (aq) + e \qquad\qquad \Delta G_1° = - FE_1° = 0$

2. $\quad H^+ \, (aq) + OH^- \, (aq) = H_2O \, (l) \qquad\qquad \Delta G_2° = RT \ln K_w$

Then $\qquad\qquad \Delta G° = \Delta G_1° + \Delta G_2° \quad$ since G is an additive property

Hence $\qquad\qquad -FE° = 0 + RT \ln K_w$

or $\qquad\qquad\qquad E° = -\dfrac{RT}{F} \ln K_w$

$$= -0.059 \log 10^{-14} \text{ at } 25°C$$

$$= 0.826 \text{ V}$$

□ Since $E° > E_1°$, H_2 is expected to act as a better electron provider, i.e. better reducing agent in alkaline medium (where the process 1 is favoured).

13.16 i. Here relevant processes are

(1) $\qquad \tfrac{1}{2}O_2 \, (g) + H_2O(l) + 2e = 2OH^- \, (aq) \qquad\qquad \Delta G_1° = -2FE_1°$

(2) $\qquad\qquad 2H^+ \, (aq) + 2e = H_2 \, (g) \qquad\qquad\qquad \Delta G_2° = 0$

(3) $H_2 \, (g) + \tfrac{1}{2}O_2 \, (g) + H_2O(l) = 2H^+ \, (aq) + 2OH^- \, (aq) \quad \Delta G_3° = -77.7 \text{ kJ mol}^{-1}$

Since (1) − (2) gives (3)

$$\Delta G_3° = \Delta G_1° - \Delta G_2°$$

Then $\qquad -77.7 = -2FE_1° - 0$

or $\qquad\qquad E_1° = \dfrac{77.7 \times 10^3 \text{ J mol}^{-1}}{2(9.65 \times 10^4 \text{ C mol}^{-1})} = 0.40 \text{ V}$

ii. The addition of Eq. (i) and $2H^+ + 2OH^- = 2H_2O$ gives the concerned half-reaction

$$\tfrac{1}{2}O_2 + 2H^+(aq) + 2e = H_2O(l) \qquad\qquad \Delta G^\circ = -2FE^\circ$$

Then $\qquad\qquad \Delta G^\circ = \Delta G_1^\circ + 2RT \ln K_w$

Hence $\qquad\qquad -2FE^\circ = -2FE_1^\circ + 2RT \ln K_w$

or $\qquad\qquad E^\circ = E_1^\circ - \dfrac{RT}{F} \ln 10^{-14} = 0.4 - 0.059 \log 10^{-14} = 1.23\ V$

□O_2. This follows from higher E° value for the half-reaction (i), and more so for (ii), compared with that for the half reaction $H^+(aq) + e = \tfrac{1}{2}H_2(g)\ (E^\circ = 0)$.

□□ Acid, because the relevant half-reactions are favoured by an acid, and are hindered by an alkali.

13.17 The given half-reactions are

(1) $\qquad Zn + 4NH_3 = Zn(NH_3)_4^{2+}(aq) + 2e \qquad \Delta G_1^\circ = -2FE_1^\circ$

(2) $\qquad\qquad\qquad Zn = Zn^{2+}(aq) + 2e \qquad\qquad \Delta G_2^\circ = -2FE_2^\circ$

(1) – (2) gives

$$Zn^{2+}(aq) + 4NH_3 = Zn(NH_3)_4^{2+}(aq) \qquad\qquad \Delta G^\circ = -RT \ln K$$

where K is stability constant for the complex ion

Then $\qquad\qquad \Delta G^\circ = \Delta G_1^\circ - \Delta G_2^\circ$

Hence $\qquad -RT \ln K = -2FE_1^\circ + 2FE_2^\circ$

or $\qquad -\dfrac{RT}{2F} \ln K = E_2^\circ - E_1^\circ$

or $\qquad -\dfrac{0.059}{2} \log K = 0.763 - 1.03 \quad$ at 25°C

whence $\qquad\qquad K = 1.12 \times 10^9$

13.18 The emf of the given cell with transfer E is related to that for the corresponding cell without transfer E' by

$$E = 2t_{Cl^-}E' \quad \text{by Eq. (13.8b) and (13.9)}$$

$$= 2(1 - t_{H^+})E'$$

or $\qquad\qquad t_{H^+} = 1 - \dfrac{E}{2E'}$

Now $\qquad\qquad E' = \dfrac{RT}{F} \ln \dfrac{(a_\pm)_{right}}{(a_\pm)_{left}}$

$$= 0.059 \log \dfrac{0.02}{0.01} = 0.0178\ V \text{ at 25°C}$$

Then $\qquad\qquad t_{H^+} = 1 - \dfrac{0.0062}{2 \times 0.0178} = 0.826$

Liquid junction potential

$$= E - E^\circ$$

$$= 0.0062 - 0.0178 = -0.0116\ V$$

Since $E < E'$, the anode side of the liquid junction is positive.

13.19 For the corresponding cell without transfer

Anode reaction $\quad\quad\quad\quad Cl^- (m_1) = \frac{1}{2} Cl_2 (P_1) + e$

Cathode reaction $\quad\quad \frac{1}{2} Cl_2 (P_2) + e = Cl^- (m_2)$

Net reaction $\frac{1}{2} Cl_2 (P_2) + Cl^- (m_1) = \frac{1}{2} Cl_2 (P_1) + Cl^- (m_2)$

Then by Nernst equation

emf without transfer

$$E' = \frac{RT}{F} \ln \frac{P_2^{\frac{1}{2}} m_1}{P_1^{\frac{1}{2}} m_2} \quad \text{ignoring deviation from ideality}$$

$$= 0.059 \log \frac{(1)^{\frac{1}{2}} (0.001)}{(0.01)^{\frac{1}{2}} (0.01)} = 0$$

The changes at the liquid junction due to transfer processes are

$$t_{Na^+} Na^+ (m_1) = t_{Na^+} Na^+ (m_2)$$

$$t_{Cl^-} Cl^- (m_2) = t_{Cl^-} Cl^- (m_1)$$

The emf associated with these changes, which is the liquid junction potential E_j, is then given according to the Nernst equation by

$$E_j = \frac{RT}{F} \ln \frac{m_1^{t_{Na^+}}}{m_2^{t_{Na^+}}} + \frac{RT}{F} \ln \frac{m_2^{t_{Cl^-}}}{m_1^{t_{Cl^-}}}$$

$$= (2t_{Na^+} - 1) \frac{RT}{F} \ln \frac{m_1}{m_2} \quad \text{putting } t_{Cl^-} = 1 - t_{Na^+}$$

$$= (2t_{Na^+} - 1) \, 0.059 \log \frac{0.001}{0.01}$$

$$= 0.059 \, (1 - 2t_{Na^+})$$

Then the emf of the cell with transfer

$$E = E' + E_j = 0 + E_j$$

$$= E_j$$

i. $\quad\quad\quad\quad E = E_j = 0.059 \, (1 - 2 \times 0.4) = 0.0118$ V

ii. Here $\quad\quad t_{Na^+} = 0$, because the entire current is carried by Cl^-

Then $\quad\quad\quad E = E_j = 0.059$ V

iii. Here $\quad\quad t_{Na^+} = 1$

Then $\quad\quad\quad E = E_j = 0.059 \, (1 - 2 \times 1) = -0.059$ V

Note:

1. The given concentration cell is neither of pure electrode concentration type nor of pure electrolyte concentration type, it is one of mixed type.

2. For the corresponding cell without transfer emf is zero, though the cell reaction is not nil.

3. Liquid junction potential can be calculated independently without any knowledge of the emf of the cell with transfer and of one without transfer.

4. Mere change of membrane causes E and E_j in (iii) to have sign opposite of those in (ii).

13.20 For the electrode $Ag|AgI(s)|I^-(aq)$, the electrode process is

$$Ag(s) + I^-(aq) = AgI(s) + e$$

Since I^- ion is involved in this electrode process, $Ag|AgI(s)|I^-(aq)$ is reversible wrt I^- ion. Also this electrode is reversible wrt Ag^+ ion with which I^- ion can establish rapid equilibrium. However, since the electrode potential is only indirectly affected by Ag^+ ion (through removal of I^- ion), only the reversibility wrt I^- ion is important.

For the $Ag|Ag^+(aq)$ electrode, the electrode process is

$$Ag(s) = Ag^+(aq) + e$$

Then the electrode is reversible wrt Ag^+ ion directly.

☐ The electrode process for the electrode $Ag|AgI(s)|I^-(aq)$ may be thought to involve the following two steps

i. $\qquad Ag(s) = Ag^+(aq) + e$

ii. $Ag^+(aq) + I^-(aq) = AgI(s)$

The process (i) gives rise to the potential of both $Ag|Ag^+(aq)$ and $Ag|AgI(s)|I^-(aq)$ electrodes. So the latter electrode might as well be treated as $Ag|Ag^+(aq)$ electrode and we can write

$$E_{Ag|AgI|I^-} = E^{\circ}_{Ag|Ag^+} - \frac{RT}{F} \ln a_{Ag^+}$$

$$= E^{\circ}_{Ag|Ag^+} - \frac{RT}{F} \ln (K_{th} AgI|a_{I^-})$$

$$= \left(E^{\circ}_{Ag|Ag^+} - \frac{RT}{F} \ln K_{th} AgI \right) + \frac{RT}{F} \ln a_{I^-}$$

$$= E^{\circ}_{Ag|AgI|I^+} + \frac{RT}{F} \ln a_{I^-}$$

Then $\qquad E^{\circ}_{Ag|AgI|I^-} = E^{\circ}_{Ag|Ag^+} + \frac{RT}{F} \ln K_{th} AgI$

Note: $E^{\circ}_{Ag|Ag^+|I^-} > E^{\circ}_{Ag|Ag^+}$ (or $E^{\circ}_{I^-|AgI|Ag} < E^{\circ}_{Ag^+|Ag}$), since $\frac{RT}{F} \ln K_{th} AgI$ is negative. This is quite likely because the tendency for the process (i) is enhanced due to removal of Ag^+ ion by the process (ii).

In the $Ag|AgI|I^-$ electrode, $Ag(s)$ is usually coated with $AgI(s)$. But this is not essential; an iodide solution saturated with AgI will serve the purpose.

13.21 $E = 0$ for (i), (ii) and (iii) but not for (iv), because for the first three cells, there is no essential difference between the two electrodes (considering each electrode reversible wrt Pb^{2+} ion at same concentration).

$E = 0$ signifies that no net process will occur in the cell when the circuit is closed.

☐ $E^{\circ} = E^{\circ}_{Pb^{2+}/Pb} - E^{\circ}_{Cl^-|PbCl_2|Pb} = -\frac{RT}{F} \ln K_{th} PbCl_2$

$$= -0.059 \log 10^{-4}$$

$$= 0.236 \text{ V.}$$

13.22 When one faraday of electricity passes through the cell from left to right the following processes will occur

first anode process $\frac{1}{2}H_2\,(P) = H^+\,(m_1) + e$

first cathode process $AgCl(s) + e = Ag(s) + Cl^-\,(m_1)$

second anode process $Ag(s) + Cl^-\,(m_2) = AgCl\,(s) + e$

second cathode process $H^+\,(m_2) + e = \frac{1}{2}H_2(P)$

Net process $H^+\,(m_2) + Cl^-\,(m_2) = H^+\,(m_1) + Cl^-\,(m_1)$

The emf (E_1) associated with this process is given, according to Nernst equation by

$$E_1 = \frac{RT}{F}\ln\frac{(a_{H^+}\cdot a_{Cl^-})_2}{(a_{H^+}\cdot a_{Cl^-})_1}$$

$$= 2\frac{RT}{F}\ln\frac{(a_\pm)_2}{(a_\pm)_1}$$

□ The given cell is one without transfer, because it involves no liquid junction (i.e. interface between two miscible electrolyte solutions) across which an additional potential can develop due to difference in mobilities of the cations and anions in the two solutions.

□□ To find transport numbers of H^+ and Cl^- ions from emf measurements, we need knowledge of emf (E_2) of the following cell with transfer

$$Pt|H_2(P)|\,HCl\,(m_1):HCl(m_2)|H_2\,(P)|\,Pt$$

For this cell $E_2 = 2t_{Cl^-}\dfrac{RT}{F}\ln\dfrac{(a_\pm)_2}{(a_\pm)_1}$ by Eq. (13.8b)

Then $t_{Cl^-} = \dfrac{E_2}{E_1}$

$$t_{H^+} = 1 - t_{Cl^-} = 1 - \frac{E_2}{E_1}$$

13.23 When the cell produces electrical energy electricity passes through the cell from left to right and the net result is the transfer of H^+ and Cl^- ions from the higher concentration (m_2) to the lower concentration (m_1) (vide Section 13.5). Under open circuit condition, such transfer (called diffusion) will occur directly across the liquid junction from m_2 to m_1 with H^+ ion, having higher mobility, leading the march. As a consequence, an electrical potential difference will develop at the liquid junction that will make the H^+ and Cl^- ions move ultimately with the same speed. The left side (i.e. the anode side) of the liquid junction potential thus formed will therefore be positive.

□ This cannot be done directly through the use of salt bridge which can eliminate the liquid junction potential only incompletely. However, we can achieve this indirectly using the following cell which involves no liquid junction

$Pt|H_2(P)||HCl(m_1)|AgCl(s)|Ag|\,AgCl(s)|HCl(m_2)|H_2(P)|Pt$, $m_2 > m_1$. The emf of this cell is twice that of the cell concerned, vide Eq. (13.9).

13.24 Here $\dfrac{dE}{dT} \approx \dfrac{\Delta E}{\Delta T} = \left(\dfrac{1.0486 - 1.0477}{298.15 - 293.15} + \dfrac{1.0494 - 1.0486}{303.15 - 298.15)}\right)\Big/2$

$$= 0.00017\ V/K$$

$$\Delta H = -nF \left[E° - T \left(\frac{\partial E}{\partial T} \right)_P \right] \quad \text{by Eq. (13.13)}$$

$$= -(1) (96485 \text{ C mol}^{-1}) [1.0486 \text{ V} - (298.15 \text{ K}) (0.00017 \text{ V/K})]$$
$$= -96284 \text{ J mol}^{-1}$$

The accuracy of the calculated ΔH

$$= \pm 96485 \sqrt{(0.001)^2 + (0.01 \times 0.00017)^2}$$
$$= \pm 96.5 \text{ J mol}^{-1}$$

13.25 Here, the cell reaction is

$$Zn + Cu^{2+} (aq) = Zn^{2+} (aq) + Cu, \text{ for 2F of electricity}$$

Then

$$E = E° - \frac{RT}{2F} \ln \frac{m_{Zn^{2+}}}{m_{Cu^{2+}}}$$

As electricity is drawn from the cell $m_{Cu^{2+}}$ decreases and $m_{Zn^{2+}}$ increases and hence E decreases. When $E = 1.0$ V,

$$1.0 \text{ V} = [0.34 \text{ V} - (-0.76 \text{ V})] - \frac{0.059 \text{ V}}{2} \log \frac{m_{Zn^{2+}}}{m_{Cu^{2+}}}$$

or

$$\frac{m_{Zn^{2+}}}{m_{Cu^{2+}}} = 2453$$

Let x be the amount of Cu^{2+} that is converted into Cu when the cell potential drops to 1.0 V. Then

$$\frac{m_{Zn^{2+}}}{m_{Cu^{2+}}} = \frac{1.0 \text{ mol} + x}{1.0 \text{ mol} - x} = 2453$$

whence $\quad\quad x = 0.999$ mol

Quantity of electricity drawn

$$= \text{charge carried by 0.999 mol of } Cu^{2+} \text{ ions}$$
$$= 2 \times 0.999 \text{ or 1.998 faradays}$$

☐ The cell can deliver electricity till $E = 0$, when

$$0 = 1.1 - \frac{0.059}{2} \log \frac{m_{Zn^{2+}}}{m_{Cu^{2+}}}$$

or

$$\frac{m_{Zn^{2+}}}{m_{Cu^{2+}}} = 1.94 \times 10^{37}$$

Then, virtually entire 1 mol of Cu^{2+} is discharged which produces 2F of electricity. Therefore, the required time is 2×96485C/0.3A or 6.43×10^5s.

Note: When a chemical cell is exhausted its E is zero but $E°$ is not.

13.26 i. $\quad\quad \Delta G° = -nFE = -nF (E°_{Cu^{2+}|Cu} - E°_{Zn^{2+}|Zn})$
$$= -(1) (96485 \text{ C mol}^{-1}) (0.34 \text{ V} + 0.76 \text{ V})$$
$$= -1.06 \times 10^5 \text{ J mol}^{-1}$$

$$\Delta S° = nF \left(\frac{\partial E°}{\partial T} \right)_T$$
$$= 0$$

$$\Delta H° = \Delta G° + T\Delta S°$$
$$= \Delta G° + 0$$
$$Q° = T\Delta S°$$
$$= 0$$

ii. The direct reaction occurs irreversibly. Yet $\Delta G°$, $\Delta S°$ and $\Delta H°$ will be same as in (i), because G, S and H are state functions and hence their changes are independent of the path followed. However, irreversibility of the process makes $Q° \neq T\Delta S°$ and hence different from that in (i). Here $Q° = \Delta H°$, provided the reaction occurs isobarically.

iii. Here the process is reversible and reverse of that occurring in (i). Hence, $\Delta G°$, $\Delta S°$, $\Delta H°$ and $Q°$ will all be numerically same as in (i) but opposite in sign.

iv. Here no electrical work is involved. The process is irreversible but reverse of that occurring in (ii). Hence the result will be numerically same as in (ii) but opposite in sign, provided the discharged cell returns (isobarically) to the same state as with the cell before charging.

Note: The result $\Delta H° = \Delta G°$ found with the well-familiar Daniel cell having negligible temperature coefficient of emf led to the faulty generalisation that in any galvanic cell, the electrical energy produced would be equal to the chemical energy consumed.

13.27 Anode reaction $\qquad H_2(g) = 2H^+ (aq) = 2e$

Cathode reaction $\quad \frac{1}{2}O_2(g) + 2H^+ (aq) + 2e = H_2O (l)$

Net cell reaction $\qquad H_2(g) + \frac{1}{2}O_2(g) = H_2O (l)$

$$E° = -\frac{\Delta G°}{2F} = -\frac{\mu_{H_2O(l)}°}{2F}$$

$$= \frac{237.13 \times 10^3 \text{ J mol}^{-1}}{2 \times 9.6485 \times 10^4 \text{ C mol}^{-1}} = 1.23 \text{ V}$$

☐ Yes, because the net cell reaction for an alkaline electrolyte, as given below, is same as that for the acidic electrolyte.

Anode reaction $\qquad H_2(g) + 2OH^- (aq) = 2H_2O (l) + 2e$

Cathode reaction $\quad \frac{1}{2}O_2(g) + H_2O(l) + 2e = 2OH^- (aq)$

Net cell reaction $\qquad H_2(g) + \frac{1}{2}O_2(g) = H_2O (l)$

☐☐ Because an internal combustion engine converts heat of combustion into mechanical energy where the efficiency of conversion, determined by the Carnot's formula, is liable to be much less than 100%. This can be understood if we remember that such a conversion amounts to transformation of random molecular motion (that causes thermal energy) into less random macroscopic motion (associated with mechanical energy), which is difficult to achieve. But this does not happen with transformation of free energy (due to organised electronic motion) into mechanical energy through a fuel cell.

13.28 Anode reaction $\qquad C_2H_6 (g) + 4H_2O (l) = 2CO_2(g) + 14H^+ (aq) + 14e$

Cathode reaction $\quad \frac{7}{2}O_2 (g) + 14H^+ (aq) + 14e = 7H_2O (l)$

Cell reaction $\qquad C_2H_6 (g) + \frac{7}{2}O_2 (g) = 2CO_2 (g) + 3H_2O (l)$

$$\Delta G° = 2\mu°_{CO_2(g)} + 3\mu°_{H_2O(l)} - \mu°_{C_2H_6(g)}$$

$$[2(-394.36) + 3(-237.13) - (-32.82)] \text{ kJ mol}^{-1}$$

$$= -1467 \text{ kJ mol}^{-1}$$

$$E° = -\frac{\Delta G°}{nF}$$

$$= -\frac{(-1467 \times 10^3 \text{ J mol}^{-1})}{(14)(96485 \text{ C mol}^{-1})} = 1.09 \text{ V}$$

□ The main problem with such fuel cells is the low rate of electrochemical oxidation of the fuel under ordinary condition. To make such reactions sufficiently fast either high temperature or a suitable catalyst is required.

□□ Coal (the cheapest fuel) can be used in fuel cells only indirectly by converting it into a gaseous fuel like CO. However, due to technological and operational difficulties, such cells are uneconomical.

13.29 The required cell is

$$Ag|Ag^+(aq)\|Cl^-(aq)|AgCl(s)|Ag$$

Anode reaction $\quad\quad Ag = Ag^+(aq) + e$

Cathode reaction $\quad\quad AgCl(s) + e = Ag + Cl^-(aq)$

Net cell reaction $\quad\quad AgCl(s) = Ag^+(aq) + Cl^-(aq)$

i. $\quad\quad E° = -\dfrac{\Delta G°}{F} = -\dfrac{\mu°_{Ag^+(aq)} + \mu°_{Cl^-(aq)} - \mu°_{AgCl^+(s)}}{F}$

$$= -\frac{(77.11 - 131.23 + 109.79) \text{ kJ mol}^{-1}}{96485 \text{ C mol}^{-1}}$$

$$= \frac{55670 \text{ J mol}^{-1}}{96485 \text{ C mol}^-} = 0.577 \text{ V}$$

ii. $\quad\quad K = e^{-\frac{\Delta G°}{RT}}$

$$= \exp\left[\frac{-55670 \text{ J mol}^{-1}}{(8.314 \text{ J K}^{-1} \text{ mol}^{-1})(298 \text{ K})}\right] = 1.75 \times 10^{-10}$$

iii. Solubility of AgCl = [solubility product of AgCl]$^{\frac{1}{2}}$ $m°$

$$= K^{\frac{1}{2}}m° = (1.75 \times 10^{-10})^{\frac{1}{2}} (1 \text{ mol kg}^{-1})$$

$$= 1.32 \times 10^{-5} \text{ mol kg}^{-1}$$

iv. Standard Gibbs free energy of formation of AgCl (aq) is

$$\mu°_{AgCl(aq)} - \mu°_{Ag(s)} - \tfrac{1}{2}\mu°_{Cl_2(g)} = \mu°_{Ag^+(aq)} - \mu°_{Cl^-(aq)} - \mu°_{Ag(s)} - \tfrac{1}{2}\mu°_{Cl_2(g)}$$

[which corresponds to the change $Ag(s) + \tfrac{1}{2}Cl_2(g) = AgCl(aq)$]

$$= (77.11 - 131.23 - 0 - 0) \text{ kJ mol}^{-1}$$

$$= -54.12 \text{ kJ mol}^{-1}$$

Note: By convention, the standard Gibbs free energy of formation of AgCl (s) equals $\mu°_{AgCl(s)}$ and that of AgCl (aq) equals $\mu°_{AgCl(aq)}$ at 25°C. Again,

$$\mu°_{AgCl(aq)} - \mu°_{AgCl(s)} = \mu°_{Ag^+(aq)} + \mu°_{Cl^-(aq)} - \mu°_{AgCl(s)} = -RT\ln K_{th}$$

13.30 Anode reaction $\frac{1}{2}H_2$ (1 atm) $= H^+(m) + e$

Cathode reaction $AgCl(s) + e = Ag(s) + Cl^-(m)$

Net reaction $\frac{1}{2}H_2$ (1 atm) $+ AgCl(s) = H^+(m) + Cl^-(m) + Ag(s)$

$$E = E^\circ - \frac{RT}{F}\ln(a_{H^+} \cdot a_{Cl^-})$$

$$= E^\circ - \frac{2RT}{F}\ln(m_{\pm}\gamma_{\pm})$$

$$= E^\circ - 2 \times 0.059\,[\log m + \log\gamma_{\pm}] \quad \text{at } 25°C \ (\text{taking } m_{\pm} = m)]$$

Then from the given data

$$0.5205 = E^\circ - 0.118\,[\log 0.0032 + \log 1], \text{ for the first solution}$$

$$0.3420 = E^\circ - 0.118\,[\log 0.1238 + \log\gamma_{\pm}], \text{ for the second solution}$$

whence $\quad E^\circ = 0.815$ V and $\gamma_{\pm} = 0.366$

$$\Delta G^\circ = -nFE^\circ$$

$$= -(1)\,(96485 \text{ C mol}^{-1})\,(0.815 \text{ V})$$

$$= -78635 \text{ J mol}^{-1}$$

Note: Here the general procedure of finding E° is to plot the quantity $E + \frac{2RT}{F}\ln m$

against a function of molality, generally $\sqrt{m}$, and then to extrapolate the resulting curve to $m = 0$ when the ordinate gives E°. m is formal concentration.

13.31 a. Anode reaction $\qquad\qquad 2Ag(s) + SO_4^{2-}\,(0.1 \text{ m}) = Ag_2SO_4(s) + 2e$

Cathode reaction $\qquad\qquad 2H^+\,(0.2 \text{ m}) + 2e = H_2$ (1 atm)

Cell reaction $\quad 2Ag(s) + 2H^+\,(0.2 \text{ m}) + SO_4^{2-}\,(0.1 \text{ m}) = Ag_2SO_4(s) + H_2$ (1 atm)

b. $$E = E^\circ + \frac{0.059}{2}\log(a_{H^+}^2 \cdot a_{SO_4^{2-}}), \quad \text{at } 25°C$$

$$= E^\circ + \frac{0.059}{2}\log\left\{m_{H^+}^2 \cdot m_{SO_4^{2-}} \cdot \gamma_{\pm}^3\right\}$$

i. $$E = -0.627 + \frac{0.059}{2}\log\left\{(0.2)^2\,(0.1)\cdot(1)^3\right\} = -0.698 \text{ V}$$

ii. $$E = -0.627 + \frac{0.059}{2}\log\left\{(0.2)^2\,(0.1)\cdot(0.7)^3\right\} = -0.711 \text{ V}$$

The negative sign of the emf thus calculated indicates that the cell reaction as written above is not spontaneous.

c. $$2Ag^+\,(aq) + e = 2Ag$$

$$Ag_2SO_4(s) = 2Ag^+\,(aq) + SO_4^{2-}\,(aq)$$

Adding, $\quad Ag_2SO_4(s) + 2e = 2Ag(s) + SO_4^{2-}\,(aq)$

Then $\qquad E^\circ_{SO_4^-|Ag_2SO_4|Ag} = E^\circ_{Ag^+|Ag} + \frac{RT}{2F}\ln K_{th}(Ag_2SO_4)$, [vide problem 13.20]

or $\qquad\qquad \log K_{th} = \frac{2}{0.059}(E^\circ_{SO_4^{2-}|Ag_2SO_4|Ag} - E^\circ_{Ag^+|Ag})$

$$= \frac{2}{0.059}(-E^\circ_{Ag|Ag_2SO_4|SO_4^{2-}} - E^\circ_{Ag^+|Ag})$$

$$= \frac{2}{0.059}[-(-0.627) - 0.799]$$

whence $K_{th} = 1.48 \times 10^{-6}$

Note: Solubility product of a sparingly soluble electrolyte can be calculated from the E°'s of two electrodes—one reversible wrt the cation and the other wrt anion of the electrolyte concerned.

In the present problem E° of the Cell $= E^{\circ}_{Ag|Ag_2SO_4|SO_4^{2-}}$.

13.32 A suitable cell of the desired type is

$$H_2 \text{ (1 atm)} | NaOH \text{ (aq)} | Ag_2O(s) | Ag$$

The cell reaction may be represented as

$$H_2 \text{ (1 atm)} + Ag_2O(s) + H_2O \text{ (l)} = 2H^+ \text{ (in NaOH soln)} + 2Ag(s) + 2OH^- \text{ (aq)}$$

$$E = E^{\circ} - \frac{RT}{2F} \ln a_{H^+}^2 \, a_{OH^-}^2$$

$$= E^{\circ} - \frac{RT}{F} \ln K_w \qquad\qquad (I)$$

where E° is the standard potential of the electrode $OH^- | Ag_2O | Ag$ (taking $E^{\circ}_{H^+|H_2} = 0$). As desired, E of the cell is independent of the concentration of the electrolyte (NaOH solution) involved. From the measurement of emf of the stated cell, we can find K_w using the relation (I) from the knowledge of E°.

Note: Determination of K_w (similar to solubility product) is based on E°'s of two electrodes—one reversible wrt H^+ ion and the other reversible wrt OH^- ion. Since E° is zero for the first electrode, only knowledge of E° of $OH^- | H_2$ (g) as the second electrode will serve the purpose.

13.33 For the first cell

$$E_1^{\circ} = E^{\circ}_{H_2|OH^-} - E^{\circ}_{Cl^-|AgCl|Ag}$$

For the second cell

$$E_2^{\circ} = E^{\circ}_{H_2|H^+} - E^{\circ}_{Cl^-|AgCl|Ag}$$

Then $E_1^{\circ} - E_2^{\circ} = E^{\circ}_{H_2|OH^-} - E^{\circ}_{H_2|H^+}$

$$= -\frac{RT}{F} \ln K_w \qquad \text{vide problem 13.15}$$

$$= -(0.059 \text{ V}) \log K_w \text{ at } 25^{\circ}C$$

$$= (1.0486 - 0.2224) \text{ V}$$

whence $K_w = 9.93 \times 10^{-15}$

Note: The cell $Pt|H_2(g)| NaOH \text{ (aq)}, NaCl \text{ (aq)} | AgCl(s) | Ag$ has same E° as the less simple cell $Pt|H_2(g)| NaOH \text{ (aq)} || NaCl \text{ (aq)} | AgCl(s) | Ag$, though the two cells differ in E even for some concentrations of NaOH and NaCl.

13.34 The relevant half-reactions are

(1) $I_2 \text{ (s)} + 2e = 2I^- \text{(aq)}$ $\Delta G_1^{\circ} = -2FE_1^{\circ}$

(2) $I_3^- \text{ (aq)} + 2e = 3I^- \text{ (aq)}$ $\Delta G_2^{\circ} = -2FE_2^{\circ}$

(1) – (2) gives $I_2 \text{ (s)} + I^- \text{ (aq)} = I_3^- \text{ (aq)}$ $\Delta G^{\circ} = -RT \ln K$

Then $\Delta G^{\circ} = \Delta G_1^{\circ} - \Delta G_2^{\circ}$

Hence $\qquad -RT \ln K = -2FE_1^\circ + 2FE_2^\circ$

or $\qquad\qquad \log K = \dfrac{2F}{2.303RT}(E_1^\circ - E_2^\circ)$

$$= \dfrac{2}{0.059 \text{ V}}(0.535 \text{ V} - 0.536 \text{ V})$$

or $\qquad\qquad K = 1.081$

Here $\qquad\qquad K = \dfrac{C_{I_3^-}}{C_{I^-}} = \dfrac{C_{I_3^-}}{0.5 - C_{I_3^-}} = 1.081$

or $\qquad\qquad C_{I_3^-} = 0.260$

Then the solubility of iodine in KI

$$= C_{I_3^-} + C_{I_2 (aq)}$$
$$= (0.260 + 0.00133) \text{ M}$$
$$\approx 0.260 \text{ M}$$

Note: In KI solution, the dissolved iodine exists almost entirely as I_3^- ion.

13.35 a.i. The relevant half reactions are

(1) $\qquad Cu^{2+} (aq) + 2e = Cu(s)$, $\qquad\qquad \Delta G_1^\circ = -2FE_1^\circ$

(2) $\qquad Cu^{2+} (aq) + e = Cu^+ (aq)$ $\qquad\qquad \Delta G_2^\circ = -2FE_2^\circ$

(1) – (2) gives $Cu^+ (aq) + e = Cu(s)$ $\qquad\qquad \Delta G^\circ = -FE^\circ$

Then $\qquad\qquad \Delta G^\circ = \Delta G_1^\circ - \Delta G_2^\circ$

Hence $\qquad\qquad -FE^\circ = -2FE_1^\circ + FE_2^\circ$

or $\qquad\qquad E^\circ = 2E_1^\circ - E_2^\circ$

$$= 2(0.34 \text{ V}) - 0.16 \text{ V}$$
$$= 0.52 \text{ V}$$

[**Note:** $E^\circ \neq E_1^\circ - E_2^\circ$, because E, unlike G, is not additive]

ii. In terms of E° of $Cu^+ (aq) + e = Cu$,

$$\mu_{Cu^+ (aq)}^\circ = FE^\circ, \text{ by convention (vide Section 13.3)}$$
$$= (9.65 \times 10^4 \text{ C mol}^{-1})(0.52 \text{ V})$$
$$= 5.02 \times 10^4 \text{ J mol}^{-1}$$

Similarly from E° of $Cu^{2+} (aq) + 2e = Cu$

$$\mu_{Cu^{2+} (aq)}^\circ = 2FE^\circ$$
$$= 2(9.65 \times 10^4 \text{ C mol}^{-1})(0.34 \text{ V})$$
$$= 6.56 \times 10^4 \text{ J mol}^{-1}$$

iii. $\qquad\qquad E^\circ = E_2^\circ - \dfrac{RT}{F} \ln K_{sp} CuI \quad \text{(vide problem 13.20)}$

$$= 0.16 \text{ V} - 0.059 \text{ V} \log (4 \times 10^{-12})$$
$$= 0.83 \text{ V}$$

iv. 2 × (2) – (1) gives

$$Cu + Cu^{2+} (aq) = 2Cu^+ (aq) \quad \Delta G^\circ = -RT \ln K$$

Then $\qquad\qquad \Delta G^\circ = 2\Delta G_2^\circ - 2\Delta G_1^\circ$

Hence $\qquad -RT \ln K = -2FE_2^\circ + 2FE_1^\circ$

or $\qquad \log K = \dfrac{2F}{2.303RT}(E_2^\circ - E_1^\circ)$

$\qquad\qquad = \dfrac{2}{0.059\text{ V}}(0.16\text{ V} - 0.34\text{ V})$ at 25°C

whence $\qquad K = 7.91 \times 10^{-7}$

b. i. Yes, because $E°$ for the half-reaction Cu^+ (aq) $+ e = Cu$ is much greater than that for Cu^{2+} (aq) $+ e = Cu^+$ (aq), when K for the reaction $2Cu^+$ (aq) $= Cu + Cu^2$ (aq) becomes very high [vide a(iv)].

ii. Although K for the reaction $Cu + Cu^{2+}$ (aq) $= 2Cu^+$ (aq) can be computed, a cell with such simply net reaction cannot be set up due to instability of Cu^+ (aq). A closely corresponding cell can, however, be set up with adequate stabilization of Cu^+, e.g. $Cu|CuCl$ (soln in conc HCl$||CuCl_2$ (aq soln)$|Cu$ with net reaction $Cu + Cu^{2+}$ (aq) $= 2Cu^+$ (as chloro complex).

iii. Cu^{2+} ion can liberate I, from KI through the reaction Cu^{2+} (aq) $+ 2I^-$ (aq) $= CuI$ (s) $+ \frac{1}{2}I_2$ (s) but not through the reaction Cu^{2+} (aq) $+ 2I^-$ (aq) $= Cu + I_2(s)$ and Cu^{2+} (aq) $+ I^-$(aq) $= Cu^+$ (aq) $+ \frac{1}{2}I_2$ (s). Because $E°$ of the half reaction $\frac{1}{2}I_2$ (s) $+ e = I^-$ (aq) is lower than that of Cu^{2+} (aq) $+ I^-$ (aq) $+ e = CuI$ (s), though it is higher than both $E_{Cu^{2+}/Cu}^\circ$ and E_{Cu^{2+}/Cu^+}°.

13.36 The given half-reactions are

(1) $\quad MnO_4^-$ (aq) $+ 8H^+$ (aq) $+ 5e = Mn^{2+}$(aq) $+ 4H_2O(l)$ $\quad \Delta G_1^\circ = -5FE_1^\circ$

(2) $\quad MnO_2$ (s) $+ 4H^+$(aq) $+ 2e = Mn^{2+}$(aq) $+ 2H_2O(l)$ $\quad \Delta G_2^\circ = -2FE_1^\circ$

(1) – (2) gives

$\qquad MnO_4^-$ (aq) $+ 4H^+$ (aq) $+ 3e = MnO_2$ (s) $+ 2H_2O(l)$ $\quad \Delta G° = -3FE°$

Then $\qquad \Delta G° = \Delta G_1^\circ - \Delta G_2^\circ$

Hence $\qquad -3FE° = -5FE_1^\circ + 2FE_2^\circ$

or $\qquad E° = \dfrac{5E_1^\circ - 2E_2^\circ}{3} = \dfrac{5 \times 1.51\text{ V} - 2 \times 1.23\text{ V}}{3} = 1.70\text{ V}$ ●

☐ Under ordinary condition MnO_4^- ion can under upto five equivalent reduction occurring only in successive steps each involving not more than two equivalent reduction. The given value of $E°$ for the half-reaction (1) represents the average of $E°$'s of half-reactions for reduction of different oxidation states to Mn^{2+} obtained on weighting each according to the number of equivalents involved. $E°$ for the half-reaction Tl^{3+} (aq) $+ 2e = Tl^+$ (aq) is less than E_1° but greater than E_2°. Therefore, Tl^+ can reduce MnO_4^- but only upto MnO_2.

Note: In predicting the product of reduction of MnO_4^-, it is safe to think separately in terms of half-reaction for reduction of each oxidation state to Mn^{2+}.

13.37 a. During titration, the potential of the Fe^{3+}/Fe^{2+} couple (given by Eq. (I)) increases and that of the Ce^{4+}/Ce^{3+} couple (given by Eq. (II)) decreases such that they become equal at each stage of the titration at equilibrium. It is this equilibrium potential which is measured in the potentiometric titration and hence it corresponds to either of the couples. This is inevitable, as the two couples share the same inert electrode in the measuring device

$$E = E^{\circ}_{Fe^{3+},Fe^{2+}} - \frac{RT}{F} \ln \frac{C_{Fe^{2+}}}{C_{Fe^{3+}}} \tag{I}$$

$$= E^{\circ}_{Ce^{4+},Ce^{3+}} - \frac{RT}{F} \ln \frac{C_{Ce^{3+}}}{C_{Ce^{4+}}} \tag{II}$$

[Here molar concentration, instead of activity, is used as is customarily done in case of a titration].

☐ (I) + (II) gives

$$E = \frac{E^{\circ}_{Fe^{3+},Fe^{2+}} + E^{\circ}_{Ce^{4+},Ce^{3+}}}{2} - \frac{RT}{2F} \ln \frac{C_{Fe^{2+}} C_{Ce^{3+}}}{C_{Fe^{3+}} C_{Ce^{4+}}}$$

$$= \frac{E^{\circ}_{Fe^{3+},Fe^{2+}} + E^{\circ}_{Ce^{4+},Ce^{3+}}}{2} \quad \text{at the equivalence point}$$

$$\left(\text{when } \frac{C_{Fe^{2+}}}{C_{Fe^{3+}}} = \frac{C_{Ce^{4+}}}{C_{Ce^{3+}}} \right)$$

which is average of E°'s of the two couples.

For two couples 1 and 2 involving n_1 and n_2 electron transfer per ion

$$E^{\circ} = \frac{n_1 E_1 + n_2 E_2}{n_1 + n_2}$$

b. i. 1%

ii. Substituting $E = \frac{1}{2}\left(E^{\circ}_{Fe^{3+},Fe^{2+}} + E^{\circ}_{Ce^{4+},Ce^{3+}}\right)$ in Eq. (I)

$$\frac{1}{2}(0.77 + 1.61)\,V = 0.77\,V - 0.059\,V \log \frac{C_{Fe^{2+}}}{C_{Fe^{3+}}}$$

whence $\quad \frac{C_{Fe^{2+}}}{C_{Fe^{3+}}} = 7.61 \times 10^{-8}$

Therefore, % of Fe^{2+} remaining unreacted

$$= \frac{C_{Fe^{2+}}}{C_{Fe^{2+}} + C_{Fe^{3+}}} \times 100$$

$$= 7.6 \times 10^{-8} \times 100$$

$$= 7.6 \times 10^{-6}$$

iii. By Eq. (I) and (II)

$$E^{\circ}_{Ce^{4+},Ce^{3+}} - \frac{RT}{F} \ln \frac{100}{1} = E^{\circ}_{Fe^{3+},Fe^{2+}} - \frac{RT}{F} \ln \frac{C_{Fe^{2+}}}{C_{Fe^{3+}}}$$

Then $\quad 1.61\,V - 0.59\,V \log 100 = 0.77\,V - 0.059\,V \log \frac{C_{Fe^{2+}}}{C_{Fe^{3+}}}$

whence $\quad \frac{C_{Fe^{2+}}}{C_{Fe^{3+}}} = 5.8 \times 10^{-13}$

Therefore, % of Fe^{2+} remaining unreacted

$$= \frac{C_{Fe^{2+}}}{C_{Fe^{2+}} + C_{Fe^{3+}}} \times 100$$

$$= 5.8 \times 10^{-13} \times 100$$

$$= 5.8 \times 10^{-11}$$

c.

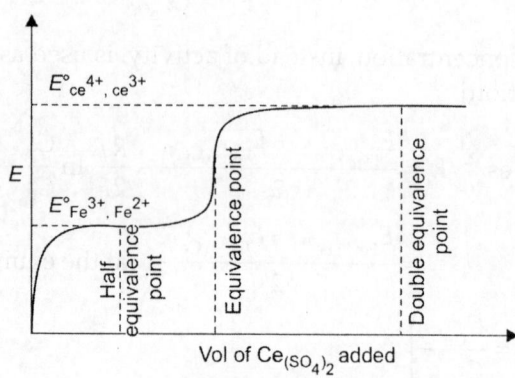

Fig. 13.1

Note:

1. In the potentiometric titration what is measured is the equilibrium potential of the electrode formed by the experimental solution containing the two couples reacting directly. It is entirely different from the potential of the cell formed by these two couples (connected through a salt bridge) without any direct reaction between them.

2. Although the relevant reaction is not absolutely complete at the equivalence point (due to reversible nature of chemical reaction), it is essentially complete, and more so after the equivalence point, for the estimation purpose.

13.38 Anode reaction $\qquad$ $Pb(s) + SO_4^{2-}(aq) = PbSO_4(s) + 2e$

Cathode reaction $\qquad$ $PbO_2(s) + 4H^+(aq) + SO_4^{2-}(aq) + 2e = PbSO_4(s) + 2H_2O(l)$

Net cell reaction $\quad$ $Pb(s) + PbO_2(s) + 4H^+(aq) + 2SO_4^{2-}(aq) = 2PbSO_4(s) + 2H_2O(l)$

Then emf of the cell $\quad E = E° + \dfrac{RT}{2F} \ln a_{H^+}^4 \cdot a_{SO_4^{2-}}^2$

$$= E° + \frac{0.059 \text{ V}}{2} \log a_\pm^6$$

Hence $\qquad \log a_\pm = \dfrac{E - E°}{3 \times 0.059 \text{ V}}$

$$= \frac{(2.016 - 2.04) \text{ V}}{3 \times 0.059 \text{ V}}$$

whence $\qquad a_\pm = 0.732$

Note: Here $PbSO_4$ is produced at the anode (due to oxidation) and also at the cathode (due to reduction).

13.39 a. Anode reaction $\quad$ Fe(s) + 2OH$^-$(aq) = FeO(s) + H$_2$O(l) + 2e

Cathode reaction $\quad$ Ni$_2$O$_3$(s) + H$_2$O(l) + 2e = 2NiO(s) + 2OH$^-$(aq)

Net cell reaction $\quad$ Fe(s) + Ni$_2$O$_3$(s) = FeO(s) + 2NiO(s)

b. Here $\Delta G^\circ = \mu^\circ_{FeO(s)} + 2\mu^\circ_{NiO(s)} - \mu^\circ_{Fe(s)} - \mu^\circ_{Ni_2O_3(s)} \neq 0$, because μ° is zero for the element Fe(s) but has different value for different compounds. Hence

$$E^\circ = -\frac{\Delta G^\circ}{2F} \neq 0$$

Again, since $\quad \Delta G^\circ = -RT\ln K \neq 0, K \neq 1$.

c. Since all the species appear in the cell reaction in pure state $\Delta G = \Delta G^\circ$, and hence $E = E^\circ$.

d. Since $E = E^\circ$, E will not be affected by addition of H$_2$SO$_4$ (though it neutralises some KOH, that does not appear in the cell reaction).

e. One advantage of the alkali storage cell is that its emf is stable because the cell reaction involves only the pure species in the solid state. But it has much lower emf than the acid storage cell.

UNIVERSITY QUESTIONS

13.1 What do you mean by a reversible electrochemical cell? Give an example.

(Calcutta BSc(H), 1999)

13.2 a. Distinguish between cell potential and electromotive force.

(Allahabad BSc, 2009)

b. Explain why emf of an electrochemical cell cannot be measured with the help of a voltmeter. (Delhi BSc, 2009)

13.3 What do you understand by the liquid junction potential? How does it arise?

(Madurai BSc, 2000)

13.4 A salt bridge can minimise the liquid junction potential. Explain.

(Kalyani BSc(H), 2003)

13.5 a. What is the parameter of utmost importance when an electrolyte is chosen for construction of a salt bridge? (Calcutta BSc(H), 1995)

b. For the potentiometric titration of AgNO$_3$ against KCl, what type of salt bridge should be used and why? (Kalyani BSc(H), 2003)

13.6 Devise a chemical cell without a liquid junction where the overall reaction will be

$$\tfrac{1}{2}H_2 (g, P_{H_2}) + AgCl(s) = AgCl(s) + H^+ (a_{H^+}) + Cl^- (a_{Cl^-})$$

and discuss the physico-chemical principle for the measurement of E° of such a cell.

(Calcutta BSc(H), 1993)

13.7 Derive an expression for the liquid junction potential across the electrolyte boundary of the following cell:

Ag(s)|AgCl(s)|HCl(a_1)|HCl(a_2)|AgCl(s)|Ag(s) $\qquad$ (Jadavpur BSc(H), 2013)

13.8 Explain what is implied by reversibility of an electrode with respect to some ion. Illustrate your answer with reference to H$^+$ and Cl$^-$ reversible electrodes.

(Calcutta BSc(H), 1995)

13.9 Define standard hydrogen electrode. What is the necessity of choosing such electrode? (Burdwan BSc(H), 2010)

13.10 What is a reference electrode? Give an example of a reference electrode and write an expression for its electrode potential. Comment on suitability of the electrode cited by you. (Burdwan BSc(H), 2000)

13.11 Justify/criticise

a. Cell emf is an extensive property. (Burdwan BSc(H), 1998)

b. Any cell whose emf is known is suitable as a standard cell.

(Burdwan BSc, 1999)

c. The equation $E = E° + \dfrac{RT}{nF} \ln Q$, where Q represents activity terms, demands that plot of E vs T would be a straight line. (Burdwan BSc(H), 1999)

13.12 Explain whether Q and E, in $E = E° + \dfrac{RT}{nF} \ln Q$, will be affected if the overall cell reaction is multiplied by x. (Burdwan BSc(H), 2009)

13.13 Write the SI unit of RT/F and its value at 300 K. (Burdwan BSc(H), 2006)

13.14 The standard electrode potentials for

$$Zn^{2+} + 2e \to Zn,\ E°_{298} = -0.76\ V$$

and $$Co^{2+} + 2e \to Co,\ E°_{298} = -0.28\ V \text{ at } 25°C$$

a. Which of the two metals is more noble?

b. If a solution containing one gm mole/litre of both $ZnSO_4$ and $CoSO_4$ is electrolysed with Pt electrodes at 25°C. What would be the approximate concentration of cobalt ions when deposition of Zn starts.

(Calcutta BSc(H), 1994)

13.15 a. With smooth Pt electrodes, decomposition potentials of H_2SO_4 and HNO_3 are found to be the same in their aqueous 1 m solution, although the acids are different. Explain.

b. Since the reversible emf of a voltaic cell with hydrogen and oxygen electrodes is 1.23 V, electrolysis of water from acidic or alkaline solution would require a minimum of 1.23 V. Comment. (Burdwan BSc(H), 1996)

13.16 An aq solution of aluminium sulphate cannot be used for electrolytic preparation of metallic aluminium. Explain. (Calcutta BSc(H), 1970)

13.17 Set up cells in which the following chemical reactions or changes would occur.

a. $\frac{1}{2}Cl_2 + Br^- = \frac{1}{2}Br_2 + Cl^-$ (Burdwan BSc(H), 2012)

b. $\frac{1}{2}H_2 + AgI = H^+ + I^- + Ag$ (Calcutta BSc(H), 2005)

c. $2H_2(g) + O_2(g) = 2H_2O(l)$ (Calcutta BSc(H), 2011)

d. $Hg + PbO(s) = Pb + HgO(s)$ (Calcutta BSc(H), 1977)

e. $Cd(Hg) + Hg_2SO_4 + \frac{8}{3}H_2O = CdSO_4 \cdot \frac{8}{3}H_2O + 2Hg$ (Calcutta BSc(H), 1997)

f. $Ag^+(aq) + Cl^-(aq) = AgCl(s)$ (Calcutta BSc(H), 2007)

g. $HCl + NaOH = NaCl + H_2O$ (Calcutta BSc(H), 1979)

h $CuSO_4(c_1) \to CuSO_4(c_2),\ c_2 < c_1$ (Calcutta BSc(H), 2003)

13.18 a. Write down the conditions under which the relation $\Delta G = -nFE$ holds.

(Burdwan BSc(H), 2005)

b. The reaction in a Daniel cell is represented as

$$Zn(s) + Cu^{2+}(aq) = Zn^{2+}(aq) + Cu(s)$$

If G°_{298} values for $Zn(s)$ and $Cu(s)$ are zero and those for $Zn^{2+}(aq)$ and $Cu^{2+}(aq)$ are -35.14 kCal mol^{-1} and 15.66 kCal mol^{-1} respectively, calculate E° of the cell and discuss the spontaneity of the above reaction under standard state.

(Calcutta BSc(H), 1997)

13.19 Consider the cell $Pt|H_2(g)|HCl(aq)|AgCl(s)|Ag$. Write down the cell reaction. Relate cell potential with ionic strength of the medium. Suggest a linear plot in order to obtain the standard emf of the cell and the Debye–Huckel constant.

(Burdwan BSc(H), 2009)

13.20 Design cells without liquid junction that can be used to determine the activity coefficients of aqueous solutions of (i) NaOH and (ii) H_2SO_4. Write down the cell reactions and hence find out the respective equations relating the emf to the mean ionic activity coefficient. (Burdwan BSc(H), 2008)

13.21 a. Cu/Cu^+ and Na/Na^+ electrodes are prepared in similar ways. Discuss.

(Burdwan BSc(H), 1996)

b. The quilibrium constant for the reaction $2Cu^+ = Cu^{2+} + Cu(s)$ at 25°C is 1.646×10^6

i. Construct a cell in which the reaction could occur.

ii. Calculate the standard emf of the cell at 25°C.

iii. Evaluate the standard potential of the Cu^+/Cu electrode at 25°C.

$[E^{\circ}_{Cu^{2+},Cu^+} = 0.153$ V at 25°C] (Calcutta BSc(H), 2004)

13.22 Consider the following half equations

$Fe^{2+}(aq) + 2e = Fe(s), \quad E^{\circ} = -0.44$ V

$4H^+(aq) + O_2(g) + 4e = 2H_2O(l), \quad E^{\circ} = +0.23$ V

Comment on the spontaneity of the corrosion reaction

$Fe(s) + 2H^+ + \frac{1}{2}O_2(g) = Fe^{2+}(aq) + H_2O(l)$ (Burdwan BSc(H), 2000)

13.23 Consider the half cell reaction $AgCl(s) + e = Ag(s) + Cl^-(aq)$. If $\mu^{\circ}_{AgCl} = -109.21$ kJ mol^{-1} and $E^{\circ} = 0.222$ V for this half cell, calculate $\mu^{\circ}_{Cl^-(aq)}$. (Burdwan BSc(H), 2009)

13.24 Given $\quad Fe^{3+}(aq) + e = Fe^{2+}(aq), \; E^{\circ}_{298} = 0.77$ V

$Fe^{2+}(aq) + 2e = Fe(s), \; E^{\circ}_{298} = -0.44$ V at 25°C

Calculate E° and ΔG° for the reaction $Fe^{3+}(aq) + 3e = Fe(s)$ and hence indicate $\mu^{\circ}_{Fe^{3+}(aq)}$ at 25°C. (Calcutta BSc(H), 1994)

13.25 Standard electrode potentials of Fe^{3+}/Fe^{2+} and $Fe(CN)_6^{3-}/Fe(CN)_6^{4-}$ couples are 0.77 V and 0.36 V respectively. Find the stability of $(FeCN)_6^{3-}$ complex relative to $Fe(CN)_6^{4-}$ complex. (Burdwan BSc(H), 1998)

13.26 Derive a cell in which the reaction is $AgBr(s) \rightleftharpoons Ag^+ + Br^-$. E° at 25°C for $Br^-/AgBr(s)/Ag$ and Ag^+/Ag electrodes are 0.071 V and 0.799 V respectively. Calculate K_{sp} of AgBr at 25°C. (Calcutta BSc(H), 2013)

13.27 Set up a reversible electrochemical cell in which the following overall reaction takes place

$$H^+ (a = 0.1) + OH^- (a = 0.1) \rightleftharpoons H_2O(l)$$

If emf of the cell is 0.708 V at 25°C, calculate the ionic product of water at 25°C.

(Calcutta BSc(H), 2000)

13.28 What are the required characteristics of a galvanic cell so that it can serve as a useful power source? (Jadavpur BSc(H), 2012)

13.29 Between 0°C and 60°C, the potential of the cells,

$Pb(s)|PbSO_4(s)|H_2SO_4(aq\,a)|PbSO_4(s)|PbO_2(s)|Pb(s)$, in 1 m H_2SO_4 is given by E (in volt) = $1.91737 + 56.1 \times 10^{-6}t + 108 \times 10^{-8}t^2$, where t is the Celsius temperature

   i. Calculate ΔG, ΔH and ΔS for the cell reaction at 25°C.

   ii. For the half cells at 25°C

$$PbO_2(s) + SO_4^{2-} + 4H^+ + 2e = PbSO_4(s) + 2H_2O, \quad E° = 1.6849\ V$$

$$PbSO_4(s) + 2e = Pb(s) + SO_4^{2-} \quad E° = -0.3553\ V$$

   Calculate the mean ionic activity coefficient of H_2SO_4 in 1 m H_2SO_4 at 25°C. Assume that the activity of water is unity. (Burdwan BSc(H), 2010

13.30 For a galvanic cell, ΔH of the unit cell reaction involving two-electron transfer is -65.7 kJ at 325 K. Calculate the emf of the cell at this temperature, given that it decreases linearly with increase of temperature at the rate of 2.2 mV K^{-1}.

(Jadavpur BSc(H), 2012)

13.31 The emf of a galvanic cell, having an exothermic cell reaction, is measured to be zero at a particular temperature. Find out the sign of temperature coefficient of the emf given justification. (Jadavpur BSc(H), 2013)

13.32 Describe the weston standard cell indicating the electrode reactions. The emf of this cell is 1.01530 V at 20°C and 1.01807 V at 25°C. Calculate ΔG, ΔH and ΔS for the cell reaction at 25°C. (Calcutta BSc(H), 1982)

13.33 The emf of the cell:

$Pt|Hg(l)|Mercurous\ nitrate\ (a_1 = 0.01)||Mercurous\ nitrate\ (a_2 = 0.10)|Hg(l)|Pt$

at 25°C is 0.0295 V. Find the molecular formula of mercurous nitrate.

(Calcutta BSc(H), 2000)

13.34 Find the cell reaction and calculate the potential of the following cell with transference:

$Pb(s)|PbSO_4(s)|CuSO_4(m_1 = 0.2, \gamma_\pm = 0.110):CuSO_4(m_2 = 0.02, \gamma_\pm = 0.320)|PbSO_4(s)|Pb$

Given that the transport number of Cu^{2+} is 0.37.

Is the cell reaction spontaneous as written? (Calcutta BSc(H), 1988)

13.35 The concentration cell $Ag|AgCl(s)|KCl\ (0.50\ N)|K_xHg|KCl\ (0.05\ N)|AgCl(s)|Ag$ has emf of -0.107 V. Formulate the corresponding cell with transference, for which the emf is found to be -0.05357 V. Evaluate the transference number of Cl^- ion.

(Burdwan BSc(H), 2011)

13.36 Give an example of a metal-insoluble salt like electrode which can be used as a pH-electrode. Explain its working principle and mention its limitations.

(Jadavpur BSc(H), 2013)

13.37 When a hydrogen electrode [$P(H_2) = 1$ atm] and a calomel electrode ($E_{cal} = 0.280$ V) are immersed in a solution at 25°C, a potential of 0.664 V is obtained. Calculate pH and the hydrogen ion activity. (Burdwan BSc(H), 2008)

13.38 The emf of the cell (at 25°C), Glass electrode || buffer solution | calomel electrode was 0.0232 V when buffer's pH was 2.5. The emf was increased to 0.112 V when another buffer solution was used. What was the pH of the second buffer?

(Calcutta BSc(H), 2013)

13.39 100 mL of a 0.1 M Fe^{2+} solution was titrated with 0.02 M $KMnO_4$ solution in 0.5 M H_2SO_4. Calculate the equilibrium constant at 25°C.

[Given $E°_{Fe^{3+},Fe^{2+}} = 0.771$ V and $E°_{MnO_4^-,Mn^{2+}} = 1.51$ V] (Burdwan BSc(H), 2011)

13.40 The electrode potential values of Cu^{2+}/Cu^+ and I_2/I^- suggest iodometric titration of Cu^{2+} ion is not possible but in practice, it is found otherwise. Explain
[$E°$ for the systems: $Cu^{2+}/Cu = 0.34$ V, $Cu^{2+}/Cu^+ = 0.15$ V, $Cu^{2+}/CuI = 0.87$ V, $I_2/I^- = 0.54$]   (Burdwan BSc(H), 2006)

13.41 A cell is set up with an aqueous solution containing Ce^{4+} and Ce^{3+} in the proportion of 80% and 20% coupled with an aqueous solution containing Fe^{3+} and Fe^{2+} in the proportion of 10% and 90%. The observed emf at 25°C is 0.93 V. Find the equilibrium constant for the reaction
$$Ce^{4+} + Fe^{2+} = Ce^{3+} + Fe^{3+} \quad at\ 25°C$$
Can this reaction be used for quantitative estimation?   (Calcutta BSc(H), 1995)

13.42 In a titration of 50 ml 0.1 N Fe^{2+} solution by 0.1 N Ce^{4+} solution, calculate the potentials when (i) 49.9 ml (ii) 50.1 ml of Ce^{4+} be added. Also calculate the concentrations of various reactants and products at the equivalence point of the above titration.
[Given $E°_{Fe^{3+},Fe^{2+}} = 0.77$ V and $E°_{Ce^{4+},Ce^{3+}} = 1.51$ V at 25°C]   (Burdwan BSc(H), 2000)

13.43 Show that, in a potentiometric titration, the equivalence point is denoted by most rapid change of the indicator electrode potential with the volume of the titrant added.   (Jadavpur BSc(H), 20120

13.44 In a potentiometric titration indicate the essential characteristic to be considered for the choice of an electrode. Explain with an example.
Show with a precipitation titration as an example that for better and convenient titration, the reversal of roles of titre and titrant requires change of electrodes.
(Calcutta BSc(H), 1995)

KEY TO UNIVERSITY QUESTIONS

13.1 Vide Section 13.2.

13.2 a. Cell potential, i.e. the potential difference between the electrodes of a cell, is a variable quantity depending on the strength of the electric current passing through the cell. The highest value of cell potential, that happens at zero current, is the emf which is a fixed quantity of a particular cell of fixed composition.
  b. Vide problem 13.2(ix).

13.3 By the liquid junction potential, associated with an electrochemical cell, we mean a potential difference across the interface of two electrolyte solutions in contact.
  ☐ This arises from the difference in mobilities of the cations and anions of the electrolytes that results in different initial rate of their interdiffusion occurring across the interface of the two electrolyte solutions. Vide problem 13.23.

13.4 Vide problem 13.2(vii).

13.5 a. The relevant parameter is the ionic mobility which has to be at least nearly same for the constituent ions of an electrolyte to be used in salt bridge.
  b. A salt bridge containing NH_4NO_3 is the suitable one, because NH_4NO_3 is unreactive to both $AgNO_3$ and KCl.

13.6 Vide problem 13.30.

13.7 For 1 F of electricity passing through the cell from left to right, the changes at the liquid junction due to transfer processes are
$$t_{H^+}.H^+(m_1) \rightarrow t_{H^+}.H^+(m_2)$$
$$t_{Cl^-}.Cl^-(m_2) \rightarrow t_{Cl^-}.Cl^-(m_1)$$

The emf associated with these changes, which is the liquid junction potential E_j, is then given according to the Nernst equation, by

$$E_j = t_{H^+} \frac{RT}{F} \ln \frac{a_1^{\frac{1}{2}}}{a_2^{\frac{1}{2}}} + t_{Cl^-} \frac{RT}{F} \ln \frac{a_2^{\frac{1}{2}}}{a_1^{\frac{1}{2}}}$$

$$= \frac{1}{2}(t_{H^+} - t_{Cl^-}) \frac{RT}{F} \ln \frac{a_1}{a_2}$$

13.8 The reversibility of an electrode with respect to some ion implies that this ion is directly involved in the reversible electrode process concerned, or can remain in rapid equilibrium with any such ion.

□ Vide problems 13.15 and 13.20 for illustration.

13.9 A standard hydrogen electrode consists of hydrogen gas bubbled at one bar pressure over a Pt-wire dipped in an aqueous solution containing hydrogen ion at unit mean ionic activity.

□ Vide Section 13.3.

Note: By IUPAC convention, all standard electrode potentials are reduction potentials, i.e. one associated with a reduction process.

13.10 Reference electrodes are ones of preassigned potentials.

□ The standard hydrogen electrode serves as the basic reference electrode whose potential has been conventionally taken to be zero at 25°C, while the calomel and silver-silver chloride electrodes, which are more easy to prepare and handle compared to the hydrogen electrode, are often used as the auxiliary reference electrodes with their potentials based on that of standard hydrogen electrode. For expression of electrode pot, vide Section 13.4(c).

□□ Both the auxiliary reference electrodes are thermodynamically equivalent to the chlorine-chloride electrode, but are preferred to the latter because they donot involve any gas.

13.11 a. Vide problem 13.1(v).

b. For a cell to be used as a standard cell, knowledge of its emf is not enough. Its emf should also be stable at a particular temperature and have low temperature coefficient. Therefore, the given statement is not justified.

c. Vide problem 13.1(x).

13.12 Multiplication with x affect the extent of cell reaction and hence Q but not E which is an intensive quantity.

Note: $\ln Q$ is proportional to n.

13.13 For the consistency of units in the Nernst equation for cell emf

$$E = E° + \frac{RT}{nF} \ln Q$$

RT/F must have the same unit as E, i.e. volt in SI (n and $\ln Q$ having no units).

At 300 K, $\quad \dfrac{RT}{F} = \dfrac{(8.314 \text{ JK}^{-1}\text{mol}^{-1})(300 \text{ K})}{(96485 \text{ C mol}^{-1})}$

$$= 0.02585 \text{ volt}$$

13.14 a. Since $E°_{Co^{2+}/Co} > E°_{Zn^{2+}/Zn}$, Co is more noble, i.e. less tendency to be oxidized.

b. On subjecting the solution to electrolysis only Co^{2+} will discharge till the potential $E_{Co^{2+}/Co}$ drops to the discharge potential of Zn^{2+}. Taking this to be $E^{\circ}_{Zn^{2+}/Zn}$ (ignoring the overpotential), the required concentration of Co^{2+} would be given by

$$E^{\circ}_{Co^{2+}/Co} + \frac{RT}{2F} \ln C_{Co^{2+}} = E^{\circ}_{Zn^{2+}/Zn}$$

(ignoring the activity coefficient effect)

Then $= -0.28 \text{ V} + \dfrac{0.059 \text{ V}}{2} \log C_{Co^{2+}} = -0.76 \text{ V}$

whence $C_{Co^{2+}} = 5.3 \times 10^{-17} \text{ M}$

Note: Virtually entire Co^{2+} is discharged when Zn^{2+} starts discharging.

13.15 a. Beause the process of electrolysis happening with both the solutions is essentially same which is decomposition of water through the following most likely half-reactions

$$2H^+ (aq) + 2e = H_2(g) \text{ at the cathode}$$

$$H_2O (l) = 2H^+ (aq) + \tfrac{1}{2}O_2 (g) + 2e \text{ at the anode}$$

Note: The two solutions will not give the same products of electrolysis at all concentrations.

b. In real electrolysis, the electrode processes occur irreversibly. For hydrogen electrode, the reversible potential associated with the reduction corresponds to the equilibrium $2H^+ (aq) + 2e = H_2(g)$. To disturb this equilibrium to the right, the potential of the electrode (metal) has to be lowered (i.e. made more negative) so that $\bar{\mu}_e$ is enhanced tending e to be consumed. Similarly, to disturb the equilibrium $H_2O (l) = 2H^+ (aq) + \tfrac{1}{2}O_2 (g) + 2e$ related to the oxygen electrode, to the right the potential of the electrode has to be raised (i.e. made more positive). The applied emf for electrolysis of water must therefore exceed the sum of the potentials corresponding to the two half-reaction equilibrium, i.e. 1.23 V.

Note: The given statement is in agreement with the thermodynamic principle that a reversible change of a system will produce greater work than when the change is brought about irreversibly, or in reverse order when consumption of work is concerned as in the present problem where electrical work is consumed.

13.16 Because Al^{3+} ion has much higher discharge potential (above 1.7 V) than H^+ ion even considering the overpotential which cannot be enhanced using alkaline medium due to its reaction with Al.

Note: Although Na^+ ion has higher discharge potential than Al^{3+} ion, the discharge of Na^+ is possible in alkaline medium over Hg that dissolves Na, leading to lowering of discharge potential.

13.17 The relevant cells are

a. $Pt|Br_2 (l)| Br^- (aq)\|Cl^- (aq)|Cl_2 (g)|Pt$

Anode reaction $\qquad\qquad Br^- (aq) = \tfrac{1}{2}Br_2 (l) + e$

Cathode reaction $\qquad \tfrac{1}{2}Cl_2 (g) + e = Cl^- (aq)$

Net reaction $\qquad \tfrac{1}{2}Cl_2 (g) + Br^- (aq) = \tfrac{1}{2}Br_2 (l) + Cl(aq)$

b. $Pt|H_2 (g)| HI(aq)|AgI(s)|Ag$ [vide problem 13.30]

c. Vide problem 13.9(iii).

d. $Pt|Hg(l)|HgO(s)|OH^-(aq)|PbO(s)|Pb$ vide problem 13.9(vii).

Anode reaction $Hg(l) + 2OH^-(aq) = HgO(s) + H_2O(l) + 2e$

Cathode reaction $PbO(s) + H_2O(l) + 2e = Pb(s) + 2OH^-(aq)$

e. Vide problem 13.6

f. Vide problem 13.29.

g. Vide problem 13.9(iv).

h. Vide problem 13.9(viii).

13.18 a. Vide Section 13.3.

b. For the cell reaction

$$\Delta G^\circ = G^\circ_{Zn^{2+}(aq)} + G^\circ_{Cu(s)} - G^\circ_{Zn(s)} - G^\circ_{Cu^{2+}(aq)}$$

$$= (-35.14 \times 4.18 \times 10^3 + 0 - 0 - 15.66 \times 4.8 \times 10^3) J\ mol^{-1}$$

$$= -212344\ J\ mol^{-1}$$

Since ΔG° is negative, the cell reaction will occur spontaneously under standard state

$$E^\circ = -\frac{\Delta G^\circ}{nF}$$

$$= -\frac{-212344\ J\ mol^{-1}}{(2)(96485\ C\ mol^{-1})} = 1.10\ V$$

13.19 Here $\qquad E = E^\circ - \dfrac{2RT}{F} \ln(m\gamma_\pm)$ vide problem 13.30

$$= E^\circ - \frac{2RT}{F} \ln m + \frac{2RT \times 2.303}{F} A\sqrt{I}$$

(by Debye–Huckel limiting law, where I is ionic strength of the medium) which is the desired relation. This may be written as

$$E + \frac{2RT}{F} \ln m = E^\circ + \frac{4.606RT}{F} A\sqrt{m} \quad \text{since } I = m$$

The plot of $E + \dfrac{2RT}{F} \ln m$ versus $\sqrt{m}$ will be linear having intercept = E° and slope $= \dfrac{4.606RT}{F} A$ wherefrom A can be obtained.

13.20 i. A relevant cell is $Pt|Na(Hg)|NaOH(m)|H_2(1\ atm)|Pt.$

Here cell reaction is

$2Na(Hg) + 2H_2O(l) = 2Na^+(aq) + 2OH^-(aq) + H_2(1\ atm)$ vide problem 13.9(vi)

Then $\qquad E^\circ = E^\circ - \dfrac{RT}{2F} \ln a^2_{Na^+} a^2_{OH^-}$

$$= E^\circ - \frac{RT}{F} \ln[(m)(m)\gamma^2_\pm]$$

$$= E^\circ - \frac{2RT}{F} \ln(m\gamma^2_\pm)$$

ii. A relevant cell is $pt|H_2 (1\ atm)|H_2SO_4\ (m)\| Ag_2SO_4\ (s)|Ag$.

Here cell reaction is

$$Ag_2SO_4(s) + H_2(1\ atm) = 2Ag(s) + 2H^+(aq) + SO_4^{2-}\ (aq)\ \text{vide problem 13.31}$$

Then

$$E = E^\circ - \frac{RT}{2F} \ln(a_{H^+}^2 \cdot a_{SO_4^{2-}})$$

$$= E^\circ - \frac{RT}{2F} \ln[(2m)^2\ (m)\ \gamma_\pm^3]$$

13.21 a. To prepare Cu/Cu^+ electrode, $Cu^+(aq)$, that readily disproportionates, has to be stabilised through complex formation, e.g. using concentrated HCl. The problem is similar with Na/Na^+ electrode where high reactivity of the Na toward H_2O has to be reduced through formation of amalgam.

b. i. Vide problem 13.35[b(ii)]

ii.
$$E^\circ = \frac{RT}{nF} \ln K$$

$$= \frac{0.059\ V}{n} \log K\ \text{at 25°C}$$

$$= \frac{0.059\ V}{(1)} \log (1.646 \times 10^6) = 0.367\ V$$

iii. Vide problem 13.35[a(i)].

13.22 Considering the two half-reactions

1. $\qquad Fe^{2+}(aq) + 2e = Fe(s) \qquad\qquad\qquad \Delta G_1^\circ = -2FE_1^\circ$

2. $\ 4H^+(aq) + O_2(g) + 4e = 2H_2O(l) \qquad\quad \Delta G_2^\circ = -4FE_2^\circ$

We have the corrosion reaction, from $\frac{1}{2} \times (2) - (1)$

$$Fe(s) + 2H^+(aq) + \tfrac{1}{2}O_2(g) = Fe^{2+}(aq) + H_2O(l)$$

Then ΔG° of this reaction will be

$$\Delta G^\circ = \tfrac{1}{2}(-4FE_2^\circ) - (-2FE_1^\circ)$$

$$= 2F(E_1^\circ - E_2^\circ)$$

$$= 2\,(96485\ C\ mol^{-1})\,(-0.44 - 0.23)\ V$$

$$= -129290\ J\ mol^{-1}$$

Since ΔG° is negative corrotion will occur spontaneously under standard state. It will be favoured more by higher acidity of the medium and higher oxygen pressure, the accompanying ΔG being then more negative.

13.23
$$E^\circ = \frac{-\Delta G^\circ}{F} = \frac{-(\mu_{Ag(s)}^\circ + \mu_{Cl^-(aq)}^\circ - \mu_{AgCl(s)}^\circ)}{F}$$

Then
$$0.222\ V = \frac{-[0 + \mu_{Cl^-(aq)}^\circ - (-109.721 \times 10^3\ J\ mol^{-1})]}{96485\ C\ mol^{-1}},\ \text{at 25°C}$$

Whence $\quad \mu_{Cl^-(aq)}^\circ = -131141\ J\ mol^{-1}$

13.24 The specified half-reactions are

1. $\qquad Fe^{3+}(aq) + e = Fe^{2+}(aq) \qquad\qquad\qquad \Delta G_1^\circ = -FE_1^\circ$

2. $\qquad Fe^{2+}(aq) + 2e = Fe(s) \qquad\qquad\qquad\quad \Delta G_2^\circ = -2FE_2^\circ$

(1) + (2) gives

$$Fe^{3+}(aq) + 3e = Fe(s) \qquad\qquad \Delta G° = -3FE°$$

Then $\qquad \Delta G° = \Delta G_1° + \Delta G_2°$

Hence $\qquad -3FE° = -FE_1° - 2FE_2°$

or $\qquad E° = \dfrac{E_1° + 2E_2°}{3}$

$$= \dfrac{0.77\text{ V} + 2(-0.44\text{ V})}{3} = -0.037\text{ V}$$

$$\Delta G° = -3FE°$$
$$= -3(9.65 \times 10^4\text{ C mol}^{-1})(-0.037\text{ V})$$
$$= 1.07 \times 10^4\text{ J mol}^{-1}$$

$$\mu°_{Fe^{3+}(aq)} = -\Delta G°$$

13.25 Considering the following processes

1. $\qquad Fe^{3+} + e = Fe^{2+} \qquad\qquad \Delta G_1° = -FE_1°$

2. $\qquad Fe(CN)_6^{3-} + e = Fe(CN)_6^{4-} \qquad \Delta G_2° = -FE_2°$

I. $\quad Fe^{3+} + 6CN^- = Fe(CN)_6^{3-} \qquad\qquad \Delta G_I° = -RT \ln K_I$

where K represents stability constant

II. $\quad Fe^{2+} + 6CN^- = Fe(CN)_6^{4-} \qquad\qquad \Delta G_{II}° = -RT \ln K_{II}$

We find that (II) − (I) ≡ (2) − (1)

Then $\qquad\qquad \Delta G_{II}° - \Delta G_I° = \Delta G_2° - \Delta G_1°$

Hence $\quad -RT \ln K_{II} + RT \ln K_I = RT \ln Kr$

(where Kr is the stability of $Fe(CN)_6^{3-}$ relative to $Fe(CN)_6^{4-}$)

$$= -FE_2° + FE_1°$$

Therefore $\qquad$ 0.059 V log Kr = −0.36 V + 0.77 V

Whence $\qquad\qquad Kr = 8.9 \times 10^6$

[The ions are all in aqueous solution]

13.26 The relevant cell is $Ag|Ag^+(aq)\|Br^-(aq)|AgBr(s)|Ag.$ $\quad$ vide problem 13.29

☐ $\qquad E°_{Br^-|AgBr|Ag} = E°_{Ag^+|Ag} + \dfrac{RT}{F}\ln K_{th}\ (AgBr)$ $\quad$ vide problem 13.20

or $\qquad$ 0.071 V = 0.799 V + 0.059 V log K_{th} at 25°C

whence $\qquad K_{th} = 4.58 \times 10^{-13}$

13.27 The relevant cell is $Pt|H_2(P)|OH^-\ (a = 0.1)\|H^+\ (a = 0.1)|H_2(P)|Pt.$

☐ Treating the cell as concentration cell with $a_{H^+} = \dfrac{K_w}{0.1}$ for the left electrode and $a_{H^+} = 0.1$ for the right electrode

$$E = \dfrac{RT}{F}\ln\dfrac{0.1}{K_w/0.1}$$

Then $\qquad$ 0.807 = 0.059 V log $\dfrac{10^{-2}}{K_w}$ at 25°C

whence $\qquad K_w = 10^{-14}$

13.28 For a galvanic cell to serve as an efficient power source, the most important criterian is that its internal resistance has to be sufficiently low so that the dissipation of power (mostly as heat) is least. This can be achieved using electrolytes of high conductivity and making interelectrode distance short. Further, for ready power supply the electrode processes need to be quite fast.

Note: The much used lead storage battery amply fulfills these criteria. Here, during charging $PbSO_4$ is reduced to Pb and oxidised to PbO_2 on the Pb electrodes (vide problem 13.38), instead of decomposition of H_2O into H_2 and O_2. This is due to higher discharge potential (overvoltage) for libration of H_2 and O_2 on Pb.

13.29 i.
$$\Delta G = -nFE \quad \text{Here } n = 2$$

$$\Delta S = nF\left(\frac{\partial E}{\partial T}\right)_P$$

$$= nF\left(\frac{\partial E}{\partial t}\right)_P \quad \text{since } \frac{\partial E}{\partial T} = \frac{\partial E}{\partial t}$$

$$\Delta H = \Delta G + T\Delta S$$

ii.
$$E = E° + \frac{RT}{F}\ln a_{H^+}^2 a_{SO_4^{2-}} \quad \text{vide problem 13.38}$$

$$= E° + 0.059 \text{ V} \log[m_{H^+}^2 \cdot m_{SO_4^{2-}} \cdot \gamma_{\pm}^3] \quad \text{at 25°C}$$

$$= (1.6849 \text{ V} + 0.3553 \text{ V}) + 0.059 \text{ V} \log[(2)^2(1)\gamma_{\pm}^3]$$

for H_2SO_4 of unit molality, etc.

13.30
$$E = -\frac{\Delta H}{nF} + T\left(\frac{\partial E}{\partial T}\right) \quad \text{by Eq. (13.13)}$$

$$= -\frac{(-65.7 \times 10^3 \text{ J mol}^{-1})}{(2)(96485 \text{ C mol}^{-1})} + (325 \text{ K})(-2.2 \times 10^{-3} \text{ V K}^{-1})$$

$$= -0.37 \text{ V}$$

The negative value of E thus calculated implies that the measured emf will be +0.37 V corresponding to the spontaneous reaction which is reverse of the reaction concerned.

13.31 From the relation $\Delta H = -nFE + nFE\left(\frac{\partial E}{\partial T}\right)_P$, it follows that for an exothermic cell reaction, i.e. one with negative ΔH, $\left(\frac{\partial E}{\partial T}\right)_P$ will be negative for E to be zero. Here $E = 0$ is justified. But $E°$ is unlikely to be so, because this corresponds to K (equilibrium constant) = 1 which is unusual.

13.32 Vide problem 13.6.

For calculation of ΔG, ΔH and ΔS [vide problem 13.26].

Here
$$\frac{\partial E}{\partial T} = \frac{\Delta E}{\Delta T}$$

13.33 Let $Hg_x(NO_3)_x$ be the molecular formula of mercurous nitrate.

Then for the given concentration cell x electrons will be transferred for the unit cell reaction represented as

$$\underset{(a_2\,=\,0.1)}{Hg_x(NO_3)_x} \rightarrow \underset{(a_1\,=\,0.1)}{Hg_x(NO_3)_x}$$

Hence
$$E = \frac{RT}{xF}\ln\frac{a_2}{a_1}$$

$$= \frac{0.059\text{ V}}{x}\log\frac{a_2}{a_1} \quad \text{at } 25°C$$

or
$$x = \frac{0.059\text{ V}}{E}\log\frac{a_2}{a_1}$$

$$= \frac{0.059\text{ V}}{0.0295\text{ V}}\log\frac{0.1}{0.01} = 2$$

13.34 For two faradays of electricity passing through the cell from left to right, the following processes will occur in the cell

Anode reaction $\quad Pb + SO_4^{2-}(m_1) = PbSO_4(s) + 2e$

Cathode reaction $\quad PbSO_4(s) + 2e = Pb + SO_4^{2-}(m_2)$

Transfer process $\quad t_{Cu^{2+}}Cu^{2+}(m_1) = t_{Cu^{2+}}Cu^{2+}(m_2)$

$$t_{SO_4^{2-}}SO_4^{2-}(m_2) = t_{SO_4^{2-}}SO_4^{2-}(m_1)$$

Net cell reaction

$$SO_4^{2-}(m_1) + t_{Cu^{2+}}Cu^{2+}(m_1) + t_{SO_4^{2-}}SO_4^{2-}(m_2) = SO_4^{2-}(m_2) + t_{Cu^{2+}}Cu^{2+}(m_2) + t_{SO_4^{2-}}SO_4^{2-}(m_1)$$

or
$$t_{Cu^{2+}}Cu^{2+}(m_1) + t_{SO_4^{2-}}SO_4^{2-}(m_1) = t_{Cu^{2+}}Cu^{2+}(m_2) + t_{SO_4^{2-}}SO_4^{2-}(m_2)$$

$$\text{putting } t_{SO_4^{2-}} = 1 - t_{Cu^{2+}}$$

$$E = t_{Cu^{2+}}\frac{RT}{2F}\ln\frac{(a_{Cu^{2+}}\cdot a_{SO_4^{2-}})_1}{(a_{Cu^{2+}}\cdot a_{SO_4^{2-}})_2}$$

$$= t_{Cu^{2+}}\frac{RT}{F}\ln\frac{(m_\pm\gamma_\pm)_1}{(m_\pm\gamma_\pm)_2}$$

$$= (0.37)(0.059\text{ V})\log\frac{(0.2)(0.110)}{(0.02)(0.320)} \quad \text{since } m_\pm = m$$

$$= 0.0117$$

☐ Since E is positive, the cell reaction as written above is spontaneous.

13.35 This is similar to problem 13.22 differing in that here the electrodes are reversible wrt the anion (Cl^-).

13.36 An example of the desired electrode is silver-silver oxide electrode for which the reduction half-reaction is

$$Ag_2O + H_2O(l) + 2e = Ag + 2OH^-(aq)$$

and
$$E = E° - \frac{RT}{F}\ln a_{OH^-}$$

☐ Being reversible wrt OH^- ions, this electrode can be used for pH measurement provided the experimental solution is not acidic (enough to dissolve Ag_2O) and is free from reducing agent (e.g. glucose) and complexing agents (e.g. NH_3).

13.37
$$E_{obs} = E_{Cal} - E_{Hydrogen}$$
$$= E_{Cal} + 0.059\ pH \quad \text{by Eq. (13.6)}$$

or
$$pH = \frac{E_{obs} - E_{cal}}{0.059}$$

$$= \frac{0.664\ V - 0.280\ V}{0.059\ V} = 6.51$$

Then
$$a_{H^+} = 10^{-pH} = 10^{-6.51} = 3.09 \times 10^{-7}$$

Note: We write $E_{obs} = E_{cal} - E_{hydrogen}$ (and not $E_{obs} = E_{hydrogen} - E_{cal}$). This follows from the order of standard potentials, $E^\circ_{cal} > E^\circ_{hydrogen}$, remembering that E_{obs} is positive.

13.38 Vide problem 13.12.

13.39 The involved red-ox reaction is

$$MnO_4^- (aq) + 5Fe^{2+} (aq) + 8H^+ (aq) = Mn^{2+} (aq) + 5Fe^{3+} (aq) + 4H_2O (l)$$

which involves 5-electron transfer per MnO_4^- ion. If the reaction were carried out through an electrochemical cell, the standard emf of the cell E° would be related to the equilibrium constant K of the reaction by

$$\frac{RT}{5F} \ln K = E^\circ = E^\circ_{MnO_4^-,Mn^{2+}} - E^\circ_{Fe^{3+},Fe^{2+}}$$

or
$$\frac{0.059\ V}{5} \log K = 1.51\ V - 0.771\ V$$

whence
$$K = 4.24 \times 10^{62}$$

13.40 Vide problem 13.35[b(iii)].

13.41 Using the given data at 25°C.

$$E_{Ce^{4+},Ce^{3+}} = E^\circ_{Ce^{4+},Ce^{3+}} - 0.059\ V \log \tfrac{20}{80}$$

$$E_{Fe^{3+},Fe^{2+}} = E^\circ_{Fe^{3+},Fe^{2+}} - 0.059\ V \log \tfrac{90}{10}$$

[by Eqs (I) and (II) in problem 13.37(a)]

Now
$$E_{obs} = E_{Ce^{4+},Ce^{3+}} - E_{Fe^{3+},Fe^{2+}}$$

Then
$$0.934 = (E_{Ce^{4+},Ce^{3+}} - E_{Fe^{3+},Fe^{2+}}) - 0.059\ V \log \tfrac{20}{80} + 0.059\ V \log \tfrac{90}{10}$$

$$= 0.059\ V \log K - 0.059\ V \log \tfrac{20}{80} + 0.059\ V \log \tfrac{90}{10}$$

whence
$$K = 1.61 \times 10^{14}$$

☐ Since the reaction has so high value of its equilibrium constant, it can be effectively used for quantitative estimation.

13.42 Since the two solutions have the same normality, one will be equivalent to the other of equal volume.

i. This corresponds to a stage of titration (0.2%) before the equivalence point where Fe^{2+} (but not Ce^{4+}) ions remain in significant extent. Here the potential can be conveniently expressed with reference to the iron couple as

$$E = E^\circ_{Fe^{3+},Fe^{2+}} - \frac{RT}{F} \ln \frac{C_{Fe^{2+}}}{C_{Fe^{3+}}}$$

$$= 0.77\ V - 0.059\ V \log \frac{50 - 49.9}{49.9} = 0.929\ V$$

ii. This corresponds to a stage of titration (0.2%) after the equivalence point where Ce^{4+} (but not Fe^{2+}) ions remain in significant extent. Here the potential can be conveniently expressed with reference to the cerium couple as

$$E = E^{\circ}_{Ce^{4+},Ce^{3+}} - \frac{RT}{F} \ln \frac{C_{Ce^{3+}}}{C_{Ce^{4+}}}$$

$$= 1.61 \text{ V} - 0.059 \text{ V} \log \frac{50}{50.1 - 50} = 1.45 \text{ V}$$

☐ Fraction of Fe^{2+} ions (and also Ce^{4+} ions) remaining unreacted at the equivalence point

$$= 7.6 \times 10^{-8} \quad \text{vide problem 13.37[b(ii)]}$$

Then at the equivalence point

$$C_{Fe^{2+}} = C_{Ce^{4+}} = \tfrac{1}{2} \times 0.1 \times 7.6 \times 10^{-8} \text{ N} \quad \text{the factor } \tfrac{1}{2} \text{ is due to double dilution}$$

$$C_{Fe^{3+}} = C_{Ce^{3+}} \approx \tfrac{1}{2} \times 0.1$$

13.43 Let us consider this with reference to the titration of Fe(II) with Ce(IV), supposing that initially iron is present almost entirely as Fe(II) and Cerium as Ce(IV). At any stage of the titration before the equivalence point let the fraction of Fe(II) reacted be x. Then the potential at this stage will be

$$E = E^{\circ}_{Fe^{3+},Fe^{2+}} - \frac{RT}{F} \ln \frac{C_{Fe^{2+}}}{C_{Fe^{3+}}}$$

$$= E^{\circ}_{Fe^{3+},Fe^{2+}} - \frac{RT}{F} \ln \frac{x}{1-x}$$

Therefore, near the equivalence point $\ln \dfrac{x}{1-x}$ and hence E will rise rapidly to a large value as $x \to 1$.

13.44 To carry out a titration potentiometrically, the electrode should necessarily be reversible with respect to the ion(s) involved in the relevant reaction. For example, in the HCl–NaOH titration, where the net reaction is $H^+ + OH^- = H_2O$, the hydrogen electrode will serve the purpose because H^+ ions are involved directly and OH^- ions indirectly in the associated electrode process.

☐ Considering the potentiometric titration of $AgNO_3$ with KI, where the net reaction is $Ag^+ (aq) + I^- (aq) = AgI (s)$, Ag electrode is the appropriate one to be used because Ag^+ ions are directly involved in the electrode process. But in the reverse titration, it is better and more convenient to choose the Ag/AgI electrode because the associated electrode process involves I^- ions directly and Ag^+ ions indirectly (vide problem 13.20).

14

Electrolytic Conductance

Electrical energy and thermal energy flow through matter following similar empirical rule but different mechanism as it appears from the fact that a good thermal conductor is not always a good electrical conductor, e.g. diamond (which has the highest room-temperature thermal conductivity of any substance but is poor in electrical conductivity).

14.1 CONDUCTION OF ELECTRICITY

Conduction of electricity means flow of electric charge. Since charge requires some material particle to carry it, the flow of electricity cannot occur through a mechanism analogous to radiation of heat. It happens either through migration of electrons, or through migration of charged molecules (called ions). Then, for a substance to conduct electricity, it must contain mobile charged particles—which are electrons in case of metals or like substances called electronic conductors while these are ions in case of electrolytic conductors or simply electrolytes. The conducting behaviours of electrolytes in solution have some interesting features. The conductance data regarding electrolytes are usually referred to their aqueous solutions. Electrolytes are called strong or weak according as they are (practically) completely or partially dissociated in their aqueous solutions of ordinary concentrations.

14.2 OHM'S LAW AND CONDUCTANCE

The flow of electric charge per unit time through any cross section of a conductor is called electric current which is a macroscopic quantity characteristic of the conductor as a whole. But current density, which is the flow of electric charge per unit time through unit cross section of the conductor perpendicular to the direction of current, is a microscopic quantity characteristic of a point within the conductor.

Electrolytic conductors, like electronic conductors, usually obey Ohm's law, i.e.

$$\text{current} \propto \text{voltage applied}$$

the proportionality constant is called conductance (inverse of resistance) of the conductor concerned.

The conductance of an electrolytic solution is measured using a conductivity cell which usually consist of two thin parallel inert-metal plates of equal size held fixed at certain distance apart. The distance between the electrodes divided by the area of each electrode is called cell constant of the cell. The conductance is measured with alternating current and it depends on the cell constant of the cell used.

14.3 DIFFERENT TYPES OF CONDUCTIVITIES

Ohm's law in terms of current density is

$$j \propto E, \text{ where } E \text{ is the applied field or potential gradient}$$
$$= kE \tag{14.1}$$

The proportionality constant k is called specific conductivity (also specific conductance) of the conducting species, which is inverse of specific resistance or resistivity, and is given by

$$k = \frac{1}{r} \cdot \frac{l}{a} \tag{14.2}$$

where r is the resistance of the electrolyte placed between two parallel electrodes each of area a and separated by a distance l.

Then specific conductivity, or simply conductivity, of an electrolyte in a solution (or in melt) is its conductance when unit volume of it is placed between two parallel electrodes each of unit area and set unit distance apart. It is obtained by multiplying measured conductance $(1/r)$ with the cell constant (l/a) of the cell used.

It is more useful to think conductivity on concentration basis by defining a quantity called equivalent conductivity (also equivalent conductance) which is the conductance of one gram equivalent of electrolyte in a solution placed between two large parallel electrodes set unit distance apart. It follows from this definition that equivalent conductivity (Λ) and specific conductivity (k) will be related as

$$\Lambda = kV \tag{14.3a}$$

where V is the volume of the solution containing 1 g equivalent of the electrolyte.
In conventional unit

$$\Lambda = \frac{1000 \, k}{C} \tag{14.3b}$$

where k is in siemens (S)/cm, i.e. ohm $(\Omega)^{-1}\text{cm}^{-1}$ and C is normality.

At infinite dilution (i.e. when further dilution has no significant effect) each of the constituent ions of an electrolyte contributes independently to Λ of the electrolyte. This contribution of an ion is called its equivalent ionic conductivity represented by λ. For a binary electrolyte

$$\Lambda = \lambda_+ + \lambda_- \tag{14.4}$$

where λ_+ corresponds to cation and λ_- to anion. This is Kohlrausch's law of independent migration of ions.

If c in Eq. (14.3b) is in molarity, then the corresponding Λ woud represent a quantity called molar conductivity (denoted by Λ_m) and the relevant λ values are called molar ionic conductivities (denoted by λ_m). Although IUPAC prefers the term *molar conductivity* to *equivalent conductivity*, the latter is still in use due to simplicity of the expression involving it. This is at the cost of clarity with IUPAC recommendation.

14.4 CONDUCTIVITY AND IONIC MOBILITY

In an electric field E an ion, due to its charge Ze, experiences a force of magnitude ZeE that causes it to move with an acceleration in the direction of the field. However, as the ion moves, it experiences a frictional (retarding) force due mainly to the viscosity of the medium; this latter force increases with the speed of the ion following Stokes law

(Section 7.4). Ultimately (within 10^{-13} s) these two opposing forces are mutually balanced when the ion attains a constant terminal speed, called *drift speed* of the ion which varies proportionally with the applied field strength. The drift speed of an ion in the direction of an applied field of unit strength is called the mobility (u) of the ion.

Under an electric field (E) the current density (j) within the solution of a binary electrolyte is given by

$$j = v_+ n_+ Z_+ e + v_- n_- Z_- e \qquad (14.5)$$

due to cation   due to anion

[where v is the drift speed, n is the number density and Ze is the charge of the type of ion indicated by the suffix.]

$$= \frac{FC}{1000} (v_+ + v_-) \quad \text{for strong electrolytes}$$

[where $n_+ Z_+ e = n_- Z_- e = \dfrac{FC}{1000}$ in CGS unit, c is normality of the electrolyte and F is Faraday constant.]

$$= kE \quad \text{by Ohm's law}$$

Then
$$k = \frac{FC}{1000}(u_+ + u_-) \quad \text{since } v = Eu \qquad (14.6)$$

or
$$\Lambda = \frac{1000\, k}{C} Fu_+ + Fu_- = \lambda_+ + \lambda_- \quad \text{by Kohlrausch's law}$$

As ionic mobility u and hence λ are characteristic properties of an ion (in dilute solution), if follows that

$$\left.\begin{array}{l} \lambda_+ = Fu_+ \\ \lambda_- = Fu_- \end{array}\right\} \qquad (14.7)$$

i.e. ionic conductivity is proportional to the ionic mobility.

In case of a weak electrolyte having degree of dissociation α

$$n_+ Z_+ e = n_- Z_- e = \frac{FC\alpha}{1000}$$

and hence the above expressions for k and Λ would have to be modified accordingly by multiplying the rhs with α.

14.5 VARIATION OF CONDUCTIVITY WITH DILUTION

As expected from the expressions of k and Λ, the dilution will always result in decrease of sp conductivity but increase in equivalent conductivity of electrolyte. In dilute solution following relations hold

for strong electrolytes, $\Lambda = \Lambda^\circ - b\sqrt{c}$ \qquad (14.8)

(due to Kohlrausch)

for weak electrolytes, $\dfrac{1}{\Lambda} = \dfrac{1}{\Lambda^\circ} + \dfrac{\Lambda c}{K(\Lambda^\circ)^2}$ \qquad (14.9)

(due to Ostwald)

where Λ° is the equivalent conductivity at infinite dilution and K is the dissociation constant of the electrolyte. Then Λ° can be obtained from the intercept at $C = 0$ of the plot of Λ against $\sqrt{c}$ for strong electrolytes and $1/\Lambda$ against Λc for weak electrolytes.

Again, for a mixture of electrolytes (strong or weak)

$$k = \frac{\sum C_i \lambda_i}{1000} \tag{14.10}$$

where C_i is normality of the ith ion and λ_i is its equivalent ionic conductivity.

14.6 DEPENDENCE OF CONDUCTANCE ON NATURE OF THE SOLVENT

If interionic forces are negligible (as in very dilute solution), the mobility u_i of the ion i in an electrolyte solution is expected to fall in proportion with the dragging force due to viscosity η of the solvent. Then, by Stokes law

$$u_i \propto \frac{Z_i}{\eta r_i}$$

where r_i is the effective radius of the (solvated ion)

Now, in very dilute solution of an electrolyte

$$\Lambda^\circ = F(u_+ + u_-)$$

Then for a particular electrolyte (where Z_i's are fixed) assuming r_i's to be independent of the solvent

$$\Lambda^\circ \eta = \text{constant} \tag{14.11}$$

which is Walden's rule. This rule fails for electrolytes having cations of smaller size due to their high polarising power that makes r_i appreciably different in different solvent.

If, however, the solution is not very dilute, the interionic forces have to be considered. Here it is useful to apply Debye–Huckel concept that in solution each ion is surrounded by a spherically symmetrical atmosphere of opposite charge in absence of an applied electric field and no net force is exerted on the ion by its atmosphere. In the presence of a field, as the ion moves in a particular direction, the atmosphere becomes asymmetric (inertial effect) with higher charge density behind the central ion producing a dragging effect on the latter (asymmetry effect). Additional lowering of ionic mobility arises also from carrying of solvent molecules by each ion and its atmosphere (electrophoretic effect) which, in effect, amounts to swimming of the ions upstream against the solvent flow. Electrophoretic retardation is in addition to viscous retardation in absence of ionic atmosphere.

The Steady state motion of an ion arises when the electric driving force equals the sum of frictional, asymmetric and electrophoretic retardations. From such considerations Onsager obtained an equation of the following form

$$\Lambda = \Lambda^\circ - (A + B\Lambda^\circ)\sqrt{C} \tag{14.12}$$

It contains an empirical parameter Λ° (which is equivalent conductivity at a particular temperature at infinite dilution, without consideration of solvent effect), and the theoretical parameters A and B both involving charge type of the electrolyte and dielectric constant of the solvent, apart from temperature. A also involves the viscosity of the solvent. For a $1 - 1$ electrolyte, the above expression becomes

$$\Lambda = \Lambda^\circ - \left[\frac{82.4}{(\epsilon_r T)^{\frac{1}{2}} \eta} + \frac{82.0 \times 10^4}{(\epsilon_r T)^{\frac{3}{2}}} \Lambda^\circ \right] \sqrt{C}$$

where ϵ_r is relative permittivity (dielectric constant) of the medium.

14.7 CONDUCTIVITY AT HIGH FIELD AND HIGH FREQUENCY

Ohm's law is valid for electrolytes, if the applied electric field is of rather low strength or of low frequency. An increase in conductivity is observed when the applied field is of high strength of the order of 10^5 volt/cm (Wien effect) or of high frequency 10^6 Hz (Debye–Falkenhagen effect). Such effects can be explained on the basis of the ionic atmosphere model.

On application of high fields, the ions move so quickly that it effectively loses its atmosphere; the atmosphere does not have time to be reformed to retard the motion of the ion (i.e., the asymmetry effect disappears). With increase in field strength, the equivalent conductivity of an electrolyte will therefore increase and approach the value for infinite dilution when the drift velocity in the direction of the applied field exceeds the random thermal velocity of the ions.

At high frequency of the field, the ions change their direction of motion so quickly that their motion cannot be followed by their sluggish atmosphere. The ions move as if they have no atmosphere, and consequently the conductivity increases. Here asymmetry effect and electrophoretic effect both disappear. These happens when the period of oscillation of the applied field is less that the relaxation time of the ionic atmosphere (this is the time required for the atmosphere to form about an ion).

14.8 TRANSPORT NUMBER

From conductance measurement alone, we can determine only the sum of ionic conductivities of the positive and negative ions. To get the individual ionic conductivity, we require independent additional knowledge of a quantity, the transport number defined below.

Of the total current passing through an electrolyte, the fraction current by a particular type of ion is called the transport number of that type of ion. For a mixture of electrolytes, the transport number t_i of the ion i is given by

$$t_i = \frac{C_i u_i}{\sum C_i u_i} = \frac{C_i \lambda_i}{\sum C_i \lambda_i} \tag{14.13a}$$

in dilute solution when the contribution of the ion i to the current density is proportional to $C_i \lambda_i$, by Eq. (14.5).

For a single electrolyte, C_+ and C_- (for C is normality)

$$\left.\begin{array}{l} t_+ = \dfrac{u_+}{u_+ + u_-} = \dfrac{\lambda_+}{\lambda_+ + \lambda_-} \\[3mm] t_- = \dfrac{u_-}{u_+ + u_-} = \dfrac{\lambda_-}{\lambda_+ + \lambda_-} \end{array}\right\} \tag{14.13b}$$

The most effective method for the determination of transport number (t_+) of an ion M^+ in an electrolyte MX is that based on the direct determination of the mobility of that ion from the measurement of the displacement l cm of a moving boundary formed by a solution of this electrolyte (of normality c) with solution of another electrolyte NX of same normality) in a tube of cross sectional area a cm^2 due to passage of Q coulomb of electricity. Here

$$t_+ \cdot \frac{Q}{F} = \frac{lac}{1000} \tag{14.14}$$

amount of electricity   (amount of M^+ ion carrying $t_+ \cdot \frac{Q}{F}$ faraday)
carried by M^+ ion

The older method of transfer number determination is that based on the measurement of the secondary effect of electric current using Hittorf principle that during electrolysis of an electrolyte with inert electrodes a loss of electrolyte occurs in the neighbourhood of either electrode. The number of g equivalents of electrolyte lost around each electrode per faraday of electricity passing is equal to the transport number of the ion migrating away from the electrode concerned. If, however, the electrolyte is produced around an electrode (e.g. $CuSO_4$ is produced around the anode when $CuSO_4$ is electrolysed with Cu electrodes), there will be a net gain of electrolyte which should necessarily be equal to the corresponding loss around the other electrode.

The determination of transport number from emf measurement (discussed in chapter 13) of suitable concentration cells and of the corresponding double cells is cumbersome which offers precise result only when the concentration of the electrolytes on either side of the liquid junction do not appreciably differ.

14.9 APPLICATION OF CONDUCTANCE MEASUREMENT

Measurement of conductance offers simple method for the determination of thermodynamic quantities like dissociation constant of electrolytes, solubility product of sparingly soluble electrolytes, ionic product of water etc.

Titration of an electrolyte by another can be conveniently done conductometrically. Any reaction which is either associated with replacement of one kind of ion by another of different mobility or with a change in ionic concentration of the reaction mixture will be associated with change in conductance at different rates before and after equivalence point of the titration. The equivalence point corresponds to the break in the titration curve obtained on plotting conductance of the reaction mixture against the volume of the titrant added.

AUXILIARY PROBLEMS

14.1 Which of the following substances belong to the categories—electrolyte, non-electrolyte or none.
Cl_2, HCl, NH_3, CH_3Cl, Na, NaH, NaHg and Al_2O_3.

14.2 Diamond, unlike graphite, is a bad conductor of electricity though, it is a good conductor of heat. Explain.

14.3 Na conducts electricity in the solid and liquid states but not in the vapour state while NaCl conducts electricity in the liquid vapour and dissolved states but not in the solid state. Explain.

14.4 Is electric current (often quoted using an arrow) a vector quantity? What about current density?

14.5 Find the dimensions of resistance, resistivity and conductivity. Also mention their SI and conventional units.

14.6 Electrolytic conductors, like electronic conductors, obey Ohm's law, but not for electric field of any strength. Explain.

14.7 What should be the nature of the plot of conductance values of an electrolyte of varying concentration measured in a particular cell against respective specific conductance values? Does this plot give any information about the cell? How essential is the information for any absolute measurement in the cell?

14.8 How are the measured conductance, sp conductance and equivalent conductance of an electrolyte solution related to one another? State the effects of dilution on them. Are these effects in accordance with the reaction you suggest?

14.9 With dilution k of an electrolyte decreases while its Λ increases. But in general, the variations of k and Λ with concentration C of an electrolyte do not follow the simple relations $k \propto c$ of and $\Lambda \propto \frac{1}{c}$. Explain.

14.10 Relate molar conductivity (Λm) of an electrolyte with (i) its equivalent conductivity (Λ aq) (ii) molar ionic conductivity (λ_m) of its constituent ions.

14.11 In dilute solution ionic conductivity $\propto$ ionic mobility. What is the proportionality constant? Do you expect that an ion with higher charge will always have higher mobility?

14.12 On which of the following factors will the drift speed and mobility of an ion in a solution depend?
(i) applied electric field (ii) ionic charge (iii) ionic strength (iv) temperature (v) pressure.

14.13 Account for the following:
a. Alternating current is used in measuring the conductance of an electrolyte solution.
b. It is necessary that the electrodes of a conductivity cell be electroplated.
c. The conductance of an electrolyte solution increases with increase in the strength and frequency of the applied electric field beyond certain limit.
d. The ionic conductivity increases in the order $K^+ < Na^+ < Li^+$ in the fused electrolyte but in reverse order in aqueous solution.
e. Proton, the smallest ion exhibits highest equivalent ionic conductivity in aqueous solution, despite its highest polarising power.

14.14 Justify or criticise the following:
a. Sp conductance = measured conductance × cell constant. Hence sp conductance should depend on the cell constant of the cell used.
b. Sp conductance of a solution is the conductance of unit volume of it.
c. Molar conductivity of an electrolyte is the conductivity per mol of it.
d. The degree of dissociation of a weak electrolyte increases with dilution. Therefore, its conductivity should increase with dilution.
e. The total number of ions does not change with dilution in case of strong electrolytes. Therefore, the equivalent conductivity of such electrolytes should not change with dilution.
f. The equivalent conductivity (Λ) of a binary electrolyte can always be expressed as $\Lambda = \lambda_+ + \lambda_-$ where λ_+ and λ_- are the characteristic equivalent ionic conductivities of the cation and anion of the electrolyte.
g. Kohlrausch's law for independent migration of ions has justification only for binary electrolytes (e.g. Na_2SO_4) and not for electrolytes which produce more than two types of ions (e.g. $NaHSO_4$).
h. With dilution Λ for weak electrolytes continuously increases and hence for such electrolytes $\Lambda°$, the value of Λ in the limit of zero concentration, cannot have a definite value.
i. $\Lambda°$ of a strong electrolyte will be greater than that of a weak electrolyte.

14.15 Arrange the following substances in the order of their Λ° values.

HCl, NaCl, NaOH, C_6H_5OH, CH_3COOH, CH_3COONa.

14.16 What is the significance of Λ/Λ° in case of strong and weak electrolytes? Which substance in the following pairs will have higher value of Λ.

(i) H_2SO_4 and Na_2SO_4, each at normal concentration.

(ii) CH_3CO_2H and CH_3CO_2Na, each at normal concentration.

14.17 a. Deduce Ostwald dilution law.

b. How will you verify it graphcally?

c. The law holds satisfactorily with CH_3CO_2H but not with $CHCl_2 \cdot CO_2H$. Explain.

d. How can $\Lambda^\circ_{CHCl_2CO_2H}$ be obtained from conductance measurement?

14.18 Calculate k/Λ for a 0.01 N KCl solution. From your calculation is it justified to state that k is less than Λ?

14.19 True or false.

a. Transport number of an ion, like its ionic conductivity, is a characteristic property of the ion.

b. Transport number of an ion cannot be zero.

c. For a particular electrolyte solution, the plot of transport number of cation versus that of anion will be linear.

d. During electrolysis of an electrolyte with inert electrodes, there must be a loss in amount of the electrolyte around each electrode.

14.20 Dilution of the electrolyte solution will result in higher ionic mobility and hence higher transport number of both cation and anion. Justify or contradict.

14.21 The transport number determined by the Hittorf method is usually apparent which differ significantly from the true transport number. Explain. Sometimes this apparent transport number seems to be abnormally low (nearing zero). Illustrate this with appropriate example. Is this likely with moving boundary method?

14.22 Establish the relation $k = \sum C_i \lambda_i /1000$ mentioning the assumption(s) involved.

14.23 In connection with conductometric titration, explain the following:

a. Conductance and not resistance is plotted.

b. The titre is much concentrated than the solution to be titrated.

c. During potentiometric titration data points near the equivalence point are important, while in a conductometric titration, the same is not true.

14.24 Draw the conductance curve for the following titrations:

a. HCl with NaOH and reverse

b. HAc with NaOH and reverse

c. HCl + HAc (in equal concentration) with NaOH

d. H_2OX with NaOH

e. H_2SO_4 with NaOH

If the titrating solution is diluted, how would the titration curves change? Would equivalence point be affected due to this dilution?

14.25 A $BaCl_2$ solution is placed between two parallel electrodes 8 cm apart and is subjected to a potential difference 4 volts by connecting the electrodes to an external battery. What will be the electrical force on a Ba^{2+} ion if the interionic attraction and the dipolar effects of the solvent molecules be disregarded?

[Dielectric constant of the medium $= 80$]

14.26 An electric current is passed through a 0.01 N KCl solution in a conductivity cell, where electrodes are each of area 1 cm² and 1 cm apart, by applying a potential difference of 5 V between the electrodes. The ionic mobilities at infinite dilution of K^+ and Cl^- are 7.62×10^{-4} cm²V⁻¹s⁻¹ and 7.91×10^{-4} cm²V⁻¹s⁻¹ respectively. Ignoring interionic forces, calculate

  a. Cell constant
  b. Conductivity
  c. Conductance
  d. Current density

14.27 If the equivalent conductivities of solutions with varying concentration of NaAc, NaCl and HCl are plotted against $\sqrt{C}$ (C is normality of the electrolyte), the intercepts obtained are 91, 128 and 425 respectively in Ω^{-1} cm²eq⁻¹ unit at 25°C. If the resistance of a 0.02 N KCl solution of sp conductivity 0.00276 Ω^{-1}cm⁻¹ is 380 Ω in certain cell and that of a 0.01 N HAc in the same cell is 6434 Ω, calculate the degree of dissociation (α) of HAc in the solution at 25°C stating the approximation(s) involved.

14.28 In a saturated solution of $BaSO_4$ at 25°C, Ba^{2+} and SO_4^{2-} ions take 151 sec and 121 sec respectively to cover a distance of 1 cm across which a potential difference of 10 V is applied. The conductance of the solution in certain cell is 2.2×10^{-6} Ω^{-1}. Using the same cell the conductance is found to be 0.82×10^{-6} Ω^{-1} with water and 1.4×10^{-3} Ω^{-1} with a 0.01 N KCl solution at the same temperature. Find the conductivity of the KCl solution at 25°C. Mention the assumption involved.
  $[K_{sp}(BaSO_4) = 10^{-10}$, in g equivalent L⁻¹ unit]

14.29 At 25°C, the resistance of certain conductivity cell is 4.4×10^6 Ω when filled with water, 10^2 Ω when filled with 0.02 N KCl and 1.02×10^5 Ω when filled with saturated solution of AgCl. Given that at 25°C $\Lambda^\circ_{KCl} = 149.8$, $\Lambda^\circ_{AgCl} = 138.4$, $u^\circ_{H^+} = 3.63 \times 10^{-3}$ and $u^\circ_{OH^-} = 2.05 \times 10^{-3}$, all in conventional units. With necessary assumption, calculate

  a. cell constant
  b. solubility of AgCl at 25°C
  c. ionic product of water at 25°C
  Comment on the result.

14.30 It is found that the resistance of certain conductivity cell is 2340 Ω when filled with 0.001 N HCl, 7900 Ω when filled with 0.001 N NaCl, and 8250 Ω when filled with 0.001 N $NaNO_3$. The equivalent conductivity of $NaNO_3$ is 121 Ω^{-1}cm²eq⁻¹. Find:

  a. conductivity of 0.001 N $NaNO_3$
  b. cell constant
  c. resistance of the same cell when filled with 0.001 HNO_3
  d. equivalent conductance of 0.001 N HNO_3.

14.31 The resistance of a dilute solution of NaCl is 40 Ω and that of a NaOH solution is 100 Ω in the same cell. If these two solutions are mixed up in 1:2 volume ratio, what will be the resistance of the resulting mixture measured in the same cell? What assumption you have taken to solve this problem?

14.32 25 ml of 0.1 N HCl was taken in a conductivity cell of cell constant 1 cm⁻¹. The conductance of the solution changed from A siemens to B siemens when 25 ml of 0.1 N NaOH was added to it. Argue that A − B is positive and hence find an expression for the equivalent conductivity of NaCl in terms of B.

14.33 In the titration of 0.01 N HCl with NaOH of same strength, find the conductivity of the resulting solution (a) before 25% of the equivalence point (b) at the equivalence point (c) after 25% of the equivalence point. Given $\Lambda_{NaOH} = 249$, $\Lambda_{HCl} = 426$ and $\Lambda_{NaCl} = 126$, all in conventional unit.

14.34 The results of measurement of equivalent conductivity at infinite dilution in $\Lambda^{-1}cm^2equiv^{-1}$ at 25°C for the following pairs of electrolytes are found as

Electrolyte	Λ°	Electrolyte	Λ°	Electrolyte	Λ°
KCl	149.9	KNO_3	145.0	KOH	271.6
LiCl	115.0	$LiNO_3$	110.1	LiOH	236.7

Generalise, the results in the form of a law and state, how it helps in obtaining the equivalent conductivity of a weak electrolyte at infinite dilution.

14.35 In a Hittorf method experiment, an exactly 0.200 molal solution of $CuSO_4$ was electrolysed between copper electrodes. After the experiment, 18.217 g of the cathode solution was found to contain 0.554 g of $CuSO_4$. Also 0.020 g of silver was deposited during the experiment in a silver coulometer connected in series. Calculate the transport number of the Cu^{2+} and SO_4^{2-} ions (Ag = 107.9).

14.36 A solution containing 2.84% of $CuSO_4$ was electrolysed using a Pt cathode and Cu anode. After electrolysis, the cathode solution was found to weigh 54.7 g and to contain 0.409 g of Cu. The increase in weight of the cathode was 0.408 g. Calculate the transport number of Cu^{2+} ion (Cu = 63.6).

What is harm if current is passed for a long time in transport number determination by Hittorf method?

14.37 In an electrolysis of $CuSO_4$ solution in a Hittorf transference cell with Cu electrodes, it is found that the anode solution has gained 0.004 mol of $CuSO_4$ and that the gain in weight of anode solution is same as the loss in weight of anode. Calculate:

a. transport number of SO_4^{2-}

b. number of mol of $CuSO_4$ lost by the cathode compartment

c. quantity of electricity passed through the cell

14.38 A Hittorf cell fitted with Ag–AgCl electrodes was filled with HCl solution that contained 0.3856×10^{-3} g HCl per g of water. A current of 2 mA was passed for exactly 3 h. After electrolysis, the cathode solution weighed 51.7436 g containing 0.0267 g of HCl. Find out t_{H^+}.

14.39 Account for the following:

a. In an electrolysis at a fixed potential, current can be increased by stirring the electrolyte solution.

b. A concentrated solution of cadmium iodide can exhibit an anion transport number of greater than 1.

14.40 The equivalent conductivity of LiCl at infinite dilution is 115.03×10^{-4} $\Omega^-m^2mol^{-1}$. The transport number of the cation is 0.336. Calculate the velocity of the cation if 6 V is applied across the electrodes 4 cm apart.

14.41 In a moving boundary experiment with 0.1 N KCl solution enclosed between two indicator electrolytes LiCl and CH_3CO_2K in a capillary of radius 1 mm, the cationic boundary shifts by 75.5 mm when a steady current of 10 mA is passed for 8 min. What will be the simultaneous shift of the anionic boundary?

14.42 Calculate the transport number of Cl^- ion in 0.01 N NaCl solution. To this solution an equal volume of 0.01 N HCl is added. Find the change in transport number of Cl^- ion, if any. In what volume ratio, the two solutions are to be mixed so that the transport numbers of Na^+ and H^+ ions are equal in the mixture?

[λ_{H^+} = 349.8, λ_{Na^+} = 50.1 and λ_{Cl^-} = 76.3, all in $\Omega^{-1}cm^2equiv^{-1}$]

14.43 A 0.1 M $MgCl_2$ solution is mixed with an equal volume of 0.1 M $MgSO_4$ solution. Calculate:

    a. conductivity of the resulting mixture.

    b. transport number of Mg^{2+} ion in the mixture.

    c. distance a Mg^{2+} ion would move in 100 sec if the mixture were placed in a tube of 5 cm^2 cross section and a current of 0.1 A were flowing.

[$\lambda_{Mg^{2+}}$ = 53, λ_{Cl^-} = 76.3 and $\lambda_{SO_4^{2-}}$ = 79.8, all in $\Omega^{-1}cm^2equiv^{-1}$]

ANSWERS

14.1 Compounds are called electrolytes or non-electrolytes according as they can exhibit appreciable electrical conductivity or not, in molten form or in solution.

Electrolytes: HCl, NH_3, NaH, Al_2O_3 [HCl and NH_3 are bad conductors of electricity in pure state, but good conductors when dissolved in polar solvent like water while NaH and Al_2O_3 are good conductors of electricity in molten form].

Non-electrolytes: CH_3Cl.

None: Cl_2, Na, NaHg (because these are not compounds.

Note:

    i. A substance which is not an electrolyte is not necessarily a non-electrolyte.

    ii. Any non-conuctor of electricity (which may be an element or a compound) is not a non-electrolyte, e.g. Cl_2.

    iii. A non-electrolyte may not be a non-conductor of electricity, e.g. Na which is an electronic conductor.

14.2 Because diamond, unlike graphite, contains no free electrons which are essential for electrical conductivity of solids. In diamond, the valence electrons become immobile in forming bonds responsible for the crystal structure. Yet it exhibits good thermal conductivity through wave motion associated with lattice vibration.

Note: A conductor of electricity is also a conductor of heat but the reverse is not necessarily true.

14.3 Because the electrical conductivity of Na is due to mobility of valence electrons which requires atoms to be in contact with one another. For NaCl, the conductivity is due to mobility of ions which is arrested in the solid state.

14.4 No. Electric current has been defined to specify the amount of charge flow regardless of its direction as the latter is of no concern. However, it is often quoted using an arrow which really means the sense of charge flow (more precisely the direction of current density).

☐ The current density is a vector quntity indicating the rate of charge flow through unit cross section of a conductor in the direction of the electric field involved.

Note: Electric current is a macroscopic quantity characteristic of a conductor as a whole while current density is a microscopic quantity characteristic of a point within the conductor.

14.5 $\qquad$ Resistance $= \dfrac{\text{voltage}}{\text{current}}$, by Ohm's law

$$= \frac{\text{energy}}{\text{charge} \times \text{current}} \equiv \frac{\text{mass} \times \text{speed}^2}{\text{current}^2 \times \text{time}}$$

Then the dimensions of resistance will be $\dfrac{(m)(l/t)^2}{(I^2)(t)}$ or $ml^2t^{-3}A^{-2}$.

The dimensions of resistivity will be the dimensions of resistance × area/length, i.e. $(ml^2t^{-3}A^{-2})$ $(l^2)/(l)$ or $ml^3t^{-3}A^{-2}$.

The dimensions of conductivity will be the dimensions of 1/resistivity, i.e. $m^{-1}l^{-3}t^3A^2$.

☐ The units are tabulated below:

Physical quantity	SI unit	Conventional unit
resistance	Ohm	Ohm
resistivity	Ohm meter	Ohm centimeter
conductivity	Siemens meter^{-1}	mho centimeter^{-1}

14.6 It appears from Eq. (14.6) that Ohm's law will be valid for an electrolytic conductor provided the ionic mobility of the constituent ions is independent of the applied field strength. This requirement is not fulfilled for quite high strength or high frequency of the applied field when the ion atmosphere of a moving ion loses its spherical shape (asymmetry effect) with consequent dragging effect on the motion of the ion. The analogous phenomenon is unlikely in case of electronic conductors where electrons serve as the mobile charged particles.

14.7 If conductance of an electrolytic solution is measured in a particular conductivity cell of cell constant K.

$$\text{measured conductance} = \text{specific conductance}/K$$

Then the plot of measured conductance against specific conductance should be linear.

☐ This plot gives the information about K of cell which is inverse of the slope of the line obtained from this plot.

☐☐ Knowledge of K is essential for obtaining absolute values of different electrical quantities from the measurement of conductance in a cell. For example, calculation of specific conductance of an electrolyte requires multiplication of the measured conductance with K of the cell used.

Note: For a particular cell, the plot of measured conductnce against sp conductance gives the same line whether the electrolyte is strong or weak.

14.8 The required relation is

measured conductance × cell constant = sp conductance $= \frac{1}{1000} \times$ eq conductance × normality

☐ With dilution (i.e. decrease in normality), the measured conductance and specific conductance of an electrolyte decrease, while its equivalent conductance increases.

☐☐ Yes. Because, although the measured conductance and specific conductance decrease with dilution, the decrease does not bear a proportional relation with concentration but occurs to a less extent, so that the equivalent conductance increases with dilution (vide next problem).

Note: The above relation has meaning with a solution containing only one electrolyte where the solvent has no significant contribution to the conductance.

14.9 k refers to a fixed volume (unit volume) of an electrolyte while Λ refers to a fixed amount (one equivalent) of an electrolyte. Due to dilution k increases mainly due to the fall in ionic concentration that has dominating effect over the opposing effect of the accompanying increase in ionic mobility. But Λ increases due to the increase in ionic mobility, even if there is no increase in the total amount of ions due to dilution.

However, the simple relation $k \propto C$ is unlikely because in general, the ionic concentration and the ionic mobility are not proportional to the concentration (C) of the electrolyte. As a consequence, $\Lambda \propto \frac{1}{C}$ will also be unlikely as $\Lambda = 1000 \, k/C$.

Note:

i. If $k \propto C$ were true Λ would have been independent of C.

ii. Λ is not proportional to k, though the measured conductance is proportional to k.

14.10 For an electrolyte $M_{v_+} X_{v_-}$ that dissociates completely into M^{Z+} and X^{Z-} ions in solution

   a. $\Lambda_m = v_+ Z_+ \Lambda_{aq} = v_- Z_- \Lambda_{aq}$

   b. $\Lambda_m = v_+ \lambda_{m^+} + v_- \lambda_{m^-}$ at infinite dilution

14.11 Taking ionic conductivity as the equivalent ionic conductivity (λ_{aq}), the proportionality constant would be the Faraday constant (F), by Eq. (14.7) following Ohm's law and Kohlrausch's law. But taking this as the molar ionic conductivity (λ_m) the proportionality constant would be ZF, since $\lambda_m/\lambda_{eq} = Z$, the charge of the ion.

   ☐ No. Because the effective force experienced by an ion in an electric field is not determined only by ionic charge [vide problem 14.12]. For example Cu^{2+} has lower mobility than Ag^+ in aq solution due to the adverse effect of higher size of Cu^{2+}.

14.12 Drift speed depends on all the given factors. It depends on the coulombic force determined by (i) and (ii); it depends on interionic forces which depend on (iii); it depends on the degree of ionic solvation which depends on (iv); it depends on viscosity of the solvent which depends on (iv) and (v).

   Ionic mobility depends on all the factors except (i), provided the applied electric field is not of very high strength.

14.13 a. Because a direct current would lead to electrolysis (that affects concentration of the electrolyte) and polarisation, i.e. modification of the immediate surroundings of the electrodes (that makes applied emf considerably different from the reversible emf).

   b. To make the electrode surface rough (through deposition of metal) so that the current density is significantly reduced. This serves as a precautionary measure against polarisation.

   c. Vide Section 14.7.

   d. For ions having the same charge, the ionic mobility and hence the ionic conductivity increase with decrease in size of the moving ion due to lower resistance involved in moving through a medium. For the given ions, the order of the ionic size in the fused state is reverse of that for the solvated ions (in aqueous solution) due to higher degree of solvation of the smaller ions having higher polarising power.

e. Because here conduction occurs through an unusual mechanism that involves rapid shifting of H^+ from H_3O^+ to H_2O in a long chain (by making and breaking of hydrogen bond) instead of actual migration of a single highly solvated H^+.

Note: OH^- ion has the highest equivalent ionic conductivity of all anions in aqueous solution due to a mechanism involving shifting of H^+ from H_2O to OH^- producing in effect the transfer of OH^- through water. But neither H^+ nor OH^- has the highest molar ionic conductivity. For example $Fe(CN)_6^{4-}$ has much higher λ_m than OH^- and even H^+.

14.14 a. The given statement is not justified because the conductance of a solution, measured in any cell, is proportional to its specific conductance.

b. The given statement will hold only if the conductance is measured in a cell where the area (a) of each electrode and their separation (l) are both unity. Only the condition, al = unity, is not enough. Vide Eq. (14.2).

c. The given statement is not in harmony with the definition of molar conductivity as the conductivity per unit molar concentration, i.e. $\Lambda_m = k/C$, in SI unit.

d. Not justified, because with dilution ionic concentration falls (though total number of ions increases due to higher degree of dissociation) with consequent fall in conductivity.

e. Not justified. Here dilution results in change (increase) in equivalent conductivity due to increase in ionic mobility, total number of ions remaining unchanged (ignoring the existence of ion-pairs).

f. Not justified. The given relation will hold only if the electrolyte is completely dissociated (when Λ will correspond with one equivalent of ionised electrolyte just as λ with one equivalent of an ion) and the interionic forces are negligible (when λ_+ and λ_- will be independent of each other). These necessitate the solution to be extremely dilute.

g. Not justified. For electrolytes producing more than two types of ions, the Kohlrausch's law implies that $\Lambda = \sum \lambda_i$, where λ_i is the equivalent ionic conductivity of the ith ion produced by the electrolyte.

h. Although Λ for weak electrolytes continuously increases with dilution due to increase in degree of dissociation, after certain stage, it does not change effectively, and $\Lambda°$ refers to that stage, called infinite dilution. Then the given statement is not justified from practical view point.

i. In dilute solution all electrolytes, weak and strong, undergo complete dissociation when

$$\Lambda° = \lambda_+° + \lambda_-°$$

Then $\Lambda°$ of an electrolyte is determined only by $\lambda_+° + \lambda_-°$ which are characteristics of the constitutional ions (irrespective of the nature of their parent electrolytes). Therefore, the given statement is not justified.

14.15 Remembering that $\Lambda° = \lambda_+° + \lambda_-°$ and λ is higher for an ion of lower size together with the abnormally high values of $\lambda_{H^+}°$ (highest of all cations) and $\lambda_{OH^-}°$ (highest of simple anions), the order of $\Lambda°$ will be as under

$$HCl > CH_3COOH > C_6H_5OH > NaOH > NaCl > CH_3COONa$$

14.16 The equivalent conductivity Λ of an electrolyte solution is expected to vary with number of gram equivalents (N) of cations and anions in the solution and their mobilities u_+ and u_-.

$$\Lambda \propto N\,(u_+ + u_-) \quad \text{(vide Section 14.4)}$$

Due to interionic forces u varies with ionic concentration in a complicated way. However, for weak electrolytes, the ionic concentration is quite low so that it does not appreciably affect u. Then for such electrolytes Λ/Λ° will signify their degree of dissociation, as Λ° corresponds to complete dissociation. For strong electrolytes, Λ/Λ° will have no such significance because they are always dissociated virtually completely leading in general to ionic concentrations enough to affect u; here Λ/Λ° may be regarded as a measure of change in ionic mobility due to variation of ionic concentration.

(i) H_2SO_4. Here both the substances dissociate completely leading to the same total number of ions, but $\lambda_{H^+} > \lambda_{Na^+}$.

(ii) CH_3CO_2Na. Because CH_3CO_2Na, unlike CH_3CO_2H is completely dissociated leading to much higher ionic concentration that has been dominating over higher conductivity of H^+ than Na^+.

Note: $\Lambda^\circ_{CH_3CO_2H} > \Lambda^\circ_{CH_3CO_2Na}$, because Λ° corresponds to infinite dilution where all electrolytes dissociate completely.

14.17 a. Let us consider an electrolyte MA with degree of dissociation α when its formal concentration is C in a solution

$$MA \rightleftharpoons M^+ + A^-$$
$$C(1-\alpha) \qquad \alpha C \qquad \alpha C$$

Dissociation constant

$$K = \frac{C_M \cdot C_{A^-}}{C_{MA}} \quad \text{for dilute solution ignoring activity coefficient effect}$$

$$= \frac{\alpha^2 C}{1-\alpha}$$

$$= \frac{\Lambda^2 C}{\Lambda^\circ\,(\Lambda^\circ - \Lambda)} \quad \text{putting } a = \Lambda/\Lambda^\circ \text{ for a weak electrolyte}$$

or $\qquad \dfrac{1}{\Lambda} = \dfrac{1}{\Lambda^\circ} + \dfrac{C\Lambda}{K(\Lambda^\circ)^2}$

which is Ostwald dilution law in most useful form.

b. The law can be verified graphically from $1/\Lambda$ vs $C\Lambda$ plot which will be linear if the law holds.

c. As pointed out in its derivation, the law will hold only for weak electrolyte like CH_3CO_2H and not for moderately strong electrolyte $CHCl_2CO_2H$.

d. Obviously $\Lambda^\circ_{CHCl_2 \cdot CO_2H}$ cannot be obtained from the intercept of $1/\Lambda$ vs $C\Lambda$ plot at $C = 0$, based on Eq. (14.9).

A reasonably accurate value of $\Lambda_{CHCl_2 \cdot CO_2H}$ can be obtained using Kohlrausch's law through the following expression

$$\Lambda_{CHCl_2 \cdot CO_2H} = \lambda_{H^+} + \lambda_{CHCl_2 \cdot CO_2^-}$$

$$= \Lambda_{HCl} + \Lambda_{CHCl_2CO_2Na} - \Lambda^\circ_{NaCl}$$

Since HCl, $CHCl_2CO_2Na$ and NaCl are all strong electrolytes, their $\Lambda°$'s can be found fairly accurately from the intercept of the Λ vs $\sqrt{C}$ plot at $C = 0$, based on the Onsager Eq. (14.12).

14.18
$$\frac{k}{\Lambda} = \frac{1}{\text{Volume of solution containing one equivalent of electrolyte}}$$

$$= \frac{0.01\,\text{equiv}/L}{1000\,\text{cm}^3/L} \text{ in conventional unit, by Eq. (14.3b)}$$

$$= 10^{-5}\,\text{equivalent cm}^{-3}$$

☐ No. Because k has dimensions $(m^{-1}l^{-3}t^3A^2)$ different from $\Lambda\,(m^{-1}t^3A^2n^{-1})$ and hence the relative order based on their numerical values will have no significance, being not independent of the system of units chosen. Here $k/\Lambda < 1$ with conventional system of units while $k/\Lambda > 1$ with SI (when $k/\Omega = 10$ equivalent m^{-3}).

14.19 a. False. Because, like ionic conductivity, the transport number of an ion, by its very definition, cannot be independent of other ions with which it is associated.

b. Normally true. However the exceptions to the given statement can arise in proper circumstances. One such example is offered by the electrolysis of an aqueous NaCl (say) in a cell divided into two parts by a membrane which is not permeable to Na^+ ions (say). Here t_{Na^+} is zero.

c. The statement will be true provided the current is carried almost exclusively by the ions of the electrolyte (i.e. conductance due to ions of the solvent is insignificant), when $t_+ + t_- \approx 1$.

d. Normally true with the loss in amount of the electrolyte around each electrode equal to the amount of the ions migrating out of the corresponding region of the electrodes. However, under the circumstances where the current is carried entirely only by one type of ion, the loss in electrolyte will occur only around that electrode wherefrom migration of ion occurs.

14.20 Dilution, that reduces interionic forces, is likely to enhance the mobility of both cation and anion but not at same rate. This must occur in compliance with the restriction that $t_+ + t_- = 1$ (ignoring the conductance due to the solvent) which demands that transport number of cation (t_+) and that of anion (t_-) cannot both increase.

14.21 This is not unlikely considering the transport of water molecules to the regions around the electrodes by the migrating ions due to their inevitable hydration. This can significantly affect the otherwise change in concentration of the electrolyte (ΔC) around the electrodes due to electrolysis. Then the transport number determined by the Hittorf method on the basis of the observed ΔC is likely to be apparent.

☐ Abnormal values of transport number are often found with electrolytes that can form complex ions. Mention may be made of Cadmium iodide which involves the following ionic equilibrium at ordinary concentrations

$$2CdI_2 \rightleftharpoons Cd^{2+} + CdI_4^{2-}$$

If Cd^{2+} and CdI_4^{2-} ions are present in comparable amounts and have comparable speeds, the fall in concentration of the electrolyte around anode may be negligible corresponding to the transport number of Cd^{2+} ions apparently zero.

☐☐ No. Because in moving boundary method, the transport number of an ion is determined from direct measurement of its speed. This is unlike Hittorf method which is based on the change in concentration of the electrolyte in the neighbourhood of the electrodes (a secondary affect of the migration of ions) and which can therefore lead to abnormal result with electrolyte forming complex (as with CdI_2).

Note: In dilute solution the complex anion CdI_4 dissociates and the normal transport number of Cd^{2+} is obtained.

For highly concentrated solution of cadmium iodide, the transport number of Cd^{2+} becomes apparently negative.

14.22 For a single electrolyte,

$$\Lambda = \lambda_+ + \lambda_- \text{ in dilute solution}$$

$$= \frac{k \times 1000}{C} \text{ where } C \text{ is electrolyte concentration in normality}$$

or $$k = \frac{C(\lambda_+ + \lambda_-)}{1000}$$

$$= \frac{C_+\lambda_+ + C_-\lambda_-)}{1000}$$

where C_+ corresponds to cations and C_- to anions. Here $C_+ = C_- = C$ due to full ionisation of the electrolyte in dilute solution.

For a mixture of electrolytes, we are to consider the contributions of all types of ions (instead of only two types for a single electrolyte), and accordingly, we have the relation

$$k = \frac{\sum C_i\lambda_i}{1000}$$

where C_i is the concentration of the ith ion in normality and λ_i is its equivalent ionic conductance.

Note: k is an additive property. In a mixture of electrolytes, the relation $\Lambda = \dfrac{k \times 1000}{C}$ can be applied to each of the electrolytes individually considering k and C due only to the chosen electrolyte. Often it is convenient to write $k_{solution} = k_{solvent} + k_{solute}$. Here $k_{solvent}$ refers to the solvent ions and k_{solute} to solute ions.

14.23 a. Because the conductance of a solution varies with its ionic concentration in a simpler way than the resistance.

  b. This is required to minimise the change in conductance of the reaction system due to dilution during the titration.

  c. Because in a potentiometric titration, the titration curve exhibits a large change of its slope at the equivalence point (as electrode potential is a logarithmic function of ionic concentration). But a conductometric titration curve is generally more or less linear with a prominent, but not large, change of its slope at the equivalence point. This difference accounts for the given statement concerning the principle of locating the equivalence point in the two titrations.

14.24 a. Here the relevant reaction is

  $$H^+ + Cl^- + Na^+ + OH^- \rightarrow Na^+ + Cl^- + H_2O$$

  where H^+ ions are replaced by Na^+ ions having lower mobility. Hence conductance of the reaction mixture gradually decreases upto the equivalence point 0 (Fig. 14.1a). After this, the conductance increases due to increase in ionic concentration arising from excess of added NaOH. The fall in conductance is faster [determined by $(u_{H^+} - u_{Na^+})$] than the rise in conductance [determined by $(u_{Na^+} + u_{OH^-})$]. For NaOH vs HCl titration, Fig. 14.1b corresponds to the replacement of OH^- ions by Cl^- ions of lower mobility upto equivalence point. After the equivalence point, the conductance rises at higher rate (determined by $u_{H^+} + u_{Cl^-}$).

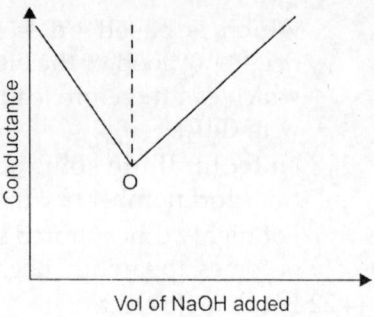

Vol of NaOH added

Fig. 14.1a

Note: Figs 14.1a and 14.1b look similar, differing only in relative slopes.

b. Here the reaction may be represented as

$$HAc \; (\rightleftharpoons H^+ + Ac^-) + Na^+ + OH^-$$
$$\rightarrow Na^+ + Ac^- + H_2O$$

In HAc vs NaOH titration, initially H^+ ions (resulting from partial dissociation of HAc) are replaced by Na^+ ions having lower mobility. Due to this and also due to suppression of dissociation, the conductance

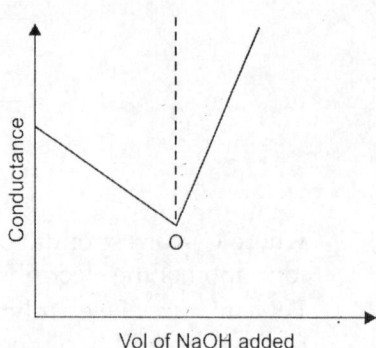

Vol of NaOH added

Fig. 14.1b

of the reaction mixture slightly decreases at the initial stage. However, after appreciable addition of NaOH, the conductance of the reaction mixture increase due to formation of highly ionised Na^+Ac^- upto the equivalence point O (Fig. 14.2a). After equivalence point, the conductance further increases but at higher rate (as $u_{Ac^-} < u_{OH^-}$).

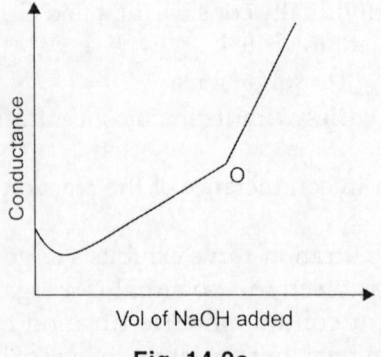

Vol of NaOH added

Fig. 14.2a

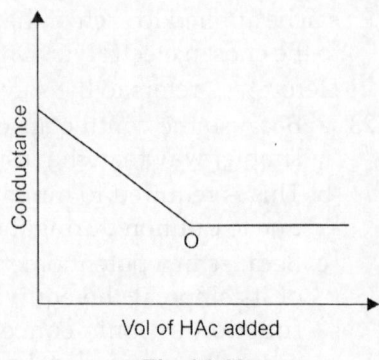

Vol of HAc added

Fig 14.2b

In NaOH vs HAc titration, the conductance decreases due to replacement of OH^- ions by Ac^- ions of lower mobility upto equivalence point. After this, the conductance remains almost unchanged since the unreacted acetic acid remains almost undissociated.

Note: Figs 14.2a and 14.2b differ widely. This is unlike the problem 14.2(a).

c. The conductance diagram (Fig. 14.3) exhibits two equivalence points—O corresponding to the preferred neutralisation of the strong acid HCl and O' corresponding to the subsequent neutralisation of the weak acid HAc. The deviation from linearity of the conductance curve is remarkable around O (because the neutralisation of HAc begins before completion of the neutralisation of HCl) and also around O' (due to appreciable hydrolysis of NaAc).

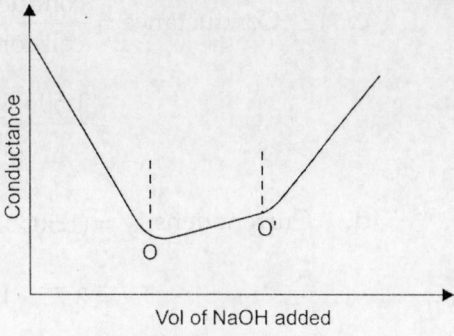

Fig. 14.3

d. $H_2C_2O_4$ is a dibasic acid. It exhibits two equivalence points. The one that appears first corresponds to the first dissociation (which is nearly complete) and the other corresponds to the much weaker second dissociation. The diagram is similar to Fig. 14.3 [Here $pK_1 = 1.2$ and $pK_2 = 4.2$].

e. In the low concentration region of titrimetric importance H_2SO_4 dissociates completely as

$$H_2SO_4 \rightarrow 2H^+ + SO_4^{2-}$$

Hence, although H_2SO_4 is a dibasic acid, it exhibits only one equivalence point, like HCl when titrated with NaOH.

Note: H_2SO_4 cannot be treated like $H_2C_2O_4$ as a mixture of two acids of different strengths.

The conductometric titration curve will remain almost unchanged in shape on diluting the titrating solution. Simply it will lie below that of the concentrated solution due to fall in conductance caused by dilution.

Then dilution will hardly affect the equivalence point.

14.25 Force on an ion (Ba^{2+})

$$= \frac{\text{Ionic charge} \times \text{potential gradient}}{\text{Dielectric constant of the medium}}$$

$$= \frac{(2 \times 4.8 \times 10^{-10} \text{ statcoulomb}) \left(\dfrac{4/300 \text{ statvolt}}{8 \text{ cm}} \right)}{80}$$

$$= 2.0 \times 10^{-14} \text{ dyn}$$

14.26 a.   Cell constant $= \dfrac{\text{distance between the electrodes}}{\text{area of each electrode}}$

$$= \frac{2.0 \text{ cm}}{1.0 \text{ cm}^2} = 2.0 \text{ cm}^{-1}$$

b. By Eq. (14.6)

Conductivity (k)

$$= \frac{(9.65 \times 10^4 \text{ C equiv}^{-1}) (0.01 \text{ equiv L}^{-1}) (7.62 + 7.91) \times 10^{-4} \text{ cm}^2 \text{V}^{-1}\text{S}^1}{((1000 \text{ cm}^3\text{L}^{-1}))}$$

$$= 1.50 \times 10^{-3} \ \Omega^{-1}\text{cm}^{-1}$$

c. $\text{Conductance} = \dfrac{\text{conductivity}}{\text{cell constant}}$

$$= \dfrac{1.50 \times 10^{-3} \Omega^{-1} \text{cm}^{-1}}{2.0 \text{ cm}^{-1}} = 0.75 \times 10^3 \Omega^{-1}$$

d. $\text{Current density} = (1.50 \times 10^{-3} \Omega^{-1} \text{cm}^{-1}) \left(\dfrac{5 \text{ V}}{2 \text{ cm}} \right)$ by Eq. (14.1)

$$= 3.75 \times 10^{-3} \text{ A cm}^{-2}$$

Note: In $\Lambda = \dfrac{k \times 1000}{C}$, the factor 1000 really implies 1000 cm^3/L

14.27 For each of the given electrolytes, being strong, the intercept is equal to its Λ° by Eq. (14.8). Then

$$\Lambda^\circ_{\text{HAc}} = \Lambda^\circ_{\text{HCl}} + \Lambda^\circ_{\text{NaAc}} - \Lambda^\circ_{\text{NaCl}}$$

$$= (425 + 91 - 128) \, \Omega^{-1} \text{cm}^2 \text{ equiv}^{-1}$$

$$= 644 \, \Omega^{-1} \text{cm}^2 \text{ equiv}^{-1}$$

Now, for same conductivity cell, the conductivities of HAc and KCl solutions will be related as

$$k_{\text{HAc}} = k_{\text{KCl}} \cdot \dfrac{r_{\text{KCl}}}{r_{\text{HAc}}} \quad \text{by Eq. (14.2)}$$

Then $\qquad \Lambda_{\text{HAc}} = \dfrac{1000 \, k_{\text{KCl}} r_{\text{KCl}}}{C r_{\text{HAc}}} \quad \text{by Eq. (14.3b)}$

$$= \dfrac{(1000 \text{ cm}^3 \text{L}^{-1}) \, (0.00276 \Omega^{-1} \text{cm}^{-1}) \, (380 \, \Omega)}{(0.01 \text{ equiv L}^{-1}) \, (6434 \, \Omega)}$$

$$= 16.3 \, \Omega^{-1} \text{cm}^2 \text{equiv}^{-1}$$

$$\alpha = \dfrac{\Lambda_{\text{HAc}}}{\Lambda^\circ_{\text{HAc}}}$$

[ignoring conductivity of H$_2$O and dependence of ionic mobility on ionic concentration in HAc solution]

$$= \dfrac{16.3 \, \Omega^{-1} \text{cm}^2 \text{equiv}^{-1}}{644.0 \, \Omega^{-1} \text{cm}^2 \text{equiv}^{-1}} = 25.3 \times 10^{-3}$$

14.28 For same conductivity cell, the conductivities due to the solutes in KCl and BaSO$_4$ solution will be related as

$$k_{\text{KCl}} = k_{\text{BaSO}_4} \left(\dfrac{1}{r_{\text{KCl soln}}} - \dfrac{1}{r_{\text{H}_2\text{O}}} \right) \Big/ \left(\dfrac{1}{r_{\text{BaSO}_4 \text{ soln}}} - \dfrac{1}{r_{\text{H}_2\text{O}}} \right) \quad \text{by Eq. (14.2)}$$

Now $\qquad k_{\text{BaSO}_4} = \dfrac{C_{\text{BaSO4}} \Lambda_{\text{BaSO}_4}}{1000} = \dfrac{\sqrt{K_{\text{sp BaSO}_4}} \cdot F(u_{\text{Ba}^{2+}} + u_{\text{SO}_4^{2-}})}{1000}$

Then, for KCl solution, where conductance due to H_2O is negligible

$$k_{KCl\,soln} = k_{KCl} = \frac{\sqrt{K_{sp\,BaSO_4}}\,F(u_{Ba^{2+}} + u_{SO_4^{2-}})}{1000\,r_{KCl\,soln}\left(\dfrac{1}{r_{BaSO_4\,soln}} - \dfrac{1}{r_{H_2O}}\right)}$$

$$= \frac{\sqrt{10^{-10}}\,\text{equiv L}^{-1})\,(9.65 \times 10^4\,\text{C equiv}^{-1})\left(\dfrac{1\,cm}{151S} + \dfrac{1\,cm}{121S}\right)\Big/\dfrac{10\,V}{1\,cm}}{(1000\,cm^3 L^{-1})\,(1.4 \times 10^{-3}\Omega^{-1})^{-1}\,(2.2 \times 10^{-6} - 0.82 \times 10^{-6})\,\Omega^{-1}}$$

$$= 1.45 \times 10^{-3}\,\Omega^{-1}cm^{-1}$$

☐ Here, it has been implicitly assumed that the solutions are prepared with water of same conductance as given.

Note: in the present problem (unlike the previous one) r, k and Λ all differ significantly for solute and solution which are indicated by the suffix.

14.29 For calculation on the basis of the given data at infinite dilution, we are to ignore the interionic forces (apart from the assumption involved in the previous problem).

a. For the KCl solution, where the conductance due to water can be ignored

$$\Lambda = \frac{K_{cell} \times 1000}{rC} \quad \text{by Eqs (14.2) and (14.3b)}$$

or

$$K_{cell} = \frac{rC\Lambda}{1000}$$

$$= \frac{(10^2\,\Omega)\,(0.02\,\text{equiv L}^{-1})\,(149.8\,\Omega^{-1}cm^2\text{equiv}^{-1})}{(1000\,cm^3 L^{-1})} = 0.30\,cm^{-1}$$

b. For the AgCl solution, the conductance due to water should not be ignored. Here solubility of the electrolyte,

$$C = \frac{1000\,K_{cell}\left(\dfrac{1}{r_{soln}} - \dfrac{1}{r_{solvent}}\right)}{\Lambda}$$

$$= \frac{\left(1000\,cm^3 L^{-1}\right)(0.30\,cm^{-1})\left(\dfrac{1}{1.02 \times 10^5\,\Omega} - \dfrac{1}{4.4 \times 10^6\,\Omega}\right)}{(138.4\,\Omega^{-1}cm^2\text{equiv}^{-1})}$$

$$= 2.0 \times 10^{-5}\,\text{equiv L}^{-1}$$

c, Conductivity of water is due to H^+ and OH^- ions. Here

$$\frac{K_{cell}}{r} = \frac{C_{H^+}\lambda_{H^+} + C_{OH^-}\lambda_{OH^-}}{1000} \quad \text{by Eq. 14.10}$$

Ionic product of water

$$= \frac{C_{H^+}}{C^\circ} \cdot \frac{C_{OH^-}}{C^\circ}$$

[where C° is the standard concentration, i.e. 1 mol L^{-1}]

$$= \left[\frac{1000\,K_{cell}}{r(\lambda_{H^+} + \lambda_{OH^-})C^\circ}\right]^2, \quad \text{since } C_{H^+} = C_{OH^-}$$

$$= \left[\frac{1000 \, K_{cell}}{rF(u_{H^+} + u_{OH^-})C^\circ} \right]^2$$

$$= \left[\frac{(1000 \, cm^3 L^{-1})(0.30 \, cm^{-1})}{(4.4 \times 10^6 \, \Omega)(9.65 \times 10^4 \, C \, mol^{-1})(3.63 + 2.05)10^{-3} \, cm^2 s^{-1} V^{-1})(1 \, mol \, L^{-1})} \right]^2$$

$$= 1.5 \times 10^{-14}$$

Ignoring of interionic forces is quite justified with AgCl solution and water (where the ionic concentration is sufficiently low) but not with 0.02 N KCl solution. As a result, the calculated K_{cell} is somewhat higher than its true value with consequent higher values of the concentration of AgCl and K_w.

14.30 a. Conductivity of $NaNO_3$ solution

$$= \Lambda \frac{C}{1000}$$

$$= (121 \, \Omega^{-1} cm^2 equiv^{-1}) \left(\frac{0.001 \, equiv \, L^{-1}}{1000 \, cm^3 L^{-1}} \right)$$

$$= 1.21 \times 10^{-4} \, \Omega^{-1} cm^{-1}$$

b.
$$K_{cell} = kr$$

$$= (1.21 \times 10^{-4} \, \Omega^{-1} cm^{-1})(8250 \, \Omega)$$

$$= 0.998 \, cm^{-1}$$

c.
$$\Lambda = \frac{1000 \, K_{cell}}{rC}$$

$$\propto \frac{1}{r}, \text{ for fixed } C \text{ and } K_{cell}$$

Now $\Lambda_{HNO_3} = \Lambda_{HCl} + \Lambda_{NaNO_3} - \Lambda_{NaCl}$, by Kohlrausch's law

Then
$$\frac{1}{r_{HNO_3}} = \frac{1}{r_{HCl}} + \frac{1}{r_{NaNO_3}} - \frac{1}{r_{NaCl}}$$

$$= \frac{1}{2340 \, \Omega} + \frac{1}{8250 \, \Omega} - \frac{1}{7900 \, \Omega}$$

$$= 4.22 \times 10^{-4} \, \Omega^{-1}$$

Therefore $r_{HNO_3} = 2369.7 \, \Omega$

d.
$$\Lambda_{HNO_3} = \frac{\gamma_{HNO_3}}{r_{HNO_3}} \Lambda_{NaNO_3} \text{ since } \Lambda \propto \frac{1}{r}$$

$$= \frac{(8250 \, \Omega)}{(2369.6 \, \Omega)} \cdot (121 \, \Omega^{-1} cm^2 equiv^{-1})$$

$$= 421.3 \, \Omega^{-1} cm^2 equiv^{-1}$$

Note: If HNO_3, HCl, $NaNO_3$ and NaCl were not of same strength, then the relevant equation would be

$$\frac{1}{(rC)_{HNO_3}} = \frac{1}{(rC)_{HCl}} + \frac{1}{(rC)_{NaNO_3}} - \frac{1}{(rC)_{NaCl}}$$

14.31 For any electrolyte

$$\Lambda = \frac{1000\,k}{C} = \frac{1000\,K_{cell}}{rC}$$

Then, ignoring the variation of Λ with C,

$$r \propto \frac{1}{C} \text{ for a particular cell}$$

Now, on mixing, the concentrations of NaCl and NaOH will be reduced by factors of $\frac{1}{3}$ and $\frac{2}{3}$ respectively. Then in the mixture

$$r_{NaCl} = 40\,\Omega \times 3 = 120\,\Omega$$

$$r_{NaOH} = 100\,\Omega \times \tfrac{3}{2} = 150\,\Omega$$

and

$$\frac{1}{r_{mix}} = \frac{1}{r_{NaCl}} + \frac{1}{r_{NaOH}} \quad \text{by Ohm's law}$$

$$= \frac{1}{120\,\Omega} + \frac{1}{150\,\Omega}$$

whence $\qquad r_{mix} = 66.7\ \Omega$

14.32 25 ml of 0.1 N HCl $\equiv$ 25 ml of 0.1 N NaOH

The two solutions will then completely neutralise one another forming a solution of NaCl of strength $\frac{0.1}{2}$ or 0.05 N (strength becomes half due to double dilution). In this reaction H^+ ions are replaced by Na^+ ions of lower mobility. Due to this and also due to dilution, the conductance of the solution will decrease, i.e. A–B will be positive.

Equivalent conductivity

$$\Lambda = \frac{1000\,k}{C} = \frac{1000}{C}\,\frac{K_{cell}}{r}$$

$$= \frac{(1000\ cm^3 L^{-1})(1.0\ cm^{-1})}{(0.05\ equiv\ L^{-1})} \cdot B$$

14.33   a. In the resulting mixture, before 25% of the equivalence point (i.e. when 75% of HCl is neutralised)

$$C_{HCl} = \left(0.01 \times \frac{25}{100}\right)\left(\frac{100}{175}\right) N = 0.00143\ N$$

$$C_{NaCl} = \left(0.01 \times \frac{75}{100}\right)\left(\frac{100}{175}\right) N = 0.00429\ N$$

$$k_{mix} = k_{HCl} + k_{NaCl}, \quad k \text{ solvent being insignificant}$$

$$= \frac{(C\Lambda)_{HCl} + (C\Lambda)_{NaCl}}{1000}$$

$$= \frac{(0.00143\ equiv\,L^{-1})(426\,\Omega^{-1}cm^2 equiv^{-1}) + (0.00429\ equiv\ L^{-1})(126\,\Omega^{-1}cm^2 equiv^{-1})}{(1000\ cm^3 L^{-1})}$$

$$= 1.15 \times 10^{-3}\,\Omega^{-1}cm^{-1}$$

b. At the equivalence point, when HCl and NaOH completely react with one another forming a solution of NaCl of strength $\frac{0.01}{2}$ or 0.005 N

$$k_{mix} = \frac{(C\Lambda)_{NaCl}}{1000}$$

$$= \frac{(0.005 \text{ equiv } L^{-1})(126\Omega^{-1}cm^2equiv^{-1})}{(1000 \text{ cm}^3L^{-1})}$$

$$= 0.63 \times 10^{-3} \Omega^{-1}cm^{-1}$$

c. After 25% of the equivalence point

$$C_{NaCl} = \left(0.01 \times \frac{100}{225}\right) N = 0.00444 \text{ N}$$

$$C_{NaOH} = \left(0.01 \times \frac{25}{100}\right)\left(\frac{100}{225}\right) N = 0.00111 \text{ N}$$

$$k_{mix} = k_{NaCl} + k_{NaOH} = \frac{(C\Lambda)_{NaCl} + (C\Lambda)_{NaOH}}{1000}$$

$$= \frac{(0.00444 \times 126 + 0.00111 \times 249)}{1000} \text{ or } 0.836 \times 10^{-3}\Omega^{-1}cm^{-1}$$

Note:

1. In conductometric titration, conductance (which is directly measured) is plotted instead of k (which has to be calculated) with no difference in shape of the corresponding graphs (as conductance $\propto k$) which is of prime importance.
2. Here k_{mix} is calculated considering the dilution effect which is avoided in the usual conductometric titrations using concentrated titrants. As expected, k_{mix} is lower in (b) than in (a) and (c).

14.34 The difference between the $\Lambda°$ values for each pair of the given electrolytes containing a common ion is same (34.9). This suggests the following general law. Cation and anion of an electrolyte contribute independently to Λ of the electrolyte at infinite dilution (Kohlrausch's law) i.e.

$$\Lambda = \lambda_+ + \lambda_-$$

where λ_+ and λ_- are the contributions (called ionic conductivity) of cation and anion to Λ.

☐ This law is very important in determining Λ of a weak electrolyte in terms of suitable strong electrolytes of which Λ can be readily found. Thus Λ of HAc can be calculated from the knowledge of Λ values of the strong electrolytes—HCl, NaAc and NaCl using the relation

$$\Lambda_{HAc} = \Lambda_{HCl} + \Lambda_{NaAc} - \Lambda_{NaCl}$$

14.35 During electrolysis, per faraday of electricity, 1 equiv of Cu^{2+} will be liberated at the cathode, and $t_{Cu^{2+}}$ equiv of Cu^{2+} will migrate in (and $t_{SO_4^{2-}}$ equiv of SO_4^{2-} will migrate out of the cathode region); the net result is the loss of $(1 - t_{Cu^{2+}})$ or $t_{SO_4^{2-}}$ equivalent of Cu^{2+} and hence $t_{SO_4^{2-}}$ equivalent of $CuSO_4$ in the neighbourhood of cathode.

Now, 0.554 g of $CuSO_4 = \frac{0.554}{(63.6 + 96.0)/2}$ or 0.0694 equiv of $CuSO_4$

Then after electrolysis, the cathode solution contains $(18.217 - 0.554)$ or 17.663 g of water and 0.00694 equiv of $CuSO_4$.

Before electrolysis, 17.663 g of water was associated with $\frac{2 \times 0.2}{1000} \times 17.663$ or 0.00706 equiv of $CuSO_4$.

Then $\qquad t_{SO_4^{2-}} = \dfrac{\text{loss in equiv of } CuSO_4 \text{ occurring with cathode soln}}{\text{amount of electricity (in faraday) passed}}$

$$= \frac{0.00706 - 0.00694}{0.020/107.9} = 0.65$$

$$t_{Cu^{2+}} = 1 - t_{SO_4^{2-}}$$

[assuming that the current is carried entirely by Cu^{2+} and SO_4^{2-}]

$$= 1 - 0.65 = 0.35$$

14.36 $\qquad 0.409 \text{ g Cu} = \dfrac{0.409 \times (63.6 + 96.0)}{63.6}$ or 1.026 g $CuSO_4$

Then after electrolysis, the cathode solution contains $(54.7 - 1.026)$ or 53.674 g water. Before electrolysis $(100 - 2.84)$ or 97.16 g of water was associated with 2.84 g of $CuSO_4$. So 53.674 g of water was associated with $\frac{2.84 \times 53.674}{97.16}$ or 1.569 g of $CuSO_4$.

Then $\quad t_{SO_4^{2-}} = 1 - t_{Cu^{2+}} = \dfrac{(1.569 - 1.026)/159.6/2}{0.408/63.6/2}$ by Hittrof principle

whence $\qquad t_{Cu^{2+}} = 0.47$

☐ The Hittrof method is based on the measurement of the change in concentration of electrolyte in the neighbourhood of an electrode resulting from the flow of current. The passage of current for a long time will lead to greater inaccuracy in the result due to greater diffusion of the electrolyte from the more concentrated to the less concentrated regions.

Note:

i. The composition of the cathode solution after electrolysis is given in terms of Cu in the present problem while it is in terms of $CuSO_4$ in the previous problem.

ii. The cathode in the present problem serves the purpose of Ag coulometer in the previous problem.

14.37 a. Per faraday of electricity passing through the solution

$$\text{loss in weight of anode} = 1 \text{ g equiv of Cu} = \frac{63.6}{2} \text{ g}$$

gain in weight of anode solution = wt of Cu dissolved + wt of SO_4^{2-} migrated

$$\text{in} -wt \text{ of } Cu^2 \text{ migrated out}$$

$$\equiv t_{SO_4^{2-}} \text{ equivalent of } CuSO_4,$$

$$\equiv t_{SO_4^{2-}} \left(\frac{63.6 + 96}{2} \right) \text{g}$$

Then, from the given condition

$$t_{SO_4^{2-}} \left(\frac{63.6 + 96}{2} \right) = \frac{63.6}{2}$$

whence $\qquad t_{SO_4^{2-}} = 0.40$

b. Here electrolysis does not result in any net change of the Hittrof cell as a whole. Hence loss in amount of $CuSO_4$ in the cathode compartment will be equal to the gain in the anode compartment, i.e. 0.004 mol.

c. Per faraday of electricity, the gain in anode solution is $t_{SO_4^{2-}}$ equiv, i.e. 0.40 equiv or 0.2 mol of $CuSO_4$ (by Hittorf principle). Then the quantity of electricity passed is 0.004/0.2 or 0.02 F.

Note: $t_{Cu^{2+}}$ found in the last three problems is not same due to difference in the concentration of $CuSO_4$ used. This is in keeping with the fact that transport number of an ion is not a characteristic property of the ion.

14.38 Here cathode reaction is

$$AgCl + e = Ag + Cl^-$$

Then, considering the migration of H^+ and Cl^- ions the cathode solution will gain t_{H^+} equiv of HCl per faraday of electricity passing through the cell.

Now, after electrolysis the cathode contains (51.7436 − 0.0267) or 51.7169 g of water and 0.0267 g of HCl.

Before electrolysis 51.7169 g of water was associated with $51.7169 \times 0.3856 \times 10^{-3}$ or 0.0199 g of HCl.

Then

$$t_{H^+} = \frac{\text{gain in equiv of HCl occurring with cathode solution}}{\text{amount of electricity (in faraday) passed}}$$

$$= \frac{(0.0267 - 0.0199)/(36.45) \text{ equiv}}{(2 \times 10^{-3}\,A)\,(3 \times 60 \times 60S)/(9.648 \times 10^5 C\,\text{equiv}^{-1})} = 0.833$$

Note: In the present problem, the cathode solution gains electrolyte which is determined by the transport number of the cation (that migrates toward the cathode), whereas in the previous problem, the anode solution gains electrolyte which is determined by the transport number of the anion (that migrates toward the anode).

14.39 a. Because stirring favours migration of the ions of the electrolyte.

b. This is due to the formation of the complex ion CdI_4^{2-}

$$2CdI_2 \rightleftharpoons Cd^{2+} + CdI_4^{2-}$$

During electrolysis of a concentrated solution of CdI_2, where CdI_4^{2-} has predominating concentration over the simple I^- ion, the anode solution may exhibit a rise in the formal concentration of cadmium when the transport number of Cd^{2+} ion becomes apparently negative. This correponds to an anion transport number greater than 1, taking $t_+ + t_- = 1$.

14.40

$$t_{Li^+} = \frac{\lambda_{Li^+}^{\circ}}{\Lambda_{LiCl}^{\circ}} \quad \text{by Eq. (14.13b)}$$

$$= \frac{Fu_{Li^+}^{\circ}}{\Lambda_{LiCl}^{\circ}} = \frac{Fv_{Li^+}^{\circ}/E}{\Lambda_{LiCl}^{\circ}}$$

or

$$v_{Li^+} = \frac{t_{Li^+}\Lambda_{LiCl}^{\circ} \cdot E}{F}$$

$$= \frac{(0.336)\,(115.03 \times 10^{-4}\,\Omega^{-1}m^2\,mol^{-1})\,(6\,V/4 \times 10^{-2}\,m)}{(9.65 \times 10^4\,C\,mol^{-1})}$$

$$= 6.0 \times 10^{-6}\,m\,s^{-1}$$

14.41 The transport number t_i of an ion i involved in a moving boundary is given by

$$t_i = \frac{lac}{1000} \cdot \frac{F}{Q} \quad \text{by Eq. (14.14)}$$

$$= \frac{\pi l r^2 cF}{1000\,it}$$

Then $\quad t_{K^+}$

$$= \frac{3.14\left(\dfrac{75.5}{10}\,\text{cm}\right)\left(\dfrac{1}{10}\,\text{cm}\right)^2 (0.1\,\text{equiv L}^{-1})(9.65 \times 10^4\,\text{C equiv}^{-1})}{(1000\,\text{cm}^2\text{L}^{-1})\left(\dfrac{10}{1000}\,A\right)(8 \times 60\,\text{s})}$$

$$= 0.476$$

$$t_{Cl^-} = 1 - 0.476 = 0.524$$

The shift in anionic boundary involving Cl^-

$$= \text{shift in cationic boundary} \times \frac{t_{Cl^-}}{t_{K^+}}$$

$$= \left(\frac{75.5}{10}\,\text{cm}\right) \times \frac{0.524}{0.476} = 8.31\,\text{cm}$$

14.42 In NaCl solution, $t_{Cl^-} = \dfrac{\lambda_{Cl^-}}{\lambda_{Na^+} + \lambda_{Cl^-}}$ by Eq. (14.13b)

$$= \frac{76.3}{50.1 + 76.3} = 0.604$$

☐ In the resulting mixture

$$C_{Na^+} = C_{H^+} \doteq 0.005\,\text{N},\ C_{Cl^-} = 0.01\,\text{N}$$

$$t_{Cl^-} = \frac{C_{Cl^-}\lambda_{Cl^-}}{C_{Na^+}\lambda_{Na^+} + C_{H^+}\lambda_{H^+} + C_{Cl^-}\lambda_{Cl^-}} \quad \text{by Eq. (14.13a)}$$

$$= \frac{0.01 \times 76.3}{0.005 \times 50.1 + 0.005 \times 349.8 + 0.01 \times 76.3} = 0.276$$

Then $\quad \Delta t_{Cl^-} = 0.276 - 0.604 = -0.328$

[**Note:** On mixing, C_{Cl^-} and $\sum C_i$ do not change, yet t_{Cl^-} decrease as $\lambda_{H^+} > \lambda_{NA^+}$].

☐☐ The condition for t_{Na^+} and t_{H^+} to be equal is that in the resulting mixture

$$\frac{C_{Na^+}}{C_{H^+}} = \frac{\lambda_{H^+}}{\lambda_{Na^+}} \quad \text{by Eq. (14.13a)}$$

$$= \frac{349.8}{50.1}$$

Then the required volume ratio, in which NaCl and HCl solutions (which are of equal strength) are to be mixed, is 349.8:50.1 or 6.98:1.

14.43 In the resulting mixture,

$$C_{Mg^{2+}} = 0.1\,\text{M} = 0.2\,\text{N}$$

$$C_{SO_4^{2-}} = \tfrac{1}{2} \times 0.1\,\text{M} = 0.1\,\text{N}$$

$$C_{Cl^-} = \tfrac{1}{2} \times 0.2\,\text{M} = 0.1\,\text{N}$$

a. Conductivity of the mixture

$$= (C_{Mg^{2+}}\lambda_{Mg^{2+}} + C_{SO_4^{2-}}\lambda_{SO_4^{2-}} + C_{Cl^-}\lambda_{Cl^-})/1000 \quad \text{by Eq. (14.10)}$$

$$= \frac{(0.2 \text{ equiv L}^{-1})(53\Omega^{-1}_{.}\text{cm}^2 \text{ equiv}^{-1}) + (0.1 \text{ equiv L}^{-1})(79.8\Omega^{-1}\text{cm}^2\text{equiv}^-) + (0.1\text{equiv L}^{-1})(76.3\Omega^{-1}\text{cm}^2\text{equiv}^{-1})}{(1000 \text{ cm}^3\text{L}^{-1})}$$

$$= 26.2 \times 19^{-3}\Omega^{-1}\text{cm}^{-1}$$

b.
$$t_{Mg^{2+}} = \frac{C_{Mg^{2+}}\lambda_{Mg^{2+}}}{C_{Mg^{2+}}\lambda_{Mg^{2+}} + C_{SO_4^{2-}}\lambda_{SO_4^{2-}} + C_{Cl^-}\lambda_{Cl^-}}$$

$$= \frac{(0.2 \text{ equiv L}^{-1})(53\Omega^{-1}\text{cm}^2\text{equiv}^{-1})}{26.2 \text{ cm}^2\text{L}^{-1}} = 0.404$$

c. Distance moved by Mg^{2+} ion

$$= \frac{1000\, t_{Mg^{2+}} \cdot it}{FaC_{Mg^{2+}}} \quad \text{by Eq. (14.14)}$$

$$= \frac{(1000 \text{ cm}^3\text{L}^{-1})(0.404)(0.1 \text{ A})(100 \text{ s})}{(9.65 \times 10^4 \text{ C equiv}^{-1})(5 \text{ cm}^2)(0.2 \text{ equiv L}^{-1})} = 0.0419 \text{ cm}$$

UNIVERSITY QUESTIONS

14.1 Is there any difference between velocity and mobility of an ion? On what factors do the mobility of an ion depend?   (Burdwan BSc(H), 1993)

14.2 Explain Wien and Debye-Falkenlagen effects.   (Burdwan BSc(H), 2010)

14.3 Justify/Criticise. Asymmetry effect is more important in solvents with high dielectric constant like water.   (Burdwan BSc(H), 1998)

14.4 Define specific and equivalent conductance. Relate them and obtain their units. Explain the effect of dilution on them. Does the relation derived above properly explain this dependence? Justify your answer.   (Calcutta BSc(H), 1996)

14.5 Explain the significance of the conductance ratio (Λ_c/Λ_o) for (i) weak and (ii) strong electrolytes.   (Calcutta BSc(H), 1996)

14.6 If k is the specific conductance and c the concentration, justify that (i) in general, the relation $k \propto c$ is untenable and (ii) only for strong electrolytes at very small c, the above proportionality may hold.   (Burdwan BSc(H), 1996)

14.7 Do you find any inconsistency in the statement 'at 25°C the specific conductivity of a 0.01 N KCl solution is less than the equivalent conductivity of a 0.01 N KCl solution'.   (Calcutta BSc(H), 1978)

14.8 a. The equivalent conductance of a 0.01 N $CaCl_2$ is 120.36 ohm^{-1}cm^{2}g equiv^{-1}. What will be the value of molar conductance in SI unit?   (Burdwan BSc(H), 2000)

b. The equivalent conductivity of an aq solution of K_2SO_4, $Al_2(SO_4)_3$, $24H_2O$ is x. What is its molar conductivity value?   (Burdwan BSc(H), 2012)

14.9 Explain why ionic conductivity at infinite dilution increaes in the order Li^+, Na^+, K^+, Rb^+.   (Burdwan BSc(H), 2011)

14.10 Give suitable reasons behind abnormally high conductivity value for both H^+ and OH^- ions in solution.   (Burdwan BSc(H), 2012)

14.11 Measurement of resistances of 0.1 molar solutions of acetic acid and of formic acid in different conductivity cells do not give us any idea as to the relative strengths of the two acids. Further even if the same conductivity cell be used, the respective resistance values do not unequivocally tell us the sequence of the strengths of the two acids. (Calcutta BSc(H), 1976)

14.12 a. Explain how do you proceed to determine equivalent conductance at infinite dilution for NaCl and CH_3COOH? (Burdwan BSc(H), 2009)

   b. Even without diluting a moderately concentrated salt solution, Λ_o can be directly measured. (Burdwan BSc(H), 1996)

14.13 The equivalent conductances at infinite dilution of $AgNO_3$, KNO_3 and KCl solutions are 133.36, 144.96 and 149.86 at s cm^2 (g-equiv)$^{-1}$ at 25°C. Calculate

   a. $\Lambda_{o\,(AgCl)}$

   b. The quantity $\Lambda_{o\,(NaCl)} - \Lambda_{o\,(NaNO_3)}$

   State the law used. (Calcutta BSc(H), 2010)

14.14 Arrange the following electrolytes in the order of their Λ_o values in water, with proper justification

   a. LiCl, HCl, KCl (Calcutta BSc(H), 2011)

   b. KCl, HCl, NaOH (Calcutta BSc(H), 2008)

   c. NaCl, NaOH, HCl, CH_3COOH (Jadavpur BSc(H), 2001)

14.15 a. Explain the effect of medium viscosity on the conductance and hence deduce the Walden's rule. (Jadavpur BSc(H), 2013)

   b. Explain why Walden's rule is not applicable to cations of small size.
   (Burdwan BSc(H), 2011)

14.16 Write an equation which describes the variation of Λ_o with temperature.
   (Burdwan BSc(H), 2002)

14.17 Find the units of A and B in the Onsager equation $\Lambda = \Lambda_o - (A - B\Lambda_o)\sqrt{C}$. How is the ratio A/B related to the viscosity of the medium? (Burdwan BSc(H), 1998)

14.18 Which one among DC and AC is used in the measurement of conductance of an electrolytic solution and why? (Jadavpur BSc(H), 2003)

14.19 Explain why in conductometric titration

   a. the strength of the titre must be at least ten times that of the solution to be titrated. (Jadavpur BSc(H), 2002)

   b. conductance is plotted against volume of titre, but resistance is never plotted.
   (Jadavpur BSc(H), 2003)

14.20 Draw the conductometric titration curves for the following titrations:

   a. CH_3COOH with NaOH (Burdwan BSc(H), 2011)

   b. NH_4OH with HCl (Jadavpur BSc(H), 2001)

   c. KCl with $AgNO_3$ (Burdwan BSc(H), 2010)

   d. $MgSO_4$ with $Ba(OH)_2$ (Jadavpur BSc(H), 2001)

   e. HCl + CH_3COOH with NaOH (Burdwan BSc(H), 2012)

   f. Oxalic acid with NaOH (Jadavpur BSc(H), 2013)

14.21 Will the conductometric titration curves for titration of oxalic acid and sulphuric acid (both dibasic) with NaOH be same? Give reasons. (Calcutta BSc(H), 1997)

14.22 How can you determine the individual strengths of NH_4Cl and NaCl in a mixture by conductometric titration? Explain the nature of the titration curves.
   (Jadavpur BSc(H), 1998)

14.23 Justify/criticise. Ionic mobility of Mg^{2+} is twice that of Na^+.

(Burdwan BSc(H), 1998)

14.24 Explain the following:

a. Transport numbers of K^+ and Cl^- ions are nearly equal in KCl.

(Burdwan BSc(H), 1999)

b. While ionic mobilities increase with temperature, both the transport number t_{H^+} and t_{Cl^-} in aqueous solution of HCl approach 0.5 as temperature is increased.

(Calcutta BSc(H), 2008)

14.25 Which of the following quantities must be same for $CaCl_2$ (aq) and NaCl (aq) at the same temperature and pressure?

$$\lambda^{\circ}_{Cl^-}, t^{\circ}_{Cl^-}, u^{\circ}_{Cl^-}$$

Give reasons for your answer. (Calcutta BSc(H), 2000)

14.26 How does a plot of t_+ vs t_- for a given electrolyte AB in a given solvent look? What is the significance of such a plot? (Burdwan BSc(H), 1998)

14.27 In between two platinum electrodes 8.0 cm apart and with a potential difference of 4.0 volt is placed a very dilute solution of $AgNO_3$. The Ag^+ ions and NO_3^- ions are found to move with velocities 3.2×10^{-4} and 3.7×10^{-4} cm per sec respectively.

Find the equivalent conductance of $AgNO_3$ and the transport number of the anion.

(Calcutta BSc(H), 1978)

14.28 0.5 N NaCl is placed between two electrodes 1.5 cm apart and having an area of each 3.0 sq cm offered a resistance of 25.0 ohms. Calculate the equivalent conductance. (Andhra BSc, 2004)

14.29 A conductivity cell of cell constant 1 cm^{-1} shows a resistance of 6667 ohms when filled with 0.001 M KCl solution at 25°C. The same cell records a resistance of 2353 ohms when filled with 0.001 M HCl solution at 25°C.

a. Calculate equivalent conductance values for KCl and HCl.

b. Calculate ion conductance of H^+ ion assuming that K^+ and Cl^- ions have same mobilities. Consider the solutions to be very dilute so that the condition of infinite dilution may be applied.

c. How far will the H^+ ion move in 10 sec when a potential difference of 2 volts is applied between two electrodes placed 2 cm apart.

d. Calculate t_{H^+} and t_{Cl^-} in the HCl solution. What will be the effect of temperature on these values? (Burdwan BSc(H), 2000)

14.30 Find the equivalent conductance of $AgNO_3$ from the following statement:

The resistance of a conductivity cell was found to be 700 ohms and 800 ohms when filled with 0.01 N KCl and 0.01 N $AgNO_3$ solution respectively, and the equivalent conductance of KCl is 150 $ohm^{-1}cm^2equiv^{-1}$.

How does the value of equivalent conductance depend on temperature?

(Burdwan BSc(H), 1994)

14.31 The specific conductivity of 0.1 M solution of NaOH is 0.0221 $ohm^{-1}cm^{-1}$. When an equal volume of 0.1 M HCl solution is added, the specific conductivity falls to 0.0056 $ohm^{-1}cm^{-1}$. Find Λ for NaCl. (Calcutta BSc(H), 2003)

14.32 The specific conductance of a solution containing 0.2 M NaCl and 0.1 M NaX (a strong electrolyte) is 0.0382 $ohm^{-1}cm^{-1}$. Calculate the equivalent ion conductance of X^-. (Given equivalent ion conductances of Na^+ and Cl^- are 50 and 76 respectively in the usual unit). (Jadavpur BSc(H), 2001)

14.33 The conductivity value of water at 25°C is 5.54×10^{-8} ohm^{-1}cm^{-1}. Calculate the degree of dissociation and ionic product of water if the ionic molar conductivity values of H$^+$ and OH$^-$ ions are estimated to be 349.8 and 197.8 ohm^{-1}cm^{2}mol^{-1} respectively. (Burdwan BSc(H), 2012)

14.34 What will be the value of Λ for a 0.001 M aqueous solution of ammonia? Given, $K_b = 1.8 \times 10^{-5}$ and $\Lambda_o = 238$ ohm^{-1}cm^{2}aq^{-1}. (Calcutta BSc(H), 1997)

14.35 Equivalent conductances at infinite dilution of HCl, NaCl and CH$_3$COONa are 426.2, 126.5 and 91 ohm^{-1}cm^{2}aq^{-1} respectively at 25°C. A conductance cell filled with 0.01 M KCl has a resistance of 257.3 ohms at 25°C. The same cell filled with 0.2 N acetic acid has a resistance of 508.6 ohms. Calculate the dissociation constant of the acid (specific conductance of 0.01 M KCl at 25°C is 1.41×10^{-3} ohm^{-1}cm^{-1}). (Calcutta BSc(H), 2001)

14.36 The ionic mobilities of CH$_3$COO$^-$ and H$^+$ ions at infinite dilution are 4.25×10^{-4} and 3.63×10^{-3} cm^2 volt^{-1}sec^{-1} at 25°C. If the specific conductance for a 0.1 N CH$_3$COOH solution be 5.20×10^{-4} ohm^{-1}cm^{-1} at the same temperature, find out the apparent as well as the thermodynamic dissociation constant of CH$_3$COOH at 25°C. (Assume limiting Debye–Huckel equation to be valid with $A = 0.509$). (Jadavpur BSc(H), 2001)

14.37 The tr number of Na$^+$ in NaCl is 0.385 and the equivalent conductance at infinite dilution of NaCl is 126.5 ohm^{-1}cm^{2}g-eq v^{-1} at 25°C. Estimate the distance travelled in 1 h by Na$^+$ ion in a very dilute solution kept in a cell with electrodes 5 cm apart when a potential difference of 3 volt is applied between the electrodes. (Jadavpur BSc(H), 2013)

14.38 A saturated aqueous solution of SrSO$_4$ shows a specific conductance of 1.5×10^{-4} ohm^{-1}cm^{-1} at 25°C. The solubility is 0.5×10^{-3} mol lit^{-1} and the equivalent conductance $\Lambda_{SrSO_4} = 140$ ohm^{-1}cm^{2}equiv^{-1}. Calculate the approximate value of specific conductance of water at the given temperature. (Burdwan BSc(H), 1997)

14.39 The specific conductance of a saturated solution of AgCl is 1.70×10^{-6} ohm^{-1}cm^{-1} and that of water is 1.50×10^{-7} ohm^{-1}cm^{-1}, both at 25°C. The mobilities of Cl$^-$ and Ag$^+$ ions are 5.6×10^{-4} and 6.8×10^{-4} cm^{2}volt^{-1}sec^{-1} respectively. Calculate the apparent as well as thermodynamic solubility product of AgCl. [Use limiting Debye–Huckel equation with $A = 0.509$]. (Jadavpur BSc(H), 2002)

14.40 In a transport experiment in 1.0 N KCl solution, the boundary between KCl and BaCl$_2$ (following) solutions is found to sweep through a volume of 0.1205 ml in 167.5 s with a current of 0.142 A. Calculate t_{Cl^-}. (Jadavpur BSc(H), 2012)

14.41 Suggest a physicochemical experiment to show that in a concentrated solution, cadmium iodide exists as Cd[CdI$_4$] and not CdI$_2$. (Jadavpur BSc(H), 1994)

14.42 Cite an example of an ion showing negative transport number in an electrolyte solution. (Burdwan BSc(H), 2008)

14.43 A AgNO$_3$ solution containing 0.00735 gm of AgNO$_3$ per gm of water is electrolysed between Ag-electrodes. At the end of the experiment 0.075 gm of Ag is found to be lost from the anode plate, also the anode solution is found to contain 23.14 gm of water and 0.255 gm of AgNO$_3$. Calculate the transport number of Ag$^+$ ion. [Ag = 108] (Jadavpur BSc(H), 2003)

14.44 During the electrolysis of a solution of potassium chloride between platinum electrodes, 0.0137 g of Cl$^-$ was lost from the anode compartment and 0.0857 g of silver was deposited in a silver coulometer connected in series with the cell. Find the transport number of K^+ and Cl^- ions.   (Delhi BSc, 2011)

14.45 A 2 molal solution of $FeCl_3$ is electrolysed between Pt-electrodes. After electrolysis, the cathode compartment solution weighing 20 gm is found to be 1.15 molal in $FeCl_3$ and 1 molal in $FeCl_2$. Calculate the transport number of Fe^{3+} and Cl^- ions.
                                                         (Calcutta BSc(H), 1999)

KEY TO UNIVERSITY QUESTIONS

14.1 Mobility of an is ion simply a particular case of its velocity when the applied electronic field is of unit strength.

☐ Vide problem 14.12.

14.2 Vide Section 14.7.

14.3 For solvents with higher dielectric constant, the interionic forces are lower and hence the asymmetry effect is less important. Then the given statement is not justified.

14.4 Vide Section 14.3.

For k, the conventional unit 'mho cm^{-1}' (i.e. $ohm^{-1}cm^{-1}$) and the SI unit 'Siemens m^{-1}' (i.e. $ohm^{-1}m^{-1}$) follows from Eq. (14.2).

For Λ, the conventional unit 'mho $cm^2 equiv^{-1}$ and the SI unit 'siemens $m^{-1}equiv^{-1}$, follow from the Eq. (14.3a).

For the last two parts of the question vide problem 14.8.

14.5 Vide problem 14.16.

14.6 Vide problem 14.9.

14.7 Vide problem 14.18.

14.8 For an electrolyte $M_{v_+} X_{v_-}$ dissociating completely into M^{Z-} and X^{Z-} ions in solution
$$\Lambda_m = v_+ Z_+ \Lambda_{eq} = v_- Z_- \Lambda_{eq}$$

a. $CaCl_2$ dissociates as
$$CaCl_2 = Ca^{2+} + 2Cl^-$$
Then $\Lambda_m = (1)\,(2)\,(120.36\ ohm^{-1}cm^2 equiv^{-1})$
$$= 240.72\ ohm^{-1}cm^2 mol^{-1}$$

b. Representing the dissociation of the given double salt as
$$KAl(SO_4)_2 \cdot 12H_2O = K(H_2O)_6^+ + Al(H_2O)_6^{3+} + 2SO_4^{2-}$$
$$\equiv KAl(H_2O)_{12}^{4+} + 2SO_4^{2-}$$
$$\Lambda_m = v_- Z_- \Lambda_{eq} = (2)\,(2)\ x$$
$$= 4x$$

14.9 Vide problem 14.13(d).

14.10 Vide problem 14.13(e)

14.11 Of the two acids having same formal concentration, one that produces higher concentration of H^+ ion is of higher strength. Now, for a particular concentration of an acid, its conductance depends not only on its ionic concentration but also on ionic mobilities and cell constant of the conductivity cell used. Here lies the explanation of the given statement.

14.12 a. For NaCl, which is a strong electrolyte, $\Lambda°$ can be obtained from the intercept at $C = 0$ of the nearly linear plot of Λ against $\sqrt{C}$ based on the Eq. (14.8), due to Kohlrausch.

For CH_3CO_2H, which is a weak electrolyte, $\Lambda°$ can be obtained from the intercept at $C = 0$ of the nearly linear plot of $1/\Lambda$ against $C\Lambda$ based on Eq. (14.9) due to Ostwald.

The method used for NaCl is useless for CH_3CO_2H (being a weak electrolyte) because Eq. (14.8) does not reasonably hold even for dilute CH_3CO_2H solution of practical importance. However, $\Lambda°$ for CH_3CO_2H can be obtained from $\Lambda°$'s of strong electrolytes HCl, NaCl and CH_3CO_2Na as in problem 14.27.

b. This is possible by determining Λ with electric field of high frequency (much above 10^6 Hz), when asymmetry effect and electrophoretic effect both disappear, i.e. ions move as if they had no atmosphere as in an extremely dilute solution Λ thus obtained will therefore be identical with $\Lambda°$ corresponding to infinite dilution.

14.13 a.
$$\Lambda°_{AgCl} = \lambda°_{Ag^+} + \lambda°_{Cl^-} \text{ by Kohlrausch's law}$$
$$= \Lambda°_{AgNO_3} + \Lambda°_{KCl} - \Lambda°_{KNO_3}$$
$$= (133.36 + 149.86 - 144.96) \text{ s cm}^2 \text{ (g-equiv)}^{-1}$$
$$= 138.26 \text{ s cm}^2 \text{ (g-equiv)}^{-1}$$

b.
$$\Lambda°_{NaCl} - \Lambda°_{NaNO_3} = \lambda°_{Cl^-} - \lambda°_{NO_3^-}$$
$$= \Lambda°_{KCl} - \Lambda°_{KNO_3}$$
$$= (149.86 - 144.96) \text{ or } 4.90 \text{ s cm}^2 \text{ (g-equiv)}^{-1}$$

vide problem 13.34.

14.14 Vide problem 14.13(d) and 14.15

a. HCl > KCl > LiCl

b. HCl > NaOH > KCl

c. HCl > CH_3COOH > NaOH > NaCl

14.15 Vide Section 14.6.

14.16 Walden's rule implies significant temperature variation of $\Lambda°$ as a consequence of temperature variation of the viscosity of the solvent involved. The most electrolytes in aqueous solution, the temperature dependence of $\Lambda°$ can be expressed as

$$\Lambda°_t = \Lambda°_{25°C} [1 + 0.02 (t - 25)]$$

This is harmony with the fact that the viscosity of water decreases by about 2% per degree rise in temperature.

Note: Pressure and temperature affect $\Lambda°$ oppositely as expected from their opposite effect on viscosity.

14.17 Dimensionally
$$A \equiv \frac{\Lambda°}{\sqrt{C}}$$

$$B \equiv \frac{1}{\sqrt{C}}$$

Then SI unit of A is siemens $m^{-1} \text{equiv}^{-1}/\sqrt{\text{equiv}/m^3}$ or simens $m^{\frac{1}{2}} \text{ equiv}^{-\frac{1}{2}}$.

SI unit of B is $m^{\frac{3}{2}} \text{equiv}^{-\frac{1}{2}}$.

☐ A cares for electrophoretic effect while B does for relatively simpler asymmetry effect so that A/B bears an inverse relation with viscosity of the medium. This is consistent with the fact that the electrophoretic retardation is lower with more viscous solvents, where the motion of both the ion and its atmosphere is slowed down.

14.18 Vide problem 14.13(a).

14.19 Vide problem 14.23.

14.20 (a) Vide problem 14.24(b).

(b) Here the reaction may be represented as

$$NH_4OH (\rightleftharpoons NH_4^+ + OH^-) + H^+ + Cl^- \rightarrow NH_4^+ + Cl^- + H_2O$$
$$\text{(aq NH}_3)$$

Regarding the conductometric titration curve the situation is similar to that in (a).

(c) In the relevant reaction

$$K^+ + Cl^- + Ag^+ + NO_3^- \rightarrow K^+ + NO_3^- + AgCl(s)$$

Cl^- ions are replaced by NO_3^- ions of slightly lower mobility. The conductometric titration curve will then have the shape as indicated in Fig. 14.4.

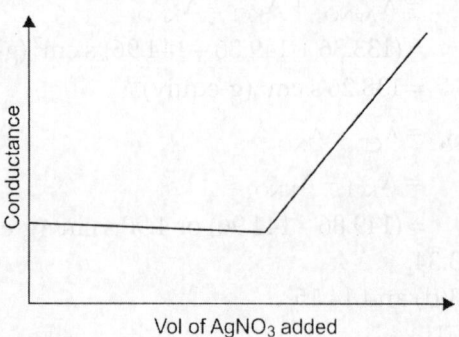

Vol of AgNO$_3$ added

Fig. 14.4.

d. In the relevant reaction

$$Mg^{2+} + SO_4^{2-} + Ba^{2+} + 2OH^- \rightarrow Mg(OH)_2(s) + BaSO_4(s)$$

the ions are removed. Then conductance of the reaction system will decrease up to equivalence point [at a rate proportional to $(\lambda_{Mg^{2+}} + \lambda_{SO_4^{2-}})$]. After equivalence point the conductance will rise at a higher rate [proportional to $(\lambda_{Ba^{2+}} + \lambda_{OH^-})$] as indicated in Fig. 14.5.

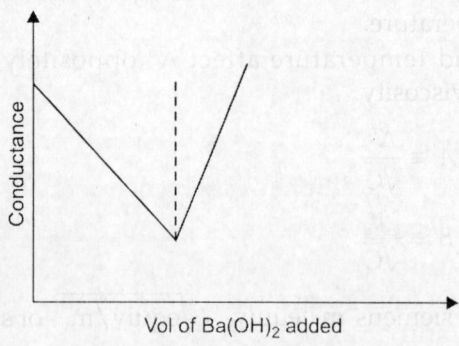

Vol of Ba(OH)$_2$ added

Fig. 14.5

e. Vide problem 14.24(c).

f. Vide problem 14.24(d).

14.21 No. Vide Problem·14.24(e).

14.22 This can be done by titrating the given mixture separately with NaOH and AgNO$_3$ solutions.

The first titration involves the reaction

$$NH_4^+ + Cl^- + Na^+ + OH^- \rightarrow Na^+ + Cl^- + NH_3 + H_2O$$

where NH_4^+ ions are replaced by Na^+ ions of lower mobility. Hence the conductance of the system will decrease up to the equivalence point and then increase. The equivalence point is determined by the concentration of NH$_4$Cl.

The second titration involves the reaction

$$Cl^- + Ag^+ + NO_3^- \rightarrow AgCl(s) + NO_3^-$$

where Cl$^-$ ions (from NH$_4$Cl and NaCl) are replaced by NO_3^- ions of slightly lower mobility. Hence conductance of the system will slightly decrease up to the equivalence point which is determined by the total concentration of NH$_4$Cl and NaCl (Fig. 14.6).

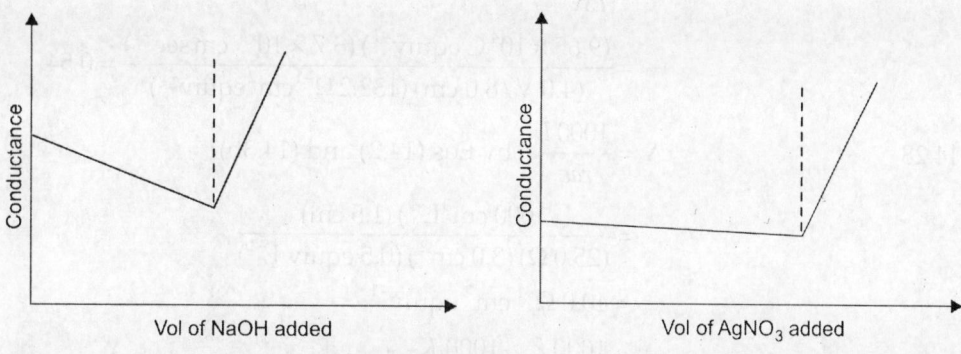

Fig. 14.6

14.23 Ionic mobility refers to the motion of an ion with no net force acting on it. It is determined by a number of component forces apart from that due directly to the applied electric field.. Then, although the latter force is double with Mg^{2+} compared with Na$^+$ (the charge of Mg^{2+} being double), the given statement should not hold because the other component forces do not change by the same proportion.

14.24 a. Because mobilities of K$^+$ and Cl ions are nearly equal, when

$$\frac{t_{K^+}}{t_{Cl^-}} = \frac{u_{K^+}}{u_{Cl^-}} \text{ by Eq. (14.36)}$$

$$\simeq 1$$

b. Vide problem 14.20.

Rise of temperature, like dilution, cannot increase both t_+ and t_- simultaneously.

Here t_{H^+} (> 0.5) decreases and t_{Cl^-} (< 0.5) increases, when both approach 0.5.

Note: Quite frequently the transport number approaches 0.5 when temperature is raised. However, this is not always true, e.g. in case of KCl. The temperature coefficient of t_+ and t_- is largely determined by the thermal stability of the relevant solvated ions.

14.25 $u^o_{Cl^-}$ and $\lambda^o_{Cl^-}$. Because u^o and λ^o, but not t^o, are characteristic properties of an ion irrespective of the other ions with which it is associated.

14.26 If the conductance due to the solvent ions is ignored, the transport numbers of cation and anion of an electrolyte in a solution will be related as $t_+ + t_- = 1$. This corresponds to a linear t_+ vs t_- plot as indicated in Fig. 14.7.

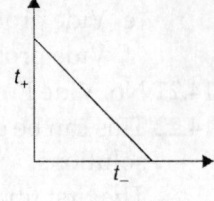

Fig. 14.7

The restriction $t_+ + t = 1$ signifies that both t_+ and t_- cannot increase (or decrease) simultaneously.

14.27
$$\Lambda = \lambda_+ + \lambda_- = F(u_+ + u_-) \text{ by Eq. (14.7)}$$
$$= \frac{F(v_+ + v_-)}{E}$$
$$= \frac{(9.65 \times 10^4 \text{ C equiv}^{-1})(3.2 + 3.7) \times 10^4 \text{ cm sec}^{-1}}{(4.0 \text{ V}/8.0 \text{ cm})}$$
$$= 133.2\ \Omega^{-1} \text{ cm}^2 \text{ equiv}^{-1}$$
$$t_- = \frac{Fv_-}{E\Lambda} \quad \text{[vide problem 14.40]}$$
$$= \frac{(9.65 \times 10^4 \text{ C equiv}^{-1})(3.7 \times 10^{-4} \text{ cm sec}^{-1})}{(4.0 \text{ V}/8.0 \text{ cm})(132.2\ \Omega^{-1}\text{cm}^2\text{equiv}^{-1})} = 0.54$$

14.28
$$\Lambda = \frac{1000\ l}{rac} \quad \text{by Eqs (14.2) and (14.3b)}$$
$$= \frac{1000 \text{ cm}^3\text{L}^{-1})(1.5 \text{ cm})}{(25.0\ \Omega)(3.0 \text{ cm}^2)(0.5 \text{ equiv L}^{-1})}$$
$$= 40.0\ \Omega^{-1} \text{ cm}^2 \text{ equiv}^{-1}$$

14.29 a.
$$\Lambda = \frac{1000\ k}{C} = \frac{1000\ K_{Cell}}{C\ r} \quad \text{[vide problem 14.29]}$$

Then
$$\Lambda_{KCl} = \frac{(1000 \text{ cm}^3\text{L}^{-1})(1 \text{ cm}^{-1})}{(0.001 \text{ equiv L}^{-1})(6667\ \Omega)} = 145\ \Omega^{-1} \text{ cm}^2\text{equiv}^{-1}$$
$$\Lambda_{HCl} = \frac{(1000 \text{ cm}^3\text{L}^{-1})(1 \text{ cm}^{-1})}{(0.001 \text{ equiv L}^{-1})(2353\ \Omega)} = 425\ \Omega^{-1} \text{ cm}^2\text{equiv}^{-1}$$

b.
$$\lambda_{H^+} = \Lambda_{HCl} - \lambda_{Cl^-} \quad \text{by Kohlrausch's law}$$
$$= \Lambda_{HCl} - \frac{\Lambda_{KCl}}{2} \quad \text{assuming } u_{K^+} = u_{Cl^-} \text{ when } \lambda_{K^+} = \lambda_{Cl^-}$$
$$= \left(425 - \frac{145}{2}\right) \text{ or } 352.5\ \Omega^{-1}\text{cm}^2\text{equiv}^{-1}$$

c. Distance moved by the H⁺ ion
$$= \text{drift speed } (Eu) \times \text{time } (t) = \frac{E\lambda_{H^+}t}{F}$$
$$= \frac{(2 \text{ V}/2 \text{ cm})(352.5\ \Omega^{-1} \text{ cm}^2 \text{ equiv}^{-1})(10S)}{(9.65 \times 10^4 \text{ C equiv}^{-1})} = 0.0365 \text{ cm}$$

d.
$$t_{H^+} = \frac{\lambda_{H^+}}{\Lambda_{HCl}}$$

$$= \frac{352.5\ \Omega^{-1}cm^2equiv^{-1}}{425\ \Omega^{-1}cm^2equiv^{-1}} = 0.829$$

$$t_{Cl^-} = 1 - t_{H^+}$$

$$= 1 - 0.829 = 0.171$$

On raising temperature t_{H^+} will decrease and t_{Cl^-} will increase. Vide question 14.26.

14.30
$$\Lambda_{AgNO_3} = \frac{r_{KCl}}{r_{AgNO_3}} \cdot \Lambda_{KCl} \quad (C \text{ and } K_{cell} \text{ being fixed}) \text{ vide problem 14.30}$$

$$= \frac{(700\ \Omega)}{(800\ \Omega)} (150\ \Omega^{-1}cm^2equiv^{-1})$$

$$= 131\ \Omega^{-1}cm^2equiv^{-1}$$

☐ Vide question 14.16.

14.31 Vide problem 14.32.

Here the resulting mixture is a solution of NaCl of concentration $\frac{0.1}{2}$ or 0.05 N

$$\Lambda_{NaCl} = \frac{1000\ k}{C}$$

$$= \frac{(1000\ cm^3 L^{-1})(0.0056\ \Omega^{-1}cm^{-1})}{(0.05\ equiv\ L^{-1})} = 112\ \Omega^{-1}cm^2equiv^{-1}$$

Note: The given value of k_{NaOH} is of no use.

14.32 By Eq. (14.10)
$$k = \frac{C_{Na^+}\lambda_{Na^+} + C_{Cl^-}\lambda_{Cl^-} + C_{X^-} \cdot \lambda_{X^-}}{1000}$$

Then $0.0382\ \Omega^{-1}cm^{-1} = \dfrac{[(0.2+0.1)\,(50) + (0.2)\,(76) + (0.1)\lambda_x]\,\Omega^{-1}cm^2L^{-1}}{1000\ cm^3L^{-1}}$

whence
$$\lambda_{X^-} = 80$$

14.33 Vide problem 14.29(c)
$$k = \frac{C_{H^+}\lambda_{H^+} + C_{OH^-}\lambda_{OH^-}}{1000}$$

Ionic product of water
$$K_W = \frac{C_{H^+}}{C^\circ} \cdot \frac{C_{OH^-}}{C^\circ}$$

$$= \left[\frac{1000\ k}{(\lambda_{H^+} + \lambda_{OH^-})C^\circ}\right]^2$$

$$= \left[\frac{(1000\ cm^3L^{-1})(5.54 \times 10^{-8}\ \Omega^{-1}cm^{-1})}{(349.8 + 197.8)\Omega^{-1}cm^2mol^{-1})(1\ mol\ L^{-1}}\right]^2 = 1.02 \times 10^{-14}$$

Degree of dissociation

$$= \frac{C_{H^+}}{C_{H_2O}} = \frac{\sqrt{K_w}\, C^\circ}{C_{H_2O}}$$

$$= \frac{(1.02 \times 10^{-14})^{\frac{1}{2}} (1\ \text{mol L}^{-1})}{\left(\frac{1000}{18}\ \text{mol L}^{-1}\right)} \qquad \text{taking density of water} = 1\ \text{g/mL}$$

$$= 0.018$$

14.34 Let a be the degree of dissociation of NH_3

$$\underset{C(1-\alpha)}{NH_3} + H_2O \overset{K}{\rightleftharpoons} \underset{\alpha C}{NH_4^+} + \underset{\alpha C}{OH^-}$$

Then

$$K_b = \frac{C_{NH_4^+} \cdot C_{OH^-}}{C_{NH_3}} = \frac{\alpha^2 C}{1-\alpha}$$

or

$$\alpha = \sqrt{\frac{K_b}{C}(1-\alpha)}$$

$$= \sqrt{\frac{1.8 \times 10^{-5}}{0.001}(1-\alpha)}$$

$$= 0.134, \text{ if } \alpha \text{ is ignored compared to 1}$$

Using this value of α, we can get its more accurate value as under

$$\alpha = \sqrt{0.018(1-0.134)} = 0.125$$

Now

$$\Lambda = \alpha\Lambda^\circ$$

$$= (0.125)\,(238\ \Omega^{-1}\,\text{cm}^2\,\text{equiv}^{-1})$$

$$= 29.7\ \Omega^{-1}\text{cm}^2\text{equiv}^{-1}$$

14.35 Vide problem 14.27.

14.36 The thermodynamic or activity dissociation constant K_a of CH_3CO_2H is defined as

$$K_a = \frac{a_{H^+}\, a_{CH_3CO_2^-}}{a_{CH_3CO_2H}}$$

$$= \frac{C_{H^+}\, C_{CH_3CO_2^-}}{C_{CH_3CO_2H}} \cdot \frac{\gamma_{H^+}\, \gamma_{CH_3CO_2^-}}{\gamma_{CH_3CO_2H}}$$

where γ is the activity coefficient

If α is the degree of dissociation of the acid having total concentration C then

$$K_a = \frac{C\alpha^2}{1-\alpha} \cdot \frac{\gamma_{H^+}\, \gamma_{CH_3CO_2^-}}{\gamma_{CH_3CO_2H}}$$

$$= K_{app} \cdot \frac{\gamma_{H^+}\, \gamma_{CH_3CO_2^-}}{\gamma_{CH_3CO_2H}}$$

Taking $\gamma_{CH_3CO_2H} = 1$, the ionic strength αC being low, and applying Debye–Huckel limiting law

$$\log \gamma_{H^+} = \log \gamma_{CH_3CO_2^-} = -A\sqrt{\alpha C} \quad \text{where } A \text{ is Debye–Huckel constant}$$

Then

$$\log K_a = \log K_{app} - 2A\sqrt{\alpha C}$$

Now $\quad \alpha = \dfrac{\Lambda_{CH_3CO_2H}}{\Lambda^{\circ}_{CH_3CO_2H}} = \dfrac{1000\, k/C}{F(u^{\circ}_{H^+} + u^{\circ}_{CH_3CO_2})}$

$$= \dfrac{(1000\ \text{cm}^3\text{L}^{-1})\,(5.20 \times 10^{-4}\,\Omega^{-1}\text{cm}^{-1})/(0.1\ \text{equiv}\ \text{L}^{-1})}{(9.65 \times 10^4\, \text{C}\ \text{mol}^{-1})\,(3.63 \times 10^{-3} + 4.25 \times 10^{-4})\ \text{cm}^2\text{V}^{-1}\text{s}^{-1}}$$

$$= 0.0133$$

Then $\qquad K_{app} = \dfrac{C\alpha^2/C^{\circ}}{1-\alpha}$

$$= \dfrac{(0.1\ \text{mol}\ \text{L}^{-1})\,(0.0133)^2\,/(1\ \text{mol}\ \text{L}^{-1})}{1 - 0.0133} = 1.78 \times 10^{-5}$$

$$\log K_a = \log(1.78 \times 10^{-5}) - 2(0.509)\sqrt{(0.0133)\,(0.1)}$$

whence $\qquad K_a = 1.63 \times 10^{-5}$

14.37 Vide question 14.29(c).

Here $\qquad \lambda_{Na^+} = t_{Na^+}\Lambda_{NaCl}$

14.38 $\qquad k_{H_2O} = k_{SrSO_4\ \text{soln}} - k_{SrSO_4}$

$$= k_{SrSO_4\ \text{soln}} - \dfrac{C\Lambda^{\circ}_{SrSO_4}}{1000}$$

(assuming Λ to be independent of concentration (C))

$$= (1.5 \times 10^{-4}\,\Omega^{-1}\text{cm}^{-1}) - \dfrac{(2 \times 0.5 \times 10^{-3}\,\text{equiv}\ \text{L}^{-1})\,(140\,\Omega^{-1}\text{cm}^2\text{equiv}^{-1})}{(1000\ \text{cm}^3\,\text{L}^{-1}}$$

$$= 10 \times 10^{-4}\,\Omega^{-1}\text{cm}^{-1}$$

Note: k_{H_2O} thus calculated is approximate due to the assumption involved.

14.39 Solubility of AgCl, $C = \dfrac{1000\, k_{AgCl}}{\Lambda_{AgCl}}$

$$= \dfrac{1000\,(k_{\text{sat AgCl soln}} - k_{H_2O})}{F(u_{Ag^+} + u_{Cl^-})}$$

$$= \dfrac{(1000\ \text{cm}^3\text{L}^{-1})\,(1.70 \times 10^{-6} - 1.50 \times 10^{-7})\,\Omega^{-1}\text{cm}^{-1}}{(9.65 \times 10^4\,\text{C}\ \text{mol}^{-1})\,(6.8 \times 10^{-4} + 5.6 \times 10^{-4})\ \text{cm}^2\text{V}^{-1}\text{s}^{-1}}$$

$$= 1.29 \times 10^{-5}\ \text{mol}\ \text{L}^{-1}$$

Apparent or concentration solubility product

$$K_{sp\ (AgCl)} = \dfrac{C_{Ag^+}}{C^{\circ}} \cdot \dfrac{C_{Cl^-}}{C^{\circ}}\ \text{in saturated AgCl solution}$$

$$= (1.29 \times 10^{-5})\,(1.29 \times 10^{-5})$$

$$= 1.66 \times 10^{-10}$$

Thermodynamic or activity solubility product

$$K_{th} = K_{sp} \cdot \gamma_{Ag^+} \cdot \gamma_{Cl^-}$$

Then using Debye–Huckel limiting law

$$\log K_{th} = \log K_{sp} - 2A\sqrt{C}$$

$$= \log(1.66 \times 10^{-10}) - 2(0.509)\sqrt{1.29 \times 10^{-5}}$$

whence $\qquad K_{th} = 1.64 \times 10^{-10}$

14.40
$$t_{Cl^-} = \frac{vc}{1000} \cdot \frac{F}{it} \quad \text{by Eq. (14.14)}$$

(where v is the volume that the boundary swept)

$$= \frac{(0.1205 \text{ cm}^3)(1.0 \text{ equiv L}^{-1})(9.65 \times 10^4 \text{C equiv}^{-1})}{(1000 \text{ cm}^3\text{L}^{-1})(0.142A)(167.5\text{s})} = 0.489$$

14.41 The relevant experiment is the one involved in the determination of transport number by Hittorf method. The observed transport number, zero and even negative, for Cd^{2+} found with concentrated solution of cadcium iodide suggests that the latter exists mostly as $Cd[Cd_4]$ and not CdI_2. Vide problems 14.21 and 14.39(b).

14.42 True transport number cannot be negative. However, the apparent transport number as found with Hittrof method can be negative. An example is offered by cadmium iodide in concentrated solution.

14.43 Vide problem 14.35.

In the present problem

the amount of electricity passed $\equiv$ amount of Ag dissolved

14.44
$$t_{K^+} = \frac{\text{loss in amount of Cl}^- \text{ in anode compartment}}{\text{amount of electricity (in faraday) passed}}$$

$$= \frac{0.0137/35.45}{0.0857/107.9} = 0.486$$

$$t_{Cl^-} = 1 - t_{K^+}$$
$$= 1 - 0.486 = 0.514$$

14.45 Let the cathode compartment contains x g of water. Now, per faraday of electricity used in the electrolysis, the cathode compartment will lose 1 mol of Fe^{3+} (due to the cathode reaction $Fe^{3+} + e + Fe^{2+}$) and gain $t_{Fe^{3+}}$ equivalent or $\frac{1}{3}t_{Fe^{3+}}$ mol of Fe^{3+} ion due to its migration towards cathode, i.e. a net loss of $\left(1 - \frac{1}{3}t_{Fe^{3+}}\right)$ mol of Fe^{3+}. Then from the given data

$$1 - \tfrac{1}{3}t_{Fe^{3+}} = \frac{2 \times \dfrac{x}{1000} - 1.15 \times \dfrac{x}{1000}}{1 \times \dfrac{x}{1000}}$$

whence $\qquad t_{Fe^{3+}} = 0.45$

Then $\qquad t_{Cl^-} = 1 - t_{Fe^{3+}} = 1 - 0.45 = 0.55$

Note:

1. Unlike the previous two problems, here a new ion (Fe^{2+}) is formed during the electrolysis which is a measure of total amount of electricity passed. But $t_{Fe^{2+}} = 0$ as Fe^{2+} is not supposed to go outside the cathode compartment.

2. Regarding electrode process, 1 equiv of $Fe^{3+} \equiv 1$ mol of Fe^{3+}. But regarding conduction of electricity, 1 equiv of $Fe^{3+} \equiv \frac{1}{3}$ mol of Fe^{3+}

3. The given weight of the cathode solution is unnecessary as the result is independent of x.

15

Chemical Kinetics and Catalysis

Mere information that 'a reaction can occur spontaneously' is of no use unless it occurs at an appreciable rate determined by its mechanism. A chemical reaction rarely occurs in a single step, called elementary reaction. Usually it occurs through a series of elementary reactions, called its mechanism, and due to consequent involvement of intermediate (s) it does not strictly follow a definite stoichiometric equation for the overall reaction from the beginning to end of the reaction. The sum of the coefficients of the reactants in the simplest stoichiometric representation of a reaction is called its stoichiometry.

15.1 TERMS AND DEFINITIONS

Rate of Reaction

For a chemical reaction having a definite stoichiometric equation, such as $v_A A + v_B B = v_C C + v_D D$, the rate ($v$) at any instant is broadly defined as

$$v = \frac{1}{V} \frac{d\xi}{dt} \tag{15.1a}$$

where ξ is the advancement of the reaction at time t and V is volume of the reaction system. The rate of a reaction is an intensive quantity; this is unlike $d\xi/dt$ (called rate of conversion).

More specifically
$$v = -\frac{1}{v_A} \frac{d[A]}{dt} = -\frac{1}{v_B} \frac{d[B]}{dt} \quad \left. \begin{array}{c} \\ \\ \\ \\ \end{array} \right\} \text{if } V \text{ is constant} \tag{15.1b}$$
$$= \frac{1}{v_C} \frac{d[C]}{dt} = \frac{1}{v_D} \frac{d[D]}{dt}$$

or more generally
$$v = \frac{1}{v_J} \cdot \frac{\delta[J]}{dt}$$

where v_J is the stoichiometric coefficient of a substance J involved in the reaction as a reactant (with negative v_J) or a product (with positive v_J).

The Eq. (15.1b) is most useful as it relates the rate of consumption of any reactant with the rate of formation of any product.

Order of Reaction

An expression for reaction rate, if it contains no quantity related to any intermediate, represents what is called a rate law or rate expression of the reaction concerned.

If the rate law of a reaction $v_A A + v_B B + ... \rightarrow$ products is of the following form (as a product of concentrations)

$$\text{rate} \propto [A]^\alpha [B]^\beta ...$$
$$= k[A]^\alpha [B]^\beta ...$$

(15.2)

then such a reaction (which is easy to handle mathematically) is characterised by some order; the reaction is said to be of (partial) order α wrt A, β wrt B ..., and of overall order (or simply order) $\alpha + \beta + ...$ The order can exist also wrt a product, if the latter appears in the rate expression of the above restricted form.

The proportionality constant k in the rate expression is called the rate constant. If the conditions for a given reaction are such that one or more of the concentration factors are constant (or nearly so) during the progress of the reaction, these factors may be included in the constant k. The reaction is then said to be of pseudo or kinetic order which is equal to the sum of the exponents of those concentration factors which change during the run. Such a situation arises if one of the reactants is present in large excess over the other(s) so that during the run, the concentration of this reactant changes only in negligible proportion. This is a common occurrence with a catalysed reaction where the catalyst concentration remains unchanged, and of course with a heterogeneous reaction where the concentration of a pure solid or liquid species involved in such reaction is absorbed in k.

The order of a reaction wrt a reactant or product may be an integer, a fraction or zero, and positive or negative. Although theoreticlly, there is no limitation as to the value of the order, it has been found experimentally that the reactions having order higher than three hardly exist.

The concept of order is useful in the kinetic classification of chemical reactions for their systematic comparative study.

Molecularity

The molecularity of an elementary reaction is the number of chemical species (molecules or ions) involved in each act leading to the chemical reaction, i.e. involved in minimum amount of the reaction.

The term molecularity has meaning only with an elementary reaction. For multi-step reactions, the term 'minimum amount of reaction' and hence molecularity, has no precise meaning due to involvement of intermediate(s).

Molecularity, by concept, is always a positive integer, never exceeding three. The reactions are called unimolecular, bimolecular or termolecular according as one, two or three species are involved in the chemical act.

Reaction Mechanism

A sequence of elementary reactions that accounts for the overall reaction is called a mechanism of the latter. A reaciion of particular stoichiometry can occur through more than one mechanism depending on the prevailing conditions. To illustrate, let us consider the gaseous reaction $A + B + C \rightarrow P$ (products), which can have the following three mechanisms involving the intermediate I.

$$A + B \xrightarrow{k_1} I, \text{slow} \qquad A + B \underset{k_{-1}}{\overset{k_1}{\rightleftharpoons}} I, \text{fast} \qquad A + B \underset{k_{-1}}{\overset{k_1}{\rightleftharpoons}} I, \text{slow}$$

$$C + I \xrightarrow{k_2} P, \text{fast} \qquad C + I \xrightarrow{k_2} P, \text{slow} \qquad C + I \xrightarrow{k_2} P, \text{fast}$$

$$\text{(i)} \qquad\qquad\qquad \text{(ii)} \qquad\qquad\qquad \text{(iii)}$$

For (i)

Overall reaction rate, v = rate of the slowest step, by bottleneck principle

$$= k_1 [A] [B] \tag{15.3a}$$

which corresponds to first order in A, first order in B and zero order in C.

For (ii)

Here also $\qquad\qquad v$ = rate of the slowest step

$$= k_2 [I] [C]$$

Now, assuming an equilibrium situation with A, B and I (being involved in fast steps)

$$[I] = \frac{k_1}{k_{-1}} [A][B]$$

Then $\qquad\qquad v = \dfrac{k_1 k_2}{k_{-1}} [A][B][C] \tag{15.3b}$

which corresponds to first order in each reactant.

For (iii)

Here, as an approximation (steady-state approximation), we can write

$$\frac{d[I]}{dt} = k_1 [A][B] - k_{-1} [I] - k_2 [I][C] = 0$$

(considering $[I]$ to be small and virtually constant during the reaction)

whence $\qquad\qquad [I] = \dfrac{k_1 [A][B]}{k_{-1} + k_2 [C]}$

Then $\qquad\qquad v = k_2 [I][C]$

$$= \frac{k_1 k_2 [A][B][C]}{k_{-1} + k_2 [C]} \tag{15.3c}$$

which does not correspond to any order.

It is important to note that Eq. (15.3c) reduces to (15.3a) when $k_{-1} \ll k_2[C]$ and to Eq. (15.3b) when $k_{-1} \gg k_2[C]$. Again, in mechanism (i) k_{-1} is ignored compared to k_2 (and not k_1). The relative magnitude of k_{-1} compared to k_1 does not affect the result.

It should be remembere that although a particular mechanism leads to a particular order, the reverse is not necessarily true. However, a knowledge of order is helpful in predicting a mechanism using the following guidelines:

a. If order of a reaction wrt each reactant is integer and equal to its coefficient in the simplest stoichiometric equation, the reaction is likely, but not necessarily, to occur in a single step (van't Hoff rule).

b. If overall order of a reaction is an integer but not equal to the stoichiometry (of its simplest representation), the reaction does not occur in a single step; it occurs in a sequence of elementary steps of which one is fairly slow compared to the others and this constitutes the rate-limiting step.

c. If overall order is a simple fraction $\left(\text{such as } \frac{1}{2}, \frac{2}{3}, \frac{4}{5} \text{ etc.}\right)$, then regarding the mechanism, same thing is indicated as in (b), but the simple fractional order additionally indicates the intervention of a radical resulting from the dissociation of a molecule.

d. If the overall order is a complex fraction or the rate law involves sum of terms in the denominator (when order does not exist), the reaction occurs in a sequence of steps of comparable rates.

15.2 EFFECT OF EQUILIBRIUM ON KINETIC PARAMETERS

Let us discuss this for the reaction $A = B + C$. If the reaction occurs in a single step, it is fairly justified to express the forward reaction rate (v_f) and backward reaction rate (v_o) according to the law of mass action as follows [vide problem (15.2)].

$$v_f = k_f [A] \text{ and } v_b = k_b [B] [C], \text{ for an ideal system}$$

But in general for a multi-step reaction, the law of mass action does not hold. Here we should write

$$v_f = [A]^\alpha [B]^\beta [C]^\gamma$$
$$v_b = [A]^{\alpha'} [B]^{\beta'} [C]^{\gamma'}$$

where α, β and γ are the orders wrt A, B and C respectively for the forward reaction, and the corresponding quantities for the backward reaction are indicated by prime. Then at equilibrium, when $v_f = v_b$

$$\frac{k_f [A]^\alpha [B]^\beta [C]^\gamma}{k_b [A]^{\alpha'} [B]^{\beta'} [C]^{\gamma'}} = 1$$

Again, even if the reaction does not occur in a single step, the equilibrium constant (K) will be given by

$$\frac{[B][C]}{[A]} = 1 \quad \text{for ideal reaction systems}$$

or

$$\frac{K[A]}{[B][C]} = 1$$

The sufficient condition for these kinetic and thermodynamic restrictions to hold is

$$\frac{k_f [A]^\alpha [B]^\beta [C]^\gamma}{k_b [A]^{\alpha'} [B]^{\beta'} [C]^{\gamma'}} = \left(\frac{K[A]}{[B][C]} \right)^n \quad \text{where } n \text{ is a rational number}$$

So that

$$\frac{k_f}{k_b} = K^n \tag{15.4}$$

$$\alpha' = \alpha - n$$
$$\beta' = \beta + n$$
$$\gamma' = \gamma + n$$

Then the orders for forward and backward reactions are not independent of each other, though their values for reaction in one direction cannot be obtained from those for the reverse direction (because n is not definitely known).

15.3 CHARACTERISTICS OF THE REACTIONS OF DIFFERENT ORDER

A. Zero Order reaction

For a zero order reaction, the rate law is

$$-\frac{dc}{dt} = k, \text{ constant}$$

where c is the concentration of the reactant wrt which the reaction is of zero order and k is apparent rate constant that includes concentrations of the other reactants which are kept constant during the run. Then

$$\int_{c_0}^{c} dc = -\int_{0}^{t} k\,dt$$

or
$$c = c_0 - kt \tag{15.5}$$

From this equation, the following characteristics of a zero order reaction are apparent:

    i. c will decrease linearly with t.

    ii. The reaction will be absolutely complete (when $c = 0$) in a finite period, $t = c_0/k$.

    iii. Zero order rate constant depends on the unit of reactant concentration.

    iv. Time for half completion of the reaction $t_{\frac{1}{2}}$ (i.e. when $c = \frac{1}{2}c_0$) is $c_0/2k$, i.e. $t_{\frac{1}{2}} \propto c_0$.

B. First Order Reaction

For a unimolecular first order reaction $A \rightarrow$ products, the rate law is

$$-\frac{dc}{dt} = kc \quad \text{where } c = [A] \text{ at time } t$$

Then
$$\int_{c_0}^{c} \frac{dc}{c} = -\int_{0}^{t} k\,dt$$

whence
$$k = \frac{1}{t}\ln\frac{c_0}{c} = \frac{1}{t}\ln\frac{a}{a-x} \tag{15.6a}$$

or
$$[A] = [A]_0 e^{-kt} \tag{15.6b}$$

where $a = c_0 = [A]_0$ and x is the decrease in reactant concentration in time t.

From Eq. (15.6a), the following characteristics of a first order reaction are apparent:

    i. $\ln c$ will decrease linearly with t.

    ii. The reaction will never be absolutely complete in a finite period, because $c = 0$ only when $t = \infty$.

    iii. First order rate constant does not depend on the unit of reactant concentration.

    iv. $t_{\frac{1}{2}} = \dfrac{\ln 2}{k}$, i.e. $t_{\frac{1}{2}}$ is independent of initial reactant concentration.

C. Second Order Reaction

For a bimolecular second order reaction $A + B \rightarrow$ Products, the rate law is

$$-\frac{d[A]}{dt} = -\frac{d[B]}{dt} = k[A][B]$$

or
$$\frac{dx}{dt} = k(a-x)(b-x)$$

where $a = [A]_o$, $b = [B]_o$, x is the decrease in concentration of each reactant in time t.

Then
$$\int_{0}^{x} \frac{dx}{(a-x)(b-x)} = \int_{0}^{t} k\,dt$$

whence
$$k = \frac{1}{(a-b)t}\ln\frac{b(a-x)}{a(b-x)} \quad \text{for } a \neq b \tag{15.7a}$$

For $\qquad A \equiv B$ or $a = b$

$$\frac{dx}{dt} = k(a-x)^2$$

which on integration gives

$$kt = \frac{1}{a-x} - \frac{1}{a}$$

or $\qquad\qquad k = \frac{1}{at}\left(\frac{x}{a-x}\right)$

(15.7b)

From Eqs (15.7a) and (15.7b), the following characteristics of a second order (bimolecular) reaction are apparent:

i. For $a \neq b$, $\ln\dfrac{b(a-x)}{a(b-x)}$ or $\ln\dfrac{a-x}{b-x}$ vs t plot will be linear and for $a = b$, $\dfrac{x}{a-x}$ or $\dfrac{1}{a-x}$ vs t plot will be linear.

ii. The reaction will never be absolutely complete in a finite period, because $[B] = 0$ happens at $t = \infty$, for $a > b$.

iii. Second order rate constant depends on the unit of reactant concentration.

iv. For a second order reaction involving a single reactant, $t_{\frac{1}{2}} = \dfrac{1}{ka}$, i.e. $t_{\frac{1}{2}} \propto \dfrac{1}{a}$.

In kinetic investigations, the course of a reaction is ordinarily followed by measuring a suitable physical quantity (e.g. reactant or product concentration, optical density, conductance etc.) at definite time intervals. From the data thus obtained, the order and rate constant of the reaction are calculated. To illustrate, let us consider the reaction between ethyl acetate and NaOH.

$$CH_3CO_2C_2H_5 + OH^- \rightarrow CH_3CO_2^- + C_2H_5OH$$

The course of this reaction can be followed by removing a definite volume (V) of the reaction mixture at definite intervals and titrating the unreacted alkali with an acid of known normality (s). If T_o, T_t and T_∞ are the titre values at time o, t and ∞ respectively, then treating the reaction as second order, k will be given by

$$k = \frac{V}{ST_\infty t} \ln \frac{T_t(T_o - T_\infty)}{T_o(T_t - T_\infty)}$$

(15.7c)

We can readily arrive at this expression if we remember that

$$a \propto T_o$$

$\left(\text{i.e.}\, [OH^-]_o\right)$

$x \propto (T_o - T_t)$, the change in titre value in time t

$b \propto (T_o - T_\infty)$, b being the highest value of x

15.4 COMPLEX REACTIONS

Most chemical reactions are not characterised by any order; they are called complex reactions. Depending on the complexity of their mechanism, they are often given special names such as opposing and reversible reactions, consecutive reactions, parallel reactions, chain reactions etc.

A. Opposing and Reversible Reactions

Two elementary reactions are said to be opposing if the reactants and products of one interchange for the other, and a reversible reaction is one that comprises one or more pairs of opposing reactions. The simplest example is offered by the reversible isomerisation $A \underset{k_{-1}}{\overset{k_1}{\rightleftharpoons}} B$ where both forward and backward processes are of first order. Let initially only A is present with initial concentration $[A]_0 = a$. If x and x_e are the concentrations of B at time t and at equilibrium ($t = \infty$) respectively, then net rate of forward reaction

$$\frac{dx}{dt} = k_1(a - x) - k_{-1}x$$

and at equilibrium $\quad \dfrac{dx}{dt} = k_1(a - x_e) - k_{-1}x_e = 0$

whence $\quad k_{-1} = \dfrac{k_1}{x_e}(a - x_e)$

Then $\quad \dfrac{dx}{dt} = \dfrac{k_1 a}{x_e}(x_e - x)$

$$\left.\begin{aligned} &= (k_1 + k_{-1})(x_e - x) \\ \text{Integration gives} \quad k_1 + k_{-1} &= \frac{1}{t}\ln\frac{x_e}{x_e - x} \end{aligned}\right\} \tag{15.8}$$

At any time t, $\quad [B] = \dfrac{k_1[A]_0}{k_1 + k_{-1}}\left[1 - e^{-(k_1 + k_{-1})t}\right]$

$$[A] = \frac{k_{-1}[A]_0}{k_1 + k_{-1}}\left[1 + \frac{k_1}{k_{-1}}e^{-(k_1 + k_{-1})t}\right]$$

The variation of $[A]$ and $[B]$ with t is shown in Fig. 15.1. $t = \dfrac{0.693}{k_1 + k_{-1}}$ corresponds to the time for half-reaction ($t_{\frac{1}{2}}$) when both $[A]$ and $[B]$ suffer half of their maximum possible change.

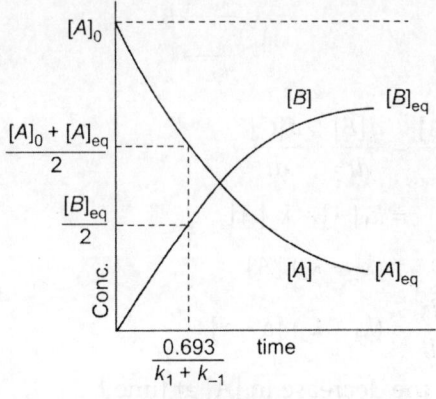

Fig. 15.1

B. Consecutive Reactions

These comprise elementary reactions such that the product of one is the reactant for the next. The simplest case is one involving only two consecutive first order reactions.

$$A \xrightarrow{k_1} B \xrightarrow{k_2} C$$

If initially only A is present at concentration $[A]_o$, then at time t

$$[A] = [A]_o e^{-k_1 t} \tag{15.9a}$$

$$[B] = \frac{k_1 [A]_o}{k_2 - k_1} \cdot (e^{-k_1 t} - e^{-k_2 t}) \tag{15.9b}$$

$$[C] = [A]_o \left(1 - \frac{k_2}{k_2 - k_1} e^{-k_1 t} + \frac{k_1}{k_2 - k_1} e^{-k_2 t}\right) \tag{15.9c}$$

when $k_1 \ll k_2$ $[C] = [A]_o (1 - e^{-k_1 t})$ and $\dfrac{d[C]}{dt} = k_1 [A]_o e^{-k_1 t} = k_1 [A]$

and when $k_2 \ll k_1$ $[C] = [A]_o (1 - e^{-k_2 t})$

Then it follows that the overall process (of formation of C) will have a simple first order kinetics corresponding to the step of lower rate constant, which forms the rate-determining step in this case.

During the reaction $[A]$ gradually decreases, $[C]$ increases, but $[B]$ passes through a maximum value

$$[B]_{max} = [A]_o \left(\frac{k_1}{k_2}\right)^{\frac{k_2}{k_2 - k_1}}$$

vide Fig. 15.3.

C. Parallel or Side Reactions

These are chemical reactions which occur simultaneously involving some common reactant(s). They are called geminal or concurrent according as they involve all the reactants common or not.

$$A + B \begin{array}{c} \nearrow C \\ \searrow D \end{array} \qquad A \begin{array}{c} \overset{+B}{\nearrow} C \\ \underset{+B^1}{\searrow} D \end{array}$$

Geminal   Concurrent

Let us consider the simplest case of two parallel first order reactions

$$A \begin{array}{c} \overset{k_1}{\nearrow} B \\ \underset{k_2}{\searrow} C \end{array}$$

Here $-\dfrac{d[A]}{dt} = \dfrac{d[B]}{dt} + \dfrac{d[C]}{dt}$

$$= k_1 [A] + k_2 [A]$$

$$= (k_1 + k_2)\, [A]$$

or $\dfrac{dx}{dt} = (k_1 + k_2)\, [a - x] \tag{15.10a}$

where $a = [A]_o$ and x is the decrease in $[A]$ at time t.

This on integration gives

$$k_1 + k_2 = \frac{1}{t} \ln \frac{a}{a - x} \tag{15.10b}$$

or $[A] = [A]_o e^{-(k_1 + k_2)t}$

Again, at any instant t

$$\frac{[B]}{[C]} = \frac{k_1}{k_2} \qquad (15.10c)$$

From the last two equations together with the restriction $[A] + [B] + [C] = [A]_c$, we have

$$[B] = \frac{k_1[A]_o}{k_1 + k_2}\left[1 - e^{-(k_1 + k_2)t}\right]$$

$$[C] = \frac{k_2[A]_o}{k_1 + k_2}\left[1 - e^{-(k_1 + k_2)t}\right]$$

The most distinguishing characteristic of parallel reactions (of geminal type) is that their products are formed in a fixed molar ratio (provided they are of same order). Again the fastest of the parallel reactions is the rate determining one for the overall reaction, which is contrary to the consecutive reactions.

If the parallel reactions considered above were reversible

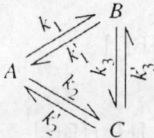

then $[B]/[C]$ would vary with time from its initial value k_1/k_2 to final value K_1/K_2 at thermodynamic equilibrium, where $K_1 = k_1/k_1'$ and $K_2 = k_2/k_2'$. This means the formation of the products is initially kinetically controlled but ultimately thermodynamically controlled. Often it so happens that $K_1/K_2 > 1$ but $k_1/k_2 < 1$. Here the formation of B is favoured thermodynamically but that of C is favoured kinetically. In synthetic organic chemistry, this kinetic control of products is often taken advantage of by making the rate of the desired parallel reaction much faster by suitable choice of catalyst and temperature.

D. Chain Reactions

A chain reaction is one that involves intermediates which are consumed and regenerated through a series of self-repeating steps.

A well-known example of such reaction occurring in gas phase is

$$H_2 + Br_2 = 2HBr$$

which follows the rate law (in the temperature range 200 to 300°C)

$$\frac{[HBr]}{dt} = \frac{k[H_2][Br_2]^{\frac{1}{2}}}{1 + k'[HBr]/[Br_2]} \qquad (15.11a)$$

The mechanism of the reaction given below, provides a good example of illustrating different types of elementary processes involved in a chain reaction.

Chain initiation $\qquad\qquad Br_2 \xrightarrow{k_1} 2Br$

Chain propagation $\qquad \begin{cases} Br + H_2 \xrightarrow{k_2} HBr + H \\ H + Br_2 \xrightarrow{k_3} HBr + Br \end{cases}$

Chain inhibition $\qquad\qquad H + HBr \xrightarrow{k_4} H_2 + Br$

Chain termination $\qquad\qquad Br + Br \xrightarrow{k_5} Br_2$

termination is facilitated by a third body that includes the wall of the reaction vessel.

Application of steady-state approximation on the short-lived intermediates Br and H gives

$$\frac{d[Br]}{dt} = 2k_1[Br_2] - k_2[Br][H_2] + k_3[H][Br_2] + k_4[H][HBr] - 2k_5[Br]^2 = 0 \quad \text{(A)}$$

$$\frac{d[H]}{dt} = k_2[Br][H_2] - k_3[H][Br_2] - k_4[H][HBr] = 0 \quad \text{(B)}$$

(A) + (B) gives

$$[Br] = \sqrt{k_1[Br_2]/k_5} \quad \text{(C)}$$

Also from (B)

$$[H] = \frac{k_2[Br][H_2]}{k_3[Br_2] + k_4[HBr]}$$

$$= \frac{k_2[H_2]\sqrt{k_1[Br_2]/k_5}}{k_3[Br_2] + k_4[HBr]} \quad \text{Substituting (C)}$$

Now

$$\frac{d[HBr]}{dt} = k_2[Br][H_2] + k_3[H][Br_2] - k_4[H][HBr] \quad \text{(D)}$$

(D) − (B) gives

$$\frac{d[HBr]}{dt} = 2k_3[H][Br_2]$$

$$= \frac{2k_2\sqrt{\dfrac{k_1}{k_5}}[H_2][Br_2]^{\frac{1}{2}}}{1 + \dfrac{k_4}{k_3} \cdot \dfrac{[HBr]}{[Br_2]}} \quad \text{(15.11b)}$$

This rate expression is of the same form as Eq. (15.11a) which is experimentally found. $[HBr]/[Br_2]$ appearing in the denominator is a consequence of the inhibition step in which HBr reacts with H in competition with Br_2.

In the above example of straight chain reaction, one chain carrier yields only one other chain carrier. But there are chain reactions where one chain carrier yields more than one carrier. These are called branched chain reactions whose rate increases very rapidly often leading to an explosion. An example of this type is the reaction $2H_2 + O_2 = 2H_2O$. In the low pressure region, the following mechanism is plausible.

Initiation $\qquad\qquad H_2 + O_2 \rightarrow HO_2 + H$

Propagation $\qquad\quad \begin{cases} H_2 + HO_2 \rightarrow H_2O + HO \\ H_2 + HO \rightarrow H_2O + H \end{cases}$

Branching $\qquad\qquad \begin{cases} O_2 + H \rightarrow HO + O \\ H_2 + O \rightarrow HO + H \end{cases}$

Termination $\qquad\quad H + H \xrightarrow{\text{wall}} H_2$

A non-chemical (nuclear) instance of chain branching is offered by the nuclear-fission bomb where a neutron is absorbed by a ^{235}U nucleus causing it to undergo fission and releasing several neutrons in the process.

The chain length in a chain reaction is the number of cycles that a chain carrier can participate in from its formation to its destruction. This is equal to the rate of overall reaction divided by the rate of the initiation reaction.

The usual characteristics of chain reactions by which they can be recognised are:

i. much greater reaction rate than that expected from the activation energy of the initiation process.

ii. abnormal effect of change in reactant concentration or pressure on the reaction rate.

iii. quite great increase or decrease of reaction rate in presence of small amounts of foreign substances (according as the latter favour generation or removal of the chain carriers).

iv. significant effect of change in shape of the reaction vessel on the chain reaction occurring in gas phase.

v. fairly high length of chain for reactions brought about by light.

15.5 EFFECT OF TEMPERATURE ON REACTION RATE

The rates of most chemical reactions increase with rise in temperature. This is due to temperature variation of the rate constant (k) that can be effectively represented by

$$\frac{d\ln k}{dT} = \frac{E_a}{RT^2} \tag{15.12a}$$

where E_a is a characteristic constant, called energy of activation of the reaction. Integration gives

$$\ln k = -\frac{E_a}{RT} + \ln A \text{ (constant), provided } E_a \text{ is independent of temperature}$$

or $\qquad k = Ae^{-E_a/RT} \tag{15.12b}$

The Eq. (15.12b), rather than Eq. (15.12a) is known as Arrhenius equation and A as pre-exponential factor. This equation is quantitatively useless for low values of E_a, particularly at high temperature, mainly due to variation of E_a with temperature.

According to Arrhenius, E_a represents the minimum energy, in excess of the average molecular energy of the reactants (at a particular temperature), which the reactant molecules must possess for participating in the reaction concerned. That is, only those molecules, called active molecules, which fulfill this energy requirement can take part in a chemical reaction. Arrhenius suggested that in every reaction system, the active molecules are in equilibrium with the normal molecules (but this is not required according to the modern theories of reaction rate). A rise in temperature favours the formation of active molecules from normal molecules and thereby enhancing the reaction rate.

It is important to note that the minimum energy required for a reactant molecule to participate in a particular reaction is really the threshold energy E_T and not E_a which are related as

$$E_T = E_a + \overline{E} \tag{15.12c}$$

where $\overline{E}$ is the average energy of the reactant molecules.

For any chemical system, the energy of activation E_a^f for forward reaction and E_a^b for backward reaction are not likely to be same. If the reaction occurs at constant pressure $E_a^f - E_a^b = \Delta H$ (enthalpy of the reaction), while if it occurs at constant volume $E_a^f - E_a^b = \Delta U$ (energy of the reaction). In course of a reaction, the energy of the reaction system first increases and then decreases after passing through a maximum as shown in Fig. 15.2.

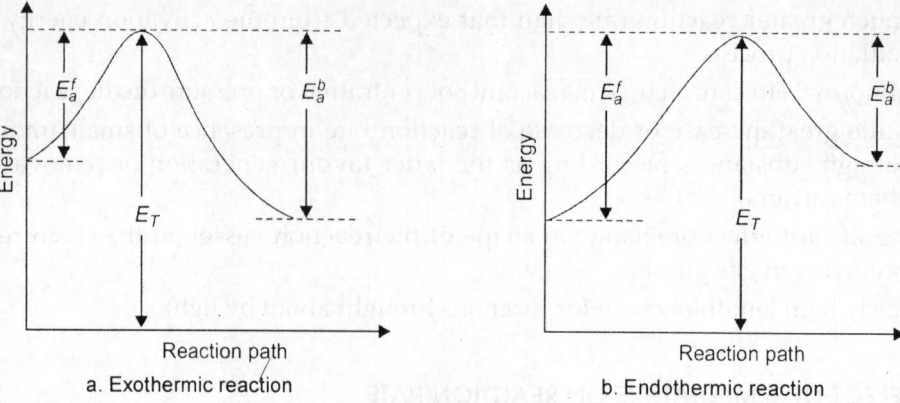

a. Exothermic reaction   b. Endothermic reaction

Fig. 15.2

15.6 EFFECT OF CATALYST ON REACTION RATE

A catalyst is a substance that affects (enhances) the rate of a reaction but remains unchanged in mass and chemical composition (not necessarily in the same physical form) at the end of the reaction. The phenomenon is known as catalysis which is called homogeneous or heterogeneous according as the reactants, products and catalyst are in same phase or not. A catalyst should be distinguished from a *sensitiser* which is a substance that is used in small amount to initiate a reaction where it is consumed. For example, $(CH_3)_2Hg$ sensitises the decomposition of C_3H_8 through a chain process initiated by CH_3 radical coming from $(CH_3)_2Hg$.

The catalytic effect is attributed to the change of reaction mechanism to one usually of lower activation energy. This results from interaction between the catalyst and the reaction system with the formation of some unstable intermediate compound(s) (in homogeneous catalysis) or adsorption complex (in heterogeneous catalysis). Here we consider only homogeneous catalysis in liquid-phase reactions of which two important types are acid-base catalysis (which is very common) and enzyme catalysis (which is of vital importance in biological processes).

A. Acid-Base Catalysis

There are many chemical reactions which are catalysed by acids or bases or by both. Some reactions are catalysed by any acid (e.g. hydrolysis of orthoesters like ethyl orthocarbonate) or by any base(e.g. decomposition of nitramide into N_2O and H_2O) and the phenomenon is called general acid or base catalysis. If the catalysis is effective only with certain acids or bases, then it is said to be specific acid or base catalysis. The most common acid catalyst in aqueous solution is the hydronium ion and the most common base catalyst is hydroxyl ion. Often enzymes appear to act as complex acid-base catalysts.

The essential feature of catalysis by an acid or a base is the transfer of a proton between the substrate and the catalyst such that the substrate acts as a base in acid catalysis or as an acid in base catalysis (in Bronsted sense of acid and base). The resulting protonated substrate (in case of acid catalysis) then reacts with either the solvent or a solute, and accordingly the mechanism is said to be protolytic or prototropic. In a protolytic mechanism the catalysis may be specific or general according to the circumstances, while in prototropic mechanism there is mostly general catalysis. This is shown below in case of acid catalysis. Similar procedure can be applied in case of base catalysis also.

The protolytic acid catalysis in aqueous solution may be represented by

$$S + BH^+ \underset{k_{-1}}{\overset{k_1}{\rightleftharpoons}} SH^+ + B$$

$$SH^+ + H_2O \xrightarrow{k_2} P + H_3O^+$$

$$B + H_3O^+ \rightleftharpoons BH^+ + H_2O$$

where BH^+ is the general acid catalyst and S is the substrate (the reactant with which the catalyst acts directly). Application of the steady-state approximation to the species SH^+ gives

$$k_1[S][BH^+] - k_{-1}[SH^+][B] - k_2[SH^+] = 0$$

$[H_3O^+]$ being absorbed in k_2. Then

$$\text{rate, } v = k_2[SH^+]$$

$$= \frac{k_1 k_2 [S][BH^+]}{k_{-1}[B] + k_2}$$

If $k_2 \gg k_{-1}[B]$ $v = k_1[S][BH^+]$

which corresponds to catalysis of general type

Again, when $k_2 \ll k_{-1}[B]$

$$v = \frac{k_1 k_2}{k_{-1}} \cdot \frac{[S][BH^+]}{[B]}$$

$$= \frac{k_1 k_2 K}{k_{-1}} \cdot [S][H_3O^+]$$

Obviously, this is a case of specific catalysis (due to H_3O^+) in spite of the fact that the initial proton transfer was from the solute species BH^+. Here

$$k_{obs} = \frac{k_1 k_2 K}{k_{-1}} [H_3O^+]$$

$$= k_{H_3O^+} [H_3O^+] \quad \text{where } k_{H_3O^+} \text{ is the catalytic coefficient for } H_3O^+ \text{ ion}$$

Then, for specific hydrogen ion catalysed reaction, the rate constant will be proportional to $[H_3O^+]$ and the plot of $\log k$ vs pH will be linear [but this simple relation does not hold in case of general acid catalysis].

The mechanism of most prototropic acid catalysis may be represented as

$$S + BH^+ \underset{k_{-1}}{\overset{k_1}{\rightleftharpoons}} SH^+ + B$$

$$SH^+ + B \xrightarrow{k_2} P + BH^+$$

Here the steady-state equation is

$$k_1[S][BH^+] - k_{-1}[SH^+][B] - k_2[SH^+][B] = 0$$

Then $v = k_2[SH^+][B]$

$$= \frac{k_1 k_2 [S][BH^+]}{k_{-1} + k_2}$$

which corresponds to general acid catalysis.

Sometimes SH⁺ reacts with a species R not acting as a base

$$SH^+ + R \xrightarrow[+H_2O]{k_2} P + H_3O^+$$

$$B + H_3O^+ \underset{}{\overset{K}{\rightleftharpoons}} BH^+ + H_2O$$

Here
$$[SH^+] = \frac{k_1[S][BH^+]}{k_{-1}[B] + k_2[R]}$$

and
$$v = k_2[SH^+][R]$$

$$= \frac{k_1k_2[S][BH^+][R]}{k_{-1}[B] + k_2[R]}$$

when
$$k_2[R] \gg k_{-1}[B], \quad v = k_1[S][BH^+]$$

which corresponds to general acid catalysis

But when
$$k_2[R] \ll k_{-1}[B],$$

$$v = \frac{k_1k_2[S][BH^+][R]}{k_{-1}[B]}$$

$$= \frac{k_1k_2K}{k_{-1}}[S][H_3O^+]$$

which corresponds to specific catalysis (due to H_3O^+)

B. Enzyme Catalysis

Enzymes mostly consist of protein molecules (certain RNA molecules also act like enzyme as catalyst). They may be associated with non-protein substances (known as coenzymes) that have effective roles in the enzyme catalysis. Some enzymes are catalytically inactive in the absence of certain metal ions.

Enzyme catalysed reactions usually begin with a step in which enzyme E combines with substrate to form a complex ES. On this basis Michaelis and Menten were first able to explain the kinetic behaviour of many enzyme catalysed reactions through the following simple mechanism

$$E + S \underset{k_{-1}}{\overset{k_1}{\rightleftharpoons}} ES$$

$$ES \xrightarrow{k_2} E + P$$

Application of steady-state approximation on the short-lived species ES gives

$$k_1[E][S] - k_{-1}[ES] - k_2[ES] = 0$$

Now the total concentration of enzyme $[E]_o$ is related to the concentration of the free enzyme [E], which is generally unknown, by

$$[E]_o = [E] + [ES]$$

Elimination of [E] between the last two equations gives

$$[ES] = \frac{k_1[E]_o[S]}{k_{-1} + k_2 + k_1[S]}$$

Then overall reaction rate,

$$v = k_2[ES]$$

$$= \frac{k_1 k_2 [E]_o [S]}{k_{-1} + k_2 + k_1 [S]}$$

$$= \frac{k_2 [E]_o [S]}{K_m + [S]}, \quad \text{where } K_m = \frac{k_{-1} + k_2}{k_1}$$

This equation is known as Michaelis–Menten equation and K_m as Michaelis constant.

When $\qquad\qquad [S] \ll K_m$

$$v = \frac{k_2}{K_m} [E]_o [S]$$

i.e. the reaction is of first order in substrate and enzyme each.

On the other hand, when $[S] \gg K_m$, v becomes maximum and is given by

$$v_{max} = k_2 [E]_o$$

i.e. the reaction is now zero order in S. Because in the limit of high concentration of S virtually all the enzyme is in the form of ES whose concentration, and hence the reaction rate, does not change with further change in $[S]$.

The quantity, $k_2 = v_{max}/[E]_o$ is called the turnover number of the enzyme which is the maximum number of moles of the substrate converted into product per unit time per mole of enzyme. This ranges from 10^{-2} to $10^6 s^{-1}$.

In terms of v_{max} Michaelis–Menten equation may be rewritten as

$$\frac{v}{v_{max}} = \frac{[S]}{K_m + [S]}$$

and $\qquad\qquad \dfrac{1}{v} = \dfrac{K_m}{v_{max}} \cdot \dfrac{1}{[S]} + \dfrac{1}{v_{max}}$

or more appropriately $\qquad \dfrac{1}{v_o} = \dfrac{K_m}{v_{max}} \cdot \dfrac{1}{[S]_o} + \dfrac{1}{v_{max}}$

[where the suffix o refers to the value at the initial stage.]

Because disregarding of the reverse reaction between E and P in the assumed mechanism has its proper justification only at the early stages of the reaction when $[P]$ is negligible.

K_m and v_{max}, and hence k_2, can be calculated from the slope (K_m/v_{max}) and intercept $(1/v_{max})$ of the linear $1/v_o$ vs $1/[S]_o$ plot.

15.7 THEORY OF REACTION RATES

Attempts to interpret Arrhenius equation in a sophisticated way led to the development of two elementary approaches—collision theory and activated complex theory. Both the theories are based on the supposition that the occurrence of a chemical reaction requires energetic collision between reactant molecules. Their essential difference lies in defining the effective collision, i.e. one that leads to the desired chemical reaction.

According to the collision theory, the rate of a bimolecular reaction $A + B \rightarrow$ products can be expressed as

$$\text{rate} = Z_{AB} e^{-\frac{E_a}{RT}} \qquad\qquad\qquad (15.13a)$$

where Z_{AB} is the frequency of bimolecular collision between A and B molecules given by Eq. (2.12b); $e^{-\frac{E_a}{RT}}$ represents the fraction of collisions occurring with impact kinetic energy not less than E_a, the Arrhenius activation energy for the reaction. Then

$$k = \pi\sigma^2 \left(\frac{8RT}{\pi\mu}\right)^{\frac{1}{2}} e^{-\frac{E_a}{RT}}$$

$$= \sigma^2 \left(\frac{8\pi RT}{N_a\mu}\right)^{\frac{1}{2}} e^{-\frac{E_a}{RT}} \tag{15.13b}$$

This indicates the temperature dependence of A in Eq. (15.12b) suggesting that it would be more appropriate to write Arrhenius equation as

$$k = aT^{\frac{1}{2}} e^{-\frac{E_a'}{RT}}$$

where a and E_a' are temperature independent quantities for the reaction. The new activation energy E_a' is still defined in the form of Eq. (15.12a) though it is now given by

$$E_a' = E_a + \tfrac{1}{2} RT \tag{15.13c}$$

Vide problem (15.51).

According to the activated complex theory, an elementary reaction proceeds through the formation of certain critical configuration, called activated complex $X^{\neq}$, that undergoes unimolecular decomposition to products. This requires $X^{\neq}$ to have certain minimum potential energy E_a relative to the reactants. Pictorially $X^{\neq}$ corresponds to the maximum in the energy profile of the reaction following a lowest energy path (resembling Fig. 15.2). $X^{\neq}$ is regarded as behaving like an ordinary molecule except that it has one mode of loose vibration of unusually low frequency ($\nu^{\neq}$) with respect to which it is unstable. Then a bimolecular reaction may be represented as

$$A + B \xrightleftharpoons{K^{\neq}} X^{\neq} \rightarrow \text{Products}$$

occurring at a rate equal to $\nu^{\neq}c^{\neq}$ (using the symbol $\neq$ to designate activated complex and its properties) with the rate constant given by

$$k = k\frac{RT}{N_A hc^{\circ}} K^{\neq}$$

In general for a reaction of molecularity n

$$k = k\frac{RT}{N_A h(C^{\circ})^{n-1}} K^{\neq} \tag{15.14a}$$

This is Eyring equation where k is the transmission coefficient which represents the probability of $X^{\neq}$ passing across the potential barrier to form products. In many cases $k \approx 1$.

Conventional treatment of $K^{\neq}$ using classical thermodynamics gives

$$\left. \begin{aligned} k &= \frac{RT}{N_A h(C^{\circ})^{n-1}} e^{-\Delta G^{\circ\neq}/RT} \quad (\text{taking } k = 1) \\[2mm] &= \frac{RT}{N_A h(C^{\circ})^{n-1}} e^{\Delta S^{\circ\neq}/R} e^{-\Delta H^{\circ\neq}/RT} \end{aligned} \right\} \tag{15.14b}$$

[Such treatment of $K^{\neq}$ is, however, not completely justified since $X^{\neq}$ does not behave really as an ordinary molecule].

For ideal-gas reaction system

$$k = e^{n} \frac{RT}{N_A h (C^\circ)^{n-1}} e^{\frac{\Delta S^{\circ \neq}}{R}} e^{-\frac{E_a}{RT}}$$ (15.14c)

where $\Delta H^{\circ \neq} = E_a - nRT$.

For liquid phase reactions

$$k = e \cdot \frac{RT}{N_A h (C^\circ)^{n-1}} e^{\frac{\Delta S^{\circ \neq}}{R}} e^{-\frac{E_a}{RT}}$$ (15.14d)

Since $\qquad \Delta H^{\circ \neq} = E_a - RT$

The quantities $\Delta G^{\circ \neq}$, $\Delta H^{\circ \neq}$ and $\Delta S^{\circ \neq}$ are calld the standard Gibbs free energy of activation, the standard enthalpy of activation and the standard entropy of activation respectively. Often the adjective 'standard' is omitted from these terms and the superscript o from their symbols, the implication remaining intact.

Simple collision theory gives satisfactory value of k only for reactions involving simple molecules. Because it fails to consider that apart from frequency of collision and energy of activation there are other rate-determining factors (like orientation of colliding molecules etc.), collectively called the probability factor (P), which are important particularly when the reactant molecules are not simple. This arises from the treatment of reactant molecules as hard spheres (that neglects internal degrees of freedom). The activated complex theory takes care for all the rate-determining factors automatically through use of partition function, though it is too sophisticated to apply for quantitative purposes due to the lack of knowledge regarding the structure of the activated complex in most cases. For a bimolecular reaction $A + B \rightarrow$ Products, the rate constant, according to this theory, is given by

$$k = \frac{RT}{N_A h} \frac{q^{\neq}}{q_A q_B} e^{-\frac{E_0}{RT}}$$ (15.15)

where q_A and q_B denote the molecular partition functions (vide Section 21.4) of A and B and $q^{\neq}$ that of the activated complex excluding the loose vibration; E_o is the activation energy of the reaction at absolute zero.

15.8 REACTIONS IN SOLUTION

Frequently the empirical rate laws found for reaction in the liquid phase are same as those found in the gas phase with nearly same rate constant. For such reactions solvent mere serves as a medium. But the theory of liquid-phase reactions is far more difficult to develop than that of gas-phase reactions because of strong intermolecular interactions in the liquid state. Except for diffusion-controlled reactions, it is hardly possible to calculate the rate constants for liquid reactions from molecular properties. Collision theory is of no use because there is no unequivocal way of calculating collision frequencies.

For a bimolecular reaction in the gas phase, the upper limit of the reaction rate is set by the collision frequency. In the liquid phase, this is determined by the first encounters between reactant molecules undergoing Brownian motion through the solution. Because having once met through diffusion, the two reactant molecules will remain close to each other for a considerable period (being surrouned by a cage of solvent molecules) enough to acquire the necessary energy of activation from their immediate surroundings.

15.9 REACTIONS INVOLVING IONS—KINETIC SALT EFFECT

The rate of ionic reactions varies much with the ionic strength of the solution in which they occur. This is called kinetic salt effect.

Activated complex theory together with Debye–Huckel limiting law leads to the following expression of rate constant for the reaction between the ions A and B in a solution of ionic strength I

$$\log k = \log k_o + 2AZ_A Z_B \sqrt{I} \tag{15.16}$$

$$= \log k_o + 1.02 Z_A Z_B \sqrt{I} \quad \text{for aqueous solution at 25°C}$$

where Z_A and Z_B are the charges of A and B respectively, A is Debye–Huckel constant, and k_o is the rate constant at infinite dilution when $I = 0$. With increase in I, k will increase or decrease according as $Z_A Z_B$ is +ve or –ve. The Eq. (15.16) was first derived by Bronsted and Bjerrum and is known after their names.

The salt effect represented by the above equation is of primary type which deals with the direct effect of ionic strength on the reaction rate. This should be distinguished from the secondary salt effect that concerns with the indirect effect of ionic strength on the reaction rate through its influence on reactant concentrations.

15.10 UNIMOLECULAR REACTIONS

Unimolecular gas reactions are interesting because of their following kinetic features:

i. Although molecules in such reactions apparently undergo chemical change independently, they need prior activation.

ii. Although activation occurs through bimolecular collisions, such reactions are first order at high pressure (though second order at low pressure).

These were first explained by Lindemann assuming collisional activation of the reactant molecules for necessary transformation of molecular energy (from translational to vibrational form) and a time gap between activation and reaction of the molecules, which led him to the following mechanism

$$A + A \underset{k_{-1}}{\overset{k_1}{\rightleftharpoons}} A^* + A$$

$$A^* \xrightarrow{k_2} \text{Products}$$

where A and A^* respectively denote the normal molecule and activated molecule of the reactant.

Application of steady-state approximation on A^* gives

$$\frac{d[A^*]}{dt} = k_1[A]^2 - k_{-1}[A^*][A] - k_2[A^*] = 0$$

whence $\qquad [A^*] = \dfrac{k_1[A]^2}{k_2 + k_{-1}[A]}$

Then rate of reaction $\quad v = k_2[A^*]$

$$= \frac{k_2 k_1[A]^2}{k_2 + k_{-1}[A]} \tag{15.17}$$

At high pressure, when

$$k_{-1}[A] \gg k_2$$

$$v = \frac{k_2 k_1}{k_{-1}}[A], \quad \text{which corresponds to first order}$$

At low pressure, when

$$k_{-1}[A] \ll k_2$$

$$v = k_1[A]^2, \quad \text{which corresponds to second order}$$

Unimolecular reactions in liquid solutions also follow the Lindemann mechanism represented as

$$A + M \underset{k_{-1}}{\overset{k_1}{\rightleftharpoons}} A* + M \quad \text{where } M \text{ represents a solvent molecule}$$

$$A* \xrightarrow{\ k_2\ } \text{Products}$$

However, in solution, there is no chance for second order kinetics be observed due to high $[M]$ that makes rate of deactivation of $A*$ (by solvent) much greater than that of its reaction.

15.11 TERMOLECULAR REACTIONS

Termolecular reactions are only limited few because of very low probability of termolecular collisions. Notable examples are offered by NO in its reaction with H_2, O_2, Cl_2, Br_2. A striking feature of such reactions is that their rate constants fall with rise of temperature. The direct explanation for this is to take the energy of activation zero and attribute the temperature effect entirely to the frequency factor that falls rapidly with rise of temperature due to lower probability of energy transfer in a termolecular collision at higher temperature. But difficulty arises in defining a triple collision.

Alternatively, the so-called termolecular reactions may be thought to proceed through bimolecular steps as given below for the reaction $2NO + O_2 \rightarrow 2NO_2$

$$2NO \rightleftharpoons N_2O_2 \quad \text{fast}$$
$$N_2O_2 + O_2 \rightarrow 2NO_2 \quad \text{slow}$$

With rise in temperature, the formation of N_2O_2, being exothermic, will be disfavoured with consequent lowering of overall reaction rate.

Other well investigated examples of simpler termolecular reactions are combination of atoms or small radicals in presence of a third body which carries away the energy liberated on bond formation. Here also the alternative mechanism involving bimolecular steps are equally likely.

AUXILIARY PROBLEMS

15.1 A chemical reaction cannot occur at a definite rate even at fixed temperature and pressure. Comment.

15.2 For a non-ideal reaction system equilibrium constant is meaningfully defined only in terms of activities of the species involved. But the rates for all reactions are customarily expressed in terms of concentrations, instead of activities, of the species. Is it justified?

15.3 On which of the following factors does the rate of a reaction depend?
(i) Pressure (ii) temperature (iii) reaction medium (iv) nature of the radiation to which the system is exposed (v) presence of a foreign substance (vi) shape of the reaction vessel.

15.4 Do the rate and rate constant of a reaction depend on its stoichiometric representation? Does $t_{\frac{1}{2}}$ do?

15.5 The rate of the elementary reaction $Br_2 \rightarrow 2Br$ can be expressed as

$$-\frac{d[Br_2]}{dt} = k[Br_2] \text{ or } \frac{d[Br]}{dt} = k'[Br]$$

which is the rate constant of the reaction k or k'?

15.6 For the reaction $H_2 + Br_2 \rightarrow 2HBr$

$$\frac{d[HBr]}{dt} = \frac{k[H_2][Br_2]^{\frac{1}{2}}}{1 + k'[HBr]/[Br_2]}$$

what can be said about the order and rate constant of this reaction?

15.7 The reactions of molecularity or order 3 are uncommon, and those of higher molecularity are unknown. Why?

15.8 Is the term 'molecularity' meaningful with (i)unimolecular reactions (ii) S_N1 reactions?

15.9 Cite example (if any) of the following:
(i) negative order of a reaction
(ii) order exceeding molecularity, and reverse.

15.10 What is the condition for a reaction to occur in a single step?

15.11 Discuss about the order, rate constant, molecularity and mechanism of the reaction $A + B + C \rightarrow$ Product, for the following rate laws
(I) rate $= k[A][B][C]$, (II) rate $= k_1[A][B]$, (III) rate $= k_1[A][B] + k_2[B][C]$

15.12 The reaction $2NO + 2H_2 \rightarrow 2H_2O + N_2$ is of order 3. Is the reaction possible in a single step? Suggest a mechanism.

15.13 (a) For a chemical system, the orders of forward and backward reactions are not independent of each other. Comment.

(b) For the reaction $A \rightarrow B + C$, $-\frac{d[A]}{dt} = k[A]$

Suggest two possible rate laws for the reverse reaction.

15.14 The law of mass action leads to the relation $K = \frac{k_f}{k_b}$. Comment.

15.15 Are the following statements true or false?
(a) The mechanism of a reaction cannot always be unambiguously decided from its rate law alone.
(b) For any multi-step reaction, there must be a rate-determining step.
(c) The incompleteness of a chemical reaction is due to simultaneous occurrence of the reverse reaction.

15.16 Under what condition, the rate of an nth order reaction can be expressed as

$$-\frac{dc}{dt} = kc^n$$

What is the kinetic requirement for such a reaction to be absolutely complete within a finite period? Using this rate law show that

$$\frac{t_{\frac{3}{4}}}{t_{\frac{1}{2}}} = 2^{n-1} + 1$$

15.17 Derive an expression representing the progress of a reaction which is same for the same order, i.e. independent of k and c_o. Mention its importance.

15.18 While it is expected that a large amount of a substance would take a larger time for its half-decomposition, the dependence of half-life on the initial concentration does not indicate so in general. Explain.

15.19 Account for the following characteristics of common zero order reactions:

(i) they cannot occur in a single step.

(ii) they can proceed to completion in a finite period.

15.20 For a reaction $A + B \rightarrow C$, rate $= k[A]^{\frac{1}{2}}[B]^{\frac{1}{2}}[C]^{-1}$, i.e. overall order is zero. Can it proceed to completion?

15.21 Find the relation between $t_{\frac{1}{2}}$ and t_{av} for a first order reaction. Which one will be greater? Justify your answer.

15.22 For the bimolecular reaction $A + B \rightarrow$ Product, with $[A]_o = a$ and $[B]_o = b$,

$$k = \frac{1}{(a-b)t} \ln \frac{b(a-x)}{a(b-x)}.$$ Therefore, when $a = b$, k will assume an absurd value of infinity. Point out the fallacy.

15.23 Find the expression of $t_{\frac{1}{2}}$ involving initial reactant concentration for the following reactions each of overall order two:

(i) $A \rightarrow$ Product, with $[A]_o = a$.

(ii) $A + B \rightarrow$ Product, having order 2 wrt A and o wrt B with $[A]_o = [B]_o = a$.

(iii) $A + B \rightarrow$ Product having order 1 wrt each of the reactants with $[A]_o = [B]_o = a$.

Will the value of $t_{\frac{1}{2}}$ be same in all three cases?

15.24 Consider the reaction $A \rightarrow$ Product. How much of A will remain unreacted at time $t = 2t_{\frac{1}{2}}$ if the reaction is of (i) zero order (ii) first order?

15.25 For the reaction $A + B \rightarrow$ Product with $[A]_o = [B]_o$. How much of A will remain unreacted at $t = 3t_{\frac{1}{2}}$ if the reaction is (i) first order in both A and B (ii) first order in A but zero order in B (iii) zero order in both A and B.

15.26 What will be $t_{\frac{1}{2}}$ for a simple first order reaction (i) $A \xrightarrow{k} B$ and for a reversible first order reaction (ii) $A \underset{k'}{\overset{k}{\rightleftharpoons}} B$. $t_{\frac{1}{2}}$ of (ii) is lower than that of (i), though the average rate of net forward reaction is lower with (ii). Explain.

15.27 (a) A reversible reaction $A \rightleftharpoons B$ may be regarded as a specific case of the consecutive reaction $A \rightarrow B \rightarrow C$ with C same as A. Comment.

(b) For the first order processes (i) $A \rightarrow B$, (ii) $A \rightarrow B \rightarrow C$ and (iii) $A \rightarrow C$, would the time for half reduction in the amount of A be same?

15.28 The overall rate of a multi-step reaction is determined by its slowest step. This is known as bottle-neck principle. Verify the validity of this principle for the reaction $A \xrightarrow{k_1} B \xrightarrow{k_2} C$ involving first order steps.

15.29 Consider the following two mechanisms involving first order steps:

(I) $A \xrightarrow{k_1} B \xrightarrow{k_2} C$ (II) $A \underset{k_{-1}}{\overset{k_1}{\rightleftharpoons}} B \xrightarrow{k_2} C$, where $k_2 \ll k_1, k_{-1}$.

Show that in both cases the overall reaction $A \to C$ follows a first order kinetics. How do they differ?

15.30 Consider the reaction (I) in the previous problem.

(a) Find the expressions for $[B]_{max}$ and the corresponding times when (i) $k_1 \neq k_2$ (ii) $k_1 = k_2$, assuming that the system initially contains only A.

(b) In which of the following two cases $[B]_{max}$ will be higher? (i) $k_1 = 10k_2$ (ii) $k_2 = 10k_1$.

Draw concentration versus time curve. Point out the important difference between the two diagrams and state how useful is this.

15.31 Are the following reactions possible?

(i) $A \to B \to C$ (ii) $A \underset{C}{\overset{B}{\diagdown\diagup}}$ (iii) $A \underset{k_2}{\overset{k_1}{\underset{k_3}{\rightleftharpoons}}} \overset{B}{\underset{C}{\diagdown}}$

15.32 For an elementary reaction, the ratio of rate constants for reactions in the forward and backward directions is equal to the equilibrium constant. Show that this holds at equilibrium also for a complex reaction, such as $A \underset{k_{-1}}{\overset{k_1}{\rightleftharpoons}} B \underset{k_{-2}}{\overset{k_2}{\rightleftharpoons}} C$.

15.33 Consider the following reactions involving first order steps

(i) $A \underset{k_{-1}}{\overset{k_1}{\rightleftharpoons}} B$ (ii) $A \overset{k_1 \nearrow B}{\underset{k_2 \searrow C}{}}$, $k_1 \gg k_2$

(a) Can both be treated as simple first order reaction?

(b) In (ii) which parallel reaction will determine the overall reaction rate? Is your answer contradictory to the bottle-neck principle?

15.34 Show what $t_{\frac{1}{2}}$ is same for the following two first order reactions?

(i) $A \overset{k_1 \nearrow B}{\underset{k_2 \searrow C}{}}$ (ii) $A \overset{k_2 \nearrow B}{\underset{k_1 \searrow C}{}}$

How can one be differentiated from the other?

15.35 Following mechanism has been proposed for the gaseous reaction:

$2N_2O_5 \longrightarrow 4NO_2 + O_2$ having rate law, rate $= k[N_2O_5]$

(i) $N_2O_5 \underset{k_{-1}}{\overset{k_1}{\rightleftharpoons}} NO_2 + NO_3$

(ii) $NO_2 + NO_3 \xrightarrow{k_2} NO_2 + NO + O_2$

(iii) $NO + NO_3 \xrightarrow{k_3} 2NO_2$

Show that it is consistent with the stoichiometry and rate law of the reaction with

$k = \dfrac{k_1 k_2}{k_{-1} + 2k_2}$. Does this mechanism correspond to a chain reaction?

15.36 The thermal decomposition of ethane, stoichiometrically represented by $C_2H_6 \longrightarrow C_2H_4 + H_2$, is supposed to occur through the following chain mechanism.

$$C_2H_6 \xrightarrow{k_1} 2CH_3$$
$$CH_3 + C_2H_6 \xrightarrow{k_2} CH_4 + C_2H_5$$
$$C_2H_5 \xrightarrow{k_3} C_2H_4 + H$$
$$H + C_2H_6 \xrightarrow{k_4} H_2 + C_2H_5$$
$$H + C_2H_5 \xrightarrow{k_5} C_2H_6$$

Find the order of the reaction and the chain length (which is defined as the number of cycles that a radical can participate in from its formation to its destruction).

15.37 The photochemical H_2–Cl_2 reaction occurs at a constant rate when carried out in a jar inverted over water. What can be said about the order and mechanism of the reaction?

15.38 Explain the following in connection with H_2–Br_2 reaction:
  (i) The chain initiation process is $Br_2 \rightarrow 2Br$, and not $H_2 \rightarrow 2H$
  (ii) The chain inhibition process is $H + HBr \rightarrow H_2 + Br$, and not $Br + HBr \rightarrow Br_2 + H$.
  (iii) The chain termination process is $Br + Br \rightarrow Br_2$, rather than $H + Br \rightarrow HBr$ or $H + H \rightarrow H_2$.

15.39 Account for the following notable differences between H_2–Br_2 and H_2–Cl_2 reactions:
  (i) Inhibition by product occurs in case of H_2–Br_2 reaction (through the reaction $H + HBr \rightarrow H_2 + Br$) but not in case of H_2–Cl_2 reaction.
  (ii) O_2, even in small concentration, has a noticeable effect on H_2–Cl_2 reaction though not on H_2–Br_2 reaction.
  (iii) Chain length is greater with H_2–Cl_2 reaction.

15.40 Although H_2–Cl_2 and H_2–Br_2 reactions proceed through a chain mechanism, H_2–I_2 reaction is simply a bimolecular one. Explain.

14.41 (a) The chain mechanism of H_2–O_2 reaction, unlike that of H_2–Cl_2, involves chain branching through the following steps
$$H + O_2 \rightarrow OH + O$$
$$O + H_2 \rightarrow OH + H$$
Chain branching is disfavoured at very high pressure. Why?
  (b) H_2–O_2 reaction can occur explosively only above certain pressure depending on temperature. Why?

15.42 Account for the following
  (a) Generally in course of a reaction between molecules, the energy of the reaction system first increases and then decreases after passing through a maximum.
  (b) Carbon burns in air spontaneously only above certain temperature, called ignition temperature, though the process is exothermic.

15.43 True or false.
  (a) A slow reaction cannot be exothermic.
  (b) On raising temperature, the rate of a reaction will rise or fall according as the reaction is endothermic or exothermic.

15.44 (a) According to Arrhenius, the rate constant $k = Ae^{-\frac{E_a}{RT}}$. Would A and E_a depend on T and nature of the reaction.
  (b) Which reaction would be affected more with temperature—one with higher or lower E_a?
  (c) Arrhenius equation fails for reactions involving free radicals. Why?

15.45 Mention the basic difference between the activated complex and the intermediate in a reaction. Can both be treated thermodynamically?

15.46 (a) The energy of activation is not same as the energy of the activated complex. Justify.

(b) What is the significance of energy of activation from the view point of (i) collision theory (ii) activated complex theory?

15.47 What makes the collision theory ineffective and activated complex theory effective? For what type of reaction would the two theories lead to the same result?

15.48 The simple $H + H \rightarrow H_2$ reaction, unlike $H_2 + I_2 \rightarrow 2HI$, is extremely slow. Explain this from the view point of both collision theory and transition state theory.

15.49 Combination of atoms, but not radicals with several bonds (e.g. CH_3), needs the presence of a third body. Why? Would you consider this third body as catalyst?

15.50 Set up a relation between E_a and $\Delta H^{\circ \ddagger}$ of a reaction using necessary approximations.

15.51 According to the activated complex theory, the rate constant of an elementary reaction can be expressed as

$$k = aT^m e^{-\frac{E_0}{RT}}$$

where a and m are temperature-independent constants, and so is E_0, the activation energy of the reaction at absolute zero. Connect these parameters with the Arrhenius parameters (A and E_a). Would the activation energy increase or decrease with rise of temperature?

15.52 What is the dimension of A in the Arrhenius equation $k = Ae^{-\frac{E_a}{RT}}$? Mention its SI units for unimolecular and bimolecular reactions.

15.53 Justify or criticise the following:

(i) The activation energy of a reaction may be zero but never negative.

(ii) For any reaction, the activation energy cannot be less than the heat of the reaction.

15.54 For nuclear disintegrations $\dfrac{d\ln k}{dT} = 0$. Hence, by Arrhenius equation $\dfrac{d\ln k}{dT} = \dfrac{E_a}{RT^2}$, the activation energy $E_a = 0$ for such processes. Comment.

15.55 Consider the following two mechanisms of the reaction $A + B + C \rightarrow$ Products

$$A + B \xrightarrow{k_1} I, \text{slow} \qquad\qquad A + B \underset{k_{-1}}{\overset{k_1}{\rightleftharpoons}} I, \text{fast}$$

$$I + C \xrightarrow{k_2} \text{Products, fast} \qquad I + C \xrightarrow{k_2} \text{Products, slow}$$

$$\text{(i)} \qquad\qquad\qquad\qquad\qquad \text{(ii)}$$

Find the expressions of activation energy of the overall reaction for the two mechanisms. Would the deduced expressions provide the values in compliance with the Arrhenius concept of activation energy as the minimum energy to be supplied to the reactant molecules for their reaction to occur?

15.56 The reaction $2NO + O_2 \rightarrow 2NO_2$, whose rate falls with rise of temperature, can proceed through a one-step termolecular mechanism or through a set of bimolecular steps as represented below.

1. $2NO \underset{k_{-1}}{\overset{k_1}{\rightleftharpoons}} N_2O_2$, fast

2. $N_2O_2 + O_2 \xrightarrow{k_2} 2NO_2$, slow

What can be said about the activation energy for the two mechanisms. For which mechanism, the reaction will be faster?

15.57 Lindemann theory of unimolecular gas reactions assumes:

(i) collisional activation of the reactant molecules.

(ii) a time-gap between the activation of a molecule and its decomposition.

Justify these assumptions.

15.58 Unimolecular gas reactions occur, according to Lindemann, through the following mechanism

$$A + A \underset{k_{-1}}{\overset{k_1}{\rightleftharpoons}} A^* + A$$

$$A^* \xrightarrow{k_2} Products$$

where A and A^* respectively denote the normal molecule and the activated molecule of the reactant. Find the order and rate constant of the reaction at high and low pressure. Interpret the variation of order with pressure. Would the activation energy of the reaction be different at high and low pressure.

15.59 Account for the following:

(a) The Lindemann mechanism agrees qualitatively with the fact that the (apparent) first-order rate constant of a unimolecular reaction falls with fall in the reactant pressure but quantitative disagreement is noticeable.

(b) A diatomic molecule cannot decompose by a unimolecular reaction.

15.60 Justify or criticise the following statements.

(a) A catalyst has effect only on those reactions which are thermodynamically possible.

(b) A catalyst cannot affect the equilibrium constant, and hence equilibrium composition of a reaction system.

(c) For homogeneous catalysis, the rate constant of a catalysed reaction would be proportional to the molar concentration of the catalyst.

(d) A catalysed reaction involves more number of steps, and hence more complicated than the uncatalysed one, hence it is unlikely to be faster than the latter.

(e) A catalyst changes the mechanism of a reaction into one of lower activation energy.

(f) The classification of catalysts as positive and negative is misleading.

15.61 CO reacts with H_2 in presence of a catalyst as

$$CO + 3H_2 \xrightarrow{Ni} CH_4 + H_2O$$

$$CO + 2H_2 \xrightarrow{Cu} CH_3OH$$

Would you conclude from such reactions that a catalyst determines the product of a reaction?

15.62 The reaction $N_2 + 3H_2 \rightarrow 2NH_3$ is known to be catalysed by Fe. Remembering that a catalyst acts selectively, would you expect Fe to catalyse the reverse reaction also?

15.63 The following two mechanisms have been proposed for the overall reaction $2A \rightleftharpoons A_2$.

$$2A \underset{k_{-1}}{\overset{k_1}{\rightleftharpoons}} A_2^* \qquad\qquad A + M \underset{}{\overset{K}{\rightleftharpoons}} AM$$

$$A_2^* + M \xrightarrow{k_2} A_2 + M \qquad\qquad AM + A \xrightarrow{k} A_2 + M$$

$$\text{(I)} \qquad\qquad\qquad\qquad \text{(II)}$$

Find the expression for the overall rate of the reaction according to the two mechanisms.

15.64 Consider the following reaction scheme

$$A + M \xrightleftharpoons[k_{-1}]{k_1} A^* + M$$

$$A^* \xrightarrow{k_2} \text{Products}$$

Show that the observed rate constant (k) for the decomposition of A is given by

$$k = k_1 k_2 [M]/(k_2 + k_{-1}[M])$$

Discuss the nature of the plot of k vs $[M]$. Express the observed activation energy (E) at high $[M]$ in terms of activation energies of the elementary steps.

15.65 Can prototropic mechanism of acid catalysis lead to specific catalysis?

15.66 A reaction is catalysed both by H_3O^+ and OH^-. Write down the expression for the first-order rate constant (k) of the catalysed reaction. Draw $\log k$ vs pH diagram and interprete the intercept thereof. To what pH does the minimum of such curve correspond?

15.67 Does enzyme catalysis belong to homogeneous or heterogeneous category? What makes enzyme much more efficient as catalyst compared to other catalysts?

15.68 In connection with enzyme catalysis account for the following facts which do not follow from the Michaelis–Menten mechanism.

(i) Sometimes the reaction rate does not attain a limiting value as the substrate concentration is increased, but passes through a maximum.

(ii) Generally the enzyme-catalysed reaction rate passes through a maximum as the pH is varied.

(iii) Frequently the enzyme-catalysed reaction rate passes through a maximum as the temperature is raised.

15.69 (a) What is the significance of Michaelis constant?

(b) The rate of an enzyme-catalysed reaction is 1.2×10^{-3} mol $L^{-1}s^{-1}$ when the substrate concentration is 0.12 mol L^{-1} and it is half of its maximum value when the substrate concentration is 0.036 mol L^{-1}. Find the maximum rate of this reaction.

15.70 The reaction $N_2O_5 + NO \rightarrow 3NO_2$ follows a rate law of the form

$$\text{rate} = k P_{N_2O_5}^x P_{NO}^y$$

Following kinetic data are available:

(i) When initial pressures of N_2O_5 and NO were 100 and 1 torr respectively, P_{NO} were found to vary linearly with time.

(ii) When initial pressures of N_2O_5 and NO were 1 and 100 torr respectively, $\log P_{N_2O_5}$ were found to vary linearly with time (in hr) with $|\text{slope}| = 0.35$. Find

(a) x and y

(b) half-life for the gas mixture (i)

(c) half-life for the gas mixture (ii)

(d) half-life when initial pressure of N_2O_5 and NO each is 50 torr.

15.71 Using the data available in the previous problem suggest a mechanism of the reaction $N_2O_5 + NO \rightarrow 2NO_2$.

15.72 Consider the reaction A → Products which can follow a first-order or second-order kinetics but with same rate constant. In which case the reaction rate will be higher for a particular concentration of A?

15.73 If $k_1 = k_2 = k_3$ for three reactions, being respectively uni-, bi- and ter-molecular, when concentration is expressed in molarity, what will be the relation between these rate constants if the concentration is expressed in mol/mL?

15.74 The rate constant (k) of a third order reaction is 0.06 concentration being expressed in mol/L and time in min. Find the value of k if concentration is expressed in molecules/mL and time in sec.

15.75 The rate constant of a gaseous reaction is 7.5×10^{-2} L mol^{-1}s^{-1} at 25°C. Find its value in atm^{-1}s^{-1}. What is the order of this reaction?

15.76 The rate constant (k) for the reaction $2A\,(g) = 3B\,(g) + C\,(g)$ is 4.5×10^{-5} mL mol^{-1} min^{-1} at 25°C. If A is initially at 1 atm and 25°C, find (a) the initial rate of reaction (b) initial rate of formation of B (c) time required for 50% decomposition of A (d) time required for 100% decomposition (e) fraction of A remaining unreacted at time $2t_{\frac{1}{2}}$.

15.77 The decomposition of a gas is a second order reaction. Using the gas at 200 mm initial pressure at a certain temperature, if 25% of the gas decomposes in 30 min, in what time will 50% of the gas decompose? Also find the value of the rate constant.

15.78 A gas decomposes by 10% in time 25 min and at that time the rate of the reaction is 15 in some unit. In the same unit, the rate of decomposition is 10 when 40% of the gas is decomposed. Find (i) order (ii) rate constant and (iii) $t_{\frac{1}{2}}$ of the reaction.

15.79 The rate law for the reaction $A + B + C \rightarrow$ Products is

$$\text{rate} = k[A]^x\,[B]^y\,[C]^z$$

It is found that:

(i) reaction rate is doubled, when [A] is doubled, [B] is halved and [C] is doubled or halved.

(ii) reaction rate is four times when [A] and [B] are both doubled.

Find x, y and z.

15.80 For the reaction $A + B + C = X + Y$, it is experimentally found that starting with each of the reactants at 0.01 M the time of half reaction is double of that found with each of the initial reactant concentration at 0.02 M. If the rate of formation of X when the reactants are in 0.01 M is 5×10^{-4} mol L^{-1}mol^{-1}, what is the rate constant and half time for the reaction in this case.

15.81 Rate $\left(\frac{d[A]}{dt}\right)$ of a reaction $2A + B \rightarrow$ Products is given below as a function of different initial concentrations of A and B:

[A] mol L^{-1}	[B] mol L^{-1}	Initial rate mol L^{-1}min^{-1}
0.1	0.1	0.25
0.2	0.1	0.50
0.1	0.2	0.25

Find (i) order (ii) rate constant (iii) $t_{\frac{1}{2}}$ of the reaction for $[A]_0 = 0.2$ and $[B_0] = 0.1$.

15.82 The hydrolysis of ethyl acetate in HCl has the following rate law

$$-\frac{d\,[\text{ester}]}{dt} = k\,[\text{ester}]\,[\text{HCl}], \text{ with } k = 0.1 \text{ L mol}^{-1}\text{h}^{-1} \text{ at room temperature}$$

If [ester] and [HCl] are both initially in 0.01 M, find the time required for half of the ester to be hydrolysed (disregarding the reverse reaction).

15.83 The inversion of sucrose obeys a rate law of the following form

$$-\frac{d[\text{sucrose}]}{dt} = k[\text{sucrose}]^x [H^+]^y$$

At room temperature, the time for half conversion of sucrose, which is independent of sucrose concentration, has the value at pH 4 which is ten times of what found at pH 5. Find x and y.

15.84 Methyl acetate is hydrolysed in 1 N HCl at room temperature. Aliquots of equal volume were removed at intervals and titrated with a standard NaOH solution when following data were obtained

t/min	0	5	15	25	∞
time value/mL	12	13.5	15.7	17.2	20

Find order (assuming it to be simple integer), rate constant and time for half of ester to react.

15.85 When ethyl acetate is saponified by NaOH, the progress of the reaction can be followed by titrating the unreacted alkali against a standard acid at different intervals. Using equal concentration of ester and alkali, the following results were obtained:

t/min	0	5	25	55	120	∞
titre value/mL	16	10.24	4.32	2.31	1.10	0

Find the order of the reaction. What fraction of ester will be decomposed in 30 min? Can rate constant be calculated from the given data?

15.86 A solution initially contains ethyl acetate at 0.0005 M and NaOH at 0.0008 M. After 10 min, a 25 mL aliquot of this mixture requires for its neutralisation 33 mL of a 0.0005 M HCl. Find the rate constant for this saponification reaction which is of first order wrt each reactant. At what time 50% of the ester will react?

15.87 The reaction $C_2H_5NO_2 + OH^- \rightarrow C_2H_5OH + NO_2^-$ obeys the following rate law

$$\text{rate} = k[C_2H_5NO_2][OH^-]$$

With initial concentration of $C_2H_5NO_2$ at 0.001 and NaOH at 0.3 M, 1 min is required for 1% of $C_2H_5NO_2$ to react. How long would it take for 50% of $C_2H_5NO_2$ to react? Also calculate the same, if the initial concentration of each reactant is 0.001 M.

15.88 The reaction $2A \rightarrow B$ is followed with the following observed data:

t/min	0	10	20	30	40
$[A]$/mol L^{-1}	0.312	0.223	0.160	0.114	0.082

Find (i) order (ii) rate constant and (iii) $t_{\frac{1}{2}}$ of the reaction.

15.89 Following data were obtained on the inversion of sucrose:

t/min	0	10	20	30	∞
angle of rotation	24.09°	20.40°	17.10°	14.16°	−10.74°

Calculate the average life of a sucrose molecule.

15.90 The dimerisation of butadiene proceeds homogeneously in the gas phase. At certain temperature, the total pressure P (at constant volume) observed on different time t is given in the following table.

t (min)	0	21	49.5	77.5	103.5
P (torr)	632	557	498	465	442.5

Establish whether the reaction is of second order. Also find the rate constant.

15.91 At 600°C, the decomposition of N_2O into N_2 and O_2 follows first order kinetics. The total pressure P exerted during the decomposition in a fixed volume at different time are given below:

t (hr)	26.5	62.5	∞
P (kg cm^{-2})	41.5	45.6	56.1

Find the average life of a N_2O molecule.

15.92 Decomposition of N_2O, initially containing some N_2 is studied with the following kinetic data:

t (hr)	0	26.5	62.5	∞
P (kg cm^{-2})	41.1	45.2	49.3	59.8

(i) Find initial composition of the reaction system.
(ii) Show that the reaction follows first-order kinetics and find the rate constant.
(iii) At what time, the total pressure will rise by 25%.

15.93 The decomposition of NH_3 catalysed by tungsten surface was studied at 1100°C, and the following data were obtained for the time of half decomposition with different initial pressures of NH_3

P_o (mmHg)	300	200	100
$t_{\frac{1}{2}}$ (min)	8.6	5.7	2.9

Find the order of the reaction and the rate constant. How will you account for the reaction order?

15.94 (a) The decomposition of a gas changes from zero order at the start to first order near the end. Suggest its probable rate law.

(b) The Pt-catalysed decomposition of HI is of zero order with rate constant $k = 500$ torr/sec at high pressure at 100°C. But at low pressure, the reaction is of first order with rate constant $k' = 50$/sec. At what pressure HI would decompose at the rate 100 torr/sec at 100°C?

15.95 Predict the effect of addition of KCl on the following reactions involved in their key steps with physical interpretation thereof:

(i) $Co(NH_3)_5Br^{2+} + Hg^{2+} \rightarrow$ Product

(ii) $S_2O_8^{2-} + I^- \rightarrow$ Product

(iii) $Co(NH_3)_5Br^{2+} + OH^- \rightarrow$ Product

15.96 (a) Consider the following reactions in aqueous medium

(i) $CH_3CO_2C_2H_5 + OH^- \rightarrow CH_3CO_2^- + C_2H_5OH$

(ii) $H_2O_2 + H^+ + I^- \rightarrow HIO + H_2O$

Ionic strength has little effect on (i) but profound effect on (ii). Explain.

(b) The reaction (ii) has rate constant (k) 12.18 L^{2}mol^{-2}min^{-1} when the ionic strength is 0.0525 mol kg^{-1}. Find the value of k at zero ionic strength.

15.97 The rate constants for the reaction of benzene with atomic oxygen are 1.44×10^7 at 300 K, 3.03×10^7 at 341 K and 6.90×10^7 at 392 K, all in L mol^{-1}s^{-1}. Express the rate constant for this reaction as a function of temperature in the given range..

15.98 The rate constant for a bimolecular decomposition of a gas at 1000 K increases by 10% per degree rise of temperature around 1000 K. Find energy of activation of the reaction. What fraction of gas molecules will possess sufficient energy to react at 1000 K?

15.99 Using collision theory, calculate the rate constant for the bimolecular reaction $2HI \rightarrow H_2 + I_2$ at 500 K. The molecular diameter of HI is 0.35 nm and the activation energy is 88.1 kJ mol^{-1}. Also calculate the rate of reaction if the reactant molecules were all activated at the same temperature at 1 atmosphere.

15.100 For the reaction $H_2 + I_2 \rightarrow 2HI$, the rate constants at 302°C is 2.45×10^{-4} L mol^{-1}s^{-1} and at 508°C is 0.950 L mol^{-1}s^{-1}. Calculate E_a and $\Delta H^{\neq}$ at 508°C.

15.101 The rate constant for the gas-phase decomposition of ozone in the low pressure region is given by $k = 4.6 \times 10^{12}e^{-(1230\ K)/T}$L mol^{-1}s^{-1}. Find (i) $\Delta H^{\neq}$ (ii) $\Delta S^{\neq}$ (iii) $\Delta G^{\neq}$ at 25°C. Here $\Delta S^{\neq}$ comes out negative. What does this imply?

15.102 For the first order reaction $2N_2O_5 (g) \rightarrow 4NO_2 (g) + O_2 (g)$

$$\log (t_{\frac{1}{2}} / \sec) = -13.6 + \frac{5463.21}{T / K}$$

Find:

(i) rate constant preexponential factor and activation energy.

(ii) fraction of N_2O_5 decomposed per second, and also the time required for total pressure to be double of the initial at 25°C.

(iii) error in rate constant and energy of activation, if the error in the determination of $t_{\frac{1}{2}}$ is 1.0%.

Mention the assumption(s) you have taken.

15.103 The acid-catalysed esterification reaction $RCO_2H + R'OH \rightarrow RCO_2R' + H_2O$ follows the rate law

$$\frac{d[\text{ester}]}{dt} = k[H_3O^+][RCO_2H][R'OH]$$

At pH = 2 and initial concentration $[RCO_2H]_0 = [R'OH]_0 = 0.01$ M, the time for 30% completion of the reaction is found to be 30 min at 25°C and 60 min at 15°C. Find

(i) apparent rate constant

(ii) true rate constant at 25°C

(iii) energy of activation

ANSWERS

15.1 This is expected with a multi-step reaction due to involvement of intermediate(s), unless the reaction system is at equilibrium.

Note: Frequently, the stoichiometry of the reaction is not considerably affected due to very low concentration of the intermediate(s).

15.2 This is justified only on practical ground considering low accuracy of most kinetic data, except for ionic reactions in solution due to high non-ideality of the system manifested in their salt effect.

Note: Even for an elementary reaction $aA + bB = cC + dD$, it is not absolutely justified to write

$$r_f = k_f a_A^a a_B^b$$
$$r_b = k_b a_C^c a_D^d$$

Such expressions should be corrected by multiplying right hand sides with a common factor Y. According to the activated complex theory $Y = \frac{1}{\gamma^{\neq}}$, where $\gamma^{\neq}$ is the activity coefficient of the activated complex.

15.3 Factors (i), (ii) and (iii) are common to all chemical reactions, arising out of their effect on the reactant activity.

The factor (iv) arises only when the reacting substances absorb the available radiation and get activated.

(v) arises only if the foreign substance acts as a catalyst/sensitiser/inhibitor.

(vi) is important only for chain reactions where the generation or removal of active intermediates is catalysed by the walls of the reaction vessel. For same volume, a spherical vessel has the least surface area and hence least catalytic activity.

15.4 According to the conventional definition (introduced by IUPAC through Eq. (15.1a), both rate and rate constant of a reaction must depend on its stoichiometric representation. However, the answer is negative with often-used practical definition of reaction rate as the rate of change of concentration of a specified reactant or product.

□ No. $t_{\frac{1}{2}}$, which refers to half-completion of a reaction possible in a given system, is not connected to stoichiometry of the reaction. It has relevance only with the practical rate constant.

15.5 Following the discussion on the previous problem, here the conventional rate constant will be k and not k'.

However, both k and k' may be considered as practical rate constants, the former wrt Br_2 and the latter wrt Br.

Note: For the given stoichiometric representation, $k' = 2k$.

15.6 The given rate expression, being not in multiplicative form [vide Eq. (15.2)], does not correspond to any order and rate constant. However, for the following restrictions

(A) When $C_{HBr} \ll C_{Br_2}$, (i.e. in the initial stage)
$$\text{rate} = kC_{H_2}C_{Br_2}^{\frac{1}{2}}$$

(B) When $C_{HBr} \gg C_{Br_2}$
$$\text{rate} = \frac{k}{k'} C_{H_2} C_{Br_2}^{\frac{3}{2}} C_{HBr}^{-1}$$

Then under the condition (A), the reaction corresponds to the order 1 wrt H_2 and $\frac{1}{2}$ wrt Br_2, with rate constant k.

But under the condition (B), the reaction corresponds to the order 1 wrt H_2, $\frac{3}{2}$ wrt Br_2 and -1 wrt HBr, with rate constant k/k'.

Note: The order of a reaction is not independent of the prevailing condition. A reaction that corresponds to an order under some condition may not do it under some other condition.

15.7 Because a reaction of molecularity n needs a simultaneous collision between n reactant molecules whose probability falls rapidly with increasing n virtually to a zero value for $n > 3$.

Regarding order of a reaction, it is determined by the prevailing reaction mechanism which involves one or (usually) more elementary reactions. Since an elementary reaction is unlikely to have a molecularity (or order) exceeding 3, we expect the same for the order.

15.8 (i) Yes. Because although a unimolecular reaction follows a multi-step mechanism, only one step (that leads to decomposition of an activated molecule) is of chemical nature.

(ii) No. Because an S_N1 reaction follows a mechanism involving more than one chemical step.

Note: The adjective 'unimolecular' is misfit for the so-called S_N1 reactions as a whole. It has meaning only with the unimolecular step involving the substrate.

15.9 (i) An example is offered by the reaction $Hg_2^{2+} + Tl^{3+} \rightarrow 2Hg^{2+} + Tl^+$ occurring in aqueous solution. Here the reaction is of order -1 wrt Hg^{2+} corresponding to the rate law

$$\text{rate} = k\frac{[Hg_2^{2+}][Tl^{3+}]}{[Hg^{2+}]}$$

and the proposed mechanism

$$Hg_2^{2+} \rightleftharpoons Hg^{2+} + Hg \quad \text{fast}$$
$$Hg + Tl^{3+} \rightarrow Hg^{2+} + Tl^+ \quad \text{slow}$$

(ii) True order of a reaction is always equal to its molecularity (if it exists). However, the kinetic order may be (I) greater or (II) less than the molecularity. An example of (I) is offered by a unimolecular gas reaction which is of kinetic order two at low pressure. An example of (II) is offered by any bimolecular reaction, e.g. $H_2(g) + I_2(g) = 2HI(s)$, with one of the reactants taken in large excess when the reaction is kinetically of order one.

15.10 The necessary, though not sufficient, condition for a reaction to occur in a single step is that the order of the reaction wrt each reactant must be integer and equal to its coefficient in the simplest stoichiometric equation representing the reaction. For example, if the order of the reaction $A + B + C \rightarrow$ Products is one wrt each of the reactants, the reaction may occur in a single step or it can follow a multi-step mechanism like (ii) considered in Section 15.1.

Note: A particular mechanism implies only a particular rate law but the converse is not necessarily true.

15.11 (I) Order is one wrt each reactant.

Rate constant is k.

Mechanism may involve a single step or multiple steps as in the mechanism (ii) considered in Section 15.1.

Molecularity is 3 for single-step-mechanism but meaningless for multi-step-mechanism.

(II) Order is one wrt A, one wrt B and zero wrt C.

Rate constant is k_1.

Mechanism involves multiple step as in the mechanism (i) considered in Section 15.1.

Molecularity is meaningless.

(III) Here the reaction occurs obeying simultaneously the following two rate laws each associated with some specific order and relevant mechanism

$$\text{rate} = k_1 C_A C_B$$
$$\text{rate} = k_2 C_B C_C$$

15.12 For the given reaction, the order (3) is not equal to the stoichiometry (4). Hence it cannot occur in a single step. A probable mechanism is

$$2NO + H_2 \rightarrow N_2 + H_2O_2, \quad \text{slow}$$
$$H_2O_2 + H_2 \rightarrow 2H_2O, \quad \text{fast}$$

(Vide mechanism (i) in Section 15.1).

15.13 (a) Vide Section 15.2.

(b) If α, β and γ are the orders of the forward reaction wrt A, B and C respectively, and α', β' and γ' are the corresponding orders of the backward reaction, then

$$\alpha' = \alpha - n \quad \text{where } n \text{ is a rational number}$$
$$\beta' = \beta + n$$
$$\gamma' = \gamma + n$$

[Vide Section 15.2]. From the given rate law $\alpha = 1$, $\beta = 0$ and $\gamma = 0$, then

$$\alpha' = \tfrac{1}{2}, \beta' = \tfrac{1}{2} \text{ and } \gamma' = \tfrac{1}{2}, \text{ for } n = \tfrac{1}{2}$$
$$\alpha' = 0, \beta' = 1 \text{ and } \gamma' = 1, \text{ for } n = 1$$

Therefore, the two possible rate laws for the backward reaction are

$$\text{rate} = k_1 C_A^{\frac{1}{2}} C_B^{\frac{1}{2}} C_C^{\frac{1}{2}}$$

and $\qquad \text{rate} = k_2 C_B C_C$

15.14 Even if the reaction rate is expressed in terms of activities of the reacting species, the relation $K = k_f/k_b$, suggested by the law of mass action holds rigorously only if no intermediate is involved as with one-step reactions. For multi-step reactions, the general relation is $K^n = k_f/k_b$ (where n is a rational number) of which $K = k_f/k_b$ is only a special case with $n = 1$ (vide Section 15.2).

15.15 (a) True, vide note on problem 15.10.

(b) False, for reactions involving elementary steps of comparable rate constants as in a chain reaction.

(c) Partially true. A reaction will occur incompletely (in a finite period) even if there is no reverse reaction, unless the order of the reaction wrt a reactant is less than one.

Note: When the order of a reactant is not less than one (e.g. a first order decomposition reaction), the reaction will take infinite time for its completion. This, in effect, amounts to incompleteness of the reaction.

15.16 The required condition is that the initial concentrations of the reactants are to be in the mole ratio in which they react (e.g. $a:b$ for reaction $aA + bB \rightarrow$ Products).

☐ Integrating the given rate expression, we have

$$\int_0^t k\,dt = -\int_{c_0}^c \frac{dc}{c^n}$$

(where c_0 is the lowest common factor between the initial concentrations of the reactants involved)

or $\qquad t = \dfrac{1}{k(n-1)}\left(\dfrac{1}{c^{n-1}} - \dfrac{1}{c_0^{n-1}} \right) \quad \text{for } n \neq 1$

and $\qquad t = \dfrac{1}{k} \ln \dfrac{c_0}{c} \quad \text{for } n = 1$

Then the reaction will be absolutely complete, i.e. $c = 0$, for a finite value of t only if $n < 1$.

☐☐ Putting $c = \tfrac{1}{2}c_0$ in the expressions of t, we have

$$t_{\frac{1}{2}} = \frac{2^{n-1} - 1}{k(n-1)c_0^{n-1}} \quad \text{for } n \neq 1$$

and $\qquad t_{\frac{1}{2}} = \dfrac{\ln 2}{k}$ $\qquad\qquad$ for $n = 1$

i.e. $\qquad t_{\frac{1}{2}} = \dfrac{K}{c_0^{n-1}}$

where $\qquad K = \dfrac{2^{n-1} - 1}{k(n-1)}$ $\qquad\qquad$ for $n \neq 1$

$$K = \dfrac{\ln 2}{k} \qquad\qquad\qquad \text{for } n = 1$$

If $t'_{\frac{1}{2}}$ is the half-life for initial concentration $c_0/2$.

$$t'_{\frac{1}{2}} = \dfrac{K}{(c_0/2)^{n-1}} = \dfrac{2^{n-1}K}{c_0^{n-1}} = 2^{n-1}t_{\frac{1}{2}}$$

Then $\qquad \dfrac{t_{\frac{3}{4}}}{t_{\frac{1}{2}}} = \dfrac{t_{\frac{1}{2}} + t'_{\frac{1}{2}}}{t_{\frac{1}{2}}} = 2^{n-1} + 1$

Note: k involved in the expression $-\dfrac{dc}{dt} = kc^n$ will be the true rate constant only if the stoichiometric coefficients are all equal. Vide problem 15.81.

15.17 Let us consider two dimensionless variables α and τ defined as

$$\alpha = \dfrac{c}{c_0} \qquad \text{the fraction of reaction to be completed}$$

$$\tau = kc_0^{n-1}t$$

Substituting these in the integrated rate equation for the nth order reaction, we have the following desired expression

$$\dfrac{1}{\alpha^{n-1}} - 1 = (n-1)\tau \qquad\qquad \text{for } n \neq 1$$

$$\ln \alpha = -\tau \qquad\qquad\qquad \text{for } n = 1$$

☐ Such as expression, being unique for each order, is important in determining the order of a reaction. The method consists in finding out the theoretical α versus $\log \tau$ curve for appropriate order (to be determined) that has the same shape as the experimental α versus $\log t$ curve. The two curves will coincide when the experimental curve is shifted along the $\log t$ axis by an amount $\log (kc_0^{n-1})$, since $\log \tau = \log t + \log (kc_0^{n-1})$.

Note: The experimental α vs t curve does not have the same shape as the theoretical α vs τ curve for the concerned order. Hence such curves will not serve our purpose.

15.18 Such expectation is fulfilled only if the order of decomposition of the substance concerned is less than one. We can readily understand this when the order is zero for which the reaction rate is constant. For first order reaction, the half-life ($t_{\frac{1}{2}}$) is independent of the initial reactant concentratin (c_0), because here the rate of decomposition falls proportionally with the amount of decomposition. Again, for reaction of higher order $t_{\frac{1}{2}}$ is lower, because here the rate of decomposition falls in lower proportion than the amount of decomposition.

15.19 (i) Because, for a reaction occurring in a single step, the overall order and stoichiometry must be equal, which is not possible if the order of the reaction is zero.

(ii) If a reaction is of zero order wrt a reactant A (say), it will occur at a constant rate if other reactants (if any) are maintained at fixed concentration. Then in course of (a finite) time A will be totally exhausted when the reaction is said to be complete.

15.20 No. Because here back reaction does not stop, the order wrt C being not zero.

Note: Zero order reactions cannot all proceed to completion. On the contrary, the reactions of higher order (but less than one) can go to completion.

15.21 For a first order reaction

$$t_{\frac{1}{2}} = \frac{\ln 2}{k}, \text{ from } \int_{C_0}^{\frac{C_0}{2}} \frac{dc}{c} = -\int_0^{t_{\frac{1}{2}}} k\, dt$$

$$t_{av} = \int_0^\infty \frac{dc}{c} t$$

$$= -\int_0^\infty k e^{-kt} t\, dt$$

[Since, for a first order reaction $dc = -kc\,dt = -kc_0 e^{-kt} dt$.]

$$= \frac{1}{k}$$

Then the required relation is

$$t_{\frac{1}{2}} = \ln 2 \cdot t_{av}$$

Obviously $\qquad t_{av} > t_{\frac{1}{2}}$

This is justified considering the lower rate (and hence higher duration) of the reaction in the second half.

15.22 The expression $k = \dfrac{1}{(a-b)t} \cdot \ln \dfrac{b(a-x)}{a(b-x)}$ follows from integration of the rate expression $\dfrac{dx}{dt} = k(a-x)(b-x)$. Here the integration is based on the assumption that $a \neq b$. Hence the given expression of k is not applicable when $a = b$. However, the differential rate expression holds whether a and b are equal or not.

Note: In deriving the expression of k for $a = b$, this is put in the differential equation which holds whether a and b are equal or not. However $k = \dfrac{1}{(a-b)t} \ln \dfrac{b(a-x)}{a(b-x)}$ leads to the same result under the limit $b \to a$ vide university question 15.31(a).

15.23 In all three cases the rate expression is same $\left(-\dfrac{dc}{dt} = kC^2 \right)$ and hence the expression for $t_{\frac{1}{2}}$ will be same (initial reactant concentrations being same) which is

$$t_{\frac{1}{2}} = \frac{1}{ka}$$

☐ But $t_{\frac{1}{2}}$ will not have the same value in the three cases k being not same. Because (i) differs from (ii) and (iii) in nature of the reaction, and (ii) differs from (iii) in reaction mechanism.

15.24 (i) For a zero order reaction,

$$\int_{C_0}^{C} dc = -\int_{0}^{2t_{\frac{1}{2}} = C_0/k} kdt$$

or $\qquad C = 0$

Alternative method: Since the reaction rate is independent of the reactant concentration, the first half of such reaction will occur in same time as the second half. Then the amount of A remaining at $t = 2t_{\frac{1}{2}}$ will be zero.

(ii) For a first order reaction

$$\int_{C_0}^{C} \frac{dc}{C} = -\int_{0}^{2t_{\frac{1}{2}} = \frac{2\ln 2}{k}} kdt$$

$$C = \frac{C_0}{4}$$

Alternative method: In each $t_{\frac{1}{2}}$ of a first order reaction, the amount of the reactant A is reduced to half. Then at $t = 2t_{\frac{1}{2}}$, the amount of A will be reduced to $\left(\frac{1}{2}\right)^2$ or $\frac{1}{4}$ fraction of the starting amount.

15.25 (i) Here the relevant rate law is

$$-\frac{dC}{dt} = kC^2$$

Then $\qquad \int_{C_0}^{C} \frac{dC}{C^2} = -\int_{0}^{3t_{\frac{1}{2}} = \frac{3}{ka}} kdt$

whence $\qquad C = \frac{C_0}{4}$

Then the amount of A will be reduced to $\frac{1}{4}$ fraction of the starting amount.

(ii) The relevant rate law is

$$-\frac{dC}{dt} = kC$$

Following the alternative method used in the previous problem 15.24(ii), the amount of A will be reduced to $\left(\frac{1}{2}\right)^3$ or $\frac{1}{8}$ fraction of the starting amount.

(iii) The relevant rate law is

$$-\frac{dC}{dt} = k$$

Here the amount of A will be reduced to zero at $t = 2t_{\frac{1}{2}}$, and obviously at $t = 3t_{\frac{1}{2}}$ also [vide problem 15.24(i)].

15.26 (i) Here $t_{\frac{1}{2}}$, the time required for the concentration of A to be reduced to half, will be

$$t_{\frac{1}{2}} = \frac{\ln 2}{k}$$

(ii) Here the reversible reaction may be treated as a simple first order reaction (i.e. without any reverse reaction) with the initial concentration of A equal to its equilibrium concentration in actual reversible reaction and the rate constant equal to $k + k'$ [from Eq. (15.8)]. Then defining $t_{\frac{1}{2}}$ as the time required for the possible change in concentration of A to go half-way, we have

$$t_{\frac{1}{2}} = \frac{\ln 2}{k + k'}$$

□ Because A suffers a lower reduction in concentration in (ii) than in (i).

15.27 (a) There is nothing wrong with thinking so. But this is of no use. Because from the general kinetic equations, one cannot arrive at the corresponding equations for the specific case considered.

(b) $t_{\frac{1}{2}}$ will be same for (i) and (ii) [due to the same rate constant for the step $A \rightarrow B$], but not for (iii).

15.28 Vide Section 15.4B.

15.29 (I) $$\frac{d[C]}{dt} = k_2 [A]_0 e^{-k_2 t} = k_2 [A]$$

(II) $$\frac{d[C]}{dt} = \frac{k_1 k_2}{k_{-1}} [A] \qquad \text{vide Section 15.1, mechanism (ii)}$$

□ (I) and (II) differ in rate constant.

15.30 (a) Here we have

$$\frac{d[A]}{dt} = -k_1 [A]$$

$$\frac{d[B]}{dt} = k_1 [A] - k_2 [B]$$

$$\frac{d[C]}{dt} = k_2 [B]$$

Integration of the first equation gives

$$[A] = [A]_0 e^{-k_1 t}$$

Substituting this in the second equation, we have

$$\frac{d[B]}{dt} + k_2 [B] = k_1 [A]_0 e^{-k_1 t} \qquad (x)$$

or $$\frac{d\left([B]e^{k_2 t}\right)}{dt} = k_1 [A]_0 e^{(k_2 - k_1)t} \qquad \text{multiplying both sides with } e^{k_2 t}$$

Integrating under the given condition

$$\int_0^{[B]} d([B]e^{k_2 t}) = k_1 [A]_0 \int_0^t e^{(k_2 - k_1)t} dt \qquad (y)$$

(i) When $k_1 \neq k_2$, (y) gives

$$[B] = \frac{k_1 [A]_0}{k_2 - k_1} (e^{-k_1 t} - e^{-k_2 t})$$

The maximum concentration of B is given by

$$\frac{d[B]}{dt} = \frac{k_1 [A]_0}{k_2 - k_1} \left(-k_1 e^{-k_1 t} + k_2 e^{-k_2 t}\right) = 0$$

whence $$t = \frac{\ln \frac{k_2}{k_1}}{k_2 - k_1}$$

which is the time required for $[B]$ to be maximum.

[The other solution $t = \infty$ corresponds to minimum J.

Again when $[B]$ attains the maximum value $[B]_{max}$, we have from (X)

$$[B]_{max} = [A]_0 \frac{k_1}{k_2} e^{-k_1 t}$$

$$= [A]_0 \frac{k_1}{k_2} e^{-\frac{k_1}{k_2 - k_1} \ln \frac{k_2}{k_1}}$$

$$= [A]_0 \frac{k_1}{k_2} \left(e^{\ln \frac{k_1}{k_2}} \right)^{\frac{k_1}{k_2 - k_1}}$$

$$= [A]_0 \frac{k_1}{k_2} \cdot \left(\frac{k_1}{k_2} \right)^{\frac{k_1}{k_2 - k_1}}$$

$$= [A]_0 \left(\frac{k_1}{k_2} \right)^{\frac{k_2}{k_2 - k_1}}$$

(ii) When $k_1 = k_2$, (y) gives

$$[B] = [A]_0 k_1 t e^{-k_1 t}$$

[This follows also from the expression of $[B]$ for $k_1 = k_2$ in the limit $k_2 \to k_1$ by Ɛ Hospital's rule vide question 15.31a.]

Then for maximum concentration of B

$$\frac{d[B]}{dt} = [A]_0 k_1 (1 - k_1 t) e^{-k_1 t} = 0$$

whence $\quad t = \dfrac{1}{k_1}$

Then $\quad [B]_{max} = \dfrac{[A]}{e}$

(b) $[B]_{max}$ is higher in case of (i), because here the formation of B occurs at a faster rate than its decomposition.

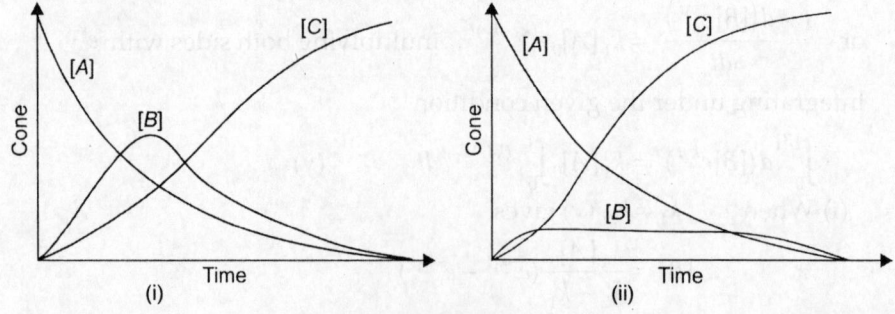

Fig. 15.3

The diagram (ii) has an important difference from (i) in that it has an almost horizontal part of low height in the $[B]$ versus t curve. This corresponds to an almost constant concentration of the intermediate (B) in the major part of the reaction, which forms the basis of the steady state approximation useful for a reaction involving unstable intermediate.

Note: $[B]_{max}$ is determined by the relative values of the rate constants while the time (t_{max}) required to reach $[B]_{max}$ is determined by their absolute values. Due to such difference t_{max} for higher $[B]_{max}$ may not be higher

15.31 (i) and (iii) are possible. But (ii), where each single-step process is opposed by a multi-step process, is prohibited by the principle of microscopic reversibility (which states that when a reaction system is at equilibrium, each elementary step and its exact reverse in a mechanism must occur at the same rate).

Note: The principle of microscopic reversibility introduces essentially the same idea as does the activated complex theory with same activated state for both forward and reverse processes. It imposes certain restriction on the rate constants, apart from the thermodynamic restrictions (vide Section 15.2), e.g.

$$k_1 k_2 k_3 / k_{-1} k_{-2} k_{-3} = 1 \text{ for (iii)}.$$

15.32 For the given reaction system attaining equilibrium

$$\frac{C_B}{C_A} = \frac{k_1}{k_{-1}}, \frac{C_C}{C_B} = \frac{k_2}{k_{-2}}$$

and hence

$$\frac{C_C}{C_A} = \frac{k_1 k_2}{k_{-1} k_{-2}}$$

At the beginning

$$-\frac{dC_A}{dt} = k_1 C_A, \text{ if only } A \text{ is initially present}$$

$$-\frac{dC_C}{dt} = k_{-2} C_C, \text{ if only } C \text{ is initially present}$$

Hence

$$\frac{k_1}{k_{-2}} \neq K \text{ (for the overall reaction), unless } k_{-1} = k_2$$

At equilibrium

$$\frac{dC_B}{dt} = k_1 C_A - k_{-1} C_B + k_{-2} C_C - k_2 C_B = 0$$

or

$$C_B = \frac{k_1 C_A + k_{-2} C_C}{k_{-1} + k_2}$$

$$-\frac{dC_A}{dt} = k_1 C_A - k_{-1} C_B = k_1 C_A - \frac{k_{-1}(k_1 C_A + k_{-2} C_C)}{k_{-1} + k_2}$$

$$= \left(\frac{k_1 k_2}{k_{-1} + k_2} \right) C_A - \left(\frac{k_{-1} k_{-2}}{k_{-1} + k_2} \right) C_C$$

i.e. the rate constant of the forward reaction is $\dfrac{k_1 k_2}{k_{-1} + k_2}$ and that of the reverse reaction is $\dfrac{k_{-1} k_{-2}}{k_{-1} + k_2}$ They have the ratio $\dfrac{k_1 k_2}{k_{-1} k_{-2}}$ which is equal to the equilibrium constant.

Note: For the elementary reversible reaction $A \underset{k_b}{\overset{k_f}{\rightleftharpoons}} C$ k_f and k_b donot change during the reaction and k_f / k_b is always equal to K.

For a complex reaction, even at equilibrim $k_f / k_b = K$ will hold provided the elementary steps add to the overall reaction.

15.33 (a) Yes.

(i) Can be treated in the specified way considering the rate constant equal to $(k_1 + k_{-1})$ and initial concentration of A equal to its equilibrium concentration in the actual reversible reaction [from Eq. (15.8)].

(ii) Can be treated similarly considering the rate constant equal to $(k_1 + k_2)$ and the initial concentration of A equal to its actual value [from Eq. (15.10a)].

Note: The difference in the nature of the two treatment lies in the initial reactant concentration.

(b) $A \rightarrow B$ having higher rate constant k_1.

□ No. Bottle-neck principle is not applicable to parallel reactions which cannot be regarded as multi-step reactions. Here $A \rightarrow B$ and $A \rightarrow C$ are two different reactions and not two steps of a reaction.

15.34 From Eq. (15.10b), we have the same $t_{\frac{1}{2}}$ for (i) and (ii) as given below

$$t_{\frac{1}{2}} = \frac{\ln 2}{k_1 + k_2}.$$

□ Differentiation between (i) and (ii) can be done from the difference in the molar ratio of B and C in the reaction system as given below

$$\frac{[B]}{[C]} = \frac{k_1}{k_2} \text{ for (i)}$$

$$\frac{[B]}{[C]} = \frac{k_2}{k_1} \text{ for (ii)}$$

15.35 The required constancy follows from the result that 2(i) + (ii) + (iii) gives

$$2N_2O_5 \rightarrow 4NO_2 + O_2$$

Applying steady state approximation principle on the intermediates NO_3 and NO, we have

$$\frac{d[NO_3]}{dt} = k_1[N_2O_5] - k_{-1}[NO_2][NO_3] - k_2[NO_2][NO_3] - k_3[NO][NO_3] = 0 \quad (A)$$

$$\frac{d[NO]}{dt} = k_2[NO_2'][NO_3] - k_3[NO][NO_3] = 0 \quad (B)$$

(A)–(B) gives

$$k_1[N_2O_5] - (k_{-1} + 2k_2)[NO_2][NO_3] = 0 \quad (C)$$

Now $\qquad \dfrac{d[N_2O_5]}{dt} = k_1[N_2O_5] - k_{-1}[NO_2][NO_3]$

$$= k_1[N_2O_5] - \left(\frac{k_{-1}k_1}{k_{-1} + 2k_2}\right)[N_2O_5] \quad \text{using [C]}$$

$$= \frac{2k_1k_2}{k_{-1} + 2k_2}[N_2O_5]$$

Then reaction rate

$$\frac{1}{2}\frac{d[N_2O_5]}{dt} = \frac{k_1k_2}{k_{-1} + 2k_2}[N_2O_5]$$

□ Since the overall stoichiometric equation is found exactly from the addition of properly multiplied elementary steps, the given mechanism does not correspond to a chain reaction.

Note: Steady state approximation principle is valid in major part of the reaction provided the intermediates are quite unstable [vide problem 15.30(b)].

Reaction involving free radicals frequently, but not always, follow a chain mechanism.

15.36 Here the mechanism involves three intermediates (two in the previous problem) for which the steady-state equations are

$$\frac{d[CH_3]}{dt} = 2k_1[C_2H_6] - k_2[CH_3][C_2H_6] = 0 \tag{A}$$

$$\frac{d[C_2H_5]}{dt} = k_1[CH_3][C_2H_6] - k_3[C_2H_5] + k_4[H][C_2H_6] - k_5[H][C_2H_5] = 0 \tag{B}$$

$$\frac{d[H]}{dt} = k_3[C_2H_5] - k_4[H][C_2H_6] - k_5[H][C_2H_5] = 0 \tag{C}$$

(A) + (B) + (C) gives

$$[H] = \frac{k_1[C_2H_6]}{k_5[C_2H_5]}$$

Substitution of this in (C) gives

$$k_3k_5[C_2H_5]^2 - k_1k_5[C_2H_6][C_2H_5] - k_1k_4[C_2H_6]^2 = 0$$

or $\quad\quad\quad\quad k_3k_5[C_2H_5]^2 - k_1k_4[C_2H_6]^2 \approx 0$, considering $k_5[C_2H_5] \ll k_4[C_2H_6]$

whence $\quad\quad\quad\quad\quad [C_2H_5] = \left(\frac{k_1k_4}{k_3k_5}\right)^{\frac{1}{2}}[C_2H_6]$

Then rate of reaction

$$-\frac{d[C_2H_6]}{dt} \approx \frac{d[C_2H_4]}{dt}$$

[the concentration of CH_4 (a byproduct) being low and of intermediates being still lower.]

$$= k_3[C_2H_5]$$

$$= \left(\frac{k_1k_3k_4}{k_5}\right)^{\frac{1}{2}}[C_2H_6]$$

Therefore the overall reaction is of first order.

☐ From the given theoretical definition

$$\text{Chain length} = \frac{\text{rate of the overall reaction}}{\text{rate of the initiation step}} \quad \text{(experimental definition)}$$

$$= \frac{(k_1k_3k_4/k_5)^{\frac{1}{2}}[C_2H_6]}{k_1[C_2H_6]}$$

$$= \left(\frac{k_3k_4}{k_1k_5}\right)^{\frac{1}{2}}$$

Note:

(i) In this problem, as in any chain reaction, the overall stoichiometric equation cannot be obtained from the elementary steps involved in the mechanism. This does not happen with the previous problem.

(ii) A simple order, even if integral, does not necessarily imply the simplicity of the reaction mechanism.

15.37 Here the molecules of HCl formed (which are equal in number to the molecules of $H_2 + Cl_2$ reacted) readily dissolve keeping the concentration of the reacting gases and hence the reaction rate constant irrespective of the mechanism followed. Then under the given experimental condition, the reaction is kinetically of zero order. But nothing can be said regarding the mechanism, or even the true order, of the reaction. However, the resemblance of Cl_2 with Br_2 suggests the H_2–Cl_2 reaction to have a chain mechanism as is found with the well-investigated H_2–Br_2 reaction.

15.38 (i) Because the Br–Br bond is much weaker, and hence easier to break, than the H–H bond.

(ii) Because the H + HBr reaction is much faster than the endothermic Br + HBr reaction of higher activation energy.

(iii) All three reactions, involving recombination of atoms, require a third body to remove the liberated energy (vide problem 15.48). However, the Br + Br reaction serves as the most efficient chain termination process because it is much less exothermic than the other two.

15.39 (i) Because the reaction H + HBr has rate comparable with the chain propagating step H + Br_2, but the corresponding reaction H + HCl (slightly endothermic) has much lower speed than the reaction H + Cl_2 (exothermic).

(ii) Because in the reaction system H_2–Cl_2 unlike H_2–Br_2, the reactions of O_2 with the existing free radicals occur significantly relative to other relevant elementary processes. To be specific, let us consider the reaction H + O_2. It has speed considerable when compared with H + Cl_2 but not so when compared with much faster H + Br_2 reaction.

(iii) Because the step ($X + H_2 \rightarrow HX + H$, X = Cl, Br) immediately following the initiation step ($X_2 \rightarrow 2X$) is much faster with chlorine than with bromine.

Note: At ordinary temperature, the reaction Br + $H_2 \rightarrow$ HBr + H is so slow that most of the Br atoms recombine to Br_2 molecules leading to a low chain length. However, with rise in temperature, the chain length increases due to increase in reaction rate of this step, being endothermic (unlike the reaction Cl + $H_2 \rightarrow$ HCl + H, which is exothermic).

15.40 In reality, both types of mechanisms are followed in all three cases with varying degree. In H_2–I_2 reaction, the bimolecular mechanism is favoured most as the activation energy for the bimolecular process $H_2 + Cl_2$, $H_2 + Br_2$ and $H_2 + I_2$ falls fairly in this order (210 $\rightarrow$ 189 $\rightarrow$ 171 kJ mol^{-1}). On the other hand, although the energy required to dissociate the halogen molecules decreases in the order $Cl_2 \rightarrow Br_2 \rightarrow I_2$, the activation energy of the chain propagating step $X + H_2 \rightarrow$ HX + H increases in the same direction (25 $\rightarrow$ 72 $\rightarrow$ 140 kJ mol^{-1}). The chain mechanism is favoured in case of chlorine and bromine due to rather low activation energy of the $X + H_2$ reaction.

15.41 (a) This follows from the consideration that debranching, which involves interaction between free radicals, requires three-body collision (vide problem 15.48) whose probability is appreciable only at very high pressure (when molecular density is sufficiently high).

(b) For such a reaction to be explosively fast, the rate of formation of the radicals must exceed the rate of their destruction. This requirement is not fulfilled below certain pressure, because at lower pressure the radicals diffuse more quickly to the walls of the container.

Note: Such pressure limit is analogous to the temperature limit for ignition.

15.42 (a) Because the interaction between the reactant molecules begins with repulsion between their electrons (being nearer to each other), and upto certain stage, such repulsion predominates over the nucleus-electron attraction between them, but after that stage reverse happens.

Note: Such energy profiles are characteristic of molecular reaction but not necessarily of ionic reactions like $NH_4^+ + NO_2^- \rightarrow N_2 + H_2O$.

(b) Ignition temperature is a consequence of the energy of activation inevitable for a molecular reaction, even if it is exothermic.

Note: The chemical process associated with burning occurs even at temperatures below the ignition temperature, but without any visible effect due to low rate of reaction at lower temperature.

15.43 (a) False. A simple example is offered by the reaction $H + H \rightarrow H_2$ which is slow but exothermic.

Note: There is no necessary relation between the reaction rate (a kinetic parameter) and the heat of reaction (a thermodynamic parameter). However, in general, the exothermic reactions are quite fast and the endothermic reactions are slow under ordinary conditions.

b. False. On raising temperature, the rate of a reaction (even if exothermic) will rise in case of a single-step reaction where $E_a \geq 0$, but not necessarily for a multi-step reaction (vide problem 15.56).

Note: With rise in temperature, the equilibrium constant of an exothermic reaction decreases, but the rate constant usually increases.

15.44 (a) E_a is defined by equation $d\ln k/dT = E_a/RT^2$ (by analogy with Van't Hoff equation $d\ln K/dT = \Delta H/RT^2$, pulling $K = k_f/k_b$ and $\Delta H = E_a^f - E_a^b$). This on integration, assuming E_a to be constant, gives $k = Ae^{-\frac{E_a}{RT}}$. Then it appears that A (which corresponds to an integration constant) would not depend on T and also on k i.e. on nature of the reaction. Evidently E_a cannot be independent of the nature of reaction.

However, theoretically both A and E_a should depend significantly on T and nature of reaction.

(b) The reaction with higher E_a. Because here $\dfrac{d\ln k}{dT} \left(= \dfrac{dk/k}{dT} \right) = \dfrac{E_a}{RT^2}$ is higher, i.e. k changes in higher proportion, though $k = Ae^{-\frac{E_a}{RT}}$, and hence dk/dT is lower.

(c) The Arrhenius equation $k = Ae^{-\frac{E_a}{RT}}$ is based on the assumption that E_a of a reaction is independent of temperature. The workability of this assumption is poor, even for short range of temperature, when E_a is quite low. Arrhenius equation fails for free-radical reactions because their E_a is very low.

15.45 An intermediate, even if it is unstable, is an actual chemical species that corresponds to a minimum in the energy profile for a multi-step reaction.

An activated complex is a hypothetical species corresponding to the maximum in the energy profile for a single-step reaction. It may be treated as an intermediate-like chemical species except that it has one loose vibration wrt which it is unstable.

☐ Strictly speaking an activated complex cannot be treated using classical thermodynamics, since it does not really behave as an ordinary chemical species. However, statistical thermodynamics can be used with due exclusion of the loose vibration wrt which the activated complex is unstable.

Note: It will be more appropriate to view the activated complex as one stage in a continuous process leading from reactants to products.

15.46 (a) Vide Section 15.5.

(b) According to Arrhenius, the energy of activation (E_a) represents the minimum energy, in excess of the average molecular energy of the reactants, which the reactant molecules must possess for participating in the reaction concerned. But he was silent regarding the nature of the energy involved.

(i) According to the collision theory, E_a (for any elementary bimolecular reaction) represents the minimum kinetic energy of impact of the colliding reactant molecules required for their chemical reaction.

Note: E_a has no relevance with the total KE of the colliding molecules nor even with the relative KE in any direction. It is significantly related to the impact KE, i.e. the KE involved in pressing together the colliding reactant molecules (along the line of their centres at the moment of collision).

(ii) According to the activated complex theory, E_a (for any elementary reaction) represents the minimum potential energy that the activated complex must possess in order that the reaction can occur.

Note: Regarding E_a, the two theories signify different nature of energy but not different amount of energy for a reaction where the potential energy results only from conversion of KE.

15.47 In essence, both collision theory and the activated complex theory equate the rate of a bimolecular reaction to the rate of a particular type of molecular collision, called the effective collision. In the simple collision theory, the effective collision is defined as the collision between the reactant molecules with certain minimum impact kinetic energy (the activation energy) treating the molecules as hard spheres (with no internal degrees of freedom). On the other hand, according to the activated complex theory, the effective collision is one that gives rise to the formation of a particular configuration of atoms (called activated complex) with certain minimum potential energy (relative to the reactant molecules that collide) through distribution of energy among various degrees of freedom (external as well as internal) in an appropriate manner. The disregarding of internal degrees of freedom in the collision theory makes it ineffective while their due consideration in the activated complex theory makes the latter effective.

☐ The reaction between atoms which are devoid of internal degrees of freedom.

Note: Before the advent of the activated complex theory, the simple collision theory was modified empirically to explain the abnormally higher values of rate constants predicted by it (often higher by a factor of 10^5 or more). This was done by introducing a correction factor (analogous to compressibility factor in gas equation), called the steric factor (P), based on the need for kinetic consideration of proper orientation of the reacting molecules, quantum mechanical restriction regarding spin change and peculiarity of atomic combination (vide problem 15.48). But the collision theory utterly failed in explaining the abnormally fast reactions.

P-factor would have served well if we could calculate it. This is not altogether impossible if the internal degrees of freedom are properlly considered. But the result would be uselessly complex mathematically.

15.48 According to the collision theory, the slowness of such atomic reaction is attributed to the low value of the P-factor while according to the activated complex theory, it is due to very low value (almost zero) of the transmission coefficient of the Eyring equation on which this theory is based.

Such phenomenon can be more clearly explained in terms of the liberated heat of the reaction which, if not removed by a third body, will cause the product molecule (H_2) to dissociate (in one-half a period of vibration). Only species containing bonded atoms (such as H_2 and I_2) can effectively serve as third body due to their ability to remove energy in form of vibration (apart from translation and rotation). This makes $H_2 + I_2 \rightarrow 2HI$ faster than $H + H \rightarrow H_2$.

Note: The recombination of atoms in the absence of a third body is possible through removal of the liberated energy in form of emitted radiation.

15.49 For combination of atoms, the liberated heat must have to be removed, otherwise it will dissociate the product molecule. Such reactions would then require the presence of a third body containing bonded atoms which can effectively remove the liberated heat. This is not essential for the combination of radicals with several chemical bonds because of distribution of the liberated energy among the bonds making no bond vibration intolerably vigorous.

□ Yes, because here the third body facilitates the reaction but remains chemically unaffected.

15.50 We have
$$E_a = RT^2 \frac{d\ln k}{dT} \quad \text{by Eq. (15.12a)}$$

$$= RT^2 \frac{d}{dT} \ln \left(\frac{RT}{N_A h} \frac{K^{\neq}}{(C^{\circ})^{n-1}} \right) \quad \text{by Eq. (15.14a)}$$

$$= RT + RT^2 \frac{d\ln k^{\neq}}{dT}$$

$$= RT + \Delta U^{\circ\neq}$$

$$= RT + \Delta H^{\circ\neq} - \Delta(PV)^{\neq}$$

In gas phase, for ideal gas reaction system
$$\Delta(PV)^{\neq} = \Delta n^{\neq} RT$$
$$= (1-n)RT$$

where $\Delta n^{\neq}$ is the change in number of moles in going from reactants to activated complex. Then
$$E_a = \Delta H^{\circ\neq} + nRT$$

in liquid phase, where $\Delta V^{\neq}$ can be ignored
$$E_a = \Delta H^{\circ\neq} + RT$$

15.51 We have
$$E_a = RT^2 \frac{d\ln k}{dT}$$

$$E_o + mRT \quad \text{substituting the given expression for } k$$

Again substituting $E_o = E_a - mRT$ in the given equation
$$k = aT^m e^m e^{-\frac{E_a}{RT}}$$

Comparing this with the Arrhenius equation, $k = Ae^{-\frac{E_a}{RT}}$

$$A = a\,(eT)^m$$

□ Since m is usually negative, it follows from the expression of E_a that E_a will generally decrease with increase in T.

15.52 Since the exponential factor in the Arrhenius equation is dimensionless A will have the dimensions and units same as the rate constant. The SI unit of A is S^{-1} for a unimolecular order reaction and m^3 $mol^{-1}s^{-1}$ for a bimolecular order reaction.

Note: A is sometimes called frequency factor. But this name is justified only for first order reactions.

15.53 (i) According to the activated complex theory

$$E_a = E_o + mRT \quad \text{vide problem 15.51}$$

Since m is usually negative, it follows from this relation that E_a for an elementary reaction (e.g. $H + H \rightarrow H_2$) may be zero which corresponds to the state where all the reactant molecules possess the necessary energy to react (and hence need no supply of energy). Obviously a negative value of E_a has no meaning with an elementary reaction. But this is not so with a multi-step reaction where E_a for the overall reaction is determined by the E_a's for more than one elementary step, e.g. E_a is negative for the reaction

$$2NO + O_2 \rightarrow 2NO_2 \quad \text{(vide problems 15.55 and 15.56)}$$

(ii) The heat of reaction (ΔH) at constant pressure is related to the activation energy (E_a) for the forward and backward reactions as

$$\Delta H = E_a^f - E_a^b$$

Then, considering the possibility of the negative value of activation energy, the given statement is not justified.

15.54 The given statement is controversial. The concept of activation energy (E_a) as the energy barrier for a process no doubt holds for nuclear disintegrations. But here, unlike ordinary chemical reactions, E_a is so much higher than the thermally available kinetic energy that a tunnelling mechanism (vide note on problem 18.89) is followed (with k independent of T) where Arrhenius equation loses its significance.

15.55 For mechanism (i), the rate constant (k) for the overall reaction is k_1, by Eq. (15.3a). Then, by the Arrhenius Eq. (15.12b), the required energy of activation will be equal to that for the first step of which the rate constant is k_1.

For the mechanism (ii), $k = k_1 k_2 / k_{-1}$, by Eq. (15.3b). Then applying Arrhenius Eq. (15.12b) to the rate constants for the overall reaction and for each step, we have

$$Ae^{-\frac{E_a}{RT}} = \left(\frac{A_1 A_2}{A_{-1}}\right) e^{-\frac{(E_{a,1} + E_{a,2} - E_{a,-1})}{RT}}$$

Then for the overall reaction, $E_a = E_{a,1} + E_{a,2} - E_{a,-1}$ (and E_a may be negative as in the reaction $2NO + O_2 \rightarrow 2NO_2$).

□ For (i), the predicted activation energy that corresponds to a particular elementary step, would comply with the Arrhenius concept of activation energy. But this would not happen with (ii) where k is determined by more than one temperature dependent quantity, viz k_1, k_{-1} and k_2.

15.56 Since NO has strong affinity for O_2 and the probability of energy transfer is negligible in a termolecular collision, E_a for the termolecular mechanism is expected to be zero.

For the three-step mechanism

$$k = k_1 k_2 / k_{-1}$$

and hence $\qquad E_a = E_{a,1} - E_{a,-1} + E_{a,2}$ (vide problem 15.55)

$$= \Delta U + E_{a,2}$$

The negative value of ΔU for the association reaction (1) is likely to make E_a negative.

☐ For termolecular reactions, A is more important than E_a in determining k. Then, considering that A, for similar reactions, decreases rapidly as we go from unimolecular to termolecular reaction (where $A \propto T^{-3}$), the mechanism involving bimolecular steps is expected to be faster.

Note:

(i) k is higher when bimolecular steps are involved though they are greater in number than that in case of termolecular mechanism. This is similar to a catalytic process due to lowering of E_a.

(ii) Fall of k due to rise of T, as in the concerned reaction, does not necessarily imply that E_a is negative.

15.57 (i) Even a unimolecular reaction requires activation of the reactant molecules, which ordinarily consists in the transfer of translational kinetic energy of the molecules into internal energy of the existing chemical bonds. This reuires collision of the molecules with each other or with the walls of the container.

Note: The mere fact that molecules are moving with high speed (i.e. high KE) does not make them unstable. The situation resembles that of fast moving cars; their KE will not wreck them unless they happen to collide when their translational KE transforms into internal energy of their parts.

(ii) The activated molecules decompose only when enough energy is accumulated into a particular vibration through a process of distribution of the internal energy of activation that takes some time. Then, the time-gap between the activation of a molecule and its decomposition is inevitable.

15.58 Vide Section 15.10.

☐ At very low pressure, the time between two successive molecular collisions is so large that every activated molecule decomposes (before it can be deactivated by collision with inactive molecules). This makes the rate of reaction equal to the rate of formation of activated molecules, which is proportional to the square of the molecular concentration. The reaction thus becomes kinetically of the second order. At high pressure, the time between two successive collisions is so short that the activated molecules mostly undergo deactivation instead of decomposition, i.e. only a small fraction of the activated molecules present at any instant decomposes. This fraction may be supposed to be more or less constant. Now, according to the Maxwell distribution law, the number of activated molecules (i.e. ones having energy above certain critical value) will be a definite fraction of the total number of reactant molecules irrespective of the pressure [vide Eq. (2.7a)]. Then the fraction of molecules decomposing in a given time will also be irrespective of pressure, i.e. molecular concentration. This corresponds to the first order kinetics.

□□ Yes. This is due to difference in rate constants at high and low pressure. We can justify this considering the Lindemann mechanism as a combination of two simple mechanisms of which one predominates at very low pressure and the other at very high pressure.

Note: For unimolecular gas reactions, the activation energy is provided in bimolecular collisions yet the reaction is of first order at high pressure, though it is of second order at low pressure.

For unimolecular reactions in liquid solution, the application of the Lindemann theory would lead only to a first order rate expression similar to that for the gaseous reaction at high pressure.

15.59 (a) Because the Lindemann theory of unimolecular reaction is based on the faulty assumption that all the activated molecules have the same probability of decomposition independent of the internal energy they acquire by collisional activation.

(b) Vide problem 15.57(ii).

A key idea in the Lindemann mechanism of unimolecular reaction is the considerable time-gap between the activation and decomposition of the reactant molecules. For a diatomic molecule, that contains only one chemical bond, the time-gap is virtually zero (the distribution of internal activation energy being out of question) so that the rate constant for the unimolecular step tends to infinity. In this situation, the bimolecular activation process will serve as the rate determining step and hence, the overall reaction will have second-order kinetics.

15.60 (a) Thermodynamically, a reaction is possible, i.e. has tendency to occur, only if the products have lower free energy (G) than the reactants at same temperature and pressure. A catalyst cannot create this tendency because, being unchanged in mass and chemical composition at the end of the reaction, it has virtually* no contribution to G of the reaction system. The given statement is thus justified.

[**Note:** * Frequently a catalyst undergoes a physical change with only insignificant ΔG. For example, a platinum gauze catalyst used in the oxidation of NH_3 becomes considerably rough after prolonged use. Evidently the catalyst is somehow involves in the mechanism of the reaction, but is regenerated in a somewhat different form at the end of the reaction.

Although a catalyst has no significant effect on $\Delta G_{T,P}$ of a reaction, it influences the tendency of a thermodynamically allowed raction by changing its mechanism to one of lower free energy of activation].

(b) The statement that a catalyst cannot affect the equilibrium constant (K) is justified by the relation $\Delta G_T^\circ = -RT \ln K$, ΔG° being not affected by the catalyst. But a homogeneous catalyst can change the equilibrium composition (x_i) of a reaction system through change of activity coefficients (γ_i) of the reactants and the products unless they happen to offset each other. This follows from the fact that K is a function of the activities $a_i = \gamma_i x_i$ at equilibrium.

[**Note:** Since a catalyst is usually present in small amounts, its effect on the equilibrium composition is usually negligible].

(c) The given statement is usually, but not always, true. Because depending on the complexity of the reaction system, the rate expression in presence of a catalyst may contain terms independent of the catalyst, terms involving various positive powers of the catalyst concentration, and even terms inversely proportional to the catalyst concentration. Here, the first term is due to the uncatalysed reaction and each of the other terms is connected to different mechanistic scheme involving the catalyst (vide problem 15.11). It may so happen that the mechanism of the catalysed reaction has rate constant different from the uncatalysed one but does not vary with catalyst concentration (as in case of hydrolysis of acetamide catalysed by an acid at high concentration of the latter). [In case the given statement holds, the term catalytic coefficient is used to represent the rate constant per unit concentration of the catalyst].

(d) The given statement is not justified. Despite a greater number of steps involved in a catalysed reaction, it is easier for the reaction system to follow the catalysed mechanism on energetic ground, just as it is easier to climb a roof-top with a ladder.

(e) The given statement does not always hold. There are few reactions where the catalysed mechanism has higher E_a than the uncatalysed one [vide J.A. Campbell, J.Chem. Edu., 61, 40 (1984)]. Considering this, it is better to say that a catalysed mechanism has lower free energy of activation in compliance with the activated complex theory.

Note: A catalysed mechanism may have higher value of A (in Arrhenius equation) than the uncatalysed one.

(f) If the definition of a catalyst (vide Section 15.6) is rigidly followed the term negative catalyst will appear awkward. Because, so-called negative catalysts are used up while influencing (retarding) the rate of a reaction. They should be better called inhibitor (behaving opposite to sensitiser).

Note: The strict adherence to the definition of catalysis will not justify, the name autocatalysis given to certain phenomenon where the amount of a catalyst varies from the beginning to the end of a reaction. Further, we should be careful not to use the adjective 'catalysed' to any reaction of accelerated rate. Thus, we should not say that 'a reaction is catalysed by light', light being not a substance.

15.61 Of the different reactions possible with the same reactants, that reaction will be favoured for which $\Delta G^{\ne}$ of the catalysed mechanism will be lowest. Since $\Delta G^{\ne}$ is determined jointly by the reactants, products and the catalyst, it is not unlikely that the final products are different with Ni and Cu, i.e. determined by the catalyst used.

Note: For the same reactants with different catalyst, the products may not be different. Thus $CO + H_2$ reaction gives the same product if Pd is used, instead of Ni, as catalyst.

15.62 For the specified reactants and products, $\Delta G^{\ne}$ for a particular catalyst is fixed and hence the selectivity of the catalytic action is out of question. A remarkable fact regarding a catalyst is that it affects rate constant (k) but not equilibrium constant $K = (k_f/k_b)^n$, (where n is a constant) of a reaction. This implies that a catalyst cannot change only k_f or only k_b, it must change both the rate constants and that too only proportionally. Therefore, for the given reaction, if the forward process is catalysed by Fe, the backward process will also be.

Note: Although Fe accelerates the dissociation of NH_3 along with its formation (thereby facilitating the attainment of equilibrium), the time required for production of a specified amount of NH_3 will be shortened by the catalytic effect of Fe.

15.63 For mechanism I (and also for II), M is involved in the reaction but regenerated at the end of the reaction and hence M acts as catalyst. It is likely that the second step, which involves M, will be fast: Applying steady state approximation to A_2^* we have

$$k_1[A]^2 - k_{-1}[A_2^*] - k_2[M][A_2^*] = 0$$

whence
$$[A_2^*] = \frac{k_1[A]^2}{k_{-1} + k_2[M]}$$

Then, overall reaction rate

$$= k_2[M][A_2^*]$$

$$= \frac{k_1 k_2[M][A]^2}{k_{-1} + k_2[M]}$$

For mechanism II, the first step is likely to be fast (being a combination reaction involving M). Here

$$[AM] = K[A][M] \text{ vide mechanism (ii) in Section 15.1}$$

Then, overall reaction rate

$$= k[AM]$$

$$= kK[M][A]^2$$

Note: For both the mechanisms, the kinetic order of the reaction is same (which is two) but the rate constant is different.

15.64 Considering that the first step is slow, we have, on applying steady state approximation to A^*.

$$k_1[M][A] - k_{-1}[M][A^*] - k_2[A^*] = 0$$

whence
$$[A^*] = \frac{k_1[M][A]}{k_2 + k_{-1}\{M\}}$$

Then overall reaction rate

$$= k_2[A^*]$$

$$= \frac{k_1 k_2[M][A]}{k_2 + k_{-1}[M]}$$

whence
$$k = \frac{k_1 k_2[M]}{k_2 + k_{-1}[M]}$$

☐ $\quad k = k_1[M] \quad$ when $[M]$ is low

i.e. $\quad k \propto [M]$

and $\quad k = \dfrac{k_1 k_2}{k_{-1}} \quad$ when $[M]$ is high

$$= \text{constant}$$

Then k vs $[M]$ plot will give a curve as shown in the Fig. 15.4.

☐☐ At high $[M] \quad k = \dfrac{k_1 k_2}{k_{-1}}$

Then $\quad\quad\quad\quad E = E_{a,1} + E_{a,2} - E_{a,-1}$

vide problem (15.55)

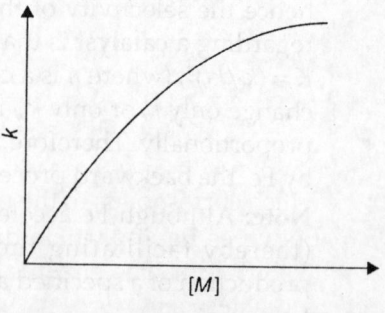

Fig. 15.4

Note: Here the step involving M (the catalyst) is considered to be slower (unlike the previous problem). This is not unlikely because here this step is bimolecular while the decomposition of A^* is unimolecular which is likely to be faster. But the situation will be different if $[M]$ is quite high when $k_{-1}[M] > k_2$.

k depends on nature of M (as k_1 and k_{-1} do so) but not on $[M]$ when $[M]$ is high, while it depends on both when $[M]$ is low.

15.65 Yes. Vide section 15.6(A).

15.66 The required expression is

$$k = k_o + k_{H_3O^+}[H_3O^+] + k_{OH^-}[OH^-] \tag{A}$$

where k_o is the rate constant for the uncatalysed reaction, $k_{H_3O^+}$ is the catalytic coefficient for H_3O^+ and k_{OH} is that for OH^-.

☐ In acid medium, when $[H_3O^+] \gg [OH^-]$, (A) reduces to

$$k \approx k_{H_3O^+}[H_3O^+], \text{ since usually } k_o \ll k_{H_3O^+}$$

or $$\log k = \log k_{H_3O^+} - pH$$

In alkaline medium, when $[H_3O^+] \ll [OH^-]$

$$k \approx k_{OH^-}[OH^-]$$

$$= k_{OH^-}\frac{K_w}{[H_3O^+]}$$

or $$\log k = \log(k_{OH^-} \cdot K_w) + pH$$

☐ Then the log k vs pH curve will be as shown in Fig. 15.5.

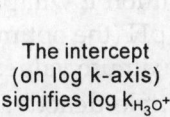
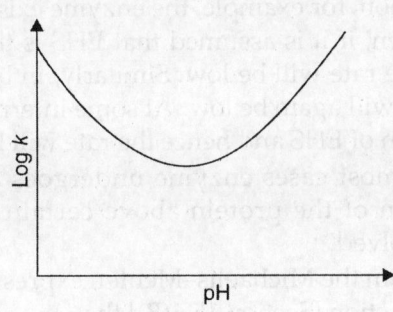

The intercept (on log k-axis) signifies log $k_{H_3O^+}$

Fig. 15.5

☐☐ Differentiating (A) wrt $[H_3O^+]$ and equating the result to zero, we have

$$[H_3O^+] = \left(\frac{k_{OH^-} \cdot K_w}{k_{H_3O^+}}\right)^{\frac{1}{2}}$$

or $$pH = \tfrac{1}{2}\log\left(\frac{k_{H_3O^+}}{k_{OH^-} \cdot K_w}\right)$$

which corresponds to the minimum of the log k vs pH curve.

15.67 Enzymes have particle diameter falling in the dimensional range of colloids $(3 - 10^3 \text{ nm})$. The enzyme catalysis is, therefore, midway between homogeneous and heterogeneous catalysis.

□ The high efficiency of an enzyme as catalyst is attributed to the fact that the elementary steps responsible for catalytic activity occur through concerted mechanism permitted by the flexibility of the structure of enzyme molecules.

Note: Catalytic activity of enzyme is highly specific. Many enzymes exhibit stereochemical specificity.

15.68 (i) This can be explained in terms of a mechanism in which a second substrate molecule becomes attached to the enzyme-substrate complex.

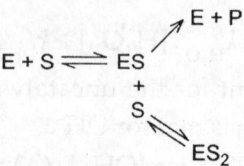

At high substrate concentration, ES_2 is formed in large amounts, and if it is unreactive, the rate will be low.

(ii) Because the active centre of the enzyme can exist in three states of ionization

$$E \rightleftharpoons EH \rightleftharpoons EH_2$$

Obviously the form EH bears one more positive charge than E, and EH_2 bears one more positive charge than EH (for simplicity the charges are omitted here). The enzyme–substrate complex can also exist in three different states of ionization

$$ES \rightleftharpoons EHS \rightleftharpoons EH_2S$$

In acid solution, for example, the enzyme exists largely as EH_2 and the complex as EH_2S. Then, if it is assumed that EHS is the only form that can give rise to products, the rate will be low. Similarly, in basic solution E will predominate, and the rate will again be low. At some intermediate pH (the optimum pH) the concentration of EHS and hence the rate will have maximum value.

(iii) Because in most cases enzyme undergoes appreciable deactivation due to denaturation of the protein above certain temperature depending on the enzyme involved.

15.69 (a) It follows from the Michaelis–Menten expression for the rate (v) of an enzyme-catalysed reaction [Section 15.6(B)] that

$$\frac{v}{v_{max}} = \frac{[S]}{K_m + [S]} \qquad \qquad \text{[A]}$$

whence $v = \frac{1}{2}v_{max}$ for $[S] = K_m$. Then K_m has the significance that it equals the substrate concentration for which the rate of the enzyme-catalysed reaction falls to half of the maximum rate.

(b) $$v_{max} = \frac{K_m + [S]}{[S]} \cdot v \quad \text{by Eq. (A)}$$

$$= \frac{(0.036 \text{ mol L}^{-1}) + (0.12 \text{ mol L}^{-1})}{(0.12 \text{ mol L}^{-1})} (1.2 \times 10^{-3} \text{ mol L}^{-1}\text{s}^{-1})$$

$$= 1.56 \times 10^{-3} \text{ mol L}^{-1}\text{s}^{-1}$$

15.70 (a)
$$\frac{dP_{NO}}{dt} = k\,P_{N_2O_5}^x\,P_{NO}^y$$

$$= k_{app}\,P_{NO}^y \quad \text{for condition (i) where } P_{N_2O_5} \text{ is nearly constant}$$

Since p_{NO} varies linearly with time, $y = 0$

Again
$$\frac{dP_{N_2O_5}}{dt} = k\,P_{N_2O_5}^x, \quad \text{since } y = 0$$

Since $\log P_{N_2O_5}$ varies linearly with time, $x = 1$ and $k = 0.35\ \text{hr}^{-1}$

(b) For gas mixture (i), since the reaction is kinetically of zero order

$$t_{\frac{1}{2}} = \frac{(P_{NO})_o}{2k_{app}} = \frac{(P_{NO})_o}{2k(P_{N_2O_5})_o}$$

$$= \frac{(1\ \text{torr})}{2(0.35\ \text{hr}^{-1})(100\ \text{torr})} = 1.43 \times 10^{-2}\ \text{hr}$$

(c) For gas mixture (ii), since the reaction is kinetically of first order

$$t_{\frac{1}{2}} = \frac{\ln 2}{k}$$

$$= \frac{\ln 2}{0.35\ \text{hr}^{-1}} = 1.98\ \text{hr}$$

(d) $t_{\frac{1}{2}}$ will be same as in (c), since k and overall order are same and, the order being one, $t_{\frac{1}{2}}$ is independent of the initial concentration.

15.71 Since the order of the reaction is zero wrt NO, the reaction should follow a multi-step mechanism, where NO will be involved after the slowest step, as given below

$$N_2O_5 \xrightarrow{k_1} NO_2 + NO_3 \quad \text{slow}$$
$$NO + NO_3 \xrightarrow{k_2} 2NO_2 \quad \quad \text{fast}$$

15.72 For first-order kinetics, the reaction rate $v_1 = kC$.

For second-order kinetics, the reaction rate $v_2 = kC^2$.

Here the relative order of rate is not independent of the unit of C chosen.

For conventional unit of concentration C, i.e. mol L^{-1}

$$v_1 > v_2 \quad \text{if } C < 1$$
$$v_2 > v_1 \quad \text{if } C > 1$$

Note: $v_1 = v_2$, for $C = 1$ in chosen unit.

15.73 Let k_1', k_2' and k_3' be the rate constants of the respective reactions in the new units. Then

$$k_1' = k_1\ \text{time}^{-1}$$

$$k_2' \equiv k_2\,\frac{L}{\text{mol time}}$$

$$= k_2\,\frac{10^3\ \text{mL}}{\text{mol time}}$$

$$k_3' \equiv k_3\,\frac{L^2}{\text{mol}^2\ \text{time}}$$

$$= k_3\,\frac{10^6\ \text{mL}^2}{\text{mol}^2\ \text{time}}$$

Then numerically

$$k_1' : k_2' : k_3' = k_1 : 10^3 k_2 : 10^6 k_3$$
$$= 1 : 10^3 : 10^6, \quad \because k_1 = k_2 = k_3$$

This is the required relation.

15.74
$$k = 0.06 \frac{L^2}{mol^2 \ min}$$

$$\equiv 0.06 \frac{10^6 \ mL^2}{(6.022 \times 10^{23} \ molecules)^2 \ (60 \ s)}$$

$$\equiv 2.76 \times 10^{-45} \frac{mL^2}{molecules^2 \ s}$$

Then the value of k in new unit is 2.76×10^{-45}.

15.75 For an nth order reaction $\dfrac{dc}{dt} = kC^n$. Hence k will have the unit $(L \ mol^{-1})s^{-1}$ when C in mol L^{-1} and time in sec. Then from the given unit of k, $n - 1 = 1$ or $n = 2$

Here
$$k = 7.5 \times 10^{-2} \frac{L}{mol \ s}$$

$$= 7.5 \times 10^{-2} \frac{RT \ (in \ L\text{-}atm)}{atm \ s}, \text{ using the ideal gas equation } P = cRT$$

$$= 7.5 \times 10^{-2} \ (0.082 \times 298) \ atm^{-1}s^{-1}$$

$$= 1.8 \times 10^{-2} \ atm^{-1}s^{-1}$$

15.76 Initial concentration of A,

$$[A]_o = \frac{P_o}{RT}$$

$$= \frac{(1 \ atm)}{(0.082 \ L\text{-}atm \ K^{-1}mol^{-1})(298 \ K)} = 0.041 \ mol \ L^{-1}$$

The reaction is of second order as implied by the unit of k.

(a) Initial rate of reaction

$$-\frac{1}{2}\frac{d[A]}{dt} = k[A]_o^2$$

$$= [4.5 \times 10^{-5} \times 10^{-3} \ L \ mol^{-1}min^{-1}) (0.041 \ mol \ L^{-1})^2$$

$$= 7.5 \times 10^{-11} \ mol \ L^{-1}min^{-1}$$

(b) Initial rate of formation of B

$$= 3 \times \text{initial rate of reaction}$$

$$= 3 \times 7.5 \times 10^{-11} \ mol \ L^{-1} \ min^{-1}$$

(c) The required time

$$t_{\frac{1}{2}} = \frac{1}{2k[A]_o}, \quad \text{since} -\frac{d[A]}{dt} = 2k[A]^2$$

$$= \frac{1}{2(4.5 \times 10^{-5} \times 10^{-3} \ L \ mol^{-1}min^{-1})(0.041 \ mol \ L^{-1})}$$

$$= 2.8 \times 10^8 \ min$$

(d) ∞, since a secopnd order reaction is not complete in a finite period of time.

(e) $\qquad -\int_{[A]_0}^{[A]} \dfrac{d[A]}{[A]^2} = 2k\int_0^{2t_{\frac{1}{2}}=\frac{1}{k[A]_0}} dt,$ the reaction being of second order

whence $\qquad [A] = \frac{1}{3}[A]_0$

or $\qquad \dfrac{[A]}{[A]_0} = \frac{1}{3}$

Note: By rate constant, we generally mean conventional rate constant and not the practical rate constant; the two rate constants are different in the problem concerned (vide problems 15.4 and 15.5). Here conventional rate constant is k and practical rate constant is $2k$ wrt A.

15.77 $\qquad\qquad k = \dfrac{1}{at}\left(\dfrac{x/a}{1-x/a}\right)$ by Eq. (15.7b)

$$= \dfrac{1}{(200 \text{ mm})(30 \text{ min})}\left[\dfrac{25/100}{1-25/100}\right]$$

$$= 5.55 \times 10^{-5} \text{ mm}^1 \text{ min}^{-1}$$

Time required for 50% decomposition

$$t_{\frac{1}{2}} = \dfrac{1}{ka}$$

$$= \dfrac{1}{(5.55\times 10^{-5} \text{ mm}^{-1}\text{min}^{-1})(200 \text{ mm})} = 90.1 \text{ min}$$

Alternatively, $t_{\frac{1}{2}}$ can be calculated directly by the relation

$$t_{\frac{1}{2}} = \dfrac{1}{ka} = \dfrac{t(1-x/a)}{x/a}$$

Note: Calculation of $t_{\frac{1}{2}}$ by the alternative method does not require any knowledge of a and k.

15.78 (i) The rate (v) of an nth order reaction is

$$v = k(a-x)^n$$

$$= ka^n\left(1-\dfrac{x}{a}\right)^n$$

Then, using the given date

$$\left.\begin{array}{l} 15 = ka^n\left(1-\frac{10}{100}\right)^n \\ 10 = ka^n\left(1-\frac{40}{100}\right)^n \end{array}\right\} \Rightarrow \dfrac{15}{10} = \left(\dfrac{90}{60}\right)^n$$

whence $n = 1$.

(ii) For a first order reaction

$$k = \dfrac{1}{t}\ln\dfrac{1}{1-x/a} \quad \text{by Eq. (15.6a)}$$

$$= \dfrac{1}{(25 \text{ min})}\ln\dfrac{1}{1-\frac{10}{100}} = 4.22\times 10^{-3} \text{ min}^{-1}$$

(iii) $\qquad\qquad t_{\frac{1}{2}} = \dfrac{\ln 2}{k} = \dfrac{0.693}{4.22\times 10^{-3}\text{ min}^{-1}} = 164.2 \text{ min}$

15.79 Here the reaction rate does not vary with variation in change of concentration of c, then $Z = 0$. Also it follows from the given data that

$$2^x \left(\tfrac{1}{2}\right)^y = 2 \quad \text{from (i)}$$

and
$$2^x \cdot 2^y = 2^2 \quad \text{from (ii)}$$

Then
$$\left.\begin{array}{l} x - y = 1 \\ x + y = 2 \end{array}\right\} \Rightarrow \begin{array}{l} x = \tfrac{3}{2} \\ y = \tfrac{1}{2} \end{array}$$

and

15.80 The half life $(t_{\frac{1}{2}})$ of an nth order reaction is related to the initial reactant concentration (C_0) by

$$t_{\frac{1}{2}} \propto \frac{1}{C_0^{n-1}} \quad \text{vide problem 15.16}$$

Then for two different initial concentrations

$$\frac{t_{\frac{1}{2}}}{t_{\frac{1}{2}}'} = \left(\frac{C_0'}{C_0}\right)^{n-1}$$

or
$$n = 1 + \frac{\log(t_{\frac{1}{2}}/t_{\frac{1}{2}}')}{\log(C_0'/C_0)}$$

$$= 1 + \frac{\log 2}{\log \frac{0.02}{0.01}}, \quad \text{from the given data}$$

$$= 2$$

Since the reaction is of second order

$$k = \frac{\text{rate}}{C^2}$$

$$= \frac{5 \times 10^{-4} \text{ mol L}^{-1}\text{min}^{-1}}{(0.01 \text{ mol L}^{-1})^2} = 5.0 \text{ L mol}^{-1}\text{min}^{-1}$$

$$t_{\frac{1}{2}} = \frac{1}{kC_0}$$

$$= \frac{1}{(5.0 \text{ L mol}^{-1}\text{min}^{-1})(0.01 \text{ mol L}^{-1})} = 2.1 \text{ min}$$

15.81 (i) For the same initial concentration (0.1 mol L^{-1}) of A, the reaction rate does not vary with change in concentration of B. Hence order wrt B is zero. Again, for the same initial concentration (0.1 mol L^{-1}) of B, the rate becomes double when the concentration of A is doubled. Hence the order wrt A is one.

(ii)
$$k = \frac{\text{reaction rate}}{[A]} \quad \text{since order wrt } B \text{ is zero}$$

$$= \frac{0.25 \text{ mol L}^{-1}\text{min}^{-1}}{0.1 \text{ mol L}^{-1}} = 2.5 \text{ min}^{-1}$$

(iii)
$$t_{\frac{1}{2}} = \frac{\ln 2}{2}$$

$$= \frac{\ln 2}{2.5 \text{ min}^{-1}} = 0.28 \text{ min}$$

Note: For $[A] = 0.2$ mol L^{-1} and $[B] = 0.1$, k and hence $t_{1/2}$ will be same as found above. Here k is the practical, and not conventional rate constant. Vide problem (15.76).

15.82 The given rate law may be rewritten as

$$-\frac{d[\text{Ester}]}{dt} = k'\,[\text{Ester}]$$

where $k' = k\,[\text{HCl}]$

= constant, HCl being a catalyst, [HCl] does not change

This expression corresponds to the first rate law. Then the time $(t_{\frac{1}{2}})$ for half of the ester to be hydrolysed is given by

$$t_{\frac{1}{2}} = \frac{\ln 2}{k'} = \frac{\ln 2}{k\,[\text{HCl}]}$$

$$= \frac{\ln 2}{(0.1\,\text{L mol}^{-1}\text{h}^{-1})\,(0.01\,\text{mol L}^{-1})} = 693\,\text{h}$$

Note: In calculating $t_{\frac{1}{2}}$ only the kinetic order is of importance. Here [HCl] is useful but [Ester] is not. though ester is reactant and HCl is catalyst, because the reaction is of first order where $t_{\frac{1}{2}}$ is independent of the reactant concentration.

15.83 Here $-\dfrac{d[\text{sucrose}]}{dt} = k'[\text{sucrose}]^x$, where $k' = k\,[\text{H}^+]^y = $ constant at constant pH. Since $t_{\frac{1}{2}}$ is constant at constant pH, it follows from this rate equation that $x = 1$ and hence

$$t_{\frac{1}{2}} = \frac{\ln 2}{k'} = \frac{\ln 2}{k\,[\text{H}^+]^y}$$

Then from the given data at pH = 4 and 5, we have

$$10 = \left(\frac{10^{-4}}{10^{-5}}\right)^y$$

whence $\qquad y = 1$

15.84 The relevant chemical reaction is

$$CH_3CO_2CH_3 + H_2O \xrightarrow{\text{HCl}} CH_3CO_2H + CH_3OH$$

During the reaction [HCl] remains constant (since HCl acts as a catalyst) and [H$_2$O] remains practically constant (since H$_2$O is present in large excess due to its function as solvent apart from reactant). Then, if the reaction is of first order wrt ester, the rate law in integrated form will be

$$k = \frac{1}{t}\ln\frac{a}{a-x}$$

$$= \frac{1}{t}\ln\frac{T_\infty - T_o}{T_\infty - T_t}$$

Since $\qquad \dfrac{x}{a} = \dfrac{T_t - T_o}{T_\infty - t_o}$

where T_o, T_t and T_∞ are the titre values at time o, t and ∞.

From the given data, we have

for $t = 5$ min; $\dfrac{1}{t} \ln \dfrac{T_\infty - t_o}{T_\infty - T_t} = \dfrac{1}{5 \text{ min}} \ln \dfrac{20 - 12}{20 - 13.5} = 0.0415 \text{ min}^{-1}$

for $t = 15$ min $\dfrac{1}{t} \ln \dfrac{T_\infty - t_o}{T_\infty - T_t} = \dfrac{1}{15 \text{ min}} \ln \dfrac{20 - 12}{20 - 15.7} = 0.0414 \text{ min}^{-1}$

for $t = 25$ min $\dfrac{1}{t} \ln \dfrac{T_\infty - t_o}{T_\infty - T_t} = \dfrac{1}{25 \text{ min}} \ln \dfrac{20 - 12}{20 - 17.2} = 0.0420 \text{ min}^{-1}$

It appears from above calculations that $\dfrac{1}{t} \ln \dfrac{T_\infty - T_o}{T_\infty - t_t}$ is nearly constant for different values of t. Hence, the reaction is of first order wrt ester (as assumed).

The required rate constant will be the average values of $\dfrac{1}{t} \ln \dfrac{T_\infty - T_o}{T_\infty - t_t}$, then

$$k = \dfrac{(0.0415 + 0.0414 + 0.0420) \text{ min}^{-1}}{3} = 0.0416 \text{ min}^{-1}$$

$$\text{Half life} = \dfrac{\ln 2}{k} \quad \text{since the reaction is kinetically of first order}$$

$$= \dfrac{0.693}{0.0416 \text{ min}^{-1}} = 16.7 \text{ min}$$

Note: k thus calculated represents the apparent rate constant which is equal to the true rate constant multiplied by $[H_2O]^\alpha$ where α (the order wrt H_2O) is unknown.

15.85 The relevant chemical reaction is

$$CH_3CO_2C_2H_5 + OH^- \rightarrow CH_3CO_2^- + C_2H_5OH$$

If the reaction is of order one wrt ester and also wrt OH^-, i.e. total order two, then under the given condition, the rate law is integrated form will be

$$ka = \dfrac{1}{t} \cdot \dfrac{x}{a - x} \quad \text{by Eq. (15.7b)}$$

$$= \dfrac{1}{t} \cdot \dfrac{T_o - T_t}{T_t - T_\infty}$$

Since $\dfrac{x}{a} = \dfrac{T_o - T_t}{T_t - T_\infty}$

From the given data, we have

for $t = 5$ min, $\dfrac{1}{t} \cdot \dfrac{T_o - T_t}{T_t - T_\infty} = \dfrac{1}{5 \text{ min}} \cdot \dfrac{16 - 10.24}{10.24 - 0} = 0.01125 \text{ min}^{-1}$

for $t = 15$ min $\dfrac{1}{t} \cdot \dfrac{T_o - T_t}{T_t - T_\infty} = \dfrac{1}{25 \text{ min}} \cdot \dfrac{16 - 4.32}{4.32 - 0} = 0.1093 \text{ min}^{-1}$

for $t = 25$ min $\dfrac{1}{t} \cdot \dfrac{T_o - T_t}{T_t - T_\infty} = \dfrac{1}{55 \text{ min}} \cdot \dfrac{16 - 2.31}{2.31 - 0} = 0.1077 \text{ min}^{-1}$

Since $\dfrac{1}{t} \cdot \dfrac{T_o - T_o}{T_t - T_\infty}$ for different value of t is fairly constant, the order of the reaction is two (as assumed). The average value of ka will be

$$ka = \dfrac{(0.1125 + 0.1093 + 0.1077) \text{ min}^{-1}}{3} = 0.110 \text{ min}^{-1}$$

☐ The fraction of ester decomposed

$$\frac{x}{a} = \frac{kat}{1+kat}$$

$$= \frac{(0.11\ \text{min}^{-1})(30\ \text{min})}{1+(0.11\ \text{min}^{-1})(30\ \text{min}^{-1})} = 0.77$$

☐☐ k cannot be calculated (from ka) without the knowledge of a.

Note: The hydrolysis of ester is of order one wrt ester in both acid and alkaline medium, though the total order is different in the two cases.

15.86 25 ml of reaction mixture $\equiv$ 33 mL of 0.0005 M HCl.

Then concentration of NaOH in the reaction mixture after 10 min $(a - x)$ will be

$$\frac{33 \times 0.0005}{25} \quad \text{or } 0.00066\ \text{M}$$

Change in concentration of NaOH, $x = (0.0008 - 0.00066)$ or 0.00014 M

$$k = \frac{1}{(a-b)t}\ln\frac{b(a-x)}{a(b-x)} \qquad \text{by Eq. (15.7a)}$$

$$= \frac{1}{(0.0008\ \text{M} - 0.0005\ \text{M})(10\ \text{min})}\ln\frac{(0.0005)(0.00066)}{(0.0008)(0.0005 - 0.00014)}$$

$$= 45.4\ \text{L mol}^{-1}\text{min}^{-1}$$

☐ Here the required time is $t_{\frac{1}{2}}$ of the reaction that corresponds to half-completion of the maximum possible amount of the reaction, i.e. $x = b/2$, and is given by

$$t_{\frac{1}{2}} = \frac{1}{k(a-b)}\ln\frac{b(a-b/2)}{a(b-b/2)}$$

$$= \frac{1}{k(a-b)}\ln\frac{2a-b}{a}$$

$$= \frac{1}{(45.5\ \text{M}^{-1}\text{min}^{-1})(0.0008\ \text{M} - 0.0005\ \text{M})}\ln\frac{2 \times 0.0008 - 0.0005}{0.0008}$$

$$= 23.3\ \text{min}$$

Note: $t_{\frac{1}{2}}$ of a reaction of given order depends on composition of the reaction system.

15.87 Here, the concentration of NaOH (0.3 M) being much higher, the given rate law may be rewritten as

$$-\frac{d[C_2H_5NO_2]}{dt} = k'[C_2H_5NO_2]\ \text{where}\ k' = k[\text{NaOH}] = (0.3\ \text{M})\,k$$

This corresponds to first-order kinetics. Hence

$$k' = \frac{1}{t}\ln\frac{1}{a-x} = \frac{1}{t}\ln\frac{1}{1-x/a}$$

$$= \frac{1}{(1\ \text{min})}\ln\frac{1}{1-\frac{1}{100}} = 0.01\ \text{min}^{-1}$$

$$t_{\frac{1}{2}} = \frac{\ln 2}{k'} = \frac{0.693}{0.01\ \text{min}^{-1}} = 69.3\ \text{min}$$

☐ Here the second-order kinetics will be followed for which

$$t_{\frac{1}{2}} = \frac{1}{ka} \text{ since the reactant concentrations are in stoichiometric ratio}$$

$$= \frac{0.3\,M}{k'a}$$

$$= \frac{0.3\,M}{(0.01\,\text{min}^{-1})\,(0.001\,M)} = 3.0 \times 10^4 \text{ min}$$

15.88 (i) For the first interval $\Delta[A]/[A]$ is $\dfrac{0.312 - 0.223}{0.312}$ or 0.2852. Similarly, for the second, third and fourth intervals, the values of $\Delta[A]/[A]$ are 0.2825, 0.2875 and 0.2807 respectively. Since the fraction of A reacted in each interval is practically same and the intervals are all of equal duration, the reaction is of first order.

(ii) Here the conventional rate law is

$$-\tfrac{1}{2}\frac{d[A]}{dt} = k[A], \text{ where } k \text{ is conventional rate constant}$$

or

$$-\frac{d[A]}{dt} = k'[A], \text{ where } k' = 2k \text{ is practical rate constant}$$

Then

$$k' = \frac{1}{t} \ln \frac{a}{a-x} = \frac{1}{t} \ln \frac{1}{1-x/a} \quad \text{by Eq. (15.6a)}$$

$$= \frac{1}{10\,\text{min}} \ln \frac{1}{1-0.2852} \quad \text{considering the first interval}$$

$$= 3.3 \times 10^{-2} \text{ min}^{-1}$$

and

$$k = \tfrac{1}{2}k' = \frac{3.3 \times 10^{-2}}{2} \text{ or } 1.65 \times 10^{-2} \text{ min}^{-1}$$

(iii)

$$t_{\frac{1}{2}} = \frac{\ln 2}{k'}$$

$$= \frac{\ln 2}{3.3 \times 10^{-2} \text{ min}^{-1}} = 21 \text{ min}$$

Note: $t_{\frac{1}{2}} \neq \dfrac{\ln 2}{k}$. [Vide problem (15.76)]

15.89 Representing sucrose by S, $\Delta S/[S]$ in each interval is given in terms of angle of rotation (α) by

$$\frac{\Delta[S]}{[S]} = \frac{\Delta\alpha \text{ in the interval concerned}}{\Delta\alpha \text{ in time } \infty}$$

Then

for the first interval

$$\frac{\Delta[S]}{[S]} = \frac{24.09 - 20.40}{24.09 - (-10.74)} = 0.10594$$

for the second interval

$$\frac{\Delta[S]}{[S]} = \frac{20.40 - 17.10}{20.40 - (10.74)} = 0.10597$$

for the third interval

$$\frac{\Delta[S]}{[S]} = \frac{17.10 - 14.16}{17.10 - (10.74)} = 0.10560$$

As in the previous problem, $\Delta S/[S]$ for each interval is virtually same and the intervals are all of equal duration. Therefore, the reaction is kinetically of first order (first order wrt sucrose). Then

$$k = \frac{1}{t} \ln \frac{1}{1-x/a} \quad \text{as in the previous problem}$$

$$= \frac{1}{10 \text{ min}} \ln \frac{1}{1-0.1059} \quad \text{considering the first interval}$$

$$= 1.12 \times 10^{-2} \text{ min}^{-1}$$

The average lifetime $= \frac{1}{k} = \frac{1}{1.12 \times 10^{-2} \text{ min}^{-1}} = 89.3 \text{ min}$

Note: If time intervals are not equal, the order has to be decided showing the constancy of k through its calculation as done in problem 15.84.

15.90 Here it is convenient to represent the dimerisation of butadiene as

$$C_4H_6 \rightarrow \tfrac{1}{2}C_8H_{12}$$

If the reaction is of second order, then

$$k = \frac{1}{at} \cdot \frac{x}{a-x} \quad \text{Eq. (15.7b)}$$

$$= \frac{1}{P_o t} \cdot \frac{2(P_o - P_t)}{2P_t - P_o}$$

Because

$$\frac{x}{a} = \frac{P_o - P_t}{P_o - P_\infty}$$

$$= \frac{P_o - P_t}{P_o - P_o/2}, \quad \text{assuming that initially only } C_4H_4 \text{ is present}$$

$$= \frac{P_o - P_t}{P_o/2}$$

whence

$$\frac{x}{a-x} = \frac{2(P_o - P_t)}{2P_t - P_o}$$

Our task is to calculate k for each P_t and to judge its constancy.

15.91 It is convenient to represent the decomposition of N_2O as

$$N_2O \rightarrow N_2 + \tfrac{1}{2}O_2$$

Since the reaction is of first order

$$k = \frac{1}{t} \ln \frac{1}{1-x/a}$$

$$= \frac{1}{t} \ln \frac{1}{3(1 - P_t/P_\infty)}$$

Because

$$\frac{x}{a} = \frac{P_t - P_o}{P_\infty - P_o}$$

$$= \frac{P_t - \tfrac{2}{3}P_\infty}{P_\infty - \tfrac{2}{3}P_\infty}$$

(assuming that the reaction system initially contains only N_2O)

$$= \frac{3P_t}{P_\infty} - 2$$

To find the rate constant, we are to calculate it for each P_t and then to take average.

$$\text{Average life} = \frac{1}{k} \quad \text{since the reaction is of first order}$$

15.92 (i) $\quad p_{N_2O} + p_{N_2} = P_o$ where p denotes the initial partial pressure

$$\tfrac{3}{2}p_{N_2O} + p_{N_2} = P_\infty$$

whence $\quad p_{N_2O} = 2(P_\infty - P_o)$

Then, the mole fraction of N_2O

$$p_{N_2O}/P_o = 2(P_* - P_o)/P_o$$
$$= 2(59.8 - 41.1)/41.1 = 0.91$$

(ii) For first-order kinetics

$$k = \frac{1}{t} \ln \frac{a}{a-x}$$

$$= \frac{1}{t} \ln \frac{P_\infty - P_o}{P_\infty - P_t}$$

Because $\quad \dfrac{x}{a} = \dfrac{P_t - P_o}{P_\infty - P_o} \quad \text{but } P_\infty \neq \tfrac{3}{2}P_o$

whence $\quad \dfrac{a}{a-x} = \dfrac{P_\infty - P_o}{P_\infty - P_t}$

Now

for $P_t = 45.2 \quad k = \dfrac{1}{26.5 \text{ h}} \ln \dfrac{59.8 - 41.1}{59.8 - 45.2} = 9.33 \times 10^3 \text{ h}^{-1}$

for $P_t = 49.3 \quad k = \dfrac{1}{62.5 \text{ h}} \ln \dfrac{59.8 - 41.1}{59.8 - 49.3} = 9.23 \times 10^3 \text{ h}^{-1}$

k thus calculated is fairly constant showing that the reaction follows first-order kinetics.

$$\text{Average crate constant} = \frac{(93.3 + 92.3)\,10^{-3}}{2} \quad \text{or } 9.28 \times 10^{-3} \text{ h}^{-1}$$

(iii) The required time

$$t = \frac{1}{k} \ln \frac{P_\infty - P_o}{P_\infty - 1.25 P_o}$$

$$= \frac{1}{9.28 \times 10^{-3} \text{ h}^{-1}} \ln \frac{59.8 - 41.1}{59.8 - 1.25 \times 41.1} = 85.9 \text{ h}$$

Note: Calculation of k requires either P_o or P_∞ for pure N_2O (in the previous problem), but both P_o and P_∞ for impure N_2O.

15.93 For an nth order reaction

$$t_{\frac{1}{2}} \propto \frac{1}{C_o^{n-1}} \quad \text{vide problem 15.16}$$

or $\quad n = 1 + \dfrac{\log\,(t_{\frac{1}{2}}/t_{\frac{1}{2}}')}{\log\,(C_o'/C_o)}$

By this relation, n can be found out for each pair of the given $t_{\frac{1}{2}}$'s and then taking their average.

Alternatively, n can be found from $P_o/t_{\frac{1}{2}}$ value which are

P_o (mm)	300	200	100
$P_o/t_{\frac{1}{2}}$ (mm/min)	$\dfrac{300}{8.6}$	$\dfrac{200}{5.7}$	$\dfrac{100}{2.9}$
	$= 34.9$	$= 35.1$	$= 34.5$

It appears that $P_o/t_{\frac{1}{2}}$ is nearly constant, which corresponds to zero order.

$$k = \tfrac{1}{2}\frac{P_o}{t_{\frac{1}{2}}} \quad \text{the order being zero}$$

$$= \tfrac{1}{2}\,\frac{34.9+35.1+34.5}{3} \quad \text{or } 17.4 \text{ mm/min}$$

[**Note:** This alternative method is effective when order of the reaction is integer. If $t_{\frac{1}{2}}$ is independent of P_o the order will be one. Again if $P_o t_{\frac{1}{2}}$ is constant, the order will be two].

☐ At very high pressure, the surface of the catalyst is occupied almost completely, so that the reaction rate, which is determined by the surface density of the reactant, does not practica.iy change with further increase in pressure.

15.94 (a) Since the pressure of a gas falls during its decomposition, the given situation corresponds to a change in reaction order from zero at high pressure to one at low pressure. A probable rate law is

$$\text{rate} = \frac{k_1 P}{1+k_2 P}$$

Note: A rate law of this form corresponds to a catalysed reaction and not to any uncatalysed reaction.

(b) Here the rate law is assumed to be of the form as in (a).

At low pressure, when $k_2 P \ll 1$

$$\text{rate} = k_1 P$$

Then from the given data

$$k_1 = k' = 50 \text{ sec}^{-1}$$

At high pressure, when $k_2 P \gg 1$

$$\text{rate} = \frac{k_1}{k_2} = k$$

or $$k_2 = \frac{k_1}{k} = \frac{50 \text{ sec}^{-1}}{500 \text{ torr}\cdot\text{sec}^{-1}} = 0.1 \text{ torr}^{-1}$$

Then the required pressure is given by

$$100 \text{ torr sec}^{-1} = \frac{(50 \text{ sec}^{-1})P}{1+(0.1 \text{ torr}^{-1})P}$$

whence $\qquad P = 2.5$ torr

Note: Uncatalysed decomposition of HI follows second order kinetics. This is not unlikely in view of the fact that in presence of a catalyst, a reaction follows a different mechanism.

15.95 (i) Here both the reacting ions have positive charge. Since the product of their charges is positive, it follows from the Bronsted–Bjerrum Eq. (15.16) that the rate constant will increase with increase in ionic strength due to addition of KCl. The underlying physical reason is that the collisions between ions of like charges are enhanced by their atmospheres which tend to reduce electrostatic repulsion between them, the effect being higher at higher ionic strength.

(ii) The effect is qualitatively same as in (i), because although the reacting ions bear negative charges, the product of ionic charges is positive differing only in magnitude from that in (i).

(iii) Here the reacting ions being oppositely charged, the product of ionic charges is negative. Hence the effect of ionic strength will be opposite of that in (i).

15.96 (a) For (i) the product of charges of the reacting species being zero, the ionic strength will have practically no effect on the rate constant (by Bronsted–Bjerrum equation).

For (ii) the rate-determining step is the bimolecular reaction $H_3O_2^+ + I^-$, where the reacting species are oppositely charged. Then the rate constant for this reaction will be remarkably lowered with increase in ionic strength.

Note: In applying Bronsted–Bjerrum equation, we need to consier the charges of the reacting species involved only in the rate-determining step. Failure to recognise this step will lead to unreliable conclusion.

(b)
$$\log \frac{k}{k_o} = 1.02 Z_A Z_B \sqrt{I} \quad \text{by Eq. (15.16)}$$

$$= 1.02 (+1)(-1)\sqrt{0.0525}$$

whence
$$\frac{k}{k_o} = 0.5838$$

Therefore
$$k_o = \frac{k}{0.5838} = \frac{12.18 \text{ L}^2 \text{mol}^{-2}\text{min}^{-1}}{0.5838}$$

$$= 20.86 \text{ L}^2 \text{ mol}^{-2}\text{min}^{-1}$$

15.97 The desired function is provided by the Arrhenius equation in the following convenient form

$$\ln k = a - \frac{b}{T}$$

b is obtained using the relation $b = (\ln k_2 - \ln k_1) / \left(\dfrac{1}{T_1} - \dfrac{1}{T_2} \right)$ separately for three

temperature ranges – (300 K, 341 K), (341 K, 392 K) and (300 K, 392 K) and averaging the values thus found.

a is obtained using the relation $a = \ln k + b/T$ separately for 300 K, 341 K and 392 K (using the average value of b), and averaging the values thus found.

15.98
$$E_a = RT^2 \frac{d \ln k}{dT}$$

$$= RT^2 \frac{dk / k}{dT}$$

$$= (8.314 \text{ JK}^{-1}\text{mol}^{-1})(1000 \text{ EK})^2 \left(\frac{10}{100} \text{ K}^{-1} \right) = 8.314 \times 10^5 \text{ J mol}^{-1}$$

□ According to the collision theory concept of activation energy for a bimolecular reaction [vide problem 15.46(b)], the required fraction will be $e^{-\frac{E_a}{RT}}$. Because, this corresponds to the energy restricted to translational kinetic energy in one component for each of the two colliding molecules or in two components for a single molecule, i.e. two dimensional distribution of translational energy [vide problem 2.42(iv)].

15.99

$$k = \sigma^2 \left(\frac{8\pi RT}{N_A \mu} \right)^{\frac{1}{2}} e^{-\frac{E_a}{RT}} \quad \text{by Eq. (15.13b)}$$

Here

$$E_a = E_a' \text{ (experimental)} - \tfrac{1}{2} RT \quad \text{by Eq. (15.13c)}$$

$$= (88.1 \times 10^3 \text{ J mol}^{-1}) - \tfrac{1}{2}(8.314 \text{ JK}^{-1}\text{mol}^{-1})(500 \text{ K})$$

$$= 86021 \text{ J mol}^{-1}$$

Then

$$k \, (m^3 s^{-1}) = (0.35 \times 10^{-9} \text{ m})^2 \left[\frac{8(3.14)(8.314 \text{ JK}^{-1}\text{mol}^{-1})(500 \text{ K})}{(127.91 \times 10^{-3} \text{ kg mol}^{-1}} \right]^{\frac{1}{2}}$$

$$\times \exp\left[-\frac{86021 \text{ J mol}^{-1}}{(8.314 \text{ JK}^{-1}\text{mol}^{-1})(500 \text{ K})} \right]$$

□ When all the reactant molecules are activated, $e^{-\frac{E_a}{RT}}$ equals one and the rate is determined only by the frequency of binary collision. Then

$$\text{rate (mols per sec)} = \sigma^2 \left(\frac{8\pi RT}{N_A \mu} \right)^{\frac{1}{2}} C_{HI}^2 \quad \text{where } \mu \text{ is mass of each molecule}$$

$$= \sigma^2 \left(\frac{8\pi RT}{N_A \mu} \right)^{\frac{1}{2}} \left(\frac{P}{kT} \right)^2$$

$$= (0.35 \times 10^{-9} \text{ m})^2 \left[\frac{8(3.14)(8.314 \text{ JK}^{-1}\text{mol}^{-1})(500 \text{ K})}{(127.91 \times 10^{-3} \text{ kg mol}^{-1})} \right]^{\frac{1}{2}}$$

$$\times \left[\frac{1.013 \times 10^5 P_a)}{(1.381 \times 10^{-23} \text{ JK}^{-1})(500 \text{ K})} \right]^2$$

Note: When all the reactant molecules are activated the reaction rate and hence k does not become infinity. This is true even in case of unimolecular reactions (vide problem 15.59).

15.100

$$E_a = \frac{R \ln \dfrac{k_2}{k_1}}{\dfrac{1}{T_1} - \dfrac{1}{T_2}} \quad \text{by Eq. (15.12b)}$$

$$= \frac{(8.314 \text{ JK}^{-1} \text{ mol}^{-1}) \ln\left(\dfrac{0.950}{2.45 \times 10^{-4}} \right)}{\dfrac{1}{(302 + 273) \text{ K}} - \dfrac{1}{(508 + 273) \text{ K}}}$$

$$= 149823 \text{ JK}^{-1}\text{mol}^{-1}$$

$$\Delta H^{\circ\neq} = E_a - nRT \quad \text{vide problem 15.50}$$
$$= E_a - 2RT \quad \text{since the reaction is of order } n = 2$$
$$= 149824 - 2\,(8.314\,\text{JK}^{-1}\text{mol}^{-1})\,(508 + 273)\,\text{K}$$
$$= 136837\,\text{JK}^{-1}\text{mol}^{-1}$$

15.101 (i) Comparing the given expression of k with the Arrhenius Eq. (15.12b), we have

$$E_a/R = 1203\,K$$

or
$$E_a = 1203\,K \times R$$
$$\Delta H^{\neq} = E_a - 2RT$$

(since the reaction is of order two as implied by the unit of k)

$$= (1203\,K - 2 \times 298\,K)\,R$$
$$= (607\,K)\,(8.314\,\text{JK}^{-1}\text{mol}^{-1})$$
$$= 5.05\,\text{kJ mol}^{-1}$$

(ii) A in Arrhenius equation is

$$A = e^n \frac{RT}{N_A h C_o^{n-1}}\, e^{\frac{\Delta S^{\neq}}{R}} \quad \text{by Eq. (15.14c)}$$

Then

$$\Delta S^{\neq} = R\ln \frac{A N_A h C_o}{e^2 RT} \quad \text{since } n = 2$$

$$= (8.314\,\text{JK}^{-1}\text{mol}^{-1})\ln \frac{(4.6 \times 10^{12} \times 10^{-3}\,\text{mol}^{-1}\text{s}^{-1})(6.02 \times 10^{23}\,\text{mol}^{-1}) \times (6.63 \times 10^{-34}\,\text{JS})(10^3\,\text{mol m}^{-3})}{(2.72)^2\,(8.314\,\text{JK}^{-1}\text{mol}^{-1})(298\,K)}$$

$$= -19.15\,\text{JK}^{-1}\text{mol}^{-1}$$

(iii)
$$\Delta G^{\neq} = \Delta H^{\neq} - T\Delta S^{\neq}$$
$$= 5.05\,\text{kJ mol}^{-1} - (298\,K)\,(-19.15 \times 10^{-3}\,\text{kJK}^{-1}\,\text{mol}^{-1})$$
$$= -0.657\,\text{kJ mol}^{-1}$$

☐ A negative value of $\Delta S^{\neq}$ indicates that the activated complex is more organised than the sum of its separate components. In other words, $\Delta S^{\neq}$ = negative reflects the probability of collision between properly oriented reactant molecules, which is more negative than that associated with collisions in general. In bimolecular gas phase reactions, $\Delta S^{\neq}$ is usually negative due to transformation of translational and rotational degrees of freedom into vibrational degrees of freedom which do not usually have significant contributions to entropy.

Note: $\Delta S^{\neq}$ is related to the steric factor (P); the more positive the $\Delta S^{\neq}$, the larger is P. If $\Delta S^{\neq}$ is positive, the activated complex is more probable and the reaction is faster than normal. If $\Delta S^{\neq}$ is negative, the activated complex is less probable and the rate is slower. Some times $\Delta S^{\neq}$ is virtually zero as in case of combination of atoms; this implies that the activated complex is similar in structure to the product.

15.102 (i) Since the reaction is of first order

$$k = \frac{0.693}{t_{\frac{1}{2}}} \quad \text{where} \quad \frac{d[N_2O_5]}{dt} = k[N_2O_5]$$

$$= \frac{0.693}{\text{Antilog}\left[-13.16 + \dfrac{5463.21}{25 + 273}\right]} = 4.65 \times 10^{-6}\,\text{sec}^{-1}$$

Then the conventional rate constant

$$= \tfrac{1}{2}k \text{ which refers to conventional rate } -\tfrac{1}{2}\frac{d[N_2O_5]}{dt}$$

$$= \frac{4.65 \times 10^{-6}}{2} \text{ or } 2.32 \times 10^{-6} \text{ sec}^{-1}$$

Again

$$t_{\frac{1}{2}} = \frac{0.693}{k} = \frac{0.693}{Ae^{-\frac{E_a}{RT}}}$$

Then

$$\log t_{\frac{1}{2}} = \frac{\ln t_{\frac{1}{2}}}{2.303} = \frac{1}{2.303}\ln\frac{0.693}{A} + \frac{E_a}{2.303R}\cdot\frac{1}{T}$$

Comparing this with the given expression for $t_{\frac{1}{2}}$ we have

$$\frac{1}{2.303}\ln\frac{0.693}{A} = -13.16 \text{ or } A = 0.693 \times e^{2.303 \times 13.16}\text{s}^{-1}$$

and

$$\frac{E}{2.303R} = 5463.21 \text{ K, or } E = 2.303\,(5463.21 \text{ K})\,(8.314 \text{ JK}^{-1}\text{mol}^{-1})$$

(ii) The required fraction of $N_2O_5 = -\dfrac{d[N_2O_5]}{[N_2O_5]}\Big/dt = k$

To find the required time, it will be convenient to write the chemical equation as follows with the concentration of each species indicated under it

$$N_2O_5 \rightarrow 2NO_2 + \tfrac{1}{2}O_2 \quad \text{Total conc}$$
$$a-x \qquad\quad 2x \qquad \tfrac{1}{2}x \qquad a+\tfrac{3}{2}x$$

or

$$\frac{x}{a} = \frac{2}{3}$$

Then

$$t = \frac{1}{k}\ln\frac{1}{1-x/a}$$

(iii)

$$k = \frac{0.693}{t_{\frac{1}{2}}}$$

or

$$\ln k = \ln 0.693 - \ln t_{\frac{1}{2}}$$

or

$$\frac{dk}{k} = -\frac{dt_{\frac{1}{2}}}{t_{\frac{1}{2}}}$$

Then the error in k will be negative of that in $t_{\frac{1}{2}}$ i.e. -1.0%.

Again,

$$\ln t_{\frac{1}{2}} = \ln\frac{0.693}{A} + \frac{E_a}{RT}$$

or

$$\frac{dt_{\frac{1}{2}}}{t_{\frac{1}{2}}} = \frac{dE_a}{RT}$$

Then error in $E_a = RT \times$ fractional error in $t_{\frac{1}{2}}$.

15.103 (i) Rewriting the given rate law, we have

$$\frac{d(\text{Ester})}{dt} = k'[RCO_2H][R'OH]; \text{ where } k' = k[H_3O^+]$$

This corresponds to second-order kinetics. Then the apparent rate constant k' is given by

$$k' = \frac{1}{at}\left(\frac{x/a}{1-x/a}\right) \quad \text{since } [RCO_2H]_o = [R'OH]_o$$

$$= \frac{1}{(0.01 \text{ M})(30 \text{ min})}\left[\frac{30/100}{1-30/100}\right]$$

$$= 1.43 \text{ M}^{-1}\text{min}^{-1}$$

(ii) True rate constant

$$k = \frac{k'}{[H_3O^+]}$$

$$= \frac{1.43 \text{ M}^{-1}\text{min}^{-1}}{10^{-2}\text{M}} = 1.43 \times 10^2 \text{ M}^{-2}\text{min}^{-1}$$

(iii) Energy of activation (E_a) is given by

$$\frac{E_a}{R}\left[\frac{1}{T_1}-\frac{1}{T_2}\right] = \ln\frac{k_2'}{k_1'}$$

$$= \ln\frac{t_1}{t_2}$$

since the time taken for a definite fraction of a reaction varies inversely with the rate constant $\left[\int_0^t kdt = -\int_{c_o}^{\phi C_o}\frac{dc}{C^n} = \text{constant for fixed } n \text{ and } \phi\right]$

or

$$E_a = \frac{R\ln\frac{t_1}{t_2}}{\frac{1}{T_1}-\frac{1}{T_2}}$$

$$= \frac{8.31 \text{ JK}^{-1}\text{mol}^{-1}\ln\frac{1}{2}}{\frac{1}{298 \text{ K}}-\frac{1}{188 \text{ K}}} = 32.3 \text{ kJ mol}^{-1}$$

UNIVERSITY QUESTIONS

15.1 For the elementary reaction $A + B \xrightarrow{k} 2C$, express $d[A]/dt$ and $d[C]/dt$ in terms of

(i) the reaction rate (r).

(ii) rate constant (k) and the molar concentration of A and B.

(Calcutta BSC(H), 2013)

15.2 The rate of a reaction is 1.0×10^{-5} mol dm^{-3}s^{-1}. Calculate the rate in molecule cm^{-3}s^{-1} unit. (Burdwan BSc(H), 2002)

15.3 Comment on the following statements:

(i) The order of a reaction is always integral.

(ii) The stoichiometry of a reaction indicates the order of the reaction.

(Calcutta BSc(H), 1994)

15.4 What is "bottle neck principle" in chemical kinetics?   (Burdwan BSc(H), 2011)

15.5 Can a reaction rate be independent of concentration? If yes, cite an example. Does such a reaction go to completion in a finite time?   (Calcutta BSc(H), 1996)

15.6 A zero order reaction cannot be a single-step reaction. Justify.

(Calcutta BSc(H), 2012)

15.7 What is the order of the reaction $A + 2B = 2C$, given that

C_A	C_B	Rate
0.10	0.10	0.12
0.10	0.20	0.24
0.20	0.20	0.24

(Burdwan BSC(H), 2007)

15.8 Write the differential rate law for the (elementary) reaction $2A + B \rightarrow P$ and turn it into a pseudo first-order and pseudo second-order rate law. Give an example of pseudo first-order reaction. What is the unit of its rate constant? How does it differ from the second-order rate constant?   (Burdwan BSc(H), 1993)

15.9 The rate constant of a reaction has the unit mol $dm^{-3}sec^{-1}$. What is the order of the reaction.   (Burdwan BSc(H), 2005)

15.10 The reaction in aqueous solution involves the following two elementary steps:

Step 1 $Hg_2^{2+} \underset{k_{-1}}{\overset{k_1}{\rightleftharpoons}} Hg^{2+} + Hg$ rapid equilibrium

Step 2 $Hg + Tl^{3+} \overset{k_2}{\longrightarrow} Hg^{2+} + Tl^{1+}$ slow

Apply the rate determining step approximation to show that the reaction is of negative order with respect to Hg^{2+}.   (Calcutta BSc(H), 2008)

15.11 Deduce the rate law expressions for the gas phase reaction of H_2 and I_2 to form HI at 400°C corresponding to each of the following mechanisms:

(i) $H_2 + I_2 \overset{k}{\longrightarrow} 2HI$ (single step)

(ii) $I_2 \underset{k_{-1}}{\overset{k_1}{\rightleftharpoons}} 2I$

$I + H_2 \underset{k_{-2}}{\overset{k_2}{\rightleftharpoons}} IH_2$

$IH_2 + I \overset{k_2}{\longrightarrow} 2HI$ (slow step)

Can the derived rate laws distinguish the given two mechanisms? Comment.

(Calcutta BSc(H), 2007)

15.12 Show that the ratio $t_{\frac{1}{2}} / t_{\frac{3}{4}}$ of any nth order reaction ($n \neq 1$), with equivalent initial concentration of the reactants, can be written as a function of n alone.

(Calcutta BSC(H), 2006)

15.13 A unimolecular gaseous reaction shows second order kinetic at low pressure. Explain using Lindemann mechanism. Indicate when such a reaction would follow first order kinetics.   (Calcutta BSc(H), 2008)

15.14 (a) Write down the overall rate expression for the gas phase reaction $H_2 + Br_2 = 2HBr$. Explain clearly that chain inhibition would involve the reaction $H + HBr \rightarrow H_2 + Br_2$ and not $Br + HBr \rightarrow Br_2 + H$.   (Burdwan BSc(H), 1996)

(b) What will be the order of the eaction in the initial stage.

(Jadavpur BSc(H), 2013)

(c) The kinetics of two thermal gaseous reactions $H_2 + I_2 \rightarrow 2HI$ and $H_2 + Br_2 \rightarrow 2HBr$ are found to be different. Account for the difference from the view point of mechanism.   (Burdwan BSc(H), 1994)

15.15 Find out the order of reaction $A \rightarrow$ products in the following cases:

(i) $t_{\frac{1}{2}} \times a =$ constant (a being the initial concentration of the reactant).

(ii) The reaction is complete in time $2t_{\frac{1}{2}}$. (Calcutta BSc(H), 2009)

(iii) On doubling a, $t_{\frac{1}{2}}$ is doubled. (Calcutta BSc(H), 2010)

15.16 The reaction $A \rightarrow$ products gives a linear plot of $1/[A]$ vs t, of intercept 1000 L mol^{-1} and slope 3×10^{-2} L mol^{-1}s^{-1}. What is the order of the reaction? Calculate $t_{\frac{1}{2}}$ of this reaction. (Calcutta BSc(H), 2008)

15.17 (a) A reaction $A \rightarrow$ Products, is of zero order with respect to A. Two experiments are performed with initial concentrations $[A]_o = 0.1$ M and 0.05 M respectively. Sketch the plots to show the variation of $[A]$ with t. (Burdwan BSc(H), 2008)

(b) Show that a zero order reaction can lead to completion.

(Burdwan BSc(H), 2008)

15.18 A chemical reaction is known to be of zero order with $k = 5 \times 10^{-8}$ mol L^{-1}s^{-1}. How long does it take to change the concentration from 4×10^{-2} mol L^{-1} to 2×10^{-4} mol L^{-1}? (Burdwan BSc(H), 1995)

15.19 Show that for a reaction obeying the rate equation $-\dfrac{dc}{dt} = kC^n$, the time for completion is always infinite except for $n < 1$. (Burdwan BSc(H), 1999)

15.20 What is meant by third order reaction? Give one example. A third order reaction is 50% complete in 100 s. Calculate the time for 75% and 100% completion. (Burdwan BSc(H), 2000)

15.21 The rate constant of a reaction is 0.05 s^{-1}. Calculate the time for half reaction when the initial concentration is 0.01 M. (Burdwan BSc(H), 2001)

15.22 For a first order process $A \rightarrow$ Products, show that the number of molecules of A present at a time t is given by

$$N_t = N_o \left(\frac{1}{2}\right)^{\frac{t}{t_{\frac{1}{2}}}}$$

where N_o is the number of molecules at $t = o$ and $t_{\frac{1}{2}}$ is the time for half-decomposition. (Burdwan BSc(H), 1994)

15.23 Prove that the average life expectancy of the molecules decomposing by a first-order reaction is equal to $1/k$ [k is rate constant). (Calcutta BSc(H), 2004)

15.24 Show that for a first-order reaction, the time required for 99.9% completion of the reaction is almost 10 times the time for 50.0%. Explain. (Calcutta BSc(H), 2007)

15.25 At 25°C the half-life period for the decomposition of N_2O_5 is 2.05×10^4 s and is independent of the initial concentration of N_2O_5.

(i) What is the order of the reaction.

(ii) What length of time is required for 80% of N_2O_5 to decompose.

(Calcutta BSc(H), 2004)

15.26 Decomposition of a substance A is studied at two different initial concentrations a_1 and a_2 where $a_1 = 3a_2$. If the observed half-lives t_1 and t_2 follow the ratio $t_2:t_1 = 2:1$, find the order of decomposition process. (Burdwan BSc(H), 1997)

15.27 The decomposition of ammonia catalysed by tungsten surface was studied at 1100°C and the following data were obtained for the time of half decomposition $(t_{\frac{1}{2}})$ with different initial pressures (P_o) of ammonia

P_o/mmHg	300	200	100
$t_{\frac{1}{2}}$ / min	8.6	5.7	2.9

Find the order of the reaction and the rate constant. (Calcutta BSc(H), 1977)

15.28 A gas decomposes according to second order kinetics. When the initial pressure is 500 torr, 40% decomposition occurs in 30 min. Find out the time required for 75% decomposition of the gas and the value of the rate constant.

(Calcutta BSc(H), 2014)

15.29 Decomposition of a gas at an initial pressure of 600 mmHg was studied in a closed vessel at a certain temperature. The gas was found to be 50% decomposed in 30 min and 75% decomposed in 90 min. Find the order and rate constant.

(Jadavpur BSc(H), 2013)

15.30 Sucrose is hydrolysed to glucose and fructose in presence of H^+ as catalyst. At any initial concentration of sucrose, the half-lives at pH = 5 and pH = 4 are 500 min and 50 min respectively. Find out the value of exponents a and b in the rate law expession

$$-d[\text{Sucrose}]/dt = k[\text{Sucrose}]^a [H^+]^b \qquad \text{(Calcutta BSc(H), 2013)}$$

15.31 (a) Suppose $[A]_o$ and $[B]_o$ are the initial concentrations of the reactants A and B respectively in a second order elementary reaction, $A + B \rightarrow$ Product(s). Find out the integrated rate question for this reaction when (i) $[B]_o = [A]_o$ and (ii) $[B]_o \neq [A]_o$. Show that the second result may directly be reduced to the first under the limit $[B]_o \rightarrow [A]_o$. (Burdwan BSc(H), 2009)

(b) When will the reaction follow a first order kinetics. (Jadavpur BSc(H), 2013)

15.32 A solution containing equal concentrations of ethyl acetate and NaOH is 25% saponified in 5 min. What will be the percent saponification after 10 min?

(Vidyasagar BSc, 2002)

15.33 Methyl acetate was hydrolysed to acetic acid and methanol using 1(N) HCl as catalyst. Aliquots of equal volume were removed at intervals and titrated with a solution of NaOH.

Time (min)	0	5	15	∞
NaOH (cm³)	24.0	27.0	31.4	40.0

Show that it is a first order reaction and evaluate the average life period of the reaction in minutes. (Calcutta BSc(H), 2012)

15.34 In kinetic study of a reaction in aqueous medium, it is customary to stop the reaction by adding **large** excess of **cold** water. Explain the underlined words.

(Burdwan BSc(H), 1994)

15.35 In the reaction $(COOH)_2 \rightarrow CO + CO_2 + H_2O$, the following results were obtained:

Time (min)	0	300	450	600	1200
Vol of KMnO₄ (mL) required to titrate oxalic acid	22.0	17.0	15.0	13.4	7.9

Show that the reaction is of first order, and determine the value of the velocity constant. (Bangalore BSc, 1978)

15.36 A dilute solution of cane sugar was hydrolysed into glucose and fructose by addition of dilute HCl. The progress of the reaction was followed in a polarimeter tube by observing the angle of rotation at different times. The results are:

t (min)	0	10	20	30	80	∞
Angle of rotation (degree)	32.4	28.8	25.5	19.6	10.3	14.1

Show that the reaction is of first order and calculate the half-life. (Delhi BSc, 1982)

15.37 The composition of the gas phase reaction $2A \to B$ was monitored by measuring the total pressure as a function of time. The following are the results:

Time (sec)	0	100	200	300	400
Pressure (mmHg)	400	322	288	268	256

What is the order of the reaction, and what is the value of the rate constant? At what time will the reaction be 99.99% complete? (Calcutta BSc(H), 1990)

15.38 (a) State the Arrhenius equation showing the variation of k with temperature. (Burdwan BSc(H), 2000)

(b) What is the feature that distinguishes pre-exponential factor in Arrhenious equation from that obtained in collision theory rate law? (Calcutta BSc(H), 2007)

(c) What is the unit of pre-exponential factor for a first order reaction? (Calcutta BSc(H), 2000)

15.39 What value of rate constant (k) is predicted by Arrhenius equation for k by $T \to \infty$? Is this result physically reasonable? (Burdwan BSc(H), 2009)

15.40 What are the Arrhenius parameters for a reaction that follows the equation

$$k = aT^m e^{-\frac{E_0}{RT}}$$

If k is rate constant for a second order reaction, find the unit of a. (Burdwan BSc(H), 2008)

15.41 The rate constant for a reaction depends on temperature as $k = B\sqrt{T}\, 10^{-\frac{C}{T}}$. Express the energy of activation in terms of C. (Burdwan BSc(H), 2000)

15.42 If all the collisions were effective, a $10°$ rise in temperature would have increased the rate of a simple bimolecular reaction by about 1% at ordinary temperature. Explain. (Burdwan BSc(H), 1998)

15.43 A reaction with activation energy $E = 20$ kJ mol^{-1} is taking place at temperature $27°C$. Calculate the percentage of increase in rate constant if temperature is increased by $1°C$. (Burdwan BSc(H), 2008)

15.44 The first order rate constant (k) of a reaction $A \to$ Products follows the equation

$$\log k\ (\text{min}^{-1}) = 33.91 - \frac{18000}{T}$$

(a) How long will it take for one mole of A to decompose 75% at $227°C$?

(b) Calculate the energy of activation.

(c) Calculate the frequency factor (A) in Arrhenius equation $k = Ae^{-\frac{E}{RT}}$.

(Burdwan BSc(H), 2009)

15.45 Calculate the activation energy of a reaction whose rate at $27°C$ gets doubled for $10°$ rise in temperature. (Mysore BSc, 2009)

15.46 For a $10°C$ rise in temperature, the rate constant doubles for reaction I, trebles for reaction II. If the two reactions have comparable pre-exponential factors, what is the ratio of their activation energies? (Calcutta BSc(H), 2010)

15.47 The values of k observed for the result $2HI \to H_2 + I_2$ at $356°C$ and $443°C$ are 3.02×10^{-5} L mol^{-1}s^{-1} and 2.53×10^{-3} L mol^{-1}s^{-1} respectively. Calculate the activation energies for the forward reaction (E_f) and reverse reaction (E_r), if $\Delta H = 16.32$ kJ mol^{-1}. (Jadavpur BSc(H), 2013)

15.48 A certain first order reaction is 20% complete in 15 minutes at 27°C, but for the same extent of reaction at 37°C only 5 minutes are required. Calculate activation energy of the reaction.   (Calcutta BSc, 2014)

15.49 A first order reaction is 20% complete in 10 minutes at 25°C and 40% complete in 12 minutes at 40°C. Calculate the energy of activation of the reaction. Find also the value of k at very high temperature.   (Burdwan BSc, 2012)

15.50 (a) The activation energy of a reaction may be zero. Comment.

(Burdwan BSc(H), 1998)

(b) The rate constant of a reaction is experimentally found to decrease with increase in temperature. Is there any contradiction with Arrhenius rate equation? Justify your answer.   (Calcutta BSc(H), 2007)

15.51 The reaction $2NO + O_2 \xrightarrow{k} 2NO_2$ proceeds through the following steps

$$2NO \underset{k_{-1}}{\overset{k_1}{\rightleftharpoons}} 2N_2O_2 \text{ (fast)}$$

$$N_2O_2 + O_2 \xrightarrow{k_2} 2NO_2 \text{ (slow)}$$

Find the order of the reaction. The overall rate constant (k) is found to decrease with rise in temperature. Explain.   (Calculate BSc(H), 2003)

15.52 (a) Justify/criticise. Rate of a reaction increases with temperature primarily because of increased kinetic energy of reactant molecules.   (Burdwan BSc(H), 1997)

(b) Rate constant for the reaction $3CH_2 \rightarrow C_3H_6$ is almost independent of temperature. Explain.   (Burdwan BSc(H), 1996)

15.53 Explain the concept of activation energy in the light of the theory of transition state.

(Calculate BSc(H), 1989)

15.54 Transition state complex does not always decompose into products. Justify or criticise with a suitable example.   (Burdwan BSc(H), 2010)

15.55 Write down the expression for the rate constant of a reaction according to the transition state theory and explain the terms involved. How is the enthalpy of activation related to the energy of activation (E_a)?   (Calcutta BSc(H), 2010)

15.56 What is the entropy of activation? How is it related to the frequency facor? Why is it usually negative?   (Burdwan BSc(H), 2010)

15.57 Justify/criticise. A high positive value of $\Delta S^{\neq}$ means a high energy of activation.

(Burdwan BSc(H), 1998)

15.58 Two reactions have the same energy of activation but the entropies of activation differ by 10 SI units. Compare the rate constants for the two reactions.

(Burdwan BSc(H), 2010)

15.59 A substance decomposes by two competing paths with rate constants k_1 and k_2. If at 10°C the value of $k_1/k_2 = 10$ and at 40°C, $k_1/k_2 = 0.1$, what is the difference of activation energies for these two reactions?   (Jadavpur BSc(H), 2000)

15.60 The elementary bimolecular gas phase reaction $CO + O_2 \rightarrow CO_2 + O$ has an activation energy 214.2 kJ mol^{-1} at 2700 K. Estimate the rate constant for the reaction by applying the collision theory of reaction rate at 2700 K. The molecular diameters of O_2 and CO are 3.6×10^{-10} m and 3.7×10^{-10} m respectively. Also estimate the probability (or steric) factor for the reaction if the observed rate constant is 3.45×10^5 Ls^{-1}mol^{-1} at 2700 K.   (Calcutta BSc(H), 2009)

15.61 The frequency factor for the conversion of cyclopropane to propylene at 500°C is 1.48×10^{15} min^{-1} and its energy of activation is 65 kCal mol^{-1}. Calculate the rate constant, $\Delta S^{\neq}$ and $\Delta H^{\neq}$ for the reaction at 500°C. Express your answer in SI unit.   (Jadavpur BSc(H), 2000)

15.62 (a) Consider the simplest consecutive reaction involving first order steps $A \xrightarrow{k_1} B \xrightarrow{k_2} C$, where only A is present initially. Show schematically the variation of $[A]$, $[B]$ and $[C]$ with time.   (Burdwan BSc, 1997)

(b) Show that the time required for the formation of maximum amount of B is independent of the initial conc of A.   (Burdwan BSc(H), 2007)

(c) $[C]$ at any time t is given by

$$[C] = [C]_o \left(1 - \frac{k_2 e^{-k_1 t}}{k_2 - k_1} + \frac{k_1 e^{-k_2 t}}{k_2 - k_1} \right)$$

Explain the significance of the result when $k_1 \ll k_2$.   (Burdwan BSc(H), 1995)

(d) Assume a steady-state approximation for the above problem to obtain expressions for $[B]$ and $[C]$. Explain clearly when this approximation is valid.
                                                                (Burdwan BSc(H), 1995)

15.63 Consider the following parallel reaction $A \xrightarrow{k_1} B,\ A \xrightarrow{k_2} C$

(i) Show that $[A] = [A]_o e^{-(k_1 + k_2)t}$

$$t_{\frac{1}{2}} = \frac{0.693}{k_1 + k_2}$$

$$[B]/[C] = \frac{k_1}{k_2} \text{ at any point of time}$$

(ii) If the activation energies for the two reactions are E_1 and E_2 respectively, show that the observed activation energy (E_a) for the disappearance of A is given by

$$E_a = \frac{k_1 E_1 + k_2 E_2}{k_1 + k_2}$$

(iii) If $[B]_o = [C]_o = 0$ and $k_1/k_2 = 2$, plot $[A]$, $[B]$ and $[C]$ as functions of time on the same graph.   (Burdwan BSc(H), 2005)

15.64 For a complex reaction, the observed rate constant is given by $k_{obs} = k_1 + 2k_2 + \frac{1}{2}k_3$, where k_1, k_2 and k_3 are respective rate constants of the elementary steps in the proposed mechanism. Find an expression of the overall activation energy (E_{obs}) in terms of the respective activation energies E_1, E_2 and E_3 of the elementary steps.

15.65 How can you experimentally distinguish the following mechanisms:

(i) $A \xrightarrow{k_1} B + C$ (ii) $A \overset{k_2 \to B}{\underset{k_3 \to C}{\diagdown}}$ (Calcutta BSc(H), 2007)

15.66 Draw a schematic diagram showing energy as a function of reaction coordinate and indicate on the plot the following:

(i) Energy of activation for forward and reverse reaction.

(ii) Role of a positive catalyst.

(iii) A thermodynamically controlled process and a kinetically controlled process.
                                                                (Burdwan BSc(H), 1999)

15.67 Slowest step in a multi-step reaction is the rate-determining step. Discuss, giving one example.   (Burdwan BSc(H), 1997)

15.68 Consider the parallel reaction $A \overset{k_1}{\underset{k_2}{\rightarrow}} \begin{matrix} B \\ C \end{matrix}$

In an experiment, it was found that 60% decomposition of A takes place in 20 minutes and analysis of product showed that B and C are in 3:1 ratio. Calculate k_1 and k_2. (Calcutta BSc(H), 2014)

15.69 A gas phase reaction at a particular temperature proceeds by parallel reaction

$$A \overset{k_1 \to B}{\underset{k_2 \searrow C}{}} \qquad \begin{matrix} k_1 = 3.75 \text{ s}^{-1} \\ k_2 = 4.65 \text{ s}^{-1} \end{matrix}$$

(i) Calculate the maximum percentage of yield of C.

(ii) Calculate the half-life time of A. (Calcutta BSc(H), 2008)

15.70 The kinetics of the reaction $A^+ + B^- \overset{k}{\longrightarrow}$ Products is studied in (i) pure water and (ii) 1 M KCl solution. In which case will the value of k be greater? Will the energy of activation be significantly different in the two cases? (Burdwan BSc(H), 1994)

15.71 Predict with reason the effect of increase in ionic strength on the rate constant for each of the following reactions:

(i) $S_2O_8^{2-} + I^-$ (Calcutta BSc(H), 2008)

(ii) $CH_3COOC_2H_5 + OH^-$ (Calcutta BSc(H), 2008)

(iii) Inversion of cane sugar. (Burdwan BSc(H), 1995)

15.72 (a) A catalyst is a substance which modifies the rate of a chemical reaction without taking part in it. How far is the statement justified. (Burdwan BSc(H), 1995)

(b) Does a catalyst affect the free energy change (ΔG) of a reaction? (Burdwan BSc(H), 2008)

(c) Heat of reaction remains the same whether a catalyst is used or not. Comment. (Burdwan BSc(H), 1999)

(d) A catalyst that increases the rate of a forward reaction also increases the rate of the backward reaction to the same extent (proportion). Explain. (Burdwan BSc(H), 1995)

(e) A positive catalyst usually makes the reaction proceed via a path having lower energy of activation. Explain. (Burdwan BSc(H), 1994)

(f) A negative catalyst acts by making the reaction to proceed by an alternative path of higher energy. Justify/criticise. (Burdwan BSc(H), 1999)

15.73 A decomposition reaction proceeds as follows

$$A + M \underset{k_{-1}}{\overset{k_1}{\rightleftharpoons}} B + C + M$$

$$C + A \overset{k_2}{\longrightarrow} 2B$$

Find $d[B]/dt$, assuming a steady-state approximation for C. (Burdwan BSc(H), 1998)

15.74 The hydrolysis of a substrate is simultaneously catalysed by H_3O^+ and OH^- ions and also occurs spontaneously. The reaction is first order with respect to all species. Write down the expression for the rate constant. Hence show that the rate is maximum when

$$[H_3O^+] = \left(\frac{k_{OH^-}}{k_{H_3O^+}} \cdot K_w \right)^{\frac{1}{2}} \qquad \text{(Calcutta BSc(H), 2008)}$$

15.75 The rate constant of a reaction catalysed by H_3O^+ and OH^- ions, and also occurring spontaneously is

$$k = k_o + k_{H_3O^+}[H_3O^+] + k_{OH^-}[OH^-]$$

where k_o = rate constant for the uncatalysed reaction.
What should be the plot of $\log k$ vs pH for
(i) sufficiently acidic solution
(ii) sufficiently basic solution
(iii) solutions when there is no catalysis in either the acidic or basic region
(Burdwan BSc(H), 1992)

15.76 H_3O^+ acts as a homogeneous catalyst in the reaction of inversion of sucrose. At pH 3 the reaction in a aqueous solution proceeds with a constant half-life of 50 minutes. What value of the half-life would you expect at pH = 4?
(Calcutta BSc(H), 1997)

15.77 At a given temperature, the rate of hydrolysis of an ester, catalysed by strong acid is almost doubled when pH is changed from 0.80 to 0.50. Justify, whether this is an example of homogeneous catalysis or not.   (Calcutta BSc(H), 2009)

15.78 A certain reaction can proceed in the absence as well as in the presence of a catalyst. The rate constants for the two mechanisms are k_a and k_c respectively. If $\Delta S_a^{\neq}$ is 41.84 JK mol^{-1} greater than $\Delta S_c^{\neq}$ and $\Delta H_a^{\neq}$ is 20.92 kJ mol^{-1} greater than $\Delta H_c^{\neq}$, show which rate constant is larger at 298 K and in what proportion?   (Jadavpur BSc(H), 2013)

15.79 (a) Using the Michaelis–Menten mechanism of enzyme catalysis show that the initial rate of such reaction is first order with respect to substrates at low [s] and zero order at high [S].   (Burdwan BSc(H), 2010)
(b) Comment on the order of the reaction as [S] is varied.   (Burdwan BSc(H), 2006)
(c) Under what condition an enzyme catalysed reaction follows second order kinetics?   (Nagpur BSc, 2000)

15.80 Show that the Michaelis–Menten rate law (v_0) for an enzyme catalysed reaction is given by

$$\frac{1}{v_0} = \frac{K_m}{v_{max}} \cdot \frac{1}{[S]_o} + \frac{1}{v_{max}}$$

where $v_{max} = \lim v_0$ and the other terms have their usual significance.
What is turnover number and what are its typical values?
(Burdwan BSc(H), 1993)

15.81 Show how do you calculate the turnover number and Michaelis constant.
(Burdwan BSc(H), 2010)

15.82 The slope and intercept of the plot of $1/v_0$ against $1/[S]_0$ are 3.5×10^2 s and 5×10^4 mol^{-1} Ls respectively, where v_0 and $[S]_0$ are the initial rate and the initial substrate concentration of an enzyme catalysed reaction obeying Michaelis–Menten kinetics. Estimate the K_m and the turnover number when the initial enzyme concentration $[E]_0 = 2.5 \times 10^{-9}$ mol L^{-1}.   (Calcutta BSc(H), 2008)

KEY TO UNIVERSITY QUESTIONS

15.1 (i) $-d[A]/dt = r$
$d[C]/dt = 2r$

(ii) $-d[A]/dt = k[A][B]$
       $d[C]/dt = 2k[A][B]$

15.2   rate $= 1.0 \times 10^{-5}$ mol dm^{-3}s^{-1}
                $\equiv 1.0 \times 10^{-5}$ (6.022 $\times 10^{23}$ molecules) (10 cm)$^{-3}$s^{-1}
                $= 6.022 \times 10^{15}$ molecules cm^{-3}s^{-1}

15.3 (i) True only for an elementary reaction.

(ii) There is no necessary connection between the stoichiometry and order of a reaction. However, the given statement is true for an elementary reaction.

15.4 Vide problem 15.28.

15.5 Yes. An example is provided by the iodination of acetone in aqueous solution.

$$I_2 + CH_3COCH_3 \xrightarrow[\text{or base}]{\text{acid}} ICH_2COCH_3 + HI$$

The rate of iodination, when catalysed by an acid or base, has been found to be independent of the concentration of iodine, but is proportional to the concentration of acetone and acid or base used. This happens because I_2 is involved after the rate-determining step in the (acid-catalysed) mechanism as represented below

$$CH_3-\overset{\overset{\displaystyle O}{||}}{C}-CH_3 + H_3O^+ \rightleftharpoons CH_3-\overset{\overset{\displaystyle {}^+O-H}{||}}{C}-CH_3 + H_2O \quad \text{fast}$$

$$H_2O + CH_3-\overset{\overset{\displaystyle {}^+O-H}{||}}{C}-CH_3 \longrightarrow CH_2{=}\overset{\overset{\displaystyle OH}{|}}{C}-CH_3 + H_3O^+ \quad \text{slow}$$

$$CH_2{=}\overset{\overset{\displaystyle OH}{|}}{C}-CH_3 + I_2 + H_2O \longrightarrow ICH_2-\overset{\overset{\displaystyle O}{||}}{C}-CH_3 + H_3O^+ + I^- \quad \text{fast}$$

Such a reaction would go to completion because the rate of the reaction does not fall with time, provided [CH$_3$COCH$_3$] is kept constant.

15.6 Vide problem 15.19(i).

15.7 Vide problem 15.81.

15.8 The differential rate law is

rate $= k[A]^2[B]$

Pseudo first-order rate law, rate $= k_1[B]$, where $k_1 = k[A]^2$

Pseudo second-order rate law, rate $= k_2[A]^2$, where $k_2 = k[B]$

An example of pseudo first-order reaction is the acid catalysed hydrolysis of an ester which is of first order wrt ester.

The rate constant of a pseudo first-order reaction, like that of a true first order reaction has the unit time^{-1}.

The rate constant of a second order reaction has the unit conc^{-1} time^{-1}, which unlike pseudo first-order reaction, is not free from unit of concentration.

15.9 Vide problem 15.75.

15.10 Overall reaction rate $\approx$ rate of the slowest step

$= k_2[Hg][Tl^{3+}]$

Now using the equilibrium condition for the first step

$$\frac{[Hg^{2+}][Hg]}{[Hg_2^{2+}]} = \frac{k_1}{k_{-1}}$$

or
$$[Hg] = \frac{k_1 [Hg_2^{2+}]}{k_{-1} [Hg^{2+}]}$$

Then
$$v = \frac{k_2 k_1}{k_{-1}} \frac{[Hg_2^{2+}][Tl^{3+}]}{[Hg^{2+}]}$$

which corresponds to a negative order wrt Hg^{2+}.

Note: In general, the appearance of a product in the rate law is avoided by studying the reaction much before the arrival of equilibrium. But here, it is unavoidable because the formation of intermediate (Hg) is not possible without the formation of a product (Hg^{2+}).

15.11 (i) The rate law is $v = k[H_2][I_2]$

(ii) $v \simeq$ rate of the slowest step
$$= k_3 [IH_2][I]$$

Now, using the equilibrium condition for the first two steps, we have

$$\frac{[I]^2}{[I_2]} = \frac{k_1}{k_{-1}}$$

and
$$\frac{[IH_2]}{[I][H_2]} = \frac{k_2}{k_{-2}}$$

whence, by multiplication

$$[IH_2][I] = \frac{k_1 k_2}{k_{-1} k_{-2}} [H_2][I_2]$$

Then
$$v = \frac{k_1 k_2 k_3}{k_{-1} k_{-2}} [H_2][I_2]$$

☐ No. Because the two mechanisms of the reaction ($H_2 + I_2 = 2HI$) correspond to the same rate law.

Note: The two mechanisms can be distinguished by detecting the intermediates, most effectively by spectroscopic means.

15.12 Vide problem 15.16.

15.13 Vide Section 15.10.

15.14 (a) Vide problem 15.38(ii).

(b) Vide problem 15.6.

(c) Vide problem 15.40.

15.15 For an nth order reaction $t_{\frac{1}{2}} \propto \dfrac{1}{a^{n-1}}$

(i) The given condition, i.e. $t_{\frac{1}{2}} \propto \frac{1}{a}$ corresponds to $n = 2$.

(ii) The required condition is $t_{\frac{1}{2}} \propto a$ which corresponds to $n = 0$.

(iii) The given condition, i.e. $t_{\frac{1}{2}} \propto a$ corresponds to $n = 0$ [which is same as in (ii)]

15.16 The given informations correspond to a second order reaction.

☐ Here
$$\frac{1}{[A]} - \frac{1}{[A]_o} = kt$$

From the linear plot of $1/[A]$ vs t, we have

$$k = \text{slope} = 3 \times 10^{-2} \text{ L mol}^{-1} \text{s}^{-1}$$

and $$1/[A]_o = \text{Intercept} = 1000 \text{ L mol}^{-1}$$

Then, $$t_{\frac{1}{2}} = \frac{1}{k[A]_o}, \quad \text{the order of the reaction being two}$$

$$= \frac{1000 \text{ L mol}^{-1}}{3 \times 10^{-2} \text{ L mol}^{-1}\text{s}^{-1}} = 3.3 \times 10^4 \text{ s}$$

15.17 (a) Since the reaction is of zero order wrt A

$$[A] = [A]_o - kt \quad \text{by Eq. (15.5)}$$

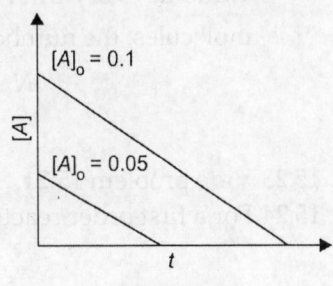

Then $[A]$ vs t plot will be linear with intercept $[A]_o$ and slope k as shown in Fig. 15.6.

(b) If a reaction is of zero order wrt to a reactant A, then the reaction rate will not change with time, provided the concentrations of the other reactants are kept unchanged. Then, under this condition, the reaction is expected to be complete, i.e. $[A] = 0$ in a finite period. This follows

Fig. 15.6

also from the equation in (a) that gives a finite value of $t = [A]_o/k$, for $[A] = 0$.

Note: A zero order reaction does not necessarily lead to completion in a finite period.

15.18 For a zero order reaction

$$-\frac{\Delta c}{\Delta t} = k$$

or $$\Delta t = -\frac{\Delta c}{k}$$

$$= -\frac{(2 \times 10^{-4} - 4 \times 10^{-2}) \text{ mol L}^{-1}}{(5 \times 10^8 \text{ mol L}^{-1}\text{s}^{-1})} = 7.96 \times 10^5 \text{ s}$$

15.19 Vide problem 15.16.

15.20 A third order reaction is one whose rate can be expressed as

$$\text{rate} \propto C^3$$

where C represents the concentration of a rate determining reactant.
☐ Vide Section 15.11.
☐☐ For an nth order reaction

$$\frac{t_{\frac{3}{4}}}{t_{\frac{1}{2}}} = 2^{n-1} + 1 \quad \text{(vide problem 15.16)}$$

Then for a third order reaction, the time for 75% completion $(t_{\frac{3}{4}})$ will be

$$t_{\frac{3}{4}} = (2^{3-1} + 1) \, t_{\frac{1}{2}}$$

$$= 5 \, (100 \text{ s}) = 500 \text{ s}$$

An nth order reaction does not go to completion in a finite period, if $n \geq 1$. Then the required time $= \infty$.

15.21 The given unit of rate constant indicates that the reaction is of first order. Then

$$t_{\frac{1}{2}} = \frac{\ln 2}{k}$$

$$= \frac{\ln 2}{0.05\,s^{-1}}$$

Note: The given reactant concentration is of no use.

15.22 By very definition of $t_{\frac{1}{2}}$ the number of molecules will be reduced to half of its initial value in every interval of duration $t_{\frac{1}{2}}$. Obviously, starting with N_o number of molecules, the number of molecules N_t remaining after time $t = mt_{\frac{1}{2}}$ will be

$$N_t = N_o \left(\tfrac{1}{2}\right)^m$$

$$= N_o \left(\tfrac{1}{2}\right)^{t/t_{\frac{1}{2}}}$$

15.23 Vide problem 15.21.

15.24 For a first order reaction

$$t = \frac{1}{k} \ln\left(\frac{1}{1-x/a}\right) \qquad \text{by Eq. (15.6a)}$$

Then

$$\frac{t_{99.9\%}}{t_{50\%}} = \frac{\ln\left(1 - \dfrac{99.9}{100}\right)}{\ln\left(1 - \dfrac{50}{100}\right)} = 9.97 \approx 10$$

15.25 (i) Since $t_{\frac{1}{2}}$ is independent of the initial concentration, the reaction is of first order.

(ii)

$$t = \frac{1}{k} \ln\left(\frac{1}{1-x/a}\right) \qquad \text{the reaction being of first order}$$

$$= \frac{t_{\frac{1}{2}}}{\ln 2} \cdot \ln\left(\frac{1}{1-x/a}\right)$$

$$= \frac{(2.05 \times 10^4\,s)}{\ln 2} \ln\left(\frac{1}{1-\frac{80}{100}}\right) = 1.65 \times 10^4\,s$$

15.26 For an nth order reaction, $t_{\frac{1}{2}} \propto \dfrac{1}{a^{n-1}}$, then for two different initial concentrations, we have

$$\frac{(t_{\frac{1}{2}})_2}{(t_{\frac{1}{2}})_1} = \left(\frac{a_1}{a_2}\right)^{n-1}$$

Substituting of the given data gives

$$2 = 3^{n-1}$$

or

$$n = 1 + \frac{\log 2}{\log 3} = 1.63$$

15.27 Vide problem 15.93.

15.28

$$k = \frac{1}{at} \left(\frac{x/a}{1-x/a}\right) \qquad \text{by Eq. (15.7b)}$$

$$= \frac{1}{(500 \text{ torr})(30 \text{ min})} \left(\frac{\frac{40}{100}}{1 - \frac{40}{100}} \right) = 4.44 \times 10^{-5} \text{ torr}^{-1}\text{min}^{-1}$$

□
$$t = \frac{1}{ka} \left(\frac{x/a}{1 - x/a} \right)$$

$$= \frac{1}{(4.44 \times 10^{-5} \text{ torr}^{-1}\text{min}^{-1})(500 \text{ torr})} \left(\frac{\frac{75}{100}}{1 - \frac{75}{100}} \right) = 135 \text{ min}$$

15.29 For an nth order reaction

$$\frac{t_{\frac{3}{4}}}{t_{\frac{1}{2}}} = 2^{n-1} + 1$$

Then by substituting the given data, we have

$$\frac{90}{30} = 2^{n-1} + 1 \quad \text{whence } n = 2$$

$$k = \frac{1}{at} \left(\frac{x/a}{1 - x/a} \right)$$

$$= \frac{1}{(600 \text{ mm})(30 \text{ min})} \left(\frac{0.5}{1 - 0.5} \right) = 5.55 \times 10^{-5} \text{ mm}^{-1}\text{min}^{-1}$$

15.30 Vide problem 15.83.

15.31 (a) Vide Section 15.3(c).

For (i)
$$kt = \frac{1}{[A]_o - x} - \frac{1}{[A]_o}$$

For (ii)
$$kt = \frac{1}{[A]_o - [B]_o} \ln \frac{[B]_o ([A]_o - x)}{[A]_o ([B]_o - x)}$$

Considering the rhs of the second expression as the ratio of two functions f and g, we can write

$$kt = \frac{\ln[B]_o + \ln([A]_o - x) - \ln[A]_o - \ln([B]_o - x)}{[A]_o - [B]_o}$$

$$= \frac{f([B]_o)}{g([B]_o)}, \text{ regarding } [A]_o \text{ and } x \text{ as fixed}$$

For $[B]_o = [A]_o$ both f and g vanish making f/g undefined.
But in the limit $[B]_o \rightarrow [A]_o$, f/g approaches a finite value. Now

$$\lim_{[B]_o \rightarrow [A]_o} \frac{f([A]_o)}{g([B]_o)} = \lim_{[B]_o \rightarrow [A]_o} \frac{f'([A]_o)}{g'([B]_o)} \quad \text{by L' Hospital's rule}$$

$$= \frac{1}{[A]_o - x} - \frac{1}{[A]_o}$$

(b) This will happen when either reactant is present in large excess of the other. Thus under the condition $[A] \gg [B]$, rate $= k[A][B]$ reduces to

$$\text{rate} = k'[B] \quad \text{where } k' = k[A] \simeq \text{constant}$$

which corresponds to a first-order kinetics.

15.32 Considering that the saponification reaction is of second order

$$ka = \frac{1}{t}\left(\frac{x/a}{1-x/a}\right) \quad \text{by Eq. (15.7b)}$$

$$= \frac{1}{(5\ \text{min})}\left(\frac{25/100}{1-25/100}\right) = \frac{1}{15}\ \text{min}^{-1}$$

For $t = 10$ min

$$\frac{1}{15}\ \text{min}^{-1} = \frac{1}{(10\ \text{min})}\left(\frac{x/a}{1-x/a}\right)$$

whence $x/a = 0.4$ i.e. 40%.

15.33 Vide problem 15.84.

  □ Average life period $= 1/k$ vide problem 15.21.

15.34 On addition of large excess of cold water, the reaction practically stops due to lowering of reaction rate arising out of dilution effect (on reactant concentration) and thermal effect (on k).

15.35 Vide problem 15.84.

  Note: k thus found is the true rate constant. This is unlike problem 15.84.

15.36 Vide problem 15.89.

15.37 Vide problem 15.90.

15.38 (a) $$\frac{d\ln k}{dT} = \frac{E_a}{RT^2} \quad \text{or commonly,}\quad k = Ae^{-\frac{E_o}{RT}}$$

  Note: The two equations are not equivalent, the integrated form presupposes E_a to be temperature independent.

  (b) The feature that A actually depends on T and also on the nature of a reaction. Vide problem 15.44(a).

  (c) Vide problem 15.52.

15.39 At the limit $T \to \infty$, the Arrhenius equation $k = Ae^{-\frac{E_o}{RT}}$ predicts k to be a constant (A) which is independent of the nature of the reaction and temperature as supposed by Arrhenius in deriving this equation (vide problem 15.44).

  □ No. Because A actually increases with T. University question 15.38(b).

15.40 Vide problem 15.51.

  □ For a bimolecular reaction $m = \frac{1}{2}$ (by simple collision theory Section 15.7).

  Then a will have the dimension of $A/T^{\frac{1}{2}}$. Therefore, the required unit of a will be $m^3\ mol^{-1}s^{-1}K^{-\frac{1}{2}}$ in SI unit.

15.41 $$E_a = RT^2\frac{d\ln k}{dT}$$

$$= CR + \tfrac{1}{2}RT \quad \text{substituting the given expression for } k$$

15.42 If all the collisions were effective, we would have ·

$$k = A$$

$$\propto \sqrt{T}, \quad \text{by collision theory for a bimolecular reaction}$$

Then per degree rise of T, the rise of k, and hence rate $= \dfrac{d\ln\sqrt{T}}{dT} \times 100\%$.

Therefore, for a 10° rise in T, the required %

$$= \frac{d\ln\sqrt{T}}{dT} \times 10 \times 100$$

$$= \frac{1000}{2T} \simeq 1 \text{ at ordinary } T \text{ (around 500 K)}$$

15.43 The required percentage, $\dfrac{d\ln k}{dT} \times 100$

$$= \frac{E_a}{RT^2} \times 100$$

$$= \frac{(20 \times 10^3 \text{ J mol}^{-1}) \times 100}{(8.31 \text{ JK}^{-1}\text{mol}^{-1})(300 \text{ K})^2} = 2.67 \text{ K}^{-1}$$

15.44 (a) at 227°C

$$\log\left(\frac{k}{\min^{-1}}\right) = 33.91 - \frac{18000 \text{ K}}{(227 + 273) \text{ K}}$$

whence $\quad k = 8.13 \times 10^{-3} \text{ min}^{-1}$

$$t = \frac{1}{k} \ln \frac{1}{1 - x/a}$$

$$= \frac{1}{8.13 \times 10^{-3} \text{ min}^{-1}} \ln \frac{1}{1 - \frac{75}{100}} = 170.6 \text{ min}$$

(b) Comparing the given equation with the Arrhenius equation in the following form

$$\log k = \log A - \frac{E_a}{2.303 \, RT}$$

We have $\quad \dfrac{E_a}{2.303 \, R} = 18000 \text{ K}$

or $\qquad E_a = 2.303 \, R \, (18000 \text{ K}) = 2.303 \, (8.314 \text{ JK}^{-1}\text{mol}^{-1}) \, (1800 \text{ K})$

(c) By the same procedure as in (b)

$$\log(A/\min^{-1}) = 33.91$$

whence $\qquad A = 8.13 \times 10^{33} \text{ min}^{-1}$

Note: A has the unit of k.

15.45 From the given condition

$$\frac{E_a}{RT^2} \times 10K = 1 \quad [\text{vide problem 15.98}]$$

or $\qquad E_a = \dfrac{RT^2}{10k}$

$$= \frac{(8.314 \text{ JK}^{-1}\text{mol}^{-1})(300k)^2}{10K} = 74826 \text{ J mol}^{-1}$$

15.46 For reaction I

$$\frac{E_a}{RT^2} \times 10k = 1, \text{ assuming } A \text{ to be same for I and II}$$

and for reaction II

$$\frac{E_a'}{RT^2} \times 10k = 2$$

Then

$$\frac{E_a}{E_a'} = \frac{1}{2}$$

15.47 For the forward reaction

$$\ln \frac{k_f'}{k_f} = \frac{E_f}{R}\left(\frac{1}{T} - \frac{1}{T'}\right)$$

or

$$E_f = \frac{R \ln \dfrac{k_f'}{k_f}}{\dfrac{1}{T} - \dfrac{1}{T'}}$$

$$= \frac{(8.314 \times 10^{-3}\, \text{kJK}^{-1}\text{mol})\ln \dfrac{2.53 \times 10^{-3}\ \text{L mol}^{-1}\text{s}^{-1}}{3.02 \times 10^{-5}\ \text{L mol}^{-1}\text{s}^{-1}}}{\dfrac{1}{(356 + 273)\ K} - \dfrac{1}{(443 + 273)\ K}}$$

$$= 190.59\ \text{kJ mol}^{-1}$$

☐ $$E_f - E_r = \Delta H$$

or $$E_r = E_f - \Delta H$$

$$= (190.59 - 16.32) \text{ or } 174.27\ \text{kJ mol}^{-1}$$

15.48 Vide problem 103(iii).

15.49 First, we are to calculate k at each temperature by the relation $k = -\frac{1}{t}\ln(1 - \frac{x}{a})$. Then the required quantities are to be calculated using the following relations

$$E_a = \frac{R \ln(k_2 / k_1)}{(1/T_1) - (1/T_2)}$$

k at very high temperature = $A = k e^{\frac{E_a}{RT}}$.

15.50 (a) Vide problem 15.53(i).

(b) No. For a multi-step reaction, the application of Arrhenius equation shows that E_a for such reaction is determined by E_a's of the elementary reactions, and it may so happen that E_a for the overall reaction becomes negative (vide problem 15.55) when k will decrease with increase in temperature.

15.51 Rate of the overall reaction

$$v = \text{rate of the slowest step}$$
$$= k_2 [N_2O_2]\,[O_2]$$

Now, considering the pre-equilibrium before the final step, we have

$$\frac{[N_2O_2]}{[NO]^2} = \frac{k_1}{k_{-1}}$$

or

$$[N_2O_2] = \frac{k_1}{k_{-1}}[NO]^2$$

Then

$$v = \frac{k_1 k_2}{k_{-1}}[NO]^2\,[O_2]$$

Hence the reaction is of second order wrt NO and first order wrt O_2.
☐ Vide problems 15.55 and 15.56.

15.52 (a) Justified. Because the temperature-variation of reaction rate is determined primarily by activation energy of the reaction which comes largely from the molecular kinetic energy that increases with temperature.

(b) Since the given reaction involves free radical (CH_2) the activation energy of the reaction would be close to zero so that $k \approx A$. The given statement is justified because the temperature-variation of A is quite low (vide university question 15.42).

15.53 Vide problem 15.46(b)(ii).

15.54 This is justified by quantum mechanics which suggests that the probability of transition state complex to get out of the potential barrier, despite its possibility on energetic ground, may be very low. This corresponds to low value of transmission coefficient in Eyring equation (on which the activated complex theory is based) e.g. in the reaction $H + H \rightarrow H_2$. Vide problem 15.48.

15.55 Vide Eq. (15.15).

☐ Vide problem 15.50.

15.56 This is similar to ΔS involved in the elementary process of formation of activated complex from the reactants. Vide Section 15.7.

☐ Vide problem 15.101.

15.57 $\Delta S^{\neq}$ is related only to A in the Arrhenius equation $k = Ae^{-\frac{E_a}{RT}}$ (by equation like 15.14c). Since $\Delta S^{\neq}$ has no necessary connection with E_a, the given statement is not justified.

15.58 For two reactions, I and II, having the same energy of activation, we have

$$\frac{k_{II}}{k_I} = e^{[(\Delta S^{\neq})_{II} - (\Delta S^{\neq})_I]/R} \quad \text{by Eqs (15.14c) or (15.14d)}$$

$$= \exp\frac{10 \text{ JK}^{-1}\text{mol}^{-1}}{8.3 \text{ JK}^{-1}\text{mol}^{-1}} \quad \text{from the given data}$$

$$= 3.3$$

15.59 If k and k' refer to rate constants at T and T' respectively, then

$$\frac{k_1/k_1'}{k_2/k_2'} = \frac{e^{-\frac{E_1}{R}\left(\frac{1}{T} - \frac{1}{T'}\right)}}{e^{-\frac{E_2}{R}\left(\frac{1}{T} - \frac{1}{T'}\right)}} \quad \text{by Arrhenius equation}$$

Then $\qquad \ln\frac{k_1/k_1'}{k_2/k_2'} = \frac{1}{R}\left(\frac{1}{T} - \frac{1}{T'}\right)(E_2 - E_1)$

or $\qquad E_2 - E_1 = \dfrac{R\ln\left(\dfrac{k_1}{k_2} \Big/ \dfrac{k_1'}{k_2'}\right)}{\dfrac{1}{T} - \dfrac{1}{T'}}$

$$= \frac{(8.3 \text{ JK}^{-1}\text{mol}) \ln\left(\frac{10}{0.1}\right)}{\dfrac{1}{(10+273)\text{ K}} - \dfrac{1}{(40+283)\text{ K}}} = 112872 \text{ J mol}^{-1}$$

15.60 $\qquad k = \sigma^2 \left(\dfrac{8\pi RT}{N_A\mu}\right)^{\frac{1}{2}} e^{-\frac{E_a}{RT}} \quad \text{by Eq. (15.13b)}$

Here $\quad\quad\quad E_a = E_a'\,(\text{experimental}) - \frac{1}{2}RT \quad$ by Eq. (15.13c)

$\quad\quad\quad\quad\quad\quad = 214.2 \times 10^3 \text{ J mol}^{-1} - \frac{1}{2}(8.314 \text{ JK}^{-1}\text{mol}^{-1})(2700 \text{ K})$

$\quad\quad\quad\quad\quad\quad = 202976 \text{ J mol}^{-1}$

$$N_A\mu = \frac{M_{CO}M_{O_2}}{M_{CO^+}M_{O_2}}$$

$$= \left(\frac{28 \times 32}{28 + 32}\right)10^{-3} \text{ kg mol}^{-1}$$

$$= 14.9 \times 10^{-3} \text{ kg mol}^{-1}$$

Then $\quad\quad k = \left(\frac{3.6 + 3.7}{2} \times 10^{-10} \text{ m}\right)^2 \left[\frac{8(3.14)(8.31 \text{ JK}^{-1}\text{ mol}^{-1})(2700 \text{ K})}{14.9 \times 10^{-3} \text{ kg mol}^{-1}}\right]^{\frac{1}{2}}$

$$\times \exp\left[-\frac{202976 \text{ J mol}^{-1}}{(8.31 \text{ JK}^{-1}\text{ mol}^{-1})(2700 \text{ K})}\right]$$

$$= 9.67 \times 10^{-20} \text{ m}^3 \text{ s}^{-1} (\text{molecule}^{-1})$$

$$\text{Steric factor} = \frac{k_{\text{observed}}}{k_{\text{calculated}}}$$

$$= \frac{3.45 \times 10^5 \text{ Ls}^{-1}\text{ mol}^{-1}}{(9.67 \times 10^{-20} \text{ m}^3\text{s}^{-1})(6.02 \times 10^{23} \text{ mol}^{-1})(10^3 \text{ Lm}^{-3})}$$

$$= 5.93 \times 10^{-3}$$

Note: Steric factor is meaningful only when it is less than one (as in this problem). Otherwise, it would be rediculous. Vide problem 15.47.

15.61 $k \quad\quad\quad k = Ae^{-\frac{E_a}{RT}}$

$$= 1.48 \times 10^{15}\,(60 \text{ s})^{-1} \times \exp\left[-\frac{(65 \times 10^3\text{cal mol}^{-1})(4.18 \text{ J cal}^{-1})}{(8.31 \text{ JK}^{-1}\text{ mol}^{-1})(500 + 273) \text{ K}}\right]$$

$$= 3.74 \times 10^{-2}\text{s}^{-1}$$

$$\Delta S^{\neq} = R\ln\frac{AN_A h}{eRT} \quad \text{by Eq. (15.14c), with } n = 1$$

$$\Delta H^{\neq} = E_a - RT$$

$$= (65 \times 10^3 \text{ cal mol}^{-1})(4.18 \text{ J cal}^{-1}) - (8.31 \text{ JK}^{-1}\text{ mol}^{-1})(500 + 273) \text{ K}$$

$$= 265.28 \text{ kJ mol}^{-1}$$

15.62 (a) Vide problem 30(b).

(b) Vide problem 30(a).

(c) Vide Section 15.4(a).

(d) Considering steady-state approximation for B i.e. $\dfrac{d[B]}{dt} = 0$, we have

$$\frac{d[C]}{dt} = -\frac{d[A]}{dt}$$

$$= k_1[A]$$

$$= k_1[A]_o\,e^{-k_1 t}$$

Integration gives

$$\int_0^{[C]} d[C] = k_1 [A]_o \int_0^t e^{-k_1 t} dt$$

whence $\qquad [C] = [A]_o (1 - e^{-k_1 t})$

Here we arrive at this expression readily without getting into the lengthy expression given in (c).

The steady state approximation is valid provided $k_1 \ll k_2$ when $[B] \approx k_1 [A]_o / k_2$ is sufficiently small and $[B]$ vs t curve exhibits an almost horizontal part of low height. Vide problem 30(b)(ii).

15.63 (i) Vide Section 15.4(c)

At any instant

$$\frac{[B]}{[C]} = \frac{\int_0^t k_1 [A] dt}{\int_0^t k_2 [A] dt}$$

assuming that the parallel reactions are of first order (or same order of any other value)

$$= \frac{k_1}{k_2}$$

(ii) Here $\qquad k = k_1 + k_2$

Then $\qquad \dfrac{dk}{dT} = \dfrac{dk_1}{dT} + \dfrac{dk_2}{dT}$

Since $\qquad \dfrac{1}{k}\dfrac{dk}{dT} = \dfrac{E}{RT^2}$

we have $\qquad \dfrac{kE}{RT^2} = \dfrac{k_1 E_1}{RT^2} + \dfrac{k_2 E_2}{RT^2}$

whence $\qquad E = \dfrac{k_1 E_1 + k_2 E_2}{k}$

$$= \dfrac{k_1 E_1 + k_2 E_2}{k_1 + k_2}$$

(iii) At any instant $[B]/[C] = 2$ and $[A] + [B] + [C] = $ constant

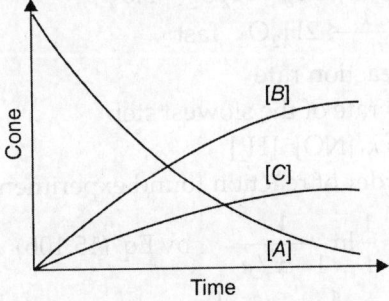

Fig. 15.7

15.64 Here
$$k_{obs} = k_1 + 2k_2 + \tfrac{1}{2}k_3$$

Then
$$E_{obs} = \frac{k_1 E_1 + 2k_2 E_2 + \tfrac{1}{2}k_3 E_3}{k_1 + 2k_2 + \tfrac{1}{2}k_3} \quad \text{vide previous problem}$$

15.65 At any ionstant
$$[B]/[C] = 1 \text{ for (i)}$$
$$[B]/[C] = k_2/k_3 \text{ for (ii)}$$
$$\neq 1 \text{ unless } k_2 = k_3$$

Then (i) and (ii) can be distinguished experimentally by determining $[B]/[C]$.

15.66 With reference to the reaction $A{\overset{\nearrow B}{\underset{\searrow C}{}}}$ the relevant energy profile is given in Fig. 15.8.

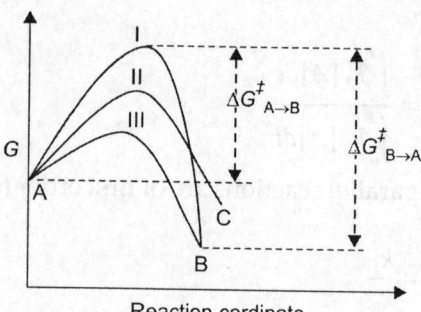

Fig. 15.8

I refers to the reaction $A = B$ with free energy of activation indicated for the forward $(A \rightarrow B)$ and reverse $(B \rightarrow A)$ processes.

II refers to the reaction $A = C$.

III refers to the catalysed $A = B$ reaction.

Here C will be formed in preference to B because $\Delta G^{\ddagger}_{A \rightarrow C} < \Delta G^{\ddagger}_{A \rightarrow B}$, and the product is said to be kinetically controlled (i.e. by rate constant). However, in presence of a catalyst (or by some other means) which is supposed to make $\Delta G^{\ddagger}_{A \rightarrow B} < \Delta G^{\ddagger}_{A \rightarrow C}$ the formation of B will be preferred where $\Delta G_{A \rightarrow B} < \Delta G_{A \rightarrow C}$, and the product is said to be thermodynamically controlled (i.e. by equilibrium constant).

15.67 The given statement is justified. We now illustrate this considering the reaction $2NO + 2H_2 \rightarrow 2H_2O + N_2$ as an example, which is found to be of total order 3.

Since the order differs from stoichiometry (4) of the reaction, it must occur following a multi-step mechanism such as the following one
$$2NO + H_2 \xrightarrow{\ k_1\ } N_2 + H_2O_2 \quad \text{slow}$$
$$H_2O_2 + H_2 \xrightarrow{\ k_2\ } 2H_2O \quad \text{fast}$$

Accordingly, overall reaction rate
$$= \text{rate of the slowest step}$$
$$= k_1 [NO]^2 [H_2]$$

This agrees with the order of reaction found experimentally.

15.68
$$k_1 + k_2 = \frac{1}{t} \ln \frac{1}{1 - x/a} \quad \text{by Eq. (15.10b)}$$

$$= \frac{1}{20 \text{ min}} \ln \frac{1}{1 - \frac{60}{100}} = 0.0458 \text{ min}^{-1} \tag{A}$$

Again $\qquad \dfrac{k_1}{k_2} = \dfrac{[B]}{[C]}$ by Eq. (15.10c)

$$= 3 \qquad\qquad\qquad\text{(B)}$$

From (A) and (B) $\quad k_1 = 3.43 \times 10^{-2}\ \text{min}^{-1}$

$$k_2 = 1.14 \times 10^{-2}\ \text{min}^{-1}$$

15.69 (i) During the reaction $\dfrac{[B]}{[C]} = \dfrac{k_1}{k_2} = $ constant, but $[B]$ and $[C]$ increase till they become maximum at the end of the reaction.

Then, the maximum % of yield of C

$$= \frac{[C]}{[B]+[C]} \times 100 = \frac{k_2}{k_1 + k_2} \times 100$$

$$= \frac{4.65\ \text{s}^{-1}}{(3.74 + 4.65)\ \text{s}^{-1}} \times 100 = 55.4$$

(ii) Half life of $\quad A = \dfrac{\ln 2}{k_1 + k_2} = \dfrac{0.693}{(3.74 + 4.65)\ \text{s}^{-1}} = 0.0826\ \text{s}$

15.70 Since the reacting ions are oppositely charged, it follows from the Bronsted–Bjerrum Eq. (15.16) that the rate constant will decrease with increase in ionic strength due to addition of KCl. Then k will be greater in (i) than in (ii).

☐ The activation energy in the two cases will not differ significantly. The lower value of k in (ii) results essentially from reduction in frequency of collision between the reacting ions due to reduction in electrostation attraction between them by their atmosphere, the effect being greater at greater ionic strength.

15.71 (i) Vide problem 15.95(ii).

(ii) Vide problem 15.96(a).

(iii) Inversion of cane sugar results from its hydrolysis (catalysed by H^+ and OH^- ions). This involves cane sugar molecule in the rate-determining step. Hence the effect of ionic strength will be same as in (ii).

15.72 (a) The given statement is unlikely to hold. Because the catalytic effect is due to change of the reaction mechanism to one of lower free energy of activation. For this to happen the catalyst and the reaction system should have appreciable interaction of chemical nature between them.

(b) A catalyst does not virtually affect ΔG of a reaction. Because it remains unchanged in mass and chemical composition at the end of the reaction, though it may undergo physical change with insignificant ΔG.

(c) Just as ΔG of a reaction is virtually same in presence or absence of catalyst, the same is true for the heat of reaction (ΔU or ΔH). [But $\Delta H^{\neq}$ of a reaction will not be same in the presence and absence of a catalyst].

(d) Vide problem 15.62.

(e) Vide problem 15.60(e).

(f) The mechanisms of action of the so-called negative catalyst and positive catalyst are not just reverse. The former retards the rate of a reaction by reacting with the intermediate(s) or the substance that accelerates the rate of the reaction.

15.73 Considering steady-state approximation for C, we have

$$k_1[M]\,[A] - k_{-1}'[M]\,[B]\,[C] - k_2[A]\,[C] = 0$$

whence $$[C] = \frac{k_1[M][A]}{k_{-1}[M][B] + k_2[A]}$$

$$\frac{d[B]}{dt} = 2k_2[C][A]$$

$$= \frac{2k_1k_2[M][A]^2}{k_{-1}[M][B] + k_2[A]}$$

15.74 Vide problem 15.66.

15.75 Vide problem 15.66.

The plot of $\log k$ vs pH should be linear in all three cases as shown in Fig. 15.9.

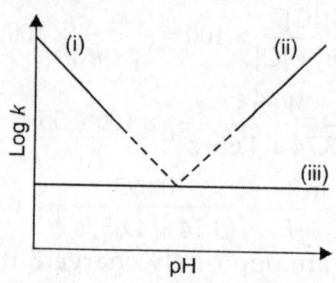

Fig. 15.9

(iii) Corresponds the uncatalysed reaction.

15.76 Since H_3O^+ acts as a homogeneous catalyst, the reaction may be supposed to be of first order wrt H_3O^+. Again constant $t_{\frac{1}{2}}$ for constant $[H_3O^+]$ (at pH = 3) implies the same wrt sucrose. Then the rate law will be

$$-\frac{d[\text{sucrose}]}{dt} = k[H_3O^+][\text{sucrose}]$$

$$= k'[\text{sucrose}]$$

Accordingly $$t_{\frac{1}{2}} = \frac{\ln 2}{k'} = \frac{\ln 2}{k[H_3O^+]}$$

Then $t_{\frac{1}{2}}$ (at pH = 4) $= t_{\frac{1}{2}}$ (at pH = 3) $\times \dfrac{10^{-3}\,M}{10^{-4}\,M}$

$$= (50 \text{ min}) \times 10 = 500 \text{ min}$$

15.77 Here $[H_3O^+]$ changes by a factor $10^{-0.5}/10^{-0.8} \simeq 2$. Then the reaction rate is proportional to $[H_3O^+]$. This justifies that H_3O^+ acts as a homogeneous catalyst in the given reaction. Vide problem 15.60(c)

15.78 $$\frac{k_c}{k_a} = e^{(\Delta S_c^{\ddagger} - \Delta S_a^{\ddagger})/R - (\Delta H_c^{\ddagger} - \Delta H_a^{\ddagger})/RT} \quad \text{by Eq. (15.14b)}$$

Now $(\Delta S_c^{\ddagger} - \Delta S_a^{\ddagger})/R - (\Delta H_c^{\ddagger} - \Delta H_a^{\ddagger})/RT$

$$= \frac{1}{R}\left[-(\Delta S_a^{\ddagger} - \Delta S_c^{\ddagger}) + (\Delta H_a^{\ddagger} - \Delta H_c^{\ddagger})/T\right]$$

$$= \frac{1}{8.314 \text{ JK}^{-1}\text{mol}^{-1}}\left[-41.84 \text{ JK}^{-1}\text{ mol}^{-1} + \frac{20.92 \times 10^3 \text{ J mol}^{-1}}{298 \text{ K}}\right] = 3.41$$

Therefore $k_c > k_a$

and $$k_c{:}k_a = e^{3.411}{:}1$$

15.79 Vide Section 15.6(B).
15.80 Vide Section 15.6(B).
15.81 Vide Section 15.6(B).

15.82   Turnover number $= \dfrac{1}{[E]_o / v_{max}}$ vide Section 15.6(B)

$$= \dfrac{1}{E_o \times \text{intercept}}$$

$$= \dfrac{1}{2.5 \times 10^{-9} \text{ mol L}^{-1})(5 \times 10^4 \text{ mol}^{-1} \text{ Ls})} = 8.0 \times 10^3 \text{ s}^{-1}$$

$K_m = v_{max} \times \text{slope}$
$ = \text{slope} / \text{intercept}$
$ = 3.5 \times 10^2 \text{ s} / 5 \times 10^4 \text{ mol}^{-1}\text{Ls}$
$ = 7.0 \times 10^{-3} \text{ mol L}^{-1}$

Adsorption and Surface Catalysis

A curious fact about a multiphase multicomponent system is that often the concentration of a component in the interface region differ much from its value in the phases in contact.

16.1 ADSORPTION

When a species in a heterogeneous system is so distributed that it is concentrated or depleted in the interfacial region, the phenomenon is called adsorption or absorption according as the enrichment/depletion is appreciable or not. The species which is absorbed is called the adsorbate and the underlying material is the adsorbent or substrate. Usually the adsorbate has higher concentration at the interface than in the bulk. This is called positive adsorption. The less familiar opposite situation is called negative adsorption.

Adsorption occurs because the molecules at the interface, unlike those in the bulk, experience an unbalanced intermolecular force (due to their asymmetric surroundings) which the system tends to reduce.

Adsorption is mostly an exothermic process. It is called physisorption or chemisorption according as the forces responsible for it are of physical or chemical type. Chemisorption, unlike physisorption, is ordinarily a slow process, difficultly reversible and highly selective going not beyond the formation of a monolayer of adsorbed molecules (just as only a limited member of atoms are combined in a molecule by chemical forces).

An expression representing the variation of surface (interface) coverage by the adsorbed species with pressure or concentration of the absorbate at a chosen temperature is called the adsorption isotherm.

16.2 GIBBS ADSORPTION ISOTHERM

Gibbs, on thermodynamic ground, deduced the following relation between interfacial tension and composition of the adsorbed species at the interface

$$d\gamma = -\sum \Gamma_i d\mu_i \tag{16.1a}$$

where Γ_i is the surface excess of the ith species (i.e. concentration of the ith species at the interface in excess of its bulk concentration in the two phases involved).

For a two-component system, if the interface (supposed to be planar) is so located that the adsorption of one of the components, say 1 (which is usually the solvent) is zero, i.e. $\Gamma_{1,1} = 0$, then the adsorption $\Gamma_{2,1}$ of component 2 relative to that of 1 is given by

$$d\gamma = -\Gamma_{2,1} d\mu_2$$

or

$$\Gamma_{2,1} = -\left(\frac{\partial \gamma}{\partial \mu_2}\right)_T \tag{16.1b}$$

This is Gibbs (relative) adsorption isotherm. (The term *adsorption isotherm* is justified on the ground that the chemical potential μ is a function of pressure or concentration for an ideal mixture).

Quantitatively this equation is of little practical importance in case of adsorption on solid surface (due to difficulty in the measurement of γ) which is an important step for reactions involving solid catalysts.

16.3 LANGMUIR ADSORPTION ISOTHERM

Langmuir first theoretically deduced an isotherm for adsorption of a gas on a solid on the basis of the following structural model: (i) the surface of a solid contains a fixed number of evenly distributed equivalent adsorption sites and the surface is flat on a microscopic scale. (ii) each site can hold only one adsorbate molecule (i.e. adsorbate forms only a unimolecular layer), (iii) there is no interaction between the adsorbed molecules on different sites (i.e. ability of a molecule to be adsorbed at any unoccupied site or to leave an occupied site is independent of the fraction of sites occupied).

Let θ be the fraction of N adsorption sites occupied by the adsorbate A at equilibrium

$$A\,(g) + M\,\text{(surface)} \underset{k_d}{\overset{k_a}{\rightleftharpoons}} AM\,\text{(surface)}$$

Then $\qquad \underset{\text{rate of adsorption}}{k_a P\,(1 - \theta)\,N} = \underset{\text{rate of desorption}}{k_d \theta N} \qquad\qquad$ (16.2a)

or $\qquad\qquad\qquad \theta = \dfrac{KP}{1 + KP} \qquad\qquad\qquad\qquad$ (16.2b)

(where $\qquad\qquad\qquad K = \dfrac{k_a}{k_d}$)

which is Langmuir adsorption isotherm. In terms of volume (V) of the adsorbed gas (at constant temperature and pressure), this becomes

$$\left.\begin{array}{c} \dfrac{V}{V_m} = \dfrac{KP}{1 + KP} \\[3mm] \dfrac{1}{V} = \dfrac{1}{V_m} + \dfrac{1}{V_m KP} \end{array}\right\} \qquad (16.3)$$

or

where V_m is the value of V corresponding to complete surface coverage (i.e. $\theta = 1$)

Langmuir isotherm fails in many cases of strong adsorption. This is particularly due to non-uniformity of actual surfaces, which either pre-exists or developed by the interaction between the adsorbed molecules. In some cases, the following empirical Freundlich isotherm is more appropriate.

$$\theta \text{ or } V = kP^{\frac{1}{a}} \qquad\qquad\qquad (16.4)$$

where k and a ($a > 1$) are constants for the system. Because this isotherm really corresponds to a modified Langmuir isotherm based on several kinds of adsorption sites having different heat of adsorption.

16.4 BET ISOTHERM

This is the most widely used isotherm, due to Brunauer, Emmett and Teller, as represented below

$$\left.\begin{aligned}\frac{V}{V_m} &= \frac{cy}{(1-y)(1-y+cy)} \\[2mm] \frac{y}{V(1-y)} &= \frac{1}{V_mc} + \frac{c-1}{V_mc} \cdot y\end{aligned}\right\}$$

(16.5)

or

where $y = P/P^*$, P^* being the vap pr of the bulk liquid adsorbate at the experimental temperature.

Here V is the volume of the material adsorbed at a given pressure, V_m is the V for complete monolayer coverage and c is a constant for the system at constant temperature which is a measure of the tendency to form monolayer adsorption compared to multilayer adsorption. This isotherm can be derived following Langmuir's treatment in generalised form assuming that the enthalpy of adsorption of the first chemisorbed layer is of much higher value than those for succeeding physically adsorbed layers happening with a liquefied substrate.

From the measured value of V_m, the surface area of the adsorbate can be computed from the knowledge of the surface area occupied by a single adsorbed molecule (estimated from the density of the liquefied adsorbate assuming the molecules to be spherical and that they are close packed in the liquid.

16.5 HETEROGENEOUS (SURFACE) CATALYSIS

A key feature of heterogeneous catalysis is adsorption (equivalent to formation of intermediate in homogeneous catalysis). In most cases the heterogeneous catalysis involves a liquid or gaseous reaction mixture and a solid catalyst, and may be thought to proceed through the following steps:

   (i) Diffusion of reactants to the catalyst surface.
  (ii) Adsorption (usually chemisorption) of at least one reactant on the surface.
 (iii) Chemical reaction of adsorbed molecules individually or mutually on adjacent sites (Langmuir–Hinshelwood mechanism) or with fluid-phase molecules (Eley–Rideal mechanism).
  (iv) Desorption of products.
   (v) Diffusion of products from the catalyst surface.

Depending on the situation, any one of these steps may be the slowest and therefore be rate-determining, but this is commonly the step (iii).

A. Surface Catalysed Unimolecular Reactions

For the gaseous reaction $A \rightarrow C + D$, let θ_A be the fraction of catalyst surface (sites) occupied. Then

reaction rate, $v = k\theta_A$

(assuming the rate to be proportional to the surface coverage)

$$= k \cdot \frac{K_A P_A}{1 + K_A P_A + K_C P_C + K_D P_D}$$

(16.6a)

assuming Langmuir isotherm to hold for A, C and D. Vide problem 16.7.

If products are adsorbed to the negligible extent

$$v = \frac{K_A P_A}{1 + K_A P_A} \tag{16.6b}$$

Then
$$v = \begin{cases} k K_A P_A & \text{at low pressure} \\ k & \text{at high pressure} \end{cases}$$

At low pressure, the reaction is of first order, θ being proportional to P. But at high pressure, the reaction is of zero order, θ being unchanged with P due to complete coverage of the surface.

B. Surface Catalysed Bimolecular Reactions

Let us consider the gaseous reaction $A + B \rightarrow P$ (products) when only one of the reactants, say A, is adsorbed (Eley–Rideal mechanism).

reaction rate, $v = k\theta_A P_B$

[where P_B is the partial pressure of the non-adsorbed gas B]

$$= \frac{k K_A P_A P_B}{1 + K_A P_A} \tag{16.7}$$

Then
$$\begin{aligned} v &= k K_A P_A P_B & \text{at low pressure} \\ &= k P_B & \text{at high pressure} \end{aligned}$$

At low pressure, when the surface is incompletely covered, the reaction is of second order. But at high pressure, when the surface is almost completely covered, the reaction is of first order where the rate-determining step happens to be the collision of B with adsorbed A.

When each of the reactants (but no product) is adsorbed (Langmuir–Hinshelwood mechanism)

$$v = k\theta_A \theta_B$$

$$= \frac{k K_A K_B P_A P_B}{(1 + K_A P_A + K_B P_B)^2} \tag{16.8a}$$

If one of the reactants, say A, is very strongly adsorbed, $K_A K_P \gg 1$ and then

$$v = k K_B P_B / K_A P_A \tag{16.8b}$$

Here v decreases with increase in P_A. In general a surface catalysed reaction is inhibited by a strongly adsorbed substance (reactant, product or a foreign material), called catalyst poison.

AUXILIARY PROBLEMS

16.1 Account for adsorption from (i) molecular standpoint (ii) thermodynamic standpoint.

16.2 State with reasons whether the following statements are true or false?
   (a) Adsorption is an exothermic process.
   (b) Physisorption is a fast process.
   (c) Chemisorpion does not go beyond monolayer formation.

16.3 The amount of nitrogen adsorbed on iron is quite large at 78 K; it decreases sharply as the temperature is raised, but increases again toward a maximum value at ~ 778 K and subsequently falls off. Explain.

16.4 How will you distinguish between physisorption and chemisorption?

16.5 Can any gas adsorb appreciably on any solid? On what factor does this depend?

16.6 Langmuir isotherm holds only in case of chemisorption, comment. Show that Langmuir isotherm is a particular case of BET isotherm.

16.7 Two gases, A and B, compete for adsorption on the same surface according to Langmuir isotherm. Show that the fraction of surface covered by A molecules is

$$\theta_A = \frac{K_A P_A}{1 + K_A P_A + K_B P_B}$$

where the terms have their usual significance.

16.8 If a diatomic gas adsorbs as atoms the Langmuir isotherm becomes

$$\theta = \frac{\sqrt{KP}}{(1 + \sqrt{KP})}$$

Establish this.

16.9 Suppose that ozone adsorbs on a particular surface in accordance with Langmuir isotherm. Discuss, how could you use the pressure dependence of the fractional coverage to distinguish between adsorption (i) without dissociation (ii) with dissociation into $O_2 + O$ and (iii) dissociation into $O + O + O$?

16.10 From Gibbs adsorption isotherm, with suitable assumption, arrive at the following expression for surface pressure (π)

$$\pi\sigma = n^\circ RT$$

16.11 Gibbs adsorption isotherm (16.1a) is of central importance in understanding various surface phenomena. Show that it leads to (i) Langmuir isotherm (ii) Freundlich isotherm.

16.12 Predict the effect of (i) NaCl (ii) Na-stearate on surface tension of water (γ). In which case the effect on γ will be more pronounced? Would sucrose increase or decrease γ of water?

16.13 Mention the essential requirement(s) for a solid to act as an efficient catalyst.

16.14 Are the following statements true or false?
(a) Surface catalysis is due to concentration of the reactants on the surface.
(b) Physisorption has no significant role in heterogeneous catalyses.
(c) Heterogeneous catalysis, unlike homogeneous one, involves no formation of intermediate.

16.15 A good catalyst should have moderate value for enthalpy of adsorption of the reatants. Why?

16.16 Active centres of a catalyst are formed by atoms present on the surface peak. Justify or criticise.

16.17 Ethyl alcohol is dehydrogenated on metals like Ni but dehydrated on alumina. Explain.

16.18 Catalytic activity of Ni surface on hydrogenation of ethylene is not same for 100, 110 and 111 planes. Explain.

16.19 Rise of temperature often adversely affects the catalytic activity of a solid. Explain.

16.20 Catalysis are frequently used not in the form of pure substance but in the form of complicated multicomponent systems. Explain this citing example.

16.21 Explain the action of a (i) catalyst promoter (ii) Catalyst poison. Can same substance act both as promoter and poison?

16.22 For cars equipped with catalytic converters (to remove pollutants from the exhaust) the gasoline to be used must be lead-free. Why?

16.23 Decomposition of ammonia on molybdenum is retarded by nitrogen, but as the surface becomes saturated with nitrogen, the rate of decomposition does not fall to zero. Why?

16.24 Decomposition of HI is of second order in absence of a catalyst. But when catalyed by Au, it is of first order at low pressure and of zero order at high pressure.

16.25 Explain the following decomposition of NH_3 using Langmuir isotherm.

(i) $-\dfrac{dP_{NH_3}}{dt}$ = constant, on tungsten surface

(ii) $-\dfrac{dP_{NH_3}}{dt} = k\dfrac{P_{NH_3}}{P_{H_2}}$ on platinum surface

16.26 Below certain pressure, the hydrogenation of C_2H_4 on Cu is of order +1 wrt each reactant. But the reaction between CO and O_2 over Pt is of order +1 wrt O_2 and of order −1 wrt CO. Explain on the basis of Langmuir–Hinshelwood mechanism.

16.27 For 0.1 M aqueous solution of some amino acid $d\gamma/dc = -4.0 \times 10^{-4}$ N m^{2}mol^{-1} at 25°C. Find surface excess of the acid and cross-sectional area of its molecules.

16.28 For aqueous solution of saturated aliphatic acids at 18°C, the surface tension obeys the relation $\gamma = [730 - 29.9 \log (ac + 1)$ dyn cm^{-1} where a is a constant and c is the acid concentration. Find $\Gamma_{2,1}$ as a function of c. Which gas–solid adsorption isotherm does the expression resemble?

16.29 The data below give the volume of H_2 (at STP) adsorbed on 1.02 g of Cu at 0°C at different pressure.

P/torr	38	114	152
V/cm³	1.22	1.31	1.36

(a) Explain whether the given data fit Langmuir isotherm better than Freundlich isotherm (density of liquid H_2 is 0.0708 g m^{-3}].

(b) Apply Langmuir isotherm to calculate
   (i) volume of H_2 that would be required for a complete monolayer formation
   (ii) the specific area of Cu
   (iii) the pressure at which 50% of the surface will be covered at equilibrium.

16.30 The desorption of certain amount of a gas is found to take 25 min at 1900 K and 2.5 min at 2000 K. How long would it take for the same amount of desorption at 2100 K?

16.31 The half-life of adsorbed oxygen atom on tungsten surface is 0.35 s at 2550 K and 3.50 s at 2360 K. Find
   (i) the activation energy for desorption.
   (ii) the residence time of a oxygen atom on tungsten surface at 2400 K.

16.32 The chemisorption of H_2 on Mn becomes 10% faster when temperature is raised from 500 K to 700 K. Calculate the activation energy for this adsorption.

16.33 The adsorption of a gas on a solid surface fits Langmuir isotherm. It is found that 15% coverage of the surface occurs at 4.5 atm at 190 K and at 35.5 atm at 273 K. Find
   (i) the standard enthalpy of adsorption.
   (ii) the standard entropy of adsorption at 273 K.
   Would the result change with increase in specific area of the solid?

ANSWERS

16.1 (i) Vide Section 16.1.

(ii) Thermodynamically, adsorption occurs spontaneously because it reduces free energy (G) of the interface in a system making ΔG negative for the system as a whole.

16.2 (a) The given statement is usually true. Because ΔS of a system usually happens to be negative when a species is adsorbed due to loss of translational freedom (i.e. of configurational randomness) of the species. Then for $\Delta G_{T,P}$ ($= \Delta H - T\Delta S$) of the system to be negative, a criterian for spontaneity, ΔH has to be negative.

Exception arises when a species is adsorbed in dissociated form and the dissociated species has high translational mobility on the surface. Thus H_2 adsorbs on glass endothermically. Here ΔS for the net process $\frac{1}{2}H_2$ (g) $\rightarrow$ H(glass) is positive and $T\Delta S$ dominate the small positive ΔH making $\Delta G_{T,P}$ negative.

(b) The given statement is usually true. Since physisorption involves weak forces of van der Waals type, the adsorbed molecules are only slightly distorted so that activation energy is low and the process becomes quite fast, unless the adsorbent is highly porous.

(c) The statement is true. Because chemisorption involves strong short-range forces of chemical nature which, being specific, acts selectively between adsorbate and adsorbent. Chemisorption is therefore restricted not to go beyond monolayer formation, just as only a limited number of atoms are combined in a molecule by chemical forces.

Note: Sometimes a physically adsorbed layer is formed on the top of an underlying chemisorbed layer. Here the same system displays physisorption at one temperature and chemisorption at some higher temperature.

16.3 Since adsorption is mostly an exothermic process, it is likely (by Chatelier principle) that the extent of any particular adsorption (physical or chemical) at equilibrium should decrease with rise in temperature. This is actually found with the given example which involves physisorption (having low activation energy) occurring significantly only in the low temperature region and chemisorption (having higher activation energy) only in high temperature region. Here the data has definite meaning only in the short range around 78 and 778 K, but appears puzzling between 78 and 778 K which includes mostly the regions where neither physisorption nor chemisorption occurs appreciably.

16.4 This can be done directly by examination of the infrared spectra of the adsorbed species. In case of chemisorption, unlike physisorption, the IR spectra of adsorbed species differ significantly from that of the species in solution.

16.5 Yes. This happens when adsorption is caused by physical forces of van der Waals type which are non-specific in nature.

□ The extent of such adsorption of a gas depends on (increases with) liquefiability of the gas which is largely determined by its critical temperature. Lower temperature, higher pressure and roughness of the surface will of course favour the adsorption of a gas, be it physical or chemical.

16.6 Langmuir isotherm is not limited only to chemisorption, it holds also to physisorption, provided the adsorption does not go beyond the formation of a monolayer of adsorbed molecules.

☐ When c is a large and $y = P/P* \ll 1$, BET isotherm (16.5) reduces to

$$\frac{V}{V_m} = \frac{cy}{1+cy}$$

$$= \frac{KP}{1+KP} \quad \text{where } K = \frac{c}{P*}$$

which is Langmuir isotherm applicable only to monolayer adsorption.

16.7 At some stage of adsorption let θ_A and θ_B be the fraction of the surface covered by A and B respectively, when the fraction bare would be $1 - \theta_A - \theta_B$. At equilibrium, we have for A

$$k_{a,A} P_A (1 - \theta_A - \theta_B) = K_{d,A} \theta_A \quad \text{by Eq. (16.2a)}$$

whence $\dfrac{\theta}{1 - \theta_A - \theta_B} = K_A P_A \quad \text{where } K_A = k_a/k_d$

Similarly $\dfrac{\theta_B}{1 - \theta_A - \theta_B} = K_B P_B$

Then $\dfrac{K_A P_A}{1 + K_B P_B} = \dfrac{\theta_A}{1 - \theta_A}$

whence $\theta_A = \dfrac{K_A P_A}{1 + K_A P_A + K_B P_B}$

Note: In case, the adsorption is not competitive (found only in limited cases) the isotherms would be same as in case of adsorption of each gas on separate surfaces.

16.8 The adsorption of a diatomic gas (A_2) as atoms may be represented as

$$A_2 \,(g) + 2M \,(\text{surface}) \underset{k_d}{\overset{k_a}{\rightleftharpoons}} 2AM \,(\text{surface})$$

At equilibrium

$$k_a P \,[N \,(1 - \theta)]^2 = k_d (N\theta)^2$$

$$\underset{\text{rate of adsorption}}{} \quad \underset{\text{rate of disorption}}{}$$

or $\dfrac{\theta}{1 - \theta} = \sqrt{\dfrac{k_a P}{k_d}}$

$$= \sqrt{KP} \quad \text{where } K = k_a / k_d$$

whence $\theta = \dfrac{\sqrt{KP}}{(1 + \sqrt{KP})}$

16.9 The adsorbed species are all same in (i) and also in (iii), but not in (ii). As a consequence, the isotherms in (i) and (iii) will have the following less complicated forms

For (i) $\dfrac{\theta}{1 - \theta} = (KP)$

For (iii) $\dfrac{\theta}{1 - \theta} = (K'P)^{\frac{1}{3}}$

Similar to that found in the previous problem.

Then the adsorption will correspond to (i) or (ii) according as $\theta/(1-\theta)$ varies linearly with $P^{\frac{1}{2}}$ or $P^{\frac{1}{3}}$. If neither happens then (ii) is likely.

16.10 From Gibbs (relative) adsorption isotherm (16.1b), we have

$$\frac{n^{\sigma}}{\sigma} = \Gamma_{2,1} = -\frac{1}{RT}\left(\frac{\partial\gamma}{\partial\ln c_2}\right)_T$$

for an ideal dilute solution where $d\mu_2 = RT\ln c_2$.

[σ is interfacial area].

Again, for dilute solution of a surfactant, the surface tension may be supposed to decrease linearly with concentration (c_2), i.e.

$$\gamma = \gamma^{\circ} - Kc_2 \text{ where } \gamma^{\circ} \text{ is the surface tension of the pure solvent}$$

Then

$$\frac{n^{\sigma}}{\sigma} = \frac{Kc_2}{RT}$$

$$= \frac{(\gamma^{\circ} - \gamma)}{RT} = \frac{\pi}{RT}$$

where $\pi = \gamma^{\circ} - \gamma$ is called the surface pressure

Hence

$$\pi\sigma = n^{\sigma}RT$$

Note: $\pi\sigma = n^{\sigma}RT$ corresponds to the two-dimensional perfece gas equation.

16.11 (i) In case of adsorption of a gas on solid, we have from Gibbs adsorption isotherm (16.1a) applied to a one-component system of surface area σ

$$\frac{n^{\sigma}}{\sigma} = -\left(\frac{\partial\gamma}{\partial\mu}\right)_T$$

where μ is chemical potential of the adsorbent and n^{σ} is its amount present in the surface

or

$$n^{\sigma} = -\sigma\left(\frac{\partial\gamma}{\partial\mu}\right)_T$$

$$= -\frac{\sigma}{RT}\left(\frac{\partial\gamma}{\partial\ln P}\right)_T$$

for an ideal gas since $\mu = \mu^{\circ}(T) + RT\ln P$

Now expressing n^{σ} in terms of volume of the gas adsorbed (V) as

$$n^{\sigma} = \frac{P*V}{RT*}$$

and expressing $d\gamma$ as a kind of chemical potential through

$$d\mu' = \frac{RT*}{P*}d\gamma$$

(where $T*$ and $P*$ are chosen temperature and pressure)

we have

$$-\frac{\sigma}{RT}\left(\frac{\partial\mu'}{\partial\ln P}\right)_T = V$$

This is equivalent to Langmuir isotherm. Because on putting $d\mu' = -\frac{RT}{\sigma}\left(\frac{V_m}{V_m - V}\right)dV$ in this expression, we can arrive at the Langmuir isotherm in its most familiar form $\frac{V}{V_m} = \frac{KP}{1+KP}$. Here it is easier to show the reverse.

[From $\quad \dfrac{KP}{1+KP} = \theta = \dfrac{V}{V_m}$

we have $\quad P = \dfrac{\theta}{K(1-\theta)} = \dfrac{V}{K(V_m - V)}$

$$\dfrac{dP}{dV} = \dfrac{1}{K(V_m - V)} + \dfrac{V}{K(V_m - V)^2} = \dfrac{V_m}{K(V_m - V)^2}$$

$$d\mu' = -\dfrac{RT}{\sigma} \cdot V d\ln P = -\dfrac{RT}{P\sigma} V dP = -\dfrac{RT}{\sigma}\left[\dfrac{K(V_m - V)}{V}\right]V\left[\dfrac{V_m}{K(V_m - V)^2}\right]dV$$

$$= -\dfrac{RT}{\sigma}\left(\dfrac{V_m}{V_m - V}\right)dV]$$

(ii) Here, on experimental ground, we take

$$d\mu' = -a\dfrac{RT}{\sigma}dV$$

where dV is the change in adsorbed volume and a is a constant for the system.

or $\qquad \left(\dfrac{\partial\mu'}{\partial\ln P}\right)_T = -a\dfrac{RT}{\sigma}\left(\dfrac{\partial V}{\partial\ln P}\right)_T$

Now $\qquad \left(\dfrac{\partial\mu'}{\partial\ln P}\right)_T = -\dfrac{RT}{\sigma}V,\ $ shown in (i)

Then $\qquad a\dfrac{dV}{V} = d\ln P$

Integration gives $V = kP^{\frac{1}{a}}$, which is Freundlich isotherm.

16.12 NaCl will accumulate more in the interior of the solution (compared with the surface layer) where the oppositely charged Na^+ and Cl^- ions can interact more to cause greater decrease in free energy of the system. Then, as a consequence of the resulting negative adsorption of NaCl, surface tension of water will be enhanced by NaCl as it appears from the Gibbs adsorption isotherm. Because in creating new surface by water molecules they are to do additional work against the associated ion-dipole attractions.

However, positive adsorption happens with Na-stearate due to hydrophobic nature of the alkyl radical in the stearate ion with consequent lowering of surface tension. Here it is less difficult to enlarge the surface compared with pure solvent because more solute will come to the new surface causing further lowering of free energy.

□ The rate of change of γ with increasing concentration of the salt is expected to be greater with Na-stearate, because of its greater tendency to be accumulated at the interface.

□□ Regarding the effect on surface tension, sucrose is expected to behave similar to NaCl. Because sucrose molecules, though neutral, contain polar $-OH$ groups through which they form H-bonds with H_2O molecules, and for doing it to a greater extent, they accumulate in the interior of the solution.

Note: Removal of oily dirt particles from solid surfaces by detergents is mainly due to the ability of the latter to lower surface tension of water.

16.13 For effective catalysis at least one reactant is required to be adsorbed fairly strongly (i.e. $|\Delta H|$ of adsorption sufficiently high) so that it can change to a distorted molecular form in which it readily undergoes the desired reaction. Very strong adsorption will, however, adversely affect the reaction rate. Vide problem 16.15.

16.14 (a) Hardly true. The enhanced rate of reaction on a surface is primarily due to lower activation energy of the surface reaction than that of the homogeneous reaction. This is consistent with the fact that in certain cases, different surfaces give rise to different products of reaction from same reactants due to difference in the activation energy involved.

(b) Usually true, $|\Delta H|$ in physisorption being quite small. Only in few cases the role of physisorption cannot be ignored, such as in recombination of radicals, where a third body facilitates the reaction by removing the liberated heat.

(c) Apparently true, considering an intermediate as a chemical species having independent existence. However, the anwer becomes opposite, if we take, as is often done, an adsorbed species as equivalent to an intermediate, considering significant distortion of a molecule after its adsorption (particularly in chemisorption).

16.15 Because the situation $|\Delta H_{ads}|$ = very small virtually corresponds to an uncatalysed reaction.

Again, the situation $|\Delta H_{ads}|$ = very high virtually corresponds to no reaction involving adsorbed reactant molecules which, being very tightly held at their adsorption sites, have little tendency to react.

Note: When $|\Delta H_{ads}|$ = very high, activation energy for the surface catalysed reaction is greater than that for the homogeneous reaction.

16.16 An active centre of a catalyst, where a chemical reaction is most likely to occur, is generally a group of a definite number of adsorption sites arranged in proper geometrical manner required for a particular transformation of reactant molecules to occur. Due to this geometrical restriction, surface atoms of a peak may not be suitable in forming active centres though such atoms have relatively higher adsorption power due to their lower coordination number. Therefore, the given statement is not justified.

16.17 Dehydrogenation of C_2H_5OH possibly involves adsorption through attachment of one H-atom to one site and another H-atom to a neighbouring site of Ni as represented below:

Alumina,, which contains oxide groups and also hydroxyl groups at the surface, dehydrates C_2H_5OH possibly through attachment with oxygen and hydrogen atoms of the latter as represented below.

The different products with nickel and alumina are due to different nature of adsorption and different configuration of the active centre.

16.18 This is due to different configuration of the active centres for the given planes.

16.19 This is attributed to sintering of the catalyst surface that reduces the number of active centres.

16.20 Because in this way, we can enhance the activity and selectivity of a catalyst (through generation of more effective centres) and impart greater thermal and mechanical stability to it.

Catalysts are often used in the following forms:

(i) Mixed catalyst, e.g. a mixture of ZnO and Cr_2O_3 which is a better catalyst than either oxide used separately for reduction of CO by H_2 to yield CH_3OH. The activity and selectivity of such catalysts depend on their composition.

(ii) Promoted catalyst where the activity of a catalyst is increased by addition of a small amount of some other substance(s), called promoter, which by itself has no significant catalytic activity. An example of such catalyst is Fe promoted by Al_2O_3 and K_2O used in the synthesis of NH_3 by Haber process. It is interesting to note that Al_2O_3 alone can increase the activity of Fe but K_2O does the reverse.

(iii) Supported catalyst where a catalytically active substance (usually a metal) is deposited on a catalytically inactive porous body. An example of this type of catalyst is platinised asbestos. (Pt deposited on asbestos).

16.21 (i) There may be different mechanisms of promoter action just as there are various mechanisms of catalysis. The increase in activity and selectivity of a catalyst may be due to change in nature of the active centres or their number or both due to change in spacing of the lattice. The higher activity may also be due to change in adsorption power of the catalyst.

(ii) Inhibition of a surface catalysed reaction by a substance, called catalyst poison, is due to its preferential adsorption on the active centres, which are usually only a small fraction of the total number of adsorption sites of the catalyst. Then, it is not unlikely to observe this often with a minute amount of the catalyst poison (insufficient to occupy all the adsorption sites).

☐ Yes. An example is offered by K_2O which, along with Al_2O_3, serves as a promoter of the Fe catalyst used in the Haber synthesis of NH_3. But K_2O acts as a catalyst poison for Fe in absence of Al_2O_3.

Note: The term 'catalyst promoter' and 'catalyst poison' have significance only with surface catalysis. Although they have no relevance to homogeneous catalysis, this is not so with enzyme catalysis (being intermediate between homogeneous and heterogeneous catalysis) where the effects of coenzymes correspond to promoting action and the effects of acid or base within certain range of pH corresponds to catalyst poisoning [vide Section 15.6(B) and problem 15.68(ii)].

16.22 Because lead has poisoning effect on the catalytic converters used in cars.

16.23 Here the active centres are not all of same type. N_2 is not able to be adsorbed on all the different types of active centres.

Note: Selectvity in poisoning makes it possible that a catalyst which is poisoned for one reaction remains unpoisoned for some other reaction. For example, Pd is poisoned by quinoline–sulphur for reduction of aldehydes by H_2 but not for reduction of acid chlorides. $BaSO_4$ is the inert catalyst support.

16.24 The decomposition of HI in gas phase follows practically a simple bimolecular mechanism, and hence it is of second order.

But the same reaction occurring on Au surface (where only HI is adsorbed) follows a unimolecular mechanism and hence it is of first order at low pressure and zero order at high pressure. Vide Section 16.5(A).

Note: The order of a unimolecular reaction varies with pressure, whether it occurs in gas phase or on the surface, though this variation is different in the two cases. This is unlike bimolecular reaction.

16.25 (i) The given rate law follows from Eq. (16.6b) as $K_{NH_3} P_{NH_3} \gg 1$ (unless P_{NH_3} is very low) due to strong adsorption of NH_3 on tungsten.

(ii) Following Eq. (16.6a), we have

$$-\frac{d P_{NH_3}}{dt} = \frac{K_{NH_3} P_{NH_3}}{K_{H_2} P_{H_2}} \quad \text{since } K_{H_2} P_{H_2} \gg 1 \text{ due to strong adsorption of } H_2$$

$$= \frac{k P_{NH_3}}{P_{H_2}}$$

Note: The reaction rate, and even rate law, depends on the power and selectivity of adsorption of the reactants and products.

16.26 For the reaction between H_2 and C_2H_4 on Cu, we have from Eq. (16.8a).

$$v = k K_{H_2} K_{C_2H_4} P_{H_2} P_{C_2H_4}$$

Since $K_{H_2} P_{H_2} \ll 1$ and also $K_{C_2H_4} P_{C_2H_4} \ll 1$ due to weak adsorption of both H_2 and C_2H_4 on Cu.

Again, for the reaction between CO and O_2, we have from Eq. (16.8b)

$$v = \frac{k K_{O_2} P_{O_2}}{K_{CO} P_{CO}} \quad \text{since CO is strongly adsorbed on Pt.}$$

Note: Here CO (a reactant) poisons the catalyst, while in problem 16.25(ii), the poisoning happens with H_2 (a product).

16.27   Surface excess $= -\dfrac{C}{RT}\left(\dfrac{\partial \gamma}{\partial C}\right)_T$ Vide problem 16.10

$$= -\frac{(0.1 \times 10^3 \text{ mol m}^{-3})}{(8.31 \text{ JK}^{-1} \text{ mol}^{-1})(298 \text{ K})}(-4 \times 10^{-4} \text{ Nm}^2 \text{ mol}^{-1})$$

$$= 1.62 \times 10^{-5} \text{ mol m}^{-2}$$

Assuming that the interface is entirely occupied by the solute molecules, cross-sectional area of the molecules

$$= \frac{1}{\text{number of molecules per unit interfacial area}}$$

$$= \frac{1}{(1.62 \times 10^{-5} \text{ mol m}^{-2})(6.02 \times 10^{23} \text{ mol}^{-1})}$$

$$= 1.02 \times 10^{-19} \text{ m}^2$$

16.28
$$\Gamma_{2,1} = -\frac{C}{RT}\left(\frac{\partial\gamma}{\partial C}\right)_T$$

$$= \frac{C}{RT}\cdot\frac{\dfrac{29.9}{2.303}\cdot 9}{1+aC}$$

$$= \frac{12.98}{RT}\cdot\frac{aC}{1+aC}$$

☐ The expression of $\Gamma_{2,1}$ resembles the following gas–solid Langmuir adsorption isotherm

$$\theta = \frac{KP}{1+KP} \quad \text{by Eq. (16.2b)}$$

$\Gamma_{2,1}$ and θ both represent surface concentration of the adsorbent, differing only in form.

16.29 (a) Langmuir isotherm (16.3) demands a constant value of $\Delta P/\Delta(P/V)$ $(= V_m)$ while Freundlich isotherm (16.4) demands a constant value for $\Delta\log V/\Delta\log P$ $(=1/a)$. To examine fulfilment of such demand, we draw up the following table

$P/$torr	38	114	152
$\log(P/$torr$)$	1.58	2.06	2.18
$\log(V/$cm$^3)$	0.086	0.117	0.133
P/V (torr cm^{-3})	31.15	87.02	111.8

Here, we have

$$\left.\begin{array}{l}\Delta P/\Delta(P/V)=1.36 \\ \Delta\log V/\Delta\log P=0.065\end{array}\right\} \text{ for the pressure range 38–114 torr}$$

$$\left.\begin{array}{l}\Delta P/\Delta(P/V)=1.53 \\ \Delta\log V/\Delta\log P=0.130\end{array}\right\} \text{ for the pressure range 114–152 torr}$$

$$\left.\begin{array}{l}\Delta P/\Delta(P/V)=1.41 \\ \Delta\log V/\Delta\log P=0.078\end{array}\right\} \text{ for the pressure range 38–152 torr}$$

It appears that with variation of pressure $\Delta P/\Delta(P/V)$ varies less than $\Delta\log V/\Delta\log P$. Hence the given data fit Langumuir isotherm better than Freundlich isotherm.

(b) (i)
$$V_m = \text{Average value of } \Delta P/\Delta(P/V)$$
$$= \frac{(1.36+1.53+1.41)\text{ cm}^3}{3} = 1.43\text{ cm}^3$$

(ii) The number of H_2 molecules present in V_m volume of the gas at STP is

$$N_{H_2} = \frac{PV_m N_A}{RT} = \frac{(1.00\text{ atm})(1.43\times10^{-3}\text{ L})(6.02\times10^{23}\text{ mol}^{-1})}{(0.0821\text{ L atm K}^{-1}\text{ mol}^{-1})(273\text{ K})} = 3.84\times10^{19}$$

Cross-sectional area of a H_2 molecule

$$A = \pi\left(\frac{3V}{4\pi}\right)^{\frac{2}{3}} \text{ where } V = \text{Volume of a molecule} = \frac{M}{eN_A}$$

$$= \pi\left(\frac{3M}{4\pi eN_A}\right)^{\frac{2}{3}}$$

$$= \pi \left[\frac{3(2.02 \text{ g mol}^{-1})}{4\pi (0.0708 \text{ g cm}^{-3})(6.02 \times 10^{23} \text{ mol}^{-1})} \right]^{\frac{2}{3}}$$

$$= 1.58 \times 10^{-15} \text{ cm}^2$$

Surface area of Cu

$$= NA$$

$$= (3.84 \times 10^{19})(1.58 \times 10^{-15} \text{ cm}^2) = 6.67 \times 10^4 \text{ cm}^2$$

$$\text{SP area of Cu} = \frac{\text{surface area}}{\text{mass}} = \frac{6.67 \times 10^4 \text{ cm}^2}{1.02 \text{ g}} = 6.54 \times 10^4 \text{ cm}^2 \text{g}^{-1}$$

(iii) The pressure P' at which $\theta = \frac{1}{2}$ is

$$P' = \frac{1}{K} \quad \text{by Eq. (16.2b)}$$

Then

$$P' = P\left(\frac{V_m}{V} - 1 \right) \quad \text{by Eq. (16.3)}$$

$$\text{Average } P' = \frac{\left[38\left(\frac{1.43}{1.22} - 1 \right) + 114\left(\frac{1.43}{1.31} - 1 \right) + 152\left(\frac{1.43}{1.35} - 1 \right) \right] \text{torr}}{3} = 8.26 \text{ torr}$$

16.30 Assuming that the temperature dependence of the desorption rate follows Arrhenius-like equation

$$E_d = \frac{R \ln(t'/t)}{\frac{1}{T'} - \frac{1}{T}}$$

where E_d is the activation energy for desorption and t is the time required for desorption at temperature T.

Then

$$\frac{R \ln\left(\frac{t''}{t} \right)}{\frac{1}{T''} - \frac{1}{T}} = \frac{R \ln\left(\frac{t'}{t} \right)}{\frac{1}{T'} - \frac{1}{T}}$$

or

$$t'' = t \exp \left[\frac{\left(\frac{1}{T''} - \frac{1}{T} \right) \ln\left(\frac{t'}{t} \right)}{\frac{1}{T'} - \frac{1}{T}} \right] \quad \begin{array}{l} \text{where } t'' \text{ corresponds to } T'' = 2100 \text{ K} \\ t' \text{ corresponds to } T' = 2000 \text{ K} \\ t \text{ corresponds to } T = 1900 \text{ K} \end{array}$$

$$= (25 \text{ min}) \exp \left[\frac{\left(\frac{1}{2100 \text{ K}} - \frac{1}{1900 \text{ K}} \right) \ln\left(\frac{2.5}{25} \right)}{\frac{1}{2000 \text{ K}} - \frac{1}{1900 \text{ K}}} \right] = 1.1 \text{ min}$$

16.31 (i) Since $t_{\frac{1}{2}} = (\ln 2/k_d)$, we have following Arrhenius equation

$$E_d = \frac{R \ln(t'_{\frac{1}{2}}/t_{\frac{1}{2}})}{(1/T') - (1/T)}$$

$$= \frac{(8.314 \text{ J K}^{-1} \text{ mol}^{-1}) \ln(3.50/0.35)}{(1/2360 \text{ K}) - (1/2550 \text{ K})} = 605.9 \text{ kJ mol}^{-1}$$

(ii) $t_{\frac{1}{2}}$ and hence the average residence time $\dfrac{1}{k_d} = \dfrac{t_{\frac{1}{2}}}{\ln 2}$ at any temperature can be calculated from the knowledge of the same at two other temperatures following the similar procedure as in the previous problem.

16.32 The activation energy for adsorption

$$E_a = \frac{R\ln(k_a'/k_a)}{(1/T)-(1/T')}$$

$$= \frac{(8.314\ \text{JK}^{-1}\ \text{mol}^{-1})\ln 1.10}{(1/500\ \text{K})-(1/700\ \text{K})} = 1.39\ \text{kJ mol}^{-1}$$

16.33 (i) According to Langmuir isotherm (16.2b)

$$KP = \frac{\theta}{1-\theta}$$

Then $\ln K + \ln P = $ constant, for constant θ

Differentiation wrt T gives

$$\left(\frac{\partial \ln P}{\partial T}\right)_{\theta} = -\left(\frac{\partial \ln K}{\partial T}\right)_{\theta}$$

$$= -\frac{\Delta H_{ad}^{\circ}}{RT^2} \quad \text{using van't Hoff equation}$$

Integration givesn

$$\Delta H_{ad}^{\circ} = \frac{R\ln(P/P')}{(1/T)-(1/T')}$$

$$= \frac{(8.314\ \text{JK}^{-1}\ \text{mol}^{-1})\ln(4.5/35.5)}{(1/190\ \text{K})-(1/273\ \text{K})} = -10.73\ \text{kJ mol}^{-1}$$

(ii)
$$\Delta G^{\circ} = -RT\ln K$$

$$= RT\left(\ln P + \ln\frac{1-\theta}{\theta}\right)$$

$$= (8.314\ \text{JK}^{-1}\ \text{mol}^{-1})(273\ \text{K})\ln\frac{35.5\ \text{atm}}{1.0\ \text{atm}} + \ln\frac{1-(15/100)}{(15/100)}$$

$$= 12.05\ \text{kJ mol}^{-1}$$

$$\Delta S^{\circ} = \frac{\Delta H^{\circ}-\Delta G^{\circ}}{T}$$

$$= \frac{-10.73\ \text{kJ mol}^{-1}-12.05\ \text{kJ mol}^{-1}}{273\ \text{K}} = -0.833\ \text{kJ K}^{-1}\ \text{mol}^{-1}$$

☐ No, provided the Langmuir isotherm remains valid. In fact with increase in specific area of the adsorbent, there is a growing chance for interaction between the adsorbed molecules which has been ignored in Langmuir isotherm.

UNIVERSITY QUESTIONS

16.1 Explain the term adsorption. Why it is caused? (Guru Nanak Dev BSc, 2004)

16.2 Distinguish between adsorption and absorption. Discuss the factors which affect the adsorption of a gas on a solid adsorbent. (Baroda BSc, 2010)

16.3 Distinguish between physical and chemical adsorption.   (Delhi BSc, 2000)

16.4 What signs of ΔH and ΔS in the case of physical adsorption are expected?
                                                                        (Punjab BSc, 2001)

16.5 Air containing gasoline vapours when passed through a bed of activate carbon, the gasoline vapours are removed from the air and the temperature of the carbon bed rises. Explain this phenomenon.   (Jadavpur BSc(H), 2012)

16.6 The enthalpy of adsorption of a compound on a surface is -125 kJ mol^{-1}. Explain whether this is physisorption or chemisorption.   (Calcutta BSc(H), 1994)

16.7 Adsorption of a gas on a solid is an exothermic process. Justify or criticise.
                                                                        (Calcutta BSc(H), 2006)

16.8 What is an adsorption isotherm? Write the equation describing Langmuir isotherm.
                                                                        (Burdwan BSc(H), 2001)

16.9 What are the limitations of Langmuir's theory of adsorption. Under what conditions Langmuir adsorption isotherm reduces to Freundlich adsorption isotherm.   (Jadavpur BSc(H), 2013)

16.10 5.0 g of a catalyst absorbs 400 cm^3 of N_2 at STP to form a monolayer. What is the surface area per gram if the area occupied by a molecule of N_2 is 16Å.
                                                                        (Madras BSc, 2008)

16.11 How does the Langmuir θ vs P plot look? Explain the nature of $1/\theta$ vs $1/P$ plots (θ is the fraction of surface covered and P is the pressure). (Burdwan BSc(H), 1998)

16.12 From the following data for adsorption of N_2 gas on a solid surface, calculate the constant involved in the Langmuir isotherm; $P = \infty$, $V = 180$ cc/gm; $P = 3.5$ atm, $V = 100$ cc/gm (V = Volume of the gas adsorbed).   (Burdwan BSc(H), 1996)

16.13 In an experiment involving adsorption of CO on charcoal at 272 K, the equilibrium pressures (P) and the corresponding volumes (V) of gas adsorbed per gm of charcoal (corrected to 1 atm) were recorded. The plot of P/V against P was linear with a slope of 0.009 cm^{-3} and an intercept of 1.2 k Pa cm^{-3}. Assuming simple monolayer adsorption, calculate the adsorption equilibrium constant.
                                                                        (Calcutta BSc(H), 2013)

16.14 If a diatomic gas is adsorbed as atom on the surface of a catalyst, then show that the fractional saturation (θ) of the surface by the gas is $\theta = \dfrac{\sqrt{KP}}{1+\sqrt{KP}}$ where the terms have their usual significance.   (Jadavpur BSc(H), 2013)

16.15 77.1 × 10^{-6} g of the alcohol, $C_{16}H_{31}OH$ produced a compact film of monolayer on a Langmuir film balance occupying an area of 402.7 sq cm. Find out the cross-section of the molecule.   (Jadavpur BSc(H), 2011)

16.16 What is Gibbs surface excess? Justify that it can be both positive and negative.
                                                                        (Calcutta BSc(H), 1996)

16.17 Starting from Gibbs surface tension equation $d\gamma = -\sum \Gamma_i d\mu_i$ where the terms have their usual significance, obtain an equation showing the variation of surface tension with concentration of a surface active agent and discuss the implication of the equation.   (Burdwan BSc(H), 1995)

16.18 Show that the Gibbs adsorption equation leads to a two-dimensional perfect gas equation for small concentration of surface-active substance.
                                                                        (Burdwan BSc(H), 1993)

16.19 Explain how the surface tensions of NaCl, acetic acid and sodium dodecyl sulphonate vary with conc in water. (Calcutta BSc(H), 2003)

16.20 Calculate the surface excess of a solute per unit area of 25°C if the conc of the solution is 5×10^{-5} mol L^{-1} and $(\partial\gamma/\partial c)_T = -1.35 \times 10^8$ dyn cm^2 mol^{-1}. Express your answer in mol m^{-2}. (Burdwan BSc(H), 1997)

16.21 For dil soln, the surface tension often varies linearly with solute molar conc (C) as $\gamma = \gamma^\circ - bc$, where γ° denotes the surface tension of the pure solvent and b is a constant. Show that the surface excess, $\Gamma_2 = (\gamma^\circ - \gamma)/RT$. (Calcutta BSc(H), 2014)

16.22 The surface tension of aq solution follows a linear dependence on conc of a solute and is reduced by six units at 0.02 M at 27°C. Calculate the Gibbs surface excess for the soln at 0.0005 M. (Calcutta BSc(H), 1996)

16.23 At 292 K, the surface tension of solution of butyric acid in water (γ) can be represented accurately by the equation

$$\gamma = \gamma_o - a \ln(1 + bc)$$ where c is the conc of butyric acid

γ_o is the surface tension of water, while a and b are constants. Set up the expression for the excess conc of solute per sq cm of surface as function of c. (Calcutta BSc(H), 2004)

16.24 Derive Freundlich adsorption isotherm from the Gibbs adsorption isotherm applied to a gas. (Arunachal BSc, 2011)

16.25 Explain how a catalyst increases the speed of a reaction. (Sambalpur BSc, 2010)

16.26 Explain the physical concepts underlying the principle of catalytic activity by solid surface for gaseous reactants. (Burdwan BSc(H), 1991)

16.27 Explain the following with examples: (i) Negative catalyst (ii) Catalyst poison (iii) Catalyst promoter. (Purvanchal BSc, 2003)

16.28 The decomposition of PH_3 on tungsten surface is found to be of first order at low pressures and zero order at high pressures. Give the possible reason. (Calcutta BSc(H), 2013)

KEY TO UNIVERSITY QUESTIONS

16.1 Vide Section 16.1 and problem 16.1.

16.2 Vide Section 16.1 and problem 16.5.

16.3 Vide Section 16.1 and problem 16.4.

16.4 ΔH and ΔS both should be negative for physical adsorption but not necessarily for chemical adsorption (where a species is absorbed in dissociated form). Vide problem 16.2(a).

16.5 This is due to adsorption of gasoline by carbon, the process being exothermic.

16.6 The given enthalpy of adsorption is quite high typical of chemisorption.

16.7 Vide problem 16.2(a).

16.8 Vide Sections 16.1 and 16.3.

16.9 Vide Section 16.3.

☐ The Langmuir adsorption isotherm, in its common form is

$$\theta = \frac{KP}{1 + KP} \quad \text{[Eq. (16.2b)]}$$

At low pressure, when $KP \ll 1$, it takes the following form

$$\theta = KP$$

Again, at high pressure when $KP \gg 1$, it becomes

$$\theta = 1$$

The last two equations for the limiting pressures suggest the following adsorption isotherm for the intermediate pressures

$$\theta = kP^{\frac{1}{a}}, \text{ the Freundlich isotherm}$$

where k and $\frac{1}{a}$ are characteristic constants for the system at the prevailing temperature and pressure, both lying between 0 and 1.

Then it follows that the Freundlich isotherm is a special case of the Langmuir isotherm, applicable only over a limited range of pressure within which $\frac{1}{a}$ and k may be regarded as constant.

Note:

1. The Freundlich isotherm is frequently employed (as an interpolation formula) in connection with adsorption from solution by replacing pressure with concentration.

2. If $\frac{1}{a}$ were equal to 1, the Freundlich isotherm would be equivalent to the Nernst distribution law.

16.10 Vide problem 16.29(b)(ii).

16.11 Langmuir isotherm $\theta = \dfrac{KP}{1 + KP}$ takes the following form

$$\theta = KP \text{ in the low pressure region}$$
$$\theta = 1 \text{ in the high pressure region}$$

Then the θ vs P plot will be almost linear in the low pressure region and also in the high pressure region, but not in the other regions.

□ Since $\dfrac{1}{\theta} = \dfrac{1}{KP} + 1$, $\dfrac{1}{\theta}$ vs $\dfrac{1}{P}$ plot will be linear in the whole range of pressure.

16.12

$$K = \dfrac{1}{P\left(\dfrac{V_m}{V} - 1\right)} \text{ by Eq. (16.3)}$$

$$= \dfrac{1}{(3.5 \text{ atm})\left[\left(\dfrac{180 \text{ cc}}{100 \text{ cc}}\right) - 1\right]} = 0.36 \text{ atm}^{-1}$$

16.13 Assuming Langmuir isotherm

$$\dfrac{P}{V} = \dfrac{P}{V_m} + \dfrac{1}{V_m K} \text{ by Eq. (16.3)}$$

Then P/V against P plot will be linear with

$$\text{Slope} = \dfrac{1}{V_m} = 0.009 \text{ cm}^{-3}$$

$$\text{Intercept} = \dfrac{1}{V_m K} = 1.2 \text{ kPa cm}^{-3}$$

Therefore

$$K = \dfrac{\text{Slope}}{\text{Intercept}}$$

$$= \dfrac{0.009 \text{ cm}^{-3}}{1.2 \text{ kPa cm}^{-3}} = 7.5 \times 10^{-3} \text{ kPa}^{-1}$$

16.14 Vide problem 16.8.

16.15 Vide problem 16.27.

16.16 Vide Section 16.2.

Due to tendency of a system to lower its free energy (G), the surface excess (Γ) of a component of the system may be both positive and negative depending on its interaction with other component(s). For a binary solution a solute will have Γ positive or negative according as it has low or high interaction with the solvent. Vide problem 16.2.

16.17 Vide Section 16.2.

The chemical potential (μ_2) of a surfactant in a binary solution is

$$\mu_2 = \mu_2^\circ + RT \ln C_2 \quad \text{assuming the solution to be ideal and dilute}$$

Then from Eq. (16.1b)

$$\Gamma_{2,1} = -\frac{1}{RT}\left(\frac{\partial \gamma}{\partial \ln C_2}\right)_T$$

which is the required relation.

The implication of this equation is as follows. If a solute accumulates at the interface, its surface excess ($\Gamma_{2,1}$) becomes positive and so $\left(\dfrac{\partial \gamma}{\partial \ln C_2}\right)_T$ becomes negative. That is, surface tension of the system decreases due to dissolution of a substance when it accumulates at the interface relative to the bulk.

16.18 Vide Section 16.10.

16.19 Surface tension of water is enhanced by NaCl due to negative surface excess of NaCl for aqueous solution. But the effect is opposite with other two substances due to their positive surface excess arising from hydrophobic nature of the alkyl group that grows with increasing size of the organic ions. Then such effect will be more intense in the following order

$$\text{NaCl} < \text{acetic acid} < \text{sodium dodecyl sulphonate}$$

Vide problem 16.12.

16.20 Vide problem 16.27.

16.21 Vide problem 16.10.

16.22 $$\Gamma_2 = \frac{\gamma^\circ - \gamma}{RT} \qquad \text{vide previous question}$$

$$= \frac{\left(\dfrac{6 \times 0.0005}{0.02}\right) \text{dyn cm}^{-1}}{(8.314 \times 10^7 \text{ erg K}^{-1} \text{ mol}^{-1})(300 \text{ K})}$$

$$= 6.01 \times 10^{-12} \text{ mol cm}^{-2}$$

16.23 Vide problem 16.28.

16.24 Vide problem 16.11(ii).

16.25 A catalyst increases the speed of a reaction through a change of reaction mechanism into one involving lower free energy of activation.

16.26 Vide Section 16.5.

16.27 (i) The term 'negative catalyst' is often used for designation of a substance that decreases the rate of a reaction when added in small amounts. Such substances function through removal of reaction intermediates or through destruction of some substance that accelerates the reaction rate. For example H_3O^+ acts as a negative catalyst for decomposition of H_2O_2 which is accelerated by BaO_2. This fact has been taken advantage of in the preparation of H_2O_2 by addition of BaO_2 to an acid and not the reverse.

Note: The classification of catalysts as positive and negative is misleading. Vide problem 15.60(f).

(ii) Vide problem 16.21(ii)

(iii) Vide problem 16.20.

16.28 Decomposition of PH_3 on tungsten surface (on which only PH_3 is appreciably adsorbed) follows a unimolecular mechanism, and hence it is of first order at low pressure and zero order at high pressure. Vide Section 16.5(A).

Electric and Magnetic Properties of Molecules

Electric properties, and to less extent, the magnetic properties of molecules determine many of the bulk properties of matter.

17.1 DIPOLE MOMENT AND POLARIZATION

If in a molecule, the centroid of its +ve charge (due to atomic nuclei) is separated from that of its −ve charge (due to electrons) by a distance l the molecule is said to be polar, behaving as an electric dipole of dipole moment ql (in magnitude), where q is the magnitude of either charge. The dipole moment is a vector quantity pointing from the −ve to the +ve charge.

   As electrons and nuclei in any molecule are, to some extent, mobile, when a molecule (polar or non-polar) is placed in an electric field, there will be a small displacement of the electrical charge centres. The molecule is then said to be polarized. The polarization of a substance is usually measured by the induced dipole moment per unit volume of the substance in the direction of the applied field. The dipole moment m of a molecule induced by a field of intensity E is given by

$$m \propto E$$
$$= \alpha E \tag{17.1}$$

where E represents the effective field acting on the molecule and not necessarily the overall applied field, α is a constant called polarizability of the molecule.

17.2 CLAUSIUS–MOSSOTTI EQUATION

Let a uniform field of strength E_0 is produced by two charged parallel plates where the intervening medium comprises a non-polar substance. The intensity of the effective electric field acting on each molecule of the medium can be obtained by assuming a unit positive charge enclosed in a small spherical cavity containing too few molecules. It is the resultant of (1) the field E_0 due to charges on the plates (2) the field $-4\pi I$ due to the charges induced on the surfaces of the medium in contact with the plates (3) the field $+\frac{4}{3}\pi I$ due to the charge induced on the surface of the spherical cavity and (4) the field caused by molecules within the cavity which is zero for a liquid or gas whose molecules are randomly oriented in the absence of an external field. These considerations lead to the following relation

$$\frac{D-1}{D+2} \cdot \frac{M}{\rho} = \frac{4}{3}\pi N_A \alpha \tag{17.2}$$

   In SI, $\quad \dfrac{D-1}{D+2} \cdot \dfrac{M}{\rho} = \dfrac{N_A \alpha}{3\epsilon_0} \quad$ where ϵ_0 is vacuum permittivity

This is Clausius–Mossotti equation which relates dielectric constant (D) and density (ρ) of the medium (which are macroscopic properties) with its polarizability (α) which is a microscopic property. The quantity on the left-hand side of this equation is called the molar polarizability or molar polarization of the medium, denoted by P_m, i.e.

$$P_m = \frac{D-1}{D+2} \cdot \frac{M}{\rho} \tag{17.3}$$

17.3 DEBYE EQUATION

Ordinarily, Clausius–Mosotti equation does not hold for polar molecules. Debye modified this equation to be applicable to polar molecules on the basis of the idea that such molecules are preferentially oriented in the applied field producing a net additional dipole moment in the direction of the field, and this has been shown (using Boltzmann distribution) to be equal to $\mu^2 F/3kT$ per molecule, μ being the permanent dipole moment of the molecules. This corresponds to a contribution $\mu^2/3\ kT$ to the polarizability called orientation polarizability (α_O). Then, for polar molecules, the appropriate equation should be

$$P_m = \frac{D-1}{D+2} \cdot \frac{M}{\rho}$$

$$= \frac{4}{3}\pi N_A (\alpha_D + \alpha_O)$$

$$= \frac{4}{3}\pi N_A \alpha_D + \frac{4}{3}\pi N_A \cdot \frac{\mu^2}{3kT} \tag{17.4}$$

where α_D is the distortion polarizability that arises from the distortion of the molecules by the applied field; α_D actually consists of two parts—electronic polarizability (which corresponds to the field-induced distortion of the electron cloud) and nuclear or atomic polarizability (which corresponds to distortion of the nucleus).

The above equation, known as Debye equation, will hold accurately only for non-associated dilute gases.

Maxwell showed that the dielectric constant D measured with a capacitor and the refractive index n measured by refraction experiment are related by

$$D = n^2$$

provided the same frequency is used in both measurements. If the radiation used in the measurement of n has the frequency of order higher than 10^{11} Hz (microwave region) orientation polarization will not arise (dipole being unable to orient with such frequency). Again, with visible light, the frequency is so high that only electron and not nuclei can move with such frequency and hence only the electron polarizability will arise. Under this condition, the Clausius–Mossotti equation will lead to the Lorentz–Lorenz equation

$$R_m = \frac{n^2-1}{n^2+2} \cdot \frac{M}{\rho} \tag{17.5}$$

where R_m is a scalar quantity called molar refraction which is a constant for a particular substance at all temperatures for a particular wavelength of light (usually in the visible region) used for measurement of the refractive index (n). R_m is largely an additive and to some extent a constitutive (i.e. structure-based) property of the atoms in the molecule.

17.4 DIPOLE MOMENT AND MOLECULAR STRUCTURE

The dipole moment of a chemical bond is called bond moment which is equal to the partial charge on either of the bonded atoms (usually due to the difference in their electronegativities) multiplied by the bond distance.

The dipole moment of a molecule may be thought of as the vector sum of the individual bond moments and the orbital moments of the unshared electrons. Then, the knowledge of dipole moment can provide information regarding the nature of the chemical bonds and their arrangement in the molecule. For this purpose a consistant set of bond moments and group moments is useful, though not completely justified.

The absence of any permanent dipole moment in a molecule (containing no unshared electrons) indicates that either all its bonds are non-polar (i.e. without any bond moment, e.g. H_2) or the molecular structure has certain symmetry elements so that the bond moments vectorially add to zero. Thus carbon dioxide O=C=O molecule is non-polar. Here all rotations around the bond axis are symmetry operations, as are end-to-end rotation and end-to-end reflection. The symmetry criterion is more important than the question of whether the atoms in the molecule are the same or not.

The observed dipole moment of a molecule provides an excellent test of its electronic structure by comparing it with the theoretical value $\int \psi * \hat{\mu}\psi \, d\tau$ calculated from the best wave function ψ. The dipole moment operator is $\hat{\mu} = \sum_i q_i r_i$, where r_i is the displacement vector of charge q_i from the origin. The summation is taken over all the nuclei and electrons in the molecule.

17.5 MAGNETIC PROPERTIES

A magnetic field and an electric current always exist together. The microcurrents (due to electronic motion inside the atoms) within a magnet is similar to a current flowing in a closed circuit. Then electric properties (due to static distribution of electric charges) and magnetic properties (due to electric currents) may be analogously treated (Table 17.1).

The temperature-independent part of χ_m is mostly due to diamagnetic effect which arises from the effect of external magnetic field on the orbital motion of the electrons. On the other hand, the temperature-dependent part of χ_m is connected to paramagnetic effect. The paramagnetism arises from intrinsic magnetic moments of the molecules (almost entirely due to spin and orbital angular momentum of electrons) through their alignment in an applied magnetic field. Below certain temperature some paramagnetic solids transform to a strongly magnetised phase called ferromagnetic, due to locking of electron spins into a parallel arrangement over large domains, while antiparallel spin locking results in weakly magnetized antiferromagnetic phase. The transition temperature is known after Curie for the first case and after Neel for the second case.

The diamagnetic effect appears in all substances. But if the molecules of the substance have the intrinsic magnetic moment, it is overridden by the more powerful paramagnetic effect.

For ions and molecules, the contributions due to electron spin are dominant. Then a measurement of permanent magnetic moment of a molecule can inform us regarding the number of unpaired electrons, the molecule contains. The magnetic moment for n unpaired

Table 17.1

Electric properties/their interrelations	Analogous magnetic properties/their interrelations
1. Electric dipole moment (μ)	1. Magnetic moment ($\mu_{(m)}$) where pole is counterpart of charge
2. Polarization (I) which is induced electric dipole moment per unit volume of a substance in the direction of the applied electric field E	2. Magnetization $I_{(m)}$ which is induced magnetic moment per unit volume in the direction of applied magnetic field (H)
3. Electric susceptibility	3. Magnetic susceptibility

$$\chi = \frac{I}{E} \text{ or } \frac{I}{\epsilon_0 E} = \epsilon_r - 1$$

where ϵ_r is dielectric constant or the permittivity of the medium relative to that of vacuum (ϵ_0), χ is a measure of the ease with which a substance can be electrically polarized. χ is always +ve

$$\chi_{(m)} = \frac{I_{(m)}}{H} \text{ or } \frac{I_{(m)}}{\mu_0 H} = \mu_r - 1$$

where μ_r is the permeability of the medium relative to that of vacuum (μ_0). $\chi_{(m)}$ is a measure of the ease with which a substance can be magnetized. $\chi_{(m)}$ may be +ve (when the substance is paramagnetic) or may be –ve (when the substance is diamagnetic)

4. Debye equation

4. Curie law in terms of molecular parameters

$$P_m = \frac{N_A}{3\epsilon_0}\left(\alpha + \frac{\mu^2}{3kT}\right)$$

$$\chi_m = N_A \mu_0 \left(\alpha_{(m)} + \frac{\mu_{(m)}^2}{3kT}\right)$$

where $P_m = \frac{\epsilon_r - 1}{\epsilon_r + 2} \cdot V_m$ is molar polarizability (which is proportional to the electric susceptibility for a non-polar substance) V_m is molar volume of the substance and α is polarizability of its molecules

$\epsilon_0 = 8.854 \times 10^{-12} \text{CV}^{-1}\text{m}^{-1}$ (or $\text{C}^2\text{N}^{-1}\text{m}^{-2}$)

where $\chi_m = \chi_{(m)} V_m$ is the molar magnetic susceptibility and $\alpha_{(m)}$ is magnetizability

$\mu_0 = 4\pi \times 10^{-7} \text{JC}^{-2}\text{m}^{-1}\text{s}^2$ (or NA^{-2})

electrons is $\sqrt{n(n+2)}\,\mu_B$, where m_B is the magnetic moment ($eh/4\pi m_e$) of an unpaired electron in the field direction, which is taken as the unit of magnetic moment, called Bohr magneton (9.2741×10^{-24} Joule/tesla).

$$\left. \begin{aligned} \chi_m \text{ (due to spin)} &= \frac{N_A \mu_0}{3kT} \mu_B^2 n(n+2) \\ \\ &= \frac{(1.5714 \text{ cm}^3 \text{ K mol}^{-1})}{T/K} \cdot n(n+2) \end{aligned} \right\} \tag{17.6}$$

AUXILIARY PROBLEMS

17.1 What do you mean by a polar molecule? Can a molecule AB be polar if the two bonded atoms A and B have same electronegativity?

17.2 Atoms are non-polar but a molecule may be either polar or non-polar. Discuss.

17.3 Can a molecule of an element be polar? Discuss with example (if any).

17.4 Even if a chemical bond is non-polar, it cannot be 100% covalent. Comment.

17.5 A polar molecule is more polarizable than a non-polar one. Comment.

17.6 Arrange N_2, O_2 and CH_4 in the order of their polarizability. Give reasons for your answer.

17.7 Although oxygen have much higher electronegativity than carbon, CO molecule has a very low dipole moment. Further, the negative end of the dipole is on the less electronegative C atom. Explain.

17.8 Find the dimensions of dipole moment (μ). Both (distortion) polarizability and μ^2/kT have the dimension of volume. Justify.

17.9 Molar polarization P_m of a dielectric is defined by

$$P_m = \frac{D-1}{D+2} \cdot \frac{M}{\rho}$$

Since the dielectric constant (D) and density (ρ) depend on temperature, P_m of a non-polar substance should also do the same. Comment.

17.10 On which of the following does P_m of a non-polar substance depend?
Pressure, temperature and field strength.

17.11 What does P_m of a dielectric signify?

17.12 Justify or criticise the following statements regarding Clausius–Mosotti equation:
(i) it holds accurately only for dilute gases.
(ii) it can be used only for non-polar substances.

17.13 Define monopole and monopole moment.

17.14 State the possibility of the following:
(a) electric dipole moment of an atom.
(b) electric moment of a nucleus.
(c) electric quadrupole moment of a nucleus.
(d) ellipsoidal nucleus.

17.15 An n-pole is defined to be an array of n point charges with an n-pole moment but no lower moment. Can a molecule form a tripole? Which of the following molecules will have (i) dipole moment (ii) quadrupole moment (ii) none?
H_2, O_3, H_2O, CO_2

17.16 H_2 has no dipole moment. Does this imply that H–H bond has no ionic character?

17.17 What multipoles are represented by the molecules CH_4, CH_3Cl and $CH_3 \cdot CH_3$?

17.18 For which of the following molecules, the dipole moment will be zero?

O_2, O_3, P_4, H_2N—⬡—NH_2 O_2N—⬡—NO_2

17.19 Arrange the following molecules in the order of their dipole moment:

Cl—⬡—Cl   Cl—⬡—Cl (with Cl below)   Cl—⬡—CH_3 H_3C—⬡—CH_3

17.20 Which substance will have higher dipole moment in the following pairs and why?
(a) HCl and HI
(b) NaCl and NaI
(c) SO_2 and CO_2
(d) H_2O_2 and C_2H_2
(e) NH_3 and NF_3
(f) H_2O (g) and H_2O (s)

(g) [naphthalene structure] and [azulene structure]

17.21 Which of the following two structures is more likely for the polar molecule ClF_3:
(a) One having three equitorial F atoms.
(b) One having two axial and one equitorial F atoms

17.22 What type of motion will be observed with an electric dipole in (i) uniform electric field (ii) non-uniform electric field.

17.23 The electric field at the mid point between two equal +ve charges is zero. If a small dipole is placed at this point, would any force act on it?

17.24 What will be the electric field at a distance r from (i) a H_2O molecule along the bisector of its bond angle (ii) a CO_2 molecule along its molecular axis? Comment on the expressions derived.

17.25 Write down the expression for potential energy (V) of interaction between two collinear dipoles at a distance r apart. From this expression, would you expect V for CO_2 molecules (having no dipole moment) to be zero.

17.26 Find the potential energy of interaction between two collinear CO_2 molecules when they are separated by a distance r?

17.27 P_m's of $CHCl_3$ (f.p. 209 K) are found to be 48.0, 48.5 and 29.8 all in cm^3/mol at 233 K, 213 K and 203 K respectively. How would you account for this temperature effect?

17.28 The dipole moments of HCl, HBr and HI are 3.44, 2.64 and 1.00 respectively in 10^{-30} c m unit. Arrange them in order of increasing polarizability. If equilibrium, internuclear distances are 127, 141 and 161 pm respectively, find the partial charges on the halogen atoms and examine whether they are in order of the electronegativities of the halogens.

17.29 In which compound, CH_3Cl or CCl_4, would you expect a higher change on chlorine, and why?

17.30 In which compound, CH_4 or CHF_3, would you expect the more polar C–H bond and why?

17.31 The dipole moments of CH_3Cl and $CHCl_3$ should be same. Justify or criticise.

17.32 Calculate the dipole moment of a H_2O molecule assuming that all the electrons in the molecule circulate symmetrically about O-atom taking O–H bond length to be 9.60×10^{-11} m and bond angle 104°. Compare the calculated dipole moment with its actual value 1.85 D, and explain the discrepancy (if any).

17.33 The dipole moments of benzaldehyde, chlorobenzene, p-chlorobenzaldehyde and m-chlorobenzaldehyde are 3, 1.6, 2 and 2.6 debye respectively. Find (a) the angle, the dipole moment of benzaldehyde makes with the 1,4-axis of the benzene ring, (b) the mole fractions of the cis and trans conformers of m-chlorobenzaldehyde in their equilibrium mixture. Is the procedure you adopt equally valid in case of O-chlorobenzaldehyde?

17.34 The density of $SiHBr_3$ (mol wt 269) is 2.69 g cm^{-3} at 25°C, its refractive index is 1.578 and dielectric constant is 3.57. Estimate its dipole moment (neglecting atom polarisability).

17.35 The dielectric constant of $SO_2(g)$ is 1.0099 at 273 K and 1.0057 at 373 K, at 1 atm. Estimate the dipole moment of SO_2. If S–O bond moment is 9.32×10^{-30} cm, estimate the OSO angle.

17.36 For CH_3Cl (g), molar polarization (P_n) is 90 cm/mol at 250 K. With rise in temperature P_m decreases till it falls to the lowest value 50 cm^3/mol. Calculate for CH_3Cl (a) distortion polarizability (b) dipole moment (c) dielectric constant at 500 K and 20 atm.

17.37 The refractive index of water is normally found to be 1.33. Calculate (a) distortion polarizability of H_2O molecule (b) induced dipole moment in an applied field of strength 1.0 kV/cm, taking necessary assumption(s)

17.38 The plates of an evacuated capacitor hold one electronic change per square micrometer. Calculate (a) the applied electric field (b) the field that result when the intervening space between the plates is filled with CH_4 gas of dielectric const 1.000886 at STP (c) the electric susceptibility of CH_4 (d) volume of a CH_4 molecule.

17.39 Calculate (i) the lowest and (ii) the highest potential energy of interaction between two H_2O molecules ($\mu = 1.85$ D) 5.0 nm apart. Can average interaction energy be also calculated from the given data?

17.40 A HCl molecule ($\mu = 1.03$ D) is aligned by an external electric field of strength 5.0 V/cm. An argon atom (polarizability 1.66×10^{-24} cm^3) slowly approaches toward it from the negative side of the dipole. At what separation, HCl molecule will will turn over pointing towards the Ar atom?

17.41 A pure liquid paraffin has density 0.66 g cm^{-3} and refractive index 1.38. If atomic refractions are 1.1 and 2.42 for H and C respectively, find:

(a) the formula of the paraffin.

(b) the refractive index in vapour state of density 0.66 mg cm^{-3} at certain temperature and pressure.

17.42 Compute the molar refraction of CH_2BrI from the following values of molar refractions in certain unit.

$CH_3I = 19.5$, $CH_3Br = 14.5$, $HBr = 9.9$, $CH_4 = 6.8$.

17.43 A nucleus cannot have intrinsic electric dipole moment. Can it have magnetic dipole moment (which is counterpart of electric dipole moment)?

17.44 In dealing with the magnetism of molecular species only its electrons are important. Why?

17.45 Paramagnetic effect should always be temperature-dependent. Comment.

17.46 What is the magnetic analogue of orientation polarization caused by an electric field? How does it originate?

17.47 Can diamagnetic effect depend on temperature?

17.48 A substance having both diamagnetism and paramagnetism always appears to be paramagnetic. Why?

17.49 Electric susceptibility is always +ve. But magnetic susceptibility may be +ve or −ve. Explain.

17.50 What will be the dimensions of magnetic susceptibility?

17.51 Explain the occurrence of Curie temperature and Neel temperature? Why the ferromagnetism and antiferromagnetism are observed with limited substances?

17.52 Classify the following substances as diamagnetic or paramagnetic: C (graphite), polyethylene, N_2, NO, O_2, S (rhombic), Na, NaCl, Ca, Pt, Cu, $CuSO_4 \cdot 5H_2O$.

17.53 Calculate the strength of (i) the electric field (E) and (ii) the magnetic field (B) at a point 1.0×10^{-10} m away from a proton, measured along the axis of its spin. The magnetic moment of the proton is 1.4×10^{-16} amp-m^2.

17.54 Consider a gas at 300 K in a magnetic field of strength 10^4 gauss. If the molecules of the gas are monatomic having a magnetic dipole moment of 1.0×10^{-23} A-m^2, find their (i) thermal energy (ii) magnetic energy. Will the energy of the gas be affected in presence of an electric field?

17.55 Molar magnetic susceptibility* of O_2 is 4.1×10^{-2} cm³/mol at 300 K and 1 atm. Find:

(a) the number of unpaired electrons present in a O_2 molecule.

(b) the permanent magnetic moment of O_2.

(c) the induced magnetic moment per cm³ at the specified temperature and pressure in a magnetic field of strength 1 A/m.

Is the magnetic data on O_2 in agreement with the theory of valency?

[* Here magnetic susceptibility refers to H and not B]

17.56 Estimate spin-only molar susceptibility of MnF_2 crystal. Compare this with the experimental value 0.1462 cm³/mol at 295 K and account for the discrepancy observed.

ANSWERS

17.1 Vide Section 17.1.

☐ Yes. Because electronic imbalance leading to formation of partial charges can arise from the difference in size of the bonded atoms (even when they have same electronegativity), the electronic charge cloud in the region of overlapped orbitals being closer to the smaller atom (and not at the centroid of the molecule).

17.2 The given statement is justified. Because, atomic orbitals (unlike molecular orbitals) being mono-centric, the centroid of the (negative) electronic charge in an atom always lies at the (positive) nucleus. But in a molecule, the centroids of negative and positive charges may not coincide.

17.3 Yes, as in the non-linear O_3 molecule where the electron density on the central O-atom being different from that on the two end atoms, two polar O-O bonds are formed.

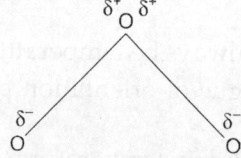

Note: Here dipole moment is low.

17.4 By 100% covalent bond, we mean one where the bond-forming electrons are 50–50 shared by the bonded atoms. This does not happen with any real bond. Even for the simplest diatomic molecule H_2 the bond will invariably have some ionic character due to a definite (though very small) probability of the ionic structure like $\ddot{H}^- H^+$ (that does not violate any rule)

17.5 The given statement is not justified. Because polarizability of a molecule is essentially determined by its size and not by its polarity (measured by its dipole moment). However, a molecule (irrespective of polarity) is more polarizable, more polar is its bonds.

17.6 Polarizability increases in the order $O_2 < N_2 < CH_4$.

☐ Becauae polarizability is higher for a molecule with larger size due to larger, and hence floppier, electron cloud. Although O_2 and N_2 molecules have almost same size, the latter contains less tightly held π-electrons forming triple bond (compared to double bond in O_2).

Note: The polarizability of a molecule is determined by its size and not necessarily by the number of electrons it contains.

17.7 For CO molecule, the outermost electron configuration is $(\sigma_1)^2 (\sigma_1{}^*)^2 (\pi)^4 (\sigma_2)^2$. O-orbitals having lower energy contribute more to the bonding MO's, whereas C-orbitals more to the antibonding MO's. Here eight out of ten electrons (six from O and four from C) occupy bonding MO's where they are held closer to O than to C and thereby tending to neutralise the effect of greater nuclear charge of the oxygen core. This makes dipole moment of CO very low.

☐ This arises from occupation of antibonding MO $(\sigma_1{}^*)$ to which C-orbitals have major contribution.

17.8 Dimensions of μ = dimensions of (charge) (displacement) = tll.

☐ LHS of the Debye Eq. (17.4) has the dimension of vol/mol. Then α_D and μ^2/kT will have the same dimension of volume, since N_A has the dimension mol^{-1}.

17.9 The given statement is not justified on the basis of the Clausius–Mosotti Eq. (17.2) where the RHS is temperature independent. T affects both D and ρ keeping P_m unchanged.

17.10 Pressure, unless it is low enough for the Clausius–Mossotti equation to hold.

17.11 It follows from the Clausius–Mossotti equation that P_m of a dielectric signifies the actual volume of its molecules per mol, provided the dielectric is non-polar and the molecules are spherical.

Note: The molar polarisation P_m does not represent the induced dipole moment per mol.

17.12 (a) The Clausius–Mosotti equation is based on the assumption of a small spherical cavity containing too few molecules. This assumption holds strictly only for dilute gases. Hence the given statement is justified.

(b) The Clausius–Mossotti equation holds also for polar substances, provided the frequency of the applied field is very high such that the molecules cannot orient quickly enough to follow it when the orientation polarization has no chance to occur. Therefore, the given statement is not justified.

17.13 A monopole is a point charge, and the monopole moment is the net charge in an array of several point charges.

17.14 (a) Vide problem 17.2.

(b) An intrinsic electric dipole moment of a nucleus is not possible.

(c) Possible through arrangement of two positive and two negative charges like the positive and negative signs for a d_{z^2} orbital (Fig. 18.7).

(d) Possible. This is justified on the ground that the existence of quadrupole moment in a nucleus inevitably makes the nuclear charge distribution somewhat non-spherical.

17.15 The definition of n-pole demands $n = 2^m$ with $m = 0, 1, 2, 3, ...$ Then formation of a molecular tripole ($n = 3$) is not tenable.

☐ (i) O_3 and H_2O (ii) CO_2 (iii) H_2.

Note: Formation of an n-pole does not happen with any value of n.

17.16 No. Because the absence of dipole moment in H_2 merely implies that there is no net charge separation happenning with the H–H bond. But it does not exclude the possibility of the formation of transient dipoles (resulting from fluctuation of the instantaneous electronic charge density in the molecule) that oscillate ($H^+H^- \leftrightarrow H^-H^+$) imparting some ionic character to the H–H bond.

17.17 CH_4 forms an octupole represented as

CH_3Cl forms a dipole

$CH_3 CH_3$ forms a quadrupole

17.18 O_2, P_4, and

Dipole moment arises in O_3 due to bond polarity and in p-diaminobenzene due to the NH_2-group moment not acting along the bond connecting NH_2-group to the benzene ring.

17.19

Here the group moments, being acted along the connecting bond, are cancelled in the p-compounds. In m-compounds, the dipole moment is lower with dimethyl benzene due to lower moment of the CH_3-group.

17.20 (a) $HI < HCl$. Because Cl-atom, having higher electronegativity attains higer partial charge than I-atom.

(b) $NaCl < NaI$. Because the opposite charges of same amount (1 unit) are separated by a greater distance in NaI.

(c) $CO_2 < SO_2$. Because the bond moments are cancelled in CO_2 molecule (linear) but not in SO_2 molecule (angular).

(d) $C_2H_2 < H_2O_2$. For the same reason as in (c).

(e) $NF_3 < NH_3$. Because the resultant bond moment (although somewhat greater in NF_3) acts in the same direction as the lone pair moment of N in case of NH_3 but in the opposite direction in case of NF_3.

(f) $H_2O\,(s) < H_2O\,(g)$, because of H-bonding that tends to reduce the partial charges in $H_2O\,(s)$.

(g) Because of charge separation leading to the stable

structure in which both the rings contain $6\,(4n + 2)$ π-electrons.

17.21 The structure (b) is more likely. Because a molecule of structure (a) (trigonal planar) will be non-polar.

17.22 (i) The dipole will experience a torque (but no net force) tending to align it along the field by rotation.

(ii) The dipole will experience a net force tending to move it in the direction of the field.

17.23 The (net) force on a dipole in a non-uniform electric field E is the dipole moment multiplied by the field gradient $(\partial E/\partial z)$ along the dipole moment. Here, at the mid point $dE/dz \neq 0$ though $E = 0$. Therefore, a net force would act on the dipole.

17.24 (i) H_2O molecule forms a dipole which can be represented by two partial charges q and $-q$ separated by a distance l along the bisector of its bond angle.

The field E generated by the dipole is the vector sum of the fields generated by each partial charge. Then

$$E = \frac{q}{(r+l/2)^2} - \frac{q}{(r-l/2)^2} \quad \text{in GSU}$$

$$= \frac{q}{r^2}\left[\frac{1}{(1+x)^2} - \frac{1}{(1-x)^2}\right] \quad \text{where } x = \frac{l}{2r}$$

$$= \frac{q}{r^2}[(1 - 2r + 3x^2 - ...) - (1 + 2x + 3x^2 + ...)]$$

$$= -\frac{q}{r^2} \cdot 4x = -\frac{q}{r}\cdot\frac{2l}{r} \quad \text{considering only the leading terms}$$

$$= -\frac{2\mu}{r^2} \tag{I}$$

(ii) CO_2 molecule forms a quadrupole

$$E = -\frac{q}{(r+l)^2} + \frac{2q}{r^2} - \frac{q}{(r-l)^2}$$

$$= \frac{q}{r^2}\left[-\frac{1}{(1+x)^2} + 2 - \frac{1}{(1-x)^2}\right] \quad \text{where } x = \frac{l}{r}$$

$$= \frac{q}{r^2} - (1 - 2x + 3x^2 - ...) + 2 - (1 + 2x + 3x^2 + ...)]$$

$$= -\frac{q}{r^2}\cdot 6x^2 = \frac{6q}{r^2}\cdot\frac{l^2}{r^2}$$

$$= -\frac{6\mu_{\text{quadrupole}}}{r^4} \tag{II}$$

□ It appears from the expressions (I) and (II) that the electric field falls off more rapidly with distance in the direction

dipole → quadrupole

17.25

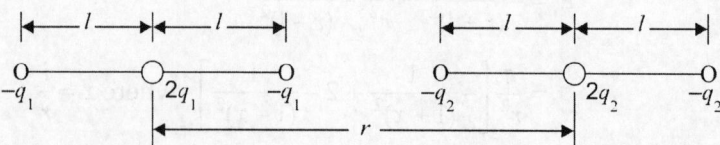

For two dipoles arranged in this way, the potential energy of interaction (v) between them is the sum of the interactions between four pairs of charges

$$V = \frac{q_1 q_2}{r} - \frac{q_1 q_2}{r+l} - \frac{q_1 q_2}{r-l} + \frac{q_1 q_2}{r}$$

$$= \frac{q_1 q_2}{r} \left(1 - \frac{1}{1+x} - \frac{1}{1-x} + 1 \right) \quad \text{where } x = \frac{l}{r}$$

$$= \frac{q_1 q_2}{r} [1 - (1 - x + x^2 - ...) - (1 + x + x^2 + ...) + 1]$$

$$= -\frac{q_1 q_2}{r} \cdot 2x^2; \text{ for } r \gg l, \text{considering only the leading terms}$$

$$= -\frac{q_1 q_2}{r} \cdot \frac{2l^2}{r^2}$$

$$= -\frac{2\mu_1 \mu_2}{r^3}$$

[**Note:** V is not independent of the relative orientation of the two dipoles. For fixed r, V will be least when each partial charge in one dipole is at least distance from the opposite partial charge in the other dipole, i.e. corresponding to the arrangement

O———————O
+   −

O———————O
−   +

□ The above expression for V holds only for dipoles. It is not applicable to CO_2 which forms quadrupole and not dipole. Hence one should not expect V of CO_2 to be zero from this expression notwithstanding the fact that the dipole moment (μ) of CO_2 is zero.

17.26

Considering all all 9 coulombic interactions for this representation

$$V = q_1 q_2 \left(\frac{1}{r} - \frac{2}{r-l} + \frac{1}{r-2l} \right) - 2 \left(\frac{1}{r+l} - \frac{2}{r} + \frac{1}{r-l} \right) + \left(\frac{1}{r+2l} - \frac{2}{r+l} + \frac{1}{r} \right)$$

$$= \frac{q_1 q_2}{r} \left(1 - \frac{2}{1-x} + \frac{1}{1-2x} \right) - 2 \left(\frac{1}{1+x} - 2 + \frac{1}{1-x} \right) + \left(\frac{1}{1+2x} - \frac{2}{1+x} + 1 \right) \quad \text{where } x = \frac{l}{r}$$

$$= \frac{q_1 q_2}{r} \cdot 24 x^4 = \frac{24 q_1 q_2 l^4}{r^5} \quad \text{considering only the leading terms}$$

$$= \frac{24 \, (\mu_1)_{\text{quadrupole}} \, (\mu_2)_{\text{quadrupole}}}{r^5} \quad \text{in GSU} .$$

Note: Here V is positive. However, V may be negative, for example, when one quadrupole molecule lies along the perpendicular of the other. The quadrupole–quadrupole interaction is too weak to have any importance in ordinary cases.

17.27 This follows from the Debye Eq. (17.4). With fall in temperature P_m rises slightly above fp but falls abruptly below fp due to disappearance of the temperature dependent orientation polarizability ($\mu^2/3\,kT$) on solidification, molecular rotation being prevented in solids.

17.28 The polarizability of a molecule usually increases with its size. Therefore the required order of polarizability is

$$HCl < HBr < HI$$

☐ Partial charge on Cl atom in HCl

$$= \frac{\text{dipole moment of HCl}}{\text{internuclear distance in HCl}}$$

$$= \frac{3.44 \times 10^{-30} \; C \cdot m}{127 \times 10^{-12} \; m} = 2.71 \times 10^{-20} \; C$$

Partial charge on Br atom in HBr

$$= \frac{2.64 \times 10^{-30} \; C \cdot m}{141 \times 10^{-12} \; m} = 1.87 \times 10^{-20} \; C$$

Partial charge on I atom in HI

$$= \frac{1.00 \; C \cdot m}{161 \times 10^{-12} \; m} = 0.621 \times 10^{-20} \; C$$

The calculated partial charges on the halogen atoms are in the order of electronegativities of the concerned halogen atoms.

Note: The major factor that determines the order of bond moments is the difference in electronegativities of the bond forming atom and not the bond distance.

17.29 C atom has lower electronegativity in CH_3Cl than in CCl_4 due to less number of (more electronegative) Cl atoms. Then in CH_3Cl, the Cl atom will have higher share of electrons forming C–Cl bond and hence a higher charge.

17.30 C atom has higher electronegativity in CHF_3 than in CH_4. This makes C–H bond more polar in CHF_3.

17.31 For our purpose, we can write

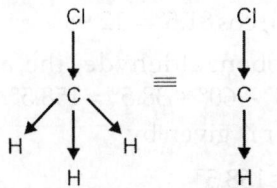

(Since moment of CH_3 gr $\equiv$ moment of C–H bond (to make CH_4 non-polar)

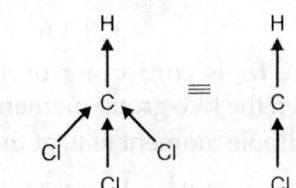

(Since moment of CCl_3 gr $\equiv$ moment of C–Cl bond to make CCl_4 non-polar)

Considering that C–Cl bond is more polar in CH_3Cl, the dipole moment of CH_3Cl will be greater than that of $CHCl_3$.

Note: Here, the purpose being qualitative, C–H bond which is quite less polar than C–Cl bond has not been considered.

17.32 For the given condition, the distribution of charge (in electronic unit) in H_2O molecule can be represented as

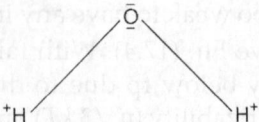

Then dipole moment of H_2O

$$= 2\cos\left(\frac{\angle HOH}{2}\right) \times O\text{–}H \text{ bond moment}$$

$$= 2\cos\left(\frac{\angle HOH}{2}\right) \times \text{charge on H atom} \times O\text{–}H \text{ bond length}$$

$$= 2\cos\frac{104°}{2}(1.60 \times 10^{-19}\,C)(9.60 \times 10^{-11}\,m)$$

$$= 18.91 \times 10^{-30}\,cm$$

$$= 18.91/3.33 \text{ or } 5.68\,D$$

The dipole moment of H_2O thus calculated in much higher than its actual value (1.85 D). Because in this calculation H-atoms are supposed to bear a unit charge which is much greater than their actual partial charge.

17.33 (a) Let θ be the angle of benzaldehyde dipole moment with the 1,4 axis of the benzene ring. Assuming that the dipole moment μ of p-chlorobenzaldehyde is the vector sum of the moment μ_1 of –CHO gr (i.e. of benzaldehyde) and the moment μ_2 of –Cl gr (i.e. of chlorobenzene), we have

$$\mu^2 = \mu_1^2 + \mu_2^2 + 2\mu_1\mu_2 \cos(180 - \theta)$$

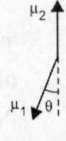

or $\qquad \cos\theta = \dfrac{\mu_1^2 + \mu_2^2 - \mu^2}{2\mu_1\mu_2} = \dfrac{3^2 + 1.6^2 - 2^2}{2(3)(1.6)} = 0.782$

$\therefore \qquad\qquad \theta = 38.5°$

(b) For the *cis* conformer of *m*-chlorobenzaldehyde, the angle between the two group moments is

$$180° - (60° + 38.5°) = 81.5°$$

Then dipole moment μ_C of this isomer is given by

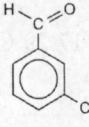

$$\mu_c^2 = \mu_1^2 + \mu_2^2 + 2\mu_1\mu_2 \cos 81.5°$$

$$= 3^2 + 1.6^2 + 2(3)(1.6)\cos 81.5° = 12.98$$

For the *trans* conformer of *m*-chlorobenzaldehyde, the angle between the two group moments is $180° - 60° + 38.5° = 158.5°$.

Then dipole moment of μ_t of this isomer is given by

$$\mu_t^2 = \mu_1^2 + \mu_2^2 + 2\mu_1\mu_2 \cos 158.5°$$

$$= 3^2 + 1.6^2 + 2(3)(1.6)\cos 158.5° = 2.63$$

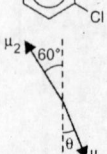

The observed μ^2 for *m*-chlorobenzaldehyde is the weighted average for its *cis* and *trans* conformer. If x is the mole fraction of the *cis* conformer then

$$\mu^2 = x\mu_c^2 + (1-x)\mu_t^2$$

or

$$x = \frac{\mu^2 - \mu_t^2}{\mu_c^2 - \mu_t^2}$$

$$= \frac{2.6^2 - 2.63}{12.98 - 2.63} = 0.40$$

[**Note:** Determination of dipole moment is based on Debye equation which contains μ^2. Hence the average value of μ will be given by the expression used above and not by $\mu = x\mu_C + (1-x)\mu_t$.

□ No, because of appreciable interaction between the two substituent groups due to their close proximity to each other in the *ortho* compound.

17.34 From combination of Eqs (17.4) and (17.5)

$$\frac{D-1}{D+2} \cdot \frac{M}{\rho} = \frac{n^2-1}{n^2+2} \cdot \frac{M}{\rho} + \frac{4}{9}\pi\frac{N_A\mu^2}{kT}, \text{ neglecting atom polarizability}$$

or

$$\mu = \left[\frac{9kT}{4\pi N_A}\left(\frac{D-1}{D+2} - \frac{n^2-1}{n^2+2}\right)\frac{M}{\rho}\right]^{\frac{1}{2}}$$

$$= \left[\frac{9(1.38\times10^{-16}\text{ erg K}^{-1})(298\text{ K})}{4(3.14)(6.02\times10^{23}\text{ mol}^{-1})}\right.$$

$$\left.\left(\frac{3.57-1}{3.57\ 2} - \frac{1.578-1}{1.578+2}\right)\frac{269\text{ g mol}^{-1}}{2.69\text{ g cm}^{-3}}\right]^{\frac{1}{2}}$$

$$= 0.79\times10^{-18}\text{ esu-cm} = 0.79\,D$$

Note: Here $D \neq n^2$, and hence P_m given by the Clausius–Mossotti equation will not be equal to R_m. If n is measured with light in the visible region (as is usually done),

$$R_m = \frac{n^2-1}{n^2+2} \cdot \frac{M}{\rho} \text{ will correspond to electron polarization.}$$

17.35 From Debye Eq. (17.4)

$$\frac{D-1}{D+2} \cdot \frac{RT}{P} = \frac{4}{3}\pi N_A\alpha_D + \frac{4}{9}\pi N_A\frac{\mu^2}{kT}$$

assuming SO_2 gas to behave ideally.

For two different temperatures T_1 and T_2 at constant pressure, it gives

$$\left(\frac{D_1-1}{D_1+2}\cdot T_1 - \frac{D_2-1}{D_2+2}\cdot T_2\right)\frac{R}{P} = \frac{4\pi N_A}{9k}\left(\frac{1}{T_1} - \frac{1}{T_2}\right)\mu^2$$

Then substituting the given data

$$\left[\frac{1.0099-1}{1.0099+2}(273\text{ K}) - \frac{1.0057-1}{1.0057+2}(373\text{ K})\right]\frac{(8.314\times10^7\text{ erg K}^{-1}\text{ mol}^{-1})}{(1.013\times10^6\text{ dyn cm}^{-2})}$$

$$= \frac{4(3.143)(6.022\times10^{23}\text{ mol}^{-1})}{9(1.381\times10^{-16}\text{ erg K}^{-1})}$$

Therefore
$$\mu = 1.623 \times 10^{-18} \text{ esu-cm}$$
$$= 3.336 \times 1.623 \times 10^{-30} \text{ C·m}$$
$$= 5.414 \times 10^{-30} \text{ C·m}$$

□ The bond angle θ of SO_2 is related to its dipole moment μ and bond moment μ' by

$$\mu = 2\mu' \cos\frac{\theta}{2}$$

or
$$\theta = 2\cos^{-1}\left(\frac{\mu}{2\mu'}\right)$$

$$= 2\cos^{-1}\frac{5.41 \times 10^{-30} \text{ C·m}}{2(9.32 \times 10^{-30} \text{ C·m})}$$

$$= 146°21'$$

17.36 (a) According to the Debye Eq. (17.4), P_m is lowest at $1/T = 0$. Then
$$50 \text{ cm}^3 \text{ mol}^{-1} = \tfrac{4}{3}\pi N_A \alpha_D$$
$$= \tfrac{4}{3}(3.14)(6.02 \times 10^{23} \text{ mol}^{-1})\, \alpha_D$$

whence
$$\alpha_D = 1.98 \times 10^{-23} \text{ cm}^3$$

(b) For two different temperatures T_1 and T_2

$$(P_m)_1 - (P_m)_2 = \frac{4\pi N_A \mu^2}{9k}\left(\frac{1}{T_1} - \frac{1}{T_2}\right)$$

or
$$\frac{4\pi N_A \mu^2}{9k} = \frac{(P_m)_1 - (P_m)_2}{\dfrac{1}{T_1} - \dfrac{1}{T_2}}$$

$$= \frac{(90 - 50) \text{ cm}^3 \text{ mol}^{-1}}{\dfrac{1}{250 \text{ K}} - 0 \text{ K}^{-1}} = 10^4 \text{ cm}^3 \text{ K mol}^{-1}$$

Then
$$\mu = \left(\frac{10^4 \times 9\text{K}}{4\pi N_A}\right)^{\frac{1}{2}}$$

$$= \frac{(10^4 \text{ cm}^3 \text{ K mol}^{-1}) \times 9\,(1.38 \times 10^{-16} \text{ erg K}^{-1})}{4(3.14)(6.02 \times 10^{23} \text{ mol}^{-1})}$$

$$= 1.28 \times 10^{-18} \text{ esu·cm} = 1.28\text{D}$$

(c)
$$\frac{D-1}{D+2}\cdot\frac{RT}{P} = \frac{4}{3}\pi N_A \alpha_D + \frac{4}{9K}\pi N_A \frac{\mu^2}{T} \qquad \text{vide problem 17.35}$$

Then, using the given data

$$\frac{D-1}{D+2}\cdot\frac{(8.31 \times 10^7 \text{ erg K}^{-1} \text{ mol}^{-1})(500 \text{ K})}{(20 \times 1.01 \times 10^6 \text{ dyn cm}^{-2})} = 50 \text{ cm}^3 \text{ mol}^{-1} + \frac{10^4 \text{ cm}^3 \text{ K mol}^{-1}}{500 \text{ K}}$$

from (b)

whence
$$D = 1.1056$$

17.37 (a) From Eqs (17.4) and (17.5)

$$\frac{n^2-1}{n^2+2}\cdot\frac{M}{\rho} = \frac{4}{3}\pi N_A \alpha_D \qquad \text{neglecting atom polarizability}$$

or $\alpha_D = \dfrac{3(n^2-1)}{4\pi N_A\,(n^2+2)} \cdot \dfrac{M}{\rho}$

$= \dfrac{3(1.33^2-1)}{4(3.14)(6.02\times10^{23}\ \text{mol}^{-1})(1.33^2+2)} \cdot \dfrac{(18\ \text{g mol}^{-1})}{(1\ \text{g cm}^{-3})}$, taking $\rho_{H_2O} = 1\,\text{g cm}^{-3}$

$= 1.46\times10^{-24}\ \text{cm}^3$

(b) Induced dipole moment

$= \alpha_D E$

$= (1.46\times10^{-24}\ \text{cm}^3)(1.0\times10^3\times\dfrac{1}{300}\ \text{statvolt cm}^{-1})$ by Eq. (17.1)

$= 4.87\times10^{-24}\ \text{esu-cm} = 4.87\ \mu D$

17.38 (a) Applied field $E_0 = \dfrac{\text{charge per unit area of either plate}}{\text{vacuum permittivity}}$

$= \dfrac{1.602\times10^{-19}\ \text{C}/(10^{-6}\ \text{m})^2}{8.854\times10^{-12}\ \text{CV}^{-1}\text{m}^{-1}}$

$= 1.809\times10^4\ \text{V m}^{-1}$

(b) Field in presence of CH_4

$E = \dfrac{E_0}{D}$

$= \dfrac{1.809\times10^4\ \text{V m}^{-1}}{1.000886}$

$= 1.807\times10^4\ \text{V m}^{-1}$

(c) Electric susceptibility

$= D-1$

$= 1.000886-1 = 0.000886$

(d) For non-polar spherical molecules (considering CH_4 molecules to be so), the RHS of Eq. (17.2) represents the actual volume of 1 mol of gas molecules. Then Volume of a CH_4 molecule

$= \dfrac{D-1}{D+2} \cdot \dfrac{RT}{PN_A}$ in GSU

$= \dfrac{1.000886-1}{1.000886+2} \cdot \dfrac{(8.314\times10^7\ \text{erg K}^{-1}\ \text{mol}^{-1})(273\ \text{K})}{(1.013\times10^6\ \text{dyn cm}^{-2})(6.022\times10^{23}\ \text{mol}^{-1})}$

$= 1.098\times10^{-23}\ \text{cm}^3$

17.39 From the result found in problem 17.25, it follows that

(i) the lowest potential energy of interaction

$= -\dfrac{2\mu^2}{r^3}$

corresponding to relative orientation of 0° of the dipoles in collinear position

$= \dfrac{-2(1.85\times10^{-18}\ \text{esu-cm})^2}{(5.0\times10^{-9}\times10^2\ \text{cm})^3}$

$= -5.48\times10^{-17}\ \text{erg}$

(ii) the highest potential energy of interaction

$$= \frac{2\mu^2}{r^3}$$

corresponding to relative orientation of 180° of the dipoles in collinear position

□ No, because the average value depends on temperature which is not given in this problem.

Note: In the given problem $r \gg l$. If, however, r is comparble to l, then the most favourable (a) and most disfavourable (b) orientations of the dipoles will be as under

Corresponding to the
lowest potential energy

Corresponding to the
highest potential energy

17.40 A molecule of HCl, due to its dipole moment (μ), will possess an energy ($-\mu E$) in an electric field (E). It will turn over at a distance r when the energy required ($2\mu E$) for this is just provided by the dipole-induced dipole interaction between HCl and Ar molecules [given by Eq. (3.19)], i.e.

$$2\,\mu E = \frac{\alpha\mu^2}{r^6} \quad \text{in GSU}$$

or $$r = \left(\frac{\alpha\mu}{2E}\right)^{\frac{1}{6}}$$

$$= \left[\frac{(1.66 \times 10^{-24}\ \text{cm}^3)\,(1.03 \times 10^{-18}\ \text{esu-cm})}{2\,(5.0 \times \frac{1}{300}\ \text{Statvolt cm}^{-1})}\right]^{\frac{1}{6}}$$

$$= 1.93 \times 10^{-7}\,\text{cm}$$

17.41 (a) Let the formula of the paraffin be $C_x H_{2x+2}$.

Then molar mass, $M = 12x + (2x+2)\,1.008$

$$= 14.016x + 2.016$$

molar refraction, $R_m = 2.42x + (2x+2)\,1.1$

$$= 4.62x + 2.2$$

Then by Eq. (17.5)

$$4.62x + 2.2 = \frac{1.38^2 - 1}{1.38^2 + 2} \cdot \frac{14.016x + 2.016}{0.66}$$

whence $x = 5$ (nearest integer)

(b) Since M and R_m are constant for a particular substance

$$\frac{n^2 - 1}{n^2 + 2} \cdot \frac{1}{\rho} = \text{constant}$$

Then from the given data

$$\frac{n^2 - 1}{n^2 + 2} \cdot \frac{1}{0.66 \times 10^{-3}\,\text{g cm}^{-3}} = \frac{1.38^2 - 1}{1.38^2 + 2} \cdot \frac{1}{0.66\,\text{g cm}^{-3}}$$

whence $n = 1.00$

Note: For calculation of n it is not required to calculate R_m and M.

17.42 Since R_m is an additive property

$$R_m(CH_2) = R_m(CH_3Br) - R_m(HBr) = 14.5 - 9.9 = 4.6$$

$$R_m(H) = \frac{[R_m(CH_4) - R_m(CH_2)]}{2} = \frac{(6.8 - 4.6)}{2} = 1.1$$

$$R_m(Br) = R_m(HBr) - R_m(H) = 9.9 - 1.1 = 8.8$$

Then
$$R_m(CH_2BrI) = R_m(CH_3I) + R_m(Br) - R_m(H)$$
$$= 19.5 + 8.8 - 1.1 = 27.2$$

Alternatively

$$R_m(CH_2BrI) = R_m(CH_3Br) + R(CH_3I) - R_m(CH_4)$$
$$= 14.5 + 19.5 - 6.8 = 27.2$$

Note: For alternative method (which finds ready application to simple compounds) the given value of $R_m(HBr)$ is unnecessary.

17.43 Yes, provided the nucleus possesses spin.

17.44 Because the nuclear contribution (if any) to the magnetism is insignificant compared to that originating from spin and orbital angular momenta of electrons.

17.45 No. Paramagnetism will be temperature-independent if it involves the interaction of the excited state with the ground state of the molecule under the influence of the applied magnetic field.

17.46 Temperature-dependent paramagnetism. It originates from spin and orbital angular momenta of electrons with predominating effect of spin.

17.47 Diamagnetism results from the change of the orbital motion of the electrons by the applied magnetic field with consequent appearance of an opposing magnetic field (by Lenz's law). Being an intra-atomic effect, determined only by the size and shape of the molecular orbitals, diamagnetism is normally temperature-independent. However, at very high temperature, it is likely to be temperature-dependent due to formation of excited molecules.

17.48 Because electron spin, which is mainly responsible for paramagnetism, has dominating effect over the orbital motion of the electron which determines the diamagnetism. This happens because within a molecule there are strong internal electric fields (directed along the chemical bonds) which greatly control the orbital angular momenta of the electrons (compared to the external magnetic field) without influencing electron spin.

17.49 Because the electric susceptibility is determined only by orbital motion of the electrons, but the magnetic susceptibility is determined jointly by the spin and orbital motion of the electrons, with their effects mutually opposing.

17.50 The answer depends on how magnetic susceptibility (χ) is defined. For its normal definition as

$$\chi = \frac{\text{magnetic dipole moment per unit volume }(I_m)}{\text{magnetic field strength }(H)}$$

χ will be dimensionless.

If, however, χ is defined as

$$\chi = \frac{I_{(m)}}{B_0} = \frac{I_{(m)}}{\mu_0 H}$$

(where B_0 is magnetic induction for vacuum and μ_0 is vacuum permeability) then χ will have the dimension of μ_0^{-1} (i.e. $\mu_0\chi$ is dimensionless).

17.51 Ferromagnetism arises from locking of electron spins into a parallel arrangement over large domains, and antiferromagnetism from locking into antiparallel arrangement. Then occurrence of a critical temperature, called Curie or Neel temperature, is quite likely above which the thermal motion in the material more than compensates the interaction (energy responsible for spin locking and thereby causing transformation) into a simple paramagnetic phase.

☐ Because spin locking becomes possible only for certain interatomic distances and certain orbital radii.

17.52 Polyethylene, NO, O_2, Na, Pt and $CuSO_4 \cdot 5H_2O$ are paramagnetic because they contain unpaired electron in the molecule.

Note: Polyethylene, NO and O_2 are paramagnetic but non-conductor of electricity (due to absence of free electron).

17.53 (i) The strength of electric field (E) at distance r from a point electric charge q in vacuum is

$$E = \frac{q}{4\pi \, \epsilon_0 \, r^2} \quad \text{in SI}$$

$$= \frac{1.602 \times 10^{-19} \, C}{4\pi \, (8.854 \times 10^{-12} \, CV^{-1} m^{-1}) \, (1.0 \times 10^{-10} \, m)^2}$$

$$= 1.44 \times 10^{11} \, V \, m^{-1}$$

(ii) The strength of magnetic field (B) due to a magnetic dipole of moment $\mu_{(m)}$ in vacuum is

$$B = \frac{\mu_0 \mu_{(m)}}{2\pi r^3} \quad \text{in SI}$$

$$= \frac{(4\pi \times 10^{-7} \, NA^{-2}) \, (1.4 \times 10^{-26} \, A \cdot m^2)}{2\pi (1.0 \times 10^{-10} \, m)^3}$$

$$= 2.8 \times 10^{11} \, NA^{-1} \, m^{-1} \text{ or } T$$

Note: $H = \dfrac{B}{\mu_0} = \dfrac{\mu_{(m)}}{2\pi r^3}$.

17.54 (i) Average thermal energy per molecule

$$= \tfrac{3}{2} kT$$

$$= \tfrac{3}{2} (1.38 \times 10^{-23} \, JK^{-1}) \, (300 \, K) = 6.21 \times 10^{-21} \, J$$

(ii) Magnetic energy per molecule

$$= 2\mu_{(m)} B$$

$$= 2(1.0 \times 10^{-23} \, A \cdot m^2) \, (1.T) = 2.0 \times 10^{-23} \, J$$

☐ Here the molecules do not have any electric polarity and hence their energy will not be affected in presence of an electric field unless the field strength is very high.

Note: The thermal energy of a molecule is much greater than the magnetic energy. Then the energy exchange in random molecular collision will significantly hamper the alignment of the magnetic dipoles with the external magnetic field.

17.55 (a) Substituting the given data in Eq. (17.6)

$$4.1 \times 10^{-2} \, cm^3 mol^{-1} = \left(\frac{1.571 \, cm^3 K \, mol^{-1})}{300 \, K} \right) n(n+2)$$

whence $\qquad n = 2$, the nearest integer

(b) The permanent magnetic moment of a O_2 molecule

$$= \sqrt{n(n+2)}\,\mu_B \quad \text{where } \mu_B \text{ is Bohr magneton}$$
$$= \sqrt{2(2+2)} \cdot (9.274 \times 10^{-24}\ JT^{-1})$$
$$= 26.231\ JT^{-1}$$

(c) Induced moment

$$\grave{I}_{(m)} = H\chi_{(m)}$$
$$= \frac{H\chi_m}{V_m}$$
$$= H\chi_m \frac{P}{RT}$$
$$= \left(\frac{1.0\ JT^{-1}}{m^3}\right)(4.1 \times 10^{-2} \times 10^{-3}\ L\ mol^{-1}) \times \frac{(1\,atm)}{(0.082\ L\ atm\ K^{-1}\ mol^{-1})(300\ K)}$$
$$= 1.67 \times 10^{-6}\ JT^{-1}\ per\ m^3$$
$$= 1.67 \times 10^{-12}\ JT^{-1}\ per\ cm^3$$

☐ The number of unpaired electrons (2) calculated from the given magnetic data is in agreement with the molecular orbital theory of valency though not with the less sophisticated valence bond theory.

Note: The fractional value of n provided by the exact solution of the Eq. (17.6) is somewhat less than 2. This is likely to be due to some orbital contribution to χ_m.

17.56 MnF_2 crystal is paramagnetic due to its constituent Mn^{2+} ions each containing 5 unpaired electrons. For $n = 5$, the Eq. (17.6) gives

$$\chi_m = \left(\frac{1.571\ cm^3\ K\ mol^{-1}}{295\ K}\right)(5)(5+2)\ \text{at 295 K}$$

$$= 0.1864\ cm^3\ mol^{-1}$$

☐ x_m thus calculated is higher than the experimental value in the following percentage

$$\frac{0.1864 - 0.1462}{0.1462} \times 100 = 27.5\%$$

This can be attributed to considerable locking of electron spins into antiparallel arrangement (i.e. antiferromagnetic effect) that impairs the validity of the Eq. (17.6).

UNIVERSITY QUESTIONS

17.1 An external electric field is incident upon a polarisable fluid dielectric medium. Explain the effect of field on the medium and give qualitative description of the various contributions that give rise to the effect.   (Burdwan BSc(H), 1994)

17.2 Explain why the polarizability of a molecule decreases at high frequency.
          (Calcutta BSc(H), 2014)

17.3 Arrange the following molecules in order of increasing polarizability with reason. HCl, HF, HI, HBr   (Burdwan BSc(H), 1997)

17.4 Explain the origin of dipole moment in a molecule with special emphasis on the fact that some molecules having heteronuclear atoms may have zero dipole moment.   (Burdwan BSc(H), 1991)

17.5 Define dipole moment of a molecule. How is it expressed? Which of the molecules CH_4, CH_3Cl, CH_2Cl_2, $CHCl_3$ and CCl_4 possess dipole moment? Why? Arrange them in order of their dipole moment. (Calcutta BSc(H), 1987)

17.6 What will be the direction of dipole moment in H_2O molecule? Give symmetry arguments as far as practicable. (Burdwan BSc(H), 1992)

17.7 (a) Explain with reasons, why para-dichlorobenzene is non-polar but para-dihydroxybenzene is polar. (Calcutta BSc(H), 2006)

(b) Which one of the following pair will have higher dipole moment. *p*-dinitro-benzene and *p*-dihydroxy benzene. Give reasons for your answer.
(Jadavpur BSc, 2012)

17.8 The dipole moment of water molecule is 1.84 D. Calculate O–H bond moment, the bond angle being 105°. (Calcutta BSc(H), 2001)

17.9 Calculate the percentage ionic character of H–Cl bond if the distance between the two atoms is 1.275 Å and its dipole moment is 1.03 D. (Sambalpur BSc, 2006)

17.10 The dipole moment of ortho-xylene is 0.693 D. Find the dipole moment of toluene. (Calcutta BSc(H), 2014)

17.11 The dipole moments of chlorobenzene and nitrobenzene are respectively 1.7 D and 3.9 D. Calculate the dipole moment of *m*-dichlorobenzene and *o*-chloronitro-benzene. (Calcutta BSc(H), 1985)

17.12 Find the dipole moment (μ) for $CHCl_3$ using vector addition of bond moments. Given the dipole moment of CH_3Cl is 6.24 D. The experimental value of μ for $CHCl_3$ is 3.37 D. Explain the anomaly. (Burdwan BSc(H), 2006)

17.13 What are the molar polarisation and molar refraction of a substance? (Calcutta BSc(H), 1995)

17.14 Write down an expression for molar polarisation of a substance in terms of experimentally measurable parameters. (Calcutta BSc(H), 1992)

17.15 Polarisability of CCl_4 is independent of temperature whereas that of $CHCl_3$ changes with temperature. Explain. (Burdwan BSc(H), 2009)

17.16 Orientational polarizability of polar molecules is inversely proportional to temperature. Justify. (Jadavpur BSc(H), 2012)

17.17 Find the dimension of μ^2/kT, the terms have their usual meanings. (Calcutta BSc(H), 2005)

17.18 The refractive index of benzene at 298 K for light of wavelength 600 nm is 1.498. Density of benzene is 0.844×10^3 kg m^{-3}. Estimate the polarizability of benzene assuming atomic polarisation as 1% of electronic polarization.
(Jadavpur BSc(H), 2011)

17.19 The molar polarization of the vapour of a compound was found to vary linearly with T^{-1}, and is 75.74 cm^3 mol^{-1} at 320 K and 71.43 cm^3 mol^{-1} at 421.7 K. Calculate the distortion polarisability and dipole moment of the molecule.
(Calcutta BSc(H), 2013)

17.20 The molar polarisation of *cis*-dichloroethylene (MW = 97) at 302 K is 93.13 cc and at 403.8 K is 74.35 cc. Calculate its dipole moment value. Also determine its refractive index if its density be 1.28 gm/cc at 302 K, assuming atomic polarization is 5% of electronic polarization. (Burdwan BSc(H), 2012)

17.21 Calculate the molar refraction of allyl chloride at a temperature at which its density is 0.938 g cm^{-3}. The experimentally observed value of refractive index at this temperature is 1.3715. (Delhi BSc, 2004)

17.22 The refractive index of gaseous normal alkane C_xH_{2x+2} is 1.00139 at STP. If the atomic refractions for H and C are 1.10 and 2.42 cc/mol respectively, find the molecular formula of the alkane. (Calcutta BSc(H), 1984)

17.23 The refractive index of CCl_4 for D line at 20°C is 1.457 and its density is 1.595 g cm^{-3}. Calculate the atomic refraction of chlorine if the atomic refraction of carbon is 2.42. (Jadavpur BSc(H), 2012)

17.24 Differentiate diamagnetism, paramagnetism and ferromagnetism with reference to:
  (i) magnetic permeability
  (ii) specific magnetic susceptibility
  (iii) lines of force through the medium (Calcutta BSc(H), 2008)

17.25 What is Bohr magnetron? Find its value in SI unit. (Jadavpur BSc(H), 2011)

17.26 A sample of an organometallic compound was placed in a susceptibility balance. The following data were observed:
  $F = 24.7$ dyn,; $A = 0.18$ cm^2, $B = 4000$ G
  Find the magnetic susceptibility of the compound. (Jadavpur BSc(H), 2011)

17.27 For O_2(g), $\chi_M = 3449.0 \times 10^{-6}$ cc per mole at 25°C. Assuming diamagnetic molar susceptibility to be negligible compared to χ_M, calculate the internal magnetic moment of this molecule. (Calcutta BSc(H), 2007)

17.28 State briefly how the number of unpaired electrons in a molecule can be found out from values of magnetic moment. (Calcutta BSc(H), 2002)

KEY TO UNIVERSITY QUESTIONS

17.1 Vide Sections 17.1 and 17.2.

17.2 Because above certain frequency of the order of 10^{11} Hz, orientation polarizability will not arise, the dipoles being unable to orient with such frequency. Again for frequency of the order of 10^{14} Hz or higher, even nuclear polarizability will not arise because nuclei being relatively massive, cannot move with so high frequency.

17.3 Vide problem 17.28.

17.4 Vide Section 17.4.

17.5 All the molecules except for the first and the last. For CH_4 and CCl_4, the bond moments vectorially add to zero due to their symmetrical (perfectly tetrahedral) structure.
  □ The order of dipole moment is $CHCl_3 < CH_2Cl_2 < CH_3Cl$ because the moment of the more polar C–Cl bond increases in this order. Vide problems 17.29 and 17.30.

17.6 The direction of dipole moment in H_2O molecule will be along its two fold symmetry axis (C_2) that bisects H–O–H bond angle. Because the component of the dipole associated with one OH bond in H_2O molecule in the direction perpendicular to the C_2 axis is cancelled by an equal but opposite component of the second OH bond.

17.7 (a) Because in *p*-dichlorobenzene, the two C–Cl bond moments act along the same line (1,4-axis of benzene ring) in opposite directions and hence cancell each other. But in *p*-dihydroxybenzene, the two OH group moments do not act along the same line and hence cannot cancel each other.

  (b) *p*-dihydroxybenzene will have higher dipole moment, because the two group moments in this compound, unlike *p*-dinitrobenzene, are not collinear.

17.8 Dipole moment of $H_2O = 2\cos\left(\dfrac{\angle HOH}{2}\right) \times$ O–H bond moment etc.

17.9 Percentage ionic character of H–Cl bond

$$= \frac{\text{Observed dipole moment of HCl}}{\text{Hypothetical dipole moment if H–Cl bond were fully ionic}} \times 100$$

$$= \frac{\text{Observed dipole moment}}{\text{Electronic charge} \times \text{bond distance}} \times 100$$

$$= \frac{(1.03 \times 10^{-18} \text{ esu-cm})}{(4.803 \times 10^{-10} \text{ esu})(1.275 \times 10^{-8} \text{ cm})} \times 100 = 16.82$$

Note: Any chemical bond is likely to have some ionic character. This is not unjustified even with homonuclear diatomic molecules (e.g. H_2) which possess transient, but no net dipole moment. Vide problem 17.4.

17.10 In o-xylene, the two CH_3-group moments, each equivalent to the dipole moment of toluene (μ), act at angle $\theta = 60°$ between them. The resultant moment, i.e. dipole moment of o-xylene (μ') is then given by

$$\mu' = 2\mu\cos(\theta/2), \text{ ignoring any interaction between the two } CH_3\text{-groups}$$

or $\qquad \mu = \dfrac{\mu'}{2\cos(\theta/2)}$

$$= \dfrac{0.693\,D}{2\cos(60°/2)}$$

17.11 Vide problem 17.33.

17.12 Vide problem 17.31.

17.13 Vide Section 17.2, Section 17.3 and problem 17.11.

17.14 Equation (17.4).

17.15 It is due to the temperature-dependent orientation polarizability that arises with $CHCl_3$, the molecules being polar unlike CCl_4.

17.16 Orientation polarizability arises from the tendency of the molecular dipoles to be oriented along the direction of the applied electric field. This tendency is reduced by the thermal (random) molecular motion that increases with temperature. This justifies the given statement qualitatively.

17.17 Vide problem 17.8.

17.18 For benzene (which is nonpolar), $\mu = 0$. Then combination of Eqs (17.4) and (17.5) gives

$$4\mu N_A \alpha_D = \underbrace{\frac{n^2 - 1}{n^2 + 2} \frac{M}{\rho}}_{\substack{\text{electronic} \\ \text{polarization}}} + \underbrace{\frac{n^2 - 1}{n^2 + 2} \frac{M}{\rho} \times \frac{1}{100}}_{\text{atomic polarization}}$$

or $\qquad \alpha_D = \dfrac{1.01}{4\pi N_A} \cdot \dfrac{n^2 - 1}{n^2 + 2} \dfrac{M}{\rho}$

$$= \frac{1.1}{4(3.143)(6.022 \times 10^{23} \text{ mol}^{-1})} \times \frac{1.498^2 - 1}{1.498^2 + 2} \times \frac{78.108 \text{ g mol}^{-1}}{0.844 \times 10^3 \times 10^3 \text{ g}}$$
$$\frac{}{(10^2)^3 \text{ cm}^3}$$

$$= 3.619 \times 10^{-24} \text{ cm}^3$$

17.19 First calculate μ by Eq. (17.4), vide problem 17.36. Then find α_D using the calculated value of μ and the given data at any one temperature.

17.20 For calculation of μ vide previous problem. Refractive index (n) is given by

$$P_m = \underbrace{\frac{n^2-1}{n^2+2}\cdot\frac{M}{\rho}}_{\substack{\text{electronic}\\\text{polarization}}} + \underbrace{\frac{n^2-1}{n^2+2}\frac{M}{\rho}\times\frac{5}{100}}_{\text{atomic polarization}} + \frac{4\mu N_A\mu^2}{9kT}$$

17.21 Substituting the given data for allyl chloride ($CH_2= CH - CH_2Cl$) in Eq. (17.5)

$$R_m = \frac{1.3715^2-1}{1.3715^2+2}\times\frac{62.584\text{ g mol}^{-1}}{0.938\text{ g cm}^{-3}}$$

$$= 15.146\text{ cm}^3\text{ mol}^{-1}$$

17.22
$$R_m = \frac{n^2-1}{n^2+2}\cdot\frac{RT}{P}\quad\text{by Eq (17.5) assuming ideal gas equation to hold}$$

Then using the give data

$$[2.42x + (2x+2)\times 1.1]\text{ cc·mol}^{-1} = \frac{1.00139^2-1}{1.00139^2+2}\cdot\frac{(8.314\times10^7\text{ erg K}^{-1}\text{mol}^{-1})(298\text{ K})}{(1.013\times10^6\text{ dyn cm}^{-2})}$$

whence $x = 4$, the nearest integer.

Then the molecular formula of the alkane is C_4H_{10}.

17.23 Let x be the atomic refraction of Cl. Then by Eq. 17.5

$$2.42 + 4x = \frac{1.457^2-1}{1.457^2+2}\cdot\frac{153.81\text{ g mol}^{-1}}{1.595\text{ g cm}^{-3}}$$

whence $\qquad x = 5.96\text{ cm}^3\text{ mol}^{-1}$

17.24

Type of magnetism	Magnetic susceptibility $\chi_{(m)}$	Magnetic permeability (μ_r)	Concentration of lines of force of the applied field H
Diamagnetism	Negative, small (usually of the order of 10^{-6} c.g.s) independent of magnetic field intensity and temperature	$\mu_r < 1$	Less in a diamagnetic substance than in vacuum that causes the substance to move to the regions of lower field intensity
Paramagnetism	Positive, small (usually of the order of 10^{-3} c.g.s), independent of the magnetic field intensity and usually decreases with increasing temperature (almost linearly with $1/T$)	$\mu_r > 1$	Greater in a paramagnetic substance than in vacuum that causes the substance to move to the regions of higher field intensity
Ferromagnetism	Positive, large (usually of the order of 10^2 c.g.s) dependent on the magnetic field intensity, temperature and previous history Above a particular temperature (called Curie temperature) $\chi_{(m)}$ decreases with increasing temperature like a paramagnetic substance. But below this temperature the rate of decrease is much higher	$\mu_r \gg 1$	Much greater than in a paramagnetic substance

17.25 Vide Section 17.5.

17.26 Here a cylindrical sample of the experimental substance is suspended vertically between the poles of an electromagnet such that at its upper end, the magnetic field (B) is negligible. The force (F) acting on the substance is

$$F = \int_O^B B\chi \, A \, dx \, \frac{dB}{dx} = \tfrac{1}{2}\chi \, AB^2$$

where dx is the length of the cylinder in the field B and A is the cross-sectional area of the specimen having susceptibility $\chi_{(m)} = I_{(m)}/B$ relative to the surrounding atmosphere (air). Then

$$\chi_{(m)} = \frac{2F}{AB^2}$$

$$= \frac{2(24.7 \times 10^{-5} \text{ N})}{(0.18 \times 10^{-4} \text{ m}^2)(4000 \times 10^4 \text{ T})^2}$$

$$= 171.5 \text{ Nm}^{-2}\text{T}^{-2}$$

Multiplication of this with μ_0 gives the dimensionless value of $\chi_{(m)}$ (vide problem 17.50).

17.27
$$\chi_m = N_A\mu_0 \left(\alpha_{(m)} + \frac{\mu_{(m)}^2}{3kT} \right)$$

[which is magnetic analogue of the Debye equation $P_m = \dfrac{N_A}{3 \, \epsilon_0}\left[\alpha + \dfrac{\mu^2}{3kT} \right]$].

The part of χ_m which is proportional to $\alpha_{(m)}$ is temperature-independent. It is the counterpart of the electrical distortion polarization connected to diamagnetism. If it is neglected, then

$$\mu_{(m)} = \left(\frac{3kT\chi_m}{N_A\mu_0} \right)^{\frac{1}{2}}$$

$$= \left[\frac{3(1.381 \times 10^{-23} \text{ JK}^{-1})(298.15 \text{ K})(3449.0 \times 10^{-6} \times 10^{-6} \text{ m}^3 \text{ mol}^{-1})}{(6.022 \times 10^{23} \text{ mol}^{-1})(4\pi \times 10^{-7} \text{NA}^{-2})} \right]^{\frac{1}{2}}$$

$$= 7.502 \times 10^{-24} \text{ A m}^2$$

17.28 The magnetic moment of a molecule containing n unpaired electrons is $\sqrt{n(n+2)}\,\mu_B$ where μ_B is Bohr magneton. Then n can be found out from the knowledge of magnetic moment of the molecule concerned.

18

Quantum Mechanics

Classical (Newtonian) mechanics cannot account for many properties of microscopic particles. This is due to their significant wave-like character (originally suggested by de Broglie in 1924). In 1926 Erwin Schrodinger and Werner Heisenberg independently developed a new kind of mechanics, called *quantum mechanics*. This is called so because this embodies the key note of *quantization of energy* which has been simply taken on postulational basis to explain the peculiar spectral distribution of black-body radiation, photoelectric effect, compton effect and line spectra of gases.

18.1 BLACK-BODY RADIATION

A hot solid produces continuous radiation (i.e. of all wavelengths) whose spectral distribution (i.e. the energy radiated with each range of wavelength) at a particular temperature usually depends on the material of the solid but it is independent of the latter with the radiation inside an isothermal hollow body, called *black-body radiation*. For black-body radiation, the density (e_v) of radiant energy per unit frequency range depends only on the region of frequency (v) and temperature (T) of its source as follows

$$e_v = \frac{8\pi h v^3}{c^3} \cdot \frac{1}{e^{\frac{hv}{kT}} - 1} \tag{18.1a}$$

In terms of wavelength (λ), this equation takes the following form

$$e_v = \frac{8\pi h c}{\lambda^5} \cdot \frac{1}{e^{\frac{hc}{\lambda kT}} - 1} \tag{18.1b}$$

[Since $v = \frac{c}{\lambda}, |dv| = \frac{c}{\lambda^2}|d\lambda|$ and $e_v |dv| = e_\lambda |d\lambda|$]

The Eq. (18.1a) was first deduced by Max Planck on the basis of the assumption that radiation is emitted and absorbed by oscillating dipoles in the solid that can only have energies that are integral multiples of hv, where h is a constant (later called Planck's constant). c is the speed of light (Fig. 18.1).

Initially the curve (a) rises quickly but after passing through the maximum, it falls off comparatively slowly. The reverse happens with the curve (b).

18.2 PHOTOELECTRIC EFFECT

Following Planck's idea that the energy of an oscillator is quantized, Einstein considered electromagnetic radiation to be made up of particles (photons) each having an energy hv, where v is frequency of radiation and h is Planck's constant.

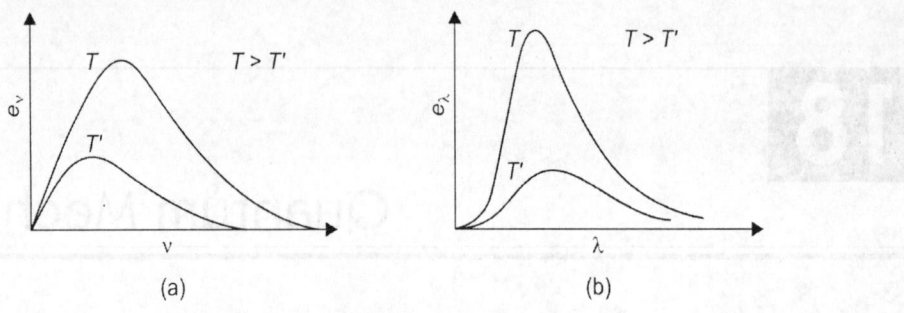

(a) (b)

Fig. 18.1

Einstein viewed photoelectric effect (the emission of electron from a surface caused by an electromagnetic radiation) as knocking off an electron from the emitting surface by a photon of incident radiation to explain the instantaneous nature of the process, the threshold frequency of the radiation and the dependence of kinetic energy of photoelectrons on the frequency (and not intensity) of the radiation involved. The maximum kinetic energy ($K \cdot E_{max}$) of the photoelectrons is given by

$$K \cdot E_{max} = h\nu - \phi \tag{18.2}$$

where ϕ is photoelectric work function which determines the threshold frequency (ϕ/λ).

18.3 COMPTON EFFECT

Scattering of electromagnetic radiation by electron accompanied by recoil of the latter is called the Compton effect. Compton observed this by allowing a beam of x-ray (high frequency radiation) to impinge on a graphite block (which contains essentially free electrons). Here emission of electron is caused by radiation which suffers scattering instead of absorption. This is unlike photoelectric effect.

Compton held that the observed effect arises from elastic collision of the photons of the incident light with the free electrons in the scattering block as shown in Fig. 18.2.

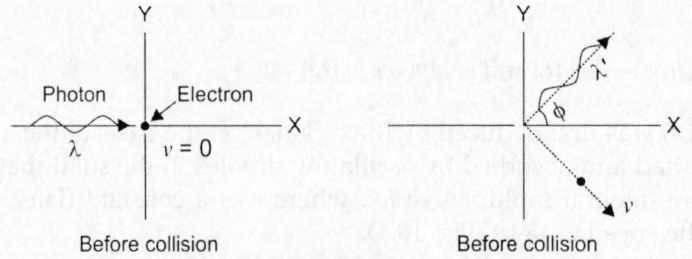

Before collision Before collision

Fig. 18.2

Here
$$\frac{hc}{\lambda} = \frac{hc}{\lambda'} + (m - m_0)c^2$$

and
$$\Delta\lambda = \lambda' - \lambda = \frac{h}{m_0 c}(1 - \cos\phi) \tag{18.3}$$

where m_0 is the mass of electron at rest and m that at speed v, $\Delta\lambda$ is the Compton shift and $h/m_0 c$ is the Compton wavelength of electron.

18.4 LINE SPECTRA

While hot solids produce continuous spectrum, hot gases produce line spectra. The simplest spectrum is that of hydrogen atoms. Here the frequency of the spectral lines $\bar{v}$ (in wave number) can be expressed in the following form

$$\bar{v} = R_H \left(\frac{1}{n_1^2} - \frac{1}{n_2^2} \right)$$

where n_1 and n_2 are integers such that $n_1 \geq 1$ and $n_2 > n_1$ and R_H is a constant called *Rydberg constant*.

To explain the above expression Niels Bohr suggested a model of H-atom where its electron moves only in circular orbits round the nucleus such that the angular momentum of the electron is an integral multiple (n) of $\hbar \, (= h/2\pi)$. This leads to only some restricted values of energy, i.e. quantization of energy of a H-like atom expressible in the following form

$$E = -\frac{Z^2 m e^4}{2n^2 \hbar^2} \tag{18.4}$$

where Z is nuclear charge, m is mass of electron and e is its charge.

Bohr further assumed that radiant energy is emitted or absorbed by a H-atom only when its electron jumps from one orbit to another of lower or higher energy with energy difference equal to hv where v is the frequency of the emitted or absorbed radiation corresponding to wave number $\bar{v}$.

18.5 WAVE-PARTICLE DUALITY

Light can behave as waves or particles (photons) depending on the situation. This is hinted by the following relation between the wavelength λ of light and the momentum p of its photons

$$\lambda = \frac{h}{p} \tag{18.5}$$

which follows from Einstein's special theory of relativity (according to which, the energy of a photon, $mc^2 = hv = hc/\lambda$).

de Broglie suggested that, if electromagnetic radiation has a dual (wave-particle) nature, the same might be true of matter (which were classically thought to consist of particles) with the relevant wavelength given by the same expression. This hypothesis was verified by Davisson and Germer who observed diffraction (a wave-like character) of electrons by a nickel crystal.

18.6 UNCERTAINTY PRINCIPLE

The wave-particle duality of matter makes it impossible to know simultaneously the exact values of particular pairs of observables, called *complementary observables*, such as momentum and position, energy and time. The limitation regarding this was formulated by Heisenberg through his uncertainty principle which is often stated thus. The uncertainty in position (Δq) and the uncertainty in momentum (Δp) of a particle will be such that

$$\Delta p \Delta q \geq \frac{h}{4\pi} \tag{18.6a}$$

Here Δq and Δp must refer to the same coordinate axis, Δp is the root mean square (rms) deviation of the momentum from its mean value and Δq is the rms deviation of the position from the mean position.

Suppose that we attempt a precise measurement of the position of a particle using a microscope. The accuracy of this measurement is $\pm \lambda$, the wavelength of the light used to illuminate the particle. To be visible, the particle will have to collide with a photon. Now a photon of wavelength λ has a momentum h/λ of which an unknown fraction is transferred to the observed particle. Thus the result of locating the particle within a distance $\Delta q \simeq \pm \lambda$ is to produce an uncertainty in its momentum $\Delta p \simeq h/\lambda$. This rather crude reasoning shows that the product of these two uncertainties is $\Delta p \Delta q \simeq \Delta h$,, i.e. of the order of h.

In terms of energy (E) and time (t), the uncertainty principle has the form

$$\Delta E \Delta t \geq \frac{h}{4\pi} \tag{18.6b}$$

It should be kept in mind that the uncertainty principle is one way of describing the wave-particle duality of all physical entities; it does not in any way depend on the design and sensitivity of the measuring device. The uncertainties involved in Eqs (18.6a) and (18.6b) are inherent in the system.

18.7 SCHRÖDINGER EQUATION

Classical mechanics describes the motion of a body by a precise trajectory following Newton's law, $F = m \cdot d^2x/dt^2$. But a trajectory has no quantum mechanical significance. In wave mechanics (Schrodinger's approach to quantum mechanics), the quantity which is analogous to trajectory is a mahematical function, called *wavefunction* that contains all the informations we hope to learn about a system. The heart of wave mechanics is an equation containing wavefunction, which is postulated by Schrodinger. The time-independent Schrodinger equation for a particle of mass m moving in one dimension in the x direction is

$$-\frac{h^2}{8\pi^2 m}\frac{d^2\psi(x)}{dx^2} + V(x)\psi(x) = E\psi(x) \tag{18.7a}$$

where $\psi(x)$ is the wavefunction which is a function of particle's coordinate, $V(x)$ is the potential energy function of the particle and E is the total energy of the particle.

In three-dimensional form, the Schrodinger equation is

$$-\frac{h^2}{8\pi^2 m}\nabla^2\psi(x,y,z) + V(x,y,z)\,\psi(x,y,z) = E\psi(x,y,z) \tag{18.7b}$$

or simply
$$\nabla^2\psi + \frac{2m}{\hbar^2}(E-V)\psi = 0$$

where ∇^2 (the Laplacian operator) is

$$\nabla^2 = \frac{\partial^2}{\partial x^2} + \frac{\partial^2}{\partial y^2} + \frac{\partial^2}{\partial z^2}$$

The time-dependent Schrodinger equation is

$$H\psi = i\hbar\frac{\partial\psi}{\partial t} \tag{18.8}$$

where
$$H = -\frac{h^2}{2m}\nabla^2 + V$$

For a one-particle one-dimensional system this is

$$-\frac{\hbar^2}{2m}\frac{\partial^2 \psi(x,t)}{\partial x^2} + V(x,t)\psi(x,t) = i\hbar\frac{\partial \psi(x,t)}{\partial t}$$

In case of a conservative system (for which the potential energy is a function of coordinate and not necessarily of time), we can write $\psi(x, t) = \psi(x)\psi(t)$ and then

$$-\frac{\hbar^2}{2m}\frac{1}{\psi(x)}\frac{\partial^2 \psi(x)}{\partial x^2} + V(x) = i\hbar\frac{1}{f(t)}\frac{df(t)}{dt} = E$$

Then $\quad -\dfrac{\hbar^2}{2m}\dfrac{1}{\psi(x)}\dfrac{\partial^2 \psi(x)}{\partial x^2} + V(x) = E$ which is time-independent Schrödinger equation

and $\qquad\qquad i\hbar\dfrac{1}{f(t)}\dfrac{df(t)}{dt} = E$

So the time-dependent wavefunction for a conservative one-dimensional system will be

$$\psi(x, t) = \psi(x)e^{-\frac{iEt}{\hbar}}$$

18.8 INTERPRETATION OF WAVEFUNCTION

Max Born successfully interpreted the wavefunction ψ in terms of the location of the particle it describes. He suggested that ψ^2 or $\psi^*\psi$ (where ψ^* is a complex conjugate of ψ) at a point is the probability density for a system described by ψ to be located at that point. This cannot be derived, but this is an effective postulate of quantum mechanics.

The form of Schrödinger wave equation and Born's interpretation of wavefunction impose the following restrictions:

 (i) ψ and its first derivative must both be continuous everywhere.
 (ii) ψ must be single-valued.
 (iii) ψ must always be finite. It may be zero but not so everywhere.
 (iv) ψ must be normalized (or at least normalizable).
   Normalization of ψ is done by multiplying it with a constant N, called normalization constant, such that

$$N^2 \int_{-\infty}^{\infty} \psi^*(x)\psi(x)dx = 1$$

for one-dimensional system.

18.9 OPERATORS AND OBSERVABLES

Observable properties of a quantum mechanical system can be derived from its wavefunction by means of appropriate operators. An operator ($\hat{\Omega}$) species a mathematical operation to be carried out on a function (f), called operand, to obtain a new function (g)

$$\hat{\Omega}f = g$$

If $\hat{\Omega}f = \omega f$, where ω is a constant then f is called an eigenfunction of $\hat{\Omega}$ with eigenvalue ω. Thus in the time-independent Schrödinger equation in one dimension

$$\left[-\frac{\hbar^2}{2m}\frac{d^2}{dx^2} + V(x)\right]\psi = E\psi$$

ψ is an eigenfunction and E (the energy) is an eigenvalue of the operator $-\dfrac{\hbar^2}{2m}\dfrac{d^2}{dx^2} + V(x)$, called Hamiltonian operator denoted by $\hat{H}$.

A quantum mechanical operator corresponding to a physical quantity (observable) can be obtained from the classical expression of the latter in terms of coordinates (q), momentum (p_q), time (t) and energy (E), by its conversion to an operator using the following rules:

Classical observable (for one-dimensional system)	Quantum mechanical operator	Operation
x	$\hat{x}$	Multiplication by x
p_x	$\hat{p}_x$	$\dfrac{\hbar}{i}\dfrac{\partial}{\partial x}$
t	$\hat{t}$	Multiplication by t
E	$\hat{E}$	$i\hbar\dfrac{\partial}{\partial t}$

The sum of two operators that operate on a function f is defined by

$$(\hat{\Omega}_1 + \hat{\Omega}_2)f = \hat{\Omega}_1 f + \hat{\Omega}_2)f$$

The product of two operators is defined by

$$\hat{\Omega}_1\hat{\Omega}_2 f = \hat{\Omega}_1 [\hat{\Omega}_2 f]$$

If $\hat{\Omega}_1\hat{\Omega}_2 = \hat{\Omega}_2\hat{\Omega}_1$, the operators $\hat{\Omega}_1$ and $\hat{\Omega}$ are said to commute with each other. This happens if eigenfunctions of $\hat{\Omega}$ are also the eigenfunctions of $\hat{\Omega}_2$. Here the observables represented by Ω_1 and Ω_2 can in principle be simultaneously measured precisely. If, however $\hat{\Omega}_1$ and $\hat{\Omega}_2$ do not commute, simultaneous measurements of the corresponding observables can be done only with limited accuracy given by the uncertainty principle. The commutator of two operators $\hat{\Omega}_1$ and $\hat{\Omega}_2$ is denoted by $[\hat{\Omega}_1, \hat{\Omega}_2]$ and defined by $[\hat{\Omega}_1, \hat{\Omega}_2] = \hat{\Omega}_1\hat{\Omega}_2 - \hat{\Omega}_2\hat{\Omega}_1$.

An operator $\hat{\Omega}$, is said to be linear if for any two functions f_1 and f_2

$$\hat{\Omega}(C_1 f_1 + C_2 f_2) = C_1\hat{\Omega}f_1 + C_2\hat{\Omega}f_2$$

where C_1 and C_2 are arbitrary numbers (real or complex).

It is a postulate of quantum mechanics that every physically observable quantity can be represented by a particular liner operator, called Hermitian operator, that satisfies the condition

$$\int\limits_{\text{all space}} \psi_1^*\hat{\Omega}\psi_2 d\tau = \int\limits_{\text{all space}} \psi_2(\hat{\Omega}\psi_1)^* d\tau$$

where $d\tau$ represents the differential of all coordinates; and ψ_1 and ψ_2 are two state functions of the system.

Another important assumption is that the expectation value $\langle\Omega\rangle$, sometimes called the average value of an observable for the system in state ψ corresponding to an operator $\hat{\Omega}$ is given by

$$\langle\Omega\rangle = \int\limits_{\text{all space}} \psi^*\hat{\Omega}\psi\, d\tau$$

If ψ is an eigenfunction of $\hat{\Omega}$ with eigen value ω, the expectation value is

$$\langle \Omega \rangle = \int \psi^* \hat{\Omega} \psi \, d\tau = \int \psi^* \omega \psi \, d\tau = \int \psi^* \psi \, d\tau = \omega$$

But, if ψ is not an eigenfunction of the operator concerned, we can still write it as a linear combination of eigen functions ψ_i of the operator, i.e. $\psi = \sum C_i \psi_i$. For simplicity, suppose $\psi = C_1 \psi_1 + C_2 \psi_2$, then

$$\langle \Omega \rangle = \int (C_1 \psi_1 + C_2 \psi_2)^* \hat{\Omega} (C_1 \psi_1 + C_2 \psi_2) d\tau$$

$$= \int (C_1 \psi_1 + C_2 \psi_2)^* (C_1 \omega_1 \psi_1 + C_2 \omega_2 \psi_2) d\tau$$

$$= C_1^* C_1 \omega_1 \int \psi_1^* \psi_1 d\tau + C_2^* C_2 \omega_2 \int \psi_2^* \psi_2 \, d\tau + C_1^* C_2 \omega_2 \int \psi_1^* \psi_2 d\tau + C_2^* C_1 \omega_1 \int \psi_2 \psi_1^* \, d\tau$$

$$= |C_1|^2 \omega_1 + |C_2|^2 \omega_2 \tag{18.9}$$

In general, $\langle \Omega \rangle = \sum |C_i|^2 \omega_i$.

Here we have made use of the fact that the individual wavefunctions are normalised and the fact that $\int \psi_m^* \psi_n d\tau = 0$ for $m \neq n$ (such wavefunctions are said to be orthogonal to one another). The orthogonality is essential for the expectation values of the quantum mechanical operators to be real.

Then, when the wavefunction is not an eigenfunction of the operator of interest, the measured value of Ω in a single observation will be one of the eigenvalues of the operator $\hat{\Omega}$, but we cannot predict which one, though the average value of all such measurements is well defined. The probability that a particular eigenvalue ω_i is measured is equal to $|C_i|^2$, where C_i is the coefficient of the eigenfunction ψ_i in the expression of ψ in terms of these eigenfunctions.

18.10 PARTICLE IN A BOX

The simplest application of the Schrödinger equation is in the treatment of the translational motion of a particle. This offers an illustration of the standard quantum mechanical procedure which consists of three steps—setting up the wave equation, solving it to obtain the wavefunction and then using the latter to derive observables of the system.

Since a truly free particle is mere an imagination, a useful quantum mechanical treatment concerns the motion of a particle confined by impenetrable walls—the so-called particle in a box. Let us consider the simpler situation of a one-dimensional system where the particle moves freely only within a length a in the x direction with its potential energy $V(x)$ as under

$$V(x) = 0 \text{ for } 0 < x < a$$
$$V(x) = \infty \text{ for } x \leq 0 \text{ and } x \geq a$$

Here the Hamiltonian operator

$$\hat{H} = -\frac{\hbar^2}{2m} \frac{d^2}{dx^2}$$

where m is particle's mass. Then Schrödinger equation ($\hat{H}\psi = E\psi$) will be

$$-\frac{\hbar^2}{2m} \frac{d^2 \psi(x)}{dx^2} = E\psi(x)$$

The general solution of this equation is

$$\psi(x) = A\sin\left(\frac{2mE}{\hbar^2}\right)^{\frac{1}{2}} x + B\cos\left(\frac{2mE}{\hbar^2}\right)^{\frac{1}{2}} x$$

or

$$\psi(x) = Ce^{i\sqrt{\frac{2mE}{\hbar^2}}\cdot x} + De^{-i\sqrt{\frac{2mE}{\hbar^2}}\cdot x}$$

(18.10)

Now for the above restrictions of potential, the Born's interpretation of $\psi(x)$ implies that

$$\psi(x) = 0 \quad \text{at } x = 0$$
$$\psi(x) = 0 \quad \text{at } x = a$$

where the probability of finding the particle is zero. These two boundary conditions will be satisfied only if $B = 0$ and

$$\left(\frac{2mE}{\hbar^2}\right)^{\frac{1}{2}} a = n\pi$$

where n is an integer (quantum number), i.e. the energy E of the particle is quantized.

Further, the normalization condition, $\int \psi^*(x)\psi(x)dx = 1$, requires $A = \left(\frac{2}{a}\right)^{\frac{1}{2}}$. Then the allowed discrete values of energy and the corresponding wavefunctions are

$$E_n = \frac{n^2 h^2}{8ma^2}$$

(18.11)

$$\psi_n(x) = \left(\frac{2}{a}\right)^{\frac{1}{2}} \sin\left(\frac{n\pi x}{a}\right)$$

(18.12)

On the contrary, the energy of a free particle is not quantized. Because, although the motion of this particle can still be described by the Schrodinger equation, ψ does not have to fulfill any boundary conditions and hence no restriction arises on E. Here the wavefunction must involve at least one arbitrary constant in the following form [from Eq. (18.10)] setting one of the arbitrary constant to zero

$$\psi(x) = A\sin\left(\frac{2mE}{\hbar^2}\right)^{\frac{1}{2}} x$$

(18.13)

or

$$\psi(x) = B\cos\left(\frac{2mE}{\hbar^2}\right)^{\frac{1}{2}} x$$

Alternatively

$$\psi(x) = Ce^{i\sqrt{\frac{2mE}{\hbar^2}}\cdot x} \quad \text{for motion in } +x \text{ direction}$$

or

$$\psi(x) = De^{-i\sqrt{\frac{2mE}{\hbar^2}}\cdot x} \quad \text{for motion in } -x \text{ direction}$$

It should be noted that the above derivation of Eq. (18.12) is questionable, as the requirement of continuity of $d\psi/dx$ (demanded by the Schrödinger equation) is hardly fulfilled for an infinite potential jump at the boundaries. However, this does not happen with real systems involving only finite potential barriers.

The wavefunctions and the probability densities are represented diagramatically in Fig. 18.3(a) and (b). The Fig. 18.3 of a wavefunction, which is similar to that for a standing wave, is connected to de Broglie wavelength (λ) through the relation

$$a = n \cdot \frac{\lambda}{2}$$

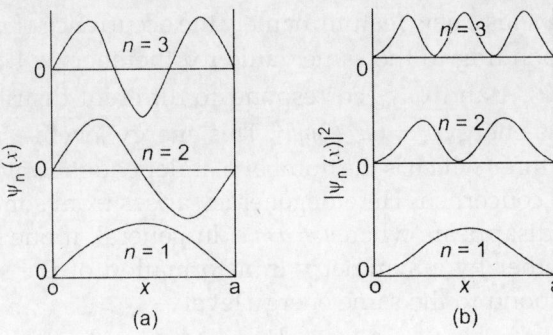

Fig. 18.3

The number of nodes (i.e. the points at which the wavefunction passes through zero) equals $n - 1$.

In Fig. 18.3(b), unlike Fig. 18.3(a), the curves meet the x-axis tangentially. In both (a) and (b), the peaks are of same height which is independent of n. For a very high value of n separation between the consecutive peaks virtually disappears, when the probability of finding the particle is same at all points, as is expected classically for constant $V(x)$.

Treatment of the particle in a one-dimensional box can be easily extended to a three-dimensional box in the form of a rectangular parallelopiped of sides a, b and c in the x, y and z directions respectively, where the potential energy $V(x, y, z)$ is zero within the box and infinity outside it. Here the wavefunction can be put as the product of three functions, each depending on just one coordinate.

$$Y(x, y, z) = X(x)\, Y(y)\, Z(z)$$

With this, Schrodinger equation for a particle of energy E separates into the following three simple differential equations

$$-\frac{\hbar^2}{2m}\frac{d^2X}{dx^2} = E_x X$$

$$-\frac{\hbar^2}{2m}\frac{d^2Y}{dy^2} = E_y Y$$

$$-\frac{\hbar^2}{2m}\frac{d^2Z}{dz^2} = E_z Z \quad \text{where } E_x + E_y + E_z = E$$

These are all similar to that found with a one-dimensional box. Then, we have the following wavefunction in case of a three-dimensional box

$$\psi_{n_x n_y n_z}(x, y, z) = \left(\frac{8}{abc}\right)^{\frac{1}{2}} \sin\frac{n_x \pi x}{a} \sin\frac{n_y \pi y}{b} \sin\frac{n_z \pi z}{c} \tag{18.14}$$

specified by a set of three quantum numbers n_x, n_y, n_z.
The allowed energy levels are

$$E_{n_x n_y n_z} = \frac{h^2}{8m}\left(\frac{n_x^2}{a^2} + \frac{n_y^2}{b^2} + \frac{n_z^2}{c^2}\right) \tag{18.15a}$$

In case of a cubical box $a = b = c$, we have

$$E_{n_x n_y n_z} = \frac{h^2}{8ma^2}(n_x^2 + n_y^2 + n_z^2) \tag{18.15b}$$

The last equation reveals a new feature, namely the occurrence of more than one distinct wavefunction corresponding to the same value for the energy. For example, the three eigenfunction $\psi_{1,2,1}$, $\psi_{2,1,1}$ and $\psi_{1,1,2}$ correspond to different distributions in space, but they all have the same energy $E = 6h^2/8ma^2$. This energy level is said to be degenerate, and the degeneracy is three which is the number of independent wavefunctions associated with the energy level concerned. Here degeneracy arises from some symmetry element of the system which disappears when $a \neq b \neq c$. In general, if one wavefunction can be transformed into another by a symmetry transformation of the system, then the two wavefunctions correspond to the same energy level.

18.11 SIMPLE HARMONIC OSCILLATOR

A particle is said to undergo a simple harmonic motion if it always moves under the action of a restoring force (F) which is proportional to the displacement from its equilibrium position

$$F = -kx; \quad \text{where } k \text{ is the force constant}$$

This corresponds to a potential energy $V = \frac{1}{2}kx^2$ which implies the motion of the particle in a parabolic well potential (that differs much from the infinite square well potential in case of *particle in a box*). Then the Schrödinger equation for the particle is

$$-\frac{\hbar^2}{2m}\frac{d^2\psi}{dx^2} + \frac{1}{2}kx^2\psi = E\psi$$

This equation provides the acceptable wavefunctions only for certain discrete values of E

$$E_v = \left(v + \tfrac{1}{2}\right)h\nu \tag{18.16}$$

$$v = 0, 1, 2, \ldots, \text{ and } \nu = \frac{1}{2\pi}\sqrt{\frac{k}{m}}$$

The well-behaved solutions to the above Schrödinger equation will be of the form e^{-ax^2} times a polynomial of degree v in x.

The first three wavefunctions are

$$\psi_0 = \left(\frac{2a}{\pi}\right)^{\frac{1}{4}} e^{-ax^2}$$

$$\psi_1 = \left(\frac{2a}{\pi}\right)^{\frac{1}{4}} 2a^{\frac{1}{2}}xe^{-ax^2} \quad \left[\text{where } a = \left(\frac{\pi}{h}\right)\sqrt{km}\right]$$

$$\psi_2 = \left(\frac{2a}{\pi}\right)^{\frac{1}{4}} (4ax^2 - 1)e^{-ax^2}$$

These wavefunctions and the corresponding probability densities are graphically represented in Fig. 18.4(a) and (b) along with the potential energy.

In both the Figs 18.4(a) and (b), the curves penetrate through the potential barrier represented by broken lines) significantly and meet the x-axis tangentially. The height of the end peaks (which is greater than those of the intermediate peaks) is lower with higher

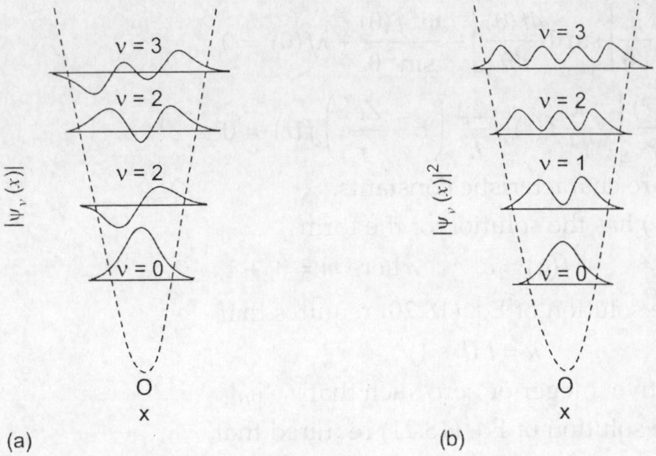

(a)　　　　　　　　　　　　　　　　(b)

Fig. 18.4

energy levels. But in Fig. 18.4(b), the area under the curve (which is a measure of total probability of finding the oscillator) is same for all energy levels.

18.12 TWO-PARTICLE ROTOR

Simplest example is the hydrogen-like atomic species which is classically thought to consist of an electron of charge $-e$ and a nucleus of charge $+Ze$, each moving about their common centre of mass following Coulomb's law. Such rotational motion may be treated as that of a single particle of charge $-e$ and reduced mass μ

$\left(\text{where } \dfrac{1}{\mu} = \dfrac{1}{m_{\text{electron}}} + \dfrac{1}{m_{\text{nucleus}}} \approx \dfrac{1}{m_{\text{electron}}}\right)$ in the electrostatic field of the nuclear charge

$+Ze$. The potential energy V of such a particle at a distance r from the nucleus is

$$V = -\frac{Ze^2}{r} \left(-\frac{Ze^2}{4\pi\,\epsilon_0\,r}\text{ in SI}\right)$$

The situation is then similar to that of a particle in a three-dimensional box with a spherically symmetrical potential well having infinite depth at the centre. At $r = 0$, $V = -\infty$ and at $r = \infty$, $V = 0$.

Then the Schrodinger equation for a hydrogen-like atom is

$$\nabla^2\psi + \frac{2\mu}{\hbar^2}\left(E + \frac{Ze^2}{r}\right)\psi = 0 \tag{18.17}$$

This transforms to the following polar form in spherical coordinates (r, θ, ϕ)

$$\frac{1}{r^2}\frac{\partial}{\partial r}\left(r^2\frac{\partial\psi}{\partial r}\right) + \frac{1}{r^2\sin\theta}\frac{\partial}{\partial\theta}\left(\sin\theta\frac{\partial\psi}{\partial\theta}\right) + \frac{1}{r^2\sin^2\theta}\frac{\partial^2\psi}{\partial\phi^2} + \frac{2\mu}{\hbar^2}\left(E + \frac{Ze^2}{r}\right)\psi = 0 \tag{18.18}$$

With the new coordinate system, it is possible to write

$$\psi = f(r)\,f(\theta)\,f(\phi)$$

Then the Schrodinger equation separates into the following three simple equations

$$\frac{1}{f(\phi)}\cdot\frac{d^2 f(\phi)}{d\phi^2} + m^2 = 0 \tag{18.19}$$

$$\frac{1}{\sin\theta}\frac{d}{d\theta}\left[\sin\theta\frac{df(\theta)}{d\theta}\right] - \frac{m^2 f(\theta)}{\sin^2\theta} + \lambda f(\theta) = 0 \tag{18.20}$$

$$\frac{1}{r^2}\frac{d}{dr}\left[r^2\frac{df(r)}{dr}\right] - \frac{\lambda}{r^2}f(r) + \frac{2\mu}{\hbar^2}\left(E + \frac{Ze^2}{r}\right)f(r) = 0 \tag{18.21}$$

where m and λ are characteristic constants.

The Eq. (18.19) has the solution of the form

$$f(\phi) = e^{imx} \quad \text{where } m = 0, \pm 1, \pm 2, \ldots$$

An acceptable solution of Eq. (18.20) requires that

$$\lambda = l\,(l+1)$$

where l is a positive integer or zero such that $l \geq |m|$.

An acceptable solution of Eq. (18.21) required that

$$E = -\frac{Z^2 \mu e^4}{2n^2 \hbar^2} \tag{18.22}$$

where n is a +ve integer such that $n \geq \lambda$, i.e. $n \geq l + 1$.

Since lowest value of l is zero, the lowest value of n is 1.

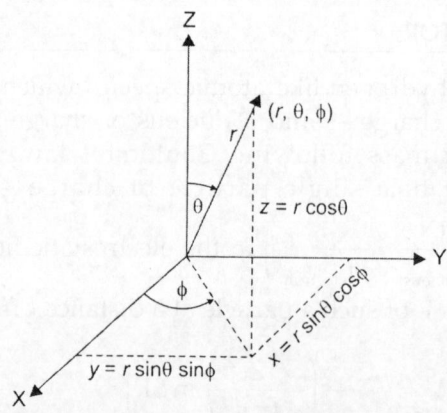

Fig. 18.5: Spherical coordinate system

Then it appears that the wavefunctions provided by the Schrodinger equation for a hydrogenic atom are characterised by three integral numbers n, l and m which are called principal quantum number, azimuthal or angular-momentum quantum number and magnetic quantum number respectively such that

$$n = 1, 2, 3, \ldots$$

$$l = 0, 1, 2, 3, \ldots$$

$$m = -l, (-l+1), \ldots, (l-1), l$$

n specifies quantization of energy through Eq. (18.22), l specifies quantization of total angular momentum which is equal to $\sqrt{l(l+1)} \cdot \hbar$, and m determines the components of angular momentum in some specified direction (usually taken as z-axis) equal to $m\hbar$ whereby specifying space quantization of angular momentum. The absence of l and m in the expression for E implies that the nth energy level of hydrogen atom will be n^2-fold degenerate (n^2 being the sum of all subshells permitted by n).

In case of rigid rotor r is constant and $V = 0$, when the Eq. (18.18) reduces to

$$\frac{1}{r^2 \sin\theta} \frac{\partial}{\partial\theta} \left(\sin\theta \frac{\partial\psi}{\partial\theta} \right) + \frac{1}{r^2 \sin^2\theta} \frac{\partial^2\psi}{\partial\phi^2} + \frac{2\mu E\psi}{\hbar^2} = 0 \qquad (A)$$

which provides the following expression for energy

$$E = \frac{l(l+1)\hbar^2}{2\mu r^2} \qquad (18.23a)$$

An expression of this of this form holds in case of diatomic (and also polyatomic liner) molecules for which moment of inertia $I = \mu r^2$, and E is usually expressed in terms of rotational quantum number J (instead of l), as

$$E = \frac{J(J+1)\hbar^2}{2I} \qquad (18.23b)$$

Although E depends only on the quantum number J, the wavefunction depends also on another quantum number m. Since m can have integral values ranging from $-J$ to J, the rotational levels are $(2J + 1)$-fold degenerate.

In case of particle-in-a-box (where $V = 0$, having negligible $1/r$ and no change in angular variables) the Eq. (A) reduces to

$$\frac{d^2\psi}{dr^2} + \frac{2\mu E\psi}{\hbar^2} = 0$$

which leads to Eq. (18.11) (since $\mu \approx m_e$).

18.13 ATOMIC ORBITALS

An atomic orbital is a one-electron wavefunction that describes an electron in an atom. It replaces the faulty idea of electron orbit in the old quantum theory of Bohr.

A group of orbitals with the same value of n is called a shell of the atom. The shells are denoted by letters K, L, M, ..., corresponding to $n = 1, 2, 3, ...$. Again, within a particular shell, the orbitals having same value of l form a subshell denoted by s, p, d, f, ..., corresponding to $l = 0, 1, 2, 3, ...$

All hydrogenic atomic orbitals are conveniently thought to consist of two parts

$$\psi_{n,l,m} = R_{n,l} Y_{l,m}$$

R is called the radial wavefunction which is of the form

$$R_{n,1} = N_{n,1} e^l L_{n,1}(e) \, e^{e/n}$$

where $N_{n,1}$ is a normalization constant, $L_{n,1}$ is a polynomial in e, $e = Zr/a_0$, and $a_0 = \dfrac{\hbar^2}{\mu e^2}$ is the Bohr radius which is equal to the radius of the first Bohr orbit of the hydrogen atom.

$Y_{1,m}$ is the angular part of the wavefunction which has the form

$$Y_{1,m} = N'_{l,m} L'_{l,m} (\sin\theta, \cos\theta) \, e^{im\phi}$$

where $N'_{l,m}$ is a normalization constant, $L'_{l,m}$ is a polynomial whose highest term is of the form $(\cos\theta)^a \cdot (\sin\theta)^b$ where $a + b = l$ and $b = |m|$.

However, instead of working with such imaginary wavefunctions (unless $m = 0$), it is often more convenient to use real wavefunctions formed by taking appropriate linear combination of complex wavefunctions, so that the imaginary part is replaced by a trigonometrical function of ϕ.

It should be remembered that ψ, R and Y must be separately normalized. For a particular wavefunction, one can readily find out n from the factor $e^{e/n}$ or from the term containing highest power of e (the power being equal to $n-1$), l from the factor e^l or from the degree of the polynomial L' ($\sin\theta$, $\cos\theta$), and m from $\sin(m\phi)$ or from $\cos(m\phi)$ where the value is negative or positive accordingly. The wavefunctions for the first few hydrogenic orbitals are given in Table 18.1.

Both R and Y are important in connection with chemical bonding which requires appropriate overlapping of the two orbitals of the bond-forming atoms. R determines

Table 18.1: Real hydrogenic wavefunctions

Orbital	n	l	m	$R_{n,l}$		$Y_{l,m}$
$1s$	1	0	0	$2\left(\dfrac{Z}{a_0}\right)^{\frac{3}{2}} e^{-e}$		$\left(\dfrac{1}{4\pi}\right)^{\frac{1}{2}}$
$2s$	2	0	0	$\dfrac{1}{2\sqrt{2}}\left(\dfrac{Z}{a_0}\right)^{\frac{3}{2}}(2-e)e^{-\frac{e}{2}}$		$\left(\dfrac{1}{4\pi}\right)^{\frac{1}{2}}$
$2p_z$	2	1	0	$\dfrac{1}{2\sqrt{6}}\left(\dfrac{Z}{a_0}\right)^{\frac{3}{2}} e e^{-\frac{e}{2}}$	Common	$\left(\dfrac{3}{4\pi}\right)^{\frac{1}{2}}\cos\theta$
$2p_x$	2	1	1	$\dfrac{1}{2\sqrt{6}}\left(\dfrac{Z}{a_0}\right)^{\frac{3}{2}} e e^{-\frac{e}{2}}$		$\left(\dfrac{3}{4\pi}\right)^{\frac{1}{2}}\sin\theta\cos\phi$
$2p_y$	2	1	-1	$\dfrac{1}{2\sqrt{6}}\left(\dfrac{Z}{a_0}\right)^{\frac{3}{2}} e e^{-\frac{e}{2}}$		$\left(\dfrac{3}{4\pi}\right)^{\frac{1}{2}}\sin\theta\sin\phi$
$3s$	3	0	0	$\dfrac{2}{81\sqrt{6}}\left(\dfrac{Z}{a_0}\right)^{\frac{3}{2}}(27-18e+2e^2)e^{-\frac{e}{3}}$		$\left(\dfrac{1}{4\pi}\right)^{\frac{1}{2}}$
$3p_z$	3	1	0	$\dfrac{4}{81\sqrt{6}}\left(\dfrac{Z}{a_0}\right)^{\frac{3}{2}}(6-e)e e^{-\frac{e}{3}}$	Common	$\left(\dfrac{3}{4\pi}\right)^{\frac{1}{2}}\cos\theta$
$3p_x$	3	1	1	$\dfrac{4}{81\sqrt{6}}\left(\dfrac{Z}{a_0}\right)^{\frac{3}{2}}(6-e)e e^{-\frac{e}{3}}$		$\left(\dfrac{3}{4\pi}\right)^{\frac{1}{2}}\sin\theta\cos\phi$
$3p_y$	3	1	-1	$\dfrac{4}{81\sqrt{6}}\left(\dfrac{Z}{a_0}\right)^{\frac{3}{2}}(6-e)e e^{-\frac{e}{3}}$		$\left(\dfrac{3}{4\pi}\right)^{\frac{1}{2}}\sin\theta\sin\phi$
$3d_{z^2}$	3	2	0	$\dfrac{4}{81\sqrt{30}}\left(\dfrac{Z}{a_0}\right)^{\frac{3}{2}} e^2 e^{-\frac{e}{3}}$		$\left(\dfrac{5}{16\pi}\right)^{\frac{1}{2}}(3\cos^2\theta-1)$
$3d_{zx}$	3	2	1	$\dfrac{4}{81\sqrt{30}}\left(\dfrac{Z}{a_0}\right)^{\frac{3}{2}} e^2 e^{-\frac{e}{3}}$		$\left(\dfrac{15}{4\pi}\right)^{\frac{1}{2}}\sin\theta\cos\theta\cos\phi$
$3d_{yz}$	3	2	-1	$\dfrac{4}{81\sqrt{30}}\left(\dfrac{Z}{a_0}\right)^{\frac{3}{2}} e^2 e^{-\frac{e}{3}}$	Common	$\left(\dfrac{15}{4\pi}\right)^{\frac{1}{2}}\sin\theta\cos\theta\sin\phi$
$3d_{x^2-y^2}$	3	2	2	$\dfrac{4}{81\sqrt{30}}\left(\dfrac{Z}{a_0}\right)^{\frac{3}{2}} e^2 e^{-\frac{e}{3}}$		$\left(\dfrac{15}{16\pi}\right)^{\frac{3}{2}}\sin^2\theta\cos2\phi$
$3d_{xy}$	3	2	-2	$\dfrac{4}{81\sqrt{30}}\left(\dfrac{Z}{a_0}\right)^{\frac{3}{2}} e^2 e^{-\frac{e}{3}}$		$\left(\dfrac{15}{16\pi}\right)^{\frac{3}{2}}\sin^2\theta\sin2\phi$

how close the two atoms must come (i.e. the bond length) and Y determines the most suitable orientation of the orbitals (i.e. the bond angle).

Radial Wavefunctions

The radial parts of the wavefunctions in Table 18.1 are diagramatically represented in Fig. 18.6(a). At large value of r, all approach zero (due to exponential factor). Also R is zero at the nucleus, unless $l = 0$. In between these two points, the number of nodes (at which $R = 0$) is $n - l - 1$. In

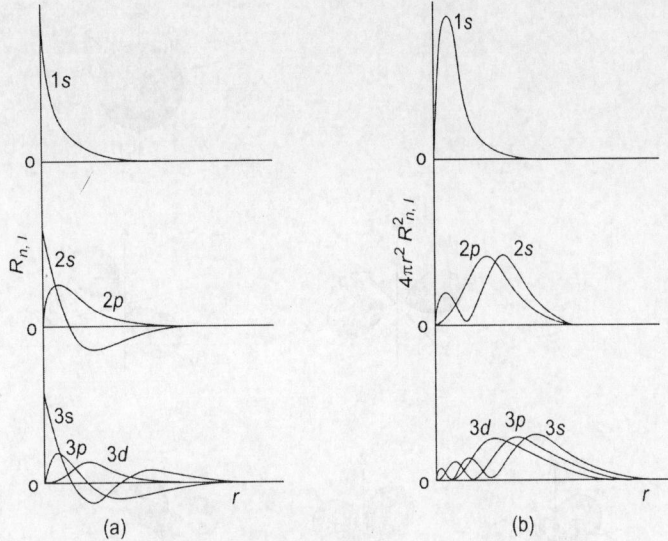

Fig. 18.6: (a) Radial wavefunctions (b) radial distribution functions (Not to scale)

each case R^2 at a particular point is proportional to the probability density for the electron to be found at that point (the proportionality constant is determined by Y). But in connection with bond formation, we need to know the probability of finding the electron anywhere in a spherical layer of thickness dr around the nucleus at a distance r from it. This is equal to $R^2 4\pi r^2 dr$. Here $4\pi r^2 R^2$ is called the *radial distribution function* represented in Fig. 18.6(b).

In the Figs 18.6(a) and 18.6(b), the number of nodes is same but the number of maxima is different; the latter equals $n - l$ in case of Fig. 18.6(b). When there are two or more maxima, the value of r corresponding to the highest value of radial distribution function is called the most probable radius of the orbital concerned which signifies the most probable distance of an electron from the nucleus when it occupies the orbital concerned.

This is lower than the mean radius of the orbital which is equal to $\int_0^\infty r \cdot R^2 r^2 dr$ (which signifies the mean radius of the relevant atom).

Further, it appears from the Fig. 18.6(b) that the most probable distances of s-, p- and d-orbitals with same n differ only slightly in order $s > p > d$. But the ability to penetrate to the nucleus is most with a ns-electron and this falls through np to nd-electrons.

Angular Part of Wavefunction

The shape of an orbital is best represented diagramatically by a surface of constant probability density that encloses a major fraction of the probability of finding the orbital electron. For real hydrogenic orbitals of lowest energy, such diagrams are shown in Fig. 18.7. An s-orbital is spherically symmetrical. p- and d-orbitals consist of distorted ellipsoids which do not touch each other. The number of nodal planes associated with an orbital is same as the number of angular nodes which is just equal to l of the orbital.

In dealing with chemical bonding qualitatively, p-orbitals are often pictured simply as a pair of mutually touching spherical lobes which actually represents the shape of the angular part of this orbital (and not of the entire orbital). Vide Fig. 18.7.

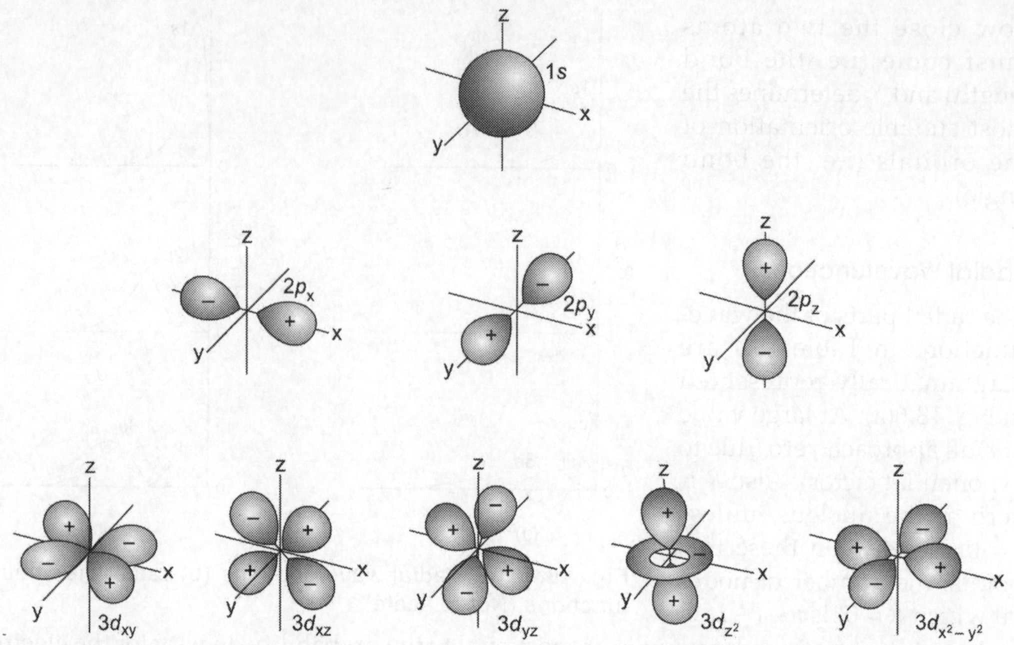

Fig. 18.7: Contours of hydrogenic orbitals. The indicated signs are those of the wavefunctions

AUXILIARY PROBLEMS

18.1 Quantum theory begins with the Planck's theory of black-body radiation. What is meant by such radiation?

18.2 In what important respect(s) does the radiation inside an isothermal hollow body differ from that emitted by its outer surface? How do you account for it?

18/3 A perfect black-body absorbs all radiation of any wavelength falling upon it and hence it will appear black in colour. Comment.

18.4 The cavity radiation, like an enclosed gas, exerts pressure on the cavity walls. But this pressure, unlike the gas pressure, is independent of volume. Explain.

18.5 Only a part of the spectral distribution curve for black-body radiation can be explained on the basis of the classical concept of energy as a continuous variable. Why?

18.6 An iron rod glows first dull red, then orange, then white as its temperature is raised. Explain.

18.7 Whereas a hot solid or liquid produces a continuous spectrum of radiation, a hot gas produces a discontinuous spectrum consisting of lines. Explain. Can any other geometrical figure, instead of line, appear in the spectrum?

18.8 To explain the line spectrum of atomic hydrogen Bohr developed a theory by imposing the restriction of quantization on angular momentum rather than on energy, though ultimately, it leads to quantization of energy which is the main concern. Why?

18.9 Bohr's idea of electron-orbits of fixed energy is contradictory to the classical theory of electrodynamics (according to which an accelerated charged body always radiates energy). How will you reconcile this?

18.10 (a) What features of the Bohr model of H-atom are not tenable to quantum mechanics?

   (b) How does the Bohr ground state of H-atom differ from that predicted by quantum mechanics?

   (c) Is there any experimental distinction between the two?

18.11 Bohr assumed quantization of angular momentum but in a faulty way. In spite of this, Bohr's theory gives the following correct expression for the energy (E) of a hydrogen atom as provided by Schrödinger model $E = -\dfrac{me^4}{2n^2\hbar^2}$. How?

18.12 It is our experience that a pendulum can be made to swing with any amplitude and hence with any energy. Does this contradict Planck's idea of quantization of energy?

18.13 What is Planck constant? What is its value in SI unit? If Planck constant were 6.6 Js, would quantum phenomena be more or less conspicuous in everyday life than they are now? Why?

18.14 Would you expect quantum effects to be generally more important at the low-frequency or high-frequency end of the electromagnetic spectrum. Why?

18.15 What is the dimension of kT/h?

18.16 The frequency of an atom emitting visible light is 10^{15} cycles/sec, while that of a big pendulum is 1 cycle/sec. Predict which one possesses the larger characteristic action over unit time. Compare this with respect to kT at a temperature of 27°C.

18.17 Mention a phenomenon that establishes the quantum nature of light. If light transfers energy by means of separate quanta, why do we not precise a faint light as a series of tiny flashes?

18.18 In photoelectric effect, what makes you sure that each electron absorbs only one photon?

18.19 How does photoelectric work function differ from ionization potential of a metal? Which one is greater?

18.20 Which phenomenon provides more conclusive evidence in favour of quantum nature of light—photoelectric effect or Compton effect?

18.21 Compton effect is observed when a beam of x-ray falls on a block of graphite. But with aluminium photoelectric effect is observed instead. Explain.

18.22 Compton shift, like Raman shift, does not depend on the incident wavelength. But the former, unlike the latter, depends on the angle between the incident and scattered radiation and is always positive. Explain.

18.23 What is Compton wavelength? What is the highest value of the compton shift?

18.24 de Broglie theory fails in case of macroscopic bodies. Comment.

18.25 Calculate the de Broglie wavelength of the electron accelerated from rest through a potential difference of (i) 1 kV (ii) 100 kV.

18.26 State with reason whether quantum mechanics needs to be applied to the following cases:

   (i) Two balls of mass 1 g joined through a spring of length 1 cm.

   (ii) Two H-atoms in H_2 molecule.

   (iii) A cricket ball delivered with a velocity of 120 km/hr?

18.27 Wave-like and particle-like aspects of matter are not contradictory but complementary. Comment.

18.28 Calculate the speed to which a stationary H-atom should be accelerated due to absorption of a photon of frequency 2.5×10^{14} Hz. Can a H-atom absorb a photon whose energy exceeds its binding energy?

18.29 A worm of mass 1.0 g emits light of wavelength 600 nm with a power of 0.1 W in the backward direction. What speed will it attain after 30 days, if it starts from rest and faces no resistance?

18.30 Uncertainty principle is often stated in terms of uncertainty in coordinate position (Δq) and uncertainty in momentum (Δp), but not uncertainty in velocity though velocity is a simplier quantity than momentum. Why? Does the value of $\Delta p \Delta q$ depend on the value of p?

18.31 Two bodies A and B move freely with uniform velocity. The velocity of A is twice that of B. What will be uncertainty in momentum and the uncertainty in position for A and B?

18.32 Suppose we attempt a precise measurement of the position of a particle through a microscope using light of wavelength λ. Here momentum of the particle and hence the accuracy in the measurement depend on λ. Then, would the uncertainty in position (Δq) and the uncertainty in momentum be given by the relation

$$\Delta p \Delta q \geq \frac{h}{4\pi}$$

18.33 Is violation of the uncertainty principle possible?

18.34 Discuss the possibility of the following relations between uncertainty in position and momentum for a particle

$$\Delta p_x \Delta x = 0$$
$$\Delta p_y \Delta x = 0$$

18.35 Suppose a body is stationary (i.e. no uncertainty in position), then the uncertainty in its velocity would be infinity which is contradicting to the fact that it is stationary. Resolve this paradox.

18.36 In general, spectral lines arising due to transitions from the ground state of an atom are sharp. Explain.

18.37 How is the stationary Bohr orbit of an atom defined in terms of de Broglie waves?

18.38 Calculate the energy of an electron in a Bohr orbit of a He^+ ion if the circumference of the orbit is three times the de Broglie wavelength of the electron.

18.39 Define classically the state of a system and comment on it.

18.40 How is state defined in quantum mechanics? What is stationary state. Verify whether the function $e^{-2\pi i \frac{Et}{h}} \psi(x)$ represents a stationary state or not.

18.41 Explain whether ψ, $-\psi$ and $2i\psi$ represent the same state, given that ψ is real.

18.42 If ψ_1 and ψ_2 are eigenfunctions of H, will $C_1\psi_1 + C_2\psi_2$ also be an eigenfunction of H? Can $C_1\psi_1 + C_2\psi_2$ be a state function?

18.43 Show that if the energies E_1 and E_2 of two independent wavefunctions $\psi_1(x)$ and $\psi_2(x)$ of a system are additive, then the composite function is the product of the two functions.

18.44 An acceptable wavefunction ψ of a system must be single-valued. Why?

18.45 Can ψ be zero or infinity?

18.46 An acceptable wavefunction is required to be normalizable. Why? Can normalization constant be –ve, zero, infinity or imaginary?

18.47 Normalize the following functions:

(i) a constant c in the range $[-L, L]$, (ii) $\sin(n\pi x/a)$ in the range $[0, a]$, (iii) $e^{-r/a}$ in three-dimensional space (iv) $xe^{-r/2a}$ in three-dimensional space.

18.48 The wavefunction $\left[\int_0^\infty x^n e^{-ax} dx = \dfrac{n!}{a^{n+1}}\right]$ for the motion of a particle in a ring is of the

form $\psi = Ne^{im\phi}$. Find the normalization constant N.

18.49 Following is an unnormalized excited state wavefunction of the H-atom

$$\psi = \left(2 - \frac{r}{a_0}\right) e^{-\frac{r}{2a_c}}$$

Normalize this. What is the angular part of the normalized wavefunction?

18.50 The radial wavefunction for the ground state of a H-atom is $\psi = Ne^{-r/a_0}$. Determine the normalization constant N.

18.51 State, citing reasons, whether the following are the acceptable wavefunctions over the indiated intervals:

(i) A constant $\qquad\qquad$ $[-\infty, +\infty]$

(ii) $\sqrt{x}$ $\qquad\qquad$ $[-1, 1]$

(iii) $\dfrac{1}{x}$ $\qquad\qquad$ $[-1, 1]$

(iv) $\ln x$ $\qquad\qquad$ $[-1, 1]$

(v) e^x $\qquad\qquad$ $[0, \infty]$

(vi) e^{-x} $\qquad\qquad$ $[0, \infty]$

(vii) e^{-x} $\qquad\qquad$ $[-\infty, +\infty]$

(viii) e^{-x^2} $\qquad\qquad$ $[-\infty, +\infty]$

(ix) $e^{-|x|}$ $\qquad\qquad$ $[-\infty, +\infty]$

(x) $\sin x$ $\qquad\qquad$ $[0, 2\pi]$

(xi) $\sqrt{\sin x}$ $\qquad\qquad$ $[0, 2\pi]$

(xii) $\sin^{-1} x$ $\qquad\qquad$ $[-1, 1]$

(xiii) $\tan x$ $\qquad\qquad$ $[0, 2\pi]$

(xiv) $e^{i\theta}$ $\qquad\qquad$ $[0, 2\pi]$

18.52 State the reasons whether the following ψ_x vs x diagram correspond to acceptable wavefunctions.

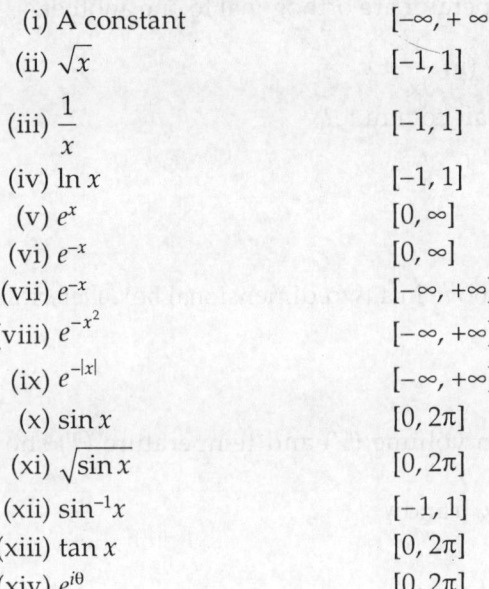

(i) $\qquad$ (ii) $\qquad$ (iii) $\qquad$ (iv) $\qquad$ (v)

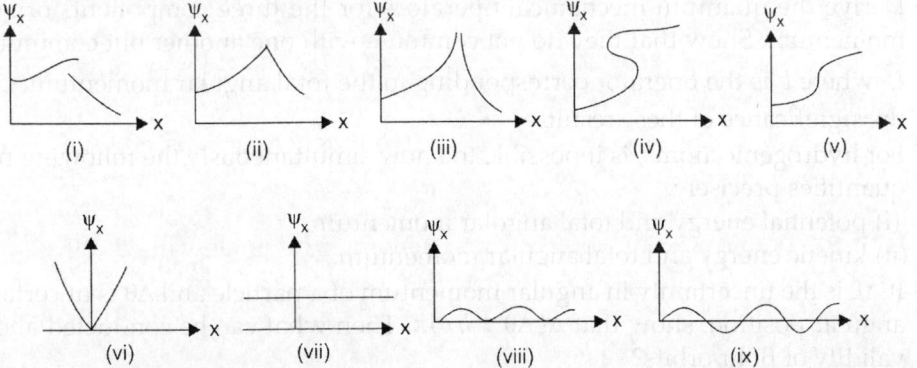

(vi) $\qquad$ (vii) $\qquad$ (viii) $\qquad$ (ix)

18.53 What are the eigenfunctions and eigenvalues of the operator d/dx?

18.54 Identify which of the following functions are eigenfunctions of the operator d/dx and d^2/dx^2: (i) e^{ikx} (ii) $\cos kx$ (iii) k (iv) kx (v) e^{-ax^2}. Give the corresponding eigenvalue, if any. Comment on the results.

18.55 Wavefunction of a free particle is $\psi = A\exp\left[\dfrac{i}{\hbar}(Px - Et)\right]$. Find out the operators for total energy E and momentum p.

18.56 A state $\psi = Ne^{-\alpha x^2}$ satisfies an eigenvalue equation $H\psi = E\psi$ with $H = -\dfrac{d^2}{dx^2} + x^2$. Find possible values of α and eigenvalue E.

18.57 Verify whether the operator ∇^2 is linear.

18.58 Show that $\hat{p}_x$ and $\hat{H}_x$ are hermitian operators.

19.59 Justify or criticise the following:
(a) Eigenvalues of a hermitian operator are all real quantities.
(b) Eigenfunctions of a hermitian operator are orthogonal to one another.

18.60 Find the operator $\hat{A}^2$, if $\hat{A}$ is (i) $\dfrac{\hbar}{i}\dfrac{\partial}{dx}$ (ii) $\dfrac{\partial}{\partial x} + x$.

18.61 Do the operators in the following pair commute?
(i) x and d^2/dx^2

(ii) $\dfrac{d}{dx}$ and $\displaystyle\int_0^x (\ ) \, dx$

(iii) $\cos$ and $\sin^{-1}$

18.62 Consider a particle constrained to move in a two-dimensional box. Determine:
(i) $[\hat{x}, \hat{p}_y]$
(ii) $[\hat{x}, \hat{p}_x]$
Comment on the result.

18.63 For a state function dependent on volume (V) and temperature (T) show that operators $\partial/\partial V$ and $\partial/\partial T$ commute.

18.64 Show that for any three operators α, β and γ
(a) $[\alpha + \beta, \gamma] = [\alpha, \gamma] + [\beta, \gamma]$
(b) $[\alpha, \beta\gamma] = [\alpha, \beta]\,\gamma + \beta\,[\alpha, \gamma]$
(c) $[\alpha\beta, \gamma] = [\alpha, \gamma]\,\beta + \alpha\,[\beta, \gamma]$

18.65 Evaluate $[\hat{x}^n, \hat{p}_x]$.

18.66 Derive the quantum mechanical operators for the three components of angular momentum. Show that they do not commute with one another but commute with $\hat{L}^2$ where $\hat{L}$ is the operator corresponding to the total angular momentum. Discuss the significance of these results.

18.67 For hydrogenic atoms, is it possible to know simultaneously the following pairs of quantities precisely:
(i) potential energy and total angular momentum.
(ii) kinetic energy and total angular momentum.

18.68 If ΔL is the uncertainty in angular momentum of a particle and $\Delta\theta$ is uncertainty in angular position, show that $\Delta L\Delta\theta \geq h/4\pi$. Then what can be conducted about the validity of Bohr orbits?

18.69 Estimate the minimum uncertainty in the speed of an electron in a hydrogen atom, taking its radius equal to that of the first Bohr orbit.

18.70 With what fundamental accuracy, we can locate the position of (i) an electron and (ii) a bullet of mass 50 g, if each has speed 400 m/s accurate to 0.01%? Comment on the result. Also find the corresponding accuracy in finding kinetic energy of the electron in this case.

18.71 (a) A particle of mass m is in a state described by the wavefunction $\psi = c\,e^{ikx}$. Find the velocity of the particle. Is the particle free or bound? Can you predict the location of the particle?

(b) If $\psi = C_1 e^{ikx} + C_2 e^{-ikx}$, find average momentum. What is the probability that the particle will exhibit the momentum $\hbar k$ in a single experiment? What would be the form of ψ so that this probability is 90%? Also evaluate kinetic energy of this particle.

18.72 Find the average momentum of a particle described by the wavefunction $\psi = \sin kx$ in the range $-\infty$ and $+\infty$.

18.73 Find an expression for ψ and E of a particle in a one-dimensional box of length $2a$ within the coordinates 0 to $2a$. Also do the same for the coordinates $-a$ to $+a$. Can $(x-a)(x+a)$ be a state function in either case?

18.74 Derive the energy levels of a particle in a box (PIB) from de Broglie relation. Is the expression for energy levels in keeping with the uncertainty principle?

18.75 Can a PIB be at rest?

18.76 From the expression $E = \dfrac{n^2 h^2}{8ma^2}$ for PIB, it appears that $E = 0$ when $a = \infty$. Is this justified?

18.77 Explain the following on the basis of PIB situation:

(a) The translational motion of a molecule may be adequately treated by the method of classical mechanics.

(b) The confinement of electron in the nucleus by electrostatic attraction is not possible.

18.78 Wavelength of a particle confined to a one-dimensional box of length a is

$$\psi = \left(\frac{2}{a}\right)^{\frac{1}{2}} \sin\left(\frac{n\pi x}{a}\right)$$

(A) Find the probability that the particle will be found between $0.49a$ and $0.51a$, for $n = 1$ and $n = 2$.

(B) For $n = 1$ and $a = 10$ nm, calculate the probability that the particle is (i) between 9.90 and 10.00 nm (ii) between $a/3$ and $2a/3$ (iii) in the right half of the box.

(C) What are the most likely locations of the particle in the box in the state (i) $n = 3$ and (ii) $n = 1$?

(D) Show that the wavefunctions of such a system are orthogonal.

(E) Verify that the given wavefunctions satisfies the relation $H\psi = E\psi$ and hence find E.

18.79 For the particle in a one-dimensional box of length a, find the uncertainty in (i) position Δx (ii) momentum Δp_x (iii) energy ΔE. Comment on the result. What is the lowest value of $\Delta p_x \Delta x$?

18.80 The life time of an excited state of an atom is 10^{-9} s. Calculate the uncertainty in the energy of this state. What will be the width of the corresponding spectral line at

600 nm? Also find the minimum limit of accuracy with which we can know the frequency of the radiated energy by the atom?

18.81 Find the frequency of absorption for the first electronic transition of 1,3-butadiene treating its π-electrons on the basis of the particle in a box model. The bond length is 154 pm for C–C and 135 pm for C = C, and radius of C-atom is 77 pm.

18.82 The energy levels of a particle in a cubic box of edge length a (like a one-dimensional box of length a) are given by

$$E = \frac{n^2 h^2}{8 m a^2}$$

What are the first three values of n?

18.83 A He molecule is confined in a cubic box $a = 1$ m. If its energy equals the average thermal molecular energy (assuming ideal behaviour) at 300 K, what will be the value of n and de Broglie wavelength? Can He molecule be treated as a classical particle?

18.84 Consider a particle in a cubic box. State the possibility of an energy level having an energy (i) three times (ii) four times (iii) five times that of the lowest level, mentioning the degeneracy of the possible energy level in each case. What is the lowest energy level having maximum possible degeneracy? Is degeneracy limited only to cubic boxes?

18.85 Consider PIB with $a:b:c = 1:2:3$. What is the lowest energy level? Find its degeneracy. What is the highest degeneracy possible with this situation?

18.86 Find the degeneracy of the first four energy levels of a particle in a box of dimensions $a = b = 2c$.

18.87 The wavefunction for the state of lowest energy of a one-dimensional harmonic oscillator is $\psi = Ae^{-Bx^2}$. Write the appropriate Schrodinger equation for the system and hence show that the total energy E of the lowest state is $\frac{1}{2}hv$. Also find A and B in terms of reduced mass (μ) and force constant k.

18.88 The probability of finding an oscillator in the lowest energy state $\psi = Ae^{-Bx^2}$ is most at its equilibrium position. But it is our experience with a pendulum (that undergoes harmonic oscillation) that its bob expends most of its time at the turning points. Does the quantum mechanical principle fail in case of pendulum? Explain.

18.89 Justify the following:

(a) Energy of a harmonic oscillator cannot be zero.

(b) Kinetic energy of a harmonic oscillator may be negative.

18.90 Calculate the minimum energy and minimum excitation energy of the (a) pendulum of length 1 m on the earth's surface (b) the balance-wheel of a watch having frequency 5 Hz (c) the O–O bond of force constant $(k) = 1177$ Nm^{-1} in O_2 molecule.

18.91 Show that for a harmonic oscillator, the uncertainty in position (Δx) and uncertainty in momentum (Δp_x) are related by

$$\Delta p_x \Delta x \geq \frac{h}{4\pi}$$

18.92 A node is the position at which the wavefunction passes through zero. Then, how do you explain the passage of electron across a node? How does the energy associated with a wavefunction vary with the number of nodes in case of PIB and harmonic oscillator?

18.93 Wavefunctions of H-atom are usually expressed in spherical coordinates while those of PIB in Cartesian coordinates? Why?

18.94 A hydrogenic orbital has the form $\psi_{lnm} =$ (constant) $r^4 \sin 4\theta \sin^4 \phi$. What are the values of n, l and m for this orbital?

18.95 Designate the following hydrogen orbitals (as ns, np etc.):

(i) $\psi = \dfrac{1}{4\sqrt{2\pi}} \left(\dfrac{Z}{a_0}\right)^{\frac{3}{2}} e e^{-\frac{e}{2}} \cos\theta$

(ii) $\psi = \dfrac{1}{4\sqrt{2\pi}} \left(\dfrac{Z}{a_0}\right)^{\frac{3}{2}} e e^{-\frac{e}{2}} \sin\theta\cos\phi$

(iii) $\psi = \dfrac{1}{4\sqrt{2\pi}} \left(\dfrac{Z}{a_0}\right)^{\frac{3}{2}} (2-e)e^{-\frac{e}{2}}$

(iv) $\psi = \dfrac{1}{4\sqrt{2\pi}} \left(\dfrac{Z}{a_0}\right)^{\frac{3}{2}} e e^{-\frac{e}{2}} \sin\theta\sin\phi$

18.96 Arrange the orbital $1s$, $2s$, $3s$, $3p$, $3d$ for H-atom in order of energy. Would the order you suggest be same for atoms of other element? What is the degeneracy of the M shell of H-atom.

18.97 How does hydrogenic $2s$ wavefunction differ from $1s$ wavefunction? Indicate their angular part (if any).

18.98 What is orbital angular momentum of an electron in a $2p$-orbital? How many angular and radial nodes this orbital will have? Show diagramatically the node (s) in $2p_x$ orbital. Do the 'signs' on the the two sides of a node make any sense in the electronic charge distribution? Mention the usefulness (if any) of marking the regions of a wavefunction with +ve and –ve sign.

18.99 The three real hydrogenic p-orbitals are designated as p_x, p_y and p_z. Rationalise such designation. Show with justification that p_z is an eigenfunction of $\hat{L}_z$, but p_x and p_y are not. Can a linear combination of p_x and p_y be an eigenfunction of $\hat{L}_z$?

18.100 Real d-orbitals are customarily designated as $d_{x^2-y^2}$, d_{z^2}, d_{xy}, d_{yz} and d_{xz}. Given that an atom has no preferred direction in space, how do you justify such irregular designation (unlike p-orbitals)?

18.101 The ground state wavefunction of a H-atom is

$$\psi = \left(\dfrac{1}{\pi a_0^3}\right)^{\frac{1}{2}} e^{-r/a_0} \quad \text{where } a_0 \text{ is the Bohr radius}$$

(a) At what point the probability of finding the electron is most? At what distance does the probability of finding the electron at a point in the H-atom falls to 50% of its highest value.

(b) At what distance the probability of finding the electron in a spherical layer of thickness dr will be most (such distance is called the most probable radius of the orbital). Is there only one or more than one radius, where this probability will fall to 50% of its highest value?

(c) Calculate the average value of the distance of the electron from the nucleus (called the mean radius of the orbital).

(d) Calculate the probability that the electron will be found within a volume of 1.0 pm^3 located (i) at the nucleus (ii) at a distance a_0 from the nucleus (volume of a H-atom is of the order of 10^7 pm^3). Comment on the result.

(e) Calculate (i) the mean kinetic energy and (ii) the mean potential energy of the electron.

18.102 True or false?

(a) The smallest allowed value of the atomic quantum number is zero.

(b) ψ cannot be maximum or minimum at infinity where $\psi = 0$

(c) $\left|\psi^2\right|$ has a maximum at the origin for the hydrogen 1s-state.

(d) $4\pi r^2 \left|\psi\right|^2$ has a maximum at the origin for hydrogen 1s-state.

18.103 In a hydrogen atom the various quantum numbers are related according to $n > l \geq |m|$, whereas for the quantum number for a particle in a box, there is no such restriction. Explain.

ANSWERS

18.1 Ideal blackbody radiation is the one whose distribution of frequencies depends only on its temperature* as it happens with the radiation in equilibrium with the interior walls of an evacuated cavity.

[* Strictly speaking, it is not the radiation to which the temperature applies, but the matter in equilibrium with the radiation.]

18.2 (i) The radiation inside a hollow body has higher intensity than that emitted by its outer surface.

(ii) In the first case, unlike the second, the distribution of frequencies does not depend on the material, and the size and shape of the body (cavity).

☐ Such differences are attributed to:

(i) the reflection of radiation by the interior surface of the body

(ii) the thermal equilibrium of the radiation with the interior surface where the photons in the system can adjust their number to maintain a constant total energy.

18.3 The given statement is justified provided the temperature of the body is not so high that it becomes self-luminous.

18.4 The pressure of cavity radiation at equilibrium is determined by the rate of emission of radiation which, at a particular temperature, depends only on the surface density of atoms (the emitter) in the cavity and hence independent of the volume of the cavity. On the contrary, the gas pressure at a paticular temperature is determined largely by the bulk density of the gas molecules, which depends on the volume of the gas, the number of gas molecules being fixed.

18.5 Only that part of the spectral distribution curve for blackbody radiation can be explained classically which corresponds to very long wavelength region. Because only for such region, the gap between consecutive energy levels of the relevant oscillators narrows down virtually to the classical expectation of energy continuum.

Note: The Raleigh–Jeans law for blackbody radiation, $e_\lambda = 8\pi kT/\lambda^4$, works (closely) only at long wavelengths. Because it is based on the law of equipartition of energy which assumes the classical concept of energy as a continuous variable.

18.6 The given colour change happens, because the radiation of higher wavelength is emitted in decreasing proportion at high temperature, as

$$\lambda_{max}T = \text{constant}, \quad \text{Wien displacement law}$$

where λ_{max} is the maximum wavelength for cavity radiation at temperature T. This is a consequence of the Planck Eq. (18.1b) for cavity radiation.

18.7 A spectral line results from the transition of the constituent particles of a system from one energy level to another. For solids and liquids, the number of accessible energy levels (that determines the number of spectral lines) is much greater and the gap between the successive energy levels (that determines the frequency of the lines) is much too small compared with gases. Because the constituent particles interact strongly in solids and liquids (due to their close proximity), but only negligibly in gases. As a consequence, the spectral lines merge together forming a continuous spectrum in case of solids or liquids, but not in case of strongly heated gases where almost free atomic species are generated.

☐ The lines in the spectrum are the images of the slits used in the usual experimental setup. Therefore, with such set-up, no other geometrical figure can appear in the spectrum.

18.8 Because in case of rotational motion, the idea of quantization can be expressed in a rather simpler way through angular momentum than through energy.

Note: Angular momentum and h have the same unit. Bohr's theory is based on this fact.

18.9 The contradiction disappears at large quantum number which corresponds to a macroscopic body, when the gap between the energy levels virtually disappears with consequent occurrence of continuous radiation.

Note: Quantum physics must reduce to classical physics at large quantum numbers. This is Bohr's correspondence principle. In fact, Bohr framed his theory of the hydrogen atom with this principle in mind.

18.10 (a) (i) 'The assumption of definite trajectory of electron' has no quantum mechanical significance.

    (ii) 'The assumption that the angular moment of the electron is an integral multiple of $\hbar$' has to be rectified to $\sqrt{l(l+1)} \cdot \hbar$ (where l is zero or an integer) on quantum mechanical ground.

  (b) (i) The Bohr ground state possesses orbitan angular momentum ($n\hbar$, with $n = 1$), while the one predicted by quantum mechanics (i.e. actual ground state) does not (since $l = 0$).

    (ii) The two ground states have quite different distribution of electron

  (c) The two models can be distinguished experimentally by examining magnetic properties in the ground state.

18.11 In his theory of H-atom, Bohr made the faulty assumption that the orbiting electron in a H-atom follows circular paths with angular momentum equal to $n\hbar$, $n = 1, 2, 3, ...$ This leads to the following expression for the energy (E) of a H-atom

$$E = -\frac{me^4}{2n^2\hbar^2}$$

This energy expression happens to be same as that predicted by quantum mechanics. The interesting point is that in the above expression due to Bohr n was in connection with angular momentum with its faulty designation as $n = 1, 2, 3, ...$ that suits (fortunately) with energy of the atom. This explains why the above expression becomes as effective as the one found quantum mechanically with n as positive integer.

Note: The essential part of Bohr's theory is not the orbits themselves (which are faulty) but their energies that determine the frequencies of the spectral lines with which this theory is concerned.

18.12 No. Because here the quantum of energy involved hv is ordinarily so small (since the frequency of oscillation v is small and the Planck constant is very small), that the energy change of a pendulum seems to occur continuously.

18.13 The energy of a photon of an electromagnetic radiation is proportional to the frequency of the latter. The relevant proportionality constant is called the Planck constant (h).

☐ In SI unit, $h = 6.626 \times 10^{-34}$ JS.

☐☐ If h were 6.6, which is very high, the quantum phenomena, such as discontinuous nature of energy change of a system, would be more conspicuous in every day life (e.g. in case of oscillation of a pendulum), the energy quanta being then large.

18.14 High-frequency end.

☐ Because, at high frequency (v), the energy quantum (hv) is high.

18.15
$$\frac{kT}{h} = v \cdot \frac{kT}{hv}$$

Both kT and hv have the same dimensions of energy [kT represents the energy of a molecule per degree of vibrational freedom and hv represens the energy per photon]. Then kT/h will have the dimension of frequency (v) i.e. t^{-1}.

18.16 Here, the action is determined by the quantum of vibrational energy (hv) that can be transferred at a time. This is large in case of an atom due to high value of the frequency (v).

☐ Compared to kT ($1.38 \times 10^{-23} \times 300$ J), hv is much higher in case of atom ($6.63 \times 19^{-34} \times 10^{15}$ J), but much less in case of pendulum ($6.63 \times 10^{-34} \times 1$ J).

Note: Although the energy content of a big pendulum is much too higher (due to very high quantum number arising from large mass), the ability to transfer energy is lower with pendulum than an atom.

18.17 The quantum nature of light is manifested when it interacts with matter, e.g. in the phenomenon of *photoelectric effect*.

☐ Even a weak light (i.e. of low intensity) involves interaction of many photons per second. Consequently, due to persistence of vision, the energy transfer by successive photons seems to occur continuously.

18.18 This lies in the verification of Einstein's photoelectric Eq. (18.2), which is based on the supposition that each electron absorbs only one photon. The plot of maximum kinetic energy of the photoelectron versus the frequency (v) of the incident radiation (of ordinary intensity) is really found to be linear with the slope equal to the Planck constant h.

18.19 The ionization potential, more appropriately ionization energy (E), of a metal is the minimum energy required to remove an electron from a free gaseous atom (in the ground state) of the metal.

The photoelectric work function of a metal is the minimum energy (ϕ) required to remove an electron from a bound atom in its crystal where the constituent atoms are combined together through metallic bonds.

□ $\phi > E$ because ϕ relates to bound atoms.

18.20 Compton effect. Because it involves elastic collision of photon with solid surface, which is a more concrete evidence of the particle character of photon than mere emission of electron by photon as in the photoelectric effect.

18.21 Graphite contains essentially free electrons which are, therefore, easily emitted by absorbing only a part of the energy of the incident photons with scattering of the latter.

Electrons in aluminium are bound fairly tightly whose photoemission is possible only by absorption of high energy photons. Such electrons can however behave like free electrons but with effective mass much greater than that of an electron. Because here the ionic core in which the electron is bound recoils as a whole during collision with a photon. As a consequence, the Compton shift for such electrons is immesurably small [vide Eq. (18.3)].

Note: The photoelectric effect may be regarded as a particular case of the Compton effect with immesurable Compton shift.

18.22 Because, both Compton shift and Raman shift involve scattering of light.

□ Because Compton shift, unlike Raman shift, involves emission of electron along with scattering of light.

18.23 Vide Section 18.3.

□ By Eq. (18.3), the highest value of Compton shift is $2h/m_0c$ which corresponds to head-on collision ($\phi = 180°$).

18.24 According to de Broglie, matter which were classically thought to consist of particles, is likely to possess certain wave-like character, and the relevant wavelength (λ) is given by

$$\lambda = h/mv$$

However, for macroscopic bodies, the mass being high, λ is much too smaller (even when their velocity is low) than the dimensions of any physical system. Hence, wave phenomena, like diffraction, are not observed with macroscopic bodies and it seems that the de Broglie theory fails for macroscopic bodies.

18.25 (i) de Broglie wavelength

$$\lambda = h/p$$

$$= \frac{h}{\sqrt{2m\,KE}} \quad \text{for } v = \left(\frac{2KE}{m}\right)^{\frac{1}{2}} \ll c$$

$$= \frac{6.63 \times 10^{-34} \text{ JS}}{\sqrt{2(9.11 \times 10^{-31} \text{ kg})(1.00 \times 10^3 \times 1.60 \times 10^{-19} \text{ J})}}$$

$$= 3.9 \times 10^{-11} \text{ m}$$

(ii) Here v is of the order of c (from preliminary survey).

Then for precise calculation of λ, we are to use the following relation

$$\lambda = \frac{h}{p} = \frac{ch}{\sqrt{(KE + m_0c^2)^2 - m_0^2c^4}}, \quad \text{since } KE = \sqrt{m_0^2c^4 + p^2c^2} - m_0c^2$$

$$= \frac{(3.00 \times 10^8 \text{ ms}^{-1})(6.63 \times 10^{-34} \text{ JS})}{\sqrt{[(100 \times 10^3 \times 1.60 \times 10^{-19} \text{ J}) + (9.11 \times 10^{-31} \text{ kg})(3.00 \times 10^8 \text{ ms}^{-1})^2]^2 - (9.11 \times 10^{-31} \text{ kg})^2 (3.00 \times 10^8 \text{ ms}^{-1})^4}}$$

$$= 2.02 \times 10^{-12} \text{ m}$$

Note: The relations used above follow from $m = \dfrac{m_0}{\sqrt{1 - \frac{v^2}{c^2}}}$.

Thus $\qquad E = \dfrac{m_0 c^2}{\sqrt{1 - \frac{v^2}{c^2}}}$, multiplying both sides by c^2

Then $\qquad E^2 = m_0^2 c^4 + \dfrac{E^2 v^2}{c^2} = m_0^2 c^4 + c^2 (mv)^2 = m_0^2 c^4 + c^2 p^2$

$$KE = E - E_0 = \sqrt{m_0^2 c^4 + p^2 c^2} - m_0 c^2$$

18.26 Quantum mechanics is applicable to all systems, but need be applied only to microscopic system like (ii) [and not (i) and (iii)] which exhibit appreciable wave-like character, i.e. which have de Broglie wavelength not too small compared to the dimensions of any physical system. In case of (i) and (iii), this does not happen due to high mass of the systems.

18.27 The behaviours of radiation and matter pertain to wave-like aspect in some phenomena and to particle-like aspect in all other phenomena. Then the two aspects are complementary. They are not contradictory as they simply carry two extreme views regarding the mode of action of radiation and matter.

Note: Being opposite in nature, the particle-like and wave-like aspects are not both operative in the same experimental situation. It should also be kept in mind that these two aspects serve only to account for a phenomenon and not to predict its possibility of occurrence. Thus the fact that light can have particle-like property explains the photoelectric effect due to ultraviolet radiation but not the inability of infrared radiation to bring about such phenomenon. Both the radiations, however, exhibit their particle-like property in their emission or absorption by atoms and molecules.

18.28 Speed of the H-atom after absorption of photon

$$= \frac{\text{Momentum of the photon absorbed } (h\nu / c)}{\text{Mass of the H-atom}}$$

$$= \frac{\left(\dfrac{(6.63 \times 10^{-34} \text{ Js}) (2.5 \times 10^{14} \text{ Hz})}{3.00 \times 10^8 \text{ ms}^{-1}} \right)}{(1.67 \times 10^{-27} \text{ kg}}$$

$$= 0.331 \text{ m}$$

[**Note:** Here the solution is based on the conservation of momentum and not of KE, the collision of photon being inelastic.]

☐ Yes. Here the energy is excess of binding energy will appear as KE of the emitted electron.

18.29 As in the previous problem, the solution is based on the law of conservation of momentum

$$\text{Speed of the worm} = \frac{\text{Total momentum of the emitted photons}}{\text{Mass of the worm}}$$

$$= \frac{\text{Total energy of the emitted photons}}{\text{Speed of light} \times \text{mass of the worm}}$$

$$= \frac{\text{Power of the worm} \times \text{Period elapsed}}{\text{Speed of light} \times \text{mass of the worm}}$$

$$= \frac{(0.1 \text{ W}) (30 \times 24 \times 60 \times 60 \text{ s})}{(3 \times 10^8 \text{ ms}^{-1})(1 \times 10^{-3} \text{ kg})} = 0.864 \text{ ms}^{-1}$$

Note: Here wavelength of light if of no use. This is unlike the previous problem. Use has been made of the relation $P = E/C$ which holds only in case of photon.

18.30 According to uncertainty principle, it is possible only within some limited accuracy to know simultaneously the values of particular pairs of observables (called complementary observables, such as position and momentum) of a physical entity due to wave-particle duality. The principle is often stated conveniently in terms of the simple quantities—position and momentum. Instead of momentum, if velocity (a simpler quantity) is used, the product of uncertainties in the two complementary observables will not be independent of mass. This is not unlikely in view of the fact that the de Broglie wavelength associated with a material particle is determined entirely by its momentum and not velocity.

☐ Yes, but cannot be lower than $h/4\pi$. Vide problem 18.79.

18.31 As implied by the meaning of uncertainty (vide Section 18.6), the given two bodies will have zero uncertainty in momentum. Also they have the same uncertainty, infinity, regarding position.

Note: Uncertainty should not be confused with experimental error.

18.32 Yes. The uncertainties Δp and Δq involved in the given expression represent the ones inherent in the system. These, unlike experimental errors, are not determined by the measuring device. However, the experimental errors give some hint regarding uncertainties. In the given measuring device, the inaccuracy of position is comparable to λ while that of momentum to h/λ, so that the product of the two becomes independent of λ and is of the order of h which is in crude agreement with the uncertainty principle.

18.33 No. Because the uncertainty principle essentially represents the wave-particle duality which is firmly established.

18.34 According to the uncertainty principle, uncertainties are involved in knowing simultaneously the position and the momentum of a particle, only if they refer to the same coordinate axis. Then $\Delta P_x \Delta x \left(\geq \frac{h}{4\pi} \right)$ cannot be zero because both Δx (the uncertainty in position) and ΔP_x (uncertainty in momentum) refer to the same coordinate axis (which is x). But $\Delta P_y \Delta x$ is zero because Δx and ΔP_y do not refer to the same coordinate axis.

18.35 The paradox will be resolved if the wavefunction of the body is supposed to be a linear combination of very large number of eigenfunctions so that it has high value only at a particular location where the position of the body is virtually fixed and hence certain.

Regarding momentum of the body, in any single experiment, the observed value will correspond to any one of the component wavefunctions and hence very uncertain. Again, considering the large number of wavefunctions, the average value of the observed momentum (which is a vector quantity) can be practically zero even for a short span of time imparting a static appearance to the body.

Then in the given situation, the uncertainty principle regarding position (having no uncertainty) and momentum (having extreme uncertainty) is not violated.

18.36 This is due to life-time broadening, a consequence of the uncertainty principle. If an atomic species lives for a time t, then its energy instead of being exactly E, may be anywhere in a range ΔE around E, where $\Delta E \gg \dfrac{h}{4\pi t}$. Since an atom in the ground state is long-lived (compared to excited states), it will be characterised by short ΔE, and hence narrow, i.e. sharp, spectral lines arising due to transitions from this state.

18.37 According to the Bohr's theory, the angular momentum (mvr) of an electron in a stationary orbit of radius r is

$$mvr = n\frac{h}{2\pi} \quad \text{where } n \text{ is an integer}$$

or $$2\pi r = n\frac{h}{mv}$$

$$= n\lambda \quad \text{where } \lambda \text{ is de Broglie wavelength}$$

Then the Bohr orbit is such that its circumference is an integral multiple of the relevant de Broglie wavelength of the orbiting electron.

18.38 From the relation $2\pi r = n\lambda$, we have for the given condition, $n = 3$ from Eq. (18.4)

$$E = -\frac{z^2}{n^2} \cdot \frac{2\pi^2 me^4}{h^2}$$

$$= -\frac{2^2}{3^2} \times 13.6 \text{ eV, since } z = 2 \text{ for He}^+$$

$$= -6.04 \text{ eV}$$

Note: $\dfrac{2\pi^2 me^4}{h^2} = 13.6$ eV is the ionization potential of the hydrogen atom (the minimum energy needed to remove the electron from a H-atom).

18.39 In classical mechanics, the state of a system is defined by specifying all the forces acting and all the positions and velocities (or momenta) of the constituent particles. But the very knowledge needed to specify the classical–mechanical state of a system is unattainable. Because according to the uncertainty principle, simultaneous specification of position and momentum of a particle is not possible.

18.40 In quantum mechanics, the state of a system is defined by a mathematical function, called state function which is a function of the coordinates of the particles of the system and also of time, such that it satisfies time-dependent Schrödinger equation.

☐ A state represented by wavefunction in Schrödinger equation is called stationary if the probability density calculated from the wavefunction is independent of time.

☐☐ For the given wavefunction, the probability density is

$$\left(e^{-2\pi i\frac{Et}{h}}\psi(x)\right)^* e^{-2\pi i\frac{Et}{h}}\psi(x) = |\psi(x)|^2$$

Since the probability density is independent of time, the given function represents a stationary state.

Note: The solution of time-independent Schrödinger equation is a stationary state function but the converse is not necessarily true.

18.41 The state of a system is represented by a wavefunction (ψ) such that the properties of the system at the concerned state can be found by certain mathematical operation on ψ. The given wavefunctions which are proportional to one another represent the same state because they provide eigenvalue for the same operator.

18.42 No, unless $E_1 = E_2$.

[Here $H\psi_1 = E_1\psi_1$ and $H\psi_2 = E_2\psi_2$.

Then $H\psi = H(C_1\psi_1 + C_2\psi_2) = E_1C_1\psi_1 + E_2C_2\psi_2 \neq$ constant $x\psi$, if $E_1 \neq E_2$]

□ But even if $C_1\psi_1 + C_2\psi_2$ is not an eigenfunction, it may be a state function if ψ_1 and ψ_2 both satisfy time-independent Schrödinger equation.

[If ψ_1 and ψ_2 are state functions (not necessarily eigenfunctions), then

$$H\psi_1 = i\hbar\frac{\partial\psi_1}{\partial t}$$

and

$$H\psi_2 = i\hbar\frac{\partial\psi_2}{\partial t}$$

Then

$$\begin{aligned}
H\psi &= H(C_1\psi_1 + C_2\psi_2) \\
&= C_1H\psi_1 + C_2H\psi_2 \\
&= C_1\left(i\hbar\frac{\partial\psi_1}{\partial t} + C_2\,i\hbar\frac{\partial\psi_2}{\partial t}\right) \\
&= i\hbar\left(\frac{\partial}{\partial t}C_1\psi_1 + \frac{\partial}{\partial t}C_2\psi_2\right) \\
&= i\hbar\left[\frac{\partial}{\partial t}(C_1\psi_1 + C_2\psi_2) = i\hbar\frac{\partial\psi}{\partial t}\right]
\end{aligned}$$

18.43 Here the energy E of the composite function ψ is

$$E = E_1 + E_2$$

Then

$$\frac{E\psi}{\psi} = \frac{E_1\psi_1}{\psi_1} + \frac{E_2\psi_2}{\psi_2}$$

or

$$\frac{H\psi}{\psi} = \frac{H\psi_1}{\psi_1} + \frac{H\psi_2}{\psi_2}$$

Since additivity of energies implies no interaction and hence no change in potential energy, we can write

$$H = -\frac{\hbar^2}{2m}\frac{\partial^2}{\partial x^2}$$

Then

$$\frac{1}{\psi}\left(\frac{\partial^2\psi}{\partial x_1^2} + \frac{\partial^2\psi}{\partial x_2^2}\right) = \frac{1}{\psi}\frac{\partial^2\psi_1}{\partial x_1^2} + \frac{1}{\psi_2}\frac{\partial^2\psi_2}{\partial x_2^2}$$

This relation will hold if

$$\psi = K\psi_1\psi_2$$
$$= \psi_1\psi_2 \text{ for } K = 1$$

18.44 Because Max Born's interpretation of ψ implies that, only a single value of probability density ($\psi^*\psi$) at each point in space is physically meaningful.

18.45 ψ cannot be infinite at any point. Because this would correspond to complete localisation of the system described by ψ, which would be inconsistent with its wave properties.

ψ can be zero but not everywhere, because the system it describes must be somewhere.

18.46 Because Max Born's interpretation of ψ demands that an acceptable ψ must be normalized, i.e. total probability of finding the system in the entire space must be unity.

□Normalization constant may be −ve or imaginary. But it cannot be zero or infinity due to the restriction that ψ cannot be zero everywhere or infinity anywhere.

18.47 Normalization constant N is given by

$$N^2 = \frac{1}{\displaystyle\int_{\text{all range}} \psi^* \psi \, d\tau} \quad \text{(vide Section 18.8)}$$

$$= \frac{1}{I}$$

(i)
$$I = \int_{-L}^{L} C^2 dx = 2C^2 L$$

Then
$$N = \frac{1}{C\sqrt{2L}}$$

(ii)
$$I = \int_0^a \sin^2 \frac{n\pi x}{a} dx = \frac{a}{2}$$

$$\left[\because \int \sin^2 \theta \, d\theta = \tfrac{1}{2} \int (\sin^2 \theta + \cos^2 \theta) \, d\theta = \tfrac{1}{2} \int d\theta, \text{ by symmetry} \right.$$

$$\left. \text{Then } \int_0^a \sin^2 \frac{n\pi x}{a} dx = \int_0^{n\pi} \sin^2 \theta \cdot \frac{a}{n\pi} d\theta = \frac{n\pi}{2} \cdot \frac{a}{n\pi} = \frac{a}{2} \right]$$

Then
$$N = \sqrt{\frac{2}{a}}$$

(iii)
$$I = \int_0^\infty e^{-\frac{2r}{a}} r^2 dr \int_0^\pi \sin\theta d\theta \int_0^{2\pi} d\phi$$

since the volume element in three dimension is $d\tau = r^2 dr \sin\theta d\theta d\phi$ (Fig. 18.5).

$$= \left(\frac{a^3 \cdot 2!}{2^3} \right)(2)(2\pi) = \pi a^3$$

Then
$$N = \frac{1}{\sqrt{\pi a^3}}$$

(iv)
$$I = \int_0^\infty r^2 \cdot r^2 e^{-\frac{r}{a}} dr \int_0^\pi \sin^3 \theta \, d\theta \int_0^{2\pi} \cos^2 \phi d\phi \quad \because x = r\cos\phi\sin\theta$$

$$= (4! a^5)\left(\tfrac{4}{3}\right)(\pi) = 32\pi a^5$$

$$\left[\int_0^\pi \sin^3 \theta = -\int_1^{-1}(1 - \cos^2 \theta) \, d\cos\theta = -\int_1^{-1}(1 - x^2)dx = \tfrac{4}{3} \right]$$

Then
$$N = \frac{1}{\sqrt{32\pi a^5}}$$

Note:

1. Only the +ve values of N is reported, as the −ve value of N does not imply a different state.

2. In normalization, one must be careful about the dimension of the situation.

18.48 Since ψ is normalized

$$1 = \int_0^{2\pi} \psi^* \psi\, d\phi \quad \text{since motion is described only by the coordinate } p$$

$$= N^2 \int_0^{2\pi} d\phi = 2\pi N^2$$

Then $\qquad N = \dfrac{1}{\sqrt{2\pi}}$

18.49 Let the normalized wavefunction be $N\psi$. Then

$$1 = N^2 \int \psi^* \psi\, d\tau$$

$$= N^2 \int_0^\infty \left(2 - \frac{r}{a_0}\right)^2 r^2 e^{-r/a_0}\, dr \int_0^\pi \sin\theta\, d\theta \int_0^{2\pi} d\phi$$

$$= N^2 \int_0^\infty \left(4r^2 - \frac{4r^3}{a_0} + \frac{r^4}{a_0^2}\right) e^{-r/a_0}\, dr \int_0^\pi \sin\theta\, d\theta \int_0^{2\pi} d\phi$$

$$= N^2 \int_0^\infty \left(4 \cdot \frac{2!}{\left(\frac{1}{a_0}\right)^3} - 4 \cdot \frac{3!}{a_0 \left(\frac{1}{a_0}\right)^4} + \frac{4!}{a_0^2 \left(\frac{1}{a_0}\right)^5}\right)(2)(2\pi)$$

$$= 32\pi a_0^3 N^2$$

Then $\qquad N = \dfrac{1}{\sqrt{32\pi a_0^3}}$

[Note: Here the wavefunction is taken to be three dimensional (because it represent an orbital) which is not mentioned unlike problem 18.47(iii)].

□ Although the given wavefunction contains no angular variables (θ and ϕ), it has

an angular part which is $\left[\int_0^\pi \sin\theta\, d\theta \cdot \int_0^{2\pi} d\phi\right]^{-\frac{1}{2}} = \dfrac{1}{\sqrt{4\pi}}$. This is in keeping with the

necessity that the total wavefunction and its radial and angular parts must be separately normalised.

18.50 Since the given function represents a radial wavefunction, its normalization is determined only by r (and not θ and ϕ unlike the previous problem) through the following restriction

$$1 = \int_{\text{all range}} \psi^* \psi = N_2 \int_0^\infty e^{-2r/a_0} r^2\, dr = N^2 \frac{2!}{\left(\frac{2}{a_0}\right)^3}$$

Then $\qquad N = \dfrac{2}{\sqrt{a^3}}$

18.51 An acceptable wavefunction is one that conforms to the restrictions (mentioned in Section 18.8) imposed by the form of Schrödinger equation and Born's interpretation of wavefunction.

(i) No, being not normalizable.

(ii) No, being not normalizable.

(iii) No, because $\frac{1}{x} = \infty$ at $x = 0$ within the given range.

(iv) No, because of the discontinuity of the first derivative $\left(\frac{1}{x}\right)$ of $\log x$ at $x = 0$ within the given range.

(v) No, because $e^x = \infty$ at $x = \infty$.

(vi) Yes. Here e^{-x} is normalisable through not normalised.

(vii) No, because $e^{-x} = \infty$ at $x = -\infty$.

(viii) Yes. Here e^{-x^2} is normalizable, though not normalized.

(ix) No, because the first derivative of $e^{-|x|}$ is not continuous at $x = 0$ (although $e^{-|x|}$ is continuouus at $x = 0$).

(x) Yes, for the same reason as in (viii).

(xi) No, being not normalizable.

(xii) No, being not single-valued.

(xiii) No, because $\tan x = \alpha$ at $x = \frac{\pi}{2}$.

(xiv) Yes. Here $e^{i\theta}$ is normalizable.

18.52 (i) No, because ψ is not continuous.

(ii) No, because the slope (i.e. the first derivative) of ψ is not continuous at the point of intersection of the two parts of the curve.

(iii) No, because ψ is infinite at certain value of x.

(iv) No, because ψ is not single-valued.

(v) No, because the slope of ψ is infinite (i.e. discontinuous) at certain value of x.

(vi) No, because $\psi \to \infty$ as $x \to \pm \infty$.

(vii) No, because $\int_{-\infty}^{+\infty} \psi^2 dx \neq 1$.

(viii) Yes, because of agreement with Schrodinger equation and Born's interpretation of wavefunction.

(ix) Yes, for the same reason as in (viii).

18.53 Let ψ be an eigenfunction of the operator d/dx with eigenvalue k. Then

$$\frac{d\psi}{dx} = k\psi$$

or

$$\frac{d\psi}{\psi} = kdx$$

Integration gives

$$\ln\psi = kx + \ln a \quad \text{where } \ln a \text{ is the integration constant}$$

or

$$\psi = ae^{kx}$$

ψ is different for each different value of k (but not for different a).

18.54 (i) $\dfrac{de^{ikx}}{dx} = ik\,e^{ikx}$ and $\dfrac{d^2 e^{ikx}}{dx^2} = -k^2 e^{ikx}$.

Then e^{ikr} is an eigenfunction of both the operators with eigenvalue ik (imaginary) and $-k^2$ (real) respectively.

(ii) $\dfrac{d\cos kx}{dx} = -k\sin kx \neq \text{constant} \times \cos kx$ and $\dfrac{d^2}{dx^2} \cos kx = -k^2 \cos kx$.

Then $\cos kx$ is not an eigenfunction of d/dx but in eigenfunction of d^2/dx^2 with eigenvalue $-k^2$.

(iii) $\dfrac{dk}{dx} = 0 = 0 \times k$ and $\dfrac{d^2 k}{dx^2} = 0 = 0 \times k$.

Then k is an eigenfunction of both the operators with same eigenvalue, zero.

(iv) $\dfrac{d(kx)}{dx} = k \neq$ constant $\times kx$ and $\dfrac{d^2(kx)}{dx^2} = 0 = 0 \times kx$.

Then kx is not an eigenfunction of d/dx but it is an eigenfunction of d^2/dx^2 with eigenvalue 0.

(v) $\dfrac{de^{-ax^2}}{dx} = -2ax\,e^{-ax^2}$ and $\dfrac{d^2 e^{-ax^2}}{dx^2} (4a^2x^2 - 2a)e^{-ax^2}$.

Then e^{-ax^2} is not an eigenfunction of both the operaters.

Note: A particular function may be an eigenfunction of more than one operator with even same eigenvalue. Again eigenvalue may be real (including zero) and also imaginary. However, eigenvalues of quantum mechanical operations are always real.

18.55 $$\dfrac{\partial \psi}{\partial t} = -\dfrac{iE}{\hbar}\psi$$

or $$-\dfrac{\hbar}{i}\dfrac{\partial}{\partial t}\psi = E\psi$$

Hence the operator for E is $-\dfrac{\hbar}{i}\dfrac{\partial}{\partial t}$.

Again $$\dfrac{\partial}{\partial x}\psi = \dfrac{iP}{\hbar}\psi$$

or $$\dfrac{\hbar}{i}\dfrac{\partial}{\partial x}\psi = P\psi$$

Hence the operator for P is $\dfrac{\hbar}{i}\dfrac{\partial}{\partial x}$.

18.56 Here $$H\psi = -\dfrac{d^2\psi}{dx^2} + x^2\psi = [(1 - 4\alpha^2)x^2 + 2\alpha]\psi.$$

Now for ψ to be an eigenfunction of H, the quantity within [ ] must be independent of x. This is possible only if
$$1 - 4\alpha^2 = 0$$
whence $\alpha = \frac{1}{2}$ and $E = 1$.

18.57 ∇^2 survives the test of linearity with any identity relation, such as the following
$$e^{ik}(x + y + z) = \cos k(x + y + z) + i\sin k\,(x + y + z)]$$
Here $$\nabla^2 e^{ik}(x + y + z) = -3k^2 e^{ik}(x + y + z)$$
$$\nabla^2 \cos k(x + y + z) + \nabla^2 i\sin k(x + y + z) \equiv 3k^2[\cos k(x + y + z) + i\sin(x + y + z)]$$
$$\equiv 3k^2 e^{ik}(x + y + z)$$

18.58 $$\int_{-\infty}^{+\infty}\psi_1^*\hat{P}_x\psi_2 dx = \int_{-\infty}^{+\infty}\psi_1^*\dfrac{\hbar}{i}\dfrac{\partial\psi_2}{\partial x}dx$$

where ψ_1 and ψ_2 are two state functions of a system

$$= \dfrac{\hbar}{i}\int_{-\infty}^{+\infty}\psi_1^* d\psi_2$$

$$= \dfrac{\hbar}{i}\left[|\psi_1^*\psi_2|_{-\infty}^{+\infty} \int_{-\infty}^{+\infty}\psi_2 d\psi_1^*\right]$$

$$= -\dfrac{\hbar}{i}\int_{-\infty}^{+\infty}\psi_2 d\psi_1^* \qquad \because |\psi_1^*\psi_2|_{-\infty}^{+\infty} = 0$$

(due to existence of nodal point at the extremity)

$$= -\frac{\hbar}{i} \int_{-\infty}^{+\infty} \psi_2 \left(\frac{\partial}{\partial x} \right) \psi_1^* dx$$

$$= \int_{-\infty}^{+\infty} \psi_2 \left(\frac{\hbar}{i} \frac{\partial \psi_1}{\partial x} \right)^* dx$$

$$= \int_{-\infty}^{+\infty} \psi_2 (\hat{P}_x \psi_1)^* dx$$

☐ $$\hat{H}_x = -\frac{\hbar^2}{2m} \frac{\partial^2}{\partial x^2} + V(x)$$

Now $$\int_{-\infty}^{+\infty} \psi_1^* \frac{\partial^2}{dx^2} \psi_2 dx = \left[\left| \psi_1^* \frac{\partial}{\partial x} \psi_2 \right|_{-\infty}^{+\infty} - \int \frac{\partial}{\partial x} \psi_2 d\psi_1^* \right]$$

$$= -\int \frac{\partial \psi_2}{\partial x} \frac{\partial \psi_1^*}{\partial x} dx$$

$$= -\left[\left| \psi_2 \frac{\partial \psi_1^*}{\partial x} \right|_{-\infty}^{+\infty} - \int \psi_2 \frac{\partial^2 \psi_1^*}{\partial x^2} dx \right]$$

$$= \int \psi_2 \left(\frac{\partial}{\partial x^2} \psi_1 \right)^* dx \qquad \frac{\partial^2}{\partial x^2} \text{ is a real operator}$$

Then $\frac{\partial^2}{dx^2}$, and hence $\frac{\hbar^2}{2m} \frac{\partial^2}{\partial x^2}$, is a Hermitian operator.

Therefore, V being a real quantity, $\hat{H}$ is a Hermitian operator.

18.59 (a) For a Hermitian operator $\hat{\Omega}$

$$\int \psi_1^* \Omega \psi_2 d\tau = \int \psi_2 (\hat{\Omega} \psi_1)^* d\tau \qquad (I)$$

w here ψ_1 and ψ_2 are any two state functions of the system.

Then, when $\psi_1 = \psi_2 = \psi$, an eigenfunction, the eigenvalue ω will be

$$\omega = \int \psi^* \hat{\Omega} \psi \, d\tau = \int \psi (\hat{\Omega} \psi)^* d\tau = \omega^*$$

This requires ω to be real.

(b) It follows from (I) that when ψ_1 and ψ_2 are eigenfunctions

$$\omega_2 \int \psi_1^* \psi_2 d\tau = \omega_1 \int \psi_2 \psi_1^* d\tau$$

or $(\omega_2 - \omega_1) \int \psi_1^* \psi_2 d\tau$.

Then $\int \psi_1^* \psi_2 d\tau. = 0$, if $\omega_1 \neq \omega_2$

i.e. ψ_1 and ψ_2 are orthogonal to one another, if they have different eigenvalues.

Note: If $\omega_1 = \omega_2$, eigenfunctions of a hermitian operator may not be orthogonal to one another.

18.60 (i)
$$\hat{A}^2 \psi = \frac{\hbar}{i}\frac{\partial}{\partial x}\left(\frac{\hbar}{i}\frac{\partial}{\partial x}\right)\psi$$

$$= \frac{\hbar^2}{i^2}\frac{\partial^2 \psi}{\partial x^2} = -\hbar^2\frac{\partial^2 \psi}{\partial x^2}$$

Then $A^2 = -\hbar^2\dfrac{\partial^2}{dx^2}$.

(ii) $\hat{A}^2\psi = \left(\dfrac{\partial}{\partial x}+x\right)^2\psi = \left(\dfrac{\partial}{\partial x}+x\right)\left(\dfrac{\partial}{\partial x}+x\right)\psi$

$$= \left(\frac{\partial}{\partial x}+x\right)\left(\frac{\partial \psi}{\partial x}+x\psi\right)$$

$$= \frac{\partial^2 \psi}{\partial x^2}+x\frac{\partial \psi}{\partial x}+\psi+x\frac{\partial \psi}{\partial x}+x^2\psi$$

$$= \frac{\partial^2}{\partial x^2}+2x\frac{\partial}{\partial x}+x^2+1)\psi$$

Therefore $\qquad \hat{A}^2 = \dfrac{\partial^2}{dx^2}+2x\dfrac{\partial}{\partial x}+x^2+1$

Note: Here algebraic squaring gives

$$\left(\frac{\partial}{dx}+x\right)^2 = \frac{\partial^2}{\partial x^2}+2x\frac{\partial}{\partial x}+x^2$$

$\hat{A}^2$ implies two consecutive operations with $\hat{A}$.

18.61 (i)
$$\frac{d^2}{dx^2}x\psi = \frac{d}{dx}\left(\psi+x\frac{d\psi}{dx}\right)$$

$$= \frac{d\psi}{dx}+x\frac{d^2\psi}{dx^2}+\frac{d\psi}{dx}$$

$$\neq x\frac{d^2\psi}{dx^2}$$

Then x and $\dfrac{d^2}{dx^2}$ do not commute.

(ii) Since $\qquad \int_0^x d\psi = \int_0^x\dfrac{d\psi}{dx}dx = \dfrac{d}{dx}\int_0^x\psi dx, \dfrac{d}{dx}$ and $\int_0^x()\,dx$ commute

(iii) $\qquad \cos\sin^{-1}\psi = \cos\cos^{-1}\sqrt{1-\psi^2} = \sqrt{1-\psi^2}$

$$\sin^{-1}\cos\psi = \sin^{-1}\sin\left(\frac{\pi}{2}-\psi\right) = \frac{\pi}{2}-\psi$$

Then $\cos\sin^{-1}\psi \neq \sin^{-1}\cos\psi$ and hence $\cos$ and $\sin^{-1}$ do not commute.

18.62 (i)
$$[\hat{x},\hat{P}_y]\psi = x\frac{\hbar}{i}\frac{\partial\psi}{\partial y}-\frac{\hbar}{i}\frac{\partial}{\partial y}(x\psi)$$

$$= \frac{\hbar}{i}x\frac{\partial\psi}{\partial y}-\frac{\hbar}{i}x\frac{\partial\psi}{\partial y}--\frac{\hbar}{i}\psi\frac{\partial x}{\partial y} = 0$$

Then $\qquad [\hat{x},\hat{P}_y] = \hat{0}$

where $\hat{0}$ is the zero operator which is multiplication by zero

(ii)
$$[\hat{x}, \hat{p}_x]\psi = \frac{\hbar}{i} x \frac{\partial \psi}{dx} - \frac{\hbar}{i} x \frac{\partial \psi}{dx} - \frac{\hbar}{i} \psi \frac{\partial x}{dx}$$

$$= -\frac{\hbar}{i}\psi$$

Then $\quad [\hat{x}, \hat{p}_x] = -\frac{\hbar}{i}\hat{I}$

where $\hat{I}$ is the identity operator which is multiplication by one.

☐ Then $\hat{x}$ and $\hat{p}_y$ commute, while $\hat{x}$ and $\hat{p}_x$ do not. This is in accordance with the uncertainty principle according to which there is no uncertainty of knowing simultaneously the position and momentum in the different directions but not in the same direction.

18.63 For any state function $f(V, T)$

$$df = \frac{\partial f}{\partial V} dV + \frac{\partial f}{\partial T} dT \tag{I}$$

Since the differential of a state function is an exact differential, we have, by applying Euler criterion of exactness on (I)

$$\frac{\partial}{\partial T}\frac{\partial f}{\partial V} = \frac{\partial}{\partial V}\frac{\partial f}{\partial T}$$

Then $\dfrac{\partial}{\partial V}$ and $\dfrac{\partial}{\partial T}$ commute.

18.64 (a)
$$[\alpha + \beta, \gamma] = (\alpha + \beta)\gamma - \gamma(\alpha + \beta) = \alpha\gamma + \beta\gamma - \gamma\alpha - \gamma\beta$$
$$= \alpha\gamma - \gamma\alpha + \beta\gamma - \gamma\beta = [\alpha, \gamma] + (\beta, \gamma]$$

(b)
$$[\alpha, \beta\gamma] = \alpha\beta\gamma - \beta\gamma\alpha = \alpha\beta\gamma - \beta\alpha\gamma + \beta\alpha\gamma - \beta\gamma\alpha$$
$$= [\alpha, \beta]\gamma + \beta[\alpha, \gamma]$$

(c)
$$[\alpha\beta, \gamma] = \alpha\beta\gamma - \gamma\alpha\beta = \alpha\beta\gamma - \alpha\gamma\beta + \alpha\gamma\beta - \gamma\alpha\beta = \alpha[\beta, \gamma] + [\alpha, \gamma]\beta$$

18.65
$$[\hat{x}^n, \hat{P}_x] = [\hat{x}\cdot\hat{x}^{n-1}, \hat{P}_x) = x[\hat{x}^{n-1}, \hat{P}_x] + [\hat{x}, \hat{P}_x]x^{n-1} \quad \text{by problem 18.64 (c)}$$
$$= x[\hat{x}\cdot\hat{x}^{n-2}, \hat{P}_x) + i\hbar\, x^{n-1} \quad \text{by problem 18.62 (ii)}$$
$$= x\cdot x[\hat{x}^{n-2}, \hat{P}_x] + x(\hat{x}, \hat{P}_x)x^{n-2} + i\hbar x^{n-1}$$
$$= x^2[\hat{x}^{n-2}, \hat{P}_x] + i\hbar x^{n-1} + i\hbar x^{n-1}$$
$$= ni\hbar x^{n-1}, \text{ since } x^n[\hat{x}°, \hat{P}_x] = \hat{0}$$

18.66 The angular momentum vector L for a classical particle is defined as the vector cross product of its position $r\ (= ix + jy + kz)$ and linear momentum p

$$L = r \times p = \begin{vmatrix} i & j & k \\ x & y & z \\ p_x & p_y & p_z \end{vmatrix}$$

$$= i(yp_z - zp_y) + j(zp_x - xp_z) + k(xp_y - yp_x)$$

where i, j, k are unit vectors in the x, y and z directions, and P_x, P_y and P_z are components of P. Then the components of the classical angular momentum vector of a single particle are

$$L_x = yp_z - zp_y$$
$$L_y = zp_x - xp_z$$
$$L_z = xp_y - yp_x$$

The quantum mechanical operators for the three components of angular momentum are obtained from these equations by replacing classical quantities with their corresponding quantum mechanical operators, such as $p_x = \dfrac{\hbar}{i}\dfrac{\partial}{\partial x}$.

$$\hat{L}_x = \frac{\hbar}{i}\left(y\frac{\partial}{\partial z} - z\frac{\partial}{\partial y}\right)$$

$$\hat{L}_y = \frac{\hbar}{i}\left(z\frac{\partial}{\partial x} - x\frac{\partial}{\partial z}\right)$$

$$\hat{L}_z = \frac{\hbar}{i}\left(x\frac{\partial}{\partial y} - y\frac{\partial}{\partial x}\right)$$

☐

$$\hat{L}_x\hat{L}_y = \frac{\hbar}{i}\left(y\frac{\partial}{\partial z} - z\frac{\partial}{\partial y}\right)\frac{\hbar}{i}\left(z\frac{\partial}{\partial x} - x\frac{\partial}{\partial z}\right)$$

$$= -\hbar^2\left(yz\frac{\partial^2}{\partial z\partial x} + y\frac{\partial}{\partial x} - yx\frac{\partial^2}{\partial z^2} - z^2\frac{\partial^2}{\partial y\partial x} + zx\frac{\partial^2}{\partial y\partial z}\right)$$

$$\hat{L}_y\hat{L}_x = -\hbar^2\left(zy\frac{\partial^2}{\partial x\partial z} - z^2\frac{\partial^2}{\partial x\partial y} - xy\frac{\partial^2}{\partial z^2} + xz\frac{\partial^2}{\partial z\partial y} + x\frac{\partial}{\partial y}\right)$$

Then $$[\hat{L}_x\hat{L}_y] = \hat{L}_x\hat{L}_y - \hat{L}_y\hat{L}_x$$

$$= -\hbar^2\left(y\frac{\partial}{\partial x} - x\frac{\partial}{\partial y}\right)$$

(since multiplication and differentiation are each commutative)

$$= -\frac{\hbar}{i}\hat{L}_z \neq \hat{0}$$

Similarly $$[\hat{L}_y\hat{L}_z] = -\frac{\hbar}{i}\hat{L}_x$$

$$[\hat{L}_z\hat{L}_x] = -\frac{\hbar}{i}\hat{L}_y$$

Then $\hat{L}_x$, $\hat{L}_y$ and $\hat{L}_z$ do not commute with one another.

Again $$[\hat{L}^2, \hat{L}_x] = [\hat{L}_x^2 + \hat{L}_y^2 + \hat{L}_z^2, \hat{L}_x]$$

$$= [\hat{L}_x^2, \hat{L}_x] + [\hat{L}_y^2, \hat{L}_x] + [\hat{L}_z^2, \hat{L}_x] \text{ vide problem 18.64(a)}$$

$$= \hat{0} \quad + \quad \hat{0} \quad + \quad \hat{0} \qquad \text{vide problem 18.64(c)}$$

Then $\hat{L}_x$ commutes with $[\hat{L}^2$, and $\hat{L}_y$ and $\hat{L}_z$ also do the same.

☐☐ The results imply that only the magnitude of the vector L^2 and one of the three components of angular momentum (arbitrarily taken as the z-component) can be specified simultaneously (unlike classical expectation of knowing all three components simultaneously). The magnitude of L^2 (the eigenvlue of $\hat{L}^2$) and L_z (the eigenvalue of $\hat{L}_z$) can have only some specified values

$$\hat{L}^2\psi = l(l+1)\hbar^2\psi \qquad\qquad l = 0, 1, 2, \ldots$$

$$\hat{L}_z\psi = m\hbar\psi \qquad\qquad m = -l, -l+1, \ldots, l-1, l$$

where l is the angular momentum quantum number and m is the magnetic quantum number.

Note: Since the spin angular momntum of an elementary particle has no analog in classical mechanics, we cannot construct the corresponding operators like orbital angular momentum operators. However, by analogy with the orbital angular momentum operators $\hat{L}^2, \hat{L}_x, \hat{L}_y, \hat{L}_z$, there are spin angular momentum operators $\hat{S}^2, \hat{S}_x, \hat{S}_y, \hat{S}_z$, such that

$$\hat{S}^2\psi = S(S+1)\hbar^2\psi \qquad\qquad S = 0, \tfrac{1}{2}, 1, \tfrac{3}{2}, ...$$
$$\hat{S}_z\psi = m_s\hbar\psi \qquad\qquad m_z = -S, -S+1, ..., S-1, S$$

where S is the spin quantum number and m_s is spin quantum number for the z component of the spin.

18.67 Possible, if the operators corresponding to each pair of observables commute. Here it is convenient to use the polar coordinates, the system being of spherical symmetry. For such system

$$\hat{L}_z = \frac{\hbar}{i}\left(x\frac{\partial}{\partial y} - y\frac{\partial}{\partial x}\right) = \frac{\hbar}{i}\frac{\partial}{\partial\phi}$$

$$\left[\frac{\partial}{\partial\phi} = \frac{\partial x}{\partial\phi}\frac{\partial}{\partial x} + \frac{\partial y}{\partial\phi}\frac{\partial}{\partial y} + \frac{\partial z}{\partial\phi}\frac{\partial}{\partial z}\right.$$

$$\left.\frac{\partial x}{\partial\phi} = -y, \frac{\partial y}{\partial\phi} = x, \frac{\partial z}{\partial\phi} = 0\text{ etc.}\right]$$

(i) For a system of spherical symmetry, since $\hat{V}(r)$ has no variable in common with $\hat{L}_z$, $\hat{V}$ will commute with $\hat{L}_z$, and hence also with $\hat{L}_x$ and $\hat{L}_y$ as they are related to each other by a simple transformation of coordinates. As a consequence V will commute with $\hat{L}^2 = \hat{L}_x^2 + \hat{L}_y^2 + \hat{L}_z^2$.

(ii) Here $\qquad \hat{K} = -\frac{\hbar^2}{2\mu}\nabla^2$

$$= -\frac{\hbar^2}{2\mu}\left(\frac{1}{r^2}\frac{\partial}{\partial r}r^2\frac{\partial}{\partial r} + \frac{1}{r^2\sin\theta}\frac{\partial}{\partial\theta}\sin\theta\frac{\partial}{\partial\theta} + \frac{1}{r^2\sin^2\theta}\frac{\partial^2}{\partial\phi^2}\right)$$

For the same reason as in (i), $\hat{L}_z$ commutes with the first and second parts of $\hat{K}$ and also with the third part since $\frac{\partial}{\partial\phi}$ commutes with $\frac{\partial^2}{\partial\phi^2}$. Then $\hat{K}$ will commute also with $\hat{L}_x$ and $\hat{L}_y$, and hence with $\hat{L}^2$.

Note: $\hat{L}^2$ will commute with $\hat{V}$ and $\hat{K}$ each, provided the system is spherically symmetrical.

18.68 Here $\Delta L = m\Delta vr = \Delta Pr$ and $\Delta\theta = \Delta x/r$, then

$$\Delta L\Delta\theta = \Delta P\Delta x \geq \frac{h}{4\pi} \quad\text{by uncertainty principle}$$

This invalidates the idea of Bohr orbit of definite radius (r).

18.69 Minimum uncertainty in speed

$$= \frac{h}{4\pi m\Delta x} \quad\text{where } \Delta x \text{ is the uncertainty in electron position}$$

$$= \frac{6.63\times10^{-34}\text{ JS}}{4\times3.14\,(9.11\times10^{-31}\text{ kg})\,(2\times52.9\times10^{-12}\text{ m})}$$

$$= 5.5\times10^{-5}\text{ ms}^{-1}$$

Here Δx has been taken to be twice the Bohr radius (52.9 Pm), electron being confined to a linear region of this length.

18.70 Fundamental accuracy, i.e. minimum uncertainty in position $\Delta x = \dfrac{h}{4\pi m \Delta v}$

(i) $\Delta x = \dfrac{6.63 \times 10^{-34} \text{ JS}}{4 \times 3.14 \,(9.11 \times 10^{-31} \text{ kg}) \,(400 \times \frac{0.01}{100} \text{ m/s})} = 0.145 \text{ m}$

(ii) $\Delta x = \dfrac{6.63 \times 10^{-34} \text{ JS}}{4 \times 3.14 \,(50 \times 10^{-3} \text{ kg}) \,(400 \times \frac{0.01}{100} \text{ m/s})} = 2.64 \times 10^{30} \text{ m}$

☐ Here Δx is measurable in case of electron, but it is too small to be measured in case of bullet. Then, the uncertainty principle has no practical significance for a macroscopic object like bullet. This is in harmony with the Bohr correspondence principle (vide problem 18.9).

[Note: Don't put $\Delta \theta$ in %].

☐☐Since $\qquad$ KE $= \tfrac{1}{2} mv^2, \dfrac{d(\text{KE})}{\text{KE}} = 2 \dfrac{dv}{v}$

Then, accuracy in finding KE $= 2 \times$ accuracy in finding v

$$= 2 \times 0.01 \text{ or } 0.02\%$$

18.71 (a) $\qquad \hat{p}\psi = \dfrac{\hbar}{i} \dfrac{d}{dx} (ce^{ikx}) = \hbar k \psi$

Then $\qquad p = \hbar k$, and hence $v = \dfrac{p}{m} = \dfrac{\hbar k}{m}$

☐ Since v does not depend on location of the particle, the particle is free.
☐☐ Probability density of the particle at any point

$$= \psi^* \psi$$
$$= C^2, \text{ which is independent of } x.$$

This means the particle has an equal probability of being found anywhere, i.e. location of the particle cannot be predicted.

Note: Here momentum of the particle is fixed but location is completely uncertain. This is an illustration of the uncertainty principle.

(b) Here, unlike (a), ψ is not an eigenfunction of $\hat{p}$. Hence the observed momentum in a single experiment will be $\hbar k$ (corresponding to the wavefunction $C_1 e^{ikx}$) or $-\hbar k$ (corresponding to $C_2 e^{-ikx}$), and the average momentum is

$$\langle p_x \rangle = C_1^2 \hbar k - C_2^2 \hbar k \quad \text{by Eq. (18.9)}$$

☐ The required probability $= C_1^2$

☐☐ If $\qquad C_1^2 = \frac{90}{100}$, i.e. $C_1 = \pm \sqrt{0.9}$

then $\qquad C_2^2 = 0.1$, i.e $C_2 = \pm \sqrt{0.1} \,(\because C_1^2 + C_2^2 = 1)$

Then the required wavefunction

$$\psi = \sqrt{0.9}\, e^{ikx} \pm \sqrt{0.1}\, e^{-ikx}$$

$$\square\square\square \qquad \hat{K}_x = \frac{\hat{p}_x^2}{2m} = -\frac{\hbar^2}{2m}\frac{d^2}{dx^2}$$

Now the given function is an eigenfunction of $\hat{K}_x$ (through not of $\hat{p}_x$) with eigenvalue $\hbar^2 k^2 / 2m$. Then KE of the particle is $\hbar^2 k^2 / 2m$ (which is always observed in any single experiment).

Note: For any single experiment

$$K_x = \frac{p_x^2}{2m}$$

For average $\qquad \langle K_x \rangle = \langle p_x^2 \rangle / 2m$

$$\neq \langle p_x \rangle^2 / 2m$$

18.72 Here ψ is not normalized. Let N be the normalization constant.

Then $\qquad \langle p_x \rangle = N^2 \int_{-\infty}^{\infty} \psi^* \hat{p}_x \psi \, dx$

$$= \frac{\int_{-\infty}^{\infty} \psi^* \hat{p}_x \psi \, dx}{\int_{-\infty}^{\infty} \psi^* \psi \, dx} = \frac{\frac{\hbar}{i}\int_{-\infty}^{\infty} \psi^* \left(\frac{d\psi}{dx}\right) dx}{\int_{-\infty}^{\infty} \psi^* \psi \, dx}$$

For the given wavefunction

$$\int_{-\infty}^{\infty} \psi^* \frac{dy}{dx} dx = k \int_{-\infty}^{\infty} \sin kx \cos kx \, dx$$

$$= 0 \quad \text{(since the integral is an odd function of } x)$$

Therefore $\qquad \langle p_x \rangle = 0$

18.73 For a particle in a one-dimensional box of length L within the coordinates 0 to L

$$\psi = \left(\frac{2}{L}\right)^{\frac{1}{2}} \sin\left(\frac{n\pi x}{L}\right)$$

$$E = \frac{n^2 h^2}{8mL^2}$$

For the given problem within the coordinates 0 to 2a, $L = 2a$, then

$$\psi = \left(\frac{2}{2a}\right)^{\frac{1}{2}} \sin\left(\frac{n\pi x}{2a}\right)$$

$$E = \frac{n^2 h^2}{8m(2a)^2}$$

$\square$ The situation with coordinates $-a$ to a can be transformed to the situation with coordinates 0 to 2a (the length of the box being same) simply by shifting the origin to $-a$. Then in this case

$$\psi = \left(\frac{2}{2a}\right)^{\frac{1}{2}} \sin\left[\frac{n\pi(x-a)}{2a}\right] \qquad \text{(I)}$$

$$E = \frac{n^2 h^2}{8m(2a)^2}$$

E is same in the two cases, since E depends on length of the box but not on change of the origin.

□□ $(x - a)(x + a)$ can be a state function in the second case (but not in the first case) because this function becomes zero at the extreme position of the particle $x = -a$ and $x = a$, though this is not an eigenfunction (since $H\psi \neq E\psi$). This becomes possible because any function can be expressed in terms of orthonormal set (I).

18.74 As expected for a standing wave, the de Broglie wavelength (λ) for PIB will be related to the length (a) of the box as follows

$$a = n \cdot \tfrac{1}{2}\lambda \quad \text{where } n = 1, 2, ...$$

or

$$\lambda = \tfrac{2a}{n}$$

Then according to the de Broglie relation

$$p = \frac{h}{\lambda} = \frac{nh}{2a}$$

Since PIB has only KE, the permitted energy levels will be

$$E = \frac{p^2}{2m} = \frac{n^2 h^2}{8ma^2}$$

□ From this expression

$$\frac{h}{2} = \sqrt{2mE} \cdot a = P \cdot a \quad \text{for } n = 1$$

If P is considered to be the limit of uncertainty with which momentum can be known, and a likewise for position, the produce of these two uncertainties is of the order of h. Then the expression for energy levels is in keeping with the uncertainty principle.

18.75 No. Because, if PIB is at rest (as is allowed classically) its energy (being wholly kinetic) will be zero corresponding to $n = 0$. This will lead to $\psi = 0$ everywhere which is not permitted by Born's interpretation of ψ (vide problem 18.45).

18.76 No. For $a = \infty$, the PIB cannot be treated as bound, and this limits the applicability of the give expression of E.

Note: E cannot be zero even for a free particle, because if $E = 0$, ψ [by Eq. (18.13)] will be zero everywhere, which is absurd.

18.77 (a) Since h^2/m is very small for a normal molecule (of the order of 10^{-30} CGS unit), the separation of successive translational energy levels [given by Eq. (18.11)] is so small for reasonable values of n, that the distribution of energy may be regarded as virtually continuous. It is for this reason that ordinarily, the translational motion of a molecule may be adequately treated classically.

(b) From Eq. (18.11), it follows that when a is very low, of the order of nuclear size (10^{-12} cm), the kinetic energy (E) for a submicroscopic particle like electron ($m \simeq 10^{-27}$ gm) is very high (of the order of 10^9 eV). It is for this reason that the confinement of electron in the nucleus by electrostatic attraction (coulombic energy being only of the order of 10^5 eV) is not possible.

Note: The existence of heavy particle like proton (which has KE less than electron) in the nucleus is not unlikely.

18.78 The probability (P) of finding a particle in a one-dimensional box in the range x_1 and x_2 is

$$P = \int_{x_1}^{x_2} \psi_n^2 dx$$

$$= \frac{x_2 - x_1}{a} - \frac{1}{2\pi}\left(\sin\frac{2\pi x_2}{2} - \sin\frac{2\pi x_1}{a}\right) \tag{I}$$

$$\approx \psi_n^2 \, \Delta x \text{ at } x = \frac{(x_1 + x_2)}{2} \tag{II}$$

when the range $\Delta x = x_2 - x_1$ is small.

A. For $n = 1$ $\psi_1^2 = \dfrac{2}{a}\sin^2\dfrac{\pi x}{a}$

$$= \frac{2}{a}\sin\frac{\pi}{2}, \text{ putting } x = 0.5a$$

$$= \frac{2}{a}$$

Then $P \approx \dfrac{2}{a} \times 0.02a = 0.04$ by relation (II)

For $n = 2$ $\psi_2^2 = \dfrac{2}{a}\sin^2\dfrac{2\pi x}{a} = \dfrac{2}{a}\sin^2\pi = 0$

Then $P \approx 0$

B. (i) $\psi_1^2 = \dfrac{2}{a}\sin^2\dfrac{\pi x}{a}$

$$= \frac{2}{10}\sin^2\pi, \text{ putting } a = 10\text{nm and } x \approx 10 \text{ nm}$$

$$= 0$$

Then $P \approx 0$

(ii) $P = \dfrac{\frac{a}{3}}{a} - \dfrac{1}{2\pi}\left[\sin\left(2\pi \times \tfrac{2}{3}\right) - \sin\left(2\pi \times \tfrac{1}{3}\right)\right]$,   by relation I

$$= 0.605$$

(iii) $P = \tfrac{1}{2}$; from symmetry of particle's motion

Since the probability in either half is half of the total probability ($= 1$).

C. (i) Most likely locations correspond to the antinodes of ψ_3, i.e. maxima or minima of ψ_3 where $\dfrac{d\psi_3}{dx} = 0$. This happens when $\cos\dfrac{3\pi x}{a} = 0$ or $\dfrac{3\pi x}{a} = (2m+1)\dfrac{\pi}{2}$. Then the most likely locations of the particle are $x = a/6$, $a/2$, $5a/6$ (remembering that $0 < x < a$).

(ii) For $n = 1$, ψ has only one antinode that occurs at $a/2$ which is therefore the most likely location of the particle.

Note: Classically the particle has same probability at all positions between 0 and a within which its potential energy does not change.

D. For two wavefunctions ψ_m and ψ_n of the given system

$$\int_0^a \psi_m \psi_n dx = \frac{2}{a} \int \sin\frac{m\pi x}{a} \sin\frac{n\pi x}{a} dx$$

$$= \frac{1}{a}\int_0^a \cos\frac{(m-n)\pi x}{a} dx - \frac{1}{a}\int_0^a \cos\frac{(m+n)\pi n}{a} dx$$

$$= \frac{a}{(m-n)\pi}\left|\sin\frac{(m-n)\pi}{a}\right|_0^a - \frac{a}{(m+n)\pi}\left|\sin\left(\frac{m+n}{a}\right)\pi\right|_0^a = 0$$

Therefore ψ_m and ψ_n are orthogonal to one another when $m \neq n$.

E. Here $\qquad \hat{H} = -\frac{\hbar^2}{2m}\frac{d^2}{dx^2}$ since the energy of the system is wholly kinetic

Then $\qquad \hat{H}\psi = \frac{\hbar^2}{2m}\frac{n^2\pi^2}{a^2}\psi \quad$ for the given wavefunction

$$= E\psi$$

whence $\qquad E = \frac{n^2\hbar^2}{8ma^2}$

18.79 (i) Normally the uncertainty in position Δx means the standard deviation of x, which is the root-mean-square deviation from the expectation value $\langle x \rangle$

i.e. $\qquad (\Delta x)^2 = \langle(x - \langle x\rangle)^2\rangle \quad [= \langle x^2 - 2x\langle x\rangle + \langle x\rangle^2\rangle$

$$= \langle x^2\rangle - \langle x\rangle^2 \qquad = \langle x^2\rangle - 2\langle x\rangle^2 + \langle x\rangle^2 = \langle x^2\rangle - \langle x\rangle^2]$$

Now $\qquad \langle x \rangle = \int_0^a \psi_n^*(n\psi_n)dx = \frac{2}{a}\int_0^a x\sin^2\left(\frac{n\pi x}{a}\right)dx$

$$= \frac{1}{a}\int_0^a x\left[1 - \cos\left(\frac{2n\pi x}{a}\right)\right]dx = \frac{a}{2}$$

Alternatively $\langle x \rangle = \frac{a}{2}$

by symmetry of the system [the probability of finding the particle at equal distance on either side of $x = a/2$ is equal.]

$$\langle x^2 \rangle = \frac{2}{a}\int_0^a x^2 \sin^2\left(\frac{n\pi x}{a}\right)dx$$

$$= \frac{1}{a}\int_0^a x^2 dx - \frac{1}{a}\int_0^a x^2\cos\left(\frac{2n\pi x}{a}\right)dx$$

$$= \frac{1}{a}\cdot\frac{a^3}{3} + \frac{1}{2n\pi}\int_0^a 2x\sin\left(\frac{2n\pi x}{a}\right)dx$$

(integration by parts, the first part being zero)

$$= \frac{a^2}{3} - \frac{a}{(2n\pi)^2}\left|2x\cos\frac{2n\pi x}{a}\right|_0^a$$

(integration by parts, the second part being zero)

$$= \frac{a^2}{3} - \frac{2a^2}{(2n\pi)^2}$$

Then

$$\Delta x = \left[\frac{a^3}{3} - \frac{2a^2}{(2n\pi)^2} - \frac{a^2}{4}\right]^{\frac{1}{2}}$$

$$= \frac{a}{2n\pi}\left[\frac{(n\pi)^2}{3} - 2\right]^{\frac{1}{2}}$$

(ii)

$$\langle p_x \rangle = \int_0^a \psi_n^* \frac{\hbar}{i}\frac{d}{dx}\psi_n dx$$

$$= \frac{2\hbar}{ia}\int_0^a \sin\left(\frac{n\pi x}{a}\right)\cos\left(\frac{n\pi x}{a}\right)dx \quad \text{for the given wavefunction}$$

$$= \frac{\hbar}{ia}\int_0^a \sin\left(\frac{2n\pi x}{a}\right)dx = 0$$

Alternatively, $\langle p_x \rangle = 0$, by symmetry. For each value of n, the momentum of the particle in either direction is numerically same and has equal probability.

$$\langle p_x^2 \rangle = -\frac{2\hbar^2}{a}\int_0^a \sin\left(\frac{n\pi x}{a}\right)\frac{d^2}{dx^2}\sin\left(\frac{n\pi x}{a}\right)dx$$

$$= \frac{2n^2\pi^2\hbar^2}{a^3}\int_0^a \sin^2\left(\frac{n\pi x}{a}\right)dx$$

$$= \frac{n^2\pi^2\hbar^2}{a^3}\int_0^a\left[1 - \cos\left(\frac{2n\pi x}{a}\right)\right]dx$$

$$= \frac{n^2h^2}{4a^3}\int_0^a dx - \frac{n^2h^2}{4a^3}\int_0^a \cos\left(\frac{2n\pi x}{a}\right)dx$$

$$= \frac{n^2h^2}{4a^3}|x|_0^a$$

$$= \frac{n^2h^2}{4a^2}$$

Alternatively

$$\langle p_x^2 \rangle = 2m\langle E \rangle$$

$$= 2m\frac{n^2h^2}{8ma^2}$$

$$= \frac{n^2h^2}{4a^2}$$

Therefore

$$\Delta P_x = \left[\langle P_x^2 \rangle - \langle P_x \rangle^2\right]^{\frac{1}{2}}$$

$$= \frac{nh}{2a}$$

(iii)

$$\Delta E = \left(\langle E^2 \rangle - \langle E \rangle^2\right)^{\frac{1}{2}}$$

Now

$$\langle E^2 \rangle = \frac{2}{a}\int_0^a \sin\left(\frac{n\pi x}{a}\right)\frac{\hbar^4}{4m^2}\frac{d^4}{dx^4}\sin\left(\frac{n\pi x}{a}\right)dx$$

$$= \frac{2}{a} \frac{\hbar^4}{4m^2} \left(\frac{nx}{a} \right)^4 \int_0^a \sin^2 \left(\frac{n\pi x}{a} \right) dx$$

$$= \left(\frac{n^2 h^2}{8ma^2} \right)^2 = \langle E \rangle^2$$

Then $\qquad \Delta E = 0$

Alternatively, here $\hat{E} = \hat{K} = -\dfrac{\hbar^2}{2m} \dfrac{d^2}{dx^2}$, energy being entirely kinetic. Since ψ is an eigenfunction of $\hat{E}$ the uncertainty in E is zero.

☐ The results are supported by the uncertainty principle which demands that

$$\Delta P_x \cdot \Delta x \text{ or } \Delta E \cdot \Delta t \; \geq \frac{h}{4\pi}$$

Here $\qquad \Delta P_x \Delta x = \dfrac{h}{4\pi} \left[\dfrac{(n\pi)^2}{3} - 2 \right]^{\frac{1}{2}} > \dfrac{h}{4\pi}$

Regarding uncertainty in energy, $\Delta E = 0$, happening at all time, corresponds to $\Delta t = \infty$.

☐☐ The lowest value of $\Delta P_x \Delta x$, which corresponds to $n = 1$, is $\dfrac{h}{4\pi} \left(\dfrac{\pi^2}{3} - 2 \right)^{\frac{1}{2}}$.

Note: Here $\Delta P_x \Delta x$ is not fixed. It increases with n (i.e. energy of the particle) and is always greater than $h/4\pi$. However, in most calculations of uncertainty $\Delta P_1 \cdot \Delta x$ is taken to be $h/4\pi$.

18.80 Minimum uncertainty in energy

$$\Delta E = \frac{h}{4\pi t} \quad \text{vide problem (18.36)}$$

$$= \frac{6.63 \times 10^{-34} \text{ JS}}{4 \times 3.14 \, (10^{-9} \text{s})} = 5.28 \times 10^{-26} \text{ J}$$

☐ The width $\Delta\lambda$ of the spectral line is given by

$$\Delta E = h\Delta v = \frac{hC\Delta\lambda}{\lambda^2}$$

or $\qquad \Delta\lambda = \dfrac{\Delta E \lambda^2}{hc}$

$$= \frac{(5.28 \times 10^{-26} \text{ J}) (600 \times 10^{-9} \text{m})^2}{(6.63 \times 10^{-34} \text{ Js}) (3.00 \times 10^8 \text{ m/s})} = 9.56 \times 10^{-14} \text{ m}$$

☐☐ Minimum accuracy of knowing the frequency of the radiant energy

$$\frac{\Delta v}{v} \times 100 = \left| \frac{\Delta v}{\lambda} \right| \times 100$$

$$= \frac{9.56 \times 10^{-14} \text{ m} \times 100}{600 \times 10^{-9} \text{ m}} = 1.59 \times 10^{-5}\%$$

18.81 In butadiene, the electronic energy levels of the π electrons, which are assumed to move along a straight line, are

$$E_n = \frac{h^2 n^2}{8ma^2} \quad \text{by Eq. (18.11)}$$

Since, by Pauli exclusion principle, each of the energy levels can hold not more than two electrons (with opposite spin), the four π electrons in the lowest energy state fill the first two levels with $n = 1$ and $n = 2$. Then the first electronic transition will occur from the level with $n = 2$ to the level with $n = 3$, and the accompanying energy change (ΔE) is given by

$$\Delta E = (3^2 - 2^2)\, h^2/8\,ma^2$$

Here $a = (2 \times 135 + 154 + 2 \times 77)$ pm or 578 pm

[a includes two C=C bonds, one C–C bond and radii of two C-atoms.]

The required frequency of absorption

$$= \frac{\Delta E}{h} = \frac{(3^2 - 2^2)\,(6.63 \times 10^{-34}\ \text{JS})}{8\,(9.11 \times 10^{-31}\ \text{kg})\,(578 \times 10^{-13}\ \text{m})^2} = 1.36 \times 10^{15}\ \text{s}^{-1}$$

Note: The calculated frequency is very close to the experimental result. How useful is the simple model PIB in understanding diverse complex phenomena!

18.82 Here $n^2 = n_x^2 + n_y^2 + n_z^2$ Vide Eq. (18.15b)

w here n_x, n_y and n_z are all integers (not zero). Then the first three values of n are

$$n = \sqrt{1^2 + 1^2 + 1^2} = \sqrt{3}$$

$$= \sqrt{2^2 + 1^2 + 1^2} = \sqrt{6}$$

$$= \sqrt{2^2 + 2^2 + 1^2} = 3$$

18.83 Here $\dfrac{n^2 h^2}{8\,ma^2} = E = \tfrac{3}{2}\,kT$

Then $n = (12k\ mT)^{\frac{1}{2}} \cdot \dfrac{a}{h}$

$$= \left[12(1.38 \times 10^{-23}\ \text{J/K}) \left(\frac{0.004\ \text{kg/mol}}{6.02 \times 10^{23}\ \text{mol}^{-1}}\right) (300\ \text{K})\right]^{\frac{1}{2}} \frac{(1\,\text{m})}{(6.63 \times 10^{-34}\ \text{JS})}$$

$$= 2.74 \times 10^{10}$$

de Broglie wavelength

$$\frac{h}{p} = \frac{h}{\sqrt{2mE}} = \frac{2a}{n}$$

$$= \frac{2\,(1\,\text{m})}{2.74 \times 10^{10}} = 7.3 \times 10^{-11}\ \text{m}$$

Here $\dfrac{E_{n+1} - E_n}{E_n} = \dfrac{2n+1}{n^2} <<< 1$, since n is very high

Then, the translational energy of the molecule is essentially continuous. Hence the translational motion of the molecule can be treated classically.

Note: This is an illustration of Bohr's correspondence principle (vide note on problem 18.9).

18.84 The energy of a particle in a cubical box is

$$E_{n_x n_y n_z} = \frac{h^2}{8\,ma^2}\,(n_x^2 + n_y^2 + n_z^2)\quad \text{by Eq. (18.15b)}$$

In the lowest level $n_x = n_y = n_z = 1$,

$$E_{111} = \frac{3h^2}{8\,ma^2}$$

(i) From the given condition $n_x^2 + n_y^2 + n_z^2 = 9$.

This is possible with $(n_x, n_y, n_z) = (1, 2, 2)$, $(2, 1, 2)$ and $(2, 2, 1)$, and hence the degeneracy is three.

(ii) Here $n_x^2 + n_y^2 + n_z^2 = 12$. This is possible with $n_x = n_y = n_z = 2$ and hence degeneracy of the energy level is one.

(iii) Here $n_x^2 + n_y^2 + n_z^2 = 15$, which is not possible (n_x, n_y and n_z being integer)

☐ The maximum degeneracy is six that happens when $n_x \neq n_y \neq n_z$.

The lowest energy level having this degeneracy is with

$$n_x^2 + n_y^2 + n_z^2 = 1^2 + 2^2 + 3^2 = 14$$

☐☐ No. Vide problem 18.85 and 18.86.

18.85 The energy levels are

$$E_{n_x n_y n_z} = \frac{h^2}{8\,m}\left(\frac{n_x^2}{a^2} + \frac{n_y^2}{b^2} + \frac{n_z^2}{c^2}\right)$$

The lowest energy level corresponds to

$$\frac{n_x}{a} = \frac{n_y}{b} = \frac{n_z}{c} = 1$$

☐ Its degeneracy is one.

☐☐ The highest degeneracy is six and this happens when $\dfrac{n_x}{a} \neq \dfrac{n_y}{b} \neq \dfrac{n_z}{c}$.

18.86 Here the energy levels are

$$E = \frac{h^2}{8\,mc^2}\left(\frac{n_x^2}{4} + \frac{n_y^2}{4} + n_z^2\right) \quad \text{by Eq. (18.15a)}$$

$$= \frac{h^2}{32\,mc^2}\left(n_x^2 + n_y^2 + 4n_z^2\right)$$

Following this relation, the first four energy levels and their degeneracy are stated below:

• The first energy level is nondegenerate corresponding to the wavefunction $\psi_{1,1,1}$ with $n_x = n_y = n_z = 1$.

• The second energy level is of degeneracy two corresponding to $\psi_{2,1,1}$ and $\psi_{1,2,1}$.

• The third energy level is nondegenerate corresponding to $\psi_{2,2,1}$.

• The fourth energy level is of degeneracy two corresponding to $\psi_{3,1,1}$ and $\psi_{1,3,1}$.

18.87 Here the appropriate Schrödinger equation is

$$\left[\frac{-\hbar^2}{2\mu}\frac{d^2}{dx^2} + \tfrac{1}{2}kx^2\right]\psi = E\psi$$

Putting $\psi = Ae^{-Bx^2}$, we have

$$\left[-\frac{\hbar^2}{2\mu}(4B^2x^2 - 2B) + \tfrac{1}{2}kx^2\right]\psi = E\psi$$

or $\left(\dfrac{\hbar^2 B}{\mu} - E\right)\psi + \left(\tfrac{1}{2}k - \dfrac{2\hbar^2 B^2}{\mu}\right)x^2\psi = 0$

Since E is independent of x, it follows from this equation that

$$\tfrac{1}{2}k - \frac{2\hbar^2 B^2}{\mu} = 0, \text{ or } B = \frac{1}{2\hbar}\sqrt{k\mu}$$

and

$$E = \frac{\hbar^2}{\mu}B$$

$$= \frac{\hbar^2}{\mu}\cdot\frac{1}{2\hbar}\sqrt{k_1\mu}$$

$$= \tfrac{1}{2}\hbar\sqrt{\frac{k}{\mu}} = \tfrac{1}{2}\frac{h}{2\pi}\sqrt{\frac{k}{\mu}} = \tfrac{1}{2}h\nu$$

$$A = \frac{1}{\left[\displaystyle\int_{-\infty}^{\infty} e^{-2Bx^2}\,dx\right]^{\frac{1}{2}}}, \text{ for normalization}$$

$$= \left(\frac{2B}{\pi}\right)^{\frac{1}{4}} = \left(\frac{2}{h}\sqrt{k\mu}\right)^{\frac{1}{4}}$$

18.88 No. The height of the end peaks (compared with the intermediate peaks) in $\psi_v^2(x)$ vs x plot increases toward the classical turning points with increase in quantum number (v) as indicated in Fig. 18.4, and when v is very high (as in case of a pendulum) the quantum mechanical expectation merges with the classical expectation (in accordance with the Bohr's correspondence principle).

18.89 (a) Because if $E = 0$ for a harmonic oscillator, then at the equilibrium position (where $PE = 0$) KE and hence momentum would be zero, i.e. position and momentum would be simultaneously known precisely, in violation of the uncertainty principle.

Alternatively: For $E = 0$, the wavefunction provided by the Schrödinger equation for harmonic oscillator woul not be acceptable, as they become ∞ at $x = \pm \infty$.

(b) The harmonic oscillator wavefunctions fall off exponentially to zero at infinite distances. This means there is a definite probability of finding the system in zones outside the classical turning points at finite distances. For such zones, the potential energy exceeds the total energy (since at the classical turning points the potential energy equals the total energy), i.e. kinetic energy is negative.

Note: The phenomenon of the existence of a system in a classically forbidden zone of negative kinetic energy is called quantum mechanical tunnelling effect. In particle-in-a-box situation, the potential barrier has been imagined to be infinitely large to avoid this effect. But in reality, this is not possible as is evidenced by the common phenomenon of radioactivity where α-particles (He^{2+}), β-particles (e^-) etc. are coming out of atomic nucleus though they have KE less than the nuclear binding energy. The tunnelling effect is appreciable only with light particles which have significant wave property. Electron tunnelling forms the basis of the scanning tunnelling microscope that is used to get pictures of the surface atoms of a solid. In chemistry, tunnelling has remarkable role, e.g. in dealing with electron-transfer reactions and anomalous isotope effects for H^+ (that tunnels more readily) compared to D^+ in some reactions.

18.90 For a harmonic oscillator, the minimum energy E_{min} is $\frac{1}{2}hv$ and minimum excitation energy ΔE is hv, by Eq. (18.16). Then

(a)
$$E_{min} = \frac{1}{2} \cdot \frac{h}{2\pi} \left(\frac{g}{L}\right)^{\frac{1}{2}}$$

$$= \frac{6.63 \times 10^{-34} \text{ JS})}{2 \times 2 \times 3.14} \left(\frac{0.980 \text{ m/s}^2}{1 \text{ m}}\right)^{\frac{1}{2}}$$

$$\Delta E = 2E_{min}$$

(b)
$$E_{min} = \frac{1}{2}hv = \frac{1}{2}(6.63 \times 10^{-34} \text{ JS})(5 \text{ Hz})$$

$$\Delta E = 2E_{min}$$

(c)
$$E_{min} = \frac{1}{2} \cdot \frac{h}{2\pi} \left(\frac{k}{\mu}\right)^{\frac{1}{2}} = \frac{h}{4\pi}\left(\frac{2k}{m_0}\right)^{\frac{1}{2}} \quad \text{since } \frac{1}{\mu} = \frac{1}{m_0} + \frac{1}{m_0}$$

$$= \frac{(6.63 \times 10^{-34} \text{ JS})}{4 \times 3.14} \left[\frac{2(1177 \text{ N.m}}{\left(\dfrac{0.016 \text{ kg/mol}}{6.02 \times 10^{23} \text{ mol}^{-1}}\right)}\right]^{\frac{1}{2}}$$

$$\Delta E = 2E_{min}$$

18.91 Here
$$\langle x \rangle = \int_{-\infty}^{\infty} x\psi^2 dx$$

$$= 0, \text{ since } x\psi^2 \text{ is an odd function of } x$$

Symmetry consideration leads to the same result

$$\langle x^2 \rangle = \frac{2}{k}\left\langle \tfrac{1}{2}kx^2 \right\rangle$$

$$= \frac{2}{k}\langle V \rangle = \frac{2}{k}\langle E_K \rangle$$

$$= \frac{2}{k} \cdot \frac{1}{2}\langle E \rangle$$

$$= \frac{1}{k}\left(v + \tfrac{1}{2}\right)hv \qquad \text{by Eq. (18.16)}$$

Then
$$\Delta x = \sqrt{\langle x^2 \rangle - \langle x \rangle^2} \quad \text{vide problem 18.79}$$

$$= \sqrt{\left(v + \tfrac{1}{2}\right)\frac{kv}{k}}$$

$$\langle p_x \rangle = 0, \text{ by the same argument as with } \langle x \rangle$$

$$\langle p_x^2 \rangle = 2m\langle E_K \rangle = 2m \cdot \tfrac{1}{2}\langle E \rangle$$

$$= m\left(v + \tfrac{1}{2}\right)hv$$

Then
$$\Delta p_x = \sqrt{\langle p_x^2 \rangle - \langle p_x \rangle^2} = \sqrt{m\left(v + \tfrac{1}{2}\right)hv}$$

Therefore
$$\Delta p_x \Delta x = \left(v + \tfrac{1}{2}\right)hv\sqrt{\frac{m}{k}}$$

$$= \left(v + \tfrac{1}{2}\right)\frac{h}{2\pi}$$

$$\geq \frac{h}{4\pi} \quad \text{since } v \geq 0$$

Note: Here, like PIB (problem 18.79), $\Delta P_x \Delta x$ is in conformity with the uncertainty principle. However, here, unlike PIB, the lowest value of $\Delta P_x \Delta x$ is $h/4\pi$.

18.92 We can explain this by assuming that the electron is smeared over the entire space around some point(s), called node(s), where the density of smearing is zero. Electron is not really passing across a node.

☐ The energy (E) associated with a wavefunction increases with the number of nodes (n). In case of PIB $E \propto (n+1)^2$, while in case of harmonic oscillator $E \propto \left(n + \frac{1}{2}\right)$.

18.93 Because the wavefunction can be conveniently expressed in spherical coordinates in case of spherically symmetrical system of H-atom but not in case of non-spherical PIB system.

18.94 Here $\psi_{nlm} \propto e^4 \sin^4\theta \sin 4\phi$, since $r \propto e$, vide Section 18.13.

Then $\qquad n - 1 = $ highest power of $e = 4$, whence $n = 5$

$\qquad l = 4$, corresponding to the factor e^4 or $\sin^4\theta$

$\qquad m = -4$, corresponding to $\sin 4\phi$

Therefore the given expression of the orbital will correspond to $\psi_{5,4,-4}$.

18.95 (i) Here $\qquad n - 1 = $ highest power of $e = 1$, whence $n = 2$

$\qquad l = 1$, corresponding to the factor e' or $\cos\theta$

$\qquad m = 0$, corresponding to $\cos 0\phi$

Then the orbital is $\psi_{2,1,0}$, i.e. ψ_{2p_z}.

(ii) The orbital is $\psi_{2,1,1}$, i.e. ψ_{2p_x}, proceeding as in (i)

(iii) Here $\qquad n - 1 = 1$ or $n = 2$

$\qquad l = 0$

$\qquad m = 0$

Then the orbital is $\psi_{2,0,0}$, i.e. ψ_{2s}

Alternatively: Since ψ is devoid of θ and ϕ, it must correspond to an s orbital.

(iv) The orbital is $\psi_{2,1,-1}$, i.e. ψ_{2p_y}

Here $m = -1$, corresponds to $\sin\phi$.

18.96 By Eq. (18.22), the required order of the orbitals is $1s < 2s = 2p < 3s = 3p < 3d$.

☐ N. Because in case of atoms, which are not of H-like, E is not determined by the principal quantum number n alone according to the Eq. (18.22).

☐☐ Since the energy of a H-atom is independent of l and m, and depends only on n, each shell is n^2-fold degenerate. Then for M shell, which corresponds to $n = 3$, the degeneracy is 3^2.

18.97 Apart from energy (which is greater with ψ_{2s}), the two wavefunctions differ in the following important respects:

(i) ψ_{2s} exhibits a radial node while ψ_{1s} does not.

(ii) At $r = \infty$, ψ_{2s} exhibits a maximum while ψ_{1s} exhibits a minimum.

(iii) Radial distribution function exhibits two maxima and two minima for ψ_{2s} but only one maximum and one minimum for ψ_{1s}. Vide Fig. 18.6(b).

☐ Although ψ_{1s} and ψ_{2s} contain no angular variables θ and ϕ, they have angular part which is same despite their differences) and is equal to

$$\left[\frac{1}{\int_0^\pi \sin\theta d\theta \int_0^{2\pi} d\phi} \right]^{\frac{1}{2}} = \frac{1}{2\sqrt{\pi}}$$

18.98 Orbital angular momentum

$$= \sqrt{l(l+1)}\, \hbar$$
$$= \sqrt{l(l+1)}\, \hbar \text{ for } 2p \text{ where } l = 1$$

☐ Number of angular nodes $= l = 1$.

Number of radial nodes $= n - l - 1$, (total number of nodes being $n - 1$)
$$= 2 - 1 - 1 = 0$$

☐☐ No. The 'signs' have relevance to the wavefunction ψ of the atomic system (with reference to the coordinate axes chosen), which have no physical significance unlike $\psi^*\psi$ (a +ve real quantity) [Schrodinger had the delusive belief that, for a charged particle, the charge distribution is connected to its wavefunction].

☐☐☐ The 'signs' of the atomic orbitals can be used qualitatively to predict bond formation through overlap of the atomic orbitals involved. If the overlap is +ve, bond formation is possible. Thus no bond is formed through interaction of s-orbital of an atom A with p_x- or p_y-orbital of an atom B (taking AB as z-axis), as shown in Fig. 18.8(a). Because here net overlap is zero (+ + overlap is neutralised by the equal + – overlap). But bond formation is possible through interaction between s-orbital and p_z-orbital [Fig. 18.8(b)] due to +ve overlap.

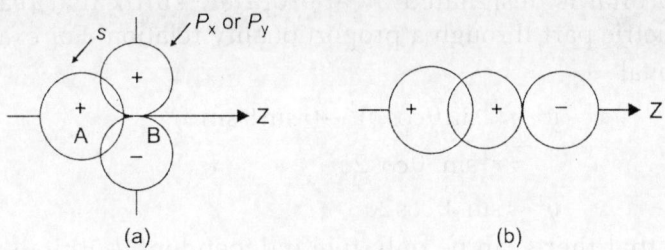

(a)   (b)

Fig. 18.8

18.99 The designations p_x, p_y and p_z are connected to the fact that their angular parts have maximum magnitude along x-, y- and z-axis respectively. Or, that the trigonometric parts of these orbitals are proportional to the respective x, y and z projections of the orbitals (as vector).

Since p_z is independent of ϕ, it will be an eigenfunction of $\hat{L}_z = \dfrac{\hbar}{i}\dfrac{\partial}{\partial\phi}$ (vide problem 18.67) with eigenvalue zero.

p_x and p_y obtained originally from the Schrodinger equation are of the form $f(r,\theta)e^{i\phi}$ and $f(r,\theta)e^{-i\phi}$ which are of course eigenfunctions of $\hat{L}_z$ having eigenvalue $\hbar$ and $-\hbar$ respectively. But due to difference in eigenvalue, the real p_x and p_y defined as

$$p_x \,(\text{real}) = \tfrac{1}{2} f(r,\theta)(e^{i\phi} + e^{-i\phi})$$

$$p_y \,(\text{real}) = \tfrac{1}{2} f(r,\theta)(e^{i\phi} - e^{-i\phi})$$

will not be eigenfunctions of $\hat{L}_z$. However, p_x (real) $+ ip_y$ (real) and p_x (real) $- ip_y$ (real), which give back original p_x and p_y, must be eigenfunctions of $\hat{L}_z$.

Note: Any linear combination of two eigenfunctions (having same eigenvalue) will also be an eigenfunction of the corresponding operator. But even if two functions are not eigenfunction, certain linear combination of them can be an eigenfunction.

18.100 The angular part Y_l^m of original complex d-orbitals and of the real orbitals obtained from appropriate linear combination of the former are:

Complex form   *Real form*

$$Y_2^2 = \left(\frac{15}{32\pi}\right)^{\frac{1}{2}} \sin^2\theta\, e^{2i\phi} \qquad d_{z^2} = Y_2^0 = \left(\frac{5}{16\pi}\right)^{\frac{1}{2}} (3\cos^2\theta - 1)$$

$$Y_2^1 = \left(\frac{15}{8\pi}\right)^{\frac{1}{2}} \sin\theta\cos\theta\, e^{i\phi} \qquad d_{xz} = \frac{1}{\sqrt{2}}(Y_2^1 + Y_2^{-1}) = \left(\frac{15}{4\pi}\right)^{\frac{1}{2}} \sin\theta\cos\theta\cos\phi$$

$$Y_2^0 = \left(\frac{5}{16\pi}\right)^{\frac{1}{2}} (3\cos^2\theta - 1) \qquad d_{yz} = \frac{1}{\sqrt{2}i}(Y_2^1 - Y_2^{-1}) = \left(\frac{15}{4\pi}\right)^{\frac{1}{2}} \sin\theta\cos\theta\sin\phi$$

$$Y_2^{-1} = \left(\frac{15}{8\pi}\right)^{\frac{1}{2}} \sin\theta\cos\theta\, e^{-i\phi} \qquad d_{x^2-y^2} = \frac{1}{\sqrt{2}}(Y_2^2 + Y_2^{-2}) = \left(\frac{15}{16\pi}\right)^{\frac{1}{2}} \sin^2\theta\cos 2\phi$$

$$Y_2^{-2} = \left(\frac{15}{32\pi}\right)^{\frac{1}{2}} \sin^2\theta\, e^{-2i\phi} \qquad d_{xy} = \frac{1}{\sqrt{2}i}(Y_2^2 - Y_2^{-2}) = \left(\frac{15}{16\pi}\right)^{\frac{1}{2}} \sin^2\theta\sin 2\phi$$

Each d-orbital is designated by appropriate suffix that readily reflects its trigonometric part through a proportionality relation. For example, in case of $d_{x^2-y^2}$ orbital

$$x^2 - y^2 = (r\sin\theta\cos\phi)^2 - (r\sin\theta\sin\phi)^2$$
$$= r^2\sin^2\theta\cos 2\phi$$

i.e. $\qquad x^2 - y^2 \propto \sin^2\theta\cos 2\phi$

The fact that there can be only five independent d-orbitals ($l = 2$) makes it impossible to have a complete set of three orbitals $d_{z^2-y^2}$, $d_{y^2-z^2}$ and $d_{x^2-z^2}$ as with d_{xy}, d_{yz} and d_{xz}. In such a situation, it is best to use a linear combination of $d_{y^2-z^2}$ and $d_{x^2-z^2}$.

$$[d_{y^2-z^2} + d_{x^2-z^2}] \propto [(\sin^2\theta\sin^2\phi - \cos^2\theta) + (\sin^2\theta\cos^2\phi - \cos^2\theta)]$$
$$\propto (3\cos^2\theta - 1)$$
$$\propto d_{z^2}$$

Then $d_{y^2-z^2} + d_{x^2-z^2}$ is equivalent to d_{z^2}. The orbital d_{z^2} is symmetrical around z-axis. It consists of two lobes along z-axis with a doughnut-shaped part resulting from fusion of the lobes of $d_{y^2-z^2}$ along y-axis and lobes of $d_{x^2-z^2}$ along x-axis.

18.101 (a) At any point at a distance r from the nucleus,

Probability density, $\psi^2 \propto e^{-2r/a_0}$

Then the probability of finding the electron will be most at a point $r = 0$ where ψ^2 is highest.

☐ The probability will fall to 50% of its highest value when

$$e^{-2r/a_0} = 0.5$$

or $\qquad r = -\frac{1}{2}a_0 \ln 0.5$

$$= 0.346 a_0$$

(b) The most probable radius (r) corresponds to the maximum in the radial distribution function, $f = 4\pi r^2 R^2$ (R is radial wavefunction) where $df/dr = 0$. Here $f \propto r^2 e^{-2r/a_0}$ and then

$$\frac{df}{dr} \propto \left(2r - \frac{2r^2}{a_0} \right) e^{-2r/a_0} = 0 \text{ for } r = a_0 \text{ and } \infty$$

Obviously $r = a_0$ is the most probable radius (the other value of r corresponds to a minimum of f).

☐ Two radii will fulfil the required condition, because, here the highest value of f corresponds to a maximum [unlike (a)].

Note: In case of (a), ψ^2 has no maximum (though it has a minimum). If the procedure in (b) is applied to (a), the result will be $r = \infty$ which corresponds to a minimum.

(c) The mean radius of the orbital is

$$\langle r \rangle = \int \psi^* r \psi d\tau$$

$$= \frac{1}{\pi a_0^3} \int_0^\infty r^3 e^{-2r/a_0} dr \int_0^\pi \sin\theta \, d\theta \int_0^{2\pi} d\phi$$

$$= \frac{1}{\pi a_0^3} \left(\frac{3! a_0^4}{2^4} \right) (2)(2\pi)$$

$$= \frac{3}{2} a_0$$

[**Note:** The Bohr radius a_0 corresponds to the most probable radius and not the mean radius. The probability of finding the electron at a point is most at $r = 0$ and not at $r = a_0$].

(d) The probability (P) that the electron will be found within a small volume ΔV is

$$P = \int_{\text{volume}} \psi^2 d\tau$$

$$\simeq \psi^2 \Delta V, \text{ if } \Delta V \text{ is small}$$

(i) at the nucleus

$$P = \left(\frac{1}{\pi a_0^3} \times 1 \right) \Delta V$$

$$= \frac{1}{3.14 \, (53 \, pm)^3} \times 1 pm^3 = 2.1 \times 10^{-6}$$

(ii) at a distance $r = a_0$

$$P = \frac{1}{\pi a_0^3} \cdot \frac{1}{e^2} \cdot \Delta V$$

$$= \frac{1}{\pi a_0^3} \cdot \frac{1}{e^2} \times 1 pm^3$$

$$= \frac{2.1 \times 10^{-6}}{2.72^2} = 2.8 \times 10^{-7}$$

☐ It appears from above calculations that there is a non-zero probability of finding the electron at $r = 0$. In spite of this, the electron avoids destruction at the nucleus. This paradox will disappear if we remember that as $r \to 0$, the electron assumes tremendous velocity which warrants consideration of relativistic effect which the given wavefunction has not taken care of.

(e) (i) Here $\hat{K} = -\dfrac{\hbar^2}{2\mu} \nabla^2$

where μ is the reduced mass which in case of H-atoms is virtually equal to the mass of electron m_e, and

$$\nabla^2 = \frac{1}{r^2}\frac{\partial}{\partial r}\left(r^2 \frac{\partial}{\partial r}\right) + \frac{1}{r^2 \sin\theta}\frac{\partial}{\partial\theta}\left(\sin\theta \frac{\partial}{\partial\theta}\right) + \frac{1}{r^2 \sin^2\theta}\frac{\partial^2}{\partial\phi^2}$$

For the given wavefunction, which is independent of θ and ϕ

$$\nabla^2\psi = \frac{1}{r^2}\frac{d}{dr}\left(r^2 \frac{d\psi}{dr}\right)$$

$$= \frac{1}{r^2}\left(2r \frac{d\psi}{dr} + r^2 \frac{d^2\psi}{dr^2}\right)$$

$$= \frac{2}{r}\frac{d\psi}{dr} + \frac{d^2\psi}{dr^2}$$

$$= \left(\frac{1}{\pi a_0^3}\right)^{\frac{1}{2}}\left(-\frac{2}{a_0 r} + \frac{1}{a_0^2}\right)e^{-r/a_0}$$

Then $\langle K \rangle = \int \psi^* \left(-\dfrac{\hbar^2}{2m_e}\nabla^2\right)\psi d\tau$

$$= -\frac{\hbar^2}{2m_e}\left(\frac{1}{\pi a_0^3}\right)\int_0^\infty \left(-\frac{2}{a_0 r} + \frac{1}{a_0^2}\right)e^{-2r/a_0}r^2 dr \int_0^\pi \sin\theta\, d\theta \int_0^{2\pi} d\phi$$

$$= \frac{\hbar^2}{2m_e a_0^2}$$

[**Note:** Although ψ does not contain θ and ϕ, they must be considered in $d\tau$ because ψ is total wavefunction]

(ii) $\langle V \rangle = \int \psi^* \hat{V}\psi d\tau$

$$= \int \psi^* \left(-\frac{e^2}{r}\right)\psi d\tau$$

$$= -\frac{e^2}{\pi a_0^3}\int_0^\infty \frac{1}{r}e^{-2r/a_0}r^2 dr \int_0^\pi \sin\theta\, d\theta + \int_0^{2\pi} d\phi$$

$$= -\frac{e^2}{a_0} \quad \text{(in Gaussian form)}$$

18.102 (a) True for azimuthal (or angular-momentum) quantum number and magnetic quantum number but not for principal quantum number.

(b) False. At infinity ψ may be minimum (as in case of ψ_{1s}) or maximum (as in case of ψ_{2s}).

(c) False. At the origin $|\psi^2|$ for 1s-state of hydrogen is highest but not maximum.

(d) False. For 1s-state of hydrogen although $4\pi r^2 |\psi|^2$ has a maximum, it occurs not at the nucleus (origin) but at certain distance (most probable radius) away from the latter.

18.103 For PIB, the wavefnction of the system can be separated into three independent parts with reference to each of the coordinates used. But in case of H-atom, although $\psi = f(r)\, f(\theta)\, f(\phi)$ (in spherical coordinate), $f(\phi)$ and $f(\theta)$ contain a common constant m, called magnetic quantum number while $f(\theta)$ and $f(r)$ contain another common constant l, called azimuthal quantum number such that $l \geq |m|$. Further $f(r)$ points to another constant n called principal quantum number such that $n > l$.

UNIVERSITY QUESTIONS

18.1 Define a perfect black-body. Write three characteristic features of intensity distribution of radiation emitted by a black body. (Calcutta BSc(H), 2000)

18.2 Express mathematically the Planck distribution law for black-body radiation and show that, under particular condition, it reduces to the Rayleigh–Jeans law. (Calcutta BSc(H), 2007)

18.3 Derive Wien's displacement law that $\lambda_{max}\, T$ is a constant from Planck distribution and deduce an expression for the constant. (Jadavpur BSc(H), 2013)

18.4 The peak in the sun's emitted energy occurs at about 480 nm. Estimate the temperature of its surface assuming as a black-body like emitter. (The second radiation constant, $C_2 = hc/k = 1.439$ cm K). (Calcutta BSc(H), 2006)

18.5 What is photoelectric effect? Explain it by using quantum theory of light. (Jadavpur BSc(H), 2011)

18.6 What is threshold frequency in photoelectric effect? How is this frequency related to the frequency of the incident radiation? (Burdwan BSc(H), 2011)

18.7 The work junction for Cs is 3.43×10^{-19} J. What is the kinetic energy of an electron liberated by radiation of 550 nm? What is stopping voltage? How many electrons are generated if the total energy absorbed at 550 nm is 1.00×10^{-3} J? (Burdwan BSc(H), 2011)

18.8 Explain what you understand by Compton effect. Determine the scattering angle for which the shift in wavelength would be maximum. (Calcutta BSc(H), 2013)

18.9 A photon has an energy of 1 eV. Estimate its momentum. (Calcutta BSc(H), 2006)

18.10 Give two examples where light behaves as particles. (Burdwan BSc(H), 2005)

18.11 Cite the example of an experiment which reveals the wave property of matter. (Burdwan BSc(H), 2006)

18.12 What is wave-particle duality of matter? Explain how the concepts of matter wave leads to Bohr's theory in a natural way. (Jadavpur BSc(H), 2011)

18.13 Explain in terms of de Broglie theory the classical behaviour of macroscopic particles. (Burdwan BSc(H), 2000)

18.14 Show that the error in the de Broglie wavelength (λ) is related to the error in velocity (v) by the relation $d\lambda\, (-\lambda/v)\, dv$. (Burdwan BSc(H), 2011)

18.15 Calculate the wavelength associated with an electron moving with a velocity of 1×10^8 cm sec^{-1}. (mass of electron $= 9.1 \times 10^{-28}$ g)   (Delhi BSc, 2005)

18.16 Calculate the de Broglie wavelength of an electron accelerated from rest through a potential difference of 100 kV.   (Jadavpur BSc(H), 2013)

18.17 Determine the wavelength of a cricket ball having a mass 4.0×10^{-2} kg and velocity 30 m sec^{-1}. Comment on the result.   (Nagpur BSc, 2003)

18.18 (a) State Heisenberg's uncertainty principle for position and momentum. What is the value of $\Delta y \Delta P_x$?   (Calcutta BSc(H), 2008)

   (b) Write the minimum value of $\Delta x \Delta P_x$.   (Burdwan BSc(H), 2006)

18.19 A microparticle can be found anywhere with equal probability within a length L. Find roughly its uncertainty in momentum. If the latter is the minimum possible value of its momentum, find also the minimum kinetic energy.

                    (Calcutta BSc(H), 2006)

18.20 An electron has a speed of 3.0×10^4 cm sec^{-1} accurate to 0.01%. Find out the uncertainty in the position of the electron.   (Vidyasagar BSc(H), 2002)

18.21 A moving ball weighing 200 g is to be located within 0.2Å. What is the uncertainty in velocity? Comment on your result.   (Lucknow BSc, 2002)

18.22 An excited atom gives up excess energy by emitting a photon of characteristic frequency. The average period that elapses between the excitation of the atom and the time it radiates is 1.0×10^{-8} s. Find the inherent uncertainty in the frequency of the photon.   (Calcutta BSc(H), 2014)

18.23 (a) How is a state defined in quantum mechanics?   (Burdwan BSc, 1998)

   (b) What are stationary states? Show that the function $e^{-2\pi i Et/h} \psi(n)$ represent a stationary state.   (Burdwan BSc, 2000)

18.24 State the conditions of 'acceptibility of wave function' in quantum mechanics with explanation.   (Calcutta BSc(H), 2003)

18.25 Determine whether each of the following functions is acceptable or not as state function over the indicated intervals:

   (a) (i) e^{-x} $[0, \infty]$ (ii) e^{-x} $[-\infty, \infty]$ (iii) $e^{-|x|}$ $[-\infty, \infty]$ (iv) $\sin^{-1}x$ $[-1, 1]$

                    (Calcutta BSc(H), 2002)

   (b) (i) $1/x$ $[0, \infty]$ (ii) $e^{-x} \cos x$ $[0, \infty]$ (Calcutta BSc(H), 2004)

   (c) xe^{-x} $[0, \infty]$ (Calcutta BSc(H), 2014)

18.26 What is normalization? Why is it necessary? What is its significance?

                    (Calcutta BSc(H), 2001)

18.27 Determine whether each of the following functions is normalisable or not over the indicated intervals:

   (i) e^x $[0, \infty]$ (ii) $\sin x$ $[0, 2\pi]$

   Normalise the function that can be normalised.   (Calcutta BSc(H), 2007)

18.28 A normalised wave function $F(f) = Ae^{im\phi}$ is defined in the range $0 \leq \phi \leq 2\pi$. Find A using the conditions for acceptability of a wavefunction. Obtain the allowed values that the quantum number m may take up.   (Calcutta BSc(H), 2008)

18.29 Normalize following functions:

   (i) $\cos(n\pi x/a)$ over the interval $-a \leq x \leq a$.

   (ii) e^{-r} over the interval $0 \leq r \leq \infty, 0 \leq \theta \leq \pi, 0 \leq \phi \leq 2\pi$.   (Burdwan BSc(H), 2002)

18.30 Prove that the function $\psi_n(x) = (2a)^{-\frac{1}{2}} e^{\frac{i\pi nx}{a}}$, $n = 0, \pm 1, \pm 2, \ldots$ over the interval $-a \leq x \leq a$ are each normalised and mutually orthogonal. (Calcultta BSc(H), 2013)

18.31 Classify the following operators as linear or non-linear

$$\frac{d^2}{dx^2}, (\ )^2, \int (\ )dx, \exp$$ (Calcultta BSc(H), 2004)

18.32 Show that the function $\psi = ae^{x/a}$ is an eigenfunction of the operator d/dx. What is the eigenvalue? (Burdwan BSc(H), 2012)

18.33 f is an eigenfunction of the operator xd/dx with eigenvalue n. Express f in terms of x. (Burdwan BSc(H), 2008)

18.34 Test whether the following functions are eigenfunctions of the operation d^2/dx^2. Write down also the corresponding eigenvalues of:
(a) $\sin x$ (b) $\sin x \cos x$, (c) $\sin mx + \cos nx$ (Burdwan BSc(H), 2011)

18.35 Find B that makes $\exp(-ax^2)$ an eigenfunction of the operator $\left(\frac{d^2}{dx^2} - Bx^2\right)$. Determine the eigenvalue. (Calcultta BSc(H), 2008)

18.36 Verify that $f = x^n$ is not an eigenfunction of d/dx. How do you modify the operator d/dx so that the said function will be an eigenfunction? (Burdwan BSc(H), 2009)

18.37 Test that e^{ikx} is an eigenfunction of the momentum operator and find the eigenvalue. (Burdwan BSc(H), 2005)

18.38 Prove that the function $f(x) = 3x^2 - 1$ is an eigenfunction of the operator

$$\hat{\theta} = -(1-x^2)\left(\frac{d^2}{dx^2}\right) + 2x\left(\frac{d}{dx}\right)$$

Find the corresponding eigenvalue. (Burdwan BSc(H), 2012)

18.39 Define a Hermitian operator. Confirm that the operator $\frac{\hbar}{i}\frac{d}{dx}$ is Hermetian. (Calcutta BSc(H), 2012)

18.40 Prove that Hermitian operators have real eigenvalues. (Jadavpur BSc(H), 2011)

18.41 If ψ_1 and ψ_2 are energy eigenfunctions with same eigenvalue E, prove that any linear combination $C_1\psi + C_2\psi_2$ is also an eigenfunction with same eigenvalue. (Calcutta BSc(H), 2003)

18.42 (a) If ψ_1 and ψ_2 are energy eigenfunctions with eigenvalues E_1 and E_2, verify whether the function $\psi = C_1\psi_1 + C_2\psi_2$ is an eigenfunction.
(b) Find the relation between C_1 and C_2 if ψ is normalised. Given ψ_1 and ψ_2 are normalized and mutually orthogonal.
(c) Find the value of $\langle E \rangle$ using the above wavefunctions (i.e. ψ_1, ψ_2 and ψ). (Burdwan BSc(H), 2006)

18.43 Construct the operator $\hat{A}^2$ from:
(i) $\hat{A} = \hat{x}\hat{P}_x$; $\hat{x}$ and $\hat{P}_x$ are position and momentum operators respectively. (Burdwan BSc(H), 2006)

(ii) $\hat{A} = \frac{d}{dx} + x$. (Calcutta BSc(H), 2012)

18.44 Two operators A and B have simultaneous eigenfunction with different eigenvalues. Show that A and B commute. (Calcutta BSc(H), 2012)

18.45 Evaluate the following commutators:

(i) $[\hat{x}, \hat{P}_x]$ (Calcutta BSc(H), 2010)

(ii) $\left[\dfrac{d^2}{dx^2}, \hat{x}\right]$ (Jadavpur BSc(H), 2003)

(iii) $[\hat{x}, \hat{H}], \hat{H} = -\dfrac{\hbar^2}{2m}\dfrac{d^2}{dx^2} + V(x)$ (Calcutta BSc(H), 2008)

Comment on the results.

18.46 Given $[\hat{x}, \hat{P}_x] = i\hbar$, evaluate $\hat{A}$ and $[\hat{x}, \hat{A}]$ where $[\hat{x}, \hat{P}_x^2] = \hat{A}$. (Calcutta BSc(H), 2008)

18.47 Consider a particle constrained to move in a two-dimensional box. Determine $[\hat{x}, \hat{P}_y]$ and interpret the result. (Calcutta BSc(H), 2002)

18.48 Consider a particle of mass m confined in an infinitely deep potential well, where $V = 0$, for $|x| < L$. The wavefunction is of the form

$$\psi(x) = A \sin kx + B \cos kx$$

Apply boundary conditions on the wavefunction to deduce that

$k = \dfrac{n\pi}{2L}, n = 1, 2, 3, \ldots$ with $A = 0$ for odd n, and $B = 0$ for even n.

(Calcutta BSc(H), 2012)

18.49 (a) The wavefunction for a particle in a box $(1 - d)$ of length L is given by

$$\psi = A\cos\left(\frac{n\pi x}{L}\right), -\frac{L}{2} \le x \le \frac{L}{2}.$$

Find A. (Burdwan BSc(H), 2005)

(b) Applying the Hamiltonian operator for the system (with zero pot energy), find the energy of the particle. Plot ψ vs x for $n = 0$ and $n = 1$.

(Burdwan BSc(H), 2006)

18.50 The normalised wavefunction of a particle in one dimensional box is given by

$$\psi_n = \sqrt{\frac{2}{L}} \sin\left(\frac{n\pi x}{L}\right), 0 \le x \le L$$

(a) (i) Find expression for E.

(ii) Why $n = 0$ is not permitted. (Calcutta BSc(H), 2003)

(b) Find $\langle E^2 \rangle > \langle E \rangle^2$. (Calcutta BSc(H), 2014)

(c) Find the value of ΔP_x where $(\Delta P_x)^2 = \langle p_x^2 \rangle - \langle p_x \rangle^2$. Hence determine the minimum possible value of Δx for the third level. (Calcutta BSc(H), 2010)

18.51 (a) What does 'zero' in 'zero-point energy' signify? (Burdwan BSc(H), 2002)

(b) Show that the zero-point energy of a particle moving in a one dimensional box freely is in accordance with Heisenberg's uncertainty principle.

(Jadavpur BSc(H), 2013)

18.52 (a) For a particle in a 1-D box of length L, apply de Broglie hypothesis to obtain energy levels. (Calcutta BSc(H), 2014)

(b) Show that (i) length of the box is integral multiple of half wavelengths (ii) number of nodes in the wavefunction ψ_n is $(n - 1)$. (Burdwan BSc(H), 2008)

18.53 Find the expressions of permitted energies when

(i) an electron is in a box of 0.01 nm length.

(ii) a 10 g marble is in a box of 10 cm length (mass of electron = 9.1×10^{-31} kg). Comment on the result. (Calcutta BSc(H), 2014)

18.54 An electron is confined in a one-dimensional box of length L. What should be the length of the box to make its zero-point energy equal to its rest mass energy (m_0c^2). Express the result in terms of the Compton wavelength. (Calcutta BSc(H), 2003)

18.55 Construct Hamiltonian operator for a free particle. (Burdwan BSc(H), 2005)

18.56 For a particle in a box of infinite barrier, what would happen to its energy if barrier is removed? (Calcutta BSc(h), 2006)

18.57 For a particle in a one-dimensional box with $0 \le x \le L$

(a) Show that $\langle x \rangle = \frac{L}{2}$ for any state. (Calcutta BSc(H), 2008)

(b) Calculate $\langle x^2 \rangle$ and comment on its value as $n \to \infty$. (Calcutta BSc, 2006)

(c) Where is the particle most likely to be found at the lowest energy level? What is the probability for the particle to be found in the left half of the box? (Burdwan BSc(H), 2005)

(d) Determine the probability of finding the particle in the region $0 \le x \le L/4$. For what value of n, this probability is a maximum? (Calcutta BSc(H), 2013)

(e) Show that $\langle P \rangle = 0$ for all stationary states. Does it mean that the particle is at rest? Explain. (Calcutta BSc(H), 2007)

18.58 Show that $\psi_i(x)$ and $\psi_j(x)$ representing the wavefunctions corresponding to two different states of a particle confined in a one-dimensional box are orthogonal. (Calcutta BSc(H), 2005)

18.59 Estimate the wavelength of light absorbed when a p_i electron of butadiene is excited from the highest occupied energy level to the lowest vacant energy level. For the sake of simplicity assume that the p_i electrons of butadiene move in a one-dimensional box of length 7.0 A. [$m_e = 9.1 \times 10^{-28}$ g]. (Calcutta BSc(H), 2004)

18.60 The terms 'state' and 'energy level' are not synonymous in quantum mechanics. For the particle in a cubic box, consider $E \le 9 h^2/8\, ma^2$. How many states and levels lie within this range? (Calcutta BSc(H), 2012)

18.61 Find the lowest kinetic energy of an electron trapped in a cubical box of edge 10^{-8} cm. What is the degeneracy of the level that has an energy 3 times that of the lowest level? [$m_e = 9.1 \times 10^{-28}$ g, $h = 6.626 \times 10^{-27}$ erg sec]. (Calcutta BSc(H), 2006]

18.62 For a particle in a cubical box, write down the energy equations for the condition: $n_x + n_y + n_z = 4$, and indicate the degeneracy, if any. (Calcutta BSc(H), 2000)

18.63 What is the degree of degeneracy in connection with particle in a cubic box, if the three quantum numbers n_x, n_y and n_z can have the value 1, 2 and 3. (Burdwan BSc(H), 2012)

18.64 For a particle of mass m in a cubic box of length L with zero potential inside and infinite potential on the walls and outside, show that if the cube is distorted in z direction only by an amount ΔL and if ΔL is small, then the energy charges approximately by

$$\Delta E = \frac{n_z^2 h^2}{4\, mL^3} \cdot \Delta L$$

State explicitly the assumption you make in deriving this result. (Calcutta BSc(H), 2012)

18.65 A cubic box with each side measuring 10 Å (with zero potential inside and infinite potential outside) contains a system of 4 quantum particles (fermions). Find the degeneracy of the lowest energy state (configuration) of the system.

(Calcutta BSc(H), 2003)

18.66 Show that, in a rectangular box with infinitely high potential walls of sides $a = L$ and $b = 2L$, there is a degeneracy between the states ($n_x = 1, n_y = 4$) and ($n_x = 2, n_y = 2$). Write down the wavefunction for the state ($n_x = 1, n_y = 0$) and comment on the result.

(Calcutta BSc(H), 2007)

18.67 Benzene may be regarded as a square box of 4 Å edge length containing 6π-electrons. Find the expression for the energy of the ground state of the system. Calculate the minimum energy required to promote one electron to the lowest unoccupied energy level.

(Calcutta BSc(H), 2013)

18.68 There is no difference between a particle in a one-dimensional box and a simple harmonic oscillator because both execute a to and fro motion on a straight line segment. Comment.

(Burdwan BSc(H), 1999)

18.69 (a) Setup the Hamittonian operator for a harmonic oscillator.

(Burdwan BSc(H), 2006)

(b) Write down time-independent Schrodinger equation of a one-dimensional harmonic oscillator.

(Calcutta BSc(H), 2006)

(c) What are characteristic features of linear harmonic oscillator obeying quantum mechanical laws.

(Calcutta BSc(H), 2001)

18.70 Normalize the wavefunction for harmonic oscillator $\psi_1 = A \cdot x e^{-\alpha x^2/2}$ (for state $n = 1$).

(Calcutta BSc(H), 2014)

18.71 For harmonic oscillator of lowest energy level $\psi = Ae^{-Bx^2}$, where $B = \frac{\sqrt{k\mu}}{2\hbar}$, show that energy of the lowest state is $\frac{1}{2}\hbar\omega$, where $\omega = \left(\frac{k}{\mu}\right)^{\frac{1}{2}}$.

(Calcutta BSc(H), 2014)

18.72 What is zero-point energy of a particle executing simple harmonic oscillation?

(Calcutta BSc(H), 2013)

18.73 Explain the correspondence principle using 1-D SHO as an example.

(Jadavpur BSc(H), 2013)

18.74 For a simple harmonic oscillator $\langle x \rangle = 0, \langle P \rangle = 0 \langle x^2 \rangle = \dfrac{\hbar}{(mk)^{\frac{1}{2}}}(v + \frac{1}{2})$ and $\langle p^2 \rangle = \hbar(mk)^{\frac{1}{2}}(v + \frac{1}{2})$. Show that a simple harmonic oscillator obeys the uncertainty principle by computing Δx and ΔP.

(Calcutta BSc(H), 2004)

18.75 Zero-point energy of an oscillator is a consequence of uncertainty principle. Explain.

(Calcutta BSc(H), 2010)

18.76 Calculate the zero-point energy of a harmonic oscillator containing a particle of mass 5.16×10^{-26} kg and force constant 285 Nm^{-1}.

(Calcutta BSc(H), 2004)

18.77 For a harmonic oscillator consisting of a particle of mass 1.33×10^{-25} kg, the difference in adjacent energy levels is 4.82×10^{-21} J. Calculate the force constant.

(Calcutta BSc(H), 2014)

18.78 The force constant for the molecule $^{79}Br-^{79}Br$ is 240 Nm^{-1}. Calculate its zero-point energy.

(Calcutta BSc(H), 2008)

18.79 Construct the Hamiltonian operator and thereby the Schrodinger equation for hydrogen atom in polar form. Separate the variables into radial and angular parts.

(Burdwan BSc(H), 2008)

18.80 What is an orbital? What does the statement 's-orbitals are spherically symmetric' signify? (Burdwan BSc(H), 2005)

18.81 What is the meaning of nodes of wavefunctions? How do you explain the passage of electron across the node? Show diagrammatically the node in $2P_x$ orbital. Do the signs on the two sides of a node make any difference in the electronic charge distribution? (Burdwan BSc(H), 2000)

18.82 The radial wavefunction for $2s$ orbital of a hydrogen atom is given by

$$R_{2,0} = N\left(2 - \frac{r}{a_0}\right)e^{-r/2a_0} \quad N \text{ is a constant}$$

Determine the number and location of node(s) in the $2s$ wavefunction. (Calcutta BSc(H), 2002)

18.83 For $1s$ state of H-atom $\psi_{1s} = b_0\, e^{-r/a_0}$
   (i) Find the normalisation constant b_0. (Calcutta BSc(H), 2002)
   (ii) Evaluate the probability density for $1s$ electron at the nucleus. (Calcutta BSc(H), 2004)

18.84 The normalised radial wavefunction $(R_{1,0})$ of a H-atom in the ground state may be expressed as $R_{1,0}\ 2\left(\frac{1}{a_0}\right)^{\frac{3}{2}} e^{-r/a_0}$, where a_0 is defined as the smallest Bohr orbit radius. Find an expression for the electronic charge content confined within the spherical shell with inner radius r and the thickness of the shell being dr. (Burdwan BSc(H), 2012)

18.85 Calculate the most probable radius (r_{mp}), at which an electron will be found when it occupies a $1s$ orbital of a hydrogenic atom of atomic number z.

$$\psi_{1s} = \frac{1}{\sqrt{\pi}}\left(\frac{Z}{a_0}\right)^{\frac{3}{2}} e^{-Zr/a_0}$$

(Calcutta BSc(H), 2013)

18.86 The wavefunction for $2P_z$ orbital of H-atom is

$$\psi_{2p_z} = \tfrac{1}{4}(2\pi)^{-\frac{1}{2}}\left(\frac{1}{a_0}\right)^{\frac{5}{2}} r e^{-r/2a_0}\cos\theta, \text{ where } a_0 = \text{molar radius}$$

where a_0 = molar radius. Find the average value of the electron-nucleus separation in the $2p_z$ state of H-atom. $\left[\int_0^\infty x^n e^{-bx}dx = \dfrac{n!}{b^{n+1}}\right]$ (Calcutta BSc, 2014)

18.87 Find the radial distribution function for the following hydrogenic orbital

$$\psi = A r\cos\theta \exp\left(-\frac{r}{2a_0}\right)$$

where A and a_0 are constants. (Calcutta BSc(H), 2014)

18.88 The radial part of the $2s$ and $2p$ orbitals of hydrogen atom are given (in au) as below
$$R_{2,0} = N_1(2 - r)e^{-\frac{r}{2}} \text{ and } R_{2,1} = N_2 r e^{-\frac{r}{2}}$$

where N_1 and N_2 are constants.
Depict graphically the plots of the radial distribution functions against r and comment. (Calcutta BSc, 2013)

18.89 Calculate the expectation value of the potential energy for a H-atom in the ground state. Show that the average kinetic energy is equal to the total energy with change in sign. $\left(\text{Given: } \psi_{1s} = (\pi a_0^3)^{-\frac{1}{2}} e^{-r/a_0}\right)$ (Burdwan BSc(H), 2012)

18.90 Find out the ground-state electronic energy of the hydrogen atom. Given that the radial wavefunction $R(r) = 2a_0^{-\frac{3}{2}} e^{-r/a_0}$ for $n = 1$ and $l = 0$.   (Jadavpur BSc(H), 2012)

18.91 Indicate the difference in the shape of $3s$ and $4s$ orbitals. Is there any possibility of f-orbitals in the third quantum shell?   (Burdwan BSc(H), 2007)

18.92 Find the degeneracy of the energy level for H-atom with $n = 2$.
     (Burdwan BSc(H), 2005)

KEY TO UNIVERSITY QUESTIONS

18.1 A perfect black-body is one that absorbs all the electromagnetic radiation of any frequency falling upon it. Experimentally the most nearly perfect blackbody is not a substance at all, but a cavity with a pin hole on one of its walls.

☐ The intensity distribution of blackbody radiation, i.e. the distribution of radiation density, depicted in Fig. 18.1(a) reflects the following characteristic features of this radiation:

  (i) At a particular temperature, the intensity varies with the frequency of the radiation passing through a maximum.

  (ii) At higher temperature, the maximum intensity occurs at higher frequency.

  (iii) At higher temperature, the intensity for a particular frequency is higher and the distribution is narrowed.

18.2 The Planck distribution law can be expressed mathematically by the Eq. (18.1b). For high wavelength, when

$$\frac{hc}{\lambda kT} \ll 1$$

$$e^{\frac{hc}{\lambda kT}} - 1 = \left(1 + \frac{hc}{\lambda kT} + \ldots\right) - 1 \approx \frac{hc}{\lambda kT}$$

Under this condition Eq. (18.1b) reduces to the Raleigh–Jeans law

$$e_\lambda = \frac{8\pi kT}{\lambda^4}$$

18.3 Blackbody spectrum has a maximum at wavelength λ_{max} for which $de_\lambda/d_\lambda = 0$. Then, λ_{max} can be obtained by differentiating Eq. (18.1b) and putting the result to zero. Thus

$$5\lambda^4 (e^{\frac{hc}{\lambda kT}} - 1) + \lambda^5 e^{\frac{hc}{\lambda kT}} \left(-\frac{hc}{kT\lambda^2}\right) = 0$$

or $$1 - e^{-x} = \frac{x}{5}, \quad \text{where } x = \frac{hc}{\lambda_{max} kT}$$

whence $$x = 4.965$$

Then $$\lambda_{max} T = \frac{hc}{4.965 k} = \text{constant}$$

This is Wien's displacement law.

Note: The equation $1 - e^{-x} = \frac{x}{5}$ cannot be solved for x analytically. It has to be solved numerically using a hand calculator.

18.4
$$T = \frac{hc}{4.965 \, k \, \lambda_{max}} \quad \text{by Wien's displacement law}$$

$$= \frac{1.439 \text{ cm K}}{4.965 \, (480 \times 10^{-7} \text{ cm})} = 6038 \text{ K}$$

18.5 Vide Section 18.2.

18.6 The minimum frequency for a radiation to be effective in causing photoelectric effect is called the threshold frequency of the emitting substance.

☐ The threshold frequency (v threshold) is related to the frequency (v) of the incident radiation (v) through the maximum kinetic energy ($K \cdot E_{max}$) of the photoelectrons as

$$K \cdot E_{max} = hv - hv_{threshold} \quad \text{by Eq. (18.2)}$$

Note: KE_{max} refers to the surface atoms of the emitter.

18.7
$$K \cdot E_{max} = \frac{hc}{\lambda} - \phi \quad \text{by Eq. (18.2)}$$

$$= \frac{(6.63 \times 10^{-34} \text{ Js}) \, (3.00 \times 10^8 \text{ m/s})}{(550 \times 10^{-9} \text{ m})} - 3.43 \times 10^{-19} \text{ J}$$

$$= 3.62 \times 10^{-19} \text{ J} - 3.43 \times 10^{-19} \text{ J} = 0.19 \times 10^{-19} \text{ J}$$

$$\text{Stopping voltage} = \frac{K \cdot E_{max}}{e} \quad \text{where } e \text{ is the charge of an electron}$$

$$= \frac{0.19 \times 10^{-19} \text{ J}}{(1.60 \times 10^{-19} \text{ C})} = 0.12 \text{ V}$$

Number of electrons emitted = Number of photons absorbed

$$= \frac{\text{Energy absorbed}}{\text{Energy of each photon } (hC/\lambda)}$$

$$= \frac{1.00 \times 10^{-3} \text{ J}}{3.62 \times 10^{-19} \text{ J}} = 2.76 \times 10^{15}$$

18.8 Vide Section 18.3.

☐ Vide problem 18.23.

18.9 Momentum of the photon $= \dfrac{\text{Energy of the photon}}{\text{Speed of light}}$

$$= \frac{1eV \times 1.602 \times 10^{-19} \text{ J/eV}}{3.00 \times 10^8 \text{ m/s}} = 5.34 \times 10^{-28} \text{ kg m s}^{-1}$$

18.10 Photoelectric effect and Compton effect.

18.11 Diffraction, e.g. diffraction of an electron beam by nickel crystal (grating) as was first evidenced by Davisson and Germer.

18.12 Matter, which were classically thought to consist of particles, has some wavelike character. This leads to an ambiguous perspective of matter, called wave-particle duality.

☐ In a stationary state of H-atom, corresponding to a definite energy, the circumference ($2\pi r$) of the orbiting electron should necessarily be an integral multiple (n) of the de Broglie wavelength (λ) of the electron (to avoid destructive interference), i.e.

$$2\pi r = n\lambda$$

$$= n\frac{h}{mv}$$

Then, angular momentum of the electron, $mvr = n.\frac{h}{2\pi}$ which is a prime assumption of Bohr's theory. So Bohr's theory results automatically from the concept of wave character of matter.

18.13 Vide problem 18.24.

18.14 De Broglie wavelength

$$\lambda = \frac{h}{mv}$$

Then
$$\ln \lambda = \ln \frac{h}{m} - \ln v$$

Differentiation wrt v gives

$$\frac{1}{\lambda}\frac{d\lambda}{dv} = -\frac{1}{v}$$

or
$$d\lambda = -\frac{\lambda}{v}dv$$

18.15
$$\lambda = \frac{h}{mv}$$

$$= \frac{(6.63\times10^{-27}\ \text{erg}\cdot\text{sec})}{(9.1\times10^{-28}\ \text{g})(1\times10^{8}\ \text{cm}\,\text{sec}^{-1})} = 7.3\times10^{-8}\ \text{cm}$$

18.16 Same as problem 18.25(ii).

18.17
$$\lambda = \frac{h}{mv} = \frac{6.63\times10^{-34}\ \text{JS}}{(4.0\times10^{-2}\ \text{kg})(30\ \text{ms}^{-1})} = 5.52\times10^{-33}\ \text{m}$$

□ λ thus calculated is too low to be measured, i.e. wave character is undetectable. Then de Broglie theory has no practical significance for a macroscopic body like cricket ball.

18.18 (a) Vide Section 18.6.

□ Vide problem 18.34.

(b) The minimum value of $\Delta x\Delta P_x$ is not same for all systems (vide note on problem 18.79), but cannot be less than $h/4\pi$.

18.19 Here uncertainty in position, $\Delta x = L$.

Then uncertainty in momentum

$$\Delta P_x = \frac{h}{4\pi\Delta x},\ \text{roughly}$$

$$= \frac{h}{4\pi L}$$

$$\text{Minimum KE} = \frac{(P_x)^2_{\min}}{2\,m}$$

$$= \frac{(h/4\pi L)^2}{2\,m}\ \text{from the given assumption}$$

18.20 Vide problem 18.70.

18.21 Min uncertainty in velocity

$$\Delta v = \frac{h}{4\pi \, m \, \Delta x}$$

$$= \frac{6.63 \times 10^{-27} \text{ erg sec}}{4 \times 3.14 \, (200 \text{ g}) \, (0.2 \times 10^{-8} \text{ cm})} = 1.32 \times 10^{-21} \text{ cm/sec}$$

☐ Vide problem 18.70.

18.22 Similar to problem 18.80.

Here minimum uncertainty in frequency $\Delta v = \frac{\Delta E}{h}$.

18.23 Vide problem 18.40.

18.24 Vide Section 18.8.

18.25 (a) Vide problem 18.51.

(b) (ii) Acceptable, for the same reason as in (a) (i).

(c) Not acceptable because $xe^{-x} = \infty$ at $x = \infty$.

18.26 Vide Section 18.8 and problem 18.46.

18.27 (i) e^x is not normalizable, because $e^x = \infty$ at $x = \infty$.

(ii) Sin x is normalizable, because it is a continuous single-valued function having always a finite value (zero only at $x = 0$, $x = \pi$ and $x = 2\pi$) in the given range.

Here the normalisation constant

$$= \left[\frac{1}{\int_0^{2\pi} \sin^2 x \, dx} \right]^{\frac{1}{2}}$$

$$= \frac{1}{\sqrt{x}}$$

18.28 Vide problem 18.48.

☐ $\qquad Ae^{im\phi} \equiv A \, (\cos m\phi + i \sin m\phi)$

Since addition of 2π to the coordinate ϕ brings us back to the same point in space, the condition for acceptability of a wavefunction that it must be single-valued, i.e. $F(\phi) = F(2\pi + \phi)$ will allow m to take up only the values $0, \pm 1, +2, \dots.$

18.29 (i) Let N be the normalisation constant. Then

$$1 = N^2 \int_{-a}^{a} \cos^2 \left(\frac{n\pi x}{a} \right) dx$$

$$= \frac{N^2}{2} \left[\int_{-a}^{a} \cos^2 \left(\frac{2n\pi x}{a} \right) dx + \int_{-a}^{a} dx \right]$$

$$= \frac{N^2}{2} \, (0 + 2a)$$

Then $\qquad N = \frac{1}{\sqrt{a}}$

(ii) Vide problem 47(iii).

18.30 Let ψ_m and ψ_n be two wavefunctions of the given set, then

$$\int_{-a}^{a} \psi_m^* \psi_n \, dx = \frac{1}{2a} \int_{-a}^{a} e^{i\pi(n-m)\frac{x}{a}} dx$$

$$= 1 \text{ for } m = n, \text{ but } 0 \text{ for } m \neq n$$

Therefore ψ_m and ψ_n are each normalised and mutually orthogonal.

18.31 An operator $\hat{\Omega}$ is said to be linear if for any two functions f_1 and f_2

$$\hat{\Omega}(c_1 f_1 + c_2 f_2) = c_1 \hat{\Omega} f_1 + c_2 \hat{\Omega} f_2$$

where c_1 and c_2 are arbitrary constants.

$\dfrac{d^2}{dx^2}$ and $\int( )dx$ which fulfil this condition are linear operators, but $( )^2$ and exp which does not, are non-linear operators.

18.32 $$\frac{d\psi}{dx} = a \cdot \frac{1}{a} e^{\frac{x}{a}} = \frac{1}{a}\psi = \text{constant} \times \psi$$

Therefore $y = ae^{x/a}$ is an eigenfunction. The eigenvalue is $1/a$.

18.33 Here $$x\frac{df}{dx} = nf$$

or $$\frac{df}{f} = n\frac{dx}{x}$$

Integration gives

$$\ln f = n \ln x + \ln a \quad \text{where } a \text{ is an arbitrary constant}$$

Then $$f = ax^n$$

18.34 (a) $$\frac{d^2(\sin x)}{dx^2} = -\sin x$$

Then $\sin x$ is an eigenfunction of d^2/dx^2 with eigenvalue -1.

(b) $$\frac{d^2}{dx^2}(\sin x \cos x) = \frac{d}{dx}(\cos_x^2 - \sin_x^2) = -4\sin x \cos x$$

Then $\sin x \cos x$ is an eigenfunction with eigenvalue -4.

(c) $$\frac{d^2}{dx^2}(\sin mx + \cos nx) = -(m^2 \sin mx + n^2 \cos nx) \neq \text{constant} \times (\sin mx + \cos nx)$$

Then $\sin mx + \cos nx$ is not an eigenfunction.

18.35 $$\left(\frac{d^2}{dx^2} - Bx^2\right)e^{-ax^2} = (4a^2x^2 - 2a - Bx^2)e^{-ax^2}$$

$$= [(4a^2 - B)x^2 - 2a]e^{-ax^2}$$

$$= \text{constant} \times e^{-ax^2} \text{ for } e^{-ax^2} \text{ to be an eigenfunction}$$

This requires $4a^2 - B = 0$ or $B = 4a^2$.

The corresponding eigenvalue is $-2a$.

18.36 $$\frac{df}{dx} = \frac{dx^n}{dx} = nx^{n-1} \neq \text{constant} \times x^n$$

Then $f = x^n$ is not an eigenfunction of d/dx.

□ Since $x\dfrac{df}{dx} = nf$, the required modification of the operator $\dfrac{d}{dx}$ is to $x\dfrac{d}{dx}$.

Note: The purpose of the modified operator $x\dfrac{d}{dx}$ is also served by $\dfrac{d^{m+1}}{dx^{m+1}}$ $(m \geq n)$ of which $f = x^n$ is an eigenfunction with eigenvalue zero.

18.37
$$\hat{p}_x e^{ikx} = \frac{\hbar}{i}\frac{d}{dx}e^{ikx} = \hbar k\, e^{ikx}$$

$$= \text{constant} \times e^{ikx}$$

Then e^{ikx} is an eigenfunction of $\hat{p}_x$ with eigenvalue $\hbar k$.

18.38
$$\hat{\theta}f(x) = -(1-x^2)\frac{d^2(3x^2-1)}{dx^2} + 2x\frac{d}{dx}(3x^2-1)$$

$$= -(1-x^2)6 + 12x^2$$

$$= 6(3x^2-1)$$

Then $f(x) = 3x^2 - 1$ is an eigenfunction of $\hat{\theta}$.

The corresponding eigenvalue is 6.

18.39 Vide Section 18.9.

□ Vide problem 18.58.

18.40 Vide problem 19.59(a).

18.41 Vide problem 18.42.

18.42 (a) Similar to the previous problem

(b) $\left.\quad 1 = \displaystyle\int (c_1\psi_1 + c_2\psi_2)^*(c_1\psi_1 + c_2\psi_2)\, d\tau = \left|c_1^2\right| + \left|c_2^2\right| \quad\right\}$ vide Eq. (18.9)

(c) $\langle E \rangle = |c_1|^2 E_1 + |c_2|^2 E_2$

18.43 (i)
$$\hat{A}^2\psi = \hat{x}\,\frac{\hbar}{i}\frac{\partial}{\partial x}\left(\hat{x}\,\frac{\hbar}{i}\frac{\partial\psi}{\partial x}\right)$$

$$= -\hbar^2\left(x\frac{\partial^2\psi}{\partial x^2} + \frac{\partial\psi}{\partial x}\right)$$

Therefore $\quad \hat{A}^2 = -\hbar^2\left(x\dfrac{\partial^2}{\partial x^2} + \dfrac{\partial}{\partial x}\right)$

(ii) Same as problem 18.60(ii).

18.44
$$\hat{A}\hat{B}\psi = \hat{A}\omega_B\psi = \omega_B\hat{A}\psi = \omega_B\omega_A\psi \quad \text{where } \omega \text{ is the eigenvalue}$$

$$\hat{B}\hat{A}\psi = \hat{B}\omega_A\psi = \omega_A\hat{B}\psi = \omega_A\omega_B\psi$$

Since multiplication of the two variables (ω_A and ω_B) is alwayc commutative

$$\hat{A}\hat{B} = \hat{B}\hat{A} \text{ i.e. } \hat{A} \text{ and } \hat{B} \text{ commute.}$$

18.45 (i) Vide problem 18.62(ii).

(ii) $\qquad \left[\dfrac{d^2}{dx^2}, x\right] = \dfrac{2d}{dx} \quad$ vide problem 18.61 (i)

(iii) Since $\hat{x}$ and $V(x)$ always commute (as their product simply implies multiplication), the given commutator is equivalent to the following

$$\left[x, -\frac{\hbar^2}{2m}\frac{d^2}{dx^2}\right] = \frac{\hbar^2}{2m}\left[\frac{d^2}{dx^2}, \hat{x}\right]$$

$$= \frac{\hbar^2}{2m} \cdot 2\frac{d}{dx}$$

$$= \frac{\hbar^2}{m}\frac{d}{dx}$$

18.46 $\hat{A} = [\hat{x}, \hat{P}_x^2] = [\hat{x}, \hat{P}_x \cdot \hat{P}_x]$

$$= [\hat{x}, \hat{P}_x]\hat{P}_x + \hat{P}_x[\hat{x}, \hat{P}_x] \quad \text{vide problem 18.64(b)}$$

$$= 2i\hbar\hat{P}_x$$

$[\hat{x}, 2i\hbar\hat{P}_x] = 2i\hbar[\hat{x}, \hat{P}_x]$

$$= 2i\hbar \cdot i\hbar\hat{I}$$

$$= -2\hbar^2\hat{I}$$

18.47 Vide problem 18.62.

18.48 Here the particle can move freely within a length $2L$ with its potential energy $V(x)$ as under

$$V(x) = 0 \qquad\qquad \text{for } -L < x < L$$
$$V(x) = \infty \qquad\qquad x \le -L \text{ and } x \ge L$$

Since the probability of finding the particle is zero at $V(x) = \infty$, the following boundary conditions are imposed on the wavefunction from Born's interpretation of the latter

$$\psi(x) = 0 \qquad\qquad \text{at } x = -L$$
$$\psi(x) = 0 \qquad\qquad \text{at } x = L$$

Applying these on the given wavefunction, we have

$$\psi(x) = B\cos kL - A\sin kL = 0 \qquad\qquad\qquad\qquad (I)$$
$$\psi(x) = B\cos kL + A\sin kL = 0$$

Then $\begin{vmatrix} \cos kL & -\sin kL \\ \cos kL & \sin kL \end{vmatrix} = 0$

i.e. $2\cos kL \cdot \sin kL = \sin 2kL = 0$

Therefore $2kL = n\pi \quad \text{where } n = 1, 2, 3, \ldots$

or $k = \dfrac{n\pi}{2L}$

Putting this in Eq. (I), we have

$$A = 0 \text{ for odd } n, \text{ and } B = 0 \text{ for even } n$$

18.49 (a) $A = \left[\dfrac{1}{\displaystyle\int_{-\frac{L}{2}}^{\frac{L}{2}} \cos^2(n\pi x/L)\,dx}\right]^{\frac{1}{2}}$

$$= \left(\frac{2}{L}\right)^{\frac{1}{2}}$$

(b)
$$H\psi = -\frac{\hbar^2}{2m}\frac{d^2}{dx^2}\psi$$

$$= +\frac{\hbar^2}{2m}\left(\frac{n\pi}{L}\right)^2 \cdot \psi \quad \text{for } \psi = A\cos\left(\frac{n\pi x}{L}\right)$$

Then
$$E = \frac{\hbar^2}{2m}\left(\frac{n\pi}{L}\right)^2$$

$$= \frac{n^2 h^2}{8\,mL^2}$$

□

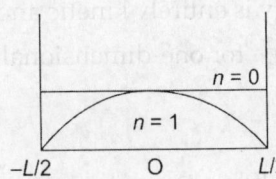

Fig. 18.9: ψ vs x plot

18.50 (a) (i) Similar to question 18.49(b).

(ii) $n = 0$ is not permitted because it gives $\psi(x) = 0$ everywhere, which has the absurd meaning that the particle is lost.

(b) Vide problem 18.79(iii).

(c) Vide problem 18.79(ii).

18.51 (a) 'Zero' in zero-point energy signifies the lowest energy (which is not really zero) expected at the absolute zero of temperature.

(b) Vide problem 18.74.

18.52 (a) Vide problem 18.74.

(b) (i) For a particle in a 1-D box of length L, the wavefunction $\psi_n = \left(\frac{2}{L}\right)^{\frac{1}{2}}\sin\left(\frac{n\pi x}{L}\right)$
(where n is an integer) is zero for only $(n + 1)$ points at $x = 0, \frac{1}{n}L, \frac{2}{n}L, \dots \frac{n}{n}L$ which are separated by half-wavelength $(\lambda/2)$ Then $L = n\cdot(\lambda/2)$.

(ii) Nodes of a wavefunction are points where it passes through zero and not simply zero. Then the number of nodes of the wavefunction ψ_n for a particle in a box will be $n - 1$ and not $n + 1$, the points at $x = 0$ and L (where ψ_n is simply zero) being excluded.

18.53 The expression of permitted energies (E) will follow from Eq. (18.11)

(i) $E = \dfrac{(6.63 \times 10^{-34} \text{ JS})^2 n^2}{8(9.1 \times 10^{-31}\text{ kg})(0.01 \times 10^{-9}\text{ m})^2} = 9.11 \times 10^{-13} n^2$

(ii) $E = \dfrac{(6.63 \times 10^{-34}\text{ JS})^2 n^2}{8(10 \times 10^{-3}\text{ kg})(10 \times 10^{-2}\text{ m})^2} = 8.29 \times 10^{-65} n^2$

□ Here E is entirely translational kinetic energy. Since E is very small in case of (ii), the separation of successive translational energy levels is so small for reasonable values of n, that the distribution of such energy may be regarded as virtually continuous. Then the translational motion of a macroscopic body may be adequately treated by the method of classical mechanics (vide problem 18.83). This is unlike electron where the gap between energy levels is significant due to low mass of the electron.

18.54 For the given condition, the length of the box L is given by

$$m_0 c^2 = \frac{h^2}{8\,m_0 L^2}$$

or

$$L = \frac{1}{\sqrt{8}}\frac{h}{m_0 c} \quad \text{(taking mass of electron as its rest mass)}$$

$$= \frac{1}{\sqrt{8}} \times \text{Compton wavelength}$$

18.55 For a free particle, the energy is entirely kinetic and is given by

$$E = \tfrac{1}{2} m v_x^2 \quad \text{for one-dimensional motion in x-direction}$$

$$= \frac{p_x^2}{2m}$$

Then the Hamiltonian operator

$$\hat{H} = \left(\frac{h}{i}\frac{d}{dx}\right)^2 \Big/ 2m$$

$$= -\frac{h^2}{2m}\frac{d^2}{dx^2}$$

18.56 Here quantization of energy will disappear, because the wavefunction does not have to fulfil any boundary conditions. Then, like a classical particle, it will have a continuous range of energies, but its energy cannot be zero (contrary to a classical particle), otherwise the wavefunction will be zero everywhere.

18.57 (a) Vide problem 18.79.

(b) Vide problem 18.79.

$\langle x^2 \rangle \to \frac{a^2}{3}$ as $n \to \infty$.

Then for very high value of n, $\langle x^2 \rangle$ is independent of state.

(c) Vide problem 18.78(c)(ii).

□ Vide problem 18.78(B)(iii).

(d) Probability $\quad P = \int_0^{\frac{L}{4}} \psi_n^2 dx$

$$= \int_0^{\frac{L}{4}} \frac{2}{L} \sin^2\left(\frac{n\pi x}{L}\right) dx$$

$$= \left| \frac{x}{L} - \frac{1}{2n\pi}\sin\frac{2n\pi x}{L} \right|_0^{\frac{L}{4}}$$

$$= \tfrac{1}{4} - \frac{1}{2n\pi}\sin\frac{n\pi}{2}$$

□ $n = 3$ which corresponds to the highest −ve value of $\dfrac{1}{2n\pi}\sin\dfrac{n\pi}{2}$.

(e) Vide problem 18.79.

□ Since momentum p is a vector quantity, $\langle p \rangle = 0$ does not necessarily mean that the particle is at rest.

18.58 Vide problem 18.78(D).

18.59 Vide problem 18.81.

18.60 Vide problem 18.84(i)

18.61 Vide problem 18.84.

18.62 Since n_x, n_y and n_z are all integer not less than 1

$$n_x^2 + n_y^2 + n_z^2$$
$$= 2^2 + 1^2 + 1^2 = 6$$

The corresponding energy equation is

$$E = \frac{6h^2}{8ma^2} \quad \text{by Eq. (18.15b)}$$

□ Degeneracy is three. Vide problem 18.84(i).

18.63 The degeneracy is six. Vide problem 18.84.

18.64
$$E = \frac{h^2}{8m}\left(\frac{n_x^2}{L^2} + \frac{n_y^2}{L^2} + \frac{n_z^2}{L^2}\right) \quad \text{from Eq. (18.5b) replacing } a \text{ by } L$$

$$\Delta E = \frac{h^2 n_z^2}{8m}\left[\frac{1}{L^2} - \frac{1}{(L + \Delta L)^2}\right]$$

(assuming that distortion of the cube does not affect its lengths along x- and y-axis)

$$= \frac{h^2 n_z^2}{8m} \cdot \frac{2L\Delta L}{L^4} \quad \text{since } \Delta L \ll L$$

$$= \frac{h^2 n_z^2}{4mL^3} \cdot \Delta L$$

18.65 For a particle in a cubic box, the energy levels are

$$E_{n_x n_y n_z} = \frac{h^2}{8ma^2}(n_x^2 + n_y^2 + n_z^2)$$

The lowest level, which corresponds to $n_x = n_y = n_z = 1$ is non-degenerate. Hence, by the Pauli exclusion principle, it can hold not more than two fermion particles (i.e., ones with half-integral spin which include electrons, protons, neutrons and some atomic nuclei). Then in the lowest energy state of the given system, two particles will go to the level with $n_x^2 + n_y^2 + n_z^2 = 3$ and the other two to the next energy level with $n_x^2 + n_y^2 + n_z^2 = 6$ for which (n_x, n_y, n_z) of the wavefunction may be $(1, 1, 2)$ $(1, 2, 1)$ or $(2, 1, 1)$. Hence the required degeneracy is three.

18.66 Here energy levels are

$$E_{n_x n_y} = \frac{h^2}{8m}\left(\frac{n_x^2}{a^2} + \frac{n_y^2}{b^2}\right) \quad \text{from Eq. (18.15a) putting } c = 0$$

$E_{n_x n_y}$ is same for two states $(n_x = 1, n_y = 4)$ and $(n_x = 2, n_y = 2)$ with $a = L$ and $b = 2L$. Hence there is a degeneracy between the given two states.

□ For the state $(n_x = 1, n_y = 0)$, the wavefunction is

$$\psi_{n_x n_y} = \left(\frac{2}{ab}\right)^{\frac{1}{2}} \sin\frac{n_x \pi x}{a} \sin\frac{n_y \pi y}{b} = 0$$

Such a wavefunction is not acceptable because it is always zero and hence contradictory to the Born's interpretation of wavefunction.

18.67 The energy of a particle in a square box of length a is

$$E_{n_x n_y} = \frac{h^2}{8ma^2}(n_x^2 + n_y^2)$$

Here first energy level corresponds to $n_x = n_y = 1$ which is non-degenerate.

Second energy level corresponds to $(n_x = 1, n_y = 2)$ or $(n_x = 2, n_y = 1)$ and is therefore doubly degenerate.

Then in the ground state two electrons belong to the first level and four electrons to the second level. Therefore

energy of the ground state $= \dfrac{h^2}{8ma^2}[2(1^2 + 1^2) + 2(1^2 + 2^2) + 2(2^2 + 1^2)]$

Now, nearst vacant energy level corresponds to $n_x = 2, n_y = 2$.

Promotion of an electron to this level from the second level will involve least energy. Then

required energy for promotion $= \dfrac{h^2}{8ma^2}[(2^2 + 2^2) - (1^2 + 2^2)]$

$$= \frac{3(6.63 \times 10^{-34} \text{ JS})^2}{8(9.1 \times 10^{-31} \text{ kg})(4 \times 10^{-10} \text{ m})}$$

18.68 The given statement is not justified. Because, although in both cases the motion is to and fro, in case of particle in a box problem, the particle moves freely within an infinite square well potential while in case of simple harmonic oscillator, the particle always moves under the action of a force in a parabolic well potential.

18.69 (a) The energy (E) of a harmonic oscillator is

$$E = \underbrace{\frac{P_x^2}{2m}}_{\substack{\text{kinetic} \\ \text{energy}}} + \underbrace{\tfrac{1}{2}kx^2}_{\substack{\text{potential} \\ \text{energy}}} \quad \text{for motion in x-direction, } k \text{ is force constant}$$

The corresponding Hamiltonian operator is

$$\hat{H} = \left(\frac{\hbar}{i}\frac{d}{dx}\right)^2 \Big/ 2m + \tfrac{1}{2}k\hat{x}^2$$

$$= -\frac{\hbar^2}{2m}\frac{d^2}{dx^2} + \tfrac{1}{2}k\hat{x}^2$$

(b) The required Schrodinger equation is

$$-\frac{\hbar^2}{2m}\frac{d^2\psi}{dx^2} + \tfrac{1}{2}kx^2\psi = E\psi \quad (\text{from } \hat{H}\psi = E\psi)$$

(c) The notable characteristic features of quantum mechanical harmonic oscillator, unlike the classical ones, are the following:

(i) their lowest energy (zero-point energy) is positive, i.e. not zero.

(ii) their kinetic energy may be negative. (Vide problem 18.89)

18.70 Normalization requires

$$1 = \int_{-\infty}^{\infty} \psi^*\psi \, dx$$

$$= A^2 \int_{-\infty}^{\infty} x^2 e^{-\alpha x^2} dx$$

$$= A^2 \tfrac{1}{2} \left(\frac{\pi}{\alpha^3} \right)^{\frac{1}{2}}$$

whence

$$A = \left(\frac{4\alpha^3}{\pi} \right)^{\frac{1}{4}}$$

18.71 See solution of problem 18.87 which is based on the relation

$$\left(\frac{\hbar^2 B}{\mu} - E \right) \psi + \left(\tfrac{1}{2}k - \frac{2\hbar^2 B^2}{\mu} \right) x^2 \psi = 0$$

Substitution of $B = \dfrac{\sqrt{k\mu}}{2\hbar}$ (given) in this relation gives

$$\frac{\hbar^2}{\mu} B\psi = E\psi$$

Hence

$$E = \frac{\hbar^2 B}{\mu}$$

$$= \frac{\hbar^2}{\mu} \cdot \frac{1}{2\hbar} \sqrt{k\mu}$$

$$= \tfrac{1}{2}\hbar\sqrt{\frac{k}{\mu}}$$

$$= \tfrac{1}{2}\hbar\omega$$

Note: We can arrive at $E = \tfrac{1}{2}\hbar\omega$, even if $B = \sqrt{k\mu}/2\hbar$ is not given as it follows from solution of the problem 18.87.

18.72 Schrodinger equation leads the following expression for the energy of a harmonic oscillator

$$E = \left(v + \tfrac{1}{2} \right)hv \quad v = 0, 1, 2, \ldots, v \text{ is frequency of the oscillator}$$

Then the zero-point energy, which corresponds to $v = 0$, is $\tfrac{1}{2}hv$.

18.73 Vide problem 18.88.

18.74 Vide problem 18.91.

18.75 Vide problem 18.89(a).

18.76 Zero-point energy

$$\tfrac{1}{2}hv = \frac{h}{4\pi}\sqrt{k/m} \quad \text{by Eq. (18.16)}$$

$$= \frac{6.63 \times 10^{-34} \text{ JS}}{4 \times 3.14} \sqrt{\frac{285 \text{ Nm}^{-1}}{5.16 \times 10^{-26} \text{ kg}}}$$

$$= 3.92 \times 10^{-21} \text{ J}$$

18.77 The difference in adjacent energy levels

$$\Delta E = hv = \frac{h}{2\pi}\sqrt{k/m} \quad \text{by Eq. (18.16)}$$

Then, force constant

$$k = \left(\frac{2\pi\Delta E}{h}\right)^2 m$$

$$= \left[\frac{2 \times 3.14\,(4.82 \times 10^{-21})}{(6.63 \times 10^{-34}\ \text{JS})}\right]^2 (1.33 \times 10^{-25}\ \text{kg})$$

$$= 277\ \text{Nm}^{-1}$$

18.78 Zero-point energy of the molecule

$$\tfrac{1}{2}hv = \tfrac{1}{2}h \cdot \frac{1}{2\pi}\sqrt{k/\mu} \quad \text{where m is the reduces mass of the molecule}$$

$$= \frac{h}{2\sqrt{2}\pi}\sqrt{k/m} \quad \text{since } \frac{1}{\mu} = \frac{1}{m_{Br}} + \frac{1}{m_{Br}}$$

$$= \frac{6.63 \times 10^{-34}\ \text{JS}}{2\sqrt{2} \times 3.14}\sqrt{\frac{240\ \text{Nm}^{-1}}{79 \times 1.66 \times 10^{-27}\ \text{kg}}} = 3.19 \times 10^{-21}\ \text{J}$$

Note: Here the two-particle problem has been treated as a one-particle problem (18.76), simply by replacing mass (m) with reduced mass $\mu = m/2$.

18.79 Vide Section 18.12.

□ Using polar (spherical) coordinates, the wavefunction of hydrogen atom (which has spherical symmetry) can be written as a product of three functions each having only one variable, i.e.

$$\psi = f(r)\,f(\theta)\,f(\phi)$$

Then Eq. (18.18) becomes

$$\frac{\sin^2\theta}{f(r)}\frac{d}{dr}\left[r^2\frac{df(r)}{dr}\right] + \frac{\sin\theta}{f(\theta)}\frac{d}{d\theta}\left[\sin\theta\frac{df(\theta)}{d\theta}\right] + \frac{2\mu}{\hbar^2}\left(E + \frac{Ze^2}{r}\right)(r^2\sin\theta) = -\frac{1}{f(\phi)}\frac{d^2 f(\phi)}{d\phi^2}$$

Here lhs is a function of r and θ only, and the rhs is a function of φ only, so neither can depend on the other's variable (s). Then each side must be equal to some common constant m^2, and considering each side of this equation, we have on rearrangement

$$\frac{1}{f(\phi)} \cdot \frac{d^2 f(\phi)}{d\phi^2} + m^2 = 0$$

and

$$\frac{1}{f(r)}\frac{d}{dr}\left[r^2\frac{df(r)}{dr}\right] + \frac{2\mu}{\hbar^2}\left(E + \frac{Ze^2}{r}\right)r^2 = \frac{m^2}{\sin\theta} - \frac{1}{\sin\theta}\frac{1}{f(\theta)}\frac{d}{d\theta}\left[\sin\theta\frac{df(\theta)}{d\theta}\right]$$

By the same argument as above, we can split this equation into following two equations, equating each side to some common constant λ (say)

radial: $\dfrac{1}{r^2}\dfrac{d}{dr}\left[r^2\dfrac{df(r)}{dr}\right] - \dfrac{\lambda}{r^2}f(r) + \dfrac{2\mu}{\hbar^2}\left(E + \dfrac{Ze^2}{r}\right)f(r) = 0$

angular: $\dfrac{1}{\sin\theta}\dfrac{d}{d\theta}\left[\sin\theta\dfrac{df(\theta)}{d\theta}\right] - \dfrac{m^2 f(\theta)}{\sin^2\theta} + \lambda f(\theta) = 0$ where $m^2 = -\dfrac{1}{f(\phi)}\dfrac{d^2 f(\phi)}{d\phi^2}$

18.80 A wavefunction for a single electron that pictures the probability of finding the electron at various parts in space is called an orbital.

□ The significance of the given statement is that the relevant wavefunction has the same value at fixed distances from the nucleus in all directions.

18.81 Vide problems 18.92 and 18.98.

18.82 Number of nodes $= n - 1 = 2 - 1 = 1$.

The location of the node is given by

$$2 - \dfrac{r}{a_0} = 0$$

or $\qquad\qquad r = 2a_0$

18.83 (i) $\qquad 1 = b_0^2 \displaystyle\int_{-\infty}^{\infty} r^2 e^{-2r/a_0}\, dr \int_0^{\pi} \sin\theta d\theta \int_0^{2\pi} d\phi$ vide problem 18.47(iii)

$$= b_0^2 \pi a_0^3$$

Then $\qquad\qquad b_0 = \dfrac{1}{\sqrt{\pi a_0^3}}$

(ii) Probability density

$$\psi_{1s}^2 = b_0^2 e^{-2r/a_0}$$

$$= b_0^2 \text{ at the nucleus, i.e. at } r = 0$$

$$= \dfrac{1}{\pi a_0^3}$$

18.84 The probability of finding the electron in a thin layer of thickness dr and radius r is R^2 (the probability density at a point at distance r) times $4\pi r^2 dr$ (the volume of the layer).

Therefore, the required charge content

$$= 4\pi r^2 R^2 e\, dr$$

where e is the charge of an electron.

18.85 Vide problem 18.101(b).

18.86 Average electron-nucleus separation

$$\langle r \rangle = \int \psi * r \psi d\tau$$

$$= \dfrac{\pi}{8a_0^5} \int_0^{\infty} r^2 \cdot r \cdot r^2 e^{-r/a_0} \int_0^{\pi} \cos^2\theta \sin\theta d\theta \int_0^{2\pi} d\phi$$

$$= \dfrac{\pi}{8a_0^5} \int_0^{\infty} r^5 e^{-r/a_0}\, dr \left(-\int_1^{-1} x^2 dx\right) \int_0^{2\pi} d\phi$$

$$= \dfrac{\pi}{8a_0^5} \cdot \dfrac{5!}{\left(\frac{1}{a_0}\right)^6}\left(\tfrac{1}{3}\right)(2\pi)$$

18.87 Let angular part of the wavefunction, $Y = B\cos\theta$

Then for normalisation of Y,

$$1 = B_2 \int_0^\pi \cos^2\theta \sin\theta \, d\theta \int_0^{2\pi} d\phi$$

$$= B^2 \left(\tfrac{2}{3}\right)(2\pi)$$

or $\qquad B = \sqrt{\tfrac{3}{4\pi}}$

Then the radial part of the given wavefunction

$$R = \frac{\psi}{\sqrt{\tfrac{3}{4}}\theta \cos\theta}$$

$$= \sqrt{\tfrac{4\pi}{3}} \cdot A r e^{-r/2a_0}$$

Therefore, the radial distribution function is

$$4\pi r^2 R^2 = 4\pi r^2 \frac{4\pi}{3} A^2 r^2 e^{-r/a_0}$$

18.88 Vide Fig. 18.6(b).

The number of maxima exhibited in the diagram is two with $2s$ but only one with $2p$. In case of $2s$, r corresponding to the higher peak represents the most probable radius of the orbital. Also it appears from the diagram that the most probable radius is greater with $2s$ than with $2p$ (though only slightly). Because the ability to approach (penetrate) to the nucleus is more with a $2s$-electron (which is pertinent to the additional maximum that provides shielding effect on nuclear attraction) than with a $2p$-electron.

Note: The phrase 'most probable radius' appears awkward with p-orbital which is not spherical. Actually, this phrase signifies the most probable distance of an electron from the nucleus when it occupies the orbital concerned.

18.89 Vide problem 101(e).

Here $\qquad \dfrac{PE}{KE} = -\dfrac{2m_e e^2 a_0}{\hbar^2} = -2$

Since $\qquad a_0 = \dfrac{\hbar^2}{m_e e^2}$

(from $\dfrac{Ze^2}{r^2} = \dfrac{mv^2}{r}$ and $mvr = n\hbar$ with $n = 1$ for the first Bohr orbit of H atom)

Then, total energy $E = KE + PE = -KE$

18.90 Answer is same as that of the problem 101(e). But for the given problem, since the radial wavefunction is given, the expression for $\langle K \rangle$ and $\langle V \rangle$ would be free from angular variables, i.e.

$$\langle K \rangle = -\frac{\hbar}{2m_e} \left(2a_0^{-\frac{3}{2}}\right)^2 \int_0^\infty \left(-\frac{2}{a_0 r} + \frac{1}{a_0^2}\right) e^{-2r/a_0} r^2 dr$$

$$\langle V \rangle = \left(2a_0^{-\frac{3}{2}}\right)^2 e^2 \int_0^\infty \frac{1}{r} e^{-2r/a_0} r^2 dr$$

The total energy $= -KE$ as in the previous university question.

18.91 Although 3s- and 4s-orbitals are both spherical in shape, they differ in the number $(n - 1)$ of nodes (where the wavefunction possess through zero) which is 2 for 3s but 3 for 4s.

☐ In the third quantum shell $(n = 3)$, there is no possibility of f-orbitals $(l = 3)$, due to the restriction $l \leq (n - 1)$.

18.92 The energy of a H-atom is independent of the quantum number l and m, i.e. depends only on n. Then the degeneracy of the energy level with $n = 2$ will be equal to the number of orbitals associated with this level, i.e. $n^2 = 2^2$.

19

Molecular Spectroscopy

Spectroscopy is the study of the electromagnetic radiation associated with the transition of matter from one energy state to another. Compared to the atomic spectra, the molecular spectra are likely to be much more complicated because of involvement of transition between molecular energy states due to rotational and vibrational motions which are not possible with an atom. The separation between different energy states are as under:

Nuclear spin states < electron spin states < rotational states < vibrational states < electronic states

Analysis of molecular spectra provides valuable informations regarding molecular parameters.

19.1 GENERAL ASPECTS OF SPECTRA

Frequency of Spectral Lines

Whether the spectra originate from emission, absorption or scattering due to molecules, each spectral line arises from a specific molecular transition from one energy state E_1 to another of different energy E_2 corresponding to a frequency of radiation given by

$$\left.\begin{array}{c}|E_1 - E_2| = h\nu \\ = hc\bar{\nu}\end{array}\right\} \tag{19.1}$$

where ν and $\bar{\nu}$ denotes the frequency in Hertz ($\sec^{-1}$) and wavenumber (inverse of wavelength) respectively, h is Planck constant and c is speed of light.

Spectral Selection Rules

Optical spectra result mainly from interaction of the electrons and nuclei of molecular species with the electric field of light which consists of oscillating electric and magnetic fields at right angle to each other. Such an interaction leading to transition of a molecule between two stationary states ψ_m and ψ_n will be probable or not according as the following integral is non-zero or not

$$\mu_{mn} = \int \psi_m^* \hat{\mu} \psi_n d\tau \tag{19.2}$$

($\hat{\mu} = \sum_i q_i r_i$ where r_i is the displacement vector of charge θ_i from the origin.)

Here $\hat{\mu}$ is the electric dipole moment operator. The quantity μ_{mn} is called the *transition dipole moment* which measures a sort of average molecular dipole that the electric field of the light acts upon during the actual process of transition of the molecule from state

ψ_m to ψ_n. The restriction $\mu_{mn} \neq 0$ (in compliance with the conservation of angular momentum in a photonic process) gives rise to the allowed change in quantum number(s) of the system, called the specific selection rules. Of course such selection rules will be of any value provided some gross selection rule is obeyed (that can be readily judged on structural ground), i.e. the transition must involve change in dipole moment that permanently exists in the molecule (for emission or absorption) or is induced by the field (for scattering).

If $\mu_{mn} = 0$ for any pair of states, the electric field of the light cannot perturb the molecule and then transition by the electric dipole mechanism will be forbidden. But here less effective mechanisms involving magnetic dipole/electric quadrupole might be operative, though the resulting spectra will generally be too weak to have any importance.

Intensity of Spectral Lines

In any spectrum, the lines occur with a variety of intensities. The intensity of a spectral line resulting from the transition of the molecule from some initial level of energy E_i will be proportional to the number of molecules (N_i) that have the energy E_i (assuming that the probability of transition is same for each energy level). This number, called population of the energy level, is given by the Boltzmann distribution

$$N_i \propto g_i \, e^{-E_i/kT} \tag{19.3}$$

where g_i is the degeneracy of the ith level.

However, the intensity of spectral lines is not independent of the transition frequency which determines the transition probability. The absorption of radiation is a stimulated process, but emission may be spontaneous or stimulated. Einstein showed that

$$\frac{\text{rate of spontaneous emission}}{\text{rate of stimulated emission}} = \frac{AN}{BIN} = \frac{8\pi h (v/c)^3}{I} \tag{19.4}$$

where A is the coefficient of spontaneous emission and B that of stimulated emission due to the radiation of intensity I and frequency v equal to the transition frequency, and N is the population of the upper state. A and B measure the probabilities of the relevant transitions. The coefficients of stimulated absorption and emission are equal. It follows from the above relation that the spontaneous emission becomes relatively more important than the stimulated emission at high frequencies (visible or ultraviolet) but not at low frequencies (microwave or radio frequency).

Width of Spectral Lines

In a spectrum, different lines have different width (continuous short range of wavelength or frequency). If the temperature (T) is not very low, this can be attributed primarily to the Doppler shift ($\Delta\lambda$) given by

$$\Delta\lambda = \frac{2\lambda}{c} \left(\frac{2kT\ln 2}{m} \right)^{\frac{1}{2}} \tag{19.5}$$

where m is mass of the molecule and λ is the wavelength of the relevant radiation. But at extremely low temperature (when Doppler effect is negligible) the width of spectral line is attributed to lifetime broadening, a consequence of uncertainty principle. According to this principle, the energy of an excited molecule having lifetime τ is not fixed at certain value E but varies within a range dE around E where $dE \leq h/4\pi\tau$. This relation will hold provided τ is determined essentially by the intrinsic deactivation (which cannot be controlled) and not by collisional deactivation (which can be minimised at low concentration and temperature.

Spectral Regions

The region in which a spectral line will appear depends on the energy states involved in the molecular transition allowed by some quantum mechanically based selection rules. The types of changes responsible for spectrum in different regions are given in Table 19.1.

Table 19.1

Spectral region	Origin
1. Radiofrequency region (10^5–10^9 Hz)	1. Changes of nuclear spin state of an elementary species with odd mass number or odd number of protons in a magnetic field (i.e. nuclear magnetic resonance)
2. Microwave region (10^9–10^{11} Hz) and Far infrared region (10^{11}–10^{12} Hz)	2. Change of electron spin state for a species with unpaired electron(s) in a magnetic field (i.e. electron spin resonance) or Transition between rotational energy levels of gas molecules (i.e. molecular rotation)
3. Infrared region (10^{12}–10^{14} Hz)	3. Change in vibrational and rotational energy states together (i.e. molecular vibration)
4. Visible and UV region (10^{14}–10^{15} Hz)	4. Change in electronic energy state accompanied by changes in vibrational and rotational energy states. This involves transition of valence electrons
5. Vacuum ultraviolet, x-ray and γ-ray region (10^{16}–10^{21} Hz)	5. Transition of inner electrons to higher levels, or dissociation or ionization of molecules

19.2 ROTATIONAL SPECTRA

A rotational spectrum arises from transition between rotational energy levels of a polar molecule. This will happen (with considerable intensity) only if the rotational motion entails a change in dipole moment of the molecule, which is the gross selection rule for rotational transition.

The rotational energy E_J of a linear (gas-phase) molecule, treated as a rigid rotor, can be expressed (on quantum mechanical ground) as

$$E_J = \tfrac{1}{2}\frac{M^2}{I} = J(J+1)\frac{h^2}{8\pi^2 I}, J = 0, 1, 2, \ldots$$

$$= BhJ(J+1) \tag{19.6}$$

where $B = \dfrac{h}{8\pi^2 I}$ is the rotational constant of the molecule, I is its moment of inertia about the axis of rotation and J is the rotational quantum number. Each energy level is $(2J + 1)$-fold degenerate.

[If we take $B = \dfrac{h}{8\pi^2 Ic}$ then the energy expression will change to $E_J = BhcJ(J + 1)$].

A molecule can undergo rotational transition only between two adjacent rotational energy levels, i.e. the specific selection rule for rotational transition is $\Delta J = \pm 1$, + sign applies to absorption and – sign to emission of radiation. The frequency of radiation ν associated with a rotational transition is then given by

$$\nu = \frac{E_{J+1} - E_J}{h} = 2B(J+1) \tag{19.7a}$$

for absorption transition $J \rightarrow J + 1, J = 0, 1, 2, \ldots$

$$\nu = \frac{E_J - E_{J-1}}{h} = 2BJ \tag{19.7b}$$

for emission transition $J \rightarrow J - 1, J = 1, 2, \ldots$

Then for each value of J, a line of frequency ν will appear in the rotational spectrum and there will be a constant separation of $2B$ between the adjacent lines. These lines appear in the microwave or far-infrared region of spectrum.

However, in the experimental rotational spectrum, the spacing between consecutive lines varies, though only slightly, from one part of the spectrum to another. This is due to centrifugal distortion of the molecule that causes the rotational energy levels closer together with increasing rotational motion. When corrected for this effect, the expression for rotational energy levels becomes

$$E_J = BhJ(J+1) - DhJ^2(J+1)^2 \tag{19.8}$$

where D is centrifugal distortion constant. For a diatomic molecule D is approximately given by

$$D = \frac{4B^3}{\nu_2^2} \tag{19.9}$$

where ν_e is the equilibrium bond frequency. Then considering the centrifugal distortion, the frequency of the spectral line for any rotational transition $J \rightarrow J + 1$ will be

$$\nu = \frac{E_{J+1} - E_J}{h} = 2B(J+1) - 4D(J+1)^3 \tag{19.10}$$

Since D is much smaller than B, ν will only slightly decrease with increase in J.

19.3 VIBRATIONAL SPECTRA

A vibrational spectrum arises from transition between vibrational energy levels of a molecule. This will happen only if the molecular vibration involves a change in dipole moment of the molecule (not necessarily polar), which is the gross selection rule for vibrational transition.

The simplest molecular vibration is that observed with a diatomic molecule. Assuming this vibration to be simple harmonic, the vibrational energy levels of the molecule are given by

$$E_v = \left(v + \tfrac{1}{2}\right)h\nu \qquad v = 0, 1, 2, \ldots \tag{19.11}$$

Here v is the vibrational quantum number and ν is the fundamental vibrational frequency of the molecule.

For harmonic vibration only those vibrational transitions are allowed for which the change in vibrational quantum number is unity (which is the specific selection rule), i.e.

$$\Delta v = \pm 1$$

The frequency of radiation associated with the transition $v \to v + 1$ (for absorption) or $v \to v - 1$ (for emission) will then be

$$v = \frac{|\Delta E_v|}{h}$$

which is same as the vibrational frequency of the molecule and is independent of the vibrational level from which transition occurs. Then, unlike rotational spectrum, the vibrational spectrum of a diatomic molecule will contain only one line of frequency v which lies in the infrared region.

However, the vibrational spectra of even diatomic molecules do not really consist of a single line, instead (i) they contain a number of apparently broad lines with appreciable separation between them, and (ii) each such broad line really consists of a large number of closely spaced components and is often called a band. The anharmonicity of oscillation gives rise to (i) while the rotational transitions accompanying a vibrational transition lead to (ii).

Effect of Anharmonicity

If the molecular vibrations were harmonic, the potential energy (V) of a diatomic molecule would have been given by the simple expression

$$V = \tfrac{1}{2}k(r - r_e)^2$$

where r is the distance between the bonded atoms, r_e is the equilibrium internuclear distance corresponding to the minimum potential energy and k is the face constant. According to this equation, the potential energy curve will be parabolic as shown by broke line in Fig. 19.1. However, the actual potential energy curve, as shown by solid line, is not parabolic, though it is nearly so at the lower part. This is due to the anharmonic character of oscillation in real molecules. Here, unlike harmonic oscillator, V cannot exceed a certain value (D) corresponding to dissociation of the molecule. D signifies spectroscopic (equilibrium) dissociation energy D_e or chemical dissociation energy D_o according as it is measured from the minimum potential energy or ground vibrational state ($v = 0$) of the molecule. The two are related approximately by

$$D_e = D_o + \tfrac{1}{2}hv_e \tag{19.12}$$

where v_e is the equilibrium vibration frequency of the molecule.

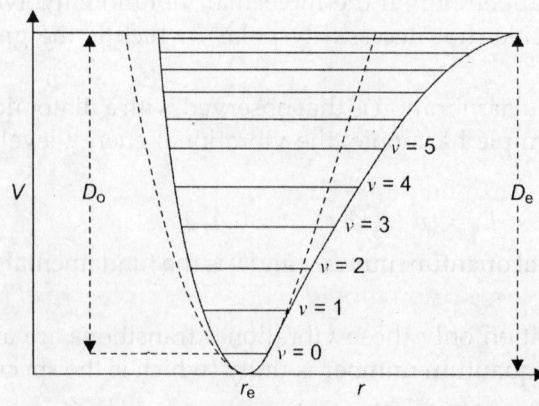

Fig. 19.1

The real potential energy curve for a diatomic molecule fits well with the following empirical function suggested by Morse

$$V = D_e \left[1 - e^{-\alpha(r - r_e)}\right]^2 \tag{19.13}$$

where $\alpha = \pi v_e \left(\dfrac{2\mu}{D_e}\right)^{\frac{1}{2}}$. When the Morse function is used, the solution of Schrodinger equation gives the following vibrational energy levels

$$E_v = \left(v + \tfrac{1}{2}\right) h v_e - x_e \left(v + \tfrac{1}{2}\right)^2 h v_e \tag{19.14}$$

where $x_e = h v_e / 4D_e$ is called the anharmonicity constant. Then, unlike the harmonic oscillator, the energy levels for a real diatomic molecule are not equally spaced, but get closer with increasing quantum umber and ultimately merge (when the molecule dissociates) as is indicated in Fig. 19.1 by horizontal lines. Therefore, vibrational transitions from different levels will occur at different frequencies that decrease with increase in vibrational quantum number (v) of the level involved. However, at ordinary temperature the molecules are mostly in their vibrational ground state and the vibrational transition $v = 0 \to v = 1$, called the fundamental transition, will be the dominating one producing the most intense band. But at sufficiently high temperature (or if the vibrational frequency of the molecule is low) the population of higher vibrational levels becomes appreciable and then transitions originating from these levels will arise producing weak bands, called combination bands or more often hot bands (due to their usual origin).

Another consequence of anharmonicity is the occurrence of weaker overtone transitions with $|\Delta v| > 1$ in addition to fundamental transition where $|\Delta v| = 1$. The frequencies for fundamental (also called first harmonic), first overtone (second harmonic), second overtone (third harmonic), ... corresponding to $0 \to 1$, $0 \to 2$, $0 \to 3$, ..., vibrational transitions are

$$\left.\begin{aligned}
v_1 &= (1 - 2x_e) v_e \\
v_2 &= (1 - 3x_e) 2v_e \\
v_3 &= (1 - 4x_e) 3v_e
\end{aligned}\right\} \tag{19.15}$$

Since $x_e \ll 1$, the frequencies of the fundamental and overtone bands are approximately v_e, $2v_e$, $3v_e$.

Vibration–Rotation Spectra

The vibrational transitions actually occur between rotational energy levels belonging to different vibrational levels, normally obeying the selection rule $\Delta v = \pm 1$ and $\Delta J = \pm 1$. Here $\Delta J = 0$ is also possible but only in some exceptional circumstances, e.g. in case of NO that possesses electronic angular momentum about the bond axis. Consequently, molecules do not usually exhibit pure vibrational spectra, instead they exhibit vibration–rotation spectra.

In order to predict the form of vibration–rotation spectra in the simple case of diatomics, let us use the following expression for the energy levels of a vibrating rotating molecule treating it as a rigid rotor and harmonic oscillator

$$E_{v,J} = \left(v + \tfrac{1}{2}\right) h v + J(J + 1) Bh \tag{19.16}$$

(Here the suffix e of v_e has been dropped).

For any vibrational transition, the spectrum will consist of three sets of lines designated as P-, Q- and R-branch according as $\Delta J = -1, 0$ or $+1$. Assuming B to be unchanged in vibrational transitioon, the frequencies of such lines will be

$$\left. \begin{array}{ll} v_P = v - 2BJ & \text{for P-branch} \\ v_Q = v & \text{for Q-branch} \\ v_R = v + 2B(J+1) & \text{for R-branch} \end{array} \right\} \tag{19.17}$$

In fact, the rotational constant of the vibrationally excited state B_1 is slightly smaller than that of the ground vibrational state B_0 because of anharmonicity of the vibration. As a result (actually found), the Q-branch (if any), instead of being a single line, consists of a series of closely spaced lines, and the lines in P- and R-branch (instead of having equal spacing $2B$) slightly diverge and converge in the respective cases as J increases. This follows from the following expressions of frequency

$$\left. \begin{array}{l} v_P = v - (B_1 + B_0)J + (B_1 - B_0)J^2 \\ v_Q = v + (B_1 - B_0)J(J+1) \\ v_R = v + (B_1 + B_0)(J+1) + (B_1 - B_0)(J+1)^2 \end{array} \right\} \tag{19.18}$$

Vibrational Spectra of Polyatomic Molecules

A polyatomic molecule having x atoms has $3x - 5$ or $3x - 6$ normal modes of vibration according as the molecule is linear or not (vide Section 2.5). Each normal mode may be supposed to behave like an independent harmonic oscillator with frequency given by

$$v_i = \frac{1}{2\pi}\left(\frac{k_i}{\mu_i}\right)^{\frac{1}{2}}$$

where v_i is the frequency of the normal mode i, k_i is the force constant and μ_i is the reduced mass for this mode. Not all normal modes of vibrations, but only those which result in a change of dipole moment of the molecule can exhibit vibrational spectra obeying the selection rule $\Delta v_i = \pm 1$.

However, the spectra predicted from the above simple consideration have no remarkable resemblance with the actual spectra of large molecules which are very difficult to analyse. The experimental spectra of polyatomic molecules are very useful to identify them and also to detect the presence of certain groups of atoms in the molecule as they have quite characteristic absorption bands.

19.4 RAMAN SPECTRA

When a beam of light passes through a medium, a certain amount of it is absorbed, a certain amount is transmitted and usually a certain amount is scattered (i.e. distributed in different directions apart from the direction of incident light). Scattering occurs mostly at the same frequency as the incident light, and this is known as Rayleigh scattering. But frequently, a small part of the scattered light differs in frequency from that of the incident light. This is known as Raman scattering or Raman effect, and the frequency differences as Raman shifts or Raman frequencies. Raman scattering is usually weak and the Raman shifts may be quite small ($10–4000$ cm^{-1}). Hence the phenomenon may be obscure unless the incident light beam is of high intensity and quite monochromatic (as with laser).

We can understand Raman effect as follows. When a photon of energy $h\nu$ of incident light collides with a molecule, it is scattered, instead of being absorbed, unless $h\nu$ fits the energy gap between two suitable energy levels of the colliding molecule. If the collision is not elastic, the energy of the photon will either increase or decrease by an amount $h\nu'$ which is the difference between two rotational or vibrational) energy levels of the molecule. The frequency ν'' of the photon after Raman scattering becomes

$$\nu'' = \nu \pm \nu'$$

The Raman shift, being the characteristic of the scattering substance, is completely independent of the frequency of the incident light. Hence, it is conveniently studied with visible or ultraviolet light. The Raman spectra consist of closely spaced lines in addition to parent (incident) line of high intensity. Raman lines are called *Stokes* or *anti-Stokes* according as they have lower or higher frequency than the parent line. The Raman spectra are exhibited only by the molecules whose polarizability changes during their rotation or vibration. This is gross selection rule for Raman scattering.

For rotational Raman transitions, the specific selection rule in case of liner molecules is usually

$$\Delta J = 0, \pm 2$$

$\Delta J = \pm 1$ is also possible but only in exceptional circumstances (where there is electronic angular momentum about the internuclear axis). $\Delta J = 0$ corresponds to Raleigh scattering. Raman transitions with $\Delta J = +2$ give rise to Stokes lines and those with $\Delta J = -2$ to antistokes lines. Their frequencies are given by

$$\nu_{Stokes} = \nu - 2B\,(2J + 3), J = 0, 1, 2, \dots \tag{19.19a}$$

$$\nu_{anti\text{-}Stokes} = \nu + 2B\,(2J - 1), J = 2, 3, \dots \tag{19.19b}$$

where ν is the frequency of the parent line. Then both stokes and antistokes lines will have same spacing equal to $4B$. But the first two lines adjacent to the parent line are separated from it by $6B$. These predictions are amply realised.

For vibrational Raman effect, the specific selection rule is the same as that for vibrational absorption or emission spectra, viz.

$$\Delta v = \pm 1$$

In case of vibration–rotation Raman spectra, the selection rule for vibrational transition is accompanied by that for totational transition. Then such spectra are expected to consist of branches designated as O-branch (for $\Delta J = -2$), P-branch (for $\Delta J = -1$), Q-branch (for $\Delta J = 0$), R-branch (for $\Delta J = +1$) and S-branch (for $\Delta J = +2$). It is important to note that diatomic molecule, which do not normally possess Q-branch in their vibrational absorption or emission spectra, have it in their Raman spectra. Expectedly, anharmonicity makes the real vibration–rotation Raman spectra more complicated, stokes lines for $\Delta v = 1$ and anti-stokes lines for $\Delta v = -1$ have their closely spaced branch structure.

Raman spectra are as useful as infrared spectra in the elucidation of molecular structure. Vibrational Raman spectra, together with infrared spectra, provide valuable information in assigning frequencies to the various normal modes of vibration of polyatomic molecules from the presence or absence of a given line in both the spectra. As a rule (exclusion rule), if a molecule has a centre of symmetry, only those vibrational modes which are symmetric with respect to the centre can be Raman-active and only those which are antisymmetric with respect to the centre can be infrared-active. Then such molecules cannot have any mode active to both (though there may be modes inactive to both with molecules of higher symmetry, e.g. SF_6).

19.5 ELECTRONIC SPECTRA

Electronic spectrum of a molecular species results from transition between its electronic energy levels corresponding to differential electronic arrangement. Each electronic transition yields a number of bands due to accompanying vibrational transitions and each band is made up of a number of closely spaced lines due to accompanying rotational transitions. For a given electronic transition, the whole set of bands is called a band system. Even for a diatomic molecule, the electronic spectra may consist of several band systems.

Only those electronic transitions are allowed which involve states of like multiplicity (i.e. states with equal number of unpaired electrone). Violation of this rule happens only with molecules exposed to strong magnetic fields or ones containing an atom having high nuclear charge which exerts such a field. However, unlike pure vibrational transition, there is no simple restriction regarding the change in vibrational quantum number in an electronic transition. Here the change in vibrational quantum as high as ten is not uncommon though the usual selection rule for rotational transition is obeyed.

Franck–Condon Principle

In any electronic transition from a given vibrational level, the change of vibrational quantum number can be predicted by a rule known as the Franck–Condon principle. According to it, the positions and velocities of the atomic nuclei in a molecule remain virtually unchanged during an electronic transition. Therefore, an electronic transition may be represented by a vertical line drawn between the potential energy curves for the two electronic states of the molecule involved. The most probable transitions will be those starting from the extreme positions of the nuclei for any given vibrational level (except the lowest vibrational level for which the transition takes place from the mid point of the line of vibration) where the nuclei spend their maximum time. Two such transitions between electronic states A and B from the vibrational level $v' = 1$ of A are shown in Fig. 19.2. Here the transitions will be most intense with $v'' = 0$ and 4. Clearly, the change of vibrational quantum number will depend on the relative shapes and

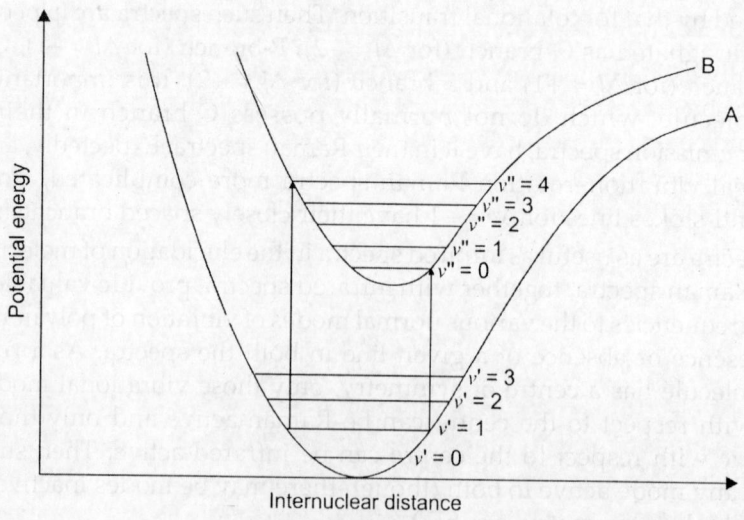

Fig. 19.2

positions of the upper and lower potential energy curves. This explains why there is no simple selection rule for vibrational transitions accompanying electronic transitions. Also we can understand the more complex nature of electronic spectra for emission (that can occur from quite a large number of vibrational levels) than for absorption (that occurs mostly from the ground vibrational level).

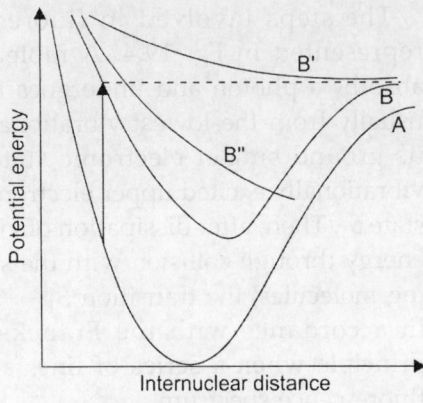

Fig. 19.3

Dissociation and Predissociation

If the molecular potential energy curves for the two electronic states A and B are related as shown in Fig. 19.3, the molecule will dissociate due to the transition shown and of course due to those involving absorption of higher energy. The electronic spectra in such cases will therefore, consist of a number of bands which converge and ultimately pass into a continuous region showing no band structure due to vibration. From the position of the convergence limit where the continuous part of the spectrum commences, the bond dissociation energy can be calculated.

If the potential energy curve for the upper electronic state has no minimum such as B′, then every possible transition to this state from lower electronic state will result in dissociation of the molecule. Then the electronic absorption spectrum for such cases will show no band structure at all.

If, however, the potential energy curve B″ intersects the curve B as shown in Fig. 19.3, the electronic transition to B near the point of intersection will lead to dissociation of the molecule through internal conversion to B″ state. [But this will not happen if the transition occurs to a vibrational level much below or much above (up to certain limit) the region of intersection]. In this situation, the electronic spectrum contains an intermediate blurred region. Such a spectrum is called predissociation spectrum.

Deactivation of Electronically Excited species

An electronically excited molecule often undergoes non-radiative deactivation to lower electronic state by transformation of its excitation energy into thermal energy of the surrounding molecules through collission. An electronic transition of this type is called internal conversion or intersystem crossing according as it happens without or with change in multiplicity of the state.

The other mode of deactivation is through less common radiative transition—fluorescence and phosphorescence.

Fluorescence

The emission of electromagnetic radiation by a species due to its transition between electronic states of same spin multiplicity is called *fluorescence*. Such transitions being allowed, the first order rate constant for fluorescence is very high (10^6–10^9 sec^{-1}) and hence the fluorescence ceases immediately after the exciting source is removed.

The steps involved in fluorescence is represented in Fg. 19.4. A molecule first absorbs a photon and undergoes transition usually from the lowest vibrational state of its ground singlet electronic state S_0 to a vibrationally excited upper electronic singlet state S_1. Then, after dissipation of vibrational energy through collision with the surrounding molecules, the transition $S_1 \rightarrow S_0$ occurs in accordance with the Franck–Condon principle when a series of lines appear as fluorescence spectrum.

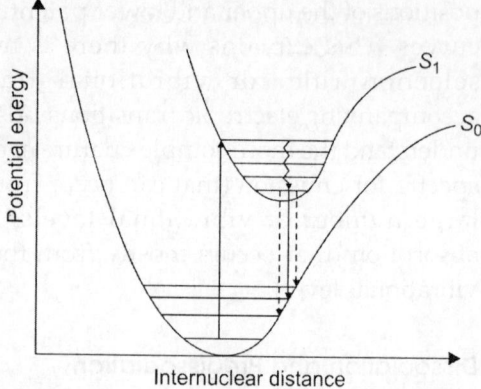

Fig. 19.4: ⌇⌇➔ represents a non-radiative process → and ----➤ represent radiative processes of strong and weak intensities

Phosphorescence

The emission of electromagnetic radiation by a species due to its transition between electronic states of different spin multiplicity (usually, Triplet → Singlet) is called phosphorescence. Such transitions being formally forbidden, phosphorescence (unlike fluorescence) does not cease immediately after the exciting radiation is cut off.

The sequence of events leading to phosphorescence is represented in Fig. 19.5. Here intersystem crossing is a vital step.

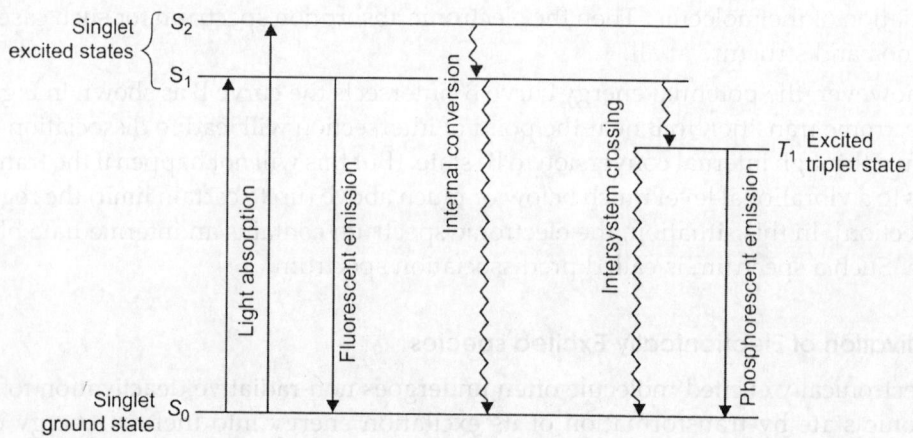

Fig. 19.5: Schematic diagram in terms of electronic energy levels, known as Jablonski diagram representing fluorescence and phosphorescence

Importance of Electronic Spectra

Electronic spectra of gases, though very complicated, can be used, at least for simple molecules, to calculate the length and force constant of a bond from the analysis of the vibrational bands in such spectra, particularly when the vibrations are not Raman or infrared active.

For polyatomic molecules, the electronic spectra are so complicated that they are useful mostly for qualitative purpose in elucidating the molecular structure from the position and intensity of a band due to a group of atoms.

For organic compounds, electronic absorption spectra usually arise due to promotion of electrons of different groups from n (nonbonding), σ or π orbitals in the ground state to antibonding orbitals (σ^* or π^*) in the excited state as given in Table 19.2.

Table 19.2

Groups	Electronic transition	Band position	Band intensity
Single bonded group like C–C, C–H	$\sigma \to \sigma^*$	Vacuum ultraviolet (< 180 nm)	Strong
Single bonded grs containing non-bonding electrons	$n \to \sigma^*$	Ultraviolet (> 180 nm)	Strong
Unsaturated grs like C=C, C=O, –N=N–, –N=O	$\pi \to \pi^*$	Above 160 nm	Strong
Unsaturated grs containing non-bonding electrons	$n \to \pi^*$	Around 280 nm	Weak

With increase in number of unsaturated groups (called chromophores) in the molecule absorption occurs at longer wavelength and the effect becomes greater when the groups are in conjugation.

AUXILIARY PROBLEMS

19.1 The translational motion of a molecule has no spectroscopic importance. Why?

19.2 NMR spectra, unlike other kinds of spectra, can be obtained only in presence of an externally applied magnetic field. Explain.

19.3 Electronic energy levels are calculated using Born–Oppenheimer idea that electronic and nuclear motion can be treated independently. Is this justified.

19.4 Which spectral line will be broader—one due to short-lived or long-lived excited state?

19.5 In which case the width of spectral will be less affected by temperature—electronic transition or rotational transition?

19.6 Show that the moment of inertia of a diatomic molecule AB composed of atoms of masses m_A and m_B and bond length R is equal to μR^2 where μ is the reduced mass of the molecule.

19.7 Possession of dipole moment is a necessary but not sufficient condition for a molecule to have any rotational spectrum due to a particular rotation. Comment.

19.8 Rotation of a polar molecule cannot produce any electric field since the fields due to the equal and opposite charges in the molecule will be mutually cancelled. Comment.

19.9 Can energy of a rigid rotor be zero? Justify your answer.

19.10 With increasing rotation would the gap between the rotational energy levels of a rigid rotor increase or decrease? What will be the degeneracy of the level with $J = 7$?

19.11 A non-polar molecule cannot emit radiation due to its rotation. Comment. Discuss whether determination of equilibrium vibration frequency of a diatomic molecule is possible from the study of its rotational spectrum.

19.12 Consider a charged particle in a box in ground state ($n = 1$). Are the transitions to $n = 2, 3, 4, 5$ all allowed?

19.13 Spontaneous emissions occur in insignificant proportion compared to stimulated emissions. Comment.

19.14 The vibrational potential energy $V(r)$ of real diatomic molecules due to their anharmonic vibration can be effectively expressed by the following empirical function of internuclear distance (r) suggested by Morse

$$V(r) = D_e [1 - e^{-\alpha(r - r_e)}]^2$$

where D_e and α are characteristic constants for a molecule (suffix refers to equilibrium situation)

(a) show that for small displacements from the equilibrium position, the above function is approximated to a simple harmonic potential.

(b) Calculate $V(r_e)$, $V(\alpha)$ and α.

(c) Comment on the values of V at $r = 0$ and $r = \infty$.

19.15 Is dissociation of a molecule possible if it acts as a harmonic oscillator?

19.16 How many normal modes of vibration CO_2 and O_3 can have? Arrange them in order of their frequency in case of CO_2. Are these mutually independent?

19.17 Find the reduced mass of (i) O_2 and (ii) CO_2 for the symmetric stretching mode. From the result you find predict whether O_2 or CO_2 will have higher frequency of vibration. Which mode of CO_2 will have highest reduced mass?

19.18 Strength of a chemical bond should be measured by its force constant and not frequency. Justify.

19.19 State giving reasons which bond in each of the following pairs will have higher vibrational frequency, higher strength and higher reactivity?

(a) C=C stretching and C≡C stretching

(b) C–H stretching and C–D strecthing

19.20 The radiation emitted or absorbed by a diatomic molecule will have same frequency as the molecule. Comment.

19.21 What is meant by the equilibrium vibrational frequency of a diatomic molecule? Is this same as the lowest frequency of the lines in its vibrational spectrum?

19.22 Rotational constant (B) can be calculated from pure rotational spectra and also from vibration–rotation spectra. Which method (i) is more convenient (ii) gives more accurate result.

19.23 Is rotational constant in vibrationally excited state greater or smaller than in ground vibrational state for a real diatomic molecule?

19.24 Define spectroscopic dissociation energy (D_e) and chemical dissociation energy (D_o). How are they related? Establish a relation between D_e and x_e. Also show that $v_1/v_2 = D_o/D_e$, where v_1 is the frequency of the spectral line due to fundamental transition and v_e is the equilibrium vibrational frequency of the molecule.

19.25 Vibrational spectra of molecules in condensed phase, unlike gas phase, consist of broad lines with no rotational structure. Explain.

19.26 A molecule can gain or lose energy from an electromagnetic radiation of frequency v only if hv equals the difference between two molecular energy levels. Comment.

19.27 Raman effect was not evident before the invention of laser. Justify.

19.28 Raman effect is conveniently studied using incident radiation in the visible or ultraviolet region. Can infrared radiation be used?

19.29 For Raman spectrum, the optical systems are made of glass or quartz, and water can be used as a solvent. Are these permitted for IR spectrum?

19.30 Experimental substances are taken as concentrated solution for Raman spectrum, but usually as dilute solution for IR spectrum. Why?

19.31 The usual selection rule is $\Delta J = \pm 2$ for rotational Raman transition but $\Delta J = \pm 1$ for ordinary rotational transition. Justify.

19.32 Normally Raman spectra are simplier than IR spectra. Why?

19.33 Justify, citing example that a given normal mode may be (a) IR active but Raman-inactive (b) Raman-active but IR-inactive (c) IR and Raman-active (d) IR- and Raman-inactive.

19.34 Give example (if any) of each of the following where vibrational modes of a molecule are all (a) IR-active (B) Raman-active (c) IR-inactive (d) Raman-inactive (e) both IR- and Raman-active.

19.35 Represent diagramatically the different vibrational modes of CO_2 and H_2O. Which of them are IR-active and which are Raman-active.

19.36 Which of the following molecules will have a vibrational mode that is only IR-active or only Raman-active?

NH_3, NF_3, CH_4, C_2H_4 and C_6H_6.

19.37 Predict the structure of NO_2^+ ion from electronic theory of valency and examine whether your prediction is in agreement with its following spectroscopic data. It exhibits one vibrational frequency at 1400 cm^{-1} which is Raman-active but IR-inactive and two frequencies at 2360 and 540 cm^{-1} which are IR-active but Raman-inactive. To which vibrational modes do these frequencies correspond?

19.38 Unlike pure vibrational transition, there is no simple restriction regarding the change in vibrational quantum number in an electronic transition, though the usual selection rule for rotational transition is obeyed. Explain.

19.39 How can you find information about molecular vibrational and rotational constants if the molecule is neither Raman-active nor IR-active?

1940 Why electronic spectra are much more complex for emission than for absorption? Which one is useful in studying the vibrational characteristics of the ground electronic state?

19.41 The electronic spectra of liquid or solid samples have no vibrational structure. Why?

19.42 (a) Comment on the following transitions: (i) $n \rightarrow n^*$ (ii) $\sigma \rightarrow \pi^*$.

(b) Which transition will occur at lower wavelength—$n \rightarrow \pi^*$ or $\pi \rightarrow \pi^*$?

(c) What is the nature of transition (i.e. $n \rightarrow \pi^*$, $\pi \rightarrow \pi^*$ etc.) associated with minimum energy change in (i) C_2H_4 (ii) HCHO and (iii) C_6H_6.

19.43 Which compound will exhibit electronic absorption maximum at longer wavelength—2-butene or 2,4-hexadiene? Jutify your answer.

19.44 Fluorescent radiation has generally a lower frequency than the incident radiation. Why?

19.45 The intensity of fluorescence in solution depends on the nature of the solvent. Explain. In which solvent the fluorescence will be less intense—H_2O or $SOCl_2$?

19.46 Unlike fluorescence, the phosphorescence does not cease immediately after the exciting radiation is cut off. Why?

19.47 The intensity of phosphorescence increases when the phosphorescent molecule contains a heavy atom. Why?

19.48 A vital step in phosphorescence is intersystem crossing which occurs quite rapidly though it involves a change in spin multiplicity. How is this possible?

19.49 The phosphorescence is most intense with solid samples and is seldom observed in liquids. Explain.

19.50 The phosphorescence spectrum of a substance is observed at longer wavelength than its fluorescence spectrum. Why?

19.51 Raman effect has similarity with fluorescence in that in both cases light is produced that has frequency different from that of the incident light. In what important respects do they differ?

19.52 How does fluorescence/phosphorescence differ from chemiluminescence and luminescence in a gas-discharge tube or television screen?

19.53 Sometimes a molecule dissociates when it is excited to an electronic state though the energy is insufficient for the dissociation in that state. How?

19.54 Both phosphorescence and predissociation phenomena involve crossing of potential energy curves. In what important respect does the nature of crossing differ in the two cases?

19.55 Based on moment of inertia about the principal axes a, b and c of the molecules, labelled as $I_a \leq I_b \leq I_c$, following classification can be made (assuming rigid molecular structure).

Moment of inertia	Class of molecule
$I_a = 0$, $I_b = I_c$	Linear
$I_a = I_b = I_c$	Spherical top
$I_a = I_b > I_c$	Prolate $\Big\}$ Symmetric top
$I_a = I_b < I_c$	Oblate
$I_a \neq I_b \neq I_c$	Asymmetric top

Classify the following molecules into the above categories:

H_2O, NH_3, BCl_3, HCN, CH_4, C_6H_6, CH_2Cl_2, CH_3Cl, $CHCl_3$ and SF_6.

19.56 Which of the following molecules can show (i) microwave spectrum (ii) infrared spectrum (iii) Raman spectrum and (iv) none.

He, H_2, HD, O_2, HCl, CH_4, CH_3Cl, CH_2Cl_2, CO_2, H_2O, H_2O_2, C_2H_2, C_2H_6 and SF_6.

19.57 Certain transition occurs at 500 mμ. If there are 1000 molecules in the ground state, what is the approximate equilibrium population of the excited state at 300 K.

19.58 The wavenumbers of $J = 0$ to $J = 1$ transition for ^{35}Cl and $D^{35}Cl$ are 20.878 and 10.784 cm^{-1} respectively. Estimate the isotope effect on bond length. The atomic masses of ^1H, D and ^{35}Cl are 1.0078 u, 2.0140 u and 34.9688 u respectively.

10.59 Should rotational constant B change due to rotational transition? If B changes by 16% in a fundamental transition of a diatomic molecule, what will be the change in bond length?

10.60 At 300 K, the rotational spectrum of CO consists of a series of equivalent lines spaced 3.8423 cm^{-1} apart. What is (i) the moment of inertia and (ii) the bond length of the molecule? Which transition will be more intense—$J = 7 \rightarrow J = 8$ or $J = 8 \rightarrow J = 9$? Which transition will be most intense?

[Given H = 1.008, C = 12, O = 15.999 and absolute mass of H-atm = 1.6734 × 10^{-24} gm.

19.61 Three consecutive lines in the microwave spectrum of $H^{79}Br$ are observed at 84.544, 101.355 and 118.112 cm^{-1}. Assign the lines to their appropriate rotational transition. Also calculate the rotational constant and the centrifugal distortion constant.

19.62 The line of lowest frequency observed in the rotational spectrum of HBr appears at 16.939 cm^{-1}. Two other lines are observed at 84.544 cm^{-1} and 118.112 cm^{-1}. To what rotational transitions are these lines due? Does HBr molecule behave as rigid rotor? Calculate equilibrium vibration frequency of HBr.

19.63 In the rotational spectrum of HF, the wavenumber of the transitions follow the following equation

$$\bar{v}(cm^{-1}) = 41.122\,(J+1) - 8.52 \times 10^{-3}\,(J+1)^3$$

Calculate the vibrational frequency and force constant of the H–F bond.

19.64 If the spacing between the rotational spectral lines for $^{12}C^{16}O$ be 3.8423 cm^{-1}, find that for $^{13}C^{16}O$. Also find the gap between the lines of maximum intensity for P and R branch of $^{12}C^{16}O$ at same temperature.

19.65 For $^{12}C^{16}O$, Q branch centres at 2143 cm^{-1} and the gap between the lines of maximum intensity for P and R branches is 60 cm^{-1} at 27°C. Find rotational constant, bond length and zero point energy.

19.66 In the spectrum of CO due to vibration–rotation interaction, the wavenumber of the fundamental vibration transition follows the following equation

$$\bar{v}\,(cm^{-1}) = 2140.28 + 3.813\,J - 0.0175\,J^2$$

Calculate the equilibrium vibration frequency equilibrium rotational constant and the constant which represents the change of rotational constant with vibrational levels.

19.67 For the fundamental absorption bond of HI, the first two lines in the P-branch appear at 2217 and 2204 cm^{-1} and the first two lines in the R branch at 2242 cm^{-1} and 2254 cm^{-1}. Find the moment of inertia and the internuclear distance of HI for the state $v = 0$ and $v = 1$.

19.68 The fundamental vibration frequency of $H^{35}Cl$ is found to be 8.667×10^{13} s^{-1}. What would be the separation between the infrared spectral lines for $H^{35}Cl$ and $H^{37}Cl$ if the force constants for the bonds are assumed to be the same?
[At mass of $^{35}Cl = 58.06 \times 10^{-27}$ kg, $^{37}Cl = 61.38 \times 10^{-27}$ kg, H = 1.673×10^{-27} kg].

19.69 The fundamental and first overtone transitions of $^{14}N^{16}O$ are centred at 1876 cm^{-1} and 3724 cm^{-1} respectively. Calculate the equilibrium vibration frequency, the force constant, anharmonicity constant and exact zero point energy of the molecule.
Also find the intensity of the hot band relative to that of the fundamental at 27°C.

19.70 The IR spectrum of $H^{35}Cl$ shows a very intense absorption at 2886 cm^{-1}, a weaker one at 5668 cm^{-1} and a very weak one at 8347 cm^{-1}. Calculate the equilibrium vibration frequency, anharmonicity constant, equilibrium dissociation energy and dissociation energy level.

19.71 The fundamental frequency and chemical dissociation energy of H_2 molecule are 4395 cm^{-1} and 4.4763 ev. Find spectroscopic dissociation energy of H_2. Also find the chemical and spectroscopic dissociation energies of D_2 molecule if it has the same force constant.

19.72 The fundamental vibration frequencies are 2885 cm^{-1} for HCl, 1990 cm^{-1} for DCl, 2990 cm^{-1} for D_2 and 3627 cm^{-1} for HD. Find the heat of the reaction HCl + $D_2 \rightarrow$ DCl + HD, when all the species involved are at their zero point energy levels. Mention the assumption involved.

19.73 When a certain compound is irradiated with radiation of wavelength 436 nm, Raman lines are found at 440, 445 and 451 nm. Calculate the Raman frequencies. At which wavelengths the absorption in the infrared region may be expected?

19.74 (a) Which of the following quantities are related to half-life of the emitting species position, width and intensity of the lines in its emission spectra?

(b) A substance fluoresces at a wavelength of 400 nm with a half-life of 1.00×10^{-8} sec. Also it phosphoresces at 500 nm. The transition probabilities for stimulated emission in the two cases are in the ratio $10^5:1$. What is the half-life of the phosphorescent state?

ANSWERS

19.1 Because, ordinarily the translational energy levels of a molecule are extremely closely spaced (virtually continuous) as it appears from the expression of energy for particle in a box. [vide problem 18.77(a)].

19.2 An atomic nucleus cannot have an intrinsic electric dipole moment (though it can have an electric quadrupole moment) and hence it cannot interact appreciably with an electric field. But some nuclei, having intrinsic spins and hence magnetic dipole moments, can interact with the magnetic field in electromagnetic radiation. NMR spectroscopy is based on this, while all other kinds of spectroscopy are based on the interaction of electric dipole moment with electromagnetic radiation. NMR spectra can be obtained only in presence of a magnetic field of high flux density to make nuclear-spin energy level separation necessarily equal to a quantum of the concerned radiation.

Note: Proton and neutron each have spin $\frac{1}{2}$, and the spin of an atomic nucleus may be thought of as the resultant of the spins of the constituent protons and neutrons. All nuclei with odd number of proton or neutron have spin with spin quantum number I equal to odd integral multiple of $\frac{1}{2}$ (i.e. $\frac{1}{2}, \frac{3}{2}$ etc.) and those with odd number of both proton and neutron have I equal to even integral multiple of $\frac{1}{2}$ (i.e. 1, 2, 3 etc.). The other nuclei (i.e. containing even number of both proton and neutron, such as ^{12}C and ^{16}O) do not have spin and hence do not interact with electromagnetic radiation and no NMR spectrum.

19.3 This is amply justified on the ground that the nuclei (being of much greater mass) move far more slowly than the electrons so that for the given purpose the nuclear motion may be ignored.

19.4 One due to short-lived excited state because of greater uncertainty in its energy.

19.5 Electronic transition. Because it takes place in shorter time than rotational transition (due to faster motion of electron than nucleus). The same is reflected by Eq. (19.5), λ being lower with electronic transition.

19.6 Let us consider the rotation of a diatomic molecule AB about an axis passing through its centre of mass and perpendicular to its bond.

$$A \bullet \xleftarrow{\quad r_A \quad} C \xrightarrow{\quad r_B \quad} \bullet B \qquad r_A + r_B = R$$

Here

$$\mu R^2 = \frac{m_A m_B}{m_A + m_B}(r_A + r_B)^2 = \frac{m_A m_B}{m_A + m_B}(r_A^2 + 2r_A r_B + r_B^2)$$

$$= \frac{m_A m_B r_A^2}{m_A + m_B} + \frac{(m_A r_A)^2}{m_A + m_B} + \frac{(m_B r_B)^2}{m_A + m_B} + \frac{m_A m_B r_B^2}{m_A + m_B} \qquad \because m_A r_A = m_B r_B$$

(considering moment about c at equilibrium)

$$= m_A r_A^2 + m_B r_B^2$$

Note: Through this relation, a two-particle problem can be looked as a one-particle one.

19.7 The necessary condition for a molecule to have a rotational spectrum due to rotation about any of its axes is that its dipole moment must change during this rotation. This requirement may not be fulfilled even with a polar molecule as in case of rotation of a polar symmetric top molecule, like CH_3Cl, about its symmetry axis along which dipole moment is directed. Then the given statement is justified.

19.8 The given statement is false. Because here the fields produced due to equal and opposite charges of the dipole act in the same direction.

19.9 Yes. This is justified by the uncertainty principle (vide problem 18.68). When energy, and hence angular momentum, of the rigid rotor is definite (equal to zero), its orientation in space is completely uncertain.

Note: But energy of a harmonic oscillator cannot be zero. Vide problem 18.89(a).

19.10 Would increase, as it appears from

$$E_{j+1} - E_J = 2Bh(J+1) \quad \text{by Eq. (19.6)}$$
$$\text{Degeneracy of } E_J = 2J + 1$$
$$= 2 \times 7 + 1 \quad \text{for } J = 7$$

Note: Gap between consecutive energy levels of a system may increase, decrease (as in anharmonic oscillators) or remain same (as in harmonic oscillators). Vide university question 19.3.

19.11 A non-polar molecule cannot radiate through involvement of electric dipole (which is absent). However, some feeble emission of radiation might happen through involvement of magnetic dipole/electric quadrupole. Vide Section 19.1.

☐ Yes. This is possible using Eq. (19.9). Vide problems 19.62 and 19.63.

19.12 For effective transition of a particle between two stationary states ψ_m and ψ_n, the transition dipole moment μ_{mn} need be nonzero. For the given one-particle one-dimensional problem, we have, using wavefunctions for PIB [Eq. (18.12)].

$$\mu_{mn} = \frac{2q}{a} \int_0^a x \sin\frac{m\pi x}{a} \sin\frac{n\pi x}{a}$$

[by Eq. (19.2) where q is charge of the particle]

$$= \frac{2q}{a} \int_0^a \frac{x}{2}\left[\cos\frac{\pi}{a}(m-n)x - \cos\frac{\pi}{a}(m+n)x\right]$$

$$\text{Now} \quad \int_0^a x\cos\frac{\pi}{a}(m-n)x = \left|-\frac{a}{\pi(m-n)}x\sin\frac{\pi(m-n)}{a} + \frac{a^2}{\pi^2(m-n)^2}\cos\frac{\pi(m-n)}{a}x\right|_0^a$$

$$= \frac{a^2}{\pi^2(m-n)^2}[\cos(m-n)\pi - 1]$$

Similarly

$$\int_0^a x\cos\frac{\pi}{a}(m+n)x = \frac{a^2}{\pi^2(m+n)^2}[\cos(m+n)\pi - 1]$$

Then

$$\mu_{mn} = qa\left[\frac{\cos(m-n)\pi - 1}{\pi^2(m-n)^2} - \frac{\cos(m+n)\pi - 1}{\pi^2(m+n)^2}\right]$$

So, for μ_{mn} to be nonzero m and n must not be both even or both odd. Hence the allowed transitions are $1 \to 2$ and 4, but not $1 \to 3$ and 5.

Note: The spectral selection rule for the particle-in-a-box is $\Delta n = \pm 1, \pm 3, \pm 5, ...$, for charged particles.

19.13 The given statement is justified only for emission at low frequencies like microwave or radiofrequency. Vide Eq. (19.4).

19.14 (a) For small displacement

$$V(r) \simeq D_e \left[\alpha (r - r_e) \right]^2$$
$$= \tfrac{1}{2} k (r - r_e)^2 \quad \text{where } k = 2 D_e \alpha^2$$

which corresponds to simple harmonic potential.

(b) $V(r_e) = 0$, putting $r = r_e$ in the given expression
$V(\infty) = D_e$, putting $r = \infty$ in the given expression

$$\alpha = \left(\frac{k}{2 D_e} \right)^{\frac{1}{2}} = \pi v_e \left(\frac{2\mu}{D_e} \right)^{\frac{1}{2}}$$

where $v_e = \dfrac{1}{2\pi} \left(\dfrac{k}{\mu} \right)^{\frac{1}{2}}$ is the equlibrium vibration frequency of the molecule.

(c) $V(r_e) = 0$ corresponds to minimum potential at equilibrium
$V(\infty) = D_e$ corresponds to maximum potential at the dissociated state

19.15 No. Because k of harmonic oscillators (unlike the anharmonic ones) cannot be zero however high be their vibrational energy.

19.16 CO_2 and O_3 have same atomicity (x) equal to 3, but different number of normal vibrational modes due to different molecular shape.

For CO_2 molecule having linear structure, number of normal modes of vibration
$$= 3x - 5$$
$$= 3 \times 3 - 5 = 4$$

For O_3 molecule having non-linear structure, number of normal modes of vibration
$$= 3x - 6$$
$$= 3 \times 3 - 6 = 3$$

☐ For CO_2 four normal modes may be represented as under

Here $v_2 > v_1 > v_3 = v_4$. Lower is the frequency of a mode higher is the ease with which it can be excited. v_3 and v_4 differ only in that the atomic nuclei move in mutually perpendicular planes and are therefore degenerate.

☐☐ Yes, they are defined as such.

Note: The representation of the normal modes of CO_2 done above is not the only way of doing it, but one of the several ways allowed by the molecular symmetry. What concerns us is the number of modes and not the way of their representation (just like number of components of a macroscopic system). In representing normal modes, it should be kept in mind that they are independent synchronous motion of atoms in a molecule that do not result in displacement of centre of mass of the molecule or rotation of the molecule as a whole.

19.17 For a normal mode, the reduced mass of a molecule is the one that is swung about by the vibration and is, in general, a complicated function of the masses of the vibrating atoms in the molecule. In case of CO_2, the reduced mass (μ) for symmetric stretching mode is, however, determined simply by the masses of the O-atoms as under (there being no contribution from the C-atom which is stationary in this mode)

$$\frac{1}{\mu} = \frac{1}{m_0} + \frac{1}{m_0}$$

or $\qquad \mu = \frac{m_0}{2}$

This is same as for O_2.

☐ CO_2 has higher force constant (k) (C=O being a stronger bond than O=O) and hence higher frequency $v = \frac{1}{2\pi}\left(\frac{k}{\mu}\right)^{\frac{1}{2}}$.

☐☐ μ of CO_2 is highest for the bending modes which is given by

$$\frac{1}{\mu} = \frac{1}{m_c} + \frac{1}{2m_0}$$

Because here the motions of the two O-atoms are alike and hence the molecular vibration amounts to a vibration of two particles, one of mass m_c and another of mass $2m_0$. This is unlike asymmetric stretching where μ is given by

$$\frac{1}{\mu} = \frac{1}{m_c} + \frac{1}{m_0} + \frac{1}{m_0}$$

Note: μ and k of a molecule depend on its normal mode. Again, k is a characteristic constant of a molecule only for equilibrium state corresponding to minimum potential energy to which k is usually referred (sometimes indicated using the suffix e).

19.18 A chemical bond between two atoms signifies a force that holds the atoms together, and hence its strength (i.e. effectiveness) should be linked directly to the force constant (k) and not the frequency (v) of the bond. Because v, unlike k, depends also on the reduced mass, apart from chemical nature of the bonded atoms.

19.19 (a) In the given two bonds, which involve like atoms, C≡C having higher bond order will have higher strength. Again, this implies higher k and hence higher vibrational frequency $v = \frac{1}{2\pi}\sqrt{\frac{k}{\mu}}$, μ being same for both C≡C and C=C. But regarding reactivity (which is a kinetic factor), there is no unique answer. For example, C≡C has higher reactivity than C=C in reaction with H_2, while the reverse happens with Br_2.

Note: Higher bond order does not necessarily imply higher strength, if the bonds involve unlike atoms. For example, O–H (bond energy $\approx$ 461 kJ mol^{-1}) has higher bond strength than O=O (bond energy $\approx$ 402 kJ mol^{-1}), though the latter has higher bond order.

(b) Here the bond strength, and hence k, is same for the given two bonds (due to same chemical nature of the bonded atoms and same bond order). But v is higher for C–H due to its lower reduced mass. C–H has higher reactivity due to lower activation energy (due to higher zero-point energy) in the (substitution) reaction, compared to C–D.

Note: (a) and (b) differ in question of reactivity. This happens due to difference in the type of reaction—addition in case of (a) but substitution (less complicated) in case of (b).

19.20 The given statement is not justified even for a hypothetical diatomic molecule behaving as a harmonic oscillator for which vibrational energy levels are equally spaced, due to various rotational transitions accompanying a vibrational transition.

19.21 Equilibrium vibration frequency of a molecule is the hypothetical frequency (v_e) for a harmonic oscillator $\left(v_e = \frac{1}{2\pi}\sqrt{\frac{k_e}{\mu}}\right)$ corresponding to the force constant (k_e) of the concerned molecule at its minimum potential energy of oscillation.

☐ No. Vide Eq. 19.15.

Note: For molecules having same anharmonicity constant (e.g. H_2 and D_2), $v_1 \propto v_2$.

19.22 (i) One using vibration–rotation spectra. Because such spectra lie in the infrared region which is experimentally quite convenient. This is unlike microwave region of pure rotational spectra.

(ii) One using pure rotational spectra. Because the microwave region of such spectra can be more thoroughly analysed due to sufficiently high resolution in this region of high wavelength.

19.23 In the vibrationally excited state, the bond length (and hence moment of inertia) is higher and hence rotational constant is lower, due to anharmonicity of vibration for a real molecule.

19.24 For definition, vide Section 19.3.

☐ $$D_e - D_0 = E_{v=0}$$

Then, assuming harmonic oscillation of the molecule

$$D_e - D_0 = \tfrac{1}{2}hv_e \quad \text{by Eq. (19.11)}$$

But, considering anharmonicity of oscillation

$$D_e - D_0 = \tfrac{1}{2}hv_e\left(1 - \tfrac{1}{2}x_e\right) \quad \text{by Eq. (19.14)}$$

☐☐ For dissociation energy level

$$D_e = \left(v + \tfrac{1}{2}\right)hv_e - x_e\left(v + \tfrac{1}{2}\right)^2 hv_e \quad \text{by Eq. (19.14)}$$

Since $\dfrac{dE}{dv} = 0$ for dissociation energy level,

$$[1 - (2v + 1)x_e]\,hv_e = 0$$

or $\qquad v + \tfrac{1}{2} = \dfrac{1}{2x_e}$

Then from the expression of D_e

$$D_e = \frac{hv_e}{4x_e}$$

☐☐☐ $\qquad \dfrac{v_1}{v_e} = 1 - 2x_e \quad \text{by Eq. (19.15)}$

$$= 1 - \frac{2hv_e}{4D_e} = 1 + \frac{4(D_0 - D_e)}{4D_e} \quad \text{by Eq. (19.12)}$$

$$= \frac{D_0}{D_e}$$

Note: $x_e = \dfrac{hv_e}{4D_e} = \dfrac{hv_1}{4D_0}$.

19.25 In vibrational spectra of molecules reasonably sharp rotational lines are found only in gas phase at sufficiently low pressure (below 10 P_a). Here lines being of low frequency, their width is determined dominantly by collisionally controlled lifetime broadening (which is least at low pressure due to low frequency of collision) and Doppler broadening.

In condensed phase Doppler broadening is not considerable (due to low translational motion of the molecules). Here rapid non-radiative collisional deactivation of rotational states (due to their low energy gap) leads to the given observation.

19.26 The given statement is not justified. Because even if the energy of a photon ($h\nu$) of electronegative radiation equals the difference between two molecular energy levels $|E_f - E_i|$, the molecule will exchange energy with the radiation only if some gross spectral selection rule is complied. Again, energy exchange can occur even if $h\nu \neq |E_f - E_i|$, as in Raman scattering.

19.27 The Raman scattering is usually so weak and the Raman shift is so small (10–4000 cm^{-1}) that the confirmation of Raman effect needs careful experiment with sufficiently monochromatic light beam of high intensity as with laser. The given statement is, therefore, quite justified.

19.28 Raman effect (the scattering of radiation at different wavelengths) originates from vibrational–rotational transition of molecules that gives rise to spectra in the infrared region. Then infrared radiation is unsuitable for studying Raman effect because of its absorption, instead of scattering, by a molecule. Although any radiation which is not absorbed by a molecule can in principle be used for this purpose, one in the visible or ultraviolet region is experimentally convenient.

19.29 No. Because glass and quartz appreciably absorb infrared radiation. Again water is opaque to infrared.

Note: Experimentally, Raman spectroscopy is advantageous over IR spectroscopy regarding selection of incident light and necessary equipment.

19.30 To make the Raman lines sufficiently intense. Because for same concentration, Raman lines have far less intensity than the corresponding IR lines.

19.31 This is due to the difference between the two transitions in their gross selection rule, which refers to dipole moment in case of ordinary rotational transition but to polarizability in case of Raman transition. For Raman transition $\Delta J = \pm 2$ is justified by the fact that the polarizability of a molecule returns to its initial value twice in every complete radiation. Also this is in harmony with the conservation of angular momentum in Raman scattering which may be thought to involve two photons—one coming in and the other going out.

19.32 Because the overtone and combination bands are normally too weak to be observed in Raman spectra.

19.33 Justification lies in the difference between the Raman transition and ordinary IR transition in their gross selection rule. Examples for different cases are cited below:

(a) Asymmetric stretching mode and also the bending modes of CO_2.

(b) Symmetric stretching mode of CO_2. The simplest example is provided by a homonuclear diatomic molecule (like H_2).

(c) Each mode in a molecule with no symmetry elements (e.g. CHFClBr) or with a few symmetry elements, other than centre of symmetry (e.g. HCl, H_2O).

(d) Examples are limited only to molecules of quite high symmetry. Mention may be made of the purely bending mode (T_{2u}) of SF_6 where only asymmetric bonds are involved.

Note: (a) and (b) are in accordance with the exclusion rule for centosymmetric molecules.

19.34 (a) For a molecule as mentioned with problem 19.33(c).

(b) Same as in (a).

(c) Only for a homonuclear diatomic molecule.

(d) For no molecule. Because every molecule has at least one Raman-active mode.

(e) Same as (a).

19.35 The three normal modes of H_2O are represented below

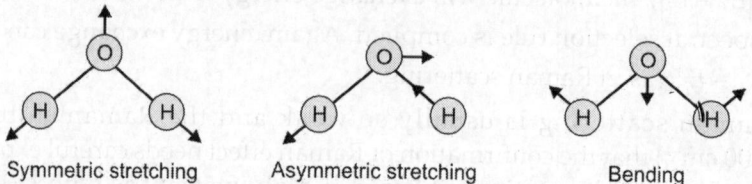

Symmetric stretching   Asymmetric stretching   Bending

For H_2O, all three modes are IR-active and also Raman-active.

For CO_2, vide problems 19.16 and 19.33(a) and (b).

19.36 All excepting NH_3.

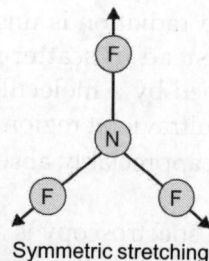

Symmetric stretching

For NF_3 (planar, unlike NH_3) purely stretching (symmetrical) mode, that involves change in polarizability but not dipole moment, is only Raman-active.

For CH_4 (tetrahedral), purely stretching symmetrical and purely bending modes are only Raman-active.

For C_2H_4 and C_6H_6, both being centosymmetric, the exclusion principle provides the answer.

Note: Fulfilment of the requirement in the question does not necessarily imply that relevant molecule is centosymmetric.

19.37 The following centosymmetric linear structure (as with CO_2) is suggested by electronic theory of valency (Lewis and VSEPR).

$$[:\ddot{O}=N=\ddot{O}:]^+$$

This structure is supported by the exclusion rule for a centosymmetric molecule in that none of the given frequencies of the vibrational modes is both infrared- and Raman-active. Further, with the given three frequencies, four possible vibrational modes can be correlated considering double degeneracy of one of them (the bending one).

Then, with the CO_2 spectrum in mind, the Raman-active frequency 1400 cm^{-1} (that involves no change in dipole moment) corresponds to the symmetric stretching mode. Of the infrared-active frequencies (that involves change in dipole moment), the one with larger frequency, i.e. 2360 cm^{-1} corresponds to the asymmetric stretching mode, while the other, i.e. 540 cm^{-1} corresponds to the bending mode.

19.38 The explanation regarding the unusual change in vibrational quantum number in an electronic transition lies in the Franck–Condon principle (vide Section 19.5). Here the compliance of usual selection rule for rotational transition is a consequence of the conservation of angular momentum in emission or adsorption of a photon which has spin 1.

Note: Electronic transitions where the vibrational quantum number changes by as much as 10 are not uncommon. This can be understood readily from potential energy diagrams with a vertical line connecting them. But the underlying mathematical procedure based on Eq. (19.2) is extremely cumbersome.

19.39 From analysis of the rotational structure of the vibrational bands in the electronic spectra of the molecules concerned.

Note: Even if a molecule (homonuclear diatomic) shows no pure-rotational or vibration–rotation spectra, it will show electronic spectra containing vibrational bands with rotational structure. Because in an electronic transition, the molecular structure undergoes appreciable change necessary for vibrational or rotational transition to occur.

19.40 Because electronic transition can occur in emission from quite a large number of vibrational levels while it occurs in absorption mostly from the ground vibrational level (which is most populated) of the ground electronic state.

□ Electronic emission spectra. Because it involves transition to different vibrational levels of the ground electronic state usually from much fewer vibrational levels of the excited electronic state.

19.41 This happens mostly due to overlapping of the vibrational bands, the gap between the vibrational energy levels being appreciably narrowed due to closer proximity of the molecules in the condensed phase compared to the gas phase.

Note: In condensed phase, the electronic spectra have no vibrational structure and vibrational spectra have no rotational structure (problem 19.25), due ultimately to the narrower spacing of the relevant energy levels.

19.42 (a) (i) Irrelevant, there being no antibonding orbital (n^*) associated with n electrons since they do not form any bond.

(ii) Symmetry forbidden, i.e. not allowed from consideration of the symmetry of the relevant molecular orbitals.

(b) $\pi \rightarrow \pi^*$, since π and π^* differ in higher energy (due to higher interaction in bond formation) than n and π^*.

(c) $\pi \rightarrow \pi^*$ for C_2H_4 and C_6H_6.

$n \rightarrow \pi^*$ for HCHO due to $>C=\ddot{O}$: group.

19.43 Here the absorption maximum at longer wavelength will correspond to $\pi \rightarrow \pi^*$ transition. Considering N conjugated π-electrons in a polyene as particles in a one-dimensional box of length Nd, where d is C–C bond length in the conjugated chain, the energy levels will be

$$E_n = \frac{n^2 h^2}{8m(Nd)^2} \quad \text{by Eq. (18.11)}$$

By Pauli exclusion principle, $n = N/2$ for the ground electronic state.

Then λ_{max} absorbed, which corresponds to the lowest excitation energy, will be given by

$$\frac{hc}{\lambda_{max}} = E_{\frac{N}{2}+1} - E_{\frac{N}{2}} = \frac{(N+1)h^2}{8md^2N^2}$$

Therefore, larger λ_{max} will be exhibited by the compound with larger N, i.e. 2,4-hexadiene.

19.44 Because the potential energy curve for the exited electronic state is usually displaced to a greater internuclear distance than the ground state curve (since excited states usually have more antibonding character). Further, the electronic emission transition occurs mostly from the lowest vibrational level of the excited state to the upper vibrational levels of the ground state as a consequence of the Franck–Condon principle (vide Fig. 19.4).

19.45 Because quenching of fluorescence in solution depends on the nature of the solvents due to difference in their ability to strip electrical energy off the excited molecules concerned.

□ H_2O. Because, due to its wider spacing of vibrational energy level, it can absorb higher quantum of energy than $SOCl_2$.

19.46 Because phosphorescence transition (usually triplet → singlet), unlike fluorescence transition, as formally forbidden due to spin inversion.

19.47 With heavy atom, due to its considerable spin–orbit interaction, spin inversion is appreciably allowed.

19.48 Because intersystem crossing involves a non-radiative transition which, unlike a radiative transition, is not restricted by any selection rule.

19.49 Because in case of solids, unlike liquids, collisional dissipation of vibrational energy does not carry the molecule too quickly past the potential crossing point and the intersystem crossing step has ample time to occur.

19.50 Because the potential energy curve for a triplet state is displaced to a greater internuclear distance than the corresponding singlet state curve (to reduce interelectronic repulsion due to same spin).

19.51 Essential differences are:

(i) Fluorescence is the emission of electromagnetic radiation from a real excited state. Raman effect is the scattering of electromagnetic radiation by a molecular species without involving any real (stationary) excited state.

(ii) Fluorescence involves electronic transition accompanied by vibrational and rotational transitions. Raman effect involves vibrational or rotational transition but never electronic transition.

19.52 Generally any emission of light from a species in excited electronic states is termed luminescence. Often it is named specifically depending on the mechanism followed. Thus fluorescence/phosphorescence, where excitation is caused by light absorption, differs from chemiluminiscence and the luminescence in a gas-discharge tube (televation screen) in that the excitations in the latter two cases are caused by chemical reaction and by collision with electron respectively.

19.53 This situation, called predissociation, can arise when potential energy curve for the excited electronic state intersects another potential energy curve having no minimum as in Fig. 19.3, and transition occurs to a vibrational level near the region of intersection.

19.54 In case of phosphorescence, unlike predissociation, the intersected potential energy curves should necessarily correspond to electronic states of different spin multiplicity.

19.55 Linear: HCN.

Spherical top: CH_4 and SF_6.

Symmetric top: Prolate: CH_3Cl (here the principal axis c contains C–Cl bond

Oblate: $CHCl_3$, NH_3, BCl_3 and C_6H_6 (in $CHCl_3$, the principal axis c contains C–H bond)

Asymmetric top: CH_2Cl_2 and H_2O.

Note: Symmetry elements are not essential for a molecule to have a principal axis, though they help in deciding about the principal axes. Thus an axis of symmetry of a molecule coincides with one of its principal axes. A plane of symmetry contains two of the principal axes and is perpendicular to the third. For a molecule with no symmetry elements, the principal axes cannot be found out in a single way.

19.56 (i) Following molecules which possess either unpaired electrons or appreciable permanent dipole moment that changes during rotation.

O_2 (due to unpaired electrons), HCl (but not HD) CH_3Cl, CH_2Cl_2, H_2O, H_2O_2.

(ii) Following molecules whose dipole moment changes during vibration:

HCl, CH_4, CH_3Cl, CH_2Cl_2, CO_2, H_2O, H_2O_2, C_2H_2, C_2H_6 and SF_6.

(iii) All the given molecules, except He, whose polarizability changes during rotation or vibration.

(iv) He. Because it contains no unpaired electrons and is not able to rotate or vibrate, the molecule being monatomic.

Note: The dipole moment of HD is too low to have any spectroscopic importance.

19.57 The population of an excited state i, i.e. the number of molecules N_i in that state, is related to that in the ground state N_o by

$$N_i = N_o e^{-E_i/kT} \quad \text{by Eq. (19.3) taking } g_i = 1$$
$$= N_o e^{-hc/\lambda kT}$$

where E_i is the energy of the molecule in the excited state relative to that in the ground state

$$= (1000)\exp\left[-\frac{(6.63 \times 10^{-34} \text{ JS})(3.00 \times 10^8 \text{ mS}^{-1})}{(500 \times 10^{-9} \text{ m})(1.38 \times 10^{-23} \text{ JK}^{-1})(300 \text{ K})}\right]$$

$$= 1.9 \times 10^{-39}$$

Note:

1. The term *energy state* and *energy level* have different meaning. They can be used synonymously only when the energy levels are all non-degenerate as assumed in this problem.

2. The units m μ (which represents millimicrometer, i.e. 10^{-9} meter) and μm (which represents micrometer, i.e. 10^{-6} meter) are to be differentiated carefully.

19.58 For $J = 0$ to $J = 1$ transition

$$\bar{v} = \frac{v}{c} = \frac{2h}{8\pi^2 Ic} \quad \text{by Eq. (19.7a)}$$

$$= \frac{2h}{8\pi^2 \mu R^2 c}$$

Since $I = \mu R^2$ vide problem (19.6), where R is the bond length

Then $\qquad \dfrac{R_{DCl}}{R_{HCl}} = \dfrac{(\mu\bar{v})^{\frac{1}{2}}_{HCl}}{(\mu\bar{v})^{\frac{1}{2}}_{DCl}}$

$$= \left[\dfrac{\left(\dfrac{1.0078 \text{ u} \times 34.9688 \text{ u}}{1.0078 \text{ u} + 34.9688 \text{ u}}\right)(20.878 \text{ cm}^{-1})}{\left(\dfrac{(2.0140 \text{ u} \times 34.9688 \text{ u})}{2.140 \text{ u} + 34.9688 \text{ u}}\right)(10.784 \text{ cm}^{-1})}\right]^{\frac{1}{2}} = 0.9979$$

Therefore, the isotope effect on bond length is

$$\dfrac{R_{DCl} - R_{HCl}}{R_{HCl}} \times 100 = -0.28\%$$

Note: The isotope effect is small but measurable.

19.59 Yes, because of centrifugal distortion, B is lower for higher rotational level due to higher bond length.

$\square \qquad\qquad B \propto \dfrac{1}{I} \propto \dfrac{1}{R^2}$

Then % change of $B = -2 \times$ % change of R

or $\quad$ % change of $R = -\frac{1}{2} \times$ % change of B

$$= -\frac{1}{2} \times 16\% = -8\%$$

Note: A real molecule does not behave as rigid rotor and hence its rotation cannot be treated purely as such. The so-called rotational spectra invariably contains some vibrational flavour that can be used in evaluating important vibrational parameters. Vide problem 19.63.

19.60 (i) Spacing between spectral lines $= 2B$, where $B = \dfrac{h}{8\pi^2 Ic}$

Then $\quad I = \dfrac{h}{8\pi^2 Bc} = \dfrac{6.63 \times 10^{-27} \text{ arg-sec}}{8(3.14)^2 (3.8423 \text{ cm}^{-1}/2)(3.00 \times 10^{10} \text{ cm sec}^{-1})}$

$$= 14.57 \times 10^{-40} \text{ g} \cdot \text{cm}^2$$

(ii) Bond length $\left(\dfrac{I}{\mu}\right)^{\frac{1}{2}} = \left[I \bigg/ \dfrac{m_c m_o}{m_c + m_o}\right]^{\frac{1}{2}}$

$$= \left[14.57 \times 10^{-40} \text{ g cm}^2 \bigg/ \dfrac{12 \times 15.999}{12 + 15.999} \times \dfrac{1.6734 \times 10^{-24} \text{ g}}{1.008}\right]^{\frac{1}{2}}$$

$$= 1.13 \times 10^{-8} \text{ cm}$$

$\square$ The intensity (P) of the transition from a rotational level J is

$$P \propto (2J + 1)e^{-E_J/kT} = (2J + 1)e^{-J(J+1)BhC/kT}$$

Then for transition from two levels J and J'

$$\dfrac{P_J}{P_{J'}} = \dfrac{2J + 1}{2J' + 1} e^{[J'(J'+1) - J(J+1)]BhC/kT}$$

Then

$$\frac{P_7}{P_8} = \frac{2 \times 7 + 1}{2 \times 8 + 1} \exp \frac{[8(8+1) - 7(7+1)](3.8423 \text{ cm}^{-1}/2)(6.63 \times 10^{-27} \text{ ergs}) \times (3.00 \times 10^{10} \text{ cms}^{-1})}{(1.38 \times 10^{-16} \text{ erg K}^{-1})(300 \text{ K})}$$

$$= 1.2$$

i.e. $P_7 > P_8$.

Note: With increase in J, the pre-exponential factor increases while the exponential factor decreases.

□□ For most intense transition

$$\frac{dP}{dJ} = 0$$

i.e. $\left[2 - (2J+1)^2 \frac{BhC}{kT} \right] e^{-J(J+1)\frac{BhC}{kT}} = 0$

or $2 - (2J+1)^2 \dfrac{BhC}{kT} = 0$

or $$J = \sqrt{\frac{kT}{2BhC}} - \frac{1}{2}$$

$$= \left[\frac{(1.38 \times 10^{-16} \text{ erg K}^{-1})(300 \text{ K})}{(3.8423 \text{ cm}^{-1})(6.63 \times 10^{-27} \text{ erg} \cdot \text{s})(3 \times 10^{10} \text{ cm} \cdot \text{s}^{-1})} \right]^{\frac{1}{2}} - \frac{1}{2}$$

$$= 6.9, \text{ i.e. } 7 \text{ (nearest integer)}$$

19.61 The frequency $(\bar{\nu})$ of any spectral line in the microwave spectra (which are usually studied by absorption of radiation) due to any rotational transition $J \to J + 1$ is

$$\bar{\nu} = 2B(J+1) - 4D(J+1)^3 \quad \text{by Eq. (19.10)}$$

If the second term on the rhs is ignored

$$J \simeq \frac{\bar{\nu}}{2B} - 1$$

Then taking $2B$ equal to the spacing between the consecutive lines, we have for the given first line

$$J \simeq \frac{84.544 \text{ cm}^{-1}}{(101.355 - 84.544) \text{ cm}^{-1}} - 1$$

$$= 4, \text{ taking the nearest integer}$$

Then the first line corresponds to $J = 4$ to $J = 5$ transition. Again, since the lines are consecutive, the second line will correspond to $J = 5$ to $J = 6$ and the third line to $J = 6$ to $J = 7$ transition. Here advantage has been taken of the fact that J cannot be fractional.

□ Using the assignment of lines thus found, we have from the above expression of $\bar{\nu}$

$$84.544 \text{ cm}^{-1} = 2B(4+1) - 4D(4+1)^3 = 10B - 500D \tag{I}$$

$$101.355 \text{ cm}^{-1} = 2B(5+1) - 4D(5+1)^3 = 12B - 864D \tag{II}$$

$$118.112 \text{ cm}^{-1} = 2B(6+1) - 4D(6+1)^3 = 14B - 1372D \tag{III}$$

(II) − (I) and (III) − (II) give

$$16.81 \text{ cm}^{-1} = 2B - 364D \tag{IV}$$

$$16.76 \text{ cm}^{-1} = 2B - 508D \tag{V}$$

(IV) − (V) gives

$$D = \frac{0.05 \text{ cm}^{-1}}{144} = 3.47 \times 10^{-4} \text{ cm}^{-1}$$

whence

$$B = \frac{16.81 \text{ cm}^{-1} + 364D}{2} = 8.47 \text{ cm}^{-1}$$

Note: Calculation of B and D can be done from the frequencies of any two spectral lines. In the above procedure, the frequencies of the three lines have been used to get the better average values of B and D.

19.62 In rotational spectra, the line of lowest frequency corresponds to $J = 0 \rightarrow J = 1$ transition. Then from the given data

$$\bar{v} = 2B = 16.939 \text{ cm}^{-1}, \text{ ignoring } D$$

Since separation between any two of the given lines, much exceeds $2B = 16.939 \text{ cm}^{-1}$, the given spectral lines are not consecutive (unlike the previous problem).

For the 84.544 cm⁻¹ line

$$J \simeq \frac{\bar{v}}{2B} - 1$$

$$\simeq \frac{84.544 \text{ cm}^{-1}}{16.939 \text{ cm}^{-1}} - 1$$

$$= 4, \text{ taking the nearest integer}$$

which corresponds to $J = 4$ to $J = 5$ transition.

For the line 118.112 cm⁻¹

$$J \simeq \frac{118.112 \text{ cm}^{-1}}{16.939 \text{ cm}^{-}} - 1 = 6, \text{ taking the nearest integer}$$

whichcorresponds to $J = 6$ to $J = 7$ transition.

□ No. Because the gap between any two of the given lines is not an exact multiple of $2B = 16.939 \text{ cm}^{-1}$.

□□ To find the equilibrium vibration frequency ($\bar{v}_e$), we are to calculate first B and D as in the previous problem, and then to use the relation

$$\bar{v}_e = \left(\frac{4B^3}{D} \right)^{\frac{1}{2}} \quad \text{by Eq. (19.9)}$$

19.63 As in problem 19.61

$$\bar{v} = 2B(J + 1) - 4D(J + 1)^3$$

Comparing this expression with the given one

$$2B = 41.122 \text{ cm}^{-1}$$

$$4D = 8.52 \times 10^{-3} \text{ cm}^{-1}$$

whence $B = 20.56 \text{ cm}^{-1}$ and $D = 2.13 \times 10^{-3} \text{ cm}^{-1}$.

(Equilibrium) vibration frequency

$$\left(\frac{4B^3}{D} \right)^{\frac{1}{2}} = \left[\frac{4(20.56 \text{ cm}^{-1})^3}{2.13 \times 10^{-3} \text{ cm}^{-1}} \right]^{\frac{1}{2}} = 4039.94 \text{ cm}^{-1}$$

Force constant

$$= 4\pi^2 \mu (\bar{\nu}_e c)^2$$

$$= 4\pi^2 \cdot \frac{m_H m_F}{m_H + m_F} \cdot \bar{\nu}_e^2 c^2$$

$$= 4(3.14)^2 \frac{1.008 \times 19}{1.008 + 19} \times 1.66 \times 10^{-24} \text{ g}) (4039.94 \text{ cm}^{-1})^2 (3 \times 10^{10} \text{ cm sec}^{-1})^2$$

$$= 9.175 \times 10^5 \text{ dyn cm}^{-1}$$

Note: This problem illustrates the evaluation of vibrational parameters of a molecule from its rotational spectrum.

19.64 Since the spacing (d) between the rotational spectral lines is proportional to B which again is inversely proportional to the moment of inertia (I), we have for two different species

$$\frac{d}{d'} = \frac{I'}{I}$$

$$= \frac{\mu'}{\mu}$$

for diatomic species of same bond length, which is nearly true for the given two species

Then $\qquad d\,(^{13}C^{16}O) = d(^{12}C^{16}O) \cdot \dfrac{\mu(^{12}C^{16}O)}{\mu(^{13}C^{16}O}$

$$= (3.8423 \text{ cm}^{-1}) \cdot \left(\frac{12 \times 16}{12 + 16} \right) \bigg/ \left(\frac{13 \times 16}{13 + 16} \right) = 3.6734 \text{ cm}^{-1}$$

□ For lines of maximum intensity

$$J_{max} = \sqrt{\frac{kT}{2Bhc}} - \frac{1}{2} \quad \text{vide problem 19.60}$$

$$= \left(\frac{(1.38 \times 10^{-16} \text{ erg K}^{-1}) (300 \text{ K})}{(3.8423 \text{ cm}^{-1}) (6.63 \times 10^{-27} \text{ erg} \cdot \text{s}) (3 \times 10^{10} \text{ cm s}^{-1})} \right)^{\frac{1}{2}} - \frac{1}{2}$$

$$= 6.9, \text{ i.e. 7 (nearest integer)}$$

The required gap $= 2B\,(J_{max} + 1) + 2B\,J_{max} \quad$ by Eq. (19.17)

$$= 2B\,(2J_{max} + 1)$$

$$= d\,(2J_{max} + 1)$$

$$= (3.8423 \text{ cm}^{-1})\,(2 \times 7 + 1)$$

$$= 37.634 \text{ cm}^{-1}$$

19.65 Gap between the lines of maximum intensity for P and R branches is

$$2B\,(2J_{max} + 1) = 2B \left(\frac{2kT}{Bhc} \right)^{\frac{1}{2}}$$

$$= \left(\frac{8kTB}{hc} \right)^{\frac{1}{2}}$$

From this relation, B can be calculated using the given value of the gap.

$$\text{Bond length} = \left(\frac{I}{\mu}\right)^{\frac{1}{2}}$$

$$= \left(\frac{h}{8\pi^2 Bc\mu}\right)^{\frac{1}{2}}$$

Zero-point energy $= \frac{1}{2}h\bar{v}_e c$, where $\bar{v}_e$ is the equilibrium vibration frequency of the molecule which is supposed to be same as the frequency of the Q branch (treating the molecule as rigid rotor).

Vide problem 19.69.

19.66 The given expression corresponds to R-branch of the vibration–rotation spectrum where

$$\bar{v} = \bar{v}_e + (B_1 + B_0)(J+1) + (B_1 - B_0)(J+1)^2 \quad \text{by Eq. (19.18)}$$

$$= (\bar{v}_e + 2B_1) + (3B_1 - B_0)J - (B_0 - B_1)J^2$$

Comparing this expression with the given one

$$3B_1 - B_0 = 3.813$$
$$B_0 - B_1 = 0.0175$$

whence $\quad B_0 = 1.932 \text{ cm}^{-1}$ and $B_1 = 1.915 \text{ cm}^{-1}$

Again $\quad \bar{v}_e + 2B_1 = 2140.28$

or $\quad \bar{v}_e = (2140.28 - 2 \times 1.915) \text{ cm}^{-1}$

$$= 2136.4 \text{ cm}^{-1}$$

The constant which represents the change in rotational constant, i.e. $B_0 - B_1$ is 0.0175.

Note: Excited vibrational state has smaller rotational constant than the ground vibrational state.

19.67 Here the vibrational transition is from $v = 0 \rightarrow v = 1$.

For P-branch, the first line corresponds to $J = 1$ to $J = 2$ transition and the second line to $J = 2$ to $J = 3$ transition. But for R-branch, the first line corresponds to $J = 0$ to $J = 1$ transition and the second line to $J = 1$ to $J = 2$ transition. Then by Eq. (19.18), we have from the given wavenumbers of the lines

$$2217 \text{ cm}^{-1} - 2204 \text{ cm}^{-1} = -(B_1 + B_0)(1-2) + (B_1 - B_0)(1^2 - 2^2) = 4B_0 - 2B_1$$
$$2254 \text{ cm}^{-1} - 2242 \text{ cm}^{-1} = (B_1 + B_0)(1-0) + (B_1 - B_0)(2^2 - 1^2) = 4B_1 - 2B_0$$

whence $\quad B_0 = 6.33 \text{ cm}^{-1}$ and $B_1 = 6.16 \text{ cm}^{-1}$

For $v = 0$, moment of inertia

$$I_0 = \frac{h}{8\pi^2 BC}$$

$$= \frac{(6.63 \times 10^{-27} \text{ erg} \cdot \text{s})}{8(3.14)^2 (6.33 \text{ cm}^{-1})(10^{10} \text{ cm} \cdot \text{s}^{-1})}$$

$$= 1.328 \times 10^{-35} \text{ g} \cdot \text{cm}^2$$

For $v = 1$, moment of inertia

$$I_1 = \frac{h}{8\pi^2 B_1 C}$$

For $v = 0$, internuclear distance

$$= \left(\frac{I_0}{\mu}\right)^{\frac{1}{2}}$$

$$= \frac{1.328 \times 10^{-35} \text{ g} \cdot \text{cm}^2}{\dfrac{1.008 \times 126.9}{1.008 + 126.9} \times 1.661 \times 10^{-24} \text{ g}} = 7.99 \times 10^{-10} \text{ cm}$$

19.68 Assuming that $H^{35}Cl$ and $H^{37}Cl$ molecules behave as harmonic oscillator with same force constant, their vibrational frequencies will be related as

$$v(H^{37}Cl) = v(H^{35}Cl)\left[\frac{\mu(H^{35}Cl)}{\mu(H^{37}Cl)}\right]^{\frac{1}{2}}$$

$$= (8.667 \times 10^{13} \text{ s}^{-1})\left(\frac{1.673 \times 58.06}{1.673 + 58.06}\Big/\frac{1.673 \times 61.38}{1.673 + 61.38}\right)^{\frac{1}{2}}$$

(since for harmonic oscillation v is equal to the fundamental frequency)

$$= 8.66 \times 10^{13} \text{ s}^{-1}$$

Then the required separation of the spectral lines $= (8.667 - 8.66) \times 10^{13} \text{ s}^{-1}$.

19.69 For a diatomic molecule, the wavenumber $\bar{v}_1$ for fundamental transition and $\bar{v}_2$ for first overtone transition are related to the equilibrium frequency $\bar{v}_e$ and anharmonicity constant x_e of the molecule as

$$\bar{v}_1 = (1 - 2x_e)\bar{v}_e$$

$$\bar{v}_2 = (1 - 3x_e)2\bar{v}_e \qquad \text{vide Eq. (19.15)}$$

Then from the given data

$$1876 \text{ cm}^{-1} = (1 - 2x_e)\bar{v}_e$$

$$3724 \text{ cm}^{-1} = (1 - 3x_e)2\bar{v}_e$$

So $\qquad \dfrac{1 - 2x_e}{2(1 - 3x_e)} = \dfrac{1876}{3724}$

or $\qquad x_e = 0.00735 \text{ cm}^{-1}$

$$\bar{v}_e = \frac{1876 \text{ cm}^{-1}}{1 - 2 \times 0.00735} = 1903.33 \text{ cm}^{-1}$$

Force constant $= 4\pi^2\mu(\bar{v}_e c)^2$

Exact zero-point energy

$$= \tfrac{1}{2}hc\bar{v}_e - \tfrac{1}{4}hcx_e\bar{v}_e \quad \text{by Eq. (19.14)}$$

$$= \tfrac{1}{2}hc\bar{v}_e - \left(1 - \frac{x_e}{2}\right)$$

$$= \tfrac{1}{2}(6.63 \times 10^{-27} \text{ erg-sec})(3 \times 10^{10} \text{ cm sec}^{-1})(1903.33 \text{ cm}^{-1})\left(1 - \frac{0.00735}{2}\right)$$

$$= 1.88 \times 10^{-13} \text{ erg}$$

□ For fundamental transition $v = 0 \to v = 1$

$$\Delta E_{0 \to 1} = hc \cdot \bar{v}_e \left(1 - 2x_e\right)$$

Hot band corresponds to $v = 1$ to $v = 2$ transition

Then $\dfrac{\text{Intensity of hot band}}{\text{Intensity of fundamental}}$

$$= \frac{N_1}{N_0} = e^{-\frac{\Delta E_{0 \to 1}}{kT}}$$

$$= \mathrm{Exp}\left[\frac{-hc\bar{v}_e\left(1 - 2x_e\right)}{kT}\right]$$

$$= \mathrm{Exp}\,\frac{(-6.63 \times 10^{-27}\,\mathrm{erg\,sec})\,(3.00 \times 10^{10}\,\mathrm{cm\,sec^{-1}})\,(1903.33\,\mathrm{cm^{-1}})\,(1 - 2 \times 0.00735)}{(1.38 \times 10^{-16}\,\mathrm{erg\,K^{-1}})\,(300\,\mathrm{K})}$$

$$= 7.04 \times 10^{-5}$$

19.70 Here the first absorption is very intense, the second is weak and the third is very weak. Further the absorption frequencies are nearly in the ratio 1:2:3. Then, it is likely that the first absorption frequency $(\bar{v}_1)$ would correspond to the fundamental transition, the second $(\bar{v}_2)$ to the first overtone transition and the third $(\bar{v}_3)$ to the second overtone transition as under. [Vide Eq. (19.15)].

$$\bar{v}_1 = (1 - 2x_e)\,\bar{v}_e \tag{I}$$

$$\bar{v}_2 = (1 - 3x_e)\,2\bar{v}_e \tag{II}$$

$$\bar{v}_3 = (1 - 4x_e)\,3\bar{v}_e \tag{III}$$

(II) – (I) and (III) – (II) give

$$5668\,\mathrm{cm^{-1}} - 2886\,\mathrm{cm^{-1}} = (1 - 4x_e)\,\bar{v}_e$$

$$8347\,\mathrm{cm^{-1}} - 5668\,\mathrm{cm^{-1}} = (1 - 6x_e)\,\bar{v}_e$$

whence
$$\bar{v}_e = 2988\,\mathrm{cm^{-1}}$$
$$x_e = 0.01723$$

Equilibrium dissociation energy

$$D_e = \frac{hc\bar{v}_e}{4x_e} \quad \text{vide problem 19.24}$$

$$= 8.62 \times 10^{-12}\,\mathrm{erg}$$

Equilibrium dissociation energy level will correspond to the quantum number v given by

$$v = \frac{1}{2x_e} - \tfrac{1}{2} \quad \text{vide problem 19.24}$$

$$= \frac{1}{2 \times 0.01723} - \tfrac{1}{2}$$

$$= 28.6, \text{ i.e. } 29 \text{ (nearest integer)}$$

19.71 Spectroscopic dissociation energy (D_e) is (approximately) related to chemical dissociation energy (D_o) as

$$D_e = D_o + \tfrac{1}{2}hv_e$$

$$= D_o + \frac{h}{2}\left(\frac{\nu_1}{1 - 2x_e}\right) \quad \text{by Eq. (19.15)}$$

$$= D_o + \frac{h}{2}\left(\frac{\nu_1}{1 - \dfrac{h\nu_1}{2D_0}}\right) \quad \because x_e = \frac{h\nu_e}{4D_e} = \frac{h\nu_1}{4D_0} \quad \text{(vide note of problem 19.24)}$$

$$= D_o + \frac{h}{\dfrac{2}{\nu_1} - \dfrac{h}{D_0}}$$

$$= D_o + \frac{h}{\dfrac{2}{c\bar{\nu}_1} - \dfrac{h}{D_0}}$$

$$= 4.4763 \times 1.602 \times 10^{-12} \text{ erg} + \frac{6.626 \times 10^{-27} \text{ erg sec}}{\dfrac{2}{(3.00 \times 10^{10} \text{ cm s}^{-1})(4395 \text{ cm}^{-1})} - \dfrac{(6.626 \times 10^{-27} \text{ erg sec})}{4.4763 \times 1.602 \times 10^{-12} \text{ erg}}}$$

$$= 7.6362 \times 10^{-12} \text{ erg}$$

$$= 4.7666 \text{ eV}$$

☐ Assuming force constant to be same for H_2 and D_2

$$\frac{(\nu_e)_{D_2}}{(\nu_e)_{H_2}} = \left(\frac{\mu_{H_2}}{\mu_{D_2}}\right)^{\frac{1}{2}}$$

$$= \left(\frac{m_H}{m_D}\right)^{\frac{1}{2}}$$

$$= \frac{1}{\sqrt{2}}$$

Again, assuming x_e to be same for H_2 and D_2

$$\frac{(D_o)_{D_2}}{(D_o)_{H_2}} = \frac{(\nu_1)_{D_2}}{(\nu_1)_{H_2}} = \frac{(\nu_e)_{D_2}}{(\nu_e)_{H_2}} = \frac{(D_e)_{D_2}}{(D_e)_{H_2}} \quad \left(\because x_e = \frac{h\nu_e}{4D_e} = \frac{h\nu_1}{4D_o}\right)$$

$$= \frac{1}{\sqrt{2}}$$

Then $\qquad (D_o)_{D_2} = \dfrac{1}{\sqrt{2}}(D_o)_{H_2} = \dfrac{4.4763}{\sqrt{2}} \text{ eV} = 3.1657 \text{ eV}$

$$(D_e)_{D_2} = \frac{1}{\sqrt{2}}(D_e)_{H_2} = \frac{4.4766}{\sqrt{2}} \text{ eV} = 3.3710 \text{ eV}$$

19.72 Assuming that all the given diatomic species behave as harmonic oscillator

zero-point energy per mol $= \frac{1}{2}N_A hc\bar{\nu}$

where $\bar{\nu}$ is the vibration frequency which is same as the fundamental vibration frequency (in wavenumber) for harmonic oscillator. Then the heat of the given reaction

$$\Delta E = \frac{1}{2}N_V hc\left(\bar{\nu}_{DCl} + \bar{\nu}_{HD} - \bar{\nu}_{HCl} - \bar{\nu}_{D_2}\right)$$

19.73 Raman frequencies (in wavenumber), of the compound corresponding to the given lines are the differences in the frequencies of these lines from that of the incident radiation, i.e. $\left(\dfrac{1}{436} - \dfrac{1}{440}\right) = 2.08 \times 10^{-5}$ nm^{-1} corresponding to the first line, etc.

□ For a particular compound the Raman frequencies will correspond to the absorption in the infrared region. Hence the required wavelengths will be just inverse of the Raman frequencies, i.e. $\dfrac{1}{2.08 \times 10^{-5} \text{ nm}^{-1}} = 4.81 \times 10^4$ nm corresponding to the first line, etc.

19.74 (a) Width of the lines [position is related to the energy levels involved and intensity is related to the population of the energy level wherefrom emission occurs].

(b) By Eq. (19.4), $A \propto B\nu^3 \propto B/\lambda^3$, then

$$\dfrac{\text{Transition probability for spontaneous fluorescence}}{\text{Transition probability for spontaneous phosphorescence}}$$

$$= \dfrac{A(400 \text{ nm})}{A(500 \text{ nm})}$$

$$= \dfrac{(500 \text{ nm})^3}{(400 \text{ nm})^3} \cdot \dfrac{B(400 \text{ nm})}{B(500 \text{ nm})}$$

$$= \dfrac{125}{64}(10^5) \quad \text{from the given data}$$

$$= 1.95 \times 10^5$$

Since for spontaneous emission

$$t_{\frac{1}{2}} \propto \frac{1}{A}$$

$$t_{\frac{1}{2}} \text{ (phosphorescence)} = \dfrac{A(400 \text{ nm})}{A(500 \text{ nm})} t_{\frac{1}{2}} \text{ (fluorescence)}$$

$$= (1.95 \times 10^5)(1.00 \times 10^{-8} \text{ s})$$

$$= 1.95 \times 10^{-3} \text{ s}$$

UNIVERSITY QUESTIONS

19.1 The electronic and vibrational transitions produce the electronic and vibrational spectra. What are the transitions responsible for NMR spectrum?

(Burdwan BSC(H), 2002)

19.2 What type of molecular transitions are brought about by absorption of (i) ultraviolet-visible radiation (ii) Infrared radiation. (Madurai BSc, 2006)

19.3 Give three examples of three different model systems where (i) energy levels are equispaced, (ii) energy gap increases with increase in quantum number (iii) energy gap decreases with increase in quantum number. (Calcutta BSc(H), 2014)

19.4 The probability of all changes $\Delta J = \pm 1$ is almost the same among the rotational energy levels in a rigid diatomic molecule but all spectral lines will not be equally intense. Explain. (Burdwan BSc(H), 1993)

19.5 On what factors does the intensity of a spectral line depend? Show that the intensity of a spectral line in the rotational spectra will be maximum when

$$J = \sqrt{\dfrac{kT}{2hBc}} - \dfrac{1}{2}$$

(Jadavpur BSc(H), 2000)

19.6 (a) Plot of intensity vs rotational number (J) in rotational spectroscopy of a diatomic molecular system passes through a maximum. Justify or criticise. How and why does the plot change if one of the atoms is replaced by its heavier isotope?

   (b) Why such an observation is not met when absorption intensity is plotted against vibrational quantum number (v) in vibrational spectroscopy?

(Jadavpur BSc(H), 2012)

19.7 Rotational spectra is not observed for every molecule in the gas phase, because all of them do not rotate. Comment. (Burdwan BSc(H), 1994)

19.8 Show that the rotational spectral lines of a rigid diatomic molecule are equispaced. (Calcutta BSc(H), 2009)

19.9 The rotational spectrum of $^{79}Br^{19}F$ shows a series of equidistant lines spaced 0.71433 cm^{-1} apart. Calculate the rotational constant, the moment of inertia and bond length of the molecule. Find which transition gives rise to the most intense spectral line at 80 K. (Burdwan BSc(H), 1997)

19.10 An estimated bond length of the species CN is 0.117 nm. Predict the positions of the first three lines in the microwave spectrum of CN. (Atomic weights of C and N are 12.011 and 14.0067 respectively). (Calcutta BSc(H), 2013)

19.11 Pure rotational spectrum of $^{12}C^{16}O$ has two successive lines at 7.72 cm^{-1} and 11.58 cm^{-1}. Assign the J-values (both initial and final) for the transition. Predict the position of the next two transitions. (Burdwan BSc(H), 2006)

19.12 In the rotational spectra of H–F molecule, the wavenumbers of the lines follow the relation

$$\bar{v}_J = [41.122\,(J+1) - 8.52 \times 10^{-3}\,(J+1)^3]\,\text{cm}^{-1}$$

Find the bond length and vibrational frequency of the molecule. Show that the bond in H–F is is relatively strong. (Jadavpur BSc(H), 2001)

19.13 A rotational spectral study not only gives the bond length but also the bond strength of a diatomic molecule. Justify or criticise. (Jadavpur BSc(H), 2010)

19.14 (a) What is meant by Raman effect? How does it differ from Raleigh scattering?

   (b) The Raman shift of a given Raman-spectrum line is inependent of the value of the exciting frequency. Explain. (Calcutta BSc(H), 2014)

19.15 The wavenumber of the incident radiation in a Raman spectrometer is 20487 cm^{-1}. What is the wavenumber of the scattered stokes radiation for the $J = 0$ to $J = 2$ transition of $^{14}N_2$ for which $B = 1.99$ cm^{-1}. (Calcutta BSc(H), 2014)

19.16 Classify the following molecules in terms of spherical top, symmetric top and asymmetric top: CH_3Cl, CH_4, NH_3, H_2O. (Burdwan BSc(H), 2011)

19.17 Which of the following molecules may give the rotational spectrum?
N_2, H_2, HCl, HD, CO, CCl_4. (Burdwan BSc(H), 2012)

19.18 (a) Write the essential condition for a molecule to be infrared active.

(Burdwan BSc(H), 2006)

   (b) Indicate which of the following molecules are active in the infrared spectrum:
N_2, H_2, HF, CO, CO_2. (Burdwan BSc(H), 1995)

19.19 Find the number of normal modes of vibration of CO_2 and H_2O. State and explain which of these vibrations are IR and Raman active. (Calcutta BSc(H), 2000)

19.20 Justify or criticise the statement "Raman spectroscopy is confirmatory as well as complementary to the rotational and vibrational spectroscopy". State and explain the rule of mutual exclusion in spectroscopy. (Jadavpur BSc(H), 1999)

19.21 A linear molecule AB_2 has either 'ABB' or 'BAB' structure. Using its IR and Raman spectra together, how would you ascertain the actual structure of the molecule.
(Calcutta BSc(H), 1992)

19.22 The molecules H_2 and CO_2 are both non-polar but the former is infrared inactive while the latter gives IR spectra. Explain. (Calcutta BSc(H), 1992)

19.23 The molecule CO_2 contains C=O group but it does not show IR absorption characteristic of the keto group. Explain. (Calcutta BSc(H), 1993)

19.24 Vibrational spectra are band spectra. Explain, why band centre is usually missing in roto-vibronic spectra? Is there any exception? (Calcutta BSc(H), 2009)

19.25 How does the IR absorption spectrum of a hypothetical diatomic molecule AB look like, when it behaves as (i) harmonic oscillator (ii) anharmonic oscillator.
(Calcutta BSc(H), 2013)

19.26 A heteronuclear diatomic molecule of reduced mass 1.68×10^{-24} gm absorbs at 2880 cm^{-1}. Calculate the force constant assuming harmonic oscillator model.
(Calcutta BSc(H), 2001)

19.27 The fundamental vibration frequency of $H^{35}Cl$ is 8.67×10^{13} sec^{-1}. Calculate the fundamental vibration frequency of $D^{35}Cl$ on the assumption that the force constants of the bonds are equal. (Burdwan BSc(H), 2012)

19.28 (a) How does HCl and DCl differ in respect of vibrational and rotational spectra (assuming the molecules to behave as rigid rotor and harmonic oscillator).
(Burdwan BSc(H), 2002)

(b) Compare the zero-point vibrational energies of H_2 and HD.
(Burdwan BSc(H), 2005)

19.29 Write the energy expression of an anharmonic oscillator for a given vibrational quantum number (v) of a diatomic molecule AB. Hence find the maximum value of v (v_{max}) in terms of the anharmonicity constant. Does the energy of the v_{max} level correspond to the 'dissociation energy' of the molecule AB? Explain.
(Calcutta BSc(H), 2006)

19.30 Using the formula for the energy levels for the Morse potential

$$E_v = hv \left(v + \tfrac{1}{2}\right) - \frac{(hv)^2}{4D_e} \left(v + \tfrac{1}{2}\right)^2$$

deduce the expression of energy spacing between adjacent levels.
For $H^{35}Cl$, $D_e = 7.41 \times 10^{-19}$ J and $v = 8.97 \times 10^{13}$ s^{-1}, calculate the smallest value of v for which $E_{v+1} - Ev < 0.5 (E_1 - E_0)$. (Burdwan BSc(H), 2010)

19.31 The vibrational energy levels of F_2 molecule is given by the expression

$$E_v \, (cm^{-1}) = 215 \left(v + \tfrac{1}{2}\right) \left\{1 - 0.003 \left(v + \tfrac{1}{2}\right)\right\}$$

Find (i) the anharmonicity constant (ii) equilibrium osillating frequency and (iii) zero-point energy of the molecule. (Calcutta BSc(H), 2008)

19.32 What do you mean by a hot band? Justify the name. Comment on the intensity of such bands. (Jadavpur BSc(H), 1999)

19.33 What is the difference between overtones and hot bands in the IR-spectra?
(Calcutta BSc(H), 2008)

19.34 Both the transitions $v = 0 \rightarrow v = 2$ and $v = 1 \rightarrow v = 2$ are weak. How would you differentiate these two transition bands? How would you name these bands?
(Jadavpur BSc(H), 2010)

19.35 The fundamental and first overtone transistors of $^{14}N^{16}O$ are centred as 1876.06 cm^{-1} and 3724.20 cm^{-1} respectively. Calculate the exact zero point energy of the molecule. (Calcutta BSc(H), 2014)

19.36 The IR spectrm of H^{35}Cl shows a very intense absorption of 2886 cm^{-1}, a weaker one at 5668 cm^{-1} and very weak one at 8347 cm^{-1}. Calculate the equilibrium oscillation frequency, anharmonicity constant and force constant of HCl molecule. (Jadavpur BSc(H), 2000)

19.37 The IR spectrum of CO molecule shows the position of lines calculated from the equation
$$\bar{v}\,(cm^{-1}) = 2143.28 + 3.813\,m - 0.0175\,m^2$$
where m has its usual significance. Calculate the equilibrium value of the rotational constant and the constant which represents the change of rotational constant with vibrational levels. (Jadavpur BSc(H), 2001)

19.38 Explain why the IR frequency of C=C vibration is lower than that of C≡C vibration. (Burdwan BSc(H), 2005)

19.39 What is the physical basis of vertical transition in electronic spectroscopy? (Calcutta BSc(H), 2013)

19.40 What are the criteria that a molecule must fulfil in order to show $n \rightarrow \pi^*$ transition. (Burdwan BSc(H), 2010)

19.41 Explain precisely why $\pi \rightarrow \pi^*$ transition is generally more intense than $n \rightarrow \pi^*$. (Burdwan BSc(H), 1993)

19.42 Explain with example of formaldehyde, which one appears at lower wavelength— a $n \rightarrow \pi^*$ or a $\pi \rightarrow \pi^*$ transition? (Burdwan BSc(H), 1997)

19.43 The UV absorption band shifts to longer wavelengths as the chain length in a conjugated polyene increases. Comment, using a suitable model of your choice. (Burdwan BSc(H), 1999)

19.44 Two isomeric dienes (A) and (B) having molecular formula C$_5$H$_8$ absorbs at λ_{max} = 223 nm and λ_{max} = 178 nm respectively. Write the structures of the two isomers. (Madras BSc, 2005)

19.45 Describe all possible unimolecular photophysical processes that could take place when a molecule is excited to the lowest vibrational state of its first excited singlet state. State the distinctive features of the radiative processes in this case. (Calcutta BSc(H), 2006)

19.46 Absorption and fluorescence hold a mirror image relationship. Comment. (Burdwan BSc(H), 2011)

19.47 What are internal conversion and inter-system crossing? (Calcutta BSc(H), 2013)

19.48 Justify the differences between fluorescence and phosphorescence in terms of the (i) wavelength of emission involved and (ii) radiative lifetimes. (Calcutta BSc(H), 2014)

19.49 In case of phosphorescence, singlet to triplet transition is possible. Justify. (Jadavpur BSc(H), 2012)

19.50 Low temperature and viscous medium are suitable for observing phosphorescence. Explain. (Calcutta BSc(H), 2013)

19.51 How does the intensity of fluorescence vary with pressure and temperature? (Jadavpur BSc(H), 2009)

19.52 What happens to fluorescence when a photosensitiser transfers its absorbed energy to some other species. (Burdwan BSc(H), 1992)

KEY TO UNIVERSITY QUESTIONS

19.1 Transitions between nuclear-spin energy levels.

19.2 (i) Transition of valence electrons to higher levels.

(ii) Transition between vibrational energy levels of the same electronic state.

19.3 (i) A hytothetical diatomic molecule assumed to behave as a harmonic oscillator.

(ii) A particle confined by impenetrable walls (the so-called particle in a box). [Another example is provided by a rigid rotor].

(iii) A hydrogen-like atom that consists of two particles (an electron and a nucleus), each moving about their common centre of mass following coulomb's law. [Another example is provided by a real molecule which actually behaves as an anharmonic oscillator].

Note: In general, if the energy of a system is wholly kinetic [as in (ii)], the gap between the consecutive energy levels increases with increase in quantum number. But when potential energy is present, this trend is opposed [as in (iii)], and even nullified [as in (i)], due to tendency of the potential energy to be minimum.

19.4 Because the intensity of a spectral line resulting from the molecular transition from an energy level is proportional to the population of that level (assuming that the probability of transition is same for each energy level) which is not same for each level.

19.5 The intensity of a spectral line depends essentially on (i) the population of the initial energy level involved in the relevant molecular transition and (ii) the probability of transition.

☐ Considering that the factor (ii) is almost same for different rotational energy levels, the required relation can be established vide problem 19.60.

19.6 (a) This is justified on the ground that the intensity (P) of a rotational spectral line is largely determined by its parent rotational level of quantum number J according to the following relation

$$P \propto (2J + 1)\, e^{-E_J / kT}$$

with increase in J, the factor $(2J + 1)$ increases while the other factor decreases but at different rate so that P passes through a maximum.

☐ P_{max} corresponds to

$$J = \sqrt{\frac{kT}{2hCB}} - \frac{1}{2} \qquad \text{vide problem 19.60}$$

Therefore when one of the atoms is replaced by its heavier isotope, B will decrease (due to increase in moment of inertia) with consequent increase in J, i.e. the position of maximum will shift to higher value of J.

(b) Because, for a diatomic molecule, the vibrational energy levels, unlike rotational energy levels, are non-degenerate. Here we can write

$$P \propto e^{-E_v / kT}$$

Then P will not exhibit any extremium for any finite change of E_v or v.

19.7 Rotational spectrum is not observed for every molecule in the gas phase not because such molecules do not rotate but because all molecular rotations do not comply with the gross selection rule for rotational transition (vide auxiliary problem 19.7). Therefore, the given statement is not justified.

19.8 This follows from the expression for frequency (ν) of the rotational spectral lines according to Eq. 19.7a or 19.7b, either of which gives $|\Delta \nu| = 2B$ for any two consecutive values of J.

19.9 Vide problem 19.60.

19.10 This is similar to the previous question put in reverse way. From the given bond length (R), B can be calculated using the relation

$$B = \frac{h}{8\pi^2 \mu R^2 C}$$

The position (in wavenumber $\bar{\nu}$) of the first three lines can then be predicted from

$$\bar{\nu} = 2B(J + 1) \quad \text{by Eq. (19.7a)}$$

putting $J = 0, 1$ and 2.

19.11 Vide problem 19.61.

19.12 Vide problem 19.63.

For bond length (vide problem 19.60), we are to consider average spacing of the lines equal to $2B$.

19.13 Using the expression for the wavenumber ($\bar{\nu}$) of the rotational spectral lines

$$\bar{\nu} = 2B(J + 1) - 4D(J + 1)^3$$

B and D can be found the experimental data. On proceeding further, the bond length (R) and force constant k (which is a measure of the bond strength) can be calculated from

$$B = \frac{h}{\cdot 8\pi^2 \mu R^2 c}$$

and

$$k = 4\pi^2 \mu (\bar{\nu}_e c)^2$$

where

$$\bar{\nu}_e = \left(\frac{4B^3}{D} \right)^{\frac{1}{2}}$$

Therefore, the given statement is justified.

19.14 (a) Vide Section 19.4.

(b) The Raman shift ν of a given Raman-spectrum line is such that $h\nu$ is the difference between two rotational or vibrational energy levels of the scattering molecule. Then, ν being the characteristics of the scattering substance, is independent of the value of the existing frequency.

19.15
$$\begin{aligned}
\bar{\nu}_{stokes} &= \bar{\nu} - 2B(2J + 3) \quad \text{by Eq. (19.19a)} \\
&= 20487 \text{ cm}^{-1} - 2(1.99 \text{ cm}^{-1})(0 + 3) \quad \text{from the given data} \\
&= 20475 \text{ cm}^{-1}
\end{aligned}$$

19.16 Vide problem 19.55.

19.17 HCl (but not HD) and CO, which possess appreciable permanent dipole moment, will give rotational absorption and emission spectrum. For HD, the dipole moment is too low to have any spectroscopic importance. All the given molecules will exhibit rotational Raman spectrum because their polarizability changes with rotation.

19.18 (a) The essential condition is that the dipole moment of the molecule must change during vibration.

(b) HF, CO and CO_2 (other than symmetric stretching vibration).

19.19 Vide problem 19.35.

19.20 Any molecule will have at least one rotational and one vibrational motion that would be Raman active. This is unlike IR spectroscopy. Obviously, a molecule having IR spectra will also have Raman spectra (though the same molecular motion may not be responsible for both). Then it follows that Raman spectroscopy is complementary in the first occasion and confirmatory in the second occasion.

□ Vide Section 19.4.

19.21 The molecule will have the structure BAB (which is centosymmetric) if none of its modes is both IR- and Raman-active (by exclusion rule), otherwise the structure will be ABB.

19.22 Even if a molecule is non-polar, its dipole moment can change during vibration for some of its modes making them IR-active. For CO_2, although symmetric stretching mode is IR-inactive, the other modes are IR-active. This cannot happen with H_2 because it has only one mode of vibration which is symmetric.

19.23 In CO_2 molecule C-atom is sp hybridised whereas C in ordinary carbonyl group is sp^2 hybridised. It is due to higher s-character that the molecular orbital of CO group has strikingly higher stretching frequency in CO_2 molecule than in ordinary carbonyl group.

19.24 Because the vibrational transitions actually occur between rotational energy levels belonging to different vibrational levels, obeying the usual selection rule $\Delta v = \pm 1$ and $\Delta J = \pm 1$.

□ Because the vibrational transitions with $\Delta J = 0$ occur only rarely, e.g. in case of NO molecule having electronic angular momentum about the bond axis.

Note: $\Delta J = 0$ happens with vibrational transition, though not with pure rotational transition.

19.25 Here we are concerned only with vibrational motion of the molecule.

In case of (i), the vibrational levels are equispaced. Then considering the spectroscopic selection rule for molecular vibration $\Delta v = \pm 1$, the vibrational transitions from all levels will occur at same frequency. The resulting vibrational spectrum will therefore contain only one line of same frequency as the vibrational frequency of the molecule.

In case of (ii), the vibrational energy levels get closer with increase in quantum number and ultimately merge. Then due to the transition $\Delta v = \pm$ each vibrational level will produce one spectral line. But these lines, due to their nearly same frequency, will merge into a broad line. Further, such transitions will be accompanied by overtone transitions $\Delta v = \pm 2, \pm 3, ...$, each producing a separate broad line. Then the resulting vibrational spectrum will consist of several broad lines of which one of lower frequency (due to the fundamental transition $\Delta v = \pm 1$) will be much stronger than the others.

19.26 For constant $= 4\pi^2 \mu (\bar{v}_e c)^2$

$$= 4 (3.14)^2 (1.63 \times 10^{-24} \text{ g}) (2880 \text{ cm}^{-1})^2 (3.00 \times 10^{10} \text{ cm sec}^{-1})^2$$

$$= 294403 \text{ dyn cm}^{-1}$$

19.27 Vide problem 19.68.

19.28 (a) Here rotational spectra differ mainly in the separation (2B) between the adjacent spectral lines, which is lower with DCl due to its lower B due to higher I.

Vibrational spectra differ mainly in the position of the single spectral line which is at lower frequency (ν) for DCl due to its higher μ.

(b) Zero-point energy $\propto \nu$ by Eq. (19.11)

$$\propto \frac{1}{\sqrt{\mu}} \quad \text{assuming force constant to be same}$$

Then HD having higher μ will have lower E_0 compared to H_2.

19.29 Vide problem 19.24.

Here the energy of the v_{max} level corresponds to the spectroscopic dissociation energy (D_e), i.e. one measured from the minimum potential energy of the molecule.

19.30 From the Morse potential

$$E_{v+1} - E_v = h\nu - \frac{(h\nu)^2}{2D_e}(v+1)$$

So $$E_1 - E_0 = h\nu - \frac{(h\nu)^2}{2D_e} \quad \text{(putting } v = 0)$$

Then for $E_{v+1} - E_v < 0.5\,(E_1 - E_0)$ to hold the condition is

$$h\nu - \frac{(h\nu)^2}{2D_e}(v+1) < 0.5\left[h\nu - \frac{(h\nu)^2}{2D_e}\right]$$

or $$v > \frac{D_e}{h\nu} - \tfrac{1}{2}$$

$$> \frac{(7.41 \times 10^{-19}\ \text{J})}{(6.63 \times 10^{-34}\ \text{Js})(8.97 \times 10^{13}\ \text{s}^{-1})} - \tfrac{1}{2}$$

$$> 11.96$$

Since v is an integer, the required smallest value of v is 12.

19.31 For a real molecule, the vibrational energy levels E_v in wavenumber, can be expressed as

$$E_v = \bar{\nu}_e\left(v + \tfrac{1}{2}\right)\left[1 - x_e\left(v + \tfrac{1}{2}\right)\right] \quad \text{by Eq. (19.14)}$$

Then from the given expression

(i) anharmonicity constant $x_e = 0.003$.

(ii) equilibrium oscillation frequency $\bar{\nu}_e = 215\ \text{cm}^{-1}$.

(iii) zero-point energy

$$= \left(\frac{215\ \text{cm}^{-1}}{2}\right)\left(1 - \frac{0.003}{2}\right)$$

(putting $v = 0$ in the given expression)

$$= \underbrace{(6.63 \times 10^{-27}\ \text{erg sec})}_{=\,h}\underbrace{(3.00 \times 10^{10}\ \text{cm/sec})}_{=\,c}\left(\frac{215\ \text{cm}^{-1}}{2}\right)\left(1 - \frac{0.003}{2}\right)$$

$$= 3.21 \times 10^{-17}\ \text{erg}$$

19.32 Vide Section 19.3.

19.33 Overtone bands correspond to vibrational transitions $|\Delta v| > 1$ involving ground vibrational level ($v = 0$) whereas hot bands correspond to vibrational transition $|\Delta v| > 1$ without involving ground vibrational level.

19.34 The transition $v = 0$ to $v = 2$ gives rise to a weaker band (since $\Delta \tilde{v} > 1$) than the transition $v = 1$ to $v = 2$. Further with rise in temperature, the intensity of the band will decrease in the first case but increase in the second case.

□ The band due to $v = 0 \to v = 2$ transition is named as overtone band while that due to $v = 1 \to v = 2$ transition is named as hot band.

19.35 Vide problem 19.69.

19.36 Vide problem 19.70.

19.37 Vide problem 19.66.

19.38 Vide problem 19.19.

19.39 The relevant physical basis lies in the fact that an electron can jump from one state to another much more quickly ($\sim 10^{-18}$ sec) than the period of vibration (10^{-13} sec) of the comparatively massive atomic nuclei with the consequence that the positions and velocities of the atomic nuclei remain almost unchanged during an electronic transition.

19.40 The molecule must have π-orbital and also n (non-bonding) orbital.

19.41 Because $n \to \pi^*$ transition, unlike $\pi \to \pi^*$, is symmetry forbidden.

19.42 Formaldehyde, $H-C{\overset{=\ddot{O}:}{\underset{H}{\diagup}}}$, containing $C=\ddot{O}:$ group, can undergo $\pi \to \pi^*$ transition, and also $n \to \pi^*$ transition, though the latter is symmetry forbidden and hence weak.

$\pi \to \pi^*$ transition occurs at lower wavelength because it involves higher energy change.

19.43 The given statement can be justified considering conjugated π-electrons in a polyene as particles in a one-dimensional box. Vide problem 19.43.

19.44 The polyene with higher λ_{max} (viz. A) will be one of longer conjugated chain (shown in the previous question). Therefore the two isomers will have the following structure

$$H_2C=CH-CH = CH-CH_3 \qquad H_2C=CH-CH_2-CH=CH_2$$
$$\qquad\qquad A \qquad\qquad\qquad\qquad\qquad B$$

19.45 The photophysical processes that can take place with deactivation of the excited molecule are mostly of two well-defined types—fluorescence and phosphorescence.

□ Vide, fluorescence and phosphorescence, in Section 19.5.

19.46 The absorption and fluorescence spectra arise due to transitions mostly from the ground vibrational level ($v = 0$) of the lower and upper electronic states respectively. They contain a common band due to O–O transition. The remaining absorption bands occur at higher frequency and the remaining fluorescence bands occur at lower frequency. Then the two spectra will appear to have a mirror-image resemblance if the vibrational structure is similar for the lower and upper electronic states. Obviously, the given statement is not exact.

19.47 Vide 'deactivation of Electronically Excited Species', Section 19.5.

19.48 (i) Vide problem 19.50.

(ii) Vide problem 19.46.

19.49 Because, here, singlet $\rightarrow$ triplet transition occurs in a non-radiative way, called intersystem crossing, which is not restricted by any selection rule.

19.50 For the given condition, frequency of collision being low, the collisional dissipation of vibrational energy occurs sufficiently slowly and then the inter-system crossing step will have enough time to occur.

19.51 For a gas, collisional deactivation of an excited molecule will be favoured and hence the intensity of fluorescence will be hampered with rise in pressure and temperature. Vide Eq. (2.10a).

For a liquid or solid, the intensity of fluorescence is practically independent of pressure but is adversely affected by a rise of temperature.

19.52 The intensity of fluorescence will fall due to adsorption of the emitted radiation by the photosensitiser, but not due to deactivation of the excited state responsible for fluorescence.

Photochemistry

Photochemistry deals primarily with the chemical effects of light in the visible and ultraviolet region, and is closely related to electronic spectroscopy. Usually the chemical effect is little with radiations of lower frequency (infrared) while it is drastic in higher frequency regions (x-ray, γ-ray).

20.1 ABSORPTION OF LIGHT

When a beam of light (radiation) passes through a medium, a certain amount of light is usually absorbed. If dI is the change (decrease) in intensity that occurs when light of intensity I passes normally through a layer of thickness dx containing an absorbing species at a molar concentration c, then the probability of absorption of a photon is given by

$$-\frac{dI}{I} \propto cdx$$
$$= \alpha cdx \qquad (20.1a)$$

Integration gives

$$-\int_{I_0}^{I} \frac{dI}{I} = \alpha c \int_{0}^{l} dx \qquad \text{ignoring absorption due to solvent}$$

or
$$\log \frac{I_0}{I} = \epsilon cl \qquad (20.1b)$$

which is Lambert–Beer law. Here I_0 and I are the intensities of the light beam before and after passing through a sample of thickness l. The quantity $\epsilon = \alpha/2.303$ is called the molar absorption coefficient (formerly extinction coefficient) or molar absorptivity or molar absorbancy index (since it is an indication of the intensity of the transition responsible for light absorption) which is a characteristic constant of the absorbing species depending on the wavelength of the incident light, the solvent and the temperature; $\log \frac{I_0}{I}$ is called the absorbance (formerly and still widely, the optical density) of the same, I/I_0 is called transmittance.

ϵ has the SI unit m^2 mol^{-1}. But it is usually cited in units of M^{-1} cm^{-1} or L mol^{-1} cm^{-1} (with C in mol L^{-1} and l in cm).

Lambert–Beer law can also be written as

$$\ln \frac{I_0}{I} = \sigma nl \qquad (20.1c)$$

where n is the number of light-absorbing molecules per unit volume, and σ is called molecular cross-section for photon absorption (having the dimensions of area).

Lambert–Beer law constitutes the basis of the various colorimetric methods of analysis. For a mixture of independently absorbing substances the absorbance, being an additive property, is given by

$$\log \frac{I_0}{I} = l \sum \epsilon_i \, C_i \tag{20.1d}$$

20.2 CHARACTERISTICS OF PHOTOCHEMICAL REACTIONS

Photochemical reactions (i.e. ones caused by light) differ from the ordinary thermal or dark reactions in the following important respects.

  (i) Photochemical reactions receive their activation energy by absorption of photons of light by reactant molecules, while ordinary thermal reactions acquire their activation energy through random successive collisions between molecules.

 (ii) Photochemical reactions are much less affected by temperature than the ordinary thermal reactions.

(iii) Certain photochemical processes involve an increase in free energy of the reaction system, while the ordinary chemical reactions are always accompanied by a decrease in free energy (at constant T and P).

20.3 LAWS OF PHOTOCHEMISTRY

Photochemical reactions are governed by the following two basic laws—the first one is qualitative and the second is quantitative.

Grotthuss–Draper Law

Only radiations which are absorbed by a reaction system can cause chemical reaction.

This provides the most common activation principle (vide problem 20.8).

Stark–Einstein Law

In a photochemical reaction, the number of reactant molecules activated equals the number of photons absorbed.

Violation of this law is not unexpected, particularly when the light used is of high intensity as with a laser beam (vide problem 20.12). Even when the law holds, it will be risky to apply it to predict the amount of overall reaction per absorbed photon. This is due to (i) partial deactivation or removal of the activated (or dissociated) reactant molecules and (ii) chain propagation of the thermal reaction (with fragments of activated molecules) that follows the primary light-absorption process. Then the efficiency of the absorbed photon in causing the concerned reaction, called the overall quantum yield (ϕ), is rarely equal to one.

$$\phi = \frac{\text{number of molecules of specified substance reacted or formed}}{\text{number of photons absorbed}}$$

$$= \frac{\text{rate of reaction}}{\text{rate of absorption of photon per unit volume } (I_{abs})}$$

I_{abs} is the intensity of light absorption.

Apparently, it is less risky to decide on the primary quantum yield that concerns only with the light absorption process. It is often taken to be one (vide problem 20.16). However, the overall quantum yield is much more important, particularly because it gives a clue to the mechanism of a photochemical reaction (like reaction order).

For light of frequency ν the energy (E) per mol of photon, which is called as einstein, is given by

$$E = N_A h\nu \tag{20.2a}$$

$$= \frac{N_A hc}{\lambda} \tag{20.2b}$$

$$= \frac{0.1196}{\lambda/\text{meter}} J \cdot \text{mol}^{-1} \tag{20.2c}$$

According to Stark–Einstein law, E is the amount of energy required to activate one mole of a substance. E is ~600 kJ mol^{-1} for λ ~ 200 nm (ultraviolet) and ~120 kJ mol^{-1} for λ ~ 1000 nm (infrared region). Since it usually takes at least 140 kJ mol^{-1} for electronic excitation of molecules, photochemical reactions are initiated by ultraviolet and visible but not usually by infrared radiation.

20.4 PHOTOSTATIONARY STATE

The equilibrium of a reaction system will be disturbed in presence of light that has direct influence on the forward and/or backward reaction rate. Here the system will adjust to equalise the two rates but at a new composition depending on the intensity and wavelength of the light used. When this happens, the system is said to be in a photo-stationary state (rather than a photochemical equilibrium).

Anthracene (A) offers an example to illustrate

$$2A \underset{\text{dark}}{\overset{\text{UV light}}{\rightleftharpoons}} A_2$$

Here a photostationary state is set up when the forward reaction rate (kI_{abs}) equals the backward reaction rate $(k'C_{A_2})$ i.e.

$$kI_{abs} = k'C_{A_2}$$

where k and k' are the rate constants for the light and dark reactions (k includes the effect of deactivation of the excited molecules produced by light absorption). Then, at photo-stationary state, the concentration of dimer (C_{A_2}) would be determined not by the concentration of the reactant (A) but by the intensity of light absorption (I_{abs}), and thereby indicating the striking fact that the laws of chemical equilibrium is not applicable in this situation.

The phenomenon of vision can be explained on the basis of photostationary state. In the eye, a light sensitive substance, known as *visual purple*, is supposed to be bleached when exposed to light producing *visual yellow* but is regenerated in an ordinary thermal process leading to a stationary state. The eye becomes specially sensitive in the dark due to accumulation of 'visual purple' while removvl of this on exposure to intense light leads to dazzling.

20.5 PHOTOSENSITISATION

Sometimes a chemical reaction is caused by light which is absorbed only by a foreign substance introduced into the reaction system concerned. The phenomenon is called photosensitisation and the foreign substance involved is called a photosensitiser.

For example, the radiation (253 nm) from a mercury lamp has no chemical effect on a mixture of H_2 and O_2, but the effect is observed in presence of traces of mercury vapour which acts as a sensitiser. This happens due to formation of H-atoms by the excited mercury atom (Hg^*) through the following thermal reactions

$$Hg^* + H_2 \rightarrow Hg + 2H$$

$$Hg^* + H_2 \rightarrow HgH + H$$

H-atoms thus formed initiate a chain reaction forming H_2O. In absence of mercury photo-atomisation of H_2 and O_2 cannot occur with 253 nm radiation because the absorption continuum for H_2 and O_2 lies much below this wavelength.

An outstanding instance of photosensitisation, which is of tremendous biological importance, is the photosynthesis of carbohydrate by plants where chlorophyll (the principal constituent of green-plant pigment) acts as a sensitiser by absorbing strongly in the visible region (600 nm–700 nm) and then transferring the energy to the reactants ($CO_2 + H_2$) causing the following reaction

$$nCO_2 + nH_2O + xh\nu \rightarrow (CH_2O)_n = nO_2$$

A quantitative application of photosensitisation is found in chemical actinometers, which are devices to measure the quantity of light from the amount of chemical change brought about by it in a system of known quantum yield. The decomposition of oxalic acid sensitised by uranyl ion, is a common actinometric reaction for radiation of short wavelength (250–450 nm)

$$UO_2^{2+} + h\nu \rightarrow (UO_2^{2+})^*$$

$$(UO_2^{2+})^* + H_2C_2O_4 \rightarrow UO_2^{2+} + CO_2 + CO + H_2O$$

AUXILIARY PROBLEMS

20.1 Is Lambert–Beer law applicable only for visible region of electromagnetic radiation? Does it hold only for liquid solutions?

20.2 On which of the following factors does molar absorption coefficient (ϵ) of a species depend and why?
  (i) Concentration of the species
  (ii) Nature of the solvent
  (iii) Wavelength of the light absorbed
  (iv) Temperature

20.3 State whether the Lambert–Beer law will be valid for (i) a complex species (ii) polychromatic radiation (iii) scattering of radiation.

20.4 Justify the following statements:
  (a) Photochemical reactions are of vital importance for the existence of life on earth.
  (b) Photochemical reactions are much less affected by temperature than the ordinary thermal reactions.
  (c) Photochemical reactions can be brought about more selectively than thermal reactions.
  (d) Although infrared radiations have little chemical effect, the photography in the infrared region is an effective technique.

20.5 The presence of chlorofluorocarbons (like $CFCl_3$) in the atmosphere is dangerous. Explain.

20.6 In the photosynthetic process

$$nCO_2 + nH_2O + xh\nu \xrightarrow{\text{Chlorophyll}} (CH_2O)_n + nO_2,$$

What is the source of O_2–CO_2 or H_2O? Give evidence for your choice.

20.7 In certain photochemical processes (e.g. photosynthesis of carbohydrates by plants) $\Delta G_{T,P}$ is positive. Do they violate any thermodynamic principle? Discuss.

20.8 Grotthuss–Draper law gives only necessary condition for any radiation to be photochemically active. Justify.

20.9 The phenomenon of photosensitisation goes against Grotthuss–Draper law. Comment.

20.10 In a photosensitised process, the absorption of energy by photosensitiser follows some selection rule but transfer of energy to the reactant does not. Explain.

20.11 How does a photosensitiser differ from an ordinary sensitiser? Would you regard them as catalyst?

20.12 Is there any possibility of the following phenomena?

(a) Transition of a molecule to an excited electronic state by absorption of two photons almost simultaneously.

(b) Excitation of two molecules by a single photon.

20.13 What are the factors on which quantum yield of a photochemical reaction depend, and how? Give reasons for your answer.

20.14 What can be said about the mechanism of a photochemical reaction if its quantum yield is (i) high (ii) low? Can primary quantum yield be greater than one?

20.15 The quantum yield of the photochemical H_2–Cl_2 reaction is very high (10^4–10^6), while it is very low (of the order of 10^{-2} at ordinary temperature) with H_2–Br_2 reaction. How do you account for this?

20.16 In a photochemical process, the primary quantum yield should always be one. Comment.

20.17 The quantum yield of the photochemical decomposition of HI to $H_2 + I_2$ due to radiation of wavelength 400 nm is initially 2. Suggest a mechanism of the reaction. For the overall reaction, the elementary reactions H + I, I + HI, H + H and H + I_2 are not considered. Why?

20.18 For a photochemical reaction, is there any analogue of chemical equilibrium?

20.19 To maintain a photostationary state, light has to be passed through the system continually. What is the fate of the absorbed light?

20.20 NO_2 is decomposed by light of wavelength around 365 nm according to the equation

$$2NO_2 \rightarrow 2NO + O_2$$

with a quantum yield of 2 at the initial stage. Suggest a mechanism for this reaction. What happens to the quantum yield when an enclosed sample of NO_2 is irradiated for a long time where the reverse reaction (which occurs thermally) cannot be ignored.

20.21 A reaction responds to both red and violet light with an equal quantum yield. In which case, the amount of photochemical reaction will be greater per joule of light absorbed?

20.22 A plausible mechanism of dimerisation of anthracene (A) is

$$A + hv \longrightarrow A^*$$

$$A^* + A \xrightarrow{k_2} A_2$$

$$A^* \xrightarrow{k_3} A + hv'$$

$$A_2 \xrightarrow{k_4} 2A$$

Show that the maximum concentration of A_2 is determined only by the intensity of light absorption. Compare this with that expected in absence of light. Justify this difference.

20.23 Show that the rate of the photochemical H_2–Br_2 reaction varies proportionally with the square root of the intensity of the light absorbed.

20.24 Photolysis of acetaldehyde is supposed to occur through the following mechanism

$$CH_3CHO + hv \longrightarrow CH_3 + CHO$$

$$CH_3 + CH_3CHO \xrightarrow{k_2} CH_4 + CH_3CO$$

$$CH_3CO \xrightarrow{k_3} CH_3 + CO$$

$$CH_3 + CH_3 \xrightarrow{k_4} C_2H_6$$

Derive the expressions for the quantum yield of CO and the rate of consumption of CH_3CHO.

20.25 (a) What is the photochemical equivalence of the radiation of wavelength 800 nm?

(b) The absorption continuum of molecular iodine vapour begins at 500 nm (which is equivalent to 57.3 kcal mol^{-1}). The heat of dissociation of molecular iodine into atoms is 35.5 kcal mol^{-1} (which corresponds in energy to the wavelength 800 nm of light). What effect is expected if molecular iodine vapour is supplied with 35.5 kcal mol^{-1} of energy in form of (i) heat (ii) light of wavelength 800 nm?

(c) Also discuss the effect of irradiating the iodine vapour with light of wavelength equal to and below 500 nm.

20.26 The reaction between H_2 and O_2 cannot be brought about by radiation of wavelength 253 nm though it is equivalent to 112 kcal mol^{-1} that exceeds the thermal energy 103 kcal mol^{-1} required to initiate the reaction. However, the reaction occurs with the same radiation in presence of traces of Hg vapour. Explain.

20.27 If photodissociation of H_2 is brought about by light of wavelength 2537 Å in presence of Hg vapour, how much heat will result per mol of H_2 dissociated? [Dissociation energy of H_2 is 430.5 kJ mol^{-1} and its absorption continuum begins at 849 Å].

20.28 The photosynthetic process through chlorophyll in plants can be represented as

$$CO_2 + H_2O + xhv \rightarrow > CHOH + O_2 \qquad \Delta U = 502 \text{ kJ mol}^{-1}$$

The absorption maximum for chlorophyll is 593 nm. Find x. Is the overall process endothermic or exothermic?

20.29 In the photochemical formation of HCl from its elements, a quantum yield of 1.0×10^6 is found with light of wavelength 4800 Å. How many g of HCl would be produced per calorie of such radiation absorbed?

20.30 An uranyl oxalate actinometer is irradiated for 15 min with light of wavelength 4350 Å and oxalic acid equivalent to 12.0 mL of 0.001 M $KMnO_4$ is found to have been decomposed. The quantum yield of the actinometer being 0.58, find the rate of light incidence.

20.31 In an experiment light (400 nm) comprising of 7.5×10^{-3} mol of photons passes simultaneously through a cell containing a substance A and another cell containing a substance B, when 1.5×10^{-3} mol of A is decomposed and 10% of light is transmitted. The same amount of light when passed only through the second cell causes decomposition of 2.5×10^{-3} mol of B. Find quantum yield for the decomposition of A, if that of B is 0.5. Disregard the effect of the cell walls.

20.32 (a) Predict the possible products of photodecomposition of acetone with light of 3000 Å. The average bond energies are C–H: 414 kJ mol^{-1}, C–C: 347 kJ mol^{-1} and C=O: 732 kJ mol^{-1}.

   (b) Photodecomposition of acetone vapour is carried out in a cell of 60.0 mL capacity at 56°C with radiation of 3000 Å for 23000 sec at a rate of 85200 erg sec^{-1}. The pressure changes from 760 torr to 790 torr. Calculate (i) the number of moles of acetone decomposed and (ii) the quantum yield.

20.33 A photocell A of 1 cm length when filled with certain solution transmits a light of given wavelength by 75%. Another cell B of 2 cm length when filled with some other solution transmits 50% of the light of same wavelength. Find the percent transmission for the light passing simultaneously through both the cells.

20.34 A photocell when filled with liquid X transmits 60% of incident light of certain wavelength and when filled with liquid Y transmits only 30% of the light of the same wavelength. What would be the optical density at this wavelength if the photocell were filled with a mixture of X and Y in 1:2 volume ratio.

20.35 A 0.002 M solution of a substance transmits 75% of incident light of wavelength 500 mµ if path length is 1 cm. Calculate the extinction coefficient and percent transmission for a 0.001 M solution in a 2 cm cell, other factors remaining unchanged. Will the result vary if wavelength of the incident light is changed?

20.36 The following data for absorption of light by a sample of bromine dissolved in CCl$_4$ were obtained at a definite wavelength using a 2 cm cell.

[Br$_2$]/mol L^{-1}	0.001	0.005	0.010
% transmission	81.4	35.6	12.7

   Find (i) extinction coefficient and (ii) absorption cross-section of bromine.

20.37 Light absorption of proteins at 280 nm is due to the amino acid tryptophan having molar absorption coefficient 540 m^2 mol^{-1}. A 5.0×10^{-5} M solution of certain protein has an absorbance of 0.54 for 1 cm path length. How many tryptophan residues are present per molecule of the protein?

20.38 Certain gas at STP absorbs 50% of incident light of wavelength 500 nm for 2 cm path length. Find the percent transmission for 1 cm path length with the same gas at 0.5 atm (wavelength and temperature same).

20.39 A radiation of wavelength 400 nm and intensity 2×10^{17} photons cm^{-2} s^{-1} is passed for 10 min through a photocell of size 1 cm × 2 cm × 10 cm which is completely filled with a solution. The radiation is absorbed to the extent of 5% by the front wall, and 5% by the opposite wall of the cell, and 45% by the solution. Find (i) intensity of the transmitted radiation (ii) optical density of the solution (iii) the rate of absorption of photon per unit volume of the solution (I_{abs}) and (iv) total amount of energy absorbed by the solution during irradiation, if the radiation passes through the entire largest face of the cell. How will the corresponding quantities be affected when the radiation falls on the smallest face of the cell? Mention the assumption(s) taken.

20.40 A radiation of wavelength 400 nm and intensity 8×10^8 erg cm^2 s^{-1} enters into a photocell of length 1 cm through an area of 2 cm^2. The cell is filled with a 0.002 M solution of a substance having molar absorption coefficient 50 M^{-1} cm^{-1} and irradiated for 15 min. Find the total amount of energy absorbed by the solution and I_{abs}.

20.41 Total OD of a 0.05 M solution of a substance is 0.60 at 300 nm and is 1.10 for 0.10 M solution. Find the OD of the solvent and the extinction coefficient of the solute. The cell length is 2 cm.

20.42 A 1×10^{-3} M solution of a dye (x) shows an absorbance of 0.20 at 450 mμ and an absorbance of 0.05 at 620 mμ. A 1×10^{-4} M solution of another dye (Y) shows 0.00 absorbance at 450 mμ and an absorbance of 0.42 at 620 mμ. Calculate the concentration of each dye present together in a solution which exhibits an absorbance of 0.38 and 0.71 at 450 and 620 mμ respectively. The same cell is used in all measurements and its thickness is 1.00 cm.

20.43 A monochromatic light beam passes vertically through a 1 cm depth of a 0.02 M aqueous solution of certain salt in a cell. The accompanying OD is 0.4. But OD changes to 0.5 when depth of the solution is changed to 4 cm by addition of pure water to the cell. Calculate OD of 1 cm thickness of pure water and molar absorption coefficient of the salt (ignoring any absorption by the cell itself).

ANSWERS

20.1 No. The law is applicable for any electromagnetic radiation, provided it is absorbed by the optical system following Stark–Einstein law and there is no appreciable scattering or emission of radiation. [Vide note on problem 20.12].

☐ No. The law is applicable to any optically homogeneous phase.

20.2 ε depends on all the factors except (i) provided the concentration of the absorbing species is so low that the interactions between them can be ignored. Since ε is determined by the intensity of the transition responsible for light absorption, it will depend on the solvent (that determined the energy levels of the absorbing species), the temperature (that determines the population of the energy levels) and of course on the wavelength of the incident light whose quantum is required to fit in the gap between the appropriate energy levels of the absorbing species.

20.3 (i) No, if the complex species dissociate (or associate) appreciably in varying proportion with dilution.

(ii) No. The law is valid strictly only for monochromatic light, because the probability of photon absorption has true meaning only for a fixed wavelength.

(iii) No. The law concerns only with absorption of radiation.

20.4 (a) Because most plant and animal life on earth depends on the photosynthesis of carbohydrates by plants through the following reaction

$$nCO_2 + nH_2O + nh\nu \rightarrow (CH_2O)_n + nO_2$$

The reverse of this reaction provides energy for living beings. In absence of light, the equilibrium for this reaction lies far to the left.

Again, the protection of life on earth is possible due to the existence of ozone layer in the upper atmosphere that effectively filters the sun's rays out of harmful ultraviolet radiation being strongly absorbed by O_3.

Note: An approximate steady state of stratospheric O_3 is due to the following reaction

$$3O_2 \xrightleftharpoons[\text{UV}]{\text{UV}} 2O_3$$

Both the forward and backward processes are initiated through the formation of O-atoms by absorption of ultraviolet radiation.

(b) This is due to the different modes of activation in the photochemical and thermal reactions. In thermal reactions, the reactant molecules are activated by thermal energy through its temperature dependent distribution by means of intermolecular collisions. Therefore, the fraction of reactant molecules activated and hence the thermal reaction rate is quite dependent on temperature.

In photochemical reactions each reactant molecule is individually activated through sudden absorption of a photon of the radiation used (and not by gradual accumulation of energy from surrounding molecules). Here the fraction of molecules activated is determined by the intensity of the radiation used and not by temperature. Therefore, the photochemical reactions are expected not to be affected by temperature or only slightly affected due to the secondary thermal reactions following the primary light absorption process.

[In thermal reaction, the activation is comparable to climbing up a hill while in photochemical reaction, it is comparable to landing on the top of the hill.

Since molecules are activated in photochemical reactions not through distribution of supplied energy following a statistical principle, they do not follow Arrhenius type of rate law].

(c) Because using monochromatic light, a particular reaction can be brought about by exciting only the appropriate species in a reaction mixture. In contrast, heating to bring about a reaction thermally increases the energies (translational, rotational and vibrational) of all species.

(d) Because, although silver halide emulsions used in photofilms respond poorly to infrared radiation, they become sufficiently sensitive to such radiation in presence of certain dyes used as photosensitiser.

20.5 Because such substances produce Cl-atoms by absorbing ultraviolet radiation, $CFCl_3 + h\nu \rightarrow CFCl_2 + Cl$. Cl atoms thus formed can cause stratospheric O_3 depletion due to their catalytic action on decomposition of O_3

$$Cl + O_3 \rightarrow ClO + O_2$$
$$ClO + O \rightarrow Cl + O_2$$

20.6 H_2O. The reaction is initiated through the formation of H-atoms from H_2O

$$H_2O \xrightarrow[\text{chlorophyll}]{h\nu} 2H + O_2$$

H-atoms thus formed subsequently reduce CO_2 producing carbohydrates.

Such a mechanism is evidenced by the fact that all the liberated oxygen is $^{18}O_2$ when $H_2{}^{18}O$ is used instead of ordinary water.

20.7 According to thermodynamic principle, $\Delta G_{T,P}$ must be negative in any spontaneous process. In photochemical reactions too, this will hold provided G of the absorbed photon (that behaves as a reactant particle) is considered in calculating ΔG, otherwise the so-called free energy change (i.e. one disregarding photonic free energy) may be positive as is found with photosynthesis of carbohydrates by plants.

Note: Heating in thermal reaction amounts to exposure to infrared radiation which gradually interacts with the reactant particles. Here the radiation is treated as wave.

In contrast, the radiation is treated as shower of particles (behaving as reactant properties) in a photochemical process due to their sudden (one-time) interaction with the reactant particles.

20.8 Because the absorbed radiation may not be able to provide activation energy necessary for a chemical reaction to occur (as in case of absorption of infrared radiation). Further, the photoactivated reactants, if generated, may not undergo any chemical reaction due to their radiative or non-radiative deactivation.

20.9 Cannot be viewed as such, considering the photosensitiser as a part of the reaction system.

20.10 Because a photosensitiser takes up energy through absorption of photons but it gives up energy to its surrounding reactant molecules in non-radiative way (i.e. without emitting any photon). The latter process, unlike the former, is not restricted by any quantum mechanical selection rule.

20.11 A photosensitiser, unlike ordinary sensitiser, rarely undergoes any chemical change in bringing about an energetically possibe reaction.

Then we can regard a photosensitiser, but not an ordinary sensitiser, as catalyst. It is better to call it a photochemical catalyst.

20.12 For a high-power laser beam (that provides a very high density of photons), both (a) and, to a less extent, (b) have some probability. Liquid O_2, for example, absorbs at 630 nm wavelength where an absorbed photon excites two O_2 molecules simultaneously to the lowest-lying O_2 excited electronic state.

Note: Lambert–Beer law is not strictly valid for laser beams due to failure of Stark–Einstein law.

20.13 The quantum yield (QY) of a photochemical reaction depends on a number of factors such as (i) wavelength of the light used (ii) intensity of light (iii) reactant concentration (iv) temperature.

The manner in which QY will be affected by these factors is determined by the nature of the secondary processes that follow the primary light absorption process. However, it is usually adversely affected with increase in value of the first three factors, while it is favoured by rise of temperature.

☐ Of the two radiations that can cause same reaction, usually the one having higher wavelength will give lower QY due to greater chance of the photoactivated molecules to be deactivated by transferring smaller proportion of their energy to the other molecules with which they collide. Above certain wavelength, the QY falls rapidly with increasing wavelength.

The effect of factor (ii) is largely due to greater tendency for the (non-radiative) reversal of the initiation reaction due to higher concentration of the activated reactants.

The appearance of factor (iii) is due to collisional deactivation of the photoactivated molecules in higher proportion at higher reactant concentration. This is supported by very low QY of most reactions occurring in liquid phase.

The higher QY at higher temperature is due to increase in rate of the secondary thermal reaction with rise in temperature.

20.14 (i) Indicates that the reaction involves a chain mechanism with chain propagating steps adequately fast.

(ii) Does not exclude a chain mechanism of the reaction but with chain propagating steps quite slow.

☐ Primary quantum yield can be greater than one only in case of irradiation with a laser beam. Vide problem 20.12.

20.15 This is mainly because the chain propagating step $(X + H_2 \rightarrow HX + H)$ immediately following the initiation step $(X_2 + hv \rightarrow 2X)$ is much faster with chlorine than with bromine. At ordinary temperature, this reaction is so slow that most of the photo-generated bromine atoms recombine to produce bromine molecules leading to lower yield.

Note: With increase in temperature QY of H_2–Br_2 reaction increases markedly due to increase in rate of the step $(Br + H_2 \rightarrow HBr + H)$ in relatively higher proportion (due to higher activation energy) compared with the reaction $(Cl + H_2 \rightarrow HCl + H,$ which is exothermic).

20.16 The given statement is not true. Because the Stark–Einstein law, on which primary QY is based, is not always valid. Vide problem 20.12.

20.17 A probable mechanism of the reaction is

$$HI + hv \rightarrow H + I$$
$$H + HI \rightarrow H_2 + I$$
$$I + I \rightarrow I_2$$

Adding three steps, we have

$$2HI + hv \rightarrow H_2 + I_2$$

which is consistent with the Q·Y of 2 for the overall reaction.

☐ The elementary atomic reactions H + H and H + I (which involve simplest atom H) are not considered because they are so highly exothermic that three-body collisions are essential for their significant occurrence. The reaction I + HI (which is endothermic) is disregarded because of its high activation energy (and hence slower rate). The reaction $H + I_2$ (though of small activation energy like H + HI reaction) has little role at the initial stage of the reaction where the concentration of I_2 is low.

Note: (i) Q·Y will differ significantly from 2 near the end of the reaction when the elementary reaction $H + I_2$ cannot be ignored (ii) The decomposition of HI is not a chain reaction.

20.18 Yes. This is so-called photochemical equilibrium which requires a continuous supply of radiant energy for its maintenance. On cutting off the source of radiation, the composition of the system will no longer remain constant. Being a steady-state situation, the phenomenon should be better called photostationary state.

20.19 Heat, which is liberated by the backward thermal reaction.

20.20 A probable mechanism of the reaction is

$$NO_2 + hv \rightarrow NO_2^*$$
$$NO_2^* + NO_2 \rightarrow 2NO + O_2$$

where the asterisk indicates an activated molecule. Here absorption of one photon leads to decomposition of two NO_2 molecules, i.e. Q·Y is 2.

☐ On prolonged illumination, the system will attain a photostationary state when there will be no net chemical reaction and all the absorbed light will appear as heat. Then continuous illumination will result in decrease of Q·Y approaching zero.

20.21 Here the amount of photochemical reaction per quantum of light absorbed is same for both red and violet light. Then red light having lower value of quantum (due to lower frequency) will cause greater amount of reaction per joule of light absorbed.

Note: Although red light is less effective than violat in bringing about chemical reactions, the above unusual result follows from the assumption of equal Q·Y for both the light.

20.22 In the steady state, when $d[A^*]\,dt = 0$, we have

$$I_{abs} - k_2[A^*][A] - k_3[A^*] = 0$$

Again, in the photostationary state, when $d[A_2]/dt = 0$, we have

$$k_2[A^*][A] - k_4[A_2] = 0$$

Elimination of $[A^*]$ from the last two equations gives

$$[A_2] = \frac{k_2[A]\,I_{abs}}{k_4(k_2[A] + k_3)}$$

$$= \frac{I_{abs}}{k_4} \quad \text{for high } [A] \text{ when } k_2[A] \gg k_3$$

Then maximum concentration of A_2, that happens with high concentration of A, is independent of $[A]$ and is determined only by the intensity of light absorption.

☐ In absence of light A_2 will be in equilibrium with A, and from thermodynamic consideration, we expect

$$[A]_2 = K[A] \quad \text{where } K \text{ is the equilibrium constant}$$

☐☐ In absence of light $[A_2]$ varies with $[A]$ following law of mass action, but in presence of light $[A_2]$ is independent of $[A]$ provided $[A]$ is high. Because at high $[A]$, when $I_{abs} \ll [A]$, the formation of A_2 is determined only by the number of photons colliding with A molecules. The law of mass action holds in absence of light but not in presence of light because in the latter case the system can attain a steady or photostationary state but never an equilibrium state.

Note: The reaction is of zero order for high concentration of reactant. But the reaction rate at low reactant concentration depends on the latter in a complicated way.

20.23 Photochemical H_2–Br_2 reaction occurs like thermal reaction [Vide Section 15.4(D)] through the same chain mechanism except for the chain initiation process

$$Br_2 + h\nu \to 2Br \quad v = I_{abs} \text{ (more generally } k \cdot I_{abs})$$

where I_{abs} is the intensity of light absorption. Then by replacing the initiation rate $k_1[Br_2]$ for thermal reaction with I_{abs} in the Eq. (15.11b), we have the following rate expression for the photochemical reaction

$$\frac{d[HBr]}{dt} = \frac{2k_2\left(\dfrac{1}{k_5}\right)^{\frac{1}{2}}[H_2]\,I_{abs}^{\frac{1}{2}}}{1 + \dfrac{k_4}{k_5}\cdot\dfrac{[HBr]}{[Br_2]}}$$

20.24 Steady-state approximation for CH_3 and CH_3CO radicals gives

$$\frac{d[CH_3]}{dt} = I_{abs} - k_2[CH_3][CH_3CHO] + k_3[CH_3CO] - 2k_4[CH_3]^2 = 0 \quad \text{(A)}$$

$$\frac{d[CH_3CO]}{dt} = k_2[CH_3][CH_3CHO] - k_3[CH_3CO] = 0 \quad \text{(B)}$$

(A) + (B) gives

$$2k_4[CH_3]^2 = I_{abs} \quad \text{or} \quad [CH_3] = \left(\frac{I_{abs}}{2k_4}\right)^{\frac{1}{2}}$$

Again from (B)
$$k_3[CH_3CO] = k_2[CH_3][CH_3CHO]$$

$$= k_2 \left(\frac{I_{abs}}{2k_4}\right)^{\frac{1}{2}} [CH_3CHO]$$

Now $\qquad \dfrac{d[CO]}{dt} = k_3[CH_3CO]$

$$= k_2 \left(\frac{I_{abs}}{2k_4}\right)^{\frac{1}{2}} [CH_3CHO]$$

Then $\qquad$ Q·Y of CO $= \dfrac{d[CO]}{dt} \Big/ I_{abs} = \dfrac{k_2[CH_3CHO]}{(k_4 I_{abs})^{\frac{1}{2}}}$

$$-\frac{d[CH_3CHO]}{dt} = I_{abs} + k_2[CH_3][CH_3CHO]$$

$$= I_{abs} + k_2 \left(\frac{I_{abs}}{2k_4}\right)^{\frac{1}{2}} [CH_3CHO]$$

Note: Here it seems unusual that $-\dfrac{d[CH_3CHO]}{dt} \neq \dfrac{d[CO]}{dt}$ when we represent overall decomposition of CH_3CHO by the fixed stoichiometric equation CH_3CHO = $CH_4 + CO$. Actually no fixed stoichiometry is followed in chain reactions.
The above rate expressions for formation of CO and consumption of CH_3CHO are approximate, because they are derived ignoring the decomposition of CHO radical

$$CHO \rightarrow CO + H$$
$$H + CH_3CHO \rightarrow CH_3CO + H_2$$

20.25 (a) One einstein of radiation

$$= \frac{N_A hc}{\lambda}$$

$$= \frac{(6.02 \times 10^{23} \text{ mol}^{-1})(6.63 \times 10^{-34} \text{ Js})(3 \times 10^8 \text{ m s}^{-1})}{(800 \times 10^{-9} \text{ m})} \text{ for the given wavelength}$$

$$= 1.50 \times 10^5 \text{ J mol}^{-1}$$

Then by Stark–Einstein law

required photochemical equivalence $= \dfrac{1}{1.50 \times 10^5 \text{ J mol}^{-1}}$

$$= 6.67 \times 10^{-6} \text{ mol J}^{-1}$$

(b) (i) I_2 molecules will dissociate, the required amount of thermal energy being supplied.

(ii) I_2 molecules will not dissociate, because the wavelength of radiation used is higher than the wavelength at which the absorption continuum of I_2 begins.

(c) At 500 nm, I_2 will dissociate. However, the absorbed energy being in much excess of 35.5 kcal m ol^{-1}, one of the iodine atoms formed will not be in ground electronic state.

Absorption at wavelength shorter than 500 nm will of course lead to dissociation of I_2 molecule. Here the extra energy will be converted into translational energy of the system's particles and will appear as heat.

Note: To bring about a reaction, the necessary amount of energy depends on nature of the latter. Here the energy to be supplied in form of light is greater than that in form of heat because of certain quantum mechanical restrictions regarding interaction of matter with radiation.

20.26 Here $H_2 + O_2$ reaction is actually brought about by thermal energy which is provided by Hg in exchange of radiant energy absorbed by it. Hence regarding absorption of energy, quantum mechanical restrictions arise with Hg (the energy being radiant) but not with $H_2 + O_2$ system (the energy being thermal).

20.27 Here Hg acts as a photosensitiser

Resulting amount of heat

= radiant energy absorbed by Hg, $N_A hc/\lambda$ – thermal energy consumed in dissociation of H_2

$$= \frac{(6.02 \times 10^{23} \text{ mol}^{-1})(6.63 \times 10^{-34} \text{ Js})(3 \times 10^8 \text{ ms}^{-1})}{(2537 \times 10^{-10} \text{ m})} - 430.5 \times 10^3 \text{ J mol}^{-1}$$

$$= 4.15 \times 10^4 \text{ J mol}^{-1}$$

20.28 $$x = \frac{\Delta U}{N_A hc/\lambda}$$

$$= \frac{502 \times 10^3 \text{ J mol}^{-1}}{(6.02 \times 10^{23} \text{ mol}^{-1})(6.63 \times 10^{-34} \text{ Js})(3 \times 10^8 \text{ ms}^{-1})/(594 \times 10^{-9} \text{ m})}$$

$$= 3 \text{ (rounded up figure)}$$

The radiant energy (corresponding to $x = 3$) is in excess of $\Delta U = 502$ kJ mol^{-1} which is likely to appear as heat (vide problem 20.27). Then the overall process is exothermic.

20.29 Energy of photon $= \dfrac{N_A hc}{\lambda}$

$$= \frac{(6.02 \times 10^{23} \text{ mol}^{-1})(6.63 \times 10^{-34} \text{ Js})(3 \times 10^8 \text{ ms}^{-1})}{4800 \times 10^{-10} \text{ m}}$$

$$= 2.49 \times 10^5 \text{ J mol}^{-1}$$

1 cal $\equiv 4.18$ J

$$\equiv \frac{4.18}{2.49 \times 10^5} \text{ or } 1.68 \times 10^{-5} \text{ mol of photon}$$

Quantity of HCl formed = moles of photon absorbed × QY

$$= (1.68 \times 10^{-5})(1.0 \times 10^6) \text{ moles}$$

$$= 1.68 \times 10 \times 36.5 \text{ or } 1226.4 \text{ g}$$

20.30 1 mol of $MnO_4 \equiv \dfrac{5}{2}$ mol of oxalic acid

Then 12.0 ml of 0.001 M KMnO$_4 \equiv \dfrac{12}{1000} \times 0.001 \times \dfrac{5}{2}$ or 3.0×10^{-5} mol of oxalic acid

$$\text{Rate of light incidence} = \frac{\text{rate of decomposition of oxalic acid}}{Q \cdot Y}$$

$$= \frac{\left(\dfrac{3.0 \times 10^{-5}}{15 \times 60}\right) \text{mol s}^{-1}}{0.58 \text{ mol einstein}^{-1}}$$

$$= 5.75 \times 10^{-8} \text{ einstein s}^{-1}$$

$$= 5.75 \times 10^{-8} \times 6.02 \times 10^{23} \text{ quanta s}^{-1}$$

Note: Here it has been assumed that all the photons entering into the actinometer is absorbed.

20.31 In the second case

$$\text{moles of photons absorbed by } B = \frac{\text{moles of } B \text{ decomposed}}{Q \cdot Y \text{ for } B}$$

$$= \frac{2.5 \times 10^{-3}}{0.5} \text{ or } 5 \times 10^{-3}$$

Then, from the first experimental data, the moles of photons absorbed by A

$$= \frac{100 - 10}{100} \times 7.5 \times 10^{-3} - 5 \times 10^{-3} = 1.75 \times 10^{-3}$$

Therefore, QY for decomposition of $A = \dfrac{\text{moles of A decomposed}}{\text{moles of photons absorbed by A}}$

$$= \frac{1.5 \times 10^{-3}}{1.75 \times 10^{-3}} \text{ or } 0.86$$

20.32 (a) Energy of photon $= \dfrac{N_A hc}{\lambda}$

$$= \frac{(6.02 \times 10^{23} \text{ mol}^{-1})(6.63 \times 10^{-34} \text{ Js})(3 \times 10^8 \text{ ms}^{-1})}{3000 \times 10^{-10} \text{ m}}$$

$$= 400 \text{ kJ mol}^{-1}$$

This is greater than C–C bond energy, but less than C–H and C=O bond energies. Then the absorption of light of wavelength 3000 Å can dissociate only C–C bond of CH_3COCH_3 leading to the possible products CO and C_2H_6.

$$\underset{}{CH_3} - \overset{\overset{\displaystyle O}{\overset{\displaystyle \|}{}}}{C} - CH_3 + 2h\nu \rightarrow CO + \underset{\displaystyle \downarrow}{2CH_3}$$
$$C_2H_6$$

(b) (i) Number of moles of acetone decomposed $= \dfrac{\Delta PV}{RT}$

$$= \frac{\left(\dfrac{30}{760} \text{ atm}\right)\left(\dfrac{60}{1000} \text{ L}\right)}{(0.082 \text{ L atm K}^{-1} \text{mol}^{-1})(329 \text{ K})}$$

$$= 8.77 \times 10^{-5} \text{ mol}$$

(ii) Number of photons absorbed

$$= \frac{\text{rate of absorption of radiant energy} \times \text{duration of irradiation}}{\text{energy of photon } (Nhc/\lambda)}$$

$$= \frac{(85200 \text{ erg s}^{-1})(23000 \text{ s})}{400 \times 10^3 \times 10^7 \text{ erg mol}^{-1}} = 4.90 \times 10^{-4} \text{ mol}$$

$$\text{Quantum yield} = \frac{\text{number of molecules decomposed}}{\text{number of photons absorbed}}$$

$$= \frac{8.77 \times 10^{-5} \text{ mol}}{4.90 \times 10^{-4} \text{ mol}} = 0.179$$

20.33 Here transmitted light for the first cell is the incident light for the second cell. Then

$$\% \text{ transmission} = \frac{75}{100} \times \frac{60}{100} \times 100 = 45$$

20.34 On mixing OD ($= \in cl$) of each substance will change only due to dilution, assuming no interaction between X and Y. Then for the mixture

$$\text{OD}_{mix} = \text{O} \cdot \text{D}_X + \text{O} \cdot \text{D}_Y, \quad \text{OD being an additive property}$$

$$= \frac{1}{3} \cdot \text{OD}_{\text{pure } X} + \frac{2}{3} \cdot \text{OD}_{\text{pure } Y}$$

$$= \frac{1}{3} \log \frac{100}{60} + \frac{2}{3} \log \frac{100}{30} = 0.42$$

20.35 Extinction coefficient, $\in = \dfrac{\log \dfrac{I_0}{I}}{cl}$ by Eq. (20.1b)

$$= \frac{\log \dfrac{100}{75}}{(0.002 \text{ M})(1 \text{ cm})} = 62.5 \text{ M}^{-1} \text{ cm}^{-1}$$

$$\% \text{ transmission} = 100 \times \frac{I}{I_0} = 100 \times \text{anti log}\,[-\in cl]$$

$$= 100 \times \text{antilog}\,[-(62.5 \text{ M}^{-1} \text{ cm}^{-1})(0.001 \text{ M})(2 \text{ cm})]$$

$$= 75.0$$

☐ Yes. Because $\in$ depends on wavelength of the light used.

Note: In the above calculations absorption due to solvent has been ignored. This is not really justified. Vide problem 20.41.

20.36 (i) By Eq. (20.1b), the average value of $\in$ is

$$\in = \frac{1}{3} \left[\frac{\log \dfrac{100}{81.4}}{(0.001 \text{ M})\left(\dfrac{2}{10} \text{ cm}\right)} + \frac{\log \dfrac{100}{35.6}}{(0.005 \text{ M})\left(\dfrac{2}{10} \text{ cm}\right)} + \frac{\log \dfrac{100}{12.7}}{(0.01 \text{ M})\left(\dfrac{2}{10} \text{ cm}\right)} \right]$$

$$= 149.3 \text{ M}^{-1} \text{ cm}^{-1}$$

(ii) From Eqs. (20.1b) and (20.1c), we have

$$\sigma = 2.303 \in \cdot \frac{c}{n} = \frac{2.303 \in}{N_A} \quad \text{in SI}$$

$$= 2.303 \, (149.3 \text{ L mol}^{-1} \text{ cm}^{-1}) \cdot \frac{1000 \text{ cm}^3 \text{L}^{-1}}{6.02 \times 10^{23} \text{ mol}^{-1}} \quad \text{in conventional unit}$$

$$= 5.71 \times 10^{-19} \text{ cm}^2$$

20.37 $$[\text{tryptophan}] = \frac{\log \dfrac{I_0}{I}}{\epsilon \, l}$$

$$\doteq \frac{0.54}{(540 \text{ m}^2 \text{mol}^{-1})(10^{-2} \text{ m})}$$

$$= 10^{-1} \text{ mol m}^{-3} = 10^{-4} \text{ mol L}^{-1}$$

Then the number of tryptophan residues per protein molecule

$$= \frac{[\text{tryptophan}]}{[\text{protein}]} = \frac{10^{-4} \text{ mol L}^{-1}}{5.0 \times 10^{-5} \text{ mol L}^{-1}} = 2$$

20.38 By Lambert–Beer law

$$\log \frac{I_0}{I} \propto \epsilon \, Pl$$

since $P \propto c$ at constant temperature assuming the gas to behave ideally

Then for two different pressures and cell lengths, we have

$$\log\left(\frac{I}{I_0}\right)_2 = \log\left(\frac{I}{I_0}\right)_1 \cdot \frac{P_2 l_2}{P_1 l_1}$$

$$= \log\left(\frac{50}{100}\right) \cdot \frac{0.5 \times 1}{1 \times 2}$$

whence $$\frac{I}{I_0} = 0.84$$

Then the % transmission = 0.84 × 100 or 84.

20.39 Assuming that part of the incident light is absorbed and the rest of it is only transmitted, each wall of the cell will transmit (100 – 5) or 95% of the light falling on it and the solution transmits (100 – 45) or 55% of the light falling on it.

(i) Intensity of the transmitted radiation

= intensity of the incident radiation × fraction transmitted by the incident wall × fraction transmitted by the solution × fraction transmitted by the opposite wall

$$= (2 \times 10^{17} \text{ photons cm}^{-2}\text{s}^{-1})\left(\frac{95}{100}\right)\left(\frac{55}{100}\right)\left(\frac{95}{100}\right)$$

$$= 0.993 \times 10^{17} \text{ photons cm}^{-2}\text{s}^{-1}$$

(ii) If I_0 is the intensity of radiation entering into the solution and I is that of the radiation transmitted by the solution then

$$\text{OD} = \log \frac{I_0}{I} = \log \frac{100}{55} = 0.26$$

(iii) $$I_{abs} = \frac{(I_0 - I) \times \text{area through which radiation enters into the solution}}{\text{volume of the solution}}$$

$$= \frac{\left(2 \times 10^{17} \text{ photons cm}^{-2}\text{s}^{-1} \times \dfrac{95}{100} - 2 \times 10^{17} \text{ photons cm}^{-2}\text{s}^{-1} \times \dfrac{95}{100} \times \dfrac{95}{100}\right)(2 \text{ cm} \times 10 \text{ cm})}{1 \text{ cm} \times 2 \text{ cm} \times 10 \text{ cm}}$$

$$= 9.5 \times 10^{15} \text{ photons cm}^{-3}\text{s}^{-1}$$

(iv) Energy of a photon $= \dfrac{hc}{\lambda}$

$$= (6.63 \times 10^{-27} \text{ erg s}) \dfrac{(3.00 \times 10^{10} \text{ cm s}^{-1})}{(400 \times 10^{-7} \text{ cm})}$$

$$= 4.97 \times 10^{-12} \text{ erg}$$

Total amount of energy absorbed by the solution

$= I_{abs} \times$ volume of the solution $\times$ duration of irradiation

$\qquad \times$ energy of a photon (hc/λ)

$$= (9.5 \times 10^{15} \text{ photons cm}^{-3}\text{s}^{-1}) \, (20 \text{ cm}^3) \, (10 \times 60 \text{ s}) \, (4.97 \times 10^{-12} \text{ erg})$$

$$= 5.66 \times 10^{12} \text{ erg}$$

☐ (i) will be less due to greater path length

(ii) will be greater due to lower transmission

(iii) will be less due to lower area of incidence though I is lower, the effect of area being predominating

(iv) will be less due to lower I_{abs}.

Note: $I_{abs} \neq I_0 - I$.

20.40 Total amount of energy absorbed

$$= (I_0 - I) \, A \, t, \quad A \text{ is the area through which the radiation entering the cell}$$

$$= I_0 \, (1 - 10^{-\epsilon \, cl}) \, A t$$

$(8 \times 10^8 \text{ erg cm}^{-2}\text{s}^{-1}) \, (1 - 10^{-50 \times 0.002 \times 1}) \, (2 \text{ cm}^2) \, (15 \times 60 \text{ s})$

$$= 3.29 \times 10^8 \text{ erg s}^{-1} \times 900 \text{ s}$$

$$= 2.96 \times 10^{11} \text{ erg}$$

Energy of a photon $= \dfrac{hc}{\lambda}$

$$= 4.97 \times 10^{-12} \text{ erg} \quad \text{[from previous problem]}$$

$$I_{abs} = \dfrac{(I_0 - I) \, A}{V} \Big/ \dfrac{hc}{\lambda}$$

$$= \dfrac{3.29 \times 10^8 \text{ erg s}^{-1}}{(1 \times 2 \text{ cm}^3) \, (4.97 \times 10^{-12} \text{ erg})}$$

$$= 3.31 \times 10^{19} \text{ photons cm}^{-3} \text{ s}^{-1}$$

20.41 $\qquad OD_{soln} = OD_{solute} + OD_{solvent}$

$$= \epsilon \, cl + OD_{solvent}$$

Then from the given data

$$\left. \begin{array}{l} 0.60 = (0.50 \text{ M}) \, (2 \text{ cm}) \, \epsilon + OD_{solvent} \\ 1.10 = (0.10 \text{ M}) \, (2 \text{ cm}) \, \epsilon + OD_{solvent} \end{array} \right\} \Rightarrow \begin{array}{l} OD_{solvent} = 0.1 \\ \epsilon = 5 \text{ M}^{-1} \text{ cm}^{-1} \end{array}$$

20.42 For X,

at 450 mμ $\qquad \epsilon_X = \dfrac{OD}{cl} = \dfrac{0.20}{(1 \times 10^{-3} \text{ M}) \, (1 \text{ cm})} = 200 \text{ M}^{-1} \text{ cm}^{-1}$

at 620 mμ $\qquad \epsilon'_X = \dfrac{OD}{cl} = \dfrac{0.05}{(1 \times 10^{-3} \text{ M}) \, (1 \text{ cm})} = 50 \text{ M}^{-1} \text{ cm}^{-1}$

For Y,

at 450 mμ $\qquad \epsilon_Y = \dfrac{OD}{cl} = 0$

at 620 mμ $\qquad \epsilon'_Y = \dfrac{OD}{cl} = \dfrac{0.42}{(1 \times 10^{-4}\ M)\,(1\ cm)} = 4200\ M^{-1}\ cm^{-1}$

For the mixture of X and Y by Eq. (20.1d)

at 450 mμ $\qquad 0.38 = (1\ cm)\,[(200\ M^{-1}\ cm^{-1})\,c_X + 0]$ whence $c_X = 1.9 \times 10^{-3}\ M$

at 620 mμ $\qquad 0.71 = (1\ cm)\,[(50\ M^{-1}\ cm^{-1})\,(1.9 \times 10^{-3}\ M) + 4200\ M^{-1}\ cm^{-1}\ c_Y]$

whence $c_Y = 1.5 \times 10^{-4}\ M$

20.43 OD_{soln} for 1 cm thickness

$\qquad 0.4 = OD_{solute}\,(\epsilon\, cl) + OD_{water}$ for 1 cm thickness $\qquad$ (A)

OD_{soln} for 4 cm thickness

$\qquad 0.5 = OD_{solute} + OD_{water}$ for 4 cm thickness

$\qquad = OD_{solute} + 4 \times OD_{water}$ for 1 cm thickness $\qquad$ (B)

From (A) and (B)

OD_{water} for 1 cm thickness $= 0.025$

$\qquad OD_{solute} = 0.375$

$$\epsilon = \dfrac{OD_{solute}}{cl}$$

$$= \dfrac{0.375}{(0.02\ M)\,(1\ cm)}$$

$$= 18.75\ M^{-1}\ cm^{-1}$$

Note: Here OD_{solute} does not change due to addition of water because cl does not change. This is unlike horizontal light path where $OD_{solvent}$ does not change due to dilution.

UNIVERSITY QUESTIONS

20.1 Photochemical reactions should follow an Arrhenius type rate law. Justify or comment. (Calcutta BSc(H), 1993)

20.2 A photochemical reaction takes place according to the following mechanism:

$$HI + h\nu \xrightarrow{\ k_1\ } H + I$$
$$H + HI \xrightarrow{\ k_2\ } H_2 + I$$
$$I + I \xrightarrow{\ k_3\ } I_2$$

Prove that two moles of HI would decompose per einstein of radiation absorbed. (Burdwan BSc(H), 2011)

20.3 Explain the effect of light on an equilibrium mixture of reactants and products when the forward reaction is a photochemical one and the backward process is a dark reaction. (Calcutta BSc(H), 1998)

20.4 What do you understand by 'photostationary state'? State with reason whether it it a steady state or an equilibrium state. (Calcutta BSc(H), 2013)

20.5 The reaction $2A \rightleftharpoons A_2$ occurs both thermally and photochemically. The photochemical reaction takes place with the following steps:

(i) $A + hv \longrightarrow A^*$

(ii) $A^* + A \xrightarrow{k_2} A_2$

(iii) $A_2 \xrightarrow{k_3} 2A$

(iv) $A^* \xrightarrow{k_4} hv'$

Applying steady state approximation to A^*, show that

$$[A_2] = \frac{I_{abs}}{k_3 \left[1 + \dfrac{k_4}{k_2[A]} \right]} \text{ at photostationary equilibrium}$$

Also show that $[A_2]$ is independent of $[A]$, when A is present in large excess. Compare the result with that in the thermal equilibrium case.

(Calcutta BSc(H), 2006)

20.6 Explain why the quantum yield for a primary photochemical process is unity. Also explain why overall quantum yield for a reaction differs from unity.

(Burdwan BSc(H), 1994)

20.7 A photochemical reaction takes place according to the following mechanisms:

$$A + hv \longrightarrow A^*$$
$$A^* + M \xrightarrow{k_2} A + M$$
$$A^* \xrightarrow{k_3} \text{Product}$$

Derive an expression for the quantum yield of the reaction.

(Burdwan BSc(H), 2006)

20.8 (a) The following mechanism has been proposed for the photochemical formation of HCl:

$$Cl_2 + hv \xrightarrow{k_1} 2Cl$$
$$Cl + H_2 \xrightarrow{k_2} HCl + H$$
$$H + Cl_2 \xrightarrow{k_3} HCl + Cl$$
$$Cl \xrightarrow{k_4} \tfrac{1}{2}Cl_2 \text{ (on wall)}$$

Obtain the expression for the overall rate of formation of HCl. Comment on the quantum yield of the reaction.   (Calcutta BSc(H), 1997)

(b) Comment on the high values of quantum yield for the photochemical formation of HCl from H_2 and Cl_2.   (Burdwan BSc(H), 2000

20.9 The quantum yield for the photochemical formation of HBr from hydrogen and bromine gases is very low. Explain.   (Calcutta BSc(H), 2013)

20.10 A suggested mechanism for the photolysis of ozone in low energy light (red light) is

$$O_3 + hv \xrightarrow{k_1} O_2 + O$$
$$O_3 + O \xrightarrow{k_2} 2O_2$$
$$O + O_2 + M \xrightarrow{k_3} O_3 + M$$

(a) Derive an expression for the overall rate of disappearance of ozone.

(b) Write the expression for the overall quantum yield for the disappearance of ozone, ϕ.

(c) At low total pressure $\phi = 2$, what is the value of k_1?   (Burdwan BSc(H), 2008)

20.11 (a) State law of photochemical equivalence. Mention one exception to it.

(Calcutta BSc(H), 2014)

(b) State Grotthuss–Draper law of photochemistry explaining its significance.

(Burdwan BSc(H), 2000)

20.12 The photodissociation of gaseous HI to form normal hydrogen and iodine atoms require radiation of 4040 Å or less. (i) Determine the molar heat of dissociation of HI, (ii) if radiation of 2537 Å is used, how much energy will appear as kinetic energy of the atoms. (Calcutta BSc(H), 2009)

20.13 The quantum yield is 2 for the photolysis of gaseous HI to $H_2 + I_2$ by light of 253.7 nm wavelength. Calculate the number of moles of HI that will decomposed if 300 J of light of this wavelength is absorbed. (Jadavpur BSc(H), 1999)

20.14 A 100 watt sodium vapour lamp radiates most of its energy in the yellow D line at 589 nm. How long will such a lamp take to excite more than half the molecules of an absorbing species in a 10^{-3} mol dm^{-3} sample if all the radiant energy are absorbed by the sample? (Calcutta BSc(H), 1991)

20.15 The quantum yield is unity for the dissociation of acetone vapour at 150°C if irradiated by 254 nm radiation. How long it will take to dissociate 10^{-2} mol of acetone using 100 W monochromatic light source of 254 nm radiation?

(Jadavpur BSc(H), 2001)

20.16 In a particular experiment, an incident light (300 nm) passes simultaneously through two cells, an empty cell and then a cell containing uranyl oxalate solution. It is found that 6.201×10^{-3} mol of oxalate has been decomposed in 2 hours of time. With the same set up when the empty cell is filled up with acetone, 1.4×10^{-3} mol of acetone decomposed in 10 hours of time and in the next cell 2.631×10^{-2} mol of oxalate has been decomposed. Find the quantum yield of acetone decomposition if that of oxlate decomposition be 0.57. (Jadavpur BSc(H), 2000)

20.17 Give an example of a photochemical reaction in which energy absorbed by one species brings about changes elsewhere. Describe an important application of it in quantitative photochemistry. (Calcutta BSc(H), 1992)

20.18 Find the wavelength of light necessary to break photochemically a H–H bond if the average bond energy is 431 kJ mol^{-1}. Which substance Hg (g) or Na (g) would serve as an efficient photosensitiser if the primary absorption wavelengths are 2563.5 Å and 3302.99 Å respectively. (Burdwan BSc(H), 1994)

20.19 Radiation of wavelength 2540 Å was passed through a cell containing 10 ml of a solution of 0.0495 M oxalic acid and 0.01 M uranyl sulphate, after the absorption of 8.81×10^8 ergs of radiation, the concentration of oxalic acid was reduced to 0.0383 M. Calculate the quantum yield for the photochemical decomposition of oxalic acid at the given wavelength. (Calcutta BSc(H), 2007)

20.20 An actinometer uses a solution of $K_3[Fe(C_2O_4)_3]$ in which Fe^{3+} is reduced and the oxalate ion is oxidised. Assuming that the quantum yield is 1.24 at 310 nm, calculate the intensity of light incidence which produces 1.3×10^{-5} mol of Fe^{2+} in 36.5 min. The same light source is used to irradiate a sample of CH_2CO for a period of 15.2 min. If the quantum yield of C_2H_4 is 1.0 and that of CO is 2.0, determine the amount of each gas produced by the photochemical reaction.

(Burdwan BSc(H), 1995)

20.21 Quantum yield for the reaction to produce Fe^{2+} in a ferrioxalate actinometer is 0.94. After an irradiation session, the resulting solution of volume 57.4 ml was reacted to produce a coloured solution with molar extinction coefficient 1.11×10^4 L mol^{-1} cm^{-1}. This solution gave an absorbance value of 0.663 in a 1 cm cell. How much quanta were absorbed by the ferrioxalate solution   (Calcutta BSc(H), 2014)

20.22 What percentage of light will be transmitted through two cells put together in the path of light, if their individual transmissions are 60% and 30%.

(Kalyani BSc(H), 2006)

20.23 State and explain Lambert–Beer law. Discuss the deviation of the law from its ideal behaviour.   (Burdwan BSc(H), 1997)

20.24 What will be the maximum and minimum value of optical density. What will be the unit of molar extinction coefficient?   (Calcutta BSc(H), 1998)

20.25 A certain substance in a cell of length l absorbs 10% of incident light. What percentage of light will be absorbed in cell which is five times as long.

(Delhi BSc, 2003)

20.26 Gaseous acetone transmits 25.1% of an incident light in a cell of certain length and at a pressure of 100 mm of Hg. Assuming Beer law, calculate the pressure at which 98% of the same incident light will be absorbed.   (Calcutta BSc(H), 1994)

20.27 What is the molar extinction coefficient of a solute which absorbs 60% at 400 nm of light when light beam is passed through a 0.1 m cell consisting of 5×10^{-6} mol dm^{-3} solution? At what concentration absorption will be 99.99%.

(Jadavpur BSc(H), 1996)

20.28 A solution of a coloured compound of concentration 1.4×10^{-4} M has 20% trans-mission in a cell of path length 1 cm at 450 nm wavelength. Calculate the molar absorption coefficient (ϵ) of the substance. If the path length and the concentration are both halved, calculate the percentage transmission. Will the value of ϵ change if light of wavelength 550 nm is used?   (Calcutta BSc(H), 2005)

20.29 A mixture of dichromate and permanganate ions was analysed spectrophotometri-cally at 440 nm and 545 nm as a means for the simultaneous determination of these two species, and the observed absorbance values were 0.385 and 0.653 respectively, at each wavelength for a 1.00 cm cell. Calculate the concentrations of dichromate and permanganate in the unknown mixture.

[**Given:**

for $C_2O_7^{2-}$ $\epsilon_{440} = 370$ M^{-1} cm^{-1}

$\epsilon_{545} = 10.8$ M^{-1} cm^{-1}

for MnO_4^- $\epsilon_{440} = 92.8$ M^{-1} cm^{-1}

$\epsilon_{545} = 2350$ M^{-1} cm^{-1} (Burdwan BSc(H), 1997

KEY TO UNIVERSITY QUESTIONS

20.1 Vide problem 20.4(b).

20.2 Vide problem 20.17.

We can also establish this considering the rates of individual steps. Here

$$\frac{d[H]}{dt} = 0 = I_{abs} - k_2[H][HI] \text{ in the steady state}$$

and $-\dfrac{d[HI]}{dt} = I_{abs} + k_2[H][HI]$

Adding these two equations, we have

$$-\frac{d[HI]}{dt} = 2I_{abs}$$

Note:

(i) Ordinarily the reaction is very slow, as I_{abs} is low even with a fairly intense light source.

(ii) The reaction rate does not depend on [HI] directly, but does so indirectly, as I_{abs} depends on [HI].

20.3 Vide Section 20.4.

20.4 Vide problem 20.18.

20.5 Vide problem 20.22.

20.6 Vide Section 20.3.

20.7 Applying steady-state approximation to the short lived A^* molecules, we have

$$\frac{d[A^*]}{dt} = 0 = I_{abs} - k_2[M][A^*] - k_3[A^*]$$

whence

$$[A^*] = \frac{I_{abs}}{k_2[M] + k_3}$$

Overall reaction rate

$$k_3[A^*] = \frac{k_3 I_{abs}}{k_2[M] + k_3}$$

Then, quantum yield $= \dfrac{k_3}{k_2[M] + k_3}$

Note: QY $\to 1$ as $[M] \to 0$.

20.8 (a) Applying steady-state approximation to the atomic species, we have

$$\frac{d[Cl]}{dt} = 2k_1 I_{abs} - k_2[Cl][H_2] + k_3[H][Cl_2] - k_4[Cl][X] = 0 \tag{A}$$

$$\frac{d[H]}{dt} = k_2[Cl][H_2] - k_3[H][Cl_2] = 0 \tag{B}$$

Again

$$\frac{d[HCl]}{dt} = k_2[Cl][H_2] + k_3[H][Cl_2] \tag{C}$$

(A) + (B) gives

$$[Cl] = \frac{2k_1 I}{k_4[X]}$$

and (B) + (C) gives

$$\frac{d[HCl]}{dt} = 2k_2[Cl][H_2]$$

$$= \frac{4k_1 k_2 I[H_2]}{k_4[X]}$$

Then the quantum yield (ϕ) of the reaction

$$\phi_{HCl} = \frac{4k_1 k_2[H_2]}{k_4[X]}$$

(b) Vide problem 20.15.

20.9 Vide problem 20.15.

20.10 (a) Applying steady-state approximation to O-atom, we have

$$\frac{d[O]}{dt} = k_1 I_{abs} - k_2 [O_3][O] - k_3 [M][O_2][O] = 0$$

Again $\quad -\frac{d[O_3]}{dt} = k_1 I_{abs} + k_2 [O_3][O] - k_3 [M][O_2][O] = 2k_1 I_{abs}$

(b) $$\phi = -\frac{d[O_3]}{dt} \Big/ I_{abs} = 2k_1$$

(c) $$k_1 = \frac{\phi}{2} = \frac{2}{2} = 1$$

20.11 (a) and (b), vide Section 20.3.

20.12 (i) Molar heat of dissociation of HI

= 1 einstein of radiation of wavelength 4040 Å

$$= \frac{(6.02 \times 10^{23} \text{ mol}^{-1})(6.63 \times 10^{-34} \text{ Js})(3 \times 10^8 \text{ ms}^{-1})}{4040 \times 10^{-10} \text{ m}} \quad \text{by Eq. (20.2b)}$$

= 299.3 kJ mol^{-1}

(ii) Assuming that the resulting hydrogen and iodine atoms are in the same electronic state as in (i), the energy appearing as kinetic energy of the atoms will be the difference between 1 einstein of radiation for wavelength 2537 Å and that for 4040 Å.

20.13 Here

$$100 \text{ J of light } = \frac{100 \text{ J}}{(6.02 \times 10^{23} \text{ mol}^{-1})(6.63 \times 10^{-34} \text{ Js})(3 \times 10^8 \text{ ms}^{-1})/(253.7 \times 10^{-9} \text{ m})}$$

$$= 2.1 \times 10^{-3} \text{ mol of photons}$$

Number of moles of HI that will be decomposed

= Number of moles of photons absorbed × QY

= $2.1 \times 10^{-3} \times 2$ or 4.2×10^{-3}

20.14 Time required

$$> \frac{\frac{1}{2} \times \text{quantity of absorbing species present in the sample}}{\text{rate of supply of photon}}$$

$$> \frac{\frac{1}{2} \times \text{quantity of absorbing species}}{\text{power supplied/photonic energy } (N_A hc / \lambda)}$$

$$> \frac{\frac{1}{2} \times 10^{-3} \text{ mol dm}^{-3}}{(100 \text{ Js}^{-1})/[(6.02 \times 10^{23} \text{ mol}^{-1})(6.63 \times 10^{-34} \text{ Js})(3 \times 10^8 \text{ ms}^{-1})/(589 \times 10^{-9} \text{ m})}$$

= 1.03 s dm^{-3}

20.15 Time required $= \dfrac{\text{quantity of acetone to be decomposed}}{\text{rate of decomposition of acetone}}$

$$= \frac{\text{quantity of acetone to be decomposed}}{\text{QY} \times \text{power supplied/photonic energy}}$$

20.16 Assuming that the photons entering into the cells are all absorbed, photons absorbed in the second experiment

$$= \frac{10\,h}{2\,h} \times \text{photons absorbed in the first experiment}$$

$= 5 \times$ moles of oxalate decomposed/QY of oxalate decomposition

$= 5 \times 6.201 \times 10^{-3}$ mol/0.57

$$\text{QY of acetone decomposition} = \frac{\text{moles of acetone decomposed}}{\text{moles of photons absorbed by acetone}}$$

$$= \frac{1.4 \times 10^{-3}}{\dfrac{5 \times 6.201 \times 10^{-3}}{0.57} - \dfrac{2.631 \times 10^{-2}}{0.57}} = 0.17$$

Note: Compare this problem with the problem 20.31 where the photons entering into the cell(s) are not all absorbed.

20.17 Vide Section 20.5.

20.18 The required wavelength should necessarily be less than λ for which one einstein $(N_A hc/\lambda)$ is equal to the given bond energy. Then

$$\lambda = \frac{(6.02 \times 10^{23}\ \text{mol}^{-1})(6.63 \times 10^{-34}\ \text{Js})(3 \times 10^8\ \text{ms}^{-1})}{431 \times 10^3\ \text{J mol}^{-1}}$$

$= 2.805 \times 10^{-7}$ m or 2805 Å

So Hg having absorption wavelength (2563.5 Å) lower than 2805 Å would serve as an effective photosensitiser.

Note: λ thus calculated provides the necessary but not sufficient condition for the radiation to have chemical effect. For a radiation to cause dissociation of H_2 the wavelength must lie in the absorption continuum (which begins at ~ 850 Å for H_2). Vide problem 20.25.

20.19 Number of photons absorbed

$$= \frac{\text{energy absorbed}}{\text{photonic energy } (N_A hc/\lambda)}$$

$$= \frac{8.81 \times 10^8 \times 10^{-7}\ \text{J}}{(6.02 \times 10^{23}\ \text{mol}^{-1})(6.63 \times 10^{-34}\ \text{Js})(3 \times 10^8\ \text{ms}^{-1})/2540 \times 10^{-10}\ \text{m}}$$

$= 1.87 \times 10^{-4}$ mol

$$\text{QY} = \frac{\text{amount of oxalic acid decomposed}}{\text{amount of photons absorbed}}$$

$$= \frac{(0.0495 - 0.0383)\ \text{mol L}^{-1} \times \dfrac{10}{1000}\ \text{L}}{1.87 \times 10^{-4}\ \text{mol}} = 0.599$$

20.20 Intensity of light incidence (i.e. rate of light incidence)

$$= \frac{\text{rate of formation of Fe}^{2+}}{\text{QY}}$$

$$= \frac{(1.3 \times 10^{-5}/36.5 \times 60)\ \text{mol s}^{-1}}{1.24}$$

$= 4.79 \times 10^{-9}$ einstein s^{-1}

☐ Amount of C_2H_4 produced $=$ Intensity of light incidence $\times$ time $\times QY_{C_2H_4}$
$$= (4.79 \times 10^{-9} \text{ einstein s}^{-1})\,(15 \text{!} 2 \times 60 \text{ s})\,(1.0)$$
$$= 4.37 \times 10^{-6} \text{ mol}$$

Similarly, amount of co produced $= 8.74 \times 10^{-6}$ mol

Note: In the reaction $2CH_2CO + hv \rightarrow C_2H_4 + 2CO$, the quantum yields of the products are in the ratio of their stoichiometric coefficients. In general this is not true.

20.21 $[Fe^{2+}]$ produced $=$ [coloured complex]

$$= \frac{\text{absorbance}}{\in l} \quad \text{by Eq. (20.1b)}$$

$$= \frac{0.663}{(1.11 \times 10^4 \text{ L mol}^{-1}\text{ cm}^{-1})\,(1\text{ cm})}$$

$$= 5.97 \times 10^{-5} \text{ mol L}^{-1}$$

Quanta absorbed $= \dfrac{Fe^{2+} \text{ formed}}{QY_{Fe^{2+}}}$

$$= \frac{(5.97 \times 10^{-5} \text{ mol L}^{-1})\,(57.4 \times 10^{-3}\text{ L})}{0.94}$$

$$= 3.65 \times 10^{-6} \text{ einstein}$$

20.22 Vide problem 20.33.

20.23 Vide Section 20.1.

☐ This law demands a linear plot of $\log \frac{I_0}{I}$ against c at all concentration. However, deviation from such ideal plot is observed when the light-absorbing species suffers appreciable association or dissociation occurring in varying proportion with dilution. Even in absence of any chemical change, considerable deviation arises at higher concentrations when $\in$ cannot be treated as constant due to appreciable solute–solute interaction.

20.24 Optical density $(OD = \log \frac{I_0}{I})$ has the maximum value infinity (which corresponds to $I = 0$, i.e. complete absorption of light) and minimum value zero (which corresponds to $I = I_0$, i.e. no absorption of light). However, such values can only be approached. This follows from Lambert–Beer law [Eq. (20.1b)] according to which $OD = \infty$ and $OD = 0$ require $C = \infty$ and $C = 0$ respectively which have no real significance.

☐ Vide Section 20.1.

20.25 Here, by Eq. (20.1b)

$$\log \frac{I_0}{I} \propto l, \quad \in \text{ and } c \text{ being constant}$$

Then, for the new cell

$$\log \frac{I_0}{I} = 5 \times \log \frac{100}{100 - 10}$$

whence $\qquad \dfrac{I_0}{I} = 1.694$

So % of light absorbed $= \dfrac{I_0 - I}{I_0} \times 100 = \dfrac{1.694 - 1}{1.694} \times 100 = 41$.

20.26 Vide problem 20.38.

20.27

$$\in = \dfrac{\log \dfrac{I_0}{I}}{Cl} \quad \text{by Eq. (20.1b)}$$

$$= \dfrac{\log \dfrac{100}{100 - 60}}{(5 \times 10^{-6} \times 10^{3} \ \text{mol m}^{-3})(0.1 \ \text{m})}$$

$$= 795.8 \ \text{m}^2 \ \text{mol}^{-1}$$

☐ Here $\log \dfrac{I_0}{I} \propto c$, $\in$ and l being constant.

Then, the required concentration $= \dfrac{\log \dfrac{100}{100 \times 99.98}}{\log \dfrac{100}{100 - 60}} \times 5 \times 10^{-6} \ \text{mol dm}^{-3}$

$$= 50.26 \times 10^{-6} \ \text{mol dm}^{-3}$$

Note: Here absorption increases from 60% of ~100% when concentration increases by a factor of ~10. With further increase in concentration, the absorption will only slightly increase but never becomes 100%.

20.28

$$\in = \dfrac{\log \dfrac{I_0}{I}}{cl}$$

$$= \dfrac{\log \dfrac{100}{20}}{(1.0 \times 10^{-4} \ \text{M})(1.0 \ \text{cm})}$$

$$= 6.99 \times 10^{3} \ \text{M}^{-1} \ \text{cm}^{-1}$$

☐ Here $\log \dfrac{I}{I_0} \propto cl$. Then under the given condition

$$\log \dfrac{I}{I_0} = \left(\dfrac{1}{2}\right)\left(\dfrac{1}{2}\right) \log \dfrac{20}{100}$$

whence $\dfrac{I}{I_0} = 0.669$

Then % transmission $= 0.669 \times 100 = 66.9$

20.29 By Eq. (20.1d), we have

$0.385 = (1.00 \ \text{cm}) \ [(370 \ \text{M}^{-1}\text{cm}^{-1}C_{Cr_2O_7^{2-}}) + (92.8 \ \text{M}^{-1}\text{cm}^{-1}C_{MnO_4^-})]$ at 440 nm

$0.653 = (1.00 \ \text{cm}) \ [(10.8 \ \text{M}^{-1}\text{cm}^{-1}C_{Cr_2O_7^{2-}}) + (2350 \ \text{M}^{-1}\text{cm}^{-1}C_{MnO_4^-})]$ at 545 nm

whence $\quad C_{Cr_2O_7^{2-}} = 1.04 \times 10^{-3} \ \text{M}$

$\quad C_{MnO_4^-} = 2.73 \times 10^{-4} \ \text{M}$

21

Statistical Thermodynamics and Third Law

Classical thermodynamics deals with macroscopic (bulk) properties of matter based on some empirical laws which led to the notion of temperature (from the zeroath law), internal energy (from the first law) and entropy (from the second law) as functions of state without any reference to the molecules (particles) of which the system is composed. Quantum mechanics, on the other hand, deals with microscopic or molecular properties. Statistical thermodynamics (which is a part of a more general branch of science known as statisticl mechanics developed by Boltzmann in Germany, Maxwell in England and Gibbs in USA in the latter part of the 19th century) links between quantum mechanics and classical thermodynamics on the basis of molecular constitution of matter. Not only does it show that the laws of classical thermodynamics are consequences of the postulates of the quantum mechanics, but also it enables one to calculate macroscopic thermodynamic properties from the informations about the molecules obtained from spectroscopic measurements and from theoretical calculations of wavefunctions. The crucial steps in going from the quantum mechanics of individual molecules to the thermodynamics of macroscopic system is to recognise that the macroscopic properties are really the average behaviours of the molecules of the system. Since the number of molecules present in usual macroscopic systems is enormous (of the order of moles), the statistical thermodynamics adopts useful probabilistic approach, instead of considering the actual states of constituent molecules of the system individually. In fact, according to the uncertainty principle, one cannot know exactly the properties of individual molecules. We can understand from the probabilistic treatment why bulk thermodynamic properties (P, U, S etc.) change with time always in the same direction (i.e. irreversibly toward the equilibrium state) though mechanical properties (position, velocity etc.) of the individual molecules change reversibly. This is inevitably due to high probability of certain configuration of the entire collection of molecules.

21.1 MICROSTATE AND THERMODYNAMIC PROBABILITY

The state of a system defined in classical thermodynamics by the macroscopic properties of the system, such as energy, volume and mol numbers is called the macrostate of the system. However, by this, a system is not completely specified from molecular view point. Because each macrostate so defined comprises a large number of microstates which are characterised by the instantaneous values of the time-dependent variables like position and energy state of each particle constituting the system.For a system at equilibrium, its observable macrostate does not change with time, but its microstate changes continually (through molecular collision) such that the macrostate remains unaltered. There must, therefore, exist an enormous set of microstates which are consistent with an equilibrium macrostate. The number of distinguishable microstates (or complexions) associated with a macrostate for a specified molecular distribution into different spatial groups or energy levels is called thermodynamic probability of the macrostate for that distribution. The thermodynamic probability is proportional to the mathematical probability. The latter is obtained by dividing the former with total number of conceivable microstates of the system for the specified macroscopic properties.

To illustrate let us consider a vessel consisting of two equal parts A and B, and it contains two gas molecules denoted by α and β. Here the uniform distribution of the gas between the two parts of the vessel can be achieved through the following two microstates

$$\frac{A \qquad B}{}$$
$$\alpha \qquad \beta$$
$$\beta \qquad \alpha$$

But the non-uniform distribution in which the entire gas is concentrated in one part of the vessel (say A) can be achieved by only a single microstate represented by

$$\frac{A \qquad B}{\alpha, \beta \qquad -}$$

In this example example, the thermodynamic probability for uniform distribution is 2, while it is 1 for each of the two non-uniform distribution. But the mathematical probability is $\frac{2}{2+1+1}$ or $\frac{1}{2}$ for uniform distribution and $\frac{1}{2+1+1}$ or $\frac{1}{4}$ for each of the two non-uniform distribution. The uniform distribution is more likely to occur since it has higher probability than the non-uniform distribution.

In the above illustration, the uniform distribution has probability only twice that of the most non-uniform distribution. But it will be many times greater if the number of molecules is very large, when there will be a dominating molecular distribution that alone will account for the properties of system (vide university Q. 21.8). In general N (chemically) identical but (mentally) distinguishable gas molecules can be distributed between two halves of their container in 2^N ways which is the total number of microstates for all possible distributions. But for a specified distribution where one half of the container contains N_1 molecules and the other half N_2 molecules, the number of microstates, i.e. thermodynamic probability (Ω), will be

$$\Omega = \frac{N!}{N_1! \, N_2!} \tag{21.1}$$

and the mathematical probability $= \dfrac{N!}{N_1! \, N_2! \, 2^N}$ where $N = N_1 + N_2$. If the vessel consists of several parts and the distribution is such that first part contains N_1 molecules, second part N_2 molecules, ..., then

$$\Omega = \frac{N!}{N_1! \, N_2! \, ...} \tag{21.2}$$

$$N = N_1 + N_2 + ...$$

If the particles are are chemically dissimilar but energetically similar, the total number of microstates due to their mixing (i.e. interchange of position between like particles), e.g. in the mixing of N_X atoms of a crystalline substance X with N_Y atoms of another crystalline substance Y will be

$$\frac{N!}{N_X! \, N_Y!} \tag{21.3}$$

(where $N = N_X + N_Y$)

which is similar to Eq. (21.1)

From quantum mechanical view point, Ω is most precisely defined as the number of independent quantum states (analogous to microstates) accessible to the system of specified macroscopic properties. Often it is useful to write (when permitted)

$$\Omega = \Omega_{\text{spatial}} \, \Omega_{\text{thermal}} \tag{21.4}$$

where Ω_{spatial} is the number of distinguishable molecular distribution in space as discussed above in an over-simplified way. Ω_{thermal} is the number of distinguishable ways in which molecules of the system can be distributed over the energy states available to them. The natural tendency of mechanical energy to be transformed into thermal energy arises from the increase in Ω_{thermal} due to spreading of total energy into greater number of energy states. Similarly, a spontaneous chemical reaction owes to spreading of total energy of the reaction system over the whole range of accessible quantum states (for different categories of energy) of the reactants and products.

21.2 ENTROPY AND PROBABILITY

In thermodynamics, the quantity which is of central importance is entropy (s) which has been defined by the second law (simply on phenomenological ground) as $dS = dQ_{\text{rev}}/T$. The probabilistic ground of statistical thermodynamics helps us to understand the significance of entropy as a real physical quantity. In statistical thermodynamics, it has been assumed that each microstate of a system for a given value of energy, volume and number of particles has equal probability. Then most likely observable macrostate (virtually the dominating molecular distribution) of an isolated system at equilibrium will be one with maximum number of microstates, i.e. with maximum thermodynamic probability (Ω). Again, entropy of an isolated system becomes maximum at equilibrium. Then these two quantities, both of which are properties of the system, must be interrelated, i.e.

$$S = f(\Omega)$$

To establish the form of this function, let us consider a system consisting of two parts, 1 and 2, each having the same intensive variables. Each of Ω_1 microstates of the first part can be chosen in combination with any of the Ω_2 microstates of the second part. The total number of microstates Ω for the combined system is therefore

$$\Omega = \Omega_1 \Omega_2$$

Again, being an additive property, entropy (s) of the system must be the sum of the entropies of the two parts of the system, i.e.

$$S = S_1 + S_2$$

Then $\qquad\qquad f(\Omega) = f(\Omega_1) + f(\Omega_2)$

or $\qquad\qquad f(\Omega_1 \Omega_1) = f(\Omega_1) + f(\Omega_2)$

The most general function that satisfies this relation is

$$S = f(\Omega) = k \ln \Omega + \text{constant} \tag{21.5}$$

where k is a constant (Boltzmann constant) which is required to be equal to R/N_A (vide question 21.21). It is due to involvement of an arbitrary constant that the absolute value of S cannot be determined. Using suitable standard for measurement of S (in harmony with the third law of thermodynamics), the arbitrary constant can be avoided when

$$S = k \ln \Omega \qquad (21.6)$$

This is known as Boltzmann relation which is consistent with the Boltzmann distribution principle (vide problem 21.38).

As implied by Eq. (21.6), entropy (unlike energy) of a system is not a property of its individual molecules (though it is calculable from molecular properties), it is determined by the probabilities of states for a large collection of molecules related to their spatial and energy distribution, i.e. randomness of the system (vide problem 21.40).

21.3 BOLTZMANN DISTRIBUTION FORMULA

It deals with distribution of independent particles in a system among the different energy states available to them. The mathematical formulation was originally deduced with an equilibriated system consisting of large number (N) of identical but distinguishable particles (molecules) which can individually exist in various states with energies $\in_0$, $\in_1$, $\in_2$, ... etc. The arrangement of the system in which N_0 particles are in energy state $\in_0$, N_1 in $\in_1$, ... and N_i in $\in_i$ can be achieved in the following Ω different ways

$$\Omega = \frac{N!}{N_0!\, N_1!\, N_2! \dots N_i!}$$

or $\qquad \ln \Omega = \ln N! - \sum_i \ln N_i! \qquad (21.7)$

If Ω changes due to the changes n_i to $n_i + dn_i$ (with no change in $\in_i$) then N, total energy E and $\ln \Omega$ will change as follows

$$dN = \partial \sum_i N_i = \sum_i \partial N_i$$

$$= 0 \qquad (21.8)$$

if N is constant

$$\partial E = \partial \sum_i N_i \in_i = \sum_i \in_i \partial N_i$$

if the particles are independent, i.e. non-interacting

$$= 0 \qquad (21.9)$$

if E is constant

$$\partial \ln \Omega = -\sum_i \partial \ln N_i! \quad \text{since } \partial \ln N_i = 0 \text{ when } N \text{ is constant}$$

$$= -\sum_i \partial (N_i \ln N_i - N_i) \quad \text{using Sterling approximation that holds if } N_i > 10$$

$$= -\sum_i \left[N_i \frac{\partial N_i}{N_i} + (\ln N_i)\, \partial N_i - \partial N_i \right]$$

$$= -\sum_i (\ln N_i)\, \partial N_i$$

For most probable (i.e. dominating) arrangement of the particles Ω, and hence $\ln \Omega$, will be maximum and then

$$\partial \ln \Omega = \sum_i (\ln N_i)\, \partial N_i = 0 \tag{21.10}$$

Simultaneous solutions of Eqs (21.8), (21.9) and (21.10) requires

$$\sum_i (\alpha + \beta \in_i + \ln N_i)\, \partial N_i = 0 \tag{21.11}$$

where α and β are arbitrary constants

or
$$N_i = \frac{e^{-\beta \in_i}}{e^\alpha}$$

Now
$$N = \sum_i N_i = \frac{\sum_i e^{-\beta \in_i}}{e^\alpha}$$

Then
$$\frac{N_i}{N} = \frac{e^{-\beta \in_i}}{\sum_i e^{-\beta \in_i}} \tag{21.12}$$

Now $\beta = \frac{1}{kT}$, which follows from equating the average translational kinetic energy $\frac{1}{2\beta}$ of an ideal gas molecule per degree of freedom to $\frac{1}{2}kT$. Then Eq. (21.12) becomes

$$\frac{N_i}{N} = \frac{e^{-\in_i/kT}}{\sum_i e^{-\in_i/kT}} \tag{21.13}$$

which is the simple form of Boltzmann equation.

If several, g_j say, states correspond to he same energy value $\in_j$, then these states are said to belong to the same energy level of degeneracy g_j and in that case, the Boltzmann equation will take the following general form

$$\frac{N_j}{N} = \frac{g_j e^{-\in_j/kT}}{\sum_{j(\text{level})} g_j e^{-\in_j/kT}} \tag{21.14}$$

The derivation of Boltzmann equation thus made involves the following assumptions and approximations:
  (i) The particles constituting the system are distinguishable (not justified by the quantum theory when the particles are in motion).
  (ii) Sterling's approximation.
  (iii) Both N_i and $\in_i$ are continuous variables (provided N_i is large and $\in_i$ is small).
  (iv) Particles are independent with regard to their energy. It is, however, possible to derive Boltzmann equation using only the assumption (iv).

Alternative Derivation of Boltzmann Equation

Boltzmann equation may be derived in a straight forward way using simply the law of probability. Let us consider a system consisting of a large number of similar particles in a container of constant volume maintained at some constant temperature T. The only restriction regarding the particles is that they are independent, in the sense that the total energy of the system is a sum of their individual energies which may be $\epsilon_0, \epsilon_1, \epsilon_2, \dots$ etc. (i.e. there is no potential energy of interaction which is a complicated function of intermolecular distance). Let us withdraw from this container a particle of energy ϵ_1 and another of energy ϵ_2. Since nothing distinguishes the particles from one another except the energy they possess, the probability of withdrawal of a particle of energy ϵ_1 must be simply a function of ϵ_1 only, i.e. $p_1(\epsilon_1)$. For the second particle, it is $p_2(\epsilon_2)$. The total energy of the pair of particles withdrawn is $\epsilon_1 + \epsilon_2$. Then the probability $p_{12}(\epsilon_1 + \epsilon_2)$ of withdrawing two particles of energy ϵ_1 for one and ϵ_2 for the other will be

$$p_{12}(\epsilon_1 + \epsilon_2) = p_1(\epsilon_1)\, p_2(\epsilon_2) \quad \text{since } p_1(\epsilon_1) \text{ and } p_2(\epsilon_2) \text{ are independent}$$

Differentiation gives

$$p_2 \cdot \frac{dp_1}{d\epsilon_1} = \frac{dp_{12}}{d(\epsilon_1 + \epsilon_2)} = p_1 \frac{dp_2}{d\epsilon_2}$$

or

$$\frac{1}{p_1} \cdot \frac{dp_1}{d\epsilon_1} = \frac{1}{p_2} \cdot \frac{dp_2}{d\epsilon_2}$$

Since ϵ_1 and ϵ_2 are independent, the left-hand side and right-hand side of this equation must be equal to some constant, say $-\beta$, which is independent of ϵ_1 and ϵ_2, then

$$\frac{1}{p_i} \cdot \frac{dp_i}{d\epsilon_i} = -\beta$$

where p_i is the probability of a particle to have energy ϵ_i.

or

$$p_i = \alpha e^{-\beta\epsilon_i}$$

where α is an integration constant, which can be found from the relation

$$\sum_i \alpha e^{-\beta\epsilon_i} = \sum_i p_i = 1$$

or

$$\alpha = \frac{1}{\sum_i e^{-\beta\epsilon_i}}$$

Then

$$p_i = \frac{e^{-\beta\epsilon_i}}{\sum_i e^{-\beta\epsilon_i}}$$

Assuming that there is equal probability of withdrawing any individual particle regardless of its energy, we have

$$\frac{N_i}{N} = p_i \quad \text{where } N = \sum_i N_i$$

$$= \frac{e^{-\beta\epsilon_i}}{\sum_i e^{-\beta\epsilon_i}}$$

Barometric Distribution Formula

Boltzmann distribution equation is valid no matter whether the molecular energy (ϵ) represents total energy or any individual form of energy, e.g. translational, rotational, etc. (because in deriving this equation, no restriction was made as to the nature of the energy). When applied to potential energy in gravity field, it gives spatial distribution of each kind of particles of mass m_i in the atmosphere, such that the number particles N_1 at height h_1 and N_2 at height h_2 will be in the following ratio

$$\frac{N_2}{N_1} = e^{-(m_i g h_2 - m_i g h_1)/kT}$$

assuming acceleration due to gravity g and T to be same at different levels.

If the particles are assumed to obey perfect gas law then $N_2/N_1 = P_2/P_1$ at constant T, where P_1 and P_2 are the partial pressures at the two levels. If height is measured from sea leave, then we have putting $h_1 = 0$ and $h_2 - h_1 = h$

$$\frac{P}{P_0} = e^{-m_i g h/kT} = e^{-M_i g h/RT} \tag{21.15}$$

This is barometric formula that gives the pressure P at different height above sea-level (where the pressure is P_0) in the atmosphere for each kind of its constituents.

21.4 PARTITION FUNCTION FOR A SYSTEM OF INDEPENDENT MOLECULES

The quantity appearing in the denominator of Boltzmann Eqs (21.13) or (21.14) is called the molecular partition function q, i.e.

$$q = \sum_i e^{-\epsilon_i/kT} \tag{21.16}$$

or

$$q = \sum_{j(levels)} g_j e^{-\epsilon_j/kT} \tag{21.17}$$

In this expression, the energy ϵ_0 corresponding to the lowest level is usually taken as zero and the energies corresponding to all other levels $\epsilon_1, \epsilon_2, \ldots$, etc. are expressed relative to ϵ_0. Then, as $T \to 0$ (i.e. $\frac{1}{T} \to \infty$), $q \to g_0$, the degeneracy of the lowest level. Again, as $T \to \infty, q \to (g_0 + g_1 + \ldots + g_j)$, which is equal to the total number of energy states of the particles (molecules) and is usually infinite.

Then the molecular partition function indicates the average number of states that are thermally accessible to a molecule of the system at the prevailing temperature. At $T = 0$, only the ground level is accessible and hence $q = g_0$. At very high temperature, virtually all states are accessible, and q is correspondingly large.

Partition Function for a System of Interacting Molecules

The molecular partition function defined by Eqs (21.16) or (21.17) has meaning only with independent particles (e.g. perfect gas molecules). Because only in this case, can we define and enumerate the states of the system in terms of quantum mechanical energy states of individual molecules. But as soon as significant interactions occur between the molecules, the description of the state of the system must involve potential energy terms which are functions of intermolecular distances. To deal with such systems, we are to define partition function in a general way using the concept of ensemble.

An ensemble is a hypothetical collection of a large number of replicas of the relevant system each of which is in the same macrostate but not in same microstate. This number may be as large as we like and this is not related to the number of molecules in the actual system. Depending on the purpose, use is made of various essembles of which the following ones are important:

- *Canonical ensemble* where N, V and T are common to all the member systems which can mutually exchange energy but not matter.
- *Microcanonical ensemble* where N, V and E are common to all the member systems which are individually isolated.
- *Grand canonical ensemble* where μ, V and T are common to all the member systems which can mutually exchange matter, and of course energy.

The concept of ensemble was introduced by Gibbs to solve the absurd task of getting the value of the observed macroscopic properties by taking a time average over the changes in enormous number of microstates of a real system. The problem cannot be solved directly using Boltzmann formula for distribution of molecular energy because of its limited applicability only for independent particles. The solution lies in the postulate of statistical thermodynamics that the average of any mechanical variable over a long time in the actual system is equal to the ensemble average. Strictly speaking, this postulate holds only if the member systems are infinite in number.

According to statistical thermodynamics, even for fixed values of N, V and T, as in case of cannonical ensemble, a member system of the latter does not have a fixed energy E, rather it has a most probable energy. The probability p_i that a member i of the ensemble has energy E_i is given, in accordance with the general Boltzmann principle, by

$$p_i = \frac{e^{-E_i/kT}}{\sum_i e^{-E_i/kT}}$$

The denominator of this expression is called the canonical ensemble partition function Q, i.e.

$$Q = \sum_i e^{-E_i/kT}$$

[Q can also be expressed in terms of allowed energy levels using degeneracy, instead of quantum states, as in case of molecular partition function.]

Q and q are related as follows

$$Q = q^N$$

when the system consists of N molecules which are identical, independent and distinguishable

$$Q = \frac{q^N}{N!} \text{ when the molecules are indistinguishable}$$

Molecules are distinguishable if they are chemically different or physically differentiable by their definite locations as in a crystal. It should be noted that q depends only on V and T, but Q depends also on N (because the molecular interaction is determined by total number of molecules).

q can be computed from the knowledge of allowed energy states of the molecules obtained from the experimental spectroscopic data or theoretical consideration based on quantum mechanics. Assuming that the total energy of a molecule is a sum of independent contribution of its translational, rotational, vibrational and electronic motion denoted by the suffix t, r, v and e respectively, we can write

$$q = \underbrace{\sum e^{-\epsilon_t/kT}}_{\text{all states}} \underbrace{\sum e^{-\epsilon_r/kT}}_{\text{all states}} \underbrace{\sum e^{-\epsilon_v/kT}}_{\text{all states}} \underbrace{\sum e^{-\epsilon_e/kT}}_{\text{all states}}$$

$$= q_t q_r q_v q_e$$

Then
$$Q = \frac{1}{N!}(q_t q_r q_v q_e)^N$$

when the molecules are independent and indistinguishable
$$= Q_t Q_r Q_v Q_e$$

where
$$Q_t = \frac{q_t^N}{N!}, \ Q_r = q_r^N, \ Q_v = q_v^N \ \text{and} \ Q_e = q_e^N$$

For distinguishable molecules, $Q_t = q_t^N$.

Partition function is an extremely important characteristic of a system, because it contains all the informations needed to calculate thermodynamic properties of a system (similar to wavefunction in quantum mechanics). In this regard, Q is more general than q, because the former is not based on the assumption that the molecules are independent.

Thermodynamic Properties in Terms of Partition Function

Internal energy U of a thermodynamic system (relative to its value at $T = 0$) may be regarded as the average value of energy for a canonical ensemble based upon the system concerned, i.e.

$$U = \sum p_i E_i$$

$$= \frac{1}{Q} \sum E_i e^{-\beta E_i}$$

$$= -\frac{1}{Q}\left(\frac{\partial Q}{\partial \beta}\right)_{N,V}$$

$$= -\left(\frac{\partial \ln Q}{\partial \beta}\right)_{N,V}$$

$$= kT^2\left(\frac{\partial \ln Q}{\partial T}\right)_{N,V} \tag{21.18}$$

where $Q = \sum e^{-\beta E_i}$ the differentiation is done at constant N and V, because E_i depends on both N (when molecules of the system interact) and V (as in particle in a box situation). This expression of U leads to the following relations

$$C_V = \frac{k}{T^2}\left[\frac{\partial^2 \ln Q}{\partial\left(\frac{1}{T}\right)^2}\right]_{N,V} \tag{21.19}$$

$$P = kT\left(\frac{\partial \ln Q}{\partial V}\right)_{N,T} \tag{21.20}$$

$$H = kT^2\left(\frac{\partial \ln Q}{\partial V}\right)_{N,V} + kT\left(\frac{\partial \ln Q}{\partial \ln V}\right)_{N,T} \tag{21.21}$$

$$S = k \ln Q + \frac{U}{T} + \text{constant}$$

However, S is often expressed in the following simple form (equivalent to 21.5) dropping the arbitrary constant which (being independent of N, V and T) does not appear in the expression for the measurable quantity ΔS with which we are primarily concerned

$$S = k \ln Q + \frac{U}{T} \tag{21.22}$$

With this expression of S, we have the following relations

$$A = -kT \ln Q \tag{21.23}$$

$$G = -kT \ln Q + kT \left(\frac{\partial \ln Q}{\partial \ln V} \right)_{N,T} \tag{21.24}$$

To find contribution of each of the different types of molecular motion to the thermodynamic properties, we are to calculate partition function (Q) for that type of motion.

Translational Partition Function for an Ideal Gas

Let us consider the simple case of a molecule of mass m moving in a box of dimensions $a \times b \times c$. Using the relation (18.15a) for the energies of the allowed quantum states, the partition function (q_t) for translational motion will be given according to Eq. (21.16), by

$$q_t = \sum_{n_x=1}^{\infty} \sum_{n_y=1}^{\infty} \sum_{n_z=1}^{\infty} \exp \left[-\frac{h^2}{8mkT} \left(\frac{n_x^2}{a^2} + \frac{n_y^2}{b^2} + \frac{n_z^2}{c^2} \right) \right] \cdot$$

$$= \sum_{n=1}^{\infty} \exp \left[-\frac{h^2 n^2}{8ma^2 kT} \right] \sum_{n=1}^{\infty} \exp \left[-\frac{h^2 n^2}{8mb^2 kT} \right] \sum_{n=1}^{\infty} \exp \left[-\frac{h^2 n^2}{8mc^2 kT} \right]$$

$$= \int_0^{\infty} \exp \left(-\frac{h^2 n^2}{8ma^2 kT} \right) dn \int_0^{\infty} \exp \left(-\frac{h^2 n^2}{8mb^2 kT} \right) dn \int_0^{\infty} \exp \left(-\frac{h^2 n^2}{8mc^2 kT} \right) dn$$

for macroscopic containers, the exponents being very small ($<< 1$) unless T is very small

$$= \left(\frac{2\pi mkT}{h^2} \right)^{\frac{3}{2}} a \cdot b \cdot c$$

$$= \left(\frac{2\pi mkT}{h^2} \right)^{\frac{3}{2}} V \tag{21.25}$$

since $abc = V$, the volume of the container

The quantity $\left(\frac{2\pi mkT}{h^2} \right)^{-\frac{1}{2}}$, which has the dimensions of length, is called the thermal de Broglie wavelength which must be small compared to the average intermolecular distance for Boltzmann distribution to be applicable.

$$Q_t = \frac{q_t^N}{N!}$$

$$= \left(\frac{2\pi mkT}{h^2} \right)^{\frac{3N}{2}} \cdot \frac{V^N}{N!}$$

$$= \left(\frac{2\pi mkT}{h^2} \right)^{\frac{3N}{2}} \cdot \frac{e^N V^N}{N^N} \tag{21.26}$$

using Sterling's approximation

$$N! = \frac{N^N}{e^N}$$

Then, the translational contributions to the thermodynamic properties, such as U, C_V, S etc. for an ideal gas will be given by

$$U_t = kT^2 \left(\frac{\partial \ln Q_t}{\partial T}\right)_{N,V} = \tfrac{3}{2} NkT \qquad (21.27)$$

$$(C_V)_t = \left(\frac{\partial U_t}{dT}\right)_V = \tfrac{3}{2} Nk \qquad (21.28)$$

$$S_t = k \ln Q_t + \frac{U_t}{T}$$

$$= \tfrac{5}{2} Nk + Nk \ln\left[\left(\frac{2\pi mkT}{h^2}\right)^{\frac{3}{2}} \frac{V}{N}\right] \qquad (21.29)$$

The last equation is a form of Sackur–Tetrode equation for the entropy of an ideal monatomic gas (whose molecules can have only translational motion).

Rotational Partition Function

Let us consider the case of rigid diatomic molecules. Using the expression (18.23b) for rotational energy levels (each of the degeneracy $2J + 1$, the rotational partition function (q_r) for heteronuclear diatomic molecules will be, according to Eq. (21.17), given by

$$q_r = \sum_{J=0}^{\infty} (2J + 1)\, e^{-J(J+1)h^2/8\pi^2 IkT}$$

$$= \int_0^{\infty} (2J + 1) \exp\left[-J(J + 1) h^2 / 8\pi^2 IkT\right] dJ$$

$$= \frac{8\pi^2 IkT}{h^2} \qquad (21.30)$$

Here summation by integration will be reasonable only when J is sufficiently high so that the rotational energy levels may be treated as varying continuously. This requires the temperature to be well above $h^2/8\pi^2 Ik$ which is called characteristic rotational temperature of diatomic molecules.

For homonuclear diatomic molecules (like N_2) q_r will be half of that given by the above expression. Because here the two orientations due to 180° rotation of the molecule are indistinguishable.

Vibrational Partition Function

Let us consider the simple case of diatomic molecules behaving as harmonic oscillators. Here the energy of a vibrational state relative to that of the lowest state is

$$\epsilon_v = vh\nu \qquad v = 0, 1, 2, \dots$$

Then, the vibrational partition function will be given, according to Eq. (21.16) by

$$q_v = \sum_{v=0}^{\infty} e^{-\frac{vh\nu}{kT}}$$

$$= 1 + e^{-\frac{h\nu}{kT}} + \left(e^{-\frac{h\nu}{kT}}\right)^2 + \dots$$

$$= \frac{1}{1 - e^{-\frac{h\nu}{kT}}} \qquad (21.31)$$

when $\qquad e^{-\frac{h\nu}{kT}} \ll 1$

The quantity hv/k is called characteristic vibrational temperature (similar to rotational temperature) at which the energy of the vibrational quantum is equal to kT (which is the energy per vibrational degree of freedom of the molecule according to classical equipartition principle).

Electronic Partition Function

The electronic partition function (q_e) of a molecule is given, according to Eq. (21.17) by

$$q_e = g_0 + g_1 e^{-\epsilon_1/kT} + ... \qquad (21.32)$$

For simple molecules, the energies of the excited electronic levels (which are customarily expressed relative to that of the lowest level) are mostly much greater than kT at ordinary temperature. Then

$q_e = g_0$, the spin degeneracy of the lowest (ground) electronic level

If this relation holds, then electronic contribution to entropy(s) will be

$k \ln Q_e = k \ln q_e^N = Nk \ln g_0$ and to G will be $- NkT \ln g_0$.

Partition Function and Equilibrium Constant

For any substance

$$G = A + PV$$
$$= -kT \ln Q + PV$$
$$= - kT \ln \left(\frac{q^N}{N!} \right) + NkT \quad \text{for an ideal gas}$$
$$= - NkT \ln \left(\frac{q}{N} \right) \qquad (21.33a)$$

(using Sterling approximation.)

This expression is based on the measurement of energy of different molecular levels relative to the lowest energy level (zero-point level) taken as zero. If, however, the energies of all levels are referred to some arbitrary level of energy E_0 per mol, then the free energy per mol of any substance i for a standard pressure (1 bar) will be given by

$$\bar{G}_i^\circ = E_{oi}^\circ - RT \ln \left(\frac{q_i}{N_i} \right) \qquad (21.33b)$$

Let us consider the simple reaction

$$A \rightleftharpoons B$$

involving ideal gases. The standard free energy ΔG° of this reaction is

$$\Delta G^\circ = \bar{G}_B^\circ - \bar{G}_A^\circ$$

Then

$$K_p = \frac{(q^\circ / N)_B}{(q^\circ / N)_A} e^{-\Delta E_0^\circ/RT} \quad \text{since } \Delta G^\circ = -RT \ln K_p$$

where $\Delta E_o^\circ = \Delta E_{oB}^\circ - \Delta E_{oA}^\circ$ is the energy change due to the reaction at absolute zero.

In general for a reaction $aA + bB + ... \rightleftharpoons lL + mM + ...$

$$K_P = \frac{(q^\circ / N)_L^l \, (q^\circ / N)_M^m \cdots}{(q^\circ / N)_A^a \, (q^\circ / N)B...} e^{-\Delta E_0^\circ/kT} \qquad (21.34a)$$

$$= \left(\frac{kT}{P^\circ} \right)^{\Delta v} \frac{(q^\circ / V)_L^l \, (q^\circ / V)_M^m \cdots}{(q^\circ / V)_A^a \, (q^\circ / V)B...} e^{-\Delta E_0^\circ/kT} \qquad (21.34b)$$

Since $N = P^\circ V/kT$ and $\Delta v = (l + m + ...) - (a + b + ...)$

For non-ideal gases, liquids and solids, we are to consider interactions of many molecules which are different to calculate.

21.5 HEAT CAPACITY OF GASES

In Section 2.5, heat capacity of gases has been discussed from classical mechanical viewpoint using equipartition principle. The values thereby predicted agree well with the experimental results for monatomic gases but not for polyatomic gases as ordinary temperature.

Correct prediction of heat capacity of a system can be done on the basis of statistical thermodynamics using the relation (21.19) considering the partition function for various modes of motion involved.

For gases, contributions from translational modes always occur because translational quanta are very small. Rotational modes (for polyatomic molecules) will significantly contribute to the heat capacity only above the characteristic rotational temperature when kT is of the order of the energy separation between the rotational energy levels, as may be seen from the Boltzmann distribution. This is more so with the vibrational modes, due to higher characteristic temperature than the rotational modes. We can now understand why observed heat capacities rise with temperature and approach their equipartition values.

Heat Capacity of Solids

Here we need to consider only vibrational modes which are $3N - 3 - 3 \simeq 3N$ in number for monatomic crystal consisting of N bound atoms (3 translational modes and 3 rotational modes of the crystal being of no use).

For a crystal, each normal mode of vibration, assuming it to be harmonic, will be characterised by a partition function of the form

$$q = \frac{1}{1 - e^{-\frac{hv}{kT}}}$$

if the frequency v of vibration is such that $e^{-\frac{hv}{kT}} \ll 1$. Then the total vibrational partition function of the crystal will be the product of the terms like this, one for each of the normal modes of vibration. But the precise determination of $3N$ normal modes of vibration for a crystal of macroscopic size would be an impossible task. This is unlike gases where vibrational modes are not large in number. However, some simple approximations are effective. Thus Einstein considered a crystal consisting of independent atomic oscillators all having same frequency when the crystalline partition function (Q) will be

$$Q = q^{3N}, \text{ atoms in a crystal being localised}$$

$$= \frac{1}{\left(1 - e^{-\frac{hv}{kT}}\right)^{3N}}$$

Then by relations (21.18) and (21.19)

$$U = \frac{3Nhv}{e^{\frac{hv}{kT}} - 1} \tag{21.35}$$

$$C_V = 3Nk \left(\frac{h\nu}{kT}\right)^2 \cdot \frac{e^{\frac{h\nu}{kT}}}{\left(e^{\frac{h\nu}{kT}} - 1\right)^2} \tag{21.36}$$

This is Einstein equation for heat capacity of solids and $h\nu/k$, having the dimension of temperature, is called Einstein temperature.

According to Einstein equation, molar heat capacity of an atomic solid at high enough temperature would approach to its equipartition value $3R$ (empirical law of Dulong and Petit). Also it predicts, in agreement with experiment, that $C_V \to 0$ as $T \to 0$ which classical theory cannot explain. However, Einstein equation predicts too rapid a decrease in heat capacity in the neighbourhood (below 20 K) of absolute zero.

Debye modified Einstein's model on the ground that a crystal should be treated as a system of coupled oscillators (and not independent oscilltors as assumed by Einstein) having frequency varying continuously from zero to a maximum value ν_m characteristic of the crystal with frequency distribution as in a continuous elastic body. He deduced the following expression

$$C_V = 9Nk \left(\frac{T}{\theta}\right)^3 \int_0^{\theta/T} \frac{x^4 e^x}{(e^x - 1)^2} \, dx$$

where $x = h\nu/kT$ and $\theta = h\nu_m/k$ is the Debye temperature. Near absolute zero, this equation reduces to

$$C_V = \frac{12\pi^4}{5} Nk \left(\frac{T}{\theta}\right)^3$$

$$= aT^3 \tag{21.37}$$

This is Debye 'T-cubed' law for the heat capacity of a solid, where $a = \dfrac{12\pi^4 Nk}{5\theta^3}$ is a characteristic constant for the crystal.

21.6 THIRD LAW OF THERMODYNAMICS

The third law of thermodynamics, unlike other three laws, introduces no qualitative concept of any state parameter or state function. It provides a numerical scale for entropy (just as the second law gives a numerical scale of temperature). We now consider the different forms of this law in their chronological order.

From the measurement of emf of galvanic cells, Richards observed that ΔG (or ΔA) and ΔH (or ΔU) of cell reactions approach each other closely the lower is the temperature. This led Nernst (1906) to the hypothesis (called Nernst heat theorem) that the temperature derivatives of ΔG and ΔH during an isothermal process involving pure solids or liquids approach zero as the temperature approaches zero, provided (as was later experimentally found by Simon) substances involved are in internal equilibrium, i.e.

$$\lim_{T \to 0} \frac{\partial(\Delta G)}{\partial T} = 0 \tag{21.38a}$$

and $$\lim_{T \to 0} \frac{\partial(\Delta H)}{\partial T} = 0 \tag{21.38b}$$

or $$\lim_{T \to 0} \Delta S = 0 \quad \text{by Gibbs–Helmholtz relation} \tag{21.39}$$

This is Nernst–Simon statement of the third law.

Instead of considering difference of entropy, Planck (1912) made stronger hypothesis that the entropy of a (homogeneously) pure substance is zero at absolute zero, i.e.

$$\lim_{T \to 0} S = 0 \qquad (21.40)$$

However, in view of unattainability of absolute zero of temperature and somewhat uncertainty in the conventional entropy values (arising from the statistical concept), it is rather safe, but satisfactory to state the third law by somewhat modification of Planck's hypothesis in the following useful form (due to Lewis and Randall, 1923). Every substance (element or compound) has a finite positive entropy, but as the temperature approaches absolute zero the entropy is likely to approach zero and does so in case of a perfectly crystalline state (He is the only exception where $S \to 0$ as $T \to 0$, though it remains in liquid form around 0 K at 1 atmosphere).

Another formulation of the third law is in expressing the unattainability of absolute zero as follows: It is impossible to reduce the temperature of a system to absolute zero in a finite number of steps. This can be understood from the cooling of a paramagnetic substance following the path represented by broken lines in Fig. 21.1. It appears that the fractional lowering of temperature in each stage of successive operations of isothermal magnetisation and adiabatic demagnetisation steadily decreases.

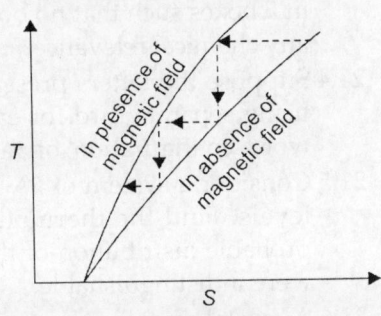

Fig. 21.1

Third-Law Entropies

The third law asserts that $S = 0$ at $T = 0$. On this basis, the entropy (called third-law entropy) S_T of a substance at any desired temperature T may be obtained by

$$S_T = \int_0^T \frac{dQ_{rev}}{T}$$

This required heat capacity (C_p) data over the whole range of temperature and enthalpy of all phase transitions in this temperature range. Since measurement of C_p cannot be carried to 0 K, Debye equation $C_p = aT^3$ is required to be used below the temperature of the lowest measurements (say T'). Then for a substance having mp T_m, the third-law entropy at a temperature T above the bp T_b may be calculated from the following expression

$$S_T = \int_0^{T'} \frac{aT^3}{T} dT + \int_{T'}^{T_m} \frac{Cp(s)}{T} dT + \frac{\Delta H_{fus}}{T_m} + \int_{T_m}^{T_b} \frac{Cp(l)}{T} dT + \frac{\Delta H_{vap}}{T_b} \int_{T_b}^{T} \frac{Cp(g)}{T} dT \qquad (21.41)$$

If there is any solid–solid phase transition, the corresponding entropy of transition would have to be included in this sum. The third-law entropies thus obtained are often called calorimetric entropies (because they are based on calorimetric measurements).

The spectroscopic entropy, which is obtained from calculation on the basis of statistical thermodynamics using spectroscopic data, is related to calorimetric entropy by

$$S_{spec} = k \ln \Omega_0 + S_{calor} \qquad (21.42)$$

$S_{spec} - S_{calor}$, i.e. $k \ln \Omega_0$ is often referred to as residual entropy, which is entropy at absolute zero (subject to some convention). Here Ω_0 is the thermodynamic probability, or degeneracy, of the lowest possible energy state of the system (for $T = 0$).

In many cases, the spectroscopic entropy value is more accurate than the calorimetric value. The entropy values found in literature are often an appropriate combination of spectroscopic and calorimetric values.

AUXILIARY PROBLEMS

21.1 Is there any difference between microstate and quantum state of a system?

21.2 Find the number of microstates for the distribution of 4 particles in 5 boxes, if the particles are (i) distinguishable (ii) indistinguishable.

21.3 Find the number of microstates for the distribution of 3 indistinguishable particles in 5 boxes such that no box contains not more than 1 particle. Does the result bear any chemical relevance?

21.4 Suppose the letters present in the word 'configuration' are all written on cards, using separate card for each letter. Find the expected frequency of obtaining the word 'configuration' on random shuffling the cards, if each shuffling requires 2 sec.

21.5 Consider a system of $2N$ distinguishable particles distributed between 2 energy levels. Find the thermodynamic and mathematical probabilities for the most probable distribution of the particles. How would the result differ if the particles were indistinguishable.

21.6 A crystal contains N_X atoms of X and N_Y atoms of Y. What will be the total number of microstates in this case? Also find the thermodynamic probability and mathematical probability of the most probable distribution when 4 atoms of the crystal X mix with 4 atoms of the crystal Y. Assume the resultant mixed crystal to be an ideal solid solution.

21.7 In statistical thermodynamics, it has been assumed that each microstate of a system for a given value of energy, volume and number of particles has equal probability. On this basis, explain the spontaneous rushing of a gas into vacuum to form uniform distribution.

21.8 A vessel contains 100 molecules of an ideal gas. Find the probability for all the molecules to be present in one-tenth volume of the vessel?

21.9 True of false.
   (a) Total molecular energy of a system is always sum of the energies of the individual molecules of the system.
   (b) Total molecular kinetic energy of a system is always sum of the kinetic energies of the individual molecules of the system.

21.10 On which of the following factors do the energy and energy levels of a system depend? (i) Volume (ii) Temperature (iii) Composition (iv) Number of moles.

21.11 Write down Boltzmann distribution formula. Does this represent an instantaneous population of the energy levels? What is the physical reason behind Boltzmann distribution?

21.12 For Boltzmann distribution to hold temperature must not be too low or pressure too high. Why? Is there any case where this formula fails even at ordinary temperature and pressure?

21.13 For a non-degenerate system, the number of molecules (N_i) present in the energy level ϵ_i may be written as $N_i = C \exp(\beta \epsilon_i)$
   (a) What is the probability of having a molecule with energy ϵ_i?
   (b) Justify that β should be negative for normal systems.

(c) Show that $\beta = -\frac{1}{kT}$.

(d) Relate C to total number of molecules, and state whether it depends on T.

(e) Show that $N_{i+1} \leq Ni$.

(f) Find $\partial N_i / \partial \beta$ in terms of energy and explain its significance.

(g) Evaluate $\sum_i \left(\dfrac{\partial N_i}{\partial \beta} \right)$.

21.14 For a degenerate system, the Boltzmann population N_i of an energy level ϵ_i of degeneracy g_i may be represented as

$$N_i \propto g_i \exp (\beta \epsilon_i)$$

(a) Is β same as in non-degenerate system?

(b) Will $N_{i+1} < N_i$ necessarily hold?

(c) What will be the value of N_o and N_i $(i \neq o)$ (i) when $T \to 0$ and (ii) when $T \to \infty$?

21.15 Are the molecular partition functions q_t, q_r, q_v and q_e extensive or intensive properties? Do they depend on N, T and V?

21.16 q_t, unlike q_r, depends on V of the system. What is expected about the dimensions of q_t and q_r? How are these quantities related to the respective canonical partition functions Q_t and Q_r? Explain the difference in such relations.

21.17 For which of the following cases will the relation $Q = \frac{q^N}{N!}$ be relevant?

(i) Ordinary He(g) (ii) He-II (l), (iii) CO_2(g) (iv) H_2O(s).

21.18 Find the dimensions of $h / \sqrt{mkT}$.

21.19 If Q_1 is the canonical partition function for 1 g of Au and Q_2 that for 10 g of Au, at 25°C and 1 atm pressure, find $\frac{\ln Q_2}{\ln Q_1}$.

21.20 Find, according to Boltzmann, the number of microstates corresponding to the macrostate state (E, V, N) of a system of ideal monatomic gas.

21.21 The Maxwell distribution of molecular speeds holds in any fluid, not just in an ideal gas. Justify this from Boltzmann energy distribution.

21.22 Derive barometric formula from Boltzmann energy distribution. Assuming that air is a mixture of N_2 and O_2 with 20 mole per cent of O_2 at sea-level where total pressure is 1 atm, find the pressure and composition at 10 km if the atmosphere has a temperature of 0°C independent of altitude.

21.23 Connect average (i) mole fraction and (ii) weight fraction of any gaseous component in the atmosphere to its mole fraction at sea-level.

21.24 Derive the relation (21.19) to (21.24).

21.25 The simplest relation between a thermodynamic quantity and the partition function arises with the work junction A, Why?

21.26 The choice of zero level of energy is arbitrary. If a constant 'C' is added to each of the system's possible energy levels which of the thermodynamic quantities—U, H, S, A and G will be affected? Justify your answer.

21.27 Using partition function, derive the ideal gas equation.

21.28 For He occupying 11.2 L at STP, calculate (a) molecular translational partition function (b) molar translational partition function. Will the results be affected due to variation of P, at constant V and T? State the assumption(s) involved and justification thereof.

21.29 Find translational contributions to $\bar{C}_P^\circ, \bar{S}^\circ$ and $\bar{G}^\circ$ for $H_2(g)$ at 25°C.

21.30 (a) For the ideal diatomic gas find expressions representing translational, rotational, vibrational and electronic contributions to C_V and S, stating their limitations with explanation thereof.

(b) At what temperature $(C_V)_t$ given by the derived expression will begin to deviate from the equipartition value.

21.31 At STP $\bar{C}_P$ is nearly same for $N_2(g)$ and $O_2(g)$ but considerably greater for $Cl_2(g)$. Explain.

21.32 An ideas gas (X) has same $\bar{C}_V$ as $N_2(g)$ at 0°C. But the two gases differ in $\bar{C}_V$ by a maximum amount of 6 cal K^{-1} mol^{-1} at high temperature. What can be inferred about the atomicity and structure of X?

21.33 Find approximate vlues of $\bar{C}_P$ for the following:

(i) H(g) (ii) $H_2(g)$ at room temperature and at its boiling point (iii) $H_2O(g)$ at room temperature and its boiling point (iv) $H_2O(l)$ at 4°C (v) Au(s) at 25°C (vi) diamond at 25°C (vii) NaCl at 25°C.

21.34 In dealing with heat capacity of a crystal only vibrational degrees of freedom is considered treating the crystal as a giant molecule. Is this justified?

21.35 Ordinarily, most solid elements have molar heat capacity around $3R$ which is the empirical law of Dulong and Petit. Justify this law and its exception due to Be, B, C and Si on the basis of Einstein's heat capacity equation. For the elements mentioned is $\bar{C}_V = 3R$ possible? What is Einstein characteristic temperature and what does it signify?

21.36 For several solid elements (e.g. alkali metals) $\bar{C}_V$ much exceeds the value $3R$ (as high at 9 cal K^{-1} mol^{-1}) at high temperature. How do you account for this?

21.37 Find an expression of chemical potential for an ideal monatomic gas and a monatomic crystal. Which crystal will have higher vapour pressure at a specified temperature—a crystal of monatomic or diatomic substance (other things being same)?

21.38 Deduce the relation $s = k \ln \Omega$ following Boltzmann distribution principle.

21.39 Does the relation $S = k \ln \Omega$ hold only for ideal gases? Is Ω an extensive or intensive quantity?

21.40 Entropy is a measure of randomness of the system. Justify.

21.41 Entropy of a system is an extensive property, but it cannot be treated as sum of entropies of the individual molecules of the system. Explain.

21.42 According to the Sackur–Tetrode equation, the translational entropy of an ideal gas depends on its molar mass. Does this imply that the translational heat capacity $(C_V)_t$, which depends on entropy, should also do the same?

21.43 The translational entropy, but not heat capacity, of an ideal gas depends on its molar mass. Explain.

21.44 State with reason whether $\bar{S}_t$ or $\bar{S}_r$ will be greater for CH_4 at STP.

21.45 Find the difference in $\bar{S}_r$ at room temperature in each of the following pairs: (i) H_2 and D_2 (ii) HD and D_2. Take suitable assumptions and approximations.

21.46 Which will have higher entropy in each of the following pairs of gaseous substances at STP:

(i) He and Ne (ii) H_2O and D_2O (iii) CO and N_2 (iv) $CH_3CH_2CH_2CH_2CH_3$ and $C(CH_3)_4$ (v) $\begin{array}{c} CH_3 \\ CH_3 \end{array} C = 0$ and $\begin{array}{c} CH_3 \\ CH_2 \\ CH_3 \end{array} O$

21.47 Why entropy of a gas increases (i) when it undergoes isothermal expansion and also (ii) when it absorbs heat at constant volume.

21.48 In adiabatic expansion of a gas $\Delta S_{rev} = 0$ but $\Delta S_{irr} > 0$. Explain.

21.49 For any substance $\Delta \bar{S}_{vap} > \Delta \bar{S}_{fus}$. Why?

21.50 $\Delta S_{freezing} > 0$ for an isolated supercooled liquid that freezes spontaneously, although the system changes to more ordered spatial arrangement of its molecules. Account for this.

21.51 The reaction $H(g) + H(g) = H_2(g)$, that occurs spontaneously, must be exothermic. Justify.

21.52 Arrange the following reactions in the order of $\Delta \bar{S}$

(i) $C(s) + O_2(g) \rightarrow CO_2(g)$ (ii) $2H_2(g) + O_2(g) \rightarrow 2H_2O(l)$

(iii) $N_2(g) + 3H_2(g) \rightarrow 2NH_3(g)$ (iv) $NH_3(g) + HCl(g) \rightarrow NH_4Cl(s)$

21.53 State with reasons, whether ΔS will be zero, +ve, −ve or none in the following processes: (i) Breaking of a graphite rod (ii) transformation of graphite into diamond (iii) combustion of graphite (iv) crystallisation of a supercooled liquid isothermally (v) crystallisation of a substance from its solution (vi) racemisation of an optically active substance (vii) conversion of ortho-H_2 to para-H_2 (viii) decomposition of O_3 to O_2 (ix) decomposition of U to Pb.

21.54 Calculate the number of microstates for the following:

(i) 1 mol of He at STP ($\bar{S}_{298}^{\circ} = 30.12$ cal K^{-1} mol^{-1} for He).

(ii) 1 mol of water relative to ice at 0°C.

(iii) a mixture of 1 mol of benzene and 1 mol of toluene at 25°C and 1 atm relative to the unmixed state (assume the mixture to be ideal).

21.55 Consider a system consisting of non-interacting, indistinguishable, identical molecules which can exist in either singlet or triplet state with energy separation ϵ. (a) Find molecular prtition function, molar partition function, $\bar{C}_V$ and $\bar{S}$ for $T = \epsilon /k$. (b) Can $\bar{U}$ be also found numerically? (c) Find the limiting values of $\bar{U}, \bar{C}_V$ and $\bar{S}$ as $T \rightarrow \infty$.

21.56 Three identical but distinguishable particles are distributed among three energy levels O, ϵ and 2ϵ. Write down the different possible distributions of the particles for total energy (i) ϵ (ii) 2ϵ and (iii) 3ϵ. (a) Find thermodynamic probability for each distribution in all three cases, mentioning the most probable distribution (b) Also find ΔS for change in energy from (i) to (ii) and also from (ii) to (iii). Interpret the result.

21.57 Consider a system of 1 mole of identical independent molecules. Each molecule has availability to it only three energy levels O, ϵ and 2ϵ. ($\epsilon = 5.52 \times 10^{-21}$ J) having degeneracy 1, 3 and 5 respectively. Calculate the following quantities (a) molecular partition function at 400 K (b) average molecular energy at 400 K (c) average number of molecules in the most populated energy level at 400 K and also at infinite temperature.

21.58 Consider a system of non-interacting particles at constant temperature which are distributed in three non-degenerate energy levels in such a way that ϵ_1, ϵ_2 and ϵ_3 levels consist of $4 \times 10^{23}, 2 \times 10^{23}$ and 1×10^{23} particles, respectively. Show that the energy levels are equispaced. If this spacing is 4.10×10^{-21} J, find the temperature of the system.

21.59 Spectroscopic observations on molecular nitrogen [vibrational frequency $(7.02 \times 10^{13}\,\text{Hz})$] show the following population of the excited vibrational levels (N_v) relative to that of the ground level (N_0) for different vibrational quantum number (v)

v	0	1	2	3
N_v/N_o	1.000	0.200	0.040	0.008

(a) Is the gas in thermodynamic equilibrium with respect to the distribution of vibrational energy?
(b) What is the vibrational temperature of the gas?
(c) Is this temperature same as the translational temperature?
(d) Find the vibrational energy of the gas.
(e) What fraction of N_2 molecules will be in the ground vibrational level at its characteristic vibrational temperature?

21.60 CO_2 molecule has rotational constant equal to 1.15×10^{10} s^{-1}. The fundamental vibrational frequencies of CO_2 are 2.00×10^{13} s^{-1}, 4.15×10^{13} s^{-1} and 7.00×10^{13} s^{-1}, the first being doubly degenerate and the other non-degenerate. Find (a) the characteristic rotational temperature of CO_2 (b) rotational contribution to $\bar{C}_V$ at 25°C (c) vibrational contribution to $\bar{C}_V$ due to the doubly degenerate vibration at 25°C.

21.61 Using the data in the previous problem, calculate the vibrational contribution to $\bar{G}$ at 25°C.

21.62 For NO molecule, the ground electronic level and the first excited electronic level have same degeneracy with energy separation of 121 cm^{-1} between them. There is no other low-lying electronic levels. Find the electronic contribution to the molar heat capacity of NO at 300 K and 500 K. Comment on the temperature dependence of the concerned heat capacity with an explanation thereof. Can you calculate electronic contribution to $\bar{G}$ from the given data?

21.63 At 298 K, kv/kT is 0.68 for $I_2(g)$ and 6.8 for HCl(g). (a) Find the vibrational heat capacity $(\bar{C}_V)_v$ of I_2 at 298 K and also at 0 K (b) At what temperature HCl(g) will have same vibrational heat capacity as I_2 at 298 K(s). (c) Comment on your result.

21.64 Use suitable Carnot's cycle to show that absolute zero cannot be attained. Would you consider this unattainability as a consequence of the second law of thermodynamics?

21.65 Justify unattainability of absolute zero on statical ground. Is negative temperature possible?

21.66 Is it possible to find absolute value of entropy?

21.67 Third-law entropies are not absolute entropies. Comment.

21.68 Is the third law of thermodynamics justified on statistical ground?

21.69 In evaluating entropy, the factors like nuclear randomness and isotope mixing are omitted? Is this justified?

21.70 Third law is applicable only to perfect crystals. True or false.

21.71 Entropy of a substance should always change with pressure at constant temperature. True or false?

21.72 For elements, the conventional values of S refer to $S = 0$ and $T = 0$, while those of H and G refer to $H° = 0$ and $G° = 0$ at $T = 298.15$ K. Why?

21.73 Does the relation $G = H - TS$ bear any meaning with usual conventional values of H, G and S?

21.74 Using the third law of thermodynamics regarding entropy establish the following:
   (a) For any substance (i) $\Delta C_p \to 0$ as $T \to 0$ and also (ii) $C_p \to 0$ as $T \to 0$.
   (b) For any substance, the coefficient of thermal expansion and coefficient of compressibility both tend to zero as $T \to 0$.

21.75 Justify the following:
   For any substance
   (i) G vs T plot has always negative slopes but this is not so with ΔG vs T plot.
   (ii) ΔH vs T plot and ΔG vs T plot have slopes of opposite sign.

21.76 It appears from the Sackur–Tetrode equation that $S_t \to -\infty$ as $T \to 0$, for fixed N and P. Does this disprove the third law of thermodynamics?

21.77 There are substances like CO, CH_3D and H_2O where S differs significantly from zero near $T = 0$ as $T \to 0$. How do you explain? Find approximate residual entropies in such cases. Apparently these point toward the failure of the third law. Comment

21.78 H_2 like CO possesses considerable residual entropy, but due to different reason. Discuss.

21.79 An average human DNA molecule has 5×10^8 dinucleotides of four different types. Assuming random choice of these four types, find the residual entropy of this DNA molecule. Comment on the result.

21.80 Considering C_p as a function of T of the form $C_p = a + bT + cT^2 + ...$ derive the relation (i) between ΔH_T and ΔH_o (ii) between ΔG_T and ΔH_o. Can ΔH_o of a reaction be positive?

21.81 Is there any latent heat of phase transformation near $T = 0$?

21.82 For a reaction at temperature (T), $\Delta G = a + bT + cT^2$
   (i) Find b.
   (ii) Find ΔH as a function of T.
   (iii) Show schematically the variation of ΔG and ΔH with T on the same plot.

ANSWERS

21.1 There is no essential difference between microstate and quantum state of a system. They imply the same only from different viewpoints.

21.2 (i) Here   number of particles, $N = 4$
             number of boxes, $l = 5$
      Then the total number of microstates $= l^N = 5^4$.
      Note: We can readily follow the result in the simplest case of distribution of 2 particles in 2 boxes where the number of microstates is 2^2, i.e. 4.
      (ii) Here also the number of particles and number of boxes are same as in (i), but the particles are indistinguishable. The total number of particles and partitions is $(N + l - 1)$ which can be arranged in $(N + l - 1)!$ ways. But out of these $(l - 1)!$ ways due to arrangement of partitions among themselves, and $N!$ ways due to

arrangement of indistinguishable particles among themselves cannot be distinguished. Then

the required number of microstates $= \dfrac{(N+l-1)!}{N!(l-1)!} = \dfrac{(4+5-1)!}{4!(5-1)!}$

Note: In the simplest case of distribution of two indistinguishable particles in two boxes, the result is $\dfrac{(2+2-1)!}{2!(2-1)!}$, i.e. 3 which corresponds to 1 homogeneous distribution and 2 heterogeneous distributions.

21.3 Here the first particle can be placed in any one of the boxes in 5 ways. For every one of these, the second particle can be placed in any one of the remaining boxes in $(5-1)$ or 4 ways. Obviously, the third particle can be placed in $(5-2)$ or 3 ways. Then the total number of possible distributions will be $5 \times 4 \times 3$. Out of these, 3! number of distributions due to arrangement of indistinguishable particles among themselves are same. Then

the required number of microstates $= \dfrac{5 \times 4 \times 3}{3!}$

[**Note:** In general, for N indistinguishable particles distributed between l boxes where no box contain more than one particle, the number of microstates will be

$$\dfrac{l!}{(l-N)!N!}, \text{ if } l > N]$$

☐ The result has chemical relevance with the Hund's rule of multiplicity.

21.4 Number of letters contained in 'configuration' = 13.

Number of different ways of shuffling the cards
$$= \text{number of ways of arranging 13 letters} = 13!$$

Assuming that every arrangement of letters has equal probability, the required frequency of obtaining the word 'configuration' will be one in 13! shuffling, i.e. once in $13! \times 2$ sec.

21.5 Required thermodynamic probability
$$= \text{number of distinguishable ways in which most probable}$$
$$\text{distribution can be achieved}$$

$$= \dfrac{(2N)!}{N!N!} \quad \text{(since most probable distribution corresponds to}$$
$$\text{equal number of particles in each energy level)}$$

Mathematical probability
$$= \dfrac{\text{Thermodynamic probability for most probable distribution}}{\text{Total number of distinguishable ways in which all possible distributions can be achieved}}$$

$$= \dfrac{(2N)!}{N!N!} \Big/ 2^{2N}$$

If the particles are indistinguishable, thermodynamic probability for uniform distribution = 1 and the mathematical probability

$$= 1 \Big/ \dfrac{(2N+1)!}{(2N)!\,1!}$$

Here the denominator is obtained following problem 21.2(ii).

Note: In case of distinguishable particles, uniform distribution has higher probability than any of the non-uniform distributions, and the difference increases with increase in N. But in case of indistinguishable particles, the uniform distribution has same probability as any of the non-uniform distributions.

21.6 Here the particles of each kind are distinguishable (their sites being so). This is unlike gas molecules.

The total number of sites is $(N_X + N_Y)$ which can be arranged in $(N_X + N_Y)!$ ways. However, the distributions due to arrangements of X atoms (which can be done in $N_X!$ ways) among themselves are all same. The same is true for Y atoms. Then, the required number of arrangements will be

$$\frac{(N_X + N_Y)!}{N_X!N_Y!}$$

21.7 For higher volume of the gas, each molecule has more positions available to it leading to more microscopic states, and hence higher thermodynamic probability of the system, that causes expansion of the gas into vacuum.

21.8 The required probability $= \dfrac{\Omega_f}{\Omega_i} = \left(\dfrac{V_f}{V_i}\right)^{100} = \left(\dfrac{1}{10}\right)^{100}$

Note: Contrary to the prediction of the second law of thermodynamics, it is not impossible for the molecules of a gas in equilibrium to come together spontaneously. However, for a large number of particles, the thermodynamic probability of such states is very low and can therefore be realised exceedingly rarely.

21.9 (a) The given statement is true only in absence of interactions between the molecules of the system.

(b) The given statement is true, even in presence of molecular interaction.

21.10 Energy will depend on all the factors except volume if there is no molecular interaction, otherwise volume will also be involved.

But the energy levels of a system for each ingredient depend only on volume (referred to particle-in-a-box situation) in absence of molecular interaction, otherwise they depend on all the factors.

21.11 Boltzmann distribution formula (21.13) represents the time average, and not the instantaneous, population of the energy levels of a system in statistical equilibrium.

☐ Its physical basis lies in the assumption that each microstate of a system for a given value of energy (E), volume and total number of particles, has equal probability. Then the probability that a molecule has a given energy (ϵ_i) will be proportional to the number of ways of distributing the rest of the energy $(E - \epsilon_i)$ among the other molecules. This number will necessarily decrease with increase in ϵ_i due to decrease in $E - \epsilon_i$. Boltzmann formula can thus be understood.

21.12 Because at very low temperature, the thermal de Broglie wavelength becomes so large that the wavefunctions of the particles overlap significantly, particularly at very high pressure and thereby affecting the independency of the particles on which Boltzmann distribution is based.

☐ The simple examples of failure of Boltzmann formula are few in number, these are liquid He II, conduction electron in metals and radiation (comprising photons).

21.13 (a) The required probability, $\dfrac{N_i}{\sum\limits_i N_i} = \dfrac{e^{\beta\epsilon_i}}{\sum\limits_i e^{\beta\epsilon_i}}$.

(b) Because only then the probability of finding a molecule will be lower at higher energy level and the total probability of all the molecules existing in different energy levels will be unity.

(c) The average kinetic energy ($\bar{\in}$) of an ideal gas molecule per degree of freedom is given by

$$\bar{\in} = \frac{\sum_i N_i \in_i}{\sum_i N_i} = \frac{\sum_i \in_i e^{\beta\in_i}}{\sum_i e^{\beta\in_i}}$$

$$= \frac{\sum_i \frac{p_i^2}{2m} e^{\beta p_i^2/2m}}{\sum_i e^{\beta p_i^2/2m}}$$

where p_i is molecular momentum per degree of freedom

$$= \frac{\frac{1}{2m}\int_{-\infty}^{+\infty} p^2 e^{\beta p^2/2m}}{\int_{-\infty}^{+\infty} e^{\beta p^2/2m} dp} \quad \text{assuming } p \text{ to vary continuously}$$

$$= -\frac{1}{2\beta}$$

Now from kinetic theory

$$\bar{\in} = \tfrac{1}{2}kT$$

Then $\qquad \beta = -\frac{1}{kT}$

(d) From the given expression

$$C = \frac{\sum_i N_i}{\sum_i e^{\beta\in_i}}$$

C must depend on T as β does so.

(e) From the given expression

$$\frac{N_i}{N_{i+1}} = \exp\left[\frac{(\in_{i+1} - \in_i)}{kT}\right], \text{ since } \beta = -\frac{1}{kT}$$

$$> 1 \text{ at finite } T, \text{ since } \in_{i+1} > \in_i$$

$$= 1 \text{ at } T = \infty$$

Then $\qquad N_{i+1} \leq N_i$

(f) Total energy of the system

$$E = \sum_i N_i \in_i$$

$$= \frac{N}{q}\sum_i \in_i e^{\beta\in_i} \quad \text{since } \frac{N_i}{N} = \frac{e^{\beta\in_i}}{q} \text{ where } q = \sum_i e^{\beta\in_i}$$

$$= \frac{N}{q}\sum_i \frac{\partial}{\partial\beta} e^{\beta\in_i}$$

$$= \frac{N}{q}\frac{\partial q}{\partial\beta}$$

Now from $\qquad \dfrac{N_i}{N} = \dfrac{e^{\beta\in_i}}{q}$

we have $\qquad \ln N_i = \ln N + \beta\in_i - \ln q$

Then
$$\frac{\partial \ln N_i}{\partial \beta} = \epsilon_i - \frac{1}{q}\frac{\partial q}{\partial \beta}$$

or
$$\frac{\partial N_i}{\partial \beta} = N_i\left(\epsilon_i - \frac{1}{q}\frac{\partial q}{\partial \beta}\right)\left[\frac{\partial \ln N_i}{\partial \beta} = \frac{1}{N_i}\frac{\partial N_i}{\partial \beta}\right]$$

$$= N_i\left(\epsilon_i - \frac{E}{N}\right) \quad \text{from the above equation}$$

which is the required relation.

$\dfrac{\partial N_i}{\partial \beta}$ signifies the energy of the molecules in the energy level i in excess of average molecular energy of the system.

(g)
$$\sum_i\left(\frac{\partial N_i}{\partial \beta}\right) = \frac{\partial}{\partial \beta}\sum_i N_i$$

$$= 0, \text{ for fixed total number of molecules}$$

This follows also from the significance of $\dfrac{\partial N_i}{\partial \beta}$ for a system of constant energy.

21.14 (a) β, which determines the distribution of energy among the molecules of the system, must be same as in non-degenerate system.

(b) Since for a degenerate system, the population N_i of an energy level i depends not only on ϵ_i but also on g_i in opposite way, the relation $N_{i+1} < N_i$ will not necessarily hold.

(c) Usually, the energy ϵ_0 corresponding to the lowest level is taken as zero and the energies corresponding to all other levels $\epsilon_1, \epsilon_2, \ldots$ are expressed relative to ϵ_0. Then

$$N_i = \frac{N g_i e^{-\epsilon_i/kT}}{g_0 + g_i e^{-\epsilon_i/kT} + \ldots}$$

Therefore

(i) When $T \to 0$
$$N_o \to N$$
$$N_i \to O$$

(ii) When $T \to \infty$

$$N_o \to \frac{N g_o}{g_o + g_i + \ldots} \quad \text{and} \quad N_i \to \frac{N g_i}{g_o + g_i + \ldots}$$

Note: The lowest energy level is not necessarily the most populated in case of degenerate systems of ordinary type and also in case of even non-degenerate systems if the energy levels are limited in number (found in some quantum systems).

21.15 q_t, q_r, q_v and q_e are all intensive properties, because none depends on number of molecules present in the system, provided the molecules do not interact (and only then the molecular partition function is meaningful).

☐ They all depend on T, rising with T (vide Sec. 21.4). Only q_t depends on V (rising with V due to lowering of energy levels). But none depend on N.

Note: q_t is intensive though it depends on V. Again, Q_t, unlike q_t, is extensive.

21.16 Both q_t and q_r are dimensionless as they are determined by the dimensionless quantities—degeneracy and $e^{-\epsilon_i/kT}$ of the energy levels.

□ The required relations (for non-interacting systems) are

$$Q_t = \frac{q_t^N}{N!}$$

$$Q_r = q_r^N$$

□□ Division by $N!$ is not required in case of rotational motion (or any other internal motions) which may be thought of as localised (distinguishable).

21.17 (i) Relevant, He molecules being indistinguishable and non-localised.

(ii) Irrelevant, Boltzmann distribution law being not obeyed.

(iii) Relevant, CO_2 molecules being indistinguishable and non-localised.

(iv) Irrelevant, H_2O molecules being localised.

21.18 From the following expression of q_t (which is dimensionless)

$$q_t = \left(\frac{2\pi mkT}{h^2} \right)^{\frac{3}{2}} V \quad \text{for an ideal gas}$$

it follows that $h/\sqrt{mkT}$ has the dimension of length.

21.19 Here $\qquad Q_1 = q^{N_1}$ and $Q_2 = q^{N_2}$

where q is molecular partition function which is independent of the amount of Au. Then

$$\frac{\ln Q_2}{\ln Q_1} = \frac{N_2}{N_1} = \frac{10}{1}$$

21.20 The number of microstates Ω (relative to that for the state at $T = 0$) is given according to Boltzmann by

$$S = k \ln \Omega$$

$$= k \ln Q_t + \frac{E}{T}$$

by Eq. (21.22) replacing U by E and Q by Q_t for an ideal monatonic gas

or $\qquad \Omega = Q_t e^{\frac{E}{kT}}$

$$= \frac{1}{N!} \left(\frac{2\pi mkT}{h^2} \right)^{\frac{3N}{2}} V^N e^{\frac{E}{kT}} \quad \text{by Eq. (21.26)}$$

$$= \frac{1}{N!} \left(\frac{2\pi m \cdot \frac{2}{3} E/N}{h^2} \right)^{\frac{3N}{2}} V^N e^{\frac{E}{kT}}$$

since $E = \frac{3}{2} NkT$ for an ideal gas.

21.21 Since total molecular kinetic energy of a system equals the sum of the kinetic energies of the individual molecules even in the presence of interaction between them, the canonical partition function for a fluid factorises into a part arising from the kinetic energy, which is same as for an ideal gas. The Boltzmann distribution law applies to this kinetic energy leading to the Maxwell distribution of speed that obviously holds in any fluid, not just in an ideal gas.

21.22 The barometric formula $P = P_0 \exp\left(-\dfrac{Mgh}{RT} \right)$ has been derived in Section 21.3.

□ $$P_{O_2} = \left(\frac{20}{100} \times 1 \text{ atm}\right) \exp\left[\frac{-9.80 \text{ m/s}^2)(32 \times 10^{-3} \text{ kg/mol})(10 \times 10^3 \text{ m})}{(8.314 \text{ JK}^{-1} \text{mol}^{-1})(273 \text{ K})}\right]$$

$$= 0.050 \text{ atm}$$

$$P_{N_2} = \left(\frac{80}{100} \times 1 \text{ atm}\right) \exp\left[\frac{-9.80 \text{ m/s}^2)(28.02 \times 10^{-3} \text{ kg/mol})(10 \times 10^3 \text{ m})}{(8.314 \text{ JK}^{-1} \text{mol}^{-1})(273 \text{ K})}\right]$$

$$= 0.238$$

$$P_{\text{total}} = P_{O_2} + P_{N_2} = (0.050 + 0.238) \text{ or } 0.288 \text{ atm}$$

$$x_{O_2} = \frac{P_{O_2}}{P_{\text{total}}} = \frac{0.050}{0.288} = 0.174$$

$$x_{N_2} = \frac{P_{N_2}}{P_{\text{total}}} = \frac{0.238}{0.288} = 0.826$$

Note:

1. The components of air appreciably fractionated to higher proportion of N_2 (the ingradient of lower molar mass) at the upper level of the atmosphere.
2. Total pressure is sometimes calculated putting M equal to the average molar mass (M_{av}) in the barometric formula. But the procedure is approximate because M_{av} is different at different levels due to appreciable fractionation.

21.23 (i) Let the number of moles of ith gaseous component in the atmosphere be N_{O_i} at sea-level and N_i at height h. Then

$$N_i = N_{O_i} e^{-M_i gh/RT} \quad \text{by Eq. (21.15)}$$

or $$N_i = \frac{N_{O_i}}{e^{M_i gh/RT}}$$

$$\simeq \frac{N_{O_i}}{1 + M_i gh/RT} \quad \text{(when } M_i gh/RT \ll 1\text{)}$$

$$= \frac{N_{O_i}}{M_i\left(\dfrac{1}{M_i} + \dfrac{gh}{RT}\right)}$$

Then, to some crude approximation, the average number of moles $\bar{N}_i$ can be expressed as

$$\bar{N}_i \propto \frac{N_{O_i}}{M_i} \tag{I}$$

for fixed h and T

or $$\bar{x}_i \propto \frac{x_{O_i}}{M_i}$$

$$= C \frac{x_{O_i}}{M_i}$$

Now $$\sum \bar{x}_i = 1$$

Hence $$C = \frac{1}{\sum x_{O_i}/M_i}$$

Therefore $$\bar{x}_i = \left(\frac{x_{O_i}}{M_i}\right) \Big/ \sum\left(\frac{x_{O_i}}{M_i}\right)$$

(ii) From Eq. (I),

$$\bar{N}_i M_i \propto N_{O_i}$$

or $$\bar{\omega}_i \propto N_{O_i}$$

where $\bar{\omega}_i$ is the average weight of the ith component in different levels
Then the average weight fraction

$$\frac{\bar{\omega}_i}{\sum \bar{\omega}_i} = \frac{N_{O_i}}{\sum N_{O_i}} = x_{O_i}$$

21.24 Derivation of Eq. (21.19):

$$C_V = \left(\frac{\partial U}{\partial T}\right)_{N,V}$$

$$= \frac{\partial}{\partial T}\left(kT^2 \frac{\partial \ln Q}{\partial T}\right)_{N,V} \quad \text{by Eq. (21.18)}$$

$$= \frac{k}{T^2}\left[\frac{\partial^2 \ln Q}{\partial\left(\frac{1}{T}\right)}\right]_{N,V}$$

Derivation of Eqs (21.20) and (21.22):

$$d\ln Q = \left(\frac{\partial \ln Q}{\partial \beta}\right)_V d\beta + \left(\frac{\partial \ln Q}{\partial V}\right)_\beta \quad (dV \text{ for a closed system where } \beta = 1/kT)$$

$$= -U d\beta + \left(\frac{\partial \ln Q}{\partial V}\right)_\beta dV \quad \text{by Eq. (21.18)}$$

or $$d(\ln Q + U\beta) = \beta dU + \left(\frac{\partial \ln Q}{\partial V}\right)_\beta dV$$

or $$dU = \frac{1}{\beta}d(\ln Q + U\beta) - \frac{1}{\beta}\left(\frac{\partial \ln Q}{\partial V}\right)_\beta dV$$

$$= kT\, d\left(\ln Q + \frac{U}{kT}\right) - kT\left(\frac{\partial \ln Q}{dV}\right)_T dV$$

Now, from classical thermodynamics
$$dU = Tds - Pdv, \text{ for reversible change in a closed system}$$
Comparing the last two equations, we have

$$P = kT\left(\frac{\partial \ln Q}{\partial V}\right)_T$$

and $$dS = kd\left(\ln Q + \frac{U}{kT}\right)$$

Then $$S = k\ln Q + \frac{U}{T} + \text{constant}$$

Adaption of suitable convention reduces this expression to the following simple form (dropping the arbitrary constant)

$$S = k\ln Q + \frac{U}{T}$$

With the above expressions for P and S, we can readily arrive at the relations (21.21), (21.23) and (21.24)

$$H = U + PV$$

$$= kT^2 \left(\frac{\partial \ln Q}{\partial T}\right)_{N,V} + kT \left(\frac{\partial \ln Q}{\partial \ln V}\right)_{N,T} \quad \text{using the Eq. (21.18)}$$

$$A = U - TS$$

$$= -kT \ln Q$$

$$G = A + PV$$

$$= -kT \ln Q + kT \left(\frac{\partial \ln Q}{\partial \ln V}\right)_{N,T}$$

21.25 Because the relevant canonical ensemble consists of systems of constant T, V and composition which are the natural variables for A.

21.26 All the given thermodynamic quantities except S will be affected due to change in partition function (Q) by a factor $e^{-\epsilon/kT}$. S, which is connected to randomisation of energy (instead of amount of energy), is not affected as the relative values of the energy levels remain same in the given change.

21.27 From the relations (21.20) and (21.26), we have the following expression for the translational contribution to the pressure of an ideal gas

$$P = \frac{NkT}{V}$$

which is ideal gas equation.

21.28 (a) Molecular translational partition function q_t is

$$q_t = \left(\frac{2\pi mkT}{h^2}\right)^{\frac{3}{2}} V \quad \text{(where } V \text{ is volume of the system)}$$

$$= \left(\frac{2\pi MkT}{N_A h^2}\right)^{\frac{3}{2}} V$$

$$= \left[\frac{2 \times 3.14) (4 \times 10^{-3} \text{ kg/mol}) (1.38 \times 10^{-23} \text{ J/K}) (273 \text{ K})}{(6.02 \times 10^{23}/\text{mol}) (6.63 \times 10^{-34} \text{ Js})}\right]^{\frac{3}{2}} (11.2 \times 10^{-3} \text{ m}^3)$$

$$= 7.57 \times 10^{-20} \text{ m}^3$$

(b) Molar translational partition function Q_t is

$$Q_t = \frac{q_t^{N_A}}{N_A!}$$

$$= \left(q_i \cdot \frac{e}{N_A}\right)^{N_A} \quad \text{using Sterling's approximation}$$

$$= \left[\frac{(7.57 \times 10^{-20} \text{ m}^3 \times 2.72)}{6.02 \times 10^{23}}\right]^{6.02 \times 10^{23}}$$

☐ The results will not be affected due to variation of P, provided V and T are constant, as it follows from the expression of q_t.

☐☐ The expressions used for calculations are based on the implicit assumptions that the molecules of the system are independent and that there is negligible probability for two molecules to have same molecular state of individual molecules.

□ □ □ These assumptions are amply justified under the given conditions of temperature (which is not very low) and pressure (which is not very high) where molecular interaction is negligible and the number of translational states of the individual molecules (measured by q_t) far exceeds the number of molecules of the system (which is of the order of 10^{23}).

Note:

1. Q_t refers to the same volume as q_t and this is the given volume and not the volume per mole.

2. N_A has unit in the expression for q_t and not for Q_t.

21.29 Using the expression (21.26), we have from Eqs. (21.21), (21.22) and (21.24)

$$(\bar{C}_p^o)_t = \frac{5}{2}N_A k = \frac{5}{2}R \quad \text{using } C_p = \left(\frac{\partial H}{\partial T}\right)_P$$

$$\bar{S}_t^o = \frac{5}{2}N_A k + N_A k \ln\left[\left(\frac{2\pi mkT}{h^2}\right)^{\frac{3}{2}}\frac{V}{N_A}\right]$$

$$= R\left\{\frac{5}{2} + \ln\left[\left(\frac{2\pi mkT}{h^2}\right)^{\frac{3}{2}}\frac{kT}{P^o}\right]\right\}$$

$$\bar{G}_t^o = -RT\ln\left[\left(\frac{2\pi mkT}{h^2}\right)^{\frac{3}{2}}\frac{kT}{P^o}\right]$$

Here $\left(\frac{2\pi mkT}{h^2}\right)^{\frac{3}{2}}\frac{kT}{P^o} = \left(\frac{2\pi MkT}{N_A h^2}\right)^{\frac{3}{2}}\frac{kT}{P^o}$

$$= \left[\frac{2\pi(28\times 10^{-3}\text{ kg/mol})(1.38\times 10^{-23}\text{ J/K (298 K)})}{(6.02\times 10^{23}/\text{mol})(6.63\times 10^{-34}\text{ Js})^2}\right]^{\frac{3}{2}}$$

$$\times\frac{(1.38\times 10^{-23}\text{ J/K})(298\text{ K})}{(10^5\,P_a)} = 5.882\times 10^{29}$$

Then $\bar{S}_t^o = (8.31\text{ JK}^{-1}\text{ mol}^{-1})\left[\frac{5}{2} + \ln(5.882\times 10^{29})\right]$

$$= 590.50\text{ JK}^{-1}\text{ mol}^{-1}$$

$$\bar{G}_t^o = -(8.31\text{ JK}^{-1}\text{ mol}^{-1})(298\text{ K})\ln(5.882\times 10^{29})$$

$$= -169.8\text{ kJ mol}^{-1}$$

Note:

1. Statistical thermodynamics and equipartition principle give the same expression of $(\bar{C}_p^o)_t$.

2. Translational motion favours entropy but disfavours free energy.

21.30 (a) For diatomic molecules

$$Q_t = \text{Same as monatomic molecules}$$

$$Q_r = q_r^N = \left(\frac{T}{\sigma\theta_r}\right)^N$$

for molecules behaving as rigid rotors where T is well above the characteristic rotational temperature θ_r (when Q_r may be computed through integration reasonably accurately) and σ is a number which is 2 for homogeneous and 1 for heterogeneous diatomics.

$$Q_v = q_v^N = \left(\frac{1}{1 - e^{-\theta_v/T}}\right)^N$$

for molecules behaving as harmonic oscillator, where T is well below the characteristic vibrational temperature θ_v so that $e^{-\theta_v/T} \ll 1$.

$$Q_e = q_e^N = g_o^N$$

for most diatomics if T is not high where g_o is the degeneracy of the lowest electronic level.

Using these expressions, we have

from Eq. (21.18) $\quad U_t = \frac{3}{2}NkT$

$$U_r = NkT$$

for $T \gg \theta_r$ when energy levels can be treated as varying continuously

$$U_v = Nk\theta_v \frac{1}{e^{\theta_v/T} - 1}$$

$$U_e = 0$$

from Eq. (21.19) $\quad (C_V)_t = \frac{3}{2}Nk$

$$(C_V)_r = Nk$$

$$(C_V)_v = Nk\left(\frac{\theta_v}{T}\right)^2 \frac{e^{\theta_v/T}}{(e^{\theta_v/T} - 1)^2}$$

$$(C_V)_e = 0$$

from Eq. (21.22) $\quad S_t = \frac{5}{2}Nk + Nk\ln\left[\left(\frac{2\pi mkT}{h^2}\right)^{\frac{3}{2}} \cdot \frac{V}{N}\right]$

$$S_r = Nk + Nk\ln\frac{T}{\sigma\theta_r}$$

$$S_v = Nk\frac{\theta_v}{T}\frac{1}{e^{\theta_v/T} - 1} - Nk\ln(1 - e^{-\theta_v/T})$$

$$S_e = Nk\ln g_o$$

(b) The required temperature (T) is one such that the energy level spacing is comparable to kT $\left(\text{i.e. } T \approx \dfrac{h^2}{8kmV^{\frac{2}{3}}}\right)$ above which Q_t and hence $(C_V)_t$, can be computed through integration reasonably accurately.

21.31 Because of considerable vibrational contribution to C_p in case of Cl_2, the Cl–Cl bond frequency being lower. But $\overline{C}_p$ is nearly same for N_2 and O_2 (though vibrational frequency is much greater with N_2), there being no significant vibrational contribution to $\overline{C}_p$ for N_2 and O_2 at STP.

21.32 At 0°C only translational and rotational contributions are to be considered. Then $\overline{C}_V$ being same for $N_2(g)$ and X (g) at 0°C, the molecules of X must be linear like N_2. But at high temperature, vibrational contribution has to be considered and the extra 6 Cal K^{-1} mol^{-1} of the equipartition value of $\overline{C}_V$ means three extra vibrational degrees of freedom which corresponds to one more atomicity. Therefore, a molecule of X will be linear triatomic.

21.33 $\bar{C}_p$ of a gas can be found out readily but approximately using equipartition principle considering full translational and rotational contributions and only partial vibrational contribution which may be taken to be 20% near room temperature and 30% at higher temperature.

(i) $\bar{C}_p = R + \bar{C}_V = R + 3R/2$, only translational contribution.

(ii) At room temperature $\bar{C}_p = R + 3R/2$ (tran) $+ 2R/2$ (rot) $+ \frac{20}{100} R$ (vib).

At the boiling temperature of H_2 which is very much lower than room temperature $\bar{C}_p = R + 3R/2$ (tran) $+ 2R/2$ (rot), there being no vibrational contribution.

(iii) At room temperature $\bar{C}_p = R + 3R/2$ (tran) $+ 2R/2$ (rot), $+ \frac{20}{100} 3R$ (vib).

At boiling temperature of H_2O which is very much higher than room temperature $\bar{C}_p = R + 3R/2$ (tran) $+ 3R/2$ (rot) $+ \frac{30}{100} 3R$ vib).

(iv) $\bar{C}_p = 18$ Cal $K^{-1}mol^{-1}$ (sp heat of water being 1 cal $K^{-1}g^{-1}$).

(v) $\bar{C}_p = 3R$ (vib), translational and rotational motion of molecules being insignificant for a solid.

(vi) Here atoms are very tightly bound and hence $\bar{C}_p$ deviates much from the usual equipartition value (given by Dulong and Petit law). Here we can consider roughly 20% of the equipartition value, i.e. $\bar{C}_p = \frac{20}{100} \times 3R$.

(vii) $\bar{C}_p = 2 \times 3R$, by Dulong and Petit law. Here the factor 2 appears because 1 mole of NaCl contains 2 moles of atoms.

21.34 Yes, because the energy due to bulk motion (which can be activated mechanically but not by heating) is not included in the internal energy.

21.35 Considering a crystal as an aggregate of atomic oscillators of same frequency v, Einstein arrived at the following equation for heat capacity of atomic solids

$$\bar{C}_V = 3R \left(\frac{hv}{kT} \right)^2 \cdot \frac{e^{hv/kT}}{(e^{\frac{kv}{kT}} - 1)^2}$$

For most solid elements, $\frac{hv}{kT} \ll 1$ at room temperature and hence $\bar{C}_V = 3R$ which is Dulong and Petit law.

In few cases like Be, B, C and Si, v is so high (due to tight binding of these light atoms) that the condition $\frac{hv}{kT} \ll 1$ is not fulfilled at room temperature.

Hence these elements provide exceptions to the Dulong and Petit law.

□ Obviously, at high temperature, when $\frac{hv}{kT} \ll 1$, the exceptional elements will have $\bar{C}_V = 3R$.

□□ The temperature $T = \frac{hv}{k}$ is called Einstein characteristic temperature of the crystal. It has the significance that at this temperature the energy per vibrational degree of freedom of an atomic oscillator of the crystal as expected from the classical principle of equipartition of energy becomes equal to one quantum of vibrational energy expected from quantum theory.

21.36 This is due to electronic contribution to $\bar{C}_V$, apart from vibrational contribution.

21.37 For an ideal monatomic gas

$$Q = \frac{q^N}{N!} \quad \text{atoms being indistinguishable}$$

Then $\qquad \ln Q = N \ln q - N \ln N + N$, by Sterling approximation

Now $\qquad A = -kT \ln Q$

Then $\qquad \mu_{gas} = \left(\frac{\partial A}{\partial N}\right)_{T,V}$

$$= -kT \ln\left(\frac{q}{N}\right)$$

For a monatomic crystal containing N atoms (which are distinguishable), we have on the basis of Einstein's model (where all $3N$ vibrational modes are assumed to have same frequency ν)

$$Q = q_v^{3N}$$

where $\qquad q_v = \dfrac{1}{1 - e^{-\frac{h\nu}{kT}}}$

Then $\qquad A = -kT \ln Q$

$$= -kT \ln q_v^{3N}$$

Then $\qquad \mu_{crystal} = \left(\frac{\partial A}{\partial N}\right)_{T,V}$

$$= -kT \ln q_v^3$$

☐ In case of diatomic molecules, rotational degrees of freedom arises with gaseous state but not with solid state. This makes μ lower (q being higher) in case of gaseous state making the vapour pressure higher.

21.38 From the second law of thermodynamics

$$dS = \frac{dQ_{rev}}{T}$$

$$= \frac{dU}{T}$$

in absence of all changes other than heating (i.e. without involvement of work) when energy levels of the system do not change

$$= k\beta \sum \epsilon_i \, dN_i \quad \text{(vide Sec. 21.3)}$$

dU being same as dE where $\beta = 1/kT$.
Then the Eq. (21.11), after rearrangement, becomes

$$dS = -k \sum \ln N_i dN_i - k\alpha \sum dN_i$$

$$= -k \sum \ln N_i dN_i \quad \text{since } \sum dN_i = 0$$

$$= kd \ln \Omega \quad \text{by Sterling approximation}$$

Then $\qquad S = k \ln \Omega + \text{constant}$

where Ω is the thermodynamic probability of the most probable distribution, or configuration of the system and this is virtually equal to the thermodynamic probability of the concerned macrostate if the system consists of large number of molecules. Even when number of molecules is not large Ω in the above expression is often referred to the macrostate concerned instead of a particular distribution.

By adapting suitable convention, the above expression can be reduced to the desired simple form (dropping the arbitrary constant).

21.39 It might seem from the derivation of $S = k \ln \Omega$ in the previous problem, that the relation holds only for non-interacting systems to which the Boltzmann distribution is limited. But this is not really so as it follows from the classical definition of S and statistical definition of Ω, vide Sec. 21.2.

□ Ω is an extensive quantity, being determined by the amount of the system.

21.40 The justification lies in the relation $S = k \ln \Omega$. A state of higher thermodynamic probability Ω, for which S is higher, is more random (disordered) because at any instant, we have less information about the exact microscopic state due to involvement of greater number of microstates.

21.41 As implied by the relation $S = k \ln \Omega$, the entropy (unlike energy) of a system is not a property of its individual molecules (though it is calculable from molecular properties). It depends on the probabilities of states having meaning only for a large collection of molecules.

21.42 It appears from the Sackur–Tetrode equation that S_t of an ideal gas depends on its molar mass (M), but ΔS_t does not. Hence $(C_V)_t = T\ (\partial S / \partial T)_{V,N}$ will be independent of M, though C_V is related to S.

21.43 The explanation lies in the fact that S is related to the dispersal (randomisation) of energy while (C_V) is related to the amount of energy (heat) absorbed by the system concerned. Then S, unlike C_V, will depend on the spacing of the energy levels of the system and hence on molar mass (imagine particle-in-a-box situation).

21.44 S_t. This is due to closer spacing of the translational energy levels compared with rotational and hence easier dispersal of translational energy.

21.45 Using the expression for S_r is problem 21.30:

(i) $\quad (\bar{S}_r)_{H_2} - (\bar{S}_r)_{D_2} = R \ln \dfrac{I_{D_2}}{I_{H_2}}$

$$= R \ln \dfrac{\mu_{D_2}}{\mu_{H_2}}$$

assuming that the bond length is same for H_2 and D_2

$$= R \ln \dfrac{m_D}{m_H}$$

$$= R \ln 2 \quad \text{taking } m_D = 2 m_H$$

(ii) $\quad (\bar{S}_r)_{HD} - (\bar{S}_r)_{D_2} = R \ln \dfrac{I_{D_2}/2}{I_{HD}}$

$$= R \ln \left(\dfrac{m_H + m_D}{m_H} \cdot \dfrac{1}{4} \right)$$

$$= R \ln \dfrac{3}{4}$$

Note: $(\bar{S}_r)_{H_2} > (\bar{S}_r)_{D_2} > (\bar{S}_r)_{HD}$. The order does not follow the molar mass due to the involvement of symmetry factor $\sigma\ (= 2)$.

21.46 (i) Ne. Because of higher molar mass that makes the spacing of energy levels lower and the ease of randomisation of energy higher with Ne.

(ii) D_2O. The reason is same as in (i).

(iii) CO. Because, althoug CO and N_2 have almost same molar mass, CO molecule has lower vibrational frequency (being less compact) and hence lower spacing of vibrational energy levels.

(iv) $CH_3CH_2CH_2CH_2CH_3$. Because of its more flexible structure than the other isomer.

(v) $\begin{matrix} CH_3 \\ CH_3 \end{matrix} {>} C = O$. The reason is same as in (iv).

21.47 (i) Because of greater number of ways in which distribution of molecules in space can be achieved at higher volume.

(ii) Because of greater number of ways in which distribution of molecules among different energy levels can be achieved due to involvement of greater number of energy levels.

21.48 For same volume expansion the increase in entropy due to spatial distribution of molecules is same whether the change occurs reversibly or irreversibly. But the decrease of energy, and hence of entropy due to energy distribution is greater in a reversible change. It is such that in a reversible adiabatic expansion the decrease in thermal entropy equals the increase in spatial entropy while it is less than the latter in an irreversible adiabatic expansion. Hence, in an adiabatic expansion $\Delta S_{rev} = 0$ but $\Delta S_{irr} > 0$.

21.49 Because $\Delta \overline{S}_{vap}$ involves greater increase in randomness, then $\Delta \overline{S}_{fus}$, gaseous state being more disordered than the condensed state of any substance.

21.50 This is due to the increase in thermal entropy of the system caused by the liberated thermal (kinetic) energy (resulting from the decrease in potential energy of the system on lattice formation) which more than compensates for the decrease in spatial entropy arising from the ordered molecular arrangement in the resulting lattice of ice.

21.51 Let us treat the reaction system as isolated. Here the decrease in spatial entropy due to formation of H_2 molecules from the combination of H-atoms has to be less than the accompanying increase in thermal entropy so that overall entropy change is positive for a spontaneous change in the isolated system. This necessitates some thermal energy to be generated by the reaction $H + H \rightarrow H_2$ which is then required to be exothermic.

21.52 For any substance, the gaseous state is most disordered. Then ΔS for a chemical reaction (in an isolated system), which increases with increase in randomness, will be greater with the reaction forming greater number of moles of the gaseous ingredients per mol of the reaction. Then the required order of $\Delta \overline{S}$ will be

$$(i) > (iii) \approx (iv) > (ii)$$

21.53 (i) $\Delta S = 0$. Because none of the state properties of the graphite rod can be regarded as affected on breaking, provided the change in the surface area is ignorable (otherwise ΔS will be +ve as in the pulvarisation of graphite into particles of atomic size, which is comparable to vaporisation).

(ii) $\Delta S = $ –ve. Because diamond has more rigid crystal structure than graphite.

(iii) $\Delta S = +ve$, for the reaction

$$C(s) + \tfrac{1}{2}O_2(g) = CO(g) \quad \text{(vide problem 21.52)}$$

(**Note:** $\Delta S = +ve$, though the process is exothermic)

However, for the reaction $C(s) + O_2(g) = CO_2(g)$, the sign of ΔS cannot be predicted from the simple considerations of the change in the number of moles of gaseous reactants and gaseous products, being numerically same. Here, what can be ascertained is that the ΔS will be small.

(iv) $\Delta S = -ve$. Because ice has orderly arrangement of H_2O molecules compared with water.

[**Note:** But ΔS will be +ve if the supercooled water is isolated vide problem 21.50].

(v) Same as (iv).

(vi) $\Delta S = 0$. Because two optically active isomers cannot be distinguished thermodynamically.

(vii) $\Delta S = -ve$. Because ortho-H_2 has higher entropy than para-H_2 due to nuclear-spin degeneracy. Vide problem 21.78.

(viii) $\Delta S = +ve$. Because, in the reaction $2O_3(g) = 3O_2(g)$, the number of moles of O_2 formed is greater than the number of moles of O_3 reacted.

(ix) In the given nuclear process, matter is appreciably converted into energy, i.e. neither mass nor energy is conserved. Consequently the laws of thermo-dynamics, and hence the concept of thermodynamic function like entropy, are not applicable to such a process.

21.54 (i)
$$\bar{S}^{\circ}_{273} = \bar{S}^{\circ}_{298} + R\ln\frac{273}{298}$$

$$= 30.12 \times 4.18\,\text{JK}^{-1}\text{mol}^{-1} + (8.31\,\text{JK}^{-1}\text{mol}^{-1})\ln\frac{273}{298}$$

$$= 125.17\,\text{JK}^{-1}\,\text{mol}^{-1}$$

Number of microstates for 1 mol

$$\Omega = \exp\left(\frac{\bar{S}^{\circ}_{273}}{k}\right) \quad \text{from } S = k\ln\Omega$$

$$= \exp\left(\frac{125.17\,\text{JK}^{-1}}{1.38 \times 10^{-23}\,\text{JK}^{-1}}\right) = 2.48 \times 10^{25}$$

(ii)
$$\Delta S_{\text{ice} \to \text{water}} = k\ln\frac{\Omega_{\text{water}}}{\Omega_{\text{ice}}}$$

$$= \frac{\Delta\bar{H}_{\text{fus}}}{T_{\text{fus}}}$$

Then for 1 mol

$$\frac{\Omega_{\text{water}}}{\Omega_{\text{ice}}} = \exp\left[\frac{\Delta H_{\text{fus}}}{kT_{\text{fus}}}\right]$$

$$= \exp\frac{80 \times 18 \times 4.18\,\text{J}}{(1.38 \times 10^{-23}\,\text{J/K})(273\,\text{K})}$$

$$= 4.94 \times 10^{24}$$

(iii)
$$\frac{\Omega_{\text{mixed}}}{\Omega_{\text{unmixed}}} = \exp\left(\frac{\Delta S_{\text{mix}}}{k}\right)$$

$$= \exp\left(\frac{-R\sum n_i \ln x_i}{k}\right)$$

Note: In (ii) $\Omega_{\text{water}}/\Omega_{\text{ice}}$, unlike $S_{\text{water}}/S_{\text{ice}}$, depends on the amount of ice transformed into water. Because, although both Ω and S depend on the amount of the substance, the former, unlike the latter, does not bear a proportionality relation with the amount. Hence $\Omega_{\text{water}}/\Omega_{\text{ice}}$ must refer to the amount of transformation.

21.55 (a)
$$q = \sum g_i e^{-\epsilon_i/kT}$$

$$= 1 + 3e^{-\epsilon/kT}$$

$$= 1 + 3e^{-1} \quad \text{for } T = \epsilon/k$$

$$= 2.10$$

$$Q = q^{N_A} \quad \text{for 1 mol}$$

$$= (2.10)^{6.02 \times 10^{23}}$$

[Here the factor $1/N!$ does not appear because the two quantum states referred, correspond to internal motion]

$$\bar{U} = kT^2 \frac{N_A}{q} \cdot \left(\frac{\partial q}{\partial T}\right)_V \quad \text{by Eq. (21.18)}$$

$$= kT^2 \frac{N_A}{q} \cdot \frac{3\epsilon}{kT^2} e^{-\frac{\epsilon}{kT}}$$

$$= \frac{3RT}{q} e^{-1} \quad \text{for } T = \frac{\epsilon}{k}$$

$$\bar{C}_V = \left(\frac{\partial \bar{U}}{\partial T}\right)_V = \frac{3R}{q} e^{-1} = 0.525R$$

$$= 4.36 \text{ JK}^{-1} \text{ mol}^{-1}$$

$$\bar{S} = k \ln Q + \frac{\bar{U}}{T} = R \ln q + \frac{3Re^{-1}}{q}$$

$$= R \ln(2.10) + 0.525R$$

$$= 10.5 \text{ JK}^{-1} \text{ mol}^{-1}$$

(b) $\bar{U}$ cannot be found numerically without numerical value of T

(c) As $T \to \infty$, $e^{-\epsilon/kT} \to 1$ and hence $q \to (1 + 3)$ or 4, then, from the above expression of $\bar{U}, \bar{C}_V$ and $\bar{S}$

$$\bar{U} = \frac{3N_A \epsilon}{q} e^{-\frac{\epsilon}{kT}} \to \frac{3N_A \epsilon}{4} \quad \text{as } T \to \infty$$

$$\bar{C}_V = \left(\frac{\partial U}{\partial T}\right)_V = \frac{3N_A \epsilon}{q} \cdot \frac{\epsilon}{kT^2} e^{-\frac{\epsilon}{kT}} \to 0 \text{ as } T \to \infty$$

This is a consequence of the limiting value of $\bar{U}$ for very high temperature

$$\bar{S} \to R \ln 4 + 0$$

Note: The limiting values of $\bar{U}, \bar{C}_V$ and $\bar{S}$ calculated above are not found with real molecules due to their unlimited number of energy levels.

21.56 (a) Only one distribution is possible for total energy ϵ and that is (2, 1, 0). The thermodynamic probability for this distribution is $\dfrac{3!}{2!1!0!} = 3$.

Two distributions are possible for total energy 2ϵ, viz (1, 2, 0) and (2, 0, 1) each of these has thermodynamic probability

$$\frac{3!}{1!2!0!} = \frac{3!}{2!0!1!} = 3$$

Then the thermodynamic probability for the state with energy 2ϵ is $3 + 3 = 6$.

Two distributions are possible also for total energy 3ϵ, viz. (0, 3, 0) and (1, 1, 1) with thermodynamic probabilities

$$\frac{3!}{0!3!0!} = 1$$

and

$$\frac{3!}{1!1!1!} = 6$$

Then the thermodynamic probability for the state with total energy 3ϵ is $1 + 6 = 7$.

The distribution (1, 1, 1) having higher thermodynamic probability is the most probable one.

(b) For the change

(i) $\rightarrow$ (ii) $\qquad \Delta S = k \ln \dfrac{\Omega ii}{\Omega i} = k \ln \dfrac{6}{3}$

(ii) $\rightarrow$ (iii) $\qquad \Delta S = k \ln \dfrac{\Omega iii}{\Omega ii} = k \ln \dfrac{7}{6}$

For the two changes, the energy change is same but entropy change is different. This can be interpreted if we remember that S is a measure of randomisation of energy and not necessarily of total energy of the system. ΔS is greater in case of (i) $\rightarrow$ (ii) change due to greater change in randomness (which is determined by thermodynamic probability) of the system, i.e. due to greater chaotic dispersal of total energy.

21.57 (a) Molecular particle function

$$q = \sum g_i e^{-\epsilon_i/kT}$$

$$= g_0 e^{-0/kT} + g_1 e^{-\epsilon/kT} + g_2 e^{-2\epsilon/kT}$$

Here $\qquad \dfrac{\epsilon}{kT} = \dfrac{5.52 \times 10^{-21}\text{ J}}{(1.38 \times 10^{-23}\text{ J/K})(400\text{ K})} = 1$

Then $\qquad q = 1 + 3e^{-1} + 5e^{-2}$

(b) Average molecular energy

$$\overline{\epsilon} = \frac{\sum N_i \, \epsilon_i}{\sum N_i}$$

$$= \frac{N_0 \, \epsilon_0 + N_1 \, \epsilon_1 + N_2 \, \epsilon_2}{N_0 + N_1 + N_2}$$

$$= \frac{\epsilon_0 + \frac{N_1}{N_0}\epsilon_1 + \frac{N_2}{N_0}\epsilon_2}{1 + \frac{N_1}{N_0} + \frac{N_2}{N_0}}$$

$$= \frac{\epsilon_0 + \frac{g_1}{g_0}e^{-(\epsilon_1 - \epsilon_0)/kT}\epsilon_1 + \frac{g_2}{g_0}e^{-(\epsilon_2 - \epsilon_0)/kT}\epsilon_2}{1 + \frac{g_1}{g_0}e^{-(\epsilon_1 - \epsilon_0)/kT} + \frac{g_2}{g_0}e^{-(\epsilon_2 - \epsilon_0)/kT}}$$

$$= \frac{\epsilon_0(0 + 3e^{-1} + 5e^{-2} \cdot 2)}{1 + 3e^{-1} + 5e^{-2}}$$

(c) At 400 K, the populations of energy levels are in the ratio $1:3e^{-1}:5e^{-2}$. Then the middle energy level is the most populated one. The average number of molecules in this energy level at 400 K is $\dfrac{3e^{-1}}{1 + 3e^{-1} + 5e^{-2}} \times 1$ mole.

At $T = \infty$, $e^{-\epsilon_i/kT} = 1$ and hence the populations of energy levels will be in the ratio 1:3:5. Then at infinite temperature, the highest energy level will be the most populated one and the average number of molecules in this energy level will be

$$\frac{5}{1 + 3 + 5} \times 1 \text{ mole}$$

Note: For degenerate systems, unlike non-degenerate one, the most populated energy level may not be the lowest level, and this may be different for different temperature. However, at very high temperature, the most populated energy level is always the one having the highest degeneracy.

21.58 Applying Boltzmann distribution law

$$\frac{N_1}{N_2} = e^{-(\epsilon_1 - \epsilon_2)/kT}$$

and

$$\frac{N_2}{N_3} = e^{-(\epsilon_2 - \epsilon_3)/kT}$$

where N_1, N_2 and N_3 are the population of the energy levels ϵ_1, ϵ_2 and ϵ_3 respectively.

Then

$$\frac{\epsilon_2 - \epsilon_1}{\epsilon_3 - \epsilon_2} = \frac{\ln \frac{N_1}{N_2}}{\ln \frac{N_2}{N_3}}$$

$= 1$, from the given data

or

$$\epsilon_2 - \epsilon_1 = \epsilon_3 - \epsilon_2$$

□

$$T = \frac{\epsilon_2 - \epsilon_1}{k \ln \frac{N_1}{N_2}}$$

$$= \frac{4.10 \times 10^{-21} \text{ J}}{(1.38 \times 10^{-23} \text{ J/K}) \ln \frac{4}{2}} = 428.7 \text{ K}$$

Note: Here T calculated from any pair of the given energy levels is same. If it were not same, we are to take the average value.

21.59 (a) Considering N_2 molecule as a harmonic oscillator, its energy levels $\epsilon_v = \left(v + \frac{1}{2}\right)hv$ will be in arithmatic progression with spacing hv. Now from the given data, the relative populations N_v/N_o of the energy levels, which are in the ratio $1:0.2:0.04:0.008$, are in geometric progression. This corresponds to the Boltzmann distribution

$$\frac{N_v}{N_o} = e^{-vhv/kT}$$

The system will then be in thermodynamic equilibrium with respect to the distribution of vibrational energy.

(b) Vibrational temperature, $T = -\dfrac{v \cdot hv}{k \ln\left(\frac{N_v}{N_o}\right)}$

$$= -\frac{(1)\,(6.63 \times 10^{-34} \text{ Js})\,(7.02 \times 10^{13} \text{ s}^{-1})}{(1.38 \times 10^{-23} \text{ J/K})\,\ln(0.20)}, \quad \text{with } v = 1$$

(c) Yes. Because the distribution of both translational energy and vibrational energy is determined by temperature of the system in accordance with the same statistical law enunciated by Boltzmann.

(d) $$\bar{U}_v = R\theta_v \cdot \frac{1}{e^{\theta_v/T} - 1}$$

where θ_v is the characteristic vibrational temperature. (Ref. problem 21.30)

$$\simeq RT \qquad T \text{ being quite high}$$

(e) $$\frac{N_o}{\sum N_v} = \frac{1}{1 + e^{-1} + e^{-2}}$$

since $hv/kT = 1$ at the characteristic vibrational temperature

$$\simeq 1 + e^{-1}$$

21.60 (a) Characteristic rotational temperature

$$= \frac{h^2}{8\pi^2 Ik} = \frac{hB}{k}, \quad \text{where } B \text{ is rotational constant}$$

$$= \frac{(6.63 \times 10^{-34} \text{ Js})\,(1.15 \times 10^{10} \text{ s}^{-1})}{(1.38 \times 10^{-23} \text{ J/K})} = 0.55 \text{ K}$$

(b) Here $(\bar{C}_V)_r$ is the same as that for a diatomic gas, the molecule being linear. Then

$$(\bar{C}_V)_r = R$$

the concerned temperature being well above the characteristic rotational temperature.

(c) For doubly degenerate vibrational frequency, the contribution to $(\bar{C}_V)_v$ will be twice that for a diatomic molecule, then

$$(\bar{C}_V)_v = 2R\left(\frac{hv}{kT}\right)^2 \cdot \frac{e^{\frac{hv}{kT}}}{\left(e^{\frac{hv}{kT}} - 1\right)^2} \qquad \text{referred to problem 21.30}$$

Here $$\frac{hv}{kT} = \frac{(6.63 \times 10^{-34} \text{ Js})\,(2.00 \times 10^{13} \text{ s}^{-1})}{(1.38 \times 10^{-23} \text{ J/K})\,(25 + 273)\text{ K}} = 3.22$$

Then $\quad (\bar{C}_V)_v = 2(8.31\ \text{JK}^{-1}\ \text{mol}^{-1})(3.22)^2 \times \dfrac{e^{3.22}}{(e^{3.22} - 1)^2}$

$$= 7.48\ \text{JK}^{-1}\ \text{mol}^{-1}$$

21.61 Considering the contribution of all the vibrational frequencies, viz. the degenerate frequency v_1 and non-degenerate v_2 and v_3

$$q_v = \left(\frac{1}{1 - e^{-hv_1/kT}}\right)^2 \left(\frac{1}{1 - e^{-hv_2/kT}}\right) \left(\frac{1}{1 - e^{-hv_3/kT}}\right)$$

$$= 1.08 \quad \text{from the given data}$$

$$\bar{G}_v = \bar{A}_v + RT = -kT \ln q_v^{N_A} + RT \quad \text{by Eq. (21.23)}$$

$$= -RT(\ln q_v - 1)$$

$$= -(8.31\ \text{JK}^{-1}\ \text{mol}^{-1})(298\ \text{K})[\ln(1.08) - 1]$$

$$= 2667\ \text{J mol}^{-1}$$

21.62 $\qquad q_e = \sum g_i e^{-\epsilon_i/kT} = g + g e^{-\epsilon/kT}$

where ϵ is the energy separation between the two levels each of degeneracy g.

$$\bar{U}_e = N_A kT^2 \left(\frac{\partial \ln q_e}{\partial T}\right)_V \quad \text{by Eq. (21.18)}$$

$$= N_A kT^2 \frac{1}{1 + e^{-\epsilon/kT}} \cdot \frac{\epsilon}{kT^2} e^{-\epsilon/kT}$$

$$= N_A \cdot \frac{\epsilon\, e^{-\epsilon/kT}}{1 + e^{-\epsilon/kT}}$$

$$(\bar{C}_V)_e = \left(\frac{\partial \bar{U}}{\partial T}\right)_V$$

$$= N_A k \left(\frac{\epsilon}{kT}\right)^2 \cdot \frac{e^{-\epsilon/kT}}{(1 + e^{-\epsilon/kT})^2} = R\left(\frac{\epsilon}{kT}\right)^2 \cdot \frac{e^{-\epsilon/kT}}{(e^{-\epsilon/kT} + 1)^2}$$

For 300 K $\quad \dfrac{\epsilon}{kT} = \dfrac{hc}{kT}\bar{v} = \dfrac{(6.63 \times 10^{-27}\ \text{ergs})(3.00 \times 10^{10}\ \text{cm s}^{-1})(121\ \text{cm}^{-1})}{(1.38 \times 10^{-16}\ \text{erg/K})(300\ \text{K})} = 0.581$

Then $\quad (\bar{C}_V)_e = (8.31\ \text{JK}^{-1}\ \text{mol}^{-1})(0.581)^2 \cdot \dfrac{e^{-0.581}}{(e^{-0.581} + 1)^2}$

$$= 0.649\ \text{JK}^{-1}\ \text{mol}^{-1}$$

For 500 K $\quad \dfrac{\epsilon}{kT} = \dfrac{hc\bar{v}}{kT} = \dfrac{(6.63 \times 10^{-27}\ \text{ergs})(3.00 \times 10^{10}\ \text{cm s}^{-1})(121\ \text{cm}^{-1})}{(1.38 \times 10^{-16}\ \text{erg/K})(500\ \text{K})}$

$$= 0.349$$

Then $\quad (\bar{C}_V)_e = (8.31\ \text{JK}^{-1}\ \text{mol}^{-1})(0.349)^2 \cdot \dfrac{e^{-0.349}}{(e^{-0.349} + 1)^2} = 0.245\ \text{JK}^{-1}\ \text{mol}^{-1}$

☐ It appears from the above expression of $(\bar{C}_V)_e$ that

$$C_V \to 0 \text{ as } T \to \infty$$

also $\qquad C_V \to 0 \text{ as } T \to 0$

beause in the low temperature region $\dfrac{e^{\epsilon.kT}}{e^{\epsilon/kT}+1} \approx e^{-\epsilon/kT}$ goes to zero more rapidly than $(\epsilon/kT)^2$ goes to infinity.

Then with rise in temperature from low value, $(\bar{C}_V)_e$ will pass through a maximum. The given temperature region where $(\bar{C}_V)_e$ falls with T, lies after this maximum.

□□ No. Because $\bar{G}_e$ unlike $(\bar{C}_V)_e$, is not independent of degeneracy, and hence requires for its calculation, the value of degeneracy which is not given.

21.63 (a)
$$(\bar{C}_V)_v = R\left(\frac{h\nu}{kT}\right)^2 \frac{e^{h\nu/kT}}{\left(e^{h\nu/kT}-1\right)^2} \qquad \text{referred to problem 21.30}$$

$$= (8.31\ \text{JK}^{-1}\text{mol}^{-1})(0.68)^2 \times \frac{e^{0.68}}{(e^{0.68}-1)^2} \quad \text{at } T = 298\ \text{K}$$

$$= 8.01\ \text{JK}^{-1}\ \text{mol}^{-1}$$

As $T \to 0$, $(\bar{C}_V)_v \to 0$, as in previous problem.

(b) The required temperature (T) is one for which
$$\frac{h\nu}{kT} = 0.68$$

Now
$$\frac{h\nu}{k \cdot 298} = 6.8$$

Then
$$T = \frac{6.8}{0.68} \times 298\ \text{K} = 2980\ \text{K}$$

(c) The temperature thus calculated is high enough for considerable dissociation of HCl, and hence it has no practical significance.

21.64 Let us consider a Carnot cycle ABCD operating between any temperature T and absolute zero. For the entire cycle

$$\oint dS = \Delta S_1 + \Delta S_2 + \Delta S_3 + \Delta S_4 = 0$$

or $\Delta S_3 = -\Delta S_1$, since $\Delta S_2 = \Delta S_4 = 0$
Now, according to the third law of thermodynamics, $\Delta S_3 = 0$ at $T = 0$ (if possible) when $\Delta S_3 = -\Delta S_1$ will not hold, since $\Delta S_1 \neq 0$. This implies that the absolute zero cannot be attained.

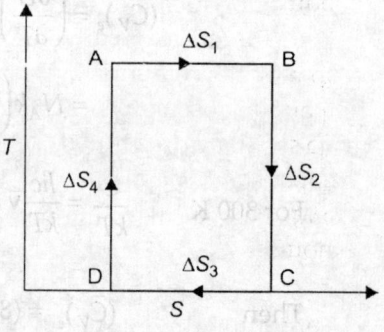

□ No. Because although we arrive at the unattainability of absolute zero using Carnot cycle (related to second law), the restriction $\Delta S_3 = 0$, which is imposed by the third law, has also been used.

21.65 For a molecular system, temperature (T) determines the number of molecules N_i in an energy level ϵ_i and N_o in the zero energy level ϵ_o according to the relation

$$\frac{N_i}{N_o} = e^{-(\epsilon_i - \epsilon_o)/kT}$$

as per Boltzmann distribution law for non-degenerate energy levels

or
$$T = \frac{\epsilon_i - \epsilon_o}{k \ln \frac{N_o}{N_i}}$$

The so-called absolute zero value of T cannot be attained because it corresponds to $N_i = 0$ which can be reached only with extraction of infinitely great amount of energy. T becomes +ve or –ve according as $N_i < N_o$ or $N_i > N_o$. For ordinary systems $N_i < N_o$ prevails because of infinitely great number of energy levels while the number of molecules is finite. However, $N_i > N_o$ is possible with some quantum systems where there is an upper limit of their internal energy (U) and the number of energy levels is finite.

To arrive at –ve temperature, instead of cooling a system (of finite number of energy levels) below absolute zero (which is impossible), we are to increase its energy (when S decreases due to decrease in randomness of molecular distribution, being concentrated at the highest energy levels, when $(\partial U/\partial S)_V$ is negative). This can be effected by absorption of light (as in lasers) or of microwave radiation (as in masers).

Note:

1. The impossibility of attainment of absolute zero temperature (affirmed by the third law) refers to cooling of a body.

2. A body with negative absolute temperature gives rise to the feeling of hotness. Because such a body, being in an extremely non-equilibrium state, will readily pass its energy to another contacting body at positive (higher) absolute temperature.

21.66 No. This is obvious from the general expression $S = k \ln \Omega$ + constant, considering the arbitrariness of the constant. Even if this constant is dropped by adapting suitable standard for measurement of S, we cannot find S absolutely because we cannot find Ω completely due to significant contribution from nuclear randomness about which nothing whatever is known. Entropies used in physical chemistry are the practical or conventional entropies which consider only translational, configurational, rotational, vibrational and electronic contributions taking them to be zero for a chosen physical state.

21.67 Yes. Because third law entropies are based on the convention that $S = 0$ for all substances in their most stable state at 0 K.

21.68 On statistical ground, entropy is formulated by the Boltzmann relation

$$S = k \ln \Omega$$

wherefrom, it is calculated considering only translational, configurational, rotational, vibrational and electronic contribution to Ω. Such contributions are likely to be zero for a substance at the lowest attainable temperature which is very close to 0 K at which every substance (except He) exists in crystalline form. The third law can be justified on this basis.

21.69 In evaluating entropy, the factors like nuclear randomness (which is uncertain) and isotope mixing (which is minor) are omitted. For chemical processes, this is justified in view of the fact that in such processes nuclei remain virtually unaffected and the isotopic composition usually remain almost constant.

21.70 The third law applies to all substances in their most stable (equilibrium) state which is not necessarily the perfect crystalline state at $T = 0$, as in case of liquid He*. Although this law does not normally hold for glassy states (because such states are not true equilibrium states), it amply holds if the phases involved are in frozen metastable equilibrium, provided this equilibrium is not disturbed in any way. Therefore, the given statement is not completely justified.

[* The peculiarity of He lies in the existence of zero-point energy which is enough in this case (due to low atomic mass) to overcome the weak interaction between He-atoms and thereby preventing them from being arranged in a regular order as in a crystal lattice. Only an external pressure above 1 atm can bring He-atoms close enough together for them to form a crystal. It is interesting to note that in the low regions of temperature near 0 K, there is no latent heat. Here phase-change does not take place by intake of heat but only by change of pressure].

21.71 The given statement is true, except at 0 K where entropy is a function of temperature only provided the system remains in internal equilibrium (by Nernst–Simon statement of the third law).

21.72 Because of the advantage that near $T = 0$, S, unlike H and G, is independent of pressure. For H and G, the conventional values are referred to $H° = 0$ and $G° = 0$ at $T = 298.15$ K (ordinary temperature) because of its practical advantage.

21.73 No quantitative meaning in general. Because the usual conventional standard for S is different from that for H and G.

Note: However, for compounds, the relation $G = H - TS$ will hold if their G, H and S are measured relative to those of their constituent elements (as in free energy of formation, enthalpy of formation and entropy of formation) which are then independent of the standard state chosen (vide University question 5.89)

21.74 (a) (i) $\Delta C_p = \dfrac{\partial \Delta H}{\partial T} \to 0$ as $T \to 0$, by Nernst heat theorem.

(ii) At constant pressure, the entropy of a substance S_T at temperature T and S_0 at $T = 0$ are related by

$$S_T = \int_0^T \frac{C_p}{T} dT + S_0 \quad \text{by the second law}$$

Now, according to the third law (Planck statement) S_T has always a finite value and $S_T \to 0$ as $T \to 0$. Then $\displaystyle\int_0^T \frac{C_p}{T} dT$ must be finite. This requires $C_p \to 0$ as $T \to 0$.

(b) By Maxwell relation

$$\left(\frac{\partial S}{\partial P}\right)_T = -\left(\frac{\partial V}{\partial T}\right)_P$$

and by the third law (Nernst–Simon statement), $\left(\dfrac{\partial S}{\partial P}\right)_T \to 0$ as $T \to 0$. Then $\left(\dfrac{\partial V}{\partial T}\right)_P$, and hence the coefficient of thermal expansion $\dfrac{1}{V}\left(\dfrac{\partial V}{\partial T}\right)_P \to 0$ as $T \to 0$.

Also $\left(\dfrac{\partial V}{\partial P}\right)_T = -\dfrac{(\partial V / \partial T)_P}{(\partial P / \partial T)_V} \to 0$ as $T \to 0$

Then the coefficient of compressibility $= -\dfrac{1}{V}\left(\dfrac{\partial V}{\partial P}\right)_T \to 0$ as $T \to 0$.

21.75 (i) any substance $\left(\dfrac{\partial G}{(\partial T)}\right)_P = -S$, if mass is fixed. Now, according to the third law S cannot be –ve. Then G vs T will always have –ve slope.

Again, $\left(\dfrac{\partial \Delta G}{\partial T}\right)_P = -\Delta S$, for any change of a closed system

According to the second law, ΔS may be +ve or –ve.

Then $\left(\dfrac{\partial \Delta G}{\partial T}\right)_P$ may be –ve or +ve, i.e. slope of ΔG vs T may not be –ve.

(ii) From Gibbs–Helmholtz relation, $\Delta G = \Delta H + T\left(\dfrac{\partial \Delta G}{\partial T}\right)_P$, we have

$$\frac{\partial \Delta G}{\partial T} = \frac{\partial \Delta H}{\partial T} + \frac{\partial \Delta G}{\partial T} + T\frac{\partial^2 \Delta G}{\partial T^2}$$

or $$\frac{\partial \Delta H}{\partial T} = -T\frac{\partial^2 \Delta G}{\partial T^2}$$

Again $\dfrac{\partial^2 \Delta G}{\partial T^2}$ has the same sign as $\dfrac{\partial \Delta G}{\partial T}$, since ΔG converges to a limiting value as $T \to 0$.

Then $\dfrac{\partial \Delta H}{\partial T}$ and $\dfrac{\partial \Delta G}{\partial T}$ have opposite sign.

21.76 Not at all. Because at very low temperature (when Boltzmann distribution is not valid), the Sackur–Tetrode equation becomes invalid.

21.77 This is due to the existence of more than one geometrical arrangement of the molecules in the crystal even near absolute zero. For example, two adjacent carbon monoxide molecules can be oriented either as (CO·CO) or as (CO·OC). However, the alternative configurations differ in energy only slightly (due to low dipole moment of CO), so that they may be regarded as equally probable. Then for a system of N molecules, the number of ways (Ω) of achieving virtually the same energy is 2^N. The entropy due to this residual disorder is

$$S = k\ln\Omega$$
$$= k\ln 2^N$$
$$= Nk\ln 2$$

In case of CH_3D (tetrahedral structure), the C–D bond can have 4 different orientations relative to neighbouring molecules. Hence, the residual entropy will be

$$S = k\ln 4^N$$
$$= NK\ln 4$$

The case of H_2O is somewhat more complex. Here, due to hydrogen bonding, each O atom is surrounded tetrahedrally by four H atoms of which two are attached by normal covalent bonds and the other two by (longer) hydrogen bonds. Then for a sample of ice containing N molecules, each of the $2N$ H atoms can be arranged in either short or long distance from an O atom with total 2^{2N} possible arrangements.

However, of the 2^4 or 16 ways of arranging four H atoms around one O atom, only six having real structural significance are acceptable. Then the number of permitted arrangements will be $2^{2N} \times \left(\frac{6}{16}\right)^N = \left(\frac{3}{2}\right)^N$ amounting to

$$S = k\ln\left(\frac{3}{2}\right)^N$$

$$= Nk\ln\left(\frac{3}{2}\right)$$

☐ The third law refers to systems in equilibrium. For the given substances, the different geometrical arrangements of the molecules are not in equilibrium, near $T = 0$. Therefore, the calculated residual entropies on statistical ground do not indicate anything regarding the failure of the third law.

21.78 The residual entropy of H_2, unlike CO, is not due to any disorder in the crystal but due to the nuclear-spin degeneracy. Ordinary hydrogen is a mixture of ortho- and para-hydrogen, which have different values of the total nuclear spin angular momentum. Ortho-hydrogen can be in any one of nine states having the same energy, while the para-hydrogen exists in a single state. As a result of mixing of the two kinds of hydrogen and the distribution of the ortho-hydrogen in nine different states, the system possesses some randomness and hence a residual entropy.

Note:

1. The third law will be applicable to a substance only if it exists in a single quantum state. Thus, the third law applies to para-hydrogen but not to ortho-hydrogen.

2. The residual entropy (S_0) is obviously zero for $\Omega = 1$. But even when $\Omega > 1$, $\overline{S}_0$ is ignorable with most substances, being of the order of 10^{-22} JK^{-1}, for Ω is as large as the Avogedro constant (N_A). For commonly known substances remarkable for their residual entropy, $\Omega = a^{N_A}$, where a is a number greater than 1.

21.79 Using the Boltzmann formula $S = Nk\ln\Omega$, we have

residual entropy per molecule $= (5 \times 10^8) (1.38 \times 10^{-23} \text{ J/K}) \ln 4$

$= 9.56 \times 10^{-15}$ J/K

☐ From the result of calculation, we conclude that even for macromolecules like DNA, the residual molecular entropy is small compared to normal entropies for macroscopic systems.

21.80 (i) $\Delta C_p = (C_p)_f - (C_p)_i = (a' + b'T + c'T^2 + ...) - (a + bT + cT^2 + ...)$

$= \alpha + \beta T + \gamma T^2 + ... \text{(say)}$

Now by Nernst heat theorem

$$\left(\frac{\partial \Delta H}{\partial T}\right)_P = \Delta C_p \to 0 \text{ as } T \to 0$$

Therefore, it follows that $\alpha = 0$ and hence

$$\left(\frac{\partial \Delta H}{\partial T}\right)_P = \beta T + \gamma T^2 + ...$$

Integration gives

$$\int_{\Delta H_0}^{\Delta H_T} d\Delta H = \int_0^T (\beta T + \gamma T^2 + ...) \, dT$$

or $\Delta H_T = \Delta H_0 + \frac{1}{2}\beta T^2 + \frac{1}{3}\gamma T^3 + ...$

which is the desired relation.

(ii) From the third law

$$\Delta S_T = \int_0^T \frac{\Delta C_p}{T}\, dT$$

$$= \int_0^T \frac{1}{T} (\beta T + \gamma T^2 + ...)\, dT$$

$$= \beta T + \tfrac{1}{2}\gamma T^2 + ...$$

Substituting ΔS_T and ΔH_T in the expression $\Delta G_T = \Delta H_T - T\Delta S_T$,

$$\Delta G_T = \left(\Delta H_0 + \tfrac{1}{2}\beta T^2 + \tfrac{1}{3}\gamma T^3 + ...\right) - T\left(\beta T + \tfrac{1}{2}\gamma T^2 + ...\right)$$

$$= \Delta H_0 - \tfrac{1}{2}\beta T^2 - \tfrac{1}{6}\gamma T^3 + ...$$

☐ Since for a spontaneous reaction ΔG_T cannot be positive at any T, it follows from the expression of ΔG_T that ΔH_o cannot be positive.

Note: At absolute zero, no endothermic process can occur.

21.81 No. Near $T = 0$, the phasechange does not take place by intake of heat but only by change of pressure.

21.82 (i) Since $\left[\dfrac{\partial \Delta G}{\partial T}\right]_P \to 0$ as $T \to 0$ (by Nernst heat theorem), we have from the given expression

$$b + 2CT = 0 \text{ at } T = 0$$

Then $\qquad b = 0$

(ii) $\qquad \Delta H = -T^2 \left[\dfrac{\partial(\Delta G/T)}{\partial T}\right]_P \quad$ by Gibbs–Helmholtz relation Eq. (5.11b)

$$= a - cT^2 \quad \text{from the given expression}$$

(iii) $\dfrac{\partial \Delta G}{\partial T}$ and $\dfrac{\partial \Delta H}{\partial T}$ have opposite sign but both tend towards zero as $T \to 0$.

Again, for a spontaneous reaction ΔG_T cannot be positive at any T (which requires C to be negative).

We then have the following diagrammatic representation (Fig. 21.2) of the variation of ΔG and ΔH with T.

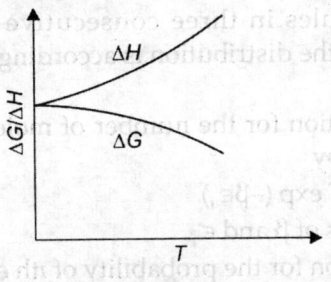

Fig. 21.2

UNIVERSITY QUESTIONS

21.1 A card is drawn from a pack of 52. What is the probability of its being an ace or a king?
(Burdwan BSc(H), 2007)

21.2 Suppose a coin is weighed so that heads come up twice as often as tails. Find the probability of getting two tails and one head in three successive tosses.
(Jadavpur BSc(H), 1996)

21.3 Calculate the possible number of ways of distributing 2 particles among 4 energy states when the particles are indistinguishable and there is no restriction on the occupancy of the energy states.
(Burdwan BSc(H), 2007)

21.4 Calculate the number of ways of arranging 5 different particles among 3 energy levels. Such that one energy level has 1 particle two have 2 each.
(Calcutta BSc(H), 2003)

21.5 Three distinguishable particles are filled at random in three different energy levels. The a-priori probability of having a particle in any one energy level out of the three is 1/3. Find the probability of two particles in one level, one particle is another level and none in the third level.
(Jadavpur BSc(H), 1999)

21.6 Define macro- and micro-states.
(Burdwan BSc(H), 1993)

21.7 A system has three distinguishable particles distributed among three compartments in all possible ways. Calculate the total number of microstates.
(Burdwan BSc(H), 2005)

21.8 The total number of microstates (W_T) of a system of N distinguishable particles, distributed in two energy states is $W_T = 2^N$. Using Stirling approximation, find the number of microstates ($W_{1:1}$) for the 1:1 distribution. Compare W_T and $W_{1:1}$ and comment.
(Calcutta BSc(H), 2008)

21.9 Calculate the mathematical and thermodynamic probability for the most probable distribution of 4 particles in two energy levels.
(Burdwan BSc(H), 1993)

21.10 "Thermodynamic processes are inherently irreversible, while mechanical processes are inherently reversible". Explain with illustration.
(Jadavpur BSc(H), 2010)

21.11 Consider 20 molecules divided equally between four non-degenerate energy levels. What is the thermodynamic probability (W) for this distribution? How does the value of W change if one molecule is removed from one level and added to another?
(Calcutta BSc(H), 2001)

21.12 State the Boltzmann formula where the energy levels are non-degenerate. Consider three successive levels with a constant gap of energy. Show that the number of molecules in the middle level is the geometric mean of the other two levels.
(Burdwan BSc(H), 1995)

21.13 The number of molecules in three consecutive levels are 1000, 100 and 10 respectively. Show that the distribution is according to Boltzmann statistics.
(Burdwan BSc(H), 2000)

21.14 The Boltzmann distribution for the number of molecules in the (non-degenerate) energy level ϵ_i is given by
$$N_i = C \exp(-\beta \epsilon_i)$$
(a) (i) Express C in terms of β and ϵ_i.
   (ii) Write an expression for the probability of ith energy level being occupied.
   (iii) Show that $N_{i+1} < N_i$. Under what condition will N_{i+1} be equal to N_i?
(Burdwan BSc(H), 1994)

(b) Is C dependent or not on temperature?

(c) Calculate C when ϵ_i's are given by $\epsilon_i = ih\nu$, $i = 0, 1, 2, 3, ...$

(Calcutta BSc(H), 2005)

21.15 Justify that the ground state vibrational level of a diatomic molecule is more densely populated? (Burdwan BSc(H), 2011)

21.16 Consider the Boltzmann population distribution (N_i) for non-degenerate energy levels $\epsilon_0, \epsilon_1, \epsilon_2, ...$ with $\epsilon_0 = 0$, and answer the following:

(i) What would be the value of N_i/N_j $(i > j)$ when $T \to \infty$?

(ii) What would be the value of N_o and N_i $(i \neq 0)$ When $T \to 0$?

(iii) Show that $W = 1$ in the limit $T \to 0$. (Burdwan BSc(H), 1996)

21.17 (a) Write down the Boltzmann distribution law for the population of the ith energy level relative to (i) the total population and (ii) the population in the ground level, in the case of degenerate and non-degenerate systems.

(b) What is partition function? Examine its value at $T = 0$ and as $T \to \infty$, and interpret the results.

(c) Evaluate the term β in $e^{-\beta\epsilon_i}$. (Burdwan BSc(H), 1992)

21.18 For a monatomic ideal gas under equilibrium at 27°C, 5% of its ground state population remains in the doubly degenerate first excited state. Find the minimum amount of energy necessary to excite one such gas particle from the ground state to the first excited state. (Jadavpur BSc(H), 2000)

21.19 Suppose that a particle is characterised by two energy levels $\epsilon_1 = 0$ and $\epsilon_2 = \epsilon$ with degeneracies of g_1 and g_2 respectively. Write an expression for the partition function and hence find the average energy of the particle. (Calcutta BSc(H), 2005)

21.20 State the conditions under which canonical ensemble partition function (Q) can be related to molecular partition function (q) as $Q = \frac{q^N}{N!}$, where N is the number of molecules in the system. Discuss the significance of partition function in statistical thermodynamics. (Jadavpur BSc(H), 1996)

21.21 The molecular partition function of an ideal monatomic gas is given by $q = \left(\frac{AT}{B}\right)^{\frac{3}{2}} V$, where A and B are constants and other terms have their usual significance. Find the expressions of molar internal energy and pressure of the gas. (Calcutta BSc(H), 2014)

21.22 How is entropy of a system related to thermodynamic probability? Comment on the expression. (Burdwan BSc(H), 2001)

21.23 Consider a system A consisting of subsystems B and C with thermodynamic probabilities $W_B = 1.0 \times 10^{20}$ and $W_C = 2.0 \times 10^{20}$. What is the number of configurations available to the system A? Compute the entropies S_A, S_B and S_C. Write down the significance of this result? (Calcutta BSc(H), 2014)

21.24 Find according to Boltzmann, the number of microstates corresponding to the state (E, V, N) of a system of ideal gas. (Burdwan BSc(H), 1999)

21.25 Calculate the relative number of microstates in water with respect to ice at 273 K (ΔH_{fus} for ice = 6.008 kJ mol^{-1} at 273 K and 1 bar). (Calcutta BSc(H), 2013)

21.26 The solidification of a supercooled liquid is spontaneous, yet it is accompanied by a decrease in entropy. Does it contradict any principle of thermodynamics? Explain. (Burdwan BSc(H), 2010)

21.27 From the probability concept of entropy, argue to show that $H + H \rightarrow H_2$ reaction must be exothermic? (Calcutta BSc(H), 2001)

21.28 Using the Boltzmann distribution law, establish equipartition of translational energy of molecules. (Jadavpur BSc(H), 2011)

21.29 Obtain the barometric formula from the Boltzmann distribution mentioning the assumption involved. (Calcutta BSc(H), 2013)

21.30 Calculate the pressure reading of a barometer on an aeroplane which is flying at an altitude of 10 km from the sea level. Assume the pressure to be 101.325 kPa at sea level and the mean temperature 243 K. Use the average molar mass of air (80% N_2 and 20% O_2). (Calcutta BSc(H), 2006)

21.31 For what increase in altitude is the earth's atmospheric pressure reduced to half? Assume, average value of temperature 250 K and the average molar mass 0.029 kg mol⁻¹. (Calcutta BSc(H), 2004)

21.32 Write down Nernst heat theorem and mention one of its exception. (Calcutta BSc(H), 2014)

21.33 State analytically the Nernst heat theorem and hence show that the heat capacity for a condensed system would remain unaltered in any transformation in the vicinity of absolute zero. (Burdwan BSc(H), 2010)

21.34 The temperature dependence of C_p is given by the expression

$$C_p = \alpha + \beta T + \gamma T^2$$

Show that for pure crystalline solid $\alpha = 0$. (Calcutta BSc(H), 2013)

21.35 Consider the following physicochemical change: Solid (0 K, P → solid (T, P). Calculate the change in entropy for the said process according to Planck formulation. Write down the expression of standard entropy in case of a liquid above the melting point and for a gas above the boiling point. (Burdwan BSc(H), 2011)

21.36 Draw a curve showing the variation of entropy when a solid (at temperature T) is heated to form vapour (at temperature T' > boiling point). How would you determine the 'absolute entropy' of the substance in the vapour phase at temperature T'? (Calcutta BSc(H), 2006)

21.37 (a) Write down the Lewis–Randall statement of the third law of thermodynamics and justify it from the concept of thermodynamic probability.

(b) Show that at absolute zero, the coefficient of cubical expansion is zero. (Burdwan BSc(H), 1992)

21.38 Consider a system for which the equation of state $\left(\frac{\partial U}{\partial V}\right)_T = -P + KTP^2$ (where K = constant) holds around room temperature. Can this equation be valid as $T \rightarrow 0$? (Calcutta BSc(H), 2008)

21.39 Entropy of a system cannot be negative. Justify. (Calcutta BSc(H), 2005)

21.40 At 25°C, the third law entropy of water is about 82 JK⁻¹ mol⁻¹ less than that of bromine at the same temperature. What does this signify? (Burdwan BSc(H), 2008)

21.41 Define the term 'residual entropy'. Carbon monoxide exhibits high value of residual entropy. Explain. (Burdwan BSc(H), 2012)

21.42 Change of Gibbs free energy for a process at low temperature at a constant pressure is given by: $\Delta G = a + bT + cT^2$. show that $b = 0$ and $\left(\frac{\partial \Delta G}{\partial T}\right)_P$ and $\left(\frac{\partial \Delta H}{\partial T}\right)_P$ have values equal in magnitude and opposite in sign and both the values tend towards zero at $T \rightarrow 0$. (Burdwan BSc(H), 2006)

21.43 Explain why C_V for N_2 is always found to be less than that of Cl_2 at ordinary temperature? (Calcutta BSc(H), 2006)

21.44 Write down the assumptions of Einstein's theory of heat capacity of solids, and indicate which of them were modified by Debye to improve the model.

(Calcutta BSc(H), 2013)

21.45 Starting from the expression of average energy of vibration with a frequency

$$v : \overline{\in} = \frac{hv}{\left(e^{hv/kT} - 1\right)},$$ arrive at an expression of C_V for an Einstein solid.

(Burdwan BSc(H), 2006)

21.46 Einstein's equation for the heat capacity of solid is given by

$$\overline{C}_V = 3R\left(\frac{hv}{kT}\right)^2 \cdot \frac{e^{hv/kT}}{(e^{hv/kT} - 1)^2}$$

where the terms have their usual meaning. Arrive at Dulong and Petit's law from the Einstein's equation. Define Einstein characteristic temperature and state its significance. (Calcutta BSc(H), 2007)

21.47 Why the value of $\overline{C}_V$ for carbon at ordinary temperature is less than $3R$?

(Burdwan BSc(H), 2006)

21.48 Deduce from Einstein equation for heat capacity $(\overline{C}_V)$ of solid the limiting value of $\overline{C}_V$ as $T \rightarrow 0$ and $T \rightarrow \infty$. (Calcutta BSc(H), 2006)

21.49 Assuming the Debye heat capacity equation to be applicable, show that the entropy of a perfect solid at very low temperature should be equal to $\frac{1}{3}C_p$ where C_p is the heat capacity at the given temperature. (Calcutta BSc(H), 2008)

21.50 The T^3 law predicts that $\lim_{T \to 0} C_V = 0$. Justify this also from the third law of thermodynamics. (Burdwan BSc(H), 1992)

KEY TO UNIVERSITY QUESTIONS

21.1 In a pack of 52 cards, there are 4 ace and 4 king cards. Then the required probability will be $\frac{4}{52} + \frac{4}{52}$.

21.2 Since 'head' has probability double of 'tail', the probability of tail is $\frac{1}{3}$ and that of head is $\frac{2}{3}$, as individual events (total probability being 1). Then the required probability of the specified joint event will be $\frac{1}{3} \times \frac{1}{3} \times \frac{2}{3}$.

21.3 Possible number of ways $= \frac{(2+4-1)!}{2!(4-1)!}$, vide problem 21.2(ii).

21.4 Required number of ways $= \frac{5!}{1!2!2!}$, by Eq. (21.2).

21.5 Thermodynamic probability $= \frac{3!}{2!1!0!}$, vide problem 21.5.

Mathematical probability $= \frac{3!}{2!1!0!}/3^3$.

21.6 Vide Section 21.1.

21.7 Total number of microstates $= 3^3$.

21.8 $\qquad W_{1:1} = \frac{N!}{\frac{N}{2}! \frac{N}{2}!}$ by Eq. (21.2) (Ω and W stand for the same quantity)

Then $\ln W_{1:1} = \ln N! - 2\ln \frac{N}{2}!$

$\qquad\qquad = N\ln N - N - 2\left(\frac{N}{2}\ln\frac{N}{2} - \frac{N}{2}\right)$, by Stirling approximation

$\qquad\qquad = N\ln 2$

or $\qquad W_{1:1} = 2^N$

i.e. $\qquad W_{1:1} = W_T$

Then, for sufficiently large value of N (when Stirling approximation holds) the thermodynamic probability for 1:1 distribution, which is the most probable (and hence dominating) one, is equal to the total number of all the microstates of the system, i.e. this dominating molecular configuration alone will account for properties of the system.

21.9 Vide problem 21.5.

21.10 Vide introduction of this chapter.

21.11 For equal distribution

$$W = \frac{20!}{5!5!5!5!}$$

On removing one molecule from one level to another, W will change (decrease) to the following

$$W = \frac{20!}{4!6!5!5!}$$

21.12 The Boltzmann formula is

$$\frac{N_i}{N} = \frac{e^{-\epsilon_i/kT}}{\sum e^{-\epsilon_i/kT}} \qquad \text{[vide Eq. (21.13)]}$$

From this, we have

$$\frac{\frac{N_1}{N_2}}{\frac{N_2}{N_3}} = \frac{e^{-(\epsilon_1-\epsilon_2)/kT}}{e^{-(\epsilon_2-\epsilon_3)/kT}} = 1$$

Since $\epsilon_2 - \epsilon_1 = \epsilon_3 - \epsilon_2$, energy levels being equispaced

Then $\qquad N_2 = \sqrt{N_1 N_3}$

21.13 Here the number of molecules (N_i and N_{i+1}) in the consecutive energy levels (ϵ_i and ϵ_{i+1}) are in geometric progression. This suggests that the distribution is in accordance with the Boltzmann statistics

$$\frac{N_i}{N_{i+1}} = e^{-(\epsilon_i-\epsilon_{i+1})/kT}$$

provided the energy levels are equispaced and non-degenerate.

Note: The problem contains no necessary information regarding energy levels.

21.14 Vide problem 21.13.

21.15 The justification lies in the Boltzmann population distribution for non-degenerate energy levels such that higher is the population, lower is the energy level.

21.16 (i) $\qquad \dfrac{N_i}{N_j} = e^{-(\epsilon_i-\epsilon_j)/kT}$

$\qquad\qquad = 1$ in the limit $T \to \infty$

(ii) $\qquad \dfrac{N_i}{N_o} = e^{-(\epsilon_i-\epsilon_o)/kT}$

$\qquad\qquad = 0$ in the limit $T \to 0$

Hence $N_i = 0$

Then $\quad N = N_0 + N_1 + N_2 + \ldots$
$\qquad\qquad = N_0$

(iii) $\qquad W = \dfrac{N!}{N_0! \, N_1! \, N_2! \ldots} = 1$

Since $N_0 = N$ in limit $T \to 0$

Note:

1. When $T \to \infty$, the population of an energy level does not depend on its energy value.
2. When $T \to 0$, molecules of a system all belong to the lowest energy level.

21.17 (a) (i) $\qquad \dfrac{N_i}{N} = \dfrac{e^{-\epsilon_i/kT}}{\sum e^{-\epsilon_i/kT}} \qquad$ for non-degenerate system

and $\qquad \dfrac{N_i}{N} = \dfrac{g_i e^{-\epsilon_i/kT}}{\sum g_i e^{-\epsilon_i/kT}}$

for degenerate system where g_i is the degeneracy of the ith energy level.

(ii) $\qquad \dfrac{N_i}{N_o} = e^{-\epsilon_i/kT}$

where ϵ_i is expressed relative to ϵ_o, the energy of the lowest level.

and $\qquad \dfrac{N_i}{N_o} = \dfrac{g_i}{g_o} e^{-\epsilon_i/kT}$

(b) Vide Section 21.4.

(c) β can be conveniently evaluated by calculating the average kinetic energy $\bar{\epsilon}$ of an ideal gas molecule for one degree of freedom which is given by

$$\bar{\epsilon} = \frac{\sum N_i \, \epsilon_i}{\sum N_i} = \frac{\sum \epsilon_i \, e^{-\beta \epsilon_i}}{\sum e^{-\beta \epsilon_i}}$$

$$= \frac{\sum \dfrac{p_i^2}{2m} e^{-\beta p_i^2/2m}}{\sum e^{-\beta p_i^2/2m}}$$

since $\epsilon_i = p_i^2/2m$ where ϵ_i and p_i are molecular energy and momentum per degree of freedom

$$= \frac{\dfrac{1}{2m} \displaystyle\int_{-\infty}^{+\infty} p^2 e^{-\beta p^2/2m} dp}{\displaystyle\int_{-\infty}^{+\infty} p^2 e^{-\beta p^2/2m} dp} \qquad \text{assuming } p \text{ to vary continuously}$$

$$= \frac{1}{2\beta}$$

Now, from kinetic theory

$$\bar{\epsilon} = \tfrac{1}{2} kT$$

Then $\qquad \beta = \dfrac{1}{kT}$

21.18
$$\frac{N_1}{N_0} = \frac{g_1}{g_0} e^{-\epsilon_1/kT} \quad \text{by Eq. (21.14)}$$

or
$$\epsilon_1 = kT \left(\ln \frac{g_1}{g_0} - \ln \frac{N_1}{N_0} \right)$$

where ϵ_1 is the energy value for the first excited energy level relative to the ground level taken as zero.

$$= (1.381 \times 10^{-23} \text{ JK}^{-1})(300.15 \text{ K}) \left(\ln \frac{2}{1} - \ln \frac{5}{100} \right)$$

$$= 1.53 \times 10^{-20} \text{ J}$$

which is the required amount of minimum energy.

21.19 Partition function $q = g_1 e^{-\epsilon_1/kT} + g_2 e^{-\epsilon_2/kT}$ by Eq. (21.17)

$$= g_1 + g_2 e^{-\epsilon/kT}$$

Now according to the Boltzmann distribution law, the probability (p_j) of finding a particle in an energy level ϵ_j of degeneracy g_j is

$$p_j = \frac{N_j}{N} = \frac{1}{q} g_j e^{-\epsilon_j/kT} \quad \text{by Eq. (21.14)}$$

Average energy of the particle

$$\bar{\epsilon} = \sum p_j \epsilon_j$$

$$= \frac{1}{q} \sum g_j \epsilon_j e^{-\epsilon_j/kT}$$

$$= \frac{0 + g_2 \epsilon e^{-\epsilon/kT}}{g_1 + g_2 e^{-\epsilon/kT}}$$

21.20 The molecules are to be independent and indistinguishable. The second part is referred to Section 21.4.

21.21 The partition function of a system (canonical ensemble) consisting of 1 mole of molecule is

$$Q = \frac{q^{N_A}}{N_A!}$$

when the molecules are independent and indistinguishable as in an ideal gas

or
$$\ln Q = N_A \ln q - \ln N_A!$$

$$= N_A \left(\tfrac{3}{2} \ln T + \ln V + \tfrac{3}{2} \ln \tfrac{A}{B} \right) - \ln N_A!$$

for the given expression of q.

Molar internal energy

$$\bar{U} = kT^2 \left(\frac{\partial \ln Q}{\partial T} \right)_V \quad \text{by Eq. (21.18)}$$

$$= \tfrac{3}{2} N_A kT$$

$$P = kT \left(\frac{\partial \ln Q}{\partial V} \right)_{N,T} \quad \text{by Eq. (21.20)}$$

$$= N_A \frac{kT}{V}$$

Note: Comparison of the above expression of P with that obtained from the imperical gas laws establishes the relation $nk = R$, the molar gas constant.

21.22 Vide Section 21.2.

21.23 The number of configuration of A, $W_A = W_B \cdot W_C = (1.0 \times 10^{20})(2.0 \times 10^{20})$

$$S_A = k \ln W_A = k \ln (1.0 \times 10^{20})(2.0 \times 10^{20})$$
$$S_B = k \ln W_B = k \ln (1.0 \times 10^{20})$$
$$S_C = k \ln W_C = k \ln (2.0 \times 10^{20})$$

□ $S_A = S_B + S_C$, which signifies that entropy is an additive property.

21.24 See problem 21.20.

21.25 See problem 21.54(ii).

21.26 On freezing, an isolated supercooled liquid suffers a decrease in configurational (spatial) entropy (and not the total entropy) due to the formation of some ordered (i.e. less random) arrangement of molecules in the resulting lattice. But this is more than compensated by the increase in thermal entropy caused by the liberated thermal energy.

□ No. Because total entropy of an isolated liquid actually increases on freezing, which is in compliance with the second law of thermodynamics.

21.27 See problem 21.51.

21.28 The average translational energy $\bar{\epsilon}$ of a gas molecule per degree of freedom is

$$\bar{\epsilon} = \frac{\sum N_i \, \epsilon_i}{\sum N_i}$$

$$= \frac{\sum \epsilon_i \, e^{-\epsilon_i / kT}}{\sum e^{-\epsilon_i / kT}} \quad \text{by Boltzman distribution law, Eq. (21.13)}$$

$$= \frac{\frac{1}{2} m \int_{-\infty}^{+\infty} p^2 e^{-p^2 / 2kT} \, dp}{\int_{-\infty}^{+\infty} e^{-p^2 / 2kT} \, dp} \quad \text{vide quesiton 21.17(c)}$$

$$= \frac{1}{2} kT$$

21.29 Vide Section 21.3, Eq. (21.15).

21.30
$$M_{air} = x_{N_2} M_{N_2} + x_{O_2} M_{O_2}$$

$$= \left(\frac{80}{100}\right)(28.02 \times 10^{-3} \text{ kg mol}^{-1}) + \left(\frac{20}{100}\right)(32 \times 10^{-3} \text{ kg mol}^{-1})$$

$$= 28.82 \times 10^{-3} \text{ kg mol}^{-1}$$

$$P = P_0 e^{-M_{air} gh / RT} \quad \text{by Eq. (21.15)}$$

$$= (101325 P_1) \exp\left[\frac{(-28.82 \times 10^{-3} \text{ kg mol}^{-1})(9.80 \text{ ms}^{-2})(10 \times 10^3 \text{ m})}{(8.314 \text{ JK}^{-1}\text{mol}^{-1})(243 \text{ K})}\right]$$

$$= 25058 \text{ Pa}$$

21.31
$$h = \frac{RT \ln \frac{P_0}{P}}{Mg} \quad \text{by Eq. (21.15)}$$

$$= \frac{(8.31 \text{ JK}^{-1} \text{ mol}^{-1})(250 \text{ K}) \ln 2}{(0.029 \text{ kg mol}^{-1})(9.80 \text{ ms}^{-2})} = 5066 \text{ m}$$

21.32 Nernst heat theorem is equivalent to the supposition that the entropy change in a reaction between pure crystalline substances approaches zero at $T = 0$.

The heat theorem implies that all crystalline substances have the same entropy at 0 K. However, this is not complied with substance like H_2 and CO. [In case of CO, the discrepancy arises from the imperfection of the crystal (vide problem 21.77). But in case of H_2, this is essentially due to nuclear spin contribution (vide problem 21.78).]

21.33
$$\lim_{T \to 0} \frac{\partial(\Delta G)}{\partial T} = 0$$

and
$$T\left(\frac{\partial \Delta H}{\partial T}\right)_V = 0$$

□
$$\Delta C_p = \frac{\partial \Delta H}{\partial T} \to 0 \text{ as } T \to 0$$

21.34
$$C_p \to 0 \text{ as } T \to 0 \quad \text{vide problem 21.74(a)(ii)}$$

Imposition of this on the given expression gives the desired result.

21.35 Vide Section 21.6, third law entropy.

21.36

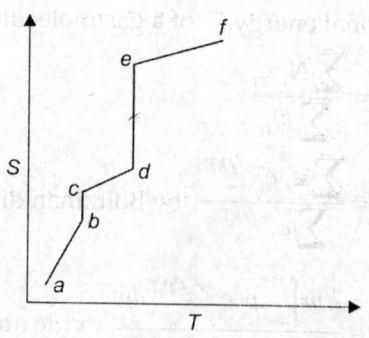

Fig. 21.3

ab corresponds to heating of the solid, bc to melting, cd to heating of the liquid, de to boiling of the liquid and ef to heating of the vapour. Note that the jump of S at the boiling point is much larger than that at the melting point, since ΔS_{vap} is much greater than ΔS_{fus} (vide problem 21.49).

□ This can be done using the expression (21.41).

21.37 (a) Vide Section 21.6 and problem 21.68.

(b) Vide problem 21.74(b).

21.38 No. Because at $T = 0$, the given equation becomes

$$\left(\frac{\partial U}{\partial V}\right)_T + P = 0$$

or
$$T\left(\frac{\partial P}{\partial T}\right)_V = 0, \text{ by Eq. (5.16), which is not possible}$$

Note: An equation of state of a real gas has no meaning at 0 K due to the phase change that invariably occurs as $T \to 0$. But even if there were no phase change, as with an ideal gas, the equation of state of the gas would not hold at $T = 0$. Thus if the ideal gas equation $PV = nRT$ were valid at $T = 0$, then coefficient of thermal

expansion $1/T$ would tend to ∞ as $T \to 0$, which is not in keeping with the third law of thermodynamics.

21.39 Statistical formulation of entropy through the Boltzmann relation $S = k \ln \Omega +$ constant implies that S has no physically fixed zero level due to arbitrariness of the involved constant. The problem disappears when this constant is dropped. Now, on the basis of the relation $S = k \ln \Omega$, S will have a lowest value of zero, i.e. no negative value, since Ω cannot be less than 1. This is also in confirmity with the third law of thermodynamics.

21.40 The given data implies higher randomisation of energy in bromine compared with water.

This is due to the higher molar mass that makes the spacings of the translational and rotational energy levels lower with bromine. Here vibrational and electronic contributions to S are of little importance as the temperature concerned is rather low.

21.41 Vide Section 21.6. Third law entropy and problem 21.77.

21.42 Vide problem 21.82.

21.43 Vide problem 21.31.

21.44 Vide Section 21.5, heat capacity of solids.

21.45 An Einstein solid is an aggregate of independent atomic oscillator each vibrating harmonically with same frequence v and average energy $\bar{\epsilon} = \dfrac{hv}{e^{hv/kT} - 1}$. Then for 1 mol of such solid having $3N_A - 6 \approx 3N_A$ vibrational modes, the total energy

$$\bar{U} = 3N_A \cdot \frac{hv}{e^{hv/kT} - 1}$$

Then
$$\bar{C}_V = \left(\frac{\partial U}{\partial T}\right)_V$$

$$= 3N_A k \left(\frac{hv}{kT}\right)^2 \cdot \frac{e^{hv/kT}}{\left(e^{hv/kT} - 1\right)^2}$$

$$= 3R \left(\frac{hv}{kT}\right)^2 \cdot \frac{e^{hv/kT}}{\left(e^{hv/kT} - 1\right)^2}$$

21.46 The given expression may be rewritten as

$$\bar{C}_V = 3Rf^2$$

where
$$f = \frac{hv}{kT}\left(\frac{e^{hv/2kT}}{e^{hv/kT} - 1}\right)$$

If T is such that $\dfrac{hv}{kT} \ll 1$, we have (using $e^x = 1 + x + \dots$ and ignoring higher terms)

$$f = \frac{hv}{kT} \cdot \frac{1 + \frac{hv}{2kT}}{\left(1 + \frac{hv}{kT}\right) - 1}$$

$$= 1 + \frac{hv}{2kT} \approx 1$$

Under this condition, $\bar{C}_V \simeq 3R$, which is the empirical law of Dulong and Petit.

☐ The quantity $h\nu/kT$, which has the dimension of temperature, is called Einstein temperature of the solid, where ν is the frequency of its constituent atoms. It signifies that at this temperature the energy per vibrational degree of freedom of the atomic oscillators of the solid, as expected from the classical principle of equipartition of energy, becomes equal to one quantum of vibrational energy expected from the quantum theory.

21.47 For carbon the vibrational quantum $h\nu$ is remarkably high so that $\bar{C}_V$ is lower than its equipartition value $3R$ at room temperature. This is essentially due to firm binding of C-atoms in the crystal.

21.48 Einstein's equation for heat capacity of solid is

$$\bar{C}_V = 3Rf^2$$

where

$$f = \frac{h\nu}{kT} \cdot \frac{e^{h\nu/2kT}}{e^{h\nu/kT} - 1}$$

$\bar{C}_V \to 3R$ (when $f \to 1$) as $T \to \infty$, vide question 21.46.

For low temperature, when $\frac{h\nu}{kT} \gg 1$

$$f \approx \frac{h\nu}{kT} \cdot \frac{e^{h\nu/2kT}}{e^{h\nu/kT}} = \frac{h\nu}{kT} \cdot e^{-h\nu/2kT}$$

Now, with decrease of temperature, the decrease of $e^{-h\nu/2kT}$ occurs more rapidly than the increase of $h\nu/kT$. Then $f \to 0$ as $T \to 0$, and therefore $\bar{C}_V \to 0$.

21.49 The entropy S of solid at temperature T' is given by

$$S = \int_0^{T'} \frac{C_p dT}{T}$$

$$= \int_0^{T'} \frac{aT^3}{T} dT \text{ by the Debye heat capacity Eq. (21.37)}$$

$$= \frac{1}{3} \cdot a(T')^3$$

$$= \frac{1}{3} C_p \text{ where } C_p \text{ is the heat capacity at temperature } T'$$

21.50 See problem 74(a)(ii).